Lizards of the World

Lizards of the World

Natural History and Taxon Accounts

GORDON H. RODDA

Photographs by Lee Grismer, Chandler Eaglestone, Robert Fisher, Glenn Gerber, Björn Lardner, Julio Lemos-Espinal, James Reardon, Robert Reed, Renee Rondeau, Susy Sanoja-Sarabia, Martin Whiting, and Elijah Wostl

Johns Hopkins University Press *Baltimore*

Printed in the United States of America on acid-free paper
9 8 7 6 5 4 3 2 1

Johns Hopkins University Press
2715 North Charles Street
Baltimore, Maryland 21218-4363
www.press.jhu.edu

Library of Congress Cataloging-in-Publication Data

Names: Rodda, Gordon H., author.
Title: Lizards of the world : natural history and taxon accounts / Gordon H. Rodda.
Description: Baltimore : Johns Hopkins University Press, 2020. | Includes bibliographical references and index.
Identifiers: LCCN 2019043071 | ISBN 9781421438238 (hardcover) | ISBN 9781421438245 (ebook)
Subjects: LCSH: Lizards—Identification. | Animals—Classification.
Classification: LCC QL666.L2 R595 2020 | DDC 597.95—dc23
LC record available at https://lccn.loc.gov/2019043071

A catalog record for this book is available from the British Library.

Special discounts are available for bulk purchases of this book. For more information, please contact Special Sales at specialsales@press.jhu.edu.

Johns Hopkins University Press uses environmentally friendly book materials, including recycled text paper that is composed of at least 30 percent post-consumer waste, whenever possible.

To the memory of Stan Rand,
a profoundly wise lizard ecologist, who listened intently to nature rather than asking it to conform to theory, promoted the discoveries of his students and colleagues in place of his own, and embraced the depth of our ignorance regarding the complexities of nature rather than trumpeting the small successes we have had in unraveling them

Contents

Lizards of the World

1 Introduction and Methods

I wish to describe for you the natural histories of all lizards. For example, in this book I describe the environment in which each species lives, what it eats, and how it reproduces. I do not help you identify lizard species or distinguish higher taxa (genera, families, etc.), nor do I report the evolutionary histories of lizard species. Instead, I focus on the natural history of living species, the macroscopic traits expressing or influencing the lizards' daily "choices," whether those choices are conscious or not. For example, a lizard may not consciously choose between defending a territory (one possible state) or roaming an undefended home range (an alternate state). A lizard might consciously choose to flee rather than remain cryptic upon the approach of a predator (two other possible traits). What large lizards choose to eat might differ from the dietary choices of small lizards, conscious or not.

Natural history incorporates the disciplines of animal behavior and ecology, but it also investigates questions between and beyond the traditional foci of either field. For example, is territoriality or seasonally expressed coloration found more often in viviparous than oviparous species? Natural history describes the situations in which each species lives; thereby it supplies the context for all subdisciplines of biology. Natural history is a specialization that collects and tabulates such observations from nature and thereby provides the fodder for the discovery of causation. Discovery of causation is a multidisciplinary process that draws on the skills of evolutionists, taxonomists, epidemiologists, ecologists, behaviorists, physiologists, and others. Some hold natural history in disrepute because it only initiates the multidisciplinary inference of causation. I say, let's get started.

There is so much left to do and time is running out. As of May 2017 (when the dataset for this work was closed to new species descriptions), there were 6528 accepted lizard species (Reptile Database: www.reptile-database.org), about twice as many as in the year 2000, with several hundred new species being described each year. Taking into account the 170 most important natural history attributes of a lizard, I found that only about one-quarter of the attributes had been addressed for an average lizard (see chapter 2 for examples, or appendix A, "Alphabetical Listing of Field Names," for formal definitions of the 170 attributes). In most cases, "addressed" means only that someone has published an anecdote or opinion on the species' modal character state. In only a small minority of cases has the question been the subject of a targeted investigation or even basic quantification. In even fewer have the facts been well established. For example, consider the most rudimentary of facts: at what time of day is a lizard active? This trait has been addressed in about 73% of lizards, but in the majority of this 73%, the inference is based on no more than collectors' impressions: it was found under a rock at 11 am; therefore, it must be nocturnal. Authors sometimes leap to conclusions based on phylogeny; if it is a *Hemidactylus* gecko it must be nocturnal. A good sample requires that individuals of all sizes and both sexes have been monitored at all times of day, without unduly disturbing the lizard and with an understanding of the individuals' differing detectabilities—often blustering males are easier to sample than females and juveniles. Many sources conveniently assume that all individuals (and sometimes all species) are equally detectable. Ample literature (e.g., Thompson et al. 1998; Mazerolle et al. 2007; McDiarmid et al. 2012; Rodda et al. 2015) demonstrates that this assumption is erroneous. The science of lizard natural history has hardly begun to address detectability and formal sampling.

To begin, I spent eight years dredging published lizard papers and books for data or opinions on the 170 most important natural history attributes (~310 character states). Naturally, eight years is not sufficient to read the entire herpetological literature; many important papers were not encountered. I encourage readers who spot an omission in this book or the accompanying database to contact me with the citation in which the missing datum may be found; I will incorporate the reference into future editions. In my tabulation of natural history attributes, I emphasized regional and monographic taxonomic works and the targeted natural history studies cited therein. I consulted only a few internet sources, primarily records from trusted institutions, such as Reptile Database and the

Red List (www.iucnredlist.org) published by the International Union for Conservation of Nature (IUCN).

In this work and accompanying database, I omitted from the list of characters highly site-specific traits, such as breeding phenology or emergence dates of hibernating species. I devoted a small number of data fields to quantifying the relative seasonality of activity or reproduction (continuous breeder v. multiple but discontinuous broods, etc.), and I recognized that some fields, such as habitat types or the number of clutches or litters per year, have a geographic component. But beyond general notes such as "viviparous at high latitudes, oviparous at low," I did not attempt to georeference attributes that vary geographically. More than 140,000 citations are given in the database, based on 5045 unique sources. Both dataset and full citations are available from the publisher as free downloads. Those database citations are not repeated in the text of this work except to support facts not otherwise referenced in the database.

This entire work is built around the database. In this chapter, I outline the methods used to aggregate and interpret the database. Chapter 2, "What Are Lizards?," describes lizards as a group, and by so doing it provides a template against which to highlight the unique features of each family and genus of lizards (chapter 5). Chapter 3, "Linkages," explores the first-order relationships between the traits documented in the database. For example, is the possession of secretory pores correlated with nocturnality? Chapter 4, "Ecological Business Models," uses the correlational structure revealed in chapter 3 to propose 14 syndromes or repeated patterns that reflect this covariation. I emphasize the predation risks associated with each of these syndromes and align those predation risks with the associated energetic benefits. The resulting balance sheets must have been positive ("in the black") for the syndromes that have persisted. To highlight the balance sheet aspect of the 14 recurring patterns, and to distinguish them from other labels that have been used for similar concepts (guilds, niches, adaptive syndromes, etc.), I call them ecological business models, or simply business models.

The bulk of this work is in chapter 5, in which each lizard family is contrasted with lizards as a whole, and each genus is described in distinction to its family. Each species is summarized in a listing at the conclusion of its genus description. This hierarchical structure accommodates phylogenetic pathways; that is, it recognizes that incremental divergence from closely related species is the course of evolution, but it does not attempt to construct a formal "phylogenetically correct" analysis of causation. Additional information is found in the database. For easier retrieval, all of the family and genus accounts are arranged alphabetically. However, if one wishes to understand the full context of a species description, read also the referenced genus, family, and lizard accounts. Common names used in the genus tables can be traced through the index. Alternate common and scientific names are found in the database.

The Interplay between Ecological and Phylogenetic Causation

The primary purpose of this work is to describe the lizards of the world, both in relation to their taxonomic lineage and their ecological associations. To what extent is an individual lizard species' natural history a product of its phylogeny versus its ecological situation? For example, is an ant-eating lizard's natural history a result of its progenitors or its ant-eating way of life? I do not have the computational skills to sort this out, but I will here briefly note a few of the conceptual obstacles.

- Every organism is the result of an iterative interaction between genotype and environment. Each species reflects the outcome after the interaction has iterated millions of times over millions of years; it is very difficult to disentangle the end result.
- The higher-level phylogeny of lizards has some profound uncertainties, which are unlikely to be definitively resolved any time soon and, in any event, are not resolved now.
- The deepest taxonomic divisions are the most ancient, and for such divisions the simplifying assumptions of evolutionary genetics are the most tenuous. Yes, the outgroups' state can provide a plausible hypothesis of the original characters of a group, but in the fullness of time the accumulation of exceptions begins to overwhelm the presumption of outgroup or ingroup stasis. This is especially true for highly labile traits, such as many behaviors cataloged in this work.
- The web of correlations among natural history traits outlined in chapter 3 (>1400 statistically significant associations) is so tangled that extant phylogenetically correct methods have difficulty analyzing them concurrently. But single-trait analyses will be distorted by the correlated factors that are omitted from the analysis.
- For the vast majority of lizards, we lack critical information on their natural history. What information we have is often erroneous.

Although it is daunting to rigorously determine the causes behind every natural history correlation, it is equally impossible to comprehend the potential multiway associations of hundreds of natural history traits among thousands of lizard species without resorting to a causal narrative framework. Therefore, I do not hesitate to suggest plausible explanations for the observed patterns when such suggestions are heuristic and consonant with ecological understanding. Nonetheless, these arm-waving suggestions should be viewed as first-approximation hypotheses rather than received knowledge. Chapters 3 and 4 are especially laden with hypotheses that are only partially tested, completely untested, or novel. There is so much yet to do.

Definitions

The process and details of assembling the database are described in appendices A–C. All database fields are listed alphabetically in appendix A. All terms defined in the text or the appendices can be searched alphabetically in the index. Most of the terms used in this text are self-explanatory, with a few exceptions.

For any continuous traits that are themselves correlated with body mass, and for which I could extract a plausible mass-free index (ontogenetic mass range, body shape, home-range area, clutch/litter count, incubation duration, longevity, clutch mass, population density), I did so. For example, home-range area is correlated with body mass—bigger lizards have bigger home ranges—and therefore I computed the standardized (mean = 0, standard deviation = 1) residuals from the regression of home-range area on body mass and used the standardized residual for each species as my metric of home-range area for that species. In a world of perfectly normal curves, one could use reported z-scores to extract the exact position of each trait in the lizard distribution; in our world of 6528 real lizards, the approximation is sufficiently close. The use of standardized residuals is especially appealing in a hierarchically structured narrative such as this, because statements in a genus account (chapter 5), which give parenthetical numerical references in the form genus-versus-family (e.g., –0.08 v. 0.33 std. resid.), enable the reader to simultaneously locate the genus's condition in relation to both family and lizards. In this home-range example, the genus, at 0.08 standardized residuals below the mean (the minus indicates less than the mean), has a body-size-adjusted home range that is near normal for lizards (larger than about 47% of species), but this is small in comparison to an average member of its family, for which the body-size-adjusted mean home range (0.33 std. resid.) is larger than about 63% of lizards. Note that these home ranges need not be absolutely large, only large for a lizard of a given size. Because these metrics are mass-adjusted, we can legitimately compare relative home-range sizes between sphaerodactylids and Komodo Dragons.

"Shape index" is the special name I gave to the mass-adjusted standardized residual of mass on snout-vent length (SVL) at adult mode (see also discussion under "Size Estimation" below). A lizard with a +1 shape index is one standard deviation heavier for a given SVL; a lizard with a –1 shape index is correspondingly lighter—about 34 percentiles—than the norm for its length.

"Enlarged toepads" indicates geckos with toes that have both toe-width scales—"lamellae"—and the ends of the digits expanded well beyond the width at the base of the digit. Taxonomists focus on the width of the lamellae, whereas this ecologist focused on the expansion of the digit.

Short-term color change, also known as metachrosis, is here divided into "any" and "advanced." "Any" denotes that the lizard is able to change its darkness, hue, or pattern within a few minutes. "Advanced" denotes that it can change its hue or pattern. A lizard exhibiting "any" but not "advanced" color change is able only to darken.

"Elongate" denotes that the lizard's forelimbs fail to meet its hindlimbs when the straight, relaxed lizard's limbs are gently pulled toward one another. Though this test is often performed and reported for skinks, it is rarely reported for other taxa: please do! One can think of this condition as either short legs or long torsos. Your choice. The standard is undefined for any lizard that is missing a pair of limbs.

"Crests" are lines of spiny dorsal scales of a size large enough to be visible to another lizard at a distance, not just visible to a taxonomist with magnification. Crests may be on head, body, or tail.

Reduced limbs have sometimes been denoted by their relative size—for which there is no agreed threshold—or by the number of digits (three or fewer). I used the latter standard.

Pores of various types (precloacal, postcloacal, femoral, etc.) are mysterious. Their function is not understood and the literature generally uses them as a taxonomic marker rather than a natural history attribute. In many taxa, they appear to be vestigial in one sex, sometimes in both sexes. I tabulate pores for species in which at least one sex appears to exhibit secretory activity. This is a very weak standard and deserves better reporting and tabulation.

Sex differences are of many types: (1) sexes differ in size, (2) sexes differ in permanent color, (3) sexes differ in seasonal color, (4) sexes differ in relative tail length, (5) sexes differ in structures such as pores, crests, head size, and so on. For most lizard species, the size difference in mass is less than twofold, but in a few cases is greater (6). I generated an index, which I call the sex differences rollup, that sums 1 point for each of these differences (aggregate scale of 0 to 6).

Sexual size dimorphism (SSD) is herein expressed on a continuum of means from –2 to 2, where individual species are credited –2 if average adult females weigh at least twice that of their male counterparts; –1 indicates females larger than males by a lesser but generally unspecified ratio, 0 represents the absence of size difference between the sexes, 1 represents males larger than females by a moderate amount, and 2 represents average adult males weighing at least twice that of their female counterparts. A genus composed of three species bearing scores of –1, 0, and 2 will have a mean score of 0.33.

"Ant specialist" (= myrmecovore) denotes a lizard whose diet is >50% ants by volume, at least at some times of year in some places. Many sources estimate rather than quantify this proportion. I do not intend this definition to imply that ant specialists are obligate ant eaters. Most eat other invertebrates opportunistically.

"Herbivore" denotes a lizard whose diet is >90% vegetable matter by volume, as least at certain times of year (many judgments are based on estimates rather than measured quantities).

"Omnivore" denotes a lizard species whose gut contents generally constituted between 1% and 90% vegetable matter, often estimated by a source rather than quantified.

I use "invertebrate eater" in place of what is commonly and erroneously termed an insectivore.

Most sources assume that foraging mode (sit-and-wait, cruise, or active foraging) is a unitary condition for each species; I allow each species to be characterized by any number of foraging modes, as reported in the literature. Werner et al. (2006), among many others, demonstrate the lability of foraging mode within a species.

"Cathemeral," or "active at any time of day," denotes a lizard that may be active by day or night depending on prevailing conditions.

Both males and females may defend space, but the data for female behavior are regrettably very thin, and for this reason I rarely report female spatial defense. For both sexes, the defense of "space" is ambiguous, because the literature is often ambiguous. A lizard may be defending personal space, mating space, a small site such as a preferred burrow or perch, a feeding territory, or a mate. Defense may be seasonal or year-round. In a few cases, the lizard may chase off other species, even some nonlizards. In the database, I distinguish these possibilities on the rare occasions when the literature reports are explicit, but for tabulations in this work I was often forced to conflate them.

I broke the spectrum of annual rainfall totals (mm/y) at the cut points shown to characterize habitats as indicated between the cut points:

55 160 400 1100 3000
hyperdry dry arid intermed. mesic wet

Elevations were similarly divided into low/medium/high with reference to cut points at 1200 and 2400 m.

"Littoral" denotes that the lizard's range is found within the geographic zone influenced by fog or other climatic phenomena associated with proximity to large water bodies, including the Great Lakes, Black Sea, and so on. This zone can be much larger than the salt spray zone used by some to denote littoral.

"Relative clutch mass" (RCM) has variously been computed as clutch mass divided by the female's pre-partum/pre-oviposition mass or post-partum/post-oviposition mass. I use the latter.

"Endemism" is based on the geographic units tabulated. I tabulate both island occurrences and political jurisdictions. In this work, endemism is always referenced to political divisions, which are countries for most of the world and states or provinces for Australia, Canada, and the United States. Thus "endemic" means one thing in the state of Georgia and another in the Republic of Georgia. Note that France maintains sprawling overseas territories; "French Overseas Territory" may be listed as the political jurisdiction for distant islands such as New Caledonia.

"Introduced" indicates that a non-native population was detected. The literature is often silent as to whether the population is "established" or persisting. A few populations that were introduced may have subsequently died out. Many more populations were no doubt introduced but failed before being detected. For most species, however, "introduced" means that the population established.

"Imperiled" is a term I used to denote a lizard species which has been deemed to be at conservation risk (rare, endangered, threatened, and similar classifications) by at least one competent authority—scientist, credible nongovernmental organization such as the IUCN, or government. I consider the judgments of the IUCN to be the most rigorous. I rank the degree of imperilment by the credibility of the source and the geographic extent deemed at conservation risk. "Globally rare" is more serious than "depleted at the edge of range." I sometimes contrast "imperiled" (at risk as denoted by any source) with the most extreme IUCN classifications, collectively: IUCN EN (endangered), IUCN CR (critically endangered), and IUCN EX (extinct).

Size Estimation

Several formal and informal standards exist for the estimation of a lizard species' geographically averaged size. The commonest standard used in field guides is said to be average or mean SVL. Some field guides report total length. Others vaguely refer to "typical" length. Maximum SVL is used by most literature reports targeting body size evolution (e.g., Meiri 2008, 2010; Feldman et al. 2015). I object to all of these standards. "Mean" is an objective measure, but in my experience, many of the literature reports are not literal population means or adult mean, but the author's subjective impression of a typical size. Furthermore, as pointed out in Rodda et al. (2015), the mean is at an unpredictable location part way down the shoulder of most size distributions, and its exact location on the shoulder is highly dependent on the degree to which juveniles or newly maturing adults are captured by the sampling method; often these smaller animals are included or excluded without explicit acknowledgment. As pointed out in many papers (Meiri 2008, 2010; Rodda et al. 2015), length is a poor metric for comparing species of differing body shape, and maxima are inherently vulnerable to differences in sample size.

My solution to these challenges is to base species size estimates on the mass of the lizard individual an unbiased sample is most likely to discover, a maximum likelihood estimator. The problem, of course, is that empirical values of this estimator are not widely available. I was able to find them for only 164 of 6528 lizard species. I estimated them for the remaining species, as outlined here and detailed in appendix B. Based on 35 species for which I had complete size and mass distributions, I identified the adult length mode. Some samples had juvenile length modes that exceeded the height of the adult mode, but the juvenile demographic bulge is easily identified and excluded by the corresponding SVL. I found that I could accurately ($r^2 > 0.99$) estimate the length mode from the mean adult length, and could do so less accurately from the maximum SVL ($r^2 = 0.93$) and maximum total length ($r^2 = 0.82$). I used the computed interspecies regressions and the available species-specific empirical size data to produce a best estimate of adult modal length for each species. These values are reported in the species listings in chapter 5. From the estimated length mode, I used either species-specific mass/length regressions or, when necessary, progressively more general mass/length regressions to estimate mean mass at modal SVL, relying heavily on the body shape-stratified familial regressions of Meiri (2010) and Feldman et al. (2015). The estimated masses for each species are given in the listings in chapter 5; the regressions used for each species are given in the database.

Analytical Methods

When describing each group of species, I often note where in the distribution of a trait a specific group appears. For example, I might note that nocturnal ground lizards have the second-highest frequency, say 14%, in mesic areas whereas a different group, say litter cryptozoids, might have the highest frequency, say 18%, in mesic areas. I use a statement such as this to draw attention to a notable ranking and to visualize the groups' relative position without asking the reader to sift through hundreds of graphs and tables. But rank-order statements such as this are not intended to make the statistical claim that the named focal group (e.g., nocturnal ground lizards) is *significantly* less endowed with the focal trait than is the first ranked group (e.g., litter cryptozoids) or significantly more endowed with the trait than the third-ranked group. Statements about relative rankings are solely descriptive.

Furthermore, I do not routinely endeavor to quantify the degree to which the itemized similarities and differences are statistically significant. However, I did so: (1) if the posited distinction is critical or unexpected, (2) if the contrasting group is identifiable and can be cleanly enumerated, and (3) when identifying first-order relationships between traits among lizards overall (chapter 3). For example, in chapter 3, I provide a fairly detailed analysis of the lizard-wide correlates of SSD (in relation to body size, sexual dichromatism, presence of vocalizations, etc.), and report tests of the statistical significance of each putative association.

I did not, however, attempt to determine cause and effect. I did not perform multiple regression analyses, classification and regression trees, or information-theoretic or phylogenetically correct models. Causal determinations usually require experimental evidence, which is well outside the scope of this work. Furthermore, I generally stopped pursuing explanatory associations at the level of first-order interactions. Many other interactions may exist, many more associations remain to be identified, and causes remain to be determined; I leave these to targeted studies, improved data quality, and the next generation of natural historians.

The species is the unit of interest in this work. I accepted the species list and classification given in Reptile Database in May 2017. Unless qualified otherwise, my use of the word "lizards" refers to lizard species rather than individuals. When I refer to "males" or "females," I am indicating males or females collectively of the focal species, not individuals. Parenthetical values in the form "(17% v. 23%)" indicate that the focal group has a mean trait frequency of 17% and the contrast group 23%. Unless otherwise qualified, the contrast group is all lizards when discussing general traits and lizard families, the family mean when discussing genera, and the genus mean when discussing species.

A special challenge facing a meta-analysis such as this work is that many facts are recorded in the literature exclusively or primarily when the positive condition is observed. For example, when lizards are seen feeding at electric lights, this is noted, but there are almost no accounts that state that a given lizard does *not* feed at electric lights. This makes it hard to frame a clean test of electric light feeding or its associations, as it is unclear which species should be treated as exemplars of lizards that do not feed at electric lights. Dangers lurk in treating all lizards not recorded at electric lights as if they do not feed there. Suppose one wished to know whether lizards feeding at electric lights are larger or smaller than lizards that do not so feed. A problem arises because collection of natural history data is biased toward larger species (see chapter 3). If one compares the size of species for which more is known (e.g., electric light use) against species for which less is known (e.g., electric light use is not documented), and one has a large sample of sizes (e.g., I report size estimates for 6524 of the 6528 species of lizards here tabulated), there is an excellent chance that a spurious but statistically significant difference in size will be discovered. In advance, I can predict its direction: known condition lizards will be statistically larger than unknown condition lizards. For this particular comparison (size of electric light users versus size of electric light nonusers), I could think of no statistically credible comparison, and therefore I performed no statistical tests and made no inferences on this topic.

In other cases, there are stronger ways to identify the negative condition groups. For example, in testing sit-and-wait lizards, I compared them to lizard species whose foraging mode has been documented but was not recorded as sit-and-wait (e.g., cruise or active foraging). In this case, the knowledge bases of the contrasted groups are fairly similar. Likewise, to compare territorial species against those not believed to be territorial, I have not used lizards not recorded as being territorial, but only those lizards that have had their social spacing behavior recorded and were found to be not territorial (e.g., intolerant but not territorial, hierarchy forming, etc.). Whenever possible I have used "control" groups that represent a similar aggregation of knowledge to the "treatment" groups, but lack the defining trait. I use the scare quotes around "treatment" and "control" because those terms imply an experiment whereas my data are descriptive rather than experimental.

Table 1.1. Cut points used for the subjective description of effect sizes. R^2 was used for comparing continuous variables. *Phi* (= square root (*Chi*-square minus *N*); Ferguson 1966) was used for nominal variables. I used Yates correction to compute *phi* for any contingency tables with a cell count <10. Standardized residuals (std. resid.) were used for variables that were mass-adjusted. I used mass ratios to compare masses.

r^2:	0.1	0.5	0.9
phi:	0.08	0.20	0.40
std. resid.:	0.2	0.4	1.0
mass ratio:	2	3	5

If I found correlations between traits for all lizards (chapter 3), those correlations were tested as appropriate to the nature of the variables involved. For example, correlations between two normally distributed interval variables were recorded as Pearson's Product-moment r^2 and the associated probability level. I used $\log_{10}$ transformations to normalize log-normal variables such as body mass or population density. Variables that are continuous but not readily normalized (e.g., mean clutch count) were ranked and tested with nonparametric tools. Associations between a nominal variable and a continuous one were tested with t-tests (variance equality tested and used for selection of t-test variant) or a Mann-Whitney nonparametric variant. I usually tested relationships between counts of nominal variables (e.g., foraging mode or diel activity pattern) using Fisher's Exact Probability tests. The latter was obtained from an online calculator (www.graphpad.com/quickcalcs/contingency1.cfm), which for two-tailed Fisher Exact tests uses the method of summing small *p* values; all other tests were performed within Microsoft Excel 2013. I report statistical probabilities for all tests, but readers should focus on the effect size (also reported), as large samples enable the statistical verification of nonetheless trivial differences.

I assessed the effect size with reference to descriptive statistics appropriate to the type of variables being compared. I divided the associations, on the basis of the cut points shown in table 1.1, into regions that I characterize as weak, moderate, strong, and very strong. Thus an effect size reflected in a mass ratio = 1.7 would be characterized as weak, and a contingency table comparing nominal variables with a *phi* = 0.41 was reported as very strong. These thresholds are arbitrary, but helpful in judging the potency of observed correlations of differing number types.

If, in the course of reading, you wish to refresh your recollection of any definitions or acronyms, they are all referenced in the index. The precise definitions used in compiling the database are given in appendix A.

Now, let's see what the average lizard looks like.

2 What Are Lizards?

To describe each lizard species is a daunting task. To keep the effort conceptually and physically manageable, I have arranged sections hierarchically. That is, after first describing lizards as a group in this chapter, each subgroup is distinguished from average lizards in later chapters. In the case of genera, the narrative accounts in chapter 5 further describe how each genus differs from its family. This chapter characterizes lizards as a group, thereby providing the template against which to contrast all first-order subgroups.

Composition

This section describes lizard groupings that are either ecological (lizards with similar ecologies grouped) or taxonomic (related species grouped). Chapters 3 and 4 focus on the distinguishing features of ecological groupings; chapter 5 is structured by taxonomic groups (family, genus, species).

Ecological

Ecologically, lizards are generalized quadrupeds very similar in size and morphology to salamanders. They differ from salamanders by virtue of their desiccation-resistant skin and eggs, which enable lizards to thrive in arid environments.

There is no widely accepted method of grouping lizards ecologically. My suggestion is embodied in the "ecological business models" explained throughout this work, especially in chapter 4. Cursory definitions for the 14 business models will help us get started (fig. 2.1); examples of each are found in the color plates.

Crack Hermit. Lizards that protect themselves from predation by spending most of their lives in a protected crevice, from which they leave primarily on momentary excursions to grab passing prey.

Diurnal Canopy Walker. Day-active lizards that protect themselves from predation by remaining within the con-

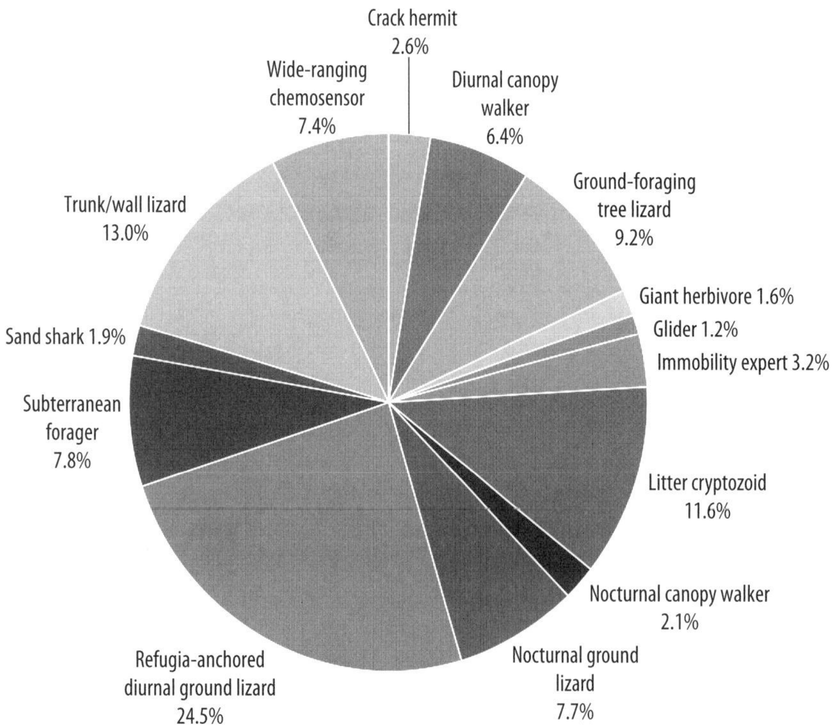

Fig. 2.1. The composition of lizard species, grouped by ecological business model. The percentage of each group is listed at the edge of wedge, based on 4912 scored species, with varying degrees of confidence in their assignment. (See color plate.)

cealment of elevated foliage at any height above ground. There is no implication that "canopy" walkers preferentially occur near the tops of vegetation. They may spend most of their time climbing just above the ground.

Giant Herbivore. Large lizards that consume readily available vegetation and minimize predation by their large size, proximity to shelter, and often by occurrence in predator-reduced environments such as high elevation, islands, or deserts.

Glider. Arboreal lizards with aerodynamic skin flaps that enable them to evade climbing predators by leaping and gliding from tree to tree.

Ground-foraging Tree Lizard. Lizards that alternate between the ground and trees to extend sight lines, capture prey, and evade predators that are nimble in only one of those microhabitats. Note that in this case "tree" is very broadly defined to include any vegetation that allows a lizard to perch above-ground (grass, shrubs, vines, trees, etc.).

Immobility Expert. Lizards with such exquisite camouflage that they can remain unseen in plain view, as long as they remain immobile, which they usually do. They eat prey that pass in front of them.

Litter Cryptozoid. Small lizards that can usually remain unseen in dense tufts of vegetation or debris on the forest floor. They consume the numerous prey inhabiting the rich duff microenvironment.

Nocturnal Canopy Walker. Night-active lizards that protect themselves from predation by darkness and the sheer complexity of vegetative branching, at any height above ground.

Nocturnal Ground Lizard. Night-active lizards that evade nocturnal ground predators by moving very slowly and quietly, often in areas of reduced predator abundance such as deserts.

Refugia-anchored Diurnal Ground Lizard. "Typical" sedentary lizards that evade predators by sprinting toward a known refuge whenever danger threatens.

Sand Shark. Specialists in loose sand, through which they "swim" until detecting a prey item on the surface above them, at which point they surface, sometimes abruptly, to make the capture.

Subterranean Forager. Limbless or functionally limbless lizards that push through the soil to find subterranean prey such as ants, termites, worms, or insect larvae.

Trunk/Wall Lizard. Lizards that take advantage of the crevices in walls and tree trunks and are able to rapidly switch sides of rocks or trunks to stalk prey or evade predators.

Wide-ranging Chemosensor. Lizards that can scent hidden prey, although foraging for such prey may be conspicuous and take them away from known refugia; therefore, these lizards are often limited to conditions ideal for high-speed flight.

Based on this sampling, the modal lizard is a refugia-anchored diurnal ground lizard, with trunk/wall lizards and litter cryptozoids in second and third place, respectively.

Taxonomic Composition

Taxonomically, lizards are the living reptiles that are not birds, turtles, crocodilians, or snakes. Snakes share their branch of life with lizards (order Squamata), but snakes have such specialized morphology and carnivorous ecology that most species of snakes are easily distinguished from lizards. Leglessness is not unique to snakes, as many lizards are also limbless (see discussion under "Morphology" below). Snakes are not covered in this work. In the above summary, no mention was made of the rhynchocephalians, another category of reptile arguably on par with turtles or crocodilians. The one living rhynchocephalian, the Tuatara, is ecologically lizard-like and therefore included in this work.

The 6528 species in this book—one rhynchocephalian and 6527 lizards (including amphisbaenians)—are often organized into 45 families in evolutionary clades following Reptile Database (accessed 2017).

- Order Rhynchocephalia
  - Suborder Sphenodontida
    - Family Sphenodontidae (1 sp.)
- Order Squamata (also includes snakes)
  - Infraorder Iguania
    - Family Agamidae (483 spp.)
    - Family Chamaeleonidae (203 spp.)
      - [12 Iguanid–sensu lato families:]
    - Family Corytophanidae (9 spp.)
    - Family Crotaphytidae (12 spp.)
    - Family Dactyloidae (416 spp.)
    - Family Hoplocercidae (19 spp.)
    - Family Iguanidae (42 spp.)
    - Family Leiocephalidae (31 spp.)
    - Family Leiosauridae (33 spp.)
    - Family Liolaemidae (306 spp.)
    - Family Opluridae (8 spp.)
    - Family Phrynosomatidae (154 spp.)
    - Family Polychrotidae (7 spp.)
    - Family Tropiduridae (136 spp.)
  - Infraorder Gekkota
    - Family Carphodactylidae (30 spp.)
    - Family Diplodactylidae (149 spp.)
    - Family Eublepharidae (36 spp.)
    - Family Gekkonidae (1120 spp.)

Family Phyllodactylidae (135 spp.)
Family Pygopodidae (46 spp.)
Family Sphaerodactylidae (214 spp.)
Infraorder Scincomorpha
Family Alopoglossidae (22 spp.)
Family Cordylidae (68 spp.)
Family Gerrhosauridae (37 spp.)
Family Gymnophthalmidae (237 spp.)
Family Lacertidae (323 spp.)
Family Scincidae (1613 spp.)
Family Teiidae (156 spp.)
Family Xantusiidae (34 spp.)
Infraorder Diploglossa
Family Anguidae (78 spp.)
Family Anniellidae (6 spp.)
Family Diploglossidae (51 spp.)
Family Xenosauridae (11 spp.)
Infraorder Dibamia
Family Dibamidae (23 spp.)
Infraorder Platynota
Family Helodermatidae (2 spp.)
Family Lanthanotidae (1 sp.)
Family Shinisauridae (1 sp.)
Family Varanidae (79 spp.)
Suborder (rank unclear; may be Superfamily) Amphisbaenia
Family Amphisbaenidae (178 spp.)
Family Bipedidae (3 spp.)
Family Blanidae (7 spp.)
Family Cadeidae (2 spp.)
Family Rhineuridae (1 sp.)
Family Trogonophidae (6 spp.)

These 45 families are a primary organizing tool for this book, and readers may find it useful to learn their names, especially of the more populous ones (table 2.1). All families are each described individually in the taxonomic accounts (chapter 5). Two families (gekkonids and skinks) are especially large (>1000 species). Just 12 families constitute 83% of the 6528 species (table 2.1). Just 14 of 568 genera account for 31% of lizard species (table 2.2).

Table 2.1. The 12 most-speciose lizard families

Family	Species	%
Scincidae	1613	24.7
Gekkonidae	1120	17.2
Agamidae	483	7.4
Dactyloidae	416	6.4
Lacertidae	323	5.0
Liolaemidae	306	4.7
Gymnophthalmidae	237	3.6
Sphaerodactylidae	214	3.3
Chamaeleonidae	203	3.1
Amphisbaenidae	178	2.7
Teiidae	156	2.4
Phrynosomatidae	154	2.4

Table 2.2. The most-speciose genera

Genus	Species	Family
Anolis	416	Dactyloidae
Liolaemus	258	Liolaemidae
Cyrtodactylus	232	Gekkonidae
Hemidactylus	144	Gekkonidae
Cnemaspis	130	Gekkonidae
Sphenomorphus	110	Scincidae
Amphisbaena	109	Amphisbaenidae
Sphaerodactylus	105	Sphaerodactylidae
Sceloporus	100	Phrynosomatidae
Ctenotus	100	Scincidae
Lerista	95	Scincidae
Varanus	79	Varanidae
Trachylepis	78	Scincidae
Emoia	77	Scincidae

Data Adequacy

Of the 1,109,760 primary cells in my data matrix (170 original fields × 6528 species), 27% of cells included some information. At least two-thirds of the most basic natural history information on lizards has yet to be collected. The best studied family, Sphenodontidae, had some answers to 71% of the questions asked; the least-documented family, Amphisbaenidae, had answers to less than 17%. Individual species ranged from 8% to 84%. Every question was answerable in several species; no species had answers to all of these basic natural history questions. Does science know more about moon rocks than lizards? Given the relative expenditure, it should.

Morphology

This section covers the physical attributes that do not differ between the sexes in certain species: size, shape, vocal ability, tail length and breakability, color, and color change.

Size

Lizards range in modal body length from 20 to 658 mm, with a median of 63 mm and a mean of 77 mm. However, length can be a poor metric because lizards that are 63 mm in length vary from ribbon-like (0.85 g) to orotund (10.2 g). The heavier lizard seems "bigger" ecologically, and it would certainly be more imposing to a competitor that weighed 0.85 grams.

Table 2.3. Lizard sizes. The last two rows do not represent deciles.

Percentile ranking	Mass (g)	Representative species
10th	1.1	*Lygodactylus* spp.
20th	1.7	*Gehyra insulensis*
30th	2.5	*Ctenotus vertebralis*
40th	3.4	*Hemidactylus ansorgii*
50th	4.6	*Liolaemus gracilis*
60th	6.3	*Plestiodon fasciatus*
70th	9.2	*Cophosaurus texanus*
80th	14.7	*Acanthodactylus schmidti*
90th	28	*Calotes emma*
95th	52	*Polychrus marmoratus*
99th	720	*Cyclura rileyi*

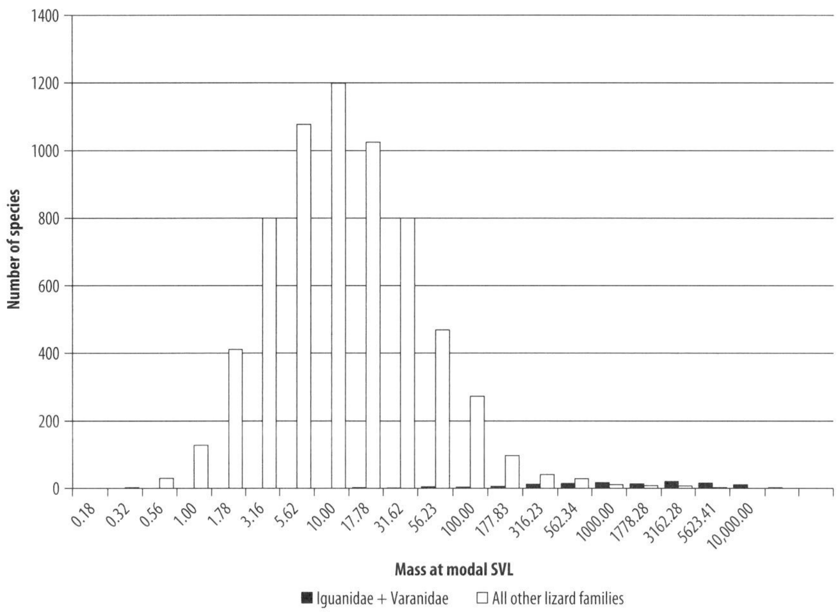

Fig. 2.2. Body mass histogram by $\log_{10}$ units.

Lizards range in mass at modal body length from 0.12 to 7000 g, with a median of 4.63 g. Masses are log-normally distributed; the mean $\log_{10}$(mass) is 0.73 or 5.37 g. A conventional metal paper clip weighs about 1 g. Some readers will be surprised at how small the average lizard is. The public tends to think of lizards such as Green Iguanas (1350 g) or Komodo Dragons (2400 g), but those are the 99th percentile of lizard size. Half of lizards weigh less than 5 g (table 2.3).

A histogram of lizard masses indicates an interesting quirk of lizard sizes (fig. 2.2). The distribution is lognormal except for an extraordinarily long "leg" out to the right. This leg is due to just two families, Iguanidae and Varanidae. When the public thinks of lizards as iguanas and monitors, they are not only contemplating the outer limits of lizard size, they are considering the two anomalous groups. Ecologists are justified in thinking of the giants (Iguanidae and Varanidae) as distinct from all other lizards. In the statistical tests of chapter 3, all tests of mass were conducted with and without the Iguanidae/Varanidae, as annotated.

Functional niches are often considered to be distinguished by ~28% differences in size (Hutchinson 1959, 1978; Roughgarden 1995), as greater size discrepancies change competitive relationships and may alter diet. Because lizards are independent organisms as soon as they hatch or are born, hatchlings/neonates effectively occupy different niches than their parents. Over their lifetime, they grow through several different weight ranges and arguably fulfill as many different niches. Expressed as the $\log_{10}$ of the mass ratio between the largest adults and the smallest hatchlings/neonates, the $\log_{10}$ ratio in lizards ranges from 0.19 (adult 1.55 times the mass of the hatchling *Micrablepharus maximiliani*) to 3.33 (adult 2000 times the mass of the hatchling *Varanus exanthematicus*). The mean ratio is 1.64 (i.e., 43.8-fold; e.g., *Urosaurus ornatus*, *Phrynocephalus versicolor*). Lizards occupy many different niches ontogenetically, some more than others.

Shape

Four common morphological specialties with behavioral-ecological overtones were singled out for special treatment: crests, elongation, limb reduction, and eye reduction. In each case the modal lizard lacks these structures, but a sizable minority possesses them. Conspicuous crests are found in 5.3% of species, but they are absent from most families and possessed by the majority of species in others (e.g., 67% of Corytophanidae). About 11.4% of lizards are elongate (nearly universal in subterranean foragers), though elongation is absent from 32 families. Limb reduction (loss of digits) occurs in 9.2% of species and often accompanies elongation and follows the same familial pattern. About 4.9% of lizards entirely lack limbs; 7% of lizards have such reduced limbs (or none) that they are functionally limbless. Eye reduction is limited to subterranean foragers (3.2% of lizard species have reduced eyes). Unlike with salamanders, there are no lizards that reside permanently in unilluminated caves and therefore have reduced eyes.

Voice

The study of vocal communication in lizards has barely begun, but it seems reasonable to suppose that lizards call to attract or repel conspecifics. Spontaneous callers probably pay a price by attracting some predators. Some stress calls might act to deter predation by startling the predator. Unfortunately, the calling sex is unknown in most cases, and behaviorists often do not know which sex the calling lizards are calling to (repel competitors or attract mates or both?). Thus the best that can be said is whether a species has been known to make sound spontaneously and whether sounds are evoked in stressful situations (usually when grabbed by a herpetologist). One has no choice but to make the false assumption that all species that have not been reported to call do not do so. Two and one-half percent of lizards are known to call spontaneously; 2.2% call when stressed, and 4.3% are known to call under some circumstances.

Tails

Tail length as a proportion of snout-vent length (SVL) ranges 0.03–4.6 and is not normally distributed in lizards, but has two modes: the low mode (0.1, the peak of 275 spp.) and the main mode (1.3, the peak of 6032 spp.). The low mode is attributable to subterranean foragers that anchor their bodies in their tunnels with a stout, often pointed, tail stub. Male lizards typically have proportionately longer tails than their female counterparts; the average ratio of male tails (expressed in relation to their SVL) to female tails similarly quantified was 1.07. That is, male tails average about 7% longer.

Five percent of lizard tails are prehensile, especially in the chameleons (88% of chamaeleonid species). Twenty-nine families (of 45) have no species with prehensile tails. Fifty-seven percent of lizard species are reported to readily break off their tails, though tail breakage is present but unreported in other species. Tail breakage is unknown in some families (e.g., Xenosauridae) and universal in others (e.g., amphisbaenians).

Color

Color can be described in relation to time (color change), sex (sexual dichromatism), and special colors, such as bright blue tails (decoying predators). The latter often has a sexual component, as adults of only one sex may display the bright blue tails. In this section, color change and blue tails are quantified in all lizards; sex differences are emphasized below under "Sexually Differing Traits."

Color Change. About 10% of lizards are documented to be able to change their color or darkness in the space of a few hours or minutes. Some kinds of color change are very strongly associated with certain clades. For example, 57.1% of chameleons have the ability to change not only their basic color, but their pattern as well. Other kinds of short-term color change are much more widely exhibited. Species in 21 of 45 families are documented to darken in certain situations.

Blue Tails. Very few lizards have bright blue tails; overall only 152, or 2.3%, of lizard species have been reported to possess bright blue tails, two-thirds of these in juveniles only. Furthermore, the phenomenon is limited to just a few families; only eight families exhibit more than 1% of species having blue tails.

Sexually Differing Traits

This section describes the traits most directly influencing the ecological relations between the sexes: size, color, and structures (horns, crests, etc.). Other sections describe more tangentially related sex differences; for example, if the sexes differ in their home-range sizes, that is described in the section on activity.

Size

Two measures of sexual size dimorphism (SSD) were tabulated: an objective measure based on reported sex-specific size measurements and a tabulation of unquantified SSD claims. The unquantified SSD claims showed two modes (males moderately larger than females and the converse), presumably due to underreporting of ambivalent data, whereas the objective size comparison exhibited a single mode near males = females. In both cases, the frequency of m>f records exceeded those of m<f ones, with strong ecological correlates associated with the differing directions of SSD (see chapters 3, 4). In contrast, the ratio of m:f relative tail lengths, though it also showed a preponderance of greater length in the males (mean m:f ratio of 1.07), was found to have few strong ecological associations (see chapter 3).

Color

It is difficult to estimate the proportion of lizards exhibiting sexual differences in color (dichromatism), as the absence of sexual dichromatism is likely underreported. Of the 1652 lizard species with an explicit claim, 82% (likely overestimate) were reported to be sexually dichromatic. Of the 6528 lizard species in total, 21% (likely underestimate) were reported to be sexually dichromatic. Of the 1149 lizard species that were documented completely enough to have a foraging mode reported (a group for which sexual size dichromatism or its absence would likely be known), 35% were documented to be sexually dichromatic.

Estimating the proportion of lizards exhibiting seasonal (i.e., "breeding") color runs into the same challenges. Assuming that seasonal color was reported whenever sexual dichromatism was reported (1162 species) yields an estimate of 17.7% showing seasonal color. Assuming that seasonal color was reported whenever foraging mode was known yields a similar estimate of 17.4%, but both of these assumptions are weak and likely underestimate the true value. Assuming that all species for which seasonal color has not been reported lack it yields a lower bound estimate of 6.4%. Seasonal color has strong ecological associations (see chapter 3).

Structures

Sexually dimorphic secretory pores are found in 15% of all lizards but are found in 69% of cordylids and 0% of most families (e.g., chamaeleonids). Half of the 344 species reported to have conspicuous crests have sexually dimorphic elaboration of the crests (generally males have the taller crests).

Sex Rollup Score

The aggregate score of sexual differences (scale of 0–6) exhibited a mean of 0.75 per species; this is likely an underestimate, as sexual differences were not reported for the lesser-known species. The highest mean was for the monotypic Sphenodontidae (4.0), which is a relatively well-known species. The 12 species of crotaphytids averaged 3.17, so the average crotaphytid differs in more than three ways between males and females. In consideration of the limitations, the sex rollup score should be used only to compare groups of species that have similar intensities of reporting. Nonetheless, sexual differences seem relatively common in lizards.

Geography

Lizards are found worldwide, with at least one species (*Zootoca vivipara*) reaching north of the Arctic Circle. The Neotropical biogeographic realm has the most species (1884 spp., Reptile Database 2017), followed closely by the Ethiopian (Africa south of the Sahara: 1388 spp.), Australasian (1274 spp.), and Oriental (1201 spp.), and followed distantly by the Palearctic (604 spp.) and Nearctic (North America: 386 spp.) The three tropical realm counts are likely to be gross undercounts.

Forty percent of lizard species occur on at least one island. For a few species, these island-record lists are exclusively freshwater islands, but most nominally insular lizards are found on saltwater islands that are large and distinct enough to have a recorded island name. More than 2600 species of lizards have colonized islands.

Nearly two-thirds of lizard species (64%) are "endemic," found in only a single political jurisdiction as defined in this work (states/provinces in the United States, Canada, and Australia). Particularly speciose jurisdictions are illustrated in figure 2.3. Roll et al. (2017) recently compiled high-quality maps showing lizard hotspots around the world.

Environment

A histogram of 619 lizard species' precipitation bands (fig. 2.4) indicates that the modal lizard (43% of species) is found in mesic areas (1100–3000 mm/y). A smaller second mode (30% of species) is represented by arid area lizards (151–400 mm/y). A few families mirror this bimodal pattern; Eublepharidae, Gerrhosauridae, Gekkonidae, and Varanidae each have one peak in the arid band and another in the mesic band (though the larger peak of varanids is in the arid), but 41 of 45 families are unimodal, primarily in the mesic band.

Dividing elevational records of lizard species into low (<1200 m), medium (1200–2400 m), and high (>2400 m)

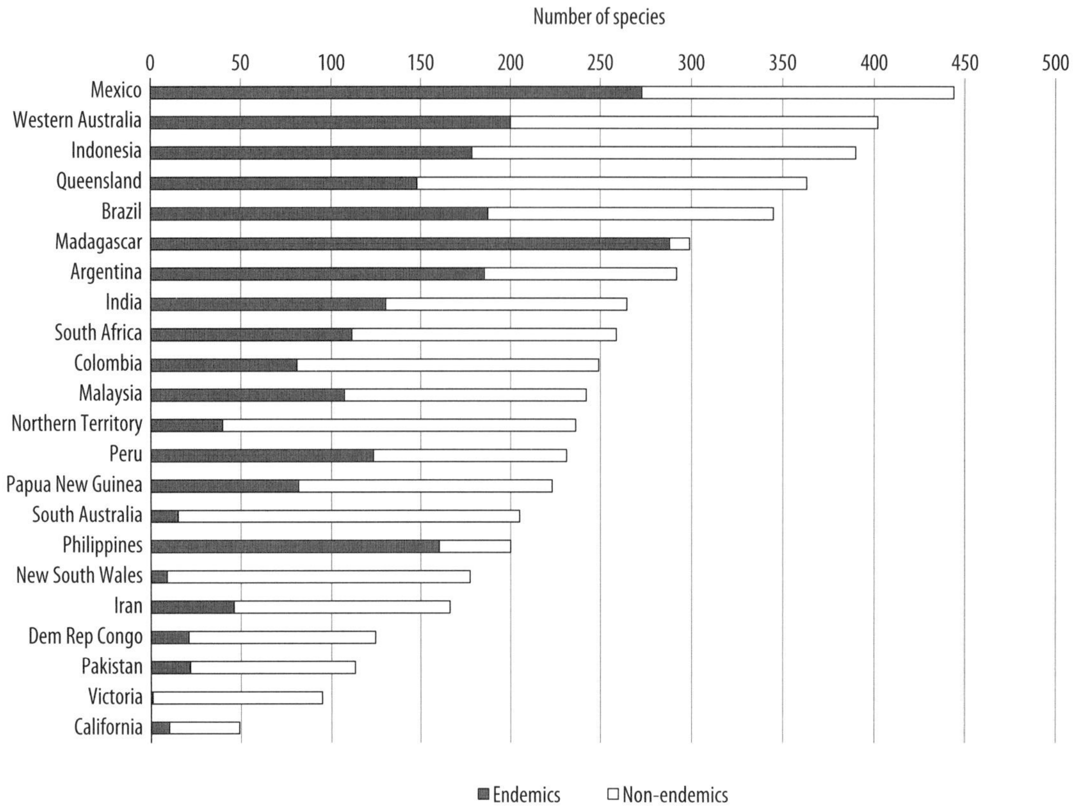

Fig. 2.3. Selected hot- and warm-spot jurisdictions.

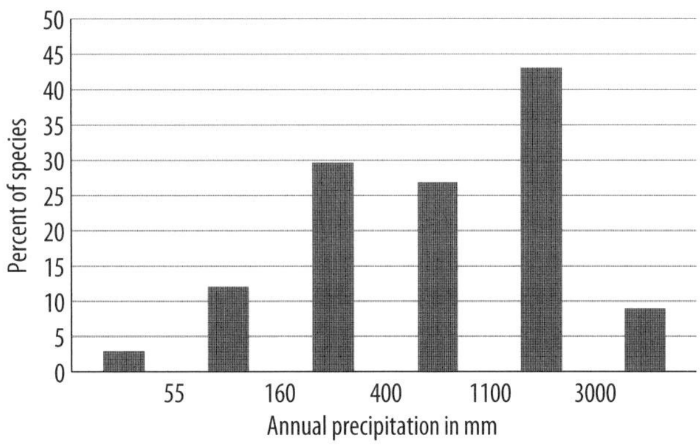

Fig. 2.4. Lizard species (n = 619) distribution by annual rainfall bands (cut points given on x-axis).

elevation bands reveals that the vast majority (82%) of lizard species are found at low elevation, which is about twice the proportion of species found at medium elevation (41%), and more than six times the proportion found in the highest band (13%). Presumably the primary driver of this low-elevation preference is heat. Partially countering this selection pressure is the benefit of higher precipitation often found at higher elevation.

Habitat

In collating natural history data on lizards, no component frustrated me more than that of habitat. There is little agreement and no consensus on which dimensions of habitat are important enough to justify consistent data collection. Appendix A offers some suggestions for approaches that might systematize the quantification of lizard habitats. Of 5393 species whose habitat was described, the keyword "littoral" was associated with 17%, "dunes" or "sand" with 22%, "rock" or "cave" with 41%, "trees" or "forest" with 77%, and "urban" or "commensal" landscapes with 19%. Without background frequencies for these habitats, it is difficult to estimate preferences, if any.

Activity

This section covers all aspects of movements and activity: seasonal inactivity such as hibernation and estivation, daily patterns of activity, movements attributable to food gathering, and social and sex influences on activity.

Inactivity

Seasonal inactivity was mentioned in 576 (9%) of 6528 species, though it is likely to be somewhat underreported. Of the seasons mentioned, the most mentioned was cold weather (511 or 89% of 576 spp.), followed by dryness (21% of 576), and rarely by wet weather (1%).

The modal lizard takes daily refuge under vegetation or surface rocks (53% of 2893 spp. documented for sleep site). A small percentage (4%) digs into loose sand, and the remainder are roughly evenly split among rock crevices (33%), elevated vegetation (26%), and dug or appropriated burrows (22%). The sum of these percentages exceeds 100% because many species exhibit more than one type of refuging behavior.

When a lizard is spotted by a predator (i.e., herpetologist), 43% of 1068 species documented first froze, apparently relying on crypsis, 11% ran away to a site not seen, 6% dove into the sand, 25% fled to a crevice, and 23% hid in vegetation (multiple tactics result in a sum >100%).

A sense of a lizard species' flight distance can be obtained from a tabulation of species reported to be conspicuously wary. This metric does not give an absolute distance, but can be used to compare groups with similar information content. These subjective assessments can be shown to be calibrated to the expectations of the herpetologist for the specific taxon involved: anoles are expected to remain still until closely approached, teiids are not. Of 860 species for which the degree of wariness evoked a comment, 60% were judged conspicuously wary and 40% were not. This upper bound can be compared to a lower bound based on the assumption that all species for which no comment was made were unwary (highly unlikely), which indicates that 8% of species are notably wary.

Activity

Diel activity was recorded in the literature for 4746 species, though it was evident that some of the records were not empirical but inferences from closely related species. Of these 4746 species, 6.8% were reported to be capable of activity at all times (= cathemeral), 1.8% were crepuscular (low light), 66.6% diurnal, and 24.7% nocturnal. The modal lizard species is diurnal.

Only about 2% of the 4746 lizards are documented to forage under electric lights, but they notably include species with all four diel activity patterns (see chapter 3).

Most lizard species are reported to bask (86% of 1358 species for which heat-gaining behavior has been documented), 9.3% are reported to gain heat indirectly (via rock, etc.), and 12.7% to be thermoconformers. The mean body temperature of active lizards was 31.3°C, varying from the anomalous Tuatara (mean = 17.8°C) to the heliophilic teiids (37.2°C).

Credible estimates of the activity or home-range areas have been obtained for only about 216 lizard species. The sample is undoubtedly biased toward the larger lizard species. The sample is a bit difficult to synthesize, as the area sizes are neither normally distributed nor independent of lizard size. Taking the $\log_{10}$ of home-range area in m^2, the mean is 2.386, which represents a home range of 243 m^2.

Foraging Movements

Though many cogent objections have been made to the somewhat cavalier application of the terms "sit-and-wait," "cruise," and "active" foraging (Perry and Pianka 1997; Perry 1999, 2007), the former conveys the useful sense that the lizard generally waits for prey to closely approach it, and the latter does the converse. A newer intermediate category—cruise—is recognized in this work, based on Butler (2005). Cruise foraging is a relatively slow movement that targets prey within striking distance of the lizard. Based on reports of 1149 species, the lizard world is nearly equally divided between sit-and-wait (46.4%) and active foragers (58.2%), with 6.6% cruising. These values add to >100% because many species exhibited multiple foraging modes.

A more quantitative approach to foraging modes is to time the proportion of time spent moving (PTM) or the number of moves per minute (MPM). These are not normally distributed, so they are presented in percentile ranks. The median MPM of 184 recorded species was about 0.645 moves per minute; the median proportion of time moving of 157 species was 0.0735, or about 7%. See the *Meroles* genus account in chapter 5 for examples of lizards that illustrate some difficulties of interpreting MPM and PTM in relation to the classical definitions of foraging modes (active v. sit-and-wait).

The surface on which a lizard was active was reported or inferred for 4893 species, though there is considerable confusion in the literature between the concept of where a lizard forages and where a herpetologist found it. Half (49.6%) of these reports indicated that the lizard had the inclination and ability to climb vegetation or vertical surfaces of rock or masonry. However, the primary active surface for each species was reported to be the ground (47.7%), elevated vegetation (32.0%), hidden in litter or loose soil (12.1%), underground (7.2%), semiaquatic (1.1%), or marine (0.02%).

Social Structuring of Spacing and Movements

It is clear that social relations have not been quantified for most lizards. Of the 427 species for which group composition was recorded, solitariness was specified in 43 (10%), 296 species (69%) were said to be found solely with a presumed mate, 87 (20%) were found in groups that were thought not to be solely relatives, and one (0.2%) was observed in a multispecies group. On a scale of 1–4, with 1 representing solitariness and 4 representing multispecies associations, the mode and median are both 2, representing mate pairs, and the mean is 2.1. Of the mate cohabitations in which the number of females appears to be specified, monogamous pairs outnumbered polygynous groups by about 2:1. About four times as many mate groups failed to mention juveniles as mentioned them. It is hard to know to what extent these group composition differences represent sampling biases. Juveniles are often ignored when present, and if females space themselves, a polygynous male would ordinarily be seen with just a single female. Nonetheless, heterosexual pairs are the group most often observed. Groups are often conspicuous and therefore are likely to be over-reported relative to other social arrangements.

Sex-based Spacing

A common observation is that male lizards chase away other males (at least seasonally) and, less often, females repel other females. The frequencies of these behaviors are difficult to establish for the usual reason that negative observations are underreported. Using the pool of 855 species for which males' social interactions are known, 15% of males were socially tolerant, 82% defended a site seasonally or under unspecified conditions, only 2.2% were reported to defend their entire home range, 1.4% were thought to defend space year round, and only 0.2% were reported to chase off females, at least at some times. Males showed hierarchical social relations in 7.7% of species, and in 7.7% of species, contrary or inconsistent social arrangements were logged.

The comparable female values for 175 species were 26% tolerance, 69% defense of a site, 0.6% defense of home range, 3.4% defense year round, 1.1% expulsion of males, 11.4% hierarchical behavior, and 3.4% inconsistent reports of social arrangements. Home-range and year-round defense may be underreported in both sexes, but the reports of them are notably scarce (cf. Stamps 1977). Social hostility, especially by males, seems to be almost exclusively associated with mating opportunities. Females are apparently more likely than males to be hierarchical, and some kind of defense of space seems to be at least seasonally present in the overwhelming majority of cases in both sexes.

Reproduction

Most lizards are oviparous (83% of 4328 species). Seventeen percent of species are viviparous and about 1% are believed to change modes geographically or contextually. Parthenogenesis has been reported to be a regular attribute of reproduction in 43 species (1% of 4328) and a rare modality in 15 species (0.3%). Evidence for temperature-dependent sex determination (TSD) has been reported in 77 lizard species. If all negative reports are taken to be negative, this represents about 1% of lizards, but the true value is likely to be several-fold larger.

Parental Care

Parental care is reported in 191 lizard species. A lower bound on the proportion of species with parental care would thus be 191 of 6528 lizards, or 3%. Of these 191, 66 species extend parental care after birth or hatching. Parental care is not easy to observe, however, and if one considers just the pool of species that have been studied enough for a foraging mode to have been noted (1149 species), 7.4% of these species are reported to exhibit some sort of parental care. Parental care is likely more difficult to observe than foraging mode (though some parental care reports are based on captive observations), so the true proportion is likely to exceed 7.4%.

Timing of Reproduction

The seasonality of reproduction may reflect the seasonality of the environment as much as the inclination of the lizard species, as many species switch their breeding pattern when introduced to areas with a different climate. Of the 798 lizard species for which reproductive seasonality has been reported, 62% have a relatively short breeding season, 27% breed over six months a year or more, and 22% breed continuously, most with seasonal modulation in output. These values add to more than 100% due to geographic variation and contradictory reports.

Incubation

Incubation duration has been reported for 1050 species, for which the mean duration is 81 days. The shortest durations reported (5 days) are for species that switch between viviparity and oviparity; the longest is for the Tuatara (405 days). However, this interval is not normally distributed and is associated with body mass, so subgroup comparisons are based on the mass-adjusted standardized residuals.

Longevity

Data on longevity are often taken from captivity, which should induce the reader to use extreme caution. If the herpetoculturist finds the ideal combination of conditions, the lizard may well survive years or decades beyond its life expectancy in the wild. Conversely, if the required conditions have not yet been discovered, the reported longevity may be no more than how long it takes the poor creature to die. For example, five species of the genus *Amphisbaena* are reported to live no longer than 0.2 years, presumably a measure of how long it takes these mysterious subterranean lizards to die in captivity (other species in the genus have survived more than 15 years).

The distribution of reported longevities is log normal; thus the mean ages cited herein are mean $\log_{10}$ values, back transformed to years. The average lizard species (5.4 g) has an average maximum reported longevity (n = 879) of 6.9 years, a remarkable value for such small vertebrates. Tuatara are documented up to 83 years and presumed to live far longer. Several *Cyclura* are documented to attain 69 years.

Clutch/Litter Attributes

The number of young in a clutch or litter is not normally distributed, as many species are constrained to clutches of one or two. The average of clutch count midpoints for 2974 species with data is 4.0 egg/young per clutch/litter. The range of brood size midpoints is 1 to 54. The number of young is associated with the size of the lizard and therefore most subsequent analyses in this work rely on standardized residuals of mass-adjusted clutch/litter count (mean of 0; standard deviation of 1). In the context of many other organisms (trees, grass, fish, frogs, turtles, crocodilians, rodents), the mean clutch count of 4.0 is remarkably low. This implies that a hatchling/neonate lizard has an exceptionally good chance of surviving to adulthood. Another group with similarly low fecundity is terrestrial salamanders, for which lizards are the dry habitat equivalent.

Many lizards breed multiple times a year, and the distribution of annual brood frequencies appears to be very poorly understood and often based on aberrant captive situations. Frequency of clutches or litters per year are not well documented, despite the wide reporting. Taking the largest number of broods per year on the 694 species for which this information is reported yields an average of 2.85 broods per year for the average lizard.

Relative Clutch Mass Ratio

Relative clutch mass (RCM), the mass ratio of the clutch/litter to the postpartum/postoviposition mass of the mother, has been recently investigated by Mesquita et al. (2015). My tabulation enumerated RCM ranges for 297 species, for which the mean was 0.226, or the clutch weighed about 23% of the mother's postoviposition mass. RCM has particular value in understanding the instantaneous physical burden placed on the fully gravid/pregnant mother. Mean RCM differs among lizard subgroups (see chapters 3 to 5).

Diet

If one characterizes each lizard species as having a single diet classification, 73% of 2860 reported lizards are primarily invertebrate eaters; 15.4% are omnivores, 4.4% are ant specialists, 4.0% are herbivores, and 3.2% are carnivores (this low value is a key distinction from the other squamate reptiles, i.e., snakes).

Although ant specialization is rare, ant eating is found in 791 species (32%) in 35 families of the 2492 species for which specific dietary items are listed. Ant eating is believed by some to be a feature of the Iguania clade (Schwenk 2000; Pianka and Vitt 2003). It is present in 100% of Iguania families and 38% of Iguania species, but also in 67% of non-Iguanian families and 28% of the non-Iguanian species.

Plant eating is slightly less common than ant eating (26% of 2492 species), but, like ant eating, is much less common as a specialization. Documentation of plant eating rarely mentioned whether the ingested material was largely leaves (requiring special bacteria and gut morphology to digest) or primarily fruits (much energy obtainable without specialized gut morphology and microbiome). It would be a reasonable first approximation to suppose that the specialists are largely folivores and the opportunists to be largely frugivores.

Populations

Population densities (individuals/ha) cannot be compared directly among species that vary greatly in body size. Mean biomass (kg/ha) is a more universal metric, but is log-normally distributed. Taking the $\log_{10}$ of 343 species' mean biomasses provides a lizard average of −0.266, which equates to 0.54 kg/ha. For a 0.2 g sphaerodactylid, this would mean 2709 individuals per hectare. For a 3000 g varanid, this would equate to 0.18 lizards per hectare. Species-mean densities range over six orders of magnitude, from *Phrynocephalus przewalskii* at 0.0002 kg/ha to *Sphenodon punctatus* at 225 kg/ha. Credible single-population density estimates (fig. 2.5) range from 0.0001 kg/ha (*Callisaurus draconoides*) to 1000 kg/ha (the six highest values are all *Sphenodon punctatus*).

Conservation Status

About half (3390 species) of lizards have been formally evaluated for the conservation threats to their population. Half (51%) of these are believed to be in no immediate danger of endangerment or serious decline. Fourteen percent

A

B

Fig. 2.5. Population densities of lizards range over seven orders of magnitude, as indicated in the text. **A,** The lowest credible estimate for a single population is for *Callisaurus draconoides* (Phrynosomatidae). **B,** The highest credible point estimate is for *Sphenodon punctatus* (Sphenodontidae). **A,** L. Grismer; **B,** J. Reardon. (For color versions of all photographs, see color plates following page 86.)

of those reported by anyone to be at risk are rated by the international body responsible for such things, the International Union for Conservation of Nature (IUCN), as extinct (EX), critically endangered (CR), or endangered (EN). An additional 28% of assessed lizards are judged by the IUCN or other competent national or international authorities to be either at lesser risk or that the risk applies to only part of the species range. An additional 8% of lizards are considered to be at risk by some competent authority but not formally classified at the national or international level. Thus half of lizard species overall have probably declined to a point where they have generated conservation concern. Lizard species that have generated conservation concern are herein referred to as "imperiled," a name that combines the worry lists of various authorities, but has no single definition.

Introductions

Only 3.3% of lizard species have been introduced by humans to new localities (average of 0.22 introductions per species, for all lizards). Introductions that demonstrably failed to result in colonization are excluded, but many literature records are ambiguous on this point. My metric for introductions is crude, both due to the paucity of hard data and to the coarse level of my geographic records. For example, a species that was introduced into two different localities within one country would count as only one introduction. Furthermore, there is substantial overlap between country and island records. For example, introductions to island nations list both the country and the island. The longer of the two lists for each species was taken to be the most accurate proxy for the number of introductions of each species. These introductions are overwhelmingly (90%) on islands. Island introductions may be easier to document (few continental range expansions can be firmly ascribed to introductions), but the 90% disproportion is so large as to suggest that islands are more often invaded.

Table 2.4. Lizards collectively accounting for half of 1484 reported introductions

Species	Number of introductions
Hemidactylus frenatus	154
Hemidactylus mabouia	140
Gehyra oceanica	103
Iguana iguana	84
Gehyra insulensis	77
Lepidodactylus lugubris	66
Gehyra mutilata	50
Varanus indicus	50
Anolis sagrei	37

In spite of the coarse reporting of introductions, it is notable that only 214 (3.3%) of 6528 species have been responsible for the introduced populations documented. One percent of lizard species are implicated in 87% of the introductions (by definition, the lizards are not responsible for the introductions—humans are). Furthermore, just nine species are involved in half the 1484 documented introductions. These species, and their introductions counts, are shown in table 2.4 (see also fig. 2.6).

Some of these species are in the pet trade, some have spread as stowaways, one is parthenogenetic (*Lepidodactylus lugubris*), and *Varanus indicus* was spread as a biocontrol agent; six of the nine are nocturnal, but they are all climb-

Fig. 2.6. A few lizard species have been transported and established in dozens of localities by humans. **A**, *Hemidactylus* spp. are the two most frequent invasives, here represented by *Hemidactylus frenatus* (Gekkonidae). **B**, Several *Gehyra* have been introduced, especially to islands; this is the Indian Ocean's *Gehyra mutilata* (Gekkonidae), nearly indistinguishable from the Pacific Ocean's highly invasive *G. insulensis* (Gekkonidae). (See also fig. 4.9B.) **C**, *Iguana iguana* (Iguanidae) is widespread in the pet trade and often released, especially in populated places; this male basking on a bridge in downtown Miami, Florida, is an example. **D**, *Varanus indicus* (Varanidae) was often transported among tropical islands in a futile attempt to control pests of coconut trees. **A**, **D**, B. Lardner; **B**, L. Grismer; **C**, G. Rodda

ing species. Taxonomically, the families and genera with high average introduction rates are almost all attributable to a single species. For example, of the 11 genera with average introduction rates of 2.0 per species or greater, eight high genus rates are due to a single species. Only the genera *Hemidactylus* (nine spp.), *Cyclura* (five spp.), and *Gehyra* (three spp.) have more than one frequent invader. Invasion appears to be highly idiosyncratic, suggesting that those frequent invaders possess some unique characteristics that enable their spread (chapter 3).

Fig. 2.7. *Sphenomorphus maindroni* (Scincidae), a typical lizard. G. Rodda

The Modal Lizard

As a memory aid to capture the essence of lizardness, the database was searched for any lizard that was found near the mode or could be at the mode (to allow for insufficient documentation of individual traits) of 12 key ecological axes cataloged: Family = Scincidae, diurnal, invertebrate eater, mass 4.5–5.6 g, a refugia-anchored diurnal ground lizard, not elongate, tail length 1.23–1.44 × SVL, males larger than females, terrestrial in microhabitat, oviparous, not parthenogenetic, and of middling subjective abundance (2.1–4.0 on scale of 0–5). The only species that qualified on all 12 counts was *Sphenomorphus maindroni* (fig. 2.7), an undistinguished, as specified, skink from the tropical island of New Guinea.

3 Linkages

This chapter dives into the correlational structure found among the 170 variables tabulated in the database. The correlations between traits are needed to begin sketching out the constraints in lizard ecology (e.g., does ant eating hinge on size?), but the relationships are only beginning to be understood and can be baffling in their collective complexity. If bewilderment is a premonition of knowledge yet to come, the contents of this chapter are a fruitful patch of research ground to fertilize. Those readers who would prefer the more straightforward and concrete tabulation of traits by species/genus/family may wish to jump directly to the taxon accounts in chapter 5.

The literature on lizard ecology was scrutinized to characterize each lizard species in about 170 fields, representing 310 conditions. For example, the field "diel activity" has a value chosen from four nominal conditions: diurnal, crepuscular (active in low light), cathemeral (may be active at any time of day), or nocturnal. In this case, the choices are mutually exclusive; in others they are not. For example, a lizard species may be both scansorial (able to climb vertical surfaces) and arboreal (forages primarily in elevated vegetation). All arboreal lizards are scansorial, but not all scansorial lizards are arboreal. Naturally there will be substantial overlap in the lists of species characterized as scansorial and arboreal. The traits scansorial and arboreal are correlated but not equivalent.

Understanding which traits are correlated with other traits is essential for understanding the deep structure of lizard ecology. One needs to know which pieces of the puzzle fit together and which do not. Chapter 4 is devoted to initiating the process of assembling the puzzle; the building blocks and their correlations are found in this chapter.

Understanding which traits are correlated with other traits is also essential for avoiding rookie errors in drawing inferences. For example, one might be tempted to infer causation from the high longevity of island iguanas, but if you understand that longevity is strongly correlated with size, and you recall that island iguanas are exceptionally large, your zeal to connect longevity to island iguanas should be tempered by the realization that their longevity might be at least partially due to their large size. Or perhaps their large size is what makes their great longevity possible. Dunham et al. (1978) provide examples of other rookie errors in reptile ecology.

Understanding which traits are correlated with other traits is also useful for improving analytical methods. For example, a recent paper on the evolution of diurnality in geckos (Gamble et al. 2015) pooled cathemeral and nocturnal lizard species together, presumably to make possible a binary analysis. The pooling, however, relies on the assumption that cathemeral and nocturnal lizards behave similarly. But a careful comparison of diel patterns in relation to other traits (see discussion below under "Time of Day") demonstrates that cathemeral lizard species rarely track the ecology of nocturnal species. Pooling cathemeral and nocturnal species thus muddies the water rather than clearing things up.

The tables in this chapter are a convenient place to look up values for ecological traits of interest (please look up families and genera in chapter 5 and individual species in the database). To take one example, suppose one was interested in the vocalizations of cruise foragers. Table 3.6 summarizes values by foraging mode and, for trait 25 ("calls spontaneously"), one can read that 18.4% of cruise foragers are reported to vocalize—more than double the proportion found in sit-and-wait foragers and nearly ninefold that of active foragers.

Finally, knowing relationships among traits is useful for understanding individual species. For example, an esteemed reptile ecologist recently posed to me the question of why *Anolis cristatellus* possesses crests. As described later in this chapter, possession of crests is strongly correlated with lizards that are arboreal, sit-and-wait foragers, males larger than females, diurnal, ant eating (but not necessarily ant specialists), characterized by high population density and sexual dichromatism. As it happens, *Anolis cristatellus* is not just one, but all of these things. The question of crests in *A. cristatellus* is thus subsumed under much broader questions of why crests are so useful for species that are arboreal, sit-and-wait foragers, males larger than females, ant eating, and so forth.

It is tempting to reach beyond the identification of lizard-wide first-order correlations to the determination of cause and effect. Evolutionary inference is beyond the scope of this work, not only due to space limitations, but also for the many conceptual challenges outlined in chapter 1 under "The Interplay between Ecological and Phylogenetic Causation." Furthermore, for selected traits, causation has already been explored in detail (e.g., Reilly et al. 2007; Losos 2009). Particularly notable causal explorations include those of viviparity (Packard 1977; Blackburn 1982; Shine 1985, 1995; Andrews 2000), diet (Schwenk 2000; Cooper and Vitt 2002; Pianka and Vitt 2003; Vitt and Pianka 2005; Costa et al. 2008), geography (Roll et al. 2017), longevity (Scharf et al. 2015), reproduction and relative clutch mass (Andrews 1979; Vitt 1981; Bauwens and Díaz-Uriarte 1997; Mesquita et al. 2015), home-range areas (Perry and Garland 2002), body size (Meiri 2008; Feldman et al. 2015), sexual size dimorphism (SSD) (Greer 2001; Cox et al. 2003), thermal biology (Cowles and Bogert 1944; Bogert 1959; Fitch 1968; Huey and Slatkin 1976; Heatwole and Taylor 1987; Huey and Bennett 1987; Huey et al. 1989; Meiri et al. 2013), and foraging modes (Perry 2007; Johnson et al. 2008).

Methods Unique to This Chapter

The statistical tests highlighted in this chapter are mostly 2 × 2 contingency tables for nominal variables in which counts of the lizard species exhibiting a certain condition, say cruise foraging, are contrasted against all other lizard species with similar information content but not the cruise-foraging trait. These two groups (cruisers and noncruisers) might, for example, be then subdivided into groups possessing or lacking crests. If crests and foraging mode are correlated, one might expect evidence of an association demonstrated by a significant lack of independence in these four counts. However, if the focal trait (cruise foraging) is near the middle of a continuum represented by conditions present on either side (sit-and-wait foraging is more sedentary, and active foraging is less so), the statistical test will be in effect asking whether cruise foragers differ from the combined pool of more sedentary and less sedentary noncruise foragers. It is as if cruise foragers were gray ink, sit-and-wait foragers white ink, and active foragers black ink. A test to distinguish cruise foragers from the others would be testing gray ink against a mix of black and white ink. Typically, no statistical difference is found between the intermediate condition (cruise foraging) and the straddling background sample ($p = 0.26$ in the case of crests), despite the discovery of strong statistical differences in crest possession between sit-and-wait foragers and the contrasting faster movers ($p < 0.0001$), and between active foragers and the contrasting slower movers ($p < 0.0001$). To fully understand whether crests and foraging mode are correlated, therefore, it is best to examine not just a single statistical test (cruise v. its contrast group), but also tests performed on related but more extreme traits (active and sit-and-wait v. their contrast groups). For this reason, the data in this chapter are often displayed in arrays of related traits.

The tables in this chapter omit nonsignificant test results unless at least two closely related tests produced significant results, or if the uniform absence of an expected statistical relationship was so striking as to warrant discussion. The reader might nonetheless find the >1400 reported tests daunting. The number has been kept to a minimum largely by performing statistical analysis only on potential correlations that had some a priori biological reason for their existence. If none was identified, a prospective test was not performed. Note that automated pairwise comparisons of the 310 conditions would produce more than 96,000 tests, and many surprising correlations would be discovered, if only by chance. The more plausible tests herein executed constitute about 1.5% of these potential first-order correlations. Many valid relationships remain for the devoted lizard natural history sleuth to uncover.

Mass

Mass and size have a pervasive influence on ecology and are vital to understanding lizard ecology. This book involves comparing and contrasting taxa of different sizes, however, and the dominance of size can mask other important relationships. Whenever possible, the size signal was factored them out of correlated analyses of continuous variables (length, ontogenetic range of body sizes, home-range area, incubation duration, clutch count, maximum longevity) by using the standardized residuals of the trait regressed on mass in place of the unadjusted trait. Biomass was used in place of absolute population density when comparing population densities of species of differing body size. For most size-related length measurements, a ratio of trait length to snout-vent length (SVL) was used in place of absolute lengths. Male:female ratios of length ratios were used to quantify sexually divergent traits. Nonetheless, one is left with many nominal variables for which no such size adjustment is readily available (table 3.1). For example, lizard species of conservation concern (table 3.1, trait 39) are 50% heavier than those that are not of conservation concern, and there is no obvious mass adjustment available to equate those two classes.

Table 3.1. Significant correlates of lizard species' mass at modal snout-vent length (SVL). Each mean corresponds to the benchmark given to its left, in units given to its right. For example, in the first trait (1), 24% of data cells have data for 1 g lizards, but 35% of data cells are filled for 100 g lizards. Statistical significance represented by p value ($0.10 < p < 0.05$) or * < 0.05, ** < 0.005, *** < 0.0005. Characterization of strength of correlation, shown beneath statistical significance, is tied to level of measurement (noncontinuous variables shown by relative strength and direction, continuous variables given as r^2; details in chapter 1). For example, in the first trait (1), the relationship was significant at the <0.0005 level, though the $r^2 = 0.07$. Log-normal continuous traits were averaged as $\log_{10}$ values but reported in original units. Sample sizes reflect the pool of species that have data appropriate to the comparison; for example, carnivores are expressed as their percentage among only those lizard species with dietary data. Reference trait number is provided in the first column to facilitate referrals in narrative. "Ig./Var." = Iguanidae and Varanidae. For definitions of traits see chapter 1 or appendix A.

Trait	Variable	Benchmarks	Means	Units (n)	Stat. sign. Strength	Notes
1	Data completeness	Mass = 1 g: Mass = 100 g:	24 35	% of cells filled (6525)	*** 0.07	$p = 3.2E{-}108$ despite tiny r^2
2	Snout-vent length (SVL)	SVL = 50 mm: SVL = 500 mm:	2.6 663	grams (6525)	*** 0.69	
3	Max. mass over min.	Mass = 1 g: Mass = 100 g:	17.7 97	ratio (40)	*** 0.39	
4	Crests	Present: Absent:	17.7 5.0	grams (6525)	*** mod+	Did dinosaurs follow suit?
5	Elongate species	Present: Absent:	2.5 5.9	grams (6525)	*** mod–	
6	Sexes differ in size	m<f: m>f:	4.7 14.1	grams (7 means)	* 0.46	See table 3.3 for details
7	M:F ratio of tail lengths	Mass = 1 g: Mass = 100 g:	1.09 1.05	ratio (716)	* 0.007	
8	Carnivores w/o Ig./Var.	Present: Absent:	24 6.3	grams (2754)	*** strong+	Diff. more w/ Ig./Var.
9	Herbivores w/o Ig./Var.	Present: Absent:	36 6.1	grams (2754)	*** v. str.+	Diff. more w/ Ig./Var.
10	Invert. eaters w/o Ig./Var.	Present: Absent:	4.6 16.2	grams (2754)	*** strong–	
11	Omnivores w/o Ig./Var.	Present: Absent:	18.4 5.3	grams (2754)	*** strong+	
12	Ant spec. w/o Ig./Var.	Present: Absent:	5.5 6.5	grams (2754)	** weak–	
13	Sit-and-wait foragers	Present: Absent:	8.0 5.2	grams (1106)	*** weak+	
14	Active foragers	Present: Absent:	5.3 8.2	grams (1106)	*** weak–	
15	Home-range area	Mass = 1 g: Mass = 100 g:	12 1018	m^2 (216)	*** 0.47	
16	Active in daytime	Present: Absent:	6.4 4.2	grams (4625)	*** weak+	
17	Active at night	Present: Absent:	3.9 6.2	grams (4625)	*** weak–	
18	Vocalization	Present: Absent:	7.3 3.6	grams (311)	* mod+	
19	Conspicuously wary	Present: Absent:	7.7 9.9	grams (827)	* weak–	
20	Sand as pred. conceal.	Present: Absent:	6.1 8.9	grams (1067)	** weak–	Diff. less w/o Ig./Var.

Table 3.1. continued

Trait	Variable	Benchmarks	Means	Units (*n*)	Stat. sign. Strength	Notes
21	Crevice as pred. conceal.	Present: Absent:	10.7 8.1	grams (1067)	* weak+	Diff. less w/o Ig./Var.
22	Male dom. hierarchy	Present: Absent:	30.2 9.8	grams (855)	*** strong+	
23	Arboreal microhabitat	Present: Absent:	6.8 5.3	grams (5173)	*** weak+	
24	Cryptozoic microhabitat	Present: Absent:	2.3 6.5	grams (5173)	*** mod–	
25	Subterranean microhabitat	Present: Absent:	2.9 6.0	grams (5173)	*** mod–	
26	Semiaquatic microhabitat	Present: Absent:	22.4 5.6	grams (5173)	*** strong+	
27	On ground microhabitat	Present: Absent:	6.9 4.8	grams (5173)	*** weak+	
28	Dug or appropriated burrows	Present: Absent:	11.6 4.8	grams (2552)	*** mod+	
29	<55 mm/y rainfall	Present: Absent:	29 7.8	grams (619)	* mod+	
30	55–150 mm/y rainfall	Present: Absent:	12.3 7.6	grams (619)	* weak+	
31	1101–3000 mm/y rain	Present: Absent:	7.0 9.0	grams (619)	* weak–	
32	1201–2400 m elevation	Present: Absent:	6.2 5.3	grams (4246)	** weak+	
33	>2400 m elevation	Present: Absent:	7.3 5.4	grams (4246)	*** weak+	
34	Number of young/event	Mass = 1 g: Mass = 100 g:	1.14 8.08	young (2974)	*** 0.26	
35	Incubation duration	Mass = 1 g: Mass = 100 g:	47 113	days (1050)	*** 0.21	
36	Endemism to one "country"	Present: Absent:	4.7 6.7	grams (6528)	*** weak–	
37	Max. longevity	Mass = 1 g: Mass = 100 g:	3.7 10.6	years (879)	*** 0.21	
38	An introduced species	Present: Absent:	11.7 5.2	grams (6525)	*** mod+	
39	At conserv. risk	Present: Absent:	7.9 5.2	grams (3390)	*** weak+	

One can use the threshold $r^2 > 0.3$ to identify those relationships for which mass has the strongest statistical relationship to continuous variables (table 3.1 trait number in parentheses):

- SVL (2) was expected to have a strong relationship to mass and does. Recall that both axes of this relationship were $\log_{10}$ transformed; thus the relationship is linear.
- Ratio of maximum mass over minimum mass increases sharply with size (3). This suggests to me that there may be a practical lower size limit for lizards, and that making proportionately small juveniles is less constrained for a big lizard than a small lizard. The absence of eyelids in many very small lizards has been related to the challenge of making extremely small eyelid scales or minimizing water loss (Greer 1983). An eyelid, like all parts of a lizard's skin, is protected by a

layer of rigid scales, but an eyelid needs to fold up very tightly as it is opened, and the scales of small lizard eyelids must therefore be microscopic to accommodate the tight folds of an eyelid tucked into the edge of an eye socket of a 0.05 g hatchling. That may not be possible for rigid scales. Relative to its body mass, a lizard's surface area is larger for a smaller lizard, promoting water loss. Vertebrate eyes seem to have a lower size limit (acuity tied to retinal area?), as the eyes of the smallest vertebrates are proportionately larger. Furthermore, transparent materials are apparently comparatively porous to water and the corneal surface can be an important pathway for water loss. The water loss from the surface of a lizard eyeball may constitute as much as 20% of total water loss in a small lizard (Reichling 1957). Larger lizards have less of their surface in their eyes, and they have a greater reservoir of body mass to hold water and address losses.

- Preponderance of m>f size dimorphism increases with size (6). Females are generally expected to be larger than males when selective pressure for high fecundity exceeds the evolutionary benefits to males of monopolizing mates through intimidation of rivals. Are larger lizards less constrained by predation pressures and therefore better able to pursue mate sequestering activities? Many selection pressures impinge on SSD.
- Home-range area increases with size (15), presumably due to basic foraging energetics; larger areas may be needed to feed more tissue.

For noncontinuous variables, the following stand out as being associated with mass differences of more than twofold:

- Crested lizards are larger (4). Crests are often envisioned as a metabolically inexpensive way to appear bigger to rivals or predators. Why would this benefit be larger when the lizard is larger?
- Elongate lizards are smaller (5). Elongation often goes along with leg reduction, which is presumably advantageous for penetrating soil or dense vegetation. Perhaps larger lizards forego locomoting through such small spaces or have more mass to overcome the physical resistance.
- Carnivores (8), herbivores (9), and omnivores (11) are larger; invertebrate eaters (10) are smaller. Digestion of cellulose-bound herbivorous matter is slow, requiring comparatively large gut, and therefore body, volumes. Carnivores eat larger prey than invertebrate eaters.
- Lizard species for which the males form dominance hierarchies (22) are larger. Large size is associated with success in hierarchical domination. Does that mean that species size has been dragged upwards by sexual size selection in hierarchical species? Or that larger lizards are more easily intimidated by their comrades? Or that larger or longer-lived species can better remember the status of their many compatriots and therefore benefit from the stability of established hierarchical relationships?
- Cryptozoids (24) and subterranean foragers (25) are smaller; semiaquatic foragers (26) are larger. Cryptozoids and subterranean foragers can push through soil, sand, and leaves more easily when they are small, but why are semiaquatic foragers larger? Perhaps larger lizard bodies retain heat longer despite the high thermal conductivity of water. Perhaps vulnerability to predation is more size-dependent in the vicinity of water?
- Those that refuge in burrows (28) are larger. As subterranean foragers are small, why are burrowers large? Penetrating soil does not seem to reward large size. The burrow association with large size might be a labeling artifact; perhaps only large lizards make excavations so sizable that they are recognizable "burrows" as opposed to just spaces under rocks?
- Lizards present in hyperdry environments (29) are larger than those that are not. Presumably moisture loss scales with relative skin area, implying that larger lizards, which have a smaller surface-to-volume ratio, suffer relatively less moisture loss.
- Lizards that have been introduced (38) are larger than those that have not. One would expect stowaways to be small and inconspicuous or transported as eggs, but deliberately transported lizards to be larger, as lizards moved intentionally at present are predominantly those in the pet trade (Kraus 2009), and people relate better to larger lizards. Both small and large lizards have been repeatedly introduced, but larger species predominate. See table 3.14 and associated discussion for other factors influencing the frequency of introduction/colonization.

The large number of correlated variables (table 3.1) speaks to the pervasiveness of size as an influence on lizard ecology and behavior. Many more factors are likely to be identified as the information content of the data set improves.

There is an interesting nonlinearity in the relationships between size and clutch/litter counts. The relationship between $\log_{10}$ mass and clutch/litter size is somewhat complicated by the non-normality of lizard clutch sizes, but if one takes clutch-count means of size-stratified samples

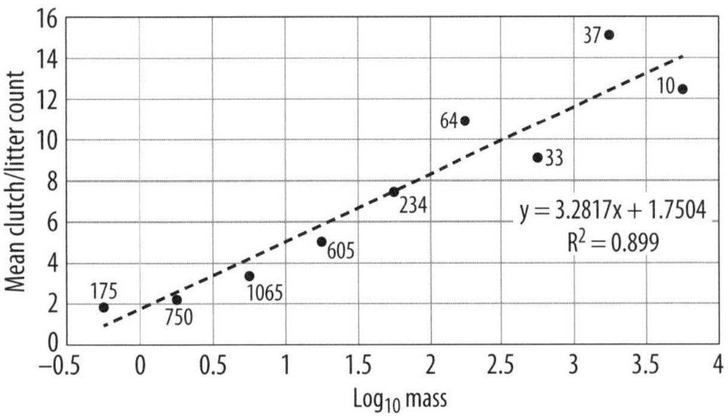

Fig. 3.1. Linear relationship between mean clutch counts for each 0.5 $\log_{10}$ units of lizard mass. Sample size of species is number associated with each point.

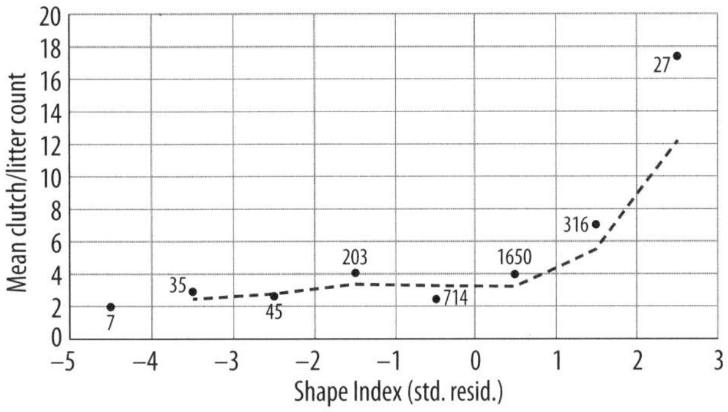

Fig. 3.2. Nonlinear relationship between mean clutch counts for each 1.0 shape index units (standardize residuals of mass on lizard length). Sample size of species is number associated with each point. Line is moving average of two points.

and relies on the central limit theorem to equivalently characterize each cohort of species grouped in 0.5 $\log_{10}$ mass units, one finds that the relationship is strongly linear (fig. 3.1). However, if one uses the same methodology to group lizard species by their shape index values (recall that both axes of shape index are $\log_{10}$ transformed), one gets a strongly nonlinear relationship to clutch/litter counts (fig. 3.2). Because shape index was created to extract the influence of size, the existence of a relationship between shape index and clutch/litter count implies the existence of an additional nonlinear effect on number of young per litter or clutch. This may address the issue of why so many relatively normally proportioned lizards seem constrained to very small clutch counts. However, once a lizard exceeds about one standardized residual of additional mass, much larger litter or clutch counts seem to become possible.

Morphology

Three somewhat disparate morphological attributes are summarized in table 3.2: crests, secretory pores, and body elongation (legs fail to overlap). Of the three, pores have the least distinctive signature, but all three have strong associations.

Statistically strong associations of crests are as follows (table 3.2 trait number):

- Large body mass (1). See discussion under "Mass" above.
- Short-term color change (2). Crests may serve to break up the lizard's outline and promote crypsis, as may color change.
- m>f SSD (3). See discussion under "Mass" above.
- Sexes differ in color (4), but none have blue tails (6). Sexual differences in color are at least loosely associated with SSD, which is also associated with large size and crests. Blue tails do not appear to be primarily sexually selected, even though adult females retain the blue associated with juveniles in many species (blue tails are a baby releaser that inhibits male aggression?). Blue tails are more strongly associated with species that are exposed suddenly (cryptozoids, sand sharks) or that depend upon flight for predator escape (wide-ranging chemosensors). None of those three groups is crested.
- Ant specialist diet (10). This association is mysterious; do the compound eyes of ants fail to recognize the "lizardness" of lizards whose outline is disrupted by crests or spines? Is that why *Moloch* and *Phrynosoma* have a disrupted outline? (For further discussion of this issue, see the *Moloch* and *Phrynosoma* accounts in chapter 5 and "Diet" below.)
- >80% sit-and-wait foraging (12), not active foraging (14). Possibly a visual versus chemical modality association of foraging modes? Perhaps a crested lizard fails to intimidate a rival focused on scent? A crest might also invalidate the longitudinal markings benefit to an active forager fleeing a predator by sprinting (transverse marks, like crests, make it easier for predators to judge a fleeing lizard's position and speed, Jackson et al. 1976, Greene 1988).
- Low movement rates (15). This clarifies that the utility of crests is tied specifically to reduced movements, not just cryptic predator concealment (see trait 23). A moving crest may elevate detection by a predator, but an immobile crested lizard may be cryptic.
- Small mass-adjusted home-range size (16). Note the association below with arboreal microhabitats, where

Table 3.2. Correlates of crests, secretory pores, and body elongation (torso longer than limbs). Conventions as in table 3.1.

Trait	Condition	Crests Stat. sign. Strength	Pores Stat. sign. Strength	Elong. Stat. sign. Strength	Units Contrast (*n*) Notes
1	Mass	17.7 *** mod+	5.1	2.5 *** mod–	grams Others w/ mass (6525)
2	Color change	44.2 *** strong+	16.2 *** weak+	0.3 *** mod–	% of focal that change All lizards (6528) See table 3.4 for details
3	Sexes differ in size	0.86 *** strong+	0.26 * strong–	–0.35 *** strong–	Score on –2 (m≪f) to 2 (m≫f) scale Others w/ SSD scores (1471) See table 3.3 for details
4	Sexes differ in color	62.2 *** strong+	24.5 ** weak+	8.0 *** mod–	% of focal w/ m:f color diff. All lizards (6528)
5	Seasonal color	11.6 *** weak+	10.9 *** weak+	4.4 * weak–	% of focal that change seasonally All lizards (6528)
6	Blue tails	0 ** weak–	2.4	4.8 *** weak+	% of focal w/ blue tails All lizards (6528)
7	Carnivores	1.3	1.0 *** weak–	2.7	% of focal that are carnivores Others w/ diet data (2860)
8	Herbivores	6.2	3.7	0.3 *** weak–	% of focal that are herbivores Others w/ diet data (2860)
9	Invertebrate eaters	65.2 * weak–	74.2	88.0 *** mod+	% of focal that are invert. eaters Others w/ diet data (2860)
10	Ant specialists	11.5 *** mod+	5.2	1.6 ** weak–	% of focal that specialize on ants Others w/ diet data (2860)
11	Omnivores	15.9	15.9	7.3 *** mod–	% of focal that are omnivores Others w/ diet data (2860)
12	Sit-and-wait foragers	82.6 *** strong+	65.3 *** mod+	20.9 *** mod–	% of focal that are sit-and-wait Others w/ foraging data (1149)
13	Cruise foragers	9.3	10.6 * weak+	7.8	% of focal that are cruisers Others w/ foraging data (1149)
14	Active foragers	18.6 *** strong–	36.5 *** mod–	84.3 *** mod+	% of focal that are active foragers Others w/ foraging data (1149)
15	Moves per minute (MPM)	31.8 *** strong–	38.7 ** mod–	59.0	Mean percentile (median = 50)
16	Home-range size (mass adj.)	–0.81 *** strong–	0.26 * mod+	0.50	Std. resid. Others w/ home ranges (216)
17	Active at any time (cathemeral)	2.6 0.055 weak–	10.8 *** weak+	14.2 *** mod+	% of focal cathemeral Others w/ activity data (4746)

Table 3.2 continued

Trait	Condition	Crests Stat. sign. Strength	Pores Stat. sign. Strength	Elong. Stat. sign. Strength	Units Contrast (*n*) Notes
18	Active in low light (crepuscular)	0.3 * weak–	1.2	4.9 *** weak+	% of focal crepuscular Others w/ activity data (4746)
19	Active during day	96.4 *** mod+	48.0 *** mod–	55.2 *** weak–	% of focal diurnal Others w/ activity data (4746)
20	Active at night	0.6 *** mod–	40.0 *** mod+	25.8	% of focal nocturnal Others w/ activity data (4746)
21	Vocalizes spontaneously	0.3 ** weak–	6.5 *** mod+	0.3 *** weak–	% of focal that calls spontaneous All lizards (6528)
22	Conspicuously wary	0.39 *** strong–	1.53 *** mod+	1.91 *** strong+	0–2 (not wary–wary) score Others w/ wariness scores (865)
23	Crypsis as primary predator concealment	67.9 *** mod+	38.0	14.7 *** mod–	% of focal that rely on crypsis Others w/ pred. conceal. data (1067)
24	Flight as primary predator concealment	5.7 * weak–	12.3	4.0 0.052 weak–	% of focal that rely on flight Others w/ pred. conceal. data (1067)
25	Sand as primary predator concealment	0 *** mod–	3.5	32.0 *** strong+	% of focal that rely on sand Others w/ pred. conceal. data (1067)
26	Crevice-burrow as predator concealment	9.3 *** mod–	35.1 ** mod+	25.3	% of focal that rely on crevice Others w/ pred. conceal. data (1067)
27	Arboreal microhabitat	77.5 *** strong+	46.7 *** mod+	5.3 *** mod–	% of focal arboreal Others w/ microhabitat data (5175)
28	Cryptozoic microhabitat	0 *** mod–	5.7 *** weak–	52.2 *** strong+	% of focal cryptozoic Others w/ microhabitat data (5175)
29	Subterranean microhabitat	0 *** weak–	2.0 *** weak–	11.3 *** weak+	% of focal subterranean Others w/ microhabitat data (5175)
30	Semiaquatic microhabitat	2.2 0.08 weak+	0.5 0.095 weak–	1.6	% of focal semiaquatic Others w/ microhabitat data (5175)
31	Terrestrial microhabitat	20.3 *** mod–	45.1 0.097 weak–	29.6 *** mod–	% of focal that forage on ground Others w/ microhabitat data (5175)
32	Spp. native on any island	54.9 *** weak+	38.2	44.6 * weak+	% of focal spp. on islands All lizards (6528)
33	Longevity (mass adj.)	–0.42 *** strong–	0.26 *** mod+	0.37 *** mod+	Std. resid. Others w/ longevity (879)
34	Species that have been introduced	9.0 *** weak+	5.5 *** weak+	2.4	% of focal introduced All lizards (6528)

Table 3.2 continued

Trait	Condition	Crests Stat. sign. Strength	Pores Stat. sign. Strength	Elong. Stat. sign. Strength	Units Contrast (*n*) Notes
35	Population density	1.39 * mod+	0.52	0.21 * weak–	kg/ha Others w/ densities (343)

home ranges are often small (except for gliders). Nongliding arboreal species often have very small home ranges (see remarks below under "Color" and "Microhabitat" and in chapter 4 under "Diurnal Canopy Walker").

- >96% diurnal (19), <1% nocturnal (20). Crests are a size-amplifying ruse or crypsis aid that works only in the light.
- Not conspicuously wary (22). Crested lizards appear to rely on crypsis rather than flight (24).
- Rely on crypsis for predator concealment (23), rather than sand (25) or crevices (26). Crests break up the quadrupedal outline (a benefit for crypsis). Crests would be a physical impediment for penetrating sand or crevices.
- >75% arboreal (27), none are cryptozoic (28) or subterranean (29), few are terrestrial (31). One can easily imagine that a crest would be a less conspicuous visual edge to a predator (but not to a close-range competitor) in the visually complex environment of arboreal foliage (v. the terrestrial environment). As noted immediately above, crests would be a physical impediment for penetrating cryptozoic or subterranean environments.
- Short longevity (33). Short longevity may reflect higher mortality (crests may have social selection benefits paired to natural selection costs in some cases—possibly not in ant specialists). For example, a large crest may enhance the social dominance of a territorial male, but its territorial patrolling (with that large flag waving on its back) may draw predators to itself and indirectly to the objects of its affection.
- High population density (35). High population density of crested species argues against the higher mortality hypothesis outlined above. However, arboreal species may achieve a comparatively higher population density (per surface area), as their numbers are spread over three dimensions instead of the two dimensions associated with terrestrial species.

The notable associations of pores are as follows (table 3.2 trait number):

- Sexes rarely differ in size (3). Why would sexual size monomorphism be associated with secretory pores? Perhaps there is a generally inverse relationship between dependence on visual and chemical senses. Pores appeal to chemosensors, but SSD appeals to more visually oriented lizards? Note that this possibility conflicts with the hypotheses in the next bullet.
- Sit-and-wait foraging (12) elevated; active foraging (14) infrequent. Sit-and-wait foragers are intrinsically visual (they wait until the prey is in view). Active foragers are associated with chemosensation. Possibly pores are used to mark key sites of sit-and-wait foragers, and active foragers simply do not defend space.
- Infrequent moves (15). The association with pores may be indirect, for example via pore-bearing species' predilection to forage by sit-and-wait (12).
- Large home-range size (16). This too conflicts with the previous two bullets, as one associates especially small home ranges with sit-and-wait foraging. Pores are especially prevalent among terrestrial lizards that defend cracks or other sites. This would appear to contradict the association of pore-bearing species with large home ranges, but perhaps the pore-bearers have large home ranges, but scent-mark only key sites.
- 40% nocturnal (20); avoids diurnality (19). Nocturnality is a recurring association with pores and suggests a reliance on nonvisual cues, such as pore secretions (though Alberts [1989] showed a possible visual component to some pore secretions).
- Vocalizes spontaneously (21). Possibly an indirect association via nocturnality, though cathemeral species are more likely than nocturnal ones to vocalize (table 3.7, trait 31).
- Conspicuously wary (22). Are pore-bearers conspicuously wary because they do not rely on conspicuous visual signals for social dominance? Is showing itself visually not part of pore-bearer display behavior?
- For predator concealment, 35% rely on crevice-burrows (26). Crevice-burrow use fits with chemically mediated spatial defense and sit-and-wait foraging. It does not fit with the association between burrow use and large size, as pore-bearers are of average size (1).

- Many arboreal (27). There is a strong negative relationship between crests and pore-bearing, but only in arboreal species (data not in table). In chapter 4, it is reported that diurnal canopy walkers have the lowest incidence of pores (2.9%) and nocturnal canopy walkers have the highest (54.4%), suggesting that pore-bearing and crest-bearing species seem to split the canopy stratum by time of day.
- Extended longevity (33). Unless the relatively retiring pore-bearing lizards live longer due to their secretive ways and subdued markings, it is difficult to see how extended longevity fits in with the other associations of pores.

The overall impression produced by this list of pore associations is that our understanding of pores in lizards is under-developed.

Highlights of elongate lizards include the following (table 3.2 trait number):

- Small size (1), the better to fit through holes.
- <1% able to change color (2). Why bother if the lizard is underground or under cover?
- m>f SSD is rare (3). Why bother if the lizard is underground or under cover?
- Only 8% have sex differences in color (4). Why bother if the lizard is underground or under cover?
- 88% have diet of invertebrates (9), few are omnivorous (11). Omnivory would ordinarily require coming out into the open.
- >80% forage actively (14), few sit-and-wait (12). This may be due to the low mobility of prey in underground and under-cover microhabitats.
- Active at any time of day: cathemeral (17). Why worry about the time of day when one is foraging underground or under cover? Diel activity sampling of elongate species may also be weak due to their often cryptozoic behavior.
- Conspicuously wary (22). It is less clear what wariness implies in a lizard that is routinely concealed. One expects a suddenly exposed cryptozoid to dive back into cover, but the same logic should apply to subterranean foragers, and those are not often described as conspicuously wary (chapter 4).
- 32% (a strong positive relationship) rely on sand swimming for escape (25) rather than crypsis (23), for which there is a moderate negative relationship. Again, these lizards are often in a nonvisual microhabitat. Elongation physically facilitates sand swimming.
- Over half are cryptozoic (28) rather than arboreal (27) or terrestrial (31). Arboreal and terrestrial are often very visual microhabitats.
- High longevity (33). Is predator pressure especially muted in cryptozoic microhabitats, thereby allowing these small slender lizards to live longer? Subsequent table 4.1 trait 70 for sand sharks mildly supports this; subterranean foragers, in contrast, are usually not deemed "elongate" because elongation is undefined in species that lack two or more limbs.

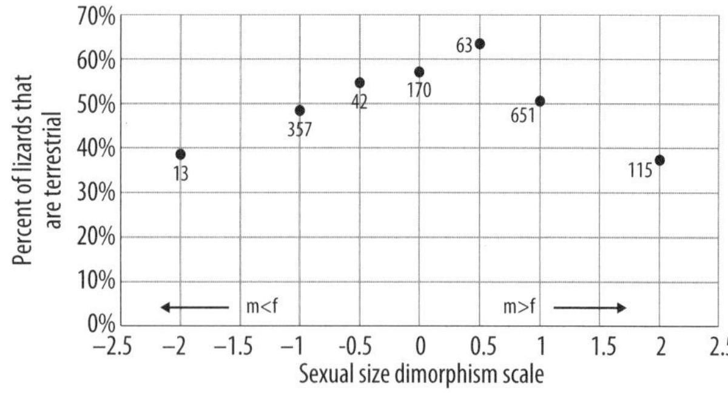

Fig. 3.3. Nonlinear relationship between proportion of lizards that are terrestrial and sexual size dimorphism (SSD), as expressed on a scale of −2 to 2. At zero on the horizontal axis, the sexes are equal in size, and the proportion of terrestrial species is near maximal. The number of lizard species contributing to each SSD class mean is shown by the adjacent number.

Sexual Size Dimorphism

The relationships of other variables to SSD are unusual in that some respond to the absolute magnitude of the m:f size discrepancy, whereas others respond to the direction of the discrepancy (m<f v. m>f). For example, terrestrial species are more frequent among those with minimal discrepancy in size, whether m<f or m>f (fig. 3.3), whereas population density is positively associated with SSD across the entire spectrum of scores (fig. 3.4). The V-shape illustrated in figure 3.3 can have a minimum at the m=f point (e.g., crests) or a maximum (terrestriality). Linear relationships can be either positive (fig. 3.4) or negative. The V-shaped functions were tested statistically by linear regression on the absolute value of the SSD score, a test which assumes the inflection point is $x = 0$. If the $x = 0$ assumption is not fully met (e.g., fig. 3.3), this test will understate fit. However, if the $x = 0$ assumption is faulty, as-yet-unrecognized phenomena are in play and warrant further investigation.

Complicating the documentation of these four possibilities (linear +, linear −, V-up, V-down) is that the points at the extreme SSD scores (2 and especially −2) are smaller samples, and they sometimes include conflicted ratings

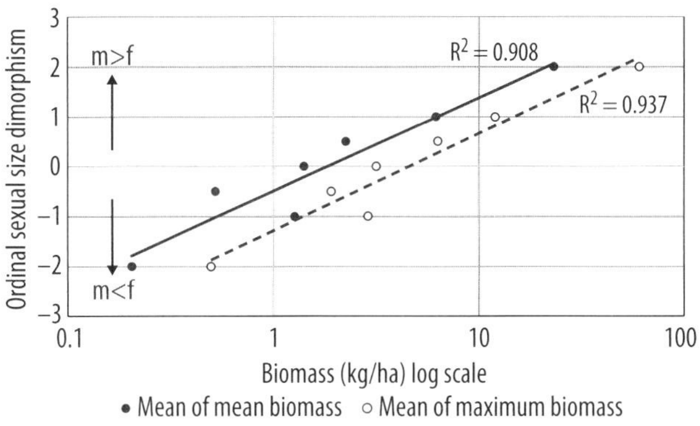

Fig. 3.4. Linear relationship between log population density and sexual size dimorphism. Note that these axes are rotated from those of the preceding figure. Because the direction of causation is unknown, viewing the relationship with alternate orientations of axes can be heuristic.

of the species (by different observers or of different populations). Treatment of conflicted responses can produce large proportional differences in small samples. For regressions using individual species as data, this heterogeneity exaggerates the influence of species at the SSD score=1 portion of the spectrum. Thus for every relationship considered, each species was evaluated with and without equal weighting for each SSD class, and if necessary broad trends were confirmed by conflating the sample into two classes in each dimension or comparing the mean SSD scores of each focal group to the aggregate of the others. If there were an appreciable fraction of conflicted scores, the samples were analyzed including or excluding conflicted species scores, with greater confidence in the analyses with conflicted scores omitted. If these two to seven analyses per trait gave consistent results, the consistent pattern is reported in table 3.3. The tabular data report sample sizes of seven (indicated in table 3.3) if based on means of SSD classes and much higher values, also indicated, if based on individual species.

The axes strongly correlated with SSD were legion: tail length, elongation, herbivorous diet, omnivorous diet, carnivorous diet, sexual dichromatism, crepuscular activity, cathemeral activity (any time of day), arboreal microhabitat, cryptozoic habits, terrestrial microhabitat, population density, conservation risk, and insularity. The other listed traits appear less important only by virtue of the strength of relationship in the traits listed above. The strength of most relationships and the lack of ready explanation for the differences between response to the magnitude of the size discrepancy (V-shaped function) and the gender direction of the size difference (linear function) suggests that further investigation of sexual size differences will be fruitful.

It may be useful to itemize traits that stand out in three regions of the spectrum (m<f, m=f, m>f). Traits most prominent in circumstances where females are much larger than males are as follows (table 3.3 trait number):

- m:f tail length ratio (8): as males get larger than females, their tails become proportionately shorter relative to those of females. This sounds like a developmental constraint (easier to shift proportions than absolute size). It does not hold in all families (notably skinks; see "Scincidae" in chapter 5).
- Invertebrate diet (13). This conflicts with the generalizations that most lizards are invertebrate eaters and most lizards are m>f. Evidently there are statistical interactions at play.
- Lizards active at any time of day (18). There is much to be learned about the unique cathemeral lizards.
- Cryptozoids (25). Cryptozoids are generally slender and small; this may put selection pressure on large female size to accommodate developing eggs (see also "Reproduction Allocation" below). Alternately, it may reflect an advantage accruing to smaller males that can locomote more easily in cryptozoic habitats and may be more nimble at finding receptive females hidden within a dense matrix. The low frequency of group living in cryptozoids (table 3.9, trait 38) and paucity of polygyny reports in cryptozoids (see "Litter Cryptozoid" in chapter 4) suggest that mate finding/guarding may be more salient than social dominance in this group.
- Mainland species (27). Mainland species are mostly at low population density and perhaps are less subjected to strong sexual selection.

Traits reaching maximum value where the sexes are equal in size are as follows:

- Lizards with secretory pores (10). This might suggest that pores are used to communicate with both sexes equally, but that is inconsistent with the unquantified observation that almost all the pores cataloged were more elaborated in one sex. Do chemical means of communication take over when sizes are equal? Or does monogamy (and therefore lack of SSD) occur at low densities, for which olfactory signals are optimal for communicating without direct contact?
- Nocturnal species (21). This suggests that if prospective mates or sexual rivals cannot easily be seen, there is less reward in towering over them.

Table 3.3. Correlates of sexual size dimorphism. Benchmarks are shown with minimum expression of the trait listed above and the maximum expression listed below, given as size relationship followed by ordinal score in parentheses. For example, the mean species mass (trait 1) of lizards is least when males are slightly or equivocally smaller than females, and species' mean mass is greatest when males are more than twice the size of females (direction of causation not determined or assumed). Correlations with $n = 7$ based on means of the 7 ordinal scores. The x-axis benchmarks given for the other traits are not applicable to trait 14, which instead compares the means of ant specialists and non–ant specialists. Other conventions as in table 3.1.

Trait	Variable	Shape of function	Min bench. Max bench.	Means	Units (*n*)	Stat. sign. Strength
1	Mass	V pointing down	m<=f (−0.5):	4.5	grams	*
			m≫f (2):	33.9	(7)	0.46
2	Shape index	Linear +	m≪f (−2):	−0.18	std. resid.	***
			m≫f (2):	0.75	(1471)	0.17
3	Elongate lizards	Linear –	m≫f (2):	0	% of spp. elongate	**
			m≪f (−2):	26	(7)	0.89
4	Tail length (× SVL)	Linear +	m≪f (−2):	1.02	ratio	*
			m≫f (2):	1.82	(7)	0.78
5	Color change (any)	V pointing down	m=f (0):	20	% changeable	*
			m≪f (−2):	62	(7)	0.61
6	Color change (advanced)	V pointing down	m=f (0):	3	% adv. change	**
			m≪f (−2):	55	(7)	0.73
7	Sexes differ in color	Linear +	m≪f (−2):	11	% dichromatic	***
			m≫f (2):	85	(7)	0.94
8	M:F ratio of tail lengths	Linear –	m≫f (2):	1.01	Ratio	***
			m≪f (−2):	1.15	(431)	0.08
9	Crests	V pointing down	m=f (0):	4	% crested	*
			m≫f (2):	32	(7)	0.64
10	Secretory pores	V pointing up	m≫f (2):	8	% w/ pores	*
			m=f (0):	32	(7)	0.52
11	Carnivores	Linear +	m≪f (−2):	0	% carnivores	*
			m≫f (2):	6	(7)	0.75
12	Herbivores	Linear +	m≪f (−2):	0	% herbivores	**
			m≫f (2):	6	(7)	0.85
13	Invert eaters	Linear –	m≫f (2):	50	% invert eaters	*
			m≪f (−2):	90	(7)	0.62
14	Ant specialists	Elev. diff.	ant spec.	0.13	mean score (−2 to 2)	*
			not ant spec.	0.41	(1126)	mod–
15	Omnivores	V pointing down (assym.)	m=f (0):	8	% omnivores	***
			m≫f (2):	33	(1126)	strong+
16	Cruise foragers	Linear –	m≫f (2):	5.4	% cruisers	**
			m<f (−1):	14.5	(548)	mod–
17	Home-range area (mass adj.)	Linear –	m≫f (2):	−0.6	std. resid.	***
			m≪f (−2):	1.4	(162)	0.07
18	Active at any time (cathemeral)	Linear –	m≫f (2):	3.5	% cathemeral	**
			m≪f (−2):	15.4	(7)	0.88
19	Active in low light (crepuscular)	Linear –	m≫f (2):	0.8	% crepuscular	**
			m<f (−1):	2.4	(6)	0.93
20	Active in daylight	Linear +	m<f (−1):	66	% diurnal	***
			m≫f (2):	96	(1371)	0.07
21	Active at night	V pointing up	m≫f (2):	0	% nocturnal	*
			m=f (0):	20	(7)	0.55
22	Reside in groups	Linear +	m≪f (−2):	4	% living in groups	*
			m≫f (2):	22	(7)	0.64

Table 3.3 continued

Trait	Variable	Shape of function	Min bench. Max bench.	Means	Units (*n*)	Stat. sign. Strength
23	Males have hierarchies	Linear +	m≪f (−2): m≫f (2):	50 97	% hierarchical (7)	* 0.61
24	Arboreal microhabitat	V pointing down	m=f (0): m≫f (2):	25 55	% arboreal (7)	* 0.74
25	Cryptozoic microhabitat	Linear −	m≫f (2): m≪f (−2):	0.5 18	% cryptozoids (7)	** 0.90
26	Terrestrial microhabitat	V pointing up	m≫f (2): m=f (0):	38 63	% terrestrial (7)	** 0.85
27	Species native on any island	Linear+	m≪f (−2): m≫f (2):	15.4 83.2	% on islands (7)	** 0.89
28	Oviparous species	Linear +	m≪f (−2): m≫f (2):	48 100	% oviparous (7)	* 0.68
29	Population density	Linear +	m≪f (−2): m≫f (2):	0.2 23.3	kg/ha (7)	*** 0.91
30	At conservation risk	V pointing down	m=f (0): m≪f (−2):	37 85	% at risk (7)	** 0.80

- Lizards foraging primarily on the ground (26), where every plant or hillock blocks vision. Perhaps there is some relationship between long sight lines (arboreal or scansorial) and the power of visual intimidation through size differences.

The hypotheses floated above concerning m=f maxima strike me as unusually weak. Let us see if that weakness is also reflected in the traits that are at minimal value where the sexes are equal in size:

- Lizard species with crests (9). If crests are used to magnify sexual size differentials, the reward may be minimal when the size difference is minimal.
- Omnivores (15). A connection between omnivory and magnitude of SSD is not obvious to me. Perhaps omnivory is favored in resource-poor environments (life is hard; eat what you can), which also tend to have low population densities and therefore fewer opportunities for social selection.
- Arboreal species (24). Potentially this arboreal minimum reflects overlap between apparent selection pressures to be strongly m>f in diurnal arboreal (especially trunk and trunk-ground) species, and m<f in some nocturnal species and gliders.
- Lizards at conservation risk (30). If strong SSD reduces effective population size relative to monomorphic species, and if low population density were associated with species at conservation risk, this association would be plausible. However (see discussion under "Imperiled Lizards" below), the second premise (endangerment ~ low population density) was not supported.

The hypotheses floated to explain the patterns associated with m=f minima also strike me as feeble.

Traits most prominent in circumstances where males are larger than females are as follows:

- Large size (1). This association is often observed. Does sexual selection for large male size drag up the mean species size? Or is there greater scope for intimidation of rivals when the species is larger?
- Stocky build (2). See preceding point.
- Long tails (4). Compared to overall body size, long tails may be relatively inexpensive to grow. If the tails are crested, they may inexpensively enhance the illusion of great size.
- Crests (9). Enhance the illusion of greater size.
- Carnivores (11), herbivores (12), omnivores (15). Carnivores must be big to consume primarily vertebrates. Herbivores (and by extension omnivores) need to be large to accommodate large lengthy guts for the digestion of plants. It is less clear in these particular cases why large size produces sexual selection for large male size. In the case of herbi-/omnivores, there is presumably less natural selection on foraging success (most of the time leaves are pretty easy to find and subdue). Does that "unshackle" sexual selection? This argument would not seem to hold for carnivores.

- Arboreal foragers (24). Arboreal dexterity would seem to be enhanced by small size, but perhaps dexterity is not as critical to reproductive success as the ability to intimidate rivals in a highly visual 3-D environment.
- Diurnally active (20). The association with diurnality underscores the likely importance of visual cues in the evolution of m>f SSD.
- Lizards in groups (22). Lizards in groups have more scope for enhancing reproductive success through mate or territory monopolization.
- Males with hierarchies (23). Hierarchical dominance belongs to the massive.
- Island species (27). Islands are likely to have higher population densities and therefore more opportunities for polygyny.
- Oviparous species (28). Oviparity tends to be more frequent in benign climates, such as maritime islands or tropical sites where population density is not constrained by abiotic factors.
- High population density (29). High population density is a common denominator in many species where social selection prevails.

In addition to the traits listed in table 3.3, traits tested pairwise for their correlations and found lacking statistical significance were mass-adjusted longevity, sit-and-wait foraging, active foraging, subterranean foraging, semiaquatic microhabitat, seasonal expression of color, wariness, and clutch size. The absence of a longevity effect goes against the conventional wisdom that strongly sexually selected species pay a survival cost (Darwin 1871). Perhaps there are countervailing or correlated factors that mask a direct effect on longevity.

Color

Table 3.4 is unusual in that the attributes portrayed side by side are not alternatives to one another, and none of them excludes any others. Thus the primary objective of this portrayal is not to compare the mean values to others on the same line, but to look for emergent linkages among traits, especially in the strength and direction of the reported correlations. Thus, for example, the direction (positive or negative) of the correlations of traits with the last two "colors"—any metachrosis and advanced metachrosis—are the same in 26 of 27 cases, though the strength of the correlation differs in 6 of 27. This concordance in direction suggests that the factors causing "any" metachrosis (typically darkening by nocturnal lizards during the day or any lizards in cold weather) may also provide the impetus for advanced metachrosis (alteration of hue or pattern in addition to darkening). The two metachrosis categories are here grouped under the name "changing colors."

The lone exception to the 26 matches in direction of the changing color correlations was secretory pores (trait 2). Among lizards expressing any metachrosis, 16.2% have pores (a weak positive relationship), whereas only 4.5% of the advanced color changers (a weak negative relationship) have secretory pores. This discrepancy suggests that darkening is especially prevalent among pore-bearing lizards. A direct comparison (not reported in table) indicates that metachrotic pore-bearing lizards are at the darkening level of metachrosis in 72% of the 157 species, whereas other metachrotic lizards (i.e., not pore-bearing) are limited to the darkening level in only 43% of the 695 species, a strong effect ($p < 0.0001$). The quantitative discrepancy in frequency between "any" and "advanced" metachrosis is also striking in nocturnal species (16), and most pore-bearing lizards are nocturnal or cathemeral (see table 3.7, trait 6).

Pores are presumed to transmit chemical signals and as such might not be expected to synergize with vision, but note the weakly positive relationship and elevated frequency of pores under the "sexes differ in color" heading in table 3.4 (trait 2, first column of means).

Another emergent pattern in table 3.4 is the parallelism between sexual dichromatism and seasonal color. Only one discrepancy in correlation direction is reported: arboreal lizards (22) are positively correlated with sexual dichromatism but weakly negatively correlated with seasonal color. No obvious association leaps out to account for this solitary divergence. These two generally parallel categories of color as here grouped as "sexual colors." Sexual colors normally act in parallel and changing colors normally align with each other, but the middle column ("blue tails") is independent of the other four.

The associations between individual enumerated traits and sexual colors, blue tails and changing colors are tabulated below. Notable associations with sexual colors (table 3.4 trait number) are as follows:

- Negative correlations with elongate (1). Elongate species tend to occur in nonvisual environments.
- Positive correlations with crests (3). Crested species tend to occur in strongly visual settings.
- m>f SSD (4), at least for dichromatism. In both size and color, the bearer may be trying to impress potential partners and intimidate rivals.
- Positive correlation with ant consumption (8) and omnivory (9), but not invertebrate eating (7). The

Table 3.4. Correlates of lizard species' colors. Any color change (= metachrosis) includes darkening, hue change, and pattern change. "Advanced color change" includes only the latter two. Other conventions as in table 3.1.

Trait	Condition	Sexes differ in color	Seasonal color	Blue tails	Any color change	Adv. color change	Units Contrast (*n*) Notes
1	Elongate lizards	4.4 *** mod–	7.8 * weak–	23.7 *** weak+	0.3 *** mod–	0 *** mod–	% elongate All lizards (6528)
2	Secretory pores	24.5 ** weak+	10.9 *** weak+	2.4	16.2 *** weak+	4.5 * weak–	% w/ pores All lizards (6528)
3	Crests	62.2 *** strong+	11.6 *** weak+	0 ** weak–	44.2 *** strong+	36.3 *** strong+	% crested All lizards (6528)
4	Sexes differ in size	0.68 *** v. str.+	0.41	0.46	0.53 *** strong+	0.59 *** strong+	Mean score (–2 to 2) Others w/ scores (1471) See table 3.3
5	Carnivores	1.3 *** weak–	0.9 ** weak–	0.8	1.5 * weak–	0.6 ** weak–	% carnivores Others w/ diet (2860)
6	Herbivores	5.6 ** weak+	4.6	0 * weak–	2.9	1.2 ** weak–	% herbivores Others w/ diet (2860)
7	Invertebrate eaters	63.9 *** strong–	62.8 *** strong–	83.2 * weak+	76.1 * weak+	78.9 * weak+	% invert. eaters Others w/ diet (2860)
8	Ant specialists	8.3 *** mod+	4.3	1.6	6.7 ** weak+	5.9 *** strong+	% ant specialists Others w/ diet (2860)
9	Omnivores	20.9 *** mod+	27.5 *** mod+	14.4	12.8 * weak–	13.4	% omnivores Others w/ diet (2860)
10	Sit-and-wait foragers	61.1 *** strong+	34.1 *** strong+ (females)	18.0 *** mod–	77.8 *** strong+	76.6 *** mod+	% sit-and-wait Others w/ foraging data (1149)
11	Cruise foragers	5.4	4.5	0 * weak–	16.9 *** mod+	21.3 *** mod+	% cruisers Others w/ foraging data (1149)
12	Active foragers	44.3 *** strong–	59.0	91.8 *** mod+	23.3 *** strong–	19.1 *** strong–	% active foragers Others w/ foraging data (1149)
13	Home-range area (mass adj.)	–0.13 * mod–	0.16	0.58 ** strong+	–0.2 * mod–	–0.6 *** strong–	Std. resid. Others w/ home ranges (216)
14	Active an any time (cathemeral)	5.6 * weak–	3.5	3.6	8.9 * weak+	2.6 *** weak+	% cathemeral Others w/ activity (4746)
15	Active in daytime	89.6 *** strong+	93.3	92.7 *** mod+	74.7 *** weak+	93.7 *** mod+	% diurnal Others w/ activity (4746)
16	Active at night	2.1 *** strong–	1.5 mod–	1.5 *** mod–	11.9 *** mod–	1.8 *** mod–	% nocturnal Others w/ activity (4746)

Table 3.4 continued

Trait	Condition	Sexes differ in color	Seasonal color	Blue tails	Any color change	Adv. color change	Units Contrast (*n*) Notes
17	Crypsis as primary pred. conceal.	43.0	34.6 * weak–	1.9 *** mod–	70.9 *** strong+	80.6 *** strong+	% rely on crypsis Others w/ pred. data (1067)
18	Flight as primary pred. conceal.	11.5	13.7	28.8 *** mod+	4.5 *** mod–	1.1 *** mod–	% rely on flight Others w/ pred. data (1067)
19	Sand as primary pred. conceal.	3.4 ** mod–	4.9	5.8	0.8 *** mod–	0.4 *** mod–	% rely on sand Others w/ pred. data (1067)
20	Crevice as primary pred. conceal.	25.5	34.6 ** mod+	40.4 * mod+	13.0 *** mod–	7.1 *** mod–	% rely on crevice Others w/ pred. (1067)
21	Veg. as primary pred. conceal.	25.5 * weak+	24.7	28.8	16.3 *** mod–	13.4 *** mod–	% rely on veg. Others w/ pred. data (1067)
22	Arboreal microhabitat	39.2 *** mod+	22.0 *** weak–	15.2 *** weak–	71.3 *** strong+	76.4 *** strong+	% arboreal Others w/ micro. (5175)
23	Cryptozoic microhabitat	6.3 *** mod–	9.3 0.07 weak–	16.6 0.09 weak+	1.0 *** mod–	0.3 *** mod–	% cryptozoic Others w/ micro. (5175)
24	Subterranean microhabitat	0 *** mod–	0 *** mod–	2.8 *** weak–	0 *** mod–	0 *** mod–	% subterranean Others w/ micro. (5175)
25	Semiaquatic microhabitat	0.6 0.08 weak–	0.2	0	0.3 * weak–	0.5	% semiaquatic Others w/ micro. (5175)
26	Terrestrial microhabitat	53.9 *** weak+	68.4 *** mod+	65.6 *** weak+	27.4 *** mod–	22.8 *** mod–	% terrestrial Others w/ micro. (5175)
27	Spp. native on any island	46.6 *** weak+	40.7	48.7 * weak+	58.6 *** mod+	56.5 *** mod+	% on islands All lizards (6528)
28	Oviparous species	88.2 *** mod+	89.8 *** weak+	86.5	91.7 *** mod+	89.2 *** strong+	% oviparous Other repro. mode (4283)
29	Longevity (mass adj.)	–0.13 ** mod–	–0.25 *** mod–	–0.10	–0.08	–0.37 *** strong–	Std. resid. Others w/ long. (879) All colors negative!
30	Population density	0.95 *** mod+	0.55	0.83	0.49	0.98	kg/ha Spp. w/ densities (343) Only dichrom. sign.
31	Spp. introduced	7.5 *** mod+	9.2 *** mod+	13.2 *** mod+	12.5 *** mod+	9.2 *** mod+	% introduced All lizards (6528) All positive!
32	At conservation risk	52.4 * weak+	51.1	45.8	74.3 *** strong+	82.6 *** strong+	% at risk Other eval. spp. (3390)

association with ant consumption may be indirect. For example, species that eat ants may be relatively immobile and therefore need to communicate without moving to the recipient. Gliders are good examples of ant specialists, some of which use color to communicate over large distances. This argument seems weak to me and does little to explain omnivores or the paucity of colorful invertebrate consumers.

- Positive correlation with sit-and-wait (10), but not active foraging (12). Sit-and-wait foragers fit the stereotype of being both relatively immobile and highly visual.
- Diurnality (15) but not nocturnality (16). Diurnality fits in with the visual modality theme.
- Small home-range size (13). Small home-range size fits in with the low mobility theme of sexual colors.
- Reliance on crevices for predator evasion (20), but not sand swimming (19). Crevice dwellers may be examples of extreme immobility, the opposite of the roaming sand swimmers.
- Arboreal (22, dichromatism only) or terrestrial (26), but not cryptozoic (23) or subterranean (24). These four distinctions seem to comport well with the visual versus nonvisual modalities. The negative relationship between seasonal color and arboreality is obscure. Seasonal colors in terrestrial lizards are often expressed exclusively or primarily latero-ventrally, suggesting a natural selection penalty (hide the color from birds flying overhead) combined with a socially or sexually selected advantage (raise the flank and flash the color to prospective mates). Perhaps this gambit does not work in an arboreal environment, where predatory birds may be peering at the lizard from below as well.
- Oviparous species (28). Perhaps this is an indirect association between benign tropical environments (which favor oviparity) and high population density promoting sexually selected colors, whereas those same selection pressures are attenuated or absent in the high latitude/high altitude environments associated with viviparity.
- Short longevity (29). Short longevity fits with the association of color with abiotically benign tropical environments and high population density on islands. No need to be iteroparous if breeding prospects are the same today as tomorrow. One might suppose that longevity would be greater on islands (unlike predator-ridden tropical mainlands). However, color may be needed to address the many sexual and social competitors present at high population densities (next bullet).
- High population density (30). High population density fits with the theme of greater sexually selected color in socially competitive environments.

One unique aspect of sexual colors is that the two sexes may have differing motivations for their expression. This is most apparent in seasonal color, for which males tend to develop color early in the pairing process, presumably to intimidate male rivals and possibly impress prospective female partners, whereas females tend to develop color after ovulation, presumably to signal to males that they have already been fertilized and are therefore no longer worthy of the males' attentions. In most species (see individual database entries), female color is a rejection signal, whereas male colors are an attraction signal. In some species, the significance of female color has yet to be firmly established (Clark et al. 2017), and initial reddening (female seasonal colors are generally reddish-orange) may excite the males (when first expressed, it may not be too late for a male to obtain a fertilization), whereas maximum expression of it repels them (Clark et al. 2017). Curiously, female-only seasonal colors are found exclusively in lizard species that have been labeled as sit-and-wait foragers. Of 24 lizard species in which only the females have seasonal color, 24 of them are described as sit-and-wait ($p < 0.0001$). Presumably sit-and-wait females are often refugia-anchored and therefore encounter greater persistence from adjacent, overly ardent suitors, whereas the more fleet-of-foot females (cruise and active foragers) can simply cruise or sprint out of the boorish male's home range.

The correlations of blue tails are fairly distinct and unique (table 3.4 trait number):

- Elongate bodies are favored (1). Elongation is associated with cryptozoids and wide-ranging chemosensors, both of which are svelte and often blue-tailed.
- Herbivory is absent (6). Herbivores, rarely blue-tailed, sit around a lot, often basking, as herbivores guts can be filled with plant material very rapidly (a few minutes per day). Blue tails would seem to be a liability for such lizards basking in the open. In contrast, cryptozoids, often blue-tailed, rarely bask conspicuously, and wide-ranging chemosensors, often blue-tailed, generally remain in refugia except when they are actively foraging.
- Active foraging is nearly universal (12); cruise foraging is absent (11); sit-and-wait foraging is avoided (10). See preceding point.
- Large home ranges strongly favored (13). This association is probably indirect, through the association with active foragers or wide-ranging chemosensors.

- Diurnality is dominant (15); nocturnality very rare (16). Blue tails would seem to offer no advantages at night.
- Crevice-burrow (20) or flight are the predator concealment tactics of choice (18); crypsis is rare (<2%; trait 17). Blue tails would invalidate crypsis. Dashing away, or into a nearby crevice, gives predators only a momentary opportunity to pounce, and pouncers are easily drawn to the conspicuous blue color.
- Many are cryptozoic (23), but not semiaquatic (25) or arboreal (22).

The microhabitat associations of blue tails are strong and consistent; blue tails are found on the ground (26) or under cover (particularly cryptozoic juveniles). Presumably the shocking brilliance of the blue tails is of startle value when a predator exposes a cryptozoic (23) or subterranean (24) lizard. Lizards that forage and rest on the surface of the ground face a different set of challenges with regard to bright blue tails. For terrestrial species, the advantage of the blue color seems to be more delicately balanced (the association—trait 26—is weak), as bright color would seem to draw predation attempts from birds. It seems that visual predators (perhaps mostly birds) prevail in the trees (22) and along streams (25), as blue tails are notably scarce (22) or absent (25) there. Are blue tails of especial value in deflecting pounces of diurnal/crepuscular mammals rather than attacks from avian predators, as suggested by the habitat and microhabitat associations? Are the lengths of sight lines crucial in determining the survival value of blue tails, as suggested by the paucity of blue tails in densely foliated habitats such as forests or riparian zones and their relative commonness in somewhat more open habitats? Perhaps lizards that are still alive can use their blue tails as effective decoys at the moment when they realize the proximity of a predator (when they are pawed out of the duff, or running for cover toward a desert shrub), but cannot do so when the predator catches them unaware (gleaned off a tree branch)? Why do so many blue-tailed lizards lose the bright color ontogenetically, and why often in one sex only? Apparently the behavior of older and especially male lizards tips the balance of benefits against having a blue tail. The ontogenetic stage at which blue tails vanish seems to vary among populations, another potentially informative trait not tabulated in the database. There is much to be learned from the context-specific frequencies of blue tails about the behavioral and ecological interactions between predators and lizard prey.

Notable correlations of color changing ability are as follows (table 3.4 trait number):

- Not elongate (1). Elongate species tend to be concealed where crypsis aids, such as color changing ability, are irrelevant.
- Crests (3). Crested species tend to be exposed and cryptic, so metachrosis is a powerful asset.
- High sexual size discrepancy (m>f or m<f) (4). Monomorphism is associated with invariant colors. This complex situation may reflect differing selection pressures at m>f than m<f. The combination does not yield to a simple hypothesis, except that perhaps sexual size differences (either direction) benefit their hosts primarily through visual modalities (e.g., by visual intimidation rather than actual mass), whereas both monomorphism and the absence of color change can be enjoyed in unilluminated or nonvisual contexts.
- Ant specialist (8). Ant specialists are generally cryptic, exposed sit-and-wait foragers. Color change may be useful both in preventing detection by predators and possibly ants, which can swarm and attack a lizard if detected.
- Sit-and-wait (10) and cruise foraging (11) but not active foraging (12). Active foragers ruin their crypsis by their movements. So there is no reason for them to bother with color change. Cruise foragers try to minimize the damage to their crypsis by moving slowing; diurnal cruise foragers often sway languidly to match the movements of windblown vegetation.
- Small home-range size (13). This may be a reflection of the negative correlation between color change and active foraging.
- Not nocturnality (16), but diurnality (15), especially for advanced metachrosis, and active at any hour (14), especially for darkening metachrosis. The diurnality/nocturnality distinction seems unsurprising for color change, but why is cathemeral (active at any hour) associated with darkening color change, but not pattern or hue? One would expect an advantage for cathemeral species to be extremely cryptic in the daytime when they are not foraging. Perhaps they are under cover when inactive. Another possibility might be that color change is metabolically expensive or reliant on rare mutations (either are consistent with the low absolute frequency of advanced color change).
- Metachrosis is positively associated with predator evasion by crypsis (70%–80% of spp.) (17), but negatively associated with the other tabulated predator evasion tactics: flight (<5% of spp.) (18), hiding in sand (19), hiding in crevices (20), or fleeing to vegetation (21). The benefits of color change to crypsis are clear, as is the lack of benefit to sprinting lizards. Hiding in

vegetation would seem to be advanced by the ability to match the background. This may be another example of the high metabolic or evolutionary costs to color change.

- Arboreality (22) is associated with color change, especially advanced metachrosis, but not cryptozoic (23), subterranean (24), or terrestrial (26) microhabitats. The association with arboreality supports the notion that hiding in vegetation is enhanced by color change.
- Oviparous species (28). Oviparity is associated with high productivity environments (consistently warm). This fits in with the idea that color change is metabolically costly (the cost is easier to bear in productive places), but it could also be an indirect reflection of greater frequencies of cryptic species resting exposed in benign environments. In colder, drier sites more species may refuge under cover.
- Short longevity (29). This fits with oviparity and is therefore equally mysterious.
- Often at conservation risk (32). It is not obvious to me why color changing ability would be correlated with high conservation risk, although many chameleons are at risk, probably due to their limited geographic ranges rather than their color changing ability.

One apparent contradiction that might be seen in the color changing list is that color changers are positively associated with cruise foraging (esp. advanced metachrosis), and yet home-range areas are the smallest in color changers (esp. advanced metachrosis). On average, cruise foragers have intermediate-sized home ranges (see discussion under "Foraging" below), but color changers (esp. advanced) seem to have especially small home ranges. Much remains to be learned about the home ranges of lizards, and though some cruise foragers, for example, chameleons, are easily sampled when they are sleeping, there are comparatively few field studies of chameleon daytime activities, home-range sizes, and geographic foraging patterns.

In reviewing the traits associated with changing colors, it is perhaps worth keeping in mind that color changes are used in a variety of situations, including visual concealment, thermoregulation, and social signaling. Darkening would appear to benefit thermoregulation and crypsis, and the parallelism between basic and advanced metachrosis argues that concealment and thermoregulation are likely to be important for advanced metachrosis as well. In this regard, note the very high mean specifically associating advanced metachrosis with crypsis (17).

A few traits show some consistency across all categories of colors. All colors show negative standard residuals for longevity (29), suggesting that whereas color may be a way of making the best of a bad situation (better to sacrifice a colorful tail than lose a life), the end result is nonetheless a relatively short life expectancy (longevity is presumed to reflect broad patterns of life expectancy). Or perhaps color is a more appropriate adaptation in environments of relative constancy or predictability (no need for high longevity to live until favorable breeding conditions reoccur)? Compared to the background frequency of 40% of lizards on islands (chapter 2), colorful lizards of all types (not significant for seasonal color) make up a disproportionate fraction of island species (27), suggesting that higher mainland predation pressure may mute the colors of mainland species.

Only sexual dichromatism is associated with high population density (30), suggesting that whereas dichromatism may emerge at high population densities, none of the other colors are particularly associated with population abundance. Perhaps commensurate with the low population densities, all of the color categories except blue tails show some association with high conservation risk (32). Color may be useful for extracting sympathy from conservation-minded humans, but it does not appear to directly facilitate sharing the planet with humans.

Diet

As dietary specializations do not lend themselves to placement on a single-axis continuum, correlations in a matrix of diet classes are not amenable to interpretation as components of a larger correlation. However, the diet table (table 3.5) conveniently contrasts mean values of different diet classes.

Carnivores are notable among diet classes for many conditions (table 3.5 trait number):

- Rarest diet class (1). Expected, as it is the highest trophic level.
- Heaviest mean mass (when Iguanidae, Varanidae included) (2). Not surprising, as greater size is required to eat larger prey.
- Lowest proportion of species with crests (5). Presumably crests are of value primarily for social/sexual reasons or for crypsis. Many carnivorous snakes hunt by sit-and-wait, but few carnivorous lizards do (*Crotaphytus collaris* is an exception). Why might social reasons for crests not apply to carnivores? Possibly because carnivores are at such low densities that social selection is attenuated. But see next bullet.
- Most male-biased SSD (9). This conflicts with the logic of the previous bullet, which reasoned that social selection was attenuated in carnivores. This

Table 3.5. Correlates of lizard species' diets. Conventions as in table 3.1.

Trait	Condition	Carnivore	Herbivore	Invert. eater	Ant spec.	Omnivore	Units Contrast (*n*)
1	Species count (no test)	3.2	4.1	72.8	4.4	15.4	% of spp.
2	Mass	135 *** v. str.+	86 *** strong+	4.8 *** mod–	5.5 *** weak–	19.7 *** mod+	Grams Others w/ diet data (2860)
3	Mass, excl. Iguanidae/ Varanidae	24 *** mod+	36 *** strong+	4.7 *** mod–	5.5	18.4 *** mod+	Grams Others w/ diet data (2754)
4	Shape index	0.68 *** strong+	1.40 *** v. str.+	–0.02 *** strong–	0.14	0.66 *** strong+	Std. resid. Others w/ diet data (2860)
5	Crests	3.3	12.0	7.1 * weak–	20.5 *** mod+	8.2	% of focal crested Others w/ diet data (2860)
6	Elongate lizards	10.9	0.9 *** weak–	15.6 *** mod+	4.7 ** weak–	6.1 *** mod–	% elongate Others w/ diet data (2860)
7	Color change	8.7 * weak–	13.7	20.1 0.055 weak+	29.1 * weak+	15.9 0.056 weak–	% color changers Others w/ diet data (2860)
8	Tail length (× SVL)	1.42	1.32	1.34 *** weak–	1.46 0.10 weak+	1.56 *** mod+	Ratio Others w/ tail length (2793)
9	Sexes differ in size	0.99 *** strong+	0.92 *** strong+	0.22 *** mod–	0.13 * mod–	0.79 *** strong+	–2 (m≪f) to 2 (m≫f) scale Others w/ SSD (1126)
10	Sexes differ in color	13.0 *** weak–	45.3 ** weak+	28.9 *** v. str.–	61.4 *** mod+	44.7 *** mod+	% dichromatic Others w/ diet data (2860)
11	Seasonal color	3.3 ** weak–	13.7	10.5 *** strong–	11.8	21.8 *** mod+	% of focal seas. color Others w/ diet data (2860)
12	Blue tails	1.1	0 * weak–	5.0 * weak+	1.6	4.1	% of focal blue tailed Others w/ diet data (2860)
13	Sit-and-wait forager	27.5 ** mod–	40.0	46.6	66.1 ** mod+	51.3	% of focal sit-and-wait Others w/ diet and foraging mode (1025)
14	Cruise forager	2.0	10.0	7.0 * weak+	5.4	2.5 * weak–	% of focal cruisers Others w/ diet and forag. (1025)
15	Active forager	70.6 *** mod+	50.0	46.4	28.6 ** mod–	46.2	% of focal active forager Others w/ diet and forag. (1025)
16	Home-range area (mass adj.)	1.01 *** strong+	–0.82 ** mod–	0.01	0.21	–0.18 0.079 mod–	Std. resid. Others w/ home ranges (212)
17	Moves per minute (MPM)	43.5	34.0	51.2	45.3	53.1	Percentile rank Others w/ MPM No assoc. w/ diet!
18	Active at any time (cathemeral)	18.0 ** weak+	1.7 ** weak–	9.1 0.08 weak+	3.1 * weak–	7.2	% cathemeral Others w/ activity (2705)

Table 3.5. continued

Trait	Condition	Carnivore	Herbivore	Invert. eater	Ant spec.	Omnivore	Units Contrast (*n*)
19	Active in low light (crepuscular)	4.5	0	2.8 * weak+	0 0.07 weak–	1.6	% crepuscular Others w/ activity (2705)
20	Active in daytime	68.5	94.8 *** mod+	66.1 *** strong–	96.1 *** mod+	86.8 *** mod+	% diurnal Others w/ activity (2705)
21	Active at night	9.0 * weak–	3.5 *** weak–	22.0 *** strong+	0.8 *** mod–	4.4 *** mod–	% nocturnal Others w/ activity (2705)
22	Conspicuously wary	25.6 ** weak+	11.6	12.6 * weak–	7.9 * weak–	16.7 * weak+	% conspic. wary Others w/ wary score (2832)
23	Crypsis for primary pred. conceal.	7.7 *** mod–	4.2 *** mod–	51.7 *** strong+	38.6	20.9 *** mod–	% rely on crypsis Others w/ pred. tactics (862)
24	Flight for primary pred. conceal.	26.9 * weak+	0	9.2 * weak–	10.5	15.7 * weak+	% rely on flight Others w/ pred. tactics (862)
25	Crevice-burrow for primary pred. conceal.	34.6	79.2 *** strong+	20.1 *** mod–	17.5	39.6 *** mod+	% rely on crevice Others w/ pred. tactics (862)
26	Veg. as primary pred. conceal.	38.5 * weak+	20.8	18.4 *** mod–	36.8 * mod+	26.9	% rely on veg. Others w/ pred. tactics (862)
27	Basks directly	90.4	100 ** weak+	81.6 *** mod–	88.3	95.8 *** mod+	% direct baskers Others w/ basking data (1044)
28	Basks indirectly	7.7	2.2 0.08 weak–	14.1 *** mod+	8.3	4.7 *** mod–	% basks indirectly Others w/ basking data (1044)
29	Thermally conforms	9.6	0 ** weak–	14.4 ** mod+	13.3	7.6 * weak–	% thermoconformers Others w/ basking data (1044)
30	Body temperature	30.9	33.1	30.0 *** weak–	29.9	32.4 *** weak+	Mean of midpoints °C Others w/ temp. data (389)
31	Group living	10.9	21.4 ** weak+	8.3 *** mod–	19.7 * weak+	19.5 *** mod+	% of focal grouped Others w/ activity (2860)
32	Males defend space	51.6 *** strong–	93.5	86.6	81.3	90.1 0.066 weak+	% of focal defensive Others w/ diet and social data (731)
33	Males hierarchical	19.4 * weak+	19.4 * weak+	5.7 ** mod–	4.2	11.0	% males hierarchical Others w/ diet and social data (731)
34	Arboreal microhabitat	27.2	13.9 *** weak–	33.9 *** weak+	49.6 *** mod+	22.6 *** mod–	% arboreal Others w/ diet and microhabitats (2818)
35	Cryptozoic microhabitat	4.3 * weak–	0.9 *** weak–	13.9 *** mod+	6.3 0.08 weak–	4.8 *** mod–	% cryptozoic Others w/ diet and microhabitat (2818)
36	Subterranean micro-habitat	3.3	0 * weak–	5.6 *** mod+	0.8 * weak–	0 *** mod–	% subterranean Others w/ diet and microhabitat (2818)

Table 3.5. continued

Trait	Condition	Carnivore	Herbivore	Invert. eater	Ant spec.	Omnivore	Units Contrast (*n*)
37	Terrestrial microhabitat	57.6	83.5 *** mod+	45.6 *** mod–	43.3 0.07 weak–	71.7 *** mod+	% terrestrial Others w/ diet and microhabitat (2818)
38	Rainfall <55 mm/y	4.5	22.2 *** strong+	1.9 * mod–	0	2.9	% in hyperdry Others w/ diet and rainfall (504)
39	Rainfall 56–150 mm/y	9.1	37.0 ** mod+	11.1 * mod–	21.4	14.4	% in dry belt Others w/ diet and rainfall (504)
40	Rainfall 401–1100 mm/y	9.1 0.051 weak–	25.9	25.1 0.098 weak–	32.1	38.5 * mod+	% in intermed. belt Others w/ diet and rainfall (504)
41	Rainfall 1101–3000 mm/y	18.2 * mod–	18.5 * mod–	44.0 * mod+	28.6	39.4	% in mesic belt Others w/ rainfall and diet (504)
42	<1200 m elevation	97.4	71.0 *** mod–	89.4 * weak+	95.9 * weak+	86.6	% found <1200 m Others w/ elev. data (1991)
43	1201–2400 m elevation	31.6	29.0 ** weak–	46.3 * weak+	46.4	44.8	% found 1201–2400 m elev. Others w/ elev. data (1991)
44	>2400 m elevation	5.3	23.0 * weak+	13.0 *** mod–	16.5	21.7 ** mod+	% found >2400 m elev. Others w/ elev. data (1991)
45	Spp. native on any island	66.3 *** mod+	38.5	42.5 *** weak–	52.8 0.068 weak+	49.2 * weak+	% on islands Others w/ diet data (2860)
46	Incubation duration (mass adj.)	0.50 *** strong+	–0.71 *** strong–	0.11 *** mod+	–0.40 *** strong–	–0.36 *** strong–	Std. resid. Others w/ incubation (859)
47	Mass-adjusted clutch/ litter count	–0.369 * mod–	0.109	0.178 ** weak+	0.172	–0.253 *** mod–	Std. resid. diff. Others w/ clutch counts (2059)
48	Endemic to a "country"	42.4	69.2 *** mod+	48.9	32.3 *** weak–	44.7	% endemic Others w/ activity (2860)
49	Longevity (mass adj.)	0.16	0.41 * strong+	–0.08 ** mod–	–0.19	0.14 * weak+	Std. resid. Others w/ longevity and diet (740)
50	Population density	0.25	12.1 *** v. str.+	0.32 *** mod–	0.50	1.59 *** mod+	kg/ha Others w/ popn. density (324)
51	Introduced	5.1	12.0 * weak+	4.9 *** mod–	8.7	12.2 *** mod+	% of focal introduced Others w/ activity (2860)
52	At conservation risk	85.5 *** mod+	64.2 0.058 weak+	53.7	30.6 *** mod–	52.6	% deemed at risk Others w/ diet and conserv. data (1792)
53	Listed by IUCN as endangered or worse	8.4	17.9 0.067 weak+	11.9	3.2 * weak–	10.5	% IUCN EN+ listed Others w/ diet and conserv. data (1792)

high level of SSD suggests otherwise. One possibility is that greater male-biased SSD is an artifact due to correlation with large body size rather than a reflection of more intense sexual or social selection in carnivores. That possibility begs the question of why would greater SSD be "caused" by larger body size. Or perhaps carnivores contest sexual relationships even at low densities.

- Least likely to show sexual dichromatism (10), seasonal color (11), or color change (7). These three color responses could be associated with reduced social/sexual selection in carnivores, or they could reflect a stronger imperative for carnivores to be cryptic to avoid alerting their vertebrate prey to their approach. Color change could support both social pressures and the cause of greater crypsis, but does not seem to have done so in most carnivores (though *Crotaphytus collaris* exhibits sexual dichromatism, seasonal color, and color change). A study of the distribution of these traits in carnivorous lizards may show greater consistency when second-order details, such as foraging mode superimposed on diet, are considered.
- Lowest sit-and-wait (13) and cruise foraging (14); highest active foraging (15). Many snake carnivores are sit-and-wait, whereas few lizard carnivores are. Presumably this reflects the ability of snakes to eat prey that are relatively large. A comparison of relative prey size among lizard carnivores of differing foraging mode would be of interest in this regard.
- Largest mass-adjusted home-range areas (16). This is probably a straightforward reflection of trophic level.
- Highest proportions of cathemeral (18) or crepuscular (19) lizard species. These associations suggest that carnivory may be aided by predatory unpredictability or low light levels. On first principles these would seem to favor the carnivore that can choose the conditions for its pounce. If so, one would expect to see this pattern mirrored in other major carnivorous taxa, such as mammals, snakes, crocodilians, and so on. Schmitz and Motani (2011) found that carnivorous dinosaurs were at least partly nocturnal, whereas their prey (herbivorous dinosaurs) were mostly cathemeral. Schmitz and Motani explicitly supported the inference that diel activity patterns were driven primarily by ecological rather than phylogenetic factors, but the pattern they observed does not mirror the lizard pattern, perhaps because lizard prey are overwhelmingly diurnal rather than cathemeral. Lizard prey are likely too small to possess the huge thermal inertia available to facilitate cathemeral foraging in herbivorous dinosaurs. Furthermore, herbivorous dinosaurs may have been homeothermic (constantly warm) and thus energetically compelled to eat through most of the night, as are modern elephants.
- Highest proportion of species said to be conspicuously wary (22). Carnivores of many taxa are conspicuously wary, presumably to facilitate ambushing. That is, the carnivorous lizard's prey remains unaware of the stealthy predator until the predator has gotten close enough to overcome the startled prey's acceleration. Furthermore, many lizard carnivores are vulnerable to human predation for their meat or skins, and thus lizard predators may need to be conspicuously wary specifically of humans.
- Highest frequency reliant on flight (24) or vegetation (26) for predator concealment. These associations are consistent with the paucity of sit-and-wait foraging among lizard carnivores.
- Lowest frequency of spatial/territorial defense (32) and highest reliance on male dominance hierarchies (tied with herbivores) (33). Presumably lizard carnivores fail to defend resource territories because their home ranges are too large for successful defense. Dominance status, in contrast, can be maintained with only occasional contact (and imposing size). The social hierarchy pressure for large size could conceivably lead to high SSD coupled with niche segregation between the sexes, but only if size assortative prey depletion was a significant driver of size-based selection. There are many unresolved questions in the complex interplay of selective forces on size.
- Highest proportion of species present at elevations below 1200 m (42) and lowest proportion above 2400 m (44). Note the difference in this condition to that of lizard herbivores (below). This altitudinal gradient suggests that lizard carnivores are better adapted to life at high temperatures (presumably endothermic carnivores replace them at higher elevation).
- Highest proportion residing in semiaquatic microhabitats (not in table; $p < 0.0001$, 7.6% v. 1.0% for noncarnivores). The reasons for the association with riparian zones deserve greater study.
- Highest proportion on islands (45). Is this a reflection of a reduction in competition from endothermic carnivores on islands? The preponderance of large islands (e.g., Indonesia, Philippines) in the island sample would argue otherwise, but a targeted comparison would be needed to distinguish among the possible contributors.
- Longest mass-adjusted incubation duration (46). The

greater incubation duration could reflect either oviposition at an earlier developmental stage or slower development. Oviposition at an earlier stage would free the mother from the minor burden of carrying early developing ova, whereas development rate would seem to be dictated primarily by soil temperature (not diet). Neither pressure seems adequate to account for the large effect size. Note that the opposite condition is present in herbivores (below).

- Lowest mass-adjusted clutch count (47). Although mass-adjusted, this metric is not trophic level adjusted. Fewer larger young may be appropriate for higher trophic levels.
- Lowest population density (50). Presumably a straightforward reflection of trophic level.
- Highest proportion said by some authority to be at conservation risk (52), but second lowest by the extreme International Union for Conservation of Nature (IUCN) ratings (53). For more on this conundrum, see the discussion under "Imperiled Lizards" below.

Though comparatively species-poor (1), herbivores have a remarkable number of distinctions, including the following (table 3.5 trait number):

- Heaviest mean mass (when Iguanidae, Varanidae excluded) (3). Selective pressure for large size to accommodate large quantities of vegetation being digested slowly does not seem sufficient to fully account for the large size of herbivores. Larger size may also reduce predation (probably the primary concern of herbivores, as vegetation is comparatively abundant); Poulin et al. (2001) demonstrated the exceptional vulnerability of small-sized lizards to bird predation. Great bulk may also enhance mating success in mating systems lacking a resource monopolization axis; prospective mates cannot enhance their attractiveness by controlling an especially good stash of leaves, but they can get bigger and thereby wow prospective mates with their caloric accomplishments. Resource monopolization can be enhanced by large size, as well, but need not be; if rapid territorial patrolling is key, the nimblest patroller may be more effective than a ponderous one.
- Heaviest body shape (4). See above.
- Shortest tails (8). Short tails in herbivores may reflect an absence of selective pressures for long tails rather that positive pressures for short tails, though in the absence of autotomy there could be direct pressure to shorten the "handle" given to predators. Long tails aid leaping from tree branch to branch, bipedal running, and propulsion through grass or other dense vegetation. None of these apply with force to large herbivores.
- Absence of blue tails (12). See earlier discussion under "Color."
- Smallest home-range size (16). Eating at a lower trophic level should minimize the area needed for sufficient food. In practice, plants have many chemical defenses and seasonal shifts in availability that in combination can induce herbivores to travel long distances to obtain seasonally preferred foods. There may be sampling biases that overlook some of these long-range forays by obligate herbivores. For example, long forays may take an herbivore out of the monitored study area.
- Lowest proportion of cathemeral species (18) and (tied with ant specialists) absence of crepuscular ones (19). Because ectothermic herbivores have low energy needs, they need to gather food for only a small fraction of their day (20–30 minutes in *Iguana iguana*), and therefore they have the option of foraging at a time of their choice. If carnivores are often cathemeral and crepuscular, it would be imprudent for an herbivore to be active at those times.
- Nearly tied with ant specialists for the highest proportion of diurnal species (20). Digestion of cellulose-bound nutrients is difficult and slow but accelerated by high temperatures. It is no surprise that most herbivores are day active (see also basking, 27).
- Least reliance on crypsis (23) or flight (24) for predator concealment; greatest reliance on crevices (25). Crypsis and flight are more difficult if you are large and corpulent.
- Highest proportion (100%) basking directly (27); lowest proportions basking indirectly (28) or conforming (29). These thermoregulatory attributes fit well with the high temperatures needed for digestion of plants.
- Highest mean body temperature (30). See above.
- Highest proportion of species living in groups (31), males defending space (32), and (tied with carnivores) showing male dominance hierarchies (33). Group living is a common practice among herbivores of many taxa, presumably because more eyes are available for predator detection. This herd mentality does not have any clean examples in adult lizards, though Burghardt (1977; Burghardt et al. 1977) reported supportive observations in juvenile *Iguana iguana* and possibly dinosaurs. If the crevices listed above are clustered,

one can get group living in herbivores due to independent attraction to a common clustered resource. In this situation, the lizards will be alerted by the fleeing actions of their peers, even if the surveillance benefits did not trigger the aggregation. In contrast to the predator evasion aspects of grouped living, the examples of herbivore males defending space or behaving hierarchically appear to be motivated primarily by mating opportunities or the potential for monopolizing mating opportunities. This too may be related to herbivory, in that herbivorous lizards are not known to defend space containing plant resources. As noted above, prime food plants tend to be dispersed and seasonally variable. Furthermore, female herbivores often cease feeding during the mating season, as their large clutches compete with bulky low-grade food for precious peritoneal space. Instead, many herbivores exhibit lek-based mating systems, whereby females choose mates based on the males' social status and their ability to protect resident females from sexually assertive subordinates.

- Lowest proportion of arboreal (34), cryptozoic (35), and (tied with omnivores) subterranean (36) species; highest terrestrial (37). These are probably more due to the herbivores great bulk and size than to proximity to key resources. As food plants are often arboreal, it is surprising that herbivores do not embrace the arboreal microhabitat (cf. *Iguana iguana*, *Corucia zebrata*), but many may be simply too large.
- Most favored in the hyperdry (38) and dry (39) rainfall belts. This is presumably due to the dearth of predators and alternate foods in those zones (see below), and arboreal opportunities are minimal in such dry places.
- Lowest proportion of species at elevations <2400 m (42, 43) and highest above that (44). This altitudinal trend seems to cut against the thrust of earlier observations noting the association of herbivores with high temperatures. The reasons are far from clear. Many liolaemid herbivores are found at high elevation, which is often very dry in the rain shadow of the Andes. Basking at high elevation (more intense radiation) in dry habitats is exceptionally effective at raising body temperature.
- Shortest incubation duration (46). Note the opposite condition in carnivores (above). One wishes to know whether the short incubation period of herbivores is due to retention of eggs in the basking mother to expedite development, or possibly accelerated development to fit with seasonal vegetation phenology. For example, herbivorous *Iguana iguana* have rapid development to complete egg development in the short period after seasonal flooding of oviposition sites and before seasonal leaf flushing of preferred food plants. The latter is critical to expedite growth of vulnerable emerging hatchlings (Rand and Greene 1982).
- Highest proportion of endemic species (48). It is far from clear why herbivores would be spotty in their distributions, though island species are normally endemic. Presumably the right combination of rich plant foods and low predation is geographically patchy. The absolutely low frequency of herbivorous species (1) is consistent with the notion that saurian (or squamate) herbivory is not an easy business model to pursue.
- Highest mass-adjusted longevity (49). High longevity could indicate either that predation is low and has limited demographic impact or that herbivores favor unpredictable environments, for which iteroparity is highly rewarded. The predominance of herbivores in the hyperdry and dry rainfall belts comports well with both associations.
- Highest population density (50). Population density estimates are biased toward sites in which the lizards are most abundant. Nonetheless, it is striking that nine of the top ten densest lizard species are plant eaters. Where the vital rates are "in the black," eating at the lowest trophic level seems to allow the achievement of higher biomass densities. The paucity of lizard herbivore species (1) suggests that conditions maintaining lizard herbivores "in the black" are geographically rare.
- Virtually tied with omnivores for the highest proportion of introduced species (51). The list of introduced species is very restricted (3.3% of species), and thus local idiosyncrasies may play a role in this association. *Iguana iguana* have been introduced at least 84 times, both as escaped pets and possible food items. Herbivores can be easy pets to care for in that they can be fed human food. Herbivores are good food items to stock on an island because they are large, fecund, and tasty. They are unlikely to be introduced as accidental stowaways.
- Highest proportion of species listed by IUCN as being at extreme conservation risk (53). Many of the most severely impacted species are large herbivores that have been constrained by their small insular ranges, high edibility, and presence on heavily developed, rat-and-mongoose-infested Caribbean Islands.

Invertebrate eaters are particularly speciose (1) but distinct from other diet classes for relatively few traits (table 3.5 trait number):

- Lowest mean mass (regardless of inclusion or exclusion of Iguanidae, Varanidae) (2,3). This fits with the notion that small prey require only small predators.
- Most slender (4). There is no reason for an invertebrate eater to be bulky.
- Nearly shortest tails (1.34 times SVL v. 1.32 for the herbivores) (8). See comments under herbivores suggesting that this is more a case of lack of selection for long tails than a positive selection for short tails.
- Highest proportion of blue tails (12). Blue tails seem to be associated with particular foraging modes (active foraging) and microhabitats (cryptozoids) rather than diet per se. Blue tails seem to be selected against for carnivores, herbivores, omnivores, and ant specialists (see those sections) but not subject to intense adverse selection in invertebrate eaters.
- Highest proportions of cathemeral (18) and nocturnal (21) species. These associations may speak to the prevalence of invertebrate prey or saurophagy throughout the 24 hours. Perhaps the diurnality of highly effective avian predators is critical. If so, the pressures on herbivores and omnivores to be diurnal allows invertebrate eaters to seem relatively nocturnal by comparison. The outlier is ant specialists (see below).
- Highest proportion of species reliant on crypsis (23); lowest on hiding in vegetation (26). Small size would appear to facilitate crypsis, but small size would seem equally useful for hiding in vegetation. This discrepancy warrants further investigation.
- Highest proportions of species that bask indirectly (28) or thermally conform (29). Unlike plant eaters, the digestive imperative for high body temperature is attenuated or lacking. That does not seem like a comprehensive explanation for the paucity of direct basking species among invertebrate eaters. Direct basking exposes invertebrate eaters of all diel persuasions to predation by diurnal birds, and invertebrate eaters are of a particularly vulnerable size (Poulin et al. 2001).
- Lowest proportion of species that live in groups (31). Group living is greatly facilitated by mutual attraction to scarce or patchy resources such as suitably creviced rock. To some degree, small body size expands the pool of suitably creviced rock, but this may not be a complete explanation for the solitariness of invertebrate eaters.
- Highest proportion of cryptozoic species (35). Small body size seems like a very strong factor in the association of cryptozoic lifestyles with invertebrate consumption.
- Highest proportion in mesic habitats (41). Invertebrate productivity is presumably shackled to primary productivity, but so is predator abundance. Do invertebrate eaters have populations less susceptible to predators than herbivores, omnivores, and carnivores?
- Largest mass-adjusted mean clutch count (47). Clutch counts are difficult to align with total reproductive effort because so many small invertebrate eaters have very short interclutch intervals. More reproductive frequency data are needed to complete the picture. That so many small lizards have the highest mass-adjusted clutch count, and possibly the shortest interclutch intervals, suggests high mortality.
- Lowest proportion of species introduced (51). Small invertebrate eaters would be good candidates for stowing away, at least in the egg stage. They do not necessarily make great pets in that provision of live prey is onerous. This association between invertebrate eating and low introduction rates supports the empirical observation that lizard introductions have become strongly associated with pets rather than stowaways (Kraus 2009).

Myrmecovores (ant dietary specialists) are notable for the following (table 3.5 trait number):

- Highest proportion of species with crests (5). As noted earlier under "Morphology" (table 3.2, trait 10), the associations of crests with ant eating could be either direct (e.g., ants are less likely to swarm and attack something that lacks the typical lizard outline as seen through the ant's compound eyes) or indirect (e.g., ant eating occurs in species that have crests for some other reason, perhaps macropredator defense). Crests or other spines (*Moloch*, *Phrynosoma*) may disrupt their saurian outline, though many saurian ant eaters lack stiff spines or crests. If stiff spines are present in *Moloch* and *Phrynosoma* as a macropredator defense, one has to posit that those two taxa uniquely share predators that are especially deterred by spines. Gape-limited snakes have been suggested as the predator targeted by this defense, yet saurophagy by snakes is hardly limited to *Moloch* and *Phrynosoma*. See the *Moloch* and *Phrynosoma* entries in chapter 5 and color plates for photographs of these taxa and additional discussion of their differences and commonalities.
- Most female-biased mean SSD (9). This suggests that lizard myrmecovores have limited sexual selection based on intermale contests (see supporting traits

32, 33). In the absence of intermale contests, one might expect m=f monomorphism rather than m<f SSD. One can imagine fecundity selection pressure in myrmecovores leading to large female size, and the clutch count evidence (47) for this is supportive. Large home ranges might reward high male mobility (large areas need to be searched for potential mates) and therefore relatively small and nimble males, and the home-range size evidence (16) partially supports that possibility.

- Highest sexual dichromatism (10) and short-term color change (7). High sexual dichromatism seems inconsistent with female-biased SSD. However, dichromatism is not exclusively a reflection of intermale contests. Are interfemale contests prevalent in ant specialists? Higher resolution social data are needed. Short-term color change is probably associated with crypsis.
- Highest proportion of sit-and-wait foragers (13) and lowest reported to forage actively (15). Ants rush frenetically about with the zeal often subjectively associated with endotherms. There is no reason for a lizard to go searching for individual ants once a trail has been located. Just sit cryptically and wait for the flow of workers to continue presenting themselves.
- Absence of crepuscular species (tied with herbivores) (19). The diel pattern of ant eating is a bit mysterious. Although there is a strong turnover in ant species at dawn and dusk, total activity is relatively uniform throughout the 24 hours (Hölldobler and Wilson 1990). The absence of crepuscular myrmecovores may reflect the inefficiencies inherent in changing targets and therefore foraging locations during the low light ant species turnover.
- Highest proportion of diurnal species (20) and lowest of nocturnal ones (21). The paucity of nocturnal ant specialists can probably not be attributed to the paucity of nocturnal ants. Either predation on ant eaters is more problematic at night (many small mammals emerge at dusk), or ant-eating lizards are less able to find and subdue their prey at night (do nocturnal ants have better defenses? Are lizard eyes poorly suited to finding ants in the dark?). This deserves more study.
- Least proportion reliant on crevices (25). The negative association of ant specialists with crevices is notable but not readily explained. The morphology of archetypical terrestrial ant specialists (wide and low) seems well matched to crevice dimensions (wide and low).
- Lowest mean body temperature (30). The nominal 0.1°C temperature difference between ant specialists and eaters of other invertebrates is trivial and unlikely to be of ecological significance.
- Lowest proportion showing male dominance hierarchies (33). This fits in with the paucity of evidence for intermale contests among ant specialists. Why so few contests? Perhaps the fairly even dispersion of females (who presumably space themselves out to match prey resource distributions, rather than refugia or mates) precludes male opportunities for mate monopolization. Mate monopolization is easiest in a planar microhabitat, and as noted immediately below, ant specialists are usually arboreal.
- Highest arboreal (34); lowest proportions that are semiaquatic (not in table) or terrestrial (37). This is a good reminder for North American herpetologists that ant specialists are mostly not like *Phrynosoma*. The aversion of ant specialists to riparian zones is mysterious. In colder climates ants may be scarce in heavily shaded sites, but that factor is unlikely to play a role in most lizard habitats.
- Absence from the driest belt (38). This too comports badly with the previous bullet's apparent aversion to riparian zones. Perhaps ants are moisture-limited and therefore rare in the hyperdry rainfall belt.
- Lowest proportion of endemic species (48). Ants can be highly habitat specific; the absence of a comparable association between endemism and ant eating suggests that many myrmecovorous lizards are able to substitute prey, at least to some degree. Invasive ants in North America have been found to be beyond the limits of suitable prey substitution for some *Phrynosoma* (Fisher et al. 2002).
- Shortest longevity (49). The mostly arboreal myrmecovores may live in a habitat that is vulnerable to heavy avian predation.
- Lowest proportion of species at conservation risk (by several different metrics) (52, 53). This is consistent with low endemism and some ability to substitute ant prey.

In some respects, the ant specialist (myrmecovore) category is unusual in being a subset rather than an equivalent class in comparison to invertebrate eaters. The directions of correlations between each of these diet classes and related axes for each of the 29 traits for which they both had significant correlations were tabulated for ant specialists and invertebrate generalists. For 10 traits, ant specialists and invertebrate eaters exhibited the same sign, but in a majority (19 of 29) they exhibited opposite signs. This suggests that ant specialists are appreciably divergent from those lizards eating other sorts of invertebrates and probably should not be lumped with them (cf. Schwenk 2000). Opportunistic consumption of ants, however, is common in invertebrate eaters (see "Diet" in chapter 2).

Omnivores are the most generalist and least distinct diet class:

- Longest tails (8). The pressures responsible for this association are unclear to me.
- Highest proportion of species showing seasonal color (11). Although not league-leading, omnivores share many traits with the strongly sexually selected herbivores: sex differences in size, sex differences in color, small home ranges, diurnality, bask directly, group living, and males defend space. Seasonal color often accompanies these traits, though whether there is a unique signature to seasonal color (which conflates male and female colors) is yet unknown.
- Highest proportion of species that have been introduced (51). Possibly the high rate of species introduction reflects the ease with which omnivores can find suitable food in a new habitat, or the ease of feeding omnivorous pets and therefore their frequent occurrence in the pet trade.

Support for the supposition that the shortness of this list is associated with dietary generalism comes from the high proportion (36/53 = 68%) of records in which omnivores have a mean value intermediate between those of invertebrate eaters and herbivores.

Idiosyncratic differences among diets are more prominent than sweeping generalizations, other than general allometry (e.g., carnivores are larger than insectivores). However, a few generalizations may be tentatively extracted from table 3.5. The strongest overall relationships with diet are found with size (2–4), SSD (9), sexual dichromatism (10, 11), diel activity (20, 21; lizards do not eat plants or ants in the dark, despite the opportunity), incubation duration (46), clutch count (47), and population density (50; plant eaters are denser). Incubation duration is notable in that there seems to be a general trophic level trend: lizards eating more plants or ants have shorter mass-adjusted incubations than those eating higher on the food chain (causation not obvious). A similar plants-n-ants similarity is expressed with regard to elevation; those lower on the food chain trend toward higher elevations (despite the challenges of digesting cellulose in colder temperatures).

Absence of an expected relationship can also be noteworthy; moves per minute (MPM) bears no statistical relationship to diet. Are stalking vertebrates and plants really that similar? Something is missing from this picture. Predation? Proportion of time moving (PTM) is not included in table 3.5 because only one of five diet types (ant specialists) showed any statistical signal (unsurprisingly, ant specialists moved slightly less of the time than other lizards). Whatever is captured by MPM and PTM does not appreciably reflect diet.

Foraging

Thirty-three notable associations of foraging mode are shown in table 3.6. Foraging modes are correlated with a large number of other natural history traits, most conspicuously: body size, body shape, relative tail length, crests, pores, sexual dichromatism, MPM, PTM, crypsis, flight, diurnality, wariness, thermal behavior, and arboreality. Some of these are probable because of the way the foraging modes are defined. For example, one is not surprised to learn that sit-and-wait foragers move less often than active foragers and have smaller home ranges. Other attributes, for example sexual dichromatism, pores, and wariness, do not have so obvious a connection.

Among the more notable traits associated with sit-and-wait foragers are the following (table 3.6 trait number):

- Highest mass (1) and most corpulent body shape (2). This body shape comports with the reduced vagility of sitting-and-waiting, as does the next bullet, but just because sedentariness permits the acquisition of greater bulk, does it select for it? One might suppose that the most contrasting foraging mode, active foraging, would select for larger body size—to reduce the number of predators that can outrun and subdue the exposed active forager—but the evidence does not support this.
- Fewest elongate body forms (3). Elongation seems to be associated with penetration of small spaces, such as is done by cryptozoids (but see "Morphology" above).
- Highest frequency of crests (6), secretory pores nearly universal (7), relatively longer male tails (10), and sex differences in color (11). All of these are associated with sexual selection. Sexual selection seems to operate at a faster clip in lizards that compete for small sites, such as crevices or mating space (see "Sexual Size Dimorphism" and "Color" above). Note that in sit-and-wait foragers, crests and pores are positively associated, whereas in arboreal lizards they are inversely related. In chapter 4, evidence is presented that, among canopy walkers, pores are strongly associated with diurnality (as in sit-and-wait foragers), but pores are scarce in nocturnal species, such as nocturnal canopy walkers.

Table 3.6. Correlates of lizard species' foraging modes: sit-and-wait, cruise, or active foraging. Note that 1–3 foraging modes were reported for each species. Conventions as in table 3.1.

Trait	Condition	Foraging			Units
		Sit-and-wait	Cruise	Active	Contrast (*n*)
1	Mass excl. Iguanidae/Varanidae	8.0 *** weak+	7.2	5.3 *** weak–	grams Others w/ mass and foraging data (1106)
2	Shape index	0.42 *** strong+	0.19	0.12 *** strong-	Std. resid. Others w/ foraging data (1149)
3	Elongate lizards	4.5 *** mod–	11.8	14.5 *** mod+	% of focal that are elongate Others w/ foraging data (1149)
4	Tail length (× SVL)	1.46 *** strong–	1.22 *** mod–	1.71 *** strong+	Ratio Others w/ foraging data (1149)
5	Tail break lizards	59.4 *** mod–	57.9	71.3 *** mod+	% of focal which break tails Others w/ foraging data (1149)
6	Crests	13.3 *** strong+	10.5	2.4 *** strong–	% of focal w/ crests others w/ foraging data (1149)
7	Secretory pores	90.0 *** mod+	23.7 * weak+	9.3 *** mod–	% of focal w/ pores Others w/ foraging data (1149)
8	Any color change	27.5 *** strong+	42.1 *** mod+	6.6 *** strong–	% of focal that color change Others w/ foraging data (1149)
9	Advanced color change	13.5 *** mod+	26.3 *** mod+	2.7 *** strong–	% of focal that change hue or pattern Others w/ foraging data (1149)
10	M:F tail length ratio	1.07 * weak+	1.07	1.04 0.05 weak–	Ratio Others w/ tail length ratios (209)
11	Sexes differ in color	46.4 *** strong+	28.9	26.9 *** strong–	% of focal that differ in color Others w/ foraging data (1149)
12	Seasonal color	64.2 *** strong–	87.5	96.6 *** strong+	% of males that change seasonally Others w/ seasonal color data (150)
13	Sexual differences rollup	1.65 *** strong+	1.61 * mod+	1.03 *** mod–	Mean score 0 (no diffs.) to 6 (all diffs.) Others w/ foraging data (1149)
14	Carnivores	2.9 ** mod–	1.7	7.5 *** mod+	% of focal that are carnivores Others w/ foraging and diet data (1025)
15	Ant specialists	7.6 ** mod+	5.0	3.3 ** mod–	% of focal that are ant specialists Others w/ foraging and diet data (1025)
16	Moves per minute (MPM)	37.9 *** strong–	47.6	63.3 *** strong+	Percentile ranks (0–100) Others w/ foraging and MPM data (139)
17	Proportion of time moving (PTM)	30.3 *** v. str.–	59.6	70.4 *** v. str.+	Percentile ranks (0–100) Others w/ foraging and PTM data (126)

Table 3.6 continued

Trait	Condition	Foraging			Units
		Sit-and-wait	Cruise	Active	Contrast (*n*)
18	Crypsis as primary pred. conceal.	40.1 *** strong+	68.9 *** strong+	9.6 *** strong–	% of focal that rely on crypsis Others w/ pred. tactics and foraging data (439)
19	Flight as primary pred. conceal.	10.5 *** strong–	6.7 * mod–	32.9 *** strong+	% of focal that rely on flight Others w/ pred. tactics and foraging data (439)
20	Home-range area (mass adj.)	0.05	0.14	0.31 ** strong+	Std. resid. Others w/ home-range data (150)
21	Active at any time (cathemeral)	10.4 0.08 weak+	13.5	6.9 * weak–	% of focal that are cathemeral Others w/ diel and foraging data (1241)
22	Active in low light (crepuscular)	1.7 0.08 weak–	9.5 ** mod+	2.8	% of focal that are crepuscular Others w/ diel and foraging data (1241)
23	Active in daytime	71.8 *** mod–	41.9 *** strong–	86.9 *** strong+	% of focal that are diurnal Others w/ diet and foraging data (1241)
24	Active at night	16.5 *** mod+	35.1 *** mod+	3.4 *** strong–	% of focal that are nocturnal Others w/ diel and foraging data (1241)
25	Calls spontaneously	7.5 *** mod+	18.4 *** mod+	2.1 *** mod–	% of focal that call spontaneously Others w/ foraging data (1149)
26	Conspicuously wary	54.0 * mod–	34.5 ** mod–	71.0 *** strong+	% conspicuously wary Others w/ foraging and wariness data (367)
27	Basks directly	82.5 ** mod–	76.7	91.8 *** mod+	% that bask directly Others w/ foraging and bask data (611)
28	Thermally conforms	18.5 *** mod+	16.7	7.4 *** strong–	% of focal that are thermoconformers Others w/ foraging and bask data (611)
29	Arboreal microhabitat	37.3 *** strong+	39.2 ** mod+	13.8 *** strong–	% arboreal Others w/ foraging and microhabitat data (1143)
30	Cryptozoic microhabitat	5.8 * weak–	6.8	9.6 ** mod+	% cryptozoic Others w/ foraging and microhabitat data (1143)
31	Terrestrial microhabitat	55.9 *** mod–	51.4 * weak–	74.3 *** strong+	% terrestrial Others w/ foraging and microhabitat data (1143)
32	Mass-adjusted clutch/litter count	–0.068	0.361 0.056 mod+	0.037	Std. resid. difference Others w/ foraging and repro. data (944)
33	Relative clutch mass	0.205	0.277	0.208	Ratio Others w/ foraging and RCM data (250)
34	Spp. native on any island	36.1 ** mod–	42.1	44.5 ** mod+	% on islands Others w/ foraging data (1149)

- Highest proportion of species where the seasonal color is exhibited by females (see "Color" above).
- Lowest proportion of crepuscular activity (22). Why would sit-and-wait lizards be averse to picking up prey as the light dims? One possibility is thermal; they may be loath to remain outside of a protective crevice as they cool—the extra food is not worth the risk. Or the prevalence of carnivores may increase in dim light (table 3.5, trait 19). The frequency of sit-and-wait foragers at night is moderate (24), suggesting that sit-and-wait is successful for species adapted to cooler temperature environments.
- highest proportion of thermoconformers (28). This tends to refute the notion that crepuscular activity is shunned because of its thermal constraints, but perhaps conformance is easier when it is sufficiently warm (i.e., during the day).
- Lowest proportion of species that are cryptozoic (30). Evidently the rate at which prey move through a cryptozoic medium such as leaf litter is too small to support sit-and-wait foraging. Possibly, the limited availability of light also constrains visual recognition of prey moving through leaf litter, but that would seem to apply to nocturnality as well (see above). Another possibility is that cryptozoic species are inadequately studied for their foraging movements.

Cruise foragers are notable for the following (table 3.6 trait number):

- Tails that are the shortest (4) and least likely to break (5). Many cruise foragers have prehensile tails, which are less likely to be breakable.
- Lowest proportion of sexual differences in color (11). The reasons for this relationship are far from clear. There is potentially a conflict between color used for crypsis (useful to cruise foragers) and color used for social signaling (usually conspicuous).
- Fewest carnivores (14). Some carnivores, perhaps especially semiaquatic carnivores, may be unrecognized cruise foragers, as cruise foraging has not been generally applied to carnivores. Or it may be as rare as indicated. Perhaps cruise foraging does not tip off invertebrate prey to the presence of the lizard, but does tip off vertebrate prey. Compound eyes do not seem to be good at detecting slow movement some distance away, but vertebrate eyes are excellent at motion detection (Fleishman 1992).
- Highest proportion relying on crypsis (18) and fewest reliant on flight (19). This pattern fits with the impression that cruise foragers are trying to remain invisible to their prey or predators.
- Highest proportion active at night (24), in low light (22), or at any time (21); lowest proportion day active (23). These generalizations rub against the image of cruise foragers as chameleons, but they would certainly fit with the notion that cruise foragers are trying to remain invisible.
- Highest proportion vocalizing spontaneously (25). Spontaneous vocalizations are also associated with cathemeral activity. Which is causing which? Do cruisers vocalize because they are cathemeral or cathemeral species vocalize because they are cruisers? Why is it important for cruisers/cathemerals to be heard? More information is needed.
- Smallest proportion of conspicuously wary species (26). This association is probably related to the cruisers' dependence on crypsis. Cryptic species appear to understand, perhaps unconsciously, that sudden flight ruins their crypsis.
- Lowest proportion that bask directly (27). Does positioning itself in direct sunlight compromise a cruiser's crypsis? Warmth would appear to benefit cruisers, but this association suggests a drawback.
- Highest proportion that are arboreal (29) and lowest proportion that are terrestrial (31). No real surprise, as cruising must be much safer within the visually confusing confines of foliage.
- Highest mass-adjusted clutch count (32). Cruising and crypsis are compatible with bulky clutch/litter counts, but are they also compelled to have many young by virtue of high mortality? For cruisers overall (i.e., chameleons may be an exception) there is no clear signal with regard to longevity.
- Highest relative clutch mass (33), though no foraging mode was significantly associated with relative clutch mass (RCM). The trend toward larger RCM in cruise foragers fits well with the large clutch counts of cruise foragers (preceding bullet), and with the generalization that many chameleons are cruise foragers.

One question that arises with regard to cruise foraging is whether it is a distinct type or simply a nondescript intermediate condition on the continuum from high movement rates to low ones. Most sources did not distinguish cruise foragers, so cruising is undoubtedly underreported and possibly distained. Support for the "nondescript intermediate" characterization comes from MPM and PTM (fig. 3.5).

Close inspection of table 3.6, however, indicates that

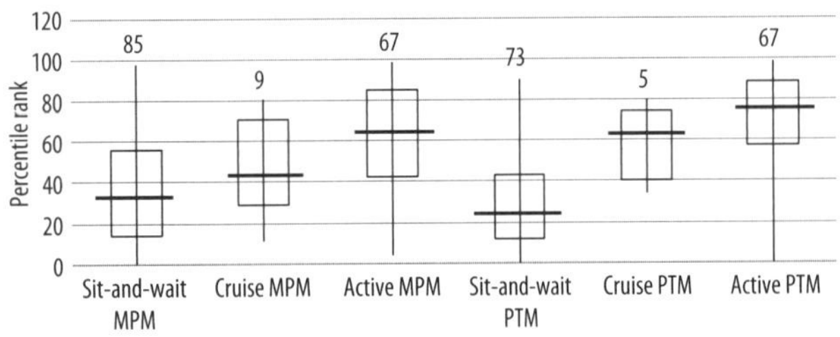

Fig. 3.5. Interquartile ranges (box) and full range (vertical line) for each of the MPM (moves per minute) and PTM (proportion of time moving) percentiles. Number above line is sample size in species. Note that these box plots are not representing confidence intervals or 95% ranges.

cruise foragers are often not in intermediate positions with regard to the traits correlated with foraging mode. For example, with regard to sexual dichromatism (11), this trait is exhibited by 92% of sit-and-wait foragers and by 78% of active foragers, but by only 29% of cruise foragers. Among the 34 traits cataloged in table 3.6, cruise foragers are intermediate in only 15 of them—indistinguishable from the 16 traits for which sit-and-wait foragers are intermediate. Active foragers are more distinct, being intermediate in only 4 of 34 traits. Using criteria of being within 30% of either sit-and-wait or active forager values (scaled to a sit-and-wait to active continuum length of 1), 11 times (of 34) cruisers are near sit-and-wait, 7 times near active, and 13 times near neither. Cruise foraging appears to be sufficiently unique to warrant recognition on par with the earlier-described foraging types, however overlapping all three types may be.

Note the high overlap in each objective metric for each of the hypothesized foraging modes (fig. 3.5). A 35th percentile score in MPM is consistent with the behavior of species that have been described as sit-and-wait, cruise, and active foraging. If one assumes (cf. this work) that each species can exhibit only one of these three foraging types, this overlap implies either that the three categories are somewhat indistinct, or that the MPM does a poor job of separating them. See the *Meroles* genus account (chapter 5) for additional complexity regarding this issue.

Active foragers are the most distinct type (table 3.6), being notable for the following:

- Lowest mass (1) and most svelte (2) and elongate (3) body forms. These fit in roughly with the physical imperatives associated with moving moderately quickly over large distances. The elongate form is not associated with high sprint speeds, however, unless the species lifts its legs (or is functionally legless) and propels itself by serpentine motion when pressed (e.g., pygopodids).
- Longest tails (4) and tails most likely to break (5). Tails can be used for propulsion and sacrificed when the alternative is capture.
- Lowest frequency of crests (6), pores (7), and color ornamentation (8, 9, 11, 12); if they exhibit seasonal color (12), it is almost exclusively males that do so. Ornaments and colors would seem to draw the attention of predators when active foragers are exposed and potentially far from a known refuge. The costs of possessing pores are much less certain. Given the chemosensory skills of many active foragers, one might expect communication between the sexes to be primarily chemical. The paucity of pores in active foragers suggests otherwise. What is the pressure selecting against pores in active foragers? No defense of space?
- Highest proportion of carnivores (14) and lowest proportion of ant specialists (15). Finding ants (a fast-moving but somewhat predictably located prey) by active searching makes no sense—moving prey should be sought by immobility; immobile prey should be sought by searching (Bell 1991, Hill 1994). The association between active foraging and vertebrate prey is less obvious. The mean size of active foragers is small (1) at 5.3 g. Thus active foragers are not obviously of a size suitable for eating vertebrates. Perhaps their exposed movements force them to possess the flight speed that is also useful for capturing other lizards.
- Largest mass-adjusted home ranges (20). Expected for active foragers, though the rationale is not clearly delineated: chemosensory prey detection does not inherently require a large search area, though herpetologists have grown accustomed to the notion that active foraging takes place over large areas. There are numerous exceptions.
- Highest proportion diurnal (23), lowest nocturnal (24). Not a given for foragers that often use chemosensory modalities, which is compatible with darkness, but a strong effect. Nocturnal ground lizards move conspicuously slowly, apparently to prevent detection by predators with good hearing. If sound-based predation is potent at night, the noisy movements of many active foragers would be a very poor fit with nocturnality.
- Highest proportion conspicuously wary (26). Active foragers are exposed and dependent on flight for their safety.

- Highest proportion that bask directly (27) and lowest of thermoconformers (28). To be able to outrun endothermic predators, an actively foraging lizard must be near its peak sprint performance, and that implies a high body temperature.
- Lowest proportion arboreal (29), highest cryptozoic (30) and terrestrial (31). Fast movement in the canopy is likely to draw the attention of sharp-eyed predatory birds. Active-foraging cryptozoids are very different from active-foraging terrestrial species, in that the cryptozoids are most likely to evade predators by being in, or diving back into, the litter or other cover, whereas the terrestrial species outrun their pursuers. This suggests that active foraging is a composite of two fairly different lifestyles. The ecological business models (chapter 4) separate these two types of active foragers.
- Highest proportion on islands (34). A first approximation would be that mammalian predation (a big threat to active foragers) is less prevalent on islands. One would like to see cryptozoids and terrestrial active foragers distinguished before embracing this hypothesis. In chapter 4, wide-ranging chemosensors (which are terrestrial active foragers) are shown to be present on islands at exactly the same frequency (40%) as are all lizards, whereas 55% of litter cryptozoids are found on islands. Apparently the insular benefit to active foragers is delivered preferentially to the cryptozoic active foragers.

Table 3.6 does not tabulate the many traits found to lack a statistically significant relationship to foraging modes: longevity; species endangerment; population density/biomass; indirect basking; sand, crevices, or vegetation use for predator concealment; limb reduction; and reproductive mode. The table also omits traits found to be only weakly correlated with a single foraging mode: number of introductions, SSD, and parental care.

Herrel (2007) proposed that active foragers tend to consume termites. Fifty-four percent of 252 lizards documented to have eaten termites were active foragers, whereas 52% of 608 lizards with dietary information that did not include termites were active foragers. The small spread was slightly greater for sit-and-wait foragers (58% v. 53%), but it too did not approach statistical significance ($p = 0.26$).

The negligible and statistically insignificant differences between sit-and-wait and active foragers in mass-adjusted clutch count (32) is notably at odds with the conventional dogma on foraging modes (Huey and Pianka 1981). The slight difference in means is in contrary direction (sit-and-wait have smaller mean clutches).

Among the statistically insignificant results, the most unexpected was the lack of association between foraging mode and RCM, as low RCM in active foragers is a central foraging-mode concept (Huey and Pianka 1981). My analysis differs from classical foraging-mode analyses in three major ways: (1) it included available data from all lizard species rather than from specific deserts or any other specified region or habitat, (2) it did not assume that each species is limited to only one foraging mode, and (3) plant eaters were often acknowledged to have active-foraging ability. Traditional analyses tend to regard all Iguania as sit-and-wait foragers; my treatment of many giant herbivores as active foragers raises mean RCM for the active class and lowers it for the sit-and-wait class. See chapter 4 for strong differences in RCM among ecological business models—differences that are not evident among foraging modes.

Time of Day (Diel Activity)

The time-of-day table (table 3.7) was compiled to assess the number of traits correlated with diel activity patterns (49 traits), identify the strongest associations (12 traits), and determine if diurnal and nocturnal conditions captured most of the variation. Not unexpectedly, diel activity patterns are correlated with most aspects of natural history, though some of the associations are not as one might expect. For example, foraging at electric lights is more prevalent among cathemeral lizards than nocturnal ones (trait 20) and is present at low levels even among diurnal species. The three traits with the strongest associations with time of day were the percent of color changing species that only darken (trait 10: nocturnals have many; diurnals has few such species), SSD (trait 11: diurnal lizards trend m>f; nocturnals m<f), and longevity (trait 47: cathemeral species are the longest-lived; diurnals the most fleeting).

Other notably strong correlations include the following (table 3.7 trait number):

- Shape index (2); cathemeral are most slender; diurnal the bulkiest. What is the advantage of being a svelte cathemeral? It seems likely to be indirect; for example, carnivores are most likely to be cathemeral, and slenderness would seem to be an advantage when capturing vertebrate prey (ask any weasel). However, many cathemerals are not carnivores; many are invertebrate eaters that feed at electric lights (how does slenderness help that?). More information is needed to sort out this tangled web.

Table 3.7. Correlates of lizard species' daily activity patterns: active at any time (cathemeral), active in low light (crepuscular), active in bright light (diurnal), active only in very low light (nocturnal). Conventions as in table 3.1.

Trait	Condition	Anytime (Cathemeral)	Low light (Crepuscular)	Day active (Diurnal)	Night active (Nocturnal)	Units Contrast (*n*)
1	Mass	5.6	5.3	7.5 *** weak+	3.9 *** weak–	grams Others w/ activity data (4746)
2	Shape index	–0.11 *** mod–	–0.03 ** mod–	0.35 *** mod+	–0.02 *** mod–	Std. resid. Others w/ activity data (4746)
3	Elongate lizards	18.8 *** mod+	24.1 *** weak+	7.5 *** weak–	9.5	% elongate Others w/ activity data (4746)
4	Limb reduction	10.5 *** mod+	4.6	1.9 *** mod–	8.4 *** mod+	% limb reduced Others w/ activity data (4746)
5	Crests	2.5 0.10 weak–	1.1 * weak–	9.4 *** mod+	0.2 *** mod-	% of focal w/ crests Others w/ activity data (4746)
6	Secretory pores	28.6 *** weak+	11.5	13.0 *** mod–	29.0 *** mod+	% of focal w/ pores Others w/ activity data (4746)
7	Tail length (× SVL)	1.214 *** mod–	1.416	1.584 *** strong+	0.992 *** strong–	Ratio Others w/ activity data (4746)
8	Tail break spp.	86.5 *** mod+	74.7 * weak+	48.6 *** strong–	87.9 *** strong+	% of spp. that break tails Others w/ activity data (4746)
9	Any color change	19.1 * weak+	13.8	16.4 *** weak+	7.1 *** mod–	% of focal that change color Others w/ activity data (4746)
10	Proportion of color changers that only darken	84 *** strong+	42	31 *** v. str. –	92 *** strong+	% that only darken Other color changers (676)
11	Sexes differ in size	–0.06 * strong–	0.05 * v. str.–	0.49 *** mod+	–0.24 *** V up	–2 (m≪f) to 2 (m≫f) score mean Others w/ SSD data (1371)
12	Sexes differ in color	21.2 * weak–	29.9	35.1 *** strong+	2.2 *** strong–	% dichromatic Others w/ activity data (4746)
13	Seasonal color	4.3 ** weak–	8.0	12.0 *** mod+	0.5 *** mod–	% that change seasonally Others w/ activity data (4746)
14	Blue tails	1.5	3.4	4.0 *** mod+	0.2 *** mod–	% w/ blue tails Others w/ activity data (4746)
15	Carnivores	7.0 ** weak+	6.1	3.1	1.7 * weak–	% carnivores Others w/ diel and diet data (2705)
16	Herbivores	0.87 ** weak–	0	5.59 *** mod+	0.87 *** weak–	% herbivores Others w/ diel and diet data (2705)
17	Invertebrate eaters	76.9	83.3 * weak+	65.9 *** strong–	93.0 *** strong+	% invert. eaters Others w/ diel and diet data (2705)

Table 3.7 continued

Trait	Condition	Anytime (Cathemeral)	Low light (Crepuscular)	Day active (Diurnal)	Night active (Nocturnal)	Units Contrast (*n*)
18	Ant specialists	1.75 * weak–	0	6.26 *** mod+	0.22 *** mod–	% ant specialists Others w/ diel and diet data (2705)
19	Omnivores	13.5	10.6	19.2 *** mod+	4.1 *** mod–	% omnivores Others w/ diel and diet data (2705)
20	Forages under electric lights	6.5 *** mod+	3.4	1.1 * weak–	2.9 *** weak+	% spp. using elect. light All lizards (6528)
21	Cruise foragers	9.2	20.6 ** mod+	3.2 *** strong–	19.5 *** mod+	% that are cruisers Others w/ foraging and microhabitat data (1241)
22	Sit-and-wait foragers	50.5 0.085 weak+	26.5 0.077 weak–	39.3 *** mod–	63.9 *** mod+	% that are sit-and-wait Others w/ foraging and microhabitat data (1241)
23	Active foragers	40.4 * weak–	52.9	57.5 *** strong+	16.5 *** strong–	% active foragers Others w/ foraging and microhabitat data (1241)
24	Rainfall 56–150 mm/y	4.8 * weak–	14.3	12.8	20.9 0.0501 mod+	% in dry zone Others w/ diel and rainfall data (567)
25	Rainfall 151–400 mm/y	33.3	28.6	28.1 * mod–	46.3 ** mod+	% in arid zone Others w/ diel and rainfall data (567)
26	Rainfall >3000 mm/y	3.2 0.071 weak–	7.1	11.8 ** mod+	1.5 * mod–	% in wet zone Others w/ diel and rainfall data (567)
27	Elevation <1200 m	95.6 *** mod+	96.8 * weak+	81.3 *** mod–	93.1 *** mod+	% of focal at low elev. Others w/ elev. and time data (3212)
28	Elevation >1200 m but <2400 m	38.9	41.9	47.5 *** mod+	24.8 *** mod–	% of focal at mid elev. Others w/ elev. and time data (3212)
29	Elevation >2400 m	6.1 *** weak–	3.2 * weak–	17.4 *** mod+	2.7 *** mod–	% of focal at high elev. Others w/ elev. and time data (3212)
30	Makes any sound	19.7 *** mod+	2.3	3.1 *** weak–	9.5 *** mod+	% that vocalize All lizards (6528)
31	Calls spontaneously	12.3 *** mod+	2.3	1.6 *** mod–	6.0 ** weak+	% that call spontaneously All lizards (6528)
32	Stress calls	10.8 *** mod+	1.1	1.6 *** mod–	4.3	% that call under duress All lizards (6528)
33	Males defend space	82.6	57.1 ** mod–	86.8 * mod+	83.3	% of spp. w/ defensive males Male sociality data (821)
34	Male hierarchies	2.9	9.5 ** mod–	8.9 * mod+	3.8	% w/ hierarchical males Male sociality data (821)

Table 3.7 continued

Trait	Condition	Anytime (Cathemeral)	Low light (Crepuscular)	Day active (Diurnal)	Night active (Nocturnal)	Units Contrast (*n*)
35	Conspicuously wary	20.3 *** mod+	32.2 *** mod+	11.8 *** weak+	3.2 *** mod–	% conspicuously wary Others w/ activity data (4746)
36	Crypsis as primary pred. conceal.	32.1 * weak–	29.6	42.4	60.7 *** mod+	% that rely on crypsis Others w/ activity and pred. conceal. data (1049)
37	Flight as primary pred. conceal.	7.1	14.8	12.0 * weak+	4.7 * weak–	% that rely on flight Others w/ activity and pred. conceal. data (1049)
38	Sand as primary pred. conceal.	11.9 * mod+	14.8	4.5 ** mod–	6.5	% that rely on sand Others w/ activity and pred. conceal. data (1049)
39	Arboreal micro-habitat	37.1	21.7 ** weak–	32.0 *** mod–	47.2 *** mod+	% arboreal Others w/ microhabitat data (4315)
40	Cryptozoic micro-habitat	22.2 *** mod+	25.3 *** weak+	7.0 *** mod–	10.4	% cryptozoic Others w/ microhabitat data (4315)
41	Subterranean microhabitat	4.1 ** weak+	2.4	0.5 *** mod–	4.9 *** mod+	% subterranean Others w/ microhabitat data (4315)
42	Terrestrial micro-habitat	34.6 *** mod–	49.4	59.3 *** strong+	37.3 *** mod–	% terrestrial Others w/ microhabitat data (4315)
43	Lives on occupied buildings	17.9 *** mod+	10.5	5.8 *** mod–	13.4 *** mod+	% of spp. living on occ. bldgs. Others w/ microhabitat data (4359)
44	Oviparous spp.	79.0 ** weak–	87.5	81.8 *** mod–	95.1 *** mod+	% oviparous Others w/ activity and repro. data (3592)
45	Mass-adjusted clutch/litter count	–0.094	0.093	0.349 *** mod+	–0.437 *** strong–	Std. resid. diff. Others w/ clutch and diel data (2636)
46	Endemic to a "country"	45.8 *** weak–	50.6	56.9 *** weak–	68.4 *** mod+	% endemic Others w/ activity data (4746)
47	Longevity (mass adj.)	0.51 *** strong+	0.49 * strong+	–0.11 *** strong–	0.16 0.09 weak+	Std. resid. Others w/ longevity data (847)
48	Species intro-duced	7.1 * weak+	8.0	4.6	2.7 ** weak–	% of spp. introduced Others w/ activity data (4746)
49	At conservation risk	53.3	55.0	52.9 ** weak+	42.5 *** mod–	% considered at risk Others evaluated for risk (2668)

- Tail length (7): diurnal tails are longest; nocturnal shortest. Most of the advantages of long tails accrue to lizards moving quickly. Nocturnal movement is rarely quick, though there may be several the reasons for this: (1) quick movement draws the attentions of sound-sensitive predators, (2) quick movement is apt to be misdirected if the lizard cannot see where it is going, and (3) in contrast to lizards that are out during the daytime (when there are almost twice as many terrestrial as arboreal lizards), nocturnal lizards are more apt to be arboreal than terrestrial; rapid movement is more difficult in trees.

- Tail breaks (8): nocturnal tails are most fragile; diurnals least so. Does this mean that predators target their pounces less precisely at night?
- Sexes differ in color (12): >1 in 3 diurnals do so; <1 in 45 nocturnals do. This does not seem surprising. Color vision is greatly reduced at night.
- Invertebrate eating (17): nearly universal in nocturnals; lowest in diurnals. See "Diet" above.
- Active foraging (23): commonest in diurnals; least in nocturnals. Active foraging is common in litter cryptozoids, but less than 2% of litter cryptozoids are known to be active at night (albeit there may be sampling difficulties). Does this reinforce the notion that making noise on the ground at night (think of a lizard moving in leaf litter) is a very bad strategy in the presence of mammalian predators?
- Terrestriality (42): lowest in cathemerals; highest in diurnals. It is not clear to me why cathemeral behavior would be strongly penalized on the ground. Presumably it is related to the penalty for being on the ground at night: the proportion of lizards that are terrestrial is about equal whether nocturnal (37%) or cathemeral (35%).
- Lives on occupied buildings (43): highest in cathemerals; lowest in diurnals. Presumably the cathemerals have the flexibility to stay out of the way of humans and predatory pets, whereas diurnal lizards' options in this regard are more limited.
- Clutches are largest in diurnal species (45), smallest in nocturnal species. These associations suggest that predation pressure is highest during the day, but additional information on brood frequency is needed to complete this picture.

The association between tail breaks and nocturnality (8) is a bit surprising given the close relationship between blue tails and diurnality (14). Expendable tails may be conceptualized as something to frustrate pouncing visual predators, but tail-breaking species are roughly twice as frequent among nocturnal and cathemeral lizards as among diurnals.

A reasonable expectation was that the two rarer and somewhat hybrid diel patterns (cathemeral and crepuscular) would have trait means intermediate between those of the numerically dominating diurnal and nocturnal samples. The 49 significant associations make possible a quantitative comparison between the different diel patterns. The expected intermediate pattern was found in 18 of the 49 analyses, but it was violated more often than honored. For example, crepuscular was the extreme condition in 17 of 49 analyses (though the sample of crepuscular species is small). Cathemeral lizards were extreme in 12 of 49 analyses. Both crepuscular and cathemeral seem sufficiently unique to warrant recognition.

Conditions in which the crepuscular mean was most extreme are as follows (table 3.7 trait number):

- Elongate (3). This implies that crepuscular lizards are often cryptozoic or otherwise dependent on penetrating narrow pathways. The association with time of day is obscure.
- Lacks secretory pores (6). This is puzzling, given the strong positive association of pores with nocturnality and the high representation among cathemeral lizards. A more plausible indirect association is with cryptozoic behavior, but why would cryptozoic behavior find secretory pores to be of little value? It is known that ants are very good at following chemical trails (Wilson 1975) and possibly would follow such trails to harass or attack resting cryptozoids. Pores are shown in chapter 4 to be exceptionally rare in diurnal canopy walkers, subterranean foragers, sand sharks, and wide-ranging chemosensors.
- Avoids eating plants (16) and ants (18). The paucity of ant eating may relate to the pores and association of cryptozoids with crepuscular activity mentioned in the preceding bullet, but this is very speculative. The avoidance of plant eating would seem to be more likely to reflect predation risks in low light (note preference of carnivorous lizards for crepuscular activity, mentioned in the discussion under "Diet" above). This inference is also very speculative.
- Cruise forager (21), not sit-and-wait (22). The association with cruise foraging may be related to the special challenge of cruise foragers to move in a manner that minimizes visual motion detection by both the lizard's prey and predators (see discussion under "Foraging" above). Low light would seem to be a favorable environment for minimizing visual detection.
- Mid elevation (28). No obvious connection.
- Lacks vocalizations (30) and stress noises (32). These negative associations are particularly striking in light of the strong positive associations of vocalizations with cathemeral lizards. Why would vocalizations be a good match for lizards that span the low light activity period, but a poor match for those that specialize in it?
- Males do not defend space (33) but do have dominance hierarchies (34). Spatial defense often entails conspicuous activity, which is not a good fit to spe-

cializing in low detection (low light, often cryptozoic behavior). Dominance hierarchies, however, can be established and maintained with infrequent and sometimes subtle interactions.

- Dives into sand (38), but does not flee (37) or remain cryptic (36). Sand diving is a good match to cryptozoic lifestyles. Flight seems to depend on high body temperatures associated with diurnality. Crypsis would seem to work in low light, but apparently there is some other issue, as yet unidentified.
- Is cryptozoic (40), but not arboreal (39). The positive association with cryptozoic behavior is mentioned often in the preceding bullets, but the aversion to arboreal microhabitats is not. Although reptile behaviorists often think of dusk activity as being crepuscular, the other half of crepuscularity occurs at dawn, when avian predators are most active in arboreal settings and lizard locomotion is most impaired by low temperatures.
- Introduced (48). This nonsignificant association has a very small effect size. More evidence is needed.
- At conservation risk (49). See remarks for preceding bullet.

Conditions in which the cathemeral mean was most extreme are as follows (table 3.7 trait number):

- Most slender (2). This is similar to the crepuscular lizards and suggests an association with cryptozoic behavior.
- Most species with limb reduction (4). Another association with narrow locomotion pathways and cryptozoic or subterranean microhabitats.
- Most color change (9). This seems to be due primarily to lizards that only darken (10), which is consistent with the cathemeral lizards' shift between day and night activity.
- Forages under electric lights (20). The capability to operate in a well-lit environment at nighttime seems like a good fit for cathemeral lizards.
- Lowest proportion in dry zone (24). No hypothesis leaps out.
- Calls spontaneously (31) and issues stress noises (32). The strong contrast with crepuscular species is striking (see discussion in the preceding bulleted list).
- Lacks male hierarchies (34). Again a very strong and nonobvious contrast with crepuscular lizards (see discussion in the preceding bulleted list).
- Exceptionally wary (35). Cathemeral lizards are exposed to all possible lizard predators. As they cannot specialize in evading a particular suite of predators, perhaps they fall back on generalized wariness?
- Forages on occupied buildings (43), but not on the ground (42). The association with occupied buildings may be driven by the advantages of foraging under electric lights. The aversion to terrestrial foraging is less clear, but may reflect vulnerability to terrestrial predators such as house cats. This may be a situation in which an analysis segregating diurnal and nocturnal terrestrial predation pressure would be rewarding.
- Viviparous species (44). Viviparity is associated with high altitudes and latitudes (low temperatures), whereas one would expect cathemeral lizards to specialize in warm climates (note positive association with low elevation belt and negative association with high elevation belt, for example). It is perhaps notable that species can be cathemeral either because they are night active in the hot summer months and day active in the spring and fall (seasonally cathemeral) or opportunistic on a daily basis (chronically cathemeral). The association with viviparity seems more likely to be manifest among the seasonally cathemeral.
- Absence of endemism (46). To the degree that cathemeral lizards are environmental generalists, they are less likely to be geographically restricted.
- Highest longevity (47). High longevity is most associated with high survival/low predation. It is not clear why lizards that are opportunistically active by time of day should experience lower predation: one untested possibility is that they switch their activity periods in response to predation pressures and thereby evade the worst predation risks. High longevity might also be promoted in climatically unpredictable environments, which reward strong iteroparity (breeding extended over many cycles), to hedge against recruitment failure during a string of reproductively adverse seasons. To some degree, these two hypotheses fit with each other, but they do not seem to fit with the cathemeral lizards' aversion to the dry rainfall belt (24).

Here are the traits that follow the convention in which the diurnal species are at one extreme, the nocturnal species are at the other, and the cathemeral and crepuscular species are intermediate (table 3.7 trait number):

- Mean mass (1). Day high; night low.
- Crests present (5). Day high; night low.
- Tail length (7) and breakage (8) seem to be the inverse of each other. The longest tails are the least breakable, and vice versa. In this case, day tails are longest and

least breakable; night tails are the shortest and most breakable.

- Color changes that do more than darken (change hue or pattern) (10). Day high; night low. Color is of low value at night.
- SSD (11). Day high; night low. Sexual selection seems to operate at a higher pitch during the day.
- Sexes differ in color (12) or show seasonal color (13). Day high; night low. Again, sexual selection seems to operate at a higher, more visual, pitch during the day.
- Blue tails (14). Day high; night low. No point having a blue tail if the predator can't see color.
- Invertebrate eater (17: day low; night high) or omnivory (19: day high; night low).
- Active foraging (23). Day high; night low. As noted above, noisy active foraging at night is a poor fit due to vulnerability to nocturnal mammalian predators.
- Arid zone occupancy (25: day low; night high) and mid (28) and high (29) elevation. The latter two are the domain of the diurnals (only diurnals can be warm enough at high elevation?).
- Aboveground microhabitat (41). Diurnals dominant above ground; nocturnals below (though the sampling of the subterranean lizards is inadequate).
- Clutch/litter count (45). Day high; night low. Nocturnal lizards are often arboreal and on small perches, where the physical cost of locomotion while transporting a large clutch may be a greater relative burden. Do nocturnal arboreal species deposit clutches more often? This clutch count comparison needs to be augmented with information on brood frequency to determine whether total reproductive output differs between night and day.

Predator Concealment Tactics

Attributes correlated with a lizard's choice of predator concealment are listed in table 3.8. Particularly strong relationships exist with foraging mode (traits 14–16), microhabitat (22–25), rainfall (26–30), and longevity (32). Particularly memorable benchmarks are listed below for specific predator tactics.

Lizards that rely on crypsis were notable for the following conditions (table 3.8 trait number):

- Crests (3). On direct observation, large crests seem visually conspicuous to humans. How does that fit with crypsis? One possibility is that sexual selection has hypertrophied small crests that were visually disruptive and therefore cryptic. Another possibility is that all crests are indeed cryptic, either to the invertebrate prey or to rapid scanning by predators against the background of foliage and branching vegetation.
- m:f differences in color (7). If the sexual differences in color are conspicuous to predators, acquisition of sexual color would seem to undermine crypsis. It is perhaps for this reason that so many sexual colors are lateral or ventral, conspicuous to conspecifics (and often erected solely in the presence of conspecifics), but mostly concealed from visual predators approaching from overhead. Another possibility is that forcing predators to adopt different search images for males and females heightens the difficulty of detecting either of them. See Henke (1998) for a fascinating example of the difficulty human predators have when forced to hold more than two search images in mind.
- Invertebrate eating (12), but not omnivory (13). Invertebrate eating can be carried out solely by sit-and-wait, whereas omnivory cannot be.
- Sit-and-wait (14) and cruise (15), but not active foraging (16). Motion conflicts with crypsis.
- Nocturnality (21). One would not expect a visual camouflage (crypsis) to have a strong association with a largely nonvisual nocturnal environment, but perhaps the species that behave cryptically during the daytime are more often active at night. It would be helpful to fully understand whether the crypsis was practiced primarily by resting or active lizards.
- Arboreal (22), but neither cryptozoic (23) nor terrestrial (25). Apropos of the preceding bullet, one would not expect terrestrial or cryptozoic lizards to be exposed during their inactive period, but arboreal ones might be (many geckos bask discreetly; some are exposed when inactive). The visually complex arboreal environment is a particularly favorable microhabitat for crypsis.
- Avoids arid (27) and intermediate (28) belts, prefers mesic (29). These associations may simply reflect the greater abundance of trees (arboreal microhabitats) in wetter environments.
- Shortest longevity (32). Wetter environments tend to be more speciose and include additional predators; these biological interactions may shorten longevity. Furthermore, wetter environments may be more predictable and therefore less supportive of selection for iteroparity. Another possibility for cryptic species is that their crypsis is driven by the local scarcity of concealing refugia; lizards that are forced to remain in an exposed location when inactive may suffer greater mortality, however good the crypsis.

Table 3.8. Correlates with lizard species' primary mode of predator concealment: crypsis, flight, sand diving, retreat to a crevice or burrow, concealment in vegetation. Conventions as in table 3.1.

Trait	Condition	Crypsis	Flight	Sand	Crevice-burrow	Veg.	Units Contrast (*n*)
1	Mass	8.2	10.1	6.1 ** weak–	10.7 * weak+	7.8	grams Others w/ pred. tactics (1067)
2	Shape index	0.45 * weak+	0.50 * weak+	–0.18 *** strong–	0.47 * weak+	0.32 0.06 weak–	Std. resid. Others w/ pred. tactics (1067)
3	Crests	20.7 *** mod+	7.0 * weak–	0 *** mod–	4.9 *** mod–	11.6	% crested Others w/ pred. tactics (1067)
4	Elongate lizards	2.4 *** mod–	2.6 0.052 weak–	38.7 *** strong+	7.2	8.7	% spp. elongate Others w/ pred. tactics (1067)
5	Any color change	61.8 *** v. str.+	15.7 *** mod–	4.8 *** mod–	19.7 *** strong–	26.9 *** mod–	% color changers Others w/ pred. tactics (1067)
6	Advanced color change	49.8 *** v. str.+	2.6 *** mod–	1.6 *** mod–	7.6 *** strong–	15.7 *** mod–	% that can change hue or pattern Others w/ pred. tactics (1067)
7	Sexes differ in color	43.0	11.5	3.4 ** mod–	25.5	25.5 * weak+	% dichromatic Others w/ pred. tactics (1067)
8	Seasonal color	13.8 * weak–	21.7	14.5	23.9 ** mod+	18.6	% that show seas. color Others w/ pred. tactics (1067)
9	Blue tails	0.2 *** mod–	13.0 *** mod+	4.8	8.0 * mod+	6.2	% blue tailed Others w/ pred. tactics (1067)
10	Carnivores	0.5 *** mod–	7.7 * weak+	0	4.2	5.4 * weak+	% that are carnivores Other w/ diet data (862)
11	Herbivores	0.3 *** mod–	0	0	8.8 *** strong+	2.7	% that are herbivores Others w/ diet data (862)
12	Invertebrate eaters	85.8 *** strong+	62.6 * weak–	63.8	57.9 *** mod–	61.3 *** mod–	% that are invert. eaters Others w/ diet data (862)
13	Omnivores	7.5 *** mod–	23.1 * weak+	31.0 ** mod+	24.5 *** mod+	19.4	% that are omnivores Others w/ diet data (862)
14	Sit-and-wait foragers	76.7 *** strong+	29.5 *** strong–	57.1	56.5	58.2	% sit-and-wait foragers Others w/ foraging and pred. data (439)
15	Cruise foragers	24.0 *** strong+	3.4 * mod–	3.6	6.1	4.9 * mod–	% cruise foragers Others w/ foraging and pred. data (439)
16	Active foragers	15.5 *** strong–	79.5 *** strong+	60.7	56.5 0.051 mod+	52.5	% active foragers Others w/ foraging and pred. data (439)
17	Home-range area (mass adj.)	0.10	0.52 * strong+	1.13	–0.06	–0.24 0.06 mod–	Std. resid. Others w/ home-range and pred. data (117)

Table 3.8 continued

Trait	Condition	Crypsis	Flight	Sand	Crevice-burrow	Veg.	Units Contrast (*n*)
18	Active at any time (cathemeral)	6.0 * weak–	5.2	17.2 * mod+	9.7	8.3	% cathemeral Others w/ activity and pred. data (1049)
19	Active in low light (crepuscular)	1.8	3.5	6.9 0.057 weak+	3.5	0.8 0.06 weak–	% crepuscular Others w/ activity and pred. data (1049)
20	Active in daylight	77.9	87.0 * weak+	63.8 * mod–	80.5	83.8 0.06 weak+	% diurnal Others w/ activity and pred. data (1049)
21	Active at night	14.4 *** mod+	4.3 * weak–	12.1	6.2 * weak–	7.1 * weak–	% nocturnal Others w/ activity and pred. data (1049)
22	Arboreal microhabitat	63.3 *** strong+	10.4 *** mod–	1.6 *** mod–	20.2 *** strong–	37.8	% arboreal Others w/ microhabitat (1061)
23	Cryptozoic micro-habitat	2.6 *** mod–	2.6	21.0 *** mod+	6.9	5.8	% cryptozoic Others w/ microhabitat (1061)
24	Subterranean micro-habitat	0 ** mod–	0	19.4 *** strong+	0 * weak–	0 0.08 weak–	% subterranean Others w/ microhabitat (1061)
25	Terrestrial microhab-itat	31.6 *** strong–	87.0 *** strong+	58.1	71.0 *** strong+	54.8	% terrestrial Others w/ microhabitat (1061)
26	Rainfall 55–150 mm/y	13.6	23.3	28.6	23.2 0.051 mod+	5.9 * mod–	% of focal in dry belt Others w/ rainfall data (223)
27	Rainfall 151–400 mm/y	17.3 ** strong–	60.0 ** strong+	35.7	37.5	33.8	% of focal in arid Others w/ rainfall data (223)
28	Rainfall 401–1100 mm/y	18.5 ** strong–	53.3 * mod+	35.7	35.7	39.7 0.08 mod+	% of focal in intermed. Others w/ rainfall data (223)
29	Rainfall 1101–3000 mm/y	60.5 ** strong+	26.7 * mod–	21.4 0.09 mod–	32.1 * mod–	42.6	% in mesic belt Others w/ rainfall data (223)
30	Rainfall >3000 mm/y	9.9	0 0.050 mod–	0	3.6 0.07 mod–	20.6 ** strong+	% in wet belt Others w/ rainfall data (223)
31	Species native on any island	46.1	37.4 * weak–	35.5 0.066 weak–	48.1	56.2 ** mod+	% on islands Others w/ pred. data (1067)
32	Longevity (mass adj.)	–0.30 ** mod–	–0.07	0.33 * strong+	0.11 * mod+	–0.08	Std. resid. Others w/ longevity data (332)

Lizards that rely on flight were notable for the following (table 3.8 trait number):

- Carnivores (10), not herbivores (11). These associations are consistent with their shape index values. Carnivores are built for speed and herbivores are built to pack in leaves.
- Active foragers (16), not sit-and-wait (14) or cruise (15). Almost by definition.
- Not cathemeral (18) or nocturnal (21). The aversion to

nocturnality is probably related to the need for heat to maximize sprint speed. Is the physiology of sprinting different in cathemeral species? Do cathemeral species have enzyme systems with broader temperature curves, or are they forced to operate under suboptimal conditions at most times?
- Absent from wet areas (30). The paucity of flight tactics in mesic areas does not seem to have been probed in the ecological literature. Wet areas offer a plethora of debris crevices and arboreal refugia that are lacking in many desert environments, but mesic sites may not have the abundance of open rock fissures characteristic of drier habitats. Where open rock fissures are sufficiently abundant, the sprinting lifestyle may be adopted without requiring precise knowledge of each crack's location. In contrast, wet environments may not so much diminish the opportunities for predator escape by flight as provide a rich panoply of alternative predator evasion options.

Lizards that hide in sand, a relatively rare condition, are exceptional for a remarkable number of traits (table 3.8 trait number):

- Lowest mean mass (1). The better to penetrate sand.
- Most slender shape (2). The better to penetrate sand.
- Absence of crests (3). The better to penetrate sand.
- Least sexual dichromatism (7). This suggests that intersexual communication takes place primarily in nonvisual environments or by nonvisual modalities. How do they do it? Pores are rare in lizards that hide in sand (table 3.2, trait 25), perhaps because they enable hostile invertebrates such as ants to trail them to their sleeping positions (see discussion under "Time of Day" above). Perhaps among sand swimmers the sexes interact only over short distances or in direct contact?
- Omnivory (13), but not carnivory (10) or herbivory (11). Omnivory is not on the table when a lizard is submerged in sand, but perhaps sandy environments have such a paucity of invertebrate prey that it pays to consume nutritious seeds when they are available or accumulate in wind-sheltered recesses. In very dry environments the only reliable source of water may be vegetation.
- Largest home range (17). This is probably dictated by the very low primary productivity of sand or dune habitats.
- Lowest proportion of diurnals (20), highest of cathemeral (18) and crepuscular (19). Deserts are infamous for their temperature swings from day to night. Excessive daytime sand temperatures may need to be avoided, as well as nighttime lows. Sand swimmers may adapt by shifting activity periods seasonally (seasonally cathemeral) or embracing moderate crepuscular temperatures.
- Cryptozoic (23) and subterranean (24) species. If a lizard can keep the small particles out of airways, sand is the easiest substrate to penetrate.
- Highest proportion in dry rainfall belt (26); lowest in areas mesic (29) or wetter (30). Wet sand is very difficult for a small lizard to penetrate. Dry sand is easy.
- Scarcity on islands (31). Large sandy areas are surprisingly rare on islands (terrestrial dunes accumulate by wind transport over long distances). The largest group of the world's islands (Southeast Asia) are too wet (sand is overgrown with trees).
- Longest longevity (32). Sand is a medium that apparently protects its saurian denizens well. Predators may also be intrinsically rare in the low productivity environments that support dunes. Longevity is associated with climatically unpredictable environments, such as the dry rainfall zone, where inhabitants must survive for a long time to ensure that they persist until there is a period favorable for recruitment.

Lizards that retreat to crevices or burrows are notable for the following (table 3.8 trait number):

- Largest size (1). Larger lizards have the power to excavate large amounts of soil. There may also be a sampling bias: small lizards may be perceived to be under rocks; larger lizards make more recognizable "burrows."
- Fattest build (2). Adding girth and mass has little penalty if one is secure in a burrow; slender lizards are more suited to speedy flight.
- Herbivory (11), but not invertebrate diet (12). Herbivores are bulky to pack in bulky foods. They need burrows to protect themselves, as they are too portly and defenseless to outrun or intimidate predators.
- Terrestrial (25) microhabitat. Other microhabitats offer other forms of concealment. The only better alternative for surface dwellers is flight, a poor option for the corpulent.
- Residence in arid belt (27). In more mesic areas, rock cracks are often filled with vegetation, and digging may be excessively difficult in damp or clayey soils. In some habitats, lizard burrowers excavate only during the seasons when soils are softer, but rely on the hardened soil layers for protection during the remainder of the year. Thus seasonally arid may be more favorable than chronically arid.

Many lizards rely on the concealment of vegetation, so veg-hiders tend to be average in many aspects, but they do stand out in a few ways (table 3.8 trait number):

- Smallest home range (17). Perhaps an indirect connection. Lizards that rely on vegetation can find shelter in very small areas, at least when the environment is wet enough to support ample plant growth.
- Least likely to be crepuscular (19). The connection is not obvious. The crepuscular sample is small.
- Avoid dry belt (26), but highest in intermediate (28) and wet (30). There is more concealing vegetation in wetter areas.
- Prevalence on islands (31). Small maritime islands often lack soil, but have ample vegetation.

Microhabitat

Microhabitat use is one of the most influential natural history attributes and is associated with about 50 other traits (table 3.9). Particularly notable correlations exist with: SSD, body mass, shape index, short-term color changing ability, relative tail length, reliance on crypsis as a primary predator defense, and home-range area. Note that the table includes both mutually exclusive microhabitat classes (arboreal/cryptozoic/subterranean/semiaquatic/terrestrial) and a nonexclusive characterization for climbing species ("climb").

The focus here is on the arboreal class; climbers are mentioned only when that category diverges from the arboreals. Based on the correlations (table 3.9), arboreal lizards are notable in relation to lizards from other microhabitats for the following (trait number):

- Lowest proportion of species that are elongate (6); highest of crests (4) and pores (5). Climbing and leaping are facilitated by long strong legs (hence not elongate). The association with crests suggests some benefit from the arboreal environment (see previous entries in this chapter), perhaps enhancement of crypsis against the jagged and visually complex background of arboreal vegetation. The benefits of pores in an arboreal environment are complicated by the interaction with time of day (see "Morphology" above). Pores seem to be beneficial primarily at night, but why are they particularly beneficial in an arboreal microhabitat? Perhaps because secretions can constitute an effective boundary marking along the linear pathways in arboreal microhabitats at night, whereas in the daytime longer-range vision guides movements, and pore secretions are frequently intermittent across the long vistas in the visually complex diurnal arboreal environment. At night, if you can sense only what is in your immediate vicinity—the bark or leaf in front of your nose—a chemical cue on the bark or leaf can be a decisive boundary signal, whereas a lizard scanning around itself in the daytime may see myriad branches and leaves, on only a tiny fraction can previously deposited chemical cues be immediately detected.
- Highest proportion of prehensile tails (10). An obvious benefit in the trees.
- Nearly greatest m>f size dimorphism (13). This needs further information. Daytime communication in the trees is often visual, enhancing the influence of larger actors. This is supported in chapter 4 by evidence that daytime canopy walkers are mostly m>f whereas nocturnal canopy walkers are mostly m<f. However, gliders and immobility experts, which may be arboreal, exhibit a female bias (i.e., m<f). Many factors appear to influence SSD in arboreal environments. See also the *Draco* account (chapter 5).
- Highest proportion of sexes differ in color (14), can change color (11), or have advanced color change (12). Color change appears particularly useful for both crypsis and communication in the trees in daylight. Sex differences in color are also reliant on daylight. It is less clear that this benefit carries over to the arboreal environment at night.
- Ant specialist diet (21). There is a mysterious interaction among arboreality, a diet of ant specialization, and diurnal activity (see "Diet" above for additional discussion of this complexity). But why is the arboreal environment favored for ant specialists? Active foraging is rarely practical in trees, as so many pathways turn out to be dead ends, active-foraging tips off visual predators, and visual predators on lizards abound in trees. Sit-and-wait foraging is an especially good match, both for the arboreal environment and for a diet of ant specialization. Ants are best taken by sit-and-wait, as they convey themselves to their predator. Immobility of the lizard is ideal for trees, as arboreal predators have difficulty detecting immobile targets.
- Sit-and-wait foraging (23). See preceding bullet.
- Lowest proportion of crepuscular species (28). Further information is needed to clarify a negative relationship between crepuscularity and arboreality. Avian foraging may be maximal at dawn and dusk.
- Highest proportion of spontaneous callers (31) or stress callers (32). Vocalization seems like a good match to the visual obstructions in an arboreal microhabitat. That visual barriers are a key is supported by

Table 3.9. Correlates with lizard species' primary microhabitat (or "climb" = scansorial = able to scale vertical surfaces). Sample size varies if shown with a plus (stated value is minimum). Other conventions as in tables 3.1 and 3.3. Microhabitat abbreviations: Arb. = arboreal; Crypt. = cryptozoic; Subt. = subterranean or fossorial; Aquat. = semiaquatic; Terr. = terrestrial (infrequently ascends vertical surfaces). "NA" indicates data not available.

Trait	Condition	Arb.	Crypt.	Subt.	Aquat.	Terr.	Climb	Units Contrast (*n*) Notes
1	Number of species (no tests)	32.0	12.1	7.2	1.1	47.7	49.6	% of all lizards of known microhabitat (4893+)
2	Mass at modal SVL	6.8 *** weak+	2.3 *** strong−	2.9 *** mod−	22.4 *** v. str.+	6.9 *** weak+	7.3 *** mod+	grams Others w/ microhabitat data (4892+)
3	Shape index	0.30 *** mod+	−0.60 *** strong−	−2.70 *** v. str.−	0.45 *** mod+	0.33 *** strong+	0.32 *** strong+	Std. resid. Others w/ microhabitat data (4892+)
4	Crests	15.0 *** strong+	0 *** mod−	0 *** weak−	12.7 0.08 weak−	2.6 *** mod−	12.2 *** mod+	% of focal w/ crests All other lizards (6528)
5	Pores	23.7 *** mod+	7.7 *** weak−	4.6 *** weak−	7.3 0.09 weak−	15.4 0.09 weak−	22.3 *** mod+	% of focal w/ pores Others w/ microhabitat data (4893+)
6	Elongate lizards	2.0 *** mod−	52.4 *** strong+	19.1 *** weak+	18.2	7.5 *** mod−	4.3 *** mod−	% of focal elongate Others w/ microhabitat data (4893+)
7	Eyes reduced	0 *** mod−	1.1 *** weak−	50.5 *** v. str+	0	0 *** mod−	0.1	% of focal eye-reduced Others w/ microhabitat data (5173)
8	Tail length (× SVL)	1.39 ** weak+	1.27 *** mod−	0.39 *** strong−	1.55 ** mod+	1.49 *** mod+	1.43 *** mod+	Ratio Others w/ microhabitat data (4738+)
9	Tail breaks	58.8	81.0 *** mod+	37.9 *** mod−				% of focal break tail Others w/ microhabitat data (4893+)
10	Tail prehensile	15.5 *** mod+	0.3 *** weak−	0 *** weak−	1.8	2.1 *** mod−	11.5 *** mod+	% of focal prehensile Others w/ microhabitat data (4893+)
11	Color change (any)	29.4 *** strong+	1.1 *** mod−	0 *** mod−	3.6 * weak−	7.6 *** mod−	24.3 *** strong+	% of focal that change color Others w/ microhabitat data (4892+)
12	Color change (advanced)	17.8 *** mod+	0.2 *** mod−	0 *** mod−	3.6	3.6 *** mod−	14.1 *** mod+	% of focal advanced Others w/ microhabitat data (4892+)
13	Sexes differ in size	0.55 *** V-down	−0.4 *** strong−	−0.6 *** strong−	0.6 weak+	0.33 V-down	0.49 *** mod+	−2 (m≪f) to 2 (m≫f) scale Others w/ SSD (1410) See table 3.3
14	Sexes differ in color	30.5 *** mod+	13.0 *** mod−	0 *** mod−	14.5 0.08 weak−	28.1 *** weak+	33.3 *** mod+	% of focal dichromatic Others w/ microhabitat data (4892+)
15	Seasonal color	5.6 *** weak−	6.2 0.07 weak−	0 *** mod−	1.8	11.6 *** mod+	9.6 ** weak+	% of focal seasonal color Others w/ microhabitat data (4892+)
16	Blue tails	1.3 *** weak−	3.8 0.09 weak+	1.1 * weak−	0	3.8 *** weak+	2.4 0.1 weak−	% of focal blue tailed Others w/ microhabitat data (4892+)
17	M:F rel. tail length	1.06 * weak−	1.03	1.21	1.08	1.09 * weak+	1.06 * weak−	Ratio Others w/ m:f ratio data (683+)

Table 3.9 continued

Trait	Condition	Arb.	Crypt.	Subt.	Aquat.	Terr.	Climb	Units Contrast (*n*) Notes
18	Carnivores	2.8	1.3 * weak–	2.5	21.2 *** mod+	3.7	4.4 *** weak+	% of focal carnivorous Others w/ diet and microhabitat (2756+)
19	Herbivores	1.8 *** weak–	0.3 *** weak–	0 * weak–	6.1	6.6 *** mod+	3.2 * weak–	% of focal herbivorous Others w/ diet and microhabitat (2756+)
20	Invertebrate eaters	77.4 *** weak+	89.3 *** mod+	96.6 *** mod+	60.6	64.3 *** mod–	71.9	% of focal invert. eater Others w/ diet and microhabitat (2756+)
21	Ant specialists	7.0 *** mod+	2.5 0.08 weak–	0.8 * weak–	0	3.8 0.07 weak–	5.4 * weak+	% of focal ant spec. Others w/ diet and microhabitat (2756+)
22	Omnivores	11.0 *** mod–	6.6 *** mod–	0 *** mod–	12.1	21.6 *** mod+	15.0	% of focal omnivores Others w/ diet and microhabitat (2756+)
23	Sit-and-wait foragers	68.8 *** strong+	35.2 * weak–	20.0	27.3	39.8 *** mod–	58.6 *** mod+	% of focal that sit-and-wait Others w/ foraging data (1123+)
24	Cruise foragers	10.1 ** mod+	5.7	20.0	0	5.1 * weak–	7.5	% of focal that cruise Others w/ foraging data (1123+)
25	Active foragers	31.9 *** strong–	72.7 ** mod+	80.0	72.7	66.5 *** strong+	46.7 *** mod–	% of focal active foragers Others w/ foraging data (1123+)
26	Home-range area (mass adj.)	–0.40 *** strong–	–0.40	NA	–0.60	0.22 *** strong+	–0.13 ** strong–	Std. resid. Others w/ home ranges (215+)
27	Active at any time (cathemeral)	7.7	17.7 *** mod+	17.6 ** weak+	13.0	4.8 *** mod–	8.0	% of focal active at any time Others w/ diet (4176+)
28	Active in low light (crepuscular)	1.2 ** weak–	5.3 *** weak+	2.7	2.2	1.8	1.3 ** weak–	% of focal active forager Others w/ diet (4176+)
29	Day active	62.7 *** mod–	52.8 *** mod–	18.9 *** mod–	80.4	78.2 *** mod+	67.2 ** weak–	% of focal diurnal Others w/ diet (4176+)
30	Night active	28.5 *** mod+	24.2	60.8 *** mod+	4.3 ** weak–	15.2 *** mod–	23.5 ** weak+	% of focal nocturnal Others w/ diet (4176+)
31	Calls spontaneously	6.7 *** mod+	3.8	1.1 * weak–	0	2.8	4.2 *** weak+	% of focal that call spontaneously Others w/ microhabitat data (4893+)
32	Stress calls	3.6 * weak+	1.6	0 *** weak–	1.8	2.8	3.9 *** weak+	% of focal that stress call Others w/ microhabitat data (4893+)
33	Conspicuously wary	6.8 *** weak–	22.6 *** mod+	1.6 *** weak–	10.9	9.7	10.5	% of focal wary Others w/ microhabitat data (4866+)
34	Crypsis for pred. conceal.	67.3 *** strong+	20.7 *** mod–	0 ** mod–	16.7 0.08 weak–	26.8 *** strong–	52.2 *** mod+	% of focal relying on crypsis Others w/ both data (1041+)

Table 3.9 continued

Trait	Condition	Arb.	Crypt.	Subt.	Aquat.	Terr.	Climb	Units Contrast (*n*) Notes
35	Flight for pred. conceal.	2.7 *** mod–	5.2	0	0	18.6 *** strong+	6.6 *** mod–	% of focal relying on flight Others w/ both data (1041+)
36	Sand for pred. conceal.	0.2 *** mod–	22.4 *** mod+	100 *** strong+	0	6.7	0.7 *** mod–	% of focal relying on sand Others w/ both data (1041+)
37	Crevice-burrow for pred. conceal.	12.0 *** strong–	31.0	0 * weak–	41.7	34.6 *** strong+	21.8 ** mod–	% of focal that run to crevice-burrow Others w/ both data (1041+)
38	Reside in social groups	10.6 *** weak+	3.7 *** weak–	0.3 *** weak–	5.5	7.3	11.7 *** mod+	% of focal grouped Others w/ microhabitat data (4893+)
39	Male defense of space	92.5 *** mod+	72.0 * weak–	50.0 * weak–	81.8	81.8 ** mod–	88.5 *** mod+	% of focal males defensive Others w/ social data (824+)
40	Male hierarchies	2.2 *** mod–	6.0	0	9.1	12.4 *** mod+	8.1	% of focal w/ male hierarchies Others w/ social data (824+)
41	Human habitat	31.7 *** mod+	16.0 * weak–	5.9 *** weak–	10	14.7 *** mod–	30.2 *** mod+	% of focal in human habitat Others w/ macro and microhabitat (4772+)
42	Spp. native on any island	57.0 *** mod+	50.6 *** weak+	24.2 *** mod–	43.6	35.1 *** mod–	42.5	% on islands Others w/ microhabitat data (5176)
43	Oviparous (v. viviparous)	90.5 *** mod+	74.0 *** weak–	66.9 *** weak–	61.0 *** weak–	82.1	88.8 *** mod+	% of focal oviparous Others w/ microhabitat and repro. data (3740+)
44	Mass-adj. clutch/ litter count	0.001	0.108 *** weak+	–0.116 *** weak–	–0.179	–0.023	0.066 * weak+	Std. resid. difference Others w/ microhabitat and clutch data (2767+)
45	Mass-adj. ontogenetic range	0.093 ** weak+	0.134 * weak+	–0.878 *** strong–	–0.184	–0.025	0.349 *** mod+	Std. resid. Other w/ microhabitat data (4893+)
46	Endemic species	65.0 *** weak+	62.7 0.07 weak+	64.8 * weak+	41.8 * weak–	54.1 *** mod–	58	% of focal endemic All lizards (6528)
47	Longevity (mass adj.)	–0	0.38 ** strong+	–0.70	NA	0.01	0.01	Std. resid. Others w/ longevity (851+)
48	Population density	0.66	0.25 * mod–	0.07	NA	0.58	0.72 ** mod+	kg/ha Others w/ densities (341)
49	Any introductions	6.2 *** weak+	1.9 ** weak–	0 *** weak	3.6	3.8 0.07 weak–	6.9 *** mod+	% of focal introduced Others w/ microhabitat data (4803+)
50	>2 introductions	2.4 *** weak+	0.3 ** weak–	0 ** weak–	0	1.1 * weak–	2.6 *** mod+	% of focal >2 intros Others w/ microhabitat data (4803+)
51	At conservation risk	59.3 *** mod+	47.8	38.6 ** weak–	54.8	46.2 *** weak–	58.1 *** mod+	% of focal at risk Others w/ conservation evaluation (2924+)

a strong tie of vocalizations to cathemeral behavior or nocturnality (table 3.7, traits 31, 32).

- Highest proportion of lizards that rely on crypsis (34). Arboreal environments provide a visually complex background ideal for crypsis.
- Highest group living (38), though climbers are a trifle higher. As many of the arboreal prey fly rather than walk to the place where a lizard can obtain them, intraspecific competition for prey is comparatively reduced in arboreal environments; the food supply is replenished continuously and from a source outside of the lizards' foraging space. This ameliorates one major drawback of group living—progressive depletion of food supplies. Climbers often utilize cliffside cracks and those are often clustered by the geology (promoting "group living" of climbing lizards).
- Highest proportion of males defending space (39). It seems likely that the defended space is relatively small, often no more than a choice refugium or good perch. Much more information is needed to document the complex geometry of lizards' spatial defense on trunks, in the canopy and on cliffs.
- Proportion living in human habitats (41). An ability to climb minimizes contact with humans and predaceous human pets.
- Highest proportion on islands (42). This is potentially an artifact of the relatively mesic and therefore tree-covered conditions found in maritime Southeast Asia and Oceania. This potential bias may be offset by the many arboreal anole species found in the drier islands of the Caribbean.
- Highest proportion of oviparous species (43), but with average clutch count (44). Arboreal lizards are often found in climatically benign tropical forests (relatively few at high latitudes), where oviparity is common. Many, such as anoles, are slender and have small clutches laid often.
- Third-smallest young (45), similar to cryptozoids but both distant to the tiny-offspring climbers. Does climbing put a lizard in the gunsights of predatory birds, for which they need to compensate by producing more (see 44) but smaller (45) young? Do gravid climbers bear a higher penalty for carrying heavy or large eggs? Cryptozoids have the highest mass-adjusted clutch count (44), climbers are second, and arboreals only average. It is hard to compile a cohesive story from these inconsistent relationships.
- Endemism (46). This fits with the high proportion found on islands.
- Species introductions (49, 50), climbers marginally higher in multiple introductions. This fits with the high proportion found on islands, as many introductions have been from island to island, facilitated by the high level of human commerce needed to support island societies.
- Proportion at conservation risk (51). This fits with the high proportion of endemics.

In almost every comparison (45 is the glaring exception), arboreal species rank more prominently than the climbers, suggesting that it is residence primarily above ground that is distinctive, rather than a more generalized ability to climb.

Based on the correlations (table 3.9), cryptozoic lizards are notable in relation to lizards from other microhabitats for the following:

- Lowest mean mass (2). As appropriate for penetrating small channels and spaces.
- Elongate form (6). This would be reasonable, both for the reason immediately above and if locomotion through the litter was sometimes serpentine.
- Shortest tails among aboveground spp. (8) and most likely to break (9), though least likely among aboveground spp. to be prehensile (10). There would appear to be no value to a cryptozoid of having a prehensile tail. If there is no reason to have a prehensile tail there is less reason to have it unbreakable. Species that push through soil are characterized by extremely short tails, which are used for anchoring in the walls of underground passages. Some cryptozoids appear to be transitioning to subterranean foragers.
- Least likely of aboveground species to change color (11, 12). As appropriate to a vision-reduced environment.
- Lowest proportion of sexes differ in color (14) among aboveground species. As appropriate to a vision-reduced environment.
- Tied with terrestrials for blue tails (16). The presence of blue tails suggests that vision is not entirely absent from the life of cryptozoids. Potentially the predators most likely to impinge on cryptozoids do so by pawing, scraping, or scratching them from the litter. The sudden appearance of a brilliant blue tail could easily have immediate startle value in this context and could help misdirect the ensuing pounce.
- Not carnivorous (18). Cryptozoids are generally too small to consume vertebrates.
- Cathemeral (27). This diel pattern seems appropriate for living in a vision-reduced environment, but the benefits of diurnality could include indirect basking under thin cover. As diurnality is moderately and negatively associated with cryptozoic behavior, one could

infer that not many cryptozoic species rely on indirect basking.

- Crepuscular (28). The association between cryptozoic and crepuscular behavior is easier to accept in the absence of the data on indirect basking presented in the preceding bullet. Crepuscular species could become nocturnal, but that would raise their exposure, if only by noise created in the litter, to the numerous nocturnal mammalian predators. Perhaps it is better for them to stay at the diel "seam" between diurnal and nocturnal risk factors.
- Conspicuously wary (33). If "conspicuously wary" implies that the exposed cryptozoid quickly dives back under the cover of litter, such behavior would seem to be an appropriate way to reduce predation.
- Reliant on sand swimming (36). Where the substrate is loose sand, sand swimming seems like an appropriate ploy and one that is compatible with their elongate form and sometimes reduced limbs.
- Largest mass-adjusted clutch count (44). Although the effect size is small, it contradicts the easy presumption that slender cryptozoids might be forced to have correspondingly small clutches.
- Least likely among aboveground spp. to be introduced (49, 50). A penchant for moving out of light and into the substrate would minimize both accidental human transport and appeal to the pet trade.

Subterranean lizards are notable in relation to lizards from other microhabitats for the following (table 3.9 trait number):

- Slender body (3). For ease of penetrating soil.
- Absence of pores (5). The absence of pores in subterranean species was unexpected, as the underground environment would seem ideal for chemosensory marking. See remarks above on the possible liability of pores to trailing by hostile ants.
- Shortest tails (8), none of which are known to be prehensile (10). Short tails make good anchors in underground passageways. Prehensile tails do not.
- Most m<f SSD (13). Apparently the natural selection advantage to subterranean lizards of small size applies with full force in males, but is offset to some degree by selection for increased fecundity in females.
- Absence of sexual dichromatism (14), seasonal color (15), and color change (11, 12). No surprise that color markings would be of little value underground.
- Active foraging (25). Life underground is so poorly documented that one cannot confidently predict that sit-and-wait foraging (for example, from side chambers of underground ant trails) is not plausible. Sampling biases may be operating on this issue.
- Least diurnality (29), most nocturnality (30), albeit with poor sampling of these traits. Given the paucity of information on the diel abundances of subterranean prey and predators, it is difficult to have firm expectations as to the diel activity of subterranean lizards. Why not be continuously active when underground? If subterranean foraging occurs often at shallow depths, diurnality might be retained underground for thermal benefits, though shallow-digging subterraneans might also be vulnerable to detection by aboveground predators (as robins find worms).
- Least likely to be conspicuously wary (33), though underreporting possible.
- Solitary (38) and not known to have male hierarchies (40). "Not known" is the operant phrase.
- Scarcity on islands (42). The paucity of subterraneans on islands could be due either to dispersal difficulties or the shortage of friable soils on many islands.
- Smallest ontogenetic range of sizes (45), indicating particularly large young. Perhaps small subterranean young are incapable of generating the force needed to wedge their way through dense soils. It is notable (table 4.1, trait 64) that sand sharks do not share this condition, perhaps because sand sharks occur only in sand and other loose media, for which much less force is needed to burrow.
- Nearly highest endemism (46: marginally behind arboreals). Many subterranean species appear to be limited in distribution by the patch size of suitable soils. If suitable soils are patchy, many endemic species are likely to arise by local adaptation.
- Absence of known introductions (49, 50). Soil is moved less often than other commercial products, and subterranean species do not make attractive pets. Curiously, a morphologically nearly identical squamate, *Indotyphlops braminus*, often called the Flower Pot Snake, is one of the most frequently introduced snakes, as it is often accidentally transported in the soil of horticultural products. Its key adaptation may be its parthenogenetic reproduction, which halves the minimum number needed to arrive at a single destination. This observation suggests that subterranean lizards are also likely to be accidentally transported in the horticulture trade, but that such colonists rarely produce progeny. The 33% of subterraneans that are viviparous (43) would seem to be especially likely to

occasionally arrive at a destination with both sexes (a male in utero), but no such events have been documented in lizards. Perhaps subterranean lizards are less suited to flower pots than is *I. braminus*.

- Least at conservation risk (51). The risk may be underestimated due to the difficulty of monitoring subterranean species. One would expect high endemism to lead to greater conservation risk.

Semiaquatic lizards are notable in relation to lizards from other microhabitats for the following reasons (table 3.9 trait number):

- Largest species (2). The only size factor unequivocally associated with aquatic existence is heat loss through conductivity upon entering cold water. However, many semiaquatic species are in warm places, where heat loss may not be a primary consideration. More information is needed.
- Longest tails (8). Tails are central to propulsion in water.
- Most m>f SSD (13). It is possible that semiaquatic forms are concentrated at the water's edge and therefore achieve high local densities (few data available). High densities are associated with m>f SSD.
- Carnivory (18), but not invertebrate (20) or ant eating (21). These fit with large size, but why semiaquatic species would be carnivorous or large is not well understood.
- None reported to cruise (24). Sample size is very small.
- Smallest home-range size (26). Presumably semiaquatic forms remain in the vicinity of the water's edge. Home-range estimation is not well resolved for species with linear distributions, and in any case, the sample size is very small.
- Diurnality (29), not nocturnality (30). Diurnality would seem appropriate for basking between cooling dives. Nocturnality of large lizards may run into risks from predatory crocodilians and snakes, especially in larger water courses.
- Lowest proportion that call spontaneously (31). Crocodilian predators are responsive to sound.
- Take cover in crevices (37) but not concealed by crypsis (34 excluding subterranean) or flight (35). The water's edge is appealing not only from the perspective of food and moisture availability, but also from the opportunities to quickly evade predators by dashing to the medium the predator is least suited for. See chapter 4 for analogous examples such as ground-foraging tree lizards. Sprinting flight and crypsis, conversely, are not well suited to the morphological requirements of swimming.
- Viviparity (43) and small litters (44). The association of semiaquatic lizards with viviparity is interesting in that virtually all semiaquatic lizards occur in tropical (53 of 55) lowlands (43 of 43), where viviparity is scarce; 8 of 43 species occur also at higher elevations. Riparian zones are often heavily shaded and imbued with moist and therefore relatively cold soil—an inauspicious environment for egg development. By carrying the developing embryos, semiaquatic lizard mothers can ensure regular if episodic warming by basking in sun spots. All else being equal (e.g., interclutch interval, development rate, etc.), small litters imply low mortality, but all else may not be equal.
- Not endemic (46). Aquatic systems are often fragmented, especially in small-order drainages, so this is somewhat unexpected. There is a body size component to endemism, with larger species less prone to smaller distributions (table 3.1, trait 36).

Terrestrial lizards are notable in relation to lizards from other microhabitats for the following (table 3.9 trait number):

- Second highest number of spp. (1), essentially tied with climbers. Many of these terrestrial lizards occur in aridland habitats where aquatic, arboreal, and leaf litter habitats have reduced availability. Is the terrestrial microhabitat intrinsically better for lizards than the alternatives, or are there simply more hectares of suitable terrestrial habitat available?
- Seasonal color (15) and blue tails (16: tied with cryptozoids). Terrestrial lizards are more often diurnal (29) than nocturnal (78% v. 15%). Colorful traits are associated with well-lit environments. Seasonal color undoubtedly has some negative consequences for detection by predators, but this may be ameliorated in terrestrial environments by proximity to burrows or other nearby refugia. Seasonal colors may also be displayed only laterally or ventrally, making them hard for birds perched or flying overhead to see.
- Herbivory (19) and omnivory (22). Warmth (diurnality) facilitates digestion of plant matter. Large size does also, but large size is more readily accommodated on the ground (heavy lizards are slow climbers, and burrows and crevices make secure refugia for large edible lizards).
- Large home ranges (26). To some degree this may be an artifact of defining home ranges by surface area (which

reflects the terrestrial microhabitat well, and other microhabitats less well). In addition, all the lizards with exceptionally large home ranges roam over the surface of the ground, as that is energetically efficient compared to climbing. Swimming is even more efficient, but truly aquatic lizards are not found, presumably because fish, snakes, crocodilians, and turtles are more highly adapted to the ectothermic aquatic niches.

- Not cathemeral (27). The scarcity of cathemeral terrestrial lizards is not immediately explicable, especially given the apparent abundance of at least seasonally cathemeral terrestrial nonlizard squamates (i.e., many snakes). Potentially the cathemeral terrestrial snakes constitute particularly effective predators on terrestrial lizards, inhibiting the lizards' cathemeral tendencies.
- Reliance on flight (35). Relative to predatory endotherms, lizards appear to be able to outrun them only when on a two-dimensional plane, such as the terrestrial microhabitat.
- Male dominance hierarchies (40). Male dominance hierarchies in lizards appear to result mostly from contests for mating with highly concentrated females. This happens most often when the females are concentrated by microhabitat features such as rock piles, crevices, or the edge of water bodies (i.e., terrestrial or semiaquatic microhabitats). In the more diffuse populations associated with arboreal or cryptozoic microhabitats, hierarchies rarely arise. Be aware that cryptozoic lizard social systems are very poorly documented, and sampling may be biased.

Although terrestrial lizards are the most speciose, they are the least distinct in terms of natural history (note the brevity of the list). Regression to the mean is more likely with a larger sample.

Reproduction Allocation

A reasonable question, asked of all organisms, is on what grounds has evolution adjusted the allocation of reproductive effort between numerous small young and a smaller number of larger young? Two mass-adjusted measures may shed light: clutch/litter count (mass-adjusted clutch count) and the size of young relative to the largest adults (mass-adjusted ontogenetic range). There is no logical requirement for these measures to be mass or size adjusted, as one can imagine organisms that scaled uniformly; for example, that progeny were always 1/10 the size of adults and clutch/litter counts that averaged, say, eight eggs in small lizards as well as in large lizards. For reasons that are not completely clear, this does not happen. Small lizards have smaller clutches and have a smaller ontogenetic range of sizes. One gets the impression that evolution cannot push the size of lizards any smaller than they already are, and so must shrink clutch counts and the range of sizes at the small end of the lizard-size spectrum.

Note that mass-adjusted ontogenetic range is the standardized residual of $\log_{10}$ of the maximum mass divided by the minimum mass; this ratio regressed on mean mass at the modal length. Therefore, lower standardized residuals reflect larger young, and higher standardized residuals the converse. In contrast, mass-adjusted clutch count is larger for more numerous young and the converse. The implicit trade-off assumes that mass-adjusted ontogenetic range and mass-adjusted clutch count are positively correlated (table 3.10, trait 7). If ontogenetic range is not mass-adjusted, and is back transformed (as it is on occasion; e.g., table 3.1, trait 3), the resultant multiplier is an intuitive value that expresses the number of young that would have the same mass as the maximum-sized adult (typically 20 to 400). When standardized, all residuals express the number of standard deviations above (positive values) or below (negative values) the mean (i.e., 34% of the population lies between –1 and zero, etc.).

The first three traits in table 3.10 are included because they are proxies for lizards that are highly visual (crests), chemosensory (pores), or tactile/cryptozoic/subterranean (elongation). The crest results (trait 1) strongly support the implicit trade-off: crested species have smaller and more young; in both cases they surpass average lizards by about 0.5 standard deviations. In contrast, the relative size of young in pored species (trait 2) is unaffected by their pore-bearing trait, yet they have a slightly reduced (0.3 std. resid.) number of young. Elongation (3) fits the assumption better; elongate species have smaller (0.5 std. resid.) and more (0.3 std. resid.) young. The difference in magnitude of the two responses in elongate lizards (0.5 v. 0.3 std. resid.) suggests that the trade-off may not always be 1:1. It also suggests that the slender body of elongate species does not just force the clutch to be fewer in number to fit in the slender body of the mother, since elongate mothers have more numerous young than other lizards.

The relationship between SSD and reproduction (table 3.10, trait 4) is complex. Ontogenetic range shows a downward-pointing V function, with maximal relative size of young in species that lack SSD and smaller young in either m≪f or m≫f species. In contrast, the clutch count relationship to SSD is linear: the most numerous young are found in m≪f species and decline in number with increasing relative male size. Thus, SSD is not reflective of a simple trade-off between number and size of young. The mechanism whereby SSD reduces the size of young is unclear and likely to be evolutionarily interesting.

Table 3.10. Correlates of reproduction allocation variables: mass-adjusted ontogenetic size range and mass-adjusted clutch count. Each standardized residual (std. resid.) value applies to the condition listed to its left in the Benchmarks column. "NA" indicates a cell that is not applicable (because it would represent a correlation of a variable with itself). Other table conventions as in table 3.1.

			Mass-adj. ontogenetic range			Mass-adj. clutch count		
Trait	Variable	Benchmarks	Std. resid.	*n*	Stat. sign. r^2	Std. resid.	*n*	Stat. sign. r^2
1	Crests	Crest: No crest:	0.539 −0.044	1089	*** strong+	0.475 −0.036	2974	*** strong+
2	Pores	Pores: No pores:	0.017 −0.004	1089		−0.076 0.018	2974	** weak−
3	Elongation	Elong.: Not elong.:	0.498 −0.054	1089	*** strong+	0.282 −0.036	2974	*** mod+
4	Ordinal SSD	m≪f: m=f: m≫f:	0.423 0.152 0.551	7	0.08 0.49	1.300 −0.058 −0.131	7	* 0.65
5	Repro. mode	Ovip.: Vivip.:	0.056 −0.163	1023	*** mod+	−0.016 0.058	2839	* weak−
6	Rel. clutch mass	−1 std. resid: 1 std. resid:	0.234 0.250	197		0.184 0.264	296	*** 0.12
7	Mass-adj. clutch	−1 std. resid: 1 std. resid:	−3.20 0.416	1068	*** 0.09	NA	NA	NA

Reproductive mode does not support the trade-off either (table 3.10, trait 5). Oviparity has a moderate positive influence on size of young (smaller when oviparous) and a very small negative influence (0.07 std. resid.) on number of young (oviparous have fewer young). This weakly suggests that viviparity is a win-win: viviparous mothers have offspring that are both more numerous and larger. This might account for the surprisingly large number of lizard species, seemingly especially skinks, that are viviparous in warm locations. This might occur, for example, if viviparous mothers are able to contribute to the welfare of their growing young through uterine nutrition, basking, or postpartum parental care after the time when oviparous mothers have oviposited and abandoned them. This suggestion implies a correlation between viviparity and parental care, for which there is a weak supportive trend but no significant correlation (Fisher's Exact $p = 0.31$; not in table).

RCM offers little support for the trade-off (table 3.10, trait 6): there is no statistical relationship between RCM and ontogenetic size range. There is a very small positive correlation ($r^2 = 0.12$; gain is 0.08 std. resid.) between RCM and clutch count: more numerous young very weakly corresponds to a higher relative clutch mass.

The direct comparison between size and number of young (table 3.10, trait 7) supports the trade-off, though the statistical relationship is conspicuously weak ($r^2 = 0.09$). Any broad generalization like the small/numerous versus large/few trade-off in young is qualified with that inevitable disclaimer, "all else being equal." As demonstrated in this section, all else is often not equal, though the broad generalization seems to hold very weakly.

Islands

Island biogeographers focus on the insular environment or individual island traits, but the present study focuses on species. The most distinct island effects are associated with small or depauperate islands (Rodda and Dean-Bradley 2002), whereas the database used in this work is not set up to distinguish lizard species of speciose or large islands (New Guinea, Borneo) from those of depauperate or small islands (1.2 ha Carrot Rock). By mixing small and large islands together, distinctions are apt to be muted if not lost. Furthermore, this work mixes together oceanic islands (overwater colonization required) and continental islands (dispersal not required), high latitude (relatively depauperate) and low latitude (speciose) islands, and freshwater (short dispersal distances) and saltwater (sometimes vast distances) islands, and it treats any species that is found on any island as insular, even though the bulk of the species' range may be continental. This heterogeneity makes it unlikely that any island differences would be detectable in this particular sample (table 3.11).

The number of significant effects (37 traits) was unexpected. Four traits stand out as particularly strong, despite the muddy sampling (table 3.11 trait number):

Table 3.11. Correlates with island residency. Lizard species that are native on any named island (freshwater, saltwater; continental, oceanic) are considered island species regardless of the extent of their mainland range. Conventions as in table 3.1.

Trait	Variable	Benchmarks	Means	Units (*n*)	Stat. sign. Strength
1	Number of species	Island spp.:	40.3	% of total	
		Mainland spp.:	59.7	(6528)	
2	Shape index	Island spp.:	0.026	Std. resid.	0.065
		Mainland spp.:	−0.018	(6525)	weak+
3	Tail breaks	Present:	44.9	% that are isl. spp.	***
		Absent:	34.2	(6528)	mod+
4	Crests	Present:	54.9	% that are isl. spp.	***
		Absent:	39.4	(6528)	weak+
5	Elongate lizards	Present:	44.6	% that are isl. spp.	*
		Absent:	39.7	(6528)	weak+
6	Any color change	Present:	58.6	% that are isl. spp.	***
		Absent:	38.1	(6528)	mod+
7	Advanced color change	Present:	56.5	% that are isl. spp.	***
		Absent:	39.2	(6528)	mod+
8	Sexes differ in size	m≪f:	15.4	% that are isl. spp.	**
		m≫f:	83.2	(7 means)	0.89
9	Sexes differ in color	Present:	46.6	% that are isl. spp.	***
		Absent:	38.6	(6528)	weak+
10	Seasonal color	Female:	55.6	% that are isl. spp.	*
		Male:	36.3	(327)	mod+
11	Blue tails	Present:	48.7	% that are isl. spp.	*
		Absent:	40.1	(6528)	weak+
12	Carnivores	Present:	66.3	% that are isl. spp.	***
		Absent:	43.9	(2860)	mod+
13	Invertebrate eaters	Present:	42.5	% that are isl. spp.	***
		Absent:	50.2	(2860)	weak−
14	Ant specialists	Present:	52.8	% that are isl. spp.	0.068
		Absent:	44.2	(2860)	weak+
15	Omnivores	Present:	49.2	% that are isl. spp.	*
		Absent:	43.8	(2860)	weak+
16	Sit-and-wait foragers	Present:	36.1	% that are isl. spp.	**
		Absent:	44.7	(1149)	mod−
17	Active foragers	Present:	44.5	% that are isl. spp.	**
		Absent:	35.4	(1149)	mod+
18	Home-range area (mass adj.)	Island spp.:	−0.197	Std. resid.	***
		Mainland spp.:	0.387	(216)	strong−
19	Active at any time (cathemeral)	Present:	48.3	% that are isl. spp.	*
		Absent:	40.9	(4746)	weak+
20	Calls spontaneously	Present:	50.6	% that are isl. spp.	*
		Absent:	40.0	(6528)	weak+
21	Conspicuously wary	Present:	46.9	% that are isl. spp.	**
		Absent:	39.5	(6501)	weak+
22	Flight as primary pred. conceal.	Present:	37.4	% that are isl. spp.	*
		Absent:	48.3	(1067)	weak−
23	Sand as primary pred. conceal.	Present:	35.5	% that are isl. spp.	0.066
		Absent:	47.9	(1067)	weak−

Table 3.11 continued

Trait	Variable	Benchmarks	Means	Units (*n*)	Stat. sign. Strength
24	Veg. as primary pred. conceal.	Present: Absent:	56.2 44.5	% that are isl. spp. (1067)	** mod+
25	Electric light users	Present: Absent:	77.7 39.7	% that are isl. spp. (6528)	*** mod+
26	Male hierarchy	Present: Absent:	40.9 54.1	% that are isl. spp. (855)	* weak–
27	Arboreal microhabitat	Present: Absent:	57.0 36.8	% that are isl. spp. (5176)	*** mod+
28	Cryptozoic microhabitat	Present: Absent:	50.6 42.3	% that are isl. spp. (5176)	*** weak+
29	Subterranean microhabitat	Present: Absent:	24.2 44.8	% that are isl. spp. (5176)	*** mod–
30	Terrestrial microhabitat	Present: Absent:	35.1 50.7	% that are isl. spp. (5176)	*** mod–
31	Human habitats	Present: Absent:	64.6 38.3	% that are isl. spp. (5393)	*** strong+
32	Mass-adjusted clutch/litter count	Island spp.: Mainland spp.:	–0.078 0.070	Std. resid. (2974)	*** weak–
33	Eggshells hard	Present: Absent:	56.6 38.7	% that are isl. spp. (6528)	*** mod+
34	Population density	Island spp.: Mainland spp.:	0.87 0.17	kg/ha (343)	*** v. str.+
35	Longevity (mass adj.)	Island spp.: Mainland spp.:	0.18 –0.18	Std. resid. (879)	*** mod+
36	At conservation risk	Present: Absent:	56.8 39.2	% that are isl. spp. (3390)	*** mod+
37	IUCN listed as endangered or worse	Present: Absent:	65.2 45.1	% that are isl. spp. (3390)	*** mod+

- Sexes differ in size (8). Insular species are more likely to be m>>f. Likely associated with their denser populations.
- Home-range areas are much smaller on islands (18). Likely associated with higher population densities.
- Lizards on islands are more likely to embrace human habitats (31). Because humans first reached most places by ship, and such settlements have thrived in their benign maritime environments, human habitats are likely more prevalent (and more often sampled) on islands.
- Island species have denser populations (34) by a factor of >5. Reduced endothermic predation has been a commonly cited cause for the higher mean lizard population densities on islands, though full or nearly full suites of predators are present on many of the islands tabulated herein.

A number of the observed correlations may be compatible with the generalization that predator pressure is reduced on islands, insular population densities are higher, and what predator pressure exists is due primarily to birds (table 3.11 trait number):

- Tail break capability is more prevalent on islands (3). Is a breaking tail more beneficial in the presence of bird predators, more beneficial under reduced total predation, or more beneficial at high population densities? Is tail breakage a measure of predatory incompetence or predator density? These same factors may raise or lower tail break frequencies within a population, a phenomenon not tabulated in this database. The evidence suggests that in at least some species, conspecifics as well as incompetent predators may pull off tails. The dynamics of tail breakage continue to be source of mystery (Schoener 1979; Schoener and Schoener 1980).

- Color change is more prevalent on islands (6, 7). Color change serves both crypsis and other uses. As there is no direct relationship between insular species and crypsis, one is tempted to posit a relationship to high population density, denser vegetation, or the benefits of social signaling.
- Sexes differ in color (9), a weak effect that may be tied to the very strong island SSD.
- Female seasonal color (10), though not male seasonal color, is more prevalent on islands. Presumably this is associated with higher population densities, as female seasonal color may be to dissuade overly ardent suitors, and perhaps desperate suitors are more numerous and harder to evade at high population densities.
- Blue tails more prevalent on islands (11), as are active foraging (17) and cryptozoids (28), the two hotbeds of blue tails.
- Lizards on islands are more likely to be cathemeral (19). Might this reflect some island niche expansion facilitated by reduced predation pressure?
- Island species are more likely to vocalize (20). This would seem plausibly due to reduced predation pressure. Subjectively, island vocalizers sound louder than many mainland lizards; this is an unquantified comparison casually reported by several sources, especially in Australia and Pacific islands.
- Mass-adjusted clutches are smaller on islands (32), perhaps because larger offspring are needed to be better competitors in the densely populated islands, in contrast to additional offspring needed to offset predation on a mainland.
- Island species are longer lived (35). This presumably reflects reduced predation pressure. It is inconsistent with the hypothesis that iteroparity is elevated in environments that are particularly unpredictable, as islands are often tropical, maritime, and assumed to be relatively predictable climatically.

Several island traits are not traceable to high population density/lower predation pressure (table 3.11 trait number):

- The nearly twofold higher electric light use (trait 25) on islands is a puzzle. Perhaps it is associated with the greater prevalence of cathemeral activity on islands (19), as cathemeral activity and electric light use are correlated (table 3.7, trait 20). Electric lights may impact a greater proportion of habitat on islands.
- The paucity of male hierarchies on islands (26) is mysterious. Perhaps it is related to the shortage of terrestrial species on islands (30) and the association between terrestriality and male hierarchies (table 3.9, trait 40).
- Island species are not protected from endangerment by their high population densities, as this benefit is more than offset (36, 37) by their restricted ranges and habitats suffering from pervasive human disturbance.

Longevity

A classic fallacy in interpreting correlations is the assumption that a preferred one of the pair of traits is the "cause" and the other "effect." Additional information is required to assess whether and which one is responsible for the correlation, or if a third factor is responsible for both conditions (see Scharf et al. [2015] for a recent, hypothesis-driven review of longevity). Do pores (table 3.12, trait 4) lengthen the life span of their bearers, or are secretory pores metabolically expensive and therefore used primarily in long-lived species for which the communication payoff is more recurring? Does herbivory (trait 10) allow lizards to live longer or does herbivory arise preferentially in habitats for which primary productivity is low and unpredictable and thus iteroparity more highly rewarded? Or is low primary productivity associated with reduced density of endothermic predators? Does parental care (trait 26) allow offspring to live longer, or does investment in hatchling survival have a larger payoff in species that produce small clutches over more years? There are many interesting puzzles to be solved here.

The largest effect sizes in table 3.12 are found in the following (trait number):

- Lizards with crests are shorter lived (3). Crests are associated with strong sexual selection (moderate negative link to longevity), ant eating (no assoc. with longevity), sit-and-wait foraging (no assoc.), diurnal activity (strong negative), crypsis (moderate negative), arboreality (no assoc.), and population density (weak positive). The net result seems to be a strong negative. Unraveling the causation will be challenging.
- Cathemeral lizards are longer lived (14). Cathemeral lizards are associated with elongation and limb reduction (moderate positive), darkening color change (no assoc.), sexual size monomorphism (no clear assoc.), forages under electric lights (no assoc.), low elevation (no assoc.), vocalizations (strong positive), conspicuously wary (strong positive), sand concealment (strong positive), cryptozoic microhabitat (strong positive), and lives on occupied buildings (no assoc.). These evidently sum to long lives for cathemerals.
- Lizards that vocalize are longer lived (17). Does

Table 3.12. Correlates with species' size-adjusted longevities (standardized residuals of maximum reported longevities regressed on mass at modal snout-vent length). Conventions as in table 3.1.

Trait	Variable	Benchmarks	Means	Units (*n*)	Stat. sign. Strength
1	Mass	1 g: 100 g:	3.7 10.6	Years (879)	*** 0.21
2	All subsequent comparisons are mass adjusted to remove indirect effect of mass				
3	Crests	Present: Absent:	–0.42 0.04	Std. resid. (879)	*** strong–
4	Secretory pores	Present: Absent:	0.26 –0.07	Std. resid. (879)	*** mod+
5	Elongate lizards	Present: Absent:	0.37 –0.03	Std. resid. (879)	** mod+
6	Advanced color change (color/pattern)	Present: Absent:	–0.38 0.05	Std. resid. (879)	*** strong–
7	Sexes differ in color	Present: Absent:	–0.13 0.09	Std. resid. (879)	** mod–
8	Seasonal color	Present: Absent:	–0.25 0.06	Std. resid. (879)	*** mod–
9	Sex differences rollup	Score=0: Score=6:	0.09 –0.23	Std. resid. (875)	* 0.004
10	Herbivores	Present: Absent:	0.41 –0.01	Std. resid. (740)	* strong+
11	Plant eating by any lizard	Present: Absent:	0.16 –0.04	Std. resid. (706)	** weak+
12	Invertebrate eaters	Present: Absent:	–0.08 0.14	Std. resid. (740)	** mod–
13	Omnivores	Present: Absent:	0.01 –0.03	Std. resid. (740)	* weak+
14	Active at any time (cathemeral)	Present: Absent:	0.51 –0.04	Std. resid. (847)	*** strong+
15	Active in low light (crepuscular)	Present: Absent:	0.49 0.00	Std. resid. (847)	* strong+
16	Active in daylight	Present: Absent:	–0.11 0.33	Std. resid. (847)	*** strong–
17	Makes any sound	Present: Absent:	0.40 –0.11	Std. resid. (119)	*** strong+
18	Conspicuously wary	Present: Absent:	0.15 –0.41	Std. resid. (296)	*** strong+
19	Crypsis as primary pred. conceal.	Present: Absent:	–0.30 0.05	Std. resid. (332)	** mod–
20	Crevice-burrow as primary pred. conceal.	Present: Absent:	0.11 –0.16	Std. resid. (332)	** mod+
21	Sand a primary pred. conceal.	Present: Absent:	0.33 –0.09	Std. resid. (332)	* strong+
22	Cryptozoic microhabitat	Present: Absent:	0.38 –0.03	Std. resid. (857)	** strong+
23	Species native on any island	Island spp.: Mainland spp.:	0.18 –0.18	Std. resid. (879)	*** mod+
24	Oviparous species	Oviparous: Viviparous:	–0.07 0.41	Std. resid. (821)	*** strong–

Table 3.12 continued

Trait	Variable	Benchmarks	Means	Units (*n*)	Stat. sign. Strength
25	Mass-adjusted clutch/litter count	−1 std. resid. long: 1 std. resid. long:	0.281 −0.096	Std. resid. (838)	*** 0.017
26	Parental care	Present: Absent:	0.42 −0.07	Std. resid. (879)	*** strong+
27	Endemism in single "country"	Present: Absent:	0.14 −0.08	Std. resid. (879)	** mod+
28	Population density	−1 std. resid.: 1 std. resid.:	0.48 1.27	kg/ha (188)	** 0.053
29	At conservation risk	Present: Absent:	0.17 −0.06	Std. resid. (595)	** weak+

vocalization reflect life in a predator-reduced environment, such as islands, or life in unpredictable environments, such as arid places, where survival or iteroparity is benefited? In reviewing longevity associations with the various rainfall belts, no compelling signal was evident. On balance, the evidence is stronger for vocalizations being enabled by low predator pressures.

- Conspicuously wary lizard species are longer lived (18). Of the associations of conspicuous wariness with other factors that are themselves correlated with longevity, the relevant factors are crests (absence is a strong positive), elongation (moderate positive), activity at any time of day (strong positive), activity in low light (strong positive), and cryptozoic behavior (strong positive). Not surprisingly, conspicuously wary lizards live longer. The exact chain of causation will likely be very hard to establish.
- Oviparous lizards are shorter lived (24). This has been noted before (e.g., Scharf et al. 2015), with the observation that viviparous species are often inactive for parts of the year (hibernation), precluding apples-to-apples comparisons of age with continuously active lizards. The sample of nonhibernating viviparous species may suffice for an apples-to-apples comparison.
- Lizards with parental care are longer lived (26), as discussed above.

A conspicuous omission from this table is any reference to foraging behavior or habitat, though all foraging modes, rainfall, and elevation bands were tested; none had a measurable association with longevity. Perhaps foraging mode and habitat are not a factor in longevity, or all foraging modes and habitats are roughly equivalent in longevity?

Population Density/Biomass

In my opinion, a significant problem with lizard population density comparisons is that so many of the available population density estimates are insufficiently precise to detect demographically critical population changes; a human population decline of 50% in a decade would be deemed apocalyptic, but in a lizard it might not be detectable. For this reason, lizard population density estimates were divided into three tiers: (1) subjective estimates, (2) objective estimates with conspicuous methodological flaws, and (3) objective estimates with fewer and less compelling flaws. Among the results portrayed in table 3.13 are comparisons of types 1 and 2 in relation to type 3. Type 3 is presumed to be the best and is lauded here as "qualified." My impression is that subjective estimates (on a scale of 0–5) are calibrated on the basis of the observer's expectations, which include adjustments for clade, habitat, insularity, visibility, and so on. Both qualified (type 3) and nonqualified (type 2) objective estimates are in units of individuals per hectare, here converted to kg/ha to facilitate comparisons among species. The ranges of the objective estimates are enormous: more than six orders of magnitude. However, the errors responsible for distinguishing type 2 from 3 are *comparatively* small, typically around 50%. This implies that types 2 and 3 will be correlated with each other over their expansive range, and trait 24 in table 3.13 confirms a relatively high r^2 of 0.51, and notes that on average, relative to the best estimates, the type 2 estimates overestimate low densities and underestimate high ones.

In contrast to the broad agreement among objective estimates, type 1 estimates show no association with the objective estimates. Trait 23 in table 3.13 documents that the r^2 between qualified absolute population biomass and subjective estimates is 0.005. Over three orders of magnitude range in objective estimates, the mean verbal

Table 3.13. Correlates of species' population densities, operationalized as mean qualified biomass density (kg/ha). Conventions as in table 3.1.

Trait	Variable	Benchmarks	Means	Units (*n*)	Stat. sign. Strength
1	Shape index	–1: 1:	0.19 0.87	kg/ha (343)	*** mod+
2	Crests	Present: Absent:	1.4 0.5	kg/ha (343)	** mod+
3	Elongate lizards	Present: Absent:	0.21 0.59	kg/ha (343)	* weak–
4	Sexes differ in size	Max. = 1 kg/ha: Max. = 10 kg/ha:	m<f (–1) m>f (1)	See fig. 3.4 (7 means)	*** 0.94
5	Sexes differ in color	Present: Absent:	0.95 0.31	kg/ha (343)	*** mod+
6	Herbivores	Present: Absent:	12.1 0.5	kg/ha (324)	*** v. str.+
7	Invertebrate eater	Present: Absent:	0.3 1.4	kg/ha (324)	*** mod–
8	Omnivores	Present: Absent:	1.6 0.4	kg/ha (324)	*** mod+
9	Ants eaten	Present: Absent:	0.58 1.13	kg/ha (269)	* weak–
10	Sit-and-wait, cruise, active foragers	Present: Absent:	0.56 0.56	kg/ha (204)	no correlation for any foraging mode
11	Home-range area (mass adj.)	kg/ha: 100 kg/ha:	1.25 –1.31	Std. resid. (125)	*** 0.35
12	Active at night	Present: Absent:	0.30 0.61	kg/ha (336)	* weak–
13	Conspicuously wary	Present: Absent:	0.77 0.33	kg/ha (139)	* weak+
14	Males have hierarchies	Present: Absent:	2.18 0.82	kg/ha max biomass (8 means)	* mod+, mean biomass n.s.
15	Cryptozoic microhabitat	Present: Absent:	0.25 0.59	kg/ha (342)	* weak–
16	Rainfall <55 mm/y	Present: Absent:	0.05 0.43	kg/ha (160)	* mod–
17	Rainfall 151–400 mm/y	Present: Absent:	0.21 0.61	kg/ha (160)	** weak–
18	Rainfall 401–1100 mm/y	Present: Absent:	0.77 0.24	kg/ha (160)	*** mod+
19	Rainfall 1101–3000 mm/y	Present: Absent:	0.97 0.28	kg/ha (160)	*** mod+
20	Human habitat: commensal or urban	Present: Absent:	1.36 0.29	kg/ha (341)	*** strong+
21	Spp. native on any island	Island spp.: Mainl. spp.:	0.87 0.17	kg/ha (343)	*** v. strong+
22	Lizard introduced	Present: Absent:	4.8 0.5	kg/ha (343)	*** strong+
23	Subjective abundance	kg/ha: 10 kg/ha:	"common" "common"	Verbal descrip. (265)	n.s. 0.005

Table 3.13 continued

Trait	Variable	Benchmarks	Means	Units (*n*)	Stat. sign. Strength
24	Nonqualified (NQ) absolute densities	kg/ha: 100 kg/ha:	0.22 39.2	kg/ha (242)	*** 0.51, NQ overestimates low and underestimates high
25	Longevity (mass adj.)	−1 std. resid.: 1 std. resid.:	0.48 1.27	kg/ha (188)	** mod+

descriptor remains "common." If the transmitter and the listener share common expectations of context, subjective estimates may not be completely worthless, but there is no obvious way of testing that possibility.

Despite the vaguely encouraging concordance between qualified and nonqualified metrics, there were only 343 species so quantified (based on 2470 population estimates). This limited the power to discover the correlates of population density. Surprisingly, many effects were nonetheless strong.

Among the most striking were the following (table 3.13 trait number):

- Sexes differ in size (4): the seven ordinal means of sexual dimorphism were associated with population density at $r^2 = 0.94$ (fig. 3.4). It apparently pays to be a relatively large male at high population densities (presumably due to elevated intermale conflict) and a small male at low densities (unclear why such males should be small, but perhaps fecundity selection on females leaves the males at the optimal survival size and females grow larger to increase the clutch count).
- Herbivores averaged 24 times the biomass density of nonherbivores (6); omnivores fourfold (8). These effects presumably reflect the greater energy available at lower trophic levels.
- Foraging modes (10) seemed to have no association with population density. Evidently the different modes are equally successful (an ideal free distribution?).
- Diel patterns were nearly irrelevant, though nocturnality was a very weak negative (12), possibly due to increased (mammalian?) predation at night.
- Population density seemed to reflect primary productivity, with mesic habitats (18, 19) showing mild positive correlations and arid habitats (16, 17) having a negative relationship.
- Lizard species residing in human habitats have nearly five times the density of their nonurbanized colleagues (20). Urbanized habitats are a complex mosaic and may provide more sunlight and water and fewer predators (Mitchell et al. 2008).
- Lizards on islands averaged a density >5 times that of their mainland colleagues (21). See "Islands" above.
- Lizard species that have been introduced had nearly 10 times the density of nonintroduced species (22). This draws attention to the importance of propagule pressure; more lizard individuals increase the likelihood of a rare accidental colonization event, and high population density appeals to commercial collectors.

Introductions

Table 3.14 will be of special interest to biosecurity risk managers, who should note also the other associations with introductions identified in chapters 4 and 5. Although few of the identified correlations are strong, they may have appreciable predictive power in aggregate. The four strong correlates all highlight certain aspects of propagule pressure; that is, the likelihood that a given lizard will enter the human transportation chain. Although one propagule rarely establishes an introduced population, anything that increases the chances of one new lizard arriving also increases the yet smaller chance that members of both sexes will arrive and find each other.

The four strong correlates of introductions were the following (table 3.14 trait number):

- Of the lizards that forage under electric lights (by definition, in human environments) 45% have been introduced, compared to only 2.7% of the other species (17). Foraging under electric lights might directly increase the probability of a lizard being collected and intentionally distributed, but it might also reflect high densities in ports and urban areas, from which ships and other vehicles accidentally transport stowaways.
- 16% of human-habitat species were introduced, compared to 1.1% of species shunning human habitats (22). See immediately above.
- Species that were introduced had a much larger ontogenetic mass range (24), indicating especially small

Table 3.14. Correlates of lizard species that have been introduced and become potentially established anywhere outside of the species' native range. Conventions as in table 3.1.

Trait	Variable	Benchmarks	Means	Units (*n*)	Stat. sign. Strength
1	Mass	Introduced: Not intro.:	11.7 5.2	grams (6525)	*** mod+
2	Crests	Present: Absent:	9.0 3.0	% introduced (6528)	*** weak+
3	Secretory pores	Present: Absent:	5.5 2.9	% introduced (6528)	*** weak+
4	Any color change	Present: Absent:	12.5 2.2	% introduced (6528)	*** mod+
5	Advanced color change (color, pattern)	Present: Absent:	9.2 2.9	% introduced (6528)	*** mod+
6	Sexes differ in color	Present: Absent:	7.5 2.2	% introduced (6528)	*** mod+
7	Seasonal color	Present: Absent:	9.2 2.9	% introduced (6528)	*** mod+
8	Blue tails	Present: Absent:	13.2 3.0	% introduced (6528)	*** mod+
9	Herbivores	Present: Absent:	12.0 6.3	% introduced (2860)	* weak+
10	Invertebrate eaters	Present: Absent:	4.9 10.9	% introduced (2860)	*** mod–
11	Omnivores	Present: Absent:	12.2 5.5	% introduced (2860)	*** mod+
12	Sit-and-wait forager	Present: Absent:	10.5 7.5	% introduced (1149)	0.077 weak+
13	Home range (mass adj.)	Introduced: Not intro.	–0.27 0.09	Std. resid. (216)	* mod–
14	Active at any time (cathemeral)	Present: Absent:	7.1 4.2	% introduced (4746)	* weak+
15	Active at night (nocturnal)	Present: Absent:	2.7 4.9	% introduced (4746)	** weak–
16	Conspicuously wary	Present: Absent:	6.9 11.6	% introduced (837)	* weak–
17	Forages under electric lights	Present: Absent:	44.7 2.7	% introduced (6528)	*** strong+
18	Arboreal microhabitat	Present: Absent:	6.2 3.5	% introduced (4803)	*** weak+
19	Cryptozoic microhabitat	Present: Absent:	1.9 4.8	% introduced (4803)	** weak–
20	Subterranean microhabitat	Present: Absent:	0 4.8%	% introduced (4803)	*** weak–
21	Terrestrial microhabitat	Present: Absent:	3.8 5.0	% introduced (4803)	0.067 weak–
22	Human habitats	Present: Absent:	16.2 1.1	% introduced (5393)	*** strong+
23	Oviparous species	Oviparous: Viviparous:	5.0 2.5	% introduced (4283)	** weak+
24	Small young as indicated by high mass-adj. ontogenetic size range	Introduced: Not intro:	0.53 –0.08	Std. resid. (1089)	*** strong+

Table 3.14 continued

Trait	Variable	Benchmarks	Means	Units (*n*)	Stat. sign. Strength
25	Endemic to single "country"	Present: Absent:	1.6 6.3	% introduced (6528)	*** mod–
26	Population density	Introduced: Not intro.:	2.3 0.36	kg/ha (343)	*** strong+
27	At conservation risk	Present: Absent:	5.1 3.7	% introduced (3390)	0.053 weak+

young, which are associated with high fecundity species (either high clutch counts or frequent clutches, the latter not well quantified).
- Lizards with high population densities (26) were more likely to be introduced. See "Population Density/Biomass" above.

Ten moderate associations of introduced species (table 3.14 trait number) speak primarily to the likely transportation pathway:

- Mass (1). Larger species are more attractive to the pet trade; they would seem to be less likely to be introduced accidentally.
- Any color change (4) or advanced color change (5). Color changing species are attractive to the pet trade, but they might also be more likely to be color-matched to their background and therefore overlooked and accidentally transported.
- Sexes differ in color (6), show seasonal color (7), or have bright blue tails (8). Such species are attractive in the pet trade.
- Not invertebrate eaters (10), as these are difficult pets to maintain.
- Omnivores (11) can often be fed some human food and are therefore easier pets to maintain.
- Small home ranges (13) facilitate housing in small cages and accidental transport (wide-roaming stowaways are apt to be detected during transport).
- Endemism (25) low. Lizards that can be found in several jurisdictions have more geographic pathways to bring them into commerce (either accidentally or intentionally) and trade in them is less likely to be effectively regulated.

Thirteen additional, but relatively weak, natural history correlates of introduced species are listed in table 3.14. Note that the surfeit of endangered species (27) is some combination of conservation translocations (authorized releases, such as Tuatara on New Zealand islets) and unauthorized releases, and is associated with the higher frequency in the pet trade of hard-to-obtain species (e.g., chameleons).

Imperiled Lizards

Many variables were examined for potential ability to predict which lizard species were likely to be at conservation risk (table 3.15). Surprisingly few factors were demonstrably associated with expressed conservation risk ("imperiled" herein), perhaps because sample sizes of many attributes were relatively small. Among the surprises was the fact that population density (trait 24) had no association, and the trend observed was in the unexpected direction (denser species more likely to be imperiled). If endangerment is more closely related to extent of range than population density within that range, range-size monitoring by occupancy models will be more predictive of extinction risk than monitoring within occupied habitat by population density estimation. IUCN accounts of lizard species (www.iucnredlist.org) emphasize limited distribution rather than low population density as a cause of conservation concern.

Human habitat use (trait 16) might seem to be a good inverse predictor of imperilment, as the world is now awash in human habitats, but the effect size was miniscule (45.6% v. 50.4% at risk). Similarly, lizards that forage under electric lights would seem to be relatively safe (trait 12), as there are many electric lights, and electric light lizards are tolerant of disturbed habitats, but no inverse association between lights and endangerment was demonstrable. These factors substantiate the concept that population density is just one aspect of rarity (Rabinowitz 1981; Rabinowitz et al. 1986).

In addition to the small size of some of these samples, a factor potentially contributing to the lack of correlates of endangerment may be the lack of concordance between different observers as to which species are at risk. In the case of carnivores (trait 6), the severe IUCN listings were appreciably different in pattern from those of other authorities' listings. The IUCN's extreme categories had

Table 3.15. Correlates of lizard species that have been considered to be at conservation risk according to any literature account (e.g., "imperiled"). Conventions as in table 3.1.

Trait	Variable	Benchmarks	Means	Units (*n*)	Stat. sign. Strength
1	Mass	Imperiled:	7.9	grams	***
		Not at risk:	5.2	(3390)	weak+
2	Any color change	Present:	74.3	% imperiled	***
		Absent:	45.0	(3390)	strong+
3	Advanced color change (hue, pattern)	Present:	82.6	% imperiled	***
		Absent:	45.8	(3390)	strong+
4	Sexes unequivocally differ in size (either m>f or m<f)	Present:	60.7	% imperiled	***
		Absent:	46.6	(939)	strong+
5	Sexes differ in color	Present:	52.4	% imperiled	*
		Absent:	47.5	(3390)	weak+
6	Carnivores	Present:	85.5	% imperiled	***
		Absent:	53.5	(1792)	mod+
7	Herbivores	Present:	64.0	% imperiled	0.058
		Absent:	54.0	(1792)	weak+
8	Ant specialists	Present:	30.6	% imperiled	***
		Absent:	55.6	(1792)	mod–
9	Active during daytime	Present:	53	% imperiled	**
		Absent:	46	(2668)	weak+
10	Active at night	Present:	42.5	% imperiled	***
		Absent:	53.0	(2668)	mod–
11	Conspicuously wary	Present:	55.8	% imperiled	*
		Absent:	48.1	(3364)	weak+
12	Forages around electric lights	Present:	41.2	% imperiled	n.s.
		Absent:	49.0	(3390)	
13	Arboreal microhabitat	Present:	59.3	% imperiled	***
		Absent:	45.9	(3031)	mod+
14	Subterranean microhabitat	Present:	38.6	% imperiled	**
		Absent:	50.9	(3031)	weak–
15	Terrestrial microhabitat	Present:	46.2	% imperiled	***
		Absent:	54.1	(3031)	weak–
16	Human habitat use	Present:	45.6	% imperiled	*
		Absent:	50.4	(3237)	weak–
17	Rainfall 1101–3000 mm/y	Present:	66.7	% imperiled	*
		Absent:	55.8	(364)	mod+
18	Rainfall >3000 mm/y	Present:	34.2	% imperiled	**
		Absent:	64.1	(364)	mod–
19	Native on any island	Island spp.:	58.1	% imperiled	***
		Mainl. spp.:	40.5	(3390)	mod+
20	Native on any island	Island spp.:	18.5	% listed as IUCN EN+	***
		Mainl. spp.:	9.0	(3390)	mod+
21	Endemic to a "country"	Present:	52.6	% imperiled	***
		Absent:	42.3	(3390)	mod+
22	Lizard introduced	Present:	56.8	% imperiled	0.053
		Absent:	48.6	(3390)	weak+
23	Longevity (mass adj.)	Imperiled:	0.17	Std. resid.	**
		Not imperiled:	–0.06	(595)	weak+
24	Population density	Imperiled:	0.58	kg/ha	n.s.
		Not imperiled:	0.40	(252)	

listed fewer carnivores than random, whereas other authorities listed carnivores at a significantly higher rate. This may in part reflect the multiple jurisdictions in which many carnivores are found (more opportunities per species for listing) and greater public focus on larger lizards. Either factor could inflate the number of authorities listing a carnivore in comparison to the listings of smaller or more range-restricted lizards.

Next Steps

This chapter flagged correlations among the natural history traits examined, and hypothesized some possible linkages, but did not attempt to ascertain which factor was cause and which effect. That inference must await a dedicated study of the evolution of the traits. Nonetheless, any overarching concept of lizard ecology must accommodate the natural history correlations, and this chapter provides a beginning by noting relationships that fit together. In the next chapter, I propose some suites of traits that accommodate this covariation and recur throughout lizards.

4 Ecological Business Models

Lizard ecologists may have been misled by endotherm envy. Due to their extremely wasteful requirement for continuous heat production, endotherms are incessantly pressed up against energetic limits. This propels endotherms into an evolutionary race to discover and exploit rich energy sources. Endotherms' continuous pressure to maintain a high level of energy intake induces many species to specialize on scarce but rich sources like seeds, termites, nectar, and insect outbreaks. Energetically desperate endotherms compete with one another to exploit and monopolize these critical food sources. Most lizards, on the other hand, have little energetic pressure. If food is scarce for a week or a month, they will be fine.

Endotherm evolutionary ecologists have traditionally focused on how their taxa minutely divide various feeding opportunities to maintain the requisite high intake levels by minimizing dietary overlap with competing species. Their focus has been on feeding niches, optimal foraging, and dietary competition. The presumption is that species need to separate themselves from other species in dietary niche space. Robert MacArthur (1972) would be proud of their focus and their conceptual calculus. But is this the problem that lizards face?

Most lizards are opportunistic generalist invertebrate feeders. For the most part, they eat the same things as their neighbors. In most mainland localities, the most successful species are able to attain population densities that are yet far below the level at which they habitually reduce prey abundances. In such situations, dietary competition may be irrelevant most of the time (Dunham 1980). On small islands, where lizard foods are often scarce and many types of food are absent, lizards routinely attain substantially higher densities. The average density-increase ratio for islands up to the size of New Guinea tabulated in the last chapter was fivefold, and it would be much larger on small islands. Leão et al. (2018) recently looked for signs of intraspecific competition in several lizard species and the Tuatara, and they found it only in the hyperdense (ca. 515 kg/ha) Stephens Island population of the Tuatara. To put that phenomenal Tuatara density in quantitative context, the average of 922 mainland lizard density estimates in my database of 1748 qualified estimates was 0.82 kg/ha, whereas the comparable figure for island density estimates ($n = 826$) was 16.52 kg/ha. No doubt intraspecific competition for food could emerge at densities of hundreds of kilograms per hectare, but hardly any lizards, even on small islands, exist under such circumstances. This suggests that most lizard populations, in most mainland localities, are usually limited more by predation than food availability. For these ectotherms at least, the MacArthurian mindset may need to be tempered. Ecologists can explore what is chronic and may be limiting: avoiding being eaten is a primary concern of lizards in most habitats, especially those on large islands or continents. Regardless of the site-specific constraints on lizard survival, ecologists need to explore the whole of lizard ecology with a wide-open mind. Both food supply and predation can be limiting on a population.

What are normally labeled antipredator defenses (tail autotomy, head hiding, venomous snake mimicry, death feigning, etc.) are mostly features that operate in the last seconds of an encounter between predator and prey. These have a low probability of success, but if your life is about to end, you might as well try anything. The more effective defenses, however, are those that reduce the probability that the predator will discover the prey at all (Rose 1981). Lizards are masters of these, and these defenses are worth enumerating because they help identify the ecological "business models" that lizards do best.

Ectothermy itself is a predator avoidance behavior, insofar as it permits life styles that lower predation risk. A xenosaurid lizard that watches for occasional prey from the safety of its rock crevice can afford the luxury of eating only occasionally. Even herbivorous lizards can remain immobile for days at a time. An old record documented a Tuatara remaining immobile for six weeks (Cree 2014). No endotherm can do that.

Lizards cluster in environments with many safe places. Although "saxicolous" is treated by some as a specific saurian lifestyle, virtually any lizard will choose to remain near crevices that permit it to get beyond the reach of a predator. Gently tapering cracks not only exclude predacious birds and mammals (which are inherently bulky), but also many snake and lizard predators. A spiny or sacrificial tail can deter a snake in a sufficiently narrow crevice. Many Australian lizards exploit the clayey outback soils that crack deeply when dried; some specialize in this habitat. Arid environments are ideal for many lizards not only because the density of endothermic predators is reduced, but also because aridland environments are rife with gently tapering cracks.

Many lizards limit their activity season to the best few weeks of the year for the prey. Ecologists tend to assume that lizards are active on the surface whenever the weather is clement, but direct observation of lizard individuals of many species demonstrate that lizards may become inactive when the temperature is fine. Perhaps the foraging rewards fail to justify the predation risk. Teiids, for example, often become inactive while the weather is suitable. Adult lizards of many species become inactive at a season when their more energetically desperate (i.e., rapidly growing) offspring are still active. In general, only ectotherms have the metabolic option to shut down while remaining warm.

Limiting activity to the best few hours of the day for evading predators is also useful. Birds are a key group of predators on small lizards, and most birds have poor night vision. Shift activity into low light or night conditions, as many lizards do, and you avoid many predators. There are hydric and other advantages to low light living, but bird-predator avoidance is undoubtedly a key reward for many nocturnal arboreal lizards.

If you must display bright colors for mating purposes, place those colors on ventral, extendable, or lateral surfaces where they are less evident to predators. Many vertebrates play this card, but few do so with the panache of lizards.

It is also helpful to reduce movements to a narrow window of time within the active time of day. A Green Iguana needs to forage 20–30 min a day for its energy needs. By resting immobile, hidden in the foliage, for the remainder of the day, detection by predators is minimized. Many active-foraging lizards simply remain hidden (in their refugia) despite good weather for the majority of their activity period (some teiids and varanids may go days without surfacing).

Lizards often reduce detectability by crypsis and immobility whenever possible. Geckos with lateral fringes that eliminate shadows (both visible light shadows and echolocation shadows), as well as lizards with body colors that perfectly mimic tree trunk lichens, are familiar and highly effective examples. Immobility can be an end in itself, not just a consequence of lack of anything to do. *Cnemaspis* geckos remain immobile and largely invisible in plain view for hours or days at a time—self-restraint that no endotherm can match. If a lizard must move its eyes for surveillance, it is ideal to develop independent turreting of eyes (e.g., chameleons, polychrotids) to eliminate the need for the lizard's head to move. If body movement is needed, lizards may move very slowly; arboreal species can move with a swaying motion that mimics vegetation blowing in the wind.

Background matching is a saurian specialty; recognize that it can occur in two ways: (1) in the classical chameleon way in which the lizard changes color to match the background, or (2) by the lizard selecting where to rest and remaining immobile on the surface that perfectly matches its body color.

Rock, vegetation, and sand offer a wealth of opportunities for concealment, through both background matching and visual obstruction. Lizards are masters at putting a leaf, limb, or rock between themselves and prying eyes.

Perhaps the ultimate protective environment is subterranean. Although direct behavioral information about the natural history of subterranean or cryptozoic lizards is notoriously rare, the large number of such species (fig. 2.1) suggest it is winning evolutionary strategy.

Small size can be a predator avoidance tactic, if it provides opportunities for enhanced concealment. The myriad species of tiny litter lizards (e.g., *Lygisaurus*, *Lygodactylus*, *Sphaerodactylus*) are good examples; by being smaller than a fallen leaf, their world is burgeoning with useful visual barriers.

Large size can also be a predator avoidance tactic—few predators will tangle with a Komodo Dragon—but this tactic works only under a highly restrictive set of evolutionary conditions (e.g., no leopards about). The vast majority of lizards avoid rather than dominate their predators and, accordingly, most are a few grams in size.

Ecological business models are my attempt to distill the patterns present in the database's array of natural history traits of lizards. Some combinations of traits recur, most likely because there are causal connections between them. For example, an emphasis on visual prey capture is often associated with sexual dichromatism, most likely because vision mediates both social relations and foraging. In my experience, natural history trait combinations can be grouped into 14 clusters, which can be a useful shorthand for summarizing common adaptive syndromes. It is similar to the concept of a functional niche (Pocheville 2015; Pianka et al. 2017). It is similar to the concept of a "guild" (Root 1967; Inger and Colwell 1977; Simberloff and Dayan 1991), though with less emphasis on feeding specialization. With varying emphases, many authors have noted this phenomenon (e.g., Stamps 1977; Grismer 2011), and the concept has been rigorously pursued for Caribbean anoles in the form of "ecomorphs" (Williams 1972; Losos 1992).

My use of "ecological business models" differs from ecomorphs in generality: ecological business models can be applied to all species of lizards in all localities, but as yet they lack the precise morphological correlates identified for anole ecomorphs (see also Vanhooydonck and Van Damme 1999). Instead, they emphasize common behavioral patterns and microhabitat strata associated with foraging and social/territorial movements. For example, there is one business model for lizards that perch low on vegetation and capture prey on the ground (or vice versa), and there is another for lizards that perch at any height on tree trunks and capture prey on tree trunks. Unfortunately, foraging behavior with this level of microhabitat resolution is rarely found in the literature. Indeed, only 13% of lizard species have the requisite behavioral observations to assign business models unequivocally. The herpetological literature often reports where a species may be found (e.g., under rocks), but not where the lizards forage. Most species are known primarily or exclusively from taxonomic studies, which do not prioritize behavioral or ecological observation.

An explicit assumption in many efforts to analyze patterns of natural history traits in lizards (e.g., Pianka et al. 2017) is that food competition selects for traits that separate a lizard's niche from those of co-occurring lizards. In contrast, my emphasis is on how the predation risk due to movement is minimized while maintaining an adequate rate of food intake and mating opportunities. My calculation is analogous to the cost-benefit computation inherent in any entrepreneur's business plan. Potentially the primary "cost" of movement is exposure to potential predators, not the time or energy postulated by many optimal foraging models. The bottom line is whether the balance sheet shows a lifestyle that combines adequate nutritive and mating benefits with sufficiently low predation risk. Each lineage must remain "in the black" throughout its evolution. Very low rates of intake can be supported if those result in very low risk of predation; high-risk foraging must be supported by high rates of intake. Many different combinations of intake and risk are possible (see discussion after enumeration of the models). The enumerated business models illustrate my first approximation of the 14 most-common winning combinations.

The ecological business models postulated in the database for each species are intended to reflect the evolutionary history of the species, acknowledging that anthropogenic environmental changes, such as the provision of outdoor electric lighting, can alter these by a species learning to exploit the novel environment. When there is an ontogenetic shift in business model, the primary characterization of the species in this work refers to adults. If there are sex differences in business model, both are given and so annotated. On the smallest islands and in a few other situations, lizards may adopt multiple business models, as noted in the database. Each adaptive node (business model) is most likely connected to adjacent adaptive modes by a continuum. Intermediate business models should all be present, though perhaps less frequent than those at the nodes. Lizards that exhibit intermediate business models may have both adjacent business models listed, though multiple models may also signify geographic variation, behavioral polyethisms, or lack of data.

This concept of ecological business models has not yet been subjected to detailed analysis or statistical testing; the models should be considered my hypotheses regarding the nonrandom structure of natural history traits. Their publication should suffice to attract the improvement of data quality and the statistical testing that is warranted. The business model hypotheses, illustrated schematically in figure 4.1, are described in the subsequent narratives. Empirical natural history data extracted from the database on the basis of my tentative model assignments are presented in table 4.1. Highlights from table 4.1 are given at the conclusion of the description of each business model, in the subsections entitled "Touchstones from the Database."

Descriptions of the 14 business models follow in alphabetical order.

Crack Hermit

Definition

Crack hermits are refuge specialists who lie at the mouth of their burrows or crevices and feed upon things that crawl past their ever-ready (cathemeral) mouths. A number of species combine this "safe" mode with more exposed foraging modes when favorably low-risk environmental conditions allow (nighttime, after heavy rain, etc.).

Implications

Because the location of these lizards is highly predictable (*Xenosaurus newmanorum*, for example, has a mean net daily movement of 39 mm; Lemos-Espinal et al. 2003), crack hermits must be assured of their safety when in their crack. A heavily armored spiny tail that stands between the crack hermit and a predator may bolster the requisite

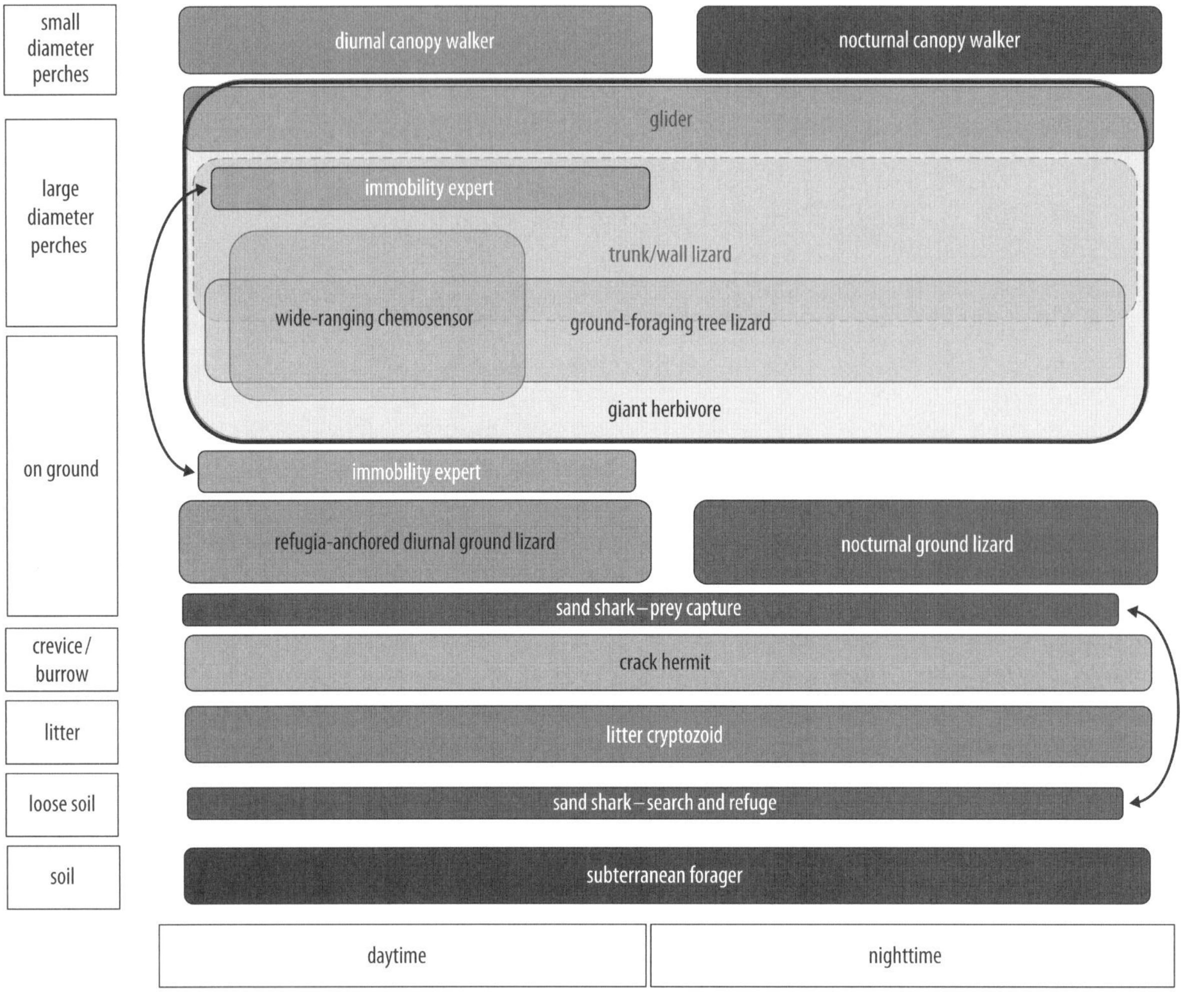

Fig. 4.1. Schematic of the 14 business models. This figure displays only two dimensions: time of day and microhabitat stratum. The other key attributes (e.g., sensory focus, territoriality, diet, activity area, movement rate, height above ground) are not represented. Species may be concentrated in subsections of the represented niche space; for example, most giant herbivores are day-active, only a few are nocturnal. (See color plate.)

assurance. Dug burrows are suboptimal for crack hermits because a predator can excavate soil and remove them; though, dug-burrow use is common in Australia, which has few native carnivores that dig well. Most crack hermits are found in deep cracks in massive rock, or at a minimum will have an escape exit for burrow-based "cracks."

Crack hermits with a good crack are in possession of a valuable resource and can be expected to defend it against competing consexuals, but to share it at mating time with potential mates, who must suspend defense of their own crack if they must leave it at mating time. Courtship and copulation can be a very hurried affair, with one or both partners returning quickly to their own cracks. Crack hermits of both sexes benefit from small size, as this increases the number of usable cracks, but female crack hermits—often viviparous, to transport the embryos out into the sunshine when they can—must be large enough to accommodate the mass of eggs or fetuses. Thus, crack hermits often exhibit m<f sexual size difference. Perhaps because the young may be born in the crack, parental attendance is common in crack hermits.

Crack hermits are likely to be clustered, as the advantages of mating proximity are prone to coincide with the patchiness of geologically optimal cracks. As the very limited sit-and-wait prey available to a crack hermit would otherwise need to be shared with crack partners, one would not expect crack cohabitation to last beyond the mating period unless the crack had multiple independent foraging positions.

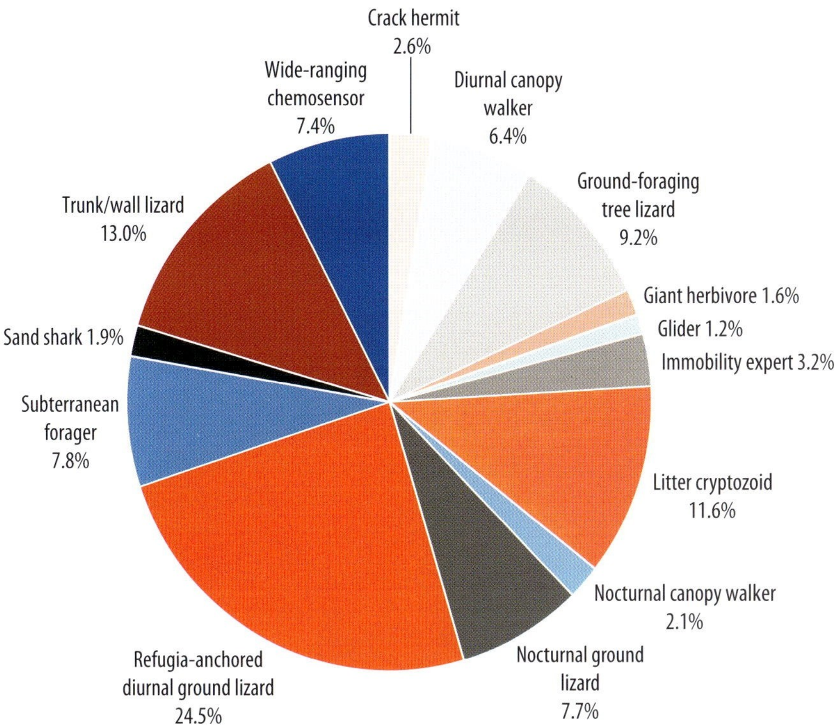

Fig. 2.1. The composition of lizard species, grouped by ecological business model. The percentage of each group is listed at the edge of wedge, based on 4912 scored species, with varying degrees of confidence in their assignment.

A

B

Fig. 2.5. Population densities of lizards range over seven orders of magnitude, as indicated in the text. **A**, The lowest credible estimate for a single population is for *Callisaurus draconoides* (Phrynosomatidae). **B**, The highest credible point estimate is for *Sphenodon punctatus* (Sphenodontidae). **A**, L. Grismer; **B**, J. Reardon

Fig. 2.6. A few lizard species have been transported and established in dozens of localities by humans. **A**, *Hemidactylus* spp. are the two most frequent invasives, here represented by *Hemidactylus frenatus* (Gekkonidae). **B**, Several *Gehyra* have been introduced, especially to islands; this is the Indian Ocean's *Gehyra mutilata* (Gekkonidae), nearly indistinguishable from the Pacific Ocean's highly invasive *G. insulensis* (Gekkonidae). (See also fig. 4.9B.) **C**, *Iguana iguana* (Iguanidae) is widespread in the pet trade and often released, especially in populated places; this male basking on a bridge in downtown Miami, Florida, is an example. **D**, *Varanus indicus* (Varanidae) was often transported among tropical islands in a futile attempt to control pests of coconut trees. **A**, **D**, B. Lardner; **B**, L. Grismer; **C**, G. Rodda

Fig. 2.7. *Sphenomorphus maindroni* (Scincidae), a typical lizard.
G. Rodda

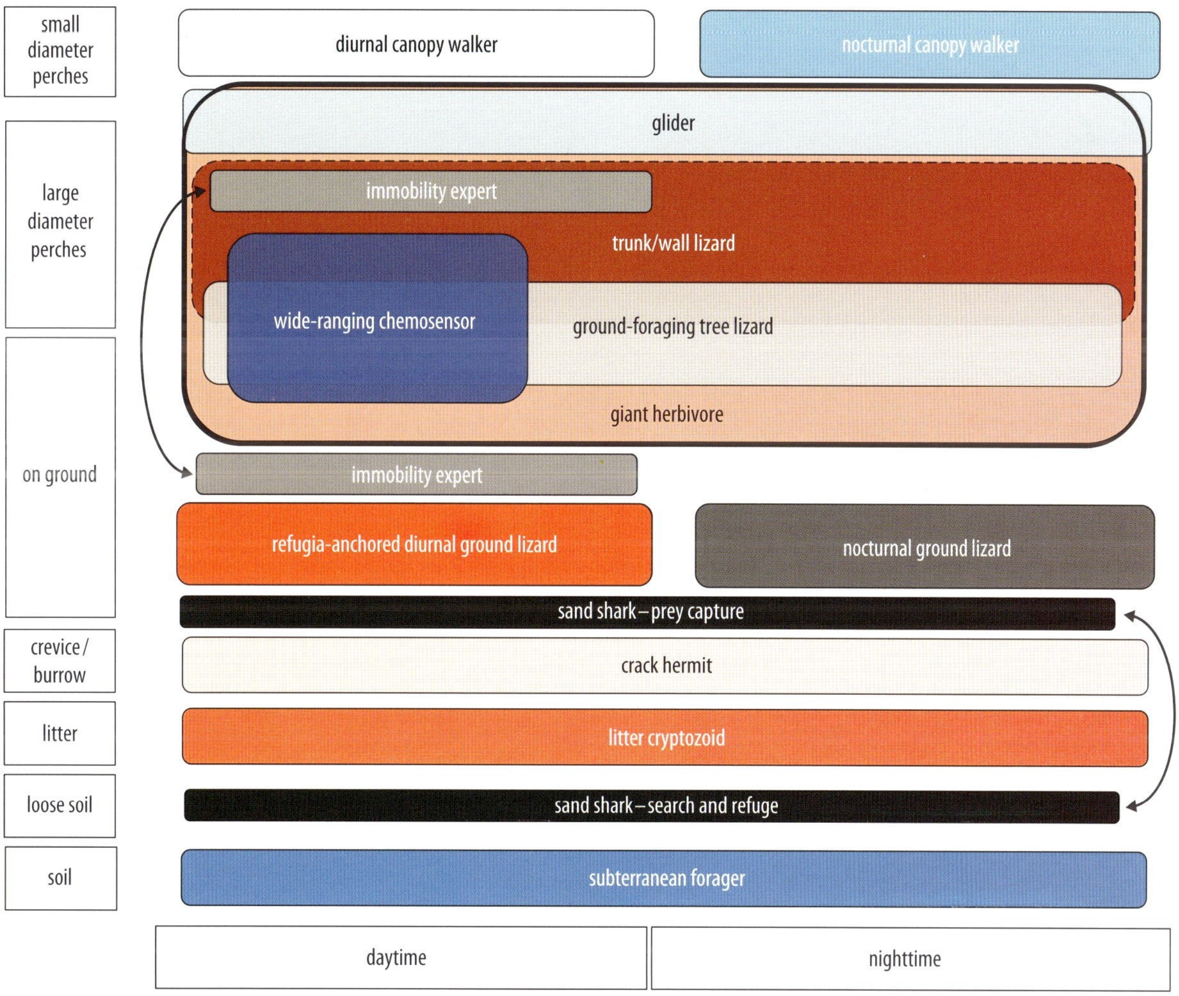

Fig. 4.1. Schematic of the 14 business models. This figure displays only two dimensions: time of day and microhabitat stratum. The other key attributes (e.g., sensory focus, territoriality, diet, activity area, movement rate, height above ground) are not represented. Species may be concentrated in subsections of the represented niche space; for example, most giant herbivores are day-active, only a few are nocturnal. Colors in this figure correspond to the wedge colors in figure 2.1.

Fig. 4.2. Crack Hermits. **A,** *Xantusia henshawi* (Xantusiidae). **B**, *Xenosaurus phalaroanthereon* (Xenosauridae). **C,** *Xenosaurus mendozai* (Xenosauridae). **D**, *Egernia striolata* (Scincidae). **A,** L. Grismer; **B**, S. Sanoja-Sarabia; **C**, J. Lemos-Espinal; **D**, M. Whiting

Fig. 4.3. Diurnal Canopy Walkers. **A**, *Anolis lineatus* (Dactyloidae). **B**, Adult male *Aphaniotis fusca* (Agamidae). **C**, Adult male *Pseudocalotes larutensis* (Agamidae). **D**, *Naultinus gemmeus* (Diplodactylidae). **E**, *Calumma brevicorne* (Chamaeleonidae). **F**, *Furcifer pardalis* (Chamaeleonidae). **A**, R. Reed; **B**, **C**, L. Grismer; **D–F**, J. Reardon

Fig. 4.4. Giant Herbivores. **A,** *Dipsosaurus dorsalis catalinensis* (Iguanidae). **B,** *Conolophus subcristatus* (Iguanidae). **C,** *Brachylophus fasciatus* (Iguanidae) **D,** *Tiliqua multifasciata* (Scincidae). **E,** *Amblyrhynchus cristatus* (Iguanidae). **F,** *Cyclura nubila caymanensis* (Iguanidae). (See also the *Iguana iguana* portrayed in figure 2.6C.) **A,** L. Grismer; **B, E,** R. Rondeau; **C,** R. Fisher; **D,** E. Wostl; **F,** G. Gerber

Fig. 4.5. Gliders. **A**, Agitated breeding male *Bronchocela cristatella* (Agamidae). **B**, Adult male *Hemidactylus craspedotus* (Gekkonidae). **C**, Adult male *Draco maximus* (Agamidae). **D**, *Draco melanopogon* (Agamidae). **A–D**, L. Grismer

Fig. 4.6. Ground-foraging Tree Lizards. **A**, *Acanthosaura armata* (Agamidae). **B**, *Calotes versicolor* (Agamidae) adult female. **C**, *Acanthosaura bintangensis* (Agamidae). **D**, *Anolis capito* (Dactyloidae). **A–C**, L. Grismer; **D**, E. Wostl

A B C D

Fig. 4.7. Immobility Experts. **A**, Adult female *Cnemaspis affinis* (Gekkonidae). **B**, *Phrynosoma platyrhinos* (Phrynosomatidae). **C**, *Naultinus manukanus* (Diplodactylidae). **D**, *Cnemaspis bidongensis* (Gekkonidae). (For other immobility experts see figure 5.3.) **A**, **B**, **D**, L. Grismer; **C**, J. Reardon

Fig. 4.8. Litter Cryptozoids. **A**, *Lygosoma bowringii* (Scincidae). **B**, newly described *Tytthoscincus martae* (Scincidae). **C**, *Delma molleri* (Pygopodidae). **D**, *Sphaerodactylus macrolepis* (male below; female above: Sphaerodactylidae). **A**, **B**, L. Grismer; **C**, J. Reardon; **D**, G. Rodda

Fig. 4.9. Nocturnal Canopy Walkers. **A**, *Aeluroscalabotes felinus* (Eublepharidae). **B**, *Gehyra mutilata* (Gekkonidae). **C**, Adult (unisexual: female) *Lepidodactylus lugubris* (Gekkonidae). **D**, *Cyrtodactylus louisiadensis* (Gekkonidae). Other nocturnal canopy walkers are featured in figure 2.6A (in the absence of electric lights) and 2.6B. **A**, **C**, L. Grismer; **B**, B. Lardner; **D**, G. Rodda

Fig. 4.10. Nocturnal Ground Lizards. **A**, *Coleonyx switaki* (Eublepharidae). **B**, *Goniurosaurus luii* (Eublepharidae). **C**, Adult female *Hemidactylus brookii* (Gekkonidae), a nocturnal ground lizard where not enticed to feed under electric lights. **D**, Adult male *Cyrtodactylus quadrivirgatus* (Gekkonidae). **A–D**, L. Grismer

Fig. 4.11. Refugia-anchored Diurnal Ground Lizards. **A**, *Petrosaurus thalassinus* (Phrynosomatidae). **B**, *Sceloporus uniformis* (Phrynosomatidae). **C**, *Sphenodon punctatus* (Sphenodontidae). **D**, *Uta stansburiana* (Phrynosomatidae). **E**, *Cnemidophorus arubensis* (Teiidae). **F**, *Carlia ailanpalai* (Scincidae). Other refugia-anchored diurnal ground lizards are featured in figure 2.5. **A**, **B**, **D**, L. Grismer; **C**, J. Reardon; **E**, R. Reed; **F**, B. Lardner

A

B

C

Fig. 4.12. Sand Sharks. **A**, *Larutia larutensis* (Scincidae, likely a sand shark). **B**, *Plestiodon reynoldsi* (Scincidae). **C**, *Eremiascincus phantasmus* (Scincidae). **A**, L. Grismer; **B**, C. Eaglestone; **C**, R. Reed

A

B

C

Fig. 4.13. Subterranean Foragers. **A**, Adult male *Dibamus tiomanensis* (Dibamidae). **B**, *Anniella geronimensis* (Anniellidae). **C**, *Bipes biporus* (Bipedidae). **A–C**, L. Grismer

Fig. 4.14. Trunk/Wall Lizards. **A**, Adult male *Cnemaspis boulengeri* (Gekkonidae). **B**, Adult male *Dasia olivacea* (Scincidae). **C**, *Cyrtodactylus pulchellus* (Gekkonidae). **D**, Adult female *Gekko gecko* (Gekkonidae). (See figure 2.6A for another species that is a trunk/wall lizard in certain places, usually in the presence of electric lights.) **A–D**, L. Grismer

Fig. 4.15. Wide-ranging Chemosensors. **A**, *Aspidoscelis gularis* (Teiidae). **B**, *Aspidoscelis hyperythra* (Teiidae). **C**, *Varanus salvator* (Varanidae). **D**, *Varanus salvator* in habitat (Varanidae). (See also the *Varanus indicus* illustrated in figure 2.6D.) **A**, E. Wostl; **B–D**, L. Grismer

Fig. 5.3. A famous case of convergent evolution. **A**, *Moloch horridus* (Agamidae). **B**, *Phrynosoma mcallii* (Phrynosomatidae). **A**, M. Whiting; **B**, L. Grismer

Table 4.1. Mean values of 76 selected traits for each of the 14 ecological business models. The units for each value are given in parentheses in the second row, under the trait description. Empty cells denote inadequate samples.

Trait number:	1	2	3	4	5	6
	Species (#)	Cells filled (%)	Mass (g)	Shape index (std. resid.)	Tail length (ratio × SVL)	Tail breaks (% of spp.)
Crack hermit	130	33	5.3	0.15	1.10	78.5
Diurnal canopy walker	313	35	9.1	0.42	1.51	21.1
Ground-foraging tree lizard	450	36	6.3	0.35	1.59	55.6
Giant herbivore	77	45	297.0	1.56	1.36	48.1
Glider	57	33	5.4	−0.33	1.54	22.8
Immobility expert	156	35	5.3	0.45	0.80	35.9
Litter cryptozoid	569	29	1.9	−0.57	1.38	79.1
Nocturnal canopy walker	103	30	3.3	0.06	0.96	96.1
Nocturnal ground lizard	379	31	3.5	0.05	0.98	91.8
Refugia-anchored diurnal ground lizard	1201	32	7.7	0.44	1.54	51.9
Subterranean forager	385	22	3.0	−2.75	0.37	38.2
Sand shark	91	29	2.3	−0.87	1.06	80.2
Trunk/wall lizard	637	32	5.6	0.26	1.25	75.7
Wide-ranging chemosensor	364	36	23.8	0.59	2.04	44.0

Trait number:	7	8	9	10	11	12
	Tail prehensile (% of spp.)	Crests (% of spp.)	Secretory pores (% of spp.)	Elongate (% of spp.)	Any color change (% of spp.)	Advanced color change (% of spp.)
Crack hermit	2.3	0.0	34.1	5.4	3.1	0.0
Diurnal canopy walker	54.3	25.6	2.9	1.6	55.0	51.1
Ground-foraging tree lizard	2.0	18.0	14.0	3.1	22.2	13.6
Giant herbivore	1.3	24.7	23.4	9.1	23.4	7.8
Glider	3.5	29.8	15.8	0.0	19.3	12.3
Immobility expert	30.1	3.8	26.3	0.0	50.6	32.1
Litter cryptozoid	0.2	0.0	6.7	46.2	0.7	0.0
Nocturnal canopy walker	19.4	1.9	54.4	2.9	11.7	2.9
Nocturnal ground lizard	1.6	0.3	29.6	5.0	6.9	0.3
Refugia-anchored diurnal ground lizard	0.7	3.2	12.6	6.6	5.2	2.4
Subterranean forager	0.0	0.0	3.9	17.9	0.0	0.0
Sand shark	0.0	0.0	3.3	75.8	0.0	0.0
Trunk/wall lizard	3.6	6.6	34.7	0.9	25.7	8.6
Wide-ranging chemosensor	3.8	0.0	4.4	5.2	1.9	0.8

Table 4.1. continued

Trait number:	13	14	15	16	17	18
	Sexes differ in size (–2 to 2 score)	Sexes differ in color (% of spp.)	Seasonal color (% of spp.)	Blue tails (% of spp.)	M:F tail length (ratio)	Sex diff. rollup (0–6 scale)
Crack hermit	–0.71	12.4	1.6	0.0	1.12	0.69
Diurnal canopy walker	0.47	44.7	11.8	0.3	1.08	1.61
Ground-foraging tree lizard	0.57	44.9	10.7	2.4	1.06	1.53
Giant herbivore	1.01	41.6	19.5	0.0	1.03	1.73
Glider	–0.25	64.9	1.8	1.8	0.94	1.79
Immobility expert	–0.65	23.7	3.2	0.0	1.22	1.16
Litter cryptozoid	–0.29	17.4	7.4	5.3	1.02	0.47
Nocturnal canopy walker	–0.53	1.9	0.0	0.0		0.73
Nocturnal ground lizard	–0.33	3.2	1.8	0.0	1.07	0.54
Refugia-anchored diurnal ground lizard	0.51	36.7	17.0	3.7	1.07	1.02
Subterranean forager	–0.70	0.3	0.0	0.3	1.24	0.10
Sand shark	–0.33	5.5	0.0	4.4		0.22
Trunk/wall lizard	0.74	23.1	3.3	1.7	1.03	1.00
Wide-ranging chemosensor	0.82	21.2	7.1	8.8	1.03	0.85

Trait number:	19	20	21	22	23	24
	Carnivores (% of spp.)	Herbivores (% of spp.)	Invert. eaters (% of spp.)	Ant spec. (% of spp.)	Omnivores (% of spp.)	Sit-and-wait foragers (% of spp.)
Crack hermit	0.0	0.0	80.3	0.0	19.7	93.3
Diurnal canopy walker	0.0	0.0	93.4	0.5	6.2	63.8
Ground-foraging tree lizard	1.1	1.1	72.7	7.0	18.1	65.5
Giant herbivore	0.0	82.2	0.0	0.0	17.8	50.0
Glider	0.0	0.0	25.0	75.0	0.0	62.5
Immobility expert	0.0	0.0	78.6	21.4	0.0	82.4
Litter cryptozoid	1.7	0.0	92.5	2.5	3.3	17.6
Nocturnal canopy walker	0.0	2.4	90.5	0.0	7.1	83.3
Nocturnal ground lizard	4.4	0.0	90.6	0.0	5.0	69.1
Refugia-anchored diurnal ground lizard	1.5	6.2	58.2	2.9	31.2	52.2
Subterranean forager	1.7	0.0	95.8	2.5	0.0	14.3
Sand shark	1.3	0.0	93.5	0.0	5.2	76.5
Trunk/wall lizard	1.7	0.3	80.9	5.9	11.2	77.1
Wide-ranging chemosensor	24.7	0.5	53.9	0.5	20.6	6.0

Table 4.1. continued

Trait number:	25	26	27	28	29	30
	Cruise foragers (% of spp.)	Active foragers (% of spp.)	Electric light use (% of spp.)	Home-range area—mass adj. (std. resid.)	Cathemeral activity (% of spp.)	Crepuscular activity (% of spp.)
Crack hermit	6.7	16.7	0.0		20.9	3.9
Diurnal canopy walker	34.0	21.3	0.0	−0.81	0.0	1.0
Ground-foraging tree lizard	1.4	38.8	5.8	−0.44	5.1	0.4
Giant herbivore	14.3	57.1	0.0	−0.73	5.2	1.3
Glider	0.0	50.0	0.0		3.5	0.0
Immobility expert	17.6	11.8	0.6	0.95	10.9	1.3
Litter cryptozoid	5.5	83.5	0.4	0.17	9.8	3.9
Nocturnal canopy walker	16.7	0.0	4.9		5.8	0.0
Nocturnal ground lizard	40.0	21.8	2.6	−0.10	8.7	0.3
Refugia-anchored diurnal ground lizard	1.9	58.4	0.3	0.13	3.2	1.6
Subterranean forager	28.6	71.4	0.0		2.3	0.3
Sand shark	0.0	70.6	0.0		8.8	3.3
Trunk/wall lizard	6.7	28.6	7.2	−0.39	9.1	1.4
Wide-ranging chemosensor	0.4	98.7	0.0	0.7	3.3	3.0

Trait number:	31	32	33	34	35	36
	Diurnal activity (% of spp.)	Nocturnal activity (% of spp.)	Calls spontaneously (% of spp.)	Stress calls (% of spp.)	Conspicuously wary (% of spp.)	Crypsis for pred. conceal. (% of spp.)
Crack hermit	38.8	20.9	4.6	6.2	23.1	4.6
Diurnal canopy walker	91.4	0.0	0.3	3.5	3.2	55.6
Ground-foraging tree lizard	76.7	9.8	2.0	3.6	9.8	8.9
Giant herbivore	89.6	1.3	6.5	2.6	26.0	1.3
Glider	77.2	15.8	1.8	0.0	8.8	15.8
Immobility expert	62.8	18.6	1.3	1.3	9.0	72.4
Litter cryptozoid	44.6	1.9	3.3	1.6	19.2	1.2
Nocturnal canopy walker	0.0	91.3	6.8	4.9	1.9	6.8
Nocturnal ground lizard	0.0	88.9	11.3	11.6	4.5	6.6
Refugia-anchored diurnal ground lizard	85.1	0.0	0.8	0.9	11.4	3.2
Subterranean forager	5.2	11.7	1.8	0.0	1.8	0.0
Sand shark	9.9	62.6	1.1	0.0	7.7	0.0
Trunk/wall lizard	43.2	38.3	6.9	4.2	7.7	3.6
Wide-ranging chemosensor	85.4	0.0	0.3	0.3	15.7	1.1

Table 4.1. continued

Trait number:	37	38	39	40	41	42
	Flight for pred. conceal. (% of spp.)	Sand for pred. conceal. (% of spp.)	Crev.-burr. for pred. conceal. (% of spp.)	Veg. for pred. conceal. (% of spp.)	Basks directly (% of spp.)	Thermo-conformer (% of spp.)
Crack hermit	0.8	0.0	10.0	0.8	84.9	15.1
Diurnal canopy walker	0.0	0.0	0.0	3.5	80.3	25.0
Ground-foraging tree lizard	1.8	0.2	6.4	8.7	78.5	20.4
Giant herbivore	0.0	0.0	27.3	7.8	100.0	0.0
Glider	0.0	0.0	0.0	0.0	100.0	22.2
Immobility expert	1.3	1.3	0.6	1.9	84.6	15.4
Litter cryptozoid	0.9	1.1	1.9	3.7	72.4	20.3
Nocturnal canopy walker	0.0	0.0	0.0	1.0	71.4	0.0
Nocturnal ground lizard	1.3	0.3	2.6	3.2	64.0	28.0
Refugia-anchored diurnal ground lizard	2.8	2.0	9.6	5.7	95.6	4.8
Subterranean forager	0.0	1.6	0.0	0.0	6.3	37.5
Sand shark	1.1	16.5	0.0	0.0	33.3	11.1
Trunk/wall lizard	0.6	0.0	5.0	7.1	79.9	22.9
Wide-ranging chemosensor	13.7	1.6	5.5	6.9	97.5	3.1

Trait number:	43	44	45	46	47	48
	Mean body temp. (°c)	Solitary (% of spp.)	Group living (% of spp.)	Males tolerate males (% of spp.)	Males defend space (% of spp.)	Males hierarchical (% of spp.)
Crack hermit	24.2	15.4	10.3	22.2	66.7	0.0
Diurnal canopy walker	29.8	25.0	15.0	7.5	90.6	3.8
Ground-foraging tree lizard	30.0	8.8	19.3	7.4	86.8	3.7
Giant herbivore	34.3	17.2	20.7	6.1	75.8	21.2
Glider	27.8	0.0	0.0	0.0	100.0	0.0
Immobility expert	33.6			40.0	60.0	0.0
Litter cryptozoid	28.8	0.0	11.1	28.6	71.4	7.1
Nocturnal canopy walker	24.6			14.3	85.7	0.0
Nocturnal ground lizard	25.7	17.2	31.0	13.9	83.3	8.3
Refugia-anchored diurnal ground lizard	32.8	9.8	21.6	9.1	87.1	13.3
Subterranean forager	27.6			80.0	20.0	0.0
Sand shark	30.8			0.0	100.0	0.0
Trunk/wall lizard	29.3	4.0	23.0	4.6	90.8	0.8
Wide-ranging chemosensor	34.6	10.0	30.0	50.6	49.4	14.1

Table 4.1. continued

Trait number:	49	50	51	52	53	54
	Hyperdry rainfall <55 mm/y (% of spp.)	Dry rainfall 56–150 mm/y (% of spp.)	Arid rainfall 151–400 mm/y (% of spp.)	Intermediate rainfall 401–1100 mm/y (% of spp.)	Mesic rainfall 1101–3000 mm/y (% of spp.)	Wet rainfall >3000 mm/y (% of spp.)
Crack hermit	0	15	31	46	8	8
Diurnal canopy walker	0	4	11	11	68	14
Ground-foraging tree lizard	0	3	25	24	45	15
Giant herbivore	20	33	23	27	23	10
Glider	0	0	0	0	0	100
Immobility expert	0	9	23	18	68	9
Litter cryptozoid	0	0	20	35	54	13
Nocturnal canopy walker	0	0	17	17	67	0
Nocturnal ground lizard	3	29	53	18	24	0
Refugia-anchored diurnal ground lizard	4	19	31	32	37	8
Subterranean forager	0	14	36	18	46	0
Sand shark	10	10	60	20	0	0
Trunk/wall lizard	5	5	15	23	66	11
Wide-ranging chemosensor	1	12	54	29	25	7

Trait number:	55	56	57	58	59	60
	Low elev. (<1200 m) (% of spp.)	Intermediate elev. (1200–2400 m) (% of spp.)	High elev. (>2400 m) (% of spp.)	Littoral habitat (% of spp.)	Dunes or sand habitat (% of spp.)	Riparian (% of spp.)
Crack hermit	78	48	5	11	9	12
Diurnal canopy walker	72	60	21	8	2	19
Ground-foraging tree lizard	91	48	11	22	15	32
Giant herbivore	97	33	6	35	38	30
Glider	100	23	3	16	0	18
Immobility expert	92	47	5	8	14	20
Litter cryptozoid	90	40	7	16	14	28
Nocturnal canopy walker	99	15	1	21	6	8
Nocturnal ground lizard	89	36	7	14	36	19
Refugia-anchored diurnal ground lizard	74	47	27	18	31	26
Subterranean forager	96	18	1	17	45	13
Sand shark	100	23	0	38	73	14
Trunk/wall lizard	94	32	4	16	4	21
Wide-ranging chemosensor	96	33	7	32	49	35

Table 4.1. continued

Trait number:	61	62	63	64	65	66
	Human habitat (% of spp.)	Oviparous (% of spp.)	Mass-adj. clutch count (std. resid.)	Mass-adj. ontogenetic mass range [− = lg. young; + = small young] (std. resid.)	Relative clutch mass (ratio)	Incubation duration mass-adjusted (std. resid.)
Crack hermit	10	34.0	−0.32	−0.68	0.232	0.05
Diurnal canopy walker	23	78.0	1.33	0.41	0.331	1.59
Ground-foraging tree lizard	38	90.7	−0.08	0.40	0.201	−0.26
Giant herbivore	22	83.6	0.43	−0.28	0.404	−0.78
Glider	27	100.0	−0.21	0.24		−0.47
Immobility expert	4	90.2	0.50	−0.10	0.340	0.08
Litter cryptozoid	18	80.5	0.17	0.15	0.181	0.16
Nocturnal canopy walker	20	97.8	−0.30	−0.40	0.244 ($n = 2$)	0.22
Nocturnal ground lizard	15	93.6	−0.30	−0.40	0.160	0.08
Refugia-anchored diurnal ground lizard	13	70.2	0.02	0.02	0.238	−0.37
Subterranean forager	8	66.5	−0.13	−0.87	0.252	−0.76
Sand shark	0	84.8	0.04	0.18	0.174	−0.48
Trunk/wall lizard	43	94.4	−0.43	−0.01	0.155	−0.05
Wide-ranging chemosensor	19	97.6	−0.26	0.13	0.215	0.25

Trait number:	67	68	69	70	71	72
	Eggs adherent (% of spp.)	Eggs laid communally (% of spp.)	Parental care (% of spp.)	Longevity mass-adjusted (std. resid.)	Population density (kg/ha)	Endemic to "country" (% of spp.)
Crack hermit	4.7	12.4	5.4	0.74	0.46	69.0
Diurnal canopy walker	1.0	1.3	0.6	−0.39	0.25	69.0
Ground-foraging tree lizard	2.7	6.7	1.8	−0.24	0.95	52.0
Giant herbivore	1.3	7.8	36.4	0.41	13.87	50.6
Glider	8.8	1.8	3.5	−0.22	0.20	54.4
Immobility expert	1.9	1.9	1.9	−0.16	0.08	71.2
Litter cryptozoid	0.2	4.7	4.2	0.01	0.18	60.3
Nocturnal canopy walker	11.7	5.8	1.9	0.61	0.17	75.7
Nocturnal ground lizard	5.3	7.7	2.9	0.22	0.60	52.8
Refugia-anchored diurnal ground lizard	0.7	3.5	3.7	−0.13	0.63	56.6
Subterranean forager	0.0	0.3	0.8	−0.92	0.02	66.8
Sand shark	0.0	0.0	2.2	0.24	0.19	51.6
Trunk/wall lizard	12.4	13.7	3.8	0.24	1.31	65.5
Wide-ranging chemosensor	0.8	2.5	5.8	−0.07	0.62	46.2

Table 4.1. continued

Trait number:	73	74	75	76
	Island species (% of spp.)	Introduced (% of spp.)	Introduced > twice (% of spp.)	Imperiled (% of spp.)
Crack hermit	19.4	1.6	0.0	68.2
Diurnal canopy walker	48.6	4.3	1.3	82.7
Ground-foraging tree lizard	53.3	9.2	2.9	42.4
Giant herbivore	67.5	30.5	8.5	91.8
Glider	84.2	1.8	0.0	11.8
Immobility expert	42.3	2.0	0.7	76.9
Litter cryptozoid	55.0	2.5	0.5	46.0
Nocturnal canopy walker	83.5	7.3	5.2	45.6
Nocturnal ground lizard	29.3	3.6	0.8	42.9
Refugia-anchored diurnal ground lizard	33.9	2.8	0.8	40.3
Subterranean forager	25.2	0.0	0.0	41.6
Sand shark	22.0	0.0	0.0	48.3
Trunk/wall lizard	59.0	8.5	4.1	54.0
Wide-ranging chemosensor	40.4	5.2	0.9	62.4

Examples

Xantusia (fig. 4.2A), xenosaurids (fig. 4.2B, 4.2C), and some cordylids are examples of full-time crack hermits; *Amalosia*, some *Cnemaspis*, *Egernia striolata* (fig. 4.2D), *Liopholis*, *Lipinia*, and *Nactus* may be opportunistic ones. Of the species assigned with full confidence to the crack hermit ecological business model, the top five species in descending order of information (filled cells) are *Euleptes europaea* (central Mediterranean sphaerodactylid), *Tiliqua adelaidensis* (Australian scincid), *Xantusia vigilis* (North American west coast xantusiid), *Xenosaurus grandis* (Mexican xenosaurid), and *Xenosaurus newmanorum* (Mexican xenosaurid).

Similar Business Models

The only similar business model is the immobility expert, which is also an extreme sit-and-wait forager, but one which is exposed and can move its preferred site at will. See also nocturnal ground lizards for a part-time variant.

Touchstones from the Database

In evaluating table 4.1 or interpreting the rankings that are highlighted in this chapter's "touchstones," recall from chapter 1 that the reported differences have not been subjected to statistical testing and represent only relative positions of the tabulated means. Crack hermits are the leading business model for the following traits (highest proportion of species unless indicated otherwise; table 4.1 trait number in parentheses):

- Most m<f sexual size difference (13). Males do not need to be large to dominate their peers, and smaller males have a larger array of suitable cracks. Female size may be constrained by selection for number of offspring.
- Sit-and-wait foragers (24). By the definition of crack hermits.
- Home-range area (28) is not listed in table 4.1 because the sample size is just one species. However, home range would be expected to be exceedingly small in crack hermits, and the one value is much smaller than any other business model (−1.31 v. lizard average of 0.0 std. resid.).
- Active at any time of day (29) or in low light (30). There is little penalty for round-the-clock vigilance.
- Coldest mean body temperature (43). Basking opportunities are presumably limited for crack hermits, even though crack hermits are average in their frequency of species that engage in direct basking when the sun position is favorable (85% v. 86%). Presumably, basking potential is a factor in their selection of home crack. Some chill may be due to the upland locations of many crack hermit species.

A B C D

Fig. 4.2. Crack Hermits. **A**, *Xantusia henshawi* (Xantusiidae). **B**, *Xenosaurus phalaroanthereon* (Xenosauridae). **C**, *Xenosaurus mendozai* (Xenosauridae). **D**, *Egernia striolata* (Scincidae). **A**, L. Grismer; **B**, S. Sanoja-Sarabia; **C**, J. Lemos-Espinal; **D**, M. Whiting

- Found in intermediate rainfall belt (52). Perhaps extreme sit-and-wait foraging is untenable in locations of very low primary productivity. In areas of very high primary productivity, small cracks are apt to be root-ridden.
- Viviparous species (62). Basking at the edge of a crack can warm the embryos as well as the mother without the risk inherent in an oviposition migration. Given their low mean active body temperature (trait 43, above), viviparity would make sense only if direct basking is regularly available.
- Smallest proportion of species on islands (71). If eggs are unavailable as a dispersal propagule, a pregnant mother is the best remaining option. If pregnant mothers rarely leave their rock crevices, accidental transport across the ocean seems unlikely.
- Highest longevity (70). Crack hermits are expected to experience low mortality due to the protection of their crevice.

Crack hermits are among the leading (i.e., ranked second or third) business models for the following traits (proportion of species elevated unless indicated otherwise; table 4.1 trait number in parentheses):

- Secretory pores (9), but not blue tails (16), crests (8), or advanced color change (12). Secretory pores are presumably used to mark possession of an occupied crack. This might prove useful if the occupant has retreated to an unobservable position, is momentarily inattentive, or is temporarily absent for purposes such as foraging or mating. Prior occupancy is a very strong determinate of contests and can be maintained for short absences by chemical cues. Blue tails, crests, or advanced color change would appear to have no strong benefit for crack hermits.
- Omnivory (23), but not carnivory (19) or myrmecovory (22). Omnivory requires the crack hermit to leave its crack, but not very far (perhaps a few meters) and not for very long (a few minutes a day). Carnivory would require lengthy excursions. Myrmecovory need not require any forays in theory, but may not be sustainable due to the behavior of the ants. Evidence for this is that both crack hermits and immobility

experts are extreme sit-and-wait foragers, both of which might be expected to take advantage of the one food that wanders reliably by. But crack hermits have 0% ant specialists, whereas the frequency of myrmecovory is the second highest in immobility experts (21.4%, behind only gliders). Direct evidence would be useful, but this indirect evidence suggests either a phylogenetic constraint (though both business models are phylogenetically distributed) or that ant feeding works differently in the two business models. The obvious difference is that crack hermits have a predictable location and immobility experts move (many observers think some *Phrynosoma* need to move regularly to avoid drawing a massed attack from the affected ants). Note, too, that the paramount ant specialists, gliders (75% of gliders are ant specialists), move their foraging sites with every glide and can evade ant attacks with a single leap (cf. diurnal canopy walkers cannot just leap away like gliders, and 99.5% of them, unlike gliders, do not specialize on ants). My guess is that crack hermits do not fear massed attacks of foraging ants because foraging ants reroute their trails to stay away from the predictable locations of crack hermits.

- Not active foragers (26). By definition of this ecological business model.
- Stress vocalizations (34) and wariness (35). Crack hermits are positively associated with spontaneous vocalizations (trait 33: 4.6% v. 3.3%), but they are league-leading only in stress vocalizations (second place). The high frequency for spontaneous vocalizations may reflect the difficulty of attracting mates when the caller's location, almost inevitably, is in a crevice out of the recipient's field of view. Conversely, spontaneous vocalizations would presumably attract unwanted attention from predators. Stress vocalizations are thought to be given for startle value—it is hard to imagine a predator that is trying to wrench a thrashing lizard from a crevice being startled by a feeble squawk—or to alert family members living nearby. In the case of crack hermits, it seems highly likely that nearby cracks are occupied by relatives, and that alarm calls might have some inclusive fitness value. They do not appear to be suited to antipredator use.
- Reliance on crevice-burrows for predator concealment (39). By ecological business model definition.
- Absence of hierarchical males (48). Opportunities for intermale interactions are probably very limited in crack hermits. When two male crack hermits do meet away from their home cracks, both are probably feeling vulnerable and might be disinclined to contest the other's presence. They have low-average levels of spatial defense overall (67% v. 82%).
- Intermediate elevations, but not low ones (56, 55). This may be an artifact of the small number of crack hermit clades. Perhaps the few clades tend to occur at intermediate elevations by chance.
- Avoidance of littoral (58) and mesic habitats (53). No reason for avoidance of littoral habitats springs to mind. Mesic habitat cracks are often filled with roots, vegetation, and soil.
- Smallest mass-adjusted clutch count (63), perhaps dictated by the requirement to fit in tight spaces, or perhaps a reflection of selection pressure for large-sized young.
- Small ontogenetic mass range (64). Low mortality reduces the selection pressure for large clutches, which may be often manifest in continued female growth and large ontogenetic body size range (e.g., diurnal canopy walkers). Low mortality is a reasonable expectation for crack hermits, as their exposure to predators is very limited. There might be selection pressure capping adult size to avoid outgrowing a well-fitting crack.
- Communal oviposition (68) and parental care (69). A third of crack hermits are oviparous; they all would presumably prefer to minimize the risk of an oviposition migration and instead use the nearest suitable oviposition site. Occupied cracks are often clustered by virtue of the geology; thus, searching gravid females are likely starting from similar origination points. These factors increase the prospects for communal oviposition. Parental care is more likely associated with viviparous crack hermits, which seem to regularly tolerate their offspring remaining in their vicinity. Neonates likely occupy narrower portions of cracks or narrower nearby cracks and therefore do not directly compete for food with their parents.

Diurnal Canopy Walker

Definition

Diurnal canopy walkers are diurnal, arboreal lizards that make frequent use of small-diameter perches (not trunks) and are incapable of crossing canopy gaps by gliding. In this case, "canopy" should be understood to include any small-diameter perch, even grass stems a short distance above the ground.

Implications

Daytime canopies are the domain of birds, which are highly effective predators on lizards. Longevity is apt to be short and therefore pressure for progeny incessant. Fortunately, canopy walkers face relatively few morphological constraints on moderate fecundity, unlike lizards that must sprint on open ground or squeeze through soil channels.

Canopies are easy to walk around, but hard to patrol, as the walkable distance between nearby twig ends may be a long and costly excursion back along the branch to the limb that connects branches leading to the two twigs. For this reason, fully arboreal territories are often exceedingly small (e.g., a fraction of a tree), or mediated by a dominance hierarchy (personal space or breeding space), rather than a feeding territory. Unfortunately, the literature on social interactions is often ambiguous as regards whether the fighting lizards were contesting turf or status (in captivity the point may be moot); discriminating data on social contests among arboreal lizards, other than a few anoles, are very scarce. One might also expect diurnal canopy walkers to rely on signaling over distances (easier than patrolling), using signals that can be concealed at other times (e.g., head bobs, dewlaps, conspicuous color in mouth lining, etc.). One would also expect avoidance of attention-grabbing signals (e.g., vocalizations) that can be localized by predators not already looking at the lizard.

The yield of prey for a stationary lizard in the canopy is likely to be low; perhaps for this reason many of the canopy species cruise their domain looking for prey, though long episodes of sitting-and-waiting—or resting (these are hard to distinguish)—are also observed. Diurnal canopy walkers are very visual and few have secretory pores. Along with an emphasis on visual modalities, they are likely to exhibit sexual dimorphism, sexual dichromatism, crests, and various ornaments, but in the case of canopy walkers, those visual attributes that might facilitate resource territorial conquests—as opposed to mating territories—are unnecessary and prone to attract predators. In the place of resource territories, diurnal canopy walkers may emphasize attributes such as color change (to facilitate crypsis, thermoregulation without having to move far, and social signaling), visually disruptive ornaments that conceal the basic quadrupedal outline, prehensile tails, and cryptic colors. To match small perch diameters and facilitate concealment, most diurnal canopy walkers are relatively small.

Examples

Examples of diurnal canopy walkers include some anoles (twig-branch ecomorphs; fig. 4.3A) and many agamids and chameleons. Of the species assigned with full confidence to the diurnal canopy walker ecological business model, the top five species in descending order of information (filled cells) are *Chamaeleo chamaeleon* (Mediterranean chamaeleonid), *Trioceros jacksonii* (East African chamaeleonid), *Chamaeleo calyptratus* (Arabian Peninsula chamaeleonid), *Anolis valencienni* (Jamaican dactyloid), and *Furcifer pardalis* (Madagascan chamaeleonid; fig. 4.3F).

Similar Business Models

Immobility experts take the immobility theme to its logical extreme; many are found in the canopy or evolved from diurnal canopy walkers—they just gave up most locomotion. Nocturnal canopy walkers shed most of the visual elaborations (and may add some sonic or chemosensory ones), but are otherwise similar to their diurnal counterparts. Many trunk/wall lizards roam out into the canopy, sometimes regularly, and do so especially on small islands; the trunk/wall lizards, however, return often to their trunks/walls to defend their territories, which in general the canopy walkers do not have. Other differences with trunk/wall lizards are highlighted in the two lists immediately below.

Touchstones from the Database

Diurnal canopy walkers are the leading business model for the following traits (highest proportion of species unless indicated otherwise):

- Tails that are prehensile (7), but not breakable (6; note contrast with trunk/wall lizards). Prehensile tails may be too valuable to be breakable.
- Lack of secretory pores (9): note contrast with trunk/wall lizards and nocturnal canopy walkers. One can imagine the utility of a marking "tool" that leaves a calling card even if the responsible lizard is on another branch at the moment. Diurnal canopy walkers do not apparently use secretory pores in this way (3%, the lowest proportion), even though nocturnal canopy walkers do (54%, the highest proportion). There seems to be some downside to secretory pores when they are not optimal. Energetic cost? Unwanted predatory attention? Nocturnal ants, snakes, or mammals might be able to find and predate exposed sleeping diurnal canopy walkers by following the lizards' secretion trails, whereas alert nocturnal canopy walkers could save themselves by timely vaulting into space.
- Darkening and advanced color change (11, 12): much higher than trunk/wall lizards; nocturnal canopy walkers darken but do not change hue or pattern. Diurnal canopy walkers are very visual and very cryptic.
- Least consumption of termites (not in table). Only 5.7% (v. 21.7% of lizards) are recorded to have eaten

Fig. 4.3. Diurnal Canopy Walkers. **A**, *Anolis lineatus* (Dactyloidae). **B**, Adult male *Aphaniotis fusca* (Agamidae). **C**, Adult male *Pseudocalotes larutensis* (Agamidae). **D**, *Naultinus gemmeus* (Diplodactylidae). **E**, *Calumma brevicorne* (Chamaeleonidae). **F**, *Furcifer pardalis* (Chamaeleonidae). **A**, R. Reed; **B**, **C**, L. Grismer; **D–F**, J. Reardon

termites. This fits well with their aversion to ants (see next list) and distinguishes them sharply from gliders (24.2% of which eat termites).

- Diurnal (31), but not cathemeral (29). Diurnal by definition, but why not cathemeral? Perhaps the many differences between diurnal and nocturnal canopy walkers are too big a gulf to bridge.
- Solitary existence (44): note contrast with trunk/wall lizards. The canopy is a comparatively uniform environment for a lizard; no reason not to spread out. Furthermore, a concentration of conspecifics would reinforce the search images and attract unwanted attention from predatory birds.
- Home-range defense (not in table) by either males (0% v. 2.2%) or females (0.0% v. 0.6%)—as opposed to the more generic "spatial" defense—is entirely absent in diurnal canopy walkers. In contrast, trunk/wall lizard males have elevated home-range defense (3.8% v. 2.2%).
- Mesic rainfall band (53). They are tied with immobility experts and barely ahead of nocturnal canopy walkers and trunk/wall lizards. Lizards in the canopy may be reliant on regular rainfall to prevent desiccation. This dominance does not extend to the wettest rainfall band, perhaps because desiccation is not a threat in a fog-enshrouded, phytotelmata-laden cloud forest. Phytotelmata are water-holding cavities in plants—such as bromeliads—that occur in terrestrial or arboreal vegetation. Heart rot in tree trunks can also create water-holding cavities.
- Medium elevation, but not low (56, 55). Possibly this is attributable to the many low-elevation sites that are not tree covered.
- Not littoral (58). This fits in with the relative paucity of low-elevation sites, but neither has an obvious explanation.
- Highest mass-adjusted clutch count (63). This fits with the notion that bird-induced mortality is high, especially on hatchlings. It also fits with their record low longevity (70) among aboveground species. The runner-up ecological business model for record low longevity is gliders, another vulnerable denizen of daytime canopies.
- Elevated ontogenetic range of masses (64), indicating relatively small hatchlings. Numerous small hatchlings, as opposed to smaller numbers of larger hatchlings, may produce the larger number of offspring needed to offset heavy bird predation.
- Slowest egg development (66). One is tempted to ascribe the long incubation period (1.59 v. 0.0 std. resid.) to the arboreal environment, except that gliders have one of the fastest development rates (–0.63 std. resid.) and adults are in the same environment. This comparison does not take into account which species oviposit aboveground versus in soil (not recorded, though the two documented *Draco* lay in soil and some diurnal canopy dwellers lay under bark—see next bullet). If anything, one would expect a gliding lizard to lay eggs at the earliest possible moment in order to reduce the mother's airframe weight, and therefore have a longer incubation period to compensate. Other arboreal lizards, like nocturnal canopy walkers (0.22 std. resid.) and trunk/wall lizards (–0.05 std. resid.), offer little support for the notion that slow development is due to the arboreal environment.
- Lack of parental care (69). Because some diurnal canopy walkers oviposit under bark, there is little the mothers can do to enhance survival (no nest guarding, wetting, or attendance to young). Others oviposit in the ground and return promptly to their arboreal perches, leaving the eggs without care.

Diurnal canopy walkers are among the leading (i.e., ranked second or third) business models for the following traits (proportion of species elevated unless indicated otherwise):

- Large size (3): mean = 9.1g (v. 5.4 g). Possibly large size is a hedge against moisture stress, as small lizards lose proportionately more water and may not have an opportunity to replenish during long spells without precipitation.
- Possesses crests (8). Crests are strongly associated with arboreality (table 3.2, trait 27), but it is less clear that crests are more advantageous in the canopy than that they are disadvantageous elsewhere. Crests are a metabolically inexpensive way to appear large, and they may break up the quadrupedal outline and thereby promote crypsis. Crypsis is a recurring association with diurnal arboreal life. Sexual selection appears to cause hypertrophy of crests in males, but this makes more sense in (often territorial) trunk/wall lizards than in the more solitary diurnal canopy walkers. Hypertrophied crests may somewhat damage crypsis, so crests should be less hypertrophied in canopy walkers than in trunk/wall lizards (database does not document degree of sexual differences in crests). Trunk/wall lizards have a lower proportion of crested species than do diurnal canopy walkers, undermining support for this argument and providing support for the notion that it is the jagged visual complexity of the foliage, rather than arboreality per se, that is the environmental feature that supports crests as an enhancement to crypsis.

- Sexes differ in color (14), show seasonal color (15), have overall sex differences (18): all frequencies at least twice those of trunk/wall lizards. These traits appear inconsistent with the hypothesized presence of territoriality in trunk/wall lizards and their absence in canopy walkers. Note however that trunk/wall lizards are more sexually dimorphic, which is consistent with the hypothesized territoriality pattern. Dichromatism may be selected against in trunk/wall lizards if it makes the lizard more conspicuous against the drab background of the trunk or wall. In contrast, lizards among the foliage have a very colorful and visually noisy background, where sex markings may be useful or communicating availability across canopy spaces.
- Invertebrate eaters (21), but not carnivores (19). These are most likely dictated by size limitations.
- Low consumption of ants (not in table): 18.8% v. 30.1% for all lizards with diet data. This is a sharp difference with gliders (79%).
- Cruise foragers (25): note sharp contrast with trunk/wall lizards. Unlike trunks, branches sway in the breeze, facilitating slow swaying cruise foraging without much loss of crypsis. On trunks, any motion stands out against the immobile background.
- Not nocturnal (33). By definition.
- Not conspicuously wary (35) and does not call spontaneously (33): note contrast with trunk/wall lizards. Both features aid crypsis, a staple of lizard canopy life.
- Rely on crypsis for predator concealment (36): note sharp contrast with trunk/wall lizards. Diurnal canopy walkers do not rely on flight (37), sand (38), or crevices (39).
- Thermoconformers (42). This is not inevitable for canopy dwellers, as there are almost always sun spots somewhere in the canopy. Presumably they avoid the sun spots either because their environment is sufficiently warm for the low level of locomotor demands placed on cryptic or cruising canopy walkers, or because birds' superb vision is more likely to be brought to bear on well-lit places near the top of the canopy, where birds often land. Remaining in the shadows may be an important part of predator avoidance in canopy dwellers. Note that diurnal canopy walkers have the lowest basking frequency (trait 41) of any diurnal aboveground business model.
- Elevation (57) high, rainfall (53, 54) mesic or wet, but not arid or intermediate (51, 52). These may merely reflect conditions optimal for tree growth.
- Relative clutch mass (RCM) (65) is high, presumably because mass-adjusted clutch count is the highest in lizards.
- Does not deposit eggs communally (68): note sharp contrast with trunk/wall lizards. It is unclear whether this merely reflects the dispersed distribution of canopy dwellers, or a more clumped distribution of oviposition sites that are on tree trunks. Data on the availability of suitable oviposition sites would be helpful in this context.
- Endemic (72). Does this reflect the extreme localization of *Anolis* species, especially insular anoles? The logic behind microhabitat partitioning in anoles has similar force for all canopy dwellers (but not gliders: see *Draco* account in chapter 5). Note that the highest endemism proportion is found in nocturnal canopy walkers. Canopy locomotion is so difficult and the morphological selection pressures high enough that diurnal canopy dwellers may often split up the ecomorph space and produce high endemism, at least on islands. Less partitioning appears to occur on mainlands.
- Short longevity (70): note contrast with trunk/wall lizards. All things being equal, the environmental predictability impinging on trunk/wall lizards and canopy walkers should not differ. Therefore, if mortality schedules were similar, size-adjusted longevity might be roughly the same between canopy walkers and trunk/wall lizards. All of the diurnal arboreal business models (diurnal canopy walker, ground-foraging tree lizard, glider, and many immobility experts) except trunk/wall lizard exhibit short longevities. These facts point in the direction of high mortality in the canopy, presumably due to avian predation. Trunk/wall lizards may have some ability to dodge avian predation by circling the trunk or hiding in bark or wall crevices. Once detected by a bird, a lizard on a twig may have fewer good options.
- Imperiled (76). High proportions of species at risk are probably associated with the high endemism and limited distributions of many canopy walkers, especially island species.

Giant Herbivore

Definition

Giant herbivores are large lizards that consume primarily vegetable matter and derive both greater digestive efficiency and reduced predator vulnerability by their large size.

Implications

Herbivores in general lack effective armament against predators (feeble dentition defending highly edible flesh), so they need to avoid predators of appropriate size. Herbivores are often extremely large (e.g., *Conolophus*, *Ctenosaura*, *Egernia*, *Hydrosaurus*, *Iguana*, *Tiliqua*) and/or occupy predator-reduced environments (desert: *Dipsosaurus*, *Laudakia*, *Sauromalus*, *Uromastyx*; island: *Amblyrhynchus*, *Brachylophus*, *Corucia*, *Cyclura*, *Gallotia*). Most are day active to thermally expedite microbial digestion of cellulose, but nocturnal herbivory is plausible in equatorial maritime environments (e.g., *Corucia*). The mating system depends on the density of suitable mates: low productivity environments favor monogamy; high concentrations (e.g., near watercourses) favor polygyny. Socially mediated tactics may adjust with density; territoriality may become lek behavior or dominance hierarchies. Sexual selection is often accompanied by sexual dichromatism, sexual dimorphism, color, crests, and other ornaments. Because herbivores are largely defenseless, juvenile mortality is often extreme (Poulin et al. 2001), leading to selection for rapid growth, large body size, and very large brood counts (e.g., *Iguana*) or large offspring sizes, especially on islands (e.g., *Amblyrhynchus*, *Brachylophus*, *Conolophus*, *Corucia*, *Tiliqua*).

Examples

See the genera mentioned in the preceding paragraph and figure 4.4. Of the species assigned with full confidence to the giant herbivore ecological business model, the top five species in descending order of information (filled cells) are *Iguana iguana* (Neotropical iguanid; fig. 2.6C), *Cyclura nubila* (Cuban iguanid), *Ctenosaura similis* (Central American iguanid), *Tiliqua rugosa* (Australian scincid), and *Cyclura pinguis* (Virgin Islands iguanid).

Similar Business Models

Refugia-anchored diurnal ground lizards of any size may be herbivorous or opportunistically herbivorous in particularly safe environments (extreme aridity, altitude), but remain in the vicinity of refugia to address the comparatively rare predation attempts. Diurnal canopy walkers differ from the few canopy herbivores only in diet (diurnal canopy walkers are overwhelmingly invertebrate eaters).

Touchstones from the Database

Giant herbivores are the leading business model for the following traits (highest proportion of species unless indicated otherwise):

- Natural history knowledge (2; 45% v. 27% of cells filled). Large lizard size attracts field studies made possible through radio telemetry, individual recognition, easy detection, and the like.
- Large body size (3; mean = 297 g) and pudgiest shape (4). Large and pudgy are associated with herbivory (chapter 3, "Diet").
- m>f sexual size dimorphism (SSD) (14). Associated with large size and herbivory (chapter 3, "Sexual Size Dimorphism").
- Seasonal color (15). Associated with strong SSD.
- Least consumption of ants (not in table; 8.8% v. 30.1%). This does not mean that 8.8% of their diet is ants, but rather that 8.8% of giant herbivore species have been known to have ants in their guts. This low level of consumption fits in with their minimal consumption of termites (see next list) and other invertebrates.
- Conspicuously wary (35). Associated with relatively defenseless herbivores, especially in places where humans consume and collect giant herbivores.
- Reliant on crevices and burrows for predator concealment (39). Nothing else will conceal a giant.
- Bask directly (41), do not conform (42). High body temperature expedites breakdown of cellulose.
- Male hierarchies (48). Male hierarchies tend to form at high population densities, which are characteristic of herbivores.
- Male giant herbivores have the highest proportion of year-round defenders (not in table; 15.2% v. 1.4%). Female giant herbivores (7.7% v. 3.4%) are also league-leading in this attribute, which is, however, inadequately documented in females. These rankings—and the male hierarchy rank—suggest an element of resource defense that is not as salient in other ecological business models. Compared to more mobile prey (insects, vertebrates), desirable plants may be more prone to depletion in a small area near a burrow or other refuge.
- Found in hyperdry (49) and dry (50) environments. These facts likely reflect reduced mammalian predation in such areas. Certainly, food plants are scarcer where primary productivity is extremely low. Potentially, animal prey are even scarcer in the dry environments, promoting a diet of poor, but available, plants. The large size of giant herbivores will give them more latitude to bridge between times of moisture availability for drinking.
- Highest RCM (65). Giant herbivores need a lot of peritoneal space for the bulky vegetation in their diet. During peak egg development, this room is often given over to heavy or numerous eggs, elevating RCM.
- Fastest egg development rate (66). Giant herbivores often undertake metabolically expensive and very predation-risky migrations to rare oviposition sites (see next bullet), presumably because they benefit

Fig. 4.4. Giant Herbivores. **A,** *Dipsosaurus dorsalis catalinensis* (Iguanidae). **B,** *Conolophus subcristatus* (Iguanidae). **C,** *Brachylophus fasciatus* (Iguanidae) **D,** *Tiliqua multifasciata* (Scincidae). **E,** *Amblyrhynchus cristatus* (Iguanidae). **F,** *Cyclura nubila caymanensis* (Iguanidae). (See also the *Iguana iguana* portrayed in figure 2.6C.) **A,** L. Grismer; **B, E,** R. Rondeau; **C,** R. Fisher; **D,** E. Wostl; **F,** G. Gerber

from the physical conditions present there. The physical conditions of interest are likely to be friable soil, high temperature, generous aeration, or optimal humidity, but especially temperature. The extraordinarily fast egg-development rate (–0.78 std. resid.) reinforces the notion that rapid egg development is important for giant herbivores. Given their large size, they may need relatively expedited development to hatch within the seasonally suitable window. They may also need to produce progeny rapidly when circumstances allow. The basic demographic challenge for giant herbivores on mainlands is overcoming predation losses through high fecundity. On islands, giant herbivores produce smaller clutches of larger young, but may still need to expedite development to match hatching time to a favorable season.

- Show parental care (69): often in the form of defending nest sites. Giant herbivore nests are often reliant on relatively scarce accumulations of loose soil. The process of digging a nest ensures the looseness of the soil and thereby gives later arriving females the option of parasitizing the efforts of their predecessors—and later diggers may inadvertently unearth previously laid eggs—unless the predecessor sticks around to guard her investment.
- High population density (71). Note the magnitude by which giant herbivores exceed all others. Associated with herbivory (chapter 3, "Diet").
- Introductions (74, 75). Two factors appear to have motivated the human dissemination of giant herbivore lizards. They are tasty and easily caught, therefore an attractive food to distribute, especially to new islands. In the modern era, iguanas have often been distributed by the pet trade. Some endangered species have been intentionally repatriated or translocated for conservation reasons.
- At conservation risk (76). Many giant herbivores have speciated on their home islands; the majority of these island endemics are now imperiled.

Giant herbivores are among the leading business models for the following traits (proportion of species elevated unless indicated otherwise):

- Few species (1). Herbivorous lizards do not appear to be very successful on mainland locations with a full suite of predators.
- Possession of crests (8), but not blue tails (16). Crests are normally associated with arboreal species. The frequency of crests in giant herbivores may reflect the metabolic advantages of using crests as an inexpensive short cut to the appearance of large size. The benefits may be both in repelling predation, especially in older juveniles, and in attaining social dominance, especially in adult males; giant herbivore crests routinely hypertrophy in males at sexual maturity. Blue tails are likely unsuitable for giant herbivores, which are slow and often immobile and basking. Many also lack tail autotomy. Bright blue tails would attract predation to these vulnerable lizards.
- Overall sex differences (18). Their high density and large size are associated with large sex differences.
- Minimal consumption of termites (not in table; 7.1% v. 21.7%). Termites are so fatty that they are eagerly consumed, even by lizards with other dietary specialties, but apparently not by giant herbivores (digestion difficult? adversely affects gastrointestinal microflora?).
- Small mass-adjusted home range (28). Perhaps home ranges can be smaller for lizards eating at a lower trophic level. This small size may contribute to their ability to defend resource-based home ranges (see social traits discussed in immediately preceding list).
- Diurnal (31). They need the sun's warmth to digest plants.
- Reliant on concealment in vegetation (40), but not flight (37) or sand (38). Vegetation is a natural background for herbivores. But they are corpulent and cannot run fast or dive into sand.
- Body temperature (43). The recurring pattern of high body temperatures appears to be for digesting cellulose. They are exceeded only by the wide-ranging chemosensors, which are reliant on flight for predator concealment and therefore need to be near optimal temperature for muscle performance. Giant herbivores rarely sprint, but they sometimes need to and may escape courtesy of their high body temperature.
- Solitary existence (44). Giant herbivores often meet up during the mating season, but they group outside of the mating season only if their refuging needs are met by clustered features such as water edges or isolated rock piles. They may temporarily group around fruit bonanzas (e.g., ripening mangoes).
- Found in littoral habitats (58). Some species will refuge in salt water, but most have no particular association with the littoral, except that many occur on islands and littoral zones may be a large proportion of small islands.
- High mass-adjusted clutch count (63). This suggests that giant herbivores are under considerable selection pressure for high fecundity. This feature fits with expeditious incubation and solicitous parental care highlighted in previous section.

- Lay eggs communally (68). Communal egg laying in most giant herbivores appears to be driven by shared environmental requirements for egg chambers rather than cooperation among lizards.
- Found on islands (73), but not endemic (72). The relationship to islands is presumed to be driven by low predation pressure on islands. Species successful on mainlands often have large ranges (plants are very widespread) and therefore low endemism.

Glider

Definition

Gliders are arboreal invertebrate-eating lizards with gliding membranes allowing them to leap and glide across canopy gaps. Some can glide across gaps as large as 50 m.

Implications

You can imagine the frustration a small lizard must feel when it sees a choice food morsel or social antagonist lingering just a few body lengths away, untouchable on a separate tree. The gliding forms are often diurnal (e.g., *Draco*, *Holaspis*) because the motivation for leaping among trees (sight of a food item, competitor, or prospective mate), and the ability to do so in a controlled manner, are enhanced by vision. Many of the diurnal forms also possess a dewlap, which may assist in controlling glides and signaling across large canopy distances. Gliders have the ability to economically defend vast arboreal territories, but is it practical for them to do so? Even if they do not repel intruders from their activity range, one would expect the activity ranges of the highly vagile gliders to be larger than those of diurnal canopy walkers. This is probably true, but documentation of glider home ranges is limited to a single species, which however does have a large home range (*Draco spilopterus* at 822 m^2 or 1.16 v. 0.0 std. resid. for lizards). This is substantially larger than the 31 m^2 (–0.81 std. resid.) recorded for the 10 documented diurnal canopy walkers.

The tremendous diversification in sexually dichromatic patagial colors suggests that, in addition to facilitating territorial defense, the elaborate colorful embellishments and refined gliding ability may serve also in mate attraction and, in some cases, intermale aerial combat. Grismer (2011) characterized one such encounter as "what can only best be described as an aerial dogfight."

Gliders often focus their movements on tree trunks, which is consistent with a territorial lifestyle. Despite their ability to glide—making them highly conspicuous to predators (which, however, except for birds, may not be able to leap canopy gaps after them)—gliders are often extremely cryptic when immobile. Thus, they often remain immobile for extended periods of time and thereby share some characteristics with immobility experts. When they do move, suddenly extending their often brightly colored flaps, the flash colors may serve a startle function, further distracting or intimidating potential predators.

Because gliding brings unwanted predatory attention, and the canopy is thick with predacious birds, one would expect gliding to be done preferentially only for the most essential outcomes, such as acquiring a mate or defeating a rival male. Courtship seems to be conducted mostly by dewlapping, which is more concealed from predators flying overhead, and may be timed preferentially to coincide with empty skies and other disruptions such as glides. The advantage of remaining immobile when not gliding provides an opportunity to forage by strict sitting-and-waiting, which may limit the energetic payoff of foraging. Many gliders respond by consuming mostly ubiquitous poor-quality prey such as ants. Diurnal canopy walkers could follow the same dietary logic, but they do not.

Gliders are often incorrectly described as imperfectly evolved fliers, but flying is a high-energy activity that would require frequent foraging and would draw unwanted attention from the pervasive bird predators. Furthermore, it would force lizards to compete energetically with the highly selective foraging of birds. Low-energy lizards are likely better off gliding than flying.

Examples

Examples of gliders include *Bronchocela cristatella*, *Draco*, *Holaspis*, *Luperosaurus kubli*, *L. sorok*, and *Ptychozoon*. Of the species assigned with full confidence to the glider ecological business model, the top five species in descending order of information (filled cells) are *Draco spilopterus* (insular Southeast Asian agamid), *Bronchocela cristatella* (Southeast Asian agamid; fig. 4.5A), *Ptychozoon kuhli* (Southeast Asian gekkonid), *Draco melanopogon* (Southeast Asian agamid; fig. 4.5D), and *Draco maculatus* (Southeast Asian agamid). The dominance of Southeast Asia among gliders is notable. Perhaps the Southeast Asian islands have a climate conducive to water conservation in lizard species with a high surface to volume ratio.

Similar Business Models

Gliding appears to be a highly derived condition, possibly an exaptation of extreme immobility (lateral fringes to eliminate side shadows also serve aerodynamically), but not easily confused with any other business model.

A

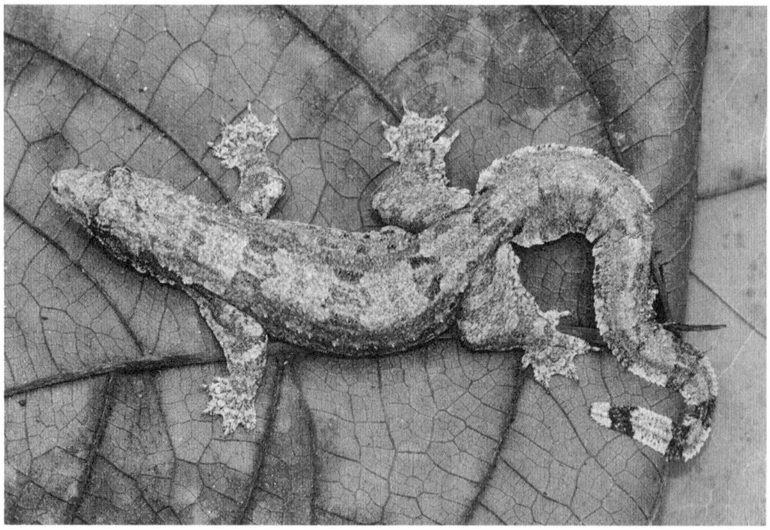

B

C

D

Fig. 4.5. Gliders. **A**, Agitated breeding male *Bronchocela cristatella* (Agamidae). **B**, Adult male *Hemidactylus craspedotus* (Gekkonidae). **C**, Adult male *Draco maximus* (Agamidae). **D**, *Draco melanopogon* (Agamidae). **A–D**, L. Grismer

Touchstones from the Database

Gliders are the leading business model for the following traits (highest proportion of species, unless indicated otherwise):

- Few species (1), though gliders are likely undersampled. See the *Draco* entry in chapter 5 for discussion of speciation in gliders. Gliding has apparently arisen on several occasions, but may require unusual circumstances to evolve.
- Crests (8). Crests are associated with several attributes of gliders: arboreal, ant eating, sit-and-wait foraging, diurnal, dependent on crypsis. Crests, including dewlaps, may have a role in flight stability or control.
- Not elongate (10). Strong or long legs are beneficial for climbing and launching.
- Tails average longer in females than in males (17). Note also the unusual m<f SSD (13). These two attributes are correlated (table 3.3, trait 8).
- Sex differences in color (16) and overall sex differences (18). Daytime canopy dwelling is associated with color signaling.
- Ant specialist diet (22). As noted above, under "Implications." Note the sharp contrast with diurnal canopy walkers. That this is a dietary specialization is supported by the observation that only an average proportion of gliders (not in table; 24.2% v. 21.7%) consume ostensibly similar termites, despite the rich nutritional rewards of termite eating.
- Not cruise foraging (25). Gliders are notably immobile when not gliding or displaying.
- Not crepuscular (30). The sample size for both gliders and crepuscular lizards is small.
- Direct basking (41). This universality contrasts noticeably with diurnal canopy walkers, perhaps driven by the coolness of the gliders' wet forest habitat. Direct basking may be done minimally, as it does not apparently result in high mean active body temperature (43).
- Social groupings (11 of 11 documented species) are m+f only (not in table), suggesting monogamy. It may not be possible for a male lizard to be polygynous when corralling mates that can glide away at the slightest provocation.
- Males defend space (47) and are not tolerant (46). More field observations are needed to clarify what exactly different gliders are defending.
- Occurrence in low elevation (56) and wet habitats (55), but not arid (52), intermediate (53), or mesic (54). Are gliders truly restricted to such a narrow range of conditions, or are wet lowland rain forests merely the habitat in which the few species of gliders evolved? The association with wet areas may reflect their vulnerability to desiccation due to their patagia and very large body surface area relative to their small body volume.
- Oviparity universal (62). Oviparity is consistent with their low-elevation, low-latitude habitats, and offloading extra weight as soon as possible to keep the airframe light.
- Island species (73). Is there selective pressure to occur on islands, or are the islands the place in which gliders happened to evolve? The islands of island gliders are mostly too large to have reduced avian predation.
- Not imperiled (76). Within their geographic ranges, gliders are found in many habitat types; they do not appear to adopt the extreme microhabitat/habitat specializations associated with diurnal canopy walkers (see *Draco* account in chapter 5).

Gliders are among the leading business models for the following traits (proportion of species in the elevated condition unless indicated otherwise):

- Svelte form (4), the lowest of above-litter ecological business models. Note divergence from immobility experts, some of which are closely related to gliders.
- Long tails (5), which infrequently break (6). If tails are important for launching and flight control, one would not want them to be prone to breakage.
- Not generalized invertebrate eater (21), carnivore (19), or omnivore (23). Gliders are more specialized on ant eating (75% of species) than any other business model (only immobility experts—21.4%—exceed 7%). Though not reported in the table, gliders are also the only aboveground business model that has never been recorded to consume plant matter (distinct from being an herbivore or omnivore). Presumably, the low energy value and great mass of plant matter is detrimental to the light bodies needed for extended gliding.
- Do not give stress calls (34). Stress calls are associated with alerting nearby kin, but the huge size of glider home ranges and the feeble intensity of stress calls greatly reduce the likelihood that any related glider would be within hearing distance.
- Reliant on crypsis (36) but not vegetation (40), sand (38), crevices (39), burrows, or running (37) for predator concealment. The lack of use of vegetation is notable and contrasts sharply with diurnal canopy walkers.

- Males not hierarchical (48). Much more field work is needed to clarify the social relations among male (and female) gliders. The few data also suggest that mainland gliders may be more conventional in their social lives than island gliders.
- Found in riparian and human habitat (61), but not in dry areas (51). Trees are probably denser and more luxuriant in riparian areas. The tolerance of human habitats is consistent with the generalization that gliders do not specialize in structural habitat types. There is a moderate positive correlation associating riparian and human habitat use among all lizards (not reported in chapter 3; Fisher's exact $p < 0.0001$). The aversion to dry areas may be due to the extremely large body surface area associated with their flight membranes (patagia) and their small body size, as noted above. Also, dry areas are infrequently forested.
- Elevated ontogenetic size range (64), indicating relatively small hatchlings. Small eggs put the least aerodynamic burden on a gravid female. Numerous hatchlings may be needed to offset the high predation losses to birds (see "Diurnal Canopy Walker" narrative for trait 64).
- Rapid egg development (66). This attribute contrasts sharply with that of diurnal canopy dwellers (–0.47 v. 1.59 std. resid.), suggesting that it is not an attribute of the arboreal environment. Why would diurnal canopy dwellers settle for the slowest egg development and gliders opt for one of the fastest? Are gliders fecundity limited? Why would they be more so than diurnal canopy dwellers? Perhaps they oviposit at an earlier developmental stage of the eggs, which makes the riddle even more difficult? *Draco* appear to oviposit in moist soil (not tabulated systematically), which may facilitate water uptake and expedite development.
- Adherent eggs (67): note similarity to trunk/wall lizards and dissimilarity to diurnal canopy walkers. This trait is attributable primarily to hard-shelled *Ptychozoon* and may not well characterize other, less-sampled, gliders.
- Not introduced (74). Stressed gliders are apt to flee by gliding, an activity so conspicuous to humans as to preclude most accidental transport. Though colorful and unique, gliders are comparatively fragile and short-lived and they need gigantic cages. They have not found a following among pet enthusiasts; nor are they worth eating, precluding most intentional transport.

Ground-foraging Tree Lizard

Definition

Feeding can be considered a two-step process: (1) prey item located, (2) prey item captured. For many species, the localization of a prey item takes place from an elevated perch, and the actual capture takes place primarily on the ground. The defensive part of this business model accentuates the benefits of fleeing terrestrial predators by leaping onto a tree and fleeing arboreal predators by diving into surface cover. From a defensive viewpoint, this business model could be called either a ground-fleeing tree lizard or a tree-fleeing ground lizard. Note also that many of the diurnal species that are conventionally considered arboreal retreat to trees when approached by potential human predators (i.e., herpetologists), but when undisturbed capture prey on the ground or in trees. Note that "tree" in this case should be interpreted broadly to include any elevated perch, including vertical rock and masonry structures. In treeless areas, the ground-foraging tree lizards routinely perch on rock walls or human structures (ruins, gravestones, wrecked automobiles, walls, etc.).

Implications

Because safety is to be found at the intersection between vertical and horizontal, ground-foraging "tree" lizards do not forage far from the tree base, and may defend "their" tree to retain exclusive use of its refuge and the modest adjacent terrestrial foraging space. Because vision is the modality that makes this business model work, these lizards are very visual in their social interactions and often resort to various displays, dewlaps, crests, sexual dimorphism, and sexual dichromatism to maintain their territories and win copulations. In some species, only one sex routinely maintains elevated perches (usually males). In general, ground-foraging tree lizards are relatively small in size, presumably because large lizards perching routinely low on tree trunks would be irresistibly rewarding to ground-based mesopredators when they descended, and larger lizards would be unable to garner sufficient food in the tiny foraging area at the base of a tree.

Examples

Ground-foraging tree lizards include many anoles (trunk-ground ecomorphs; fig. 4.6D), *Basiliscus*, some *Ctenosaura*, geckos (*Cyrtodactylus*, *Gonatodes*, *Hemidactylus*, *Hemiphyllodactylus*, *Lygodactylus*, *Phyllodactylus*, *Sphaerodactylus*, *Strophurus*, and *Tarentola*), agamids (*Acanthosaura* [figs. 4.6A, 4.6C], *Agama*, *Aphaniotis acutirostris*, *Calotes* [fig. 4.6B], *Ceratophora*, *Chlamydosaurus kingii*, a few *Cnemaspis*, *Diporiphora*, *Gonocephalus*, some *Japalura*, and *Pogona*),

Fig. 4.6. Ground-foraging Tree Lizards. **A**, *Acanthosaura armata* (Agamidae). **B**, *Calotes versicolor* (Agamidae) adult female. **C**, *Acanthosaura bintangensis* (Agamidae). **D**, *Anolis capito* (Dactyloidae). A–C, L. Grismer; D, E. Wostl

a few lacertids (*Algyroides marchi* and *Darevskia*), many *Sceloporus*, and a scattering of skinks (many *Cryptoblepharus*, some *Emoia*, *Sphenomorphus*, and *Trachylepis*). Of the species assigned with full confidence to the ground-foraging tree lizard ecological business model, the top five species in descending order of information (filled cells) are *Anolis sagrei* (Cuban dactyloid), *Hemidactylus turcicus* (Mediterranean gekkonid), *Urosaurus ornatus* (North American desert phrynosomatid), *Calotes versicolor* (southern continental Asian agamid; fig. 4.6B), and *Anolis cybotes* (Hispaniolan dactyloid).

Similar Business Models

This business model is closely allied with the refugia-anchored diurnal ground lizards, many of which scan for prey from prominences. Ground-foraging tree lizards are anchored to their arboreal refugia and in that respect are very similar to refugia-anchored ground lizards, save their ability to climb vertical surfaces. If they abandon prey capture on the ground, they become trunk/wall lizards or diurnal canopy walkers. For the many species for which direct observations of foraging are not available (and the diet information does not distinguish between ground prey and arboreal prey), it can be difficult to discriminate trunk/wall from ground-foraging tree lizard exemplars, and the differences between these two models are minimal, though ground-foraging tree lizards are almost exclusively diurnal. Some ground-foraging *Varanus* routinely ascend trees when herpetologists are about; however, varanids have vastly larger activity areas and body sizes and most are examples of wide-ranging chemosensors.

Touchstones from the Database

Ground-foraging tree lizards are the leading business model for only their reliance on defensive concealment in vegetation.

Ground-foraging tree lizards the among the leading business models for the following traits (proportion of species elevated unless indicated otherwise):

- Long tails (5). This association deserves further study; do the long tails assist with leaping from horizontal to vertical or vice versa?
- Advanced color change (12). Note divergence from refugia-anchored diurnal ground lizards. All roles of advanced color change (crypsis, thermoregulation, social signaling) find ready application in these highly visual lizards.
- Sexes differ in color (14). Note divergence from trunk/wall lizards.
- Ant specialists (22). As sit-and-wait foragers are tied to the defense of the small area around the base of a tree or cliff, it pays to take advantage of whatever invertebrate prey are at hand.
- Electric light use (27). Note divergence from refugia-anchored diurnal ground lizards. They are more similar to trunk/wall lizards in their electric light use, suggesting that scansorial abilities are more important than nocturnality in promoting electric light use.
- Males defend home ranges (not in table), as opposed to other variants of spatial defense, at the second-highest level (4.4% v. 2.2% for lizards). This contrasts with the pattern for diurnal canopy walkers (0.0%) and fits well with the pattern in trunk/wall lizards (3.8%).
- Medium (56) and high (57) elevation. Note divergence from trunk/wall lizards. The driver(s) behind these associations are obscure.
- Wet habitats (54). Does this reflect a genuine preference for wet areas, or that wet areas are consistently tree covered and therefore provide the right habitat? Unlike drier forests, wet forests are apt to have a closed canopy and therefore longer sight lines at ground level.
- Riparian (60) and human habitats (61). Note divergence from refugia-anchored diurnal ground lizards. Ground lizards are chronically vulnerable to depredation by human pets, but scansorial lizards such as ground-foraging tree lizards can readily climb out of harm's way. Relationship to riparian zones is less evident, though the riparian-human habitat connection shows up in lizards in general (see riparian bullet under "Glider," above).
- Elevated ontogenetic range of sizes (64), indicating comparatively small hatchlings (likely for the same high fecundity reasons hypothesized above under "Diurnal Canopy Walker").
- Lack parental care (69). As do most arboreal lizards.
- Short longevity (70). Note similarity to diurnal canopy walkers and divergence from trunk/wall lizards. Working the intersection between horizontal and vertical environments helps especially with nonavian predators, perhaps leaving bird predation to dominate their higher-than-average mortality.
- Introductions (74). Note divergence from refugia-anchored diurnal ground lizards. Perhaps because they are at ease on both horizontal and vertical surfaces, and require little space, ground-foraging tree lizards are easy pets and may enter a wide variety of human cargoes.
- High population density (71). Among invertebrate eaters, the mean density of ground-foraging tree lizards is exceeded only by trunk/wall lizards, whose vertical surface area is not considered in areal computations. To a lesser extent, ground-foraging tree lizards share this mensural bias. It seems doubtful that the bias fully accounts for density differences among the business models; perhaps hunting from elevated perches truly does enhance population densities to some degree.

Immobility Expert

Definition

An immobility expert is an exposed invertebrate-eating lizard that has evolved such effective crypsis and immobility that it can remain exposed throughout the day—albeit possibly in shade or partially hidden among foliage—as an extreme sit-and-wait predator. Immobility experts may be found in any aboveground stratum. Some species, immobile by day, forage slowly at night.

Implications

The requisite crypsis is effective against suspicious prey as well as intruding predators. Exquisite crypsis produces no shadows; no chromatic contrast; no recognizable out-

Fig. 4.7. Immobility Experts. **A,** Adult female *Cnemaspis affinis* (Gekkonidae). **B,** *Phrynosoma platyrhinos* (Phrynosomatidae). **C,** *Naultinus manukanus* (Diplodactylidae). **D,** *Cnemaspis bidongensis* (Gekkonidae). (For other immobility experts see figure 5.3.) **A, B, D,** L. Grismer; **C,** J. Reardon

line. The lizard must be able to subsist largely on the prey items that happen to walk in front of its mouth. Because it preys only on food items that walk in front of its mouth, often ants, an immobility expert does not need to defend any foraging space other than personal space, and in any event, the requirement for immobility largely precludes territorial patrolling or defense. Twenty-six percent of immobility experts have secretory pores, possibly to aid in mate finding, as low-density immobile animals probably find it hard to locate each other. Immobility experts are most likely to be encountered by chance rather than through distant detection, and thus these lizards tend to position themselves in low predator traffic areas to avoid chance encounters. For example, many species appear to avoid tree trunks. A few immobility experts need to relocate periodically to avoid massed attacks from the ant prey that they specialize in; such lizards may have very large activity ranges, but most immobility experts do not. This business model is less relevant at night, when visual crypsis is not advantageous; there may be nocturnal olfactory or seismic equivalents of crypsis, but such nocturnal crypsis is not evident to visual humans.

Examples

Immobility experts include some *Cnemaspis* (figs. 4.7A, 4.7D), *Phrynosoma* (figs. 4.7B, 5.3B), *Rhampholeon*, *Saltuarius*, and *Uroplatus*. Of the species assigned with full confidence to the immobility expert ecological business model, the top five species in descending order of information (filled cells) are *Phrynosoma cornutum* (southwestern North America deserts phrynosomatid), *Phrynosoma platyrhinos* (southwestern North America deserts phrynosomatid; fig. 4.7B), *Moloch horridus* (Australian agamid; fig. 5.3A), *Phrynosoma douglasii* (northwestern United States phrynosomatid), and *Uroplatus fimbriatus* (Madagascar gekkonid).

Similar Business Models

This business model has practitioners whose negligible locomotor activity is closest to that of crack hermits, but differs in requiring exquisite camouflage. Crack hermits may have bright colors and often have armored tails. Immobility experts remain exposed whereas crack hermits are normally mostly concealed, especially during daylight. Although it is a large leap from immobility expert to glider, there is a strong connection between the two, as the large lateral flanges acquired by immobility experts for eliminating shadows are likely to be a morphological substrate from which some gliding surfaces evolve. For example, the lateral fringes of *Hemidactylus cosymbotus*, a marginal glider, are transitional.

Touchstones from the Database

Immobility experts are the leading business model for the following traits (highest proportion of species unless indicated otherwise):

- Not elongate (10). The reasons for the frequently pancaked shape of many immobility experts is not well understood. Hypotheses include improved crypsis (the shadow-defeating attributes of lateral fringes), more room in the gut for low-energy foods such as ants, and improved resistance to ingestion by gape-limited snakes.
- Large mass-adjusted home-range area (28). This factor may combine with low population density (71) in some way. For example, large home ranges may reflect low availability of suitably dense prey concentrations; evenly dispersed prey might not lend themselves to foraging by extreme sit-and-wait. Large areas might need to be searched to find a good patch. Or there may be a penalty for frequent encounters with an individual predator or prey (e.g., ant colony). Some ant colonies that lose too many workers will become agitated and attack en masse. Better for the lizard to nibble at the margins of the colony and move on. A predator that rarely encounters an immobility expert may never acquire the appropriate search image for it, or may find the search image not worth pursuing because target density is too low. This could be critical for lizard prey that depend on crypsis for their safety.
- Reliant on crypsis for predator evasion (36). By the ecological business model definition.
- Found in mesic (53) habitats. Epiphyte-encrusted old growth cloud forests seem to be optimal for evolved crypsis by arboreal species. Ground-dwelling immobility experts are sometimes found in much drier climes. Exceptionally dry climates may have too few prey for sit-and-wait.

Immobility experts are among the leading business models for the following traits (proportion of species elevated unless indicated otherwise):

- Pudgy body shape (4). In the case of immobility experts (cf. giant herbivores), selection pressure for corpulent bodies does not seem driven by RCM or need for peritoneal space for bulky foods such as leaves. See the first bullet of the preceding list for some hypotheses about the shape of immobility experts.
- Tails short (5), prehensile (7), and rarely break (6). Short tails disrupt the lizard "gestalt." Prehensile tails are climbing aids for an arboreal existence (many immobility experts). If a prehensile tail is valuable for a purpose other than predator distraction, the owner is less likely to discard it.
- Darkening and advanced color change (11, 12). Probably for purposes of crypsis; a targeted study would be needed to determine whether it is used for thermoregulation or social signaling in immobility experts (prediction: rarely used for social signaling).
- Zero blue tails (16). Bright color would invalidate their crypsis. Furthermore, one should never draw a predator's attention to a tail that cannot break (6).
- m<f SSD (13). Immobility experts are expected to occur at very low densities and be immobile most of the time. These two conditions alone dictate that contact between individuals will be very infrequent and not worth contending. This implies that intermale conflict will be infrequent and unlikely to warrant the evolution of large male size. Small male size has the benefit of further reducing the number of individuals with a size-matched search image.
- Male tails longer than female tails (17). The longer male tails phenomenon is consistent with the general pattern (table 3.3, trait 8) that SSD and m:f relative tail length are inversely related (see chapter 3 discussion of that point).

- Ant specialists (22), not carnivores (19) or omnivores (23). If a lizard is totally immobile, ants are the primary prey available. Both ants (not in table; 51.6% v. 30.1%) and termites (not in table; 17.6% v. 21.7%) are taken by immobility experts overall (including both specialists and opportunists). Ant consumption overall, like ant specialization, is league-leading, but termite consumption is not, despite the rich nutritional value of termites. This reinforces the notion that ant consumption is a specialization rather than a side effect of immobility.
- Sit-and-wait forager (24), not active (26). By business model definition.
- Active at any time of day (cathemeral) (29). No reason to have a circadian cycle if chained to one spot. The low rate of food acquisition may drive extended foraging hours.
- High body temperature (43). The reason for this is far from apparent. Possibly high temperatures are tolerated to avoid the crypsis-destroying movement of shuttle thermoregulation.
- Males socially tolerant of other males (46) and not hierarchical (48) or defensive of space (47). See discussion under SSD bullet above.
- Infrequent in riparian zones (60), human habitat (61), and littoral zones (58). Human habitats have novel surfaces, materials, and colors. These do not lend themselves to crypsis evolved over the eons. The reason(s) for an aversion to riparian and littoral is/are not obvious, but may be attributable to predator densities. A parallelism between human and riparian habitats is commonly observed (see "Glider," "Ground-foraging Tree Lizard," and "Sand Shark").
- High clutch count (63) and relative clutch mass (65). Their corpulent bodies and lack of dependence on sprinting may give them clutch volume and mass options not available to more svelte lizards. Their corpulent bodies and lack of sprinting ability may elevate mortality to a level that requires high fecundity.
- Low population density (71). For reasons outline in the SSD bullet above, extreme crypsis is incompatible with high population density. Only the abysmally documented subterranean foragers have a lower mean population density.
- Endemic (72). Highly evolved crypsis is inevitably local, and would presumably result in high endemism.
- Imperiled (76). Conservation risk is probably tied to the same local adaptation that produced strong endemism (see endemism bullet above).

Litter Cryptozoid

Definition

Litter cryptozoids are lizards that remain under the leaf litter, or other dense but penetrable surface cover, at least whenever they sense the proximity of a predation threat. The penetrable surface cover can take many forms: loose dry loam in some places, dense grass in meadows and urban areas, spinifex in Australia, stones in some places, coarse woody debris in others, but most often leaf litter.

Implications

It helps to be small, or at least small enough to be concealed by the prevailing ground cover (turf, litter, etc.). In many arid environments, there are substantial hydric benefits to remaining within the litter. If basking is needed to achieve thermal requirements, the lizard must be able to either bask discreetly—in a depression surrounded by concealing leaves—or indirectly from heat transmitted through a stone or fallen leaf. Litter cryptozoids are often elongate and often have limb reduction to facilitate slithering through litter channels without displacing the leaves and drawing unwanted attention. Elongation and limb reduction likely enhance sprint speed within a dense but porous matrix.

Litter cryptozoids may retain diurnality to keep warm or avoid nocturnal predators in litter, but may also be opportunistically or routinely cathemeral where predator pressure and thermal conditions allow. Inactive litter cryptozoids may refuge under more hardened roofs than those provided by fallen leaves (under logs or rocks), but this is variable.

Life in the litter is so poorly known that it is difficult to speculate on the behavioral implications of litter existence. Diurnal litter cryptozoids often have some sexual dichromatism—and often brightly colored tails—but limited sexual dimorphism, suggesting that mate attraction may be somewhat visual, but territoriality and/or dominance relations are muted or mediated by nonvisual pathways. This is commensurate with the limits on sight lines and the apparent difficulty in patrolling a resource or mating territory in the complex, tortuous, and unstable channels through the litter.

Bright tails may act to startle or misdirect the bite of a predator that paws them out of the leaf litter. The ontogenetic loss of brightly colored tails in many refugia-anchored diurnal ground lizards suggests that juveniles of these species may restrict their movements to the leaf litter when young, perhaps to avoid evoking the territorial impulses of adults, and may be ontogenetically operating a different business model than their parents.

A B C D

Fig. 4.8. Litter Cryptozoids. **A**, *Lygosoma bowringii* (Scincidae). **B**, newly described *Tytthoscincus martae* (Scincidae). **C**, *Delma molleri* (Pygopodidae). **D**, *Sphaerodactylus macrolepis* (male below; female above: Sphaerodactylidae). **A**, **B**, L. Grismer; **C**, J. Reardon; **D**, G. Rodda

Numerical densities of litter cryptozoids can also be astronomical (e.g., >50,000/ha; Rodda et al. 2001), which would inhibit defensibility of territories or harems. Mate guarding may be a better choice than areal defense, and the very limited data suggesting pairing in litter cryptozoids is consistent with either monogamy or mate guarding. Instability of litter availability in some areas may promote short generation times (though spinifex "tangles" may have high longevity). A limitation is that litter production is often seasonal, and it can be destroyed by fire, wind, surface water movement, and other unpredictable climatic events. Accordingly, some species (e.g., *Delma grayii*) expressly avoid refuging in combustible cover (Ehmann 1992).

Examples

Many *Lygodactylus*, *Sphaerodactylus* (fig. 4.8D), *Delma* (fig. 4.8C), and some skinks (fig. 4.8B) are litter cryptozoids. Of the species assigned with full confidence to the litter cryptozoid ecological business model, the top five species in descending order of information (filled cells) are all skinks: *Scincella lateralis* (Eastern United States), *Scincella cherriei* (Central America), *Chalcides striatus* (Iberian Peninsula), *Plestiodon anthracinus* (Eastern United States), and *Ablepharus budaki* (Middle East).

Similar Business Models

The main feature distinguishing this business model from that of refugia-anchored diurnal ground lizards is that litter cryptozoids endeavor to remain undetected from their predators by staying under the litter, whereas refugia-anchored diurnal ground lizards are often highly conspicuous as they forage, display, and defend territories against conspecifics, and sometimes other species. Refugia-anchored diurnal ground lizards are routinely heliophilic, or at least conspicuous when basking, and they favor light gaps; many cryptozoic species are found in deep shade and a sizable minority have never been observed in the open. Those that have never been seen in the open could be unrecognized crack hermits. In distinction to sand sharks, litter cryptozoids search for prey while above the hard ground but below the litter; they may use visual cues to detect the prey. If the litter they inhabit rests over a loose substrate such as sand, litter cryptozoids routinely dive into the sand to escape from predators that remove the litter cover. This situation facilitates the transition to a purely subterranean forager.

Touchstones from the Database

The only attribute in which litter cryptozoids exceed all other business models is small size (mean mass = 1.9 g).

Litter cryptozoids are among the leading business models for the following traits based on proportion of species elevated unless indicated otherwise (table 4.1 trait number in parentheses):

- Number of species (569 spp.) (1). This species richness may come as a surprise to fans of big reptiles.
- Poorly known (2: low percent filled). Herpetologists tend to study subjects other than these chronically hidden lizards.
- Elongate body shape (10) and svelte body shape (4), but not prehensile tails (7). See "Implications" above.
- Blue tails (16) but not color change (11) or crests (8). See "Implications" subsection for blue tails. Color change is presumably unwarranted due to reduction in visual exposure. Crests are contraindicated, as they would displace litter and draw unwanted predatory attention.
- Few sex differences (18). This suggests (direct data minimal) that site-defense and polygyny are uncommon in litter cryptozoids.
- Active foraging (26) and large mass-adjusted home ranges (28), but not sit-and-wait (24). Evidently litter cryptozoids must search for their prey; chemosensory cues are likely to prevail over visual ones. Relative to their tiny body sizes, home ranges are comparatively large (data weak).
- Crepuscular (30) and cathemeral (29) activity. There may be no visual reason to be day active in the dark of the litter. However, night activity may be avoided to minimize predation by nocturnal mammals with good hearing.
- Conspicuously wary (35). A litter cryptozoid would be expected to rapidly seek the protection of the litter whenever exposed. Humans call this conspicuously wary.
- Alone with single mate (not in table), albeit based on only nine litter cryptozoid species with documentation. Litter cryptozoids are in heterosexual pairs (89%) at a frequency exceeded only by gliders (100%). Glider females can evade male consorts with a single leap; perhaps female litter cryptozoids have an analogous opportunity in the warren of litter channels. In both cases, males may be compelled to follow rather than defend space, for which they are fourth from lowest.
- Male tolerance (46). If spatial defense is not practical, tolerance is appropriate.
- Infrequently found in dry habitats (50). Litter production in dry habitats may be slow, and litter in such places is prone to elimination by wind or fires.
- Frequently found in intermediate rainfall band (52). Litter accumulation may be highest in places where trees are common yet litter-destroying decay is seasonally interrupted by dry spells.
- Lacks adherent eggs (67). This probably reflects the preponderance of soil-deposited eggs.

Nocturnal Canopy Walker

Definition

Nocturnal canopy walkers are nocturnal arboreal lizards that frequent small-diameter perches. Small-diameter perches may occur at any height above ground in any type of vegetation; nocturnal canopy walkers are not limited to the tops of trees.

Implications

Nighttime in the canopy is a pretty safe place, as long as you remain out at the twig ends, which is probably why so many arboreal lizards sleep there. The twig ends also provide access to a wealth of insect life, even if the density of such prey is low per twig; cool, ectothermic lizards have minimal caloric needs.

At night, vision is of relatively little use for detection at a distance in the canopy, though large-eyed species probably have usable short-range monochromatic vision under most lighting conditions. For longer distances, sound, chemical, and seismic cues likely prevail.

For reasons given in the narrative on diurnal canopy walkers, territorial defense seems uneconomic (too hard to patrol and monitor twig ends), but olfactory marking seems plausible (and unstudied) for dominance relations. If a lizard were to scent mark the base of each branch within its domain, it might be able to inexpensively establish a chemical fence for itself. Support for this undocumented possibility comes from the unsurpassed proportion (54%) of species with pores among nocturnal canopy walkers. (Note that nocturnal canopy walkers are often at-

Fig. 4.9. Nocturnal Canopy Walkers. **A,** *Aeluroscalabotes felinus* (Eublepharidae). **B,** *Gehyra mutilata* (Gekkonidae). **C,** Adult (unisexual: female) *Lepidodactylus lugubris* (Gekkonidae). **D,** *Cyrtodactylus louisiadensis* (Gekkonidae). Other nocturnal canopy walkers are featured in figure 2.6A (in the absence of electric lights) and 2.6B. **A, C,** L. Grismer; **B,** B. Lardner; **D,** G. Rodda

tracted to the wealth of insect prey around electric lights and can therefore be induced to effectively become trunk/wall lizards, for which territoriality is sometimes effective.) Nocturnal canopy walkers necessarily also travel on trunks on occasion, but they do not limit their movements to large diameter perches. To match the perch diameters, most canopy walkers are small and possess short legs.

Examples

Many arboreal gekkonids are nocturnal canopy walkers, including *Cyrtodactylus* (fig. 4.9D), *Gehyra* (figs. 2.6B, 4.9B), *Hemidactylus* (fig. 2.6A), and *Lepidodactylus* (fig. 4.9C). Of the species assigned with full confidence to the nocturnal canopy walker ecological business model, the top five species in descending order of information (filled cells) are *Lepidodactylus lugubris* (parthenogenetic Indo-Pacific gekkonid; fig. 4.9C), *Gehyra insulensis* (Pacific gekkonid), *Gehyra mutilata* (Asian gekkonid; fig. 2.6B), *Aeluroscalabotes felinus* (Southeast Asian eublepharid), and *Hemidactylus garnotii* (parthenogenetic Indo-Pacific gekkonid).

Similar Business Models

Presumably, some nocturnal arboreal lizard species could easily transition between canopy walker and trunk/wall lizard, though leg length may be matched to perch diameters and would discourage this, and the trunk environment no doubt entails higher predation risks and likely more social strife. The canopy is usually inappropriate for territorial behavior and the trunk/wall environment often facilitates territoriality, but little is known about feeding territory behavior in fully arboreal nocturnal lizards. Defense of nocturnal arboreal territorial boundaries is likely mediated by auditory and chemical cues, but there are no well-documented cases. If nocturnal canopy walkers cannot easily find a concealed place to sleep during the day, there would be strong selection pressure on them to become increasingly cryptic, perhaps eventually transitioning to an immobility expert.

Touchstones from the Database

Nocturnal canopy walkers are the leading business model for the following traits (highest proportion of species unless indicated otherwise):

- Tail breaks (6). Is this driven by avian predation on inactive (i.e., daytime) lizards?
- Secretory pores (9). This is consistent with an expectation of chemical communication at night, but more information is needed, especially to understand to what purpose the chemical communication is put. Diurnal canopy walkers are the opposite extreme, having the lowest frequency of secretory pores, highlighting the importance of nocturnality rather than arboreality.
- Not active foragers (26). The canopy environment does not facilitate active foraging. The renewal of invertebrate prey is at least partially aerial, minimizing the value of movement; the rain of aerial insects is the same here as there; why move?
- Not crepuscular (30). To some degree, this follows from the definition of the ecological business model. This association may also reflect high risk of predation at dusk, when carnivorous lizards and possible other reptiles are more active (table 3.5, trait 19), and especially dawn, when birds are hungry and lizards are cold and sluggish.
- Not thermoconformers (42). Sampling may be problematic.
- Not at medium (56) elevation. Additional information is needed, but nocturnal arboreal species have a hard time staying warm enough in any habitat, and they mostly avoid higher elevations.
- Endemic species (72). Dispersal across habitat gaps is likely to be low for canopy dwellers, leading to high endemism.

Nocturnal canopy walkers are among the leading business models for the following traits based on proportion of species elevated unless indicated otherwise (trait number in parentheses):

- Short (5) and prehensile (7) but not blue tails (16). Color has limited potential at night. Prehensile tails are a good fit to arboreal species favoring small-diameter perches. Short tails are less likely to be a handle for predators. Movement through the trees at night is comparatively deliberate and therefore may have less need for a long counterbalance.
- No sex differences in color (14) or seasonal color (15). These are consistent with the reduction in vision at night.
- Not carnivores (19) or ant specialists (22). Most canopy dwellers are too small (mean mass 3.3 g) to capture vertebrates. The paucity of nocturnal ant specialists is mysterious (see "Diet" in chapter 3). This mystery is at least consistent with the low ranking of nocturnal canopy walkers for any consumption of termites (not in table; 8.1% v. 21.7%) and ants (11.1% v. 30.1%). Breeding termites are a nutritional bonanza, rich in fat, defenseless, and flying into everything, which deepens the mystery as to why they are apparently avoided by arboreal generalists; they are also low in diurnal canopy walkers.
- Sit-and-wait foraging (24). Movement at night tends to draw unwanted predatory attention, through sound or detection of branch movements. Lizards that sit-and-wait are unlikely to be detected at night (prior to the invention of head lamps).
- Electric light use (27). Climbing nocturnal lizards are well positioned to take advantage of electric lighting. That they easily shift from a small perch (canopy walkers) to a large one (walls lit by lights) suggests few locomotor specializations for perch diameter in nocturnal lizards (cf. Caribbean anoles; see *Anolis* account in chapter 5).
- Spontaneous calling (33) and stress calls (34). Vocalizations are an obvious communication pathway for nocturnal lizards; values are also high for trunk/wall lizards and nocturnal ground lizards. Unlike those other ecological business models, however, nocturnal canopy walkers are unlikely to be territorial, suggesting that calls may serve as much to attract mates as to repel sexual competitors. Few direct observations on

the context of nocturnal vocalizations are available to evaluate this possibility.

- Not conspicuously wary (35). When humans assess a nocturnal species as not conspicuously wary, they may mean only that the lizards do not frantically dash for cover when illuminated with a headlamp. Nocturnal lizards habitually move slowly, presumably to avoid making telltale noise. They cannot be expected to have an effective evolved response to headlamps.
- Does not rely on sand (38), flight (37), or crevice-burrows (39) for predator concealment. See above bullet.
- Low body temperature (43). Unavoidable if nocturnal and active above ground in the cooling breezes.
- Do not show male hierarchies (48). Canopy dwellers are unlikely to be at high enough densities to warrant the formation of hierarchies. Most are probably solitary most of the time.
- Resident in mesic areas (53) and at low elevation (55), but not high (57) elevation areas or areas of dry (50), intermediate (52), or wet (54) rainfall. Warm mesic areas are moist enough at night to prevent desiccation. High elevations are probably too cold at night, and dry habitats would likely be desiccating, especially when exposed to the wind. The wet rainfall band might be pathogenically challenging for nocturnal arboreal lizards that cannot easily bask to repel infections. By contrast, nocturnal terrestrial lizards can more easily and safely hug the underside of a hot rock during the day.
- Oviparous (62). To be warm enough at night, nocturnal canopy dwellers avoid the high latitudes and high altitudes that promote viviparity.
- Low clutch count (63). The sharp contrast with the record-high clutch count of diurnal canopy walkers suggests that less rapid recruitment is needed at night, although uncertainty about breeding frequency could compensate for it or exacerbate this difference. Note the similarity in mass-adjusted clutch count (63) between nocturnal canopy walkers and nocturnal ground lizards. Bird predation must be almost eliminated at night, but perhaps other predators make up for the lack of birds. The second-highest longevity (70), behind only crack hermits, suggests that nocturnal canopy predation is very low.
- Slow egg development (66). Although nocturnal canopy walkers are in third place in the league tables, it is a very distant third (0.22 v. 1.59 std. resid. diurnal canopy walkers). They are nearly the same as the second-place wide-ranging chemosensors (0.25 std. resid.), suggesting that canopy dwelling is not the driving force. Many nocturnal ground lizards are gekkonids with hard-shelled eggs. Hard-shelled eggs can be deposited in aerial crevices (no need to absorb moisture from soil), but developmental speed may to some degree be sacrificed.
- Adherent eggs (67). As befits oviposition on surfaces that are mostly vertical.
- Island species (73). Table 3.11 (trait 27) indicates that arboreal species are especially common on islands. Many associations could be involved. Possibly the preponderance of forested islands in Southeast Asia are also favorable for the habitat needs of nocturnal canopy dwellers.
- Low population density (71). The input of food into the branches of tropical canopies at night is perhaps predictably low. Densities would be expected to be low in comparison to lizards eating lower on the food chain (herbivores, omnivores) or on the ground (nocturnal ground lizard, refugia-anchored diurnal ground lizard, etc.).
- Introductions (74, 75). Many of the nocturnal canopy walkers that have been introduced seem to have arrived as passive stowaways. Their climbing abilities may facilitate their inadvertent entry into human cargo. Their propensity to exploit electric lighting also draws them into contact with buildings, ports, and accidental human transport.

Nocturnal Ground Lizard

Definition

This nocturnal ground lizard business model is that of nocturnal terrestrial lizards that move slowly and quietly. There are relatively few continental species that forage out in the open at night, and perhaps none that crash about in the manner of teiids. The nighttime environment generally has a low level of background noise and rapid movement makes noise that draws unwanted predatory attention from carnivorous bats, birds, and rodents.

Implications

The nocturnal environment limits the utility of vision and largely eliminates the signaling value of color (nighttime photoreceptors cannot generally perceive color). Large size may enhance behavioral dominance, although the opportunities for dominance may be limited by low population density, and territoriality—mediated by sound or scent—is still possible, but very poorly understood and rarely documented. One nocturnal ground lizard, *Eublepharis macularius*, is reported to spread femoral pore secretions

Fig. 4.10. Nocturnal Ground Lizards. A, *Coleonyx switaki* (Eublepharidae). B, *Goniurosaurus luii* (Eublepharidae). C, Adult female *Hemidactylus brookii* (Gekkonidae), a nocturnal ground lizard where not enticed to feed under electric lights. D, Adult male *Cyrtodactylus quadrivirgatus* (Gekkonidae). A–D, L. Grismer

on the ground and over its legs when faced with subordinate males, which then retreat (Brandstaetter 1992). The nighttime environment is colder, which either compels nocturnal lizards to adopt lower-temperature metabolic pathways or retain heat from the day. Basking directly or indirectly during the day is common in nocturnal ground lizards—often abetted by crypsis and immobility—and juveniles of many species are day active. Lower environmental temperatures may constrain nighttime choices in predator evasion, and sprinting does not seem to be a common response to predator proximity.

The planar arrangement of ground foraging could promote territoriality in nocturnal ground lizards (perimeter is relatively small and easily patrolled), but there are few examples of this, presumably because the predation risk associated with patrolling movements is high, and the ability to detect conspecific intruders in a timely manner is low. Mate guarding is a possible strategy, perhaps accompanied by scent marking to repel sexual competitors. Many male practitioners of this business model possess secretory pores, though they are less frequent than in nocturnal canopy walkers. Many nocturnal ground lizards will also monitor and seize passing prey from the safety of their daytime refugia. Almost all nocturnal ground lizards are invertebrate eaters, though the larger ones capture some vertebrates. Slow nighttime ground foraging seems to be especially fruitful in moist microenvironments (e.g., beach wrack, streamside). A striking, but unquantified proportion of nocturnal ground lizards have taken up electric light feeding. Apparently the food bonanza is worth the small effort to become scansorial.

Examples

Many eublepharids, sphaerodactylids, and gekkonids are nocturnal ground lizards, including *Coleonyx* (fig. 4.10A), *Cyrtodactylus* (fig. 4.10D), *Cyrtopodion*, *Eublepharis*, *Hemidactylus* (fig. 4.10C), *Nactus*, and *Teratoscincus*. Of the species assigned with full confidence to the nocturnal ground lizard ecological business model, the top five species in descending order of information (filled cells) are *Coleonyx variegatus* (Western North American eublepharid), *Heteronotia binoei* (Australian gekkonid), *Hemidactylus brookii* (southern Asian gekkonid; fig. 4.10C), *Ptenopus garrulus* (southwest African gekkonid), and *Eublepharis macularius* (southern Asian eublepharid).

Similar Business Models

Nocturnal species that can climb well can easily shift to the trunk/wall business model, especially where electric lights enhance the profitability of vertical movement. Nocturnal species with a crepuscular bent may be able to transform to a litter cryptozoid. Those nocturnal ground lizards that gain enough nutrition from seizing prey when it passes in front of a burrow or crevice may become full-time crack hermits.

Touchstones from the Database

Nocturnal ground lizards are the leading business model for the following traits (based on highest proportion of species unless indicated otherwise):

- Cruise foraging (25). The distinction between active and cruise foraging may be muddied in a class of lizards that habitually move only slowly.
- Spontaneous (33) and stress calling (34). Associated with communication in the dark. To a lesser degree, vocalizations are associated with two-dimensional surfaces (second place is trunk/wall lizards). Two-dimensional surfaces facilitate territoriality, which may also be mediated by nighttime vocalizations. Both business models are of lizards that need to find mates that they often cannot see, for which vocalization is an obvious advantage.
- Residence in groups (45). This is a very interesting phenomenon in that it does not seem to follow from the basics of nocturnal ground foraging. It is also inconsistent with the second-place status of nocturnal ground lizards for solitariness (see next list). All nine nocturnal ground lizards that are found in groups are species that use rock refugia, suggesting that geological clumping may be responsible for what clustering occurs in nocturnal ground lizards.
- Defense of territories against members of the opposite sex (not in table). For males, the proportion is low but much higher than for lizards in general (2.8% v. 0.2%). It is more widely reported in females (14.3% v. 1.1%). This suggests a resource defense motivation, but other indications of resource defense (e.g., home-range defense, female hierarchical behavior) are lacking. Is it possible that chases in darkness sometimes occur in the absence of firm identification of the intruder's sex? Larger samples and more definitive observations on the nature of territorial behavior in nocturnal ground lizards would be valuable.

Nocturnal ground lizards are among the leading business models for the following traits (proportion of species elevated unless indicated otherwise):

- Tail breaks (6), but not blue tails (16). Blue is pointless in the dark. A number of nocturnal ground lizards have tails banded with white (see fig. 4.10), which would certainly captivate the attention of predators with monochromatic vision.
- Sexes are usually the same color (14). Color is largely pointless in the dark.
- Carnivores (19), not ant specialists (22). Though in second place for carnivores (4.4%), it is a distant second to wide-ranging chemosensors (24.7%). This seems unsurprising, given their small mean size (cf. 23.8 g for wide-ranging chemosensors). The paucity of ant specialists at night is more mysterious. Ants are reported to be just as numerous at night (see "Diet" in chapter 3). Sitting-and-waiting for nocturnal ants would seem to be compatible with typical foraging of nocturnal ground lizards, and though no nocturnal ground lizard is reported to be an ant specialist, they do consume ants at below average frequency (not in table; 18% v. 30.1%). Do lizards find ants too hard to see at night? They appear able to find termites (not in table; 30.4% v. 21.7%). Are nocturnal ants more aggressive to lizards, or are nocturnal lizards more vulnerable to retaliation because their skins are not thickened for water loss reduction? Are the lizards less capable of making critical distinctions between benign and unsuitable ants at night?
- Low body temperature (43) and thermoconformers (42), not baskers (41). All expected at night.
- Solitary living (44). Perhaps the reduced noise associated with solitary living is more essential than frequent contact with potential mates.
- Residence in dry (50) and arid (51), but not wet (54) areas. The preference for arid lands is fairly decisive, and it may be related to reduced concentrations of mammalian predators and greater availability of tapering cracks. Mammalian predators of lizards may be more abundant in wet areas (data on this point not compiled), but nighttime temperatures are apt to be lower, with fewer hot rocks holding the heat from earlier in the day. In wet areas, refugia cracks tend to be filled with vegetation and soil. Unlike nocturnal canopy walkers, nocturnal ground lizards—having

access to hot sand and rocks—tolerate higher elevations (57).

- Low clutch counts (63) and low RCM (65). Longevity (70) is slightly elevated for nocturnal ground lizards; perhaps they do not need to crank out young so rapidly. Note also the similarity (in mass-adjusted clutch count) to nocturnal canopy walkers. It is most likely that predation is reduced at night, due to the paucity of saurophagous birds. Reduced RCM is sometimes associated with a sprinting lifestyle, which is not the explanation applicable to nocturnal ground lizards, as most move conspicuously slowly.
- Low ontogenetic range of sizes (64), indicating relatively large hatchlings. Perhaps large hatchlings are better able to outgrow their vulnerability to nocturnal ground predators such as centipedes and mice.

Refugia-anchored Diurnal Ground Lizard

Definition

This refugia-anchored diurnal ground lizard business model is embraced by diurnal terrestrial lizards that sprint a short distance to safety—the quintessential ectothermic predator avoidance tactic. Activity ranges may be small for insectivores or omnivores, often a few hundred square meters or less, as the residents need to remain within sprinting distance of a refuge with a familiar location. Refugia can be crevices—of substrate or vegetation, above or under water—or turbid water. Foraging may entail either sit-and-wait or active movements based on chemosensory or visual cues. Carnivorous species may have large activity areas, especially in low productivity environments. Though this is the business model that most people intuitively associate with lizards and it is the modal lizard business model, only about 25% of lizards exhibit it.

Implications

Because sprinting is required, basking immediately adjacent to a refugium is often the activity that starts off the day; this requirement may be waived in consistently warm environments. Because activity ranges are small and these lizards routinely use elevated perches, sight lines often extend to include much of the activity area, and because the perimeter of the two-dimensional activity range is relatively short and easily patrolled, territoriality for acquisition of food or mates is often profitable. If the population density is high enough, territoriality may be replaced by a dominance hierarchy, but in either case—dominance or territoriality—resource acquisition and reproductive success may be socially mediated, often with a strong visual component: crests, color, selective pressure for intimidating appearance, and polygyny are likely.

If reproductive success can be enhanced by territoriality, features that increase monopolization of mates by dominant males (sexual dichromatism, breeding season color, SSD, scent marking, and in some cases regular perimeter patrols) will be rewarded. These mating-enhancing activities may have natural selection costs. For example, regular patrolling makes males far more vulnerable to detection by motion-detecting predators. Thus, while the reproductive success rewards may be great, mortality rates may also be fairly high, requiring high fecundity and inducing high population turnover and sometimes skewed sex ratios.

Data on scale ornamentation (striations, pointed scales, carination, etc.) were not tabulated for this work, but many refugia-anchored diurnal ground lizards seem to have very rough skin. Might that be an adaptation to deny predators that ability to extract them from their refugia?

Although there are many more refugia per hectare for small lizards than there are for large ones, the median mass of refugia-anchored diurnal ground lizards (7.7 g) is slightly elevated, perhaps in response to social pressures for larger size. Larger individuals (e.g., *Crotaphytus*, *Gambelia*) tend to have larger home ranges and feed at a slightly higher trophic level (specifically, more vertebrates consumed), with reduced or no dependence on territoriality. Stamps (1977) argued also that the form-recognition visual systems of vertebrate prey create selection pressure against conspicuous postures in carnivorous lizards, rendering impractical for them the territorial surveillance systems—patrol from high point to high point—practiced by so many of the insectivorous refugia-anchored diurnal ground lizards.

Examples

The refugia-anchored diurnal ground lizard is the most conspicuous (to humans) lizard business model. Examples are *Carlia* (fig. 4.11F), *Ctenotus*, *Lacerta*, *Sceloporus* (fig. 4.11B), *Sphenodon* (figs. 2.5B, 4.11C), *Tropidonophis*, *Uta* (fig. 4.11D), and many lacertids. Of the species assigned with full confidence to the refugia-anchored diurnal ground lizard ecological business model, the top five species in descending order of information (filled cells) are *Uta stansburiana* (Western North American phrynosomatid; fig. 4.11D), *Lacerta agilis* (Eurasian lacertid), *Crotaphytus collaris* (western North American crotaphytid), *Cnemidophorus lemniscatus* (South American teiid; see related species in fig. 4.11E), and *Gambelia wislizenii* (Western North American crotaphytid).

Similar Business Models

Ground-foraging tree lizards share many similarities with other business models, but must possess the scansorial

Fig. 4.11. Refugia-anchored Diurnal Ground Lizards. **A**, *Petrosaurus thalassinus* (Phrynosomatidae). **B**, *Sceloporus uniformis* (Phrynosomatidae). **C**, *Sphenodon punctatus* (Sphenodontidae). **D**, *Uta stansburiana* (Phrynosomatidae). **E**, *Cnemidophorus arubensis* (Teiidae). **F**, *Carlia ailanpalai* (Scincidae). Other refugia-anchored diurnal ground lizards are featured in figure 2.5. **A**, **B**, **D**, L. Grismer; **C**, J. Reardon; **E**, R. Reed; **F**, B. Lardner

skills to climb to safety. Nocturnal ground lizards, which are much less visual than refugia-anchored diurnal ground lizards, have many more constraints on territoriality. If a lizard's attachment to its refugia is tight enough, this business model can morph into that of a crack hermit, which resembles a refugia-anchored diurnal ground lizard without the surface forays or commitment to diurnality. Immobility experts share with refugia-anchored diurnal ground lizards diurnality and emphasis on vision, but they are not usually territorial or refugia-anchored. Litter cryptozoids are often close to refugia, but they rely much more on chronic concealment and avoid conspicuous visual social signaling or patrolling that would draw the attention of predators.

Touchstones from the Database

Refugia-anchored diurnal ground lizards are the leading business model for the following traits (highest proportion of species unless indicated otherwise):

- Number of species (1201 spp.) (1).
- Omnivores (23). Geographic restriction (to a small area around known refugia) favors lizards with a broad diet.
- Residence at high (57) elevation. This association is possibly due to the thermal protection offered by their refugia; they can emerge and bask when the sun shines. Potentially there are also more rock shelters at higher elevations.

Refugia-anchored diurnal ground lizards are among the leading business models for the following traits (the proportion of species is elevated unless indicated otherwise):

- Seasonal color (15). Associated with polygyny or spatial defense.
- Herbivores (20). Herbivores are bulky and often need to remain near a refuge for protection.
- Not cathemeral (29). A nonobvious association, particularly obscure in relation to species that are seasonally cathemeral (nocturnal in the heat of summer; diurnal in cooler seasons).
- Not calling spontaneously (33). Calling seems to be selected against in diurnal species. The reasons for this are a bit unclear, as many diurnal birds are frequent vocalizers. However, birds move so often that a sonically guided predator searching progressively for a calling bird is apt to be frustrated by the bird's rapid change of location, whereas the same progressively searching sonically guided predator can home in on a lizard before the lizard has a chance to move.
- Reliance on crevices-burrows (39) and sand (38) for predator concealment.
- Male hierarchies (48). Associated with polygyny, which is probably easier to carry out in areas with high density or clustered refuges. Females may also be socially intolerant (not in table; 18.5% tolerant v. lizard mean of 26%). Female spatial defense (not in table) is the highest of any ecological business model with adequate sample (76% v. lizard mean of 69%). Given the tie between refugia-anchored diurnal ground lizards and their refugia, it is not surprising that both sexes defend refugia.
- Residence in areas of dry and intermediate rainfall (50, 52), but not low elevation (56). Not a clear environmental pattern. As nocturnal species concentrate in warmer environments, daytime species might seem reduced by comparison at low elevation.
- Viviparous (62). This is consistent with high elevation (see preceding bullet) and high latitude (not tabulated in this study).
- Not imperiled (76). If species diversity is a reflection of ecological success, and conservation risk a reflection of lack of ecological success, this broad pattern seems reasonable.

Sand Shark

Definition

Sand sharks are lizards that remain underground most of the time, to emerge to capture prey on the surface or in loose surface material. These may or may not emerge suddenly or occur in sand; loose soil will do. Most seem to be present in the loose surface duff and to hunt for arthropods in or above it. The key is whether the prey consumed are on the surface of the substrate and are not subterranean at the time of consumption, though the prey are routinely detected from a subsurface position.

Implications

This business model seems to require a very loose and displaceable substrate. Sand sharks do not push through heavy soils or burrow deeply, except to refuge. Known species distributions are therefore extremely patchy and many populations are isolated and presumably prone to metapopulation extinction and speciation. Foraging sand sharks are presumably sensitive to seismic/auditory and olfactory cues, but may retain visual and quadrupedal systems, though some have reduced limbs and many are elongate. Potentially cathemeral, but many retain a daily cycle for thermoregulation in shallow sunny sand or for predator avoidance—some predators can detect shallow lizards by sound, odor, or feel. Juveniles often have brightly

A

B

C

Fig. 4.12. Sand Sharks. **A**, *Larutia larutensis* (Scincidae, likely a sand shark). **B**, *Plestiodon reynoldsi* (Scincidae). **C**, *Eremiascincus phantasmus* (Scincidae). **A**, L. Grismer; **B**, C. Eaglestone; **C**, R. Reed

colored tails, possibly for startle effect, possibly also an ontogenetic shift in business model or risk calculus.

Examples

Sand sharks include *Plestiodon reynoldsi* (fig. 4.12B) and many *Lerista* and *Ophiomorus*. Of the species assigned with full confidence to the sand shark ecological business model, the top five species in descending order of information (filled cells) are *Plestiodon reynoldsi* (Florida scincid; fig. 4.12B), males of *Phrynocephalus mystaceus* (central Asian agamid), *Scincus scincus* (skinks from the deserts of North Africa and southwest Asia), *Lerista punctatovittata* (eastern Australian skink), and *Lerista bougainvillii* (southeast Australian skink).

Similar Business Models

As indicated under litter cryptozoid, the sand shark differs in lacking most visual guidance for prey attacks, and by requiring specialized seismic and/or olfactory detectors for locating prey. However, the sand shark model does not require the extreme specializations for burrowing through resistant media that characterize fully subterranean foragers.

Touchstones from the Database

Sand sharks are the leading business model for the following traits (highest proportion of species unless indicated otherwise):

- Elongate (10: limbs fail to overlap). One might expect elongation to be even more extreme in subterranean lizards, but most subterranean foragers lack limbs entirely, which precludes the test for elongation (i.e., whether adpressed limbs overlap).
- Tail not prehensile (7). Prehensile tails are rare in nonclimbing species.
- Consumption of termites (not in table; 77.8% v. 21.7%). Termites could be taken both under and above ground (not specified); my guess is that most are taken when the termites are lying on the surface after nuptial flights—a nutritional bonanza for the lizards.
- Not cruise foragers (25). Might this be an artifact of limited information on foraging movements under the surface?
- Not reliant on crypsis (36) or color change (11). Most of the time they live in a nonvisual environment.
- Males defend space (47) and are not tolerant of other males (46: though sample size is small). Sampling is likely to be limited.
- Found in arid areas (51) and at low elevation (55), but

not mesic areas (54), riparian areas (60), human habitat (61), or high elevations (57). Loose sand is more likely to be found in low-elevation arid (not riparian) areas. Humans are disinclined to build residences around blowing sand.

- Littoral (58). Probably due to the ready availability of sand in littoral areas.
- Eggs not adherent (67) or laid communally (68). If damp sand is a good medium, there is likely to be no shortage of it and therefore no reason to deposit where other lizards lay.
- Not introduced (74). The specialized habitat requirements of sand sharks militate against accidental transportation to a suitable destination. Sand sharks are so rarely visible that they have negligible appeal to the pet trade.

Because sand sharks and subterranean foragers both move through soil, often by lateral undulation without use of limbs, they share many traits (conventions as above):

- Small size (3: means 2.3–3.0 g), slender build (4), elongate (10).
- Lack of crests (8), secretory pores (9), seasonal color (15), or prehensile tail (7).
- Lack color change (11), few sex differences (18).
- Invertebrate eaters (21). Though neither specializes in ants, they both eat substantial quantities of them; presumably, many of those eaten by subterranean foragers are larvae or pupae.
- Consumption of ants (not in table; 74.0% v. 30.1%). Some sources do not specify whether the ants taken are mobile (adults) or immobile (larvae, pupae). The latter would be taken underground, and may be inconsistent with the presumed foraging strategy of sand sharks, but a diet transitional to subterranean foragers is imaginable (direct data needed).
- Lack stress calls (34). Note, however, that all business models, including both sand sharks and subterranean foragers, have some species that call spontaneously. This implies that at least females can hear the soft calls, and that mate attraction is likely facilitated by calling in all media (soil, air, trees) and at all times of day.
- Lack direct basking (41). Not compatible with subterranean existence.
- Avoid medium (56) or high (57) elevation, where the heat loss to cool soils may be untenable.
- Avoid wet rainfall zones (54), for the same reason, and because damp soil is hard to push through.
- Sand/dune habitats (59), avoid riparian (60) and human habitats (61). Riparian habitats and human habitats are likely to have soils that are cool and well-packed.
- Rapid egg development (66). Perhaps the subterranean dwellers are uniquely knowledgeable about small-scale temperature differences in the soil and can exploit those fully to expedite development.
- Lack eggs that are adherent (67) or communally deposited (68).
- Not island species (73). Suitable soils rare or patchy; dispersal to islands very difficult.
- Rarely introduced (74, 75). See preceding bullet.

The following traits distinguish sand sharks from subterranean foragers, with the listed trait possessed by sand sharks but not subterranean foragers:

- Tail breaks (6). Sand shark often have quasi-normal tails (mean 1.06 × SVL), whereas subterranean forager tails are so short as to be unbreakable (mean 0.37 × SVL).
- Blue tails (16). Blue tails have startle value when sand sharks suddenly appear topside, by their own volition or otherwise, whereas subterranean foragers rarely appear on the surface.
- Ant consumption (not the ant specialization listed in table). Sand sharks have the second-highest frequency of ant consumption (74% v. 30.1%, second only to gliders 79%), whereas 41% of subterranean foragers consume some ants (presumably immobile larvae and pupae). It is notable that subterranean foragers in some sense prefer termites to ants; 64.5% of species consume termites, but only 40.7% of species consume ants. Perhaps ants have better defenses.
- Not cruise foraging (25). It seems that sand sharks mostly sit-and-wait for insects to bump the sand overhead, whereas subterranean foragers appear to cruise through underground tunnels searching for ants, termites, and other invertebrate larvae.
- Crepuscular (30). At dusk or dawn, the surface sand temperature may be more moderate for both predator and prey. Subterranean foragers are better buffered from temperature extremes and have no reason to be on or near the surface in low light.
- Indirect basking (not in table). Sand sharks have the highest frequency (78% v. mean of 9%) and subterranean foragers are reported to be lower (63%).
- Not thermoconformers (42). Subterranean foragers may have no way to avoid conforming to the temperature of the soil.
- Mean active body temperature (43). Sand sharks are about average (30.8°C v. mean of 31.1°C), whereas subterranean foragers are fourth from lowest (27.6°C).

A

B

C

Fig. 4.13. Subterranean Foragers. **A,** Adult male *Dibamus tiomanensis* (Dibamidae). **B,** *Anniella geronimensis* (Anniellidae). **C,** *Bipes biporus* (Bipedidae). **A–C,** L. Grismer

- Males socially intolerant (46). In contrast, the energetic cost of travel by subterranean foragers may be too great to support any sort of social sequestration of mates or food. Too little is known about the dispersion of key resources of subterranean foragers to profitably speculate about causation, but male social tolerance among subterranean foragers is the highest ecological business model.
- Male sand sharks defend space (47). This fits with their social intolerance. Subterranean foragers are more tolerant.
- Resident in hyperdry environment (49). Hyperdry environments would seem to facilitate travel in sand, but inhibit underground travel in clay or other soil types. Sand sharks do not occur in clay or other harder soil types; subterranean foragers sometimes do, at least in moister areas.
- Oviparity (62) is moderately common in sand sharks (85%), but is the second lowest value (67%) in subterranean foragers. The benefits of viviparity to subterranean foragers are far from obvious. Viviparity is usually associated with basking, which is not typical of subterranean foragers. Do pregnant subterranean foragers spend a lot of time hugging the underside of hot rocks? That should be easily testable.
- Relatively small hatchlings, as indicated by their elevated ontogenetic size range (64). Whereas sand shark hatchlings are the fourth smallest in this regard, subterranean forager hatchlings are the largest. Subterranean forager hatchlings need mass and strength to push through compacted soil; sand sharks reside only in loose substrates, and therefore do not drive through resistant media.
- Low RCM (65). Although subterranean foragers have moderately low clutch counts, the young of subterranean foragers are the largest of any business model (preceding bullet), precluding a low RCM.

Subterranean Forager

Definition

Subterranean foragers are the ultimate predator-avoiders by remaining underground and out of reach of virtually all vertebrate predators at all times. This is one of the most widely adopted lizard business models (385 species), albeit almost completely unstudied from a behavioral ecology perspective.

Implications

If a lizard is always underground, there is no payoff to eyes or limbs, and most subterranean lizards appear to be in the evolutionary process of losing both. One imagines that such lizards are very sensitive to seismic cues and olfaction. It would seem reasonable to also presume that these lizards, which appear to be perpetually in the dark, would ignore the daily alternation of day and night, though there have been recorded almost no information on diel activity cycles in obligatory subterranean forms. Not surprisingly, subterranean lizards use extant burrows of worms, ants, and termites as ready-made pathways when they can, and consume those species when they encounter them. One would expect those prey to have advanced chemical and physical defenses against such intruders, and the lizards to have highly evolved physical and chemical countermeasures, but almost nothing is known about this arms race.

Examples

Examples of subterranean forager include *Acontias*, *Amphisbaena*, *Ancylocranium*, *Bipes* (fig. 4.13C), *Cadea*, *Chirindia*, *Cynisca*, *Dalophia*, *Dibamus* (fig. 4.13A), *Grandidierina*, some *Lerista*, *Monopeltis*, and *Voeltzkowia*. Of the species assigned with full confidence to the subterranean forager ecological business model, the top five species in descending order of information (filled cells) are *Trogonophis wiegmanni* (western Mediterranean trogonophid), *Bachia trisanale* (western South American gymnophthalmid), *Acontias kgalagadi* (southern African skink), *Bipes biporus* (southern North American bipedid; fig. 4.13C), and *Acontias gariepensis* (southwestern Africa skink).

Similar Business Models and Touchstones from Database

See above narrative on sand sharks.

Trunk/Wall Lizard

Definition

Tree trunks and cliffs, and similar vertical structures such as the sides of buildings, differ from branches and foliage in that trunks, walls, and some large limbs provide highways for rapid and efficient movement in two dimensions (up or down and often around, or into crevices). To fit into the trunk/wall business model, height above ground is not crucial, as long as most movement is in a vertical plane. The actual substrate is unimportant (wood, rock, masonry, vegetation), as long as the expanse of climbable terrain is relatively easy to surveil. Trunks and cliffs are somewhat similar to the two-dimensional surface of the ground, but sight lines are often more limited. Small and medium lizards with grippy digits are especially good at scaling trunks and cliffs, and a number of species—both diurnal and nocturnal—specialize in this agreeable stratum.

Implications

Trunks are easily accessed and patrolled by climbing lizards, and they are often economical to defend. Scent marking or vocalization could be used to demark a trunk territory and would be expected at night, but that possibility is virtually unexamined. Defense may often be mediated by visual cues, which favors but does not require diurnality.

The presence of electric lights radically enhances the energetic payoff to wall existence. Anthropogenic lights can convert an otherwise nocturnal canopy walker into a trunk/wall lizard. Indeed, utilization of the trunk/wall business model among nocturnal species may have been rare prior to the invention of electric lights. Electric lights can increase lizard densities to such an unnatural level that normal territorial impulses may be counterproductive, leading to "aggressive neglect" or dominance hierarchies. In the absence of electric lights, energetic intake will be much lower, potentially requiring the lizard to search for prey over a number of trees, or to accept low-quality food or retain a small body size.

Territorial defense rewards larger body size, but territoriality is rarely practical when it includes a number of isolated trunks (the area needed to feed a larger lizard). Territorial patrolling during the day also notably increases the risk of avian predation; thus, many trunk/wall lizards opt for small body size, small territory size (with limited patrolling or, if nocturnal, scent or vocal marking), and low energy intake. Large trees can support medium-sized lizards.

Examples

Trunk/wall lizards include some *Anolis*, *Calotes*, *Cnemaspis* (fig. 4.14A), *Cyrtodactylus* (fig. 4.14C), *Dasia*, *Gekko* (fig. 4.14D), *Hemidactylus* (fig. 2.6A), and *Lamprolepis*. Of the species assigned with full confidence to the trunk/wall lizard ecological business model, the top five species in descending order of information (filled cells) are *Hemidactylus frenatus* (Indo-Pacific gekkonid; fig. 2.6A), *Lamprolepis smaragdina* (Oceania scincid), *Anolis carolinensis* (eastern North American dactyloid), *Gekko gecko* (southeastern Asian gekkonid; fig. 4.14D), and *Anolis aeneus* (Grenada dactyloid).

Similar Business Models

Nocturnal examples can easily shift to the nocturnal ground lizard business model, as both exploit large flat surfaces with limited visibility. Diurnal examples can eas-

Fig. 4.14. Trunk/Wall Lizards. A, Adult male *Cnemaspis boulengeri* (Gekkonidae). B, Adult male *Dasia olivacea* (Scincidae). C, *Cyrtodactylus pulchellus* (Gekkonidae). D, Adult female *Gekko gecko* (Gekkonidae). (See figure 2.6A for another species that is a trunk/wall lizard in certain places, usually in the presence of electric lights.) A–D, L. Grismer

ily transition to a ground-foraging tree lizard. Diurnal examples may also wander out into the foliage at times, especially if searching for suspected intruders, but they return regularly to their trunk/walls, which canopy walkers do only infrequently.

Touchstones from the Database

Trunk/wall lizards are the leading business model for the following traits (highest proportions of species unless indicated otherwise):

- Electric light use (27). Attracted by the food bonanza.
- Riparian zones (60). These often have large trees.
- Human habitat (61). Humans are good at making walls.
- Lowest mass-adjusted clutch count (63) and lowest RCM (65). This contrasts maximally with diurnal canopy walkers, which have the largest clutches and third-highest RCM. Why is the trunk/wall environment so much less demanding of number of progeny? Perhaps all those bark and cliff cracks reduce the mortality of hatchlings?
- Adherent (67) and communally laid eggs (68). These seem to be associated with risk of arboreally deposited eggs falling down. Many ground-foraging tree lizards seem to lay on the ground, but the incidence of soil oviposition is likely lower in trunk/wall lizards. Oviposition site was not tabulated in this study.

Trunk/wall lizards are among the leading business models for the following traits (proportion of species elevated unless indicated otherwise):

- Number of species (637) (1).
- Not elongate (10). Robust legs are helpful for expeditious travel on trunks or walls.
- Any color change (11: especially darkening). Trunk/

wall lizards may be of any diel pattern; those nocturnally inclined are apt to darken during the day. Those diurnally inclined are apt to use color signals.

- Secretory pores (9). Territoriality is apt to be practiced only by those species residing on one or a few adjacent trunks. Such a small area seems suitable for chemical marking.
- m>f SSD (13). Associated with two-dimensional environments and defense of space or social intolerance.
- Call spontaneously (33). Associated with cathemeral lizards, and to a lesser degree with nocturnal ones; trunk/wall lizards have many representatives in both.
- Reliance on vegetation (40) but not sand (38) for predator concealment. Much vegetation but little sand on tree trunks.
- Males defend space (47) and are not tolerant (46). Expected, given the easy defense of a tree trunk or small wall. Hierarchies are not often formed in males (48) or females (not in table). It might make sense to have a hierarchy when electric lights are present (alpha lizard gets first crack), but this phenomenon would be hard to observe and has not been reported for lizard individuals. Larger *species* often exclude smaller species at lights (Church and Lim 1961; Bolger and Case 1992). The superabundance of prey at a light may create other distortions: defense may not be worth the trouble, given that there are enough prey for all.
- Residence in hyperdry (49) but not arid (51) environments. Presumably, the association with hyperdry environments is attributable to wall but not trunk lizards.
- Residence in mesic (53) environments. A common attribute of arboreal species.
- High population density (71). Potentially misleading, as the horizontal surface area used for density calculations is not the vertical surface area of relevance to trunk/wall lizards.
- Introductions (74). This may be an artifact of the predilection of trunk/wall lizards for human habitations.

Wide-ranging Chemosensor

Definition

Wide-ranging chemosensors are fast-moving foragers that leave the vicinity of known or defended refugia to discover hidden arthropod or vertebrate prey by a combination of chemosensation and other sensory modalities. Not all wide-ranging chemosensors have absolutely large home ranges, but they forage frenetically and noisily explore potential food items detected by chemical and other cues.

Implications

Because their frequent and rapid locomotion makes them conspicuous to visual predators, they need to be fleet of foot and therefore near optimal temperature. They are usually heliophilic. They are often large enough to escape the interest of middling predators. They are never nocturnal, presumably so they can see the distant approach of predators; day activity also provides a noisier background level of sound, to minimize their exposure to predation that noisy locomotion and excavations might trigger. In the quiet of night, predators could readily use these sounds to localize and ambush them. Some wide-ranging chemosensors restrict their movements entirely to the ground (most teiids), whereas varanids may also explore prey possibilities in the trees and some lacertids routinely flee to trees when threatened. Teiids do not normally scan from elevated perches, which would enable them to monitor territorial intrusions, as least in the area within view. Many lacertid and some varanid sightings in trees likely reflect the lizard's response to the human observer rather than routine foraging behavior. Due to their often large activity areas, such areas are not defended, nor are the locations of all available refugia memorized by those species with large activity ranges, though varanids seem proficient at memorizing tree holes. Potentially, varanids' notorious cleverness may have been evolutionarily acquired by the selection benefits of memorizing vast numbers of prey pockets and refugia. Except for sacrificial blue tails, gaudy ornaments are avoided by wide-ranging chemosensors.

High-intensity active foraging has much larger reward rates than other foraging practices, which makes it possible to garner the necessary nutrients very quickly and spend the rest of the year safely concealed in burrows, tree holes, and so forth. Many wide-ranging chemosensors emerge only in optimal weather and seasons, and even under ideal conditions may remain immobile in their refugia for days or months at a time.

Although most "wide-ranging" chemosensors have large home ranges (mean is ranked second to immobility experts), some do not. Those that do not, especially perhaps lacertids, are often familiar with nearby refugia and will flee to them without hesitation when pressed. More field data are needed to understand the relationship between movement rate, foraging mode, and home-range size. The key elements may be frequency of movements, which draw predators, and absence of spatial defense (reliance on mate attraction or hierarchies), rather than the absolute size of home ranges.

Fig. 4.15. Wide-ranging Chemosensors. **A**, *Aspidoscelis gularis* (Teiidae). **B**, *Aspidoscelis hyperythra* (Teiidae). **C**, *Varanus salvator* (Varanidae). **D**, *Varanus salvator* in habitat (Varanidae). (See also the *Varanus indicus* illustrated in figure 2.6D.) **A**, E. Wostl; **B–D**, L. Grismer

Examples

Teiids (figs. 4.15A, 4.15B), some varanids (figs. 2.6D, 4.15C, 4.15D), and some lacertids are examples of wide-ranging chemosensors. Of the species assigned with full confidence to the wide-ranging chemosensor ecological business model, the top five species in descending order of information (filled cells) are *Aspidoscelis sexlineata* (North American teiid), *Ameiva ameiva* (Neotropical teiid), *Acanthodactylus boskianus* (a lacertid from the deserts of North Africa and southwestern Asia), *Aspidoscelis tigris* (southwestern North American teiid), and *Varanus griseus* (varanid from northern Africa and western Asia).

Similar Business Models

The only other business model routinely embraced by large lizards foraging actively is the giant herbivore model, which has a different foraging objective and often a very different foraging style: infrequent cautious movements rather than incessant short darts. A few lizards (e.g., *Scincus mitranus*) combine this active-foraging model with that of a sand shark, either rising or descending to attack prey that are detected by scent or sound. Many refugia-anchored diurnal ground lizards use chemosensation to locate prey, but they do not leave the vicinity of known refugia, so they do not adopt the high-risk/high-reward lifestyle denoted by "wide-ranging" and associated with frequent movements.

Touchstones from the Database

Wide-ranging chemosensors are among the most distinct of the business models and exhibit the following leading traits (highest proportion of species unless indicated otherwise):

- Tail length (5). Wide-ranging chemosensors are the only business model for which tail length average exceeds 2 × SVL (2.04). Long tails seem to facilitate bipedal running and undulation through grass and other dense but yielding media.
- Blue tails (16). The longitudinal striping of many wide-ranging chemosensors inhibits a predator's ability to precisely localize the lizard's position. The solid

blue color offers the predator an easy but misleading alternative.

- Carnivores (19). Large size, high speed, big heads. The epitome of saurian carnivory.
- Active foragers (26), not sit-and-wait (24). By the ecological business model's definition.
- Does not call spontaneously (33). They do not need any more reasons to be detectable.
- Reliance on flight (37) for predator concealment. Almost by definition.
- Body temperature (43: mean = 34.6°C). The better to sprint with.
- Not endemic (72). Highly vagile (wide-ranging) species are more likely to cross habitat discontinuities and are therefore less likely to be endemic.

Wide-ranging chemosensors are among the leading business models for the following traits (proportion of species elevated unless indicated otherwise):

- Body mass (3: mean = 23.8 g) and pudgy build (4).
- Lack crests (8). The complete absence of crests suggests that phylogeny may amplify, reinforce, or cause this association. Certainly wide-ranging chemosensors do not need any additional features to make them conspicuous. Crests are also strongly associated with arboreality, which is rare in wide-ranging chemosensors.
- Not in possession of sexually dimorphic secretory pores (9). They are lower than all aboveground ecological business models except the highly visual diurnal canopy walkers. Given their chemosensory adaptations, one might expect more use of secretory pores.
- m>f SSD (13). Associated with large size, which is typical of wide-ranging chemosensors, and high population density, which does not match most wide-ranging chemosensors.
- Omnivores (23), but not invertebrate eaters (21). Many wide-ranging chemosensors are too large for exclusively invertebrate consumption.
- Not cruise foragers (25). Cruise foraging is contingent on some level of crypsis; an excavating chemosensor cannot maintain crypsis. It might as well move actively.
- Large mass-adjusted home range (28). Note that the home ranges of wide-ranging chemosensors are not only absolutely large, but large for large lizards.
- Diurnal (31), not nocturnal (32). They need the heat to have the speed.
- Rely on sand (38) but not crypsis (36) for predator evasion.
- Bask directly (41), not thermoconformers (42). As mentioned above, they need the heat to have the speed.
- Group living (45). This trait does not have a clear association with their business model. More information is warranted. The responsible species tend to be riparian and to use burrows. Perhaps they group due to common attraction to a clumped soil resource.
- Males have hierarchies (48), but do not defend space (47). Males are socially tolerant (46). Their home ranges are too vast to defend, but they can work out their differences dyadically or hierarchically and monopolize matings in that manner. Female hierarchical behavior (not in table) is also the second highest among business models (33% v. mean of 11%), suggesting an ecological rather than sexual selection reason for these hierarchies.
- Reside in arid (51) areas. Arid areas often have long sight lines, which is beneficial for a group of lizards that uses vision to identify approaching predators. In that case, why would they not thrive in even drier areas (dry and hyperdry)? Possibly prey are too sparse in even drier areas, but further information would be most helpful.
- Sandy habitats (59), littoral areas (58). This may reflect the long sight lines in such areas.
- Oviparous (62). Wide-ranging chemosensors like it hot. Hot is suitable for oviparity.
- Slow egg development (66). Although wide-ranging chemosensors are in second place for slowest egg development, they are a very distant second (0.25 v. 1.59 std. resid.), suggesting that there are no strong selection pressures for slow development.
- Parental care (69). Almost all parental care in wide-ranging chemosensors takes the form of female guarding of either eggs or oviposition sites. All taxa show some predilection for oviposition in termite mounds, which in some species (especially *Varanus*) requires the female to return (or stay) to open the termite mound after eggs have hatched. *Ophisaurus attenuatus* is reported to raise clutch temperature when necessary by shivering thermogenesis.

Though not listed in the tables, tabulation of ant consumption (distinct from ant dietary specialization) indicates that 30% of wide-ranging chemosensors are documented to eat ants. This is something for which their tongues are said to be poorly adapted (Vitt and Pianka 2005). The mean frequency of ant eating (not specialization) in all lizards is also 30%, suggesting that wide-ranging chemosensors are not distinguishable from other lizards on the basis of ant consumption.

Table 4.2. Seven business models whose practitioners may be present on the ground during daylight. SSD = Sexual Size Dimorphism.

	Crack hermit	Giant herbivore	Ground-foraging tree lizard	Immobility expert	Litter cryptozoid	Refugia-anchored diurnal ground lizard	Wide-ranging chemosensor
Defends space?	Refuge site only	Possibly mating site	Usually	No	No	Usually	No
Predator defense	Remain in refuge	Large size; retreat to refuge	Sprint to tree or refuge	Crypsis	Avoid detection	Retreat to refuge	Flight
Primary sense	Visual	Visual	Visual	Visual	Visual + chemical	Visual	Chemical + visual
Size	Medium	Large–huge	Medium	Medium	Small	Medium	Large–huge
SSD	m<f	m≫f	m>f	m<f	m≤f	m>f	m>f
Longevity (size-adj.)	High	Medium–high	Low	Low–medium	Medium	Medium–low	Medium–low
Example	*Xenosaurus*	*Cyclura*	*Anolis*	*Phrynosoma*	*Sphaerodactylus*	*Carlia*	*Ameiva*

Table 4.3. Hypothesized risks and rewards. Business models ordered roughly by estimated predation risk from lowest to highest. Note that giant herbivores appear twice in the table (as juveniles and adults) to highlight ontogenetic change in benefits and risks. Other species experience similar ontogenetic shifts.

Business model	Relative predation risk per hour foraging	Relative net energy gain per hour foraging
Subterranean forager (data inadequate)	Extremely low?	Extremely low?
Crack hermit	Very low	Very low
Sand shark	Very low	Very low
Nocturnal canopy walker	Low	Low
Giant herbivore adult	Low	Low
Nocturnal ground lizard	Low–medium	Low–medium
Trunk/wall lizard	Low–medium	Low–medium
Immobility expert	Medium	Medium
Glider (risk from movement, not foraging)	Medium	Medium
Litter cryptozoid	Medium	Medium
Refugia-anchored diurnal ground lizard	Medium	Medium
Ground-foraging tree lizard	Medium	Medium
Giant herbivore juvenile (similar to diurnal canopy walker)	High	High
Diurnal canopy walker	High	High
Wide-ranging chemosensor (forages infrequently)	Extremely high	Extremely high

Overview of Ecological Business Models

The greatest difficulty in distinguishing among ecological business models occurs with species that are diurnal and terrestrial (seven models possible). For guidance on distinguishing these seven, see table 4.2.

The search for structure in lizard natural history traits can be organized by the tabulation of benefits (survival, growth, and reproduction) and the mortality risks. This benefit-cost exercise is somewhat akin to the financial reports prepared by businesses. Like businesses, only species that are in the black on average will persist. Populations that average in the red will be population sinks. In broad terms, this requires that benefits offset risks. Over evolutionary time, life histories consistently "in the black" will diversify and those routinely "in the red" will disappear. The net result is a very close match between benefits and costs (table 4.3).

One can wonder if classification of lizards into these business models is worthwhile, recognizing that intermediate conditions exist and a few lizards may not conform to any of the models. It seems to me that understanding the deep structure of natural history traits—their connections and independences—is the essence of knowledge, even if some species are exceptions. Dependencies are also the starting point for determining causation (Pearl and Mackenzie 2018). Furthermore, having convenient terms to summarize complex interrelations expedites the dialogue that will eventually uncover causation. A historical analogy exists: the "sit-and-wait" versus "active foraging" discussion was undoubtedly an advance in our understanding of lizard foraging, even if a fuller understanding reveals that foraging movements reside on a seamless continuum, and even the continuum fails to adequately convey the complexities of foraging movements.

Probing the correlations between traits is also useful for suggesting hypotheses about the ecology and behavior of species known only from the morphology of their fossil remains. Were crests in dinosaurs prevalent under the same conditions that promote crests in extant reptiles?

On an even more basic level, many species' microhabitats are known today only by the rock under which they were captured, which may be irrelevant to where they forage or sleep. The key to the ecological business models is understanding where each lizard captures or finds its prey. This information is vital and often missing; cataloging species by their ecological business model requires addressing this gaping hole in our knowledge. How could ecologists begin to understand a lizard's ecology without knowing how and where it forages?

To some extent, the business models are structured around the ways in which a lizard avoids predation. Immobility and refugia are key elements, but those are rarely studied, perhaps because "nobody sees predation" happen. Yet it is clear to me from the astronomical densities reached by lizards in predator-free environments that predation is a key constraint on lizard ecology in normal predator-occupied habitats. Intensive observation of known lizard individuals documents that behaviors to avoid predation trump almost any other daily lizard activity. These saurian exhibitions of fear reflect millions of years of selection; consideration of this behavioral outcome seems warranted. By obtaining the data needed to classify lizards by their business models, one can begin to understand the role of predation, even if scientists rarely get to witness the mortal deed.

Finally, business models reflect behavioral and ecological biodiversity. Scientists are understandably keen to record the world's saurian genetic diversity before it vanishes, but more is involved in biodiversity enumeration than counting scale rows and splitting taxa. Spending the time to watch how lizards forage and escape predation broadens our understanding of biodiversity conservation in a fundamental way. Moreover, it's fun.

5 Taxon Accounts

Every lizard family and genus is described in this chapter. In the case of monotypic families or genera, the duplicative accounts are merged, as indicated by the entry title. Each family is described by contrasting its attributes with those of lizards as a group (characterized in chapter 2, "What Are Lizards?"). Each genus is described by contrasting its attributes to those of its family, as characterized in this chapter. Unless explicitly described otherwise, the parenthetical quantifications in this chapter's family entries are given in the form family value v. lizard value. Similarly, the parenthetical quantifications in genus accounts are genus value v. family value. All percentage values express the proportion of species, not individuals.

The most notable features of a genus are conveyed in the first sentence for short entries, or the first paragraph for multiple paragraph accounts. Features are considered notable if they are possessed by nearly all members of the focal group in distinction from the reference group, or if the ranking of the focal group is at or very near the top or bottom of the ranked listings. For example, if 98% of a focal taxon's member were diurnal within a reference group that was only 65% diurnal, that dominance would be notable. Or, if the focal taxon was the second highest in frequency of blue tails (9% v. 2.3%), that ranking would be notable regardless of the percentage.

How closely must the focal and reference values match to be considered roughly equivalent? This depends on sample size (the number of species in the focal group), the precision of the estimate, and the magnitude of the reference value: values near 0% or 100% are more critical. For example, for a focal genus of three species, the proportions can be only 0%, 33%, 66%, and 100%, and with such large increments, 33% is a closest possible match to a reference value of 48%, even though the numerical difference is 15%. Conversely, with a large focal group (e.g., skinks) and a precisely documented feature (e.g., blue tails), a small and precise reference value (e.g., 2.3%) might elicit a definitive statement with only a small discrepancy: blue tails are notably overrepresented in skinks, even though the difference is less than 2% (4.0% v. 2.3%).

After highlighting the notable traits, subsequent sentences or paragraphs list attributes in standardized order, from morphological traits and sex differences, ascending through progressively more emergent attributes, to demography and conservation status.

Each genus narrative includes a table providing key attributes of each species, such as in the example below.

The range core is a short verbal characterization of the center of the species' distribution, not a description of the entire range. Activity pattern options are cathemeral (active at any time), crepuscular (active in low light), diurnal, or nocturnal. Diet classes are carnivores (>10% vertebrates), herbivores (>90% plants), invertebrates, ant specialists (>50%), and omnivores (1%–90% plants). In the diet column of the species tables, "ants" indicates ant specialists. Size is estimated as snout-vent length (SVL) and mass at adult size mode. The number of asterisks preceding the ecological business model name denotes the degree of confidence (0–4) in model assignment. In the example given, I was mostly, but not entirely (3 of 4 possible stars), convinced that *Acanthocercus adramitanus* is a refugia-anchored diurnal ground lizard. Ecological business

Acanthocercus spp.	English name				Range core
			Size		
	Activity	Diet	mm	g	Ecological business model
adramitanus	Yemen Blue Agama				Arabian Peninsula
	diurnal	invert.	133	82	*** refugia-anchored diurnal gr. liz.

models such as this are explicated in chapter 4. Full descriptions of all fields—whether presented in tables, narratives, or the database—are given in appendix A. Please also note that blank cells in the tables do not indicate that data do not exist for a given species; a blank indicates that the literature reviewed did not adequately document this trait.

Ablepharus

Ablepharus is a moderate radiation (10 spp.) of uniformly small (1.2 g v. 3.3 g), slender (shape index –0.71 v. –0.46 std. resid.), diurnal (100% v. 74%), litter cryptozoid (83% v. 28%) skinks from southeastern Europe to central Asia (Hungary to Kazakhstan, and Egypt and Yemen to India). As denoted by their genus name, these diminutive lizards lack movable eyelids, a trait common in small lizards (see chapter 3, "Mass"). The one assignable non-litter cryptozoid (*bivittatus*) is a refugia-anchored diurnal ground lizard (17% v. 30%).

The thin shape of these lizards is often manifest as elongation (60% v. 35%). Tails are a bit long (1.55 v. 1.33 × SVL), only half of which are documented to be breakable (50% v. 73%). Although none has a blue tail, juveniles of at least two species (*budaki*, *rueppellii*) have red tails.

Males are smaller than females (–1.0 v. –0.31 on –2 to 2 scale) and sexual colors are fairly common: sexual dichromatism (20% v. 7.6%); seasonal color (30% v. 6.2%). Thus *Ablepharus* has large sex differences for a skink (sex differences rollup score: 1.0 v. 0.28 on 0 to 6 scale).

Most eat generalized invertebrates (60% v. 84%), the remaining documented species specialize on ants (40% v. 1.6%). The diets are unknown for half of the species. Foraging mode is documented only for *grayanus* (active). Electric light use and home-range area are unknown.

All documented species are diurnal (as noted above), but *kitaibelii* is possibly crepuscular. At least four species are known to locomote without using their limbs; *pannonicus* is also reported to leap. Three species are conspicuously wary (30% v. 11%). Predator concealment is primarily by retreat to vegetation (100% v. 25%). Elevational range is known for seven species; all seven are found in the intermediate elevation band (100% v. 35%). There may be a small trend toward occurrence in riparian (43% v. 23%) and human (43% v. 14%) habitats.

Reproduction is by eggs (100% v. 64%), with remarkably large clutches (0.65 v. 0.06 std. resid.). Young are correspondingly small (ontogenetic mass range elevated: 0.50 v. 0.12 std. resid.). Incubation is extended (0.42 v. –0.40 std. resid.), which equates to 65 days (*budaki*, *chernovi*) or 75 days (*kitaibelii*), perhaps due to their high latitude and elevation.

Longevity is also lengthy (0.64 v. 0.17 std. resid.), equating to 4 years (*kitaibelii*) or 8.3 years (*rueppellii*). Endemism is low (20% v. 64%). None has been introduced (v. 2.1%). Two-thirds of evaluated species are considered imperiled (67% v. 47%), though none is on one of the lists of the most endangered (v. 14%) published by the International Union for

Ablepharus spp.	English name		Size		Range core
	Activity	Diet	mm	g	Ecological business model
bivittatus	Two-streaked Snake-eyed Skink				Azerbaijan
	diurnal		53	1.69	*** refugia-anchored diurnal gr. liz.
budaki	Budak's Snake-eyed Skink				Turkey
	diurnal	invert.	42	0.87	**** litter cryptozoid
chernovi	Chernov's Snake-eyed Skink				Turkey
			45	1.65	
darvazi	Darvaz Mountains Skink				Tajikistan
			41	1.20	
deserti	Desert Snake-eyed Skink				Uzbekistan
			52	2.54	
grayanus	Minor Snake-eyed Skink				India
	diurnal	ants	33	0.46	*** litter cryptozoid
kitaibelii	Juniper Skink				Balkans
	diurnal	invert.	46	0.78	*** litter cryptozoid (where litter) or refugia-anchored diurnal gr. liz.
lindbergi	Lindberg's Twin-striped Skink				Afghanistan
			45	1.65	
pannonicus	Middle East Snake-eyed Skink				Iran
	diurnal	ant. spec.	49	1.34	*** litter cryptozoid
rueppellii	Rueppel's Snake-eyed Skink				Jordan
	diurnal	invert.	47	0.83	**** litter cryptozoid

Conservation of Nature (IUCN). In other respects *Ablepharus* are typical skinks.

Abronia

Abronia is a speciose (29 spp.) radiation of little-known (19% v. 27% filled cells), medium-sized (10.0 g v. 10.1 g), arboreal anguids of Mexico and northern Central America (mostly Guatemala and Belize, but one reaching Honduras), living mostly in the canopies of bromeliad-laden cloud forests. It is of course difficult to study any canopy lizard in the field, and cloud forests are especially challenging. Almost all (91% v. 38%) species assigned a business model are diurnal canopy walkers, but confidence in these assignments is very low. The remainder are considered to be trunk/wall lizards (9% v. 4%), but those are also very tentative assignments.

Abronia are of slightly heavier build than most anguids (−0.11 v. −0.60 std. resid.), with typical tail lengths and tail break frequencies, but a high proportion of documented prehensile tails (66% v. 31%). The proportion of elongate species is half that of anguids (10% v. 22%). Sexual dichromatism is underrepresented (or underreported) in *Abronia* (7% v. 26%), and seasonal color is unknown (v. 4%). All *Abronia* are active (v. 71%) diurnal (v. 70%) foragers that rely on crypsis (v. 44%) for predator concealment. As with most cryptic species, they do not act conspicuously wary (3.4% v. 18%). All documented males have been reported to defend space (v. 46%), but only *anzuetoi* and *bogerti* have been documented. The close attachment of *Abronia* to individual bromeliads may give these "active" foragers a sedentary object worthy of defense.

Even more than average anguids, *Abronia* are associated with higher

Abronia spp.	English name		Size		Range core
	Activity	Diet	mm	g	Ecological business model
anzuetoi	Anzuetoi Arboreal Alligator Lizard				Guatemala
	diurnal		106	19.1	*** diurnal canopy walker
aurita	Cope's Arboreal Alligator Lizard				Guatemala
	diurnal		99	15.0	* diurnal canopy walker
bogerti	Bogert's Arboreal Alligator Lizard				Belize
	diurnal		77	2.82	*** trunk/wall lizard or diurnal canopy walker
campbelli	Campbell's Alligator Lizard				Guatemala
	diurnal		100	15.8	* diurnal canopy walker
chiszari	Chiszar's Arboreal Alligator Lizard				Mexico
	diurnal		84	3.59	* diurnal canopy walker
cuetzpali	Sierra de Miahuatlan Alligator Lizard				Mexico
	diurnal		92	11.5	* diurnal canopy walker
deppii	Deppe's Arboreal Alligator Lizard				Mexico
	diurnal	invert.	92	11.5	* diurnal canopy walker
fimbriata	Vera Paz Arboreal Alligator Lizard				Guatemala
	diurnal		102	16.9	* diurnal canopy walker
frosti	Frost's Arboreal Alligator Lizard				Guatemala
	diurnal		88	10.0	* diurnal canopy walker
fuscolabialis	Mount Zempoaltepec Alligator Lizard				Mexico
			94	12.5	** diurnal canopy walker
gaiophantasma	Brilliant Arboreal Alligator Lizard				Guatemala
	diurnal		88	10.0	* trunk/wall lizard
graminea	Sierra de Tehuacan Alligator Lizard				Mexico
	diurnal		85	8.95	* diurnal canopy walker
leurolepis	Smoothback Arboreal Alligator Lizard				Mexico
	diurnal		84	8.7	
lythrochila	Red-lipped Arboreal Alligator Lizard				Mexico
	diurnal		95	13.2	* diurnal canopy walker
martindelcampoi	Martin del Campo Alligator Lizard				Mexico
			92	11.5	** diurnal canopy walker
matudai	Matuda's Arboreal Alligator Lizard				Guatemala
	diurnal		88	10.0	** diurnal canopy walker
meledona	Jalapa Alligator Lizard				Guatemala
	diurnal		95	13.2	
mitchelli	Mitchell's Arboreal Alligator Lizard				Mexico
			84	8.7	
mixteca	Mixteca Arboreal Alligator Lizard				Mexico
	diurnal		115	25.6	* diurnal canopy walker

Abronia spp.	English name				Range core
			Size		
	Activity	Diet	mm	g	Ecological business model
montecristoi	Montecristo Arboreal Alligator Lizard				Honduras
	diurnal		76	6.0	*** diurnal canopy walker
oaxacae	Oaxacan Arboreal Alligator Lizard				Mexico
			93	12.2	
ochoterenai	Northern Chiapas Alligator Lizard				Mexico
	diurnal		91	11.2	
ornelasi	Cerro Baul Arboreal Alligator Lizard				Mexico
	diurnal		79	6.8	** diurnal canopy walker
ramirezi	Ramirez's Arboreal Alligator Lizard				Mexico
	diurnal		76	2.76	
reidi	Reid's Arboreal Alligator Lizard				Mexico
			81	7.5	** diurnal canopy walker
salvadorensis	Salvador Arboreal Alligator Lizard				Honduras
	diurnal		77	6.2	** diurnal canopy walker
smithi	Smith's Arboreal Alligator Lizard				Mexico
	diurnal		88	10.0	** diurnal canopy walker
taeniata	Bromeliad Arboreal Alligator Lizard				Mexico
			108	20.5	* diurnal canopy walker
vasconcelosii	Guatemala Arboreal Alligator Lizard				Guatemala
			97	14.2	

elevations; only two species (*chiszari*, *ramirezi*) have low elevation ranges (both <800 m). *Abronia* are rarely low (17% v. 47%), are often at medium elevation (90% v. 77%), and are equivalent to Anguidae for high elevation (41% v. 41%). They are unknown in littoral (v. 9%), sandy (v. 15%), or human (v. 11%) habitats. *Abronia* are notably underrepresented in riparian zones (0% v. 19%); perhaps in the cloud forest mists, association with high moisture habitat is irrelevant.

All known species are viviparous (v. 61%). Litter counts are low (0.24 v. 0.85 std. resid.).

Longevity is short (–0.89 v. 0.83 std. resid.), but based on only two species (*graminea*: 2.5 years; *mixteca*: 4.6 years). Endemism is elevated (83% v. 65%). None occurs on an island (v. 17%). None has been introduced (v. 3.8%). Seventy-five percent (v. 63%) of *Abronia* are of conservation concern, and 50% (v. 28%) of *Abronia* are listed by the IUCN in its most endangered categories. The climate change to come will presumably add more stress on these mountain-top lizards.

Acanthocercus

Acanthocercus is an eight-species genus of crestless (0% v. 38%) agamids found on both sides of the Red Sea from Saudi Arabia to South Africa. Most are refugia-anchored diurnal ground lizards (71% v. 43%), the remainder are ground-foraging tree lizards (29% v. 19%). Though much larger (42 g v. 15.6 g) and somewhat more robust (shape index 1.2 v. 0.8 std. resid.) than average agamids, they exhibit typical agamid color and spininess. Their ontogenetic range of sizes is among the largest in Agamidae (1.24 v. 0.69 std. resid.), indicating unusually small hatchlings. *Acanthocercus* forage for invertebrates (75% v. 46%) near or on the ground (63% v. 56%), often around rock piles.

They differ from most agamids in their fragile tails (75% reported to break). One-quarter have secretory pores (25% v. 13%). Three of the eight species are known to exhibit seasonal color (38% v. 20%) and advanced color change (38% v. 15%). Males are usually larger than females (0.90 v. 0.53 on –2 to 2 scale). All bask directly (v. 24%). *A. annectans* is unusual in living in hierarchical groups, perhaps due to high densities at their obligate rock piles. Group living is common in the genus (50% v. 12%). Their elevational profile is skewed toward higher elevations (low/medium/high, or l/m/h: 50/100/50 v. 83/52/14). Most are tolerant of human habitats (50% v. 16%); they are not recorded in littoral (v. 10%) or sandy (v. 26%) areas. *A. atricollis* has been recorded to live for 10 years (mass-adjusted: 0.15 v. –0.46 std. resid.). Only *branchi* is endemic to a country (Zambia, 13% v. 49%), and none has been introduced (v. 1.7%) or is imperiled (v. 37%).

Acanthocercus spp.	English name	Activity	Diet	Size mm	Size g	Range core	Ecological business model
adramitanus	Yemen Blue Agama	diurnal	invert.	133	82	Arabian Peninsula	*** refugia-anchored diurnal gr. liz.
annectans	Eritrean Ridgeback Agama	diurnal	invert.	118	56	Ethiopia	** refugia-anchored diurnal gr. liz.
atricollis	Black-necked Ridgeback Agama	diurnal	ants	126	68	Zambia	**** ground-foraging tree lizard
branchi	Zambia Agama	diurnal		118	56	Zambia	
cyanogaster	Blue-bellied Ridgeback Agama	diurnal	invert.	113	48.7	Ethiopia	** ground-foraging tree lizard
guentherpetersi	Peters' Ridgeback Agama	diurnal		82	18.2	Ethiopia	* refugia-anchored diurnal gr. liz.
phillipsii	Phillips' Ridgeback Agama	diurnal		77	15.1	Ethiopia	* refugia-anchored diurnal gr. liz.
yemenensis	Yemen Ridgeback Agama	diurnal		107	40.7	Arabian Peninsula	* refugia-anchored diurnal gr. liz.

Acanthodactylus

Acanthodactylus is a major radiation (41 spp.) of sand-habitat (78% v. 44%), burrow-refuging (92% v. 48%), diurnal, terrestrial lacertids found across the Saharan and Southwestern Asian deserts from Senegal to India (north to Portugal, Turkey, and Afghanistan; south to Nigeria and Yemen). Adaptations for life in the sand include the fringe of scales on their toes (for propulsion on loose material), the relatively desiccation-resistant eggshells (40% hard v. 14%), and a high mean active body temperature (35.7°C v. 33.3°C). Some taxa (e.g., *gongrorhynchatus*, *schmidti*) are among the lizards that lift diametrically opposed limbs regularly to limit heat uptake from the hot sand. Many of these diurnal species suspend activity at midday in the warm season. The only rainfall belt in which they are overrepresented is the hyperdry (<55 mm/y: 13% v. 4%; within Lacertidae only *Meroles* has higher representation in the hyperdry). Although their ecological business models are only weakly classified (and 46% unknown), half of the classified species are refugia-anchored diurnal ground lizards (v. 57%); the other half are wide-ranging chemosensors (v. 24.5%).

In many other respects, *Acanthodactylus* are typical lacertids. Body size is average (4.94 g v. 5.10 g), shape index is exactly average for the family (0.27 v. 0.27 std. resid.), tail length is only a trifle short (1.88 v. 2.06 × SVL), and virtually all tails break (98% v. 92%). Secretory pores are a bit overrepresented (7.3% v. 4.0%), as is color change (4.9% v. 2.2%). Blue tails are fairly common (39% v. 13%). Juveniles of *erythrurus* and *schreiberi* possess a red under-tail/leg color that is also shown by females, most dramatically at the beginning but not the end of the breeding season. In juveniles, at least, this red coloration serves to reduce aggression by adults (baby releasers). Juvenile and adult female *opheodurus* and *tilburyi* also exhibit red tails, but the social significance of those is not documented. *A. boskianus* tails may be red or blue. Yellow tails are also found in this genus (e.g., *micropholis*).

Although sexual size dimorphism (SSD) is shifted slightly in the m>f direction (1.07 v. 0.49 on −2 to 2 scale), the other components of sex differences (dichromatism, seasonal color, relative tail length) are average, leading to only a mildly elevated sex differences rollup (1.12 v. 0.86 on 0 to 6 scale).

Diet in *Acanthodactylus* is notable for the surfeit of ant specialists (6.7% v. 1.7%). Perhaps there aren't many alternative prey in the Sahara? Omnivory is often associated with environments with limited food, but omnivory is underrepresented (10% v. 19%) in *Acanthodactylus*. Flexibility in foraging tactics might be another sign of limited food availability; in this respect, it is notable that, because of multiple tactics per species, *Acanthodactylus* is overrepresented in all three modes: sit-and-wait (40% v. 23%), cruise (6.7% v. 2.4%), and active (93% v. 92%). Home-range area is large (0.72 v. 0.27 std. resid.), another possible sign of energetic constraints. Home-range data are based on only three species.

Acanthodactylus are of average wariness (10% v. 11%). Of the various primary predator concealment tactics, they are slightly overrepresented in reliance on flight (27% v. 22%). Perhaps surprisingly, they do not dive into sand (0% v. 7%) to escape predators (they may do so to forage), and they are slightly underrepresented in concealment by crevices/burrows (41% v. 50%) and vegetation (32% v. 42%). *A. gongrorhynchatus* closes its burrow using tail sweeps as it enters.

Compared to an average lacertid, they are substantially less frequently socially tolerant (14% v. 41%) and more prone to defend space (86% v. 59%). They are not reported to engage in male hierarchies (v. 25%), despite their relatively high population density (see below).

Aside from their notable association with sand habitats, they are not strongly related to littoral habitat (35% v. 22%), riparian zones (20% v. 28%), or human habitat (3% v. 16%), of which there is not much in this part of the world. Although sand is a common substrate, it is not universal; *ahmaddisii*, *beershebensis*, *boskianus*, and *opheodurus* are noted for their dependence on cryptobiotic crusts, which can be damaged or destroyed by grazing livestock. The causation for this association is not explicit, but these species may use the crust as a hardened roof over their burrows.

Acanthodactylus are typical in all tabulated reproductive traits except eggshell hardness (as mentioned at the beginning of this account). The hatchlings of *Acanthodactylus* are relatively large, as indicated by the low ontogenetic size range (−0.59 v. −0.05 std. resid.). Parental care is rare (2.4% v. 1.9%), but includes posthatching association of young with female *scutellatus*.

Recorded longevity is short (−0.32 v. 0.19 std. resid.). The longest absolute values are 9.3 years (*scutellatus*) and 9 years (*boskianus*). One suspects that longer values would be available if *Acanthodactylus* was of greater interest to herpetoculturalists. However, it is also notable that several observers comment on the improba-

Acanthodactylus spp.	English name				Range core
			Size		
	Activity	Diet	mm	g	Ecological business model
aegyptius	Long-footed Teethling				Egypt
	diurnal	invert.	47	2.40	* wide-ranging chemosensor
ahmaddisii	Disi's Fringe-toed Lizard				Jordan
	diurnal		66	6.7	* refugia-anchored diurnal gr. liz.
arabicus	Arabia Fringe-toed Lizard				Yemen
			64	6.3	
aureus	Western Sahara Fringe-toed Lizard				Senegal
	diurnal	invert.	56	4.03	* refugia-anchored diurnal gr. liz.
bedriagai	Bedriaga's Fringe-toed Lizard				Algeria
	diurnal	invert.	63	5.8	
beershebensis	Be'er Sheva Teethling				Israel
	diurnal	invert.	81	12.1	
blanci	White Fringe-toed Lizard				Algeria
	diurnal	invert.	74	9.4	
blanfordii	Blanford's Fringe-toed Lizard				Iran
	diurnal	invert.	55	3.9	**** refugia-anchored diurnal gr. liz.
boskianus	Bosc's Fringe-toed Lizard				Egypt
	diurnal	invert.	69	8.0	**** wide-ranging chemosensor
boueti	Bouet's Fringe-toed Lizard				Nigeria
	diurnal		54	3.80	
busacki	Busack's Fringe-toed Lizard				Western Sahara
	diurnal	invert.	62	5.7	
cantoris	Indian Fringe-toed Lizard				Pakistan
	diurnal	invert.	50	3.00	*** refugia-anchored diurnal gr. liz.
dumerilii	Duméril's Fringe-toed Lizard				Mauritania
	diurnal	invert.	47	2.50	** refugia-anchored diurnal gr. liz.
erythrurus	Spiny-footed Lizard				Morocco
	diurnal	omnivore	73	9.1	** refugia-anchored diurnal gr. liz.
felicis	South Arabia Fringe-toed Lizard				Yemen
	diurnal	invert.	56	4.10	* refugia-anchored diurnal gr. liz.
gongrorhynchatus	Saudi Fringe-toed Lizard				Saudi Arabia
	diurnal	invert.	56	4.10	** wide-ranging chemosensor
grandis	Giant Fringe-toed Lizard				Iraq
	diurnal	invert.	83	13.2	** wide-ranging chemosensor

Acanthodactylus spp.	English name	Activity	Diet	Size mm	Size g	Range core	Ecological business model
guineensis	Guinean Fringe-toed Lizard	diurnal		52	3.38	Nigeria	
haasi	Haas' Fringe-toed Lizard	diurnal	invert.	56	4.10	Saudi Arabia	
hardyi	Hardy's Fringe-toed Lizard	diurnal	invert.	55	3.89	Saudi Arabia	* wide-ranging chemosensor
harranensis	Harran Fringe-toed Lizard	diurnal		77	10.7	Turkey	* refugia-anchored diurnal gr. liz.
khamirensis	Hormozgan Sand Lizard	diurnal		44	1.99	Iran	
longipes	Long-footed Fringe-toed Lizard	diurnal	ants	54	3.71	Sahara	
maculatus	Leopard Fringe-toed Lizard	diurnal	invert.	56	4.19	Northern Sahara	* refugia-anchored diurnal gr. liz.
masirae	Masirah Fringe-toed Lizard	diurnal	invert.	47	2.40	Oman	
micropholis	Yellow-tailed Fringe-toed Lizard	diurnal	invert.	56	4.10	Afghanistan	*** refugia-anchored diurnal gr. liz.
nilsoni	Nilson's Spiny-toed Lizard			63	5.9	Iran	
opheodurus	Arnold's Fringe-toed Lizard	diurnal	invert.	57	4.40	Saudi Arabia	** wide-ranging chemosensor
orientalis	Angel's Fringe-toed Lizard	diurnal		56	4.10	Syria	
pardalis	Egyptian Fringe-toed Lizard	diurnal	invert.	61	5.4	Egypt	
robustus	Robust Fringe-toed Lizard	diurnal	invert.	64	6.3	Jordan	* wide-ranging chemosensor
savignyi	Savigny's Fringe-toed Lizard		invert.	62	5.5	Algeria	
schmidti	Schmidt's Fringe-toed Lizard	diurnal	invert.	86	14.6	Saudi Arabia	**** wide-ranging chemosensor
schreiberi	Schreiber's Fringe-toed Lizard	diurnal	omnivore	72	8.2	Turkey	* wide-ranging chemosensor or refugia-anchored diurnal gr. liz.
scutellatus	Nidua Fringe-toed Lizard	diurnal	omnivore	63	6.1	Sahara	** refugia-anchored diurnal gr. liz.

bility of *Acanthodactylus* individuals living beyond 1 year (*aegyptius, beershebensis, scutellatus*).

Population density is slightly low compared to Lacertidae (1.68 kg/ha v. 2.07 kg/ha), but high in comparison to lizards (0.54 kg/ha). Given that many lizard population estimates derive from higher productivity habitats or islands, the achievement of 1.68 kg/ha seems remarkable. Unfortunately, population density estimation is based on only two *Acanthodactylus* (*erythrurus, scutellatus*), though they yielded consistent densities (1.8 kg/ha, 1.56 kg/ha).

Much *Acanthodactylus* habitat is extensive, and few species are localized or endemic (22% v. 35%). Most species are not found on any island (15% v. 26%). None has been introduced (v. 4.6%), and a typical proportion are imperiled (48% v. 56%) or listed by the IUCN as among the most endangered (19% v. 16%).

Acanthodactylus spp.	English name				Range core
			Size		
	Activity	Diet	mm	g	Ecological business model
senegalensis	Senegal Fringe-toed Lizard				West Sahara
	diurnal	ants	52	3.38	
spinicauda	Doumergue's Fringe-toed Lizard				Algeria
			52	3.39	
taghitensis	Taghit Fringe-toed Lizard				Algeria
	diurnal		49	2.74	
tilburyi	Tilbury's Fringe-toed Lizard				Saudi Arabia
	diurnal	invert.	56	4.10	* wide-ranging chemosensor
tristrami	Lebanon Fringe-toed Lizard				Syria
	diurnal	invert.	77	10.7	* wide-ranging chemosensor
yemenicus	Yemen Fringe-toed Lizard				Yemen
	diurnal		49	2.86	

Acanthosaura (figs. 4.6A, 4.6C)

Acanthosaura is a moderate (11 spp.) genus of crested (100% v. 38%) semi-arboreal agamids from Southeast Asia (India and China to Indonesia, but especially Thailand to Vietnam). Many have distinctive but variable spines over their eyes. Though larger (39 g v. 15.6 g) and more robust (shape index 1.16 v. 0.80 std. resid.) than average agamids, they exhibit typical agamid spininess, and most forage for invertebrates (83% v. 46%) near or on the ground from arboreal perches (75% v. 43%). *A. nataliae* is an herbivore (17% v. 8%). Some *Acanthosaura* have the ability to darken (27% v. 22%), but only *lepidogaster* and *nataliae* change color or pattern (18% v. 15%). *Acanthosaura* are sexually differentiated in both body (m>f: 1.0 v. 0.5 on –2 to 2 scale) and tail size (m:f ratio of relative tail lengths 1.13 v. 1.07, the fifth-highest value in Agamidae) and color (82% v. 53%), but do not change color seasonally (v. 20%). All bask directly (v. 24%). Males defend space (100% v. 97%) but do not form hierarchies (v. 18%). They hide by crypsis (50% v. 51%) or in vegetation (50% v. 17%), but they do not use burrows for daily refugia (v. 40%). Taking crypsis to its logical extreme, *A. armata* is reported to have remained motionless for more than 24 hours. Though found at low latitude, some are seasonally inactive due to cold or dryness.

Mass-adjusted clutches are very large (2.16 v. 0.20 std. resid.), the third-highest value in Agamidae), and brood frequency is undocumented. The size of young can be inferred from the ontogenetic size ratio, which is mildly elevated (0.98 v. 0.69 std. resid.), indicating mildly small hatchlings. Incubation period is by far the longest in Agamidae (1.53 v. –0.55 std. resid.; 192 days in *armata*, 180 days in *capra*, 166 days in *crucigera*) for no obvious reason. Potentially their forested environment has few sunny patches to expedite development, or development is seasonally suspended to synchronize eclosion to the next year's rains. Longevity is short (–1.7 v. –0.5 std. resid.) for the two species documented (*armata* 2.1 years; *crucigera* 2.6 years). No *Acanthosaura* has been introduced (v. 1.7%) or is imperiled (v. 37%).

Acanthosaura spp.	English name				Range core
			Size		
	Activity	Diet	mm	g	Ecological business model
armata	Armored Pricklenape				Southern East Asia
	diurnal	invert.	112	47.4	**** ground-foraging tree lizard
bintangensis	Bintang Horned Tree Lizard				Malaysia
	diurnal		111	46.0	*** ground-foraging tree lizard
brachypoda	Fan Si Pan Spiny Lizard				Vietnam
	diurnal		93	26.7	* ground-foraging tree lizard
capra	Indochinese Spiny Lizard				Southeast Asia
	diurnal	invert.	109	44.2	** ground-foraging tree lizard
cardamomensis	Cardamom Spiny Lizard				Southeast Asia
			116	52.7	
coronata	Crowned Spiny Lizard				Vietnam
	diurnal	invert.	81	17.3	**** ground-foraging tree lizard
crucigera	Masked Horned Tree Lizard				Southern Asia
	diurnal	invert.	120	59	**** ground-foraging tree lizard
lepidogaster	Brown Pricklenape				Southeast Asia
	diurnal	invert.	87	21.5	**** ground-foraging tree lizard
nataliae	Natalia's Pricklenape				Vietnam
	diurnal	herbivore	122	62	**** giant herbivore
phuketensis	Phuket Horned Tree Agamid				Thailand
			98	31.4	
titwangsaensis	Titiwangsa Horned Tree Lizard				Malaysia
	diurnal		117	54	**** ground-foraging tree lizard

Acontias

Acontias is a major skink radiation (21 spp.) of slightly large (6.8 g v. 3.3 g), but exceedingly slender (shape index –2.27 v. –0.46 std. resid.), limbless (100% v. 16%) subterranean foragers from southern Africa (Angola and Kenya south to Cape).

Tail length is very short (0.15 v. 1.33 × SVL), as is typical of subterranean foragers (see chapter 2). Mean SSD is m<f (–1.0 v. –0.31 on –2 to 2 scale), suggesting that these lizards, while large for skinks, are small enough for fecundity selection on females to eclipse intermale-competition selection for large male size. As expected for subterranean foragers, color changes and sexual color differences are not known.

Their diet is primarily soft invertebrates (86% v. 84%), with at least four of the invertebrate eaters specializing in termites (*gariepensis*, *jappi*, *kgalagadi*, *richardi*, *rieppeli*?). *A. gariepensis* is said to forage both actively and by cruising (*kgalagadi* also cruises); it is not completely clear to me what the different foraging mode descriptors mean in a subterranean context. Similarly, insight into the diel activity patterns of *Acontias* is both limited (only three species characterized) and potentially confusing. Is nocturnal (67% v. 14.5%) a distinct category for subterranean foragers, or does it merely reflect our sampling methods? One species (*kgalagadi*) is reported to live in groups; what exactly does this mean or imply about their social interactions? Or is the requisite soft soil highly patchy in its distribution?

Acontias live at the dry end of the skink rainfall spectrum (dry: 67% v. 4%; arid: 67% v. 40%). All documented species are found exclusively at low elevations (100% v. 89%). They are overrepresented in sandy places (56% v. 25%), but either absent (0% v. 23% riparian; 0% v. 14% human habitat) or underrepresented (6% v. 21% littoral) in the other specialized habitat types.

All *Acontias* are viviparous (0% v. 64% oviparous); they have smallish litter counts (–0.38 v. 0.06 std. resid.) and light relative clutch masses (RCMs) of 0.194 (v. 0.236). Despite this, neonates are conspicuously large (ontogenetic mass range –1.08 v. 0.12 std. resid.).

Based on *meleagris* and *orientalis*, the longevity of *Acontias* is low (–0.62 v. 0.17 std. resid.), equating to 3.9 years. Population density is unknown. Endemism is average (62% v. 64%). No *Acontias* is found on islands (v. 57%). They have not been introduced (v. 2.1%). The frequency of imperiled species is low (29% v. 47%), and none is found on one of the IUCN's lists of the most endangered species (v. 14%). In all other documented respects *Acontias* are typical skinks.

Acontias spp.	English name	Activity	Diet	Size mm	Size g	Range core	Ecological business model
aurantiacus	Golden Blind Legless Skink			165	7.9	Botswana	**** subterranean forager
bicolor	Two-color Legless Skink			163	7.6	Zimbabwe	* subterranean forager
breviceps	Short-headed Legless Skink			148	6.1	South Africa	*** subterranean forager
cregoi	Cregoe's Blind Legless Skink			163	7.6	South Africa	**** subterranean forager
gariepensis	Gariep Blind Legless Skink	nocturnal	invert.	112	3.18	Botswana	**** subterranean forager
gracilicauda	Slender-tailed Legless Skink			201	12.3	South Africa	*** subterranean forager
jappi	Few-striped Legless Skink		invert.	132	4.71	Angola	*** subterranean forager
kgalagadi	Striped Blind Legless Skink	nocturnal	invert.	135	4.91	Botswana	**** subterranean forager
lineatus	Striped Legless Skink			117	3.51	South Africa	*** subterranean forager
litoralis	Coastal Legless Skink			105	2.74	South Africa	*** subterranean forager
meleagris	Cape Legless Skink	cathemeral		213	14.0	South Africa	*** subterranean forager
namaquensis	Namaqua Thin-tailed Legless Skink			154	6.6	South Africa	*** subterranean forager
occidentalis	Western Legless Skink		invert.	152	6.5	Botswana	*** subterranean forager
orientalis	Linnaeus' Lance Skink			172	8.6	South Africa	** subterranean forager
percivali	Percival's Legless Skink		invert.	193	11.2	Southern Africa	*** subterranean forager
plumbeus	Giant Legless Skink		carnivore	328	37.9	Mozambique	*** subterranean forager
poecilus	Variable Legless Skink			261	22.6	South Africa	*** subterranean forager
richardi	Richard's Legless Skink		invert.	101	2.54	South Africa	
rieppeli	Woodbush Legless Skink			146	5.9	South Africa	*** subterranean forager
schmitzi	Kataba Forest Legless Skink			135	4.94	Zambia	**** subterranean forager

Acontias spp.	English name				Range core
			Size		
	Activity	Diet	mm	g	Ecological business model
tristis	Melancholy Legless Skink				South Africa
			109	3.04	*** subterranean forager

Acratosaura spp.	English name				Range core
			Size		
	Activity	Diet	mm	g	Ecological business model
mentalis	Amaral's Colobosaur				Brazil
			52	3.05	
spinosa	Spiny Colobosaur				Brazil
			47	2.25	

Adercosaurus sp.	English name				Range core
			Size		
	Activity	Diet	mm	g	Ecological business model
vixadnexus	Yutaje Lizard				Venezuela
			49	2.49	

Adolfus spp.	English name				Range core
			Size		
	Activity	Diet	mm	g	Ecological business model
africanus	Multi-scaled Forest Lizard				Congo
	diurnal	invert.	56	4.06	** ground-foraging tree lizard
alleni	Alpine-meadow Lizard				Kenya
	diurnal		53	3.58	** refugia-anchored diurnal gr. liz.
jacksoni	Jackson's Forest Lizard				Central Africa
	diurnal	invert.	62	5.7	** ground-foraging tree lizard
masavaensis	Mount Elgon Lizard				Kenya
	diurnal		49	2.86	* litter cryptozoid

Acratosaura

Acratosaura (two spp.) are medium-sized (2.62 g v. 2.35 g) terrestrial Brazilian gymnophthalmids. Even compared to other gymnophthalmids, they are poorly known (17% v. 19% filled cells). They are less slender than the family average (0.02 v. −0.32 std. resid.). Tails are not known to break (v. 28%). They have secretory pores (50% v. 29%), but no other sex differences. Clutches are on the small side (−0.32 v. −0.11 std. resid.). They are not known to have been introduced (v. 0.8%), nor are they imperiled (v. 25%). In other respects *Acratosaura* are typical gymnophthalmids.

Adercosaurus

Adercosaurus vixadnexus is the sole member of this genus of medium-sized (2.49 g v. 2.35 g) terrestrial Venezuelan gymnophthalmids. Compared to other gymnophthalmids, it is poorly known (14% v. 19% filled cells); information is based on one specimen collected in 1995 at an elevation of 1700 m in primary forest. Tails are not known to break (v. 28%). Clutch count is unknown (v. 1.9). It is not known to have been introduced (v. 0.8%), nor is it imperiled (v. 25%).

Adolfus

Adolfus (four spp.) is a genus of diurnal invertebrate eaters (Lacertidae) found in central Africa (Cameroon to Zambia), at relatively high elevations (see below), in a relatively mesic part of Africa. They are diverse in ecological business model: two ground-foraging tree lizards (v. 8%), one litter cryptozoid (v. 2.4%), and one refugia-anchored diurnal ground lizard (v. 57%). They are of average size (3.92 g v. 5.1 g) and build (shape index 0.22 v. 0.27 std. resid.). Tails are of average length (1.90 v. 2.06 × SVL) and at least half are breakable (v. 92%). Secretory pores are present (50% v. 4%). Color change is absent (v. 2.2%). Males are often larger than females (1.0 v. 0.49 on −2 to 2 scale), and relative tail length is about 12% longer in males than females. This puts *Adolfus* at a slightly elevated point on the sex rollup spectrum (1.0 v. 0.86 on 0 to 6 scale).

Direct basking is high (75% v. 25%), possibly reflecting cool temperatures at elevation. Burrow refugia are present but underrepresented (33% v. 48%), presumably lower as a result of the half of the genus that is arboreal. The elevational range of the genus is biased high (l/m/h: 50/75/75 v. 87/61/20). Three species extend to over 3000 m (highest is *alleni* at 4500 m). They are not found in littoral (v. 22%) or sandy (v. 44%) habitats.

Egg count is on the small side (–0.20 v. 0.13 std. resid.), and incubation is relatively slow (0.13 v. –0.36 std. resid.), likely reflecting the high elevation. *A. jacksoni* lays communally (v. 1.9%).

Both evaluated species (*alleni, masavaensis*) are considered imperiled (v. 56%), but neither is among the most endangered (v. 16%). In other respects *Adolfus* are similar to other lacertids.

Aeluroscalabotes (fig. 4.9A)

Aeluroscalabotes felinus is the sole representative of this eublepharid genus, found from Thailand to Indonesia. It is the only nocturnal canopy walker within the Eublepharidae. It is larger than a typical eublepharid (20.5 g v. 13.2 g) and has a prehensile tail (v. 8.3%). The outermost two digits of this gecko's feet are opposable, reminiscent of a chameleon, presumably facilitating climbing on the small-diameter perches it frequents. It cruise forages slowly, even by the standards of nocturnal lizards. Perhaps it is this "catatonic" motion that gives rise to the reputation it holds among some local residents of Borneo that it is a harbinger of death.

The sexes differ in color (v. 11%). Unlike most eublepharid genera, it does not vocalize (v. 25%), nor is it considered to be conspicuously wary—it would have to move quickly for that. It frequents riparian zones (v. 30%). Its incubation period is short (–0.73 v. –0.28 std. resid.), and the female may remain with the clutch for a day or so after it is laid. In other respects it is a typical eublepharid.

Aeluroscalabotes sp.	English name		Size		Range core
	Activity	Diet	mm	g	Ecological business model
felinus	Cat Gecko				Malaysia
	nocturnal	invert.	100	20.5	*** nocturnal canopy walker (low)

Afroedura

Afroedura is a 27-species genus of the Gekkonidae in southern Africa (south of a line connecting Angola to Mozambique). Most species are nocturnal (96% v. 73%), terrestrial (69% v. 31%), and rock-loving. As befits crack dwellers, they are mostly flattened in profile. They are not very well known (22% v. 25% cells filled), and only one species (*africana*) is known well enough to assign an ecological business model with full confidence. The modal ecological business model is crack hermit (54%

Afroedura spp.	English name		Size		Range core
	Activity	Diet	mm	g	Ecological business model
africana	African Flat Gecko				South Africa
	nocturnal		55	3.58	**** nocturnal ground lizard
amatolica	Amatola Flat Gecko				South Africa
	nocturnal		53	3.25	** nocturnal ground lizard
bogerti	Bogert's Flat Gecko				Namibia
	nocturnal		45	2.08	** nocturnal ground lizard
broadleyi	Broadley's Flat Gecko				South Africa
	nocturnal		49	2.73	
granitica	Granite Flat Gecko				South Africa
	nocturnal		51	2.95	
haackei	Haacke's Flat Gecko				South Africa
	nocturnal		47	2.31	
halli	Inland Rock Gecko				South Africa
	cathemeral	invert.	55	3.69	** crack hermit
hawequensis	Hawequa Flat Gecko				South Africa
	nocturnal		69	6.7	** crack hermit
karroica	Karoo Flat Gecko				South Africa
	nocturnal	invert.	49	2.60	* crack hermit
langi	Lowveld Flat Gecko				South Africa
	nocturnal		39	1.37	* crack hermit
leoloensis	Leolo Flat Gecko				South Africa
	nocturnal		38	1.36	
loveridgei	Loveridge's Flat Gecko				Mozambique
	nocturnal	invert.	48	2.51	* crack hermit
major	Major Flat Gecko				South Africa
	nocturnal		64	5.5	* crack hermit

Afroedura spp.	English name				Range core
			Size		
	Activity	Diet	mm	g	Ecological business model
maripi	Maripi Flat Gecko				South Africa
	nocturnal		54	3.56	
marleyi	Marley's Rock Gecko				South Africa
			35	1.07	* crack hermit
multiporis	Multipored Flat Gecko				South Africa
	nocturnal		57	4.10	
namaquensis	Namaqua Flat Gecko				South Africa
	nocturnal		49	2.73	
nivaria	Drakensberg Flat Gecko				South Africa
	nocturnal	invert	54	3.54	
pienaari	Pienaar's Flat Gecko				South Africa
	nocturnal		47	2.41	
pondolia	Pondo Flat Gecko				South Africa
	nocturnal	invert.	56	3.75	* trunk/wall lizard
pongola	Pongola Flat Gecko				South Africa
	nocturnal		38	1.32	
rondavelica	Three Rondavels Flat Gecko				South Africa
	nocturnal		49	2.62	
rupestris	Rocky Flat Gecko				South Africa
	nocturnal		54	3.43	
tembulica	Tembo Flat Gecko				South Africa
	nocturnal		51	2.99	* nocturnal ground lizard
tirasensis	Tira Farm Flat Gecko				Namibia
	nocturnal		49	2.62	
transvaalica	Transvaal Flat Gecko				Botswana
	nocturnal	invert.	59	4.49	*** nocturnal ground lizard
waterbergensis	Waterberg Flat Gecko				South Africa
	nocturnal		42	1.77	

v. 3%); many others are nocturnal ground lizards (38.5% v. 27%), and a few are trunk/wall lizards (7.7% v. 45%). They are small even for gekkonids (2.75 g v. 3.31 g), and documented tail breakage is less common than in the family (48% v. 87%). No tails are prehensile (v. 2%). Secretory pores are the rule (89% v. 46%), and color change is not reported (v. 15%).

Male:female size dimorphism is not documented in *Afroedura*, but m:f relative tail length ratio is reported for *A. pondolia* (0.88), and the two are often inversely related (table 3.3, trait 8). This suggests that m>f size dimorphism is likely to apply, at least in *pondolia*. Fixed and seasonal color differences between the sexes are not reported (0% v. 6.6% and 0% v. 0.8%, respectively).

All documented *Afroedura* are invertebrate eaters, though dietary information is lacking for the majority of species. In addition to the nocturnal activity mode mentioned in the first paragraph, 3.8% (v. 10.8%) of *Afroedura* are active at any time. Spontaneous calling is recorded at half the family level (3.7% v. 6.5%), but underreporting for these poorly known geckos seems likely. Direct basking is reported in 7.4% of species (v. 4.9%); for many of the crack dwellers, this may indicate basking at the entrance of the crack without full body exposure. Group living is reported in a third of documented species (v. 25.3%); group living often occurs in rock pile species in places where suitable rock cracks are clumped in distribution. Branch (1998) specifies that up to 20 individuals can share the same crack in unspecified *Afroedura*. Though 69% of *Afroedura* are considered to be terrestrial, and almost all species are strongly tied to a shelter, no species is reported to refuge in a burrow (v. 10%).

Afroedura is an exception to the family generalization that most species are found at low elevation. The three elevation tiers are represented 50/50/33 (v. 92/26/2). *Afroedura* appear to avoid human habitats (7% v. 27%).

Reproduction in *Afroedura* closely matches the family template with the exception that eggs are more often adherent (44.4% v. 10.5%) and laid communally (22.2% v. 8.6%). Potentially the predilection of crack-living or arboreal lizards to lay communally is driven by inability to sense the suitability of the oviposition medium. Soil-nesting species can sense the moisture content of the soil and therefore may not need to rely on insight from previous successful hatching to determine the suitability of a site, but crack and bark nesters may. The young are slightly larger in *Afroedura* than Gekkonidae (ontogenetic mass range −0.28 v. −0.19 std. resid.).

Longevity in *Afroedura* is reported

to be short (–0.77 v. 0.15 std. resid.), though this is based on only two species in a poorly documented genus. *Afroedura* have not been introduced (v. 3.7%). Three-quarters of the evaluated species are considered imperiled (v. 47%), though none is on the IUCN's lists of most endangered (v. 11.3%). In other respects *Afroedura* are typical gekkonids.

Afrogecko

Afrogecko is a two-species genus of small (2.06 g v. 3.31 g) gekkonid geckos from southern Africa (and now introduced to at least nine South Atlantic islands). They have leaf-shaped toepads reminiscent of the phyllodactylids, to which they were once assigned. The one species with a recognizable ecological business model is *porphyreus*, which is a trunk/wall lizard. *Afrogecko* have not been reported to change color (v. 15%) or differ between the sexes in color (v. 6.6%). They do use electric lights (50% v. 3.8%) and live in groups (100% v. 25%). Males are reported to be tolerant of one another (v. 13%) and to not defend space (v. 87% defend). All are found at low elevation (v. 92%) and around human habitats (v. 27%). Incubation is on the long side (0.56 v. 0.17 std. resid.), and communal oviposition is reported (50% v. 8.6%). Introduction to multiple places is comparatively uncommon in the family, but present in the genus (50% v. 2.1%). They have not been evaluated for conservation risk.

Afrogecko spp.	English name				Range core
			Size		
	Activity	Diet	mm	g	Ecological business model
ansorgii	Ansorge's Gecko				Angola
	nocturnal		42	1.69	
porphyreus	Marbled African Gecko				South Africa
	nocturnal		48	2.50	*** trunk/wall lizard

Agama

Agama is the largest agamid genus (46 spp.) and is found in Africa south of the Sahara. Though slightly larger (22 g v. 16 g) and more robust (shape index 0.99 v. 0.80 std. resid.) than average agamids, they exhibit typical agamid form, and most forage near (ground-foraging tree lizards 21% v. 19%) or on the ground (refugia-anchored diurnal ground lizards 70% v. 43%). Another 9% (v. 7%) are trunk/wall lizards. Scales tend to be pointy, especially in the head and nape region, but they lack the profile-changing crests (crests 20% v. 38%) found in many Southeast Asian agamids.

Unlike most agamids, *Agama* tails mostly break (76% v. 17%). Secretory pores are found in half of the species (46% v. 13%), despite the complete diurnality of the genus (100% v. 98%). Short-term color change involves pattern and color (11% v. 15%).

Most species (76% v. 53%) exhibit sexual differences in color and a remarkable half change color seasonally (52% v. 20%). Most (89%) *Agama* are m>f sexually dimorphic in size (0.92 v. 0.53 on –2 to 2 scale) and rank eighth in overall sex differences among the 59 agamid genera (2.63 v. 1.61 on 0 to 6 scale). In *A. agama*, color denotes not only the bearer's sex and seasonal condition, but also its social status.

There is a high proportion of omnivores in *Agama* (42% v. 26%) and a corresponding elevation in active foraging (29% v. 17%). This does not necessarily mean that they all move frequently; *aculeata* exhibits one of the lowest recorded proportions of time moving (0th–2nd percentile); the movements of *anchietae* are so infrequent that they have not been quantifiable. Herbivory is absent (v. 8%) and invertebrate eating is underrepresented (38% v. 46%). *Agama* includes a normal frequency of ant specialists (21% v. 20%), despite the lizards' large size. Many *Agama* species that are not ant specialists nonetheless consume a high proportion of ants in their diet. Electric light foraging is conspicuously elevated (2.2% v. 0.2%). Home ranges are small (–0.38 v. 0.35 std. resid.).

All bask directly (v. 24%). Wariness is comparatively high (17% v. 6%). Compared to typical agamids, they are less dependent on crypsis (20% v. 51%) and more dependent on flight (20% v. 8%) and escape to crevices or burrows (40% v. 29%). None is said to use burrows as a daily refugium. Their mean active body temperature is high even by agamid standards (35.5°C v. 34.8°C).

Males have not been documented to defend space (v. 97%), but 82% (v. 18%) of the species documented for social interactions form hierarchies. Most species are colonial because suitable rock crevices are clumped. In establishing dominance, male *lionotus* use both teeth and their club-like tails as weapons. Although male *aculeata* are notably larger than females (usually an indicator of polygynous aspirations), they are said to be monogamous within a breeding season.

Despite residence near the Sahara, none is documented to occur in hyperdry regions, and they are less frequent (25% v. 30%) in the dry zone (56–150 mm/y) than typical agamids. They are slightly overrepre-

Agama spp.	English name	Range core	Activity	Diet	Size mm	Size g	Ecological business model
aculeata	Ground Agama	Southern Africa	diurnal	ants	88	22.5	*** ground-foraging tree lizard
africana	African Forest Agama	Ghana	diurnal	invert.	98	31.0	** ground-foraging tree lizard
agama	African Rainbow Lizard	Nigeria	diurnal	omnivore	111	46.8	*** ground-foraging tree lizard or refugia-anchored diurnal gr. liz.
anchietae	Anchieta's Agama	Southern Africa	diurnal	omnivore	89	23.2	*** refugia-anchored diurnal gr. liz.
armata	Peters' Ground Agama	Southern Africa	diurnal	omnivore	77	14.7	** refugia-anchored diurnal gr. liz.
atra	Southern Rock Agama	South Africa	diurnal	omnivore	90	24.2	**** refugia-anchored diurnal gr. liz.
bocourti	Bocourt's Agama	Senegal	diurnal		67	9.5	
boensis	Boé Agama	Guinea	diurnal		83	18.5	* refugia-anchored diurnal gr. liz.
bottegi	Somali Agama	Somalia	diurnal		98	31.0	
boueti	Bouet's Agama	Southern Sahara	diurnal	invert.	82	18.2	* refugia-anchored diurnal gr. liz.
boulengeri	Boulenger's Agama	Mauritania	diurnal		109	44.2	** refugia-anchored diurnal gr. liz.
caudospinosa	Elmenteita Rock Agama	Kenya	diurnal	invert.	117	54	** refugia-anchored diurnal gr. liz.
cristata	Crested Agama	Guinea	diurnal		102	35.8	* ground-foraging tree lizard
doriae	Benoue Agama	Nigeria	diurnal		95	28.3	*** refugia-anchored diurnal gr. liz.
etoshae	Etosha Agama	Namibia	diurnal	invert.	63	8.0	*** refugia-anchored diurnal gr. liz.
finchi	Finch's Agama	Kenya	diurnal	invert.	94	27.3	** ground-foraging tree lizard

sented in the arid (50% v. 35%) zone and moderately overrepresented in the intermediate (50% v. 12%) zone. They are absent from the wet end of the rainfall spectrum (>1100 mm/y). They are slightly overrepresented in dunes and sand (37% v. 26%).

Clutch counts are high for the lizards' size (0.81 v. 0.20 std. resid.), and young hatch at a relatively small size (0.80 v. 0.69 std. resid.). Thirteen of 15 documented species lay more than once per year. Incubation is short (−0.81 v. −0.55 std. resid.). RCM ratio is high (0.374 v. 0.336).

Population density is high for an agamid (0.65 kg/ha v. 0.16 kg/ha). Longevity is short (−1.4 v. −0.46 std. resid.), which equates to a median of 2.2 years. *Agama* are overrepresented among introductions (4.0% v. 1.7%), including one species (*A. agama*) that has been introduced nearly a dozen times (2% v. 0.6% for >2 times). Relatively few *Agama* are considered imperiled (10% v. 37%) and only 9% (v. 21%) are on one of the IUCN's lists of the most endangered. The head bobs of *lionotus* are sometimes considered offensive by Muslims (Spawls et al. 2018).

Agama spp.	English name				Range core
			Size		
	Activity	Diet	mm	g	Ecological business model
gracilimembris	Little Ground Agama				Nigeria
	diurnal		51	4.03	* refugia-anchored diurnal gr. liz.
hartmanni	Hartmann's Agama				Ethiopia
	diurnal		75	13.8	
hispida	Southern Spiny Agama				Southern Africa
	diurnal	invert.	94	27.5	*** refugia-anchored diurnal gr. liz.
hulbertorum	Elangata Wuas Agama				Kenya
	diurnal		75	13.4	*** refugia-anchored diurnal gr. liz.
impalearis	Bibron's Agama				Western Sahara
	diurnal	omnivore	106	39.7	*** refugia-anchored diurnal gr. liz.
insularis	Island Agama				Guinea
			97	30.7	** refugia-anchored diurnal gr. liz.
kaimosae	Kaimosi Agama				Tanzania
	diurnal	invert.	108	42.7	** refugia-anchored diurnal gr. liz.
kirkii	Kirk's Rock Agama				Zambia
	diurnal	ants	92	25.4	** refugia-anchored diurnal gr. liz.
knobeli	Knobel's Rock Agama				Namibia
	diurnal	omnivore	106	39.8	* refugia-anchored diurnal gr. liz.
lanzai	Puntland Agama				Somalia
	diurnal		87	21.5	*** refugia-anchored diurnal gr. liz.
lebretoni	Lebreton's Agama				Nigeria
	diurnal		111	46.0	** trunk/wall lizard
lionotus	Red-headed Rock Agama				East Africa
	diurnal	omnivore	104	37.6	** ground-foraging tree lizard or trunk/wall lizard
lucyae	Lucy's Agama				Ethiopia
			52	4.28	** refugia-anchored diurnal gr. liz.
montana	Montane Rock Agama				Tanzania
	diurnal	ants	65	8.9	** ground-foraging tree lizard
mossambica	Mozambique Agama				Zambia
	diurnal	ants	91	25.2	** ground-foraging tree lizard
mucosoensis	Mucoso Agama				Angola
			99	32.8	** trunk/wall lizard

Agama spp.	English name	Range core	Activity	Diet	Size mm	Size g	Ecological business model
mwanzae	Mwanza Flat-headed Agama	Tanzania	diurnal	ants	108	41.9	*** refugia-anchored diurnal gr. liz.
parafricana	Small Forest Agama	Ghana			82	17.8	** ground-foraging tree lizard
paragama	West African Tree Agama	Ghana	diurnal		88	22.5	** trunk/wall lizard
persimilis	Painted Agama	Ethiopia	diurnal	invert.	55	5.3	*** refugia-anchored diurnal gr. liz.
planiceps	Namibian Rock Agama	Angola	diurnal	omnivore	103	39.6	*** refugia-anchored diurnal gr. liz.
robecchii	Robecchi's Agama	Ethiopia	diurnal		107	41.5	** refugia-anchored diurnal gr. liz.
rueppelli	Rueppell's Agama	East Africa	diurnal	invert.	76	14.0	*** refugia-anchored diurnal gr. liz.
sankaranica	Sankaranian Agama	Nigeria	diurnal		70	11.1	** refugia-anchored diurnal gr. liz.
somalica	Somalia Rock Agama	Somalia	diurnal		81	17.7	** refugia-anchored diurnal gr. liz.
spinosa	Gray's Agama	Horn of Africa	diurnal	omnivore	97	30.7	** refugia-anchored diurnal gr. liz.
sylvana	Forest Agama	Ghana	diurnal		84	19.8	** trunk/wall lizard
tassiliensis	Tassili Agama	Sahara	diurnal		105	39.0	** refugia-anchored diurnal gr. liz.
turuensis	Turu Agama	Tanzania	diurnal	omnivore	104	38.3	** refugia-anchored diurnal gr. liz.
weidholzi	Weidholz's Agama	Senegal	diurnal		56	5.5	* refugia-anchored diurnal gr. liz.

Agamidae (figs. 4.3B, 4.3C, 4.5A, 4.5C, 4.5D, 4.6A, 4.6B, 4.6C, 5.3A)

The Agamidae and Chamaeleonidae, along with 12 families formerly grouped as Iguanidae, make up the infraorder Iguania, an intuitive and well-supported clade. The chameleons are highly specialized, with their turreted eyes, ballistic tongues, cruise foraging, and coiling prehensile tails, but the other Iguania species are more generalized and similar to each other, save that the agamids are restricted to the Old World (except Madagascar) and iguanids (in the broad sense) mostly to the New World (plus Madagascar). Both groups are diurnal and mostly terrestrial; they tend to be large and socially assertive. They are often spiny or crested, colorful, and omnivorous, but they are never elongate.

As currently delineated, the agamids are the third-largest family, with 483 species in the following 59 genera: *Acanthocercus, Acanthosaura, Agama, Amphibolurus, Aphaniotis, Bronchocela, Bufoniceps. Calotes, Ceratophora, Chelosania, Chlamydosaurus, Complicitus, Cophotis, Coryphophylax, Cryptagama, Ctenophorus, Dendragama, Diporiphora, Draco, Gonocephalus, Gowidon, Harpesaurus, Hydrosaurus, Hypsicalotes, Hypsilurus, Intellagama, Japalura, Laudakia, Leiolepis, Lophocalotes, Lophognathus, Lophosaurus, Lyriocephalus, Malayodracon, Mantheyus, Moloch, Oriocalotes, Otocryptis, Paralaudakia, Phoxophrys, Phrynocephalus, Physignathus, Pogona, Psammophilus, Pseudocalotes, Pseudocophotis, Pseudotrapelus, Ptyctolaemus, Rankinia, Saara, Salea, Sarada, Sitana, Stellagama, Thaumatorhynchus, Trapelus, Tympanocryptis, Uromastyx, Xenagama*.

Demonstrating that phylogeny is not ecological destiny, 10 of the 14 ecological business models are found within the Agamidae. The commonest (42.6% of species) are the refugia-anchored diurnal ground lizards, with nearly twice the representation found in average lizards (24.5%). Gliders, which have no close analogue within the other Iguania, constitute 11.4% of agamids, nearly 10-fold the representation in lizards overall (1.2%). Gliders are also ecologically distinctive from the remainder of the Agamidae. For example, unlike other agamids, the agamid gliders are not terrestrial, large, socially assertive—though they mate-guard—or omnivorous; they are ant specialists. Ground-foraging tree lizards (19.3% v. 9.2%) have about double the representation in agamids as they do in average lizards. Despite the penchant of agamids for foraging from or to sites low on tree trunks, agamids have not apparently moved up the tree trunks, except for gliders. Trunk/wall lizards are notably scarce in Agamidae (7.4% v. 13%). Giant herbivores are strongly overrepresented (6.4% v. 1.6%). Nocturnal ground lizards are nearly unknown (0.3% v. 7.7%). Litter cryptozoids are similarly scarce (1.1% v. 11.6%). Agamids are not reported to be crack hermits, nocturnal canopy walkers, subterranean foragers, or wide-ranging chemosensors.

Agamids run larger than average lizards (15.6 g v. 5.3 g), but sizes range over nearly the full spectrum, from the 10th percentile (1.15 g *Diporiphora convergens*) to the 99th (1170 g *Uromastyx aegyptia*). Agamids are portly (shape index 0.8 v. 0.0 std. resid.), long-tailed (1.86 v. 1.37 × SVL), and disinclined to break their tails (17% v. 57%), which are rarely prehensile (1.4% v. 5%). Agamids never have blue tails (v. 2.3%). Agamids are frequently ornamented with crests (38% v. 5%), and they are about twice as frequently in possession of color changing ability, either darkening (22% v. 11%) or advanced color change (15% v. 6%).

Most agamids are moderately m>f in SSD (0.53 v. 0.36 on –2 to 2 scale), with gliders, especially *Draco*, a clear m<f exception (see *Draco* account). Agamids tend to have sexual dichromatism (53% v. 21%) and reproductive-season color change is relatively common (19.7% v. 6.5%). Taking all sex differences collectively, agamids rank eighth from the most differing of the 45 families (1.61 v. 0.75 on 0 to 6 sex differences rollup score).

Despite their relatively large size, less than 1% of agamids eat vertebrates (v. 3.2%). Nor is invertebrate consumption up to the lizard proportion (46% v. 73%). Instead, agamids' large size is manifest in herbivory (8.2% v. 4.1% for lizards) and omnivory (26.1% v. 15.4%). Smaller agamids indulge in ant specialization (19.5% v. 4.4%), which is especially dominant in gliders (see *Draco* account). Agamids are resolutely sit-and-wait foragers (88% v. 46%) and avoid active foraging (17% v. 58%). They cruise forage at lizard-typical proportions (6% v. 7%). They almost never use electric lights (0.3% v. 2.0). Mass-adjusted home-range areas are moderately large (0.35 v. 0.0 std. resid.).

Agamids are nearly all diurnal (98% v. 67%), and nocturnal species have not been reported (v. 25%).

Agamids rarely vocalize spontaneously (0.2% v. 2.5%) but are capable of stress sounds (1.4% v. 2.2%). The proportion that is conspicuously wary is normal (5.8% v. 8.2%).

The mix of tactics used by agamids to evade predator detection is similar to that of other lizards, with a greater emphasis on crypsis (51% v. 43%) and a reduced dependence on flight (7.6% v. 10.8%). Their thermoregulatory behaviors are typical (e.g., direct basking 24% v. 18%). However, they maintain a higher body temperature (mean 34.8°C) than the average lizard (30.8°C), perhaps because they occupy such low latitudes.

In their social interactions, male agamids rank low on tolerance of other males (3.1% v. 14.6%), and high on defense of space (96% v. 81.5%), which may include hierarchical behavior (17.5% v. 7.7%). Female agamids are less well-documented, but exhibit the same generally pugnacious rankings.

The microhabitat characterizations of agamids are confused by their penchant for being gliders (high on tree trunks) and low on tree trunks, but not particularly in the middle. The ecological business models capture the agamid pattern better than traditional microhabitats. Although agamids are not known to be subterranean foragers, they dig or use burrows at twice the proportion of typical lizards (40% v. 22%).

Agamids are not particularly notable with regard to habitat selection; they are also similar to most lizards with regard to elevation. They are overrepresented at the dry end of the rainfall belt (hyperdry/dry/arid: 17/30/35 v. 3/12/30) and underrepresented in the wetter regions (intermed./mesic/wet: 12/20/7 v. 27/43/9). In the hyperdry (<55 mm/y) zone they are ranked second, behind only Phyllodactylidae. They are typical lizards with regard to habitat specializations such as littoral (10% v. 17%), sandy (26% v. 22%), riparian (21% v. 22%), and human (16% v. 19%) habitats.

Agamids are nearly universally oviparous (98% v. 82%). Their mass-adjusted clutch counts (0.20 v. 0.0 std. resid.) are negligibly elevated. The mass-adjusted ontogenetic range of sizes in Agamidae is high (0.69 v. 0.0 std. resid.), indicating that agamid young are unusually small, about 1/100 of their maximal size. Mean RCM of agamids (0.336 v.

0.226) is the third-highest family, for reasons that are elusive. Incubation periods are short for their size (–0.55 v. 0.0 std. resid.). There is substantial variation among genera in incubation duration (*Acanthosaura* is longest at 1.53 std. resid.; *Hydrosaurus* is shortest at –1.22), but this difference yields no obvious ecological signal. Their eggs are rarely adherent (0.8% v. 2.4%; gliders are an exception) or laid communally (0.4% v. 4.2%).

Agamid longevity is low (ranked 33 of 45 families), and their population density is low-average (0.16 kg/ha v. 0.54 kg/ha). They are of typical frequency on islands (36% v. 40%) or among endemic species (49% v. 64%). Among lizards that are introduced (1.7% v. 3.3%), they are moderately underrepresented, but *Agama*, *Calotes*, and *Stellagama* have species that are serial invaders (0.6% v. 1.1%; see those genus accounts). Agamidae is underrepresented among imperiled species (37% v. 49%). The proportion that are on one of the IUCN's lists of the most endangered is relatively low (5.1% v. 13.6%).

Agamodon

Agamodon is a small trogonophid genus (three spp.) of tiny (0.83 g v. 1.78 g), very poorly known (filled cells 13% v. 22%) subterranean foragers (100% v. 80%) found in Somalia and Yemen. Body shape is extreme (shape index –3.87 v. –3.43 std. resid.). Sex differences are unknown. One species (*arabicus*) is known to be oviparous (100% v. 50%). All species are endemic (v. 67%). None is on an island (v. 33%). In other respects *Agamodon* are typical trogonophids.

Agamura

Agamura persica is the sole representative of this gekkonid genus. It is found in the deserts of south central Asia—Afghanistan, Pakistan, Iran—and is famous for its large head and slow, straight-legged nighttime walk. It is a bit larger than a typical gekkonid (4.75 g v. 3.31 g), and its short tail (0.82 v. 1.07 × SVL) is difficult to break, and if broken does not regenerate. It lacks enlarged toepads. Its SSD of body size is not known, but the relative tail length dimorphism is strongly m>f (1.25 ratio), suggesting that body size dimorphism is most likely m<f. It basks directly (v. 4.9%) and is found at low and medium elevation (in this case, 0–1900 m). It frequents sandy areas (v. 11%) and riparian zones (v. 17%) but not human habitat (v. 27%). It reproduces like other gekkonids. It has not been introduced (v. 3.7%) and is not considered to be imperiled (v. 47%). In other respects it is a typical gekkonid.

Agamodon spp.	English name				Range core
			Size		
	Activity	Diet	mm	g	Ecological business model
anguliceps	Angled Worm Lizard				Somalia
			152	3.67	** subterranean forager
arabicus	Arabian Worm Lizard				Yemen
			102	0.25	
compressus	Flat Worm Lizard				Somalia
			83	0.62	** subterranean forager

Agamura sp.	English name				Range core
			Size		
	Activity	Diet	mm	g	Ecological business model
persica	Spider Gecko				Afghanistan
	cathemeral	invert.	60	4.75	*** nocturnal ground lizard

Ailuronyx

Ailuronyx (three spp.) are nocturnal arboreal gekkonids found in the Seychelles Islands of the western Indian Ocean. They remind me of *Phelsuma* without the gaudy colors. They are much larger than most gekkonids (16.5 g v. 3.3 g). Even the smallest species (*tachyscopaeus*), called a "dwarf," is 7.2 g. Its diet is unknown; *seychellensis*, which is twice the mass of *tachyscopaeus*, is an omnivore; *trachygaster*, which is nearly three times the mass of *seychellensis*, is a full herbivore, though of fruit, nectar, and pollen, not the leaves that characterize most giant herbivores. *A. trachygaster* specializes in the nectar and pollen of *Lodoicea maldivica* palms, an extreme specialization. Another perspective is to ask why such a dietary specialist is so large. Unlike folivores, it does not need to be large to digest vast quantities of low-energy leaves; pollen is highly proteinaceous and fruit and nectar are highly caloric. It would not seem to need to be huge to escape predation, though large birds in the Seychelles could exert considerable selective pressure. Possibly it is large to store energy to withstand long periods (e.g., after a cyclone) when fruit, nectar, and pollen are simultaneously unavailable.

Tails are short (0.90 v. 1.07 × SVL), breakable (v. 87%), and not prehensile (v. 1.8%). Only one species is documented to have secretory pores (v. 46%). One of three species shows basic color change (v. 15%). The

Ailuronyx spp.	English name				Range core
			Size		
	Activity	Diet	mm	g	Ecological business model
seychellensis	Seychelles Bronze Gecko				Seychelles
	nocturnal	omnivore	92	15.3	*** trunk/wall lizard
tachyscopaeus	Dwarf Bronze Gecko				Seychelles
	nocturnal		70	7.2	*** trunk/wall lizard
trachygaster	Giant Bronze Gecko				Seychelles
	nocturnal	herbivore	132	41.2	**** nocturnal canopy walker

Alexandresaurus sp.	English name				Range core
			Size		
	Activity	Diet	mm	g	Ecological business model
camacan	Alexandre's Forest Lizard				Brazil
			59	2.87	*** litter cryptozoid

sexes are not known to differ in color (v. 6.6%). Although their greenish cryptic body colors would suggest diurnality (e.g., *Phelsuma*), all three are believed to be strictly nocturnal. *Ailuronyx* may need to be camouflaged while sleeping semiexposed during the day. All males are reported to defend space (v. 87%).

Ailuronyx are overrepresented in littoral zones (33% v. 11%); there may be few places in the Seychelles that extend beyond the littoral. They are not reported in riparian areas (v. 17%), but many atolls and raised coral islands have no surface features that could be called riparian. They are overrepresented in human habitat (67% v. 27%), perhaps because humans pervade small islands.

The mean clutch count of *Ailuronyx* is 1.5, less than the family standard of two (−0.83 v. −0.26 std. resid.). All documented *Ailuronyx* eggs are adherent (v. 10.5%), and many are communally laid (33% v. 8.6%). Parental care is present (33% v. 1.9%), and in the reported case (*seychellensis*) may extend throughout the entire incubation period, which is extended in *Ailuronyx* (0.30 v. 0.17 std. resid.).

Longevity is short (−0.13 v. 0.15) suggesting that predation may be a larger factor than presumed in the preceding discussion. Longevity is documented for only the middle-sized species (*seychellensis*: 6.2 years), which does not seem excessively short for a gecko. This species has skin that tears easily, even during mating. Such a defense mechanism suggests that predation must be high enough to warrant such a costly defense. Population density (based on *seychellensis*) is high (1.22 kg/ha v. 0.37 kg/ha), a value not exceptional on islands. No *Ailuronyx* has been introduced (v. 3.7%) and two of three are considered imperiled (v. 47%), though none is listed on the IUCN's lists of most endangered species (v. 11.3%). In other respects *Ailuronyx* are typical gekkonids.

Alexandresaurus

Alexandresaurus camacan is the sole member of this genus of medium-sized (2.87 g v. 2.35 g) terrestrial Brazilian gymnophthalmids. Like most gymnophthalmids, it is a litter cryptozoid (v. 55%). Its tail is a bit long (2.0 v. 1.7 × SVL) and not known to break (v. 28%). It has secretory pores (v. 29%) and is elongate (v. 41%), and—unusually for small gymnophthalmids—it exhibits m>f SSD. The sexes differ in color (v. 13%). Thus the sex difference rollup score is strongly elevated (3.0 v. 0.49 on 0 to 6 scale). Clutch count is unknown (v. 1.9). This species is not known to have been introduced (v. 0.8%), nor has it been evaluated for conservation status.

Algyroides

Algyroides (four spp.) is a genus of relatively well-known (filled cells 50% v. 33%), small (2.24 g v. 5.1 g) diurnal lacertid lizards of diverse ecologies found in European countries bordering the Mediterranean from Spain to Greece. Two are thought to be ground-foraging tree lizards (50% v. 8%), two are litter cryptozoids, at least in some places (50% v. 2.4%), and one is a trunk/wall lizard (25% v. 2.4%).

In build they approximate an average lizard (shape index 0.07 v. 0.27 std. resid.). Their tails are slightly short for a lacertid (1.91 v. 2.06 × SVL) and all tails break (v. 92%). At least one (*fitzingeri*) has secretory pores.

They are almost equally prone to m>f and m<f SSD (0.25 v. 0.49 on −2 to 2 scale), but the sexes are routinely demarked by color (75% v. 33%) and seasonal color (50% v. 15%). Thus their sex differences rollup score well exceeds the lacertid norm (2.0 v. 0.86 on 0 to 6 scale).

All are conspicuously wary (v. 11%), and at least two species bask directly (50% v. 25%). Males defend space (100% v. 59%). *A. marchi* is reported to be inactive in a crevice for up to 30% of its active period.

The Iberian species (*marchi*) is found in the intermediate rainfall band (685–837 mm/y). They are overrepresented in riparian areas (100%

Algyroides spp.	English name				Range core
			Size		
	Activity	Diet	mm	g	Ecological business model
fitzingeri	Pygmy Algyroides				France
	diurnal	invert.	36	1.10	*** trunk/wall lizard
marchi	Spain Keeled Lizard				Spain
	diurnal		46	2.30	*** ground-foraging tree lizard
moreoticus	Greek Keeled Lizard				Greece
	diurnal	invert.	45	2.16	**** litter cryptozoid
nigropunctatus	Blue-throated Keeled Lizard				Balkans
	diurnal	invert.	58	4.59	** ground-foraging tree lizard or litter cryptozoid (geog. var.)

Alinea spp.	English name				Range core
			Size		
	Activity	Diet	mm	g	Ecological business model
lanceolata	Barbados Skink				Barbados
			77	9.1	
luciae	Saint Lucia Skink				Saint Lucia
			87	13.9	

v. 28%) and missing from sandy areas (0% v. 44%). Reproduction and longevity match the lacertid pattern. Population density is known from only one sample of one species (*fitzingeri*), for which it is low (0.14 kg/ha v. 2.07 kg/ha). Half of the species are endemic (v. 35%), three-quarters are found on islands (v. 26%), and none has been introduced (v. 4.6%). All evaluated species are considered imperiled (v. 56%), and one species (*marchi*) is considered to be seriously endangered by the IUCN (v. 15.5%). In other respects *Algyroides* are similar to other lacertids.

Alinea

Alinea is a two-species genus of large (11.2 g v. 3.3 g), likely extinct, skinks from the Lesser Antilles (Barbados and St. Lucia). The St. Lucian skink is the better known, being based on four specimens (three of these from Barbados), most recently collected in 1889 (filled cells for genus: 19% v. 25%). It was formally declared extinct by the IUCN in 2017; the Barbados one is considered only "possibly extinct" (as listed by the IUCN in 2017).

Aside from approximate size, available information includes only that male *lanceolata* were likely smaller than females (though, $n = 3$!), the three specimens come from riparian areas (v. 23%), and one female appears to have been viviparous with a litter of seven (0.66 v. 0.06 std. resid.). Introduced mammals are likely a reason from their extinction.

Alopoglossidae

The 22 alopoglossids (filled cells 23% v. 27%) are relatively similar to the equally mysterious 237 gymnophthalmids (19% v. 27%). Both are reclusive, relatively small, active foragers, and they reside in mesic areas of the Neotropics. Alopoglossidae includes only two genera: *Alopoglossus* and *Ptychoglossus*, both composed entirely of invertebrate-eating (100% v. 73%) litter cryptozoids (100% v. 12%). Gymnophthalmids are also almost exclusively invertebrate feeders (see Gymnophthalmidae account), but they exhibit 9 of the 14 ecological business models. Unlike gymnophthalmids, few alopoglossids are elongate (13.6% v. 11.4% for lizards, v. 41% for gymnophthalmids).

Alopoglossids run smaller than average lizards (mean of 2.5 g v. 5.3 g) and do not vary much in size (range 0.7–6.5 g). Alopoglossids are average in build (shape index −0.06 v. 0.0 std. resid.). Their tails are slightly long (1.63 v. 1.37 × SVL), and three-quarters of them are documented to break their tails (73% v. 57%), which are not known to be prehensile (0% v. 5%). They never have crests (0% v. 5.3%) or color changing ability (0% v. 10.7%), but many (23% v. 15%) have secretory pores.

About 1/4 of alopoglossids have sex differences in color (27% v. 21%), but only one species is documented to have seasonal color (4.5% v. 6.5%, likely underreported); none has a blue tail (0% v. 2.3%). None has been recorded foraging at electric lights (0% v. 2%) or vocalizing (0% v. 2.5%). They are diurnal (64% v. 67%), crepuscular (9% v. 2%), or active at any time of day (27% v. 7%). They conceal themselves from predators by vegetation (67% v. 23%) or flight (33% v. 11%) based on an inadequate sample. There are two records of thermoconforming (9% v. 2.6%) for one each of basking directly (4.5% v. 18%) and indirectly, another inadequate sample. Alopoglossids are unusually concentrated in the mesic (1101–3000 mm/y) rainfall band (100%, tied for first place among families, v. 43%). Their elevational profile indicates a great overrepresentation at intermediate elevations:

(l/m/h) 70/75/0 (v. 82/41/13). A small proportion has been recorded as residing in sand or dune areas (11% v. 22%) or around human habitations (6% v. 19%).

They are all oviparous (100% v. 82%). Median clutch counts are small (−0.17 v. 0.0 std. resid.), and their ontogenetic range of masses is low (−0.32 v. 0.0 std resid.), reflecting particularly large young (1/15 v. 1/44 of maximum size). Incubation duration is longer than the lizard average (0.29 v. 0.0 std. resid.), equating to about 79 days. One species (*Ptychoglossus festae*) has been recorded ovipositing communally (4.5% v. 4.2%). Two species with recorded RCMs were very light (0.091, 0.113).

One species (*Ptychoglossus plicatus*) has a density estimate of 0.36 kg/ha, indistinguishable from the lizard average (0.54 kg/ha). Endemism is low (45% v. 64%). Island species are rare (4.5% v. 40%). None has been introduced (v. 3.3%). Twenty-nine percent are considered at conservation risk (v. 49%), likely biased low by insufficient information. The ranks of the IUCN's most endangered species are lightly populated with alopoglossids (7.1% v. 13.6%), perhaps only partly due to lack of information.

Alopoglossus

Alopoglossus is a moderate (seven spp.) genus of average-sized (2.7 g v. 2.5 g), average-build (shape index 0.03 v. -0.06 std. resid.), litter cryptozoid (100% v. 100%) alopoglossids from northwestern South America. Unlike their sister genus *Ptychoglossus*, few species are known to break tails (14% v. 100% in *Ptychoglossus*). They are more likely to have secretory pores (43% v. 13% in *Pty.*). None is elongate (v. 20% in *Ptychoglossus*). The only documented SSD is m<f (1.0 v. 0.0 on −2 to 2 scale). Two of three species documented for activity are cathemeral (67% v. 27%). *A. festae* are vulnerable to hyperthermia when placed in human hands, suggesting a low temperature range. As far as is known, males are tolerant of each other (100% v. 100%). *A. angulatus* females oviposit in light gaps, sometimes in ant nests. In other respects they do not deviate notably from other alopoglossids.

Alopoglossus spp.	English name / Activity	Diet	Size mm	Size g	Range core / Ecological business model
angulatus	Large-scaled Forest Lizard				Northern South America
	cathemeral	invert.	54	3.42	*** litter cryptozoid
atriventris	Black-bellied Shade Lizard				Colombia
	cathemeral	invert.	47	2.22	*** litter cryptozoid
buckleyi	Buckley's Forest Lizard				Peru
			49	2.54	*** litter cryptozoid
copii	Drab Shade Lizard				Ecuador
		invert.	60	5.1	*** litter cryptozoid
festae	Western Copper Lizard				Ecuador
	diurnal		56	3.90	* litter cryptozoid
lehmanni	Lehmann's Shade Lizard				Colombia
			33	0.70	
viridiceps	Green-headed Shade Lizard				Ecuador
			55	3.74	

Alsophylax

Alsophylax (six spp., Gekkonidae) are small (0.90 g v. 3.31 g) toepadless crack hermits of the central Asian deserts (Iran to Uzbekistan, Tajikistan, Kazakhstan, Russia, Mongolia, and China). Tails are slightly longer than average (1.24 v. 1.07 × SVL). Tails are breakable (v. 87%) and not prehensile (v. 1.8%). Secretory pores are likely universal (83% v. 46%). They do not change color (v. 15%).

All species are m<f in SSD; but other sex differences are absent. With the pores and size difference, their sex rollup score is the third highest in gekkonids (1.83 v. 0.63 on 0 to 6 scale).

Their daily activity stands out as often cathemeral (33% v. 11%). They have high frequencies of spontaneous calling (67% v. 6.5%), stress calling (16.7% v. 3.7%), and conspicuous wariness (33% v. 4.4%). The latter is a bit unusual in relation to their 100% dependence on crypsis for primary predator concealment (v. 50%).

Group living is not reported in *Alsophylax* (v. 25%). They are overrepresented among the cryptozoids (20% v. 0.9%) and the terrestrials (80% v. 31%). All species use burrows as daily refugia (v. 10%). They are overrepresented in the dry zone (56–150 mm/y: 100% v. 13%) and arid zone (151–400 mm/y: 100% v. 27%). They are overrepresented in sandy areas (40% v. 11%) and around human habitats (40% v. 27%).

They are not quite typical gekkonids in terms of reproduction: clutch counts fall below the standard two (1.5 v. 2.0); when expressed in mass-adjusted terms, the clutch counts are larger than lizard average rather than smaller (0.13 v. −0.26 std. resid.), and the young are noticeably larger than average gekkonids (ontogenetic range –0.54 v. –0.19 std. resid.). No eggs are adherent (v. 10.5%), and half of the species lay communally (v. 8.6%), which is characteristic of crevice dwellers.

Longevity is slightly short (–0.19 v. 0.15 std. resid.), though only *pipiens* (3.5 years) is documented. No *Alsophylax* have been introduced (v. 3.7%). Five of six are considered imperiled (v. 47%), though none is on the most endangered species lists of the IUCN (v. 11.3%). In other respects *Alsophylax* are typical gekkonids.

Alsophylax spp.	English name				Range core
			Size		
	Activity	Diet	mm	g	Ecological business model
laevis	Smooth Pygmy Gecko				Afghanistan
	cathemeral	invert.	34	1.01	** crack hermit
loricatus	Armored Pygmy Gecko				Uzbekistan
	nocturnal		30	0.72	* crack hermit
pipiens	Squeaky Pygmy Gecko				Central Asia
	nocturnal	invert.	36	1.60	** crack hermit or nocturnal ground lizard
przewalskii	Xinjiang Pygmy Gecko				China
	nocturnal		32	0.85	** crack hermit
szczerbaki	Szczerbak's Pygmy Gecko				Turkmenistan
	cathemeral	invert.	30	0.70	*** crack hermit
tadjikiensis	Tajik Pygmy Gecko				Tajikistan
	nocturnal	invert.	31	0.76	** crack hermit

Altiphylax

Altiphylax (six spp.) are small (1.98 g v. 3.31 g), toepadless, primarily terrestrial gekkonids of the central Asian deserts (India to Afghanistan and Kyrgyzstan and possibly China). Most occur at high elevations (up to 3700 m), are nocturnal (80% v. 73%), and move slowly. It is cold up there (*levitoni* active down to air temperatures of 8°C). Of the species with assigned ecological business models, half are nocturnal ground lizards (v. 27%) and one-quarter each are crack hermits (v. 3.2%) and trunk/wall lizards (v. 45%).

Tails are not prehensile (v. 1.8%). Secretory pores are common (33% v. 46%). They do not change color (v. 15%).

The sexes are roughly equal in body size and are not reported to be dichromatic (v. 6.6%).

Electric light use stands out (50% v. 3.8%). Vocalizations have not been reported (v. 6.5%). They tend to be more wary than gekkonids (17% v. 4%).

Group living is not reported in *Altiphylax* (v. 25%). All males defend space (v. 87%). They are more terrestrial than typical gekkonids (75% v. 31%), and the others are arboreal (25% v. 67%). Some refuge in burrows (25% v. 10%). They are notably overrepresented at medium (60% v. 26%) and especially high elevation (60% v. 2%). Correspondingly, they are absent from littoral zones (v. 11%). They are overrepresented in riparian zones (60% v. 17%) and in human habitats (60% v. 27%).

They are typical gekkonids in terms of reproduction (clutch count 1.5 or –0.10 v. –0.26 std. resid.), though incubation understandably takes a little longer (0.55 v. 0.17 std. resid.). Some eggs are adherent (17% v. 10.5%) and some laid communally (17% v. 8.6%), presumably in the rock cracks they inhabit.

No *Altiphylax* has been introduced (v. 3.7%). One-third are considered imperiled (v. 47%), though none is on the most endangered species

Altiphylax spp.	English name				Range core
			Size		
	Activity	Diet	mm	g	Ecological business model
baturensis	Batura Glacier Gecko				Pakistan
	nocturnal	invert.	51	3.01	* nocturnal ground lizard or (in houses) trunk/wall lizard
levitoni	Leviton's Gecko				Afghanistan
	nocturnal	invert.	39	1.46	** trunk/wall lizard
mintoni	Minton's Swati Gecko				Pakistan
			38	1.32	
stoliczkai	Karakorum Gecko				Himalayas
	nocturnal	invert.	49	2.62	** nocturnal ground lizard
tokobajevi	Tien Shan Pygmy Gecko				Kyrgyzstan
	cathemeral	invert.	43	1.80	*** crack hermit
yarkandensis	Yarkand Bow-fingered Gecko				Pakistan
	nocturnal		46	2.22	

lists of the IUCN (v. 11.3%). In other respects *Altiphylax* are typical gekkonids.

Amalosia

Amalosia is a small (four spp.) Australian genus of fairly typical diplodactylid geckos. They are cryptically marked and often found in association with exfoliating rock, but their ecological business models are not well established (see following table). They are typical diplodactylids in size (4.58 g v. 4.15 g) and shape, though none has a prehensile tail (v. 26%) or color change (v. 5%). Sex differences are not documented. Two of four species are cathemeral (50% v. 13%). Three of four are found in littoral areas (75% v. 17%), none is found in sandy areas (v. 24%), one is found in riparian zones (25% v. 10%), and two of four are found in human habitat (50% v. 5%). One of four is reported to lay eggs in a communal site (25% v. 2.7%). *A. lesueurii* has an atypically large RCM (0.310 v. 0.225); the others are undocumented. Mass-adjusted longevity (0.69 std. resid.) is long for lizards (v. 0.0), but a little short for diplodactylids (0.90). *A. rhombifer* has been introduced to a disjunct population in the Northern Territory. In other respects they are typical diplodactylids.

Amalosia spp.	English name				Range core
			Size		
	Activity	Diet	mm	g	Ecological business model
jacovae	Clouded Velvet Gecko				Queensland
	nocturnal		56	3.52	* trunk/wall lizard
lesueurii	Lesueur's Velvet Gecko				Queensland
	cathemeral	invert.	67	5.9	** crack hermit
obscura	Slim Velvet Gecko				Western Australia
	nocturnal		57	3.61	** crack hermit
rhombifer	Zigzag Velvet Gecko				Northern Australia
	cathemeral	invert.	67	5.9	** trunk/wall lizard

Amapasaurus

Amapasaurus tetradactylus is the sole member of this genus of tiny (0.55 g v. 2.35 g) terrestrial Brazilian gymnophthalmids. Like most gymnophthalmids, it is a litter cryptozoid (v. 55%). It is on the slender side (shape index −0.54 v. −0.32 std. resid.). Its tail is a bit short (1.46 v. 1.7 × SVL) and not known to break (v. 28%). Sex differences are not documented. Clutch count is unknown (v. 1.9). This species is not known to have been introduced (v. 0.8%), nor has it been deemed imperiled (v. 25%).

Amapasaurus sp.	English name				Range core
			Size		
	Activity	Diet	mm	g	Ecological business model
tetradactylus	Five-toed Shade Lizard				Brazil
			31	0.55	*** litter cryptozoid

Amblyrhynchus (fig. 4.4E)

Amblyrhynchus cristatus is the sole member of this herbivorous iguanid genus endemic to the Galápagos Islands (Ecuador). This charismatic giant among giants (2550 g v. 609 g) is widely heralded as the world's only marine lizard. However, it is only the largest male individuals of this species that have the size and thermal mass needed for extended foraging underneath the surface of the cool Galápagoan waters. This species is strongly m>f in size, with males weighing as much as 10 times that of their mates (mean ratio not available). This indicates that almost all of the divers are adult males; the juveniles, females, and smaller males forage on foot in tide pools and among terrestrial vegetation, a microhabitat extensively used by other lizards.

The mean mass of the genus is the highest of the Iguanidae (second-highest family). The mean shape index (2.48 v. 2.11 std. resid.) is also the highest in the Iguanidae, which is the stoutest family, and presumably reflects selection pressure for a heavy compact shape. Tails rarely break (v. 57%); they need the tails to swim. Secretory pores are absent (v. 38%), as is color change (v. 21%). Sex differences overall (2.0 v. 1.9 on 0 to 6 scale) are as expected for an iguanid.

The diet of male *Amblyrhynchus* is as unique as their foraging environment: algae, especially red and green algae during good years and brown algae during El Niño events, which

Amblyrhynchus sp.	English name				Range core
			Size		
	Activity	Diet	mm	g	Ecological business model
cristatus	Marine Iguana				Galápagos Islands
	diurnal	herbivore	391	2550	**** giant herbivore

routinely induce starvation. As noted above, the diet of the smaller individuals is more diverse and may even include invertebrates.

As with many Galápagos iguanids, *Amblyrhynchus* exhibits island tameness, a feature that so endeared them to Charles Darwin. Males, however, will truculently defend their breeding space against other males (v. 90%).

Reproduction in *Amblyrhynchus* is normal, though the clutch count is remarkably small (–2.76 v. 0.45). This corresponds to 2.4 very large eggs (42 × 93 mm: min × max). Such large eggs produce the second-largest hatchlings in the family (mass-adjusted ontogenetic size range –1.47 v. –0.55 std. resid.), behind only *Brachylophus*. Given the regular occurrence of El Niño events in the Galápagos, which induces widespread starvation in *Amblyrhynchus*, large hatchling size may be beneficial to withstand periodic food stress. The paucity of predators on large lizards may be another factor promoting large initial size. Female *Amblyrhynchus* may defend their oviposition site briefly. Incubation duration is notably short (–1.10 v. –0.76 std. resid.), perhaps to complete development of a large lizard within a short seasonal window.

Longevity is about average (0.29 v. 0.16 std. resid.), corresponding to 28 years. *Amblyrhynchus* has not been introduced (v. 33%), but it is considered imperiled (v. 97%), though it is not among the IUCN's lists of the most endangered (v. 51%).

Ameiva

Ameiva (Teiidae) is a moderate genus (13 spp.) of large-sized (33.7 g v. 22.1 g), socially tolerant (100% v. 55%), flight-dependent (100% v. 77%) wide-ranging chemosensors (100% v. 97%). The genus is found from Argentina and Brazil, north to Costa Rica, the Guiana Shield countries, and islands of the Caribbean as far north as Swan Island (Honduras) and St. Vincent (Lesser Antilles).

They are robustly built (shape index 1.01 v. 0.88 std. resid.). Their tails are average in length (2.33 v. 2.30 × SVL), mostly unbreakable (7.7% v. 21.3% breakable), and never blue (v. 15%). Long tails probably aid in bipedal running (*ameiva*, *bifrontata*). At least *ameiva* exhibits the teiid trait of lifting an arm and waving it at intervals while foraging. Secretory pores are present in a normal frequency (7.7% v. 9.0%). Color change is elevated (7.7% v. 1.3% for any change; 7.7% v. 0.6% for advanced).

SSD is slightly more m>f than average among teiids (1.17 v. 0.89 on –2 to 2 scale), and sexual dichromatism is present in the majority of species (54% v. 35%). Seasonal color frequency is average (7.7% v. 9.6%). Overall, the sex differences rollup score is above average (1.62 v. 1.19 on 0 to 6 scale).

Ameiva spp.	English name				Range core
			Size		
	Activity	Diet	mm	g	Ecological business model
aggerecusans	Free-flowing Racerunner				Peru
			80	14.9	* wide-ranging chemosensor
ameiva	Giant Ameiva				Neotropics
	diurnal	omnivore	138	81	**** wide-ranging chemosensor
atrigularis	Venezuela Racerunner				Venezuela
	diurnal	omnivore	144	93	*** wide-ranging chemosensor
bifrontata	Red-snouted Ameiva				Venezuela
	diurnal	invert.	101	30.5	**** wide-ranging chemosensor
concolor	Crisnejas Racerunner				Peru
	diurnal		101	30.5	*** wide-ranging chemosensor
fuliginosa	Swan Island Racerunner				Colombia
			84	17.6	* wide-ranging chemosensor
jacuba	Emas Racerunner				Brazil
			85	18.0	** wide-ranging chemosensor
nodam	Free-flowing Ameiva				Peru
			82	15.8	** wide-ranging chemosensor
pantherina	Monagas Racerunner				Venezuela
	diurnal		118	49.7	** wide-ranging chemosensor
parecis	Rondônia Racerunner				Brazil
	diurnal		75	12.3	**** wide-ranging chemosensor
praesignis	Borriguerro Ameiva				Panama
	diurnal	invert.	163	137	*** wide-ranging chemosensor
provitaae	Red-nosed Racerunner				Venezuela
	diurnal	invert.	106	35.4	*** wide-ranging chemosensor
reticulata	Peru Racerunner				Peru
			116	46.9	* wide-ranging chemosensor

The genus is fairly evenly divided between omnivores (40% v. 33%) and invertebrate eaters (60% v. 62%). Home-range size is small, below even the lizard average (−0.37 v. 0.19 std. resid.). This belies the conventional wisdom about the large home ranges of teiids; nonetheless, the dogma that teiid home ranges are undefended seems to be born out. The home ranges of individual *A. ameiva* overlap those of five others, on average. The quantified movements of this species are surprisingly variable: 51st to 99th percentile for moves per minute (MPM); 65th to 88th percentile for proportion of time moving (PTM). Wariness is about average (8% v. 13%).

Direct basking is reported at a lower frequency that for the family (39% v. 58%). Mean active body temperature is also lower (35.0°C v. 36.9°C). Several species are noted to prefer disturbed habitats and to be heliophilic. *A. ameiva* may do better in urban areas with relatively few mongooses.

Burrows are less often noted as the daily refugium than in their family (50% v. 82%). *Ameiva* are found at the wetter end of the teiid range (intermediate 100% v. 50%; mesic 100% v. 30%). Their elevational profile is normal: (l/m/h) 100/33/8 v. 95/32/6. They are underrepresented in sandy areas (22% v. 47%) and riparian zones (11% v. 46%), but average in littoral (33% v. 38%) and human (33% v. 29%) habitats.

No *Ameiva* is parthenogenetic. Clutch count is low (−0.99 v. −0.51 std resid.), and young are smaller than average (ontogenetic mass range 0.845 v. 0.10 std. resid.). These are reflected in a RCM ratio that is the low even within teiids (0.124 v. 0.163). Incubation is longer than average (0.20 v. −0.38 std. resid.), possibly reflecting the low soil temperatures of their relatively mesic forested environment. Parental care (0.0% v. 3.2%) and communal oviposition (0.0% v. 2.6%) are absent.

Longevity is short, even for teiids (−1.06 v. −0.74 std. resid.), but population density may be very high (15.2 kg/ha v. 1.06 kg/ha), based on one island sample of *A. ameiva*. Endemism is average (54% v. 50%); insularity is average (31% v. 37%). Introductions are very frequent (23% v. 8% for >0 introductions; 7.7% v. 0.6% for >2 introductions). Only one-quarter of the species in the genus are imperiled (25% v. 41%), but all of those are listed on one of the IUCN's lists of the most endangered species (25% v. 12.1%). In other respects *Ameiva* are typical teiids.

Ameivula

Ameivula (Teiidae), formerly in *Ameiva*, is a moderate genus (10 spp.) of small-sized (8.7 g v. 22.1 g), space-defending (100% v. 45%), flight-dependent (100% v. 77%), invertebrate-eating (100% v. 62%) wide-ranging chemosensors (100% v. 97%). The genus is found primarily in Brazil, but also in Bolivia, Paraguay, and Argentina.

They are built slightly lighter than average teiids (shape index 0.60 v. 0.88 std. resid.). Their tails are exceptionally long (3.29 v. 2.30 × SVL), are unbreakable (0% v. 21.3% breakable), and are never blue (v. 15%). The tails of *jalapensis* (4.6 × SVL) and *mumbuca* (4.5 × SVL) are the longest relative

Ameivula spp.	English name				Range core
			Size		
	Activity	Diet	mm	g	Ecological business model
abalosi	Spotted Whiptail				Paraguay
	diurnal		64	7.6	*** wide-ranging chemosensor
cipoensis	Cipo Mountain Lizard				Brazil
	diurnal		62	6.8	*** wide-ranging chemosensor
confusioniba	Confusion Mountains Lizard				Brazil
	diurnal		67	8.4	**** wide-ranging chemosensor
jalapensis	Jalapao Lizard				Brazil
	diurnal		51	3.60	**** wide-ranging chemosensor
mumbuca	Tocantins Lizard				Brazil
	diurnal		52	3.80	**** wide-ranging chemosensor
nativo	Espírito Santo Lizard				Brazil
	diurnal		57	5.2	**** wide-ranging chemosensor
nigrigula	Blue Portholed Whiptail				Brazil
			73	11.0	*** wide-ranging chemosensor
ocellifera	Spix's Ocellated Whiptail				Bolivia
	diurnal	invert.	149	103	**** wide-ranging chemosensor
pyrrhogularis	Orange-throated Lizard				Brazil
			73	11.2	*** wide-ranging chemosensor
xacriaba	Peruacu Lizard				Brazil
	diurnal		61	6.3	*** wide-ranging chemosensor

lengths in lizards. Secretory pores are absent (v. 9.0%), as is color change (v. 1.3% for any change).

SSD is average (1.00 v. 0.89 on −2 to 2 scale), as is sexual dichromatism (30% v. 35%) and seasonal color (10% v. 9.6%). Overall, the sex differences rollup score is average (1.11 v. 1.19 on 0 to 6 scale).

Direct basking is common (70% v. 58%). Locomotor activity is frequent and protracted in *ocellifera*, about 90th percentile for MPM and PTM. Documented rainfall bands are intermediate (50% v. 50%) and mesic (50% v. 30%). All documented elevations are low (100% v. 95%). They are found in sandy areas (33% v. 47%) but not littoral (v. 38%), riparian (v. 46%), or human (v. 29%) habitats.

A. nativo is parthenogenetic. Clutch count for the genus is about the family average (−0.63 v. −0.51 std. resid.). Parental care (v. 3.2%), adherent eggs (v. 0.6%), and communal oviposition (v. 2.6%) are not reported. RCM ratio is average (0.164 v. 0.163).

Endemism is high (80% v. 50%), perhaps in part because Brazil is so large. No species is found on islands (v. 37%) or has been introduced (v. 8.3%). The only species evaluated for conservation status (*nativo*) is considered imperiled (v. 41%) and is endangered according to the IUCN. In other respects *Ameivula* are typical teiids.

Amphibolurus

Amphibolurus is small genus (three spp.) of large (29.5 g v. 15.6 g) agamids found in arid Australia (where agamids are known as dragons). All are refugia-anchored diurnal ground lizards (v. 43%). Though slightly more robust than average agamids (shape index 1.09 v. 0.80 std. resid.), they are morphologically notable for their long, unbreakable tails (2.0–2.8 v. 1.86 × SVL). Their crests are subtle (67% v. 38%); secretory pores are absent (v. 13%). Two of three species darken (67% v. 38%) and show sexual differences in color (67% v. 53%), but none has advanced color changing ability (v. 15%) or seasonal coloration (v. 20%). Males tend to be smaller than females (−0.25 v. 0.53 on −2 to 2 scale), despite the species' large body size. Two of three are omnivorous (67% v. 23%), though all (100% v. 88%) are said to be sit-and-wait foragers. That cannot be true when they are eating plants. All bask directly. As befits their hot environment, they sport an agamid-typical high mean body temperature (34.0°C v. 34.8°C). Social interactions are poorly known. They are overrepresented in littoral (67% v. 10%) habitats but are average in sandy areas (33% v. 26%) and riparian zones (33% v. 21%), and they are absent from human habitat (v. 16%). Mean clutch count is slightly low for an agamid (−0.20 v. 0.20 std. resid.), and multiple clutches per year are expected, as least in favorable years. RCM is notably low (0.220 v. 0.336), though based on just *muricatus*. Longevity is about average (−0.58 v. −0.46 std. resid.). None has been introduced (v. 1.7%), and one of two evaluated species is considered imperiled (50% v. 37%).

Amphiglossus

Amphiglossus is a major radiation (25 spp.) of slightly large (4.7 g v. 3.3 g), salamander-like or *Sphenomorphus*-like skinks in Madagascar. Although their filled cells (21% v. 25%) are not strikingly low, they are known primarily from their morphology

Amphibolurus spp.	English name		Size		Range core
	Activity	Diet	mm	g	Ecological business model
centralis	Centralian Lashtail Dragon				Australia
	diurnal	invert.	84	19.8	** refugia-anchored diurnal gr. liz.
muricatus	Jacky Lizard				Eastern Australia
	diurnal	omnivore	103	36.3	*** refugia-anchored diurnal gr. liz.
norrisi	Mallee Tree Dragon				Southern Australia
	diurnal	omnivore	102	35.9	** refugia-anchored diurnal gr. liz.

Amphiglossus spp.	English name		Size		Range core
	Activity	Diet	mm	g	Ecological business model
alluaudi	Brygoo's Burrowing Skink				Madagascar
			70	3.70	
andranovahensis	Andranovaho Skink				Madagascar
	nocturnal		37	0.61	
anosyensis	Red-vented Skink				Madagascar
	diurnal		56	2.00	*** litter cryptozoid
ardouini	Yellow Skink				Madagascar
			118	15.1	

Amphiglossus spp.	English name	Activity	Diet	Size mm	Size g	Range core	Ecological business model
astrolabi	Diving Skink	cathemeral		171	42.2	Madagascar	** refugia-anchored diurnal gr. liz.
crenni	Banded Burrowing Skink			127	9.8	Madagascar	
decaryi	Fort Dauphin Rock Skink			40	0.79	Madagascar	
elongatus	Elongate Burrowing Skink			81	5.4	Madagascar	
frontoparietalis	Boulenger's Madagascar Skink			63	2.68	Madagascar	
gastrostictus	O'Shaughnessy's Madagascar Skink			85	6.2	Madagascar	
johannae	Anjouan Island Skink			82	5.7	Comoros	* subterranean forager
macrocercus	Black-striped Skink			88	6.8	Madagascar	
mandady	Masaola Creeping Skink			54	1.75	Madagascar	
mandokava	Manongarivo Skink			132	20.6	Madagascar	* litter cryptozoid
melanurus	Toamasina Spotted Skink			52	1.57	Madagascar	
meva	Splendid Bicolor Skink			117	14.8	Madagascar	
ornaticeps	Gray Skink	diurnal		53	1.72	Madagascar	* refugia-anchored diurnal gr. liz.
punctatus	Eastern Madagascar Skink	diurnal		60	2.35	Madagascar	
reticulatus	Reticulate Skink	cathemeral		166	3.9.3	Madagascar	* nocturnal ground lizard
spilostichus	Tsararano Forest Skink			68	3.40	Madagascar	
splendidus	Splendid Skink			98	9.1	Madagascar	* litter cryptozoid
stylus	Masaola Rainforest Skink			56	1.29	Madagascar	** subterranean forager
tanysoma	Berara Forest Skink	nocturnal		83	3.44	Madagascar	

rather than from their ecology. Note the many blank cells in the species listing below, and keep in mind that many of the tabulated states are based on contradicted reports. The majority of species cannot be characterized in terms of diet, diel pattern, or ecological business model. Litter cryptozoid (38% v. 28%) is the most often documented ecological business model, followed by refugia-anchored diurnal ground lizard (25% v. 30%), subterranean forager (25% v. 12%), and nocturnal ground lizard (13% v. 2%). Nearly all are elongate (96% v. 35%). One-quarter of species are semiaquatic (25% v. 2%); none is found arboreally (v. 30%). *A. astrolabi* and *reticulatus* bear some ecological resemblance to the North American Mudpuppy (*Necturus*). All documented species are found in mesic (1101–3000 mm/y) rainfall areas.

Amphiglossus are relatively average in build (shape index –0.63 v. –0.46 std. resid.), tail length (1.22 v. 1.33 × SVL), and tail breakability (84% v. 73%), though *crenni* tails are somewhat prehensile (4% v. 0.7%). The only sex difference known is relative m:f tail length ratio (1.10 v. 1.03).

If the small number of nominal diel activity patterns are representative, *Amphiglossus* are overrepresented in cathemeral (29% v. 9%) and nocturnal species (29% v. 14.5%), and they are underrepresented in diurnal ones (43% v. 74%). A normal frequency of *Amphiglossus* are conspicuously wary (12% v. 11%). A few bask directly (8% v. 21%). They are underrepresented in littoral (4% v. 21%), sandy (4% v. 25%), and human (0% v. 14%) habitats, and they are overrepresented in riparian ones (33% v. 23%).

Reproduction is about evenly split between oviparous (60% v. 64%) and viviparous species (40%). The one duration-documented oviparous spe-

Amphiglossus spp.	English name		Size		Range core
	Activity	Diet	mm	g	Ecological business model
tsaratananensis	Tsaratanana Skink				Madagascar
			69	3.55	
valhallae	Valhalla Skink				Comoros
			85	12.8	

Amphisbaena spp.	English name		Size		Range core
	Activity	Diet	mm	g	Ecological business model
absaberi	Mato Grosso Worm Lizard				Brazil
			182	2.57	*** subterranean forager
acrobeles	Lance-snouted Worm Lizard				Brazil
		invert.	250	10	**** subterranean forager
alba	Red Worm Lizard				Venezuela
	cathemeral	carnivore	520	173	**** subterranean forager
albocingulata	Paraguay Worm Lizard				Paraguay
			108	0.32	** subterranean forager
amazonica	Amazon Worm Lizard				Brazil
			388	53	**** subterranean forager
anaemariae	Ana Marie's Worm Lizard				Brazil
			137	0.82	*** subterranean forager
angustifrons	South America Worm Lizard				Bolivia
		invert	256	10.0	**** subterranean forager
anomala	Barbour's Worm Lizard				Brazil
			204	4.07	*** subterranean forager
arda	Vanilla Worm Lizard				Brazil
			213	4.80	*** subterranean forager
arenaria	Bahia Sanddune Worm Lizard				Brazil
			149	1.14	*** subterranean forager
bagual	Bagual Worm Lizard				Argentina
			220	5.5	
bahiana	Bahia Worm Lizard				Brazil
			163	1.65	*** subterranean forager
bakeri	Baker's Worm Lizard				Puerto Rico
			195	3.38	**** subterranean forager
barbouri	Cuban Many-ringed Worm Lizard				Cuba
			136	0.79	**** subterranean forager
bassleri	Bassler's Worm Lizard				Peru
	cathemeral	invert.	406	64	**** subterranean forager

cies (*reticulatus*) retains its embryos so long (incubation extends development only another 14 days) that one can easily imagine a transition to viviparity. Clutch/litter count is average (0.02 v. 0.06 std. resid.).

Endemism is at typical Madagascar levels (96% v. 64%); the nonendemics are found nearby in the Comoros (two countries are present in the Comoros). All are island species (v. 57%). None has been introduced (v. 2.1%). Relatively few are considered imperiled (26% v. 47%), though one suspects ignorance about status as a contributing factor to the low frequency of imperilment. One species (*decaryi*) is listed by the IUCN on one of its lists (in this case endangered) of the most endangered (4.3% v. 14%). In other documented respects *Amphiglossus* are typical skinks.

Amphisbaena

Amphisbaena is the largest amphisbaenid genus (109 spp.) and is found throughout South America (minus Chile, plus Panama) and the Greater Antilles (plus islands of the Puerto Rican Bank). If you wish to study this genus, a proficiency in Portuguese is warranted; a remarkable 68% of species occur in Brazil. The full range of Amphisbaenidae sizes (0.12–272 g) are present in the genus *Amphisbaena*, with the paltry mean of 2.89 g (v. 2.73 g). Their shape index is correspondingly average (–3.94 v. –3.97 std. resid.). Tail breaks are slightly more frequent than in the family (23% v. 15%), possibly due to better documentation, as the tail length ratio is not longer (0.10 v. 0.11 × SVL). Most of the tails that break do so only reluctantly.

One large (64 g) species (*bassleri*) is known to use rectilinear motion when locomoting on the surface, a straight-line progression normally associated with only the largest constrictor snakes.

Because the genus *Amphisbaena* has some of the largest amphisbaenids, it has the family's only carnivores (5% v. 2.6%). Many *Amphisbaena* consume insects associated with leaf litter; it is unclear whether the lizards burrow up into the leaf litter to consume the insects, or the insects burrow down into the soil where they are consumed. *Amphisbaena* is underrepresented in diurnal species (14% v. 40%) and correspondingly overrepresented in cathemeral (29% v. 20%) and nocturnal (57% v. 40%) species, but the sampling is unconvincing. Most *Amphisbaena* are found in mesic sites (67% v. 29%). See the Amphisbaenidae account for evidence of short incubation in *A. darwinii*. In all other respects *Amphisbaena* are very similar to the other Amphisbaenidae.

Amphisbaena spp.	English name		Size		Range core
	Activity	Diet	mm	g	Ecological business model
bedai	Mato Grosso do Sul Worm Lizard				Brazil
			216	5.1	*** subterranean forager
bilabialata	Two-lipped Worm Lizard				Brazil
			229	6.4	*** subterranean forager
bolivica	False Crooked Worm Lizard				Paraguay
	nocturnal		415	70	**** subterranean forager
borelli	Chaco Forest Worm Lizard				Argentina
			147	1.10	**** subterranean forager
brasiliana	Brazil Worm Lizard				Brazil
			236	7.2	*** subterranean forager
brevis	Brazil Short Worm Lizard				Brazil
			107	0.31	* subterranean forager
caeca	Blind Worm Lizard				Puerto Rico
		invert.	217	5.2	**** subterranean forager
caiari	Rondônia Worm Lizard				Brazil
			122	0.51	* subterranean forager
camura	Crooked Worm Lizard				Paraguay
			391	55	**** subterranean forager
carlgansi	Cuban Pink Worm Lizard				Cuba
			129	0.65	**** subterranean forager
carli	Carl's Worm Lizard				Brazil
			201	3.85	* subterranean forager
carvalhoi	Carvalho's Worm Lizard				Brazil
			10	0.25	*** subterranean forager
caudalis	Cayemite Long-tailed Worm Lizard				Haiti
		invert.	201	3.85	**** subterranean forager
cayemite	Cayemite Short-tailed Worm Lizard				Haiti
		invert.	166	1.79	**** subterranean forager
cegei	Pampa Grande Worm Lizard				Bolivia
		invert.	173	2.09	*** subterranean forager
cerradensis	Cerrado Worm Lizard				Brazil
			284	15.3	*** subterranean forager
crisae	Mato Grosso Forest Worm Lizard				Brazil
			116	0.42	*** subterranean forager
cubana	Cuban Worm Lizard				Cuba
		invert.	156	1.40	**** subterranean forager
cuiabana	Cuiaba Worm Lizard				Brazil
			235	7.1	*** subterranean forager
cunhai	Cunha's Worm Lizard				Brazil
			402	61	*** subterranean forager

Amphisbaena spp.	English name	Activity	Diet	Size mm	Size g	Range core	Ecological business model
darwinii	Darwin's Ringed Worm Lizard		invert.	266	11.8	Uruguay	**** subterranean forager
dubia	Uncertain Worm Lizard	nocturnal		222	5.7	Argentina	*** subterranean forager
fenestrata	Cope's Worm Lizard		invert.	182	2.57	Virgin Islands	*** subterranean forager
filiformis	Tocantins Worm Lizard			175	2.19	Brazil	*** subterranean forager
frontalis	Alagoado Worm Lizard			212	4.74	Brazil	*** subterranean forager
fuliginosa	Black and White Worm Lizard	diurnal	invert.	423	76	Venezuela	**** subterranean forager
gonavensis	Gonave Worm Lizard		invert.	159	1.51	Hispaniola	**** subterranean forager
gracilis	Slender Worm Lizard			131	0.68	Venezuela	*** subterranean forager
hastata	Ibiraba Worm Lizard			118	0.45	Brazil	**** subterranean forager
heathi	Heath's Worm Lizard			98	0.21	Brazil	*** subterranean forager
heterozonata	False Darwin's Worm Lizard		invert.	209	4.49	Bolivia	**** subterranean forager
hiata	False Uncertain Worm Lizard			155	1.35	Argentina	*** subterranean forager
hogei	Hoge's Worm Lizard			102	0.25	Brazil	*** subterranean forager
hyporissor	Barahona Worm Lizard	nocturnal	invert.	231	6.6	Hispaniola	**** subterranean forager
ibijara	Soil King Worm Lizard			184	2.66	Brazil	*** subterranean forager
ignatiana	Holy Hint Worm Lizard			144	0.99	Brazil	*** subterranean forager
infraorbitale	Berthold's Worm Lizard			453	99	Brazil	*** subterranean forager
innocens	Innocent Worm Lizard		invert.	201	3.80	Hispaniola	**** subterranean forager
kingii	King's Worm Lizard		invert.	189	2.96	Uruguay	**** subterranean forager
kisteumacheri	Manga Worm Lizard			264	11.3	Brazil	*** subterranean forager

Amphisbaena spp.	English name				Range core
			Size		
	Activity	Diet	mm	g	Ecological business model
kraoh	Kraoh Worm Lizard				Brazil
			227	6.2	*** subterranean forager
leali	Pestel Worm Lizard				Haiti
			191	3.09	**** subterranean forager
leeseri	Urucum Worm Lizard				Brazil
			121	0.50	*** subterranean forager
leucocephala	Whiteheaded Worm Lizard				Uruguay
			259	10.6	*** subterranean forager
littoralis	Seaside Worm Lizard				Brazil
			226	6.1	* subterranean forager
lumbricalis	Earthy Worm Lizard				Brazil
			121	0.50	*** subterranean forager
manni	Mann's Worm Lizard				Hispaniola
	nocturnal		161	1.56	**** subterranean forager
maranhensis	Maranhão Worm Lizard				Brazil
			99	0.22	* subterranean forager
maximus	Four-hundred Worm Lizard				Brazil
			323	25.7	* subterranean forager
medemi	Medem's Worm Lizard				Colombia
			141	0.92	*** subterranean forager
mensae	Mesa Mountains Worm Lizard				Brazil
			102	0.25	*** subterranean forager
mertensii	Merten's Worm Lizard				Paraguay
			324	25.9	*** subterranean forager
metallurga	Minas Gerais Worm Lizard				Brazil
			147	1.10	* subterranean forager
microcephalum	Small-headed Worm Lizard				Argentina
		invert.	389	54	**** subterranean forager
miringoera	Tapirapes Worm Lizard				Brazil
			111	0.35	*** subterranean forager
mitchelli	Mitchell's Worm Lizard				Brazil
			127	0.61	*** subterranean forager
munoai	Munoa Worm Lizard				Brazil
		invert.	132	0.71	**** subterranean forager
myersi	Myers' Worm Lizard				Suriname
			114	0.99	*** subterranean forager
neglecta	Neglected Worm Lizard				Brazil
			117	0.43	*** subterranean forager
nigricauda	Black-tailed Worm Lizard				Brazil
			129	0.65	*** subterranean forager

Amphisbaena spp.	English name				Range core
			Size		
	Activity	Diet	mm	g	Ecological business model
occidentalis	Western Worm Lizard				Peru
			196	3.48	*** subterranean forager
octostegum	Duméril's Worm Lizard				Brazil
			249	9.0	*** subterranean forager
pericensis	Perico Worm Lizard				Peru
			119	0.46	*** subterranean forager
persephone	Queen Worm Lizard				Brazil
			124	0.56	*** subterranean forager
plumbea	Lead-colored Worm Lizard				Argentina
			211	4.68	**** subterranean forager
polygrammica	Werner's Worm Lizard				Peru
			171	2.02	*** subterranean forager
polystegum	Northeastern Worm Lizard				Brazil
			345	33.2	*** subterranean forager
pretrei	Rio Grande Worm Lizard				Brazil
			339	31.1	*** subterranean forager
prunicolor	Plum-colored Worm Lizard				Paraguay
			179	2.42	**** subterranean forager
ridleyi	Ridley's Worm Lizard				Brazil
			190	3.05	*** subterranean forager
roberti	Robert's Worm Lizard				Paraguay
			212	4.70	*** subterranean forager
rozei	Rose's Worm Lizard				Venezuela
			149	1.16	*** subterranean forager
sanctaeritae	Santa Rita Worm Lizard				Brazil
			127	0.61	*** subterranean forager
saxosa	Flagged Worm Lizard				Brazil
			277	13.8	*** subterranean forager
schmidti	Schmidt's Worm Lizard				Puerto Rico
			181	2.49	**** subterranean forager
scutigerum	Shielded Worm Lizard				Brazil
			366	42.3	*** subterranean forager
silvestrii	Silvestri's Worm Lizard				Brazil
			125	0.57	* subterranean forager
slateri	Slater's Worm Lizard				Peru
			109	0.32	*** subterranean forager
slevini	Slevin's Worm Lizard				Brazil
			95	0.19	**** subterranean forager
spurelli	Spurrell's Worm Lizard				Colombia
			238	7.5	**** subterranean forager

Amphisbaena spp.	English name		Size		Range core
	Activity	Diet	mm	g	Ecological business model
steindachneri	Steindachner's Worm Lizard				Bolivia
			183	2.62	*** subterranean forager
stejnegeri	Stejneger's Worm Lizard				Guyana
			199	3.64	**** subterranean forager
supernumeraria	Catimbau Worm Lizard				Brazil
			149	1.14	* subterranean forager
talisiae	Pitomba Mountains Worm Lizard				Brazil
			84	0.12	*** subterranean forager
townsendi	Townsend's Worm Lizard				Peru
			189	3.00	*** subterranean forager
trachura	Montenegro Worm Lizard				Paraguay
			266	11.8	**** subterranean forager
tragorrhectes	Oriximina Worm Lizard				Brazil
			88	0.14	*** subterranean forager
uroxena	Anchor-tailed Worm Lizard				Brazil
			126	0.58	*** subterranean forager
vanzolinii	Vanzolini's Worm Lizard				Guiana Shield
		invert.	119	0.46	**** subterranean forager
varia	Northwestern Worm Lizard				Colombia
			388	54	**** subterranean forager
vermicularis	Wagler's Worm Lizard				Brazil
			241	8.0	*** subterranean forager
wiedi	South American Worm Lizard				Brazil
			382	50	**** subterranean forager
wuchereri	Wucherer's Worm Lizard				Brazil
			314	22.9	*** subterranean forager
xera	Dry Worm Lizard				Puerto Rico
			184	2.70	**** subterranean forager

Amphisbaenidae

The 178 amphisbaenids are all limbless subterranean foragers (100% v. 7.8%), for which very little natural history information is available (filled cells 17% v. 27%). Most of the laboratory observations have been focused on how they move through the soil. Some ram directly forward, some slide a wedge underneath and push up and down, and some ram a bladelike snout forward and force it side to side. These different modes of progression are reflected in their snout shape and often feature in their common name. Field observations and other facts are almost entirely lacking. Most amphisbaenians are in the genus *Amphisbaena* (109 of 178); the other genera in the family are *Ancylocranium*, *Baikia*, *Chirindia*, *Cynisca*, *Dalophia*, *Geocalamus*, *Loveridgea*, *Mesobaena*, *Monopeltis*, and *Zygaspis*. Amphisbaenids are strikingly restricted in their distributions. They are Neotropical and Ethiopian; none is found in the Australasian, Nearctic, Oriental, or Palearctic realms. This differs from the cosmopolitan distribution of the ecologically and morphologically similar worm snakes.

Amphisbaenids are typically small (2.7 g v. 5.3 g), but cover almost the full range of lizard sizes (range: 0.12–172 g or 1st to 98th percentile). Most have reduced eyes, though the documentation of this is inadequate to state a proportion with confidence. They stand out for their slender body form (shape index –3.97 v. 0.0 std. resid.), which is consistently extreme (–5.57 to –1.85 std. resid.). All eight of the most-slender families are composed primarily of subterranean foragers, which presumably need to be slender to facilitate pushing through soil. Amphisbaenids are also consistently extreme in the shortness of their tails (mean 0.11 v. 1.37 × SVL), with genus averages ranging 0.05 (*Ancylocranium*) to 0.23 (*Zygaspis*). Most tails are simply too short to break (14.5% v. 57%), but a small minority are breakable in *Amphisbaena* and *Zygaspis*. No amphisbaenid has a prehensile tail (v. 5%), crests (v. 5.3%), color change (v. 10.7%), sexual dichromatism (v. 21%), seasonal color (v. 6.5%), or blue tail (v. 2.3%). A small number in *Amphisbaena*, *Chirindia*, and *Loveridgea* have secretory pores (3.4% in amphisbaenids v. 14.9% in all lizards). No obvious commonality unifies the pore-bearing species.

Almost all amphisbaenids (97% v. 73%) eat invertebrates, but a few of the larger ones will take vertebrates (2.6% v. 3.2%). Despite the large number of underground ant colonies, no amphisbaenids are known to be ant specialists (v. 4.4%). Nearly twice as many are recorded to eat termites (59%) as to eat ants (31%).

In my opinion, the diel cycles of lizards are poorly known, and surely among the worst must be the documentation of subterranean foragers. Nonetheless, it is striking that cathemeral (20% v. 6.8%) and nocturnal (40% v. 25%) species are overrepresented, and diurnal species are underrepresented (40% v. 67%). Why would 40% of amphisbaenids find it advantageous to be active only at night? Are they not all in the dark? Do diurnal invertebrates that are active during the day above ground burrow in and become available to amphisbaenids at night? Are there diurnal subterranean predators? If so, why are the predators diurnal? To what depth in the soil is it warmer during the day? Do amphisbaenids forage often in such superficial depths? As John Wesley Powell remarked as he embarked on the first descent of the tumultuous Colorado River through the hitherto unexplored Grand Canyon, "We may conjecture many things."

No amphisbaenid is known to vocalize (v. 2.5%) or bask (v. 18%), and none climbs trees (cf. blind snakes), forages through litter (v. 12%), forages in water (v. 1.1%), or forages on the surface (v. 48%). None (v. 3%) is found in the hyperdry (<55 mm/y) or (v. 9%) wet (>3000 mm/y) climate zones. They are overrepresented in the arid zone (151–401 mm/y: 43% v. 30%) and at low elevation (98% v. 82%). Presumably subterranean foragers are unable to achieve a temperature other than that of the soil, which may limit them to low elevation. Amphisbaenids are slightly underrepresented in littoral areas (9% v. 17%) and around humans (8% v. 18.5%), but they are slightly overrepresented in sandy or dune areas (28% v. 21.6%). Their presence in riparian zones is average (20% v. 22%).

Many amphisbaenids appear to be able to switch reproductive mode (5.6%, the second-highest family, v. 1%); the selective forces rewarding switching in the underground are not obvious. Amphisbaenids do not differ much from typical lizards in oviparity (86% v. 82%), clutch count (−0.12 v. 0.0 std. resid.), or relative size of young (−0.04 v. 0.0 std. resid.). The latter two values are poorly documented and exhibit important differences among genera: *Loveridgea* and *Zygaspis* have notably large clutch counts, whereas *Ancylocranium*, *Dalophia*, and *Monopeltis* have conspicuously small clutches (none of these genera are documented for relative size of young). Eggs are always soft (0% v. 41% hard) and never adherent (0% v. 2.4% adherent). Communal oviposition (0.6% v. 4.2%) and parental care are rare (1.1% v. 2.9%). Based on one species (*Amphisbaena darwinii*), the incubation period is anomalously short (−0.75 v. 0.0 std. resid.). RCM has not been documented.

More data are needed to ascertain whether amphisbaenids in general are fast-lived (rapid development, short longevity). Documented longevities are extremely short (−3.02 std. resid.), but that condition may reflect herpetoculturists' lack of experience maintaining them in captivity.

One species (*Zygaspis violacea*) has a mean density estimate of 0.00069 kg/ha (from three samples), but it is impossible to know if those estimates are representative. The paper reporting those estimates implies that the site was chosen because it was of high density. No one who has searched for amphisbaenids will claim that they are numerous (though many are said to be "common"), at least not in the shallow soil that has been dug for them. Can they drive though the soil faster than a human can dig after them?

Few are on islands (12% v. 40%). None has been introduced (v. 3.3%), another difference from the ecologically similar worm snakes. Twenty-three percent are considered at conservation risk (v. 49%), likely biased low by insufficient information. A typical percentage (12.7% v. 13.6%) are on one of the IUCN's lists of the most endangered.

Anadia

Anadia is a 19-species genus of the Gymnophthalmidae centered in Colombia and extending from Costa Rica to Ecuador. The mix of ecological business models (see below) and high uncertainty make it difficult to draw any generalizations about their ecology. They are about evenly split between arboreal (60% v. 8%) and terrestrial (40% v. 30%) microhabitats. *Anadia* are comparatively large (5.6 g v. 2.4 g) and robustly built (shape index 0.06 v. −0.32 std. resid.). Tails are a little short (1.60 v. 1.72 × SVL), often breakable (21% v. 28%), and frequently prehensile (16% v. 2%), at least in the arboreal species. Elongation is present in about half the species (47% v. 41%), and secretory pores are present (11% v. 29%). One species (*bitaeniata*) is sexually dichromatic. This is the only gymnophthalmid genus that reports cruise foraging (*A. brevifrontalis*, a high altitude species, >3000 m). In addition to *brevifrontalis*, at least three other *Anadia* extend above 3000 m: *altaserrania*, *bitaeniata*, *bogotensis*. Active foraging is not mentioned (v. 96%); direct basking is (21% v. 8%). Male *hobarti* are reported to defend space (v. 13%). Reproduction is unremarkable, though clutches are small (−0.40 v. −0.11 std. resid.), and they are prone to communal oviposition (11% v. 3.4%). Thirty-five percent of species are considered imperiled (v. 25%), and 17.6% (v. 11.5%) are on the IUCN's lists of the most endangered. In other respects *Anadia* are typical gymnophthalmids.

Anadia spp.	English name				Range core
	Activity	Diet	Size mm	Size g	Ecological business model
altaserrania	High Montane Anadia				Colombia
			64	6.0	
antioquensis	Antioquia Anadia				Colombia
			73	9.4	
bitaeniata	Two-banded Anadia				Venezuela
	diurnal	invert.	72	5.5	** refugia-anchored diurnal gr. liz.
blakei	Blake's Anadia				Venezuela
			74	10.0	
bogotensis	Bogota Anadia				Colombia
			58	4.49	
brevifrontalis	Merida Anadia				Venezuela
	diurnal	invert.	74	5.9	
bumanguesa	Eastern Mountain Anadia				Colombia
	diurnal		75	6.2	** litter cryptozoid
escalerae	Escalera Anadia				Venezuela
			57	4.24	
hobarti	Hobart's Anadia				Venezuela
	diurnal	invert.	72	5.3	
marmorata	Spotted Anadia				Venezuela
			77	11.4	
mcdiarmidi	McDiarmid's Anadia				Venezuela
	diurnal		54	2.06	
ocellata	Bromeliad Anadia				Panama
	diurnal	invert.	63	3.48	** diurnal canopy walker
pamplonensis	Pamplona Anadia				Colombia
			70	8.3	
pariaensis	Paria Anadia				Venezuela
	diurnal		72	8.8	
petersi	Peters' Anadia				Ecuador
			62	5.6	
pulchella	Ruthven's Anadia				Colombia
			85	15.7	
rhombifera	Diamond Anadia				Ecuador
	diurnal		58	2.65	** ground-foraging tree lizard
steyeri	Steyer's Anadia				Venezuela
	diurnal		68	4.52	
vittata	Boulenger's Anadia				Panama
	diurnal	invert.	57	2.50	* ground-foraging tree lizard

Anatololacerta

Anatololacerta (four spp.) is a genus of moderate-sized (6.0 g v. 5.1 g) refugia-anchored diurnal ground lacertids found in Turkey and Greece. In morphology, they are typical lacertids, though two species exhibit blue tails. Sex differences are few and poorly documented (sex rollup 0.25 v. 0.86 on 0 to 6 scale). All three documented species reach moderate elevation (in this case up to 1700–2100 m). Two are also found in littoral zones (v. 22%) and three in riparian zones (75% v. 28%) and human habitat (75% v. 16%). Reproduction is typical, with slightly larger clutch counts (0.42 v. 0.13 std. resid.) and slightly shorter incubation (−0.62 v. −0.36 std. resid.). Young are slightly large, as indicated by low ontogenetic mass range (−0.26 v. −0.05 std. resid.). Longevity is documented only for *anatolica*, for which it is 10 years (0.76 v. 0.19 std. resid.). All species are found on islands (v. 26%). None has been introduced (v. 4.6%). All are considered imperiled (v. 56%), but none is classified among the most endangered by the IUCN (v. 15.5%). In other respects *Anatololacerta* are similar to other lacertids.

Anatololacerta spp.	English name				Range core
			Size		
	Activity	Diet	mm	g	Ecological business model
anatolica	North Anatolia Lizard				Turkey
	diurnal		63	5.85	** refugia-anchored diurnal gr. liz.
budaki	Budak's Lizard				Turkey
	diurnal		57	4.26	* refugia-anchored diurnal gr. liz
danfordi	Danford's Lizard				Greece
	diurnal	invert.	63	5.85	** refugia-anchored diurnal gr. liz.
pelasgiana	Rhodes Island Lizard				Aegean Islands
	diurnal	invert.	73	9.0	** refugia-anchored diurnal gr. liz.

Ancylocranium spp.	English name				Range core
			Size		
	Activity	Diet	mm	g	Ecological business model
barkeri	Lindi Sharp-snouted Worm Lizard				Tanzania
		invert.	184	2.70	*** subterranean forager
ionidesi	Kilwa Sharp-snouted Worm Lizard				Tanzania
		invert.	244	8.24	** subterranean forager
somalicum	Somali Sharp-snouted Worm Lizard				Ethiopia
			154	1.30	** subterranean forager

Anelytropsis sp.	English name				Range core
			Size		
	Activity	Diet	mm	g	Ecological business model
papillosus	Mexican Blind Lizard				Mexico
			277	25.8	**** subterranean forager

Anepischetosia sp.	English name				Range core
			Size		
	Activity	Diet	mm	g	Ecological business model
maccoyi	Highlands Forest-skink				New South Wales
		invert.	51	1.03	** litter cryptozoid

Ancylocranium

Ancylocranium is small genus (three spp.) of average-sized (3.07 g v. 2.73 g) amphisbaenids from East Africa (Tanzania, Ethiopia, Somalia). They are morphologically undistinguished amphisbaenids, except for their radically short tails (0.05 v. 0.11 × SVL). All are found at low elevations; one of two habitat-characterized species (*barkeri*) occurs in sand (50% v. 28%). Two species have known clutch counts of one, which is small even for an amphisbaenid (–0.65 v. –0.12 std. resid.). No species occurs on an island (0% v. 12%), and none is considered imperiled (v. 23%).

Anelytropsis

Anelytropsis papillosus is the sole member of this genus of dibamid lizards, endemic to Mexico; all other dibamids are found in Southeast Asia. It is known from only a few specimens. It is large (25.8 g v. 2.9 g), lacks pores (v. 8.7%), and has no recorded sex differences (rollup score 0 v. 0.30 on 0 to 6 scale). It is classified as threatened in Mexico. In other respects it is a typical dibamid.

Anepischetosia

Anepischetosia maccoyi is the sole representative of a genus of small (1.03 g v. 3.3 g), elongate (v. 35%), litter cryptozoid skinks found in temperate southeastern Australia, from New South Wales to Victoria. It is comparatively well known (filled cells 39% v. 25%), though apparently not often seen.

Its shape index is notably slender (–1.27 v. –0.46 std. resid.). Sex differences are not reported. It is a thermoconformer (v. 1.5%) with a remarkably low mean active body temperature (17.0°C v. 31.1°C). Clutch count is high (0.55 v. 0.06 std. resid.), and the hatchlings are small (ontogenetic mass range 0.56 v. 0.12 std. resid.). This thermoconforming oviparous lizard retains its eggs to an advanced stage of development and then deposits them in a fairly cold environment. The incubation duration is 64 days (0.34 v. –0.40 std. resid.), suggesting that either basking directly prior to oviposition or warmer nest site choice could substantially accelerate development. The eggs are laid communally (v. 2.6%). Thus, in a sense, the nest site choice is crowd-sourced. One wonders which attribute of the nest site is evolutionarily rewarded by this arrangement: temperature? moisture conservation? predation reduction?

The species is not endemic (v. 64%); it is not found on an island (v. 57%), nor has it been introduced (v. 2.1%). It has not been evaluated for conservation status.

Anguidae

The 78 anguids are a diverse group, distributed worldwide but missing from the Australasian and Ethiopian realms, and absent from the Andes, Amazon, and most of eastern Asia. Six of the genera (*Abronia*, *Barisia*, *Coloptychon*, *Elgaria*, *Gerrhonotus*, and *Mesaspis*) exhibit a typical quadrupedal build. Five genera are effectively limbless (*Anguis*, *Dopasia*, *Hyalosaurus*, *Ophisaurus*, and *Pseudopus*). Most have osteoderms in their skin, especially dorsally and ventrally, a lateral fold to facilitate expansion and contraction of the armored body cavity (i.e., breathing), and straight transverse rows of scales, giving them the gestalt of a small-headed crocodilian. Many limbed species are colloquially known as alligator lizards. The limbless lizards are often known as glass lizards for their readily broken tails. The largest genus of anguids is *Abronia* (29 spp.), which is arboreal; the other quadrupedal genera are more or less terrestrial.

Anguids are not ecologically uniform: 1.9% are crack hermits (v. 2.6%), a third (38% v. 6%) are diurnal canopy walkers, 6% (v. 12%) are litter cryptozoids, 2% (v. 8%) are nocturnal ground lizards, a third (34% v. 25%) are refugia-anchored diurnal ground lizards, 2% (v. 8%) are subterranean foragers, 4% (v. 13%) are trunk/wall lizards, and 13% (v. 7%) are wide-ranging chemosensors. This eclecticism in business models is mirrored in the wide range of conditions found in many of the following traits.

Anguids average twice the mass of average lizards (10.1 g v. 5.3 g; range 2.5–40 g), and are somewhat slender (shape index –0.60 v. 0.0 std. resid.). As a group, their natural history is of average information content (27% v. 27% cells filled). Tails run a little long (1.66 v. 1.37 × SVL), and the extreme values are both found in limbless species: 0.88 × SVL in *Anguis fragilis* to 2.73 × SVL in *Ophisaurus mimicus*. Despite autotomy (87% v. 57%), prehensile tails are common (31% v. 5%), found in *Abronia*, *Elgaria*, and *Gerrhonotus* and absent in all other genera. Crests (0% v. 5%), secretory pores (0% v. 15%), and color change (0% v. 11%) are absent. Elongation is highly variable (22% v. 11%), being absent in some genera (*Anguis*, *Barisia*, and the limbless forms), and universal in other genera (*Coloptychon*, *Elgaria*, *Hyalosaurus*).

Mean anguid SSD is similar to that of all lizards (0.23 v. 0.36 on –2 to 2 scale), but individual genera do not conform to this pattern. *Anguis* is strongly m<f (–1 on –2 to 2 scale), whereas *Gerrhonotus* and *Pseudopus* average 1 (m>f) on the same scale. All known *Barisia* are sexually dichromatic, whereas no *Coloptychon*, *Dopasia*, *Hyalosaurus*, or *Pseudopus* are (family mean 26% v. 21% lizard mean). Seasonal color 3.8% v. 6.5%) is found only in *Anguis* and *Barisia*. Blue tails are unknown (v. 2.3%).

Almost all anguids (93% v. 73%) eat invertebrates, but a few of the larger ones will take vertebrates (7.4% v. 3.2%). The cruise (29% v. 7%) and active (71% v. 58%) foraging modes are overrepresented. Sit-and-wait foraging is comparatively rare (14% v. 46%). Electric light use is not reported (v. 2.0%). Only one anguid has home-range information (*Ophisaurus attenuatus*), and it is large (2227 m²), even when size adjusted (0.90, the third-highest family mean, v. 0.0 std. resid.).

Cathemeral (16.7% v. 6.8%) and crepuscular (13% v. 1.8%) species are overrepresented, diurnality is about average (70% v. 67%), and nocturnality unknown (v. 25%). Anguids are not known to vocalize (v. 2.5%), and they are disproportionately conspicuously wary (18% v. 8%). Their choice of predator concealment and basking behavior is typical of lizards. On average they adopt a lower active body temperature (mean = 24.5°C, third lowest, v. 31.1°C), which may reflect the many species that occur at relatively high elevation (47% v. 82% at low; 77% v. 41% at medium; 41%, the second-highest value, v. 13% at high). Anguids are also absent from the dry end of the rainfall spectrum (none in hyperdry, dry, or arid). All recorded species are found in the mesic band (100%, tied for first place, v. 43%; 1101–3000 mm/y), and they are overrepresented in intermediate (401–1100 mm/y: 50%, tied for second place, v. 27%) and wet (>3000 mm/y: 25% v. 9%). Anguids like moisture and their preference for high elevation may be moisture-driven. One might think that anguids' armored skin evolved to reduce dermal moisture loss, but if so, their skin has not allowed them to disperse into dry areas. Perhaps their armored skin gives them a degree of protection from mold.

Anguids are underrepresented in littoral (9% v. 17%), sandy (15% v. 22%), and human (11% v. 18.5%) habitats. In regard to riparian habitats, anguids are unremarkable as a family (19% v. 22%), but this conceals much variation within: *Abronia*, *Barisia*, *Dopasia*, and *Mesaspis* are large enough to be representative, but none of those are reported to occur in riparian zones. Conversely, 60% of *Anguis* species are found in riparian zones, and all of the seven

Elgaria are. What ecological attribute joins those divisions?

Group living is poorly documented in anguids, but it is found in half of documented species (50% v. 20%), as is solitary living (50% v. 10%). Males are more often tolerant (54% v. 15%) and less likely to defend space (46% v. 85%). Male hierarchies are unknown in anguids (v. 8%). Microhabitat use is fairly typical of lizards, but anguids are slightly overrepresented in the arboreal (39% v. 32%) and burrow refuging (36% v. 22%) roles.

Anguids are strongly inclined toward viviparity (61% v. 18%), presumably in association with their high elevations and latitudes. As far as is known, all *Abronia*, *Anguis*, *Barisia*, and *Mesaspis* are entirely viviparous, whereas all *Dopasia*, *Ophisaurus*, and *Pseudopus* are oviparous. When size adjusted, anguid litter/clutch counts stand out as large (0.85 v. 0.0 std. resid.). Anguid young are comparatively small, as indicated by the wide ontogenetic range of sizes (0.62 v. 0.0 std. resid.). Anguid incubation periods are short (–0.80 v. 0.0 std. resid.), with no notable variation among genera (albeit no data for seven genera). In summary, anguid young are small, numerous, and develop quickly. No anguid eggs are hard or adherent (v. 2.4%), and only 1.3% (v. 4.2%) are deposited communally. Parental care is common (17% v. 3%). Parental care often takes the form of neonates staying with their mother and is therefore likely associated with viviparity. RCMs are average (0.268 v. 0.226).

Documented longevities are long (0.83 v. 0.0 std. resid.), especially in *Anguis* (2.51 std. resid.) and *Pseudopus* (2.99 std. resid.), but not in *Abronia* (–0.89 std. resid.) Only two limbless species have had their population densities estimated (three credible times each), and both means are exceptionally high: 36 kg/ha in *Anguis fragilis* and 2.4 kg/ha in *Ophisaurus attenuatus* (v. 0.54 kg/ha in lizards).

Anguids tend not to occur on islands (17% v. 40%). They have been introduced at an arguably elevated rate (3.8% v. 3.3%), but none has been introduced more than twice (0% v. 1.1%). They are overrepresented among the imperiled lizards (63% v. 49%), especially in the Central American cloud forest group, which is rapidly losing its habitat to expanded human populations. Climate change—in the form of climate zones ascending beyond their summits—looms as the anguid terminator.

This family represents a challenge for those confident that ecology follows from phylogeny or morphology. The distinctive osteoderms might seem to offer anguids some unique advantage. But the patchy geographic distribution, widely varying morphologies, and inconsistent sex differences, behaviors, reproduction, densities, and longevities do not come together in a coherent ecological pattern.

Anguis

Anguis is a moderate (six spp.) genus of moderate-sized (18.3 g v. 10.1 g) slender-bodied (shape index –1.32 v. –0.6 std. resid.) limbless anguids from Europe and Mexico. The five (of six) species that live in Europe are not morphologically distinguishable from each other and are all considered to be wide-ranging chemosensors, often under litter; the Mexican species is thought to be a subterranean forager, as it is known only from dead on road specimens. All members of the genus are particularly vulnerable to road mortality as they resemble snakes and move slowly, hence their common name.

Half of *Anguis* are reported to break tails (v. 87%), but underreporting of this condition is likely due to confusion over species identification. No tails are prehensile (v. 31%).

Female *Anguis* are larger than their male counterparts (–1.0 v. 0.23 on –2 to 2 scale). One of six species reports seasonal color (v. 3.8%).

All are invertebrate eaters (v. 93%). All are active foragers (v. 71%). All are active at all times of day (v. 17%). Two of six are said to be conspicu-

Anguis spp.	English name		Size		Range core
	Activity	Diet	mm	g	Ecological business model
cephallonica	Peloponnese Slow Worm				Greece
			195	25.2	** wide-ranging chemosensor
colchica	Colchican Slow Worm				Eastern Europe
	cathemeral	invert.	195	24.9	*** wide-ranging chemosensor
fragilis	Slow Worm				Europe
	cathemeral	invert.	157	17.0	**** wide-ranging chemosensor
graeca	Greek Slow Worm				Greece
	cathemeral	invert.	159	14.0	*** wide-ranging chemosensor
incomptus	Plain-necked Glass Lizard				Mexico
			174	18.1	*** subterranean forager
veronensis	Italian Slow Worm				Italy
	cathemeral	invert.	159	14.0	*** wide-ranging chemosensor

ously wary (v. 33%); but underreporting may distort this apparent exact match. Only pregnant females bask (v. 15%). In general, they have a low body temperature, as do most anguids (23.8°C v. 24.5°C).

All species live in groups (v. 50%). All males defend space (v. 46%); defense may be for the breeding season only. Half of the species are considered to be cryptozoic (v. 14%).

Half of species are present in intermediate (401–1100 mm/y) and mesic (1101–3000 mm/y) bands, versus 40% and 80%, respectively. All species are found at low elevation (v. 47%). *Anguis* are often found in littoral (40% v. 9%), sandy (60% v. 15%), and human (60% v. 11%) habitats. *Anguis* are notably overrepresented in riparian zones (60% v. 19%); perhaps such areas have favorably dense herbaceous vegetation.

All are viviparous (v. 61%), and they have litters that are large, but typical of anguids (0.79 v. 0.85 std. resid.); parental care has not been recorded, though slow worms are notoriously hard to observe. Their young are a tiny fraction of maximum size (1/289 v. 1/72), as reflected in the mass-adjusted standard residuals (2.19 v. 0.62). *A. fragilis* is reported to copulate for 10 hours, perhaps another meaning of "slow worm"? RCM is above average (0.309 v. 0.268).

Longevity is very high (2.51 v. 0.83 std. resid.), which equates to 54 years, reported for multiple species, but this may reflect taxonomic confusion rather than identical longevity records. This implies low mortality prior to automobiles, or selection for great iteroparity, usually associated with infrequent opportunities for successful recruitment.

Population density, based on three estimates, averages a staggering 36 kg/ha. This may be biased by the tendency of *A. fragilis* to live in dense colonies, presumably associated with sandy soil that is easy to excavate. Nonetheless, it indicates that this species can do very well where suitable habitat is found. Endemism is low (33% v. 65%). Island species are high (33% v. 17%). *Anguis* has not been introduced (v. 3.8%). In spite of localized high densities, all *Anguis* are considered to be at conservation risk (v. 63%), though none is on one of the IUCN's lists of the most endangered (v. 28%). Possibly some species are listed due to similarity of appearance; they cannot be distinguished morphologically. Regardless, specialized habitat requirements are a significant liability in a region as fragmented and developed as Europe.

Anisolepis spp.	English name				Range core
			Size		
	Activity	Diet	mm	g	Ecological business model
grilli	Boulenger's Tree Lizard				Argentina
	diurnal		80	14.1	** diurnal canopy walker
longicauda	Argentine Longtail Iguana				Paraguay
			79	13.9	* diurnal canopy walker
undulatus	Wiegmann's Tree Lizard				Uruguay
	diurnal	invert.	72	10.1	

Anisolepis

Anisolepis is a three-species genus of moderately large (12.6 g v. 2.35 g), cryptic (100% v. 100%), diurnal (100% v. 86%), arboreal (100% v. 45%) leiosaurids found in southern South America (Paraguay to Uruguay). Two species are thought to be diurnal canopy walkers (100% v. 16%), though *grilli* is known for protracted immobility and may be an unrecognized immobility expert. They are relatively heavily built (shape index 0.67 v. 0.76 std. resid.) and have exceptionally long tails (2.79 v. 1.66 × SVL) that are not breakable (0% v. 3%) but may be prehensile (33% v. 9%). Sex differences are found in body size (–1.0 v. –1.0 on –2 to 2 scale) and relative tail length (1.16 v. 1.20 m:f ratio) but not color (0% v. 15%).

Although one species is considered an invertebrate eater, its diet has not been quantified, and there is no information on the genus's foraging mode or home-range size. *A. grilli* is found in urban areas (human habitat 33% v. 8%).

Reproduction is oviparous (100% v. 95%) but with a smaller clutch count (0.16 v. 0.68 std. resid.) than the family. *Anisolepis* has the lengthy incubation duration characteristic of the family (2.99 v. 2.81 std. resid.). One of two evaluated species is considered imperiled (50% v. 14%). In other respects *Anisolepis* are probably typical leiosaurids.

Anniellidae and *Anniella* (fig. 4.13B)

Anniella is the only genus in the amphisbaenian family Anniellidae. It is a moderate (six spp.) genus of small-sized (3.6 g v. 5.3 g; in this account, all unspecified reference values are the lizard average) limbless anniellids from California and Mexico. Morphologically, the family and genus stand out for having relatively long tails compared to other amphisbaenians (0.51 v. 0.11 × SVL); all anniellid tails are breakable (100% v. 15% for amphisbaenians). Anniellidae are also thicker than Amphisbaenidae

Anniella spp.	English name				Range core
			Size		
	Activity	Diet	mm	g	Ecological business model
alexanderae	Temblor Legless Lizard				California
			123	3.97	*** subterranean forager
campi	Southern Sierra Legless Lizard				California
	nocturnal		118	3.61	*** subterranean forager
geronimensis	Baja California Legless Lizard				Mexico
		invert.	114	3.40	**** subterranean forager
grinnelli	Bakersfield Legless Lizard				California
			115	3.42	*** subterranean forager
pulchra	Northern California Legless Lizard				Mexico
	crepuscular	invert.	135	4.94	*** subterranean forager
stebbinsi	Southern California Legless Lizard				Mexico
			104	2.69	*** subterranean forager

(–2.23 v. –3.97 shape index). Similar to Amphisbaenidae, no tails are prehensile (v. 5% in lizards). Anniellids are not reported to have crests (v. 5%), secretory pores (v. 15%), change color (v. 11%), SSD (though m<f possible), sexual dichromatism (v. 21%), seasonal color (v. 6.5%), or blue tails (v. 2.3%). All are invertebrate eaters (v. 73% in lizards, 97% in amphisbaenians). All anniellids are active foragers (v. 58% in lizards). One species each is known to be active at night (50% v. 25%) or in low light (50% v. 1.8%). One of six species is said to be conspicuously wary (17% v. 8%) but others may not be documented. The documented mean body temperature is very low (19.5°C v. 30.8°C).

As far as is known, all males are tolerant (v. 15%). All species are found in the dry zone (56–150 mm/y: 100% v. 12%). Similar to lizards (82%), most live at low elevation (80%), and about half are found at medium elevation (40% v. 41%). Half of *Anniella* live in littoral zones (v. 17%), and all are associated with sand or dunes (v. 22%). None is associated with human habitats (v. 18.5%).

All are viviparous (v. 18%). Viviparity would seem to offer few advantages to subterranean foragers with such a low mean active body temperature, but presumably development in cold soil would not be an improvement. Perhaps pregnant females bask indirectly by moving into shallow, sun-heated soil. Litters are small (–0.44 v. 0.0 std. resid.). Their ontogenetic size range is small, indicating relatively large young (–0.26 v. 0.0 std. resid.). Parental care has not been recorded (v. 3%; underreporting plausible). Mass-adjusted longevity is surprisingly high (1.02 v. 0.0 std. resid.), though based on only one species (*pulchra*: 11 years). There is one population density estimate (*pulchra*: 1 kg/ha v. 0.54 kg/ha), which is moderate, but perhaps is large for a subterranean forager (normative values unavailable). No *Anniella* has been introduced (v. 3.3%) and two-thirds (67% v. 49%) are considered imperiled. Given their restrictive ecological requirements—*geronimensis* requires coastal dunes; *campi* is found only around desert springs—habitat fragmentation and alteration are likely to be problematic.

Anolis (figs. 4.3A, 4.6D)

Anolis is the largest genus (416 spp., see table 2.2) and is found in the Neotropics and southern North America (introduced elsewhere). *Anolis* is nearly twice the size of the next-largest genus (258 spp. *Liolaemus*). This sticks in the craw of many taxonomists, who have repeatedly endeavored to break it up, only to have their suggested divisions founder on the shoals of genetically confirmed paraphyly. At least some nontaxonomists are not so sure that division is necessary, as the general gestalt of anoles is so decisive that the grouping feels natural, and the single name *Anolis* is familiar and entrenched. One is tempted to think that even anoles feel this way, as interspecific territoriality is uniquely rife within the genus; anoles apparently recognize other anoles as competitors for space/food, even if most can easily distinguish potential mates on the basis of distinctive dewlap colors. At present, *Anolis* is the sole member of the Dactyloidae. In this entry, the focus is on the ecological distinctions within the genus *Anolis*; the Dactyloidae entry focuses on the ways in which anoles differ from other lizard families.

Although anoles occur from the eastern United States to the middle of South America, the countries bordering the Caribbean have the highest *Anolis* diversity. Selected species counts:

Islands (North to South)	
Cuba	64 spp.
Hispaniola (Dom. Rep. and Haiti)	54
Puerto Rico	13
Mainland Countries (North to South)	
Mexico	58
Guatemala	21
Honduras	38
Costa Rica	39
Colombia	75
Venezuela	24

A number of ecologists have noted that anoles in the islands seem to subdivide microhabitats more finely and differently than they do on the mainland (Irschick et al. 1997; Losos 2009). There is a respectable sample of each condition: about half of *Anolis* species (196 v. 220 of 416) are found on islands. Are there gross differences between the island and mainland anoles? It should be kept in mind that our knowledge base is much better for island anoles (37% v. 23% filled cells). For example, the diets are known for 55% (107/196) of island anoles, but only 18% (40/220) of mainland anoles.

Although island anoles average 12% heavier than mainland anoles (table 5.1), the primary difference between the distributions is in variance (twofold) and kurtosis (threefold). Island anoles have a wider and more skewed range of sizes. The 65 g weight of the heaviest island anole species (*noblei*) substantially outweighs the corresponding mainland species (39 g *insignis*). There are more ecological business models represented on islands (6 v. 4), though that might be an artifact of better island sampling. More striking is the nearly sixfold difference in frequency of refugia-anchored diurnal ground lizards (mainland anoles often hide in terrestrial refugia). Conversely, trunk/wall lizards are almost fivefold more frequent on islands. Is bird predation so heavy as to render tree-trunk existence untenable on the mainland? Does a lizard need the security of a nearby ground refuge on the mainland?

Ingram et al. (2016) found that mainland anoles average larger dewlaps than island anoles. Macrini et al. (2003) found that mainland anoles have smaller toepads (in keeping with their reduced presence on tree trunks).

Mainland anoles are almost exclusively generalized invertebrate eaters (table 5.1). Island anoles are

Table 5.1. Selected differences between island and mainland anole species

Trait	Island	Mainland	*n* (island/main.)	Units
Mass, mean	4.11	3.66	196/220	grams
Variance	0.21	0.11	196/220	$\log_{10}$(mass)
Kurtosis	0.64	0.22	196/220	$\log_{10}$(mass)
Skewness	0.99	0.94	196/220	$\log_{10}$(mass)
Maximum	65	39	196/220	grams
Ecological business models	6	4	182/87	count
Diurnal canopy walker	18.7	26.4	182/87	% of spp.
Ground-foraging tree lizard	39.0	47.1	182/87	% of spp.
Refugia-anchored diurnal ground lizard	3.3	18.4	182/87	% of spp.
Trunk/wall lizards	37.9	8.0	182/87	% of spp.
Diet				
Invertebrate eaters	63	93	107/40	% of spp.
Ant specialists	16	2.5	107/40	% of spp.
Omnivores	21.5	2.5	107/40	% of spp.
Any plants eaten	58.2	6.5	91/31	% of spp.
Sexual size dimorphism	1.22	0.43	150/55	–2 to 2 scale
Electric light use	10.2	0.5	196/220	% of spp.
Home-range area				
Unadjusted	7.0	26.4	25/8	m^2
Mass-adjusted	–0.85	–0.20	25/8	std. resid.
Population density	3.8	0.22	39/9	kg/ha
Introduced species	14.3	0	196/220	% of spp.

Anolis spp.	English name				Range core
			Size		
	Activity	Diet	mm	g	Ecological business model
acutus	St. Croix Anole				US Virgin Isl.
	diurnal	invert.	53	3.23	**** ground-foraging tree lizard
aeneus	Grenadian Bush Anole				Grenada
	diurnal	ants	62	5.2	**** trunk/wall lizard (male) or diurnal canopy walker (female)
aequatorialis	Equatorial Anole				Ecuador
	diurnal	omnivore	88	14.4	** diurnal canopy walker
agassizi	Malpelo Anole				Colombia
	diurnal	invert.	91	15.9	** immobility expert
agueroi	Cabo Cruz Bearded Anole				Cuba
	diurnal	omnivore	122	38.1	* trunk/wall lizard
ahli	Escambray Blue-eyed Anole				Cuba
	diurnal		51	2.93	*** ground-foraging tree lizard

Anolis spp.	English name		Size		Range core
	Activity	Diet	mm	g	Ecological business model
alayoni	Guantanamo Twig Anole				Cuba
	diurnal		41	1.56	** diurnal canopy walker
alfaroi	Small-fanned Anole				Cuba
	diurnal		35	0.98	*** diurnal canopy walker (low)
aliniger	La Vega Anole				Hispaniola
	diurnal		54	3.38	**** trunk/wall lizard (male) and ground-foraging tree lizard (female)
allisoni	Allison's Anole				Cuba
	diurnal	ants	71	7.7	**** trunk/wall lizard
allogus	Spanish Flag Anole				Cuba
	diurnal	invert.	51	2.95	*** ground-foraging tree lizard
alocomyos	San Gerardo Anole				Costa Rica
	diurnal		48	2.45	* refugia-anchored diurnal gr. liz.
altae	High Anole				Costa Rica
	diurnal	invert.	47	2.24	*** ground-foraging tree lizard
altavelensis	Alto Velo Anole				Dominican Republic
	diurnal		43	1.77	* ground-foraging tree lizard
altitudinalis	Turquino Green-mottled Anole				Cuba
	diurnal		47	2.24	* trunk/wall lizard (male) or ground-foraging tree lizard (female)
alumina	Shiny Anole				Hispaniola
	diurnal		38	1.23	** ground-foraging tree lizard
alutaceus	Blue-eyed Grass-bush Anole				Cuba
	diurnal	invert.	39	1.28	** ground-foraging tree anole
alvarezdeltoroi	Alvarez del Toro's Anole				Mexico
			65	5.8	* refugia-anchored diurnal gr. liz.
amplisquamosus	Cusuco Big-scaled Anole				Honduras
	diurnal		45	1.99	*** ground-foraging tree lizard
anatoloros	Barinas Anole				Venezuela
	diurnal	invert.	58	4.26	
anchicayae	Lime-flanked Anole				Colombia
			54	3.54	
anfiloquioi	Oriente Anole				Cuba
	diurnal		37	1.13	* ground-foraging tree lizard
angusticeps	Cuban Twig Anole				Cuba
	diurnal	invert.	51	2.80	*** trunk/wall lizard
anisolepis	Chiapas Ornate Anole				Mexico
	diurnal		51	2.96	

much more diverse, with particularly notable increases in the percentage specializing on ants (about sixfold) and omnivores (over eightfold). The difference is equally striking in the frequency of anoles that will sometimes consume plants, a ninefold difference between islands and the mainland. Although one does not think of anoles as plant eaters, the frequency of lizard species that eat plants is 26%, but 45% of anoles eat plants. This is not attributable to size, even though herbivory is strongly associated with large size (table 3.1, traits 9, 11), and the average anole is smaller than the average lizard. However, the frequency of plant eating among island lizards is 29% and 58% among island anoles. Thus plant-eating anoles are overrepresented even within the ranks of island lizards, whereas mainland anoles show the converse pattern (6.5% v. 23%). There appears to be a strong interaction between insularity and plant eating in *Anolis*. Potentially this dietary diversity is driven by greater population pressure, as the mean island anole population density is more than 15-fold that on the mainland (table 5.1). Greater population pressure might also influence the observed expression of much greater m>f SSD on islands (island anoles need sharper metaphoric elbows) and the more compressed home ranges in the islands.

The greater use of electric lights for foraging by island anoles (table 5.1) is no doubt partly attributable to a higher density of electric lights in the islands, but surely there are enough electric lights on the mainland for anoles there to exploit them if they so choose. The 20-fold difference in frequency is unlikely to be completely attributable to greater opportunities in the islands.

Opportunity is likely to play some role in the greater frequency of introductions among island anoles (table 5.1). The appearance on an island of a new anole is easier to recognize than would be a human-transport assisted movement on the mainland. Nonetheless, there would appear to be some difference between island and mainland anoles that would account for the dramatic difference in introduction rates. Is there more emigration propagule pressure due to increased population density in the islands? Are island species more attractive to the pet trade because island anoles are easier to collect? Does the greater apparent selection pressure for SSD in island anoles also generate colors, behaviors, and morphology that are attractive to pet owners? Does the social assertiveness of island anoles bring these species more often into contact with human cargoes and accidental stowing-away events? Much remains to be learned about the natural history differences between island and mainland anoles. In general, however, it appears that from the perspective of an *Anolis*, living in the islands is not getting away from it all, but being subjected to more competition, more social conflicts, and more food stress. They may benefit from reduced predation (Andrews 1976, 1979)

No less than four species of anoles were named in awe of the size of the males' hemipenes (*apletophallus*, *bellipeniculus*, *macrophallus*, *magnaphallus*). Notably, all four are found on the mainland. Also notable, no lizard in any other genus is so consecrated. This suggests that hemipenis size may be important to at least mainland anoles (or that *Anolis* taxonomists have a "unique" perspective). Perhaps copulation opportunities in the mainland are sufficiently rare that the males are selected to have a strong copulatory lock and use the extended copulation time to maximize the number of sperm intro-

Anolis spp.	English name				Range core
			Size		
	Activity	Diet	mm	g	Ecological business model
annectens	Annex Anole				Venezuela
			65	6.0	
anoriensis	Anori Anole				Colombia
			72	8.1	
antioquiae	Antioquia Anole				Colombia
			64	5.8	
antonii	Anton Anole				Colombia
			49	2.55	
apletolepis	Furcy Anole				Hispaniola
			52	3.02	** diurnal canopy walker
apletophallus	Very Well-hung Anole				Panama
			43	1.77	
apollinaris	Boulenger's Anole				Colombia
			89	15.2	
aquaticus	Water Anole				Costa Rica
	diurnal	invert.	64	5.8	**** refugia-anchored diurnal gr. liz.
argenteolus	Guantanamo Anole				Cuba
	diurnal	ants	53	3.18	**** trunk/wall lizard
argillaceus	Bay Anole				Cuba
	diurnal		41	1.51	*** trunk/wall lizard
armouri	Armour's Anole				Hispaniola
	diurnal	invert.	62	5.3	** ground-foraging tree lizard
auratus	Grass Anole				Venezuela
	diurnal	invert.	48	2.37	*** refugia-anchored diurnal gr. liz.
bahorucoensis	Bahoruco Long-snouted Anole				Hispaniola
	diurnal	ants	45	2.00	**** ground-foraging tree lizard
baleatus	Dominican Giant Anole				Dominican Republic
	diurnal		144	62	*** trunk/wall lizard
baracoae	Baracoa Anole				Cuba
	diurnal	carnivore	131	46.6	*** trunk/wall lizard
barahonae	Barahona Anole				Hispaniola
	diurnal	omnivore	132	48.1	**** trunk/wall lizard
barbatus	Western Bearded Anole				Cuba
	diurnal	omnivore	125	40.7	*** diurnal canopy walker or trunk/wall lizard
barbouri	Hispaniola Hopping Anole				Hispaniola
	diurnal	invert.	49	2.55	* litter cryptozoid
barkeri	Barker's Anole				Mexico
	diurnal	invert.	82	11.6	** refugia-anchored diurnal gr. liz.

Anolis spp.	English name				Range core
			Size		
	Activity	Diet	mm	g	Ecological business model
bartschi	West Cuban Anole				Cuba
	diurnal	ants	64	5.7	**** trunk/wall lizard
beckeri	Becker's Short-legged Anole				Northern Central America
	diurnal	invert.	58	4.26	* diurnal canopy walker
bellipeniculus	Beautifully Hung Anole				Venezuela
			59	4.58	
benedikti	Amistad Anole				Costa Rica
	diurnal		44	1.91	
bicaorum	Utila Anole				Honduras
	diurnal		65	6.0	** trunk/wall lizard
bimaculatus	Two-spotted Tree Anole				Lesser Antilles
	diurnal	invert.	99	20.3	*** trunk/wall lizard
binotatus	Two-marked Anole				Colombia
			42	1.69	
biporcatus	Giant Green Anole				Central America
	diurnal	invert.	89	14.8	*** trunk/wall lizard
birama	Branch Anole				Cuba
	diurnal		53	3.23	* ground-foraging tree lizard
bitectus	Roof Anole				Ecuador
			50	2.78	
blanquillanus	Hummelinck's Anole				Venezuela
	diurnal	invert.	72	7.9	
boettgeri	Boettger's Anole				Peru
			59	4.42	
bombiceps	Blue-lipped Forest Anole				Western Amazon
	diurnal	invert.	65	6.1	* refugia-anchored diurnal gr. liz.
bonairensis	Bonaire Anole				Bonaire
	diurnal	ants	66	7.9	*** trunk/wall lizard
boulengerianus	Tehuantepec Anole				Mexico
			57	3.83	
brasiliensis	Brazil Anole				Brazil
	diurnal		63	5.5	* diurnal canopy walker
bremeri	Herradura Anole				Cuba
	diurnal		56	3.90	* ground-foraging tree lizard
breslini	Northwest Anole				Haiti
	diurnal		52	3.14	* ground-foraging tree lizard
brevirostris	Desert Gracile Anole				Hispaniola
	diurnal	invert.	46	2.13	** trunk/wall lizard
brunneus	Crooked Island Anole				Bahamas
	diurnal	invert.	64	5.6	** diurnal canopy walker

mitted, thereby winning the sperm competition race; at higher population densities on the islands, males may have more short (before they are interrupted) copulation opportunities or may wish to distribute their sperm among the many females with overlapping home ranges.

Anole copulation times (not systematically tabulated in this work) are highly variable among species. *A. chrysolepis*, a mainland species, is said to copulate for three hours and females to store sperm over multiple years. This would seem to diminish the scope for intermale competition leading to male sequestration of females, and *chrysolepis* is one (of three—all mainland) of the very rare anoles in which males are smaller than females and sexually dimorphic structures are not reported. Only 10 anoles have unequivocally smaller males than females; 9 of 10 are mainland species.

SSD has an interesting consequence in *bimaculatus*, for which the adult males average about 2.5 times the mass of the females, and the sexes are rarely seen together. Some have suggested that female *bimaculatus* may fear the carnivorous tendencies of the much larger males. This concern surfaces in other lizard species with extraordinary SSD; see *Bradypodion* entry for an example of males keeping their distance from the much larger females.

Anoles sometimes approximate chamaeleonid foraging patterns. Twelve anoles have prehensile tails. Slow creeping movements are particularly associated with diurnal canopy walking anoles. Like some chameleons, several anole species (e.g., *barbatus*, *porcus*) rock back and forth (up to 20 times in *barbatus*) while slowly taking a single step. This distorts quantification of PTM. *Anolis chamaeleonides* uses its long tongue to capture insects in a manner redolent of chameleons. A terrestrial anole (*agassizi*) exhibits social facilitation in movements: when one moves others nearby follow suit. Conversely, some anoles hardly move at all; *A. capito* is one of the least active lizards yet quantified (<1st percentile for both MPM and PTM); it also has a conspicuously diminutive dewlap, which it rarely uses.

Most anoles stalk moving insects. One semiaquatic species, *aquaticus*, is a rarity in that it can recognize nonmoving objects as potential prey. Other semiaquatic anoles (*barkeri*, *eugenegrahami*, *lionotus*, *maculigula*, *oxylophus*, *purpuronectes*) would be good candidates for determining if this skill is an adaptation to foraging in water. One nonaquatic anole (*bimaculatus*) will capture fish in captivity, suggesting substantial adaptability in that small island species.

Although head bobbing and dewlapping are signature activities of anoles, *porcatus* is said to lack an assertion display. *A. cristatellus*, on the other hand, is reported to continue its assertion display after dark on brightly moonlit nights.

When there is a difference in perch height between male and female anoles, it is almost always the case that males perch higher, with many females and juveniles relegated to ground level. In the trunk/wall lizard *A. distichus*, the heights are notably reversed, with juveniles having the highest perches, under which are females, with males nearest the ground. *A. distichus* is an ant specialist, and it is possible that ant prey are more often replenished from the ground level. This Hispaniolan anole would be a worthy object for foraging observations targeting this inversion. The Panamanian anole *pentaprion* may be capable of "parachuting" (gliding at more than a 45% glide angle), another observation begging for confirmation. *A. pulchellus* is

Anolis spp.	English name	Range core	Activity	Diet	Size mm	Size g	Ecological business model
calimae	Ayala's Anole	Colombia			52	3.02	
campbelli	Campbell's Anole	Guatemala			46	2.14	
capito	Pug-nosed Anole	Central America	diurnal	invert.	77	9.9	**** ground-foraging tree anole
caquetae	Caqueta Anole	Colombia			51	2.90	
carlliebi	Lieb's Anole	Mexico			42	1.70	
carlostoddi	Todd's Anole	Venezuela			49	2.55	
carolinensis	Green Anole	Southeastern United States	diurnal	invert.	66	6.2	**** trunk/wall lizard
carpenteri	Carpenter's Anole	Costa Rica	diurnal	invert.	41	1.48	* ground-foraging tree lizard
casildae	Casilda's Anole	Panama	diurnal		94	17.6	
caudalis	Cochran's Large-headed Anole	Haiti	diurnal	invert.	47	2.35	*** trunk/wall lizard
centralis	Central Anole	Cuba	diurnal	invert.	41	1.49	*** trunk/wall lizard
chamaeleonides	Short-bearded Anole	Cuba	diurnal	invert.	137	54	**** diurnal canopy walker
charlesmyersi	Myer's Anole	Costa Rica			65	6.0	
chloris	Boulenger's Green Anole	Colombia	diurnal		52	3.12	
chlorocyanus	Dominican Republic Blue-green Anole	Hispaniola	diurnal	invert.	67	6.4	**** trunk/wall lizard
chlorodius	Jacmel Anole	Hispaniola	diurnal		68	6.8	
christophei	King Christophe Anole	Hispaniola	diurnal	invert.	46	2.10	**** trunk/wall lizard
chrysolepis	Goldenscale Anole	Guyana Shield	diurnal	invert.	65	6.1	* ground-foraging tree lizard
chrysops	Guadeloupe Anole	Guadeloupe	diurnal	invert.	62	5.1	*** ground-foraging tree lizard
clivicola	Mountain Anole	Cuba	diurnal		43	1.82	*** ground-foraging tree lizard

Anolis spp.	English name				Range core
			Size		
	Activity	Diet	mm	g	Ecological business model
cobanensis	Stuart's Anole				Mexico
	diurnal		45	2.04	
coelestinus	Jeremie Anole				Hispaniola
	diurnal	invert.	68	6.9	** trunk/wall lizard or diurnal canopy walker
compressicauda	Malposo Scaly Anole				Mexico
			49	2.55	* ground-foraging tree lizard
concolor	Isla San Andres Anole				Colombia
	diurnal		62	5.3	** trunk/wall lizard
confusus	Cabo Cruz Trunk Anole				Cuba
	diurnal		47	2.26	** ground-foraging tree lizard
conspersus	Grand Cayman Blue-fanned Anole				Cayman Islands
	diurnal	omnivore	58	4.26	*** trunk/wall lizard
cooki	Cook's Pallid Anole				Puerto Rico
	diurnal		58	4.26	**** ground-foraging tree lizard or trunk/wall lizard if displaced
crassulus	Ornate Anole				Honduras
	diurnal		48	2.48	* trunk/wall lizard
cristatellus	Puerto Rico Crested Anole				Puerto Rico
	diurnal	omnivore	62	5.2	*** ground-foraging tree lizard
cristifer	Cristifer Anole				Mexico
	diurnal		72	8.1	
cryptolimifrons	Pseudo Slender Anole				Costa Rica
			42	1.60	
cupeyalensis	Cupeyal Anole				Cuba
	diurnal		32	0.71	*** refugia-anchored diurnal gr. liz.
cupreus	Copper Anole				Nicaragua
	diurnal	invert.	47	2.24	*** ground-foraging tree lizard
cuprinus	Chiapas Anole				Mexico
	diurnal		58	4.27	
cuscoensis	Cusco Anole				Peru
			54	3.54	
cusuco	Cusuco Anole				Honduras
	diurnal		42	1.69	** ground-foraging tree lizard
cuvieri	Puerto Rico Giant Anole				Puerto Rico
	diurnal	invert.	130	46	*** trunk/wall lizard
cyanopleurus	Yateras Anole				Cuba
	diurnal		37	1.17	*** ground-foraging tree lizard
cyanostictus	Santo Domingo Green Anole				Dominican Republic
			59	4.58	

reported to be capable of walking on water.

Many anoles sleep on leaves at the ends of branches, but the full range of sleep site options is practiced by some anoles. For some reason *christophei* sleeps perpendicular to stems rather than along them. *A. lineatopus* males may sleep outside of their territories, and they may sleep next to males that they would fight with if encountered during the day. Both sexes of *mccraniei* sleep in the leaf litter. *A. smaragdinus* males have overlapping territories, despite their willingness to fight overlapping males once they are in view.

Anoles exhibit the full range of social relations in their mating systems. *A. agassizi* males have two morphs, one of which is small and resembles females and is not recognized as male or repelled by full-sized males. Several species (*limifrons*, *pogus*, *lineatopus*) are thought to be monogamous, with a suggestion that *limifrons* may mate for life. *A. pogus* are unique in that their monogamy occurs not in the usual low-density situation, but in a very dense island environment (mean density of 1696 individuals per hectare).

Oviposition sites are normally solitary in anoles, but one *angusticeps* nest site was discovered with 1625 eggshells present. Female *aeneus* retain fully shelled eggs until the beginning of the rainy season, presumably to ensure better development conditions for the embryos. *A. sagrei* eggs hatch preferentially at dawn, as may be true for many diurnal lizards.

It is often asserted that island anoles routinely achieve densities of greater than 1 per m^2, but this requires one to be very selective about which square meter one chooses to measure. Credible average population density estimates exceed 1 per m^2 in only three species: *trinitatus*, 15,000/ha or 45 kg/ha; *poncensis*,

14,000/ha or 8 kg/ha; *bahorucoensis*, 12,000/ha or 18 kg/ha.

Anoles are generally innocuous and are associated with few phobias. Some species are loath to release their bite, leading a few daring fashionistas to wear live anoles as earlobe jewelry. In Jamaica, however, there are reports of *valencienni* being considered venomous.

Anolis spp.	English name	Activity	Diet	Size mm	Size g	Range core	Ecological business model
cybotes	Large-headed Anole	diurnal	ants	62	5.01	Hispaniola	**** ground-foraging tree lizard
cymbops	Cope's Veracruz Anole			42	1.69	Mexico	
damulus	Cope's Smooth Anole			47	2.27	Mexico	
danieli	Gonzalez's Anole			99	20.3	Colombia	
darlingtoni	Darlington's Anole	diurnal		66	6.3	Haiti	
datzorum	Datzs' Anole			44	1.95	Costa Rica	
delafuentei	Guamuhaya Anole	diurnal		56	3.92	Cuba	* ground-foraging tree lizard
deltae	Delta Anole			51	2.90	Venezuela	
demissus	Grand Cayemite Green Anole	diurnal		64	5.6	Haiti	* diurnal canopy walker
desechensis	Heatwole's Anole	diurnal	omnivore	50	2.78	Puerto Rico	** ground-foraging tree lizard
desiradei	Desirade Anole	diurnal	invert.	67	6.4	Guadeloupe	*** ground-foraging tree lizard
dissimilis	Odd Anole			49	2.67	Peru	
distichus	Bark Anole	diurnal	ants	37	1.10	Hispaniola	**** trunk/wall lizard
divius	Sky Blue Anole			58	4.26	Dominican Republic	** ground-foraging tree lizard
dolichocephalus	Place Negre Anole	diurnal		47	2.24	Haiti	* ground-foraging tree lizard
dollfusianus	Coffee Anole	diurnal		39	1.35	Mexico	
duellmani	Duellman's Pygmy Anole			38	1.27	Mexico	
dunni	Dunn's Anole	diurnal		51	3.00	Mexico	
eladioi	Eladio's Anole	diurnal		54	3.54	Dominican Republic	** trunk/wall lizard
elcopeensis	El Cope Anole			42	1.60	Panama	

Anolis spp.	English name				Range core
			Size		
	Activity	Diet	mm	g	Ecological business model
equestris	Knight Anole				Cuba
	diurnal	omnivore	142	60	**** trunk/wall lizard
ernestwilliamsi	Carrot Rock Anole				British Virgin Islands
	diurnal	invert.	68	6.8	*** ground-foraging tree lizard
etheridgei	Etheridge's Anole				Dominican Republic
	diurnal	invert.	40	1.45	* diurnal canopy walker (low)
eugenegrahami	Eugene's Anole				Haiti
	diurnal		61	4.91	** refugia-anchored diurnal gr. liz.
eulaemus	Good Anole				Colombia
	diurnal		81	11.3	* diurnal canopy walker
euskalerriari	Perija Anole				Venezuela
			47	2.34	
evermanni	Emerald Anole				Puerto Rico
	diurnal	omnivore	59	4.48	**** trunk/wall lizard
extremus	Barbados Anole				Barbados
	diurnal	invert.	61	4.97	**** trunk/wall lizard or ground-foraging tree lizard (esp. female)
fairchildi	Cay Sal Anole				Bahamas
	diurnal		64	5.6	** trunk/wall lizard
fasciatus	Banded Anole				Ecuador
			61	4.91	
ferreus	Comb Anole				Guadeloupe
	diurnal	invert.	96	18.7	*** trunk/wall lizard
festae	Veronica's Anole				Ecuador
	diurnal		53	3.27	
fitchi	Fitch's Anole				Colombia
			78	10.2	
forresti	Barbuda Anole				Antigua and Barbados
			47	2.24	**** ground-foraging tree lizard (male), refugia-anchored diurnal gr. liz. (female)
fortunensis	Fortuna Forest Anole				Panama
	diurnal		44	1.95	
fowleri	Fowler's Anole				Dominican Republic
	diurnal		64	5.8	* diurnal canopy walker
fraseri	Hippie Anole				Colombia
	diurnal		101	21.7	*** trunk/wall lizard
frenatus	Bridled Anole				Panama
	diurnal	invert.	118	34.1	**** ground-foraging tree lizard

Anolis spp.	English name	Activity	Diet	Size mm	Size g	Range core	Ecological business model
fugitivus	Moa Anole	diurnal		34	0.87	Cuba	*** refugia-anchored diurnal gr. liz.
fungosus	Moldy Anole	diurnal		43	1.77	Costa Rica	
fuscoauratus	Brown-eared Anole	diurnal	invert.	45	1.99	Venezuela	**** ground-foraging tree lizard
gadovii	Gadow's Anole	diurnal		67	6.4	Mexico	* refugia-anchored diurnal gr. liz.
gaigei	Gaige's Anole			85	13.0	Colombia	
garmani	Jamaican Giant Anole	diurnal	omnivore	101	21.6	Jamaica	**** trunk/wall lizard
garridoi	Escambray Twig Anole	diurnal	invert.	39	1.31	Cuba	**** diurnal canopy walker
gemmosus	Gem Anole	diurnal	invert.	58	4.30	Colombia	**** diurnal canopy walker
gibbiceps	Hook Anole			44	1.95	Venezuela	
ginaelisae	Mount Sagui Anole	diurnal		89	15.2	Panama	
gingivinus	Anguilla Bank Anole	diurnal	ants	57	4.01	Lesser Antilles	** ground-foraging tree lizard and refugia-anchored diurnal gr. liz.
gonavensis	Gonave Green Anole			40	1.41	Haiti	* ground-foraging tree lizard
gorgonae	Gorgona Anole			57	4.11	Colombia	
gracilipes	Charm Anole	diurnal	invert.	49	2.55	Colombia	**** diurnal canopy walker
grahami	Graham's Anole	diurnal	ants	60	4.62	Jamaica	**** trunk/wall lizard or ground-foraging tree lizard (esp. females)
granuliceps	Granular Anole			44	1.95	Ecuador	
griseus	Saint Vincent Tree Anole	diurnal	omnivore	121	37.0	Saint Vincent	**** trunk/wall lizard
gruuo	Tabasara Anole			47	2.24	Panama	

Anolis spp.	English name				Range core
			Size		
	Activity	Diet	mm	g	Ecological business model
guafe	Cabo Cruz Banded Anole				Cuba
	diurnal		42	1.70	** ground-foraging tree lizard
guamuhaya	Escambray Bearded Anole				Cuba
	diurnal		141	58	* trunk/wall lizard
guazuma	Sierra Anole				Cuba
	diurnal		42	1.70	**** diurnal canopy walker
gundlachi	Yellow-chinned Anole				Puerto Rico
	diurnal	omnivore	60	4.78	*** ground-foraging tree lizard
haetianus	Tiburon Stout Anole				Haiti
	diurnal	invert.	63	5.4	** ground-foraging tree lizard
haguei	Hague's Anole				Guatemala
			46	2.14	
hendersoni	Henderson's Anole				Haiti
	diurnal		45	1.98	*** refugia-anchored diurnal gr. liz. or ground-foraging tree lizard
heterodermus	Flat Andes Anole				Colombia
	diurnal	invert.	74	8.9	**** diurnal canopy walker (low)
hobartsmithi	Hobart's Anole				Mexico
	diurnal		44	1.95	
homolechis	Cuban White-fanned Anole				Cuba`
	diurnal	ants	53	3.24	**** ground-foraging tree lizard
huilae	Huila Anole				Colombia
			68	6.8	
humilis	Ground Anole				Costa Rica
	diurnal	invert.	41	1.55	**** ground-foraging tree lizard
ibague	Ibague Anole				Colombia
			47	2.30	
ibanezi	Ibanez's Anole				Costa Rica
			67	6.6	* diurnal canopy walker
imias	Imias Anole				Cuba
	crepuscular	carnivore	55	3.65	*** trunk/wall lizard
immaculogularis	White-throated Anole				Mexico
			45	1.99	
impetigosus	Pustulous Anole				Venezuela
			44	1.95	
incredulus	Turquino Emerald Anole				Cuba
	diurnal		34	0.87	* trunk/wall lizard

Anolis spp.	English name				Range core
			Size		
	Activity	Diet	mm	g	Ecological business model
inderenae	Cundinamarca Anole				Colombia
			94	17.5	
inexpectatus	Pineland Bush Anole				Cuba
	diurnal		34	0.90	** diurnal canopy walker (low)
insignis	Decorated Anole				Costa Rica
	diurnal		124	39.5	**** diurnal canopy walker
insolitus	La Palma Anole				Dominican Republic
	diurnal		42	1.67	*** diurnal canopy walker
intermedius	Intermediate Anole				Costa Rica
	diurnal		47	2.24	*** diurnal canopy walker
isolepis	Jatibonico Anole				Cuba
	diurnal		37	1.14	*** diurnal canopy walker
jacare	Jacare Anole				Colombia
	diurnal		63	5.4	** diurnal canopy walker
johnmeyeri	Merendon Blue-spotted Anole				Honduras
	diurnal		62	5.2	* diurnal canopy walker
juangundlachi	Yellow-lipped Grass Anole				Cuba
	diurnal		33	0.83	*** diurnal canopy walker (low)
jubar	Cubitas Anole				Cuba
	diurnal		51	2.89	*** ground-foraging tree lizard
kahouannensis	Kahouanne Anole				Guadeloupe
	diurnal	invert.	64	5.6	*** ground-foraging tree lizard
kemptoni	Kempton's Anole				Costa Rica
	diurnal		50	2.80	**** ground-foraging tree lizard
koopmani	Brown Red-bellied Anole				Haiti
	diurnal		39	1.37	*** ground-foraging tree lizard
kreutzi	Fortuna Anole				Honduras
	diurnal		46	2.14	
krugi	Olive Bush Anole				Puerto Rico
	diurnal	invert.	46	2.09	*** diurnal canopy walker (low)
kunayalae	Kunayala Giant Anole				Panama
			87	14.2	
laevis	Smooth Anole				Peru
			52	3.14	
laeviventris	White Anole				Central America
			57	4.03	
lamari	Lamar's Anole				Colombia
			40	1.43	

Anolis spp.	English name	Activity	Diet	Size: mm	Size: g	Range core	Ecological business model
landestoyi	Charco Azul Anole			106	25.0	Dominican Republic	
latifrons	Giant Tropical Anole	diurnal		103	23.0	Colombia	
leachii	Panther Anole	diurnal	omnivore	98	19.9	Antigua and Barbuda	*** trunk/wall lizard or ground-foraging tree lizard (esp. females)
leditzigorum	Tilaran Ground Anole	diurnal		47	2.34	Costa Rica	** refugia-anchored diurnal gr. liz.
lemniscatus	Ribboned Anole			47	2.24	Ecuador	
lemurinus	Ghost Anole	diurnal	invert.	62	5.3	Central America	**** trunk/wall lizard
leucodera	Saint Nicholas Green Anole	diurnal		64	5.8	Haiti	** trunk/wall lizard
limifrons	Slender Anole	diurnal	invert.	43	1.36	Central America	**** ground-foraging tree lizard
limon	Antioquia Dark-eyed Anole	diurnal		77	9.9	Colombia	
lineatopus	Stripe-footed Anole	diurnal	invert.	56	3.78	Jamaica	**** ground-foraging tree lizard
lineatus	Striped Anole	diurnal	invert.	70	7.5	Curacao	**** diurnal canopy walker
liogaster	Guerrero Anole		carnivore	57	3.96	Mexico	** ground-foraging tree lizard
lionotus	Lion Anole	diurnal	invert.	68	6.8	Panama	*** refugia-anchored diurnal gr. liz.
litoralis	Oriente Pallid Anole	diurnal	invert.	40	1.48	Cuba	*** diurnal canopy walker
lividus	Plymouth Anole	diurnal	invert.	59	4.53	Montserrat	*** ground-foraging tree lizard
longiceps	Navassa Anole	diurnal	invert.	69	7.0	Navassa Island	**** trunk/wall lizard (male) and ground-foraging tree lizard (female)
longitibialis	Beata Anole	diurnal	invert.	65	5.9	Hispaniola	*** ground-foraging tree lizard

Anolis spp.	English name				Range core
			Size		
	Activity	Diet	mm	g	Ecological business model
loveridgei	Honduras Giant Anole				Honduras
			94	17.5	
loysiana	Peach Anole				Cuba
	diurnal	invert.	42	1.65	*** trunk/wall lizard and immobility expert
luciae	Saint Lucia Anole				Saint Lucia
	diurnal	invert.	74	8.9	*** trunk/wall lizard and diurnal canopy walker and ground-foraging tree lizard
lucius	Cave Anole				Cuba
	diurnal	ants	56	3.79	**** trunk/wall lizard
luteogularis	Western Giant Anole				Cuba
	diurnal	omnivore	141	58	**** trunk/wall lizard
luteosignifer	Cayman Brac Anole				Cayman Islands
	diurnal	omnivore	49	2.66	*** ground-foraging tree lizard
lynchi	Lynch's Anole				Colombia
			54	3.40	
lyra	Lyre Anole				Colombia
	diurnal	invert.	64	5.8	*** diurnal canopy walker (low)
macilentus	Black-cheeked Bush Anole				Cuba
	diurnal		39	1.30	** diurnal canopy walker (low)
macrinii	Macrinius' Anole				Mexico
			78	10.2	* diurnal canopy walker
macrolepis	Big-scaled Anole				Colombia
			54	3.40	
macrophallus	Big Hemipene Anole				Guatemala
	diurnal		48	2.45	*** ground-foraging tree lizard
maculigula	Marked Anole				Colombia
	diurnal	invert.	86	13.5	** refugia-anchored diurnal gr. liz.
maculiventris	Blotched-belly Anole				Colombia
	diurnal	invert.	45	1.96	* ground-foraging tree lizard
magnaphallus	Well-hung Anole				Panama
			52	3.02	
maia	Maia's Anole				Panama
			65	6.0	* diurnal canopy walker
marcanoi	Red-fanned Stout Anole				Dominican Republic
	diurnal		56	3.82	*** ground-foraging tree lizard
mariarum	Blemished Anole				Colombia
			49	2.67	

Anolis spp.	English name				Range core
			Size		
	Activity	Diet	mm	g	Ecological business model
marmoratus	Leopard Anole				Guadeloupe
	diurnal	omnivore	70	7.5	**** ground-foraging tree lizard
marron	Jacmel Gracile Anole				Haiti
	diurnal		64	5.6	* trunk/wall lizard
marsupialis	Marsupial Anole				Costa Rica
	diurnal		42	1.60	** ground-foraging tree lizard
matudai	Matuda's Anole				Mexico
	diurnal		49	2.55	
maynardi	Maynard's Anole				Cayman Islands
	diurnal	omnivore	62	5.1	*** trunk/wall lizard
mccraniei	McCranie's Anole				Honduras
	diurnal		49	2.60	** ground-foraging tree lizard
medemi	Medem's Anole				Colombia
			47	2.24	
megalopithecus	Rueda's Anole				Colombia
	diurnal		70	7.5	* diurnal canopy walker
megapholidotus	Large-scaled Anole				Mexico
			46	2.14	* ground-foraging tree lizard
menta	Mixed Anole				Colombia
			49	2.67	
meridionalis	Meridian Anole				Paraguay
	diurnal	invert.	49	2.56	* ground-foraging tree lizard
mestrei	Red-fanned Rock Anole				Cuba
	diurnal		50	2.77	*** ground-foraging tree lizard
microlepidotus	Guerrera Oak Anole				Mexico
			45	2.07	
microlepis	Chiapas Smooth Anole				Mexico
			39	1.30	
microtus	Short-legged Canopy Anole				Costa Rica
	diurnal		89	14.9	*** diurnal canopy walker
milleri	Miller's Anole				Mexico
	diurnal		50	2.78	
mirus	San Juan White-lipped Anole				Colombia
			84	12.8	
monensis	Mona Anole				Puerto Rico
	diurnal	invert.	52	3.14	*** trunk/wall lizard (male?) and ground-foraging tree lizard (female)
monteverde	Monteverde Anole				Costa Rica
			45	2.04	

Anolis spp.	English name				Range core
			Size		
	Activity	Diet	mm	g	Ecological business model
monticola	Foothill Anole				Haiti
	diurnal		49	2.67	*** ground-foraging tree lizard
morazani	Yoro Mountains Anole				Honduras
			52	3.02	
muralla	Muralla Anole				Honduras
	diurnal		50	2.78	
nasofrontalis	Nose Anole				Brazil
			42	1.60	
naufragus	Hidalgo Anole				Mexico
			49	2.55	** ground-foraging tree lizard
neblininus	Myers' Andes Anole				Venezuela
			55	3.67	
nebuloides	False Clouded Anole				Mexico
			49	2.67	** ground-foraging tree lizard
nebulosus	Clouded Anole				Mexico
	diurnal	invert.	46	2.12	**** ground-foraging tree lizard
nelsoni	Nelson's Anole				Honduras
	diurnal	omnivore	56	3.87	** ground-foraging tree lizard
nicefori	Niceforo's Andes Anole				Venezuela
			54	3.54	
nietoi	Nieto Montes de Oca Anole				Mexico
			45	2.04	
nigrolineatus	Black-lined Anole				Ecuador
	diurnal		49	2.59	* trunk/wall lizard
noblei	Giant Anole				Cuba
	diurnal		147	65	*** trunk/wall lizard
notopholis	Scalyback Anole				Colombia
			47	2.24	
nubilus	Redonda Anole				Antigua and Barbuda
	diurnal	invert.	67	6.6	** ground-foraging tree lizard
occultus	Puerto Rico Twig Anole				Puerto Rico
	diurnal		39	1.28	**** diurnal canopy walker
ocelloscapularis	Merendon Shoulder-spotted Anole				Honduras
	diurnal		44	1.95	*** ground-foraging tree lizard
oculatus	Dominica Anole				Dominica
	diurnal	ants	76	9.1	*** ground-foraging tree lizard
oligaspis	Bahamian Anole				Bahamas
	diurnal		41	1.52	

Anolis spp.	English name	Range core	Activity	Diet	Size mm	Size g	Ecological business model
olssoni	Desert Grass Anole	Hispaniola	diurnal	invert.	46	2.18	**** diurnal canopy walker (low)
omiltemanus	Omilteme Anole	Mexico			44	1.86	** ground-foraging tree lizard
onca	Sand Anole	Colombia	diurnal	invert.	69	7.0	*** ground-foraging tree lizard
opalinus	Bluefield's Anole	Jamaica	diurnal	invert.	49	2.67	*** trunk/wall lizard
ophiolepis	Five-striped Grass Anole	Cuba	diurnal	invert.	36	1.05	**** refugia-anchored diurnal gr. liz.
oporinus	Pimienta Green Anole	Cuba	diurnal		43	1.77	* trunk/wall lizard
orcesi	Orces' Andes Anole	Ecuador			57	4.11	
ordinatus	Bahamian Brown Anole	Bahamas	diurnal	invert.	46	2.16	**** ground-foraging tree lizard
ortonii	Amazon Bark Anole	Venezuela	diurnal	ants	51	2.98	*** ground-foraging tree lizard
osa	Osa Anole	Costa Rica			50	2.78	
otongae	Otonga Anole	Ecuador			59	4.58	
oxylophus	Stream Anole	Central America	diurnal	invert.	66	6.2	*** refugia-anchored diurnal gr. liz.
pachypus	Thick Anole	Costa Rica	diurnal		47	2.24	*** refugia-anchored diurnal gr. liz.
paravertebralis	Santa Maria Anole	Colombia			45	2.04	
parilis	Ecuador Anole	Ecuador			67	6.6	
parvicirculatus	Berriozabal Anole	Mexico	diurnal		47	2.32	
paternus	Nueva Gerona Anole	Cuba	diurnal		44	1.83	**** ground-foraging tree lizard
pecuarius	Vache Green Anole	Haiti	diurnal		65	6.0	* diurnal canopy walker or trunk/wall lizard

Anolis spp.	English name				Range core
			Size		
	Activity	Diet	mm	g	Ecological business model
pentaprion	Lichen Anole				Panama
	diurnal	invert.	65	5.9	*** diurnal canopy walker
peraccae	Perraca's Anole				Colombia
	diurnal		51	2.96	* trunk/wall lizard
peruensis	Peru Anole				Peru
			49	2.67	
petersi	Peters' Giant Anole				Guatemala
	diurnal		101	21.6	* trunk/wall lizard
peucephilus	Oaxaca Pine Forest Anole				Mexico
	diurnal		42	1.63	* diurnal canopy walker
peynadoi	Northern Hispaniola Green Anole				Dominican Republic
			64	5.6	* trunk/wall lizard
phyllorhinus	Leaf-nosed Anole				Brazil
			75	9.2	
pigmaequestris	Cayo Frances Anole				Cuba
	diurnal		115	31.5	*** trunk/wall lizard
pijolense	Pijol Peak Anole				Honduras
	diurnal		53	3.25	
pinchoti	Crab Cay Anole				Colombia
	diurnal		47	2.26	* diurnal canopy walker
placidus	Placid Anole				Dominican Republic
	diurnal		43	1.73	*** diurnal canopy walker
planiceps	Orange-fanned Leaf-litter Anole				Venezuela
	diurnal	invert.	63	5.4	** ground-foraging tree lizard
podocarpus	Podocarpus Anole				Ecuador
			78	10.2	
poecilopus	Dappled Anole				Colombia
	diurnal	invert.	61	5.0	*** refugia-anchored diurnal gr. liz.
poei	Telimbela Anole				Ecuador
	diurnal		54	3.44	
pogus	Anguilla Bank Bush Anole				Lesser Antilles
	crepuscular	omnivore	51	2.90	*** ground-foraging tree lizard
polylepis	Many-scaled Anole				Costa Rica
	diurnal	invert.	49	2.65	*** ground-foraging tree lizard
poncensis	Ponce Anole				Puerto Rico
	diurnal	invert.	42	1.67	*** diurnal canopy walker (low)
porcatus	Cuban Green Anole				Cuba
	diurnal	ants	66	6.3	*** trunk/wall lizard and diurnal canopy walker

Anolis spp.	English name				Range core
			Size		
	Activity	Diet	mm	g	Ecological business model
porcus	Oriente Bearded Anole				Cuba
	diurnal	invert.	136	52	**** trunk/wall lizard
prasinorius	Cortico Anole				Dominican Republic
	diurnal		42	1.60	
princeps	First Anole				Colombia
	diurnal		106	24.9	
proboscis	Pinocchio Anole				Ecuador
	diurnal	invert.	69	7.2	** diurnal canopy walker (low)
propinquus	Neighbor Anole				Colombia
	diurnal		55	3.50	
pseudokemptoni	False Kempton's Anole				Panama
			49	2.55	
pseudopachypus	Sunrise Anole				Panama
			44	1.86	
pseudotigrinus	False Tiger Anole				Brazil
			42	1.60	
pulchellus	Puerto Rico Bush Anole				Puerto Rico
	diurnal	ants	44	1.88	*** ground-foraging tree lizard
pumilus	Cuban Spiny-plant Anole				Cuba
	diurnal	invert.	35	0.98	* diurnal canopy walker (low)
punctatus	Amazon Green Anole				Venezuela
	diurnal	invert.	74	8.9	*** trunk/wall lizard
purpurescens	Purple Anole				Colombia
			99	20.3	* diurnal canopy walker
purpurgularis	Yoro Crested Anole				Honduras
	diurnal		54	3.38	
purpuronectes	Oaxaca Water Anole				Mexico
			74	8.6	** refugia-anchored diurnal gr. liz.
pygmaeus	Chiapas Pygmy Anole				Mexico
	diurnal		34	0.92	
quadriocellifer	Cuban Eyespot Anole				Cuba
	diurnal	ants	46	2.20	**** ground-foraging tree lizard
quaggulus	Flabby Anole				Nicaragua
	diurnal		37	1.10	* refugia-anchored diurnal gr. liz.
quercorum	Gray Anole				Mexico
	diurnal		40	1.47	* ground-foraging tree lizard
radulinus	Truando Anole				Colombia
			42	1.60	

Anolis spp.	English name				Range core
			Size		
	Activity	Diet	mm	g	Ecological business model
reconditus	Blue Mountains Anole				Jamaica
	diurnal	invert.	81	11.3	* trunk/wall lizard
rejectus	Santiago Grass Anole				Cuba
	diurnal		35	0.95	**** diurnal canopy walker (low)
richardii	Grenada Tree Anole				Grenada
	diurnal	omnivore	101	21.7	** ground-foraging tree lizard or trunk/wall lizard
ricordii	Haitian Giant Anole				Hispaniola
	diurnal		145	63	*** trunk/wall lizard
rimarum	Marmelade Anole				Haiti
	diurnal		42	1.60	*** refugia-anchored diurnal gr. liz.
rivalis	Rival Anole				Colombia
			55	3.67	
roatanensis	American Flag Anole				Honduras
	diurnal		56	3.82	* ground-foraging tree lizard
rodriguezii	Rodriguez's Anole				Northern Central America
	diurnal	invert.	44	1.84	* ground-foraging tree lizard
roosevelti	Culebra Giant Anole				Puerto Rico
	diurnal	omnivore	125	40.8	*** diurnal canopy walker or giant herbivore
roquet	Martinique Anole				Martinique
	diurnal	invert.	61	5.1	**** trunk/wall lizard
rubiginosus	Juarez Mountains Anole				Mexico
			45	2.04	* ground-foraging tree lizard
rubribarbaris	Red-bearded Anole				Honduras
			47	2.24	
rubribarbus	Sagua de Tanamo Anole				Cuba
	diurnal		53	3.34	*** ground-foraging tree lizard
ruibali	Cabo Cruz Pallid Anole				Cuba
	diurnal		36	1.01	
ruizii	Ruiz's Anole				Colombia
			51	2.90	* diurnal canopy walker
rupinae	Castillon Anole				Haiti
	diurnal		50	2.78	
sabanus	Saba Tree Anole				Saba
	diurnal	invert.	54	3.42	*** ground-foraging tree lizard and trunk/wall lizard
sacamecatensis	Sacamecates Mountain Anole				Mexico
			41	1.52	

Anolis spp.	English name				Range core
			Size		
	Activity	Diet	mm	g	Ecological business model
sagrei	Brown Anole				Cuba
	diurnal	invert.	55	3.71	**** ground-foraging tree lizard
salvini	Salvin's Anole				Costa Rica
			57	3.96	
santamartae	Santa Marta Anole				Colombia
			49	2.55	
scapularis	False Brown-eared Anole				Peru
			40	1.45	
schiedii	Schiede's Anole				Mexico
			52	3.14	
schwartzi	Schwartz's Anole				St. Kitts and Nevis
	diurnal	invert.	41	1.53	**** ground-foraging tree lizard
scriptus	Silver Key Anole				Bahamas
	diurnal		64	5.6	**** ground-foraging tree lizard
scypheus	Yellow-tongued Forest Anole				Colombia
	diurnal	invert.	66	6.2	**** ground-foraging tree lizard
semilineatus	Hispaniola Grass Anole				Hispaniola
	diurnal		42	1.64	**** ground-foraging tree lizard
sericeus	Silky Anole				Mexico
	diurnal	invert.	45	1.99	*** ground-foraging tree lizard
serranoi	Serrano's Anole				Guatemala
	diurnal	invert.	70	7.5	* ground-foraging tree lizard
sheplani	Cabral Anole				Dominican Republic
	diurnal	invert.	39	1.30	**** diurnal canopy walker
shrevei	Shreve's Anole				Dominican Republic
	diurnal		52	3.14	*** ground-foraging tree lizard
sierramaestrae	Maestra Mountains Anole				Cuba
	diurnal		126	41.5	
singularis	Pourcine Anole				Hispaniola
	diurnal		48	2.46	**** trunk/wall lizard
smallwoodi	Smallwood's Anole				Cuba
	diurnal	omnivore	143	61	*** trunk/wall lizard
smaragdinus	Bahamian Green Anole				Bahamas
	diurnal	invert.	36	1.00	*** diurnal canopy walker
sminthus	Mouse Anole				Honduras
			51	2.89	* ground-foraging tree lizard

Anolis spp.	English name				Range core
			Size		
	Activity	Diet	mm	g	Ecological business model
soinii	Soini's Anole				Peru
			69	7.0	
solitarius	Solitaire Anole				Colombia
			47	2.24	
spectrum	Matanzas Anole				Cuba
	diurnal		38	1.23	**** ground-foraging tree lizard
spilorhipis	Mount Ombligo Anole				Mexico
			53	3.27	
squamulatus	Small-scaled Anole				Venezuela
			99	20.3	
stevepoei	Poe's Anole				Mexico
			46	2.14	
strahmi	Strahm's Anole				Hispaniola
	diurnal	invert.	66	6.2	*** ground-foraging tree lizard
stratulus	Barred Anole				Puerto Rico
	diurnal	ants	47	2.37	**** trunk/wall lizard
subocularis	Pacific Anole				Mexico
	diurnal		51	2.87	* ground-foraging tree lizard
sulcifrons	Grooved Anole				Colombia
			55	3.67	
tandai	Blue-throated Anole				Peru
	diurnal		63	5.5	** ground-foraging tree lizard
taylori	Taylor's Anole				Mexico
			66	6.1	* ground-foraging tree lizard
tenorioensis	Tenorio Volcano Anole				Costa Rica
	diurnal		42	1.60	
terraealtae	Les Saintes Anole				Guadeloupe
	diurnal	invert.	67	6.4	*** ground-foraging tree lizard
terueli	Yellow-fanned Pallid Anole				Cuba
	diurnal		38	1.23	* ground-foraging tree lizard
tetarii	Tetari Anole				Venezuela
			71	7.7	
tigrinus	Tiger Anole				Venezuela
			51	2.90	
toldo	Gray-banded Green Anole				Cuba
	diurnal		53	3.27	* trunk/wall lizard
tolimensis	Tolima Canyon Anole				Colombia
			52	3.14	

Anolis spp.	English name	Activity	Diet	Size mm	Size g	Range core	Ecological business model
townsendi	Cocos Anole	diurnal	invert.	44	1.95	Costa Rica	*** trunk/wall liz., ground-foraging tree lizard, and diurnal canopy walker
trachyderma	Roughskin Anole	diurnal	invert.	51	2.90	Northwest South America	*** ground-foraging tree lizard
transversalis	Banded Tree Anole	diurnal	invert.	76	9.6	Northwest South America	*** diurnal canopy walker
trinitatis	Saint Vincent Bush Anole	diurnal	invert.	58	4.25	Saint Vincent	*** trunk/wall lizard and ground-foraging tree lizard
triumphalis	Triumphant Anole	diurnal		49	2.55	Panama	
tropidogaster	Tropical Anole	diurnal		50	2.82	Colombia	
tropidolepis	Swift Anole	diurnal	invert.	52	3.01	Costa Rica	*** ground-foraging tree lizard
tropidonotus	Greater Scaly Anole	diurnal	invert.	53	3.32	Honduras	**** ground-foraging tree lizard
umbrivagus	Forest Wanderer Anole			42	1.69	Colombia	
uniformis	Lesser Scaly Anole	diurnal	invert.	37	1.18	Guatemala	**** refugia-anchored diurnal gr. liz.
unilobatus	Indigo-throated Anole	diurnal		47	2.24	Central America	*** diurnal canopy walker (low)
ustus	Belize Silky Anole			44	1.94	Mexico	
utilensis	Mangrove Anole	diurnal		52	3.02	Honduras	*** diurnal canopy walker
valencienni	Jamaica Twig Anole	diurnal	omnivore	67	6.6	Jamaica	**** diurnal canopy walker
vanidicus	Escambray Grass Anole	diurnal		36	1.01	Cuba	*** ground-foraging tree lizard
vanzolinii	Vanzolini's Anole			84	12.5	Ecuador	
vaupesianus	Williams' Anole			68	6.8	Colombia	
ventrimaculatus	Speckled Emerald Anole			67	6.4	Colombia	* diurnal canopy walker

Anolis spp.	English name				Range core
			Size		
	Activity	Diet	mm	g	Ecological business model
vermiculatus	Cuban Stream Anole				Cuba
	diurnal	omnivore	100	20.9	*** ground-foraging tree lizard
vescus	Purial Bush Anole				Cuba
	diurnal		40	1.39	
vicarius	Rio Amparrado Anole				Colombia
			43	1.77	
villai	Country Anole				Nicaragua
	diurnal		52	3.14	
viridius	Neiba Green Anole				Dominican Republic
	diurnal		69	7.0	** trunk/wall lizard or diurnal canopy walker
vittigerus	Garland Anole				Colombia
	diurnal		59	4.56	
wampuensis	Wampu River Anole				Honduras
	diurnal		46	2.14	*** ground-foraging tree lizard
wattsi	Watts' Anole				Antigua and Barbuda
	diurnal	invert.	51	2.90	*** ground-foraging tree lizard
websteri	Webster's Anole				Haiti
	diurnal	invert.	46	2.14	* trunk/wall lizard
wellbornae	Wellborn's Anole				Honduras
	diurnal		45	2.04	*** trunk/wall lizard
wermuthi	Wermuth's Anole				Nicaragua
	diurnal		48	2.45	
whitemani	Pallid Stout Anole				Hispaniola
	diurnal	invert.	57	4.11	*** ground-foraging tree lizard
williamsii	Williams' Brazil Anole				Brazil
			45	2.04	
williamsmitter-meierorum	Williams-Mittermeier Anole				Peru
			57	3.96	
wilsoni	Wilson's Anole				Honduras
	diurnal		52	3.02	** ground-foraging tree lizard
woodi	Wood's Anole				Costa Rica
	diurnal		79	10.7	**** trunk/wall lizard
yoroensis	Yoro Anole				Honduras
	diurnal		43	1.77	
zapotecorum	Zapotecan Anole				Mexico
			45	2.04	
zeus	Name of God Mountains Anole				Honduras
	diurnal		41	1.52	

Anomalopus

Anomalopus is a moderate radiation (seven spp.) of average-sized (3.95 v. 3.25 g), slender (shape index −1.42 v. −0.46 std. resid.), functionally limbless subterranean skinks from northeastern Australia (New South Wales and Queensland). Although functionally limbless, most species have remnants of all limbs, and it is notable that, although considered to be subterranean foragers, at least one species (*verreauxii*) is known to ambush aboveground prey from the mouth of its burrow. Thus these lizards may be somewhat transitional between cryptozoids and obligate subterranean foragers.

Morphology is unremarkable; sex differences or color are not reported.

One species each (see listing) is reported to be diurnal (50% v. 74%) and nocturnal (50% v. 14.5%). They are found in littoral (29% v. 21%) and sandy (29% v. 25%) areas at expected frequencies, but they are absent from riparian (v. 23%) and human (v. 14%) habitats.

Most but not all are oviparous (86% v. 64%). They have average clutch/litter counts (−0.07 v. 0.06 std. resid.). Young are particularly large (ontogenetic mass range −0.94 v. 0.12 std. resid.).

Longevity is about average (0.32 v. 0.17 std. resid.), albeit based solely on *swansoni* (8 years) and *verreauxii* (7 years). Endemism is average (57% v. 64%). No species is found on an island (v. 57%). None has been introduced (v. 2.1%). Most are imperiled (75% v. 47%), a reflection of limited distributions. None is on an IUCN list of the most endangered (v. 14%). In other respects *Anomalopus* are much like other skinks.

Anomalopus spp.	English name		Size		Range core
	Activity	Diet	mm	g	Ecological business model
brevicollis	Short-necked Worm-skink				Queensland
			74	1.25	**** subterranean forager
gowi	Speckled Worm-skink				Queensland
			87	1.81	**** subterranean forager
leuckartii	Two-clawed Worm-skink				Queensland
			112	13.2	*** subterranean forager
mackayi	Mackay's Worm-skink				Queensland
			98	9.1	* subterranean forager
pluto	Cape York Worm-skink				Queensland
			67	1.00	**** subterranean forager
swansoni	Swanson's Worm-skink				New South Wales
	diurnal	invert.	92	2.04	**** subterranean forager
verreauxii	Three-clawed Worm-skink				Queensland
	nocturnal	invert.	145	26.9	*** subterranean forager

Anotosaura spp.	English name		Size		Range core
	Activity	Diet	mm	g	Ecological business model
collaris	Collared Earless Lizard				Brazil
			42	0.64	*** litter cryptozoid
vanzolinia	Vanzolini's Earless Lizard				Brazil
			44	1.06	** litter cryptozoid

Anotosaura

Anotosaura (two spp.) are very small (0.83 g v. 2.35 g), elongate, terrestrial Brazilian gymnophthalmids that from 1933 to 2013 were known from only a single specimen. They are litter cryptozoids (v. 55%). They are very slender (−1.02 v. −0.32 std. resid.) with comparatively short tails (1.52 v. 1.72 × SVL). Tails are known to break (50% v. 28%). They have secretory pores (50% v. 29%) but no other sex differences. Unlike other gymnophthalmids, they are found in the intermediate rainfall belt (in this case, 513–811 mm/y) at low elevation (276–895 m). Reproduction is typical; clutch counts are slightly large (0.18 v. −0.11 std. resid.), with RCM low to fit their slender build (0.111 v. 0.192 ratio). Both are endemic to Brazil and have not been evaluated for conservation status. In other respects *Anotosaura* are typical gymnophthalmids.

Apathya

Apathya (two spp.) is a genus of moderate-sized (4.3 g v. 5.1 g) refugia-anchored diurnal ground lacertids found in southwestern Asia (Turkey to Iran). In morphology they are typical lacertids. Sex differences are few and poorly documented (sex rollup 0.50 v. 0.86 on 0 to 6 scale), but include sexual dichromatism in *yassujica* and seasonal color in *cappadocica*. The latter also has a bright blue tail, at least in juveniles. Both are conspicuously wary (v. 11%), and both

Apathya spp.	English name		Size		Range core
	Activity	Diet	mm	g	Ecological business model
cappadocica	Anatolia Rock Lizard				Iraq
	diurnal	invert.	64	6.1	** refugia-anchored diurnal gr. liz.
yassujica	Yassujian Lizard				Iran
			51	3.11	* refugia-anchored diurnal gr. liz

Aphaniotis spp.	English name		Size		Range core
	Activity	Diet	mm	g	Ecological business model
acutirostris	Long-snouted Shrub Lizard				Indonesia
	diurnal	invert.	55	5.2	*** ground-foraging tree lizard
fusca	Dusky Earless Lizard				Malaysia
	diurnal	invert.	60	6.8	*** diurnal canopy walker
ornata	Ornate Shrub Lizard				Borneo
	diurnal	invert.	50	3.93	* diurnal canopy walker

rely primarily on retreat to a crevice or burrow for predator concealment. Both are found up to relatively high elevation (2200, 2300 m). *A. cappadocica* is associated with riparian habitats (v. 28%). Reproduction is typical. Neither species occurs on islands (v. 26%). Neither has been introduced (v. 4.6%) or considered imperiled (v. 56%). In other respects *Apathya* are similar to other lacertids.

Aphaniotis (fig. 4.3B)

Aphaniotis is small genus (three spp.) of arboreal agamids found at low elevation in Southeast Asia. One species (*A. ornata*) is renowned for a short backward-pointing fleshy appendage on its snout. All species favor small-diameter perches (diurnal canopy walkers 67% v. 7%), and one species (*acutirostris*) is regularly found foraging from trees onto the ground (ground-foraging tree lizard 33% v. 19%). They are small (5.2 g v. 15.6 g), slender (shape index 0.58 v. 0.80 std. resid.) agamids, notable for their long (2.0–2.2 v. 1.9 × SVL), unbreakable tails (v. 17% breakable). Their crests (100% v. 38%) are subtle; secretory pores are absent (v. 13%). They do not change color quickly (v. 22%) or seasonally (v. 20%), but they do show sexual differences in color (100% v. 53%), and males are larger than females (1.00 v. 0.53 on –2 to 2 scale). Among 59 agamid genera, they are ranked seventh for overall sex differences (2.67 v. 1.61 on 0 to 6 scale). All are generalized invertebrate eaters (v. 46%). The one species characterized for thermal behavior (*fusca*) is a thermoconformer. This species has a blue mouth lining, which it displays whenever pressed (blue seems to be favored for demonstrating a willingness to bite). Social interactions are poorly known. They lay one to two eggs every few months. This puts them below the lizard and agamid averages for mass-adjusted clutch count (–0.51 v. 0.20 std. resid.). The young are slightly larger than typical agamid young (0.54 v. 0.69 std. resid.), which corresponds to young being about 1/50 of maximal mass rather than the agamid standard of 1/100. All are found on islands (v. 36%). None has been introduced (v. 1.7%) or judged to be imperiled (v. 37%). In other respects *Aphaniotis* are typical agamids.

Aprasia

Aprasia is a moderate radiation (14 spp.) of average-sized (5.0 g v. 5.2 g) pygopodids from Australia. They differ strongly from the rest of the family in that the two diet-documented species are ant larvae/pupae specialists (100% v. 8.7%), and four without diet documentation live in ant tunnels. Furthermore, most *Aprasia* have moved underground (subterranean foragers 78% v. 19%). The others are the more conventional litter cryptozoids (22% v. 68%). They are largely diurnal (85% v. 61%), despite their darkened or dimly lit microhabitat. They are the only pygopodid genus with members that are on the IUCN's lists of the most endangered (17% v. 6%).

Although they lack blue tails (v. 4.3%), *inaurita* has a red tail and *parapulchella*'s tail is pink-red. The sex rollup score for *Aprasia* is about average (0.79 v. 0.50 on 0 to 6 scale).

Although the diet class for *Aprasia* is ant specialist, they do not apparently eat adult ants, which perhaps could attack them effectively within their tunnels. Based on the usually prey-less gastrointestinal tracts of *Aprasia*, and the ability of this species to consume numerous larvae in seconds, *Aprasia* may indulge in infrequent but very rapid binge eating, perhaps to avoid triggering adverse responses from the adult ants.

Aprasia that are not diurnal are cathemeral. At least one of the species regarded as diurnal (*rostrata*) is known to occasionally remain active at night. Does diurnality inside ant tunnels persist for its thermal advantages, or are *Aprasia* moving in the direction of continuous activity (cathemerality)? Despite their normally hidden microhabitats, 14% (v. 17%) of *Aprasia* bask directly. Almost all *Aprasia* vocalize spontaneously (93% v. 67%), but they are not reported to stress call (v. 28%). Is calling necessary to find each other?

Aprasia do not noticeably differ from the family pattern in habitat use. Clutch count is low (–0.42 v. 0.44 std. resid.). Young are slightly larger than the large family average (ontogenetic mass range –2.54 v. –2.12 std. resid.). They have not been reported to oviposit communally (v. 6.5%), but it might occur unnoticed. Endemism is elevated (71% v. 51%), perhaps because of substrate specialization. Consonant with that notion is a high level of imperilment (100% v. 88%), often associated with isolated populations. As mentioned above, the genus has a high frequency of placement on the IUCN's lists of the most endangered.

Aprasia spp.	English name	Activity	Diet	Size mm	Size g	Range core	Ecological business model
aurita	Eared Worm-lizard	diurnal	ants	101	5.1	Southeastern Australia	** subterranean forager
clairae	Claire's Legless Lizard	diurnal		74	3.32	Western Australia	
haroldi	Shark Bay Worm-lizard	diurnal		91	4.46	Western Australia	
inaurita	Mallee Worm-lizard	diurnal		108	5.6	Australia	* litter cryptozoid
litorea	Gnaraloo Legless Lizard	diurnal		79	3.68	Western Australia	
parapulchella	Pink-tailed Worm-lizard	diurnal	ants	114	6.0	New South Wales	** subterranean forager
picturata	Black-headed Worm-lizard	diurnal		114	6.1	Western Australia	
pseudopulchella	Flinders Ranges Worm-lizard	diurnal		125	6.9	South Australia	** subterranean forager
pulchella	Pretty Worm-lizard	cathemeral		105	5.4	Western Australia	*** subterranean forager
repens	Sedgelands Worm-lizard	diurnal		100	5.0	Western Australia	*** subterranean forager
rostrata	Exmouth Worm-lizard	diurnal		102	5.2	Western Australia	** subterranean forager
smithi	Zuytdorp Worm-lizard	diurnal		102	5.2	Western Australia	*** subterranean forager
striolata	Striated Worm-lizard	cathemeral		109	5.7	Australia	* litter cryptozoid
wicherina	Wicherina Worm-lizard			75	3.41	Western Australia	

Archaeolacerta

Archaeolacerta bedriagae is the sole species in a genus of moderately large (8.3 g v. 5.1 g) trunk/wall lacertids found in France and Italy. It is notable for its omnivorous diet (v. 19%) and occurrence at high elevation (0–2710 m). It is relatively well known (filled cells 58% v. 33%).

In morphology, it is a typical lacertid. Tails are on the short side (1.75 v. 2.06 × SVL) and are bright blue in juveniles. Sex differences are few and poorly documented (0.0 v. 0.86 on 0 to 6 scale). Plant material is found in half the stomachs sampled. As befits their high elevation, mean active body temperature is low (30.0°C v. 33.3°C). Their social lives appear to be highly variable, or perhaps contingent on the conformation of cracks and density of the population, as males are reported as being both tolerant—up to 30 individuals sharing a crack—and intolerant—tenaciously defend against other males. Curiously, the males are singled out as being prone to suspend midday activity in the hottest months. The species is associated with littoral (v. 22%), riparian (v. 28%), and human (v. 16%) habitats. Reproduction is typical for lacertids. Hatchlings are relatively large, as indicated by reduced onto-

Archaeolacerta sp.	English name	Activity	Diet	Size mm	Size g	Range core	Ecological business model
bedriagae	Bedriaga's Rock Lizard	diurnal	omnivore	71	8.3	France	** trunk/wall lizard

genetic mass range (–0.87 v. –0.05 std. resid.). Documented longevity is short (–0.87 v. 0.19 std. resid.), corresponding to 3 years. The species occurs on many islands (v. 26%) but has not been introduced (v. 4.6%). It is subject to conservation concern (v. 56%), but it is not listed on the IUCN lists of the most endangered (v. 15.5%). In other respects *Archaeolacerta* is similar to other lacertids.

Archaius

Archaius tigris is the sole species in this genus of chamaeleonids, and it is restricted to the Seychelles Islands in the Indian Ocean. Despite its derived location, it is judged the most basal surviving chameleon (hence the genus name); the clade originated in Africa, but none of the more basal species persisted on the continent. It is large for a chamaeleonid (13.1 g v. 8.1 g) but average for a typical chameleon (see table 5.2, found in Chamaeleonidae entry). Although the sexes differ statistically in color (v. 53%), the difference is not diagnostic for individuals. This ambiguity may reflect failure to fully understand the social status of individuals. *Archaius* basks directly (v. 19%), is found in the littoral zone (v. 6%), and has small clutches (0.13 v. 2.04 std. resid.). Females may oviposit communally (v. 0.5%). Compared to typical chameleons (see table 5.2), as opposed to the family as a whole, *Archaius* has small clutches (0.13 v. 2.81 std. resid.) and rapid egg development (0.33 v. 2.43 std. resid.). It is listed as endangered by the IUCN (v. 25%), presumably due to threats from invasive mammals. In other respects it conforms to the patterns of "typical" chameleons.

Archaius sp.	English name				Range core
			Size		
	Activity	Diet	mm	g	Ecological business model
tigris	Seychelles Tiger Chameleon				Seychelles Islands
	diurnal	invert.	81	13.1	*** diurnal canopy walker

Aristelliger

Aristelliger is an eight-species genus of padless, cathemeral (38%) or nocturnal (63%) sphaerodactylids found in the West Indies (Bahamas to Hispaniola) and westward to Mexico and Honduras. Though nominally well known (filled cells 31% v. 27%), there is much uncertainty regarding their diel activity pattern and either uncertainty or geographic variation in their microhabitat choice. The result is that none can be assigned to an ecological business model with certainty. Most are likely trunk/wall lizards (71% v. 12%); others are ground-foraging tree lizards (14% v. 12%) and litter cryptozoids (14% v. 53%). Their arboreality (86% v. 29%) distinguishes them from most other sphaerodactylids. Most are known for tearing their skin easily, presumably to escape predators.

Aristelliger spp.	English name				Range core
			Size		
	Activity	Diet	mm	g	Ecological business model
barbouri	Striped Caribbean Gecko				Bahamas
	nocturnal		45	2.27	
cochranae	Cochran's Caribbean Gecko				Navassa Island
	cathemeral	invert.	54	3.95	* ground-foraging tree lizard
expectatus	Hispaniola Gecko				Hispaniola
	nocturnal		49	2.85	* litter cryptozoid
georgeensis	Weatherman Caribbean Gecko				Central America
	nocturnal	invert.	92	18.74	** trunk/wall lizard
hechti	Hecht's Caribbean Gecko				Turks and Caicos Islands
	nocturnal	omnivore	74	9.7	** trunk/wall lizard
lar	Spotted Caribbean Gecko				Hispaniola
	nocturnal	omnivore	117	37.9	* trunk/wall lizard
praesignis	Jamaica Croaking Gecko				Jamaica
	cathemeral	invert.	76	10.6	* trunk/wall lizard
reyesi	Reyes' Caribbean Gecko				Cuba
	cathemeral		55	4.11	** trunk/wall lizard

They are much larger than average sphaerodactylids (7.3 g v. 1.1 g) and moderately robust (shape index 0.40 v. –0.08 std. resid.). They often express basic color change (25% v. 6.5%), but they do not have advanced color change (v. 1.4%).

As befits their large size, they are more likely to have larger males than typical sphaerodactylids (1.00 v. 0.04 on –2 to 2 scale). Sexual dichromatism is muted (13% v. 35%) and seasonal color absent (v. 3%). Overall, the sex differences are average (0.63 v. 0.54 on 0 to 6 scale).

Although omnivory is the categorization for one in four species (25% v. 1.5%), the modal pattern is invertebrate eating (75% v. 82%). Foraging mode is unknown. Electric light use is common (25% v. 7%). Home-range sizes are unknown.

Although described as nocturnal/cathemeral, most species have never been seen active, and diel assignments are highly uncertain. It is possible that this failure to observe active lizards suggests more cryptic life styles (e.g., crack hermits, immobility experts) than is presently recognized. Consistent with cathemeral/nocturnal activity, spontaneous calling (25% v. 2%) and stress calling (13% v. 2%) are overrepresented. Most are conspicuously wary (75% v. 7%). Direct basking is unknown (0% v. 6.5%). Mean active body temperature is average (30.8°C v. 30.4°C). Social behavior is unknown.

They are known only from low elevation (v. 96%). They are found in littoral areas (38% v. 31%), but they are unrepresented in sandy areas (v. 15%), riparian zones (v. 19%), and human habitats (v. 33%).

Reproduction is typical, though their apparently fixed (at one to two) clutch size combines with their large size to indicate a fairly constrained clutch count (−0.78 v. −0.11 std. resid.). The young are conspicuously small (ontogenetic mass range 0.61 v. −0.12 std. resid.), indicating that they do not have large young to make up for the low count. Only one species (*praesignis*) is known to oviposit more than once a year, though the number of times has not been established. Incubation duration is a little short (0.14 v. 0.44 std. resid.). Communal oviposition (38% v. 11%) and adherent eggs are nearly universal (75% v. 6%), presumably to keep the eggs from falling out of trees (table 4.1, trait 67). Parental care is lacking (v. 1.4%).

Reported longevity, based solely on *lar*, is very short (−1.90 v. −0.15 std. resid.). Endemism is about average (50% v. 65%). All species occur on at least one island (100% v. 64%). Introductions are rampant (12.5% v. 5.1%), though none has been introduced to more than two islands (v. 2.3%). Many evaluated species are imperiled (40% v. 41%), though none is on any of the IUCN lists of the most endangered (v. 23%). In other respects *Aristelliger* are typical sphaerodactylids.

Aristelliger georgeensis is reputed to call more in advance of approaching storms; this trait is attributed to its forecasting ability, hence the common name.

Arthrosaura

Arthrosaura is a seven-species genus of terrestrial/cryptozoic gymnophthalmids found in northwestern South America, many in the tepui region of Venezuela and the Guiana Shield. Half of the assignable species are litter cryptozoids (v. 55%), the remainder are refugia-anchored diurnal ground lizards (v. 19%). They are about average in size (2.09 g v. 2.35 g) and build (shape index −0.14 v. −0.32 std. resid.). Only one (*reticulata*: 14% v. 41%) is marginally elongate. They have typical-length (1.67 v. 1.72 × SVL), breakable (29% v. 28%) tails. Secretory pores are common (43% v. 29%). The sexes are also likely to differ in color (29% v. 13%) and some show seasonal color (14% v. 1.7%). A third (v. 14%) are active at any time of day, the remainder (67% v. 79%) are diurnal. One species (*reticulata*) is conspicuously wary (14% v. 7%). They are overrepresented in frequency of direct basking (14% v. 8%) and thermal conformance (29% v. 4.6%), suggesting better data on thermoregulatory behavior in this genus. Mean body temperature is cool (26.0°C v. 28.8°C). One species (*kockii*) is socially documented and reported to lack any defense or intolerance. They are often associated with riparian habitats (80% v. 29%), which is reflected in their common name (creek lizards). *A. reticulata* will seek shelter under water. Endemism is low (43% v. 68%). None is considered endangered (v. 25%). In other

Arthrosaura spp.	English name				Range core
			Size		
	Activity	Diet	mm	g	Ecological business model
kockii	Orange-bellied Creek Lizard				Guyana Shield
	diurnal	invert.	54	3.42	** litter cryptozoid
montigena	Auyantepui Creek Lizard				Venezuela
			44	1.74	* refugia-anchored diurnal gr. liz.
reticulata	Yellow-bellied Creek Lizard				Venezuela
	cathemeral	invert.	54	2.03	** refugia-anchored diurnal gr. liz.
synaptolepis	Donnelly's Creek Lizard				Venezuela
			46	2.04	
testigensis	Testigos Creek Lizard				Venezuela
			35	0.85	
tyleri	Duida Creek Lizard				Venezuela
			43	1.65	
versteegii	Hoogmoed's Equivocal Lizard				Guyana
	diurnal		60	4.98	** litter cryptozoid

respects *Arthrosaura* are typical gymnophthalmids.

Asaccus

Asaccus is a moderate radiation (18 spp.) of smallish (3.6 g v. 4.7 g) trunk/wall (90% v. 44%) phyllodactylids from southwestern Asia (Turkey to Oman and Iran). One species (*zagrosicus*) may be a crack hermit (10% v. 1%). Compared to other phyllodactylids, they are notably cathemeral (56% v. 19%), arboreal (100% v. 52%), and riparian (53% v. 16%). They are poorly known (22% v. 27%).

They are slightly slender (shape index 0.32 v. 0.37 std. resid.). Their tails are comparatively long (1.21 v. 1.03 × SVL) and often prehensile (11% v. 2.2%). None is reported to change color (0% v. 11%). All possess enlarged (at the very end only) toe-pads (v. 92%). SSD is absent (0.0 v. 0.32 on –2 to 2 scale), as is seasonal color (v. 0%). Sexual dichromatism is present in about 22% (v. 7%). Taken collectively, these sex differences are minimal (0.22 v. 0.31 on 0 to 6 scale).

All documented species eat invertebrates (v. 93%). Those *Asaccus* that are not cathemeral are nocturnal (44% v. 78%). See the Phyllodactylidae account for discussion of the many controverted diel activity assessments in this group, especially *Asaccus*. A few call spontaneously (5.6% v. 11%), but none is known to give stress calls (v. 6%). No information is available on the size of social groups, if any. All documented species have males that defend space (v. 100%).

Asaccus are found in the middle of the rainfall spectrum (300–600 mm/y), based solely on *barani*. The full elevational range is occupied (0–3000 m in this case), at roughly normal proportions. They are grossly underrepresented in sandy areas (0% v. 25%) and human habitats (7% v. 40%).

Clutch count is average (–0.52 v. –0.57 std. resid.), and incubation is relatively short (–0.26 v. 0.18 std. resid.).

Endemism is exactly average (67% v. 67%). No species occurs on islands (v. 40%). None has been introduced (v. 7.4%), and only 11% (v. 29%) are considered imperiled (none severely, v. 7%). The lack of introductions may

Asaccus spp.	English name				Range core
			Size		
	Activity	Diet	mm	g	Ecological business model
andersoni	Mount Darbaste Gecko				Iran
	cathemeral		58	5.2	
barani	Harran Leaf-toed Gecko				Turkey
	cathemeral		48	2.89	
caudivolvulus	Musandam Leaf-toed Gecko				Oman
	nocturnal	invert.	54	4.27	** trunk/wall lizard
elisae	Werner's Leaf-toed Gecko				Iraq
	nocturnal	invert.	62	6.3	*** trunk/wall lizard
gallagheri	Gallagher's Leaf-toed Gecko				Oman
	nocturnal		38	1.49	** trunk/wall lizard
gardneri	Gardner's Leaf-toed Gecko				Oman
	nocturnal		60	5.7	** trunk/wall lizard
granularis	Lorestan Leaf-toed Gecko				Iran
	cathemeral		59	5.5	
griseonotus	Gray-spotted Leaf-toed Gecko				Iran
	cathemeral		60	5.7	* trunk/wall lizard
iranicus	Bushehr Leaf-toed Gecko				Iran
	cathemeral		39	1.64	
kermanshahensis	Kermanshah Leaf-toed Gecko				Iran
	nocturnal		49	3.22	* trunk/wall lizard
kurdistanensis	Kurdistan Leaf-toed Gecko				Iran
	cathemeral		55	4.44	
margaritae	Margarita's Leaf-toed Gecko				Oman
	nocturnal		52	3.64	
montanus	Mountain Leaf-toed Gecko				Oman
	nocturnal		37	1.38	* trunk/wall lizard
nasrullahi	Nasrullah's Leaf-toed Gecko				Iran
	cathemeral		59	5.5	
platyrhynchus	Flat-headed Leaf-toed Gecko				Oman
	nocturnal		54	4.27	* trunk/wall lizard
saffinae	Saffine Mountain Leaf-toed Gecko				Iraq
	cathemeral		50	3.36	
tangestanensis	Tangestan Leaf-toed Gecko				Iran
	cathemeral		51	3.62	** trunk/wall lizard
zagrosicus	Zagros Mountains Leaf-toed Gecko				Iran
	cathemeral	invert.	49	3.19	* crack hermit

relate to the absence of the features (electric light use, human habitat attraction) shown by the family.

The species *A. margaritae* honors herpetologist Margarita Metallinou (1985–2015), one of at least three herpetologists who have been trampled to death by elephants while conducting fieldwork.

Aspidoscelis (fig. 4.15A, 4.15B)

Aspidoscelis (Teiidae) is a major radiation (40 spp.) of moderately small (13.5 g v. 22.1 g), often parthenogenetic (60% of the specified species), invertebrate-eating (97% v. 62%), wide-ranging chemosensors (100% v. 97%) found primarily in Mexico, but extending southward to Costa Rica and northward to Oregon, Wisconsin, Michigan, and Maryland. Many species were formerly treated as *Cnemidophorus*.

Shape index (0.74 v. 0.88 std. resid), tail length (2.28 v. 2.30 × SVL), and tail breakage are normal (15% v. 21%). Secretory pore frequency may be slightly elevated (15% v. 9%), and blue tails are overrepresented (29% v. 15%). Color change is absent (0% v. 1.3%).

SSD (0.77 v. 0.89 on −2 to 2 scale), sexual dichromatism (29% v. 35%), and seasonal color (7% v. 10%) are average or slightly subdued for teiids.

The one species that is not an invertebrate eater (*ceralbensis*) is an island endemic and eats some fruit (omnivore 3.4% v. 33%). As fits the whiptail stereotype, the mass-adjusted home-range area of *Aspidoscelis* is well above the lizard average (0.83 v. 0.19 for Teiidae). The home ranges of female *hyperythra* and *sexlineata* are curiously larger than their male counterparts. Female *exsanguis* (males are not present) are sufficiently belligerent that they will fight to the death if confined together. *A. deppii, exsanguis, flagellicauda, sonorae*, and *tesselata* are

Aspidoscelis spp.	English name		Size		Range core
	Activity	Diet	mm	g	Ecological business model
angusticeps	Yucatan Whiptail				Guatemala
	diurnal	invert.	83	16.7	**** wide-ranging chemosensor
burti	Canyon Spotted Whiptail				Mexico
	diurnal		106	35.3	*** wide-ranging chemosensor
calidipes	Rio Tepalcatepec Valley Whiptail				Mexico
	diurnal		67	8.7	*** wide-ranging chemosensor
ceralbensis	Cerralvo Whiptail				Mexico
	diurnal	omnivore	77	13.3	*** wide-ranging chemosensor
communis	Colima Giant Whiptail				Mexico
	diurnal	invert.	108	38.2	*** wide-ranging chemosensor
costatus	Western Mexico Whiptail				Mexico
	diurnal	invert.	76	12.7	**** wide-ranging chemosensor
cozumela	Cozumel Whiptail				Mexico
	diurnal		67	8.6	*** wide-ranging chemosensor
danheimae	San Jose Whiptail				Mexico
	diurnal	invert.	57	5.1	*** wide-ranging chemosensor
deppii	Black-bellied Racerunner				Central America
	diurnal	invert.	73	11.0	**** wide-ranging chemosensor
exsanguis	Chihuahuan Spotted Whiptail				Mexico
	diurnal	invert.	82	16.3	**** wide-ranging chemosensor
flagellicaudus	Gila Spotted Whiptail				Arizona
	diurnal	invert.	83	16.7	**** wide-ranging chemosensor
gularis	Texas Spotted Whiptail				Texas
	diurnal	invert.	82	16.0	**** wide-ranging chemosensor
guttatus	Mexican Racerunner				Mexico
	diurnal		90	22.0	*** wide-ranging chemosensor
hyperythrus	Orange-throated Whiptail				Mexico
	diurnal	invert.	58	5.3	**** wide-ranging chemosensor
inornatus	Little Striped Whiptail				New Mexico
	diurnal	invert.	59	5.8	**** wide-ranging chemosensor
labialis	Baja California Whiptail				Mexico
	diurnal		55	4.5	**** wide-ranging chemosensor
laredoensis	Laredo Striped Whiptail				Mexico
	diurnal	invert.	69	9.2	**** wide-ranging chemosensor

among the most incessantly moving lizards (92nd–99th percentile for PTM, albeit only 56th–65th percentile for MPM). A similar situation is noted for *sexlineata*, which is around the 90th percentile for PTM and is less outstanding (77th percentile) for MPM. *A. tigris* has the most extreme discrepancy between the two measures: its MPM are only the 12th percentile of MPM, but the PTM is 96th–98th percentile. *A. gularis* are more deliberate in their movements, but Dixon and Lemos-Espinal (2010) declare them to be "the swiftest lizards on the planet." Foraging *laredoensis* do not move far from their refuge (net displacement over 24 hours = 2.2 m; net displacement over 1 year = 40 m), but they move rapidly (47 m in 15 minutes). The similarly sized *uniparens* will move as much as 1600 m in a morning. All *Aspidoscelis* for which comment on ontogenetic changes in activity have been logged (eight spp.) mention that the activity season of juveniles is noticeably longer than that of adults, by up to several months per year. At least *sexlineata* and *tigris* are arm wavers (pause while foraging, raise and wave one arm). The genus has a high frequency of species judged conspicuously wary (24% v. 13%).

Crypsis as a primary predator concealment tactic is low, but it is overrepresented in *Aspidoscelis* (6.7%, the highest value in Teiidae, v. 1.8%). In addition to the usual heavy teiid reliance on flight (60% v. 77%), *Aspidoscelis* retreat to crevices or burrows (20% v. 12%) and retreat to vegetation (13% v. 7%). Direct basking is common (49% v. 58%). One species is deemed a thermoconformer (*parvisocia*: 2.4% v. 0.6%). Mean active body temperature is very hot (37.8° v. 36.9°C).

Males (in the bisexual species) are usually socially tolerant (69% v.

Aspidoscelis spp.	English name				Range core
	Activity	Diet	Size mm	Size g	Ecological business model
lineattissimus	Many-lined Whiptail				Mexico
	diurnal	invert.	82	16.0	*** wide-ranging chemosensor
marmoratus	Marbled Whiptail				Texas
	diurnal	invert.	86	18.5	**** wide-ranging chemosensor
maslini	Maslin's Whiptail				Guatemala
	diurnal		68	8.8	*** wide-ranging chemosensor
maximus	Cape Region Whiptail				Mexico
			100	30.0	* wide-ranging chemosensor
mexicanus	Mexican Whiptail				Mexico
	diurnal		76	13.0	*** wide-ranging chemosensor
motaguae	Giant Whiptail				Honduras
	diurnal	invert.	119	51.0	*** wide-ranging chemosensor
neomexicanus	New Mexico Whiptail				New Mexico
	diurnal	invert.	69	9.2	**** wide-ranging chemosensor
neotesselatus	Colorado Checkered Whiptail				Colorado
	diurnal	invert.	86	18.4	**** wide-ranging chemosensor
opatae	Opata Whiptail				Mexico
	diurnal	invert.	60	6.1	**** wide-ranging chemosensor
pai	Pai Striped Whiptail				Arizona
	diurnal	invert.	64	7.3	*** wide-ranging chemosensor
parvisocius	Mexican Pygmy Whiptail				Mexico
	diurnal		64	7.6	*** wide-ranging chemosensor
pictus	Monserrate Whiptail				Mexico
	diurnal	invert.	60	6.12	*** wide-ranging chemosensor
rodecki	Rodeck's Whiptail				Mexico
	diurnal	invert.	63	7.1	*** wide-ranging chemosensor
sackii	Sack's Spotted Whiptail				Mexico
	diurnal		118	49.5	*** wide-ranging chemosensor
scalaris	Plateau Spotted Whiptail				Mexico
	diurnal		92	22.7	*** wide-ranging chemosensor
sexlineatus	Six-lined Racerunner				Arkansas
	diurnal	invert.	70	9.8	**** wide-ranging chemosensor
sonorae	Sonoran Spotted Whiptail				Arizona
	diurnal	invert.	77	13.4	*** wide-ranging chemosensor
stictogrammus	Giant Spotted Whiptail				Mexico
	diurnal	invert.	106	35.4	**** wide-ranging chemosensor

Aspidoscelis spp.	English name		Size		Range core
	Activity	Diet	mm	g	Ecological business model
tesselatus	Common Checkered Whiptail				Texas
	diurnal	invert.	84	17.2	**** wide-ranging chemo-sensor
tigris	Tiger Whiptail				Utah
	diurnal	invert.	92	23.0	**** wide-ranging chemo-sensor
uniparens	Desert Grassland Whiptail				New Mexico
	diurnal	invert.	71	10.2	**** wide-ranging chemo-sensor
velox	Plateau Striped Whiptail				Arizona
	diurnal	invert.	72	10.9	*** wide-ranging chemosensor
xanthanotus	Red-backed Whiptail				Mexico
	diurnal	invert.	102	31.6	**** wide-ranging chemo-sensor

Aspronema spp.	English name		Size		Range core
	Activity	Diet	mm	g	Ecological business model
cochabambae	Bolivia White-striped Skink				Bolivia
			69	6.4	
dorsivittatum	South America White-striped Skink				Paraguay
	diurnal	invert.	72	3.85	** refugia-anchored diurnal gr. liz.

55%). Half of the species live in discernible groups (50% v. 29%). Male hierarchies are present in typical frequency (15% v. 16%).

All habitat-documented species use burrows for daily refugia (v. 82%). *Aspidoscelis* are most often found in the arid rainfall belt (88% v. 30%) and, like the family, are absent only from the hyperdry (v. 0%) and wet (v. 0%) zones. Their elevational profile is skewed toward higher elevations: (l/m/h) 90/61/13 v. 95/32/6. They are slightly overrepresented in riparian zones (66% v. 44%) and underrepresented in human habitat (18% v. 29%). They are about average in sandy areas (53% v. 47%) and littoral zones (39% v. 38%).

Although parthenogenesis is common in the genus, it has not obviously changed the basic reproductive values. Clutch count is average (–0.46 v. –0.51 std. resid.), young are of average size (0.05 v. 0.10 std. resid.), RCM ratio is normal (0.159 v. 0.163), and incubation duration is normal (–0.54 v. –0.38 std. resid.). Parental care (4.9% v. 3.2%) and communal oviposition (2.4% v. 2.6%) are present at normal frequencies.

Longevity is a bit longer than the family average (–0.43 v. –0.74 std. resid.), perhaps because their inactive season is so long (88% of species are Nearctic). Population density is extraordinarily well documented in *Aspidoscelis*: 60 credible estimates of 10 species (perhaps conspicuous lizards are easier to survey). The mean density is low (0.23 kg/ha v. 1.06 kg/ha). Endemism is average (49% v. 50%); island species are only slightly underrepresented (29% v. 37%). Nearly 10% have been introduced (9.8% v. 8.3%), but none has been introduced more than twice (v. 0.6%). Fewer than random are imperiled (28% v. 41%), and none is on one of the IUCN's lists of the most endangered (v. 12%). In other respects *Aspidoscelis* are typical teiids.

Aspronema

Aspronema is a two-species genus of slightly large (5.0 g v. 3.3 g), long-tailed (2.00 v. 1.33 × SVL) skinks from southern South America (Bolivia to Uruguay). At least one species (*dorsivittatum*) is elongate (v. 35%). Sex differences are not reported. *A. dorsivittatum* is considered to be an invertebrate eater (v. 84%), though it is known to eat plants. *A. dorsivittatum* is found low (0–1000 m); *cochabambae* is high (2700–3900 m). *A. dorsivittatum* is found in littoral (v. 21%), sandy (v. 25%), and riparian (v. 23%) areas. Both species are viviparous (v. 36%), having average litters (0.14 v. 0.06 std. resid.). Endemism is about average (50% v. 64%). Neither species is found on an island (v. 57%), nor is either known to have been introduced (v. 2.1%). They have not been evaluated for conservation status.

Asymblepharus spp.	English name				Range core
			Size		
	Activity	Diet	mm	g	Ecological business model
alaicus	Alai False Snake-eyed Skink				Central Asia
			56	3.29	
eremchenkoi	Eremchenko's False Snake-eyed Skink				Kazakhstan
			52	1.63	
himalayanus	Himalaya Ground Skink				Himalayas
	nocturnal	invert.	53	1.66	* nocturnal ground lizard
ladacensis	Ladak Ground Skink				Himalayas
	diurnal	invert.	60	2.39	* refugia-anchored diurnal gr. liz.
mahabharatus	Mahabharat Ground Skink				Nepal
			60	4.00	
nepalensis	Nepal Ground Skink				Nepal
			39	1.01	
sikimmensis	Sikkim Ground Skink				Himalayas
	diurnal	invert.	45	1.11	* litter cryptozoid (where litter) or refugia-anchored diurnal gr. liz.
tragbulense	Tragbul Pass Skink				India
			50	2.32	

Asymblepharus

Asymblepharus is a moderate radiation (eight spp.) of small (1.96 g v. 3.3 g), terrestrial (75% v. 48%) or cryptozoid (25% v. 30%) skinks from the Himalayan region of central Asia (Uzbekistan and Pakistan to Bangladesh and China). As denoted by their genus name, these diminutive lizards lack movable eyelids, a trait common in small lizards. A diversity of ecological business models has been assigned, all with low confidence: litter cryptozoid (33% v. 28%), nocturnal ground lizard (33% v. 30%), and refugia-anchored diurnal ground lizard (33% v. 30%). Many species are elongate (50% v. 35%). Two species (*himalayanus* and *sikimmensis*) show seasonal color (25% v. 6%). Sit-and-wait foraging is widespread for a skink (50% v. 16%). Nocturnality is surprisingly common for such a cold region (33% v. 14.5%); the remainder are diurnal (67% v. 74%). Direct basking is unsurprisingly common (38% v. 21%), but thermoconforming is overrepresented as well (13% v. 2%). Mean active body temperature is a relatively normal 27.5°C (v. 31.1°C). All documented primary predator concealment tactics are retreat to vegetation (v. 25%). Social relations are inadequately documented, but suggest both male intolerance and hierarchies.

Recorded elevations are of course high: intermediate (83% v. 35%); high (67% v. 6%). The upper limits for the documented species are *himalayanus*, 4500 m; *ladacensis*, 6000 m; *mahabharatus*, 2300 m; *nepalensis*, 1580 m; *sikimmensis*, 3200 m; and *tragbulense*, 2650 m. *Asymblepharus* are overrepresented in riparian areas (40% v. 23%) and human habitats (20% v. 14%), but they are not documented in littoral (v. 21%) or sandy (v. 25%) areas.

Most documented species are understandably viviparous (67% v. 36%). Clutch/litter count is moderately large (0.47 v. 0.06 std. resid.). Young are notably small (ontogenetic mass range 1.00 v. 0.12 std. resid.). The combination of small and numerous young is somewhat uncommon in viviparous skinks, but viviparity in this case is likely induced by the high elevation and high latitude constraints on environmental warming of eggs. Some nocturnal species may find a way to bask from concealment when pregnant.

Endemism is low (38% v. 64%). No species occurs on islands (v. 57%), nor has any been introduced (v. 2.1%). This genus has not been evaluated for conservation status. In other respects *Asymblepharus* are typical skinks.

Ateuchosaurus

Ateuchosaurus is a two-species genus of average (3.0 g v. 3.3 g), often elongate (100% v. 35%), cathemeral, litter cryptozoid (100% v. 28%) skinks from extreme eastern Asia (Vietnam, China, Japan). Tails are on the short side (1.00 v. 1.33 × SVL) and breakable (100% v. 73%). Sex differences are not noted. At least one species (*chinensis*) is conspicuously wary (v. 11%). Their elevational profile is normal (l/m/h: 100/50/50 v. 89/35/6). They are found in human habitats (50% v. 14%). Both are oviparous (v. 64%), with normal to slightly numerous clutch counts (0.31 v. 0.06 std. resid.). Endemism is about average (50% v. 24%). Both species are found on islands (v. 57%). Neither species has been introduced (v. 2.1%). Neither has been evaluated for conservation status. In other respects they are probably typical skinks.

Ateuchosaurus spp.	English name				Range core
			Size		
	Activity	Diet	mm	g	Ecological business model
chinensis	Chinese Short-limbed Skink				Vietnam
	cathemeral	invert.	73	4.10	*** litter cryptozoid
pellopleurus	Ryukyu Short-limbed Skink				Japan
	cathemeral	invert.	59	2.24	*** litter cryptozoid

Atlantolacerta sp.	English name				Range core
			Size		
	Activity	Diet	mm	g	Ecological business model
andreanskyi	Atlas Dwarf Lizard				Morocco
	diurnal	omnivore	46	2.20	* refugia-anchored diurnal gr. liz.

Aurivela spp.	English name				Range core
			Size		
	Activity	Diet	mm	g	Ecological business model
longicauda	Long-tailed Whiptail				Argentina
	diurnal	invert.	67	7.4	**** wide-ranging chemosensor
tergolaevigata	Montane Whiptail				Argentina
	diurnal	invert.	52	3.42	**** wide-ranging chemosensor

Atlantolacerta

Atlantolacerta andreanskyi is the sole species in a genus of small (2.2 g v. 5.1 g) refugia-anchored diurnal ground lacertids found in alpine meadows (2000–3800 m) in the Atlas Mountains of Morocco. Tails are short (1.30 v. 2.06 × SVL) and bright blue in juveniles. Surprisingly, nearly half of the population has damaged toes. There are contradictory accounts of SSD. This may be at least in part due to differing proportions in the bodies of the two sexes: for a given SVL, males are heavier, have larger heads, and have slightly longer limbs, but no information was tabulated on how the SVL distributions of the two sexes compare. The sexes differ in color (v. 33%) but not seasonally (v. 15%). Overall, the sex differences rollup documents fairly large differences between the sexes (2.0 v. 0.86 on 0 to 6 scale).

Although *Atlantolacerta andreanskyi* is often said to be an invertebrate eater, 3%–6% of its diet is vegetation. As befits its high elevation, its mean active body temperature is low (28.5°C v. 33.3°C). It is a direct basker (v. 25%), active forager (v. 92%), and is not conspicuously wary (v. 11%). Nonetheless, individuals often remain under rocks, especially adults during cloudy weather or midday in summer. When they emerge, they may steal food from other lizards (kleptoparasitic). Maybe that is related to the extraordinary incidence of damaged toes. Males are intolerant of other males; keep away from my toes?

The species is associated with riparian habitats (v. 28%). Reproduction is oviparous; mean clutch count is slightly low (−0.09 v. 0.13 std. resid.). Hatchlings are relatively large, as indicated by reduced ontogenetic mass range (−0.53 v. −0.05 std. resid.). Females are reported to copulate only once per day, whereas documented restrictions on copulation frequency in some other lizards apply to just males. Incubation is relatively slow (0.16 v. −0.36 std. resid.), possibly due to the elevation. Females may oviposit communally (v. 1.9%). In captivity, the proximity of hatchlings is tolerated by mothers but not fathers. RCM is normal (0.237 v. 0.267 ratio).

Longevity is also normal (0.16 v. 0.19 std. resid.). The species does not occur on islands (v. 26%), nor has it been introduced (v. 4.6%). It is considered imperiled (v. 56%), but it is not on the IUCN's lists of the most endangered (v. 15.5%). In other respects *Atlantolacerta* is similar to other lacertids.

Aurivela

Aurivela is a two-species genus of small (5.0 g v. 22.1 g), wide-ranging chemosensor teiids (100% v. 97%) endemic to sandy (100% v. 47%), low elevation (100% v. 95%), dry (100% v. 10%) areas of Argentina. They have the family's highest mean active body temperature (40.8°C v. 36.9°C), albeit based solely on *longicauda*. This species has been recorded active at a remarkable body temperature of 45°C. In general, *Aurivela* differ little from archetypical teiids, except that they have low clutch counts (−0.77 v. −0.51 std. resid.).

Aurivela are slender in build (0.31 v.

0.88 std. resid.). Tails are on the long side (2.48 v. 2.30 × SVL) and may be breakable (50% v. 21%) but not blue (0% v. 15%). The only reported sex difference is m>f relative size (1.00 v. 0.89 on –2 to 2 scale). Both species are generalized invertebrate eaters (100% v. 62%) and reliant on flight (100% v. 77%) as a primary predator concealment tactic. They bask directly (v. 58%). They have not been introduced (v. 8.3%) and are not imperiled (v. 41%). In other respects *Aurivela* are typical teiids.

Australolacerta

Australolacerta australis is the sole species in a genus of average (4.6 g v. 5.1 g) refugia-anchored diurnal ground lacertids endemic to South Africa. It is poorly known (filled cells 25% v. 33%).

In morphology it is a typical lacertid. Sex differences are few and poorly documented (0.0 v. 0.86 on 0 to 6 scale). It is conspicuously wary (v. 11%). Reproduction is typical for lacertids. The species does not occur on islands (v. 26%) and has not been introduced (v. 4.6%). It is subject to conservation concern (v. 56%), but it is not listed on the IUCN lists of the most endangered (v. 15.5%). In other respects *Australolacerta* is similar to other lacertids.

Australolacerta sp.	English name		Size		Range core
	Activity	Diet	mm	g	Ecological business model
australis	Southern Rock Lizard				South Africa
	diurnal		58	4.62	* refugia-anchored diurnal gr. liz.

Bachia

Bachia (25 spp.) is the second-largest genus of gymnophthalmids (behind *Riama* 30 spp.). Like *Riama*, *Bachia* is predominantly elongate species (96% v. 41%), but unlike *Riama*, most *Bachia* are limbless or have reduced limbs. Two-thirds of *Bachia* are subterranean foragers; the remainder litter cryptozoids (v. 55%). They are found at low elevation (v. 51%) from Saint Vincent, in the Lesser Antilles, and Costa Rica to Brazil. They are a bit small (1.92 g v. 2.35 g) and very

Bachia spp.	English name		Size		Range core
	Activity	Diet	mm	g	Ecological business model
barbouri	Barbour's Earless Teju				Peru
			58	1.42	
bicolor	Two-colored Earless Teju				Venezuela
		invert.	63	1.70	*** subterranean forager
blairi	Panarica Earless Teju				Costa Rica
		invert.	54	1.21	** litter cryptozoid
bresslaui	São Paulo Earless Teju				Paraguay
			84	3.52	*** subterranean forager
cacerensis	Caceres Earless Teju				Brazil
		invert.	68	2.10	
didactyla	Two-fingered Teju				Brazil
			73	2.51	
dorbignyi	D'Orbigny's Earless Teju				Paraguay
			67	2.00	*** subterranean forager
flavescens	Guianan Earless Teju				Venezuela
	cathemeral	invert.	67	2.04	** subterranean forager
geralista	Geralista Earless Teju				Brazil
			82	3.30	
guianensis	Northwestern Earless Teju				Colombia
			54	1.21	
heteropa	La Guaira Earless Teju				Venezuela
	diurnal	invert.	55	1.25	*** litter cryptozoid
huallagana	Dixon's Earless Teju				Peru
			62	1.65	
intermedia	Noble's Earless Teju				Peru
			84	3.52	
micromela	Micro-footed Earless Teju				Brazil
			70	2.27	
oxyrhina	Cerrado Earless Teju				Brazil
			67	0.97	
pallidiceps	Cope's Earless Teju				Panama
		invert.	62	1.65	** litter cryptozoid

Bachia spp.	English name				Range core
			Size		
	Activity	Diet	mm	g	Ecological business model
panoplia	Manaus Earless Teju				Brazil
			70	2.27	*** subterranean forager
peruana	Peru Earless Teju				Peru
			86	3.74	*** subterranean forager
psamophila	Sand-loving Earless Teju				Brazil
			62	1.69	** subterranean forager
pyburni	Pyburn's Earless Teju				Colombia
			70	2.27	
remota	Tumucumaque Earless Teju				Brazil
			72	2.39	* litter cryptozoid
scaea	Madeira Earless Teju				Brazil
			59	1.46	
scolecoides	Vanzolini's Earless Teju				Brazil
			65	1.89	
talpa	Ruthven's Earless Teju				Colombia
			56	1.29	
trisanale	Stacy's Earless Teju				Northwestern South America
	diurnal	invert.	66	1.93	**** subterranean forager

Baikia spp.	English name				Range core
			Size		
	Activity	Diet	mm	g	Ecological business model
africana	West African Worm Lizard				Cameroon, Nigeria
			156	1.40	**** subterranean forager

slender (shape index –1.28 v. –0.32 std. resid.), but they have normal-length tails (1.66 v. 1.72 × SVL), which are all breakable (v. 28%). Secretory pores are common (36% v. 29%). Males tend to be smaller than females (–0.67 v. –0.12 on –2 to 2 scale). One species (*flavescens*) is sexually dichromatic, suggesting that it may rise above ground on some sexually important occasions. In keeping with the smaller male mass, the ratio of relative tail lengths is longer in males (1.17 v. 1.05 ratio). Mean body temperature is normal (28.5°C v. 28.8°C). Two species have been characterized socially (*flavescens, trisanale*); both have tolerant males. Like many subterranean foragers, they are partial to sandy habitats (36% v. 14%). They are also overrepresented in human habitats (29% v. 15%), perhaps capitalizing on disturbed or loosened soil. Population density is relatively high for gymnophthalmids (see also family account): 0.7 kg/ha v. 0.2 kg/ha. Endemism is comparatively low (48% v. 68%). Twenty-nine percent are considered imperiled (v. 25%), but none is listed on the IUCN's lists of most endangered species (v. 11.5%). In other respects *Bachia* are typical gymnophthalmids.

Baikia

Baikia is a single-species genus of central African (Cameroon, Nigeria) amphisbaenid; it is little-known, even by amphisbaenid standards (13% v. 17% cells filled). None has been collected since 1963, when it was found in grasslands.

Barisia

Barisia is a moderate (seven spp.) genus of large-bodied (22.2 g v. 10.1 g), robust (shape index 0.37 v. –0.60 std. resid.), terrestrial (86% v. 45%) or cryptozoic (14% v. 14%) anguids from Mexico. Outwardly they resemble the arboreal *Abronia*, but are uniformly refugia-anchored diurnal ground lizards (v. 34%). Their tails are a bit short for anguids (1.30 v. 1.66 × SVL), and none is prehensile (v. 31%). No *Barisia* is elongate (v. 22%). All are sexually dichromatic, however (v. 26%), and 28.6% (v. 3.8%) have seasonal color. These add up to a higher sex rollup score (1.43 v. 0.62 on 0 to 6 scale).

All *Barisia* eat invertebrates (v. 93%) by active foraging (v. 71%) during the day (v. 70%). For predator concealment, half of the species flee to a crevice or burrow (v. 33%); the other half flee to vegetation (v. 11%).

Recorded body temperatures for *Barisia* are low (23.8°C), though not especially so for anguids (v. 24.5°C). *Barisia* are overrepresented in the terrestrial microhabitat stratum (86% v. 45%) and in burrow use for refugia (75% v. 36%). None is arboreal (v. 39%). One-third of species are found in the intermediate rainfall belt (401–1100 mm/y v. 50%), one-third in mesic (1101–3000 mm/y v. 100%), and one-third in wet (>3000 mm/y v. 25%) zones. All species are

Barisia spp.	English name				Range core
			Size		
	Activity	Diet	mm	g	Ecological business model
ciliaris	Sierra Alligator Lizard				Mexico
	diurnal	invert.	115	33.9	* refugia-anchored diurnal gr. liz.
herrerae	Herrera's Alligator Lizard				Mexico
	diurnal		106	27.2	* refugia-anchored diurnal gr. liz.
imbricata	Imbricate Alligator Lizard				Mexico
	diurnal	invert.	102	17.5	** refugia-anchored diurnal gr. liz.
jonesi	Jones's Alligator Lizard				Mexico
		invert.	104	17.8	
levicollis	Chihuahua Alligator Lizard				Mexico
		invert.	122	31.5	** refugia-anchored diurnal gr. liz.
planifrons	Oaxaca Alligator Lizard				Mexico
			103	17.4	** refugia-anchored diurnal gr. liz.
rudicollis	Rough-necked Alligator Lizard				Mexico
			102	16.9	** refugia-anchored diurnal gr. liz.

Barkudia spp.	English name				Range core
			Size		
	Activity	Diet	mm	g	Ecological business model
insularis	Orissa Limbless Skink				India
	nocturnal	invert.	92	2.02	*** subterranean forager
melanosticta	Russell's Legless Skink				India
			127	4.30	**** subterranean forager

found at high (v. 41%) and medium (v. 77%) elevation belts and only 14% (v. 47%) are low. They are not found in littoral (v. 9%), sandy (v. 15%), riparian (v. 19%), or human (v. 11%) habitats.

All *Barisia* are viviparous (v. 61%). The litters are small for anguids (0.28 v. 0.85 std. resid.); more distinctively, their young are remarkably large for anguids (ontogenetic range −0.65 v. 0.62). The young are about 1/34 of maximum size, rather than the family's 1/72. No *Barisia* have been reported to exhibit parental care (v. 17%).

All are endemic to Mexico (v. 65% endemism). None is found on islands (v. 17%), and none has been introduced (v. 4%); 71% are considered imperiled (v. 63%). Nearly a third (29% v. 28%) are on one of the IUCN's lists of the most endangered.

Barkudia

Barkudia is a two-species genus of average (3.0 g v. 3.3 g), limbless (100% v. 16%), subterranean forager (100% v. 12%) skinks from India. They are very skinny (shape index −2.23 v. −0.46 std. resid.), with moderately long tails for subterranean foragers (0.50 v. 1.33 × SVL). Eyes are reduced. Sex differences and thermal and social behavior are not reported. They are known from only low elevation (v. 89%). They occur in littoral (50% v. 21%), sandy (50% v. 25%), and human (50% v. 14%) habitats but not riparian areas (v. 23%). They are oviparous (v. 64%), but clutch count is not known. Both species are endemic (v. 64%). *B. insularis* occurs on an island (v. 57%). Neither has been introduced (v. 2.1%). *B. insularis* is critically endangered (CR on IUCN's list of the most endangered; 50% v. 14%). In other respects they are probably typical subterranean skinks.

Basiliscus

Basiliscus is a small (four spp.) genus of large-sized (150 g v. 67 g), stocky (shape index 1.47 v. 1.20 std. resid.) diurnal, ground-foraging tree lizards in Corytophanidae from Central America and Colombia. Males sport flamboyant ornaments, often nuchal, dorsal, and caudal crests that may be elaborated into fins. Both sexes have corytophanid-typical long tails (2.70 v. 2.75 × SVL), which break readily in juveniles. Unlike other corytophanid genera, *Basiliscus* do not show color change (v. 22%). In keeping with their large size, *Basiliscus* are strongly m>f dimorphic in body size (1.67 v. 0.90 on −2 to 2 scale). *Basiliscus* are also overrepresented in sexual dichromatism (50% v. 22%) and seasonal color (25% v. 11%). These combine to produce an exceptionally high score on the sex differences rollup (3.0 v. 1.6 on 0 to 6 scale).

All basilisks are omnivores (v. 50%), incorporating the consumption of vertebrates. However, in this unusual case, the vertebrates

Basiliscus spp.	English name				Range core
			Size		
	Activity	Diet	mm	g	Ecological business model
basiliscus	Common Basilisk				Colombia
	diurnal	omnivore	189	238	*** ground-foraging tree lizard
galeritus	Western Basilisk				Colombia
	diurnal	omnivore	181	193	*** ground-foraging tree lizard
plumifrons	Green Basilisk				Nicaragua
	diurnal		152	109	*** ground-foraging tree lizard
vittatus	Brown Basilisk				Central America
	diurnal	omnivore	148	101	*** ground-foraging tree lizard

Bassiana spp.	English name				Range core
			Size		
	Activity	Diet	mm	g	Ecological business model
duperreyi	Eastern Three-lined Skink				Eastern Australia
	diurnal	invert.	62	2.63	* refugia-anchored diurnal gr. liz.
platynota	Red-throated Skink				Eastern Australia
	diurnal	invert.	66	3.03	* refugia-anchored diurnal gr. liz.
trilineata	Western Three-lined Skink				Southwestern Australia
	diurnal	invert.	60	2.36	** refugia-anchored diurnal gr. liz.

eaten include fish (all spp. except *vittatus*), which basilisks capture with equal representation of sit-and-wait (50% v. 80%) and active (50% v. 20%) foraging, though the sample is only two. On the basis of ca. 12 m^2 home ranges in *vittatus*, basilisks have exceptionally small activity areas (–2.05 std. resid.; comparison to other corytophanids not available). All basilisks are judged conspicuously wary (v. 56%).

Their primary predator concealment tactic is retreat to vegetation (v. 20%). All species bask directly (v. 44%). They maintain a typical tropical body temperature (32.4°C v. 30.3°C).

Male hierarchies are known in half of the species (50% v. 33%).

Three-quarters of the species are primarily arboreal (v. 89%). They generally sleep in vegetation hanging over water. If illuminated at night, *vittatus* will change sleeping site even if they have not been otherwise disturbed. They are overrepresented in littoral zones (25% v. 11%), sandy areas (50% v. 22%), riparian zones (100% v. 44%), and around humans (50% v. 33%).

They are all oviparous (v. 89%), with an average clutch count (–0.12 v. –0.22 std. resid.) and the family's smallest hatchlings (ontogenetic range 1.67 v. 1.16 std. resid.). Incubation period is somewhat short (–0.70 v. –0.52 std. resid.). They have not been observed to show parental care (v. 11%). Longevity is short (–0.41 v. –0.28 std. resid.), though *plumifrons* is recorded to live for 13 years.

Population density is unremarkable (0.50 v. 0.20 kg/ha), though difficult to quantify in lizards such as these that are found in linear home ranges along watercourses. Three-quarters of *Basiliscus* have been introduced (75% v. 33%). The eggs of *galeritus* are sought-after for human food. None is deemed to be at conservation risk (v. 25%). In other respects they are typical corytophanids.

Bassiana

Bassiana is a small genus (three spp.) of uniformly average (2.7 g v. 3.3 g), marginally elongate (100% v. 35%), diurnal (100% v. 74%), refugia-anchored ground (100% v. 30%) skinks from Australia (including Tasmania, excluding the Northern Territory). They are comparatively well known (filled cells 39% v. 25%). Seasonal color is present in all species (100% v. 6%) but somewhat variable geographically, at least in *duperreyi*. All species bask directly (100% v. 21%); their mean active body temperature is 27.5°C (v. 31.1°C). All species are described as group living (v. 16%). This extends to reproduction, at least in the sense that *duperreyi* has been found to prefer to lay their eggs communally, especially if the other eggs present have been recently deposited. They are unusual in having both genetic and environmental components to their sex determination. Given the difficulty of a mother sampling the thermal consequences of her choice of oviposition site, crowd sourcing the location may be a way of taking advantage of the assessments made by other females. All species are known to lay communally (v. 2.6%), though the oviposition site preferences of the other species have not been tested. Clutch count in *Bassiana* is relatively high (0.67 v. 0.06 std. resid.) and incubation is somewhat short (–0.53 v. –0.40 std. resid.), which equates

Bavayia spp.	English name				Range core
	Activity	Diet	Size mm	Size g	Ecological business model
crassicollis	Strand Bavayia				New Caledonia
	nocturnal	invert	71	7.1	*** nocturnal canopy walker
cyclura	Forest Bavayia				New Caledonia
	nocturnal	invert.	61	4.48	*** nocturnal canopy walker
exsuccida	Sclerophyll Bavayia				New Caledonia
	nocturnal	invert.	43	1.55	**** nocturnal canopy walker
geitaina	Gracile Bavayia				New Caledonia
	nocturnal	invert	61	4.48	** trunk/wall lizard
goroensis	Goro Bavayia				New Caledonia
	nocturnal		44	1.71	*** nocturnal canopy walker
montana	Montane Bavayia				New Caledonia
	nocturnal	invert.	64	5.2	** nocturnal canopy walker
nubila	Three Peaks Gecko				New Caledonia
	nocturnal		57	3.72	* nocturnal canopy walker
ornata	Ornate Bavayia				New Caledonia
	nocturnal		59	4.01	** trunk/wall lizard
pulchella	Pretty Bavayia				New Caledonia
	nocturnal		44	1.71	*** nocturnal canopy walker
robusta	Robust Forest Bavayia				New Caledonia
	nocturnal		69	6.5	*** nocturnal canopy walker
sauvagii	Sauvage's Bavayia				New Caledonia
	nocturnal	invert.	54	3.06	*** nocturnal canopy walker
septuiclavis	Pale-striped Bavayia				New Caledonia
	nocturnal	invert.	45	1.80	** nocturnal canopy walker

to 34 days (*duperreyi*). The young are moderately small (0.59 v. 0.012 std. resid.). Their greater number evidently drives up their collective mass, as RCM ratio in *platynota* is an awe-inspiring 0.460 (v. 0.236).

Longevity is relatively high (0.54 v. 0.17 std. resid.) based on *duperreyi* (7 years). Population density is not known. Two-thirds of species occur on islands (v. 57%). None has been introduced (v. 2.1%). All are considered imperiled (v. 47%), but none is on one of the IUCN's lists of the most endangered (v. 14%). In other respects *Bassiana* are typical skinks.

Bavayia

Bavayia (12 spp.) are the pervasive, small nocturnal arboreal geckos of New Caledonia. These diplodactylids are mostly (83% v. 25%) nocturnal canopy walkers, with the remainder (17% v. 20%) residing primarily on trunks. They are a bit smaller than a typical diplodactylid (3.55 g v. 4.15 g). Most (83% v. 26%) have prehensile tails and all have secretory pores (v. 34%). In most, if not all species, the grip of the prehensile tail is aided by adhesive scansors on the tail tip. They lack color change (v. 5.4%). SSD and dichromatism are not reported. Social interactions are undocumented. All eat invertebrates (v. 80%). They have not been reported to bask (v. 17%). They are overrepresented in littoral (27% v. 17%) and human (18% v. 5%) habitats, but they are not documented to occur in sandy areas (v. 24%) or riparian zones (v. 10%). Longevity is short for a diplodactylid (0.49 v. 0.90 std. resid.). They are all island species (v. 46%) and endemic (v. 72%) to New Caledonia. As with many remote island forms, they are well represented on the lists of species of conservation concern (64% v. 81%), including the most endangered IUCN categories (27% v. 29%). In other respects these are typical diplodactylids.

Bellatorias

Bellatorias is a small genus (three spp.) of very large (212 g v. 3.3 g), omnivorous (100% v. 13%), refugia-anchored diurnal ground (100% v. 30%) skinks found in northeastern Australia (New South Wales, Queensland, the Northern Territory) and Papua New Guinea. The genus is composed of corpulent skinks (shape index mean of 1.31 v. −0.46 std. resid.), most notably *major* (shape index 1.65 std. resid.). All three species obtain their mass with an omnivorous diet, which, at least in the case of *major*, remarkably includes the consumption of fungi. Note that the "Major Skink" is not *B. major*.

Sex differences, foraging mode, and home-range areas are poorly known. In the case of *major*, the home range is slightly small for its size (−0.24 v. −0.07 std. resid.). One species each is reported to be cathemeral (33% v. 9%), crepuscular (33% v. 2.4%), and diurnal (33% v. 74%). Perhaps they are relatively flexible in diel activity pattern. All are conspicuously wary (v. 11%). They retreat to a crevice or burrow as their primary predator concealment tactic (v.

Bellatorias spp.	English name				Range core
			Size		
	Activity	Diet	mm	g	Ecological business model
frerei	Major Skink				Queensland
	diurnal	omnivore	171	123	*** refugia-anchored diurnal gr. liz.
major	Land Mullet				Queensland
	cathemeral	omnivore	283	619	*** refugia-anchored diurnal gr. liz.
obiri	Arnhem Land Gorges Skink				Northern Territory
	crepuscular	omnivore	173	126	** refugia-anchored diurnal gr. liz.

Bipes spp.	English name				Range core
			Size		
	Activity	Diet	mm	g	Ecological business model
biporus	Ajolote				Mexico
	diurnal	invert.	194	28.0	**** subterranean forager
canaliculatus	Four-toed Worm Lizard				Mexico
			191	28.0	**** subterranean forager
tridactylus	Three-toed Worm Lizard				Mexico
			131	10.6	**** subterranean forager

43%). All bask directly (v. 21%). Two-thirds may use burrows as their daily refugium (v. 24%). At least in the case of *major*, they seem to cluster in family groups, and this species shows genetic evidence of monogamy. They are found in the littoral zones (33% v. 21%) and sandy areas (33% v. 25%), but not riparian (0% v. 23%) or human (0% v. 14%) habitats.

All are viviparous (v. 36%) with very small litter counts (–1.48 v. 0.06 std. resid.). The few young are, however, large (ontogenetic mass range –0.94 v. 0.12 std. resid.). The young may remain in the vicinity of their parents for some time. Large young in large lizards is a common pattern on islands, presumably to expedite the date on which the young get large enough to escape vulnerability to predators (offspring of small lizards may never pass this threshold). Perhaps this phenomenon operates to a degree on a continent that is comparatively bereft of predatory mammals, but has many predatory birds.

Longevity is about average (0.21 v. 0.17 std. resid.), based on *frerei* (10.3 years) and *major* (23 years). Endemism is comparatively low (33% v. 64%). Only one-third occur on islands (v. 57%). None has been introduced (v. 2.1%). All are considered imperiled (v. 47%), including placement on the IUCN's lists of the most endangered (100% v. 14%). In other respects *Bellatorias* are typical skinks.

Bipedidae and *Bipes* (fig. 4.13C)

Bipes is the only genus in the amphisbaenian family Bipedidae. It is a small (three spp.) genus of bizarre, moderately large-sized (20.3 g v. 5.3 g lizards), relatively robust (shape index –1.14 v. –3.97 for Amphisbaenidae), front-limbed subterranean foragers from Mexico. The large but short front limbs are used mole-like in scratching soil from the sides of the tunnel as *Bipes* rams its head forward and flexes it side to side. *Bipes* are sometimes called mole lizards.

Although *Bipes* tails are breakable, regrowth is minimal. Similar to Amphisbaenidae, no tails are prehensile (v. 5% for lizards). They are not reported to have crests (v. 5%), secretory pores (v. 15%), color change (v. 11%), SSD (0.0 v. 0.36 on –2 to 2 scale), sexual dichromatism (v. 21%), seasonal color (v. 6.5%), or blue tails (v. 2.3%). All are invertebrate eaters (v. 73% in lizards, 97% in amphisbaenians). One species is known to be active during daytime (v. 67% in lizards). The documented mean body temperature is normal (30.2°C v. 30.8°C for lizards). Unlike other subterranean forager families, their depth in the soil has been quantified for these three species: *biporus* is found in soil (not in litter) beneath litter at all seasons, at a mean soil depth of 36 mm (they rise closer to sunny surfaces when needed for thermoregulation); *canaliculatus* has a depth range of 25–600 mm; *tridactylus* is shallower on average (50–200 mm), though it will go deeper if necessary to avoid desiccation during the dry season. One species (*biporus*) is associated with (damp) sandy soils; one (*tridactylus*) is found in urban areas.

As far as is known, all are oviparous and have small clutch counts (–0.87 v. 0.0 std. resid.). The young are very large, as indicated by a below-average ontogenetic mass range (–1.47 v. 0.0 std. resid.), equating to a 1/17 range in *Bipes* versus 1/44 range in lizards.

Longevity is short (–1.11 v. 0.0 std. resid.), based on a single species (*biporus*, 3.3 years). None is found on islands (v. 40% of lizards), none has been introduced (v. 3.3% of lizards), and none is considered imperiled

(v. 49% of lizards). The mole-like front legs have not escaped the attentions of local residents in the area of occupancy. At least some residents believe that they are used to invade the rectums of sleeping humans and other animals.

Blaesodactylus

Blaesodactylus (six spp.) are large (15.1 g v. 3.31 g) arboreal gekkonids from Madagascar. They remind me of large *Gehyra*. They are only a little less well known than a typical gekkonid (23% v. 25% of cells filled), which in this case means that no ecological business models or daily activity patterns can be assigned with confidence. Trunk/wall lizard is the modal ecological business model (75% v. 45%); one species is possibly a ground-foraging tree lizard (25% v. 5.3%).

Blaesodactylus spp.	English name		Size		Range core
	Activity	Diet	mm	g	Ecological business model
ambonihazo	Mahajanga Gecko				Madagascar
	cathemeral		87	12.8	*** trunk/wall lizard
antongilensis	Antongil Gecko				Madagascar
	cathemeral		79	9.8	* ground-foraging tree lizard
boivini	Northern Giant Gecko				Madagascar
	cathemeral	invert.	117	29.5	*** trunk/wall lizard
microtuberculatus	Ankarana Gecko				Madagascar
			93	15.6	
sakalava	Dry Forest Tree Gecko				Madagascar
	cathemeral		81	10.6	*** trunk/wall lizard
victori	Victor's Madagascar Velvet Gecko				Madagascar
	nocturnal		100	19.1	

Tails are on the long side of gekkonids (1.27 v. 1.07 × SVL); none is said to be prehensile (v. 1.8%). Secretory pores are not reported (v. 46%), nor is color change (v. 15%), sexual dichromatism (v. 6.6%), or other sex differences (0 v. 0.63 on 0 to 6 scale). A remarkable 80% are categorized as active at any time (v. 11%); the remaining documented species is nocturnal (20% v. 73%). Direct basking is in evidence (17% v. 5%). All males defend space (v. 87%). Use of burrow refugia is not reported (0% v. 10%).

All are found exclusively at low elevation (v. 92%). Although not reported in the natural environments of littoral (v. 11%), sandy (v. 11%), or riparian (v. 17%) zones, they are present in human habitats (50% v. 27%).

Because *Blaesodactylus* are large, the fixed clutch count of two is notably small in mass-adjusted terms (−0.89 v. −0.26 std. resid.). Perhaps to compensate, the resultant hatchlings are conspicuously large (ontogenetic mass range −1.83 v. −0.19 std. resid.). Reproduction otherwise hews to gekkonid norms, except that communal oviposition is not reported (v. 8.6%) and parental care is elevated (17% v. 2%). The exemplar of parental care, *boivini*, not only protects the eggs, but the female guards the young after hatching. The incubation period for *boivini* is unusually variable in length (70–211 days). More information is needed to understand that flexibility. Does development abate when necessary to avoid hatching during inauspicious seasons?

Longevity is mildly short compared to gekkonids (−0.12 v. 0.15 std. resid.), equating to 8 years for *sakalava* and 5.2 years for *boivini*. *Blaesodactylus* have not been introduced (v. 3.7%), and only 25% of evaluated species are considered imperiled (v. 47%), though none is on one of the IUCN's lists of most endangered (v. 11%). In other respects *Blaesodactylus* are typical gekkonids.

Blanidae and *Blanus*

Blanus is the sole genus (seven spp.) of Blanidae. They are medium-bodied (4.2 g v. 5.3 g), slender (shape index = −2.97 v. 0.0 or −3.97 std. resid. for the closely related Amphisbaenidae), limbless subterranean foragers (v. 8%) from around the Mediterranean (Portugal, Spain, Morocco, and possibly Algeria; Greece, Turkey, Lebanon, Syria, and Iraq). Morphologically they are similar to amphisbaenids with a slightly higher frequency of species that will break tails (29% v. 15%), though regrowth is minimal. None has secretory pores (v. 3.4% in amphisbaenids). SSD is recorded in one species (*cinereus*): m>f. Several species are not morphologically distinguishable from *cinereus*; therefore, this relationship may apply to multiple species.

As far as is known, they are all cathemeral (100% v. 7%) eaters of invertebrates (100% v. 73%). *B. cinereus* is reported to be able to locate prey by the prey's vibrations. It is also said to be conspicuously wary (14% v. 8%). Heat is obtained resting on the underside of sun-heated rocks. Recorded species are found in the arid (100% v. 30% in lizards: 151–400 mm/y) and intermediate (100% v. 27% in lizards: 401–1100 mm/y) rainfall zones and low (100% v. 82% in lizards) and medium elevations (80% v. 41% in lizards). Few subter-

Blanus spp.	English name / Activity	Diet	Size mm	Size g	Range core / Ecological business model
alexandri	Marden Worm Lizard				Turkey
			157	4.01	**** subterranean forager
aporus	Poreless Worm Lizard				Syria
			115	1.62	*** subterranean forager
cinereus	Iberian Worm Lizard				Iberian Peninsula
	cathemeral	invert.	180	6.0	**** subterranean forager
mariae	Maria's Worm Lizard				Portugal
			180	5.97	* subterranean forager
mettetali	Morocco Worm Lizard				Morocco
		invert.	154	3.77	*** subterranean forager`
strauchi	Anatolia Worm Lizard				Turkey
	cathemeral	invert.	180	6.0	**** subterranean forager
tingitanus	Tangier Worm Lizard				Morocco
			164	4.53	*** subterranean forager

Brachylophus spp.	English name / Activity	Diet	Size mm	Size g	Range core / Ecological business model
bulabula	Fiji Banded Iguana				Fiji
	diurnal	herbivore	188	288	**** giant herbivore
fasciatus	Lau Banded Iguana				Fiji
	diurnal	herbivore	148	143	**** giant herbivore
vitiensis	Fiji Crested Iguana				Fiji
	diurnal	herbivore	192	308	**** giant herbivore

ranean foragers extend up into the intermediate elevations (thermal constraint?), but blanids do. They are regularly found in littoral zones (25% v. 17% in lizards), sandy or dune areas (75% v. 22% in lizards), and human habitat (25% v. 18.5% in lizards).

Reproductive mode (100% v. 82% oviparous) has been recorded in three species, which all have very small clutches (–0.61 v. 0.0 std. resid.). Young are especially large, as indicated by reduced ontogenetic size range (–0.60 v. 0.0 std. resid.). Phrased another way, their young are 1/19 of maximal mass (v. 1/44). Parental care is not known (v. 2.9%).

Longevity is recorded for three species (6.3–16 years), which, when size adjusted, is longer than average (1.0 v. 0.0 std. resid.). One species (*strauchi*) occurs on islands (14% v. 40%)—an unexpectedly low frequency given the circum-Mediterranean distribution. None has been introduced (v. 3.3%), also odd given the long history of commerce in that area. One of four species (25% v. 49%) evaluated for conservation status has been deemed imperiled (*strauchi*). It is not listed on one of the IUCN's lists of the most endangered (v. 14%).

Brachylophus (fig. 4.4C)

Brachylophus is a three-species genus of green arboreal giant herbivores endemic to Fiji. As Fiji is not near any landmass formerly or presently occupied by Iguanidae, this implies an extraordinary feat of overwater dispersal. *Brachylophus* mean size is the second from smallest within the Iguanidae, though they are sizable (233 g v. 609 g). Shape index (1.89 v. 2.11 std. resid.) is also the family's second smallest (in both cases surpassed only by *Dipsosaurus*). Their tails are comparatively long (2.33 v. 1.53 × SVL) and may be breakable (33% v. 38%), though the tail vertebrae lack autotomy planes. In this work, all three species are categorized as ecologically crestless, though for taxonomic purposes a line of peaked middorsal scales can be discerned. Secretory pores are absent (v. 38%). Basic color change is universal (v. 21%), though none exhibits advanced color change (v. 2.4%).

SSD is at the typical m>f level (1.0 v. 1.3 on –2 to 2 scale). Two of three species are sexually dichromatic (v. 38%); seasonal color is not reported. The m:f ratio of relative tail lengths is low, as expected (0.88 v. 0.94). The sex differences rollup score (on 0 to 6 scale) is strong but unremarkable (2.00 v. 1.88).

Compared to other iguanids, *Brachylophus* is notable for its dependence upon crypsis (100% v. 9%). These giant herbivores are known for remaining invisible high in the canopy of Fiji's tropical forests. They are not considered to be conspicuously wary (v. 29%), as they freeze surrounded by similarly green foliage; they do not flee. But they also rarely descend to a forest stratum that would allow them to be seen by humans. One species is recorded to bask directly (33% v. 55%); active body temperature is 32.0°C (v. 34.3°C).

Males defend space against other

males (v. 90%), though details of their social lives are understandably sparse.

They are found at low elevation in mesic (1101–3000 mm/y) and wet areas (up to 5000 mm/y). They are not particularly attracted to littoral (33% v. 41%), sandy (0% v. 34%), riparian (0% v. 22%), or human (33% v. 24%) habitats.

Their reproduction is typical oviparity, with a small clutch count (–1.29 v. 0.45 std. resid., second only to *Amblyrhynchus*). In lieu of large clutches, *Brachylophus* lay fairly often (ca. every 4 months), and they mate after each oviposition. The ontogenetic size range is smaller than all other iguanid genera, including *Amblyrhynchus* (–1.93 v. –0.55 std. resid.), indicating exceptionally large hatchlings. The production of few large hatchlings suggests that *Amblyrhynchus* and *Brachylophus* share similar predatory pressures. My guess is that, in both cases, these remote island lizards are vulnerable to only birds (among natural predators) and only when hatchlings are small. This might generate selective pressure to produce young that grow quickly to an invulnerable size, and having become relatively invulnerable, they suffer little subsequent mortality, allowing population survival despite the low fecundity. A major difference between *Amblyrhynchus* and *Brachylophus* is the former's vulnerability to periodic El Niño–driven starvation.

Females may defend their nest sites (for how long?), yielding a high frequency of parental care (67% v. 31%), and one that probably serves them ill in the modern era of introduced terrestrial mammals. Incubation duration is very long (1.90 v. –0.76 std. resid.), corresponding to about 220 days. This lengthy incubation may reflect the relative constancy of Fiji's tropical maritime climate (no urgency to complete development before a stressful season) and limited insolation and warming on the (pre-human) forest floor.

Longevity is unremarkable (0.18 v. 0.16 std. resid.), corresponding to 14–15 years. Population density is high (24.5 kg/ha based on six estimates for *vitiensis*). Prior to the introduction of mongoose, *Brachylophus* probably occurred at such densities throughout the Fijian archipelago. All species are considered to be imperiled (v. 97%), and all appear on the most endangered lists of the IUCN (v. 51%). For this reason, two of three species have been translocated to some of the few remaining mongoose-free islands (v. 33%), though none has been translocated more than twice (v. 9.5%). In other respects *Brachylophus* are typical iguanids.

Brachymeles

Brachymeles is a major radiation (41 spp.) of slightly small (2.51 g v. 3.25 g), elongate (88% v. 35%) or limbless skinks, found primarily in the Philippines. One species (*apus*) is found on Borneo (Malaysia), and one (*miriamae*) is found on the mainland (Thailand). Most are litter cryptozoids (80% v. 28%), the remainder are subterranean foragers (20% v. 12%). *Brachymeles* are known at an average-scincid level (filled cells 26% v. 25%). This is in some sense damning with faint praise, as ecological business models could be assigned with full confidence to only a single species. The diel activity pattern or diet is known for only 5 of 41 species (12%). Although most are believed to be litter cryptozoids, these assignments are poorly supported. At least one species (*makusog*), although unseen on the surface, is readily captured in pitfall traps, suggesting secretive surface travel.

Brachymeles are slender (shape index –1.24 v. –0.46 std. resid.), with short tails (0.85 v. 1.33 × SVL). Given that the genus is composed of subterranean foragers (mean relative tail length 0.37 × SVL: table 4.1, trait 5) and litter cryptozoids (mean relative tail length 1.38 × SVL), one might suppose that the subterranean foragers within *Brachymeles* would have tails shorter than the genus mean and the litter cryptozoids longer. This is not supported by the data (another case of phylogenetic inertia, or erroneous assignment of ecological business model?); both types have indistinguishably intermediate tails, which are short for skinks. Tails are generally breakable (93% v. 73%).

The mean SSD for the genus is very strong in the direction of males larger than females (1.50 v. –0.31 on –2 to 2 scale), an attribute unexpected in such small lizards. However, it is based on only two species (*bonitae*, *tridactylus*), and 12 other species (not described explicitly in relation to SSD) have measurements with minimal sex differences. Other sex differences are generally absent; therefore, the sex rollup score is an underwhelming 0.07 (v. 0.28 on 0 to 6 scale).

Most species are invertebrate eaters (83% v. 84%); the remainder are omnivores (17% v. 13%)—a surprisingly high frequency of plant intake for such small lizards. Although none is said to be carnivorous, ingestion of vertebrates has been noted in two species that are larger than average but not absolutely large (*boulengeri*, 3.5 g; *talinis*, 7.6 g). Daily activity patterns are a diverse mix: 20% (v. 9%) cathemeral, 40% (v. 74%) diurnal, and 40% (v. 14.5%) nocturnal. Given the difficulties in ascertaining the diel activity pattern of generally invisible lizards, these values are likely to be revised with additional observations, as is the evidence that

Brachymeles spp.	English name		Size		Range core
	Activity	Diet	mm	g	Ecological business model
apus	Bornean Slender Skink				Malaysia
	nocturnal	invert.	103	2.64	*** subterranean forager
bicolandia	Bicol Slender Skink				Philippines
			58	1.39	* subterranean forager
bicolor	Two-colored Slender Skink				Philippines
			126	9.7	* subterranean forager
boholensis	Bohol Slender Skink				Philippines
			81	3.28	* subterranean forager
bonitae	Pretty Slender Skink				Philippines
			70	2.27	* litter cryptozoid
boulengeri	Boulenger's Slender Skink				Philippines
		invert.	84	3.53	* litter cryptozoid
brevidactylus	Southern Bicol Slender Skink				Philippines
			52	1.10	* litter cryptozoid
cebuensis	Cebu Slender Skink				Philippines
			62	1.64	**** subterranean forager
cobos	Catanduañes Slender Skink				Philippines
			57	1.33	* litter cryptozoid
dalawangdaliri	Tablas Slender Skink				Philippines
			67	2.05	* litter cryptozoid
elerae	Common Slender Skink				Philippines
			61	1.60	
gracilis	Graceful Slender Skink				Philippines
	diurnal	invert	69	2.16	* litter cryptozoid
hilong	Mindanao Slender Skink				Philippines
			69	2.17	* litter cryptozoid
ilocandia	Ilokano Slender Skink				Philippines
			65	1.89	
isangdaliri	Aurora Slender Skink				Philippines
			52	1.10	
kadwa	Jessi's Slender Skink				Philippines
			106	6.4	* litter cryptozoid
libayani	Lapinig Islands Slender Skink				Philippines
			57	1.33	* litter cryptozoid
ligtas	Lubang Slender Skink				Philippines
			67	2.00	
lukbani	Lukban's Loamswimming Skink				Philippines
			73	1.20	* litter cryptozoid
makusog	Bicol Loamswimming Skink				Philippines
			102	5.8	* litter cryptozoid

only 5% of *Brachymeles* species are conspicuously wary (v. 11%). Locomotion without limbs is reported in the species that have limbs.

Foraging, predator concealment, and thermoregulatory behavior have not been observed. Their elevational profile emphasizes low elevations and is unremarkable (l/m/h: 91/26/0 v. 89/35/6). They are underreported for all habitat types—likely due in part to the paucity of observations—and absent from sandy areas (0% v. 25%).

All documented species are viviparous—a challenge for the usual assumption that viviparity is a response to high latitude or high elevation. Litter counts are average (–0.02 v. 0.06 std. resid.). Parental care is not reported (v. 3.2%) and might exist unnoticed.

Population density has been estimated for *gracilis* ($n = 3$; 0.81 kg/ha) and *schadenbergi* ($n = 1$; 0.12 kg/ha), for a mean very close to the family benchmark (0.31 kg/ha v. 0.35 kg/ha). All species are endemic to their country (v. 64%). Almost all are found on islands (98% v. 57%). None has been introduced (v. 2.1%). They are imperiled at an average level (33% v. 47%). Those listed by the IUCN on one of the most endangered lists are also of family-typical frequency (13.3% v. 14.1%). In other respects *Brachymeles* are typical skinks.

Brachymeles spp.	English name	Activity	Diet	Size mm	Size g	Range core	Ecological business model
mapalanggaon	Masbate Slender Skink			64	1.79	Philippines	
mindorensis	Mindoro Slender Skink			91	4.35	Philippines	* litter cryptozoid
minimus	Mini Slender Skink			55	0.63	Philippines	* litter cryptozoid
miriamae	Miriam's Legless Skink	cathemeral	invert.	91	1.98	Thailand	**** subterranean forager
muntingkamay	Caraballo Mountains Loamswimming Skink			71	3.80	Philippines	* litter cryptozoid
orientalis	Southern Slender Skink			96	4.98	Philippines	* litter cryptozoid
paeforum	Paef Slender Skink			56	1.32	Philippines	* litter cryptozoid
pathfinderi	Pathfinder Slender Skink			59	1.46	Philippines	
samad	Eastern Visayas Slender Skink			64	2.89	Philippines	* litter cryptozoid
samarensis	Walter's Slender Skink			59	1.49	Philippines	* litter cryptozoid
schadenbergi	Schadenberg's Slender Skink	diurnal		98	5.2	Philippines	* litter cryptozoid
suluensis	Sulu Slender Skink			61	2.47	Philippines	* litter cryptozoid
talinis	Large Striped Slender Skink	nocturnal	invert.	114	7.6	Philippines	** litter cryptozoid
taylori	Taylor's Five-toed Worm Lizard		omnivore	84	3.53	Philippines	* litter cryptozoid
tiboliorum	Western Mindanao Slender Skink			64	2.80	Philippines	* litter cryptozoid
tridactylus	Three-fingered Slender Skink			64	1.84	Philippines	* litter cryptozoid
tungaoi	Tungao's Slender Skink			89	7.1	Philippines	* litter cryptozoid
vermis	Taylor's Limbless Skink			71	1.12	Philippines	* litter cryptozoid
vindumi	Jen's Slender Skink			91	7.43	Philippines	* litter cryptozoid
vulcani	Camiguin Sur Slender Skink			67	3.16	Philippines	* litter cryptozoid

Brachymeles spp.	English name				Range core
			Size		
	Activity	Diet	mm	g	Ecological business model
wrighti	Wright's Slender Skink				Philippines
			102	5.8	** subterranean forager

Bradypodion spp.	English name				Range core
			Size		
	Activity	Diet	mm	g	Ecological business model
atromontanum	Swartberg Dwarf Chameleon				South Africa
	diurnal	invert.	64	6.8	*** diurnal canopy walker
caeruleogula	Eshowe Dwarf Chameleon				South Africa
	diurnal	invert.	57	5.2	*** diurnal canopy walker
caffer	Pondo Dwarf Chameleon				South Africa
	diurnal	invert.	64	7.1	*** diurnal canopy walker
damaranum	Knysna Dwarf Chameleon				South Africa
	diurnal	invert.	69	8.7	*** diurnal canopy walker
dracomontanum	Drakensberg Dwarf Chameleon				South Africa
	diurnal	invert.	66	7.5	*** diurnal canopy walker
gutturale	Little Karoo Dwarf Chameleon				South Africa
	diurnal	invert.	66	7.7	*** diurnal canopy walker
kentanicum	Kentani Dwarf Chameleon				South Africa
	diurnal	invert.	53	4.26	** diurnal canopy walker
melanocephalum	KwaZulu Dwarf Chameleon				South Africa
	diurnal	invert.	65	7.4	*** diurnal canopy walker
nemorale	Qudeni Dwarf Chameleon				South Africa
	diurnal	invert.	68	8.2	*** diurnal canopy walker
ngomeense	Ngome Dwarf Chameleon				South Africa
	diurnal	invert.	64	7.1	*** diurnal canopy walker
occidentale	Namaqua Dwarf Chameleon				South Africa
	diurnal	invert.	71	9.1	*** diurnal canopy walker
pumilum	Cape Dwarf Chameleon				South Africa
	diurnal	invert.	76	11.0	*** diurnal canopy walker
setaroi	Setaro's Dwarf Chameleon				South Africa
	diurnal	invert.	62	6.3	*** diurnal canopy walker
taeniabronchum	Elandsberg Dwarf Chameleon				South Africa
	diurnal	invert.	59	5.7	*** diurnal canopy walker
thamnobates	Natal Midlands Dwarf Chameleon				South Africa
	diurnal	invert.	76	10.9	*** diurnal canopy walker
transvaalense	Wolkberg Dwarf Chameleon				South Africa
	diurnal	invert.	65	7.3	*** diurnal canopy walker
ventrale	Eastern Cape Dwarf Chameleon				South Africa
	diurnal	invert.	73	9.8	*** diurnal canopy walker

Bradypodion

Bradypodion is a moderate (17 spp.) radiation of small-bodied typical chameleons (7.4 g v. 13.5 g) from far southern Africa (South Africa, Eswatini—former Swaziland—and likely Mozambique, and Namibia). A possible source of confusion is that whereas table 5.2 includes *Bradypodion* in the "typical" chameleon category, as opposed to "dwarf" chameleons, the common names used below characterize them as "dwarf" chameleons. They are small and hence dwarf "typical" chameleons. With reference to either group, their SSD is anomalous (–1.60 v. 0.14 for the family, on –2 to 2 scale). Though nearer the "typical" group in most characters, they are somewhat intermediate in reproductive characters: clutch/litter count (1.75 v. 2.81 std. resid. for "typical"); ontogenetic size range (0.30 v. 0.60 std. resid. for "typical"). In all other respects, they conform to the typical chameleons in ecological business model (diurnal canopy walker), shape index (0.49 v. 0.57 std. resid.), relative tail length (0.97 v. 1.09 × SVL), crests (29% v. 36%), microhabitat use (arboreal), and population density (0.14 v. 0.18 kg/ha).

As highlighted above, *Bradypodion* are anomalous in the high proportion of species in which males are not only smaller than females, but often much smaller (0.52–0.75 × female SVL). Therefore, mature females weigh two to five times that of mature males. Females are so much larger that small males hesitate to approach them. In contrast, Tolley and Burger (2007) emphasize that size differences cannot reliably be used for distinguishing the sexes throughout *Bradypodion*. This discrepancy may in part be attributable to the large proportion of young present in many populations (neonate to adult ratio in *occidentale* is reported to be as high as 15:1). The influx of newly mature females will therefore be of individuals about the same size as most adult males.

Seasonal color is universal in the *Bradypodion* for which color is documented (v. 18%). In combination with the strong SSD, this leads to a larger than average chamaeleonid sexual differences rollup (3.24 v. 2.16 on 0 to 6 scale).

Cruise foraging was coined to describe the foraging movements of *Bradypodion* (Butler 2005), and as far as is known, all genus members cruise (v. 58%). They also bask directly (100% v. 19%); many are considered heliophilic. For a chameleon, an unusually high proportion of males are judged tolerant (25% v. 7.5%); defense of space is less than the chamaeleonid norm (75% v. 93%). *Bradypodion* are twice as represented in littoral zones (12% v. 6%), dunes or sand (6% v. 3%), and human habitats (47% v. 23%) than other chameleons.

Almost all *Bradypodion* are viviparous (94% v. 20%), and the relative mass of fetuses is extraordinary (0.568 v. 0.444 for chameleons and 0.226 for lizards). Perhaps this wears them out, as their mass-adjusted longevity is absolutely short, though not by chameleon standards (−0.66 v. −0.83 std. resid.).

Bradypodion do not occur on islands (v. 48%), and several have been introduced (18% v. 5.4%), though not more than twice (0% v. 1.1%). All *Bradypodion* (100% v. 49%) are labeled as at conservation risk (almost all have highly restricted ranges), but they are underrepresented among the IUCN strongest categories of endangerment (12% v. 25%). Genetic distances between currently accepted species are often minimal, despite strong morphological differences. These may represent the effects of recent habitat fragmentation and several species may yet be regrouped taxonomically.

Brasiliscincus

Brasiliscincus is a small genus (three spp.) of moderately large (7.2 g v. 3.3 g), arboreal (33% v. 10%) and terrestrial (67% v. 48%) viviparous (100% v. 36%) skinks endemic to Brazil. The genus is composed of slightly hefty skinks (shape index mean of 0.20 v. −0.46 std. resid.) that are not elongate (0% v. 35%). They are not well-enough known to assign ecological business models.

Two of three species are m<f in SSD (−1.0 v. −0.31 on −2 to 2 scale), but documentation of other sex differences are lacking (1.0 v. 0.28 on 0 to 6 sex differences rollup score). Two species are invertebrate eaters (v. 84%) that forage actively, as is typical of skinks (100% v. 92%). Males defend space (100% v. 94%).

Litter count is average (0.05 v. 0.06 std. resid.), and young are of average relative size (ontogenetic mass range −0.02 v. 0.12 std. resid.). RCM is high (0.326 v. 0.236), but this is not surprising for heavily built lizards.

One species (*caissara*) is found on Sao Sebastiao Island (33% v. 57%). None has been introduced (v. 2.1%). They have not been evaluated by the IUCN for conservation status. In other respects *Brasiliscincus* are typical skinks.

Broadleysaurus

Broadleysaurus major is an active-foraging, diurnal, gerrhosaurid ground omnivore from southern Africa, from Ghana to Eritrea southward. Overall it is reasonably well known (cells filled: 44% v. 29%), but the foraging movement pattern is not well documented and therefore the assignment of ecological business model is uncertain. It is large even by gerrhosaurid standards (166 g v. 26.5 g) and heavily built (shape index 1.26 v. 0.38 std. resid.), with a short tail (1.00 v. 1.92 × SVL) that is shed only under great pressure. The skin is rough in texture and protected with osteoderms. It has secretory pores (v. 8%) but is not elongate (v. 19%). Although the sexes are not known to differ in size or color, males show seasonal color (v. 27%). *B. major* eats flowers, fruit (plants 11% by occurrence), and any vertebrate or inverte-

Brasiliscincus spp.	English name				Range core
			Size		
	Activity	Diet	mm	g	Ecological business model
agilis	Atlantic Forest Small-headed Skink				Brazil
		invert.	72	7.5	
caissara	São Paulo Small-headed Skink				Brazil
			72	7.5	
heathi	Caatinga Small-headed Skink				Brazil
		invert.	69	6.6	

Broadleysaurus spp.	English name				Range core
			Size		
	Activity	Diet	mm	g	Ecological business model
major	Rough-scaled Plated Lizard				Southern Africa
	diurnal	omnivore	186	166	* wide-ranging chemosensor

brate it can capture. It usually retreats to rock crevices, but the number of crevices an individual uses and their distribution are not reported. Males defend at least some of these crevices. It occurs from sea level to 1700 m, mostly in savanna habitat, though it tolerates human habitats as well (v. 11%). Reproduction is notable only for the small number (3.8) of eggs (−1.32 v. −0.20 std. resid.) and shortness of incubation (−0.89 v. −0.05 std. resid.), equating to about 76 days. The eggs seem unusually large (23 × 58 mm); a more objective measure, ontogenetic size range, is not available. Conversely, longevity is notably lengthy (0.89 v. −0.09 std. resid.), equating to 24 years. It has not been evaluated for conservation status by the IUCN. In other respects *Broadleysaurus* is a typical gerrhosaurid.

Bronchocela (fig. 4.5A)

Bronchocela is a moderate genus (13 spp.) of invertebrate-eating (100% v. 46%), arboreal (100% v. 43%) agamids from Southeast Asia. Though slightly larger (25 g v. 15.6 g) and more robust (shape index 1.03 v. 0.80 std. resid.) than average agamids, they have very long tails (mean 3.4 v. 1.9 × SVL) and gangly limbs and, therefore, convey a spindly impression. The two species classified as gliders expand their ribs to capture air when gliding and use their whip-like tails to launch themselves and rudder once airborne. They do not have lateral flaps or patagia.

Most *Bronchocela* (69% v. 38%) have modest nuchal crests (especially males). Secretory pores are absent (v. 13%). Short-term color change is slightly elevated (31% v. 22% for any color change, 23% v. 15% for advanced), and sexual color differences are muted (15% v. 53%). None (v. 20%) is reported to change color seasonally, though underreporting is likely. Males have substantially longer tails than females (1.19 v. 1.07 ratio, the second-highest ratio in Agamidae). The mean SSD is also slightly more male biased (0.83 v. 0.53 on −2 to 2 scale) than an average agamid, but the fascinating aspect of SSD in *Bronchocela* is that the two glider/leapers with SSD data are strongly male biased (*burmana* m>f; *cristatella* m≫f), which is contrary to the general glider pattern. Perhaps crudely gliding *Bronchocela* lack the aerial maneuverability characteristic of other gliders, and therefore females cannot select prospective mates by their dogfighting ability. Given the different morphological basis, gliding is presumably an independent innovation in *Bronchocela*. As far as is known, all bask directly (v. 24%) and hunt exclusively by sit-and-wait (v. 88%).

Observational data on this genus are inadequate on social relations, predator defense, habitat, longevity, and spacing. The two social observations are consistent with business model generalizations: the diurnal canopy walker (*smaragdina*) is

Bronchocela spp.	English name / Activity	Diet	Size: mm	Size: g	Range core / Ecological business model
burmana	Myanmar Green Crested Lizard				Myanmar
			79	16.4	** glider
celebensis	Sulawesi Garden Lizard				Indonesia
	diurnal		97	30.6	* diurnal canopy walker
cristatella	Green Crested Lizard				Southeast Asia
	diurnal	invert.	104	37.9	**** glider
danieli	Great Nicobar Forest Lizard				India
	diurnal		67	9.5	
hayeki	Sumatra Bloodsucker				Indonesia
	diurnal		95	28.6	
jubata	Maned Forest Lizard				Indonesia
	diurnal	invert.	117	55	*** diurnal canopy walker
marmorata	Philippines Garden Lizard				Philippines
	diurnal	invert.	98	31.8	* diurnal canopy walker
orlovi	Orlov's Forest Lizard				Vietnam
	diurnal		88	22.5	
rayaensis	Gunung Raya Green-crested Lizard				Thailand
			76	14.0	* diurnal canopy walker
rubrigularis	Nicobar Crested Lizard				India
			85	20.3	
shenlong	Shen Long Lizard				Malaysia
			91	25.1	
smaragdina	Chanthabun Forest Lizard				Southeast Asia
	diurnal	invert.	95	28.6	**** ground-foraging tree lizard
vietnamensis	Vietnam Forest Lizard				Vietnam
	diurnal		97	30.0	

solitary and the glider (*cristatella*) resident in a family group. Clutches are notably small (–1.06 v. 0.20 std. resid.), and multiple clutches per year are presumed. None has been introduced (v. 1.7%) and one of three evaluated species is considered imperiled (v. 37%). None is on one of the IUCN's lists of the most endangered (v. 5%).

Brookesia

Brookesia is a large (30 spp.) radiation of notably small dwarf chameleons (sensu table 5.2) from Madagascar, many with highly restricted distributions. They are among the smallest amniotes. Many have spines, crests, and unexpected flanges colored perfectly to mimic the dried leaves, twigs, and other vegetative features among which they remain cryptically immobile. Compared to other dwarf chameleons, they are slightly smaller (mean 1.48 g v. 2.04 g, though they range down to the 0.35 g *micra*), are all immobility experts (v. 98%), are slightly thinner (shape index 0.27 v. 0.32 std. resid.), have slightly longer tails (0.64 v. 0.53 × SVL), lack crests (v. 0%), are strongly m<f (–0.82 v. –0.70 on –2 to 2 scale of SSD), are 79% terrestrial (v. 83%), have small clutches (0.48 v. 0.60 std. resid.), have large young (ontogenetic size range –0.26 v. –0.16 std. resid.), are slightly shorter in mass-adjusted incubation duration (0.20 v. 0.44 std. resid.), and have the same mean population density (0.03 v. 0.03 kg/ha). In other words, they fit the profile of dwarf chameleons superbly. Most research on these tiny cryptic lizards revolves around collecting them at night, when they are sleeping exposed on low vegetation and can be found.

With so much pressure for crypsis, it is not surprising that sexual dichromatism (6.7% v. 53% in chameleons)

Brookesia spp.	English name				Range core
			Size		
	Activity	Diet	mm	g	Ecological business model
ambreensis	Amber Leaf-litter Chameleon				Madagascar
	diurnal	invert.	49	3.36	* immobility expert
antakarana	Antakarana Leaf Chameleon				Madagascar
	diurnal	invert.	48	3.17	*** immobility expert
bekolosy	Bekolosy Rainforest Chameleon				Madagascar
	diurnal	invert.	34	1.26	
betschi	Marojejy Dwarf Chameleon				Madagascar
	diurnal	invert.	39	1.91	** immobility expert
bonsi	Namoroka Leaf Chameleon				Madagascar
	diurnal	invert.	37	1.56	
brunoi	Anja Leaf Chameleon				Madagascar
	diurnal	invert.	40	1.98	* immobility expert
brygooi	Brygoo's Leaf Chameleon				Madagascar
	diurnal	invert.	47	2.98	* immobility expert
confidens	Ankarana Dwarf Chameleon				Madagascar
	diurnal	invert.	24	0.49	** immobility expert
decaryi	Spiny Leaf Chameleon				Madagascar
	diurnal	invert.	47	3.10	*** immobility expert
dentata	Toothed Leaf Chameleon				Madagascar
	diurnal	invert.	28	0.74	** immobility expert
desperata	Ambre Forest Dwarf Chameleon				Madagascar
	diurnal	invert.	29	0.85	** immobility expert
ebenaui	Northern Leaf Chameleon				Madagascar
	diurnal	invert.	47	2.98	** immobility expert
exarmata	Tsingy de Bemaraha Chameleon				Madagascar
	diurnal	invert.	26	0.64	*** immobility expert
griveaudi	Marojejy Leaf Chameleon				Madagascar
	diurnal	invert.	55	4.68	*** immobility expert
karchei	Ambavaomby Chameleon				Madagascar
	diurnal	invert.	32	1.05	* immobility expert
lambertoni	Fito Leaf Chameleon				Madagascar
	diurnal	invert.	41	2.10	
lineata	Lined Leaf Chameleon				Madagascar
	diurnal	invert.	42	2.20	** immobility expert
micra	Nosy Hara Micro Chameleon				Madagascar
	diurnal	invert.	21	0.35	*** immobility expert
minima	Minute Leaf Chameleon				Madagascar
	diurnal	invert.	21	0.37	*** immobility expert
perarmata	Antsingy Leaf Chameleon				Madagascar
	diurnal	invert.	57	5.0	*** immobility expert

Brookesia spp.	English name				Range core
			Size		
	Activity	Diet	mm	g	Ecological business model
peyrierasi	Maroantsetra Leaf Chameleon				Madagascar
	diurnal	invert.	26	0.61	** immobility expert
ramanantsoai	Ambohibooataba Leaf Chameleon				Madagascar
	diurnal	invert.	25	0.56	** immobility expert
stumpffi	Plated Leaf Chameleon				Madagascar
	diurnal	invert.	52	4.02	** immobility expert
superciliaris	East Madagascar Leaf Chameleon				Madagascar
	diurnal	invert.	44	2.60	** immobility expert
therezieni	Perinet Leaf Chameleon				Madagascar
	diurnal	invert.	47	3.10	** immobility expert
thieli	Domergue's Leaf Chameleon				Madagascar
	diurnal	invert.	42	2.20	*** immobility expert
tristis	French Mountain Leaf Chameleon				Madagascar
	diurnal	invert.	23	0.47	** immobility expert
tuberculata	Amber Mountain Leaf Chameleon				Madagascar
	diurnal	invert.	22	0.40	** immobility expert
vadoni	Iaraka River Leaf Chameleon				Madagascar
	diurnal	invert.	34	1.33	** immobility expert
valerieae	Manongarivo Leaf Chameleon				Madagascar
	diurnal	invert.	47	3.10	** immobility expert

Bufoniceps spp.	English name				Range core
			Size		
	Activity	Diet	mm	g	Ecological business model
laungwalaensis	Rajasthan Toad-headed Lizard				India/Pakistan
	diurnal	ants	59	6.7	** immobility expert

and seasonal color (0% v. 18%) are underrepresented in *Brookesia*. Correspondingly, the sex differences rollup score for *Brookesia* is subdued compared to chameleons (0.97 v. 2.16 on 0 to 6 scale).

In keeping with their crypsis, *Brookesia* go in for sit-and-wait foraging (80% v. 47%). They rarely bask directly (3.3% v. 19%). They live alone (100% v. 64%). No males are tolerant of other males (0% v. 7.5%). All *Brookesia* males are reported to defend space (v. 93%).

Brookesia are found in mesic or wetter habitats, with a greater representation in wet (>3000 mm/y) habitats than other chameleons (20% v. 5%). They are overrepresented at low elevation (100% v. 76%) and underrepresented above. They are not known to be found in littoral (v. 6%), sandy (v. 3%), or human (v. 23%) habitats. They are notably underrepresented in riparian zones (3% v. 21%). One would think riparian zones excellent for foraging *Brookesia*—are riparian zones also home to high densities of *Brookesia* predators? Birds? Snakes that prey on sleeping *Brookesia*?

They are all oviparous (v. 80%). Their mass-adjusted longevity is short even by chameleon standards (−1.16 v. −0.83 std. resid.). They have not been introduced (v. 5.4%) and are all considered imperiled (v. 95% of chameleons). Even more than chameleons, they are found in the ranks of the IUCN's most endangered (47% v. 25%). They probably need nothing more than intact habitat to survive.

Bufoniceps

The confusingly named *Bufoniceps* is but a single, small, sand-dune agamid living along the Pakistan-India border. Unlike some better-known ant-eating (v. 20%) immobility experts (v. 4.5%; e.g., *Moloch, Phrynosoma*), *Bufoniceps* is unornamented with pointed scales or crests, though it shares the short (0.6 v. 1.86 × SVL) unbreakable tail. The placement of nostrils at the top of its head and direct observation suggests that it may forage from a position within the sand from which only the eyes and nostrils protrude, negating the camouflage value of pointed scales or crests. Also, unlike some rotund ant eaters (e.g., *Phrynosoma*), it is relatively thin (shape index 0.67 v. 0.80), perhaps to facilitate movement through sand. It is not reported to have color changing ability (v. 22%) or sexual (v. 53%) or seasonal (v. 20%) differences in color or size. Unlike many sand swimmers, which dive and swim into the sand to submerge, this dune lizard shimmies laterally into the sand. It can disappear in under four seconds.

Bunopus

Bunopus (three spp.) are small (2.46 g v. 3.31 g) ground geckos of the southwestern Asian deserts (Syria to

Pakistan). Though they are nominally well known (36% v. 25% of cells filled), little is known of their behavior. All are considered nocturnal ground lizards (v. 27%), though the nocturnality is contested for each of the three species, suggesting considerable adaptability in daily activity pattern. The modal pattern is cathemeral (67% v. 11%).

All have secretory pores (v. 46%), and basic color change is present (33% v. 15%). *B. crassicauda* is reported to have minimal toepads, *tuberculatus* to lack them. Sexual size differences are very subtle but tend in the direction of m<f (–0.25 v. –0.03 on –2 to 2 scale). Sexual dichromatism is not reported (v. 6.6%).

Vocalization is present in both spontaneous (33% v. 6.5%) and especially stress (67% v. 4%) contexts. Predator concealment is primarily by retreat to vegetation (v. 19%). Mean active body temperature is high for a gekkonid (30.0°C v. 27.6°C), presumably reflecting the prevailing summer temperatures in that part of the world. Half of the documented species use burrows for daily refugia (50% v. 10%), another adaptation to that environment. They are found at both low (100% v. 92%) and medium elevation (100% v. 26%). They are overrepresented in littoral zones (50% v. 11%), sandy areas (50% v. 11%), riparian zones (50% v. 17%), and human habitats (50% v. 27%).

Reproduction in *Bunopus* is similar to that of other gekkonids, though they are not known to exhibit adherent eggshells (v. 10.5%), communal oviposition (v. 8.6%), or parental care (v. 2%).

Their longevity notably exceeds the gekkonid standard (0.36 v. 0.15 std. resid.), which corresponds to 6.0 years for *blanfordii*. *Bunopus* have not been introduced (v. 3.7%). Half of the evaluated species are imperiled (v. 47%), though none is on any of the IUCN's lists of the most endangered (v. 11%). In other respects *Bunopus* are typical gekkonids.

Bunopus spp.	English name				Range core
			Size		
	Activity	Diet	mm	g	Ecological business model
blanfordii	Blanford's Rock Gecko				Jordan
	cathemeral		48	2.52	
crassicauda	Thick-tailed Tuberculated Gecko				Iran
	cathemeral		47	2.34	** nocturnal ground lizard
tuberculatus	Baluch Desert Gecko				Iran
	nocturnal	invert.	48	2.52	**** nocturnal ground lizard

Cadeidae and *Cadea*

Cadea spp.	English name				Range core
			Size		
	Activity	Diet	mm	g	Ecological business model
blanoides	Cuban Spotted Amphisbaena				Cuba
		invert.	195	7.6	**** subterranean forager
palirostrata	White-nosed Worm Lizard				Cuba
			209	9.4	**** subterranean forager

Cadea is the only genus in the amphisbaenian family Cadeidae. It is a small (two spp.) genus of moderately large-sized (8.5 g v. 5.3 g lizards), relatively slender (shape index –2.8 v. 0.0 or –3.97 std. resid. for Amphisbaenidae), limbless subterranean foragers (v. 8%) from the country of Cuba. Tails are very short (0.05 × SVL v. 1.37 or 0.11 for amphisbaenids) and are not known to break (v. 57%). *Cadea* are oviparous, with moderately small-sized clutches in *blanoides* (–0.58 v. 0.0 std. resid. or –0.12 for amphisbaenids). Only *palirostrata* has been evaluated for conservation status, and it is judged biologically vulnerable (100% v. 49% imperiled). In other respects information on Cadeidae is lacking or undistinguished from other amphisbaenians.

Caesoris

Caesoris sp.	English name				Range core
			Size		
	Activity	Diet	mm	g	Ecological business model
novaecaledoniae	Blue-mouthed Skink				New Caledonia
	diurnal		58	3.71	* diurnal canopy walker

Caesoris novaecaledoniae is the sole representative of a genus of average (3.7 g v. 3.3 g), nonelongate (0% v. 35% elongate), diurnal canopy walker (v. 1.8%) skinks endemic to New Caledonia. It is known from only a few specimens (filled cells 25% v. 25%).

It has a blue mouth lining, wide

toepads, and a long tail (2.00 v. 1.33 × SVL). Its tail is not reported to break (v. 73%). No sex differences are known. It is found at low elevation (0–400 m) in littoral zones (v. 21%). It lays about 4.0 eggs (0.23 v. 0.06 std. resid.). It is protected in New Caledonia (v. 47% imperiled) but not listed by the IUCN (v. 14%) in the most endangered categories.

Caledoniscincus

Caledoniscincus is a moderate radiation (14 spp.) of uniformly small (2.0 g v. 3.3 g), nonelongate (v. 35% elongate), diurnal (100% v. 84%) ground (93% v. 48%) skinks endemic (93% v. 64%) to New Caledonia (*atropunctatus* also in Vanuatu). Most are tentatively refugia-anchored diurnal ground lizards (60% v. 30%); the remainder are tentatively considered to be litter cryptozoids (40% v. 28%), though documentation of foraging stratum in relation to the litter layer (under or above?) is unclear for many species. See *Calyptotis* for an example of a nearby skink genus that unequivocally forages under the litter; it may be that the absence of mammalian ground predators on New Caledonia permits a more relaxed range of foraging strata use.

Caledoniscincus spp.	English name				Range core
			Size		
	Activity	Diet	mm	g	Ecological business model
aquilonius	Northern Litterskink				New Caledonia
	diurnal		44	1.57	** litter cryptozoid
atropunctatus	Speckled Litterskink				New Caledonia
	diurnal	invert.	45	1.67	** refugia-anchored diurnal gr. liz.
auratus	Koumac Litterskink				New Caledonia
	diurnal	invert.	46	1.74	** refugia-anchored diurnal gr. liz.
austrocaledonicus	Caledonian Litterskink				New Caledonia
	diurnal	invert.	53	2.78	** refugia-anchored diurnal gr. liz.
chazeaui	Chazeau's Litterskink				New Caledonia
	diurnal		40	1.13	** litter cryptozoid
constellatus	Starry Litterskink				New Caledonia
			50	2.32	
cryptos	Cryptic Litterskink				New Caledonia
	diurnal		42	1.27	
festivus	Giant Litterskink				New Caledonia
	diurnal	invert.	61	4.33	** refugia-anchored diurnal gr. liz.
haplorhinus	Strand Litterskink				New Caledonia
	diurnal	invert.	49	2.11	** refugia-anchored diurnal gr. liz.
notialis	Brown-nosed Litterskink				New Caledonia
	diurnal		48	2.03	** refugia-anchored diurnal gr. liz.
orestes	Panie Litterskink				New Caledonia
	diurnal		56	3.29	
pelletieri	Pelletier's Litterskink				New Caledonia
			52	2.54	
renevieri	Renevier's Litterskink				New Caledonia
	diurnal		46	1.74	*** litter cryptozoid
terma	Mandjelia Litterskink				New Caledonia
	diurnal		45	1.65	** litter cryptozoid

Tails are a bit long (1.54 v. 1.33 × SVL), and most are not reported to be breakable (14% v. 73% breakable). Although none is said to differ in size between the sexes, most differ in color between the sexes (71% v. 8%). None shows seasonal color (v. 6%). All documented species are invertebrate eaters (v. 84%) that forage actively (100% v. 92%). Wariness is about average (14% v. 11%). Direct basking is common if not universal (79% v. 21%). Their elevational profile likely matches that of New Caledonia (l/m/h: 100/17/0 v. 89/35/6). They are slightly underrepresented in sandy (8% v. 25%), riparian (8% v. 23%), and human (8% v. 14%) habitats and about average in littoral zones (25% v. 21%).

All species are oviparous (v. 64%) with an average clutch count (0.32 v. 0.06 std. resid.).

None has been introduced (v. 2.1%). They are notably imperiled (58% v. 47%) and one-third are on the IUCN's lists of the most endangered (33% v. 14%). In all other documented respects *Caledoniscincus* are typical skinks.

Callisaurus (fig. 2.5A)

Callisaurus sp.	English name				Range core
			Size		
	Activity	Diet	mm	g	Ecological business model
draconoides	Zebra-tailed Lizard				Arizona
	diurnal	omnivore	79	15.4	*** refugia-anchored diurnal gr. liz.

Callopistes spp.	English name				Range core
			Size		
	Activity	Diet	mm	g	Ecological business model
flavipunctatus	False Monitor				Peru
	diurnal	omnivore	260	582	*** wide-ranging chemosensor
maculatus	Spotted False Monitor				Chile
	diurnal	carnivore	135	76	*** wide-ranging chemosensor

Callisaurus draconoides is the sole species in a genus of moderately large (15.4 g v. 11.2 g) phrynosomatids from the North American southwest, ranging from Utah to Mexico. This refugia-anchored diurnal ground lizard is comparatively well known (filled cells 65% v. 38%). It is often seen scurrying across the desert floor with its conspicuously black and white striped tail underside held visibly above its head, scorpion fashion. It exhibits m>f SSD, sexual dichromatism, secretory pores, and seasonal color, for a family-lead-tied sex difference rollup of 4.0 (v. 1.68 on 0 to 6 scale). Its diet is notably opportunistic, varying greatly from time to time and place to place. It gives stress vocalizations and is conspicuously wary. It relies on both crypsis and flight, alternately, for predator concealment. Its mean active body temperature is a sizzling 38.5°C (v. 34.2°C). It inhabits dry places (93–209 mm/y) and may sleep buried in sand or occasionally simply lying on the surface. In general it seems to favor places with firm rather than loose soil. Its elevational range is wide (0–1520 m), and it inhabits all tabulated habitats except human habitats. Clutch count is small for a phrynosomatid (0.03 v. 0.68 std. resid.), and incubation is short (–1.11 v. –0.58 std. resid.), equating to about 31 days. The young are average in relative size (ontogenetic mass range –0.24 v. –0.26 std. resid.). RCM is a light 0.154 (v. 0.300). Longevity is about average (–0.19 v. –0.25 std. resid.), equating to a maximum of 5.9 years. Population density is a thin 0.04 kg/ha (v. 0.26 kg/ha), one of the thinnest of all lizards (see chapter 2, "Populations"). It occurs on many islands in the Sea of Cortez. It is listed as threatened in Mexico, but is not listed by the IUCN. In other respects *Callisaurus* is similar to other phrynosomatids.

Callopistes

Callopistes (Teiidae) is a two-species genus of large (210 g v. 22.1 g), direct-basking (100% v. 58%), wide-ranging chemosensors (100% v. 97%) found along the west coast of southern South America (Ecuador to Chile). The resemblance to a monitor lizard referenced in the common names is probably reflective of the relatively rounded snout and large head of *flavipunctatus*, but the short neck of both species and pointed snout of *maculatus* do not fit the varanid gestalt.

Callopistes are robust in build (shape index 1.54 v. 0.88 std. resid.). Tails are slightly short (1.90 v. 2.30 × SVL) and neither breakable (0% v. 21%) nor blue (0% v. 15%). The only reported sex difference is dichromatism in *maculatus* (50% v. 35%). The larger species (*flavipunctatus*) is omnivorous (v. 33%), including in its diet both endothermic and ectothermic vertebrates; the smaller species is a carnivore (v. 3%) and is conspicuously wary (v. 13%). Both are reliant on flight (100% v. 77%) as a primary predator concealment tactic and use burrows for their daily refugium (v. 82%). *C. maculatus* is associated with sandy (v. 47%) littoral areas (v. 38%).

Clutch count is average (–0.44 v. –0.51 std. resid.). Longevity is known for both species and is very short (–1.26 v. –0.74 std. resid.). *C. maculatus* is endemic to Chile (v. 50%). They have not been introduced (v. 8.3%) and are not imperiled (v. 41%). In other respects *Callopistes* are typical teiids.

Calodactylodes

Calodactylodes (two spp.) are moderately large (8.6 g v. 3.31 g) gekkonids that are trunk/wall lizards in South Asia (one endemic to India; one endemic to Sri Lanka). These cathemeral lizards aggregate in the mouth of rock shelters from which they sally forth at night.

Though they are nominally well known (45% v. 25% of cells filled), much is yet to be learned about their complex social behavior, which appears to include male hierarchical dominance of a group of 3–11, though

Calodactylodes spp.	English name				Range core
			Size		
	Activity	Diet	mm	g	Ecological business model
aureus	Indian Golden Gecko				India
	cathemeral	invert.	70	7.0	*** trunk/wall lizard
illingworthorum	Sri Lanka Golden Gecko				Sri Lanka
	cathemeral	invert.	81	10.5	*** trunk/wall lizard

caves may house up to 50 individuals. Males, especially the dominant male, develop a bright yellow throat, recognized in their common names.

Both species have secretory pores (v. 46%), and basic color change is present (50% v. 15%). Mysteriously, males are smaller than females (–1.0 v. –0.03 on –2 to 2 scale), despite the apparent benefits of large male size for acquiring behavioral dominance and therefore outsized reproductive advantages. Males have secretory pores (100% v. 46%), as well as sexual dichromatism (50% v. 15%), resulting in an unusually strong sex differences rollup score (2.0 v. 0.63 on 0 to 6 scale). Perhaps the solution to this mystery lies in the darkness of the caves in which these lizards spend their days. These lizards penetrate the zone of complete darkness, at least in some places, though it is not known where or how they sort out their social relations. If mobility, vocalizations, or reflectance in low light rather than visual bulk is the key to achieving social dominance, males may benefit from small size. It is also possible that the lizards adopt a lekking behavior at breeding time, and females choose males (presumably on some basis other than size), rather than males sorting out their dominance among themselves.

Both species call spontaneously (100% v. 6.5%), but neither is documented to issue stress vocalizations. The vocalizations are reported to be as loud as normal human conversational speech. This suggests that energetically expensive calling behavior, rather than size, may be pivotal to their social success.

They are associated with riparian zones (50% v. 17%) and human habitat (50% v. 27%), but not littoral (v. 11%) or sandy (v. 11%) areas.

Although *Calodactylodes* are oviparous (v. 100%), they have anomalous clutch counts (mean 6.3 v. 1.9). This mean is based on the midpoint of clutch count range in *aureus* (2.0, the gekkonid standard) and *illingworthorum* (10.5). The most likely explanation in this case is that these lizards are strongly communal (100% v. 8.6%) in their oviposition behavior (hundreds of eggs seen together), and clutches from multiple females contributed to an *illingworthorum* egg cluster that was interpreted as a single clutch. The language in the original source does not specify the number of females responsible for the 6–15 egg count range. The eggs are adherent (100% v. 10.5%), as they are for many cave-dwelling lizards.

Calodactylodes have not been introduced (v. 3.7%). Half of the evaluated species are imperiled (v. 47%) and are on the IUCN's lists of the most endangered (v. 11%). In other respects *Calodactylodes* are typical gekkonids.

Calotes (fig. 4.6B)

This genus has a tortured history, both because of rampant confusion as to which species are to be included and how to distinguish them, and because of some unfortunate interactions with humans. Many species acquire red marking on the face seasonally and are therefore known colloquially as bloodsuckers. Some residents mistakenly believe the red wash reflects their penchant for human sanguinvory. According to Minton (1966), others are offended by its head bobs, which they take to be mocking Muslim prayer habits; it was considered particularly pious to kill these lizards on Fridays, though more recent documentation of this phenomenon is lacking. Others use the subordinate males relegated to resting on roads as target practice for their driving skills. However, the main cause of their decline is believed by Masroor (2012) to be the pet trade. These lizards have had it rough.

As of 2017, there were 26 species assigned to *Calotes*. These 90% (v. 43%) arboreal agamids are spread throughout southern Asia. About half will forage from trees onto the ground (45% v. 19%); the remainder are trunk/wall lizards (35% v. 7%) or diurnal canopy walkers (15% v. 7%). There is one (*minor*) nocturnal ground lizard (5%, the highest value within Agamidae, v. 0.3%), a very rare role for an agamid. Crests are common (85% v. 38%), especially well developed in the neck region and especially large in males.

Mean mass is mildly larger than the family (18.7 v. 15.6 g); shape index is correspondingly a bit high (0.93 v. 0.80 std. resid.). Tails are long (2.4 v. 1.9 × SVL) and unbreakable (0% v. 17% breakable). The tail is prehensile in *versicolor* (3.8% v. 1.4%). Seasonal color is elevated vis-à-vis agamids (31% v. 20%) but dichromatism is not (38% v. 53%). SSD, like size, is mildly elevated (0.82 v. 0.53 on –2 to 2 scale). Secretory pores are rare (4.0%

v. 12.6%). Collectively these sex differences hew to the agamid norm (1.06 v. 1.07 on 0 to 6 scale).

The eyes are able to move independently in *versicolor*. A few species (*emma*, *versicolor*) enlarge their dewlaps during the breeding season, and *mystaceus* and *versicolor* seasonally develop a gular pouch. The one species that is nocturnal (*minor*) nonetheless retains seasonal color markings. There are reports that *minor* becomes active shortly before dark (which would enhance the value of color markings), and its nocturnality is presumably an evolutionarily recent activity change. Like most nocturnal ground lizards, it moves slowly.

Stress vocalizations are usually well developed in *Calotes* (15%, the highest proportion within Agamidae, v. 1%) and take the form of an "alarm whistle," which contrasts with the inchoate squawk characteristic of many lizard stress vocalizations. Note that "whistling" is in the common name of *liolepis* and *manamendrai*. All species rely on sit-and-wait foraging (v. 88%) and crypsis (v. 51%) for predator concealment; 80% (v. 24%) are reported to bask directly. Several species are strictly limited to trunks, whereas many are eclectic in their choice of perch diameter. Social relations are largely unreported. *Calotes* are found in habitats wetter than typical agamids, including 60% (v. 20%) representation in the mesic zone (1101–3000 mm/y) and 40% (v. 7%) representation in the wet zone (>3000 mm/y). Half the species are found in human habitats (52% v. 16%).

Breeding habits are similar to most agamids, though the maximum clutch size is an impressive 69 (*versicolor*), exceeded by only five chamaeleonids and two iguanids. The mass-adjusted mean clutch count is slightly above the agamid

Calotes spp.	English name	Activity	Diet	Size mm	Size g	Range core	Ecological business model
aurantolabium	Orange-lipped Forest Lizard			66	9.3	India	* diurnal canopy walker
bachae	Vietnam Blue-crested Lizard	diurnal	invert.	79	15.9	Cambodia	** ground-foraging tree lizard
bhutanensis	Bhutan Garden Lizard			53	4.66	Bhutan	
calotes	Green Garden Lizard	diurnal	omnivore	108	42.5	India	** diurnal canopy walker
ceylonensis	Painted Lip Lizard	diurnal	invert.	82	17.9	Sri Lanka	**** trunk/wall lizard
chincollium	Chin Hills Forest Lizard	diurnal	herbivore	112	46.9	Myanmar	** ground-foraging tree lizard
desilvai	Desilva's Lizard	diurnal		70	11.0	Sri Lanka	**** trunk/wall lizard
ellioti	South Indian Hills Calotes	diurnal	invert.	64	8.3	India	* trunk/wall lizard or diurnal canopy walker
emma	Forest Crested Lizard	diurnal	invert.	95	28.1	Southern East Asia	* ground-foraging tree lizard
grandisquamis	Big-scaled Variable Lizard	diurnal	invert.	117	53.7	India	*** diurnal canopy walker
htunwini	Htunwin's Forest Lizard	diurnal		74	12.9	Myanmar	
irawadi	Ayeyarwady Forest Lizard	diurnal		82	17.8	Myanmar	
jerdoni	Jerdon's Forest Lizard	diurnal	invert.	95	28.4	India	* ground-foraging tree lizard
liocephalus	Lionhead Agama	diurnal	invert.	77	15.1	Sri Lanka	* trunk/wall lizard
liolepis	Whistling Lizard	diurnal	invert.	79	16.3	Sri Lanka	*** trunk/wall lizard
manamendrai	Manamendra-Arachchi's Whistling Lizard			59	6.7	Sri Lanka	
maria	Maria's Forest Lizard	diurnal	invert.	95	28.6	India	** ground-foraging tree lizard
medogensis	Medog Bloodsucker			64	8.3	China	
minor	Dwarf Rock Agama	cathemeral	omnivore	890	16.8	India	**** nocturnal ground lizard

average (0.37 v. 0.20 std. resid.); more notable is the small size of *Calotes* hatchlings (ontogenetic size range: 1.87, the second-highest value within Agamidae v. 0.69 std. resid.). This equates to a hatchling that is 1/267 of the species maximum size rather than the 1/100 typical of Agamidae. Thus, *Calotes* young are slightly more numerous in a clutch and much smaller than other agamids. Although most lizard hatchlings do not differ in size by sex, male *versicolor* hatchlings are larger than female hatchlings. Male *mystaceus* are reported to provide parental care by guarding the nest; this extraordinary role needs better documentation. In some lizards, the combination of communal oviposition and postoviposition receptivity combine to induce male nest guarding, but neither of these phenomena are documented for *Calotes*.

The very short mean longevity (–1.9 v. –0.46 std. resid.) also needs more complete observations, as it is based only on *emma* (0.7 years) and *versicolor* (5 years). Population density is conspicuously inconsistent (and poorly documented), with several species known from only a few specimens and others being among the densest in mainland Asian forests. Mean density (1.65 kg/ha v. 0.16 kg/ha, the third highest within Agamidae) is high for mainland lizards. Introduction rates are elevated (8% v. 1.7%), including 4% (the second-highest value in Agamidae v. 0.6%) of serial invaders (*versicolor* is found in at least 18 new island populations). *Calotes* are also imperiled at an arguably elevated rate (43% v. 37%), including 7% (v. 5%) placement on one of the IUCN's lists of the most endangered.

Calotes spp.	English name				Range core
			Size		
	Activity	Diet	mm	g	Ecological business model
mystaceus	Blue Forest Lizard				Southeast Asia
	diurnal	invert.	112	33.4	*** trunk/wall lizard or ground-foraging tree lizard
nemoricola	Nilgiri Forest Lizard				India
	diurnal	invert.	103	36.9	** ground-foraging tree lizard
nigrilabris	Dark-lipped Lizard				Sri Lanka
	diurnal	invert.	95	28.2	** ground-foraging tree lizard
nigriplicatus	Ambon Dragon				Indonesia
			100	33.0	
pethiyagodai	Pethiyagoda's Crestless Lizard				Sri Lanka
			77	14.6	* trunk/wall lizard
rouxii	Forest Blood Sucker				India
	diurnal	invert.	68	10.1	** ground-foraging tree lizard
versicolor	Common Garden Lizard				Asia
	diurnal	omnivore	112	48.1	**** ground-foraging tree lizard

Calumma spp.	English name				Range core
			Size		
	Activity	Diet	mm	g	Ecological business model
amber	Amber Mountain Chameleon				Madagascar
	diurnal	invert.	88	16.3	*** diurnal canopy walker
ambreense	Ambre Mountain Chameleon				Madagascar
	diurnal	invert.	133	49.5	*** diurnal canopy walker
andringitraense	Andringitra Chameleon				Madagascar
	diurnal	invert.	56	4.95	*** diurnal canopy walker
boettgeri	Boettger's Chameleon				Madagascar
	diurnal	invert.	52	3.92	*** diurnal canopy walker
brevicorne	Short-horned Chameleon				Madagascar
	diurnal	invert.	126	42.5	*** diurnal canopy walker
capuroni	Anosyennes Chameleon				Madagascar
	diurnal	invert.	74	10.2	*** diurnal canopy walker
crypticum	Cryptic Chameleon				Madagascar
	diurnal	invert.	86	15.3	*** diurnal canopy walker

Calumma (fig. 4.3E)

Calumma is a large (33 spp.) Madagascan radiation of typical chameleons (see table 5.2). They vary greatly in size (2.1–199 g; mean 10.5 g) and are consistent with the profile of typical chameleons in ecological business model (97% v. 98% diurnal canopy walkers), shape index (0.53 v. 0.57 std. resid.), tail length (1.14 v. 1.09 × SVL), crests (30% v. 36%), SSD (0.85 v. 0.41 on –2 to 2 scale), arboreality (97% v. 99%), clutch count (3.5 v. 2.8 std. resid.), incubation duration (2.43 v. 2.43 std. resid.), and population density (0.08 v. 0.18 kg/ha). With their frequently coiled prehensile tail and exuberant rostral appendages, they are instantly recognizable as an archetypical chameleon.

The sexes routinely differ in color (70% v. 53% in chamaeleonids), though they are underrepresented among chameleons showing seasonal color (9% v. 18%). Still, their overall sex differences rollup score is very close to the chamaeleonid mean (2.42 v. 2.16 on 0 to 6 scale).

There is one giant *Calumma* (*parsonii*) that is carnivorous as an adult. *Calumma* are more frequently sit-and-wait foragers (75% v. 46%), cruise less frequently (25% v. 58%), and never forage actively (0% v. 10.5%). Their home-range area is recorded for only three *Calumma*, but it is notably small, even by chameleon standards (–1.80 v. –1.12 std. resid.).

Calumma are slightly less pugnacious than average chameleons, as a slightly higher proportion of males are tolerant of other males (25% v. 7.5%), and fewer males defend space (75% v. 93%).

All documented *Calumma* are found in the mesic (1101–3000 mm/y) rainfall belt (v. 85%).

All are oviparous (v. 80%). Although they have relatively large clutches (see above), they do not fit the "typical" chameleon pattern of having small young (ontogenetic range –0.45 v. 0.63 std. resid.). Their young are comparatively large (compared to either the "typical" chameleons, all chameleons, or lizards). *C. parsonii* incubation can last up to two years. *Calumma* longevity is documented for only one species (*parsonii*, 14 years), but it is the longest of any chameleon genus (0.15 v. –0.83 std. resid.). No *Calumma* has been introduced, but as with so many Madagascan lizards, they are almost all im-

Calumma spp.	English name				Range core
			Size		
	Activity	Diet	mm	g	Ecological business model
cucullatum	Epaulet Chameleon				Madagascar
	diurnal	invert.	145	63	*** diurnal canopy walker
fallax	Green-spotted Chameleon				Madagascar
	diurnal	invert.	41	2.10	*** diurnal canopy walker
furcifer	Twin-snouted Chameleon				Madagascar
	diurnal	invert.	61	6.1	*** diurnal canopy walker
gallus	Pinocchio Chameleon				Madagascar
	diurnal	invert.	52	4.06	*** diurnal canopy walker
gastrotaenia	Short-nosed Chameleon				Madagascar
	diurnal	invert.	60	5.9	*** diurnal canopy walker
glawi	Glaw's Chameleon				Madagascar
	diurnal	invert.	58	5.4	*** diurnal canopy walker
globifer	Large-headed Chameleon				Madagascar
	diurnal	invert.	143	60	*** diurnal canopy walker
guibei	Antsahamanara Chameleon				Madagascar
	diurnal	invert	80	12.0	*** diurnal canopy walker
guillaumeti	Guillaumet's Chameleon				Madagascar
	diurnal	invert.	55	4.61	*** diurnal canopy walker
hafahafa	Bizarre-nosed Chameleon				Madagascar
	diurnal	invert.	88	16.4	*** diurnal canopy walker
hilleniusi	Hillenius' Short-nosed Chameleon				Madagascar
	diurnal	invert.	62	6.3	*** diurnal canopy walker
jejy	Marojejy Peak Chameleon				Madagascar
	diurnal	invert.	79	12.3	** ground-foraging tree lizard
linotum	Lined Chameleon				Madagascar
	diurnal		47	2.98	*** diurnal canopy walker
malthe	Malthe's Chameleon				Madagascar
	diurnal	invert.	106	26.9	*** diurnal canopy walker
marojezense	Sleek Chameleon				Madagascar
	diurnal	invert.	61	6.0	*** diurnal canopy walker
nasutum	Nose-horned Chameleon				Madagascar
	diurnal	invert	48	3.30	*** diurnal canopy walker
oshaughnessyi	O'Shaughnessy's Chameleon				Madagascar
	crepuscular	invert.	125	50	*** diurnal canopy walker or immobility expert
parsonii	Parson's Giant Chameleon				Madagascar
	diurnal	carnivore	224	199	*** diurnal canopy walker
peltierorum	Peltiers' Chameleon				Madagascar
	diurnal	invert.	91	17.7	*** diurnal canopy walker
peyrierasi	Marojejy Mountain Chameleon				Madagascar
	diurnal	invert.	45	2.74	*** diurnal canopy walker

Calumma spp.	English name				Range core
			Size		
	Activity	Diet	mm	g	Ecological business model
tarzan	Tarzan Chameleon				Madagascar
	diurnal		61	6.1	** diurnal canopy walker
tsaratananense	Tsaratanana Dwarf Chameleon				Madagascar
	diurnal	invert.	55	4.68	** diurnal canopy walker
tsycorne	Blunt-nosed Chameleon				Madagascar
	diurnal	invert.	101	23.8	**** diurnal canopy walker
vatosoa	Tsararano Chameleon				Madagascar
	diurnal	invert.	52	4.06	* diurnal canopy walker
vencesi	Vences' Chameleon				Madagascar
	diurnal	invert.	62	6.3	*** diurnal canopy walker
vohibola	Mathevon's Chameleon				Madagascar
	diurnal	invert.	45	2.72	*** diurnal canopy walker

Calyptommatus spp.	English name				Range core
			Size		
	Activity	Diet	mm	g	Ecological business model
confusionibus	Confusion Mountains Sand Slider				Brazil
	nocturnal		61	1.60	** subterranean forager
leiolepis	Ibiraba Sand Slider				Brazil
			57	1.34	*** subterranean forager
nicterus	Rodrigues' Sand Slider				Brazil
			59	1.46	*** subterranean forager
sinebrachiatus	Lesser Sand Slider				Brazil
			60	1.55	*** subterranean forager

Calyptotis spp.	English name				Range core
			Size		
	Activity	Diet	mm	g	Ecological business model
lepidorostrum	Cone-eared Calyptotis				Queensland
		invert.	49	1.39	
ruficauda	Red-eared Calyptotis				New South Wales
		invert.	47	1.21	** litter cryptozoid
scutirostrum	Scute-snouted Calyptotis				Queensland
		invert.	50	1.42	** litter cryptozoid
temporalis	Broad-templed Calyptotis				Queensland
		invert.	38	0.68	* litter cryptozoid
thorntonensis	Thornton Peak Calyptotis				Queensland
		invert.	35	0.53	* litter cryptozoid

periled (97% v. 95%). Thirty percent (v. 25%) are in the most endangered IUCN categories.

Calyptommatus

Calyptommatus is a four-species genus of functionally limbless subterranean foraging gymnophthalmids endemic to Brazil. They are small (1.49 g v. 2.35 g) and thin (shape index −1.26 v. −0.32 std. resid.). Sex differences are not reported. One species (*confusionibus*) is said to be nocturnal, but activity is hard to discern in subterranean foragers. They are found in sandy areas (v. 14%). *C. confusionibus* has been evaluated for conservation risk and deemed to be endangered by the IUCN. In other respects *Calyptommatus* are probably typical gymnophthalmids.

Calyptotis

Calyptotis is a minor radiation (five spp.) of diminutive (0.97 g v. 3.25 g), slender (shape index −0.85 v. −0.46 std. resid.), elongate (100% v. 35%), sexually dichromatic (100% v. 8%), litter cryptozoid (100% v. 28%) skinks found in eastern coastal Australia. Often they exhibit red coloration in their tails or tail bases.

They are moderately well known (filled cells 31% v. 25%). Their tails are slightly short (1.11 v. 1.33 × SVL) and uniformly breakable (100% v. 73%). Seasonal color is absent (v. 6% present). Their sex differences rollup score (on scale of 0 to 6) is 1.60 (v. 0.28) due to the sexual dichromatism. They are all conspicuously wary (100% v. 11%), presumably keen to dive back into the litter and hope that any predators focus on their red tail. They are not reported to bask directly (0% v. 21%). They are found exclusively at low elevation. They are associated only with riparian habitats (20% v. 23%).

All are oviparous (v. 64%) with typical to slightly large clutch counts

(0.43 v. 0.06 std. resid.). Parental care (20% v. 3%) is noted in female *scutirostrum*, which may attend her clutch for a week after oviposition, at least in captivity.

Endemism is high (80% v. 64%). None has been found on an island (v. 57%) or introduced (v. 2.1%). All are considered imperiled (v. 47%), but none is on any of the IUCN's lists of the most endangered (v. 14%). In other respects *Calyptotis* are much like other skinks.

Caparaonia

Caparaonia itaiquara is the sole member of this genus of Brazilian gymnophthalmids. It is among the least-known of the little-known gymnophthalmids (cells filled: 12% v. 19%). Beyond that in the table below, available information is that *Caparaonia* has a long tail (2.00 v. 1.72 × SVL) that is not known to break (v. 28%). This species has secretory pores (v. 29%) and is elongate (v. 41%). Other sex differences are not documented. Clutch counts are not known (v. 1.9). This species is not known to have been introduced (v. 0.8%), nor has it been evaluated for conservation status.

Capitellum

Capitellum is a small genus (three spp.) of moderately large (5.3 g v. 3.3 g), cryptozoic (v. 30%), viviparous (100% v. 36%) skinks in the Lesser Antilles. As far as is known, they are all extinct; their ecological business models are unknown. *C. mariagalantae* and *parvicruzae* have not been seen in 150 years, *metallicum* in 100 years. Behavior and ecology are thus essentially unknowable.

They were slightly more robust than an average skink (shape index 0.10 v. –0.46 std. resid.). All tails were breakable (v. 73%). They were not elongate (v. 35% elongate). They had moderately large litters (0.50 v. 0.06 std. resid.) based on *mariagalantae*. They were all endemic (v. 64%) and occurred on islands (v. 57%). None was known to have been introduced (v. 2.1%). They all occur on one of the IUCN's lists of the most endangered (v. 14%); in this case they are critically endangered. In other respects *Capitellum* were likely typical skinks.

Carlia (fig. 4.11F)

Carlia is a major radiation (44 spp.) of uniformly small (2.07 g v. 3.25 g), nonelongate (v. 35% elongate) refugia-anchored diurnal ground (96% v. 30%) skinks, found primarily in Australasia (especially Queensland and New Guinea). Several species reach across Wallace's Line into the Lesser Sundas of Indonesia, and *ailanpalai* has been repeatedly introduced in Micronesia. The only other ecological business model represented is litter cryptozoid (4.5% v. 28%).

Tails are of normal length (1.48 v. 1.33 × SVL) and breakable (96% v. 73%). *Carlia* are not blue-tailed

Caparaonia sp.	English name				Range core
			Size		
	Activity	Diet	mm	g	Ecological business model
itaiquara	Itaiquara Saxicolous Lizard				Brazil
			52	1.88	

Capitellum spp.	English name				Range core
			Size		
	Activity	Diet	mm	g	Ecological business model
mariagalantae	Marie-Galante Skink				Guadeloupe
			71	7.0	
metallicum	Lesser Martinique Skink				Martinique
			62	4.67	
parvicruzae	Lesser Saint Croix Skink				US Virgin Islands
			62	4.67	

Carlia spp.	English name				Range core
			Size		
	Activity	Diet	mm	g	Ecological business model
aenigma	Enigmatic Rainbow-skink				New Guinea
	diurnal	invert.	51	2.50	* refugia-anchored diurnal gr. liz.
ailanpalai	Curious Skink				New Guinea
	diurnal	invert.	56	4.20	**** refugia-anchored diurnal gr. liz.
amax	Bauxite Rainbow-skink				Northern Australia
	diurnal	invert.	38	0.95	** refugia-anchored diurnal gr. liz.

Carlia spp.	English name	Activity	Diet	Size mm	Size g	Range core	Ecological business model
aramia	Aramia Rainbow-skink	diurnal	invert.	50	2.34	New Guinea	** refugia-anchored diurnal gr. liz.
babarensis	Babar Rainbow-skink	diurnal	invert.	46	1.79	Indonesia	* refugia-anchored diurnal gr. liz.
beccarii	Kai Rainbow-skink	diurnal	invert.	67	5.8	Indonesia	* refugia-anchored diurnal gr. liz.
bicarinata	Yule Rainbow-skink	diurnal		43	1.36	New Guinea	** refugia-anchored diurnal gr. liz.
bomberai	Bomberai Rainbow-skink			48	2.03	Indonesia	** refugia-anchored diurnal gr. liz.
caesius	Blue-throated Rainbow Lizard			53	2.78	Indonesia	*** refugia-anchored diurnal gr. liz.
decora	Elegant Rainbow-skink	diurnal		44	1.57	Queensland	** refugia-anchored diurnal gr. liz.
diguliensis	Digul River Rainbow-skink	diurnal	invert.	49	2.13	New Guinea	** refugia-anchored diurnal gr. liz.
dogare	Sandy Rainbow-skink	diurnal		46	1.75	Queensland	** refugia-anchored diurnal gr. liz.
eothen	Milne Bay Rainbow-skink	diurnal	invert.	57	3.43	New Guinea	** refugia-anchored diurnal gr. liz.
fusca	Indonesia Brown Carlia	diurnal	invert.	57	3.57	New Guinea	** refugia-anchored diurnal gr. liz.
gracilis	Slender Rainbow-skink	diurnal	invert.	38	0.97	Western Australia	*** refugia-anchored diurnal gr. liz.
inconnexa	Whitsunday Rainbow-skink	diurnal		47	1.83	Queensland	* refugia-anchored diurnal gr. liz.
jarnoldae	Lined Rainbow-skink	diurnal	invert.	43	1.45	Queensland	*** refugia-anchored diurnal gr. liz.
johnstonei	Rough Brown Rainbow-skink	diurnal	invert.	41	1.22	Western Australia	*** refugia-anchored diurnal gr. liz.

(v. 4%). Males are usually larger than females (0.75 v. −0.31 on −2 to 2 scale). Sexual dichromatism is common (43% v. 7.6%), as is seasonal color (50% v. 6%). Hence, the mean sex differences rollup score (1.16 v. 0.28 on 0 to 6 scale) is the sixth-highest among the 155 genera of skinks. The seasonal color of *pectoralis* is a blue throat which reflects ultraviolet light and is presumably brighter to its recipient.

Although all species are nominally diurnal (v. 74%), several are known to occasionally remain active after dark (*amax*, *gracilis*, *munda*).

All *Carlia* are invertebrate eaters (v. 84%) and active foragers (v. 92%). Several species are kleptoparasitic (*ailanpalai*, *amax*, *munda*). Their primary predator concealment tactic is equally divided between retreat to a crevice or burrow (50% v. 43%) and retreat to vegetation (50% v. 25%). Direct basking is very common (64% v. 21%); many forest species prefer edges or light gaps. Mean active body temperature is slightly cool (29.9°C v. 31.1°C). Although all males defend space (100% v. 94%), 20% of males also form hierarchies (v. 5.6%).

In reference to the annual rainfall spectrum, they are unimodal in the intermediate band (401–1101 mm/y: 57% v. 34%), and some species occur from arid (151–400 mm/y) to wet (>3000 mm/y) areas. Only 4% of *Carlia* are found at elevations above 1200 m; all are found in the low elevation band (100% v. 89%). They are slightly overrepresented in riparian (30% v. 23%) and human (24% v. 14%) habitats, and they are about average in littoral (16% v. 21%) and sandy (19% v. 25%) areas.

All are oviparous (v. 64%), with clutches of average count (−0.05 v. 0.06 std. resid.); hatchlings are slightly small (ontogenetic mass range 0.31 v. 0.12 std. resid.). Incubation is slightly short (−0.53 v. −0.40 std. resid.), which equates to 46 days for an average-sized (2.75 g) species such as *rubrigularis*. Communal oviposition is slightly overrepresented (6.8% v. 2.6%); *tetradactyla* is known to oviposit in ant nests. Parental care is unknown (v. 3.2%). RCM is light (0.165 v. 0.236 ratio). At least two species hatch explosively in response to disturbance (*ailanpalai*, *schmeltzii*).

Longevity is short (−0.49 v. 0.17 std. resid.), equating to 3.8 years for *rostralis*, the longest-lived species ($n = 5$ spp. documented). Only one species (*ailanpalai*) has been documented for population density (7.7 kg/ha; $n = 18$), but this species occurs primarily on predator-reduced islands and is unlikely to be representative of the genus (mean 0.35 kg/ha for skinks).

Ninety-three percent of *Carlia* occur in the Australasian biogeographic realm (v. 46%); 14% occur in the Oriental realm (v. 24%). Endemism occurs at a typical level (66% v. 64%), and the overwhelming majority of species occur on islands (91% v. 57%). This genus is prone to introduction (4.5% v. 2.1%), including 4.5% (v. 0.7%) introduced more than twice. None is considered imperiled. In other respects *Carlia* are typical skinks.

Carlia spp.	English name			Size		Range core
	Activity	Diet	mm	g		Ecological business model
leucotaenia	Seram Rainbow-skink					Indonesia
		invert.	47	1.91		* refugia-anchored diurnal gr. liz.
longipes	Closed-litter Rainbow-skink					Queensland
	diurnal	invert.	60	4.18		*** refugia-anchored diurnal gr. liz.
luctuosa	Hooded Rainbow-skink					New Guinea
	diurnal		64	5.1		** refugia-anchored diurnal gr. liz.
munda	Shaded-litter Rainbow-skink					Northern Australia
	diurnal	invert.	40	1.11		** refugia-anchored diurnal gr. liz.
mysi	Mys' Rainbow-skink					New Guinea
	diurnal	invert.	53	2.76		*** refugia-anchored diurnal gr. liz.
nigrauris	Tindjil Rainbow-skink					Indonesia
			53	2.76		* refugia-anchored diurnal gr. liz.
pectoralis	Open-litter Rainbow-skink					Queensland
	diurnal		50	2.37		** refugia-anchored diurnal gr. liz.
peronii	Timor Rainbow-skink					Indonesia
			39	1.03		* refugia-anchored diurnal gr. liz.
pulla	North Coast Rainbow-skink					New Guinea
	diurnal		55	3.20		* refugia-anchored diurnal gr. liz.
quinquecarinata	Eastern Torres Rainbow-skink					Queensland
	diurnal		57	3.50		* refugia-anchored diurnal gr. liz.
rhomboidalis	Blue-throated Rainbow-skink					Queensland
	diurnal		52	2.56		* refugia-anchored diurnal gr. liz.
rimula	Crevice Rainbow-skink					Queensland
			37	0.87		* refugia-anchored diurnal gr. liz.
rostralis	Black-throated Rainbow-skink					Queensland
	diurnal	invert.	60	4.12		** refugia-anchored diurnal gr. liz.
rublgo	Orange-flanked Rainbow-skink					Queensland
	diurnal		43	1.38		* refugia-anchored diurnal gr. liz.
rubrigularis	Red-throated Rainbow-skink					Queensland
	diurnal	invert.	53	2.75		** refugia-anchored diurnal gr. liz.

Carlia spp.	English name				Range core
			Size		
	Activity	Diet	mm	g	Ecological business model
rufilatus	Red-sided Rainbow-skink				Western Australia
	diurnal		39	1.04	* litter cryptozoid
schmeltzii	Robust Rainbow-skink				Queensland
	diurnal		61	4.33	* litter cryptozoid
sexdentata	Macleay's Rainbow-skink				Queensland
	diurnal	invert.	58	3.70	* refugia-anchored diurnal gr. liz.
spinauris	Lelogama Four-toed Lizard				Indonesia
			39	1.03	* refugia-anchored diurnal gr. liz.
storri	Storr's Rainbow-skink				New Guinea
	diurnal	invert.	42	1.35	* refugia-anchored diurnal gr. liz.
sukur	Sukur Four-toed Skink				Indonesia
			39	1.05	* refugia-anchored diurnal gr. liz.
tetradactyla	Southern Rainbow-skink				Eastern Australia
	diurnal		54	3.01	** refugia-anchored diurnal gr. liz.
triacantha	Desert Rainbow-skink				Western Australia
	diurnal		45	1.65	* refugia-anchored diurnal gr. liz.
tutela	Maluku Rainbow-skink				Indonesia
	diurnal	invert.	47	1.85	* refugia-anchored diurnal gr. liz.
vivax	Tussock Rainbow-skink				Queensland
	diurnal	invert.	45	1.60	* refugia-anchored diurnal gr. liz.
wundalthini	Cape Melville Rainbow Skink				Queensland
			44	1.57	* refugia-anchored diurnal gr. liz.

Carphodactylidae

The 30 carphodactylids are nocturnal (86% v. 25%) or cathemeral (14% v. 7%), padless geckos from Australia. Many have spectacularly elaborated "leaf" tails, aiding in crypsis. They are diverse in diet and microhabitat, but all rely on crypsis (v. 43%); they are sit-and-wait foragers (v. 46%), occur in arid areas (v. 30%), and are oviparous (v. 82%). Included genera are *Carphodactylus*, *Nephrurus*, *Orraya*, *Phyllurus*, *Saltuarius*, *Underwoodisaurus*, and *Uvidicolus*. A plurality are immobility experts (47% v. 3.2%); other ecological business models include nocturnal ground lizards (30% v. 7.7%), crack hermits (13.3% v. 2.6%), trunk/wall lizards (6.7% v. 13%), and ground-foraging tree lizards (3.3% v. 9.2%).

Carphodactylids are large geckos (14.5 g v. 5.3 g for lizards and 3.3 g for Gekkonidae), with a moderately heavy build (shape index 0.35 v. 0.0 std. resid.). Tails are short (0.54 v. 1.37 × SVL), most of which break (87% v. 57%); none is prehensile (v. 5.0%). A few species have crests (3.3% v. 5.3%) and pores (10% v. 15%). None is elongate (v. 11%), shows any color change (v. 10.7%), sexual dichromatism (v. 21%), or seasonal color (v. 6.5%). Overall differences between the sexes are negligible (0.1 on 0 to 6 scale).

Although half of the carphodactylids are invertebrate eaters (v. 73%), 40% are large enough to take vertebrates (v. 3.2%), and 10% are omnivores (v. 15.4% in lizards and 3.8% in gekkonids). None feeds under electric lights (v. 1.4%). No carphodactylid is considered to be conspicuously wary (v. 8.2% of lizards). Carphodactylids call: 13.3% are recorded as vocalizing spontaneously (v. 2.5%), this is twice the 6.5% frequency of spontaneous calling in gekkonids. Moreover, a third of carphodactylids will call when stressed (33% v. 2.2% of lizards and 3.7% of gekkonids). The only other lizard family with a comparably ratio of stress calling to spontaneous calling is the nocturnal terrestrial eublepharid geckos (25% stress to 8% spontaneous). One has to strain to find an explanatory ecological theme that would unite just these two gecko families. They are from the same gecko clade, though other embedded clade members do not show the same pattern.

Their mean active body temperature is a chilly 22.8°C (v. 30.8°C). As carphodactylids are found at moderate latitudes and low elevations (see below), this low temperature must reflect their nocturnality and (poorly documented) basking choices. Carphodactylids are not especially poorly known (filled cells 29% v. 27%), suggesting that the absence of basking records likely reflects an absence of direct basking. Their elevational profile is skewed strongly toward low elevations: (l/m/h) 100/17/0 (v. 89/41/13). This suggests that they are averse to the cold at higher elevations, but it raises the question of why they are not warmer in mean body temperature at the elevations where they do occur.

None is recorded as living solitarily, and a third (v. 20.4%) are found in groups of more than two. Male hierarchies are not recorded, and there are insufficient data to characterize intermale social tolerance.

The carphodactylids are about evenly split between arboreal (53% v. 32%; entire genera *Orraya*, *Saltuarius*, most *Phyllurus*) and terrestrial living (43% v. 48%; entire genera: *Carphodactylus*, *Nephrurus*, *Underwoodisaurus*, *Uvidicolus*), with the remaining 3.3% cryptozoic (v. 12%; remainder of *Phyllurus*). About half (47.6% v. 22.2%) use burrows for daily refugia. Only 3% (v. 16.6%) are found in the littoral zone, 28% are associated with sand or dunes (v. 22%), and 3% (v. 18.5%) are associated with human habitats.

Almost all species lay two eggs (−0.82 v. 0.0 std. resid.), which are not adherent (v. 2.4%) or hard-shelled (v. 41%). Because clutch count appears to be phylogenetically conserved in Carphodactylidae, the standardized residuals express the degree to which a count of two diverges from the overall lizard pattern, rather than reflecting the outcome of selection on clutch number within Carphodactylidae. A few species oviposit communally (13.3% v. 4.2%). The young are conspicuously large, as indicated by strongly reduced ontogenetic mass range (−1.21 v. 0.0 std. resid.). The average carphodactylid young is 1/19 of its maximum mass (v. 1/44 for lizards), despite the large body sizes within this family. Parental care is not reported (v. 2.9%), and incubation duration (−0.07 v. 0.0 std. resid.) and RCM average about normal (0.215 v. 0.226).

Mass-adjusted longevity is on the high side (0.42 v. 0.0 std. resid.). The highest values are *Phyllurus platurus* (20 years) and *Underwoodisaurus milii* (18 years). A single credible population density estimate exists for a carphodactylid: 0.38 kg/ha for *Nephrurus levis* (v. 0.54 kg/ha mean for lizards). No carphodactylid is found on an island. Two-thirds of species are considered imperiled (v. 49%). Restricted range is characteristic of immobility experts and is cited as the reason for conservation concern with many of the carphodactylids.

Those of us not from Australasia are not personally familiar with Carphodactylidae and Diplodactylidae, two gecko families with confusingly similar names and habits. As a service to readers attempting to grasp their ecological similarities and differences, the following notes may assist with differentiating them.

- Carphodactylids are limited to mainland Australia; diplodactylids are also found in New Zealand and New Caledonia.
- No carphodactylid is found on islands; 46% of diplodactylids are insular.
- Few (30 spp.) are in Carphodactylidae; many (149 spp.) are in Diplodactylidae.
- Carphodactylids favor low-mobility lifestyles: immobility experts (47%), nocturnal ground lizards (30%), and crack hermits (13%); diplodactylids are more elevated and nocturnal: nocturnal ground lizards (35%), nocturnal canopy walkers (25%), and trunk/wall lizards (20%).
- Carphodactylids run bigger (14.5 g) than diplodactylids (4.2 g).
- Carphodactylids have very large young (−1.21 std. resid. mass range) compared to diplodactylids (−0.57).
- Carphodactylids are mostly padless (97%); only 16% of diplodactylids are padless.
- Carnivory has higher penetration in carphodactylids (40%) versus 2% among diplodactylids.
- Oviparity is universal in carphodactylids; viviparity is found in a significant minority (18%) of diplodactylids.
- In many other respects the two families (carpho- v. diplo-) are similar: nocturnal (86% v. 79%), sit-and-wait foraging (100% v. 100%), dependence on crypsis (100% v. 88%), arboreality (53% v. 64%), terrestriality (43% v. 34%), and RCM ratio (0.22 v. 0.23).

Carphodactylus

Carphodactylus is a single-species genus of large (24.9 g v. 14.5 g) carphodactylid from Queensland. It differs from the rest of the family by possessing a low middorsal ridge or crest. It has a striking dark patch under the eye, perhaps to dampen glare under difficult light conditions. It is one of the few nocturnal species that forages both on the ground and in low vegetation, suggesting that it might have exceptional visual acuity under low light. As far as is known, it is otherwise similar to the other family members.

Carphodactylus sp.	English name				Range core
			Size		
	Activity	Diet	mm	g	Ecological business model
laevis	Chameleon Gecko				Queensland
	nocturnal		114	24.9	** ground-foraging tree lizard

Celatiscincus

Celatiscincus is a two-species genus of small (1.2 g v. 3.3 g), nonelongate (v. 35% elongate), refugia-anchored diurnal ground (100% v. 30%) skinks endemic to New Caledonia. For their small size, they are moderately robust (shape index −0.38 v. −0.46 std. resid.), with moderately long tails (1.60 v. 1.33 × SVL) that are not reported (needs confirmation) to break (v. 73% breakable). The sexes differ only in color (50% v. 7.6%). They bask directly (100% v. 21%) and are associated with littoral areas (50% v. 21%). They lay about 2.5 eggs (0.33 v. 0.06 std. resid.). They have not been introduced (v. 2.1%). They are both listed on the IUCN's lists of the most endangered (v. 14%), in this case endangered, due to highly restricted ranges. In other respects *Celatiscincus* are probably typical skinks.

Celatiscincus spp.	English name	Range core	Activity	Diet	Size mm	Size g	Ecological business model
euryotis	Pines Pale-hipped Skink	New Caledonia	diurnal	invert.	39	1.07	** refugia-anchored diurnal gr. liz.
similis	Northern Pale-hipped Skink	New Caledonia	diurnal		42	1.30	** refugia-anchored diurnal gr. liz.

Celestus

Celestus is the largest (29 spp.) genus of the Diploglossidae. Its members are ecologically and morphologically similar to *Diploglossus*, and many species have wandered taxonomically back and forth between the two genera. As currently understood, roughly half the *Celestus* species are found in Hispaniola or outliers, seven in Jamaica, and nine in Central America. But the species' distributions and composition are incompletely known: three species are flagged "known from only a few specimens," two are thought extinct, six are known from one sex only, three are known from only a single locality, and one is known from only a single specimen (some species flagged for multiple reasons). Color photographs from life are readily available for fewer than half the known species. Most species are found on only a single landmass, suggesting a limited ability to disperse or colonize.

The majority of *Celestus* are litter cryptozoids (55% v. 58%), though note the low confidence in many assignments. Ground-foraging tree lizards (15% v. 9%), trunk/wall lizards (15% v. 12%), refugia-anchored diurnal ground lizards (5% v. 12%), and immobility experts (5% v. 3%) round out the diverse roster.

Celestus are of average diploglossid size (5.1 g v. 5.3 g) and build (shape index −1.01 v. −0.98 std. resid.). Though the mean mass is typical,

Celestus spp.	English name	Range core	Activity	Diet	Size mm	Size g	Ecological business model
adercus	Surprise Galliwasp	Panama			64	1.84	* trunk/wall lizard
agasepsoides	Serpentine Four-toed Galliwasp	Dominican Republic	diurnal	invert.	60	1.55	* litter cryptozoid
anelpistus	Altagracia Giant Galliwasp	Dominican Republic	nocturnal		213	217	* nocturnal ground lizard
badius	Navassa Galliwasp	Navassa Island	diurnal	invert	84	8.4	**** litter cryptozoid
barbouri	Limestone Forest Galliwasp	Jamaica			85	3.67	
bivittatus	Two-striped Galliwasp	Honduras			89	4.06	** litter cryptozoid
carraui	Puerta Planta Galliwasp	Dominican Republic			211	34.7	
costatus	Common Hispaniolan Galliwasp	Hispaniola	diurnal	omnivore	100	5.4	* ground-foraging tree lizard or litter cryptozoid
crusculus	Garman's Galliwasp	Jamaica		invert.	72	2.40	* litter cryptozoid or immobility expert
curtissi	Curtiss' Galliwasp	Hispaniola			70	2.27	* litter cryptozoid
cyanochloris	Cordilleran Galliwasp	Costa Rica			80	3.16	* ground-foraging tree lizard

the range is very large: 1.1–47 g. The six giants occur on Hispaniola (four spp.) and Jamaica (two spp.).

They are less inclined to be cathemeral (9% v. 25%) and more likely to be diurnal (82% v. 70%). Their mix of microhabitats mirrors that of the family. One species with rainfall data fell into the wet (>3000 mm/y) belt (v. 75%). For such a poorly known group, a remarkable 48% (v. 36%) are found in human habitats.

All known species are viviparous (v. 86%), though the evidence for *badius* is contradicted. Litter counts are notably large (1.11 v. 0.86 std. resid.). The ontogenetic mass range is also conspicuously high (2.31 v. 2.13 std. resid.), indicating very small young. Parental care has not been reported for *Celestus*.

They are even more imperiled than other members of the Diploglossidae: 63% v. 51% for any expression of conservation concern; 37% (v. 30%) are on the IUCN's most endangered lists. Introduced mongoose have been very hard on island species. In other respects they are typical diploglossids.

Celestus spp.	English name		Size		Range core
	Activity	Diet	mm	g	Ecological business model
darlingtoni	Darlington's Galliwasp				Dominican Republic
			70	2.27	
duquesneyi	DuQuesnay's Galliwasp				Jamaica
			89	4.06	* litter cryptozoid
enneagrammus	Huasteca Lesser Galliwasp				Mexico
			85	3.63	
fowleri	Bromeliad Galliwasp				Jamaica
	diurnal		84	8.7	* immobility expert
haetianus	Earless Galliwasp				Haiti
		invert.	79	1.86	* litter cryptozoid
hewardi	Heward's Galliwasp				Jamaica
		invert.	138	48.0	* refugia-anchored diurnal gr. liz.
hylaius	Rainforest Celestus				Costa Rica
	diurnal		86	3.74	* litter cryptozoid
macrotus	Thomas' Galliwasp				Haiti
			52	1.64	
marcanoi	Marcano's Galliwasp				Dominican Republic
			65	3.53	
microblepharis	Tiny Eyelid Galliwasp				Jamaica
			72	2.39	
montanus	Merendon Galliwasp				Guatemala
	diurnal		76	2.76	* ground-foraging tree lizard
occiduus	Jamaican Giant Galliwasp				Jamaica
		omnivore	238	46.5	
orobius	Hermit Galliwasp				Costa Rica
	diurnal		69	1.23	
rozellae	Rozella's Galliwasp				Northern Central America
	diurnal		83	3.43	* trunk/wall lizard
scansorius	Mountainside Galliwasp				Honduras
			89	4.06	* trunk/wall lizard
sepsoides	Common Four-toed Galliwasp				Hispaniola
			65	1.89	** litter cryptozoid or immobility expert
stenurus	Cope's Galliwasp				Hispaniola
	diurnal		132	8.1	*** litter cryptozoid
warreni	Haitian Giant Galliwasp				Hispaniola
	cathemeral	carnivore	213	35.3	** litter cryptozoid

Ceratophora

Ceratophora are five agamids famed for their fleshy rostral appendages, which are reduced in some species (*erdeleni*, *karu*) and most females. The appendages may be leafy in outline (*aspera*, *tennentii*) or horn-like (*stoddartii*). There are also a few bumps and spines, but only modest nuchal crests (40% v. 38%). All species are Sri Lanka endemics; of those documented, all are cruise foraging (100%, the highest value in Agamidae, v. 6%), invertebrate eaters (v. 46%) reliant on crypsis (v. 51%). Two species are small (~1.5 g) and are litter cryptozoids (40% v. 1%); three species are large (10–13 g) and are ground-foraging tree lizards (60% v. 19%).

Of the two litter cryptozoids, one (*aspera*) moves slowly and is often found in heterosexual pairs, possibly mate guarding; the other (*karu*) is solitary and moves quickly. Clutches are small (−0.27 v. 0.20 std. resid.), and laid immediately after mating in *stoddartii* (presumably after a short interval for shelling). The only incubation-documented species, *stoddartii*, has an incubation period (86 days) that is about average for lizards (0.04 v. 0.0 std. resid.), but notably long for agamids (v. −0.55). All are on the Sri Lanka Red List of endangered species (v. 37% imperiled).

Ceratophora spp.	English name				Range core
			Size		
	Activity	Diet	mm	g	Ecological business model
aspera	Rough-horned Lizard				Sri Lanka
	diurnal	invert.	38	1.66	**** litter cryptozoid
erdeleni	Erdelen's Horned Lizard				Sri Lanka
	diurnal		73	12.4	** trunk/wall lizard or ground-foraging tree lizard
karu	Karu's Horned Lizard				Sri Lanka
	diurnal		36	1.45	**** litter cryptozoid
stoddartii	Rhino Horn Lizard				Sri Lanka
	diurnal	invert.	75	13.4	**** ground-foraging tree lizard
tennentii	Leaf-nosed Lizard				Sri Lanka
	diurnal	invert.	68	9.95	**** ground-foraging tree lizard

Cercosaura

Cercosaura is a 13-species genus of relatively small (1.76 g v. 2.35 g) terrestrial gymnophthalmids centered in northwestern South America and extending throughout the continent. Most are litter cryptozoids (86% v. 55%); the remainder wide-ranging chemosensors (14% v. 2%). Tails tend to be long (1.97 v. 1.72 × SVL), breakable (39% v. 28%), and prehensile (7.7% v. 1.7%). The latter is notably unusual in terrestrial species. A few are elongate (15% v. 41%). Secretory

Cercosaura spp.	English name				Range core
			Size		
	Activity	Diet	mm	g	Ecological business model
argulus	Elegant-eyed Langelot				Northwestern South America
	cathemeral	invert.	41	1.37	** wide-ranging chemosensor
bassleri	Bassler's Ocellated Langelot				Peru
			54	3.58	
doanae	Tiffany's Langelot				Peru
	diurnal		49	2.61	** litter cryptozoid
eigenmanni	Eigenmann's Langelot				Bolivia
	diurnal	invert.	44	1.81	*** litter cryptozoid
hypnoides	Meta Langelot				Colombia
			36	0.89	
manicata	Slender Langelot				Ecuador
	diurnal	invert.	59	4.63	*** litter cryptozoid
nigroventris	Black-bellied Langelot				Venezuela
			41	1.40	
ocellata	Ocellated Langelot				Western South America
	diurnal	invert.	52	3.09	** litter cryptozoid
parkeri	Parker's Langelot				Southern South America
	diurnal		44	1.04	* litter cryptozoid
phelpsorum	Tepui Langelot				Venezuela
			49	2.49	
quadrilineata	Four-lined Langelot				Brazil
			38	1.13	
schreibersii	Schreiber's Langelot				Argentina
	diurnal	invert.	42	0.88	** litter cryptozoid

pores are common (46% v. 29%), and males are the smaller sex (–1.00 on –2 to 2 scale). Sexual dichromatism is overrepresented both in permanent (31% v. 13%) and seasonal form (15% v. 1.7%). The sex rollup score (on scale of 0 to 6) is 0.92 (v. 0.49). Although they are somewhat overrepresented among species that are conspicuously wary (15% v. 7%), *eigenmanni* renders itself conspicuous by waving an arm in teiid fashion. All documented species (*argula, manicata*) hide in vegetation as their primary predator concealment tactic. Direct basking is well represented (31% v. 8%), and the mean active body temperature is normal (28.5°C v. 28.8°C). All documented males (*argula, manicata*) are socially tolerant (v. 88%). They are partial to sandy (22% v. 14%), riparian (56% v. 29%), and human (33% v. 15%) habitats. Reproduction is normal, except that communal oviposition is common (15% v. 3.4%), and RCM (based on *eigenmanni, ocellata*) is light (0.099 v. 0.192 ratio). Ontogenetic size range is notably large (1.40 v. 0.62 std. resid.), indicating that young are especially small. *C. argula* oviposits in ant nests (along with other lizards). Endemism is relatively low (46% v. 68%), and relatively few species are considered imperiled (17% v. 25%); none is on the IUCN's lists of the most endangered (v. 11.5%). In other respects *Cercosaura* are typical gymnophthalmids.

Cercosaura spp.	English name		Size		Range core
	Activity	Diet	mm	g	Ecological business model
steyeri	Steyer's Langelot				Argentina
			42	1.47	

Chalarodon

Chalarodon is a two-species genus of small (4.3 g v. 19.8 g) Madagascar oplurids. Both are refugia-anchored diurnal ground lizards (v. 50%). They have a slender build (–0.07 v. 0.70 std. resid.) and crests (100% v. 25%). Their tails may break (50% v. 12.5%). *C. madagascariensis* is known to run bipedally. In recognition of their m>f SSD (1.0 v. 1.0 on –2 to 2 scale) and seasonal color (50% v. 25%), the genus has a relatively high sex differences rollup (2.0 v. 0.75 on 0 to 6 scale). For primary predator concealment they rely on flight (v. 17%). They are all terrestrial (v. 50%) and may refuge overnight in burrows (v. 40%). They are all found at low elevation (v. 100%) only. Both species occur in sandy areas (v. 25%). Their clutch count is low, as is characteristic of oplurids (–0.62 v. –0.62). In other respects *Chalarodon* are probably typical oplurids.

Chalarodon spp.	English name		Size		Range core
	Activity	Diet	mm	g	Ecological business model
madagascariensis	Madagascar Iguana				Madagascar
	diurnal	omnivore	74	8.5	* refugia-anchored diurnal gr. liz.
steinkampi	Steinkamp's Madagascar Iguana				Madagascar
	diurnal		53	2.19	*** refugia-anchored diurnal gr. liz.

Chalcides

Chalcides is a major radiation (31 spp.) of moderate-sized (7.5 g v. 3.3 g) Old World skinks that are scattered along the morphological continuum to limblessness: 87% are elongate or extremely so (v. 35%), 35% have limbs reduced to some degree, and one (*striatus*: 3%) has reduced eyes. None however is fully limbless and four (13%) have normal (overlapping) limbs. They are found across northern Africa (Senegal to Somalia and Nigeria to the Mediterranean) and southern Europe (Portugal to Greece) to Turkmenistan and Sri Lanka (Palearctic 81% v. 6%; Ethiopian 29% v. 20%). Some species are relatively well known (mean filled cells 34% v. 25%), though half (48%) cannot yet be characterized by ecological business model. The known species fill a diversity of ecological business models: 47% (v. 28%) litter cryptozoids, 24% (v. 12%) subterranean foragers, 18% (v. 7%) sand sharks, and 6% (v. 6%) wide-ranging chemosensors. Note that only the latter business model involves regular exposure.

Chalcides are slender (shape index –0.66 v. –0.46 std. resid.), with mildly short tails (1.04 v. 1.33 × SVL) that are breakable (100% v. 73%). Not all tails regenerate well (e.g., *chalcides*). Blue tails are relatively common (6.5% v. 4%), as befits litter cryptozoids (table 4.1, trait 16), which benefit from the startle value of brilliant color when suddenly exposed. Juvenile *Chalcides* may also be red tailed (*mauritanicus*). Many species may locomote without use of limbs and sand swimming is very common.

Chalcides are normally m<f in SSD (–1.0 v. –0.31 on –2 to 2 scale), which

Chalcides spp.	English name	Range core	Activity	Diet	Size mm	Size g	Ecological business model
armitagei	Armitage's Skink	Senegal		invert.	95	8.4	*** litter cryptozoid
bedriagai	Bedriaga's Skink	Iberian Peninsula	diurnal	invert.	77	2.88	*** subterranean forager
bottegi	Bottego's Ocellated Skink	Ethiopia		invert.	110	29.6	
boulengeri	Barbary Sand-skink	Northern Sahara	cathemeral	invert.	85	3.67	** sand shark
chalcides	Italian Three-toed Skink	Central Mediterranean	cathemeral	invert.	140	12.6	*** litter cryptozoid
coeruleopunctatus	Gomierro Skink	Spain			74	8.3	
colosii	Riffian Skink	Morocco	diurnal		102	10.1	* refugia-anchored diurnal gr. liz.
delislei	Delisle's Three-fingered Skink	Sahara	diurnal	invert.	80	3.11	** sand shark
ebneri	Ebner's Skink	Morocco	diurnal		74	4.19	
guentheri	Guenther's Cylindrical Skink	Syria	diurnal	invert.	133	11.1	*** litter cryptozoid
lanzai	Lanza's Cylindrical Skink	Morocco	diurnal	invert.	103	23.9	
levitoni	Leviton's Cylindrical Skink	Saudi Arabia			80	10.5	
manueli	Manuel's Skink	Morocco			67	3.16	
mauritanicus	Two-fingered Skink	Morocco		invert.	70	3.58	*** subterranean forager
mertensi	Algerian Three-toed Skink	Algeria			137	22.8	* subterranean forager
minutus	Small Three-toed Skink	Morocco	diurnal		92	4.39	
mionecton	Mionecton Skink	Morocco			84	3.58	* wide-ranging chemosensor
montanus	Morocco Mountain Skink	Morocco			82	11.3	
ocellatus	Ocellated Skink	Mediterranean	cathemeral	invert.	127	9.8	*** litter cryptozoid
parallelus	Chafarinas Skink	Morocco		invert.	74	2.61	

is especially notable given their relatively large size. Presumably the peritoneal-space demands of pregnancy keep females large, and intermale competition is not mediated by size. Males of *viridanus* guard their mates.

Most eat invertebrates (95% v. 84%), though omnivory is also known (5% v. 13%) and likely underreported (e.g., the invertebrate eater *ocellatus* also occasionally eats leaves and fruit). *C. boulengeri* and *mionecton* are believed to locate prey by sound/vibrations. All are active foragers (100% v. 92%), but sit-and-wait foraging is also reported (33% v. 16%). Home-range area is very small (–1.24 v. –0.07 std. resid.), though this estimate is based solely on 15 m^2 for *striatus* and *pseudostriatus*. Taxonomic confusion may have a hand in the small home-range area estimate for the large *pseudostriatus*, which was formerly confused with *chalcides* and *striatus*.

Most species are diurnal (71% v. 74%), but a substantial minority are cathemeral (29% v. 9%). Additional species are likely to be found to be cathemeral with greater sampling. Nearly a third (32% v. 11%) are conspicuously wary, which is consistent with their hidden lifestyles. All primary predator concealment tactics are used by *Chalcides*: overrepresented tactics are retreat to sand (47% v. 22%) and flight (6% v. 4%); retreat to vegetation (18% v. 25%), crevices, and burrows (24% v. 43%) and crypsis (6% v. 9%) are underrepresented. Direct basking is slightly overrepresented (29% v. 21%), as is thermoconformance (3.2% v. 1.5%). Mean active body temperature is about average (30.5°C v. 31.1°C).

The only social relations reported are male spatial defense (100% v. 94%).

Chalcides emphasize the drier end of the rainfall spectrum, but they are not reported in the hyperdry or dry (underreporting likely in the Sahara). They are overrepresented in arid (151–400 mm/y: 100% v. 40%) and intermediate (401–1100 mm/y: 50% v. 34%) zones. They are underrepresented in mesic (1101–3000 mm/y: 25% v. 40%) and absent from the wet (v. 7%) areas. They are overrepresented in medium elevations (1200–2400 m: 61% v. 35%) and about average above and below that band. They are overrepresented in littoral (38% v. 21%), sandy (66% v. 25%), riparian (35% v. 23%), and human (31% v. 14%) habitats. The consistency of overrepresentation suggests better quality reporting, but the strong association with sandy areas is likely accurate.

Despite their low latitude and warm environments, all species are viviparous (v. 36%). Litter counts are unremarkable given their body size (0.14 v. 0.06 std. resid.). Young are of normal relative size (ontogenetic mass range 0.17 v. 0.12 std. resid.). Many species have multiple litters per year (data not precise enough to estimate mean frequency). RCM is normal (0.264 v. 0.236 ratio).

Longevity is high (0.84 v. 0.17 std. resid.), which equates to 14 years in the cases of *bottegi*, *chalcides*, *colosii*, *lanzai*, *mertensi*, *ocellatus*, *polylepis*, *pulchellus*, *ragazzii*, *striatus*, and *thierryi*. It is astonishing that so many species have exactly the same maximum longevity, suggesting taxonomic confusion. Population density is moderately high (0.64 kg/ha v. 0.35 kg/ha), especially when considering the low productivity of the habitat. It is based on *striatus* ($n = 2$; 1.75 kg/ha), *sepsoides* ($n = 1$, 1.5 kg/ha), and *bedriagai* ($n = 2$, 0.1 kg/ha).

Endemism is low (35% v. 64%). Island species frequency is slightly low (42% v. 57%). Many species have been introduced (13% v. 2.1%), though none more than twice (0% v. 0.7%). The majority are imperiled (56% v. 47%), including 11% (v. 14%) that occupy the IUCN's lists of the most endangered. At least two such species (*mauritanicus*, EN; *parallelus*, EN) are unique in suffering from human collection of driftwood, which is an important cover object for them. *C. sepsoides* was apparently revered by the ancient Egyptians, who included mummified examples in their coffins. One would love to know what attribute of this relatively mundane skink so enamored the ancient Egyptians. *C. striatus*, on the other hand, is loathed by moderns for its apparent leglessness and is considered by some local residents to be venomous. In other respects *Chalcides* are typical skinks.

Chalcides spp.	English name	Range core	Activity	Diet	Size mm	Size g	Ecological business model
pentadactylus	Kuddle Poondy Skink	India			52	1.10	
polylepis	Many-scaled Cylindrical Skink	Morocco	cathemeral	omnivore	121	8.7	
pseudostriatus	Moroccan Three-toed Skink	Morocco	diurnal	invert.	159	34.9	* litter cryptozoid
pulchellus	West African Cylindrical Skink	Guinea		invert.	110	12.7	
ragazzii	Ragazzi's Bronze Skink	Horn of Africa	diurnal	invert.	106	11.2	
sepsoides	Audouin's Sand-skink	Egypt	cathemeral	invert.	94	4.71	*** sand shark
sexlineatus	Gran Canaria Skink	Spain		invert.	84	3.56	** litter cryptozoid
sphenopsiformis	Senegal Skink	Western Sahara	diurnal		84	5.9	*** subterranean forager
striatus	Western Three-toed Skink	Iberian Peninsula	diurnal	invert.	139	12.4	**** litter cryptozoid
thierryi	Thierry's Cylindrical Skink	Nigeria			121	16.3	
viridanus	West Canary Skink	Spain	diurnal	invert.	103	10.6	* litter cryptozoid

Chalcidoseps

Chalcidoseps thwaitesi is the sole representative of a genus of average (2.57 g v. 3.25 g), highly elongate (v. 35% elongate), litter cryptozoid (v. 28%) skinks endemic to Sri Lanka. It is reasonably well known (filled cells 39% v. 25%).

Although it has four digits on each limb, it locomotes without use of limbs and is very slender (shape index –1.24 v. –0.46 std. resid.). Its tail is short (0.80 v. 1.33 × SVL) and not breakable (v. 73% breakable). Sex differences are not reported. It specializes on eating termites, but it will eat a variety of invertebrates (v. 84%). It is diurnal or crepuscular and dives back into litter when disturbed. It is

Chalcidoseps sp.	English name				Range core
			Size		
	Activity	Diet	mm	g	Ecological business model
thwaitesi	Four-toed Snakeskink				Sri Lanka
	diurnal	invert	74	2.57	* litter cryptozoid

found at low and medium elevations (600–1600 m) and is associated with riparian (v. 23%) and human (v. 14%) habitats.

It lays two eggs (–0.15 v. 0.06 std. resid.).

It has not been introduced (v. 2.1%). It is considered imperiled (v. 47%), though it is not listed by the IUCN (v. 14%) in the most endangered categories.

Chamaeleo

Chamaeleo is a moderate (14 spp.) radiation of large-bodied (40.6 g v. 13.5 g) typical chameleons from Africa, the Mediterranean, and southwestern Asia (through India). Proximity to Europe and their unusual habits have probably combined to evoke greater-than-usual reporting of their natural history (49% v. 38% of cells filled). *Chamaeleo* is an ecologically diverse genus, perhaps the most diverse chamaeleonid genus, but it fits the mold of "typical" chameleons (table 5.2) in all respects except SSD. Several species within *Chamaeleo* have conflicting reports of SSD, presumably attributable to geographic variation. Taking the mean of the reports led me to a mean *Chamaeleo* estimate of –0.20 on a –2 to 2 scale, which is about midway between the typical chameleon mean of 0.41 (m>f) and the dwarf chameleon mean of –0.70 (m<f). However, *Chamaeleo* matches the "typical" chameleon mean for business model (diurnal canopy walker frequency of 86% v. 98%), shape index of 0.69 (v. 0.57 std. resid.), tail length of 1.00 (v. 1.09 × SVL), crest frequency of 29% (v. 36%), arboreal microhabitat frequency of 93% (v. 99%), clutch count (6.57, the highest in Chamaeleonidae, v. 2.81 std. resid.), small offspring size (ontogenetic size range 1.24, the highest in Chamaeleonidae, v. 0.63 std. resid.), population density of 0.15 (v. 0.18 kg/ha), and incubation duration of 1.59 (v. 2.43 std. resid.). The reproduction values are hypertrophied in comparison to average "typical" chameleons. Note that many *Chamaeleo* species have extraordinarily wide ranges in incubation duration (widest range of any chamaeleonid). Four of ten species have ranges of more than twofold (the lizard mean is 1.4-fold). *C. chamaeleon* ranges more than sixfold: 56–365 days. Several species are noted for the phenological tie of their oviposition or hatching date to the onset of rains (loosening soil? promoting insect availability?). Perhaps some of the variation in incubation period reflects developmental contingency designed to promote emergence during favorable conditions.

All *Chamaeleo* have prehensile tails (v. 88%) and advanced color change (v. 92%). They are cruise foragers (v. 58%), are found at low elevation (v. 76%), are oviparous (v. 80%), and males defend space (v. 93%).

Although *Chamaeleo* do not greatly stand out from other chameleons is their sexual distinctions (sex rollup score = 2.36 v. 2.16 on 0 to 6 scale), they are notably more frequent exhibitors of seasonal color (36% v. 18%). Seasonal color must be difficult to detect in lizards that universally exhibit advanced color change, so the importance of this distinction needs to be evaluated when the other species have been studied as thoroughly as *Chamaeleo*.

Not surprisingly, most *Chamaeleo* are invertebrate eaters (86% v. 98%); more surprisingly, several of the desert chameleons (*calyptratus*, *namaquensis*) are omnivores (14.3% v. 1.0%). Two additional species take plant material on occasion (*arabicus*, *chamaeleon*), revealing that their projectile tongues are adaptable. *C. namaquensis* sometimes uses its tongue to consume small vipers. Compared to other chameleons, *Chamaeleo* have relatively large home ranges (0.39 v. –1.12 std. resid.). Presumably this large hunting area reflects the low primary productivity in many of their habitats.

A relatively large fraction of *Chamaeleo* give stress calls (21% v. 3.4%). Given the trenchant solitariness of *Chamaeleo* (75% v. 36%), these stress calls seem unlikely to benefit kin, implying that the startle effect on predators is most likely responsible for the maintenance of this trait. *C. namaquensis* is noted for both sexes defending against both sexes; it is primarily terrestrial (defense is easier to maintain in two-dimensions), and it is reported to be capable of 5 kph sprints—a spritely pace for a chameleon.

Chamaeleo are mostly arboreal (93% v. 78%), but even some of the arboreal species dig burrows on occasion (33% v. 4%). One does not easily envision zygodactylous feet as digging tools. *Chamaeleo* are underrepresented in the mesic rainfall belt (1101–3000 mm/y; 40% v. 85%) and overrepresented in the arid (151–400 mm/y; 40% v. 10%) and intermediate (401–1100 mm/y; 40% v. 15%) belts. They are also overrepresented in littoral areas (43% v. 6%), sand/dunes environments (36% v.

3%), riparian zones (57% v. 21%), and human habitats (43% v. 23%). That they are overrepresented in all of these special habitats suggests that reporting may be more complete for *Chamaeleo* than for other lizards.

Chamaeleo have extraordinarily large clutches (as noted above), although this is not manifest in an especially high RCM (0.386 v. 0.444). Apparently the increase in numbers is offset by their small size. The pressure for fecundity is perhaps responsible, at least in part, for the high level of parental care (14.3% v. 1.0%). Note however, that several authorities have commented on evidence that suggests that female *Chamaeleo* (especially *chamaeleo*, *zeylanicus*) often die (30%–40%) promptly after oviposition, suggesting programmed senescence. See also the *Furcifer* account. If female *Chamaeleo* are terminally depleted by the act of oviposition, their penchant for remaining in the vicinity of the nest may connote exhaustion, which might also predispose them to being reported as parental-care givers. That *Chamaeleo* die young is supported by the short maximum longevity reported for these well-studied chameleons (–1.50 v. –0.83 std. resid.). Although *calyptratus* has been recorded to live for 8 years, most species of these large lizards have maxima of less than 2 years. In combination with long and variable incubation periods, this short longevity (combined with rapid maturation dictated by the short longevity) means that, for *dilepis*, the length of time between conception and hatching is longer than the period of time from hatching to maturity. Is the long time period underground extended because egg mortality is so low (and allows the eggs to safely survive seasons of poor survival), or is the time from hatching to maturity greatly accelerated because juvenile mortality is so high? Or both? The pattern in *Chamaeleo* is strikingly different from that of most lizards.

Another striking aspect of *Chamaeleo* natural history is that very few species are geographically restricted (7% v. 76% endemism). Their wide geographic ranges and apparent developmental flexibility may contribute to the high frequency with which these species have been introduced at least once (28.6% v. 5.4%),

Chamaeleo spp.	English name	Range core	Activity	Diet	Size mm	Size g	Ecological business model
africanus	African Chameleon	Central Africa	diurnal	invert.	152	71	*** diurnal canopy walker
anchietae	Angola Chameleon	Congo	diurnal	invert.	69	8.5	*** diurnal canopy walker
arabicus	Arabian Chameleon	Yemen	diurnal	invert.	174	101	** ground-foraging tree lizard or diurnal canopy walker (as veg. allows)
calcaricarens	Sahel Chameleon	Horn of Africa	diurnal	invert.	117	34.8	*** diurnal canopy walker
calyptratus	Veiled Chameleon	Yemen	diurnal	omnivore	186	122	**** diurnal canopy walker
chamaeleon	Mediterranean Chameleon	Mediterranean	diurnal	invert.	121	38.6	**** diurnal canopy walker (where trees) or immobility expert
dilepis	Flap-necked Chameleon	Southern Africa	diurnal	invert.	122	39.1	*** diurnal canopy walker (where trees) or immobility expert
gracilis	Graceful Chameleon	Ghana	diurnal	invert.	119	36.7	**** diurnal canopy walker
laevigatus	Smooth Chameleon	Africa	diurnal	invert.	108	28.7	**** diurnal canopy walker
monachus	Socotra Chameleon	Yemen	diurnal	invert.	134	50	*** diurnal canopy walker (where trees) or immobility expert
namaquensis	Namaqua Chameleon	Namibia	diurnal	omnivore	118	35.6	** refugia-anchored diurnal gr. liz.
necasi	Necas' Chameleon	Benin	diurnal	invert.	96	20.6	*** diurnal canopy walker
senegalensis	Senegal Chameleon	Ghana	diurnal	invert.	120	37.2	*** diurnal canopy walker
zeylanicus	Asian Chameleon	India	diurnal	invert.	160	60	**** diurnal canopy walker

or more than twice (14.3% v. 1.5%). *C. chamaeleon* is reported to colonize greenhouses and to breed continuously in temperate zones where they can access the year-round heat inside greenhouses. This reproductive opportunism may contribute to the high rate of introductions, as may inadvertent stowing-away within plants shipped from the greenhouses.

Herpetologists are unlikely to view chameleons as threatening or harmful creatures, but herpetologists have long known that they are not like most people. On the island of Cyprus, some residents are convinced that *C. chamaeleon* are venomous, and if bitten by them the human's death can be forestalled only by burning a donkey alive—every hair of the donkey must be incinerated. If that isn't enough of a karma burden on the chameleon, there is also a belief that an angel called upon a chameleon to warn Jesus that Judas would betray him, but the slow locomotion of the chameleon—evidently not a *namaquensis*—meant that this warning was not delivered in a timely manner.

Slow locomotion contributes to the deaths of many chameleons, especially males crossing roads in search of females during the mating season. They are also victims of desertification, deforestation, and commercial collecting. All species of *Chamaeleo* are considered imperiled (v. 95% of chameleons), but in credit to their extensive geographic ranges, 0% (v. 25%) are deemed to be in the most severe IUCN categories.

Chamaeleonidae (figs. 4.3E, 4.3F)

The 203 chamaeleonids fit comfortably within the Iguania, but they are also a specialized group instantly distinguishable from all other iguanians by their independently moving turreted eyes, projectile tongues, opposable toes, advanced color changing ability, and body that is compressed in a vertical plane. No such features can be used for instant recognition of the other iguanian families. Within the chameleons there are two ecological groups that may or may not reflect their phylogeny (recent evidence is less supportive). Klaver and Böhme (1986) declared the small chameleons to be the Brookesiinae, or dwarf, chameleons and the larger ones to be the Chamaeleoninae, or typical, chameleons. This ecological division is well supported (table 5.2), even if it does not reflect phylogeny.

Chamaeleonids are found in Africa (including Madagascar), around the Mediterranean (probably at least partially assisted by humans), and eastward to India. They are comparatively well known (38% v. 27% of cells filled). The family's mean mass, mean shape index, mean relative tail length, frequency of crests, clutch count, and ontogenetic size range are intermediate between typical and dwarf chameleons (table 5.2); the two modes represented in table 5.2 are more informative. Except for the very shortest ones, most tails are prehensile (88% of chameleons v. 5%

Table 5.2. Salient ecological attributes distinguishing the dwarf chameleons from the typical chameleons

Informal name:	Typical chameleons	Dwarf chameleons
Genera	*Archaius, Bradypodion, Calumma, Chamaeleo, Furcifer, Kinyongia, Nadzikambia, Trioceros*	*Brookesia, Palleon, Rhampholeon, Rieppeleon*
Species	149	54
Mass (g)	13.49	2.04
Diurnal canopy walkers (% of spp.)	98	2
Immobility experts (% of spp.)	0	98
Shape index (std. resid.)	0.57	0.32
Relative tail length (× SVL)	1.09	0.53
Crests (% of spp.)	36	0
Sexual size dimorphism (–2 to 2 scale)	0.41 (i.e., m ≥ f)	–0.70 (i.e., m ≤ f)
Arboreal (% of spp.)	99	17
On ground (% of spp.)	1	83
Mass-adjusted clutch count (std. resid.)	2.81	0.60
Mass-adjusted ontogenetic size range (std. resid.)	0.63	–0.16
Mass-adjusted incubation duration	2.43	0.44
Population density (kg/ha)	0.18	0.03

of lizards). None has secretory pores (v. 15%) or is elongate (v. 11.4%). Advanced color change is nearly universal (91% v. 6%).

Mean chamaeleonid SSD is broadly similar to that of all lizards (0.14 v. 0.36 on –2 to 2 scale), but individual genera do not conform to this pattern (see genus accounts). *Bradypodion* is strongly m<f (–1.60 on –2 to 2 scale), whereas *Furcifer* and *Kinyongia* are strongly m>f (average 1.50 and 1.22, respectively). About half of chameleons are sexually dichromatic (53% v. 21%), and seasonal color is about threefold more frequent in chameleons than lizards (18.2% v. 6.5%). Sexual differences in coloration are hard to detect in chameleons against the bewildering context of rapid and dramatic short-term color change; it is likely to be underreported. Blue tails are unknown (v. 2.3%). Taken with morphological differences between the sexes, overall sex differences are much larger in chameleons than in lizards (2.16 v. 0.75 on 0 to 6 scale). Among families with more than one species, only Crotaphytidae is higher.

Almost all chameleons (98% v. 73%) eat invertebrates, but 1.6% (only in *Chamaeleo* 14.3%) are omnivores (v. 15.4%). Sit-and-wait foraging is represented in chameleons at almost exactly the lizard average (47% v. 46%), though this conceals considerable variation among genera (0% in *Bradypodion* and 100% in *Kinyongia*). Cruise foraging is a chameleon specialty (58% v. 7%) but shows considerable variation among genera (100% in *Bradypodion*, *Chamaeleo*; 0% in *Kinyongia*, *Rieppeleon*). Active foraging is rare (10.5% v. 58%) and is reported only from *Brookesia* and *Chamaeleo*. Electric light use is unknown (v. 1.4%). Size-adjusted home ranges are very small (–1.12 v. 0.0 std. resid.) but are known only from *Calumma* (–1.80 std. resid.) and *Chamaeleo* (0.39 std. resid.). These lizard cruise about, but only in a relatively small and presumably familiar area.

Almost all chameleons are diurnal (98.5% v. 66.6%); all others are crepuscular (1.5% v. 1.8%). No chameleons call spontaneously (v. 2.5%), but 3.4% give stress calls (v. 2.2%), and many of the dwarf chameleons give a barely audible stress vibration (buzzing presumed to deter predation, perhaps by mimicking dangerous invertebrates). All chameleons rely on crypsis (v. 43%).

Their choice of basking behavior is typical of lizards. On average they adopt a lower active body temperature (27.0°C v. 30.8°C). Many are solitary (36% v. 10%), and only 9.1% are found in groups larger than a pair (v. 20.4%). Few males are tolerant of other males (7.5% v. 14.6%), and 92.5% of males defend some space at some time (v. 85.4%). Male chameleons are not known to form hierarchies (v. 7.7%). Female tolerance is arguably lower (13.3%, the third lowest, v. 36%). Females are the third highest in year-round defense (6.7% v. 3.4%). These rankings reinforce the chameleon's reputation for curmudgeonly solitude.

See table 5.2 for microhabitat patterns. Chameleons are not documented from the dry end of the rainfall spectrum (none in hyperdry or dry), though it is likely that *namaquensis* occurs there. They are overrepresented only in the mesic band (1101–3000 mm/y; 85% v. 43%). Chameleons are slightly overrepresented in the higher elevation bands, perhaps because such place abound in Africa. Similarly, chameleons are underrepresented in littoral (6% v. 17%) and sandy (3% v. 22%) areas and unremarkable in relation to human habitats (23% v. 18.5%). Riparian occupancy is normal (21% v. 22%).

Chameleons are like most lizards in their reproductive mode (80% v. 82% oviparous), but they have very large clutches (2.04 v. 0.0 std. resid.). Apparently chameleons are pressured to have high fecundity, as the RCMs are notably high (0.444 v. 0.226), and the mean size of young is moderately low (0.53 v. 0.0 std. resid. for ontogenetic size range), equating to young that are 1/72 of maximum size rather than 1/44. Is this responsible for the vertically compressed body shape (to stuff more eggs/young into a body that must often balance over a narrow branch)? These features associated with high fecundity are amplified in the typical chameleons (table 5.2), as is the vertically compressed body. Although chameleons may be under selective pressure for high fecundity, they have exceptionally slow mass-adjusted development prior to hatching (1.89 v. 0.0 std. resid.). This mass-adjusted incubation duration is exceeded only by leiosaurids and the Tuatara. The chameleons with the slowest development are grouped as typical chameleons (table 5.2), but there is considerable diversity among genera within each of the ecological groups. Typical chameleon incubation duration (std. resid.) ranges from 0.33 (*Archaius*; mean = 100 days) to a remarkable 4.01 (*Kinyongia*: mean = 285 days). Dwarf chameleon incubation durations range from 0.01 (*Rieppeleon*; mean = 67 days) to 0.85 std. resid. (*Rhampholeon*; mean = 108 days). Their eggshells are soft (v. 59%) and not adherent (v. 97.6%); only *Archaius tigris* is reported to lay communally (0.5% v. 4.2%). Parental care is rare (1.0% v. 2.9%).

Lifespans are short (–0.83 v. 0.0 std. resid.), and at least one population of *Furcifer labordi* is semelparous (breeds once and dies; the majority of the population survives only as eggs for most of the year). These features are consistent with high mortality, for which high fecundity is a necessary offset. Temperature-dependent sex determination (TSD) has been reported in *Chamaeleo*, *Rhampholeon*, and *Trioceros*, but it is well documented only in *Chamaeleo chamaeleon*. Do the frequent movements of chameleons in trees result in high mortality by drawing the attention of predatory birds?

Chameleons are found on islands (48% v. 40%) and introduced (5.4% v. 3.3%) in roughly the same frequency as lizards. Almost all species are at conservation risk (95% v. 49%). Their frequency in the extreme category of endangerment is nearly double that of lizards (25% v. 13.6%), probably reflecting the large number of species with highly restricted ranges in remnant forests (often high elevation) in Africa and Madagascar.

Chamaesaura

Chamaesaura (five spp.) are the outliers of the Cordylidae; functionally limbless, slender (shape index –1.20 v. 0.35 std. resid.) diurnal grass-swimmers. Their southern African range is nearly the same as the family, and many species forage by patient observation from the safety of cover, but their cover is dense tufts of grass rather than rock crevices. *Chamaesaura* are all categorized as litter cryptozoids, as they use the visual complexity of grass mats to provide concealment, but they are not hidden under dead leaves as are the classic litter species.

Chamaesaura average smaller than cordylids (6.2 g v. 10.8 g), are elongate (100% v. 7%), have much longer tails (2.68 v. 1.16 × SVL), and their tails break readily (80% v. 70% of species), but only the most distal portion of the tail is lost, as the tail is vital for propulsion through the grass mat. Unlike most cordylids, they lack secretory pores (0% v. 69%) and do not show sexual dichromatism (0% v. 37%) or seasonal color (0% v. 28%). Information on the relative sizes and proportions of the sexes is unavailable; thus they have no documented sex differences.

They all eat invertebrates (100% v. 70%). Information on foraging mode is missing or contradictory, though they rely on immobility for concealment. Home-range areas have not been estimated. In keeping with their cryptic hunting tactic, they are not conspicuously wary (close approach allowed before fleeing or diving into vegetation). One species (*tenuior*) has been reported to bask directly (v. 74%). Though they may occasional climb low vegetation, they are primarily terrestrial (100% v. 76%). They occur up to 3000 m in elevation (*tenuior*).

Chamaesaura are all viviparous (v. 75%), with the largest litters of any genus in the Cordylidae (1.07 v. –0.48 std. resid.). They have not been observed to show parental care (v. 7.4%). As far as is known, they are very short lived (–1.65 v. 0.02 std. resid.). The longest known longevity is 1.8 years for *aenea*. No *Chamaesaura* is considered to be at conservation risk (v. 79%). In other respects they match the cordylid template.

Chatogekko

Chatogekko amazonicus is a tiny (0.45 g v. 1.07 g), padless, diurnal or cathemeral litter cryptozoic (v. 53%) sphaerodactylid from the northern Amazon and the Guiana Shield. Its skin tears easily. Males are smaller than females (–1.0 v. 0.04 on –2 to 2 scale). It sits-and-waits (v. 83%), retreats to vegetation when threatened (v. 27%), does not bask directly (v. 6.5%), is partial to riparian areas (v. 19%) at low elevation (v. 96%), is below the size at which a clutch of one is of average relative size (mass-adjusted clutch 0.28 v. –0.11 std. resid.), and has a family-typical RCM (0.099 v. 0.095). It has not been

Chamaesaura spp.	English name		Size		Range core
	Activity	Diet	mm	g	Ecological business model
aenea	Coppery Grass Lizard				South Africa
	diurnal		102	5.8	** litter cryptozoid
anguina	Cape Grass Lizard				South Africa
	diurnal		108	6.6	** litter cryptozoid
macrolepis	Large-scaled Grass Lizard				South Africa
	diurnal		113	7.3	** litter cryptozoid
miopropus	Zambian Grass Lizard				Tanzania
	diurnal		106	6.3	* litter cryptozoid
tenuior	Highland Grass Lizard				Congo
	diurnal		99	5.4	* litter cryptozoid

Chatogekko sp.	English name		Size		Range core
	Activity	Diet	mm	g	Ecological business model
amazonicus	Northern Amazon Sphaero				Guiana Shield
	diurnal	invert.	26	0.45	*** litter cryptozoid

introduced (v. 5%) nor has it been evaluated for conservation status. In other respects *Chatogekko* is a typical sphaerodactylid.

Chelosania sp.	English name				Range core
			Size		
	Activity	Diet	mm	g	Ecological business model
brunnea	Chameleon Dragon				Northern Australia
	diurnal		98	31.4	** diurnal canopy walker

Chelosania

Chelosania is but a single, little-known agamid that cruises (v. 6%) slowly through the woodlands of far northern Australia. Although characterized as a "chameleon" dragon for its upright cross section, slow rocking progression, dependence on crypsis (v. 51%), erectable crests (v. 38%) and dewlap, color change (v. 15% advanced), and vaguely turreted eyes, it does not have a prehensile tail (v. 1.4%) or ballistic tongue. Nor, of course, is it descended from chameleons; the similarities are examples of convergence. No records were encountered of its diet, home range, vocalization, thermoregulation, rainfall preferences, longevity, population density, or conservation status. Males are intolerant of other males (v. 97%). It has been studied for its awesome RCM (0.695 v. 0.336), which is manifest despite the species having an utterly normal clutch count (–0.09 v. 0.20 std. resid.). Presumably the eggs are exceptionally large, an attribute unlike chameleons. The species has not been introduced (v. 1.7%).

Chioninia spp.	English name				Range core
			Size		
	Activity	Diet	mm	g	Ecological business model
coctei	Cape Verde Giant Skink				Cape Verde Islands
	cathemeral	carnivore	259	466	** refugia-anchored diurnal gr. liz.
delalandii	Delalande's Skink				Cape Verde Islands
	diurnal		75	8.6	* refugia-anchored diurnal gr. liz.
fogoensis	Santo Antao Skink				Cape Verde Islands
	diurnal		66	5.6	* refugia-anchored diurnal gr. liz.
nicolauensis	Sao Nicolau Skink				Cape Verde Islands
	diurnal	omnivore	59	3.86	* refugia-anchored diurnal gr. liz.
spinalis	Sharkskin Skink				Cape Verde Islands
			72	7.3	* refugia-anchored diurnal gr. liz.
stangeri	Cape Verde Skink				Cape Verde Islands
	diurnal	invert.	67	5.8	** refugia-anchored diurnal gr. liz.
vaillantii	Vaillant's Skink				Cape Verde Islands
	diurnal	omnivore	181	145	** refugia-anchored diurnal gr. liz.

Chioninia

Chioninia is a moderate radiation (seven spp.) of large and small (17.6 g v. 3.3 g) refugia-anchored diurnal ground skinks (100% v. 30%) endemic to the Cape Verde Islands west of Senegal. They are poorly known (filled cells 21% v. 25%).

They are fairly robust (shape index 0.48 v. –0.46 std. resid.). They have a diverse diet that often includes carrion. Most species are diurnal (83% v. 74%), but cathemeral species are well represented (17% v. 9%). *Chioninia* are found at low elevation (100% v. 89%) and are found in specialized habitats, except riparian zones (0% v. 23%), at typical frequencies: littoral (17% v. 21%), sandy (33% v. 25%), and human (33% v. 14%) habitats.

Despite the lowland equatorial habitat, all known species are viviparous (v. 36%). Their litter of two is the second-lowest mass-adjusted clutch count in Scincidae (–2.20 v. 0.06 std. resid.), second to *Corucia*. The huge *Chioninia coctei* has a litter count apparently fixed at two.

Longevity is about average for a lizard (0.07 v. 0.17 std. resid.), equating to 16 years for the only documented species (*coctei*). One species (*delalandii*) has been introduced to three other islands in the Cape Verde Islands (v. 0.7%). Half of the evaluated species are considered imperiled (v. 50% v. 47%), and one-third are on the IUCN's lists of the most endangered (33% v. 14%). In other respects *Chioninia* are typical skinks.

Chirindia

Chirindia is a small (five spp.) genus of tiny (0.5 g v. 2.7 g), threadlike (shape index –4.87 v. –3.97 std. resid.) amphisbaenids from East and southern Africa (Tanzania to South

Africa). Sixty percent (possibly all) have secretory pores (v. 3.4%). The habitat use is in keeping with the family profile. All four species for which clutch count is documented have clutches of one, which counts for a relatively large clutch given their diminutive size (0.27 v. −0.12 std. resid.). Unfortunately, there is no information on clutch frequency or longevity. None occurs on islands (v. 12%) or is imperiled (v. 23%).

Chirindia spp.	English name				Range core
			Size		
	Activity	Diet	mm	g	Ecological business model
ewerbecki	Mbanja Worm Lizard				Tanzania
		invert.	119	0.46	**** subterranean forager
langi	Lang's Round-headed Worm Lizard				Zimbabwe
		invert.	124	0.55	**** subterranean forager
mpwapwaensis	Mpwapwa Worm Lizard				Tanzania
		invert.	134	0.76	*** subterranean forager
rondoensis	Nchingidi Worm Lizard				Tanzania
			102	0.25	**** subterranean forager
swynnertoni	Swynnerton's Round-headed Worm Lizard				Mozambique
		invert.	122	0.51	**** subterranean forager

Chlamydosaurus

Chlamydosaurus is a single-species genus of agamid from woodlands of far northern Australia and adjacent New Guinea, a relatively well-known (65% v. 30%) and richly celebrated species. Its signature accoutrement is a large erectable frill that surrounds its head in both sexes, and which is displayed to shock and awe consexual rivals—who presumably possess their own and therefore should not be too dazzled—or predators. In the absence of mirrors, perhaps even the consexual rivals are mortified by the sudden enlargement of their rival's head.

Chlamydosaurus sp.	English name				Range core
			Size		
	Activity	Diet	mm	g	Ecological business model
kingii	Frilled Lizard				Northern Territory
	diurnal		185	225	**** ground-foraging tree lizard

It is an unusually large (225 g v. 16 g) agamid that is heavily built (shape index 1.67 v. 0.80 std. resid.) and in possession of a long tail (2.00 v. 1.86 × SVL). Males average 83% heavier than their mates (1.00 v. 0.53 on −2 to 2 scale), suggesting an evolutionary arms race among males for at least the appearance of large size. Males also differ in color from their mates (v. 53%), but do not change seasonally (v. 20%) or on short notice (v. 22%). Males are intolerant of other males (v. 97%).

Despite its large size, *C. kingii* eats invertebrates, foraging over a proportionately large home range (0.70 v. 0.35 std. resid.). It basks directly (v. 24%), without using its expandable frill to gain surface area. Perhaps that is not necessary, as the mean reported body temperature is a sizzling 40.0°C (v. 34.8°C). It is found in the mesic belt (1480–1657 mm/y), away from humans (v. 16%).

Females average 10.5 eggs per clutch (0.31 v. 0.20 std. resid.), deposited once or twice a year. The size of hatchlings is strikingly low (ontogenetic size range 1.75 v. 0.69 std. resid.). This equates to a hatchling that is 1/550 of maximum mass compared to the 1/100 normal for agamids. Its mass-adjusted incubation period is notably short (73 days), even by agamid standards (−1.04 v. −0.55 std. resid.). Its mass-adjusted longevity is relatively long (0.21 v. −0.46 std. resid.), and its density is moderate (0.18 kg/ha v. 0.16 kg/ha). It has not been introduced (v. 1.7%), and is not imperiled (v. 37%), though declining at the southern (more arid) margin of its range.

Chondrodactylus

Chondrodactylus (five spp.) are moderately large (9.7 g v. 3.31 g), nocturnal, mostly arboreal (80% v. 68%) gekkonids in southern Africa (especially the west coast: Angola, Namibia). They tend to have large heads and rough-skin (tuberculate); the arboreal forms are especially rough-skinned, and the ground lizard (*angulifer*) has a particularly short snout.

Tails are breakable (80% v. 87%) and moderately short (0.90 v. 1.07 × SVL). Secretory pores are absent (v. 46%).

SSD is more m>f than most gekkonids (0.75 v. −0.03 on −2 to 2 scale), as befits their larger size. Sexual dichromatism is overrepresented (20% v. 6.6%). Unusually for a m>f group, male tail length (based however on only *angulifer*) greatly exceeds that of females (1.38 v. 1.05 ratio).

These lizards are large enough to eat vertebrates; one of three is con-

Chondrodactylus spp.	English name				Range core
			Size		
	Activity	Diet	mm	g	Ecological business model
angulifer	Namib Giant Ground Gecko				Botswana
	cathemeral	carnivore	86	12.6	*** nocturnal ground lizard
bibronii	Bibron's Gecko				South Africa
	nocturnal	invert.	78	9.5	** trunk/wall lizard
fitzsimonsi	FitzSimons' Gecko				Angola
	nocturnal		73	8.1	*** trunk/wall lizard
pulitzerae	Angola Thick-toed Gecko				Angola
	nocturnal		80	10.0	** trunk/wall lizard (if commensal)
turneri	Turner's Gecko				Southern Africa
	nocturnal	invert.	75	8.7	*** ground-foraging tree lizard or trunk/wall lizard (if commensal)

sidered to be a carnivore (v. 1%), though all three species with dietary information include vertebrates in their diet. The majority (60% v. 3.8%) feed at electric lights. Vocalization is common, both spontaneously (40% v. 6.5%) and under duress (40% v. 3.7%). None is considered conspicuously wary (v. 4.4%). All rely on retreat to crevices or burrows (v. 31%) for primary predator concealment. Mean active body temperature is on the low side for a gekkonid, but unsurprising for a nocturnal gekkonid (26.5°C v. 27.6°C). Use of burrows as refugia is found in the nocturnal ground lizard, *angulifer* (25% v. 10.4%).

Solitary living is well represented in *Chondrodactylus* (50% v. 7.7%).

Chondrodactylus are overrepresented at the dry end of the rainfall spectrum: dry (56–150 mm/y: 33% v. 13%); arid (151–400 mm/y: 100% v. 27%). They are also overrepresented in sandy areas (40% v. 11%) and around human habitat (80% v. 27%), but underrepresented in the littoral zone (0% v. 11%).

Reproduction is similar to most gekkonids, though incubation is quicker (–0.38 v. 0.17 std. resid.) and eggshells are never adherent (v. 10.5%) nor laid communally (v. 8.6%). Parental care is present (20% v. 2%), in the form of clutch defense by female *angulifer*. The young are relatively large, as indicated by the reduced ontogenetic mass range (–0.42 v. –0.19 std. resid.).

One species has been introduced (*bibronii* to Florida). No species is considered imperiled (v. 47%). In other respects *Chondrodactylus* are typical gekkonids.

Christinus

Christinus (three spp.) are moderately large (7.0 g v. 3.31 g), nocturnal (100% v. 73%), arboreal (100% v. 68%) gekkonids of southern Australia and remote Lord Howe Islands east of Australia. Though they are nominally well known (40% v. 25% of cells filled), little is known of their social interactions. All are considered nocturnal canopy walkers (v. 9%), though they are unusual lizards in that they may refuge underground during the day and feed arboreally at night. Most lizards that switch strata daily feed terrestrially and refuge arboreally. Despite their terrestrial refuging, they do not do so in burrows (v. 10.4%).

All tails break (v. 87%), though the fat content of *marmoratus* tails is 33% (compared to 18% fat in body) and the owner must lament the loss of so caloric an appendage. On the other hand, this packet may be so attractive to predators that they might be satisfied capturing it without its former owner. I have seen birds too small to capture other lizard species nonetheless pursue them to remove their tails. Secretory pores are absent (v. 46%), as is color change (v. 15%). *Christinus* have enlarged toepads. *C. guentheri* has a reflective layer in its retina, producing a bright eyeshine for those observing with an electric headlamp.

The only documented species (*guentheri*) has average adult males about 35% heavier than their mates. No other sex differences are reported; the aggregated sex rollup score (on 0 to 6 scale) is therefore low compared to other gekkonids (0.33 v. 0.63).

Of the two species with documented diets, *marmoratus* is a nonspecified invertebrate eater, and *guentheri* is omnivorous, notably a nectar lapper. At good nectar sources, large numbers of *guentheri* may aggregate, raising the possibility of interesting social interactions that are however not documented. *C. guentheri* have been observed to open closed packets of sugar, suggesting that they are highly motivated and innovative at accessing it.

The mean active body temperature is low (21.6°C v. 27.6°C), presumably reflecting the high latitude environment. *C. marmoratus* is documented to remain active down to a 13.5°C body temperature. *Christinus* are overrepresented in the littoral zone (33% v. 11%) and sandy areas (33% v. 11%), though not in riparian zones (0% v. 17%).

Christinus spp.	English name				Range core
			Size		
	Activity	Diet	mm	g	Ecological business model
alexanderi	Alexander's Gecko				Southwestern Australia
	nocturnal		57	4.08	* nocturnal canopy walker
guentheri	Lord Howe Island Gecko				Lord Howe Island
	nocturnal	omnivore	91	20.5	*** nocturnal canopy walker
marmoratus	Marbled Gecko				Australia
	nocturnal	invert.	57	4.08	** nocturnal canopy walker

Christinus are typical gekkonids in reproductive mode and clutch count, though they have an extraordinary range in incubation durations: *guentheri* ranges from 91 to 273 days, with much of that variation associated with differing islands. That implied explanation does not account for the equally wide range in the mainland *marmoratus*: 70–207 days. Potentially, eggs laid late in the season must overwinter, whereas those laid early enough can complete incubation within the summer. *Christinus* eggs are often adherent (33% v. 10.5%) and laid communally (67% v. 8.6%). Parental care is absent (v. 2%). The relative size of young is noteworthy in that it is smaller than the lizard average, whereas most gekkonids are larger (ontogenetic range 0.08 v. −0.19 std. resid.). This would suggest that a small fixed clutch of small young would produce a light RCM ratio. RCM is documented only for *marmoratus*, for which it is exceptionally heavy 0.325 (v. 0.140). The eggs of *marmoratus* are relatively large (8.5 × 14 mm) for a 4 g gecko, so this species may not well represent the genus.

Longevity is documented only for *marmoratus*. It is long in absolute time (12.8 years) and when adjusted for mass (1.18 v. 0.15 std. resid.). No *Christinus* has been introduced (v. 3.7%). The evaluated species (*guentheri*) is considered imperiled (listed as vulnerable by IUCN, federal, and state authorities). In other respects *Christinus* are typical gekkonids.

Cnemaspis (figs. 4.7A, 4.7D, 4.14A)

Cnemaspis is the third-largest genus (130 spp.) of the Gekkonidae and sixth-largest overall (table 2.2). It is found primarily in the Oriental biogeographic realm (90%) and secondarily in equatorial Africa (10%). The Asian distribution spans from India to Indonesia; the richest countries are Malaysia (31), India (28), Sri Lanka (24), Indonesia (17), Thailand (16), and Vietnam (7). The genus is not yet known from Myanmar or the Philippines. The African distribution spans from Guinea to Ethiopia and south to Tanzania. Geographic sampling is incomplete, but Cameroon (six spp.), Tanzania (five), Kenya (four), and Uganda (four) top the current listings. The modal ecological business model is trunk/wall lizard (54% v. 45%). Another major group is the immobility experts (20.7% v. 4.5%), but there are also crack hermits (5.7% v. 3.2%), ground-foraging tree lizards (5.7% v. 5.3%), litter cryptozoids (3.4% v. 0.5%), nocturnal ground lizards (4.6% v. 27%), and refugia-anchored diurnal ground lizards (5.7% v. 3.0%). Perhaps equally pertinent is that the ecological business models of 43 of 130, or 33% of *Cnemaspis*, could not be estimated from the available data, and only 12% were assigned with full confidence.

Cnemaspis are small (1.58 g v. 3.31 g), but proportioned as typical gekkonids (shape index −0.02 v. 0.10). At least 72% of *Cnemaspis* are padless (v. 30%). The Asian contingent has a few species with prehensile tails (4.2% of genus v. 1.8% of family), though none of the African *Cnemaspis* do. One Asian *Cnemaspis* with an interesting tail (*caudanivea*) can fold the end of its tail forward 180° on itself while rolling and unrolling it slowly in a lateral plane, making it look like a small, white worm wiggling on a rock. This is notable because prehensile tails are generally useful only for grasping narrow perches. The African *Cnemaspis* also lack color change (0% v. 34% for Asian *Cnemaspis* v. 14.6% for gekkonids). Both groups are amply endowed with secretory pores (69% v. 46% for gekkonids).

SSD is explicitly recorded for only three (Asian) *Cnemaspis*, for which it is specified that there is none. However, morphological and color differences are present at gekkonid levels, suggesting that *Cnemaspis* are contesting social and sexual status in ways independent of size. The genus's mean sex differences rollup score comports well with the family (0.55 v. 0.63 on 0 to 6 scale).

Foraging mode observations have rarely been made of *Cnemaspis*: one was made of an African species (active) and two Asian species were noted (both sit-and-wait). The Asian *peninsularis* has been documented by MPM (1st percentile) and PTM (6th percentile), one of the least active of all lizards. *Cnemaspis* have not been recorded to use electric lights (v. 3.8%), perhaps out of deference to larger species of geckos at the lights. All known diets are invertebrate (v. 95%).

The modal *Cnemaspis* is active at any time of day (cathemeral: 48% v. 11%), with that component about twice as strong (85% v. 42%) in the African species as the Asian. The next biggest group is the diurnal species (34% of genus v. 15% of family), stronger (37% v. 15%) in the Asian group. However, even at 37%, the common name for this genus of "day" geckos is a bit of a stretch. Perhaps they should be called "all-day" geckos, though that would characterize the African species better than the Asian ones. Caution is in order for interpreting diel activity patterns in general and especially in this group; for example, *baueri, flavigaster, harimau, pemanggilensis, perhentianensis*, and *sundainsula* are said to have maximal locomotor activity (and wariness) during the day, but maximal exposure at night. Some reporters equate activity with exposure and some focus on locomotion. Just to round out the confusion, *quattuorseriata* has been described as cathemeral, crepuscular, diurnal, and nocturnal. Subjectively, this genus has more contradicted assessments of daily activity cycle than any other.

Despite their activity throughout the day, *Cnemaspis* have not been heard to vocalize spontaneously (v. 6.5%) or under duress (v. 3.7%). This is extraordinary for cathemeral geckos; perhaps their common name should be "mute geckos." They are prone to wariness (15% v. 4.4%); greater wariness is reported for the Asian species (16% v. 8%). Crypsis (35% v. 50%) and retreat to vegetation (9% v. 19%) are underrepresented as predator concealment tactics; crevices and burrows collectively are overrepresented (57% v. 31%).

Direct basking is prevalent at the gekkonid average (3.8% v. 4.9%). Solitary distributions are also gekkonid average (9% v. 8%), as is group living (27% v. 25%). Arboreality is overrep-

Cnemaspis spp.	English name		Size		Range core
	Activity	Diet	mm	g	Ecological business model
aceh	Aceh Day Gecko				Indonesia
	diurnal		31	0.78	* trunk/wall lizard
adii	Adi's Day Gecko				India
			34	1.00	
affinis	Pinang Island Rock Gecko				Malaysia
	cathemeral	invert.	46	2.22	**** immobility expert
africana	Usambara Forest Gecko				Equatorial Africa
	cathemeral	invert.	51	2.91	** crack hermit
alantika	Alantika Forest Gecko				Cameroon
	cathemeral		43	1.90	*** crack hermit
alwisi	Alwis' Day Gecko				Sri Lanka
			38	1.28	** trunk/wall lizard
amith	Amith's Day Gecko				Sri Lanka
			33	0.90	
anaikattiensis	Anaikatti Hills Gecko				India
	crepuscular		53	3.31	* refugia-anchored diurnal gr. liz.
andalas	West Sumatra Day Gecko				Indonesia
	diurnal		32	0.84	*** trunk/wall lizard
andersonii	Andaman Rock Gecko				India
			32	0.82	
argus	Lawit Mountain Day Gecko				Malaysia
	cathemeral		57	3.96	
assamensis	Assam Day Gecko				India
	diurnal	invert.	33	0.91	*** ground-foraging tree lizard
aurantiacopes	Saffron-limbed Gecko				Vietnam
	nocturnal		51	2.95	*** crack hermit
australis	Southern Day Gecko				India
			31	0.76	
barbouri	Uluguru Forest Gecko				Tanzania
	cathemeral	invert.	39	1.46	
baueri	Aur Island Rock Gecko				Malaysia
	cathemeral		57	4.10	**** immobility expert
bayuensis	Bayu Cave Rock Gecko				Malaysia
	cathemeral		42	1.78	**** immobility expert
beddomei	Beddome's Day Gecko				India
	cathemeral	invert.	46	2.22	** immobility expert
bidongensis	Bidong Island Rock Gecko				Malaysia
	cathemeral		51	2.95	** immobility expert
biocellata	Twin-spot Day Gecko				Thailand
	cathemeral		38	1.33	**** immobility expert

Cnemaspis spp.	English name				Range core
			Size		
	Activity	Diet	mm	g	Ecological business model
boiei	Boie's Day Gecko				India
			37	1.24	
boulengeri	Con Dao Round-eyed Gecko				Vietnam
	cathemeral		59	4.38	*** trunk/wall lizard
caudanivea	Hon Tre Island Rock Gecko				Vietnam
	cathemeral		43	1.87	*** immobility expert
chanardi	Chan-ard's Rock Gecko				Thailand
	diurnal		38	1.33	** trunk/wall lizard
chanthaburiensis	Chanthaburi Gecko				Southeast Asia
	cathemeral		40	1.49	
clivicola	Highland Day Gecko				Sri Lanka
			38	1.32	* trunk/wall lizard
dezwaani	De Zwaan's Day Gecko				Indonesia
	diurnal		34	1.01	* trunk/wall lizard
dickersonae	Dickerson's Forest Gecko				Congo
	cathemeral	invert.	39	1.39	
dilepis	Two-scaled Forest Gecko				Cameroon
	cathemeral		32	0.84	
dringi	Dring's Day Gecko				Borneo
			43	1.81	
elgonensis	Mt. Elgon Forest Gecko				Kenya
	cathemeral	invert.	50	2.86	** trunk/wall lizard
flavigaster	Orange-bellied Rock Gecko				Malaysia
	cathemeral		45	2.12	**** immobility expert
flavolineata	Titi Wangsa Rock Gecko				Malaysia
	cathemeral		43	1.83	
gemunu	Gemunu's Gecko				Sri Lanka
		invert.	33	0.90	* trunk/wall lizard
gigas	Giant Forest Gecko				Nigeria
	cathemeral	invert.	59	4.50	* immobility expert
girii	Giri's Day Gecko				India
			34	1.01	
goaensis	Goa Forest Gecko				India
	nocturnal	invert.	30	0.67	* trunk/wall lizard
gracilis	South-Indian Forest Gecko				India
	nocturnal	invert.	33	0.89	
grismeri	Grismer's Rock Gecko				Malaysia
	diurnal		46	2.22	
hangus	Bukit Hangus Rock Gecko				Malaysia
	diurnal		46	2.22	*** trunk/wall lizard

resented (79% v. 68%), and terrestriality underrepresented (19% v. 31%). Burrow use for daily refugia is not documented in *Cnemaspis* (v. 10.4%). All documented distributions are in the mesic (1101–3000 mm/y) rainfall belt (v. 66%). Elevational distribution matches gekkonids very closely (94/26/1 v. 92/26/2). *Cnemaspis* are underrepresented in littoral zones (6% v. 11%), in sandy areas (0% v. 11%), and around humans (20% v. 27%); and overrepresented in riparian zones (25% v. 17%). The paucity of *Cnemaspis* geckos in sandy areas may relate to the extra burden that damp, adherent sand imposes on extremely small lizards: a few wet grains are a large gritty burden for a 0.7 g lizard.

All reproductive traits in Gekkonidae match those in *Cnemaspis*, though this genus occasionally exceeds clutches of two (0.12 v. –0.26 std. resid.). Only one species has been quantified for population density: the central African *quattuorseriata* at 0.05 kg/ha (v. 0.37 kg/ha). *Cnemaspis* geckos undoubtedly attain high population densities in very small patches around suitable rocks, but average population density might be fairly sparse, as indicated by *quattuorseriata*. Other species may be more extensively common; for example, *scalpensis* is a trunk/wall lizard that is said to have more than 50 individuals present on single large trees, especially figs. *C. molligodai*, found primarily on large-diameter tree trunks, occurs in colonies of 30–40 individuals.

Most species are endemic to their host country (87% v. 71%). None has been introduced (v. 3.7%). *Cnemaspis* are overrepresented on lists of imperiled species (54% v. 47%), including the IUCN's lists of the most endangered (22% v. 11%). This overrepresentation presumably reflects the many species with very limited ranges. One extreme example is *psychedelica*, which is not only greatly restricted in distribution, but also possesses striking luminous colors attractive to the pet trade. Grismer et al. (2014) reported that in late 2013 this species was selling for about US$4000/pair on the black market, and it was likely to suffer increased smuggling pressure as tourist access improved and legal pressure for protection increased. Shortly thereafter, it was listed as endangered by the IUCN. Another species with an apparently very restricted distribution is *boiei*, which has not been detected in 150 years.

Cnemaspis spp.	English name				Range core
			Size		
	Activity	Diet	mm	g	Ecological business model
harimau	Tiger Rock Gecko				Malaysia
	cathemeral		39	1.37	** refugia-anchored diurnal gr. liz.
heteropholis	Gund Gecko				India
	cathemeral		42	1.70	** trunk/wall lizard
huaseesom	Yellow-headed Rock Gecko				Thailand
	nocturnal		41	1.58	** trunk/wall lizard
indica	Nilgiri Dwarf Gecko				India
	cathemeral	invert.	38	1.36	* nocturnal ground lizard
indraneildasii	Das' Gecko				India
			31	0.74	
jacobsoni	Jacobson's Day Gecko				Indonesia
			31	0.77	
jerdonii	Anaimalai Hill Gecko				India
	nocturnal	invert.	40	1.54	*** nocturnal ground lizard
kallima	Sri Lanka Ornate Gecko				Sri Lanka
			31	0.76	* crack hermit or nocturnal ground lizard
kamolnorranathi	Kamolnorranath's Rock Gecko				Thailand
	nocturnal		36	1.18	* trunk/wall lizard
kandambyi	Knuckles Range Epiphyte Gecko				Sri Lanka
	diurnal		26	0.48	* trunk/wall lizard
kandiana	Kandy Rock Gecko				Sri Lanka
	cathemeral		37	1.26	*** trunk/wall lizard
karsticola	Karst Rock Gecko				Malaysia
			44	1.95	
kendallii	Kendall's Rock Gecko				Borneo
	diurnal	invert.	52	3.18	**** immobility expert
koehleri	Koehler's Forest Gecko				Cameroon
	cathemeral		48	2.47	** trunk/wall lizard
kolhapurensis	Kolhapur Gecko				India
	diurnal		39	1.38	* litter cryptozoid
kottiyoorensis	Kottiyoor Day Gecko				India
			39	1.44	
kumarasinghei	Kumarasinghe's Gecko				Sri Lanka
	cathemeral		32	0.84	** trunk/wall lizard
kumpoli	Trang Province Gecko				Thailand
	cathemeral		54	3.56	
laoensis	Lao Rock Gecko				Laos
			39	1.38	

Cnemaspis spp.	English name				Range core
			Size		
	Activity	Diet	mm	g	Ecological business model
latha	Flowery Day Gecko				Sri Lanka
			31	0.73	** trunk/wall lizard
limi	Tioman Island Rock Gecko				Malaysia
	cathemeral	invert.	74	8.2	**** immobility expert
lineogularis	Stripe-throated Rock Gecko				Thailand
	diurnal		37	1.19	*** trunk/wall lizard
littoralis	Coastal Day Gecko				India
	diurnal	invert.	34	0.97	
mahsuriae	Mahsuri's Rock Gecko				Malaysia
			36	1.10	
mcguirei	McGuire's Rock Gecko				Malaysia
	cathemeral	invert.	56	3.82	**** immobility expert
menikay	Jewel Day Gecko				Sri Lanka
			29	0.60	** trunk/wall lizard
minang	Padang Day Gecko				Indonesia
	diurnal		32	0.81	** trunk/wall lizard
modiglianii	Modigliani's Day Gecko				Indonesia
			162	72	
molligodai	Molligoda's Day Gecko				Sri Lanka
			29	0.62	*** trunk/wall lizard
monachorum	Monks' Rock Gecko				Malaysia
	diurnal		35	1.01	**** immobility expert
monticola	Waynaad Gecko				India
			33	0.90	
mumpuniae	Mumpuni's Rock Gecko				Indonesia
	diurnal		53	3.31	*** trunk/wall lizard
mysoriensis	Mysore Gecko				India
	cathemeral		31	0.76	** litter cryptozoid or (when commensal) trunk/wall lizard
nairi	Ponmudi Day Gecko				India
	diurnal	invert.	41	1.61	** litter cryptozoid or (when commensal) trunk/wall lizard
narathiwatensis	Narathiwat Rock Gecko				Thailand
	cathemeral		47	2.41	
neangthyi	Neang Thy's Rock Gecko				Cambodia
	nocturnal		48	2.52	*** trunk/wall lizard or immobility expert
nigridia	Black Rock Gecko				Malaysia
	nocturnal	invert.	64	5.5	*** nocturnal ground lizard
nilagirica	Nilgiri Gecko				India
			40	1.48	

Cnemaspis spp.	English name	Activity	Diet	Size mm	Size g	Range core	Ecological business model
niyomwanae	Niyomwan's Rock Gecko	nocturnal		50	2.84	Southeast Asia	** trunk/wall lizard or crack hermit
nuicamensis	Nui Cam Hill Day Gecko	diurnal		44	1.96	Vietnam	** trunk/wall lizard or crack hermit
occidentalis	Western Forest Gecko	diurnal	invert.	50	2.80	Liberia	** ground-foraging tree lizard
omari	Omar's Gecko	cathemeral		39	1.41	Malaysia	*** trunk/wall lizard
ornata	Ornate Dwarf Gecko	diurnal		47	2.31	India	
otai	Vellore Gecko	diurnal		32	0.85	India	
pagai	Pagai Day Gecko	diurnal		32	0.84	Indonesia	
paripari	Fairy Rock Gecko	diurnal		46	2.22	Indonesia	** trunk/wall lizard
pava	Little Day Gecko			32	0.81	Sri Lanka	* trunk/wall lizard
pemanggilensis	Pemanggil Island Rock Gecko	cathemeral		64	5.5	Malaysia	**** immobility expert
peninsularis	Peninsular Rock Gecko	cathemeral	invert.	52	3.18	Malaysia	**** ground-foraging tree lizard
perhentianensis	Perhentian Day Gecko	cathemeral		43	1.87	Malaysia	**** immobility expert
petrodroma	Ondo Forest Gecko	diurnal	invert.	55	3.69	Nigeria	** trunk/wall lizard
phangngaensis	Phangnga Rock Gecko	cathemeral		40	1.55	Thailand	*** trunk/wall lizard
phillipsi	Phillips' Day Gecko			34	0.97	Sri Lanka	** trunk/wall lizard
phuketensis	Phuket Day Gecko	cathemeral		30	0.70	Thailand	**** refugia-anchored diurnal gr. liz.
podihuna	Dwarf Day Gecko	diurnal		29	0.61	Sri Lanka	*** trunk/wall lizard
pseudomcguirei	False McGuire's Rock Gecko	nocturnal		40	1.50	Malaysia	

Cnemaspis spp.	English name				Range core
			Size		
	Activity	Diet	mm	g	Ecological business model
psychedelica	Psychedelic Rock Gecko				Vietnam
	diurnal		63	5.3	** trunk/wall lizard
pulchra	Paradise Day Gecko				Sri Lanka
			33	0.91	
punctata	Spotted Day Gecko				Sri Lanka
			35	1.08	* trunk/wall lizard
punctatonuchalis	Spotted-neck Rock Gecko				Thailand
	nocturnal		45	2.09	** trunk/wall lizard
quattuorseriata	Four-lined Forest Gecko				Congo
	cathemeral	invert.	43	1.86	** ground-foraging tree lizard
rajabasa	Mount Rajabasa Gecko				Indonesia
			52	3.07	
rajakarunai	Rajakaruna's Day Gecko				Sri Lanka
	diurnal		38	1.32	* trunk/wall lizard
rammalensis	Rammale Day Gecko				Sri Lanka
			48	2.52	
retigalensis	Retigala Day Gecko				Sri Lanka
			31	0.74	
roticanai	Roti Canal Rock Gecko				Malaysia
			43	1.86	
samanalensis	Samanala Day Gecko				Sri Lanka
			37	1.20	
scalpensis	Ferguson's Day Gecko				Sri Lanka
	diurnal		34	0.99	**** trunk/wall lizard
selamatkan-merapoh	Merapoh Rock Gecko				Malaysia
	nocturnal		40	1.57	** trunk/wall lizard
shahruli	Shahrul's Rock Gecko				Malaysia
	cathemeral		36	1.10	**** immobility expert
siamensis	Thai Forest Gecko				Thailand
	cathemeral		38	1.30	**** refugia-anchored diurnal gr. liz.
silvula	Forest Day Gecko				Sri Lanka
	diurnal		31	0.76	** trunk/wall lizard
sisparensis	Sispara Gecko				India
	cathemeral	invert.	54	3.43	** refugia-anchored diurnal gr. liz.
spinicollis	Spiny-necked Forest Gecko				Nigeria
	cathemeral	invert.	49	2.62	** trunk/wall lizard
stongensis	Stong Mountain Rock Gecko				Malaysia
	nocturnal		45	2.06	*** trunk/wall lizard

Cnemaspis spp.	English name	Activity	Diet	Size mm	Size g	Range core	Ecological business model
sundagekko	Anambas Rock Gecko			58	4.24	Indonesia	
sundainsula	Sunda Island Rock Gecko	cathemeral		70	7.2	Indonesia	** crack hermit or trunk/wall liz.
tapanuli	Tapanuli Day Gecko	diurnal		32	0.85	Indonesia	* ground-foraging tree lizard
temiah	Temiah Gecko			43	1.83	Malaysia	* immobility expert
thachanaensis	Tha Chana Rock Gecko	diurnal		37	1.25	Thailand	*** trunk/wall lizard
timoriensis	Timor Gecko			34	1.01	Indonesia	
tropidogaster	Rough-bellied Day Gecko	diurnal	invert.	35	1.03	Sri Lanka	* trunk/wall lizard
tucdupensis	Tuc Dup Hill Day Gecko	cathemeral		46	2.22	Vietnam	** immobility expert
upendrai	Upendra's Day Gecko			32	0.84	Sri Lanka	
uzungwae	Uzungwa Forest Gecko	cathemeral	invert.	38	1.32	Tanzania	
vandeventeri	VanDeventer's Rock Gecko	nocturnal		41	1.67	Thailand	** trunk/wall lizard
whittenorum	Siberut Day Gecko			32	0.84	Indonesia	
wicksi	Preparis Gecko			30	0.68	India	
wynadensis	Wayanad Night Gecko	nocturnal	invert.	40	1.47	India	** nocturnal ground lizard
yercaudensis	Yercaud Gecko	nocturnal		31	0.73	India	** trunk/wall lizard

Cnemidophorus (fig. 4.11E)

Cnemidophorus (Teiidae) is a large radiation (20 spp.) of slightly small (12.3 g v. 22.1 g), flight-reliant (100% v. 77%), mostly direct-basking (85% v. 58%), mostly wide-ranging chemosensors (90% v. 97%) found primarily in northwestern South America, but extending northward to Mexico and St. Lucia and many islands just north of Venezuela, eastward to French Guiana, and southward to Brazil. Unlike the closely related *Aspidoscelis*—many of which were formerly in *Cnemidophorus*—few *Cnemidophorus* are obligate parthenogens (*cryptus*, *pseudolemniscatus*); more are either a mix of parthenogenetic and bisexual

Cnemidophorus spp.	English name	Activity	Diet	Size mm	Size g	Range core	Ecological business model
arenivagus	Green Desertification Whiptail	diurnal	omnivore	70	9.7	Venezuela	**** wide-ranging chemo-sensor
arubensis	Aruba Whiptail	diurnal	omnivore	93	24.0	Aruba	**refugia-anchored diurnal gr. liz.
cryptus	Cryptic Whiptail	diurnal	invert.	61	6.4	Venezuela	**** wide-ranging chemo-sensor
duellmani	Duellman's Whiptail	diurnal		74	11.9	Panama	* wide-ranging chemosensor
espeuti	Providencia Whiptail			81	15.3	Colombia	* wide-ranging chemosensor
flavissimus	Los Frailes Yellow Whiptail	diurnal		62	6.8	Venezuela	**** wide-ranging chemo-sensor
gaigei	Gaige's Whiptail			81	15.3	Colombia	* wide-ranging chemosensor
gramivagus	La Guanota Whiptail	diurnal	invert.	77	13.4	Colombia	**** wide-ranging chemo-sensor
lemniscatus	Rainbow Whiptail	diurnal	omnivore	63	6.6	Venezuela	**** refugia-anchored diurnal gr. liz.
leucopsammus	Blanquilla White-sand Whiptail	diurnal		75	12.2	Venezuela	*** wide-ranging chemosensor
martyris	San Pedro Martir Whiptail	diurnal		67	8.4	Mexico	*** wide-ranging chemosensor
murinus	Laurenti's Whiptail	diurnal	omnivore	112	42.2	Curacao	*** wide-ranging chemosensor
nigricolor	Starry Night Whiptail	diurnal		73	11.2	Venezuela	*** wide-ranging chemosensor
pseudolemniscatus	Cole's Whiptail	diurnal		59	5.9	Guiana Shield	*** wide-ranging chemosensor
rostralis	Tortuga Island Whiptail	diurnal		66	8.1	Venezuela	*** wide-ranging chemosensor
ruatanus	Shaky-paw	diurnal		79	14.1	Honduras	*** wide-ranging chemosensor
ruthveni	Bonaire Whiptail	diurnal		111	40.6	Bonaire	*** wide-ranging chemosensor
senectus	Gray-headed Whiptail	diurnal	omnivore	67	8.7	Venezuela	*** wide-ranging chemosensor

forms or parthenogenesis is a rare variant (*espeuti, gaigei, lemniscatus, splendidus*). The history of taxonomy has been to split up nominal species that include both parthenogenetic and bisexual forms; more revision of this group can be anticipated.

Shape index (0.71 v. 0.88 std. resid) and tail length (2.44 v. 2.30 × SVL) are average. Tail breakage (35% v. 21%) and secretory pore frequency (15% v. 9%) are elevated. Color change is absent (0% v. 1.3%).

SSD is m>f (1.11 v. 0.89 on −2 to 2 scale), and *Cnemidophorus* are colorful: sexual dichromatism (60% v. 35%), seasonal color (15% v. 10%), and blue tails (5% v. 15%) are present. These sex differences roll up to 1.67 (v. 1.19) on the 0 to 6 scale.

The diets of *Cnemidophorus* are diverse: 44% (v. 62%) invertebrate eaters; 56% (v. 33%) omnivores. Home-range area is known only for *lemniscatus*, for which it is surprisingly small (40 m^2 or −0.76 v. 0.19 std. resid.). No *Cnemidophorus* is conspicuously wary (v. 13%). Mean active body temperature is exactly average (36.9°C v. 36.9°C). Half of the species are known to have socially tolerant males (v. 55%). All use burrows as daily refugia (v. 82%). *Cnemidophorus* are found in the middle of the rainfall spectrum (arid to mesic) at low elevation (100% v. 95%). They are not strongly associated with any particular habitat type, though they are overrepresented in littoral (64% v. 38%), sandy (73% v. 36%), and human (45% v. 29%) habitats.

Clutch count is notably low (−0.87 v. −0.51 std. resid.), young are small (ontogenetic mass range 0.45 v. 0.10 std. resid.), incubation is about average (−0.44 v. −0.38 std. resid.), and RCM ratio is on the low side of the low teiid mean (0.149 v. 0.163).

Longevity is about average for teiids, based on four species (−0.64 v. −0.74 std. resid.). Population den-

Cnemidophorus spp.	English name				Range core
			Size		
	Activity	Diet	mm	g	Ecological business model
splendidus	Blue Rainbow Lizard				Venezuela
	diurnal	invert.	66	8.0	*** wide-ranging chemosensor
vanzoi	Marias Whiptail				St. Lucia
	diurnal	invert.	99	28.8	*** wide-ranging chemosensor

Coeranoscincus spp.	English name				Range core
			Size		
	Activity	Diet	mm	g	Ecological business model
frontalis	Limbless Snake-tooth Skink				Queensland
			229	16.6	*** subterranean forager
reticulatus	Three-toed Snake-tooth Skink				Queensland
		invert.	155	16.1	*** subterranean forager

Coggeria sp.	English name				Range core
			Size		
	Activity	Diet	mm	g	Ecological business model
naufragus	Satinay Sand Skink				Queensland
		invert.	112	7.2*	*** subterranean forager

sity, based on eight samples of three island species (*arubensis*, *murinus*, *vanzoi*) is very high (6.8 kg/ha v. 1.06 kg/ha). Endemism is average (55% v. 50%). Many species occur on islands (70% v. 37%). *Cnemidophorus* have not been introduced (v. 8.3%). A usual proportion is imperiled (50% v. 40%), including half that are on one of the IUCN's lists of the most endangered (50% v. 20%). In other respects *Cnemidophorus* are typical teiids.

Coeranoscincus

Coeranoscincus (Scincidae) is a two-species genus of large (16.3 g v. 3.3 g), elongate (50% v. 35%) or limbless (50% v. 12%) subterranean foragers found in coastal eastern Australia. Although considered "subterranean" they seem to spend most of their time not in soil but in the punk under rotted logs, where they may be preying primarily on earthworms, which they snag with strongly recurved teeth.

They are very slender (shape index −1.74 v. −0.46 std. resid.), with moderately short tails (0.80 v. 1.33 × SVL) that are breakable (100% v. 73%). No sex differences are known. They thermoconform (50% v. 1.5%). They are associated with sandy areas (50% v. 25%).

They are oviparous (v. 64%) with moderately small clutch counts (−0.32 v. 0.06 std. resid.) and notably large young for such large lizards (ontogenetic mass range: 20 v. 39-fold; −1.17 v. 0.12 std. resid.). With the exception of *Cyclodomorphus*, large young seem to be relatively common in the larger skinks of Australia (see *Bellatorias* account). Large young are also characteristic of subterranean foragers (table 4.1, trait 64).

C. frontalis is endemic (50% v. 64%). Neither has been introduced (v. 2.1%). Both are considered imperiled (v. 47%), but neither is on an IUCN list of the most endangered species (v. 14%). In other respects *Coeranoscincus* are probably typical skinks.

Coggeria

Coggeria naufragus is the sole representative of a genus of large (7.2 g v. 3.25 g), functionally limbless, subterranean forager (v. 12%) skinks endemic to Fraser Island on the southeastern coast of Queensland.

Although it has three digits on each limb, it locomotes without use of limbs and is slender (shape index −1.20 v. −0.46 std. resid.). Its tail is breakable (v. 73%). Sex differences are not reported. It may specialize on eating earthworms, but will eat a variety of invertebrates (v. 84%). It dives back into sand when disturbed. It is associated with sandy (v. 22%) habitats. Reproduction is not documented. It has not been introduced (v. 2.1%), nor has it been evaluated by the IUCN for conservation status.

Coleodactylus

Coleodactylus is a five-species genus of tiny (0.59 g v. 1.07 g), padless, diurnal (100% v. 62%) sphaerodactylids found in Brazil and the Guiana Shield. The modal *Coleodactylus* is a litter cryptozoid (67% v. 53%), but one may be an immobility expert (33% v. 1.2%). Tails are short (0.78 v. 1.07 × SVL) and mostly breakable (80% v. 94%). Males are smaller than females (−1.00 v. 0.04 on −2 to 2 scale). They sit-and-wait (100% v. 82%), are found in mesic areas (100% v. 45%) at low elevation (100% v. 96%), have an average RCM ratio (0.106 v. 0.095), and have a low population density (0.02 kg/ha v. 0.32 kg/ha), albeit this comparison is based solely on four samples of

Coleodactylus spp.	English name				Range core
			Size		
	Activity	Diet	mm	g	Ecological business model
brachystoma	Dwarf Pygmy Gecko				Brazil
	diurnal		29	0.59	
elizae	Eliza's Bromeliad Gecko				Brazil
	diurnal		29	0.61	* immobility expert
meridionalis	Meridian Pygmy Gecko				Brazil
	diurnal	invert.	29	0.59	
natalensis	Brazil Natal Pygmy Gecko				Brazil
	diurnal		27	0.47	**** litter cryptozoid
septentrionalis	Maraca Pygmy Gecko				Venezuela
	diurnal	invert.	30	0.71	*** litter cryptozoid

Coleonyx spp.	English name				Range core
			Size		
	Activity	Diet	mm	g	Ecological business model
brevis	Texas Banded Gecko				Texas
	nocturnal	invert.	56	3.03	**** nocturnal ground lizard
elegans	Yucatán Banded Gecko				Central America
	nocturnal	invert.	90	14.4	*** nocturnal ground lizard
fasciatus	Black Banded Gecko				Mexico
	nocturnal	invert.	64	4.70	*** nocturnal ground lizard
gypsicolus	San Marcos Barefoot Gecko				Mexico
	nocturnal	invert.	73	7.3	*** nocturnal ground lizard
mitratus	Central American Banded Gecko				Central America
	nocturnal	invert.	77	8.7	**** nocturnal ground lizard
reticulatus	Reticulate Banded Gecko				Mexico
	nocturnal	invert.	80	9.9	**** nocturnal ground lizard
switaki	Switak's Banded Gecko				Mexico
	nocturnal	invert.	73	7.3	**** nocturnal ground lizard
variegatus	Western Banded Gecko				Arizona
	nocturnal	invert.	61	4.07	**** nocturnal ground lizard

natalensis. Endemism is average (60% v. 65%). Few occur on islands (20% v. 64%). None has been introduced (v. 5%), and none is considered imperiled (v. 41%). In other respects *Coleodactylus* are typical sphaerodactylids.

Coleonyx (fig. 4.10A)

Coleonyx (eight spp.) are the eublepharids found in the New World (from Utah to Costa Rica). They have many similarities to the Old World *Eublepharis*. *Coleonyx* are nocturnal ground lizards and small for the family (6.7 g v. 13.2 g), but otherwise morphologically representative. Males tend to be smaller than females, but only *variegatus* is decisively m<f and the mean SSD score (–2 to 2 scale) is slight (–0.20 v. –0.09). No fixed sexual dichromatism is reported (v. 11%), but *Coleonyx* is the only eublepharid genus with (male) seasonal color changes (25% v. 5.6%).

Foraging in *Coleonyx* is a mix of cruising (100% v. 56%) and active foraging (50% v. 44%). Nominally nocturnal, six of eight species have annotations indicating that activity may be extended into daylight hours by rain, season, rock shelters, or other special circumstances. Males may roam more than females during the season for mate searching. Although *Coleonyx* are not reported to vocalize spontaneously (0% v. 8.3%), they have the family's highest frequency of stress calls (63% v. 25%), and the family's highest frequency of conspicuously wary species (50% v. 19%). One wonders whether stress calling is due solely to individual selection; if so, what predator is sufficiently flighty to drop its dinner because the captured gecko squawks? As discussed below, there might be an inclusive fitness benefit (kin nearby).

Males routinely defend space, as do all evaluated male eublepharids, but *variegatus* stands out by also grouping, perhaps due to clumping of preferred refugia. Members of a group are apt to be genetically related. *Coleonyx* are found in dry and arid habitats (56–400 mm/y) at low (100% v. 100%) and medium (71% v. 35%) elevations. Perhaps because they are better known (46% v. 35% of cells filled), *Coleonyx* are overrepresented in littoral zones (13% v. 7%), sandy areas (25% v. 10%), riparian zones (38% v. 30%), and human habitats (25% v. 17%).

Young diverge from the family pattern in being nearer lizard average (0.0 std. resid.) than eublepharid average in relative size (ontogenetic size range –0.17 v. –0.67 std. resid.). Eggs are sometimes adherent (13% v. 3%), but parental care is not reported (0% v. 14%). Longevity is high (1.51 v.

1.16 std. resid.). Examples of longevity and population density are given in the family account. *Coleonyx* are imperiled at a typical frequency (50% v. 60%), though none appears on the IUCN's list of most endangered species (0% v. 14%). In other respects these are typical eublepharids.

See the family account for legends associated with *variegatus*. *C. elegans* is believed by some local residents to be capable of envenoming humans at a distance by hurling its stinger (what is an upraised tail if not a stinger?).

Colobodactylus

Colobodactylus (two spp.) are exceptionally poorly known (cells filled 14% v. 19%), elongate, terrestrial Brazilian gymnophthalmids. Ecological business models cannot be assigned with the information available. They are somewhat small (1.65 g v. 2.35 g) and slender (shape index −0.62 v. −0.32 std. resid.) and have tails that are long (2.00 v. 1.72 × SVL) and breakable (50% v. 28%). Sex differences are not reported. They are associated with riparian areas. None is considered imperiled (v. 25%). In other respects *Colobodactylus* areprobably typical gymnophthalmids.

Colobodactylus spp.	English name				Range core
			Size		
	Activity	Diet	mm	g	Ecological business model
dalcyanus	Oliveira's Lizard				Brazil
			45	1.16	
taunayi	Taunay Lizard				Brazil
			56	2.34	

Colobosaura

Colobosaura (two spp.) are elongate terrestrial gymnophthalmids found in Paraguay and adjacent Brazil. *C. kraepelini* is known from only a single immature specimen (size estimated from comparison to sister taxon). They are somewhat small (1.5 g v. 2.35 g) and slender (shape index −0.66 v. −0.32 std. resid.) and have tails that are long (2.13 v. 1.72 × SVL) and breakable (50% v. 28%). Secretory pores are present (50% v. 28%). The sexes also differ in color in the one species for which both sexes have been sampled (50% v. 13%). *Colobosaura* are associated with sandy areas (50% v. 14%) and riparian zones (50% v. 29%). Reproduction is normal in the one sampled species; clutch count is slightly elevated (0.07 v. −0.11 std. resid.), and RCM is fairly high (0.260 v. 0.182 ratio). Neither species has been evaluated for conservation risk. In other respects *Colobosaura* are probably typical gymnophthalmids.

Colobosaura spp.	English name				Range core
			Size		
	Activity	Diet	mm	g	Ecological business model
kraepelini	Chaco Colobosaur				Paraguay
	diurnal		50	1.60	* refugia-anchored diurnal gr. liz.
modesta	Bahia Colobosaur				Paraguay
	diurnal		49	1.48	** refugia-anchored diurnal gr. liz.

Colobosauroides

Colobosauroides (two spp.) are poorly known (filled cells 17% v. 19%), probably elongate gymnophthalmids endemic to Brazil. Their microhabitat, diet, and diel activity pattern are unknown. They are very small (0.67 g v. 2.35 g) and slender (shape index −0.67 v. −0.32 std. resid.) and have tails that are short (1.54 v. 1.72 × SVL). Secretory pores and other possible sex differences are not documented (0% v. 28%). They are documented to occur in the annual rainfall range 1084–1515 mm/y, and the elevational range 183–800 m. Reproduction is undocumented. Neither species has been evaluated for conservation risk. In other respects *Colobosauroides* are probably typical gymnophthalmids.

Colobosauroides spp.	English name				Range core
			Size		
	Activity	Diet	mm	g	Ecological business model
carvalhoi	Carvalho's Colobosaur				Brazil
			36	0.57	
cearensis	Cunha's Colobosaur				Brazil
			40	0.78	

Coloptychon

Coloptychon is a single-species genus of little-known (17% of cells filled v. 27%), elongate (v. 22%) anguid from low-elevation forests in Costa Rica and Panama. Its long tail (2.25 v. 1.66 × SVL) breaks (v. 87%). Little else is

Coloptychon spp.	English name		Size		Range core
	Activity	Diet	mm	g	Ecological business model
rhombifer	Isthmian Alligator Lizard				Costa Rica
			95	4.82	

Colopus spp.	English name		Size		Range core
	Activity	Diet	mm	g	Ecological business model
kochii	Namib Ghost Gecko				Namibia
	nocturnal		46	2.26	**** nocturnal ground lizard
wahlbergii	Kalahari Ground Gecko				Botswana
	nocturnal	invert.	52	3.05	*** nocturnal ground lizards

Complicitus spp.	English name		Size		Range core
	Activity	Diet	mm	g	Ecological business model
nigrigularis	Kinabalu Forest Lizard				Malaysia
	diurnal		59	6.6	* diurnal canopy walker

known; no record was obtained of what stratum of the forest it inhabits.

Colopus

Colopus (two spp.) are gekkonid nocturnal ground lizards found on sandy desert plains along the west coast of southern Africa (Namibia, Botswana, and South Africa). These relatively well-known geckos (36% v. 25% filled) are slightly small (2.63 g v. 3.31 g) and have slightly shorter (nonprehensile) tails than most gekkonids (0.97 v. 1.07 × SVL). They are morphologically distinctive for largely lacking enlarged toepads (they each have two to three narrow scansors), and claws are lacking except in the hind feet of female *wahlbergii*. This sexual dimorphism is assumed to derive from the female's need to excavate a nest cavity; this logic implies that at least males do not burrow to create a daily refugium. Unspecified daily burrows are used by *wahlbergii*, but presumably they do not dig them. Neither species is reported to change color (v. 15%).

Males are documented to be shorter than females in *wahlbergii* and may be so also in *kochii*. Other dimorphic characters (secretory pores, color differences, tail length differences) are not reported.

C. wahlbergii is reported to forage by both cruise (v. 24%) and sit-and-wait (v. 81%); *kochii* forages by cruising. Neither is reported to forage at electric lights (v. 3.8%). Home-range area is undocumented. Mean active body temperature is a cool 25.5°C (v. 27.6°C). Unlike most gekkonids (mode in mesic belt: 1101–3000 mm/y), *wahlbergii* is found mostly in the arid belt (actual documented range 145–286 mm/y) and *kochii* is probably very similar. Both species are associated with sand (v. 11%), but not littoral (v. 11%), riparian (v. 17%), or human (v. 27%) habitats.

Reproduction in *Colopus* is typical of gekkonids in reproductive mode, clutch count, eggshell hardness, RCM (0.086 v. 0.140 ratio), lack of egg adhesion (v. 10.5% adherent), communal oviposition (v. 8.6%), and parental care (v. 1.9%), but mass-adjusted incubation (based on *kochii*) is nearly 10 standard deviations off the lizard mean (9.7 std. resid. compared to 0.17 for gekkonids): a remarkable 480–600 days. This extraordinary incubation interval lacks an obvious explanation, though it may help these geckos survive multiyear droughts. *C. wahlbergii* is a dietary specialist on termites; termites in the desert fly rarely and only under highly specific environmental conditions, often including rainfall. Conceivably eclosion is triggered by the same infrequent environmental conditions, providing the hatchlings a food bonanza upon emergence.

Longevity is unremarkable (0.17 v. 0.15 std. resid.); population density is undocumented. *Colopus* have not been introduced (v. 3.7%), nor have they been evaluated for conservation risk (v. 47%). In other respects *Colopus* are typical gekkonids.

Complicitus

Complicitus is a single-species genus of agamid, but as it is known from only a single male specimen, it is virtually undocumented. The specimen has a crest (v. 38%) but no secretory pores (v. 13%). It was found at medium elevation (1500 m) on Borneo.

Concinnia

Concinnia is a moderate radiation (eight spp.) of large (6.6 g v. 3.3 g), ecologically diverse, coastal skinks from Queensland (two species range into New South Wales). Although a majority are trunk/wall lizards, an unusual microhabitat for skinks (57% v. 3%), one is a refugia-anchored diurnal ground lizard (14% v. 30%), one a litter cryptozoid (14% v. 28%), and another a crack hermit (14% v. 1.7%). The tendency to ambush prey from the safety of a refuge is reported in three species (*amplus, queenslandiae, tigrinus*), suggesting that the impulse to be a crack hermit is widespread, even though all species are said to also forage actively (v. 92%).

Concinnia spp.	English name				Range core
			Size		
	Activity	Diet	mm	g	Ecological business model
amplus	Lemon-barred Forest-skink				Queensland
	diurnal	invert.	101	21.9	** refugia-anchored diurnal gr. liz.
brachysoma	Northern Bar-sided Skink				Queensland
	diurnal	invert.	66	5.6	* trunk/wall lizard
frerei	Bartle Frere Bar-sided Skink				Queensland
	diurnal	invert.	63	4.83	* trunk/wall lizard
martini	Dark Bar-sided Skink				Queensland
	diurnal	invert.	62	4.69	* trunk/wall lizard
queenslandiae	Prickly Forest Skink				Queensland
	cathemeral	invert.	74	4.28	*** litter cryptozoid
sokosoma	Stout Bar-sided Skink				Queensland
	diurnal	invert.	65	5.3	* trunk/wall lizard
tenuis	Bar-sided Skink				Queensland
	cathemeral	omnivore	70	6.7	
tigrinus	Yellow-blotched Forest-skink				Queensland
	cathemeral	invert.	76	8.7	* crack hermit

These lizards are heavy-bodied for skinks but normal for lizards (shape index 0.08 v. –0.46 std. resid.). Tails are a bit short (1.18 v. 1.33 × SVL) and universally breakable (v. 73%). *C. queenslandiae* may be elongate (13% v. 35%). No sex differences are known.

Most are strictly invertebrate eaters (87% v. 84%), though *tenuis* also takes fruit (13% v. 13% omnivores). One-quarter are acknowledged to forage by sit-and-wait (v. 16%). The majority are diurnal (63% v. 74%); the remainder are cathemeral (37% v. 9%), though there is some uncertainty about the diel pattern of all eight species. For example, five species are arguably crepuscular. This high level of uncertainty suggests circadian opportunism. The opportunism may extend beyond diel activity pattern; *C. amplus*, though nominally a refugia-anchored diurnal ground lizard that forages actively, is known to sit and wait for prey at its burrow mouth and to sleep exposed on rock faces at night. *Concinnia* seem to be very adaptable lizards.

Stress calls are surprisingly common for a skink (25% v. 0.4%) and are not associated with any obvious correlate other than cathemerality (table 3.7, trait 32). They do not call spontaneously (v. 0.1%), which is arguably an even a stronger associate of cathemerality (table 3.7, trait 31). Wariness is found at typical frequency (13% v. 11%). Direct basking is common (50% v. 21%); thermal conformance is overrepresented as well (13% v. 1.5%).

The elevational profile of *Concinnia* is normal (l/m/h: 100/50/0 v. 89/35/6). They are overrepresented in riparian (50% v. 23%) and human (38% v. 14%) habitats. They are absent from littoral (v. 21%) and sandy (v. 25%) areas. This absence from the littoral zone suggests that their presence in coastal Australia is more closely associated with coastal climates (locally much wetter in Queensland) than with the littoral zone itself. *C. queenslandiae* is associated with seeps.

All species are viviparous (v. 36%), despite their warm maritime climate. Litter sizes are about average (0.12 v. 0.06 std. resid.), but neonates are notably large (ontogenetic mass range –0.64 v. 0.12 std. resid.). See *Bellatorias* account for further discussion of this phenomenon in Australia. Parental care is found in *queenslandiae* (13% v. 3.2%), for which neonates have been observed in the company of their presumed parents.

Longevity is high (0.85 v. 0.17 std. resid.), based on 10 years in *queenslandiae*. Endemism is about average (75% v. 64%). Half of the species occur on islands (v. 57%). None has been introduced (v. 2.1%). Three-quarters of evaluated species are considered imperiled (v. 47%), though none is on any of the IUCN's lists of the most endangered (v. 14%). In other respects *Concinnia* are typical skinks.

Congolacerta

Congolacerta (two spp.) is a very poorly known (filled cells 23% v. 33%) genus of smallish (3.36 g v. 5.1 g) refugia-anchored diurnal ground lacertids found in equatorial Africa (Democratic Republic of the Congo to Uganda and Burundi). In morphology they are typical lacertids. Sex differences are few and poorly documented (0.0 v. 0.86 on 0 to 6

Congolacerta spp.	English name				Range core
			Size		
	Activity	Diet	mm	g	Ecological business model
asukului	Asukulu's Grass Lizard				Congo
	diurnal		52	3.27	
vaueresellí	Sparse-scaled Forest Lizard				Congo
	diurnal		53	3.45	* refugia-anchored diurnal gr. liz

Conolophus spp.	English name				Range core
			Size		
	Activity	Diet	mm	g	Ecological business model
marthae	Pink Land Iguana				Galápagos Islands
	diurnal	herbivore	398	2681	**** giant herbivore
pallidus	Santa Fe Land Iguana				Galápagos Islands
	diurnal	herbivore	344	1745	**** giant herbivore
subcristatus	Galápagos Land Iguana				Galápagos Islands
	diurnal	herbivore	380	2268	**** giant herbivore

scale). Neither are conspicuously wary (v. 11%). Both are found up to relatively high elevation (2400 m, 2876 m). *C. vaueresellí* is associated with mesic (1622 mm/y) habitats (v. 34%). Reproduction is not described. Neither species occurs on islands (v. 26%). Neither has been introduced (v. 4.6%). They do not appear to have been evaluated for conservation concern.

Conolophus (fig. 4.4B)

Conolophus is a three-species genus of terrestrial giant herbivores endemic to the Galápagos Islands. At a mean of 2197 g (v. 609 g), *Conolophus* is the genus with the second-largest lizards within the Iguanidae, after monotypic *Amblyrhynchus*. Shape index (2.43 v. 2.11 std. resid.) is similarly second highest. Their tails are comparatively short (1.27 v. 1.53 × SVL) and are not normally breakable (v. 38%). If broken, they do not regenerate. Secretory pores are absent (v. 38%). Color change is absent (v. 21%).

SSD is at the typical m>f level (1.2 v. 1.3 on –2 to 2 scale). None is sexually dichromatic (v. 38%) and seasonal color is not reported (v. 12%). The sex differences rollup score (on 0 to 6 scale) is an unremarkable 1.00 (v. 1.88).

Conolophus often cluster, especially females. Males defend mating space, which usually includes several burrows of value to potential mates, but do not defend their small foraging area. Home ranges are moderately small (–0.57 v. –1.06 std. resid.). Like most Galápagos lizards, *Conolophus* are not conspicuously wary. Indeed, they show island tameness.

They are not particularly attracted to littoral (0% v. 41%), sandy (0% v. 34%), riparian (0% v. 22%), or human (0% v. 24%) habitats.

Their reproduction is typical oviparity, with a notably small clutch count (–0.81 v. 0.45 std. resid.) that is exceeded only by *Amblyrhynchus* and *Brachylophus*. The ontogenetic size range is correspondingly low, indicating especially large hatchlings (–1.27 v. –0.55 std. resid.), exceeded only by *Amblyrhynchus* and *Brachylophus*. All three of these genera are found on remote islands (Galápagos, Fiji) for which the natural predators would be primarily birds of a size capable of taking only hatchlings. Presumably, they are under selective pressure to produce hatchlings that can grow quickly to an invulnerable size.

Incubation is relatively short (–0.99 v. –0.76 std. resid.), and in at least the case of *subcristatus*, they expedite incubation by ovipositing in geothermally heated soil. This may require an oviposition migration of up to 15 km and may take them up to elevations not normally occupied at the rim of the crater, from which they descend to the hot parts of the volcano. Both *pallidus* and *subcristatus* defend their nests for days to weeks (parental care 67% v. 31%). *C. marthae* and *subcristatus* live on the same island, but are associated with different volcanoes. As no recruitment for *marthae* has been detected, they may also be capable of parental care (though undocumented at present). Oviposition is considered to be communal in the case of *subcristatus* (33% v. 14%).

Longevity is the highest in the family (1.33 v. 0.16 std. resid.), corresponding to 60 years. Population density is second only to *Iguana* (31.1 kg/ha), based on six estimates for *pallidus*: (12 kg/ha) and one for *subcristatus* (81 kg/ha). They may lose weight when the rains fail (El Niño events), but are seemingly too large for predation losses in the absence of introduced mammals. All species are considered to be imperiled (v. 97%) and the nonrecruiting one (*marthae*) appears on the most endangered lists of the IUCN (v. 51%). For conservation reason *subcristatus* has been translocated to North Seymour Island (33% v. 33% for introductions), though none has been introduced more than twice (v. 9.5%). In other respects *Conolophus* are typical iguanids.

Contomastix

Contomastix (Teiidae) is a small genus (five spp.) of small-sized (6.0 g v. 22.1 g), poorly known (filled cells 26% v. 33%), flight-dependent (100% v. 77%), invertebrate-eating (100% v. 62%) wide-ranging chemosensors (100% v. 97%). In other words, fairly typical whiptails. The genus is found in Argentina, Brazil, Bolivia, and Uruguay.

Their build is light (shape index 0.50 v. 0.88 std. resid.). Tails are short (1.82 v. 2.30 × SVL), breakable at a normal frequency (20% v. 21%), and not blue (0% v. 15%). Secretory pores are absent (v. 9%); color change is unreported (v. 1.3%). Males tend to be smaller than females (–0.50 v. 0.89 on –2 to 2 scale); sexual dichromatism is underrepresented (20% v. 35%) and seasonal color absent (v. 10%). Direct basking is common (80% v. 58%). Use of burrows as a daily refugium may be low (50% v. 82%). Their elevational profile is skewed upwards (l/m/h: 50/25/25 v. 95/32/6). *Contomastix* are absent from littoral (v. 38%) and human (v. 29%) habitats, but roughly normal in sandy areas (40% v. 47%) and riparian zones (20% v. 46%).

Clutch counts are high (–0.03 v. –0.51 std. resid.); young are larger than average (ontogenetic mass range –0.38 v. 0.10 std. resid.); in combination, these suggest that RCM ratio should be exceptionally high, but direct observations of that are unavailable. Incubation duration is also unknown.

Endemism is normal (60% v. 50%). No species is found on islands (v. 40%), nor has any been introduced (v. 8.3%). Half of evaluated species are imperiled (50% v. 41%), and these species (*lacertoides*, *vittata*) are on the IUCN's lists of the most endangered (50% v. 14%). In other respects *Contomastix* are typical teiids.

Contomastix spp.	English name				Range core
			Size		
	Activity	Diet	mm	g	Ecological business model
lacertoides	Bibron's Whiptail				Uruguay
	diurnal	invert.	65	8.0	*** wide-ranging chemosensor
leachei	Jujuy Border Whiptail				Argentina
	diurnal		55	4.67	*** wide-ranging chemosensor
serrana	Cordoba Mountains Whiptail				Argentina
	diurnal		54	4.31	*** wide-ranging chemosensor
vacariensis	Spot-flanked Whiptail				Brazil
	diurnal		61	6.4	*** wide-ranging chemosensor
vittata	Cochabamba Whiptail				Bolivia
	diurnal		64	7.3	* wide-ranging chemosensor

Copeoglossum

Copeoglossum is a minor radiation (five spp.) of large (12.7 g v. 3.25 g), diurnal (100% v. 74%), terrestrial (75% v. 48%) or arboreal (25% v. 10%) skinks found in South America and the Lesser Antilles. Half of classifiable species are refugia-anchored diurnal ground lizards (v. 30%); the other half move partially up tree trunks to become ground-foraging tree lizards (v. 7%). One species is presumed extinct (*redondae*), and another is known from only a few specimens (*margaritae*), but the remainder are reasonably well known (filled cells 30% v. 25%).

Shape index is high for a skink (0.26 v. –0.46 std. resid.). *C. nigropunctatum* is marginally elongate (20% v. 35%), and their juveniles have a blue tail (20% v. 4%). Three species are characterized for SSD, and all are m<f (–1.0 v. –0.31 on –2 to 2 scale), but that is the only sex difference reported.

Diet is known from only two species (*arajara*, *nigropunctatum*), both

Copeoglossum spp.	English name				Range core
			Size		
	Activity	Diet	mm	g	Ecological business model
arajara	Caatinga Bronze Skink				Brazil
	diurnal	invert.	88	16.0	
aurae	Greater Windward Skink				Trinidad and Tobago
			87	13.9	* ground-foraging tree lizard
margaritae	Margarita Skink				Venezuela
			96	18.8	
nigropunctatum	South America Spotted Skink				Venezuela
	diurnal	invert.	91	7.4	** refugia-anchored diurnal gr. liz.
redondae	Redonda Skink				Antigua and Barbuda
			81	10.8	

Cophosaurus sp.	English name				Range core
			Size		
	Activity	Diet	mm	g	Ecological business model
texanus	Greater Earless Lizard				Texas
	diurnal	invert.	68	9.2	*** refugia-anchored diurnal gr. liz.

Cophoscincopus spp.	English name				Range core
			Size		
	Activity	Diet	mm	g	Ecological business model
durus	Cope's Keeled Water Skink				Ghana
	diurnal	invert.	52	2.66	* refugia-anchored diurnal gr. liz.
greeri	Greer's Keeled Water Skink				Ghana
	diurnal	invert.	57	3.42	* refugia-anchored diurnal gr. liz.
senegalensis	Senegal Keeled Water Skink				Senegal
			58	3.79	* refugia-anchored diurnal gr. liz.
simulans	Vaillant's Keeled Water Skink				Ghana
	diurnal	invert.	53	2.78	* refugia-anchored diurnal gr. liz.

of which are generalized invertebrate eaters (v. 84%) with a penchant for termites. Active foraging (100% v. 92% and sit-and-wait foraging (50% v. 16%) are common. Direct basking is found at a normal level (20% v. 21%), and mean active body temperature is warm (33.3°C v. 31.1°C). The active body temperature datum is based on *nigropunctatum*, which is noted to be heliophilic. *Copeoglossum* are found toward the wet end of the rainfall spectrum; they are overrepresented in the intermediate belt (401–1100 mm/y: 67% v. 34%) and slightly underrepresented in the mesic (1101–3000 mm/y: 33% v. 40%). All species are found at low elevation (l/m/h: 100/33/0 v. 89/35/6). They are absent from littoral zones (v. 21%), but representative in other habitat types: sandy areas (33% v.25%), riparian areas (33% v. 23%), and human habitats (33% v. 14%).

All species are viviparous (v. 36%), despite their tropical low-elevation climates. Their litter count is commensurate with their large size (0.17 v. 0.06 std. resid.). Neonatal size is also average (ontogenetic mass range −0.07 v. 0.12 std. resid.).

Endemism is normal (60% v. 64%). Most species occur on islands (80% v. 57%). None has been introduced (v. 2.1%). All are considered imperiled (v. 47%), and one-third of evaluated species are on one of the IUCN's lists of the most endangered (v. 14%). In other respects *Copeoglossum* are much like other skinks.

Cophosaurus

Cophosaurus texanus is the sole species in a genus of moderately small (9.2 g v. 11.2 g) phrynosomatids from the North American southwest, from Oklahoma to Mexico. This refugia-anchored diurnal ground lizard is well known (filled cells 72% v. 38%). It exhibits m>f SSD, sexual dichromatism, secretory pores, and seasonal color, for a family-lead-tied sex difference rollup of 4.0 (v. 1.68 on 0 to 6 scale). Its invertebrate diet is notably opportunistic, varying greatly from time to time and place to place. It does not issue stress vocalizations and is not conspicuously wary. It relies on flight for predator concealment. Its mean active body temperature is a sizzling 39°C (v. 34.2°C). It may be seen running with its tail curled over its head, scorpion fashion. It inhabits moderately dry places (200–910 mm/y) and may sleep buried in the sand or in rock crevices. Its elevational range is wide (0–2100 m), and it inhabits sandy and riparian areas, but not littoral or human habitats. Clutch count is low for a phrynosomatid (0.0 v. 0.68 std. resid.), and incubation is normal (−0.58 v. −0.58 std. resid.), equating to 50 days. RCM is a bit light 0.218 (v. 0.300). Longevity is near the lizard norm and is therefore a bit long for a phrynosomatid (−0.02 v. −0.25 std. resid.), equating to a maximum of 6 years. Population density is a thin 0.10 kg/ha (v. 0.26 kg/ha). It does not occur on islands. It is threatened in Mexico, but not listed by the IUCN. In other respects *Cophosaurus* is similar to other phrynosomatids.

Cophoscincopus

Cophoscincopus is a minor radiation (four spp.) of semiaquatic (100% v. 2%), diurnal (100% v. 74%) skinks of West Africa (Senegal to Togo). They are likely to be refugia-anchored diurnal ground lizards (100% v. 30%). As indicated by their common names, these average-sized (3.13 g v. 3.25 g) skinks have distinctive longitudinal keels on most scales, giving them a very unskink-like gestalt. Nonetheless, they have very average general morphology (shape index −0.08 v. −0.46 std. resid.) and normal-length (1.38 v. 1.33 × SVL) breakable (100% v. 73%) tails. They are not elongate (v. 35%), nor do they have blue tails (v. 4%). Sex differences are not noted. They are presumed to be invertebrate eaters. When threatened, they retreat to a crevice or burrow (100% v. 43%). Their elevational profile is low (l/m/h: 100/25/0 v. 89/35/6). A brood count of two (−0.16 v. 0.06 std. resid.) is specified for *durus* without explicitly indicating reproductive mode. Incubation (conceivably gestation) for that species is 60 days (−0.02 v. −0.40 std. resid.). No genus members are documented to be endemic (v. 64%), introduced (v. 2.1%), imperiled (v. 47%), or to occur on islands (v. 57%). In other respects *Cophoscincopus* are much like other skinks.

Cophotis

Cophotis is a tiny genus (two spp.) of small (5.0 g v. 15.6 g) trunk/wall (100% v. 7%) agamids, endemic to Sri Lanka. Like many agamids, they have a low crest (50% v. 38%); unlike most moderate-elevation tropical lizards, they are viviparous (v. 2%). They have prehensile tails (v. 1.4%), sexual dimorphism in relative tail length (male tails long), independently moving eyes, and are capable of advanced color change (50% v. 15%); they cruise forage (100% v. 6%), rely on crypsis (100% v. 51%), bask directly (100% v. 24%), expel other males (v. 97%), and eat only invertebrates (100% v. 46%)—all redolent of chameleons, though their body plan is more anoline than chamaeleonid. Neither species has been introduced (v. 1.7%); both are judged to have sparse populations and are listed as endangered or critically endangered on the Sri Lanka Red List).

Cophotis spp.	English name				Range core
			Size		
	Activity	Diet	mm	g	Ecological business model
ceylanica	Sri Lanka Deaf Agama				Sri Lanka
		invert.	56	5.4	** trunk/wall lizard
dumbara	Knuckles Pygmy Lizard				Sri Lanka
	diurnal	invert.	53	4.7	**** trunk/wall lizard

Cordylidae

The 68 cordylids are endemic to Africa, especially southern Africa (as far north as the Democratic Republic of the Congo and Ethiopia). Most are spiny and armored, typical quadrupedal, sedentary, rock-crevice aficionados, though one genus of five species (*Chamaesaura*) is of effectively legless lizards actively foraging within grassy mats. The distribution of ecological business models within Cordylidae is crack hermits (43% v. 2.6%), ground-foraging tree lizards (6.2% v. 9.2%), litter cryptozoids (7.7% v. 11.6%), refugia-anchored diurnal ground lizards (29.2% v. 24.5%), and trunk/wall lizards (13.8% v. 13.0%). Ten genera are found within Cordylidae at present: *Chamaesaura, Cordylus, Hemicordylus, Karusasaurus, Namazonurus, Ninurta, Ouroborus, Platysaurus, Pseudocordylus*, and *Smaug*.

The cordylids have special bone structures in their heads enabling them to wedge themselves tightly in crevices. This defensive maneuver is frequently aided by a spiny tail and osteoderms in the skin. Cordylids are larger than average lizards (10.8 g v. 5.3 g), and they have a robust but not bulky build (shape index 0.35 v. 0.0 std. resid.). Tails are a bit shorter than average (1.16 v. 1.37 × SVL), are never prehensile (v. 5.0%), and genera vary from having uniformly breakable tails to uniformly unbreakable ones within the family (mean 60% breakable v. 57% breakable). Most species have secretory pores (69% v. 15%). Only *Chamaesaura* are elongate (mean for family 7.4% v. 11. 4%). No cordylid is reported to be able to change color (v. 10.7%). None has blue tails (v. 2.3%).

Sexual differences in size mirror those of lizards exactly (0.36 v. 0.36 on −2 to 2 scale), though documentation is lacking for most cordylid genera. Males of many species have proportionately larger heads. In contrast to morphological conservatism, the sexes routinely differ in color (37% v. 21%) and seasonal color (28% v. 6.5%). In the frequency of seasonal color, they are second only to the small family Crotaphytidae. Both forms of color dichromatism exhibit strong differences among genera: *Platysaurus* are nearly universally dichromatic and often seasonally variable, whereas no *Chamaesaura* or *Cordylus* is dichromatic. The relative tail length ratio of males to females is documented for only a single cordylid (*Platysaurus attenboroughi*), for which it is m<f (0.95 v. 1.07). These factors combine in a sex differences rollup greater than the lizard average (1.41 v. 0.75 on 0 to 6 scale).

All cordylids are either invertebrate eaters (70% v. 73%) or omnivores (30% v. 15%). Ninety-four percent use sit-and-wait foraging (v. 46%) and 19% (v. 58%) active foraging. None cruises. None has been reported to feed at electric lights (v. 1.4%). Home-range size is not documented in cordylids.

Almost all cordylids are diurnal (98.5% v. 67%); the remainder are crepuscular (1.5% v. 1.8%). Most bask directly (74% v. 18%), though body temperatures are not reported. As befits lizards keen on sprinting to the security of an impregnable crevice, many (26.5% v. 8.2%) are characterized as conspicuously wary. Only five cordylids have been characterized for predator concealment behavior, and they represent a broad mix of lizard tactics.

Social behavior is barely documented in cordylids, but they exhibit typical levels of solitary residence (14% v. 10%) and group living (29% v. 20%). Five percent of cordylids are reported to occur in multispecies groups, an attribute not reported for any other family. Males in three genera (*Cordylus*, *Platysaurus*, and *Pseudocordylus*) have been uniformly described as defending space (v. 85%). Male hierarchies are known in 43% of cordylids (v. 7.7% of lizards), probably a reflection of high density living among favorable cracks.

Most cordylids reside on the ground (76% v. 48%), but some are cryptozoic (7.6% v. 12.1%), others arboreal (17% v. 32%). Despite their passion for crevices, only 10% are reported to burrow (v. 22%) for their daily refugium. As with many African taxa, they are underrepresented at low elevation (55% v. 82%) and overrepresented above (59% v. 41% at mid elevation; 21% v. 13% above 2400 m). They are underrepresented in littoral zones (9% v. 17%), sandy areas (3% v. 22%), riparian zones (1% v. 22%), and in human habitats (0% v. 18.5%). One might think that crevice lizards in a relatively arid part of the world would be keen to live in close proximity to water, for easy drinking during droughts and probably greater prey abundance. This is not supported by their virtual absence from riparian zones; there is only one riparian cordylid, *Platysaurus torquatus*, when 15 riparian species would be expected in 68 typical lizard species. Three other *Platysaurus* species (*attenboroughi*, *broadleyi*, and *mitchelli*) have distributions clustered in the vicinity of major rivers, but are not exclusively riparian, nor is *torquatus* exclusively riparian. This suggests that cordylids may have exceptional tolerance for moisture stress, or that their riparian zones are home to predators dangerous to them.

The majority are viviparous (75% v. 18%), with strong differences between genera. The largest genus (*Cordylus*, 22 spp.) is uniformly viviparous and the second-largest (*Platysaurus*, 16 spp.) uniformly oviparous. Clutch/litter size is fairly small for lizards of this size (−0.48 v. 0.0 std. resid.) and varies from a mean of one (*Ouroborus*) to eight (*Chamaesaura*). Young are larger than lizard average (ontogenetic size range −0.37 v. 0.0 std. resid.). Twenty percent of cordylids use communal egg deposition sites (v. 4.2%), and 7.4% (v. 2.9%) show parental care, which is often higher in viviparous species. RCM has not been adequately sampled for cordylids.

Longevity is unremarkable (0.02 v. 0.0 std. resid.); population density is very difficult to quantify in a species dependent on cracks. Is the sampling unit one crack? Only one species (*Cordylus cordylus*) has been characterized for density, and it is about average (0.72 v. 0.54 kg/ha). Endemism is about average (53% v. 64%). Few cordylids appear on islands (3% v. 40%). None has been introduced (v. 3.3%), and 79% are judged imperiled (v. 49%), though only 5% (v. 14%) are listed in the most severe IUCN categories.

Cordylosaurus sp.	English name				Range core
			Size		
	Activity	Diet	mm	g	Ecological business model
subtessellatus	Dwarf Plated Lizard				Namibia
	diurnal	invert.	44	1.96	** refugia-anchored diurnal gr. liz.

Cordylosaurus

Cordylosaurus subtesselatus is an active-foraging refugia-anchored diurnal ground lizard (Gerrhosauridae) from southwestern Africa, ranging from Angola to South Africa. Though it possesses the characteristic lateral fold of a gerrhosaurid, it has the gestalt of a small blue-tailed cordylid, a resemblance reflected in the confusing genus name. It is small by any reference (1.96 g v. 16 g) and of average lizard build (shape index 0.01 v. 0.38 std. resid.). It has secretory pores (v. 8%). Sex differences are not documented. It basks directly (v. 24%). It is found in sandy areas (v. 25%), often with succulent plants. Commensurate with its small size, it has a clutch of two eggs (−0.04 v. −0.20 std. resid.). It is not imperiled (v. 50%). In other respects *Cordylosaurus* is a typical gerrhosaurid.

Cordylus

Cordylus seem designed to jamb themselves in cracks; almost every scale is armored, strongly keeled, or spiny. At 22 species, this is the largest genus in the Cordylidae, and the genus extends over the family's full southern African range (as far north as Ethiopia). The majority of species (68% v. 43%) are crack hermits; others are ground-foraging tree lizards (9% v. 6%) or refugia-anchored diurnal ground lizards (23% v. 29%). They are of average size (7.3 g v. 10.8 g) and build (shape index 0.29 v. 0.35 std. resid.), with slightly shorter tails (0.84 v. 1.16 × SVL), all of which break (v. 60%). Most species have secretory pores (82% v. 69%). None is elongate (v. 7.4%).

Sex differences in size are consistently m<f for the three species documented (*angolensis, jonesii, macropholis*). In the absence of other sexual dimorphic features, the sex differences rollup is relatively muted (0.68 v. 1.41 on 0 to 6 scale).

All *Cordylus* eat invertebrates (v. 73%), which they catch by sitting and waiting (100% v. 94%). In keeping with their primary predator concealment of diving into or residing in a crack, many are judged to be conspicuously wary (36% v. 27%).

The degree of sociality in saxicolous lizards is shaped by the degree of clumping in suitable rock crevices. Solitary *Cordylus* are overrepresented (43% v. 14%), but group living is found in 29% of species, as it is in the family. Up to a dozen may share refugia in *macropholis*; up to 60 may live in a cluster of *cataphractus*. All males defend space (if only their crack). None is known to form male hierarchies, but data on social relations are very sparse.

Cordylus are overrepresented at mid elevations (75% v. 59%) and underrepresented below (42% v. 55%) and above (8% v. 21%).

All *Cordylus* are viviparous (v. 75%), and litter sizes are typical (−0.46 v. −0.48 std. resid.), about 2.1 neonates. Young are slightly larger than average (ontogenetic range −0.56 v. −0.37 std. resid.). Parental care (attendance to neonates) is documented at family-typical levels (9.1% v. 7.4%). Longevity is a bit longer than in cordylids (0.44 v. 0.02 std. resid.), and population density (0.72 v. 0.54 kg/ha for lizards) is known only from one species in the Cordylidae, appropriately enough *Cordylus cordylus*. Endemism in *Cordylus* is normal (64% v. 53%), and no cordylid has been introduced (v. 3.3% of lizards), though *jonesii* is reported to stow away in bundles of

Cordylus spp.	English name				Range core
			Size		
	Activity	Diet	mm	g	Ecological business model
angolensis	Angolan Girdled Lizard				Angola
	diurnal		77	10.5	* crack hermit
aridus	Eastern Dwarf Girdled Lizard				South Africa
	diurnal		53	2.81	*** crack hermit
beraduccii	Masai Girdled Lizard				Tanzania
	diurnal		61	4.58	*** crack hermit
cloetei	Cloete's Girdled Lizard				South Africa
	diurnal		56	3.32	** crack hermit
cordylus	Cape Girdled Lizard				South Africa
	diurnal	invert.	77	10.4	*** refugia-anchored diurnal gr. liz.
imkeae	Rooiberg Girdled Lizard				South Africa
	diurnal		55	3.17	*** crack hermit
jonesii	Jones' Girdled Lizard				Botswana
	diurnal	invert.	72	8.4	** ground-foraging tree lizard
machadoi	Machado's Girdled Lizard				Namibia
	diurnal		79	11.3	*** crack hermit
macropholis	Large-scaled Girdled Lizard				South Africa
	diurnal	invert.	63	5.2	** crack hermit or refugia-anchored diurnal gr. liz.
marunguensis	Marungu Girdled Lizard				Congo
	diurnal		77	10.4	** crack hermit or refugia-anchored diurnal gr. liz.
mclachlani	McLachlan's Girdled Lizard				South Africa
	diurnal		60	4.36	** crack hermit
meculae	Mecula Girdled Lizard				Mozambique
	diurnal		77	10.4	*** crack hermit
minor	Western Dwarf Girdled Lizard				South Africa
	diurnal		56	3.47	*** crack hermit
namakuiyus	Kaokoveld Girdled Lizard				Angola
			86	15.4	*** crack hermit
niger	Black Girdled Lizard				South Africa
	diurnal		75	9.7	*** crack hermit
nyikae	Nyika Girdled Lizard				Malawi
	diurnal		77	10.7	** crack hermit
oelofseni	Oelofsen's Girdled Lizard				South Africa
	diurnal		60	4.24	** crack hermit
rhodesianus	Zimbabwe Girdled Lizard				Zimbabwe
	diurnal	invert.	72	8.3	** refugia-anchored diurnal gr. liz. or crack hermit

Cordylus spp.	English name				Range core
			Size		
	Activity	Diet	mm	g	Ecological business model
rivae	Ethiopian Girdled Lizard				Ethiopia
	diurnal		82	13.3	** refugia-anchored diurnal gr. liz.
tropidosternum	Tropical Girdled Lizard				Zambia
	diurnal	invert.	82	13.5	** ground-foraging tree lizard or crack hermit
ukingensis	Ikinga Girdled Lizard				Tanzania
	diurnal		69	7.3	* refugia-anchored diurnal gr. liz.
vittifer	Transvaal Girdled Lizard				Mozambique
	diurnal	invert.	77	10.4	*** refugia-anchored diurnal gr. liz.

Correlophus spp.	English name				Range core
			Size		
	Activity	Diet	mm	g	Ecological business model
belepensis	Belep Islands Giant Gecko				New Caledonia
	nocturnal		81	10.7	
ciliatus	Crested Gecko				New Caledonia
	nocturnal	omnivore	102	21.9	** nocturnal canopy walker
sarasinorum	Slender Prehensile-tailed Gecko				New Caledonia
	nocturnal		106	24.3	** trunk/wall lizard

firewood. All *Cordylus* are considered imperiled (v. 79%), though only 9.5% are in the most endangered IUCN categories (v. 4.8%). In other respects these are typical cordylids.

Correlophus

Correlophus is a small (three spp.) genus of very large (17.8 g v. 4.2 g) arboreal New Caledonian diplodactylid geckos. With its wavy dorsolateral crests *C. ciliatus*, then in *Rhacodactylus*, made a splash among herpetoculturists in the 1990s when this large gecko (up to 60 g) was rediscovered after having been thought extinct. It was subsequently bred in large numbers for the pet trade. *Correlophus* foraging habits and microhabitat preferences are not well known in the wild. They are a bit bulkier than the average diplodactylid (shape index 0.45 v. 0.05 std. resid.), and most, if not all, tails are prehensile (v. 26%). Two of three have secretory pores (v. 34%), but they lack color changing ability (v. 5%).

C. ciliatus females are larger than males (–1.00 v. –0.65 on –2 to 2 scale), but other sex differences are not known (sex rollup score of 1.0 v. 0.54 on 0 to 6 scale). All known *Correlophus* are omnivores (v. 15%). All are nocturnal (v. 79%). One of three gives stress calls (v. 5%). None is known to bask (v. 17%). All are found in littoral zones (v. 17%), but none is found in sandy (v. 24%), riparian (v. 10%), or human (v. 5%) habitats. Longevity is short, especially for a diplodactylid (–0.36 v. 0.90 std. resid.). All species are endemic (v. 72%) to New Caledonia, insular (v. 46%), and of conservation concern (v. 81%), though none is listed in the most endangered categories by the IUCN (v. 29%). In other respects they are typical diplodactylids.

Corucia

Corucia zebrata, the Solomon Islands' Prehensile-tailed Skink, was a notable event among herpetoculturists when it first became widely available in the trade, but it should make an even larger impression on natural historians because it is so different from most other herbivores, and indeed most other skinks. It is huge (662 g v. 3.3 g skinks), herbivorous (v. 0.9% skinks), nocturnal (v. 14.5% skinks and 3.5% herbivores), arboreal (v. 10% skinks and 14% herbivores), and found on an island (v. 57% skinks and 39% herbivores) at low elevation (v. 89% skinks and 71% herbivores). Furthermore, it is utterly lacking in SSD (0.0 v. 0.92 for herbivores and –0.31 for skinks on –2 to 2 scale), sexual dichromatism (v. 8% skinks and 45% herbivores), and seasonal color (v. 6% skinks and 14% herbivores). The herbivore values are taken from table 3.5. That table indicates that most herbivores are characterized by strong m>f SSD (second only to carnivores), strong sexual dichromatism (second only to ant specialists), strong seasonal color (second only to omnivores), strong diurnality (second only to ant specialists), strong terrestriality (second to none), being least likely to be found at low elevation (second to none), and being least likely to be found on islands (second to none). In all these attributes, *Corucia zebrata* breaks the mold. Is there a theme holding these disparate facts together? It is also viviparous despite living on a warm tropical island,

and it has a litter size of one (–2.60, the lowest in Scincidae), v. 0.06 std. resid.) every year or two—an extraordinarily low reproductive output for so large a lizard. It lives for up to 24 years (0.50 v. 0.17 std. resid.).

The natural history of *Corucia* is not particularly well known, though filled cells are high (41% v. 25%). Heavy collecting pressure for the pet trade has made the surviving wild lizards very skittish and hard to observe in nature, which amplifies the intrinsic challenges of studying a species that is nocturnal and resides in dense canopies.

Corucia sp.	English name				Range core
			Size		
	Activity	Diet	mm	g	Ecological business model
zebrata	Prehensile-tailed Skink				Solomon Islands
	nocturnal	herbivore	289	662	**** giant herbivore

The Solomon Islands, which include Bougainville Island, politically part of Papua New Guinea, are oceanic enough to have very few indigenous mammals. Most native mammals in the Solomons are giant herbivorous rats, which do not generally bother other animals. There are several diurnal carnivorous birds, however; thus a large edible lizard would best be nocturnal—the owls of the Solomon Islands are relatively small to take so large a lizard, especially a lizard often sequestered in tree holes in dense canopies. Occupancy of the trees also partially liberates this lizard from the larger boid snakes (*Candoia)* that primarily patrol ground level in the Solomons. The large size of *Corucia* may be an evolutionary response specifically to *Candoia bibroni* predation; adults are just a bit large for predation by this semiarboreal snake. *Corucia* has the family's lowest ontogenetic mass range (–3.32 v. 0.12 std. resid.), indicating that a neonate is born at a very large size—perhaps to start life at a size resistant to *Candoia* predation. The relative lack of predation may allow *Corucia* to live for a long time and have a relatively low reproductive output.

Corucia are relatively unique among plant-eating lizards, not only in being folivorous, but in their specialization on a single species: *Epipremnum pinnatum*, an aroid vine. Most giant herbivorous lizards eat a wide variety of leaves and fruits, perhaps for better nutrition, and perhaps to minimize the metabolic impact of the chemical defenses that many plants leaves possess. By shifting their plant targets throughout the year, most herbivorous lizards can capture seasonal fruits—to be dispersed by animals, fruits must have few chemical defenses against vertebrates—and this shifting seasonal fruit diet requires relatively large home ranges. In contrast, the home range of a *Corucia* is a single tree. Keep in mind that most plants have chemical defenses against plagues of phytophagous insects, not against sparse lizards. The chemical defenses of *Epipremnum pinnatum* are not documented in the herpetological literature, but on general ecological principles it is unlikely that its chemical defenses target *Corucia*. If it does have chemical defenses against *Corucia*, *Corucia* seems to have won the arms race; that is, *Corucia* will be able to detoxify the plant's chemicals, if any. Given *Corucia*'s specialization on it, it also appears that *Epipremnum pinnatum* is nutritionally complete, or nearly so, from the perspective of low-metabolism *Corucia*. *Corucia* do not have to move far, fast, or often to be fed.

The social life of *Corucia* is not unique among lizards, but it is exceptionally rare: like some *Tiliqua* and *Egernia* skinks (see those genus accounts), *Corucia* live, at least some times, in relatively isolated cohesive family groups (male, female, offspring). The social relationship of *Corucia* parents with offspring is essentially unstudied, but at least one observation reported adults feeding the offspring, and several accounts note the delayed dispersal of young. The family group is in possession of a key resource—a tree containing *Epipremnum pinnatum*; if they were birds, one would expect the juveniles to remain with the parents for some time and share in the group defense of this resource. One might even expect the juveniles of at least one sex to remain with the parents so they might inherit the territory when the parents die. Inheritance of *Corucia* territories has not been studied, but it would be a worthy focus. Unfortunately, given the lizards' longevity, this would not be a good thesis topic.

Are trees with *Epipremnum* vines close enough together to elicit territorial challenges by neighboring males? Data on this critical subject are lacking, though *Corucia* that survived the pet craze are now widely spaced, according to anecdotal field reports. However, the absence of sex differences—size, dichromatism, and seasonal color—suggests that social interactions between male *Corucia* are relatively unimportant and likely were so even prior to the recent trade-induced decline in *Corucia* populations. Male *Corucia* are more likely to mate-guard than to contest territories or form hierarchies with neighboring males. *Corucia* does not appear to have the opportunity to become polygynous. It is likely that

Corucia will be found to be strongly monogamous, following the *Egernia* and *Tiliqua* examples.

The *Tiliqua* examples (especially *rugosa*) merit comparison with *Corucia*. The three consistent differences are *Corucia*'s nocturnality, arboreality, and insularity. The nocturnality of *Corucia* is, perhaps, a reflection of the unique suite of predators present in the Solomon Islands. Arboreality is not a good option for large *Tiliqua* lizards in the outback of Australia, as trees are comparatively few and well-defended chemically; *Egernia* are somewhat arboreal. Islands are rare in the outback. Thus those three differences (nocturnality, arboreality, insularity) may be site-specific. The general absence of sex differences is well supported by the best-studied (monogamous) *Tiliqua* (*rugosa*). Four *Tiliqua* show m<f SSD—the opposite of the usual giant herbivore condition—suggesting fecundity selection dominates on *Tiliqua* females of *adelaidensis*, *gigas*, *multifasciata*, and *nigrolutea*. However, none of those four are demonstrably monogamous and there are other differences. For example, juvenile dispersal happens within days of birth in *rugosa*; *adelaidensis* is a crack hermit in which both sexes strenuously defend their own burrows. Sex differences in color are absent in all *Tiliqua*. Thus there are tantalizing analogues between *Tiliqua* and *Corucia*, but detailed field observations of these unique skinks, such as has been done for *Tiliqua rugosa* by Michael Bull and his many academic progeny (see database for citations), will be needed to confirm or deny the hypothesis of monogamy in *Corucia*.

Does *Corucia* have other notable attributes beyond those tied directly to their unique social lives? *Corucia* is extraordinarily bulky (shape index 1.68 v. −0.46 std. resid.), perhaps to forestall snake predation or pack in quantities of leaves. Tails are not breakable (v. 73%); unbreakable tails are expected in prehensile species (v. 0.7%) and those that store much nutrition in the tail. Unlike many herbivores, *Corucia* does not have crests (v. 0%), perhaps a reflection of their scincid ancestry or the futility of living in a tree hole and keeping a crest upright. *Corucia* has not been introduced (v. 2.1%). It is considered to be imperiled (v. 47%), but is not on the IUCN's lists of most endangered species.

Coryphophylax

Coryphophylax is a tiny genus (two spp.) of tree-trunk agamids, endemic to the Andaman and Nicobar Islands of India. The small one (*brevicaudus*: 5.8 g) forages on trunks (trunk/wall lizard 50% v. 7%), and the large one (*subcristatus*: 23.6 g) forages also on the ground (ground-foraging tree lizard: 50% v. 19%). Like many agamids, they have a low crest (v. 38%) and a dewlap; unlike most agamids, they can regenerate their long tails (2.3 v. 1.9 × SVL). At least one species has advanced color change (50% v. 15%). Males are larger than females (1.0 v. 0.53 on −2 to 2 scale), and sexual dichromatism is known (50% v. 53%). They have a slightly elevated sex differences rollup score (2.0 v. 1.6 on 0 to 6 scale). They sit-and-wait (100% v. 88%) for insects (100% v. 46%). Contradictory evidence suggests that the large species has a clutch of about 11, which is a mass-adjusted 1.33 (v. 0.20 std. resid.), the sixth-highest clutch count in Agamidae. They have not been introduced (v. 1.7%), nor formally evaluated for conservation status.

Corytophanes

Corytophanes is a small (three spp.) genus of diurnal arboreal lizards that are large for lizards but small for corytophanids (24.4 g v. 67 g—family or 5.3 g—lizards) and slender for corytophanids but stocky for lizards (shape index 0.86 v. 1.20 or 0.0 std. resid.). They are found in Central America and are either immobility experts or ground-foraging tree lizards. Males are ornamented in the dorsal and especially nuchal area, with crests, fins, or "helmets." Both sexes have corytophanid-typical long tails (2.28 v. 2.75 × SVL), which do not readily break. At least some species show color change (33% v. 22%). Unlike closely related *Basiliscus*, *Corytophanes* are mildly m<f dimorphic in body size (−0.25 v. corytophanid 0.90 on −2 to 2 scale). *Corytophanes* have not been reported to be sexually dichromatic (0% v. 22%) or to show seasonal color (0% v. 11%). Thus, *Corytophanes* have a low score on the sex differences rollup (0.7 v. 1.6 on 0 to 6 scale).

All *Corytophanes* are invertebrate eaters (v. 50%), with at least *cristatus* emphasizing large prey, perhaps to minimize crypsis-disrupting movements. *Corytophanes* rely on extreme sit-and-wait (v. 80%) foraging. They seem to emphasize small-diameter perches, perhaps to increase their

Coryphophylax spp.	English name				Range core
			Size		
	Activity	Diet	mm	g	Ecological business model
brevicaudus	Short-tailed Bay Island Forest Lizard				India
	diurnal		57	5.8	* trunk/wall lizard
subcristatus	Bay Island Forest Lizard				India
	diurnal	invert.	89	23.6	*** ground-foraging tree lizard

Corytophanes spp.	English name				Range core
			Size		
	Activity	Diet	mm	g	Ecological business model
cristatus	Elegant Helmeted Basilisk				Central America
	diurnal	invert.	103	31	*** immobility expert
hernandesii	Hernandez's Helmeted Basilisk				Guatemala
	diurnal	invert.	96	25	** immobility expert
percarinatus	Keeled Helmeted Basilisk				Honduras
	diurnal		88	18.7	* ground-foraging tree lizard

field of view. Their immobility may be reflected in the liverworts, algae, and slime mold that have been found on the "helmet," no doubt enhancing their crypsis. Slow-growing vegetation is not normally found on lizards, due to their frequent shedding; the phenology and biochemistry of skin renewal of *Corytophanes* vis-à-vis plant growth seems worthy of a targeted investigation. The best-studied species, *cristatus*, is judged to be conspicuously wary (v. 56%) and to be a thermoconformer (v. 11%). It has a low mean body temperature (26°C v. 30.3°C for family).

Male hierarchies are not known (v. 33%). *Corytophanes* are found in wet areas (>3000 mm/y) in all elevation bands. Unlike *Basiliscus*, *Corytophanes* are not found in riparian zones (v. 44% for the family). They are also absent from littoral areas (v. 11%) and sandy areas (v. 22%).

Two of three are oviparous (v. 89%), with an average clutch/litter count (–0.15 v. –0.22 std. resid.) and moderately small young even for corytophanids (ontogenetic range 1.43 v. 1.16 std. resid.). Incubation period is comparatively long (0.15 v. –0.52 std. resid.). One species (*cristatus*) is documented to show nest defense parental care (v. 11%). Longevity is corytophanid-short (–0.16 v. –0.28 std. resid.). Based on one *cristatus* population density estimate, their density is sparse but unremarkable (0.08 v. 0.20 kg/ha). No *Corytophanes* has been introduced (v. 33%). *C. cristatus* is feared by some Costa Rican residents as venomous. *C. hernandesii* is specially protected in Mexico. In other respects they are typical corytophanids.

Corytophanidae

The nine corytophanids are three genera (*Basiliscus*, *Corytophanes*, *Laemanctus*) of large (67 v. 5.3 g), arboreal diurnal lizards with some sort of cranial ornamentation such as crests, fins, or casques. They range from central Mexico to Colombia and Venezuela. They were once grouped with the iguanas (sensu lato) and even today are popularly called iguanas (or basilisks). All species will sometimes run or hop bipedally. Basilisks, with their fringed toes, are often capable of running bipedally across the surface of streams, for which they are sometimes known as Jesus Christ lizards. Corytophanids are ground-foraging tree lizards (67% v. 9%), immobility experts (22% v. 3%), or diurnal canopy walkers (11% v. 6%).

They are comparatively well known (46% of cells filled v. 27%), robust in build (shape index 1.20 v. 0.0 std. resid.), and long-tailed (2.75 v. 1.37 × SVL), though not prehensile-tailed (0% v. 5%). The head ornaments differ by genus: in *Basiliscus* they resemble fins, in *Corytophanes* they appear as great dorsal extensions of the neck, and in *Laemanctus* they are transverse "casques." None has secretory pores (v. 15%), blue tails (v. 2.3%), or is elongate (v. 11%). Color change is absent in *Basiliscus*, but otherwise common (22% in family v. 11% for any change, 6% for advanced).

SSD averages 0.90 (m>f) on –2 to 2 scale (v. 0.36 for lizards) and corytophanids average 22% (v. 21%) in sexual dichromatism and 11% (v. 6.5%) with seasonal color. Combined with their routinely dimorphic crests, these sex differences sum to an elevated rollup score of 1.56 (v. 0.75 on a scale of 0 to 6).

Half of the corytophanids are invertebrate eaters (v. 73%) and half are omnivores (v. 15%). Many species are large enough to capture vertebrates, even fish, opportunistically. The majority (80% v. 46%) are regarded as sit-and-wait foragers; 20% are regarded as active foragers (v. 58%). They have not been recorded foraging at electric lights (v. 1.4%) and have the smallest mean home-range areas of any family (–2.05 v. 0.0 std. resid.). One might suppose that the three-dimensionality of corytophanid home ranges is responsible for their compactness, but such a pattern is not evident from the five next-smallest size-adjusted home range families: Sphenodontidae (–1.99), Xantusiidae (–1.31), Opluridae (–1.31), Chamaeleonidae (–1.12), and Iguanidae (–1.06), a well-mixed suite of terrestrial and arboreal families.

Corytophanids are cathemeral (11% v. 7%) or diurnal (89% v. 67%). They are not known to call spontaneously (v. 2.5%) or under duress (v. 2.2%). A majority (56% v. 8%) are judged to be conspicuously wary. Eighty percent rely on crypsis (v. 43%); the remainder hide in vegetation (v. 23%). They are overrepresented among the species that bask directly (44% v. 18%) but also overrepresented among thermoconformers (11%, the third-highest family, v. 3%). They may be overrepresented in both basking conditions because they are better known than many other lizards. Their mean body temperature is average (30.3°C v. 31.1°C).

Corytophanids are not known for living in groups, though *Basiliscus* cluster along streams (see below). All species have males that defend space (v. 85%), and a third of species form male hierarchies (v. 8%).

Most corytophanids are arboreal (89% v. 32%), the others terrestrial (11% v. 48%), and none is known to burrow (v. 22%) for a daily refugium. They are found at the wet end of the rainfall spectrum, overrepresented in mesic (1101–3000 mm/y: 71% v. 43%) and wet (>3000 mm/y: 71%, the second highest, v. 9%) areas. They have a fairly typical altitudinal distribution (100/78/11 v. 81/41/13) and are found at lizard-typical levels in littoral (11% v. 17%) and sandy (22% v. 22%) areas, but are overrepresented in human habitats (33% v. 19%). They are notably associated with riparian zones (44% v. 22%), but this conceals a key difference among corytophanid genera: *Basiliscus* are universal in riparian zones, and the other corytophanids are not reported from riparian zones.

Corytophanids are oviparous at a typical frequency (89% v. 82%) and have clutches of normal size (−0.22 v. 0.0 std. resid.) and young that are exceptionally small (ontogenetic range 1.16 v. 0.0 std. resid.). This corresponds to a hatchling that is 1/230 of maximum size rather than 1/44. The combination of average or light clutch counts with exceptionally small hatchlings matches the light RCM of the family (0.170 v. 0.226, though based solely on *Basiliscus*). Perhaps the light RCM aids gravid *Basiliscus* in running on water. Mass-adjusted incubation period is short (−0.52 v. 0.0 std. resid.), which is unusual among lizards found in densely shaded tropical areas (hot sunlit soils are often required for expedited development). Eggshells are soft (v. 59%), nonadherent (v. 97.6%) and not reported to be laid communally (100% v. 95.8%). Corytophanids exhibit parental care at higher than normal frequency (11% v. 3%).

Corytophanid longevity is mildly short (−0.28 v. 0.0 std. resid.) and population density is relatively sparse (0.20 v. 0.54 kg/ha). Corytophanids (*Basiliscus* only) have been introduced (33% v. 3.3%) but not more than twice (0% v. 1.1%). One-quarter of corytophanids are at conservation risk (v. 49%), but none is on the IUCN's lists of most endangered (v. 13.6%).

Crenadactylus

Crenadactylus (seven spp.) are minute (0.62 g v. 4.2 g), clawless, striped, terrestrial diplodactylid geckos of the Australian outback. Both the longitudinal striping and the clawlessness are probably attributable to their habit of residing within clumps of stiff spinifex (*Triodia*) grass. Claws may be incapable of gripping the hard blades, and longitudinal striping camouflages the lizard within the linear matrix. The various species are similar enough to each other that they have only recently been split from *ocellatus*. Although most traits have been documented only in *ocellatus*, they are likely to apply more broadly. At present, few data are available for the genus (17% v. 28% cells filled). *C. ocellatus* is reported to be alert and capable of feeding at all hours but remains within the *Spinifex* clump during the day. Thus is it classified as a nocturnal ground lizard, though during the daytime it might be considered a litter cryptozoid.

The diminutive size of these geckos fits well with their slender build (shape index −0.47 v. 0.05), m<f SSD and cryptozoic habits (v. 2.3%). That they are cathemeral (v. 13%) and cryptozoic fits with their being conspicuously wary (v. 3%) and not basking directly (v. 83%). They are overrepresented in littoral (29% v. 17%), sandy (57% v. 24%), and ripar-

Crenadactylus spp.	English name		Size		Range core
	Activity	Diet	mm	g	Ecological business model
horni	Central Uplands Clawless Gecko				Central Australia
			34	0.78	
naso	Northern Clawless Gecko				Western Australia
			31	0.59	
occidentalis	Western Clawless Gecko				Western Australia
			33	0.67	
ocellatus	Southwestern Clawless Gecko				Australia
	cathemeral		33	0.71	** nocturnal ground lizard
pilbarensis	Pilbara Clawless Gecko				Western Australia
			30	0.50	
rostralis	Southwest Kimberly Clawless Gecko				Western Australia
			33	0.68	
tuberculatus	Cape Range Clawless Gecko				Western Australia
			29	0.48	

ian (29% v. 10%) zones, but not in human habitats (v. 5%).

Communal egg deposition is reported (v. 2.7%). Endemism is underrepresented (57% v. 72%). *Crenadactylus* have not been introduced (v. 2%), and they have not been evaluated for conservation risk. In other respects they match typical diplodactylids.

Cricosaura

Cricosaura typica is the sole member of its xantusiid genus of small-sized (1.05 g v. 4.02 g), ant specialist (v. 7.1%) litter cryptozoids (v. 3.8%) endemic to Cuba. It is comparatively svelte (shape index −0.34 v. 0.05 std. resid.). It possesses secretory pores (v. 24%) but not color change (v. 6%). Mean active body temperature is above the family mean (29.8°C v. 27.6°C). It is found at low elevation (0–200 m). Immatures are more strictly cryptozoic (i.e., below leaf litter) than are adults. It is not associated with any of the specialty habitats (littoral, sandy, riparian, or human habitats). Reproduction is not documented. It has not been introduced (v. 3%). It is imperiled (v. 31%), but it is not on one of the IUCN's lists of the most endangered (v. 3.8%). In other respects *Cricosaura* is probably a typical xantusiid.

Crocodilurus

Crocodilurus amazonicus is the sole member of its teiid genus of very large (212 g v. 22.1 g), semiaquatic (100% v. 2.2%) refugia-anchored diurnal ground lizards (100% v. 2.6%) found in northwestern South America, from Colombia to Peru to Brazil and French Guiana. The resemblance to crocodiles is more behavioral—it is carnivorous and prefers to scull through wetlands—than morphological. It looks like the large teiid that it is.

Crocodilurus is robust in build (shape index 1.54 v. 0.88 std. resid.). Its tail is about average in length (2.06 v. 2.30 × SVL) and breakable (100% v. 21%) but not blue (0% v. 15%). The tail is much taller than it is wide, presumably to enhance propulsion through water. *C. amazonicus* does not change color (v. 1.3%) or possess secretory pores (v. 9%). No sex differences are known.

Two unusual prey are fish and Cane Toads (*Rhinella horribilis*). *Crocodilurus* basks directly (v. 58%) and uses burrows for daily refugia (v. 82%). Naturally it is found in riparian areas (v. 46%), but it is also found in littoral ones (v. 38%), though not sandy areas (v. 47%) or human habitat (v. 29%).

Its clutch count is low for its size (−1.08 v. −0.51 std. resid.), and its RCM ratio is low (0.132 v. 0.163). Longevity is average (−0.61 v. −0.74 std. resid.). It is not endemic (v. 50%), insular (v. 37%), or introduced (v. 8.3%). It is imperiled (v. 41%), but it is not on one of the IUCN's lists of the most endangered (v. 12%). In other respects *Crocodilurus* is a typical teiid.

Cricosaura sp.	English name				Range core
			Size		
	Activity	Diet	mm	g	Ecological business model
typica	Cuban Night Lizard				Cuba
	cathemeral	ants	38	1.05	** litter cryptozoid or nocturnal ground lizard

Crocodilurus sp.	English name				Range core
			Size		
	Activity	Diet	mm	g	Ecological business model
amazonicus	Crocodile Tegu				Venezuela
	diurnal	carnivore	188	212	** refugia-anchored diurnal gr. liz.

Crossobamon

Crossobamon (two spp.) are gekkonid nocturnal ground lizards found in sandy deserts of central Asia from Kazakhstan to India. These relatively well-known geckos (50% v. 25% filled) are slightly small (2.67 g v. 3.31 g) and have slightly longer tails than most gekkonids (1.14 v. 1.07 × SVL). The tails of *eversmanni* are prehensile, which is rare in gekkonids (v. 1.8%) and nocturnal ground lizards (v. 1.6%: table 4.1, trait 7). Are *eversmanni* recently derived from arboreal ancestors? Both species lack enlarged toepads (characteristic of ground lizards); 30% of gekkonids are functionally padless. *C. eversmanni* has fringed toes, often associated with propulsion in sand. Secretory pores are present in both species (v. 46%), though presence of pores in male *orientalis* is individually variable. Color change is not reported (v. 15%).

Both species of *Crossobamon* have populations in which males are smaller than females (−0.75 v. −0.03 on −2 to 2 scale), but the sexes are alike in color and lacking in seasonal color change. Their sex rollup score (on scale of 0 to 6) is slightly elevated compared to Gekkonidae (1.0 v. 0.6).

Both species are reported to sometimes be alert but immobile at their burrow entrance during daytime, though *eversmanni* may be abroad during spring days; after dark they forage abroad, albeit slowly, at least

Crossobamon spp.	English name				Range core
			Size		
	Activity	Diet	mm	g	Ecological business model
eversmanni	Comb-toed Gecko				Afghanistan
	nocturnal	invert.	52	3.21	*** nocturnal ground lizard
orientalis	Sindh Sand Gecko				India
	cathemeral	invert.	46	2.21	**** nocturnal ground lizard

in the case of *eversmanni*. This species is reported to be an active forager, as is *orientalis*, but slow active foraging might also be described as cruise foraging. Foraging in *orientalis* is reported to be both active, perhaps mostly at night, and sit-and-wait, perhaps mostly during daytime. *C. eversmanni* is reported to avoid surface activity under bright moon, but may move slowly up to 150 m from its burrow during a—presumably moonless—night. As it enters its burrow, *eversmanni* may plug the burrow entrance, using its protruding tail tip to sweep sand toward the plug. The overall foraging mode and diel activity pattern seems to be flexible in *Crossobamon*. Electric light use is not reported (v. 3.8%) and home-range size is undocumented.

Both species vocalize, though the sound is said to be spontaneous (v. 6.5%) in *eversmanni* and in response to stress in *orientalis* (v. 3.7%). *C. eversmanni* is reported to be conspicuously wary (v. 4%). Although mean active body temperature is not reported for *Crossobamon*, it is documented that *eversmanni* is active at air temperatures over the cool range 10°C–27°C. The same species is said to live in unspecified groups.

Similar to many other gekkonid nocturnal ground lizards (e.g., *Colopus*), *Crossobamon* differ from most gekkonids (mode in mesic belt: 1101–3000 mm/y) by residing in the dry rainfall belt (ca. 110 mm/y). *Crossobamon* are not reported from the medium (v. 26%) or high (v. 2%) elevation belts. Nor are they known to be found in littoral (v. 11%), riparian (v. 17%), or human (v. 27%) habitats; both species are strongly associated with sandy or dune-covered areas (v. 11%).

Crossobamon are similar to other gekkonids in most documented reproductive characters, except that *eversmanni* may have adherent eggs (v. 11%). This is another suggestion of an arboreal ancestry for this species (see table 4.1, trait 67). The young of *Crossobamon* are smaller than lizard average, unlike most gekkonids (ontogenetic range 0.32 v. –0.19 std. resid.).

Crossobamon have not been introduced (v. 3.7%), nor evaluated for conservation status.

In other respects *Crossobamon* are typical gekkonids.

Crotaphytidae

The 12 crotaphytids belong to two genera (*Crotaphytus*, *Gambelia*) of large (37 g v. 5.3 g) refugia-anchored diurnal ground lizards (100% v. 25%) of North America (including northern Mexico). All are omnivorous (100% v. 15%) and will take vertebrates when the opportunity arises. All species will sometimes run bipedally and most are found in open country that is arid (67% v. 30%) or intermediate in rainfall (67%% v. 27%). They are comparatively well known (cells filled 46% v. 27%), robust in build (shape index 1.39, third highest, v. 0.0 std. resid.), and moderately long-tailed (1.97 v. 1.37 × SVL), though not prehensile-tailed (0% v. 5%). Many species have secretory pores (58% v. 15%), but not blue tails (v. 2.3%) or crests (v. 5.3%), nor are they elongate (v. 11%). Color change is moderate (25% v. 11% for any change; 0% v. 6% for advanced).

SSD averages 0.43 (m≥f) on –2 to 2 scale (v. 0.36 for lizards), but that conceals large differences between the genera (see accounts). Crotaphytids average 75% (v. 21%) in sexual dichromatism and 100% (v. 6.5%) for seasonal color. Combined, these sex differences result in an elevated sex differences rollup score of 3.17 (v. 0.75 on a scale of 0 to 6).

Forty-six percent of crotaphytids are regarded as invertebrate eaters (v. 73%) and 46% as omnivores (v. 15%), with the remaining 9% carnivores (v. 3.2%), but the similarities among species are more striking than the differences. Most species are very generalist feeders and will take plants, invertebrates, and vertebrates as the opportunity arises. All evaluated species (100% v. 46%) are regarded as sit-and-wait foragers, but 30% are also active foragers (v. 58%) as well (especially for plants). Their foraging modes appear opportunistic. They have not been recorded foraging at electric lights (v. 1.4%) and have moderately large home-range areas (0.72 v. 0.0 std. resid.) on average, but this conceals large differences between the genera (see accounts).

They are overrepresented among lizards giving stress calls (8.3% v. 2.2%), and half of the species are reported to be conspicuously wary (v. 8.2%). Every predator concealment tactic except diving into sand is reported for the Crotaphytidae: crypsis (40% v. 43%), flight (40% v. 11%), crevices and burrows (20% v. 25%), and hiding in vegetation (20% v. 23%). Eighty-three percent have been documented to bask directly (v. 18%). None is regarded as a thermoconformer (v. 2.6%). Mean body temperature is a toasty 37.4°C (v. 31.3°C), presumably reflecting their arid environments.

The social lives of most crotaphytids are poorly known, but 33% (v. 15%) of species are socially tolerant and 67% defend space (v. 85%). Many use burrows as daily refugia (63% v. 22%). They are overrepresented in the dry (33% v. 12%), arid (67% v. 30%), and intermediate (67% v. 27%) rainfall belts. Their altitudinal distribution is the same as most lizards (see chapter 2). They are notably overrepresented in sandy/dune areas (42% v. 22%) and riparian areas (58% v. 22%), but they are absent from areas of human habitat (v. 19%).

Crotaphytids are oviparous (v. 82%), and have slightly small clutch counts (−0.36 v. 0.0 std. resid.). The size of the offspring is larger than average (ontogenetic mass range −0.38 v. 0.0 std. resid.). These two conditions nearly offset each other, such that RCMs are about average (0.299 v. 0.226), but this conceals large differences between the genera (see accounts). Mass-adjusted incubation period is short (−0.71 v. 0.0 std. resid.). Eggshells are soft (v. 59%) and nonadherent (v. 97.6%), and 8.3% of species are reported to lay eggs communally (v. 4.2%). Crotaphytids exhibit parental care at higher than normal frequency (8% v. 3%).

Corytophanid longevity is a bit short (−0.11 v. 0.0 std. resid.) and population density is relatively sparse (0.06 v. 0.54 kg/ha). Crotaphytids (*Crotaphytus* only) have been introduced (8.3% v. 3.3%) but not more than twice (0% v. 1.1%). Nearly half of crotaphytids are at conservation risk (v. 49%), and 17% are on the IUCN's lists of most endangered (v. 14%).

Crotaphytus

Crotaphytus (nine spp.) are very similar to the only other members (*Gambelia*) of their family (Crotaphytidae) in their basic ecology and many aspects of the natural history: ecological business model (100% refugia-anchored diurnal ground lizard), size (36.6 g v. *Gambelia's* 40.1 g), build (shape index 1.45 v. *Gambelia*'s 1.23 std. resid.), tail length (1.97 v. 1.98 × SVL), secretory pores (56% v. 67%), seasonal color (78% v. 67%), sex differences rollup (3.22 v. 3.00 on 0 to 6 scale), diet (very opportunistic omnivore/carnivores), foraging modes (a dab of active with a backbone of sit-and-wait), reliance on direct basking (89% v. 67%), oviparity (100% v. 100%), incubation duration (−0.67 v. −0.79 std. resid.), and longevity (0.20 v. −0.26 std. resid.). They also mostly reside in the same arid parts of western North America. If diet were destiny, these two genera would be peas in a pod.

The most striking difference between *Crotaphytus* and *Gambelia* is found in their home-range areas (here not mass-adjusted, as their mean masses are so similar). The maximum number of mean square meters for *C. collaris* is 324 m²; that for *C. reticulatus* is 700 m². The corresponding available data for *Gambelia* are *sila*, 93,600 m², and *wislizenii*,

Crotaphytus spp.	English name				Range core
			Size		
	Activity	Diet	mm	g	Ecological business model
antiquus	Venerable Collared Lizard				Mexico
	diurnal		92	35.9	*** refugia-anchored diurnal gr. liz.
bicinctores	Great Basin Collared Lizard				Nevada
	diurnal	invert.	92	35.6	*** refugia-anchored diurnal gr. liz.
collaris	Eastern Collared Lizard				Texas
	diurnal	carnivore	98	40.1	**** refugia-anchored diurnal gr. liz.
dickersonae	Tiburon Collared Lizard				Mexico
	diurnal	omnivore	91	35.4	*** refugia-anchored diurnal gr. liz.
grismeri	Sierra Los Cucapas Collared Lizard				Mexico
	diurnal	invert.	81	29.3	*** refugia-anchored diurnal gr. liz.
insularis	Angel de la Guarda Collared Lizard				Mexico
	diurnal	omnivore	93	36.4	*** refugia-anchored diurnal gr. liz.
nebrius	Sonoran Collared Lizard				Mexico
	diurnal	invert.	90	34.5	*** refugia-anchored diurnal gr. liz.
reticulatus	Reticulate Collared Lizard				Mexico
	diurnal	omnivore	110	47.2	**** refugia-anchored diurnal gr. liz.
vestigium	Baja California Collared Lizard				Mexico
	diurnal	omnivore	95	37.8	*** refugia-anchored diurnal gr. liz.

23,500 m². Expressed mass-adjusted, *Crotaphytus* home ranges are 0.02 and *Gambelia* 1.42 standardized residuals. Male *G. sila* average about 100 m net movement per 24 hours, and females average about 65 m. *Gambelia* appear to move daily from burrow to burrow, often appropriating mammal burrows as they go; burrow use for daily refugia is 50% for *Crotaphytus* but 100% for *Gambelia*. *Gambelia* are said to be 33% dependent on burrows for predator evasion, whereas no *Crotaphytus* is so described. *Crotaphytus* hide in vegetation, but *Gambelia* are not so reported. Despite the huge activity areas, however, both genera capture most prey by sitting and waiting, though sources reported that two of three *Gambelia* exhibited active foraging, but only one of seven *Crotaphytus* did. *Gambelia* are not wide-ranging chemosensors; 67% of *Gambelia* rely on crypsis v. 0% of *Crotaphytus*. The huge activity areas of *Gambelia* probably reflect the low primary productivity of their habitat, which is drier than that of *Crotaphytus*. Documented *Crotaphytus* are absent from the dry rainfall belt (56–150 mm/y), whereas 100% of documented *Gambelia* are in the dry. All *Crotaphytus* are found in the arid (151–400 mm/y) and intermediate (401–1100 mm/y) belts, whereas *Gambelia* are absent from those wetter bands. One-hundred percent of *Gambelia* species are associated with sandy/dune areas (v. 22% in *Crotaphytus*), and 100% of *Gambelia* are present in riparian zones (v. 44% of *Crotaphytus*). The most consistent theme of these habitat differences is that *Crotaphytus* and *Gambelia* use different tactics to forage and hide in their landscapes, even though they both are basically generalist sit-and-wait omnivores. *Crotaphytus* do not readily break tails, but *Gambelia* do; how does that fit in with these differences? *Crotaphytus* are more often said to be conspicuously wary (56% v. 33%)—is that because *Gambelia* freeze and rely on crypsis? Two-thirds of *Gambelia* exhibit crypsis-useful color change (v. 11% of *Crotaphytus*).

The social differences between *Crotaphytus* and *Gambelia* are few, but critical. *Crotaphytus* males are usually larger than females (mean 1.00 on −2 to 2 scale), whereas *Gambelia* tend to the converse (−0.33). The majority of *Gambelia* males are tolerant of other males (67% v. 0%). Only a third of *Gambelia* males defend space, whereas 100% of *Crotaphytus* do. Are *Gambelia* more tolerant because they occur at low density in more arid places? Population density estimates would be of interest, but are unavailable.

The two genera appear also to differ in RCM. *Crotaphytus* are very heavily laden (RCM mean of 0.45), whereas *Gambelia* are at the opposite extreme (RCM mean of 0.15). The ontogenetic mass range of *Gambelia* is lower (−0.53 v. −0.32 std. resid.), indicating that *Gambelia* have moderately larger young, but they have fewer of them; mass-adjusted clutch count of *Gambelia* is −0.45 (v. −0.33 std. resid.). Do *Gambelia* females need to be lightly loaded in the event they are caught far from a known burrow? Finer-grained movement data are needed to evaluate this possibility.

Cryptactites

Cryptactites peringueyi is the only member of this unusual gekkonid genus. It is probably a nocturnal canopy walker, and it is known only from a salt marsh in a very small area of coastal South Africa. For 80 years it was known from only two poorly labeled specimens, as no one thought to look for a gecko in a salt marsh. It is very small (0.48 g v. 3.31 g) and slender (shape index −0.19 v. 0.10 std. resid.) and lacks secretory pores (v. 46%) and color change (v. 15%). It is sexually dichromatic (v. 7%), but other sex differences have not been reported. It is not reported to vocalize (v. 7%). It stands out only for its littoral habitat (v. 11%). Because it is so small, the conventional gekkonid clutch count of two is found to be above the mass-adjusted lizard norm in this species (0.52 v. −0.26 std. resid.). It is classified as endangered by South Africa, though only “data deficient” by the IUCN. In other respects it is a typical gekkonid.

Cryptagama

Cryptagama is such an effective pebble mimic that this single-species agamid genus of immobility experts (v. 4.5%) from northwestern Australia is known from only a few specimens. Its tail is short (0.75 v. 1.86 × SVL). It sits-and-waits (v. 88%) for

Cryptactites sp.	English name				Range core
			Size		
	Activity	Diet	mm	g	Ecological business model
peringueyi	Salt Marsh Gecko				South Africa
	nocturnal	invert.	26	0.48	** nocturnal canopy walker

Cryptagama sp.	English name				Range core
			Size		
	Activity	Diet	mm	g	Ecological business model
aurita	Gravel Dragon				Northwestern Australia
	diurnal	ants	44	2.54	**immobility expert

ants (v. 20%) during the day (v. 98%). It is oviparous (v. 98%). Little else is known.

Cryptoblepharus

Cryptoblepharus is a major radiation (53 spp.) of uniformly small (1.09 g v. 3.25 g), mostly littoral (58% v. 21%), mostly ground-foraging tree skinks (75% v. 7%) from throughout the Eastern Hemisphere (Australasian: 64% v. 46%; Ethiopian: 21% v. 20%; Oriental: 15% v. 24%), with a smattering of Neotropical (2% v. 2%) and Palearctic (2% v. 6%) for good measure. They are absent only from the Nearctic (0% v. 2%). Those that are not ground-foraging tree lizards are refugia-anchored diurnal ground lizards (17% v. 30%), trunk/wall lizards (6% v. 3%), or diurnal canopy walkers (2.1% v. 1.8%). Like many very small lizards, they lack movable eyelids (possibly a water conservation measure; see "Mass" in chapter 3) and are therefore popularly deemed "snake-eyed." Although the genus as a whole has considerable ecological diversity—and on small islands great ecological diversity may prevail within a single population—many of the littoral species are geographically isolated along the coast and have speciated perhaps more than they have ecologically diverged. Nearly 50 of the currently recognized species were once deemed to be a single variable species.

In keeping with their small size, males tend to be smaller than females (–0.67 v. –0.31 on –2 to 2 scale). Blue tails are present at the family frequency (3.8% v. 4.0%). Elongation is not recorded (v. 35%). The genus is dominated by invertebrate eaters (95% v. 84%), with one species (*pulcher*) a specialist on ants (5% v. 1.6%), though it avoids worker ants and targets winged reproductives. Intertidal foraging is common; at least two species (*cognatus*,

Cryptoblepharus spp.	English name	Activity	Diet	Size mm	Size g	Range core	Ecological business model
adamsi	Adams' Snake-eyed Skink	diurnal		36	0.81	Queensland	* ground-foraging tree lizard
africanus	East-African Snake-eyed Skink	diurnal	invert.	42	1.32	Tanzania	** ground-foraging tree lizard
ahli	Ahl's Snake-eyed Skink			44	1.55	Mozambique	
aldabrae	Aldabra Snake-eyed Skink			39	1.06	Seychelles	** ground-foraging tree lizard
ater	Grande Comore Snake-eyed Skink			42	1.34	Comoros	
australis	Inland Snake-eyed Skink	diurnal		41	1.25	Australia	** ground-foraging tree lizard
balinensis	Bali Snake-eyed Skink	diurnal		39	1.05	Indonesia	** diurnal canopy walker
bitaeniatus	Europa Snake-eyed Skink			40	1.11	Mozambique	
boutonii	Mauritius Snake-eyed Skink	diurnal	invert.	38	0.93	Mauritius	** refugia-anchored diurnal gr. liz.
buchananii	Buchanan's Snake-eyed Skink	cathemeral	invert.	40	1.10	Western Australia	* ground-foraging tree lizard
burdeni	Komodo Snake-eyed Skink			43	1.45	Indonesia	** refugia-anchored diurnal gr. liz.
caudatus	Juan de Nova Snake-eyed Skink			44	1.50	Mozambique	* refugia-anchored diurnal gr. liz.
cognatus	Kindred Snake-eyed Skink	diurnal	invert.	41	1.18	Madagascar	** refugia-anchored diurnal gr. liz.
cursor	Beach Snake-eyed Skink	diurnal		39	1.04	Indonesia	** ground-foraging tree lizard
cygnatus	Swanson's Snake-eyed Skink	diurnal	invert.	40	1.11	Northern Territory	* ground-foraging tree lizard
daedalos	Dappled Snake-eyed Skink	diurnal		39	1.04	Northern Territory	* ground-foraging tree lizard
egeriae	Christmas Island Blue-tailed Skink	diurnal	invert.	45	1.67	Christmas Island (Australia)	* ground-foraging tree lizard
eximius	Fiji Snake-eyed Skink	diurnal	invert.	37	0.85	Fiji	* ground-foraging tree lizard
exochus	Noble Snake-eyed Skink	diurnal		40	1.08	Northern Territory	* ground-foraging tree lizard

Cryptoblepharus spp.	English name				Range core
			Size		
	Activity	Diet	mm	g	Ecological business model
fuhni	Fuhn's Snake-eyed Skink				Queensland
	diurnal	invert.	43	1.40	* trunk/wall lizard
furvus	Normanby Snake-eyed Skink				New Guinea
			44	1.47	* ground-foraging tree lizard
gloriosus	Glorioso Snake-eyed Skink				Comoros
			39	1.05	
gurrmul	Arafura Snake-eyed Skink				Northern Territory
	diurnal		40	1.10	* ground-foraging tree lizard
juno	Juno's Snake-eyed Skink				Western Australia
	diurnal		39	1.07	* ground-foraging tree lizard
keiensis	Maluku Snake-eyed Skink				Indonesia
			40	1.12	* ground-foraging tree lizard
leschenault	Leschenault's Snake-eyed Skink				Indonesia
			40	1.09	** ground-foraging tree lizard
litoralis	Coastal Snake-eyed Skink				Queensland
	diurnal	invert.	46	1.74	* refugia-anchored diurnal gr. liz.
megastictus	Blotched Snake-eyed Skink				Western Australia
	diurnal	invert.	38	0.92	** refugia-anchored diurnal gr. liz.
mertensi	Mertens' Snake-eyed Skink				Northern Territory
	diurnal		36	0.83	* ground-foraging tree lizard
metallicus	Metallic Snake-eyed Skink				Northern Australia
	diurnal		42	1.31	* ground-foraging tree lizard
nigropunctatus	Ogasawara Snake-eyed Skink				Japan
	diurnal	invert.	49	2.22	** ground-foraging tree lizard
novaeguineae	Bird's-head Snake-eyed Skink				New Guinea
	diurnal		37	0.90	** trunk/wall lizard
novocaledonicus	New Caledonia Shore Skink				New Caledonia
	diurnal	invert.	39	1.01	** ground-foraging tree lizard
novohebridicus	Vanuatu Snake-eyed Skink				Vanuatu
	diurnal	invert.	35	0.73	* ground-foraging tree lizard
ochrus	Pale Snake-eyed Skink				South Australia
	diurnal		40	1.11	* ground-foraging tree lizard
pannosus	Ragged Snake-eyed Skink				Eastern Australia
	diurnal		38	0.95	* ground-foraging tree lizard
plagiocephalus	Peron's Snake-eyed Skink				Western Australia
	diurnal	invert.	41	1.17	* ground-foraging tree lizard
poecilopleurus	Pacific Snake-eyed Skink				Pacific
	diurnal	invert.	42	1.30	**** ground-foraging tree lizard

voeltzkowi) collect fish that have died in drying tide pools; *africanus* and *boutonii* eat fish that may be alive at the time of ingestion. At least two arboreal species (*juno, pulcher*) are kleptoparasitic not on other lizards, but on ants: they steal the prey the ants are carting back to the nest. All are considered to be active foragers (v. 92%), and one species forages under electric lights (*poecilopleurus*). One is cathemeral (*buchananii,* 2% v. 9%); the rest are diurnal (98% v. 74%). All documented species retreat to crevices or burrows in the face of a threat (100% v. 43%). One arboreal species (*exochus*) has the remarkable habit of fleeing, not to vegetation (the arboreal staple), but by leaping off its tree and diving into soil cracks. In some places, but probably not most, *poecilopleurus* will sand swim in loose volcanic ash. Thermal behavior matches the family; mean active body temperature is 31.0°C (v. 31.1°C).

Solitary living is not reported (v. 8%), and group living occurs in half the documented species (50% v. 16%). Nonetheless, all documented males defend space (v. 94%).

Cryptoblepharus occupy a wide swath in the middle of the annual rainfall spectrum (arid to mesic), at approximately the family's frequencies in those bands. They are found exclusively at low elevation (<1200 m). As noted above, they are strongly overrepresented in the littoral zone; they are also notably common in human habitats (33% v. 14%) and present in riparian (23% v. 23%) and sandy (31% v. 25%) areas at family frequencies.

All are oviparous (v. 64%), most with a clutch of two (0.11 v. 0.06 std. resid.). Hatchlings are relatively small despite the species' small body size (ontogenetic mass range 0.52 v. 0.12 std. resid.). Eggshells are hard in at least one-third (33% v. 9%), a feature that no doubt assists in their dispersal and survival in littoral areas. Communal oviposition is very common (9.4% v. 2.6%), but parental care is unknown (v. 3.2%). RCM is exactly average (0.240 v. 0.236 ratio).

Endemism is average (68% v. 64%); island species are overrepresented (76% v. 57%). Many species have been introduced (9.4% v. 2.1%), two species (3.8% v. 0.7%) more than twice (*balinensis* as many as six times and *poecilopleurus* at least eight times). *Cryptoblepharus* are imperiled at a reduced frequency (33% v. 47%), and none is on one of the IUCN's lists of most endangered species (v. 14%). On Christmas Island (Australia), *egeriae* has declined sharply since the 1980s and may be extinct; suspicion centers on the Wolf Snake (*Lycodon capucinus*) which was introduced at the time the skink population began dropping. In other respects *Cryptoblepharus* are typical skinks.

Cryptoblepharus spp.	English name				Range core
			Size		
	Activity	Diet	mm	g	Ecological business model
pulcher	Elegant Snake-eyed Skink				Australia
	diurnal	ants	39	1.03	* ground-foraging tree lizard
quinquetaeniatus	Five-lined Snake-eyed Skink				Comoros
			40	1.08	
renschi	Blue-tailed Snake-eyed Skink				Indonesia
	diurnal	invert.	41	1.25	** ground-foraging tree lizard
richardsi	Misima Snake-eyed Skink				New Guinea
	diurnal		40	1.08	** ground-foraging tree lizard
ruber	Tawny Snake-eyed Skink				Western Australia
	diurnal		42	1.32	* ground-foraging tree lizard
rutilus	Palau Snake-eyed Skink				Palau
	diurnal		36	0.79	* ground-foraging tree lizard
schlegelianus	Schlegel's Snake-eyed Skink				Indonesia
	diurnal		38	0.97	**** refugia-anchored diurnal gr. liz.
tytthos	Pygmy Snake-eyed Skink				Northern Australia
	diurnal		35	0.73	* ground-foraging tree lizard
ustulatus	Russet Snake-eyed Skink				Western Australia
	diurnal		38	0.96	* ground-foraging tree lizard
virgatus	Striped Snake-eyed Skink				New Guinea
	diurnal	invert.	38	0.96	* ground-foraging tree lizard
voeltzkowi	Voeltzkow's Snake-eyed Skink				Madagascar
		invert.	40	1.08	** refugia-anchored diurnal gr. liz.
wulbu	Spangled Snake-eyed Skink				Northern Territory
	diurnal	invert.	36	0.81	* ground-foraging tree lizard
xenikos	Trans-Fly Snake-eyed Skink				New Guinea
	diurnal		35	0.73	** ground-foraging tree lizard
yulensis	Yule Island Snake-eyed Skink				New Guinea
	diurnal		39	1.02	* trunk/wall lizard
zoticus	Agile Snake-eyed Skink				Queensland
	diurnal		36	0.77	* ground-foraging tree lizard

Ctenoblepharys

Ctenoblepharys adspersa is the sole species in a genus of average-sized (6.4 g v. 9.4 g) liolaemids from littoral Peru. It is so poorly known (filled cells 16% v. 22%) that information is lacking on its diel pattern, diet, and ecological business model. It is heavier built than an average lizard but has the lightest average build of any liolaemid genus (0.41 v. 0.76 std. resid.). It is not known to have a breakable tail (v. 22%), secretory pores (v. 22%), or color change (v. 2%). The sexes are not known to differ in size or color (v. 29%). It is found exclusively at low elevation (0–750 m) in areas of sand and dunes (v. 32%). It is endemic to littoral Peru (v. 82%) and is not found on any islands (v. 2.3%). It is considered imperiled (IUCN, near threatened), but it is not on the IUCN's lists of the most endangered. In other respects

Ctenoblepharys sp.	English name				Range core
			Size		
	Activity	Diet	mm	g	Ecological business model
adspersa	Coastal Desert Iguana				Peru
			63	6.4	

Ctenoblepharys is probably similar to other liolaemids.

Ctenophorus

Ctenophorus is a large (29 spp.), possibly composite, genus of generalized terrestrial (100% v. 56%) agamids found in Australia. Most (83%) lack a crest (17% v. 38% crested), though many have a relatively large head. They are smaller (9.4 g v. 15.6 g) than most agamids. Commensurate with their terrestrial lifestyle and desert home, most are refugia-anchored diurnal ground lizards (89% v. 43%), and a few of the ant specialists (36% v. 20%) are immobility experts (11% v. 4.5%).

Pores are present in substantially greater frequency than in agamids (31% v. 12.6%). Most species (86% v. 53%) have sexes that differ in color, and 24% (v. 20%) change color seasonally. Sexual dimorphism score is normal (0.45 v. 0.53 on –2 to 2 scale). *Ctenophorus* exhibit the fourth-highest m:f ratio of relative tail lengths (1.13 v. 1.07), though the consequences or causes of this are obscure. Their sex differences rollup score is elevated (2.24 v. 1.61 on 0 to 6 scale).

Ant specialists and omnivores (36% v. 26%) are overrepresented, presumably due to the low productivity of their arid (71%, the second-highest rate in Agamidae v. 35%) or dry (14% v. 30%) environment. The majority (59% v. 26%) are associated with sand or dunes. This low-productivity environment may be responsible for the large home-range area (0.99 v. 0.35 std. resid.), the third-largest in agamids. All are said to sit-and-wait (100% v. 88%), though presumably the omnivores stir themselves to pursue plants.

Male social tolerance is the second highest in agamids (11% v. 3%). Male spatial defense is low (89% v. 97%).

Ctenophorus spp.	English name				Range core
			Size		
	Activity	Diet	mm	g	Ecological business model
adelaidensis	Western Heath Dragon				Western Australia
	diurnal	ants	47	3.15	** immobility expert
butleri	Butler's Dragon				Western Australia
	diurnal		43	2.44	** immobility expert
caudicinctus	Ring-tailed Dragon				Australia
	diurnal	invert.	78	15.4	** refugia-anchored diurnal gr. liz.
chapmani	Chapman's Dragon				Southwest Australia
	diurnal	ants	48	3.48	* immobility expert
clayi	Black-collared Dragon				Western Australia
	diurnal	omnivore	50	3.77	*** refugia-anchored diurnal gr. liz.
cristatus	Crested Dragon				Southwest Australia
	diurnal	invert.	93	26.3	**** refugia-anchored diurnal gr. liz.
decresii	Tawny Dragon				New South Wales
	diurnal	invert.	75	13.9	** refugia-anchored diurnal gr. liz.
femoralis	Long-tailed Sand Dragon				Western Australia
	diurnal	invert.	50	3.97	** refugia-anchored diurnal gr. liz.
fionni	Peninsula Dragon				South Australia
	diurnal	invert.	78	15.4	*** refugia-anchored diurnal gr. liz.
fordi	Mallee Dragon				Australia
	diurnal	ants	52	4.39	** refugia-anchored diurnal gr. liz.
gibba	Gibber Dragon				South Australia
	diurnal	omnivore	78	15.2	** refugia-anchored diurnal gr. liz.
isolepis	Military Dragon				Australia
	diurnal	ants	56	5.5	** refugia-anchored diurnal gr. liz.
maculatus	Spotted Dragon				Southwestern Australia
	diurnal	ants	58	6.0	** refugia-anchored diurnal gr. liz.
maculosus	Lake Eyre Dragon				South Australia
	diurnal	ants	61	7.3	*** refugia-anchored diurnal gr. liz.
mckenziei	McKenzie's Dragon				Southwestern Australia
	diurnal	omnivore	67	9.6	** refugia-anchored diurnal gr. liz.
mirrityana	Barrier Range Dragon				New South Wales
	diurnal		74	13.4	

Ctenophorus spp.	English name				Range core
			Size		
	Activity	Diet	mm	g	Ecological business model
nguyarna	Lake Disappointment Dragon				Western Australia
	diurnal	ants	65	8.8	** refugia-anchored diurnal gr. liz.
nuchalis	Central Netted Dragon				Australia
	diurnal	omnivore	94	27.9	*** refugia-anchored diurnal gr. liz.
ornatus	Ornate Dragon				Western Australia
	diurnal	omnivore	78	15.5	*** refugia-anchored diurnal gr. liz.
parviceps	Gnaraloo Heath Dragon				Western Australia
	diurnal		46	2.95	* refugia-anchored diurnal gr. liz.
pictus	Painted Dragon				Australia
	diurnal	omnivore	62	7.6	*** refugia-anchored diurnal gr. liz.
reticulatus	Western Netted Dragon				Southwestern Australia
	diurnal	omnivore	83	18.7	*** refugia-anchored diurnal gr. liz.
rubens	Russet Dragon				Western Australia
	diurnal	ants	75	13.7	** refugia-anchored diurnal gr. liz.
rufescens	Rusty Dragon				Western Australia
	diurnal		83	18.8	*** refugia-anchored diurnal gr. liz.
salinarum	Claypan Dragon				Southwestern Australia
	diurnal	omnivore	64	8.5	*** refugia-anchored diurnal gr. liz.
scutulatus	Lozenge-marked Dragon				Southwestern Australia
	diurnal	invert.	93	26.6	** refugia-anchored diurnal gr. liz.
tjantjalka	Ochre Dragon				South Australia
	diurnal		68	9.92	** refugia-anchored diurnal gr. liz.
vadnappa	Red-barred Dragon				South Australia
	diurnal		77	15.0	*** refugia-anchored diurnal gr. liz.
yinnietharra	Yinnietharra Rock Dragon				Western Australia
	diurnal		77	14.7	** refugia-anchored diurnal gr. liz.

Male hierarchy formation is the third-highest frequency in Agamidae (22% v. 17.5%). All are oviparous (100% v. 98%) and lay few eggs (−0.26 v. 0.20 std. resid.). Incubation is short for an agamid (−0.07 v. −0.55 std. resid.), perhaps reflecting high soil temperatures. RCM ratio is low (0.186 v. 0.336), suggesting that the females are not unduly burdened by their small clutches. The hatchlings are comparatively, but not absolutely, large (0.09 v. 0.69 std. resid. for ontogenetic size range). There are no reports of parental care (v. 3.5%).

Longevity is average (−0.43 v. −0.46 std. resid.). Population density is a trifle low (mean of 0.09 v. 0.16 kg/ha). None has been introduced (v. 1.7%). Half of the evaluated species are believed to be imperiled (50% v. 37%), though none is on one of the IUCN's lists of the most endangered (v. 5%). In other respects *Ctenophorus* are typical agamids.

Ctenosaura

Ctenosaura (17 spp.) is a moderate genus of diurnal iguanids centered in southern Mexico and extending south to Colombia. Within Iguanidae it is anomalous in its mainland distribution and diversity of diet, ecological business models, and sizes. Although giant herbivores are the modal pattern (universal in other iguanid genera), 59% (v. 83%), there are probably also ground-foraging tree lizards (29% v. 12%) and diurnal canopy walkers (12% v. 5%). *Ctenosaura* are of relatively small size (388 g v. 609 g)—but note wide range (82–2810 g) and average build (shape index 2.00 v. 2.11 std. resid.)—with tails of average length (1.63 v. 1.53 × SVL) that are not breakable (0% v. 38%) or only breakable in juveniles. The majority have crests (59% v. 31%), which are usually much taller in males. Secretory pores are present at typical frequency (41% v. 38%). Color change is underrepresented (12% v. 21%), but advanced color change is overrepresented (6% v. 2.4%).

Sex differences in *Ctenosaura* follow the iguanid template relatively closely, but in most cases reflect slightly greater differences. For example, the SSD score is a little greater (1.45 v. 1.30 on −2 to 2 scale), sexual dichromatism is a bit greater (47% v. 38%), and the sex rollup score (on 0

Ctenosaura spp.	English name		Size		Range core
	Activity	Diet	mm	g	Ecological business model
acanthura	Northeastern Spiny-tailed Iguana				Mexico
	diurnal	herbivore	273	873	**** giant herbivore
alfredschmidti	Campeche Spiny-tailed Iguana				Mexico
	diurnal		133	103	* diurnal canopy walker
bakeri	Utila Spiny-tailed Iguana				Honduras
	diurnal	omnivore	225	494	** giant herbivore
clarki	Michoacan Club Tail				Mexico
	diurnal		188	288	* ground-foraging tree lizard
conspicuosa	San Esteban Spiny-tailed Iguana				Mexico
	diurnal	omnivore	226	502	** giant herbivore or ground-foraging tree lizard
defensor	Yucatán Spiny-tailed Iguana				Mexico
	diurnal	omnivore	123	82	* ground-foraging tree lizard
flavidorsalis	Yellow-backed Spiny-tailed Iguana				Honduras
	diurnal	herbivore	141	124	* ground-foraging tree lizard
hemilopha	Cape Spiny-tailed Iguana				Mexico
	diurnal	herbivore	340	1678	**** giant herbivore
macrolopha	Sonora Spiny-tailed Iguana				Mexico
	diurnal	herbivore	161	181	**** giant herbivore
melanosterna	Black-shouldered Spiny-tailed Iguana				Honduras
	diurnal	omnivore	239	591	* giant herbivore
nolascensis	San Pedro Nolasco Spiny-tailed Iguana				Mexico
	diurnal	herbivore	213	418	**** giant herbivore
oaxacana	Oaxaca Spiny-tailed Iguana				Mexico
	diurnal		131	98	* ground-foraging tree lizard
oedirhina	Roatan Spiny-tailed Iguana				Honduras
	diurnal		202	358	* ground-foraging tree lizard
palearis	Motagua Spiny-tailed Iguana				Guatemala
	diurnal	omnivore	404	2810	**** diurnal canopy walker
pectinata	Mexican Spiny-tailed Iguana				Mexico
	diurnal	herbivore	293	1085	**** giant herbivore
quinquecarinata	Five-keeled Spiny-tailed Iguana				Central America
	diurnal	omnivore	152	154	* giant herbivore
similis	Black Spiny-tailed Iguana				Central America
	diurnal	herbivore	336	1276	**** giant herbivore

to 6 scale) is 2.06 (v. 1.88). Seasonal color is present at typical frequency (11.8% v. 11.9%).

Unlike other iguanid genera, *Ctenosaura* includes some omnivores (46% v. 16%), probably reflecting the smaller sizes within the genus. Many species undergo an ontogenetic shift from omnivory to herbivory. When it is very dry, *nolascensis* will eat the stems of cholla cactus. *C. similis* is the only genus member for which foraging mode has been characterized, and it is said to be both sit-and-wait and active. Although active, this species is very refuge-centric, and most activities take place within 10 m of the refuge. It seems to be particularly fond of unmortared rock walls, such as are often found in Mayan ruins. They are described as quite wary where they are hunted by humans or dogs, but quickly become habituated to human proximity where they are not hunted and can remain near a refuge (e.g., heavily visited Mayan ruins). The genus averages 24% conspicuously wary (v. 29%). The primary predator concealment tactic is retreat to crevices or burrows (75% v. 64%); the remaining 25% (v. 36%) retreat to vegetation. *C. nolascensis* exhibits island tameness.

Despite their diurnality, some *Ctenosaura* give stress calls (5.9% v. 4.8%). In *oedirhina* the stress calls given in response to human approach result in the flight of all conspecifics within hearing range.

Solitariness is slightly overrepresented (14% v. 8%), and group living is present at low frequency (14% v. 42%). All males are said to defend space (v. 90%), though hierarchies are also widely reported (50% v. 25%). As both male tactics are reported for the same species, they likely vary with population density (see table 3.13, trait 14).

About half of *Ctenosaura* are each arboreal (53% v. 36%) and terrestrial (47% v. 62%). Burrow use for daily refugia is present in a few species (18% v. 44%). *Ctenosaura* are found primarily in the mesic rainfall band (75% v. 50%) and secondarily in the intermediate band (25% v. 50%). Their altitudinal distribution matches that of the family (l/m/h: 100/18/0 v. 100/19/0). They are slightly overrepresented in human habitats (29% v.

24%), and slightly underrepresented in littoral (29% v. 41%), sandy (24% v. 34%), and riparian (8% v. 22%) zones.

Reproduction reflects their mainland existence (see also *Iguana* account). Their clutches have the largest counts after *Iguana*: 1.74 (v. 0.45 for family) std. resid. Their ontogenetic size range is the largest in the family (0.43 v. –0.55 std. resid.), indicating particularly small hatchlings. Unlike the iguanas of remote oceanic islands (*Amblyrhynchus*, *Brachylophus*, *Conolophus*), *Ctenosaura* have the full suite of lizard predators to contend with, and there is no size beyond which the hatchlings become invulnerable to predation. Presumably, this is the basis for the numerous small eggs produced by *Ctenosaura* and the primarily mainland *Iguana*.

Incubation is short (–1.06 v. –0.76 std. resid.), probably to complete development within one season. Communal oviposition occurs (6% v. 14%). In *similis* (mean 34 eggs), the females may oviposit in a much-branched network of communal tunnels from which each female digs her own side branch and egg chamber.

Average mass-adjusted longevity is comparatively short (–0.35 v. 0.16 std. resid.), corresponding to 5.6 years in *clarki* but 60 years in *similis*. Population density is known for only one sample of one species (*palearis*), a species which is unusual in being both the largest species and yet not a giant herbivore (it is a diurnal canopy walker). At 0.3 kg/ha (v. 12.4 kg/ha), this mainland density is the lowest of any iguanid genus, but may be unrepresentative.

Endemism is about average in *Ctenosaura* (71% v. 69%). Forty-seven percent of species occur on islands (v. 79%). Introductions are relatively common (24% v. 33%). Seri Indians report that they introduced *clarki* to an island to ensure food for stranded fishermen. Released pets are probably the cause of populations in Arizona (*hemilopha*, *similis*?), Florida (*pectinata*, *similis*), Texas (*pectinata*), and the Bahamas (*similis*). Six percent (v. 9.5%) of *Ctenosaura* have been introduced more than twice. All *Ctenosaura* are considered imperiled (v. 97%), and about half are listed by the IUCN as among the most endangered (47% v. 51%). In other respects *Ctenosaura* are typical iguanids.

The large, widely distributed *similis* is the best-studied species, and it encapsulates the many ways in which humans interact with these large lizards. Local male human residents in several areas assert that consumption of this species—typically oviductal eggs or follicles—is a cure for impotence. Note the many colorful common names for this species in the database, such as Witchwilly. The species is believed to be depleted over much of its range by human consumption and dog predation; laying females are particularly vulnerable and attractive to dogs. Other species that are of conservation concern due to human consumption are *bakeri*, *conspicuosa*, *flavidorsalis*, *hemilopha*, *macrolopha*, *melanosterna*, *nolascensis*, *oedirhina*, *palearis*, and *quinquecarinata*. The latter two are also collected for the pet trade.

Ctenotus

Ctenotus is the second-largest skink genus (100 spp., after *Sphenomorphus*; table 2.2). Most species are medium-sized (3.9 g v. 3.3 g; range 0.9–26 g), hyperactive, heliophilic, longitudinally striped, diurnal (97% v. 74%) terrestrial (99% v. 48%) lizards of the Australian outback (*spaldingi* extends to New Guinea). They give an impression of being more like teiids than skinks. Like teiids, they are dominated by wide-ranging chemosensors (81% v. 6%); the remainder are refugia-anchored diurnal ground lizards (17% v. 30%) and litter cryptozoids (2.3% v. 28%). *Ctenotus* are sufficiently recognizable and unique that the genus name has become a common name.

Ctenotus are somewhat more robust than the average skink (shape index –0.06 v. –0.46 std. resid.). Tails are long (2.03 v. 1.33 × SVL) and breakable (97% v. 73%). About 9% of species are elongate (v. 35%). Bright blue tails are absent (v. 4%, a distinction from teiids), though *calurus* has a pastel blue tail that it waves while foraging—perhaps to signal to a predator that it has detected it—and *leae* has a red tail. Several species (*atlas*, *coggeri*, *inornatus*) are noted to have either skin or scales that flake off easily, presumably as an aid to escaping the grip of a predator.

SSD in *Ctenotus* is m<f (–1.0 v. –0.31 on –2 to 2 scale). For comparison, the teiid average is 0.89 (i.e., m>f). Sexual dichromatism is nearly absent in *Ctenotus* (1% v. 7.6%), and most other sex differences are absent; the sex difference rollup score for the genus is 0.10 v. 0.28 (on 0 to 6 scale). In teiids it is 1.19.

Most *Ctenotus* are invertebrate eaters (67% v. 84%); most others are omnivores (30% v. 13%). A few are carnivores (3.7% v. 1.0%). Termites play a big role in the diets of *Ctenotus*; they are mentioned in nearly half (13) of 28 species with dietary information. *C. calurus*, *inornatus*, *leonhardii*, *pantherinus*, and *schomburgkii* may be considered to be termite specialists. *C. catenifer* may consume nectar. Active foraging (v. 92%) is mentioned for all species for which foraging mode is described; sit-and-wait foraging is rare (3.2% v. 16%). Electric light use is not recorded. Home-range areas and quantitative movement measures are missing, except *taeniolatus* has a PTM in the 67th percentile.

Ctenotus that are not diurnal are

Ctenotus spp.	English name	Activity	Diet	Size mm	Size g	Range core	Ecological business model
agrestis	Grassplains Ctenotus	diurnal		77	9.1	Queensland	** refugia-anchored diurnal gr. liz.
alacer	Lively Ctenotus	diurnal		59	3.82	Australia	* wide-ranging chemosensor
alleni	Ajana Ctenotus	diurnal		83	11.6	Western Australia	* wide-ranging chemosensor
allotropis	Brown-blazed Wedgesnout Ctenotus	diurnal		50	2.36	Queensland	* wide-ranging chemosensor
angusticeps	Airlie Island Ctenotus	diurnal		66	5.71	Western Australia	** refugia-anchored diurnal gr. liz.
aphrodite	Aphrodite's Ctenotus	diurnal		64	5.1	Queensland	* wide-ranging chemosensor
arcanus	Arcane Ctenotus	diurnal		80	10.5	Queensland	* wide-ranging chemosensor
ariadnae	Ariadna's Ctenotus	diurnal	invert.	57	3.50	Australia	* wide-ranging chemosensor
arnhemensis	Jabiluka Ctenotus	diurnal		51	2.47	Northern Territory	* wide-ranging chemosensor
astarte	Stony Downs Ctenotus	diurnal		73	8.0	Queensland	* wide-ranging chemosensor
astictus	Elegant Ctenotus	diurnal		48	2.07	Northern Territory	* wide-ranging chemosensor
atlas	Southern Mallee Ctenotus	diurnal	invert.	64	5.0	New South Wales	* wide-ranging chemosensor
australis	Western Limestone Ctenotus	diurnal		93	17.0	Western Australia	* wide-ranging chemosensor
borealis	White-faced Ctenotus	diurnal		105	25.3	Northern Territory	* litter cryptozoid
brevipes	Short-footed Ctenotus	diurnal		53	2.84	Queensland	* wide-ranging chemosensor
brooksi	Wedgesnout Ctenotus	diurnal	omnivore	47	1.94	Western Australia	* wide-ranging chemosensor
burbidgei	Burbidge's Ctenotus	diurnal		53	2.84	Western Australia	* wide-ranging chemosensor
calurus	Blue-tailed Finesnout Ctenotus	diurnal	invert.	45	1.58	Western Australia	* wide-ranging chemosensor
capricorni	Capricorn Ctenotus	diurnal		59	3.87	Queensland	

mostly crepuscular (2.1% v. 2.4%); the large *pantherinus* (16 g) is cathemeral (1% v. 9%). Wariness is low (5% v. 11%). Primarily predator concealment tactics are diverse. Unlike teiids, the modal behavior is retreat to a crevice or burrow (71% v. 43% for skinks; 12% for teiids). One-seventh retreat to vegetation (14% v. 25%) or rely on flight (14% v. 4% for skinks, 77% for teiids). None is cryptic (v. 9%). None relies primarily on diving into sand (v. 22%), though *calurus* is capable of sand swimming. Direct basking is underrepresented (11% v. 21%) and may be underreported. Mean active body temperature is a hot 34.4°C (v. 31.1°C). No species is considered to be a thermoconformer (v. 1.5%).

Group living is not reported in *Ctenotus* (v. 16%); all documented males defend space (v. 94%). Male hierarchies are not known (v. 5.6%).

For nighttime retreats, burrows are mentioned in 90% (v. 24%); soil cracks are specifically recorded as a refuge in *agrestis, astarte, decaneurus, delli, joanae, militaris, pulchellus,* and *rimacolus*.

Hyperdry (<55 mm/y) and dry (56–150 mm/y) are not documented as habitats for *Ctenotus*, but they are strongly overrepresented in the arid (151–400 mm/y: 88% v. 40%). They are also present in the intermediate rainfall belt (401–1100 mm/y) but are underrepresented there (13% v. 34%). All documented species are found solely at low elevation (v. 89%). *Ctenotus* are notably overrepresented in sandy areas (67% v. 25%) and underrepresented in human habitats (3% v. 14%), potentially because human settlements are rare in the outback. Their occurrence in littoral areas (29% v. 21%) and riparian zones (18% v. 23%) is normal.

All *Ctenotus* are oviparous (v. 64%) and have average clutch counts (0.04 v. 0.06 std. resid.). Hatchlings are of average relative size (ontogenetic mass range –0.07 v. 0.12 std. resid.). Incubation is on the quick side (–0.70 v. –0.40 std. resid.), perhaps due to the hot sand. Eggshells are all soft (v. 9% hard). Communal oviposition (v. 2.6%) and parental care (v. 3.2%) are not known. RCM is average (0.220 v. 0.236).

Longevity is notably short (–0.29 v. 0.17 std. resid.), equating to 8.9 years in the large (17 g) *australis* and 7 years in the more normal (5.8 g) *leonhardii*. Due to the highly unpredictable productivity of the outback environment, one might expect higher longevity to ensure that a lizard will live long enough to breed during a year of good recruitment. Population density is a bit low (0.18 kg/ha v. 0.35 kg/ha), perhaps reflecting the low productivity of their environment. Population density is exceptionally well documented in *Ctenotus*, with credible estimates for eight species (one to three estimates for each species). The density range is 0.02 to 0.70 kg/ha (i.e., consistently low or medium).

Endemism is normal (58% v. 64%). Few islands occur in the range of this genus (22% of species occur on at least one, v. 57%). No *Ctenotus* has been introduced (v. 2.1%). Most are considered to be imperiled (70% v. 47%), but none is on one of the IUCN's lists of the most endangered (v. 14%). In other respects *Ctenotus* are similar to average skinks.

Ctenotus spp.	English name				Range core
			Size		
	Activity	Diet	mm	g	Ecological business model
catenifer	Chain-striped Southwest Ctenotus				Western Australia
	diurnal		52	1.60	
coggeri	Brown-backed Ctenotus				Northern Territory
	diurnal		75	4.35	
colletti	Buff-tailed Finesnout Ctenotus				Western Australia
	diurnal	invert.	42	0.91	
decaneurus	Ten-lined Ctenotus				Northern Australia
	crepuscular		49	2.22	* wide-ranging chemosensor
delli	Darling Range Southwest Ctenotus				Western Australia
	diurnal		56	1.98	* wide-ranging chemosensor
duricola	Duricola Ctenotus				Western Australia
	diurnal	invert.	55	3.05	* wide-ranging chemosensor
dux	Fine Side-lined Ctenotus				Australia
	diurnal	omnivore	58	3.67	* wide-ranging chemosensor
ehmanni	Ehmann's Ctenotus				Western Australia
	diurnal	invert.	39	1.07	* wide-ranging chemosensor
essingtonii	Lowland Plain-backed Ctenotus				Northern Territory
	diurnal		60	4.18	* wide-ranging chemosensor
euclae	Nullarbor Ctenotus				Southwestern Australia
			49	2.18	* wide-ranging chemosensor
eurydice	Brown-backed Yellow-lined Ctenotus				Queensland
	diurnal		69	3.46	* wide-ranging chemosensor
eutaenius	Black-backed Yellow-lined Ctenotus				Queensland
	diurnal		81	10.8	* wide-ranging chemosensor
gagudju	Magela Ctenotus				Northern Territory
	diurnal		50	2.24	* wide-ranging chemosensor
gemmula	Jewelled Southwest Ctenotus				Western Australia
	diurnal		53	1.70	* wide-ranging chemosensor
grandis	Grant Ctenotus				Western Australia
	diurnal	omnivore	97	19.7	* wide-ranging chemosensor
greeri	Spotted-necked Ctenotus				Western Australia
	diurnal		56	3.38	* wide-ranging chemosensor
halysis	Chained Ctenotus				Western Australia
	diurnal		53	2.74	* wide-ranging chemosensor
hanloni	Nimble Ctenotus				Western Australia
	diurnal		65	5.5	* wide-ranging chemosensor
hebetior	Stout Ctenotus				Queensland
	diurnal		56	3.23	* wide-ranging chemosensor
hilli	Top-end Lowlands Ctenotus				Northern Territory
	diurnal	invert.	47	1.91	* litter cryptozoid

Ctenotus spp.	English name				Range core
			Size		
	Activity	Diet	mm	g	Ecological business model
iapetus	Titan Ctenotus				Western Australia
	diurnal	invert.	62	4.69	* wide-ranging chemosensor
impar	Odd-striped Ctenotus				Western Australia
	diurnal	invert.	61	4.27	* wide-ranging chemosensor
ingrami	Unspotted Yellow-sided Ctenotus				Queensland
	diurnal		75	8.4	* wide-ranging chemosensor
inornatus	Bar-shouldered Ctenotus				Australia
	diurnal	invert.	82	5.5	* refugia-anchored diurnal gr. liz. or litter cryptozoid (geog. var.)
joanae	Black-soil Ctenotus				Queensland
	diurnal		74	4.2	* wide-ranging chemosensor
kurnbudj	Kurnbudj Ctenotus				Northern Territory
	diurnal		50	2.34	* wide-ranging chemosensor
labillardieri	Common Southwest Ctenotus				Western Australia
	diurnal		65	5.4	* wide-ranging chemosensor
lancelini	Lancelin Island Ctenotus				Western Australia
	diurnal		70	3.56	* wide-ranging chemosensor
lateralis	Gravelly-soil Ctenotus				Queensland
	diurnal		77	9.1	* wide-ranging chemosensor
leae	Orange-tailed Finesnout Ctenotus				Australia
	diurnal	omnivore	55	3.14	* wide-ranging chemosensor
leonhardii	Leonhardi's Ctenotus				Australia
	diurnal	omnivore	66	5.8	* wide-ranging chemosensor
maryani	Maryan's Ctenotus				Western Australia
	diurnal		55	3.08	
mastigura	Whiptail Ctenotus				Western Australia
	diurnal		72	7.5	* wide-ranging chemosensor
mesotes	Median-striped Ctenotus				Western Australia
	diurnal		39	1.06	
militaris	Soldier Ctenotus				Western Australia
	diurnal		57	3.57	* wide-ranging chemosensor
mimetes	Checker-sided Ctenotus				Western Australia
	diurnal		73	7.9	* wide-ranging chemosensor
monticola	Atherton Ctenotus				Queensland
	diurnal		59	3.87	* wide-ranging chemosensor
nasutus	Nasute Finesnout Ctenotus				Western Australia
	diurnal	invert.	43	1.41	
nigrilineatus	Pin-striped Finesnout Ctenotus				Western Australia
	diurnal		45	1.68	

Ctenotus spp.	English name				Range core
			Size		
	Activity	Diet	mm	g	Ecological business model
nullum	Nullum Ctenotus				Queensland
	diurnal	carnivore	70	7.0	
olympicus	Olympic Ctenotus				Australia
	diurnal		64	5.2	
ora	Coastal Plains Skink				Western Australia
	diurnal		56	3.23	
orientalis	Oriental Ctenotus				Australia
	diurnal		72	7.3	
pallescens	Northwestern Wedgesnout Ctenotus				Northern Territory
	diurnal		43	1.45	* wide-ranging chemosensor
pantherinus	Leopard Ctenotus				Australia
	cathemeral	invert.	92	16.2	* wide-ranging chemosensor
piankai	Coarse Sands Ctenotus				Australia
	diurnal	omnivore	51	2.41	** refugia-anchored diurnal gr. liz.
pulchellus	Red-sided Ctenotus				Queensland
	diurnal		72	7.3	* wide-ranging chemosensor
quattuordecim-lineatus	Fourteen-lined Ctenotus				Australia
	diurnal	omnivore	60	4.20	* wide-ranging chemosensor
quinkan	Quinkan Ctenotus				Queensland
	diurnal		67	5.9	* wide-ranging chemosensor
quirinus	Arnhem Plain-backed Ctenotus				Northern Territory
	diurnal		60	4.08	* wide-ranging chemosensor
rawlinsoni	Cape Heath Ctenotus				Queensland
	diurnal		72	7.6	* wide-ranging chemosensor
regius	Pale-rumped Ctenotus				Australia
	diurnal	omnivore	68	6.1	** refugia-anchored diurnal gr. liz.
rimacolus	Fissure Ctenotus				Western Australia
	diurnal		81	10.9	* wide-ranging chemosensor
robustus	Robust Ctenotus				Australia
	diurnal	invert.	106	26.0	* refugia-anchored diurnal gr. liz. or crack hermit
rosarium	Rosy Ctenotus				Queensland
	diurnal		41	1.20	
rubicundus	Ruddy Ctenotus				Western Australia
	diurnal		90	15.2	* wide-ranging chemosensor
rufescens	Rufous Finesnout Ctenotus				Western Australia
	diurnal		42	1.33	* wide-ranging chemosensor
rutilans	Rusty-shouldered Ctenotus				Western Australia
	diurnal		50	2.32	* wide-ranging chemosensor

Ctenotus spp.	English name				Range core
			Size		
	Activity	Diet	mm	g	Ecological business model
schevilli	Black-soil Rises Ctenotus				Queensland
	diurnal		76	8.7	* wide-ranging chemosensor
schomburgkii	Barred Wedge-snout Ctenotus				Australia
	diurnal	invert.	46	1.81	* wide-ranging chemosensor
septenarius	Massive-gibber Ctenotus				Western Australia
	crepuscular		64	5.0	* wide-ranging chemosensor
serotinus	Gravel-downs Ctenotus				Queensland
	diurnal		47	1.91	* wide-ranging chemosensor
serventyi	Serventy's Ctenotus				Western Australia
	diurnal		52	2.54	* wide-ranging chemosensor
spaldingi	Straight-browed Ctenotus				Eastern Australia
	diurnal	invert.	78	9.7	* wide-ranging chemosensor
storri	Buff-striped Ctenotus				Northern Territory
	diurnal		39	1.03	** refugia-anchored diurnal gr. liz.
strauchii	Eastern Barred Wedgesnout Ctenotus				Eastern Australia
	diurnal	invert.	53	2.82	*** refugia-anchored diurnal gr. liz.
striaticeps	Stripe-headed Finesnout Ctenotus				Queensland
	diurnal		47	1.91	** refugia-anchored diurnal gr. liz.
stuarti	Stuart's Ctenotus				Northern Territory
	diurnal		50	2.24	* wide-ranging chemosensor
superciliaris	Sharp-browed Ctenotus				Western Australia
			67	6.0	* wide-ranging chemosensor
taeniatus	Ribbon Ctenotus				Queensland
			49	2.18	** refugia-anchored diurnal gr. liz.
taeniolatus	Copper-tailed Ctenotus				Queensland
	diurnal	invert.	71	7.2	* wide-ranging chemosensor
tanamiensis	Tanami Ctenotus				Western Australia
	diurnal		83	11.8	** refugia-anchored diurnal gr. liz.
tantillus	Kimberley Wedgesnout Ctenotus				Western Australia
	diurnal		44	1.47	** refugia-anchored diurnal gr. liz.
terrareginae	Hinchinbrook Ctenotus				Queensland
	diurnal		81	10.8	* wide-ranging chemosensor
uber	Spotted Ctenotus				Western Australia
	diurnal		68	6.1	** refugia-anchored diurnal gr. liz.
vagus	Uneven-striped Ctenotus				Western Australia
	diurnal		43	1.37	* wide-ranging chemosensor

Ctenotus spp.	English name				Range core
			Size		
	Activity	Diet	mm	g	Ecological business model
vertebralis	Scant-striped Ctenotus				Northern Territory
	diurnal	invert.	51	2.51	** refugia-anchored diurnal gr. liz.
xenopleura	Wide-striped Ctenotus				Western Australia
	diurnal		47	1.88	** refugia-anchored diurnal gr. liz.
youngsoni	Shark Bay Southwest Ctenotus				Western Australia
	diurnal		67	6.0	* wide-ranging chemosensor
zastictus	Hamlin Pool Ctenotus				Western Australia
	diurnal		56	3.23	* wide-ranging chemosensor
zebrilla	Southern Cape York Finesnout Ctenotus				Queensland
	diurnal		39	1.02	* wide-ranging chemosensor

Cyclodomorphus

Cyclodomorphus is a moderate radiation (nine spp.) of moderately large (8.7 g v. 3.3 g), very slender (shape index –1.18 v. –0.46 std. resid.), elongate (100% v. 35%), mostly omnivorous (78% v. 13%) viviparous (100% v. 36%) Australian skinks of diverse ecologies. They are found primarily in riparian areas (67% v. 23%). They may be litter cryptozoids (33% v. 28%), nocturnal ground lizards (33% v. 2.3%), refugia-anchored diurnal ground lizards (22% v. 30%), or ground-foraging tree lizards (11% v. 7%). Some of the litter cryptozoids are notably large to be hiding in litter, but their habitats often include dense ground-level vegetation (e.g., spinifex grass tuffs) that provides adequate cover for relatively large lizards. *Cyclodomorphus* are relatively well known (filled cells 40% v. 25%).

Cyclodomorphus are related to *Tiliqua*, and many species, like those in *Tiliqua*, sport blue tongues, which they reveal conspicuously as a predator defense. Unlike the much larger *Tiliqua* (40 g), *Cyclodomorphus* do not (literally) have large heads and huge masseter muscles capable of delivering a memorable bite. The protruding blue tongue in *Cyclodomorphus* probably conveys detection of the predator, a willingness to bite, and identity as one of the feared blue-tongue lizards. In the case of *Cyclodomorphus*, this identification may rely on mimicry of *Tiliqua* rather than actual menace to be effective.

Cyclodomorphus have a conspicuously wide ontogenetic range of masses (0.74 v. 0.12 std. resid.), indicating exceptionally small neonates. This is consistent with the large litter counts in this genus (0.73 v. 0.06 std. resid.). This pattern of many small young is a strong contrast with the closely related *Tiliqua*. Perhaps it is size mediated, as the larger *Tiliqua* hatchlings may be able to grow quickly to an invulnerable size (e.g., too big for kookaburras), whereas *Cyclodomorphus* may not have that option. A contested maximum litter count for *gerrardii* is 67; an uncontested one for *casuarinae* is 19. *Cyclodomorphus* may have many small young.

Tails are short (0.98 v. 1.33 × SVL), breakable (100% v. 73%), and may be (*gerrardii*) prehensile (11% v. 0.7%) but not blue (0% v. 4%).

Females are notably larger than males in *Cyclodomorphus*; –1.25 v. –0.31 on –2 to 2 scale. This is unexpected for an omnivore (table 3.5, trait 9), and especially unusual for large lizards (table 3.3, trait 1), but it is consistent with the apparent selection pressure for high fecundity in *Cyclodomorphus* females (see above). The sexes differ in color at a usual frequency (11% v. 8%), but do not show seasonal color (v. 6%). However the body size differences between the sexes are often large, and therefore the sex difference rollup score for the genus is the fourth highest in Scincidae: 1.56 v. 0.28 on 0 to 6 scale.

As noted above, most *Cyclodomorphus* are omnivorous; the remainder are invertebrate eaters (22% v. 84%). At least six species eat hard objects such as mollusks; *gerrardii* is specialized for this diet. It uses hard objects such as rocks for a tool to crack the snail shells and discards the shell fragments before eating the succulent body parts. All documented species are active foragers (v. 92%). None has been recorded feeding at electric lights (v. 0.4%), and home-range sizes have not been documented.

The majority of species are cathemeral (50% v. 9%); others are diurnal (25% v. 74%) or nocturnal (25% v. 14.5%), though Cogger (2014) characterizes the genus as nocturnal/crepuscular. One-third are conspicuously wary (33% v. 11%). Predator defense concealment is not well documented. *C. celatus* dives into sand; when cornered *casuarinae* mimics Australia's dangerous reptiles—varanids and elapids—by rapidly tongue-flicking and locomoting without use of limbs. The semiarboreal *gerrardii* locomotes with only front limbs when on the ground, but uses all four and its prehensile tail when climbing. Direct basking is overrepresented in the genus (33% v. 21%); thermoconforming is not known (v. 1.5%). Mean

Cyclodomorphus spp.	English name		Size		Range core
	Activity	Diet	mm	g	Ecological business model
branchialis	Common Slender Bluetongue				Western Australia
	nocturnal	omnivore	79	3.07	*** nocturnal ground lizard
casuarinae	Tasmanian She-oak Skink				Tasmania
	cathemeral	omnivore	136	11.7	*** refugia-anchored diurnal gr. liz.
celatus	Western Slender Bluetongue				Western Australia
	nocturnal	omnivore	98	5.2	*** nocturnal ground lizard
gerrardii	Pink-tongued Lizard				Queensland
	cathemeral	omnivore	191	27.0	** ground-foraging tree lizard
maximus	Giant Slender Bluetongue				Western Australia
	cathemeral	omnivore	185	25.1	** litter cryptozoid
melanops	Spinifex Slender Bluetongue				Australia
	diurnal	invert.	99	5.4	** litter cryptozoid
michaeli	Mainland She-oak Skink				New South Wales
	cathemeral	invert.	148	14.3	** nocturnal ground lizard
praealtus	Alpine She-oak Skink				New South Wales
	diurnal	omnivore	117	8.0	** refugia-anchored diurnal gr. liz. or litter cryptozoid
venustus	Saltbush Slender Bluetongue				Queensland
		omnivore	86	3.75	** litter cryptozoid

active body temperature is a warm 32.8°C (v. 31.1°C). *C. melanops* is found in the arid rainfall belt.

Cyclodomorphus are overrepresented in riparian zones (as noted in the first paragraph), as well as in littoral areas (44% v. 21%) and human habitat (33% v. 14%). They are of average frequency in sandy areas (33% v. 25%). At least two species (*branchialis*, *venustus*) are noted to use soil cracks as a daily refuge.

Based on four species, longevity is high (1.06 v. 0.17 std. resid.), equating to 39 years in *branchialis*. This fits oddly with the several lines of evidence suggesting great selective pressure for high annual fecundity (see third paragraph of this account). All five documented *Cyclodomorphus* produce one litter per year, so clutch count is probably equivalent to annual fecundity. Long-lived lizards are characterized by few young per clutch or litter (table 3.12, trait 25). Population density is comparatively high (1.30 kg/ha v. 0.35 kg/ha), albeit based on just one sample (*melanops*). Endemism is low (44% v. 64%). Few species occur on islands (44% v. 57%). None has been introduced (v. 2.1%). Most are imperiled (83% v. 47%), and an average proportion are on the IUCN's lists of the most endangered (17% v. 14%). In other respects *Cyclodomorphus* are typical skinks.

As used above in several common names, a "she-oak" is vernacular for trees of the genus *Casuarina*. However, these lizards are not especially connected to the tree. One species (*casuarinae*) is however connected to a schooner named after the tree. The schooner's voyage resulted in the first collection of this species, though the original collector died before completing the species description, as did the person who received the specimen from the deceased. One might think that the hexed name would be buried along with the unlucky herpetologists who bequeathed the specimen, but it stuck and has spread to several related lizards.

Cyclura (fig. 4.4F)

Cyclura (nine spp.) is a genus of diurnal terrestrial giant herbivores (Iguanidae) found in the West Indies. Thanks largely to the heroic exertions of members of the IUCN West Indian Iguana Specialist Group, these lizards are among the best known (filled cells 56% v. 43%).

Cyclura are giants (2109 g v. 609 g) and heavy in build (shape index 2.41 v. 2.11 std. resid.) with shortish tails (1.29 v. 1.53 × SVL), which are generally breakable (89% v. 38%), though perhaps breakable only in juveniles. The two lizard species with the heaviest absolute masses at modal SVL are in *Cyclura*: *pinguis* (7000 g) and *cornuta* (5100 g). The three species with the highest shape index values are also in *Cyclura*: *pinguis* (2.84 std. resid.), *cornuta* (2.74), and *cychlura* (2.55). It is possible that the modal size in these three species has been shifted upwards by indirect anthropogenic predation selectively on juveniles (by dogs, cats, mongoose, etc.), as these three species are imperiled by these introduced predators. All *Cyclura* have crests (v. 31%), though these are generally sexually monomorphic in *Cyclura*. Secretory pores are present (22% dimorphic v. 38%). Color change is underrepresented (11% v. 21%), and advanced color change absent (v. 2.4%).

Sex differences in *Cyclura* follow the iguanid template relatively closely. The SSD score is typical (1.43 v. 1.30 on –2 to 2 scale); sexual dichromatism is a bit lower (33%

Cyclura spp.	English name		Size		Range core
	Activity	Diet	mm	g	Ecological business model
carinata	Turks Island Iguana				Bahamas
	diurnal	herbivore	235	500	**** giant herbivore
collei	Jamaican Iguana				Jamaica
	diurnal	herbivore	315	1338	*** giant herbivore
cornuta	Rhinoceros Iguana				Hispaniola
	diurnal	herbivore	480	5100	**** giant herbivore
cychlura	Northern Bahamian Iguana				Bahamas
	diurnal	herbivore	429	3362	**** giant herbivore
lewisi	Grand Cayman Blue Iguana				Cayman Islands
	diurnal	herbivore	402	2764	**** giant herbivore
nubila	Cuban Iguana				Cuba
	diurnal	herbivore	410	2750	**** giant herbivore
pinguis	Stout Iguana				British Virgin Islands
	diurnal	herbivore	530	7000	**** giant herbivore
ricordi	Ricord's Iguana				Hispaniola
	diurnal	herbivore	351	1853	**** giant herbivore
rileyi	White Cay Ground Iguana				Bahamas
	diurnal	herbivore	272	730	**** giant herbivore

v. 38%), and the sex rollup score is 1.78 (v. 1.88) on 0 to 6 scale. Seasonal color is absent (v. 11.9%).

All species are herbivores (v. 84%), though most will take small quantities of invertebrates opportunistically, especially when young. They have been described as both cruise (50% v. 29%) and active foragers (50% v. 71%). Home-range size is small (–1.58 v. –1.06 std. resid.), though home-range measurements may fail to capture some excursions for seasonal fruit (see also comments under *Ctenosaura*). Observers of *lewisi* have suggested the possibility of airborne olfactory communication among adult males, eliciting exceptional male movements.

Despite their diurnality, some *Cyclura* give stress calls (11% v. 4.8%). One-fifth of these edible iguanas is considered to be conspicuously wary (22% v. 29%). They are smart enough to adapt to prevailing hunting practices and habituate selectively to benign humans. Their primary predator concealment tactic is retreat to a crevice or burrow (75% v. 64%) or retreat to vegetation (25% v. 36%). They bask directly in typical iguanid frequency (56% v. 55%), with a mean active body temperature of 34.8°C (v. 34.3°C).

Solitariness is not reported (v. 8%), and group living is present (50% v. 42%). All males are said to defend space (v. 90%), though hierarchies are also present (14% v. 25%).

A few *Cyclura*, especially juveniles, are arboreal (11% v. 36%), but 89% are terrestrial (v. 62%). Retreat to a daily burrow is common (67% v. 44%). Many species (but not *carinata*) readily enter salt water to seek refuge or travel among islands. Juveniles may run bipedally. They are found primarily in the intermediate rainfall band (100% v. 50%) and secondarily in the mesic band (33% v. 50%). They are found only below 300 m. They are slightly overrepresented in littoral (67% v. 41%) and sandy (78% v. 34%) habitats, and they are absent from riparian zones (v. 22%), which are relatively scarce on small low islands.

Reproduction is similar to most other iguanids, though their incubation is substantially shorter than the short family mean (–1.28 v. –0.76 std. resid.). One might suppose that very large iguanas living on islands (*Cyclura*) would follow the large-young/few-in-number pattern seen in iguanas of remote oceanic islands (*Amblyrhynchus*, *Brachylophus*, *Conolophus*), but they do not. The predator suite found in West Indian islands may have been more complete over evolutionary time (large raptors and opportunistic mammals, some now extinct) and for the past century has gained a number of key predators on terrestrial lizards (e.g., mongoose). Clutch count in *Cyclura* is low (–0.68 v. 0.45 std. resid.), and young are comparatively large (–0.19 v. –0.55 std. resid.), but they are in the middle of the iguanid pack.

Communal oviposition is common (33% v. 14%), and parental care dominates (67% v. 31%), mostly in defense of nest sites on islands where deep loose soil is in short supply. The amount of time spent guarding ranges from a few days (*carinata*, *rileyi*) or several weeks (*collei*, *nubila*) to 2 months (*cornuta*) or the hatching of the young in termitaria (*cychlura*). Mating systems in *Cyclura* seem adapted to the paucity of suitable nest sites. Males of several species (*cychlura*, *nubila*) guard burrows or optimal nest sites and use that resource to elicit copulations or acquire monogamous mates (*pinguis*). In other cases, dominant males simply guard females (*rileyi*). Subordinate males attempt forced copulations (*nubila*, *rileyi*). In situations where suitable nest sites are not close to the good places to live, females may migrate up to 1 km to nest (*cornuta, rileyi*). RCM is average (0.243 v. 0.231 ratio). An interesting aspect of incubation period in *Cyclura* is that young may remain in the nest cavity for days or weeks after hatching. This may be an opportunity to complete yolk absorption and carry out their first shed in subterranean safety. This phenomenon may be widespread but is as yet undetected in most other taxa; it is known in *Iguana*.

Average mass-adjusted longevity is normal (0.02 v. 0.16 std. resid.), corresponding to 3.2 years in *pinguis*—undoubtedly an underestimate—and 69 years in *nubila*. Population density is remarkably well known for the five species documented ($n = 2$ to 15 per sp.). Species mean densities range from 7.6 kg/ha (*cornuta*) to 51 kg/ha (*carinata*); genus mean is 15.1 kg/ha (v. 12.4 kg/ha). Under good conditions these island herbivores reach impressive densities.

Endemism is about average in *Cyclura* (56% v. 69%). All species occur on islands (v. 79%). Introductions are relatively common (56% v. 33%), often for conservation purposes (*carinata, cychlura, nubila, pinguis, rileyi*). Released pets are probably the cause of a population in Arizona (*carinata*). All *Cyclura* are considered imperiled (v. 97%), and two-thirds are listed by the IUCN as among the most endangered (67% v. 51%). The threats to these island populations are the usual: mongoose, cats, dogs, urbanization, and habitat degradation by introduced ungulates. In other respects *Cyclura* are typical iguanids.

Cynisca

Cynisca is a major (20 spp.) radiation of tiny (0.5 g v. 2.7 g), thread-like (shape index –4.81 v. –3.97 std. resid.) amphisbaenids at low elevation in West Africa (Senegal to

Cynisca spp.	English name				Range core
			Size		
	Activity	Diet	mm	g	Ecological business model
bifrontalis	Omboue Worm Lizard				Gabon
			103	0.26	**** subterranean forager
chirioi	Chirio's Worm Lizard				Guinea
			216	5.13	**** subterranean forager
degrysi	Sierra Leone Worm Lizard				Sierra Leone
			95	0.19	
feae	Fea's Worm Lizard				Senegal
			129	0.65	**** subterranean forager
gansi	Gans' Worm Lizard				Nigeria
			123	0.53	**** subterranean forager
haugi	Haug's Worm Lizard				Gabon
			98	0.21	**** subterranean forager
ivoirensis	Ivory Coast Worm Lizard				Ivory Coast
		invert.	105	0.28	**** subterranean forager
kigomensis	Kigom Worm Lizard				Nigeria
			131	0.68	**** subterranean forager
kraussi	Ghana Worm Lizard				Ghana
			124	0.55	**** subterranean forager
leonina	Archipelago Worm Lizard				Guinea
			120	0.48	**** subterranean forager
leucura	Coast Worm Lizard				Ghana
			194	3.28	**** subterranean forager
liberiensis	Liberia Worm Lizard				Sierra Leone
			122	0.52	**** subterranean forager
manei	Mane's Worm Lizard				Senegal
			102	0.25	**** subterranean forager
muelleri	Mueller's Worm Lizard				Ghana
			134	0.65	**** subterranean forager
nigeriensis	Nigerian Worm Lizard				Nigeria
		invert.	133	0.73	**** subterranean forager
oligopholis	Cassine River Worm Lizard				Guinea-Bissau
			144	0.40	**** subterranean forager
rouxae	Roux's Worm Lizard				Ivory Coast
			98	0.21	**** subterranean forager
schaeferi	Cameroon Worm Lizard				Cameroon
			150	1.18	**** subterranean forager
senegalensis	Senegal Worm Lizard				Senegal
			108	0.32	**** subterranean forager
williamsi	Williams' Worm Lizard				Ghana
			124	0.56	**** subterranean forager

Central African Republic and south to Gabon). They are associated with riparian areas (36% v. 20%). Endemism is high (80% v. 66%). Island species are low (5% v. 12%). *Cynisca* are imperiled at a slightly higher rate than amphisbaenids overall (33% v. 22%), including placement on some of the IUCN's most endangered lists (20% v. 13%). In other respects they are similar.

Cyrtodactylus (figs. 4.9D, 4.10D, 4.14C)

Cyrtodactylus is a major radiation of lizards (third-largest genus), and the largest gekkonid genus (232 spp.; table 2.2). Morphologically they are distinguished from other gekkonids primarily by their larger size (mean 7.6 g v. 3.3 g). They are poorly known (21% v. 25% of cells filled), perhaps at least in part because they are mostly nocturnal (97% v. 73%). This proportion of cells filled does not fully capture the lack of information, which is spotty. For example, information—sometimes contested—is available on the diel activity pattern in 93% of *Cyrtodactylus* but dietary information for only 16%. Information on ecological business model is lacking for 47%, and ecological business models can be assigned with full confidence in only 8%. The modal business model is trunk/wall lizard (56% v. 45%), and the second most-frequent pattern is nocturnal ground lizard (23% v. 27%). Other reported models include nocturnal canopy walker (12% v. 9%), ground-foraging tree lizard (7% v. 5%), crack hermit (0.8% v. 3.2%), and litter cryptozoid (0.8% v. 0.5%).

All *Cyrtodactylus* are found in the Oriental (82% v. 43%) and Australasian biogeographic realms (19% v. 12%). The highest concentrations of species are found in areas recently worked by Lee Grismer, though not entirely due to his efforts: Malay-

Cyrtodactylus spp.	English name				Range core
			Size		
	Activity	Diet	mm	g	Ecological business model
aaroni	Aaron's Bent-toed Gecko				Indonesia
	nocturnal		82	10.9	
adleri	Nicobar Bent-toed Gecko				India
	nocturnal		59	4.38	
adorus	Pascoe River Ring-tailed Gecko				Queensland
	nocturnal		107	22.8	
aequalis	Mon State Bent-toed Gecko				Myanmar
	nocturnal		74	8.2	
agamensis	Agam Bow-fingered Gecko				Indonesia
			45	2.10	
agusanensis	Agusan Bent-toed Gecko				Philippines
	nocturnal		85	12.2	** trunk/wall lizard
albofasciatus	Canara Gecko				India
	nocturnal		64	5.5	* nocturnal ground lizard
angularis	Angled Bent-toed Gecko				Thailand
	nocturnal	invert.	77	9.4	**** nocturnal ground lizard
annandalei	Annandale's Bent-toed Gecko				Myanmar
	nocturnal		49	2.62	
annulatus	Small Bent-toed Gecko				Philippines
	nocturnal		67	6.4	** ground-foraging tree lizard
arcanus	Dagbaru Bent-toed Gecko				New Guinea
	nocturnal		75	8.7	
astrum	Starry Bent-toed Gecko				Malaysia
	nocturnal		87	12.8	
aurensis	Aur Island Bent-toed Gecko				Malaysia
	nocturnal		80	10.3	
auribalteatus	Golden-bellied Bent-toed Gecko				Thailand
	nocturnal		79	10.1	
australotitiwangsaensis	Southern Titiwangsa Bent-toed Gecko				Malaysia
	nocturnal		95	16.6	
ayeyarwadyensis	Ayeyarwady Bent-toed Gecko				Myanmar
	cathemeral	invert.	65	5.8	** trunk/wall lizard
badenensis	Ba Den Bent-toed Gecko				Vietnam
	nocturnal		62	5.2	** trunk/wall lizard
baluensis	Kinabalu Bent-toed Gecko				Borneo
	nocturnal	invert.	81	10.7	** nocturnal canopy walker
bansocensis	Ban Soc Bent-toed Gecko				Laos
			62	5.2	
batik	Batik Bent-toed Gecko				Indonesia
	nocturnal		92	14.9	

Cyrtodactylus spp.	English name	Activity	Diet	Size mm	Size g	Range core	Ecological business model
batucolus	Besar Island Bent-toed Gecko	nocturnal		63	5.4	Malaysia	**** trunk/wall lizard
bichnganae	Bich Ngan's Bent-toed Gecko	nocturnal		81	10.6	Vietnam	
bidoupimontis	Bidoup Bent-toed Gecko	nocturnal		71	7.4	Vietnam	
bintangrendah	Bintang Lowland Bent-toed Gecko	nocturnal		91	14.6	Malaysia	
bintangtinggi	Bintang Summit Bent-toed Gecko	nocturnal		89	13.7	Malaysia	
biordinis	Guadalcanal Banded Palm Gecko	nocturnal	invert.	82	11.0	Solomon Islands	*** nocturnal canopy walker
bobrovi	Bobrov's Bent-toed Gecko			78	9.6	Vietnam	
boreoclivus	Foja Mountains Bent-toed Gecko	nocturnal		87	13.1	New Guinea	
brevidactylus	Map Bent-toed Gecko	nocturnal		72	7.8	Myanmar	
brevipalmatus	Web-footed Bent-toed Gecko	nocturnal		66	6.2	Thailand	** nocturnal canopy walker or ground-foraging tree lizard
buchardi	Xepian Bent-toed Gecko	nocturnal		56	3.82	Laos	
bugiamapensis	Bu Gia Map Bent-toed Gecko	nocturnal		66	6.0	Vietnam	
calamei	Calame's Bent-toed Gecko	nocturnal		75	8.8	Laos	** trunk/wall lizard
caovansungi	Cao Van Sung's Bent-toed Gecko	nocturnal		77	9.1	Vietnam	
capreoloides	Roe Bent-toed Gecko	nocturnal		69	7.0	Papua New Guinea	
cattienensis	Cat Tien Bent-toed Gecko	nocturnal	invert.	59	4.38	Vietnam	** ground-foraging tree lizard
cavernicolus	Niah Cave Gecko	nocturnal	carnivore	67	6.4	Malaysia	*** trunk/wall lizard
celatus	Unseen Gecko	nocturnal		41	1.61	Indonesia	
chanhomeae	Saraburi Cave Gecko	cathemeral		66	6.0	Thailand	*** trunk/wall lizard
chauquangensis	Chau Quang Gecko	nocturnal		80	10.3	Vietnam	*** trunk/wall lizard

sia (43 spp.), Vietnam (38 spp.), Thailand (35 spp.), and Indonesia (35 spp.). Species numbers decline to the west (Myanmar, 20 spp.; India, 16 spp.), east (Philippines, 9 spp.), south (Queensland, 5 spp.), and north (China, 3 spp.). Islands are not particularly speciose: Borneo (10), Mindanao (3), New Guinea (14), and Sumatra (7).

Prehensile tails are present at typical levels (2.2% v. 1.8%), but secretory pores are underrepresented (34% v. 46%) as is color change (4.7% v. 14.6%). The majority of species lack expanded toepads (56% v. 30% padless).

SSD is more female biased than in a typical gekkonid (–0.33 v. –0.03 on –2 to 2 scale), a trend which is unexpected for lizards that are above average in size. Sexual dichromatism is not reported (v. 6.6%). The mean ratio of m:f relative tail lengths is a bit short (0.98 v. 1.05), which is inconsistent with the m<f SSD (table 3.3, trait 8). Taken together, these sex differences are weaker than those of a typical gekkonid (0.35 v. 0.63 on 0 to 6 scale).

As noted above, dietary information on *Cyrtodactylus* is sparse: 97% (v. 95%) are categorized as invertebrate eaters, with all others considered to be carnivores (unsurprising given their size). There are no herbivores (v. 0.3%), ant specialists (v. 0%), or omnivores (v. 3.8%). Their distribution of foraging modes is very close to that of all gekkonids (genus is 57% sit-and-wait, 29% cruise, and 14% active). Electric light use is rare in *Cyrtodactylus* (0.4% v. 3.8%), despite the dominance of trunk/wall lizards. Size-adjusted home-range areas are exceptionally small (–0.99 v. –0.28 std. resid.), though documentation exists for only *philippinicus*, an atypical *Cyrtodactylus*. *C. philippinicus* is an uncertain ground-foraging tree lizard.

Despite their overwhelming nocturnality, few *Cyrtodactylus* call spontaneously (2.6% v. 6.5%) or under duress (2.6% v. 3.7%), potentially a reflection of inadequate information. For the same reason, it is possible that wariness (0.9% v. 4.4%) is underreported, though large nocturnal geckos are often truculent and unskittish. As a predator concealment tactic, crypsis is underreported (22% v. 50%) and hiding in vegetation dominant (44% v. 19%). Both are consistent with nocturnality.

Group living is not reported (v. 25%), nor is solitary living (v. 7.7%) or burrow refugium use (v. 10%). All documented species live in mesic areas (v. 66%; 1101–3000 mm/y). The elevational distribution (96/14/2) is very similar to that of gekkonids (92/26/2). They are overrepresented in riparian zones (22% v. 17%) but underrepresented in littoral (4% v. 11%), sandy (1% v. 11%), and human (9% v. 27%) habitats.

Mass-adjusted clutch count is low (–0.58 v. –0.26 std. resid.), though numerically it adheres to the gekkonid standard of two eggs. Mass-adjusted ontogenetic size range is very low (–0.77 v. –0.19 std. resid.), indicating exceptionally large hatchlings, which is not expected for lizards that are larger than average. Mass-adjusted incubation duration is lengthy (0.94 v. 0.17 std. resid.), but most other reproductive values are in line with the family. Adherent or communally laid eggs are expected among arboreal or trunk/wall lizards (table 4.1, traits 67, 68) but are not so found in *Cyrtodactylus* (0% v. 10.5% adherent; 0% v. 8.6% communally laid).

Mass-adjusted longevity is reported to be slightly short (–0.13 v. 0.15 std. resid.), but that may reflect sparse data (*n* = 4). *Cyrtodactylus* have not been introduced (v. 3.7%). Twenty-

Cyrtodactylus spp.	English name				Range core
			Size		
	Activity	Diet	mm	g	Ecological business model
chrysopylos	Shan State Bent-toed Gecko				Myanmar
	nocturnal		66	6.0	
collegalensis	Kollegal Ground Gecko				India
	crepuscular	invert.	50	2.88	* litter cryptozoid
condorensis	Con Dao Bent-toed Gecko				Vietnam
	nocturnal		71	7.5	
consobrinoides	Dawei Bent-toed Gecko				Myanmar
	nocturnal	invert.	45	2.12	
consobrinus	Giant Bent-toed Gecko				Malaysia
	nocturnal	invert.	106	22.1	**** trunk/wall lizard
cracens	Sinharaja Bent-toed Gecko				Sri Lanka
	nocturnal		92	15.2	
cryptus	Dark-collared Bent-toed Gecko				Vietnam
	nocturnal		75	8.8	** trunk/wall lizard
cucdongensis	Cucdong Bent-toed Gecko				Vietnam
	nocturnal		57	3.96	
cucphuongensis	Cuc Phuong Bent-toed Gecko				Vietnam
	nocturnal		86	12.4	
darevskii	Darevsky's Bent-toed Gecko				Laos
	nocturnal		81	10.6	
darmandvillei	Darmandville Bow-fingered Gecko				Indonesia
	nocturnal		71	7.4	
dati	Dat Bent-toed Gecko				Vietnam
	nocturnal		59	4.53	
deccanensis	Deccan Ground Gecko				India
	nocturnal		70	7.2	* nocturnal ground lizard
derongo	Derongo Bent-toed Gecko				New Guinea
	nocturnal		95	16.6	
deveti	Morotai Bent-toed Gecko				Indonesia
	nocturnal		84	11.9	
doisuthep	Doi Suthep Bent-toed Gecko				Thailand
	nocturnal		74	8.4	
dumnuii	Dumnui's Bent-toed Gecko				Thailand
	nocturnal		69	7.0	
durio	Durian Bent-toed Gecko				Malaysia
	nocturnal		66	6.0	* nocturnal canopy walker
edwardtaylori	Edward Taylor's Gecko				Sri Lanka
	nocturnal		86	12.4	*** trunk/wall lizard
eisenmanae	Eisenman's Bent-toed Gecko				Vietnam
	nocturnal		73	8.0	* trunk/wall lizard

Cyrtodactylus spp.	English name				Range core
			Size		
	Activity	Diet	mm	g	Ecological business model
elok	Beautiful Bent-toed Gecko				Malaysia
	nocturnal		68	6.6	*** nocturnal canopy walker or trunk/wall lizard
epiroticus	East End Bent-toed Gecko				New Guinea
	nocturnal		106	22.6	** trunk/wall lizard
equestris	Large Sandaun Bent-toed Gecko				New Guinea
			109	24.0	
erythrops	Red-eyed Bent-toed Gecko				Thailand
	nocturnal		65	5.8	
fasciolatus	West Himalayan Rock Gecko				India
	nocturnal	invert.	68	6.6	*** nocturnal ground lizard
feae	Fea's Bent-toed Gecko				Myanmar
	nocturnal	invert.	43	1.86	* trunk/wall lizard
fraenatus	Sri Lanka Bent-toed Gecko				Sri Lanka
	cathemeral	invert.	84	11.9	**** trunk/wall lizard
fumosus	Tamarind Forest Gecko				Indonesia
	nocturnal		60	4.71	** trunk/wall lizard
gansi	Gans' Bent-toed Gecko				Myanmar
	nocturnal		54	3.56	
gordongekkoi	Sendanggila Falls Gecko				Indonesia
	nocturnal		62	5.0	* trunk/wall lizard
grismeri	Grismer's Bent-toed Gecko				Vietnam
	nocturnal		77	9.4	* trunk/wall lizard
guakanthanensis	Kanthan Cave Bent-toed Gecko				Malaysia
	nocturnal		68	6.6	
gubaot	Leyte Bent-toed Gecko				Philippines
	nocturnal		81	10.6	*** ground-foraging tree lizard
gubernatoris	Eastern Himalaya Bent-toed Gecko				India
	nocturnal	invert.	47	2.41	** nocturnal ground lizard
gunungsenyumensis	Gunung Senyum Bent-toed Gecko				Malaysia
	nocturnal		63	5.3	** trunk/wall lizard
halmahericus	Halmahera Spotted Gecko				Indonesia
	nocturnal		62	5.0	
hidupselamanya	Chiku Bent-toed Gecko				Malaysia
	nocturnal		151	60.0	**** trunk/wall lizard
hikidai	Bunguran Bent-toed Gecko				Indonesia
	nocturnal		82	11.1	
hinnamnoensis	Hinnamno Bent-toed Gecko				Laos
	nocturnal		82	10.9	** trunk/wall lizard

nine percent of species are imperiled (v. 47%), and 4% (v. 11.3%) are listed by the IUCN in the most at-risk categories. In other respects *Cyrtodactylus* are typical gekkonids.

Cyrtodactylus spp.	English name	Range core	Activity	Diet	Size mm	Size g	Ecological business model
hitchi	Mekongga Mountains Gecko	Indonesia			66	6.0	** trunk/wall lizard
hontreensis	Hon Tre Bent-toed Gecko	Vietnam	nocturnal		75	8.8	** trunk/wall lizard
hoskini	Hoskin's Ring-tailed Gecko	Queensland	nocturnal		98	18.1	* trunk/wall lizard
huongsonensis	Huong Son Bent-toed Gecko	Vietnam	nocturnal		74	8.2	
huynhi	Chua Can Mountain Bent-toed Gecko	Vietnam	nocturnal		67	6.2	* trunk/wall lizard
ingeri	Inger's Bent-toed Gecko	Borneo	nocturnal		67	6.2	* nocturnal canopy walker
interdigitalis	Nao Nam Bent-toed Gecko	Thailand	nocturnal		67	6.2	**** nocturnal ground lizard
intermedius	Intermediate Bow-fingered Gecko	Southeast Asia	nocturnal	invert.	75	8.7	**** nocturnal ground lizard
inthanon	Doi Inthanon Bent-toed Gecko	Thailand	nocturnal		72	7.6	
irianjayaensis	Roesler's Mystery Bent-toed Gecko	Thailand	nocturnal		132	41.2	
irregularis	Irregular Bent-toed Gecko	Vietnam	nocturnal		74	8.2	
jaegeri	Jaeger's Gecko	Laos	nocturnal		59	4.38	** trunk/wall lizard
jambangan	Jambangan Bent-toed Gecko	Philippines	nocturnal		68	6.6	
jarakensis	Jarak Island Bent-toed Gecko	Malaysia	nocturnal		57	4.10	
jarujini	Jarujin's Bent-toed Gecko	Thailand	nocturnal		74	8.2	**** trunk/wall lizard
jelawangensis	Jelawang Bent-toed Gecko	Malaysia	nocturnal		95	16.6	* trunk/wall lizard
jellesmae	Kabaena Bent-toed Gecko	Indonesia	nocturnal		55	3.69	* ground-foraging tree lizard
jeyporensis	Patinghe Indian Gecko	India	nocturnal		52	3.18	
khammouanensis	Khammouane Bent-toed Gecko	Laos	nocturnal		62	5.0	
khasiensis	Khasi Hills Bent-toed Gecko	Himalayas	nocturnal	invert.	74	8.2	* nocturnal ground lizard

Cyrtodactylus spp.	English name	Range core	Activity	Diet	Size mm	Size g	Ecological business model
khelangensis	Lampang Bent-toed Gecko	Thailand	nocturnal		77	9.4	
kimberleyensis	East Montalivet Island Gecko	Western Australia	nocturnal		42	1.69	
kingsadai	Kingsada's Gecko	Vietnam	nocturnal		77	9.1	
klakahensis	Klakah Bent-toed Gecko	Indonesia			58	4.24	
klugei	Sudest Bent-toed Gecko	New Guinea	nocturnal		112	25.8	* trunk/wall lizard
kunyai	Kunya's Bent-toed Gecko	Thailand	nocturnal		72	7.8	
laevigatus	Komodo Bent-toed Gecko	Indonesia	nocturnal		43	1.86	
langkawiensis	Langkawi Bent-toed Gecko	Malaysia	nocturnal		81	10.6	
lateralis	Sumatra Bent-toed Gecko	Indonesia	nocturnal		70	7.2	
lawderanum	Lawder's Bent-toed Gecko	India	nocturnal	invert.	49	2.62	** nocturnal ground lizard
leegrismeri	Tenggol Island Bent-toed Gecko	Malaysia	nocturnal		75	8.7	**** trunk/wall lizard
lekaguli	Waterfall Bent-toed Gecko	Thailand	nocturnal		84	11.7	* trunk/wall lizard
lenggongensis	Lenggong Bent-toed Gecko	Malaysia	nocturnal		83	11.4	*** trunk/wall lizard
lenya	Lenya Banded Bent-toed Gecko	Myanmar			62	5.2	* trunk/wall lizard
lomyenensis	Lomyen Bent-toed Gecko	Laos	nocturnal		61	4.84	
loriae	Boulenger's Bent-toed Gecko	New Guinea	nocturnal		121	32.1	
louisiadensis	Ring-tailed Gecko	New Guinea	nocturnal	invert.	114	27.1	*** nocturnal canopy walker
macrotuberculatus	Raya Mountain Gecko	Malaysia	nocturnal		96	17.1	**** trunk/wall lizard
majulah	Singapore Bent-toed Gecko	Indonesia	nocturnal		59	4.38	
malayanus	Borneo Bent-toed Gecko	Borneo	nocturnal		99	18.7	*** trunk/wall lizard

Cyrtodactylus spp.	English name				Range core
			Size		
	Activity	Diet	mm	g	Ecological business model
malcolmsmithi	Smith's Bent-toed Gecko				India
	nocturnal		49	2.62	
mamanwa	Mamanwa Forest Gecko				Philippines
	nocturnal		75	8.7	** trunk/wall lizard
mandalayensis	Mandalay Bent-toed Gecko				Myanmar
	nocturnal		54	3.43	
markuscombaii	Comba's Gecko				Nepal
	nocturnal		61	4.84	*** trunk/wall lizard
marmoratus	Marbled Bow-fingered Gecko				New Guinea
	nocturnal		68	6.6	* nocturnal canopy walker
martini	Martin's Bent-toed Gecko				Vietnam
	nocturnal		78	9.6	
martinstolli	Stoll's Gecko				Nepal
	nocturnal		67	6.4	*** trunk/wall lizard
matsuii	Matsui's Bent-toed Gecko				Malaysia
	nocturnal	invert.	84	11.9	** ground-foraging tree lizard
mcdonaldi	McDonald's Ring-tailed Gecko				Queensland
	nocturnal		91	14.6	* trunk/wall lizard
medioclivus	Wanakipa Bent-toed Gecko				New Guinea
	nocturnal		83	11.4	
metropolis	Batu Caves Gecko				Malaysia
	nocturnal		68	6.6	
mimikanus	Mimika Bent-toed Gecko				New Guinea
	nocturnal		83	11.4	
minor	Huon Dwarf Bent-toed Gecko				New Guinea
	nocturnal		61	4.84	** nocturnal canopy walker
multiporus	Multipored Bent-toed Gecko				Laos
	nocturnal		79	10.1	
murua	Woodlark Bent-toed Gecko				New Guinea
	nocturnal		97	17.3	** trunk/wall lizard
nebulosus	Clouded Ground Gecko				India
	nocturnal	invert.	48	2.52	*** nocturnal ground lizard
nepalensis	Nepalese Rock Gecko				Nepal
	nocturnal		62	5.0	*** trunk/wall lizard
nigriocularis	Black-eyed Bent-toed Gecko				Vietnam
	nocturnal		87	12.8	** trunk/wall lizard
novaeguineae	New Guinea Giant Bent-toed Gecko				New Guinea
	nocturnal		132	41.2	* trunk/wall lizard
nuaulu	Manusela Bent-toed Gecko				Indonesia
	nocturnal		73	8.0	

Cyrtodactylus spp.	English name				Range core
			Size		
	Activity	Diet	mm	g	Ecological business model
oldhami	Oldham's Bent-toed Gecko				Myanmar
	nocturnal	invert.	65	5.8	**** nocturnal ground lizard
otai	Ota's Bent-toed Gecko				Vietnam
	nocturnal		74	8.4	
pageli	Pagel's Bent-toed Gecko				Laos
	nocturnal		73	8.0	
pantiensis	Panti Mountain Bent-toed Gecko				Malaysia
	nocturnal		64	5.7	**** ground-foraging tree lizard
papilionoides	Butterfly-marked Gecko				Thailand
	nocturnal		76	8.9	**** nocturnal ground lizard
papuensis	Papua Dwarf Bent-toed Gecko				New Guinea
	nocturnal		62	4.98	
paradoxus	Paradoxical Bent-toed Gecko				Vietnam
	nocturnal		73	7.9	** trunk/wall lizard
payacola	Swampland Bent-toed Gecko				Malaysia
	nocturnal	invert.	58	4.24	
payarhtanensis	Tenasserim Mountain Bent-toed Gecko				Myanmar
	diurnal		69	6.8	* trunk/wall lizard
peguensis	Pegu Forest Gecko				Thailand
	nocturnal	invert.	71	7.3	*** nocturnal ground lizard
petani	Farmers' Bent-toed Gecko				Indonesia
			50	2.84	
phetchaburiensis	Khao Tomo Bent-toed Gecko				Thailand
			54	3.56	** trunk/wall lizard
philippinicus	Philippine Bent-toed Gecko				Borneo
	nocturnal	invert.	79	10.1	* ground-foraging tree lizard
phongnhake-bangensis	Phong Nha-Ke Bang Gecko				Vietnam
	nocturnal	invert.	83	11.4	** trunk/wall lizard
phuketensis	Phuket Bent-toed Gecko				Thailand
	nocturnal		92	14.9	
phuocbinhensis	Phuoc Binh Gecko				Vietnam
	nocturnal		52	3.18	** trunk/wall lizard
phuquocensis	Phu Quoc Bent-toed Gecko				Vietnam
	nocturnal		71	7.4	* crack hermit
pronarus	McIlwraith Ring-tailed Gecko				Queensland
	nocturnal		115	28.1	* trunk/wall lizard
psarops	Speckle-faced Bent-toed Gecko				Indonesia
			68	6.6	
pseudoquadri-virgatus	False Lines Bent-toed Gecko				Vietnam
	nocturnal	invert.	2	7.8	

Cyrtodactylus spp.	English name	Activity	Diet	Size mm	Size g	Range core	Ecological business model
pubisulcus	Grooved Bent-toed Gecko	nocturnal	invert.	68	6.6	Borneo	** nocturnal canopy walker
puhuensis	Pu Hu Gecko	nocturnal		66	6.0	Vietnam	
pulchellus	Banded Bent-toed Gecko	nocturnal	invert.	99	18.4	Thailand	**** trunk/wall lizard
quadrivirgatus	Four-striped Forest Gecko	nocturnal	invert.	62	5.1	Malaysia	*** nocturnal ground lizard
ramboda	Ramboda Bent-toed Gecko	nocturnal		86	12.7	Sri Lanka	*** trunk/wall lizard
ranongensis	Ranong Bent-toed Gecko	nocturnal		52	3.18	Thailand	
redimiculus	Palawan Bent-toed Gecko	nocturnal		77	9.1	Philippines	* nocturnal ground lizard
rex	King Bent-toed Gecko			132	41.2	New Guinea	* trunk/wall lizard
rishivalleyensis	Rishi Valley Bent-toed Gecko	nocturnal		48	2.52	India	*** nocturnal ground lizard
robustus	Rossel Bent-toed Gecko	nocturnal		124	34.8	New Guinea	** trunk/wall lizard
roesleri	Roesler's Bent-toed Gecko	nocturnal		63	5.3	Vietnam	
rosichonariefi	Rosichonarief's Bent-toed Gecko	nocturnal		49	2.62	Indonesia	
rubidus	Red Bow-fingered Gecko	nocturnal	invert.	67	6.2	India	* trunk/wall lizard
rufford	Rufford Bent-toed Gecko			62	5.0	Laos	
russelli	Russell's Bent-toed Gecko	nocturnal		92	15.3	Myanmar	
sadleiri	Christmas Island Forest Gecko	nocturnal	invert.	83	10.5	Christmas Island (Australia)	* ground-foraging tree lizard
saiyok	Sai Yok Bent-toed Gecko	nocturnal		53	3.3	Thailand	* nocturnal canopy walker
salomonensis	Solomon Giant Gecko	nocturnal		11	25.3	New Guinea	** nocturnal canopy walker
samroiyot	Three Hundred Peaks Gecko	nocturnal		57	4.1	Thailand	
sanook	Sanook Fun Gecko	nocturnal		67	6.2	Thailand	

Cyrtodactylus spp.	English name				Range core
			Size		
	Activity	Diet	mm	g	Ecological business model
semenanjungensis	Peninsula Bent-toed Gecko				Malaysia
	nocturnal		59	4.49	*** nocturnal canopy walker
semiadii	Semiadi's Gecko				Indonesia
	nocturnal		43	1.87	
semicinctus	Kerinci Bent-toed Gecko				Indonesia
			73	8.0	
seribuatensis	Seribuat Bent-toed Gecko				Malaysia
	nocturnal		62	5.2	**** nocturnal ground lizard
sermowaiensis	DeRooij's Bent-toed Gecko				New Guinea
	nocturnal		76	9.0	
serratus	Sandaun Bent-toed Gecko				New Guinea
	nocturnal		124	34.3	** nocturnal canopy walker
sharkari	Sharkar's Gecko				Malaysia
	nocturnal		81	10.6	
slowinskii	Slowinski's Bent-toed Gecko				Myanmar
	nocturnal		87	12.8	
soba	Knuckles Bent-toed Gecko				Sri Lanka
	cathemeral	invert.	86	12.7	**** trunk/wall lizard
sommerladi	Sommerlad's Bent-toed Gecko				Laos
	nocturnal		68	6.7	** trunk/wall lizard
soni	Son's Bent-toed Gecko				Vietnam
	nocturnal		83	11.4	*** trunk/wall lizard
soudthichaki	Soudthichak's Bent-toed Gecko				Laos
			59	4.53	
speciosus	Forest Spotted Gecko				India
			48	2.52	** nocturnal ground lizard
spelaeus	Khuang Lang Cave Gecko				Laos
	nocturnal		74	8.43	* trunk/wall lizard
spinosus	Spiny Bent-toed Gecko				Indonesia
	nocturnal		69	6.8	
srilekhae	Bangalore Bent-toed Gecko				India
	nocturnal		45	2.16	*** nocturnal ground lizard
stresemanni	Stresemann's Bent-toed Gecko				Malaysia
			81	10.6	
subsolanus	Rakwana Bent-toed Gecko				Sri Lanka
	nocturnal		92	15.2	** trunk/wall lizard
sumonthai	Sumontha's Cave Gecko				Thailand
	nocturnal		68	6.7	*** trunk/wall lizard
sumuroi	Sumuroy's Gecko				Philippines
	nocturnal		70	7.2	** trunk/wall lizard

Cyrtodactylus spp.	English name		Size		Range core
	Activity	Diet	mm	g	Ecological business model
surin	Surin Island Gecko				Thailand
	nocturnal		67	6.2	
sworderi	Kota-Tinggi Forest Gecko				Malaysia
	nocturnal		67	6.2	** ground-foraging tree lizard
takouensis	Ta Kou Bent-toed Gecko				Vietnam
	nocturnal		67	6.4	** trunk/wall lizard
tamaiensis	Tamai Bent-toed Gecko				Myanmar
	nocturnal		74	8.2	
tautbatorum	Mainit Falls Gecko				Philippines
	nocturnal		59	4.38	
taynguyenensis	Central Highlands Gecko				Vietnam
	nocturnal		70	7.2	
tebuensis	Tebu Mountain Gecko				Malaysia
	nocturnal		69	7.0	
teyniei	Teynié's Bent-toed Gecko				Laos
	nocturnal		74	8.2	
thirakhupti	Thirakhupt's Cave Gecko				Thailand
	nocturnal		67	6.2	** trunk/wall lizard
thochuensis	Tho Cho Gecko				Vietnam
	nocturnal		71	7.4	
thuongae	Ba Den Mountain Gecko				Vietnam
	nocturnal		65	5.8	
tibetanus	Tibetan Thin-toed Gecko				China
	nocturnal		47	2.34	* nocturnal ground lizard
tigroides	Tiger Bent-toed Gecko				Thailand
	nocturnal		70	7.2	*** nocturnal ground lizard
timur	Banjaran Timur Gecko				Malaysia
	nocturnal		96	16.9	* nocturnal ground lizard
tiomanensis	Tioman Island Bent-toed Gecko				Malaysia
	nocturnal	invert.	75	8.7	** trunk/wall lizard
triedrus	Sir Lanka Spotted Gecko				Sri Lanka
	cathemeral	invert.	54	3.45	*** nocturnal ground lizard
trilatofasciatus	Cameron Highlands Bent-toed Gecko				Malaysia
	nocturnal		97	17.3	
tripartitus	Misima Bent-toed Gecko				New Guinea
	nocturnal		115	28.1	*** trunk/wall lizard
tuberculatus	Tuberculated Ring-tailed Gecko				Queensland
	nocturnal	invert.	120	31.5	*** nocturnal ground lizard
varadgirii	Giri's Gecko				India
	nocturnal		47	2.40	**** nocturnal ground lizard

Cyrtodactylus spp.	English name / Activity	Diet	Size mm	Size g	Range core / Ecological business model
variegatus	Moulmein Forest Gecko				Myanmar
	nocturnal	invert.	60	4.68	*** nocturnal ground lizard
vilaphongi	Kanyasone's Gecko				Laos
	nocturnal		71	7.4	
wakeorum	Wakes' Bent-toed Gecko				Myanmar
	nocturnal		55	3.69	
wallacei	Wallace's Bent-toed Gecko				Indonesia
	nocturnal		91	14.6	**** trunk/wall lizard
wangkulangkulae	Wangkulangkul's Gecko				Thailand
	nocturnal		62	5.2	
wayakonei	Kao Rao Cave Gecko				Laos
	nocturnal		76	8.9	
wetariensis	Wetar Bent-toed Gecko				Indonesia
	nocturnal		59	4.53	* nocturnal canopy walker or trunk/wall lizard
yakhuna	Blotch Bent-toed Gecko				Sri Lanka
	nocturnal	invert.	44	2.01	**** nocturnal ground lizard
yangbayensis	Yangbay Bent-toed Gecko				Vietnam
	nocturnal		78	9.8	
yoshii	Yoshi's Bent-toed Gecko				Malaysia
	nocturnal		84	11.7	** trunk/wall lizard
zhaoermii	Alpine Bent-toed Gecko				China
	nocturnal		55	3.18	* nocturnal ground lizard
ziegleri	Ziegler's Bent-toed Gecko				Vietnam
	nocturnal		76	8.9	* trunk/wall lizard
zugi	Zug's Bent-toed Gecko				Indonesia
	nocturnal		123	33.7	*** trunk/wall lizard

Cyrtopodion

Cyrtopodion is a 27-species genus of the Gekkonidae in the deserts of southwest Asia (especially Iran and Pakistan), with one species (*medogense*) extending into China and another (*scabrum*) spreading as far west as Sudan, possibly with human assistance. Most are nocturnal ground lizards (88% v. 27%), and many of these readily adapt to electric lighting, becoming trunk/wall lizards in the process (only 6% are primarily trunk/wall lizards v. 45% in family).

They are unremarkable in size (2.35 g v. 3.31 g) and build (shape index 0.04 v. 0.10 std. resid.). Most species' bodies are highlighted with tubercles, and their tails are characterized by spines at intervals. Tails are a little long (1.25 v. 1.07 × SVL), and most are breakable (93% v. 87%) but not prehensile (v. 1.8%). Secretory pores are overrepresented (56% v. 46%) and color change rare (3.7% v. 15%). Virtually all (93% v. 30%) are documented as padless. Their lack of enlarged toepads and presence of long legs and large heads are presumably responsible for the "spider gecko" moniker. They are somewhat similar in appearance and activity to the similarly named *Cyrtodactylus*, but they are more terrestrial than the mostly arboreal *Cyrtodactylus* and more disposed toward deserts, whereas *Cyrtodactylus* are found in mesic forests.

SSD has been noted, but only for its absence, in only two species (*scabrum*, *watsoni*), suggesting a general lack of sex differences in size. *Cyrtopodion* also lack reports of color differences or seasonal color.

All species with documented diets are invertebrate eaters (v. 95%) that forage by sit-and-wait (v. 81%). Electric light use is especially common (18.5% v. 3.8%). Home-range area is documented only in *scabrum*, for which it is an unremarkable 7–36 m^2 (−0.36 v. −0.28 std. resid.).

Cathemeral (18.5% v. 10.8%), crepuscular (3.7% v. 1.0%), and nocturnal (78% v. 73%) habits are overrepresented; diurnality is unknown (v. 15%). Accordingly, vocalizations are relatively frequent: spontaneous, 14.8% (v. 6.5%), and stress, 7.4% (v. 3.7%). *C. agamuroides* calls are thought to be audible by humans 100 m away. Wariness is present at gekkonid-typical frequency (3.7% v. 4.4%). All species with data conceal themselves from predators by escaping into burrows or crevices (v. 31%).

Males defend space (100% v. 87%), and none is socially tolerant (v. 13%), group-living (v. 25%), hierarchical (v. 0%), or documented to be solitary (v. 7.7%). In other words, most appear to live in spatially defended pairs.

Arboreality is uncommon (13% v. 68%); terrestriality is dominant (88% v. 31%), but for such terrestrial lizards, use of burrows for daily refugia is not widely reported (only 29%, though the family mean is even lower at 10%). *Cyrtopodion* do not stand out for residence in littoral habitats (6% v. 11%), sandy habitats (12% v. 11%), or riparian zones (24% v. 17%), but they are notably common in human habitats (53% v. 27%).

Reproduction hews to the gekkonid pattern, though incubation is notably rapid (–0.21 v. 0.17 std. resid.), probably reflecting high summer soil temperatures in this region. One species (*fortmunroi*) is said to have anomalously soft eggshells. Communal egg deposition is relatively common (18.5% v. 8.6%), and parental care is not reported (v. 1.9%). One would be eager to identify the traits that characterize the communal egg deposition sites; persistent high humidity seems likely to be the key scarce resource in this region.

Longevity is relatively short (–0.21 v. 0.15), but based on only *scabrum*. Island species are understandably rare (4% v. 45%). Introductions are gekkonid average (3.7% v. 3.7%), though their penchant for electric light feeding and human habitats puts them at greater risk of introduction. No *Cyrtopodion* is considered imperiled (v. 47%). In other respects *Cyrtopodion* are typical gekkonids.

Cyrtopodion spp.	English name				Range core
			Size		
	Activity	Diet	mm	g	Ecological business model
agamuroides	Makran Spider Gecko				Iran
	cathemeral	invert.	45	2.12	*** nocturnal ground lizard
aravallensis	Delhi Rock Gecko				India
	nocturnal		46	2.22	
baigii	Chilas Rock Gecko				Pakistan
	nocturnal		43	1.82	
battalense	Reticulate Plump-bodied Gecko				Pakistan
	nocturnal	invert.	55	3.69	*** nocturnal ground lizard
belaense	Bela Rock Gecko				Pakistan
	nocturnal		38	1.36	
brevipes	Blanford's Short-toed Gecko				Iran
	nocturnal		41	1.61	** nocturnal ground lizard
dattanense	Plump Rock Gecko				India
	nocturnal	invert.	54	3.43	** nocturnal ground lizard
fortmunroi	Fort Munro Sandstone Gecko				Pakistan
	nocturnal	invert.	45	2.12	*** trunk/wall lizard
gastrophole	Farsian Spider Gecko				Iran
	nocturnal		46	2.27	
golubevi	Sistan Spider Gecko				Iran
	nocturnal		52	3.07	
himalayanum	Himalaya Spider Gecko				India
	nocturnal		64	5.5	
hormozganum	Hormozgan Spider Gecko				Iran
	nocturnal		42	1.77	
indusoani	Soan Gecko				Pakistan
	cathemeral	invert.	48	2.52	** nocturnal ground lizard
kachhense	Kachh Gecko				Pakistan
	cathemeral	invert.	43	1.86	**** nocturnal ground lizard
kiabii	Bushehr Spider Gecko				Iran
	nocturnal		43	1.85	
kirmanense	Kerman Spider Gecko				Iran
	crepuscular		46	2.22	** refugia-anchored diurnal gr. liz.
kohsulaimanai	Sulaiman Range Gecko				Pakistan
	nocturnal	invert.	52	3.07	*** nocturnal ground lizard or trunk/wall lizard (around elec. lights)
mansarulus	Jammu Spider Gecko				India
	nocturnal		50	2.50	
medogense	Medog Spider Gecko				China
	nocturnal		37	1.19	

Cyrtopodion spp.	English name				Range core
			Size		
	Activity	Diet	mm	g	Ecological business model
montiumsalsorum	Salt Range Gecko				Pakistan
	nocturnal	invert.	43	1.87	** nocturnal ground lizard or trunk/wall lizard (around elec. lights)
persepolense	Persepolis Bent-toed Gecko				Iran
	nocturnal		46	2.22	
potoharense	Potohar Gecko				Pakistan
	nocturnal		47	2.31	* nocturnal ground lizard or trunk/wall lizard (around elec. lights)
rhodocauda	Rosy-tailed Sandstone Gecko				Pakistan
	cathemeral		55	3.69	* nocturnal ground lizard
rohtasfortai	Rohtas Gecko				Pakistan
	nocturnal		47	2.41	** nocturnal ground lizard or trunk/wall lizard (around elec. lights)
scabrum	Rough-tailed Gecko				Afghanistan
	nocturnal	invert.	51	3.04	**** nocturnal ground lizard or trunk/wall lizard (around elec. lights)
sistanense	Sistan Bent-toed Gecko				Iran
	nocturnal		50	2.84	* nocturnal ground lizard
watsoni	Quetta Rock Gecko				Pakistan
	cathemeral	invert.	48	2.49	*** nocturnal ground lizard

Dactylocnemis sp.	English name				Range core
			Size		
	Activity	Diet	mm	g	Ecological business model
pacificus	Pacific Gecko				New Zealand
	nocturnal	omnivore	82	11.2	** nocturnal canopy walker

Dactylocnemis

Dactylocnemis pacificus is the sole representative of this diplodactylid genus. Found on many islands in the north of New Zealand, this scansorial gecko is both uniquely tolerant (v. 24%) of other lizards and uniquely wary (v. 2.7%), even after long-term residence in captivity. Illustrating its tolerance for other lizards, it has been seen sharing flower heads while lapping nectar, and as many as 10 individuals have been seen simultaneously drinking from a pool of water. This implies to me an extraordinarily high population density, yet such groupings of lizards are claimed to be asocial, as the groupings do not persist beyond the transient situation that independently attracted the individuals. Most of the time they remain within the confines of their refuge.

Dactylocnemis pacificus is larger than a typical diplodactylid (11.2 g v. 4.2 g). It has a prehensile tail (v. 26%). Males are larger than females (1.0 v. −0.65 on −2 to 2 scale), but no other sex differences are known. It eats fruit, nectar, and invertebrates (v. 15% diplodactylid omnivores). It is primarily nocturnal (v. 79%), though it is not known to call spontaneously (v. 5.4%) or under duress (v. 5.4%). It will bask directly (v. 17%), but only from within the shelter of its crevice. It is found in the intermediate rainfall zone (401–1100 mm/y, v. 33%) and in littoral areas (v. 17%), but it is not reported in sandy areas (v. 24%), riparian zones (v. 10%), or human habitats (v. 5%). Like most New Zealand lizards, it is viviparous (v. 18%) and endemic (v. 72%). It has not been introduced (v. 2%) or evaluated for conservation risk. In other respects it is a typical diplodactylid.

Dactyloidae (figs. 4.3A, 4.6D)

The 416 dactyloids are presently all in the genus *Anolis*. Thus a description of the family is a description of the genus. This entry will focus on the ways in which anoles differ from other lizard families; the *Anolis* entry explores the variation within the genus. Because so much high-quality work has been done on the ecomorphology of anoles, one would be excused for thinking that anoles are particularly well documented, but the evidence is otherwise (filled cells 28% v. 27%). In comparison to other lizards, anoles have a higher proportion of species so poorly documented that a business model cannot be guessed (35% v. 25%). The proportion of *Anolis* species in which the ecological business model was assigned with full confidence—four stars in species listings—was not notably better than lizards in general (16% v. 13%). The diet class was assignable in only 35% of anoles (v. 44% of lizards). In other words, ecologists do not know what 65% of anoles eat.

Anoles are overrepresented in three ecological business models: ground-foraging tree lizards (42% v. 19%), trunk/wall lizards (28% v. 13%), and diurnal canopy walkers (21% v. 6%). Some anoles are also refugia-anchored diurnal ground lizards (8.2% v. 24.5%), immobility experts (0.4% v. 3.2%), and litter cryptozoids (0.4% v. 11.6%). They are found throughout the Neotropics and into the southern United States. They are also widely introduced around the Pacific.

Anole size range covers almost the entirety of lizards (0.71 g to 65 g; 4.5th to 96th percentile), but has a smaller than average mean (3.86 g v. 5.33 g). Their shape index is average (0.13 v. 0.0 std. resid.), and their tails are a bit long (1.86 v. 1.37 × SVL). Only 18% of anole tails are reported to break (v. 57%), though tail breakage in *Anolis* is underreported because it is expected to be nearly universal. Only 2.6% of anole tails are reported to be prehensile (v. 5.0%).

The signature morphological feature of anoles is their erectile dewlap, which varies not only in size relative to the lizard, but also in size of male dewlaps relative to female dewlaps. Losos and Chu (1998) and Harrison and Poe (2012) provide fascinating introductions to the evolution of dewlap size in male and female anoles respectively.

Anoles are never elongate (v. 11%), nor do they exhibit secretory pores (v. 15%), but crests are present in about double the lizard frequency (11% v. 5.3%). Color change (29% v. 10.7%) and advanced color change (20% v. 6%) are high and yet probably underreported. Blue tails are barely present in anoles (0.2% v. 2.3%).

Sexual dichromatism (39% v. 21%) is common in anoles, but seasonal color differences are not (0.2% v. 6.5%), perhaps because so many anoles can change color on short notice to thermoregulate or communicate to conspecifics. On a –2 to 2 scale of SSD, anoles are strongly in the m>f category (mean 1.01 v. 0.36). This puts Dactyloidae in fifth place among families, behind only groups that are much larger in body size and would therefore be expected to be more male-biased (Sphenodontidae, Iguanidae, Varanidae, and Leiocephalidae). In aggregate, anoles show many sex differences (sex rollup score 1.33 v. 0.75 on 0 to 6 scale), which puts them 13th of the 45 families.

No anoles are herbivorous (v. 4.1%), but omnivory (16.3% v. 15.4%) is present at lizard-typical frequencies and is observed in some very small species, especially small island species (e.g., 2.7 g *luteosignifer*, 2.8 g *desechensis*). Carnivory (2.0% v. 3.2%) and invertebrate eating (69% v. 73%) are present at typical rates, and ant specialization is about triple the lizard rate (12.2% v. 4.4%). Sit-and-wait foraging is present in anoles at about double the lizard rate (89% v. 46%), cruise foraging at representative frequency (8.5% v. 6.6%), and active foraging is grossly underrepresented (8.5% v. 58%). Foraging under electric lights is overrepresented, the second highest (Phyllodactylidae is the highest at 9.2%) of all families (7.2% v. 2.0%), despite the absence of nocturnal or cathemeral anoles. What is it about anoles that facilitates electric light feeding? Perhaps they are especially focused on visually detecting flying insects within a short distance of themselves, and an electric light swarm is a super-normal stimulus? Size-adjusted mean home-range area is very small (–0.69 v. 0.0 std. resid.) and averages only 9.7 m^2 (the lowest family mean) when unadjusted for size.

Anoles are almost exclusively diurnal (99.3% v. 67%), with the few other species crepuscular (v. 1.8%). A typical proportion of anoles call spontaneously (2.4% v. 2.5%), but stress calling is overrepresented (3.6% v. 2.2%). Few (4.8% v. 8%) are judged to be conspicuously wary. Although many rely on crypsis (43% v. 43%), anoles' primary predator concealment tactic is retreat to vegetation (56%, the highest family, v. 23%). This would not seem unusual for an arboreal group, but the closest families in this particular league table do not follow that mold: Varanidae (13.9%), Corytophanidae (11.1%), Iguanidae (9.5%), Lacertidae (9.3%), Phrynosomatidae (9.1%), and Alopoglossidae (9.1%). Considering all lizards, there is no statistical correlation between arboreality and vegetation as a primary predator concealment tactic (table 3.8, trait 22). Thus hiding in vegetation does not seem to be primarily or exclusively an arboreal trait. Perhaps it is related to the large number of anoles (143 of 416 species or 34%) that straddle the terrestrial-arboreal boundary, leading anoles to cross that boundary into concealing vegetation to escape predators. Ground-foraging tree lizard is the business model having the highest rank in relying on escape to vegetation as a primary predator avoidance tactic (table 4.1, trait 40).

Anoles are often thought of as typical day-active arboreal lizards, basking and chasing each other in the warm tropical sunlight. However, the data suggest a shadier reality; Dactyloidae has the second-highest proportion of species reported to be thermal conformers (16.1% v. 18.2% for largely crack dwelling Xenosauridae), far above the lizard mean of 2.6%. Anoles are average among the species that bask directly (19.5% v. 18%). Their mean body temperature of the family is mildly low (29.0°C v.

30.8°C), especially given their largely tropical distribution.

Anoles are underrepresented in social groups (12.1% v. 20.4%) and about average in solitary existence (12.1% v. 10.1%). Males are more than usually prone to defending space (94% v. 85%) and less inclined to be tolerant (6.1% v. 14.6%). Dactyloidae is the second-highest family for male defense of entire home ranges—as opposed to defense of some smaller space: refugium, personal space, mating space, and so on (6.1% v. 2.2%). Anole females are notable with regard to defense of home ranges, as they are top ranked among families (5.3% v. 0.6%), though female sociality is so poorly documented as to warrant some skepticism. Neither sex is notable for year-round defense, suggesting that their defense of their entire, often tiny, home range may be an artifact of defending mating space.

Anoles overwhelmingly occupy the arboreal microhabitat (80% v. 32%) and are underrepresented on the ground (17% v. 48%). Perhaps surprisingly, they are overrepresented among semiaquatic lizards (2.6% v. 1.1%). They almost never use burrows as daily retreats (0.7% v. 22%). They are found at the wet end of the rainfall spectrum, slightly overrepresented in mesic (1101–3000 mm/y: 49% v. 43%) and strongly in excess in the wet (>3000 mm/y: 27% v. 9%). Most reproduction occurs during the rainy part of the year, and inactivity—reproductive or locomotor—is often associated with dry periods. Anoles are also somewhat unique in that rain often stimulates locomotor activity rather than inhibits it. Anoles have a fairly typical altitudinal distribution (76% low, 54% medium, 7% high v. 81/41/13) and are found at lizard-typical levels in littoral areas (21% v. 17%); they are rare in sandy areas (2% v. 22%) and overrepresented in riparian zones (30% v. 22%) and human habitats (30% v. 19%). The latter two are likely associated with anoles' penchant for semiaquatic existence and their high representation around electric lights, respectively.

Anoles are universally oviparous (v. 82%), and almost all have clutches of one or occasionally two. Reflecting this small count, the mass-adjusted clutch count is below average (−0.61 v. 0.0 std. resid.). There is some terminological confusion over whether the clutch count of anoles is always one, with a second egg laid a day or so later, or whether it is a clutch count of two split between days. The ontogenetic mass range of anoles is high (0.35 v. 0.0 std. resid.), indicating that anole hatchlings are comparatively small, which they need not be given the low clutch count. Particularly small hatchlings are also unusual in small lizards. RCMs are thus very light (0.092 v. 0.226), as anoles lay a few small eggs often rather than many or large eggs infrequently. Mass-adjusted incubation period is short (−0.36 v. 0.0 std. resid.), presumably to ensure hatching during the optimal eclosion season, even though eggs are laid over an extended oviposition season. Eggshells vary in hardness (18% hard v. 41%), are mostly nonadherent (1% v. 2.4% adherent), and are laid communally at typical frequencies (4.6% v. 4.2%). Anoles rarely (only *oculatus*) offer parental care (0.2% v. 2.9%). Why so little parental care? Are their oviposition sites dangerous places for them to tarry? *A. oculatus* is a ground-foraging tree lizard, which may bring the females into regular proximity to their terrestrial oviposition site, and their distribution on Dominica may be relatively safe from predation. They might be defending their foraging space at the base of their tree and only inadvertently defending their nest, though this argument would appear applicable to many anoles, and parental care is not reported in the others.

Dactyloid longevity is somewhat short (−0.18 v. 0.0 std. resid.), but this does not appear to limit population density, which is high (2.24 v. 0.54 kg/ha), no doubt elevated to some degree by the many island species. Endemism and insularity are only slightly elevated (71% v. 64% endemic; 47% v. 40% on islands). Anoles have often been introduced (6.7% v. 3.3%) and are also overrepresented among those introduced more than twice (1.9% v. 1.1%). Anoles are probably prone to introduction both accidentally—eggs inadvertently stowed away in cargo, e.g., *marmoratus* oviposits in flower pots—and intentionally via the pet trade—colorful, bold, and conspicuous lizards make good pets. The majority of anoles are at conservation risk (63% v. 49%), presumably connected with the high proportion of restricted distributions, but their frequency on the IUCN's lists of the most endangered species is typical (13.4% v. 13.6%).

Dalmatolacerta

Dalmatolacerta oxycephala is the sole species in a genus of average-sized (4.8 g v. 5.1 g) trunk/wall lacertids found in the Balkans. It is reasonably well known (filled cells 50% v. 33%). It is very tolerant of cold and may be found active in proximity to snow, though it reaches to an elevation of only 1600 m.

In morphology, it is a typical lacertid. Its tail is a bit short (1.75 v. 2.06 × SVL), breakable, and bright blue in juveniles. It is one of the few lacertids with color change (v. 2.2%). The sexes differ in color (v. 33%). Other sex differences are few and poorly documented (sum to 1.0 v. 0.86 on 0 to 6 scale). Like most lacertids, it eats

Dalmatolacerta sp.	English name				Range core
			Size		
	Activity	Diet	mm	g	Ecological business model
oxycephala	Sharp-snouted Rock Lizard				Balkans
	diurnal	invert.	59	4.77	**** trunk/wall lizard

Dalophia spp.	English name				Range core
			Size		
	Activity	Diet	mm	g	Ecological business model
angolensis	Angola Worm Lizard				Angola
			268	12.1	**** subterranean forager
ellenbergeri	Zambia Worm Lizard				Zambia
			309	21.4	**** subterranean forager
gigantea	Giant Congo Worm Lizard				Congo
			513	162	**** subterranean forager
longicauda	Long-tailed Worm Lizard				Botswana
		invert.	335	29.5	**** subterranean forager
luluae	Congo Worm Lizard				Congo
			195	3.38	**** subterranean forager
pistillum	Blunt-tailed Worm Lizard				Botswana
		invert.	411	67	**** subterranean forager

invertebrates (v. 77%); it specializes in flying insects. Also like most lacertids, it is an active forager, moving so frequently (2.2 MPM) that it is in the 90th percentile for movement rate (63rd percentile for PTM). The combination of a specialization on flying prey and active foraging is an exception to the basic search principle that highly mobile prey are most easily waited for, whereas immobile prey must be actively searched for. Perhaps the agility required for capturing flying insects is present only in highly mobile lizards.

D. oxycephala is conspicuously wary (v. 11%), and when threatened is most likely to retreat to a crevice (v. 50%). It basks directly (v. 25%) and defends space (v. 59%) during the breeding season. It tolerates human habitat (v. 16%).

Reproduction is typical for lacertids, and despite its high latitude, *D. oxycephala* has a relatively short incubation period of 39 days (−0.61 v. −0.36 std. resid.). Hatchlings are comparatively large (ontogenetic mass range low: −0.54 v. −0.05 std. resid.).

Longevity is short (−0.49 v. 0.19 std. resid.), which corresponds to 3.8 years. The species is found on islands (v. 26%), and has not been introduced (v. 4.6%). It is an object of conservation concern (v. 56%), but it is not on the IUCN lists of the most endangered (v. 15.5%). In other respects *Dalmatolacerta* is similar to other lacertids.

Dalophia

Dalophia is a moderate genus (six spp.) of the largest-bodied (25.6 g v. 2.7 g), most robust (shape index −2.8 v. −4.0 std. resid.) amphisbaenids. They reside in central and southern Africa (Democratic Republic of the Congo to South Africa). They are associated with sandy (50% v. 28%) and riparian areas (50% v. 20%) but not littoral (0% v. 9%) or human (0% v. 8%) habitats. Their clutch counts of about 3.5 are particularly small for lizards this large (−0.88 v. −0.12 std. resid.). None occurs on islands (v. 12%), nor have they been evaluated for conservation status. In other respects they resemble other amphisbaenids.

Darevskia

Darevskia is a major radiation (31 spp.) of slightly small (4.4 g v. 5.1 g), mostly terrestrial (93% v. 90%), mostly diurnal (96% v. 95%), mostly saxicolous (83%) lacertids found primarily in west-central Asia (especially Transcaucasus; one species in Balkans). They are unique for the large number of parthenogenetic species (at least seven). Had they longitudinal striping, they would resemble *Aspidoscelis*, another bastion of parthenogens. They also resemble *Aspidoscelis* in their exceptional social tolerance: 85% *Darevskia* v. 77% *Aspidoscelis* (v. 41% Lacertidae v. 15% lizards). *Darevskia* are notable, also, for their frequency of male hierarchies (85%), which far exceeds that of *Aspidoscelis* (15%), Lacertidae (25%), or lizards (7.7%). Note that, like *Aspidoscelis*, the taxonomy of *Darevskia* is confused and bedeviled by hybridization.

Darevskia are notable for the lack of confidence in ecological business model assignments (see listing that follows). The modal pattern is refugia-anchored diurnal ground lizard (65% v. 57%), followed by ground-foraging tree lizard (22% v. 8%) and litter cryptozoid (13% v. 2.4%).

Tails are a bit short (1.88 v. 2.06 × SVL), and most tails are documented to break (81% v. 92%). They lack secretory pores (v. 4.0%), color change (v. 2.2%), and blue tails (v. 13%).

Although SSD is shifted appreciably in the m<f direction (−0.40 v. 0.49 on −2 to 2 scale), sex differences in color (36% v. 33%) are about average, and seasonal color (26% v. 15%) is only mildly elevated, leading to a slightly elevated sex differences rollup (1.21 v. 0.86 on 0 to 6 scale).

Diet in *Darevskia* is notable for the presence of omnivores, albeit at a frequency less than that of the family (8.3% v. 19.2%). Home-range area (based on one approximate value for *lindholmi*) is high (0.80 v. 0.27 std. resid.). *Darevskia* are not especially wary (3% v. 11%). Mean active body temperature is 29.8°C (v. 33.3°C). In keeping with their high social tolerance, only 15% (v. 59%) of males defend space at any time. One-third have been reported to use burrows as daily refugia (v. 48%).

Aside from their notable aversion to sand habitats (3% v. 44%), they are not strongly related to littoral habitat (13% v. 22%), riparian zones (40% v. 28%), or human habitat (10% v. 16%).

Darevskia are typical in all tabulated reproductive traits. Some complexity in sexual selection has been noted: *lindholmi* has peripheral as well as territorial males, and the peripheral males achieve some reproductive success.

Recorded longevity is average (0.16 v. 0.19 std. resid.). Few species are localized or endemic (26% v. 35%). No species is found on any island (v. 26%). *D. armeniaca* has been introduced (3.2% v. 4.6%), and a typical proportion are imperiled (46% v. 56%) or listed by the IUCN as among the most endangered (16.7% v. 16%). *D. praticola* (Meadow Lizard) is vulnerable to afforestation efforts, as the coniferous trees used in place of native vegetation are unusable by this lizard. In other respects *Darevskia* appear to be typical lacertids.

Darevskia spp.	English name				Range core
			Size		
	Activity	Diet	mm	g	Ecological business model
alpina	Alpine Rock Lizard				Georgia (Republic)
	diurnal		58	4.51	* refugia-anchored diurnal gr. liz.
armeniaca	Armenia Rock Lizard				Georgia (Republic)
	diurnal	invert.	59	4.80	** ground-foraging tree lizard
bendimahiensis	Bendimahi Waterfall Rock Lizard				Turkey
			55	3.95	* refugia-anchored diurnal gr. liz.
bithynica	Uludag Rock Lizard				Eastern Black Sea
	diurnal		62	5.7	
brauneri	Brauner's Rock Lizard				Eastern Black Sea
	diurnal	omnivore	67	6.9	** ground-foraging tree lizard
caspica	Mazandaran Rock Lizard				Iran
	diurnal		57	4.42	
caucasica	Caucasus Rock Lizard				Eastern Black Sea
	diurnal	invert.	55	3.95	* refugia-anchored diurnal gr. liz.
chlorogaster	Green-bellied Rock Lizard				Iran
	diurnal	invert.	60	5.0	** ground-foraging tree lizard
clarkorum	Clarks' Rock Lizard				Eastern Black Sea
	diurnal		62	5.6	* refugia-anchored diurnal gr. liz.
daghestanica	Dagestan Rock Lizard				Eastern Black Sea
			49	2.85	* refugia-anchored diurnal gr. liz.
dahli	Dahl's Rock Lizard				Eastern Black Sea
	diurnal		56	4.21	** ground-foraging tree lizard
defilippii	Alborz Rock Lizard				Iran
			51	3.11	* refugia-anchored diurnal gr. liz.
derjugini	Artvin Rock Lizard				Eastern Black Sea
	diurnal	invert.	56	4.11	* litter cryptozoid
kamii	Naharkhoran Forest Lizard				Iran
			59	4.75	
kopetdaghica	Kopet Dagh Rock Lizard				Iran
			50	2.99	

Darevskia spp.	English name				Range core
			Size		
	Activity	Diet	mm	g	Ecological business model
lindholmi	Crimea Rock Lizard				Crimea
	diurnal	invert.	61	5.4	
mixta	Ajar Rock Lizard				Eastern Black Sea
	diurnal		55	3.86	* refugia-anchored diurnal gr. liz.
parvula	Red-bellied Rock Lizard				Eastern Black Sea
	diurnal		52	3.40	* refugia-anchored diurnal gr. liz.
pontica	Black Sea Meadow Lizard				Balkans
	crepuscular	invert.	56	4.10	*** litter cryptozoid
portschinskii	River Kura Rock Lizard				Eastern Black Sea
	diurnal		56	4.06	* refugia-anchored diurnal gr. liz.
praticola	Meadow Lizard				Eastern Black Sea
	diurnal	invert.	56	4.19	** litter cryptozoid
raddei	Azerbaijan Rock Lizard				Iran
	diurnal	invert.	61	5.4	** refugia-anchored diurnal gr. liz.
rostombekowi	Rostombekov's Rock Lizard				Armenia
	diurnal		52	3.27	* refugia-anchored diurnal gr. liz.
rudis	Spiny-tailed Rock Lizard				Eastern Black Sea
	diurnal		70	8.1	* ground-foraging tree lizard
sapphirina	Lake Van Rock Lizard				Turkey
			50	2.99	* refugia-anchored diurnal gr. liz.
saxicola	Caucasia Rock Lizard				Eastern Black Sea
	diurnal	invert.	66	6.7	* refugia-anchored diurnal gr. liz.
schaekeli	Alborz Mountain Lizard				Iran
			49	2.86	
steineri	Steiner's Rock Lizard				Iran
			60	5.1	
unisexualis	White-bellied Rock Lizard				Eastern Black Sea
	diurnal	invert.	61	5.4	* refugia-anchored diurnal gr. liz.
uzzelli	Uzzell's Rock Lizard				Turkey
			52	3.38	
valentini	Valentin's Rock Lizard				Eastern Black Sea
	diurnal	invert.	66	6.8	* refugia-anchored diurnal gr. liz.

Dasia (fig. 4.14B)

Dasia is a moderate radiation (10 spp.) of large-sized (11.0 g v. 3.3 g) diurnal (100% v. 74%), arboreal (100% v. 10%) skinks from southern Asia (India and Sri Lanka to Vietnam, the Philippines and Borneo). Most are trunk/wall lizards (80% v. 3%); one (*griffini*) is a diurnal canopy walker (10% v. 2%); the other non-trunk/wall lizard (*haliana*) is a ground-foraging tree lizard (10% v. 7%). All are associated with large trees. Thus all have unusual ecological business models for skinks.

They are stoutly built (shape index 0.27 v. −0.46 std. resid.). Tails are of typical length (1.12 v. 1.33 × SVL) and breakable at typical frequencies (70% v. 73%). Only one species (*haliana*) is marginally elongate (10% v. 35%). None has a blue tail (v. 4%).

Males are consistently smaller than females (−1.0 v. −0.31 on −2 to 2 scale). This is remarkable for trunk/wall lizards (mean 0.74 on same scale: table 4.1, trait 13). Such trunk dwellers tend to be territorial, and indeed 100% of *Dasia* males (v. 94%) are known to defend space. Males of *olivacea* even defend against other lizard species. Social dominance is often accompanied by larger male size (table 3.1, trait 22), though spatial defense is less clearly tied to large size. Sexual dichromatism is typically rare in *Dasia* (10% v. 8%) and seasonal color unknown (0% v. 6%). Hence the sex differences rollup is a milquetoast 0.40 (v. 0.28 on a scale of 0 to 6).

In terms of diet, *Dasia* eat what is present on tree trunks: invertebrates (50% v. 84%), ants (25% v. 1.6%), and invertebrates plus fruit (25% v. 13%). Though all are said to forage actively (100% v. 92%), it is unclear what active foraging entails for a trunk/wall lizard or ground-foraging tree lizard. How far can a lizard go within those confines? Unfortunately data on home-range size are unavailable to clarify whether these lizards search multiple trunks or simply nose around the trunk of their primary residence. For *vittata*, locomotor activity ranges over one to three trees. Thirty percent are conspicuously wary (v. 11%), though that may mean little more than a tendency to squirrel around to the opposite side of the trunk.

Their modal primary predator concealment tactic is to hide in vegetation (67% v. 25%), but a surprising 33% take refuge in sand or loose soil (v. 22%). Direct basking is common (40% v. 21%). Mean active body temperature is average (31.5°C v. 31.1°C). For arboreal lizards *Dasia* seem remarkably prone to using burrows as daily refugia (20% v. 24%). Their elevational profile is slightly skewed toward lower elevations (l/m/h: 100/17/0 v. 89/35/6). They are slightly overrepresented in littoral (30% v. 21%), riparian (30% v.23%), and human (30% v. 14%) habitats and absent from sandy areas (0% v. 25%). The *Dasia* that dives into sand/loose soil (*haliana*) is not recorded to inhabit sandy areas, so perhaps it resides in areas with loose soil but not truly sandy soil.

All are oviparous (v. 64%). They have a clutch count almost commensurate with their size (−0.21 v. 0.06 std. resid.). Hatchlings are conspicuously large (ontogenetic mass range −0.78 v. 0.12 std. resid.). The selective forces responsible for slightly few, substantially large young have not been worked out; the selective forces might be responsible for the relatively larger females. Incubation

Dasia spp.	English name		Size		Range core
	Activity	Diet	mm	g	Ecological business model
griffini	Mindo-Palawan Tree Skink				Philippines
			93	17.0	* diurnal canopy walker
grisea	Gray Tree Skink				Borneo
	diurnal	omnivore	102	23.1	** trunk/wall lizard
haliana	Sri Lanka Tree Skink				Sri Lanka
	diurnal	invert.	74	4.24	* ground-foraging tree lizard
johnsinghi	Barred Tree Skink				India
	diurnal		74	8.3	*** trunk/wall lizard
nicobarensis	Nicobar Tree Skink				India
	diurnal		79	10.2	*** trunk/wall lizard
olivacea	Olive Tree Skink				Southern Asia
	diurnal	invert.	110	29.3	**** trunk/wall lizard
semicincta	Half-banded Tree Skink				Malaysia
	diurnal		102	23.1	* trunk/wall lizard
subcaerulea	Boulenger's Tree Skink				India
	diurnal		74	8.0	* trunk/wall lizard
vittata	Striped Borneo Tree Skink				Borneo
	diurnal	ants	78	9.7	**** trunk/wall lizard
vyneri	Vyner's Tree Skink				Borneo
	diurnal		57	3.42	* trunk/wall lizard

is notably short (–0.75 v. –0.40 std. resid.), which equates to 58 days in *olivacea*. Longevity data are lacking.

Most species occur on islands (80% v. 57%). None has been introduced (v. 2.1%). All evaluated species are imperiled (v. 47%), and they are overrepresented on the IUCN's lists of the most endangered (25% v. 14%). Perhaps it is a bad century to dwell in large trees in southern Asia. In other respects *Dasia* are typical skinks.

One of the species (*vyneri*) has a curious honoree; it is named to honor the politically astute "White Rajah" of Borneo, Charles Vyner Brooke (1875–1963), who not once but twice inveigled his way into being named the Rajah of Sarawak, a title normally reserved for hereditary native leaders.

Delma (fig. 4.8C)

Delma is a major radiation (22 spp.) of uniformly slightly small (4.1 g v. 5.2 g), functionally limbless (100% v. 100%), litter cryptozoid (95% v. 68%) pygopodids from Australia. In addition to the many litter cryptozoids, there is at least one (*impar*) crack hermit (5% v. 2.7%). *Delma* are notable for their complete absence of strongly contrasting features; in every important way, they conform to the pattern set by the average pygopodid! The only strong contrast is their uniform reliance on vegetation for primary predator concealment tactic (100% v. 25%). At least four species are documented to "leap" when pressed. Small leaping lizards.

The other differences are not only minor in importance but small in magnitude. They are a bit heavier built (shape index –1.10 v. –1.32 std. resid.). Their tails are a bit longer (2.99 v. 2.65 × SVL). They are slightly more inclined to exhibit color change (9.1% v. 4.3%). They are more often classified as invertebrate eaters (82% v. 65%). They are not nocturnal (0% v. 10%). They are somewhat

Delma spp.	English name				Range core
			Size		
	Activity	Diet	mm	g	Ecological business model
australis	Marble-faced Delma				Australia
	cathemeral		77	3.52	** litter cryptozoid
borea	Rusty-topped Delma				Australia
	cathemeral		79	3.63	** litter cryptozoid
butleri	Unbanded Delma				Australia
	diurnal	invert.	78	3.60	*** litter cryptozoid
concinna	Javelin Delma				Western Australia
	cathemeral	omnivore	88	4.20	
desmosa	Desmosa				Western Australia
	diurnal		81	3.76	** litter cryptozoid
elegans	Pilbara Delma				Western Australia
	diurnal	invert.	86	4.14	** litter cryptozoid
fraseri	Fraser's Delma				Western Australia
	cathemeral	invert.	106	5.5	*** litter cryptozoid
grayii	Side-barred Delma				Western Australia
	diurnal		103	5.2	*** litter cryptozoid
haroldi	Neck-barred Delma				Western Australia
			67	2.92	** litter cryptozoid
hebesa	Heath Delma				Western Australia
			66	2.85	* litter cryptozoid
impar	Many-lined Delma				New South Wales
	diurnal	invert.	87	4.14	*** crack hermit
inornata	Patternless Delma				Eastern Australia
	cathemeral	invert.	105	5.4	** litter cryptozoid
labialis	Single-striped Delma				Queensland
	diurnal		94	4.66	*** litter cryptozoid
mitella	Atherton Delma				Queensland
	diurnal		131	7.3	
molleri	Gulfs Delma				South Australia
	diurnal	omnivore	93	4.56	*** litter cryptozoid
nasuta	Sharp-snouted Delma				Australia
	cathemeral	invert.	85	4.06	*** litter cryptozoid
pax	Peace Delma				Western Australia
	diurnal		84	4.00	** litter cryptozoid
petersoni	Peterson's Delma				Southwestern Australia
		invert.	99	5.0	*** litter cryptozoid
plebeia	Leaden Delma				Queensland
	diurnal		102	5.2	*** litter cryptozoid
tealei	Teale's Delma				Western Australia
	diurnal		78	3.56	* litter cryptozoid

Delma spp.	English name				Range core
			Size		
	Activity	Diet	mm	g	Ecological business model
tincta	Excitable Delma				Australia
	diurnal	invert.	75	3.38	*** litter cryptozoid
torquata	Adorned Delma				Queensland
	diurnal	invert.	58	2.41	** litter cryptozoid

Dendragama spp.	English name				Range core
			Size		
	Activity	Diet	mm	g	Ecological business model
boulengeri	Sumatra Tree Lizard				Indonesia
	diurnal		63	8.0	* ground-foraging tree lizard

more prone to being conspicuously wary (14% v. 9%). Their young are the smallest of the exceptionally large hatchlings in the family (ontogenetic mass range –1.67 v. –2.12 std. resid.). None of the 83% (v. 88%) imperiled species is listed on the IUCN's lists of the most endangered (v. 6%). In all other documented respects *Delma* are typical pygopodids.

Dendragama

Dendragama is a single-species genus of agamid; it is known from only Sumatra Island in Indonesia. Both sexes have spindly legs, a long tail (2.2 v. 1.9 × SVL), a dewlap, and a medium-sized crest (100% v. 38%), which seems to be slightly better developed in the male. The sexes also differ in color, though this judgment is contested (100% v. 53%). It is found in human habitats (100% v. 16%) and riparian zones (v. 21%), but not littoral (v. 10%) or sandy (v. 26%) areas. Its elevational range covers all three belts (1200–2800 m). It lays a clutch of about four eggs, which is slightly small for its size (–0.07 v. 0.20 std. resid.). It has not been introduced (v. 1.7%), nor has it been evaluated for conservation status.

Dibamidae (fig. 4.13A)

The 23 dibamids are subterranean foragers with a strange distribution: *Anelytropsis* is a monotypic genus found in Mexico; *Dibamus* includes the remaining species spread from India and China to New Guinea. *Anelytropsis* is large (25.7 g), and *Dibamus* are small (mean 2.6 g). The family mean is 2.9 g (v. 5.3 g). Both genera are very slender (shape index –2.23 v. 0.0 std. resid.), functionally limbless, and most species have only vestigial eyes. As with most subterranean foragers, tails are very short (0.19 v. 1.37 × SVL), tails are not prehensile (v. 5%), and dibamids have no crests (v. 5.3%); they lack color change (v. 10.7%), dichromatism (v. 20.7%), and blue tails (v. 2.3%). Unlike many subterranean foragers, some dibamids have breakable tails (61% v. 57% for all lizards; 38% for subterranean foragers) and secretory pores (8.7% v. 15% for all lizards; 3.9% for subterranean foragers). The one sexually dimorphic feature of dibamids is that males have substantially longer tails than females (ratio 1.46, documented only in *Dibamus*). Accordingly, the rollup score (on 0 to 6 scale) of sex differences is minimal in dibamids (0.30 v. 0.75).

As far as is known, all dibamids are invertebrate eaters (v. 73%). They are not known to vocalize (v. 2.5%) or be conspicuously wary (v. 8.2%). Their thermoregulatory behaviors are undocumented. Dibamids are more concentrated at low elevations (l/m/h: 94/19/0) than are other lizards (82/41/13). They are underrepresented in littoral areas (6% v. 17%), dunes or sand (6% v. 22%), riparian areas (0% v. 22%), and human habitats (0% v. 18.5%).

Dibamids are oviparous (v. 82%), with a clutch count of only one recorded in two species (–0.53 v. 0.0 std. resid.). Eggshells are hard (v. 41%), nonadherent (v. 97.6%), and laid in solitary nests (v. 95.8%). Parental care is not recorded (v. 2.9%), though one can easily imagine underreporting of this and many other phenomena in subterranean foragers.

One population density estimate exists for dibamids, and it is an underwhelming 0.004 kg/ha, based on *Dibamus leucurus*. As with many subterranean foragers, endemism is high (78% v. 64%) but so too is occurrence of dibamid species on islands (65% v. 40% of lizards v. 25% of subterranean foragers). They have not been introduced (v. 3.3%). The sole *Anelytropsis* is considered imperiled, though no *Dibamus* are (25% of family members evaluated v. 49%); none is considered to be in the highest risk IUCN categories (v. 13.6%).

Dibamus (fig. 4.13A)

Dibamus is a poorly known (21% of cells filled v. 27% of lizards) genus of 22 small (2.6 g v. 2.9 g) Southeast Asian dibamids. Like other dibamids, these are subterranean foragers lacking functional limbs and possessing only vestigial eyes. Many male *Dibamus* have stylar hind limbs, which the females lack. The males' stylar hind limbs are presumed to play a role in courtship. *D. booliati* and *tiomanensis* have a remarkable predator defense that includes feigning death, combined with making their body appear wrinkled by twisting edges of the body scales into a perpendicular position, lending verisimilitude to their resemblance to a toxic earthworm found syntopically. No *Dibamus* are of conservation concern. In all other documented respects these are typical dibamids.

Dibamus spp.	English name		Size		Range core
	Activity	Diet	mm	g	Ecological business model
alfredi	Alfred's Worm Lizard				Malaysia
			106	2.82	**** subterranean forager
bogadeki	Bogadek's Blind Skink				China
			138	5.2	**** subterranean forager
booliati	Boo Liat's Worm Lizard				Malaysia
			83	1.61	**** subterranean forager
bourreti	Bourret's Worm Lizard				Vietnam
			138	5.2	**** subterranean forager
celebensis	Sulawesi Blind Lizard				Indonesia
			144	5.7	**** subterranean forager
dalaiensis	Dalai Worm Lizard				Cambodia
			101	2.52	**** subterranean forager
deharvengi	DeHarveng's Burrowing Lizard				Vietnam
			75	1.28	**** subterranean forager
dezwaani	Nias Blind Lizard				Indonesia
			97	2.32	**** subterranean forager
greeri	Greer's Burrowing Lizard				Vietnam
			71	1.12	**** subterranean forager
ingeri	Inger's Worm Lizard				Borneo
			78	1.40	**** subterranean forager
kondaoensis	Con Dao Worm Lizard				Vietnam
			89	1.91	**** subterranean forager
leucurus	White-tailed Worm Lizard				Borneo
		invert.	107	2.86	**** subterranean forager
montanus	Montane Worm Lizard				Vietnam
			102	2.60	**** subterranean forager
nicobaricum	Nicobar Worm Lizard				India
		invert.	127	4.30	**** subterranean forager
novaeguineae	New Guinea Blind Lizard				Indonesia
			127	4.30	**** subterranean forager
seramensis	Seram Blind Skink				Indonesia
			154	6.7	**** subterranean forager
smithi	Smith's Worm Lizard				Vietnam
			87	1.77	**** subterranean forager
somsaki	Khao Soi Dao Worm Lizard				Thailand
			86	1.74	**** subterranean forager
taylori	Taylor's Worm Lizard				Indonesia
			130	4.53	**** subterranean forager
tebal	Simeuleu Blind Skink				Indonesia
			105	2.77	**** subterranean forager

Dibamus spp.	English name				Range core
			Size		
	Activity	Diet	mm	g	Ecological business model
tiomanensis	Tioman Worm Lizard				Malaysia
			106	2.82	**** subterranean forager
vorisi	Voris' Worm Lizard				Indonesia
			74	1.23	**** subterranean forager

Dicrodon spp.	English name				Range core
			Size		
	Activity	Diet	mm	g	Ecological business model
guttulatum	Peru Desert Tegu				Ecuador
	diurnal	invert.	126	61	** wide-ranging chemosensor
heterolepis	Ecuador Desert Tegu				Peru
	diurnal	invert.	108	37.5	** wide-ranging chemosensor
holmbergi	Holmberg's Desert Tegu				Peru
	diurnal	herbivore	109	38.5	**** giant herbivore

Dicrodon

Dicrodon is a small teiid genus (three spp.) of gaudy, large-sized (44.4 g v. 22.1 g), poorly known (filled cells 23% v. 33%), low-elevation (100% v. 95%), terrestrial (100% v. 98%), mostly wide-ranging chemosensors (67% v. 97%). The other species is a giant herbivore (33% v. 0.6%). The genus is found in Ecuador and Peru.

Body shape is about average (shape index 1.08 v. 0.88 std. resid.). Tails are long (2.44 v. 2.30 × SVL) and neither breakable (v. 21%) nor blue (v. 15%). They are incapable of color change (v. 1.3%) and lack secretory pores (v. 9%). *D. guttulatum* is an arm waver; it lifts a front limb and periodically shakes it while foraging.

The sexes differ in size m>f (1.00 v. 0.89 on –2 to 2 scale) and color (67% v. 35%) but not seasonal color (v. 10%). Two species are invertebrate eaters (v. 62%), the other an herbivore (v. 1.1%). *D. holmbergi* is unique in storing food in its burrows (pods of *Prosopis juliflora*, a plant genus with a curiously disjunct distribution in North and South American deserts). All species use burrows as daily refugia (v. 82%). They are absent from human habitat (v. 29%, though perhaps there are few human habitations in the coastal deserts of Peru), and they are roughly representative in littoral (33% v. 38%), riparian (33% v. 46%), and sandy (67% v. 47%) habitats.

There are almost no reproductive data. *D. holmbergi* has been reported as both oviparous and viviparous. Endemism is slightly elevated (67% v. 50%). No species occurs on an island (v. 37%), has been introduced (v. 8%), or is imperiled (v. 41%). Two species are collected in large numbers: *heterolepis* for the pet trade and *holmbergi* for "medicinal" use. The latter is thought to be an aphrodisiac, perhaps in part or wholly because it consumes the pods of *Prosopis juliflora*, which are said to be an aphrodisiac. Though claims of sexually restorative powers are made for a number of herbivorous lizards (usually their gonads), this one is notable for the lizard's small size. It is not an obvious food source, though the lizards are reported to be much better tasting than the pods. In case you imagine that herpetologists invented drift fences for catching lizards, the traditional artisanal method for catching *holmbergi* entails 100–200 m long drift fences made from local vegetation. In other respects *Dicrodon* are typical teiids.

Dierogekko

Dierogekko (nine spp.) are small (1.16 g v. 4.15 g), slender (shape index –0.31 v. 0.05), striped, arboreal geckos endemic to New Caledonia. These diplodactylids are mostly (86% v. 25%) nocturnal canopy walkers; the remainder (14% v. 7%) are ground-foraging tree lizards, though assignments are highly uncertain. Eighty-nine percent (v. 34%) possess secretory pores, but none is documented to show color change (v. 5.4%). Except for pores, sex differences are not known. They are not associated with littoral (v. 17%), sandy (v. 24%), or human (v. 5%) habitats. They have not been introduced (v. 2%); all species are of conservation concern (v. 81%), and 88% of *Dierogekko* are classified by IUCN in the most endangered categories. In other respects these are typical diplodactylids.

Dierogekko spp.	English name				Range core
			Size		
	Activity	Diet	mm	g	Ecological business model
baaba	Baaba Gecko				New Caledonia
			36	0.90	
inexpectatus	Unique New Caledonian Gecko				New Caledonia
	nocturnal		38	1.03	* ground-foraging tree lizard
insularis	Islands Striped Gecko				New Caledonia
	nocturnal		39	1.15	* nocturnal canopy walker
kaalaensis	Kaala Striped Gecko				New Caledonia
	nocturnal		42	1.43	* nocturnal canopy walker
koniambo	Koniambo Striped Gecko				New Caledonia
	nocturnal		40	1.26	* nocturnal canopy walker
nehoueensis	Nehoue Nimble Gecko				New Caledonia
	nocturnal		38	1.08	* nocturnal canopy walker
poumensis	Poum Striped Gecko				New Caledonia
	nocturnal		37	0.96	* nocturnal canopy walker
thomaswhitei	Taom Striped Gecko				New Caledonia
	nocturnal		42	1.39	* nocturnal canopy walker
validiclavis	Bold-striped Bavayia				New Caledonia
	nocturnal		42	1.40	

Dinarolacerta spp.	English name				Range core
			Size		
	Activity	Diet	mm	g	Ecological business model
montenegrina	Prokletije Rock Lizard				Montenegro
	diurnal		55	3.95	
mosorensis	Mosor Rock Lizard				Balkans
	diurnal		65	6.4	*** refugia-anchored diurnal gr. liz

Dinarolacerta

Dinarolacerta (two spp.) is a lacertid genus of average (5.0 g v. 5.1 g) refugia-anchored diurnal ground lizards found in the Balkans. In morphology they are typical lacertids, although at least *mosorensis* has blue-tailed juveniles. Sex differences are few or poorly documented (0.0 v. 0.86 on 0 to 6 scale). Neither species is conspicuously wary (v. 11%). Both are found up to mildly high elevation (1660 m, 1900 m). *D. montenegrina* is associated with rock wall habitats (v. 9% "arboreal") in riparian zones (v. 28%). Reproduction is known only in *mosorensis*, which lays an average of 4.4 eggs at an advanced stage of development, and they hatch about 20 days later (–1.07 v. –0.36 std. resid.).

Longevity is relatively long (0.60 v. 0.19 std. resid.), based on nine years for *mosorensis*.

Neither species occurs on islands (v. 26%). Neither has been introduced (v. 4.6%). One of two (*mosorensis*) is imperiled (50% v. 56%), but neither is listed among the most endangered by the IUCN (v. 15.5%). In other respects they are probably typical lacertids.

Diplodactylidae (fig. 4.3D, 4.7C)

The 149 diplodactylids are geckos in 25 genera from Australia, New Caledonia, and New Zealand: *Amalosia*, *Bavayia*, *Correlophus*, *Crenadactylus*, *Dactylocnemis*, *Dierogekko*, *Diplodactylus*, *Eurydactylodes*, *Hesperoedura*, *Hoplodactylus*, *Lucasium*, *Mniarogekko*, *Mokopirirakau*, *Naultinus*, *Nebulifera*, *Oedodera*, *Oedura*, *Paniegekko*, *Pseudothecadactylus*, *Rhacodactylus*, *Rhynchoedura*, *Strophurus*, *Toropuku*, *Tukutuku*, and *Woodworthia*. Expanded toepads are present in most genera; exceptions are found primarily in *Lucasium*, *Mokopirirakau*, *Rhynchoedura*, and *Tukutuku*. They are of average information content (28% v. 27% filled, ranging in genera from 17% *Crenadactylus* to 47% *Hesperoedura*), and are ecologically diverse. Eight ecological business models are represented: crack hermits (3.9% v. 2.6%), diurnal canopy walkers (7.0% v. 6.4%), ground-foraging tree lizards (7.0% v. 9.2%), immobility experts (1.6% v. 3.2%), litter cryptozoids (0.8% v. 11.6%), nocturnal canopy walkers (25% v. 2.1%), nocturnal ground lizards (35.2% v. 7.7%), and trunk/wall lizards (19.5% v. 13%).

Diplodactylidae also show great variation in body mass (mean 4.15 g v. 5.33 g), with genus means ranging from 0.62 g *Crenadactylus* to 152 g *Hoplodactylus*. Diplodactylids are moderately variable in average build (shape index mean 0.05 v. 0.0 std. resid), ranging from –0.47 *Crenadactylus* to 1.04 *Hoplodactylus*. Tail length averages slightly below SVL length (0.85 v. 1.37 × SVL). The shortest genus mean for relative tail length ratio is 0.51 *Diplodactylus*; the longest 1.89 *Naultinus*. Almost all tails are breakable (91% v. 57%). Prehensile tails vary by genus from none to all (mean 26% v. 5%). Crests are found only in *Rhacodactylus* and *Strophurus* (mean 1.3% v. 5.3%), elongation is not reported (v. 11.4%), but secretory pores also vary by genus from none to all (mean 34% v. 15%). Basic (5.4% v. 10.7%; range 0 to 88%) and advanced (1.3% v. 6.0%; range 0 to 25%) color changing abilities are scattered through the family.

SSD is relatively fixed within each diplodactylid genus; m<f genera outnumber m>f genera such that the family mean is well onto the m<f side (−0.65 on −2 to 2 scale), though the average lizard is m>f (0.36). Many examples of m<f size dimorphism are attributed to fecundity selection, which is not likely to be the key in Diplodactylidae, as clutches in the family are uniformly small (see below). Male mobility (small males) may be more critical than male intimidation (large males). Only in *Naultinus* and *Rhacodactylus* is sexual dichromatism reported (6% of family v. 20.7% of lizards). Seasonal color is not reported in any diplodactylid (v. 6.5%); blue tails are absent (v. 2.3%). Ratios of m:f tail lengths (expressed relative to SVL) are documented only for *Naultinus* (1.03) and *Pseudothecadactylus* (0.72), yielding a low family mean (0.88 v. 1.07; see table 3.3, trait 8, for context). Due both to the paucity of information and the scarcity of strong sex differences, the mean sex rollup score for Diplodactylidae is a modest 0.54 (v. 0.75).

Keeping with the pattern of great ecological diversity, all diets except ant specialization (v. 4.4%) are represented in Diplodactylidae: carnivores (1.9% v. 3.2%), herbivores (3.7% v. 4.1%), invertebrate eaters (80% v. 73%), and omnivores (15% v. 15%). Foraging modes are very poorly documented (in only 7 of 25 genera), but all known examples are sit-and-wait (v. 46.4%). Electric light use is unknown (v. 1.4%); home-range areas are unavailable.

Diel activity patterns are diverse within Diplodactylidae: cathemeral (13% v. 6.8%), crepuscular (0% v. 1.8%), diurnal (7.6% v. 66.6%), and mostly nocturnal (79.4% v. 24.7%). Both spontaneous calling (5.4% v. 2.5%) and stress calling (5.4% v. 2.2%) are overrepresented, as is common in nocturnal and cathemeral species. Diplodactylids are not especially wary (2.7% v. 8.2%), as they rely primarily on crypsis (88% v. 43%), along with flight (8% v. 11%) and hiding in vegetation (8.0% v. 23%).

Direct basking is reported in 17% of diplodactylids (v. 18%); only 0.7% (v. 2.6%) are said to thermoconform. Mean body temperature is a very cool 24.3°C; low body temperatures are found throughout the family, not just in the high latitude forms.

Social life is almost undocumented in diplodactylids; males seem to be more tolerant than typical lizards (24% v. 15%). Most are either arboreal (64% v. 32%) or terrestrial (34% v. 48%); the remainder are cryptozoic (2% v. 12%). Burrow use for daily refugia is present in typical frequency (22% v. 22%), despite the dominance of arboreal species. All are found either in the arid (151–400 mm/y; 67%, the second-highest value, v. 30%) or intermediate rainfall belts (401–1100 mm/y; 33% v. 27%). They are concentrated at low elevations (l/m/h: 100/11/2) compared to typical lizards (82/41/13), a bias common in nocturnal species. Diplodactylids are found in littoral and sandy habitats at typical frequencies (17% v. 17% and 24% v. 22%, respectively), but are underrepresented in riparian (10% v. 22%) and human (5% v. 18.5%) habitats.

Viviparity pops up in eight genera (see accounts), but the family averages the same frequency of oviparous species as typical lizards (82% v. 82%). Clutch count is a nearly invariant two (−0.50 v. 0.0 std. resid.). Because egg/litter count is invariant, the mass-adjusted clutch count is largely a mirror of the species size; arguably it also may reflect the degree to which the clutch count is suboptimal. The mass-adjusted ontogenetic mass ratio of young, however, is free to vary (−0.57 v. 0.0 std. resid.; range of genus means −1.67 to 0.54). This indicates that diplodactylid young are larger than lizard average (1/23 of maximum mass v. 1/44). Incubation duration is a little short in every documented genus in the family (−0.22 v. 0.0 std. resid.; range −0.06 to −0.46). Eggshells are rarely hard (family mean 5% v. 41%), but occur in *Eurydactylodes*, *Mniarogekko*, and *Oedura*. Adherent eggs are found in *Strophurus* (family mean 0.7% v. 2.4%). Communally deposited eggs are scattered throughout the family (*Amalosia*, *Crenadactylus*, *Oedura*, *Rhacodactylus*); mean 2.7% v. 4.2%. Parental care is equally dispersed (*Mniarogekko*, *Naultinus*, *Oedura*, *Rhacodactylus*); mean 3.4% v. 2.9%. RCMs are average (0.225 v. 0.226), but the knowledge base is sparse (known from examples in only 6 of 25 genera: range 0.143 to 0.310).

Mass-adjusted longevity is substantially longer than average (0.90 v. 0.0 std. resid.). Only four genera have short average longevities (*Correlophus*, *Diplodactylus*, *Rhynchoedura*, *Strophurus*). No pattern that would include just those four genera springs to my mind. Conversely, some diplodactylids have high longevities: *Hoplodactylus* (2.16 std. resid. or 50 years for *H. duvaucelii*); *Woodworthia* (1.86 std. resid. or 42 years for *W. brunneus* and 37 years for *W. maculatus*). Those three long-lived species share the trait of occurring on the chilly South Island of New Zealand. New Zealand stands out with regard to longevities

Table 5.3. Mass-adjusted longevities of diplodactylids on three land masses, stratified by subjective habitat classes: treeless, forested dry (sclerophyll), forested wet. Value shown is mean (number of documented species in parentheses).

	Australia	New Caledonia	New Zealand	All
Treeless dry	–0.03 (8)	– (0)	– (0)	–0.03 (8)
Forest dry	0.52 (4)	0.37 (2)	1.73 (7)	1.15 (13)
Forest wet	1.08 (3)	0.47 (6)	1.70 (8)	1.15 (17)
All	0.34 (15)	0.44 (8)	1.70 (15)	0.90 (38)

of many taxa and should perhaps be considered separately for many life history values. The uniqueness of New Zealand is apparent from inspection of the table 5.3.

Table 5.3 is stratified by habitat type in case more arid habitats would induce greater longevity to ensure survival long enough to guarantee reproduction during a beneficially rainy year—environmental unpredictability may promote iteroparity—or have intrinsically fewer predators. No support for the environmental unpredictability hypothesis was observed here or in the lizard-wide analysis in chapter 3, as the aridland diplodactylids exhibited shorter, not longer, longevities. However, there was ample support for the uniqueness of New Zealand. Determination of cause will require additional data. Two plausible hypotheses are that high longevity evolved in the absence of predatory mammals, or colder climate slows metabolism, shortens the activity season, and thereby lengthens calendar longevity. In relation to the latter, be aware that exceptional longevities are not restricted to the temperate parts of New Zealand.

Endemism is moderately high in the diplodactylids (72% v. 64%). The southwestern Pacific domain of Diplodactylidae hints at the likelihood of island species, which are moderately overrepresented in the family (46% v. 40%). Members of *Amalosia*, *Hoplodactylus*, and *Naultinus* have been introduced (2.0% v. 3.3%)—*Hoplodactylus duvaucelii* more than twice (presumably intentional conservation translocations). Conservation concern for the diplodactylids is high, as 81% (v. 49%) are considered imperiled; 29.3% (v.13.6%) are classified in the most endangered IUCN categories.

The theme of great ecological variation runs throughout the family, perhaps in part due to the wide range of latitudes represented. They are not especially variable in moisture or elevational regimes. Nonetheless, within moderately wet low-elevation sites occurs an impressive radiation in morphologies, diets, microhabitats, activity patterns, sizes, sexual relations, and reproduction. Social relations may be correspondingly diverse but are not yet sufficiently documented.

Diplodactylus

Diplodactylus is a large (27 spp.) radiation of nocturnal terrestrial diplodactylids in Australia. Though highly variable in pattern, even within a species, most of these geckos share a short snout and a short (0.51 v. 0.85 × SVL), conspicuously enlarged tail. The variable but often bold color patterns may be largely free of selection pressure at night or the bold marks may help disrupt the lizard's outline in dim light, and the variability may create a multiplicity of search images to confound predators. Presumably, the tail is a fat storage organ useful in an energetically unpredictable desert environment, though their tails will break, thereby sacrificing their rainy day fund (instead of their lives). When cornered, at least two species (*conspicillatus*, *tessellatus*) will cover their heads with their juicy tails. When *conspicillatus* uses a spider burrow for its refugium, it uses its tail to plug the burrow entrance. Spider burrows are used as refugia by *pulcher* also; *tessellatus* refuges in cracking clay soils.

Diplodactylus are small for diplodactylids (2.67 g v. 4.15 g). About 11% of *Diplodactylus* (v. 16%) have toepads that are greatly reduced in width, perhaps functionally padless. Males are smaller than females, as is often seen in lizards of small body size (table 3.1, trait 6). Other sex differences are lacking (e.g., secretory pores 0% v. 34%), yielding a particularly low sex differences rollup score for a diplodactylid (0.15 v. 0.54 on 0 to 6 scale).

Diet has not been established for 19 *Diplodactylus*; the other eight are uniformly invertebrate eaters (v. 73%). *D. conspicillatus* and *pulcher* are termite specialists; six of eight species with dietary data report termites.

Predator concealment is known for only two species (one crypsis; one flight). Another species (*tessellatus*) combines the two, by running in spurts even when pursued. At night this could bring crypsis into play if the pursuer loses sight of the target even momentarily. Social relations are known for only *tessellatus*, which is tolerant of other males.

Body temperatures are typical for diplodactylids (25.1°C v. 24.3°C), though low temperature activity is notable; *tessellatus* is active down to 10.5°C.

Many *Diplodactylus* use burrows as

Diplodactylus spp.	English name	Range core	Activity	Diet	Size mm	Size g	Ecological business model
ameyi	Eastern Deserts Fat-tailed Gecko	Queensland			52	2.82	
barraganae	Gulf Fat-tailed Gecko	Queensland			43	1.56	
bilybara	Western Fat-tailed Gecko	Western Australia			51	2.65	
calcicolus	Eastern Stone Gecko	Southwest Australia	nocturnal		57	3.71	*** nocturnal ground lizard
capensis	Northwest Cape Diplodactylus	Western Australia	nocturnal		56	3.41	*** nocturnal ground lizard
conspicillatus	Variable Fat-tailed Diplodactylus	Australia	nocturnal	invert.	58	3.92	*** nocturnal ground lizard
custos	Kimberly Fat-tailed Gecko	Western Australia			52	2.70	
fulleri	Lake Disappointment Ground Gecko	Western Australia	nocturnal		48	2.17	*** nocturnal ground lizard
furcosus	Ranges Stone Gecko	New South Wales	nocturnal		57	3.64	*** nocturnal ground lizard
galaxias	Northern Pilbara Beak-faced Gecko	Western Australia	nocturnal		42	1.45	*** nocturnal ground lizard
galeatus	Helmeted Gecko	Central Australia	nocturnal		50	2.52	*** nocturnal ground lizard
granariensis	Wheat-belt Stone Gecko	Western Australia	nocturnal	invert.	64	5.3	*** nocturnal ground lizard
hillii	Northern Fat-tailed Gecko	Northern Territory	nocturnal	invert.	46	1.92	*** nocturnal ground lizard
kenneallyi	Kenneally's Gecko	Western Australia	nocturnal		45	1.81	*** nocturnal ground lizard
klugei	Kluge's Diplodactylus	Western Australia	nocturnal		53	3.00	*** nocturnal ground lizard
laevis	Desert Fat-tailed Gecko	Western Australia		invert.	55	3.27	** nocturnal ground lizard
lateroides	Laterite Stone Gecko	Western Australia	nocturnal		46	1.86	*** nocturnal ground lizard
mitchelli	Mitchell's Stone Gecko	Western Australia	nocturnal		64	5.1	*** nocturnal ground lizard
nebulosus	Cloud-backed Stone Gecko	Western Australia	nocturnal		49	2.37	
ornatus	Ornate Stone Gecko	Western Australia	nocturnal		52	2.76	*** nocturnal ground lizard

daily retreats (58% v. 22%). They are found only in the arid zone (151–400 mm/y; v. 67%). They are underrepresented in littoral (4% v. 17%) and human (0% v. 5%) habitats and overrepresented in sandy areas (52% v. 24%).

They are uniformly oviparous (v. 82%), with normal diplodactylid clutch characteristics. They are short-lived (−0.55 v. 0.90 std. resid.).

Diplodactylus are reported to be relatively sparse in population density (0.18 v. 0.44 kg/ha), perhaps a reflection of low primary productivity in their environment. No *Diplodactylus* has been introduced (v. 2%). Half of the evaluated species are considered imperiled (v. 81%), but none is on the IUCN's lists of most endangered (v. 29%). In other respects *Diplodactylus* are typical diplodactylids.

Diplodactylus spp.	English name		Size		Range core
	Activity	Diet	mm	g	Ecological business model
platyurus	Eastern Fat-tailed Gecko				Queensland
		invert.	50	2.48	** nocturnal ground lizard
polyophthalmus	Speckled Stone Gecko				Western Australia
	nocturnal		51	2.57	*** nocturnal ground lizard
pulcher	Fine-faced Gecko				Southwest Australia
	nocturnal	invert.	55	3.21	*** nocturnal ground lizard
savagei	Yellow-spotted Pilbara Gecko				Western Australia
	nocturnal		44	1.69	*** nocturnal ground lizard
tessellatus	Tesselated Gecko				Eastern Australia
	nocturnal	invert.	51	2.61	*** nocturnal ground lizard
vittatus	Wood Gecko				Eastern Australia
	nocturnal	invert.	51	2.57	*** nocturnal ground lizard
wiru	Desert Wood Gecko				Southwestern Australia
	nocturnal		53	2.96	*** nocturnal ground lizard

Diploglossidae

The 51 diploglossids belong to the genera *Celestus* (29 spp.), *Diploglossus* (17 spp.), and *Ophiodes* (5 spp.). The former two genera bear a strong resemblance to mildly elongate skinks or anguids; *Ophiodes* are functionally limbless and are snakelike. They are found from Mexico to Uruguay on the mainland and scattered throughout the Caribbean. Although a few species are familiar, most are secretive, scarce, or extinct; all are poorly known (23% v. 27% of cells filled). The largest group (58% v. 12%) are litter cryptozoids; an eighth are each refugia-anchored diurnal ground lizards (12% v. 25%) and trunk/wall lizards (12% v. 13%); 9% (v. 9%) are ground-foraging tree lizards. One species each is an immobility expert (3% v. 3.2%), nocturnal ground lizard (3% v. 7.7%), and sand shark (3% v. 1.9%). This family is distinguished by having particularly numerous and small young (see below).

They are slightly larger than typical lizards (6.3 g v. 5.3 g) and substantially more slender than normal for their mass (shape index –0.98 v. 0.0 std. resid.). Tails are of normal length (1.27 v. 1.37 × SVL) and breakable (98% v. 57%) but not prehensile (v. 5%). They lack crests (v. 5%) and secretory pores (v. 15%), but most are elongate (71% v. 11%). My impression is that additional species are also elongate but not so documented. No diploglossids are reported to change color (v. 11%).

SSD is strongly m>f (0.94 v. 0.36 on –2 to 2 scale), but sexual dichromatism is rare (4% v. 21%) and seasonal color not reported (v. 6.5%). Blue tails are present in typical frequency (2.0% v. 2.3%). In aggregate, sex differences are muted (0.28 v. 0.75 on 0 to 6 scale).

Diploglossids have diverse diets: carnivores (20% v. 3%), invertebrate eaters (60% v. 73%), and omnivores (20% v. 15%). The foraging mode of one species is recorded as active foraging. Electric light use is not reported (v. 1.4%). Home-range areas are unknown.

Daily activity patterns of diploglossids are poorly understood, but cathemeral is overrepresented (25% v. 7%), crepuscular is unknown (v. 2%), and nocturnal is rare (5% v. 25%). Vocalizations are not reported. Sixteen percent (v. 8%) are said to be conspicuously wary.

Direct basking is underrepresented (8% v. 18%), though body temperatures are normal given the poor sample (30.0°C v. 32°C). Social groups are reported in one species, but social documentation goes no further. Diploglossids are primarily cryptozoic (51% v. 12%) but present in all microhabitats except the semiaquatic one: arboreal (17% v. 32%), subterranean (2.4% v. 7.2%), and terrestrial (29% v. 48%). Burrow use for daily refuging is documented in 7% (v. 22%) of species but is likely underreported.

In light of their Neotropical range, one would expect diploglossids to be found in the wet end of the rainfall spectrum, and they are: mesic (1101–3000 mm/y: 25% v. 43%); wet (>3000 mm/y: 75%, the highest family, v. 9%). Their elevational distribution is like that of all lizards (l/m/h: 86/49/6). Oddly for such poorly understood lizards, they are overrepresented in human habitats (36% v. 18.5%) and underrepresented in the other types: littoral (7% v. 17%), sandy (0% v. 22%), and riparian (18% v. 22%) zones.

For a group that is almost exclusively tropical and mostly at low elevation, oviparity is surprisingly rare (15% v. 82%). Clutch/litter count is well above average (0.86, the third-highest family, v. 0.0 std. resid.), corresponding to five to nine young. The young are exceptionally small (ontogenetic size range 2.13, the highest family, v. 0.0 std. resid.), corresponding to mass ratios of 1/214 versus 1/44. Incubation duration and RCM are unknown. Parental care, which is loosely associated with viviparity, is overrepresented (7.8% v. 2.9%).

Longevity is mildly elevated (0.46 v. 0.0 std. resid.). Population density is based on a single estimate of an island species (*Diploglossus millepunctatus*) and is very high (7.0 v. 0.54 kg/ha). Given the aberrant situation in which *millepunctatus* is found—huge input of dropped fish from nesting seabirds, many anoles to eat, almost no predators on very remote Malpelo Island—no inferences about diploglossids in general can be drawn from this example.

Diploglossids have not been introduced (v. 3.3%). They are imperiled at lizard-typical frequencies (51% v. 49%) but unfortunately are strongly overrepresented on the IUCN's lists of most endangered lizards (30% v. 14%).

Diploglossus

Diploglossus is a moderate (17 spp.) genus of the Diploglossidae. Its members are ecologically and morphologically similar to *Celestus*, and many species have wandered taxonomically back and forth between the two genera. *Diploglossus* are found from Mexico to Brazil and on a few Caribbean islands—Cuba, Montserrat, Puerto Rico—and Malpelo in the Pacific. Like *Celestus*, many species of *Diploglossus* are barely documented. Note the low confidence in ecological business model assignments: litter cryptozoid (73% v. 58%), refugia-anchored diurnal ground lizard (9% v. 12%), and sand shark (9% v. 3%).

Diploglossus spp.	English name				Range core
			Size		
	Activity	Diet	mm	g	Ecological business model
atitlanensis	Torpedo Galliwasp				Honduras
	diurnal		95	4.82	* trunk/wall lizard
bilobatus	Talamancan Galliwasp				Costa Rica
	diurnal		80	3.16	** litter cryptozoid
delasagra	Cuban Galliwasp				Cuba
			102	3.82	* litter cryptozoid
fasciatus	Banded Galliwasp				Bolivia
	diurnal		131	10.6	* litter cryptozoid
garridoi	Cuban Small-eared Galliwasp				Cuba
			83	8.2	* litter cryptozoid
ingridae	Ingrid's Galliwasp				Mexico
			84	8.7	** trunk/wall lizard
legnotus	Puebla Galliwasp				Mexico
			90	10.9	* litter cryptozoid
lessonae	Brazil Galliwasp				Brazil
		invert.	133	42.2	** sand shark
microcephalus	Small-headed Galliwasp				Mexico
			47	1.16	
microlepis	Small-lipped Galliwasp				unknown!
			62	3.02	
millepunctatus	Malpelo Galliwasp				Colombia
	cathemeral	carnivore	210	34.3	** refugia-anchored diurnal gr. liz.
monotropis	Coral-mimic Galliwasp				Panama
	diurnal	carnivore	170	16.8	* litter cryptozoid
montisilvestris	Montane Forest Galliwasp				Panama
			81	7.5	
montisserrati	Montserrat Galliwasp				Montserrat
	cathemeral		140	51	* litter cryptozoid
nigropunctatus	Black-spotted Galliwasp				Cuba
			96	13.5	
owenii	Owen's Galliwasp				Mexico
			55	1.25	
pleii	Puerto Rico Galliwasp				Puerto Rico
		omnivore	102	3.87	**** litter cryptozoid

Diploglossus are slightly larger than other diploglossids (7.6 g v. 6.3 g) and have a slightly heavier build (shape index −0.66 v. −0.98 std. resid.). A smaller proportion are elongate (47% v. 71%). A larger percentage exhibit sexual dichromatism (12% v. 4%).

Diploglossus are more inclined toward carnivory (50% v. 20%) and less so toward eating invertebrates (25% v. 60%). On Malpelo Island there is an unusual situation in which the remote military base discards its food waste during the day, attracting dozens of apparently sociable carrion-eating *millepunctatus* to gorge side by side. However, other adults (evidently subordinate) and juvenile *millepunctatus* feed at night on natural food. The daily activity pattern of *Diploglossus* species is similar to diploglossids, but they are more inclined to be conspicuously wary (24% v. 16%). They bask a bit more (12% v. 8%) and have relatively low body temperature mean (28.5°C v. 30.0°C).

Unlike *Celestus*, *Diploglossus* is underrepresented in human habitat (8% v. 36% in diploglossids and 48% in *Celestus*). Also unlike *Celestus*, which is uniformly viviparous, *Diploglossus* is mostly oviparous (64% v. 14%). *D. millepunctatus* is known to copulate for 2.5 hours. Clutch count in *Diploglossus* is low in comparison to other diploglossids, but it is above the lizard average (0.28 v. 0.86 std. resid. or 0.0 for lizards). Similarly, the size of young *Diploglossus* is well above the family average but below that of lizards (ontogenetic mass range 0.77 v. 2.13 std. resid. or 0.0 for lizards). In another strong difference with the superficially similar *Celestus*, *Diploglossus* exhibits parental care (24% v. 7.8% in Diploglossidae and 0% in *Celestus*). In *nigropunctatus* both sexes of parents protect the eggs.

In a final and hopeful distinction with *Celestus*, *Diploglossus* is comparatively infrequent among the ranks of the imperiled (22% v. 51%), with only 11% (v. 30%) on the IUCN's list of the most endangered. In other respects *Diploglossus* are typical diploglossids.

Two *Diploglossus* species have interesting interactions with humans. *D. monotropis* is called the Coral-mimic Galliwasp, not because it resembles submarine animals, but because its red color banding reminds local humans of the hazardous coral snake. It enhances this mimicry by using serpentine motion (legs folded and unused) when pressed. This has led to the local conviction that this harmless lizard is venomous. Whether that misinformation leads to it being left alone or targeted for persecution is not recorded.

D. owenii is named in honor of Sir Richard Owen (1804–1892). When you feel overworked, you should consider the resume of Sir Owen, a solder/surgeon/anatomist/paleontologist who created the separate Natural History Museum, London, tutored Queen Victoria's children, discovered the fossil of *Archaeopteryx*, coined the name Dinosauria, and disagreed vociferously with Darwin over the subject of evolution.

Diplolaemus

Diplolaemus is a four-species genus of moderately large (18.9 g v. 16.4 g), diurnal (100% v. 86%), terrestrial (67% v. 50%) or cryptozoic (33% v. 5%) leiosaurids found in Chile and Argentina. They all dig refuge burrows (v. 67%) and are refugia-anchored diurnal ground lizards (v. 63%).

Diplolaemus are notably robust (shape index 0.82 v. 0.76) and in possession of relatively skinny, exceptionally short tails (1.00 v. 1.66 × SVL), which are neither breakable (v. 3%) nor prehensile (v. 9%). They are not known to change color (v. 3%), though the sexes may differ in color (25% v. 15%).

As is usual for refugia-anchored diurnal ground lizards (table 4.1, trait 47), males defend space (100% v. 100%). Both elevationally documented species are found in all elevation bands.

Although 75% of *Diplolaemus* are nominally oviparous (v. 95%), contradictory claims are made for both *darwini* and *leopardinus* (nominally "both" reproductive modes). Perhaps there is altitudinal variation in reproductive mode within these species.

One-quarter of *Diplolaemus* are judged to be imperiled (v. 14%), though none is listed on the IUCN's lists of the most endangered species (v. 0%). In other respects *Diplolaemus* are probably typical leiosaurids.

Diplometopon

Diplometopon zarudnyi is the sole member of its trogonophid genus, a medium-large (3.55 g v. 1.78 g), speckled, nocturnal (100% v. 100%)

Diplolaemus spp.	English name				Range core
			Size		
	Activity	Diet	mm	g	Ecological business model
bibronii	Bibron's Iguana				Argentina
	diurnal	invert.	95	25.6	** refugia-anchored diurnal gr. liz.
darwini	Darwin's Iguana				Argentina
		invert.	95	25.6	** refugia-anchored diurnal gr. liz.
leopardinus	Leopard Iguana				Argentina
	diurnal	invert.	67	7.6	** refugia-anchored diurnal gr. liz.
sexcinctus	Six-banded Iguana				Argentina
		invert.	95	25.6	* refugia-anchored diurnal gr. liz.

sand shark (100% v. 20%) found in the Arabian Peninsula. It can locate prey by vibrations or scent. Surface locomotion is attained by curling the tail to one side, planting the pointed tip, and straightening out. The purpose of these surface forays is unknown, perhaps to find or travel to suitable patches of loose substrate.

Diplometopon is of moderate build (shape index –3.00 v. –3.43 std. resid.). Sex differences are unknown. It lacks secretory pores (v. 17%). It is found at low elevation, in this case 0–1000 m, and it is associated with littoral (v. 33%) and sandy (v. 100%) areas. Its clutch count is unknown. It is not endemic (v. 50%) or found on an island (v. 37%). In other respects *Diplometopon* is a typical trogonophid.

Diplometopon sp.	English name	Activity	Diet	Size mm	Size g	Range core	Ecological business model
zarudnyi	Zarudnyi's Worm Lizard	nocturnal		150	3.55	Saudi Arabia	** sand shark or subterranean forager

Diporiphora

Diporiphora is a large (22 spp.) genus of generalized, diurnal (82% v. 98%), terrestrial agamids found in Australia (*bilineata* reaches New Guinea). Few (9% v. 38%) have a crest and most have long tails (2.37 v. 1.86 × SVL) and spindly limbs. They stand out from typical agamids in an inchoate assortment of ways: they often exhibit seasonal color (55%, the fourth-highest frequency in the family v. 20%), they cruise forage (25%, also fourth, v. 6%), they are cathemeral (9%, the fourth highest, v. 1.8%) and crepuscular (9%, the top frequency, v. 0.5%), they retreat to vegetation (60%, the second highest, v. 17%), and they are especially frequent in arid (67%, the fourth-highest value, v. 35%) and intermediate (35%, tied for third, v. 12%) rainfall belts. Although most are diurnal, their diurnality frequency is the fourth lowest in the family. They are smaller (7.3 g v. 15.6 g) than most agamids. Commensurate with their terrestrial lifestyle and arid home, half are refugia-anchored diurnal ground lizards (50% v. 43%), one is an immobility expert (7% v. 4.5%), and 36% (v. 19%) are ground-foraging tree lizards. They are not very well studied (filled cells 26% v. 30%), and many of the business model assignments have low confidence.

Three species are reported to avoid walking on dry leaves (which predator is listening? Kookaburra?). A low proportion of species (14% v. 22%) exhibit any rapid color change. A high proportion are invertebrate eaters (83% v. 46%), and other diets are underrepresented. All are said to sit-and-wait (100% v. 88%), and one that sits-and-waits (*superba*) also cruises with a rocking, chameleon-like motion. All bask directly (v. 24%), and they have a very high mean body temperature (37.4°C v. 34.8°C). The

Diporiphora spp.	English name	Activity	Diet	Size mm	Size g	Range core	Ecological business model
adductus	Carnarvon Dragon	diurnal		61	7.1	Western Australia	* refugia-anchored diurnal gr. liz.
albilabris	White-lipped Two-lined Dragon	diurnal		52	4.44	Western Australia	* refugia-anchored diurnal gr. liz.
ameliae	Amelia's Two-lined Dragon	diurnal		60	7.0	Queensland	
amphiboluroides	Mulga Dragon	diurnal	invert.	83	18.8	Western Australia	**** immobility expert
arnhemica	Arnhem Dragon	diurnal		59	6.3	Northern Territory	** refugia-anchored diurnal gr. liz.
australis	Tommy Roundhead	diurnal	invert.	66	9.4	New Guinea	*** ground-foraging tree lizard
bennettii	Robust Two-lined Dragon	diurnal		69	10.5	Western Australia	* refugia-anchored diurnal gr. liz.
bilineata	Two-lined Dragon	diurnal	invert.	63	7.8	Northern Territory	* ground-foraging tree lizard
convergens	Crystal Creek Two-lined Dragon	diurnal		34	1.15	Western Australia	

Diporiphora spp.	English name	Activity	Diet	Size mm	Size g	Range core	Ecological business model
lalliae	Lally's Two-lined Dragon	cathemeral		67	9.4	Western Australia	** refugia-anchored diurnal gr. liz.
linga	Pink Two-lined Dragon	diurnal		50	3.92	South Australia	** refugia-anchored diurnal gr. liz.
magna	Yellow-sided Two-lined Dragon	crepuscular		81	17.4	Northern Australia	** refugia-anchored diurnal gr. liz.
margaretae	Groote Eylandt Two-lined Dragon	diurnal		52	4.28	Western Australia	
nobbi	Nobbi	cathemeral		74	13.2	Eastern Australia	** ground-foraging tree lizard
paraconvergens	Grey-striped Western Desert Dragon	diurnal		56	5.4	Western Australia	
phaeospinosa	Brigalow Nobbi	diurnal		68	10.1	Queensland	
pindan	Pindan Two-lined Dragon	diurnal		56	5.5	Western Australia	* ground-foraging tree lizard
reginae	Plain-backed Two-lined Dragon	diurnal		65	8.6	Western Australia	
superba	Superb Two-lined Dragon	crepuscular	invert.	80	16.7	Western Australia	* diurnal canopy walker
valens	Pilbara Two-lined Dragon	diurnal		63	8.1	Western Australia	
vescus	Northern Pilbara Tree Dragon	diurnal	invert.	57	5.7	Western Australia	
winneckei	Canegrass Dragon	diurnal	omnivore	58	6.1	Queensland	* ground-foraging tree lizard

majority (62% v. 26%) are associated with sand and dunes. None has been introduced (v. 1.7%); none is considered imperiled (v. 37%).

Dipsosaurus (fig. 4.4A)

Dipsosaurus (Iguanidae) is a two-species genus of small (53 g v. 609 g), terrestrial, giant herbivores found in the western United States and northwestern Mexico. Its shape index (1.53 v. 2.11 std. resid.) is the lowest in the family. Their tails are comparatively short (1.34 v. 1.53 × SVL) and breakable (50% v. 38%). Secretory pores are present (50% v. 38%), as is basic color change (50% v. 21%). The femoral pore secretions of *dorsalis* gained some notoriety when Alberts (1989) discovered that the secretions emit light in the UV range. The benefits of this feature are unclear; it may simply brighten spots on which the secretions are deposited, elevating the chance that a passing lizard will tongue-touch that spot and thereby receive the chemical signal (though the message is unclear to humans).

SSD is at the typical m>f level (1.0 v. 1.3 on –2 to 2 scale). None is sexually dichromatic (0% v. 38%), and seasonal color is reported only in *catalinensis* (50% v. 12%). The sex differences rollup score (on 0 to 6 scale) is an unremarkable 1.00 (v. 1.88).

The home-range area of *dorsalis* is moderate (–0.23 v. –1.06 std. resid.), and *dorsalis* is considered to be wary, though some island populations show island tameness. The mean active body temperature of *dorsalis* is a blistering 41.3°C (v. 34.3°C). Males defend space (v. 90%), and both species use burrows as refugia (v. 44%). *D. dorsalis* is found at low and medium elevation (0–1520 m) in dry areas (v. 21%; in this case 93–209 mm/y). It appears partial to sandy (v. 34%) and riparian areas (v. 22%) but not littoral (v. 41%) or human (v. 24%) habitats.

Their reproduction is typical oviparity, with a notably small clutch count (–0.75 v. 0.45). Hatchlings are comparatively large, as indicated by the very small ontogenetic mass range (–0.95 v. –0.55). RCM is slightly low (0.208 v. 0.231). Incubation is relatively short (–0.93 v. –0.76 std. resid.). Parental care is not reported (v. 31%).

Longevity is moderately elevated in *dorsalis* (0.55 v. 0.16 std. resid.), corresponding to 14.6 years. Population density in *dorsalis* has been estimated 11 times (3.1 kg/ha v. 12.4 kg/ha). This relatively low-density value probably reflects the mainland predator pressure, low productivity of the desert environment, and high latitude.

Endemism is moderate (50% v. 69%). No species has been introduced (v. 33%), and none is considered to be imperiled (v. 97%). In other respects *Dipsosaurus* are typical iguanids. These small innocuous her-

Dipsosaurus spp.	English name				Range core
			Size		
	Activity	Diet	mm	g	Ecological business model
catalinensis	Catalina Desert Iguana				Mexico
	diurnal	herbivore	101	46	**** giant herbivore
dorsalis	Desert Iguana				California
	diurnal	herbivore	112	62	**** giant herbivore

Dixonius spp.	English name				Range core
			Size		
	Activity	Diet	mm	g	Ecological business model
aaronbaueri	Bauer's Leaf-toed Gecko				Vietnam
	nocturnal		37	1.23	
hangseesom	Orange-tailed Ground Gecko				Thailand
	nocturnal		40	1.47	**** nocturnal ground lizard
kaweesaki	Sam Roi Yot Leaf-toed Gecko				Thailand
	nocturnal		39	1.43	
melanostictus	Dark-sided Ground Gecko				Thailand
	nocturnal		45	2.12	**** nocturnal ground lizard
minhlei	Vinh Cuu Leaf-toed Gecko				Vietnam
			43	1.90	
siamensis	Thai Leaf-toed Gecko				Southeast Asia
	nocturnal	invert.	50	2.84	**** nocturnal ground lizard
taoi	Tao's Leaf-toed Gecko				Vietnam
			41	1.61	
vietnamensis	Vietnam Leaf-toed Gecko				Vietnam
	nocturnal	invert.	45	2.12	**** nocturnal ground lizard

bivores are the subject of a curious belief of the Seri people of Mexico that a bite from the lizard is instantly fatal. I can personally report that this is not invariably so.

Dixonius

Dixonius is an eight-species genus of small (1.78 g v. 3.31 g), nocturnal terrestrial gekkonids from the lowland forests of Southeast Asia (Vietnam to Thailand and possibly Myanmar). They are relatively poorly known (19% v. 25% of cells filled) nocturnal ground lizards (100% v. 27%). Most are documented to have secretory pores (88% v. 36%), and all have small toepads (v. 30%). SSD and sexual dichromatism have not been reported. Despite their nocturnality, vocalizations have not been reported (v. 6.5% for spontaneous calling). Despite their terrestriality, burrow use has not been reported (v. 10.4%), perhaps because size-appropriate burrows are hard to detect or suitable extant crevices suffice. Despite the paucity of information on *Dixonius*, they are overrepresented in littoral zones (50% v. 11%), sandy areas (25% v. 11%), riparian zones (25% v. 17%), and in human habitats (50% v. 27%).

Reproduction is typical for gekkonids, though incubation duration (based solely on *siamensis*) is long (0.47 v. 0.17 std. resid.), presumably reflecting the cooler soils in mesic forests. Longevity is comparatively long (1.17 v. 0.15 std. resid.), though based on only one species (10 years in *hangseesom*). None has been introduced (v. 3.7%), nor is any considered imperiled (v. 47%). In other respects *Dixonius* are typical gekkonids.

Dopasia

Dopasia is a moderate (seven spp.) genus of medium-large (14.2 g v. 10.1 g), slender (shape index −1.43 v. −0.60 std. resid.) limbless anguids from southeastern Asia. Most are found on mainlands—Nepal and India to China and Vietnam—but they are also found on the large islands of Taiwan, Borneo, and Sumatra. Many are colloquially known as glass lizards for their readily breakable tails (100% v. 87%), which are slightly longer than the anguid norm (1.77 v. 1.66 × SVL). Their ecological business models are very poorly known (20% v. 27% of cells filled), and only two species assignments have been promulgated: a nocturnal (in this case cathemeral) ground lizard (v. 2%) and a litter cryptozoid (v. 6%).

None has a prehensile tail (v. 31%), and none shows sexual dichromatism (v. 26%) or seasonal color (v. 3.8%). None is arboreal (v. 39%). They are mostly found at low elevation (83% v. 47%). Absence of reports in littoral (v. 9%), sandy (v. 15%), riparian (v. 19%), and human (v. 11%) habitats are possibly due to underreporting.

Dopasia spp.	English name	Activity	Diet	Size mm	Size g	Range core	Ecological business model
buettikoferi	Borneo Glass Lizard	diurnal	invert.	99	3.49	Borneo	* litter cryptozoid
gracilis	Elegant Glass Lizard	cathemeral	invert.	223	36.8	Southern East Asia	* nocturnal ground lizard
hainanensis	Hainan Glass Lizard			213	32.3	China	
harti	South China Glass Lizard	diurnal		214	32.6	Southeast Asia	
ludovici	Tonkin Gulf Glass Lizard			149	11.4	China	
sokolovi	Vietnam Glass Lizard			135	8.7	Vietnam	
wegneri	Western Sumatra Glass Lizard			134	8.5	Indonesia	

Dracaena spp.	English name	Activity	Diet	Size mm	Size g	Range core	Ecological business model
guianensis	Northern Caiman Lizard	diurnal	invert.	312	1019	Ecuador	* refugia-anchored diurnal gr. liz.
paraguayensis	Paraguay Caiman Lizard		invert.	331	1224	Brazil	* wide-ranging chemosensor

Dracaena

Dracaena is a small teiid genus (two spp.) of very large-sized (1117 g v. 22.1 g), semiaquatic (100% v. 2.2%), burrow-retreating (100% v. 12%), mollusk-eating lizards. One is a refugia-anchored diurnal ground lizard (v. 25%), the other a wide-ranging chemosensor (v. 97%). The genus is found in freshwater marsh habitats of interior South America (Colombia to Paraguay, Brazil, and French Guiana). Unlike *Crocodilurus*, the resemblance to a crocodilian is both behavioral and morphological (the dorsum is a rectangular array of raised bony scutes). *D. guianensis* is one of the few nominally aquatic lizards that actually forages underwater. *Dracaena* lizards are unlikely to be confused with plants bearing the same genus name.

They are heavily built (shape index 2.02 v. 0.88 std. resid.), with short necks, rounded snouts, and massive heads. Their tails are short (1.70 v. 2.30 × SVL), unbreakable (v. 21%), and not blue (v. 15%). They do not change color (v. 1.3%) or possess secretory pores (v. 9%).

The sexes differ in color in *guianensis* (50% v. 35%). The relative tail lengths are much larger in male *guianensis* (m:f ratio 1.54, the highest ratio in Teiidae, v. 1.04), though the sample is small and may be unrepresentative.

Both species are dietary specialists on mollusks (mostly underwater ones in *guianensis*; terrestrial snails in *paraguayensis*). The shells are crushed with specialized molariform teeth, and the broken shell fragments are spit out. Their primary predator concealment tactic is not the teiid-standard flight (0% v. 77%), but escape to a burrow. *D. guianensis* basks directly (v. 58%). They are associated with mesic habitats (100% v. 30%) at low elevation (100% v. 95%). Both species are riparian (v. 46%); *guianensis* is found in human habitats (50% v. 29%); they are absent from littoral (v. 38%) and sandy (v. 47%) areas.

Clutch count is exceptionally small (–1.33 v. –0.51 std. resid.). Conversely, the young are exceptionally large (ontogenetic mass range –1.88 v. 0.10 std. resid.). Incubation is very long (0.79 v. –0.38 std. resid.). Perhaps these lizard hatchlings need to be of a minimum size to generate enough bite force to crush the local mollusks or to repel the predators resident in these productive wetlands.

Longevity is slightly short (–0.84 v. –0.74 std. resid.). Neither species is endemic (v. 50%). One occurs on one island: *guianensis* on Marajo Island (50% v. 37%). Neither has been introduced (v. 8%). Both are imperiled (v. 41%), but neither is on one of the IUCN's lists of the most endangered (v. 12%). In other respects *Dracaena* are typical teiids.

Draco (figs. 4.5C, 4.5D)

Draco is the second-largest agamid genus (41 spp.), and in many ways, the most unique. It is composed entirely of gliders, and *Draco* gliders seem to be the most highly specialized gliders, though very little fieldwork has been directed at these remarkable lizards, which are found in the warm southern parts of Asia from India to Vietnam and China, south through Malaysia and the Philippines to Indonesia. As *Draco* are a radiation of insular small green lizards that use their distinctively colored dewlaps to signal conspecifics as they run through the trees, one is often tempted to think of them as Asian analogues of *Anolis*. This analogy merits further consideration.

On first principles, gliders differ from diurnal canopy walkers in several ways. The breakthrough achieved by gliders is their escape from the chains of terrestrial locomotion. Gliders are no longer forced to crawl to the next branch via retracing the original branch back to its junction with the destination branch and following that branch out to the lizard's objective. This implies that the cost of territorial patrolling is dramatically reduced. It also implies that the size of a territory can be increased at minimal energetic expense. Gliders' mass-adjusted home ranges are the largest of any agamid genus, albeit based on an inadequate sample. If territories are much larger than a portion of a tree, sight lines become limiting; it may be easy to glide to a spot three trees over (*Draco* glide up to 50 m), but would the territory owner even be aware of an intruder at that distant spot, and, having traveled three trees over, what is to stop another intruder from invading the just-vacated but now distant and invisible perch? Although the territorial behavior of gliders is very poorly understood, it seems that island *Draco* solve this conundrum in many areas not by defending huge territories but by mate guarding. *Draco* males seem to choose monogamy so that they can keep their beloved in view by gliding after her. Anecdotal reports (*volans*) suggest that a female may challenge suitors to a flying contest and accept only those that can match her aerobatics. *D. melanopogon* males are also reported to physically engage rival males while gliding after females.

Tracking a mate in this manner is facilitated by *Draco*'s routine diet of ants. Sitting and waiting next to a column of ants is a very sedentary activity, making it comparatively easy to feed and remain in line-of-sight of a mate for long periods at low energetic cost. Displaying is less essential and, in any event, would conflict with the precious immobility and crypsis upon which 100% (v. 51%) of documented *Draco* depend. This often-immobile monogamy is a very different existence from the blustering polygyny of anoles.

If this line of reasoning is correct, it sheds light on why *Anolis* is so speciose and why *Draco* is not. The comparatively pedestrian anoles cannot easily travel far, so they divide a tree into subsections for which their locomotor capabilities are optimal. For example, long legs for trunk anoles, short legs for twig anoles. Large body sizes for canopy species; small bodies for grass species. Anole aficionados have identified dozens of morphological specializations for differing strata (Beuttell and Losos 1999; Henderson and Powell 2009; Losos 2009). Adoption of these specializations allows spectacular numbers of anole species to coexist in very small areas.

Unfortunately, a similar analysis has not been done with *Draco*, but the upper limits of *Draco* specialization can be seen in the number of species on islands of differing size (fig. 5.1). Most southeastern Asian islands under about 10,000 km^2 have one to two *Draco* species. Larger islands may also be depauperate. For example, the Philippine island of Luzon (104,683 km^2) supports one *Draco*; the comparably sized island of Cuba (109,879 km^2) has 63 *Anolis* species.

With little to be gained by morphological specialization on specific perch types, *Draco* seem to have foregone most body size radiation. Inspection of the *Draco* species table below confirms that almost all *Draco* fit within the 3–8 g range; only three are appreciably larger—all species occurring on the mainland: *abbreviatus* (10.4 g), *blanfordii* (10.4 g), and *maximus* (15.7 g). In contrast, *Anolis* sizes have radiated to mean masses ranging from 0.71 g to 65.2 g (fig. 5.2). In general, it seems that *Draco* speciate by geographic or insular isolation, rather than by microhabitat specialization. Despite their superficial similarity, *Draco* are not analogues of *Anolis*, and gliders are not analogous to canopy walkers.

Draco are about the size of an average lizard but small for an agamid (5.0 g v. 15.6 g). They are very light and slender (–0.58 std. resid v. 0.80), as befits an aerialist. Their tails/rudders do not break (v. 17%). Secretory pores are absent (v. 13%); *Draco* seem to emphasize the visual modality, which fits with their universal diurnality. Eighty-five percent of species show sex differences in color (v. 53%), but seasonal coloration is limited (2% v. 20%), and short-term color change is muted (12% v. 22% for any color change).

If males obtain mates by dexterous flying rather than pugilistic encounters with consexuals, selection might reward males that are small and nimble. Of agamid genera with more than two species reported, *Draco* has the lowest mean SSD (–0.46 v. 0.53 on –2 to 2 scale); there are 16 *Draco* species exhibiting m<f for five m>f. The five with males larger are all found on the Asian mainland (*maculatus* is also reported to engage in intense intermale competition for mates). All reported males are, however, intolerant of other males (v. 3% tolerant), though none is known to be hierarchical (v. 18%).

Eighty-three percent of *Draco* are documented to bask directly

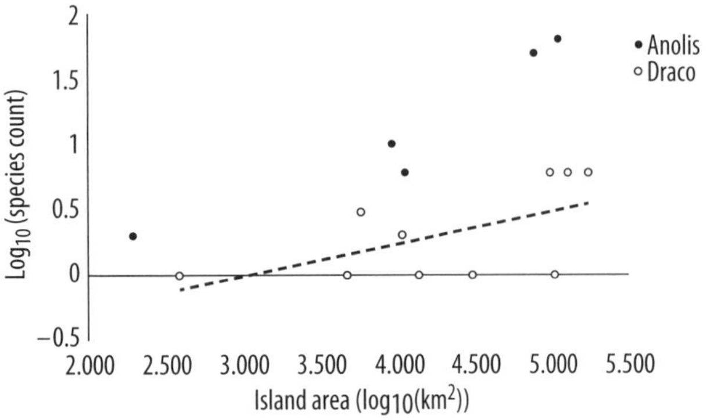

Fig. 5.1. The relationship between island area and species richness for the genera *Draco* and *Anolis*. Asian islands larger than Sulawesi omitted due to lack of analogues within the range of *Anolis*. Note that the modal species richness of *Draco* is 1 ($\log_{10}$ = 0) and the maximum is 6.

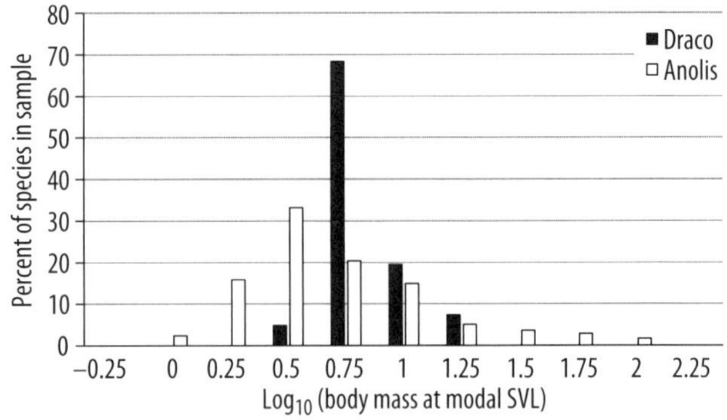

Fig. 5.2. Histogram of body masses for the genera *Anolis* and *Draco*.

(v. 24%); despite their basking, they maintain a very low mean body temperature (27.8°C v. 34.8°C). This may reflect their environment, which for all documented species (100%, tied for first place, v. 7%) is within the wettest zone (>3000 mm/y) at low elevation (v. 83%), though 22% extend to higher elevations.

Clutch count is slightly low (−0.08 v. 0.20 std. resid.), as befits a species dependent on bodily dexterity. The mass-adjusted ontogenetic size range is slightly small (0.46 v. 0.69 std. resid.), suggesting that *Draco* young are slightly large. Clutch frequency is very poorly documented. Grismer (2011) suggests that at least one species breeds continuously, but of the nine species with documentation, five are annual and four multiple brooders. Good estimates of RCM (none available) and brood frequency would be desirable for understanding the relationship between fecundity and weight loading of these aerial lizards. *Draco* are unusual for agamids in having adherent eggs (2%, fourth highest in family, v. 0.8%). Parental care (female nest defense in *sumatranus*) is also known (2% v. 3.5%) and likely to be underreported in these poorly known and hard-to-observe lizards.

Draco spp.	English name		Size		Range core
	Activity	Diet	mm	g	Ecological business model
abbreviatus	Orange-bearded Gliding Lizard				Malaysia
	diurnal	ants	104	10.4	**** glider
beccarii	Beccari's Gliding Lizard				Indonesia
	diurnal	ants	69	3.51	**** glider
biaro	Biaro Gliding Lizard				Indonesia
	diurnal	ants	67	3.33	**** glider
bimaculatus	Two-spotted Gliding Lizard				Philippines
	diurnal	ants	64	2.87	**** glider
blanfordii	Blanford's Gliding Lizard				Southern East Asia
	diurnal	ants	104	10.4	**** glider
boschmai	Flores Gliding Lizard				Indonesia
	diurnal		76	4.52	**** glider
bourouniensis	Buru Gliding Lizard				Indonesia
	diurnal		72	4.01	**** glider
caerulhians	Sangihe Gliding Lizard				Indonesia
	diurnal		71	3.75	**** glider
cornutus	Horned Gliding Lizard				Borneo
	diurnal	ants	76	4.57	**** glider
cristatellus	Crested Flying Lizard				Malaysia
	diurnal	ants	80	5.29	**** glider
cyanopterus	Mindanao Blue Gliding Lizard				Philippines
	diurnal	ants	79	5.06	**** glider
dussumieri	Western Ghats Gliding Lizard				India
	diurnal	ants	79	5.0	**** glider
fimbriatus	Crested Gliding Lizard				Malaysia
	diurnal	ants	82	5.5	**** glider

Draco spp.	English name				Range core
			Size		
	Activity	Diet	mm	g	Ecological business model
formosus	Beautiful Gliding Lizard				Thailand
	diurnal	ants	92	7.47	**** glider
guentheri	Sulu Gliding Lizard				Philippines
	diurnal	ants	75	4.38	**** glider
haematopogon	Yellow-bearded Gliding Lizard				Malaysia
	diurnal	ants	79	5.1	**** glider
indochinensis	Indochinese Gliding Lizard				Vietnam
	diurnal	ants	97	8.8	**** glider
iskandari	Tahulandang Flying Lizard				Indonesia
	diurnal	ants	65	3.04	**** glider
jareckii	Batan Gliding Lizard				Philippines
	diurnal	ants	74	4.21	**** glider
lineatus	Island Gliding Lizard				Borneo
	diurnal	ants	70	3.69	**** glider
maculatus	Spotted Gliding Lizard				Southern East Asia
	diurnal	ants	76	4.50	**** glider
maximus	Giant Gliding Lizard				Malaysia
	diurnal	ants	122	15.86	**** glider
melanopogon	Black-bearded Gliding Lizard				Borneo
	diurnal	ants	77	4.66	**** glider
mindanensis	Mindanao Orange Gliding Lizard				Philippines
	diurnal		88	6.7	**** glider
modiglianii	Enggano Gliding Lizard				Indonesia
	diurnal		66	3.18	**** glider
norvillii	Assam Gliding Lizard				India
	diurnal	invert.	96	8.4	**** glider
obscurus	Dusky Gliding Lizard				Malaysia
	diurnal	ants	85	6.1	**** glider
ornatus	White-spotted Gliding Lizard				Philippines
	diurnal	ants	78	4.97	**** glider
palawanensis	Palawan Gliding Lizard				Philippines
	diurnal		75	4.41	**** glider
quadrasi	Sibuyan Gliding Lizard				Philippines
	diurnal		71	3.77	**** glider
quinquefasciatus	Five-banded Gliding Lizard				Borneo
	diurnal	ants	92	7.5	**** glider
reticulatus	Leyte Gliding Lizard				Philippines
	diurnal	ants	91	7.31	**** glider
rhytisma	Banggai Flying Lizard				Indonesia
	diurnal	invert.	70	3.70	**** glider

Mass-adjusted longevity is very short (–0.77 v. –0.46 std. resid.), presumably reflecting high mortality in the bird-ridden trees of the wet tropics. Ninety-three percent (v. 36%) of species reside on at least one island; none has been introduced (v. 1.7%). Only 8% (v. 37%) are deemed imperiled; none of the imperiled lizards is on one of the IUCN's lists of the most endangered (v. 5%).

Draco spp.	English name				Range core
			Size		
	Activity	Diet	mm	g	Ecological business model
spilonotus	Sulawesi Gliding Lizard				Indonesia
	diurnal	ants	68	3.46	**** glider
spilopterus	Philippine Gliding Lizard				Borneo
	diurnal	ants	84	5.91	**** glider
sumatranus	Common Gliding Lizard				Borneo
	diurnal	ants	75	4.52	**** glider
supriatnai	Togian Flying Lizard				Indonesia
	diurnal		69	3.50	**** glider
taeniopterus	Barred Gliding Lizard				Southeast Asia
	diurnal	ants	72	3.96	**** glider
timoriensis	Timor Gliding Lizard				Indonesia
	diurnal		76	4.63	**** glider
volans	Java Gliding Lizard				Indonesia
	diurnal	ants	76	4.55	**** glider
walkeri	Sulawesi Black Gliding Lizard				Indonesia
	diurnal	ants	75	4.43	**** glider

Dryadosaura sp.	English name				Range core
			Size		
	Activity	Diet	mm	g	Ecological business model
nordestina	João Pessoa Eared Lizard				Brazil
			46	1.27	

Ebenavia spp.	English name				Range core
			Size		
	Activity	Diet	mm	g	Ecological business model
inunguis	Square-toed Gecko				Madagascar
	cathemeral	invert.	37	1.19	** nocturnal canopy walker
maintimainty	Dark Gecko				Madagascar
	nocturnal	invert.	27	0.49	**** nocturnal ground lizard

Dryadosaura

Dryadosaura nordestina is the sole member of this genus of elongate terrestrial Brazilian gymnophthalmids. It is slender (shape index −0.67 v. −0.32 std. resid.) and has a slightly short tail (1.53 v. 1.72 × SVL) that is not known to break (v. 28%). It lacks secretory pores (v. 29%); other sex differences are not documented. It is documented to occur over the annual rainfall range of 1120–1874 mm/y (mesic) at elevations of 0–343 m (low). Clutches are slightly large (0.26 v. −0.11 std. resid.). The reported RCM is anomalously high (0.56 v. 0.19). Are the eggs especially large (no data)? Are the young especially large (no data on ontogenetic size range)? This species is not known to have been introduced (v. 0.8%), nor has it been evaluated for conservation status.

Ebenavia

Ebenavia (two spp.) are very small (0.77 g v. 3.31 g) nocturnal gekkonids from Madagascar and nearby islands. One is arboreal (small perches), the other terrestrial. They have noticeably pointed snouts and relatively slender shape (shape index −0.18 v. 0.10 std. resid.). They have small toepads on the ends of their toes, and females (only) have claws (presumably to dig nests). Tails are a trifle short (0.99 v. 1.07 × SVL). They are not reported to have secretory pores (v. 46%). One species (*inunguis*) is documented for SSD, which is m<f, in keeping with their small size. No other sex differences have been noted. Male tolerance is also noted for one species (*inunguis*), which is consistent with small males. The apparently fixed clutch size of two is either waived or rarely respected in *Ebenavia*, presumably due to these geckos' small size; mean clutch count is 1.3 (0.14 v. −0.26 std. resid.). This species has been introduced to Mauritius, Tanzania, and possibly other insular sites. The other species (*maintimainty*) is considered endangered by the IUCN. In other respects *Ebenavia* are typical gekkonids.

Echinosaura

Echinosaura is a seven-species genus of terrestrial/cryptozoic or, arguably, semiaquatic gymnophthalmids found in northwestern South America and southern Central America (Panama to Ecuador). Ecological business models are diverse: litter cryptozoid (20% v. 55%), nocturnal ground lizard (40% v. 3%), and refugia-anchored diurnal ground lizard (40% v. 19%). *Echinosaura* are a

Echinosaura spp.	English name	Activity	Diet	Size mm	Size g	Range core	Ecological business model
brachycephala	Blunt Hedgehog Lizard	diurnal		65	6.5	Ecuador	** refugia-anchored diurnal gr. liz.
horrida	Rough Hedgehog Lizard			68	7.4	Colombia	** litter cryptozoid
keyi	Key's Lizard			67	7.0	Ecuador	
orcesi	Orces' Hedgehog Lizard			67	7.2	Ecuador	** refugia-anchored diurnal gr. liz.
palmeri	Palmer's Hedgehog Lizard	cathemeral		62	5.5	Colombia	*** nocturnal ground lizard
panamensis	Panama Hedgehog Lizard	cathemeral		63	5.9	Panama	*** nocturnal ground lizard
sulcarostrum	Furrow-snouted Hedgehog Lizard			39	1.24	Guyana	

Ecpleopus sp.	English name	Activity	Diet	Size mm	Size g	Range core	Ecological business model
gaudichaudii	Gaudichaud's Ecpleopus			40	0.80	Brazil	** litter cryptozoid

bit large (5.2 g v. 2.35 g) and relatively stout in build (shape index 0.26 v. –0.32 std. resid.). Only one (*sulcarostrum*, 14% v. 28%) is said to break its tail. Secretory pores are not reported (v. 29%); other sex differences are not documented. Two of the three poorly documented species (v. 14%) are active at any time of day, though also said to be nocturnal; the remainder (33% v. 79%) are diurnal. One species (*brachycephala*) is reported to be conspicuously wary (14% v. 7%). Two species are either cryptozoic or semiaquatic, another is terrestrial or semiaquatic, and the remaining two with data are terrestrial. They are overrepresented in riparian zones (100% v. 29%). Reproduction is normal, though the clutches are small (–0.46 v. –0.11 std. resid.). They are unusually frequent in expressions of conservation concern (50% v. 25%), and one-quarter (v. 11.5%) are classified among the most endangered categories of the IUCN. In other respects *Echinosaura* are typical gymnophthalmids.

Ecpleopus

Ecpleopus gaudichaudii is the sole member of this poorly known (filled cells 14% v. 19%) genus of elongate cryptozoic Brazilian gymnophthalmids. It is slender (shape index –0.85 v. –0.32 std. resid.). It lacks secretory pores (v. 29%); other sex differences are not documented. Clutch count is not documented. This species is not known to have been introduced (v. 0.8%), nor has it been evaluated for conservation status.

Egernia (fig. 4.2D)

Egernia is a moderate radiation (17 spp.) of large (43.7 g v. 3.3 g), robust (shape index 0.68 v. –0.46 std. resid.), viviparous (100% v. 36%), mostly diurnal (88% v. 74%) skinks of diverse ecologies endemic to Australia. They are comparatively well documented (filled cells 36% v. 25%). All documented species have some plants in their diet. Most are refugia-anchored diurnal ground lizards (65% v. 30%), but some are giant herbivores (24% v. 1%), and 12% (v. 1.7%) are crack hermits. Most species are tightly associated with crevices or burrows—100% (v. 43%) use crevices as their primary predator concealment tactic—and some have a short, heavily spined tail to repel intruders from their lairs. Several use traditional "latrines" where all or most feces are deposited in a traditional location near their refuge (*cunninghami*, *hosmeri*, *kingii*, *rugosa*, *stokesii*, and *striolata*). Many (see below) live in social groups of kin, in some cases including multiple generations.

Tails are short (0.91 v. 1.33 × SVL). Not all tails are breakable (82% v. 73%); this is especially true of the short spiny tailed species (*depressa*, *stokesii*). Two species may be elongate (*saxatilis*, *stokesii*; 12% v. 35%).

Only one species (*cygnitos*) has an explicit statement about SSD; the reported m<f is unusual for such large lizards, but it is characteristic of several ecological analogues (*Corucia*, *Tiliqua*). No other sex differences are known, yielding a very small (0.06 v. 0.28 on 0 to 6 scale) sex differences rollup score. This may reflect monogamy, which is reported for *cunninghami*, *saxatilis*, and *stokesii*.

As mentioned above, all species have some plants in their diet, though the proportion varies widely among species and populations. Twenty-three percent are considered to be herbivorous (v. 0.9%); 77% are omnivores (v. 13%), often with an ontogenetic shift toward greater plant consumption. At least one (*depressa*) capitalizes heavily on termite bonanzas (up to 86% of diet). Unusual for skinks, most *Egernia* are sit-and-wait foragers (75% v. 16%). One-quarter forage actively (v. 92%), as must all of the plant eaters. Some species (e.g., *stokesii*) eat only what is passing in front of their burrows during the day and forage actively only at night. Seven species are documented to eat vertebrates, though none does so with enough regularity to be considered carnivorous. Seabird eggs are considered a key seasonal food for *kingii* living on islands. *E. saxatilis* eats fungi. None is known to forage at electric lights (v. 0.4%). Home-range area is exceeding small (–0.87 v. –0.07 std. resid.) but poorly documented, equating to 15 m^2 in *cunninghami* and 800 m^2 in the much smaller *napoleonis*.

Direct basking is common (77% v. 21%). Mean active body temperature is warm (33.1°C v. 31.1°C). *Egernia* are not noted to live in groups of unrelated individuals (v. 16%), but several species are found in extended family groups: *saxatilis* juveniles may remain with their parents for as much as three years, during which time younger siblings will join them. *E. stokesii* aggregations of up to 17 kin have been observed; daughters seem more likely than sons to remain with their mothers and may inherit territories. Multigenerational family groups have been documented in *cunninghami*, *saxatilis*, and *stokesii*. Only 50% of male *Egernia* have been documented to defend space (v. 94%), although underreporting is likely.

Egernia are found in the middle of the annual rainfall spectrum (arid

Egernia spp.	English name				Range core
			Size		
	Activity	Diet	mm	g	Ecological business model
cunninghami	Cunningham's Skink				Eastern Australia
	diurnal	herbivore	197	191	**** giant herbivore
cygnitos	Western Pilbara Spiny-tailed Skink				Western Australia
	diurnal		95	18.3	** refugia-anchored diurnal gr. liz.
depressa	Pygmy Spiny-tailed Skink				Western Australia
	diurnal	omnivore	95	18.3	** refugia-anchored diurnal gr. liz.
douglasi	Kimberley Crevice-skink				Western Australia
	diurnal		125	44.8	** refugia-anchored diurnal gr. liz.
eos	Central Pygmy Spiny-tailed Skink				Western Australia
	diurnal		95	18.3	** refugia-anchored diurnal gr. liz.
epsisolus	Eastern Pilbara Spiny-tailed Skink				Western Australia
	diurnal		93	16.9	** refugia-anchored diurnal gr. liz.
formosa	Goldfields Crevice-skink				Western Australia
	diurnal	omnivore	89	14.6	** refugia-anchored diurnal gr. liz.
hosmeri	Hosmer's Skink				Queensland
	diurnal	omnivore	149	77.5	** refugia-anchored diurnal gr. liz. or crack hermit
kingii	King's Skink				Western Australia
	diurnal	herbivore	200	200	**** giant herbivore
mcpheei	Eastern Crevice Skink				Queensland
	diurnal	omnivore	122	41	* refugia-anchored diurnal gr. liz.
napoleonis	South-western Crevice-skink				Western Australia
	diurnal	omnivore	106	25.8	*** refugia-anchored diurnal gr. liz.
pilbarensis	Pilbara Crevice-skink				Western Australia
	diurnal	omnivore	106	26.1	* refugia-anchored diurnal gr. liz.
richardi	Bright Crevice-skink				Southwestern Australia
	diurnal	omnivore	92	16.6	** crack hermit
rugosa	Yakka Skink				Queensland
	cathemeral	herbivore	187	162	**** giant herbivore
saxatilis	Black Rock Skink				New South Wales
	diurnal	omnivore	113	13.4	* refugia-anchored diurnal gr. liz.
stokesii	Gidgee Skink				Australia
	crepuscular	omnivore	164	37.8	**** giant herbivore; juveniles may be crack hermit

Egernia spp.	English name				Range core
			Size		
	Activity	Diet	mm	g	Ecological business model
striolata	Tree Skink				Eastern Australia
	diurnal	omnivore	333	1047	*** crack hermit

Elasmodactylus spp.	English name				Range core
			Size		
	Activity	Diet	mm	g	Ecological business model
tetensis	Tete Fragile-skinned Gecko				Mozambique
	nocturnal	invert.	79	10.1	*** nocturnal ground lizard
tuberculosus	East African Fragile-skinned Gecko				Congo
	nocturnal	invert.	66	6.1	** trunk/wall lizard

to mesic: 151–3000 mm/y). They are known only from low elevation (<1200 m). Though not found in human habitats (v. 14%), they are about average in representation in littoral (18% v. 21%), sandy (12% v. 25%), and riparian (6% v. 23%) zones.

Few large young are the pattern in *Egernia*. Litter counts in *Egernia* are the fourth smallest in Scincidae (–0.97 v. 0.06 std. resid.). Neonatal size is correspondingly large (ontogenetic mass range –1.43, the second lowest in Scincidae, v. 0.12 std. resid.). Documented frequencies of pregnancies are mostly once per year but range from 0.94 per year in *stokesii* to more than 1 in *hosmeri*. RCM is average (0.230 v. 0.236). Parental care is very common (29% v. 3.2%) and often extends well beyond birth; it is documented in *depressa*, *kingii*, *napoleonis*, *saxatilis*, and *stokesii*. *Egernia* invest a lot in each neonate. Why do they bother? See the *Tiliqua* entry for further discussion of this point.

Longevity is about average (0.27 v. 0.17 std. resid.), equating to 3.3 years (*douglasi*) and 3.9 years (*depressa*) up to 26 years (*hosmeri*) and 40 years (*cunninghami*). Population density is low (0.09 kg/ha v. 0.35 kg/ha), based solely on *striolata* (n=1 sample). Endemism is about average (53% v. 64%). Island species are low (35% v. 57%). No species has been introduced (v. 2.1%). Most are imperiled (75% v. 47%), though none is on one of the IUCN's lists of the most endangered (v. 14%). In all other documented respects *Egernia* are typical skinks.

Elasmodactylus

Elasmodactylus (two spp.) are large (7.8 g v. 3.31 g) tuberculate, nocturnal gekkonids from central Africa, from the Congo through Mozambique. One is arboreal (trunk/wall); the other is terrestrial. One has fragile skin (*tetensis*); the other likely does also. *E. tetensis* is also documented for SSD (m>f), which is likely the case for *tuberculosus* as well (based on maximum sizes recorded). *E. tetensis* is documented to have secretory pores. No other sex differences are documented. *E. tuberculosus* gives stress calls. *E. tetensis* lives in groups (no details given). *E. tuberculosus* may refuge in unspecified burrows and is sometimes commensal with humans. Neither species has been introduced, nor are either considered imperiled (v. 47%). In other respects *Elasmodactylus* are typical gekkonids.

Elgaria

Elgaria is a moderate (seven spp.) genus of small-bodied (5.7 g v. 10.1 g), slender (shape index –1.2 v. –0.6), elongate (v. 31%) terrestrial anguids from coastal western North America (British Colombia to Mexico). Despite their proximity to major research universities, only about a third of the cells are filled (36%), and no ecological business models can be assigned with confidence. Most are refugia-anchored diurnal ground lizards (71% v. 8%); others are crack hermits (14% v. 2%) or litter cryptozoids (14% v. 6%).

Tails are on the long side (1.87 v. 1.66 × SVL), breakable (v. 87%), and despite their fragility are sometimes prehensile (29% v. 31%).

Sit-and-wait foraging dominates in *Elgaria* (67% v. 14%), and all foraging modes are present in the genus. Diel activity patterns are diverse, with cathemeral (29% v. 17%) and crepuscular (43% v. 13%) conspicuously overrepresented and diurnality rare (29% v. 70%). Many *Elgaria* are conspicuously wary (71% v. 18%). Despite a relatively high frequency of direct basking (43% v. 15%), mean active body temperatures are cool (25.5°C v. 24.5°C). *Elgaria* are found in diverse microhabitats but not arboreal (v. 39%). They are notably overrepresented in sand/dune (29% v. 15%) and riparian (100% v. 19%) habitats.

Elgaria are mostly oviparous (67% v. 39%), with an elevated clutch count (1.57 v. 0.85 std. resid.) and an elevated frequency of communal oviposition (14.3% v. 1.3%). The young are exceptionally small (1/282 v. 1/72 of maximum size; 2.73 v. 0.62 std. resid.). Parental care, in this case females remaining with their clutch, is about double the anguid frequency (29% v. 17%). Perhaps more memorable is that copulation durations have been recorded for two species,

Elgaria spp.	English name				Range core
			Size		
	Activity	Diet	mm	g	Ecological business model
cedrosensis	Cedros Alligator Lizard				Mexico
	diurnal	invert.	78	2.95	** crack hermit or litter cryptozoid
coerulea	Northern Alligator Lizard				Idaho
	cathemeral	invert.	98	5.1	** refugia-anchored diurnal gr. liz.
kingii	Madrean Alligator Lizard				Arizona
	cathemeral	invert.	100	5.5	** litter cryptozoid
multicarinata	Southern Alligator Lizard				California
	crepuscular	carnivore	128	10.0	** refugia-anchored diurnal gr. liz.
panamintina	Panamint Alligator Lizard				California
	crepuscular	invert.	118	8.2	** refugia-anchored diurnal gr. liz.
paucicarinata	San Lucan Alligator Lizard				Mexico
	diurnal	invert.	100	5.5	** refugia-anchored diurnal gr. liz. or litter cryptozoid
velazquezi	Central Baja California Alligator Lizard				Mexico
	cathemeral	invert.	99	5.4	** refugia-anchored diurnal gr. liz.

coerulea and *multicarinata*, and the maximums recorded have exceeded 4 hours and 24 hours, respectively. One might think that this stamina would elicit a degree of reverence by humans, but the most notable human-*Elgaria* interaction is that human parents are known to warn their (presumably misbehaving) children that *E. paucicarinata* may enter their rectums and cause disease while they sleep. No record is made as to whether this helps rambunctious children to settle down.

Population density estimates are unavailable. Many *Elgaria* occur on islands (43% v. 17%), one (*multicarinata*) has been introduced (14% v. .8%), and 29% (v. 63%) are considered imperiled, though none is on one of the IUCN's lists of the most endangered (v. 28%).

Emoia

Emoia is the fifth-largest skink genus (table 2.2) and a major radiation in Indo-Pacific islands (77 spp.). Most species are medium-sized (3.8 g v. 3.3 g; range 0.9–22 g), diurnal (100% v. 74%) terrestrial (60% v. 48%) or arboreal (36% v. 10%) lizards of islands (99% v. 57%) in Australasia (92% v. 46%) or southeastern Asia (Oriental 10% v. 24%). The overall geographic range is from Vietnam and the Ryukyu Islands to French Polynesia. Some species are found on many islands; for example, *caeruleocauda* is recorded from more than 250 named islands. Species richness declines with distance from New Guinea. The following series of islands (from west to east) illustrates this point: Sumatra (1 sp.), Java (2 spp.), Borneo (3), Sulawesi (4), Halmahera (5), Seram (4), New Guinea (34), New Britain (7), Bougainville (9), Choiseul (7), Guadalcanal (5), Vanuatu (1). The modal *Emoia* is a refugia-anchored diurnal ground lizard (47% v. 30%); others are trunk/wall lizards (18% v. 3%), ground-foraging tree lizards (13% v. 7%), diurnal canopy walkers (13% v. 2%), and litter cryptozoids (9% v. 28%). *E. atrocostata* is unusual in specializing on tide pools, into which it plunges for foraging or refuge. It also inhabits mangrove swamps but, except in the Philippines, does not frequent sandy beaches.

Emoia are of typical lizard build, but a little heavy (and sharp-snouted) for skinks (shape index −0.02 v. −0.46 std. resid.). Their tails are on the long side (1.62 v. 1.33 × SVL), and breakability is reported as fairly low (17% v. 73%). The 17% may be a gross undercount, if most reporters are taxonomists who assume that all *Emoia* have breakable tails and therefore fail to state the breakability expressly. Only 1.3% (v. 35%) of *Emoia* are elongate. Blue tails are relatively common (10.4% v. 4%).

SSD is mildly m>f (0.43 v. −0.31 on −2 to 2 scale), as expected for lizards of these ecological business models (table 4.1, trait 13). Sexual dichromatism (3.9% v. 7.6%) and seasonal color (1.3% v. 6.2%) are relatively infrequent, which is not characteristic of these ecological business models (table 4.1, traits 14, 15). The sex differences rollup score (on scale of 0 to 6) for the genus is about average (0.25 v. 0.28).

Emoia are overwhelmingly invertebrate eaters (97% v. 84%); *concolor* is known to be omnivorous (3% v. 13%). All are active foragers (v. 92%). An entomologist found that any insects dropped on the forest floor were rapidly seized by (likely kleptoparasitic) *caeruleocauda*. Electric light use is comparatively common (2.1% v. 0.4%). Home-range sizes are

Emoia spp.	English name	Activity	Diet	Size mm	Size g	Range core	Ecological business model
adspersa	Striped Small-scaled Skink	diurnal	invert.	74	8.2	Pacific Islands	** refugia-anchored diurnal gr. liz.
aenea	Bronze Emoia	diurnal		61	4.33	New Guinea	* refugia-anchored diurnal gr. liz.
aneityumensis	Anatom Treeskink			80	10.6	Vanuatu	* trunk/wall lizard
arnoensis	Micronesia Black Skink			75	8.5	Micronesia	** refugia-anchored diurnal gr. liz.
atrocostata	Seaside Skink	diurnal	invert.	63	4.8	Indo-Pacific	**** refugia-anchored diurnal gr. liz.
aurulenta	Brown's Emoia	diurnal		45	1.61	New Guinea	** refugia-anchored diurnal gr. liz.
battersbyi	Battersby's Emoia			64	5.2	New Guinea	
baudini	Baudin's Emoia			47	1.92	Indonesia	
bismarckensis	Bismarck Emoia	diurnal		61	4.33	New Guinea	* refugia-anchored diurnal gr. liz.
boettgeri	Boettger's Emoia	diurnal	invert.	64	5.2	Micronesia	*** refugia-anchored diurnal gr. liz.
bogerti	Bogert's Emoia			52	2.54	Indonesia	
brongersmai	Brongersma's Emoia	diurnal		56	3.36	New Guinea	
caeruleocauda	Pacific Bluetailed Skink	diurnal	invert.	49	2.00	Indo-Pacific Islands	**** refugia-anchored diurnal gr. liz.
callisticta	West Papua Emoia			46	1.69	Indonesia	
campbelli	Vitilevu Mountain Treeskink	diurnal	invert.	79	10.2	Fiji	** diurnal canopy walker or crack hermit
coggeri	Cogger's Emoia			41	1.20	New Guinea	
concolor	Fiji Slender Treeskink	diurnal	omnivore	81	10.8	Fiji	*** trunk/wall lizard

unknown. Only a few are conspicuously wary (5.2% v. 11.1%), a boldness which is characteristic of many island lizards. Primary predator concealment tactics are evenly split between retreat to vegetation (50% v. 25%) and retreat to a crevice or burrow (50% v. 43%).

Direct basking is about average (25% v. 21%); *submetallica* conforms (1.3% v. 1.5%). Mean active body temperature is mildly elevated (32.8°C v. 31.1°C). All documented species have males that defend space (v. 94%). The elevational profile of the genus is close to normal (l/m/h: 97/33/0 v. 89/35/6). They are found in littoral settings at a typical frequency (19% v. 21%) but are notably overrepresented in human habitat (36% v. 14%) and underrepresented in sandy areas (3% v. 25%) and riparian habitat (7.5% v. 23.0%).

All are oviparous (v. 64%) and have relatively small clutches (–0.24 v. 0.06 std. resid.). The ontogenetic mass range has been characterized for only four species but is notably large (0.75 v. 0.12 std. resid.), indicating relatively small hatchlings. Clutch frequency is not sufficiently documented to characterize, but for most species reproduction continues throughout the season of appropriate weather (usually the wet season). RCM is low, in keeping with the small clutches (0.133 v. 0.236 ratio). Incubation duration is about average for lizards of this size (–0.13 v. –0.40 std. resid.), equating to 105 days (3.3 g *guttata*) or 64 days (4.8 g *atrocostata*). Communal oviposition is about average (3.9% v. 2.6%). Eggs may hatch explosively in response to disturbance (*caeruleocauda*, *parkeri*). Parental care is absent (v. 3.2%).

Longevity is about average (–0.04 v. 0.17 std. resid.), equating to 3.5 years (*atrocostata*) and 10.8 years (*concolor*, 10.8 g). Population density is high (1.67 kg/ha v. 0.35 kg/ha), as expected for island lizards (table 3.13, trait 21). In this case, it is based on *caeruleocauda* ($n = 19$, 3.44 kg/ha) and *atrocostata* ($n = 2$, 0.81 kg/ha). Endemism is about average (57% v. 64%). Introductions are relatively common (3.9% v. 2.1%), as are introductions to more than two jurisdictions (1.3% v. 0.7). Relatively few *Emoia* are imperiled (31% v. 47%), but *Emoia* are overrepresented on the IUCN's lists of the most endangered (17% v. 14%). In other respects *Emoia* are similar to average skinks.

Some of these island systems may prove to be valuable for understanding population dynamics. For example, *guttata* (New Guinea) was undetectable in the 1930s, very common in the 1970s (>700 collected, at a time of little herpetological activity in New Guinea), and it subsequently became very rare. On Christmas Island *nativitatis* became very rare after some adverse introductions (Yellow Crazy Ants, Wolf Snake) and may be extinct as it has not been seen since 2008 (Cogger 2014).

Emoia spp.	English name		Size		Range core
	Activity	Diet	mm	g	Ecological business model
cyanogaster	Teal Treeskink				Melanesia
	diurnal	invert.	80	10.4	*** diurnal canopy walker
cyanura	Copper-tailed Skink				Pacific Islands
	diurnal	invert.	51	2.30	*** refugia-anchored diurnal gr. liz.
cyclops	Cyclops Emoia				Indonesia
			51	2.43	
digul	Digul Emoia				Indonesia
			46	1.80	
erronan	Futuna Treeskink				Vanuatu
	diurnal	invert.	42	11.4	* diurnal canopy walker
flavigularis	Yellow-throated Emoia				New Guinea
	diurnal	invert.	68	6.2	** ground-foraging tree lizard
guttata	Allison's Emoia				New Guinea
	diurnal		56	3.29	** refugia-anchored diurnal gr. liz.
impar	Azure-tailed Skink				Pacific Islands
	diurnal	invert.	48	2.04	*** refugia-anchored diurnal gr. liz.
irianensis	West Irian Mountains Emoia				Indonesia
			48	2.00	
isolata	Bellona Skink				Solomon Islands
	diurnal	invert.	53	2.80	*** refugia-anchored diurnal gr. liz.
jakati	Jakati Emoia				New Guinea
	diurnal	invert.	46	1.75	*** refugia-anchored diurnal gr. liz.
jamur	Lake Jamur Emoia				Indonesia
			47	1.83	
kitcheneri	Kitchener's Emoia				Indonesia
			37	0.89	
klossi	Kloss' Emoia				New Guinea
			74	8.3	
kordoana	Meyer's Emoia				New Guinea
			53	2.78	* trunk/wall lizard
kuekenthali	Kuekenthal's Maluku Emoia				Indonesia
			67	5.8	
laobaoense	Laobao Mangrove Skink				Vietnam
			62	4.67	
lawesi	Olive Small-scaled Skink				Samoa
	diurnal	invert.	89	14.7	** refugia-anchored diurnal gr. liz.

Emoia spp.	English name				Range core
			Size		
	Activity	Diet	mm	g	Ecological business model
longicauda	Shrub Whiptail Skink				New Guinea
	diurnal		82	11.3	** ground-foraging tree lizard
loveridgei	Loveridge's Emoia				New Guinea
			42	1.34	
loyaltiensis	Loyalty Islands Emoia				New Caledonia
	diurnal		69	6.4	**** trunk/wall lizard
maculata	Spotted Blue-tailed Skink				Solomon Islands
	diurnal	invert.	53	2.85	** ground-foraging tree lizard
maxima	Beefy Emoia				New Guinea
			43	1.41	* refugia-anchored diurnal gr. liz.
mivarti	Admiralties Emoia				New Guinea
	diurnal		50	2.32	** refugia-anchored diurnal gr. liz.
mokolahi	Tonga Robust Treeskink				Tonga
	diurnal	invert.	89	14.8	*** ground-foraging tree lizard
mokosariniveikau	Vanua Levu Slender Treeskink				Fiji
	diurnal		51	2.51	* diurnal canopy walker
montana	New Guinea Mountains Emoia				New Guinea
			59	4.01	
nativitatis	Christmas Island Whiptail-skink				Christmas Island (Australia)
	diurnal	invert.	74	7.3	** litter cryptozoid
nigra	South Pacific Black Skink				Pacific Islands
	diurnal	invert.	101	22.3	** refugia-anchored diurnal gr. liz.
nigromarginata	Vanuatu Coppery Vineskink				Vanuatu
	diurnal	invert.	66	5.5	** diurnal canopy walker
obscura	Dark New Guinea Skink				New Guinea
			52	2.67	
oribata	Blotch-necked Emoia				New Guinea
			63	4.86	
oriva	Rotuma Barred Treeskink				Fiji
	diurnal	invert.	77	9.4	* trunk/wall lizard
pallidiceps	Pale-headed Emoia				New Guinea
	diurnal		51	2.52	** refugia-anchored diurnal gr. liz.
paniai	Coastal Emoia				Indonesia
			46	1.78	
parkeri	Fiji Copper-headed Skink				Fiji
	diurnal	invert.	48	2.00	** trunk/wall lizard (if mongoose) or ground-foraging tree lizard

Emoia spp.	English name				Range core
			Size		
	Activity	Diet	mm	g	Ecological business model
physicae	Slender Emoia				New Guinea
			58	3.74	** ground-foraging tree lizard
physicina	Small Forest Emoia				New Guinea
	diurnal		46	1.74	** refugia-anchored diurnal gr. liz.
ponapea	Pohnpei Skink				Micronesia
	diurnal	invert.	44	1.54	** refugia-anchored diurnal gr. liz.
popei	Pope's Emoia				New Guinea
	diurnal		56	3.29	** litter cryptozoid
pseudocyanura	Solomons Blue-tailed Skink				New Guinea
	diurnal	invert.	57	3.45	* refugia-anchored diurnal gr. liz.
pseudopallidiceps	Side-slashed Emoia				New Guinea
			55	3.15	** litter cryptozoid
reimschiisseli	Reimschiissel's Emoia				Indonesia
			61	4.26	
rennellensis	Rennell Bluetailed Skink				Solomon Islands
	diurnal	invert.	42	1.34	** ground-foraging tree lizard
ruficauda	Red-tailed Swamp Skink				Philippines
	diurnal		48	2.02	** ground-foraging tree lizard
rufilabialis	Red-lipped Striped Skink				Solomon Islands
	diurnal	invert.	56	3.36	** refugia-anchored diurnal gr. liz.
samoensis	Pacific Robust Treeskink				Samoa
	diurnal		96	18.9	** trunk/wall lizard
sanfordi	Toupeed Treeskink				Vanuatu
	diurnal	invert.	95	18.3	*** trunk/wall lizard
schmidti	Schmidt's Skink				Solomon Islands
	diurnal	invert.	54	2.87	** refugia-anchored diurnal gr. liz.
similis	Dunn's Emoia				Indonesia
			42	0.87	
slevini	Mariana Skink				Mariana Islands
	diurnal	invert.	70	7.6	*** refugia-anchored diurnal gr. liz.
sorex	Sorex Emoia				Indonesia
			51	2.43	
submetallica	Macleay's Emoia				New Guinea
			55	3.15	* refugia-anchored diurnal gr. liz.
taumakoensis	Taumako Skink				Solomon Islands
	diurnal	invert.	51	2.51	** litter cryptozoid

Emoia spp.	English name				Range core
			Size		
	Activity	Diet	mm	g	Ecological business model
tetrataenia	Four-striped Emoia				New Guinea
			55	3.15	** diurnal canopy walker
tongana	Polynesia Slender Treeskink				Samoa
	diurnal	invert.	62	4.62	** trunk/wall lizard
tropidolepis	Strong-keeled Emoia				New Guinea
			56	3.29	** refugia-anchored diurnal gr. liz.
trossula	Fiji Barred Treeskink				Fiji
	diurnal	invert.	89	14.5	** trunk/wall lizard
tuitarere	Rarotonga Treeskink				Cook Islands
	diurnal		78	9.7	** diurnal canopy walker
veracunda	Vogelkop Emoia				New Guinea
	diurnal		47	1.92	** litter cryptozoid

Enyalioides spp.	English name				Range core
			Size		
	Activity	Diet	mm	g	Ecological business model
altotambo	Alto Tambo Woodlizard				Ecuador
	diurnal		105	33.4	
anisolepis	Rough-scaled Woodlizard				Ecuador
	diurnal		102	30.5	
azulae	Cordillera Azul Woodlizard				Peru
			78	12.7	
binzayedi	Picota Woodlizard				Peru
			97	25.3	
cofanorum	Cofan Woodlizard				Ecuador
	diurnal	invert.	111	39.9	** ground-foraging tree lizard
heterolepis	Bocourt's Woodlizard				Colombia
	diurnal		108	36.4	*** ground-foraging tree lizard
laticeps	Amazon Woodlizard				Peru
	diurnal	invert.	117	46.6	** ground-foraging tree lizard
microlepis	Small-scaled Woodlizard				Ecuador
			100	28.5	* ground-foraging tree lizard
oshaughnessyi	Red-eyed Woodlizard				Colombia
	diurnal		146	98	* diurnal canopy walker
palpebralis	Horned Woodlizard				Bolivia
	diurnal	invert.	105	32.8	* immobility expert
praestabilis	Canelos Woodlizard				Ecuador
	diurnal		101	29.2	* ground-foraging tree lizard

Enyalioides

Enyalioides (15 spp.) is a moderate genus of crested, diurnal, cryptic, arboreal hoplocercids centered in Peru and extending southeast to Brazil and north to Panama. Information is available to assign only about half of the species to ecological business models, which are predominantly ground-foraging tree lizards (71% v. 50%). Single species are also described as diurnal canopy walkers (14% v. 10%) and immobility experts (14% v. 10%).

Enyalioides are of average hoplocercid size (32.5 g v. 31.3 g) and build (0.95 v. 0.94 std. resid.). Tails are of average length (1.40 v. 1.34 × SVL), and although breakage is reported at typical frequency (60% v. 47%), many species are annotated to note that tail autotomy declines ontogenetically; the 60% value might best be thought to characterize juveniles only. Crests are nearly universal (80% recorded v. 63%). Secretory pores are normal (7% v. 10.5%). Color change (20% v. 16%) and advanced color change (13% v. 10.5%) are at typical hoplocercid frequencies. Sex differences follow the familial pattern closely.

Diet, diel activity, predator evasion, habitat use, and reproduction also closely follow the family template. Basking and thermal conformance are underreported (6.7% v. 10.5% and 0% v. 5.3%, respectively). *Enyalioides* are underrepresented in the ranks of the imperiled (14% v. 30%); none is found on the IUCN's lists of the most endangered. In other respects *Enyalioides* are typical hoplocercids.

Enyalioides spp.	English name				Range core
			Size		
	Activity	Diet	mm	g	Ecological business model
rubrigularis	Red-throated Woodlizard				Ecuador
			102	30.5	
rudolfarndti	Oxapampa Woodlizard				Peru
	diurnal		92	21.9	
sophiarothschildae	Rothchild's Woodlizard				Peru
	diurnal		106	34.1	
touzeti	Touzet's Woodlizard				Ecuador
			107	34.9	

Enyalius spp.	English name				Range core
			Size		
	Activity	Diet	mm	g	Ecological business model
bibronii	Bibron's Fathead Anole				Brazil
			95	25.3	* diurnal canopy walker
bilineatus	Two-lined Fathead Anole				Brazil
	diurnal		76	12.2	* ground-foraging tree lizard
boulengeri	Boulenger's Fathead Anole				Brazil
			93	23.7	
brasiliensis	Brazil Fathead Anole				Uruguay
			86	18.2	
catenatus	Wied's Fathead Anole				Brazil
			89	20.6	
erythroceneus	Red-flanked Fathead Anole				Brazil
			74	10.8	
iheringii	Ihering's Fathead Anole				Brazil
	diurnal	invert.	98	28.3	
leechii	Leech's Fathead Anole				Brazil
			95	25.1	* ground-foraging tree lizard
perditus	Jackson's Fathead Anole				Brazil
			76	11.9	
pictus	Tattooed Fathead Anole				Brazil
			89	20.2	

Enyalius

Enyalius (10 spp.) is a moderate genus of poorly known (19% v. 21%), relatively large (18.6 g v. 16.4 g), diurnal (100% v. 86%) leiosaurids from Brazil (and occasionally surrounding countries). Reflecting their obscurity, ecological business models can be guessed for only three species. In only one species (*leechii*) is there an appraisal of both foraging mode (active) and primary predator concealment tactic (crypsis), two traits that are rarely found together (table 3.8, trait 16). Thus what little information there is on the natural history of *Enyalius* is perplexing.

Enyalius are robust (shape index 0.81 v. 0.76) with large heads ("fathead anoles") and long tails (2.00 v. 1.66 × SVL) that are neither breakable (v. 3%) nor prehensile (v. 9%). The only species documented for SSD (*leechii*) follows the family pattern (m<f). It is also sexually dichromatic (10% v. 15%).

The only species documented for rainfall (*brasiliensis*) is found in the mesic band (1140 mm/y in this case). Two species are documented for elevational band (both low). *Enyalius* are overrepresented in riparian zones (17% v. 4%) and human habitats (17% v. 8%).

Reproduction is oviparous (v. 95%) with a slightly elevated clutch count (0.90 v. 0.68 std. resid.). Incubation is slow by lizard standards but relatively quick by leiosaurid standards (1.82 v. 2.81 std. resid.), perhaps a reflection of their low elevation. One (*bibronii*) RCM ratio is known for the genus (0.152), which is also the only datum for the family (v. 0.226 for lizards).

Endemism is relatively high for *Enyalius* (80% v. 70%), probably because the country of Brazil is so large. No *Enyalius* is considered imperiled (v. 14%), though the profound level of ignorance about this genus is likely a contributing factor. In other respects *Enyalius* are typical leiosaurids.

Epibator

Epibator is a two-species genus of large (6.5 g v. 3.3 g), diurnal (100% v. 74%), arboreal (100% v. 10%) skinks endemic to the island of New Caledonia. One species (*nigrofasciolatum*) is known to be a ground-foraging tree lizard (v. 7%); the other (*greeri*) is known with certainty from only one specimen.

They are moderately robust (shape index 0.17 v. –0.46 std. resid.), with average-length tails (1.42 v. 1.33 × SVL) that are not documented to be breakable (0% v. 73%). No sex differences are known. The better-known species basks directly (50% v. 21%) and is conspicuously wary (50% v. 11%). They are associated with littoral areas (50% v. 21%) and human habitat (50% v. 14%)..

They are oviparous (v. 64%) with average clutch counts (0.21 v. 0.06 std. resid.). Young are small (ontogenetic mass range 0.52 v. 0.12 std. resid.). Clutch frequency is unknown. Incubation is slightly short (–0.58 v. –0.40 std. resid.), based on *nigrofasciolatum* (45 days).

The maximum recorded longevity of the same species is 10 years (0.48 v. 0.17 std. resid.). Neither has been introduced (v. 2.1%). *E. greeri* is considered imperiled (v. 47%), but neither is on an IUCN list of the most endangered species (v. 14%). In other respects *Epibator* are probably typical skinks.

Epibator spp.	English name				Range core
			Size		
	Activity	Diet	mm	g	Ecological business model
greeri	Greer's Tree Skink				New Caledonia
			53	2.78	
nigrofasciolatum	Green-bellied Tree Skink				New Caledonia
	diurnal	invert.	89	15.0	*** ground-foraging tree lizard

Eremias

Eremias is a major radiation (36 spp.) of burrow-using (100% v. 48%), diurnal (100% v. 95%), terrestrial (100% v. 90%) lacertids found predominantly in dry (70% v. 30%) areas of central Asia (especially the "——stans," but ranging from Romania to North Korea, north to Russia and south to Pakistan). Most are refugia-anchored diurnal ground lizards (67% v. 57%). The remainder are categorized as wide-ranging chemosensors (33% v. 25%), though none is assigned with high confidence and two-thirds are unknown.

Eremias are exactly average in size (5.10 g v. 5.10 g) and build (shape index 0.27 v. 0.27 std. resid.), with tails that are a bit short (1.78 v. 2.06 × SVL). SSD, in this case m>f, is recorded for only one species (*persica*) and sexual dichromatism for only *lalezharica* (v. 33%).

Ant specialists, an uncommon feature in Lacertidae (1.7%), show up in this genus (*pleskei*). All *Eremias* are active foragers (v. 92%), albeit with home ranges that may not be large (–0.10 v. 0.27 std. resid., based on three species). The mix of primary predator concealment tactics is perhaps unusual for active foragers: retreat to crevices or burrows (36% v. 50%), retreat to vegetation (36% v. 42%), sand diving (18% v. 7%), and flight (9% v. 22%).

Direct basking is rare (3% v. 25%), and mean active body temperature is moderate (29.0°C v. 33.3°C).

All *Eremias* males defend space (v. 59%). *Eremias* are restricted to the dry (88% v. 30%) and arid (38% v. 40%) rainfall belts (56–400 mm/y). *Eremias* are underrepresented in littoral (4% v. 22%), riparian (13% v. 28%), and human (0% v. 16%) habitats, but they are routinely found in sandy areas (61% v. 44%).

Eremias is the only lacertid genus in which viviparity appears, and it occurs at least five times, apparently only in species that are hard to spell (*multiocellata*, *przewalskii*, *stummeri*,

Eremias spp.	English name				Range core
			Size		
	Activity	Diet	mm	g	Ecological business model
acutirostris	Point-snouted Racerunner				Afghanistan
	diurnal	invert.	67	6.9	** wide-ranging chemosensor
afghanistanica	Afghan Racerunner				Afghanistan
			57	4.42	
andersoni	Anderson's Racerunner				Iran
	diurnal		38	1.32	
argus	Mongolia Racerunner				Mongolia
	diurnal	invert.	64	6.1	* refugia-anchored diurnal gr. liz. or wide-ranging chemosensor
arguta	Steppe Racerunner				Central Asia
	diurnal	invert.	72	8.5	** wide-ranging chemosensor
aria	Aria Racerunner				Afghanistan
			53	3.51	
brenchelyi	Ordos Racerunner				China
			63	5.9	* refugia-anchored diurnal gr. liz. or wide-ranging chemosensor

szczerbaki, yarkandensis). Cold soil temperatures probably promote viviparity in this genus. Clutch/litter count is normal (–0.10 v. 0.13 std. resid.). Incubation is relatively short (–0.57 v. –0.36). Hard eggshells appear in Lacertidae only in *Acanthodactylus* and *Eremias* (50% v. 14%). Did the hard shells of *Eremias* play a role in promoting viviparity? Perhaps viviparity and hard shells are different responses to the cold, dry climate? (There is no viviparity reported for *Acanthodactylus*, but its climate is dry and hot). Parental care is not reported (v. 1.9%).

Recorded longevity is above average (0.30 v. 0.19 std. resid., n = 3 spp.), which equates to 11 years in *argus*. Credible population density estimates (eight) are available for three species; they average a low 0.12 kg/ha (v. 2.07 kg/ha). Noncredible estimates are available for nine species (see database). The low productivity of these dry temperate lands probably depresses population density. An average number of species are endemic (33% v. 35%). No species is found on any island (v. 26%) or introduced (v. 4.6%). A typical proportion are imperiled (46% v. 56%), and some are listed by the IUCN as among the most endangered (6.7% v. 16%). In other respects *Eremias* appear to be typical lacertids.

Eremias spp.	English name				Range core
			Size		
	Activity	Diet	mm	g	Ecological business model
buechneri	Kaschar Racerunner				China
			57	4.26	
cholistanica	Cholistan Racerunner				Pakistan
			47	2.51	
fasciata	Sistan Racerunner				Afghanistan
	diurnal	invert	56	4.10	** refugia-anchored diurnal gr. liz.
grammica	Reticulate Racerunner				Central Asia
	diurnal	invert.	70	7.9	*** refugia-anchored diurnal gr. liz.
intermedia	Aralo-Caspian Racerunner				Turkmenistan
	diurnal		55	4.00	* refugia-anchored diurnal gr. liz. or wide-ranging chemosensor
isfahanica	Ghomishloo Racerunner				Iran
	diurnal		58	4.58	* refugia-anchored diurnal gr. liz.
kavirensis	Maranjaab Racerunner				Iran
			69	7.6	
kokshaaliensis	Kokshaal Racerunner				China
			50	2.99	
kopetdaghica	Kopet Dagh Racerunner				South of Caspian Sea
			64	6.1	
lalezharica	Lalehzar Racerunner				Iran
	diurnal		60	5.1	
lineolata	Striped Racerunner				Afghanistan
	diurnal		45	2.23	
montana	Mountain Racerunner				Iran
	diurnal		52	3.38	
multiocellata	Multi-ocellated Racerunner				Central Asia
	diurnal	invert.	63	6.0	
nigrocellata	Black-ocellated Racerunner				Turkmenistan
			64	6.1	
nikolskii	Kirghiz Racerunner				Central Asia
	diurnal		64	6.1	
papenfussi	Papenfuss' Racerunner				Iran
			54	3.66	
persica	Caspian Steppe Racerunner				Black and Caspian Sea areas
	diurnal	invert.	81	12.1	** wide-ranging chemosensor
pleskei	Pleske's Racerunner				Black and Caspian Sea areas
	diurnal	ants	53	3.50	

Eremias spp.	English name		Size		Range core
	Activity	Diet	mm	g	Ecological business model
przewalskii	Gobi Racerunner				Mongolia
	diurnal	invert.	79	11.6	
quadrifrons	Alachan Racerunner				China
			81	12.2	
regeli	Tajik Racerunner				Turkmenistan
			60	5.0	
scripta	Sand Racerunner				Turkmenistan
	diurnal	invert.	57	4.26	** wide-ranging chemosensor
strauchi	Strauch's Racerunner				S Caspian Sea
	diurnal	invert.	67	7.0	* refugia-anchored diurnal gr. liz.
stummeri	Tien Shan Racerunner				Central Asia
			69	7.8	
suphani	Suphan Racerunner				Turkey
	diurnal		62	5.6	
szczerbaki	Szczerbak's Racerunner				Kyrgyzstan
			57	4.26	
velox	Central Asia Racerunner				Central Asia
	diurnal	invert.	72	8.6	** refugia-anchored diurnal gr. liz.
vermiculata	Variegated Racerunner				Mongolia
			61	5.3	
yarkandensis	Yarkend Racerunner				China
			53	3.51	

Eremiascincus (fig. 4.12C)

Eremiascincus is a moderate radiation (14 spp.) of slightly large (5.2 g v. 3.3 g), sand-swimming (100% v. 22%), nocturnal (67% v. 15%) or cathemeral (33% v. 9%) skinks of diverse ecologies found in Australia, New Guinea, and adjacent parts of Indonesia. The modal ecological business model is sand shark (50% v. 7%). Nocturnal ground lizards are also common (30% v. 2%); the remainder are litter cryptozoids (20% v. 28%). These business models suggest a general aversion to being seen and a predilection for loose substrates.

Eremiascincus are morphologically typical of skinks with regard to build (shape index −0.12 v. −0.46 std. resid.), tail length (1.44 v. 1.33 × SVL), tail breakability (57% v. 73%), and elongation (36% v. 35%). They differ from average skinks in their complete absence of blue tails (v. 4%). Sex differences are not reported.

The diet of *Eremiascincus* is exclusively invertebrates (v. 84%), though a few vertebrates are taken occasionally. Compared to most skinks they are notably committed to sit-and-wait foraging (100% v. 16%), though two-thirds are also said to forage actively (67% v. 92%). *E. pardalis* ambushes prey from the mouth of its burrow during the day but goes abroad at night. *Eremiascincus* do not feed at electric lights (v. 0.4%) and home-range size is unknown (see last paragraph).

Staying out of sight would seem to limit direct basking (0% v. 21%); *richardsonii* basks indirectly. It is the only genus member for which mean active body temperature has been recorded (26.3°C v. 31.1°C). This low temperature likely reflects *richardsonii*'s nocturnality. Social relations are poorly understood; the only documented males (*douglasi*) defend space (v. 94%).

Burrow use is widely reported (57% v. 24%). *E. fasciolatus* and *richardsonii* are found in the arid rainfall belt (151–400 mm/y). The three elevationally documented species are all low (<1200 m). *Eremiascincus* are overrepresented in littoral (36% v. 21%), sandy (73% v. 25%), and riparian (36% v. 23%) habitats. They are found in human habitats at average frequency (9% v. 14%).

All are oviparous (v. 64%), though *pardalis* lays eggs at such an advanced stage of development that viviparity looms. Clutch number is about average for lizards of this size (0.15 v. 0.06 std. resid.). Hatchlings are comparatively small (ontogenetic mass range 0.69 v. 0.12 std. resid.). RCM, based on three species, is comparatively light (0.168 v. 0.236). Parental care is present (7% v. 3.2%): *richardsonii* hatchlings may remain with the mother for an unspecified period.

Reported maximum longevity (based on *pardalis*) is short (−0.23 v. 0.35 std. resid.), equating to 4 years. Because hidden lizards are not a favorite of herpetoculturists, it is likely that additional sampling would raise longevity estimates. Population density is moderately high (0.65 kg/ha v. 0.35 kg/ha), based on one estimate of *richardsonii*. Endemism is low (29%

Eremiascincus spp.	English name				Range core
	Activity	Diet	Size mm	Size g	Ecological business model
antoniorum	Antoni Timor Skink				Indonesia
			57	3.57	
brongersmai	Brown-sided Bar-lipped Skink				Western Australia
	nocturnal	invert.	85	6.3	** nocturnal ground lizard
butlerorum	Sumba Skink				Indonesia
			45	1.65	
douglasi	Orange-side Bar-lipped Skink				Northern Territory
	nocturnal	invert.	73	4.03	** nocturnal ground lizard
emigrans	Sumba Bar-lipped Skink				New Guinea
			59	2.24	
fasciolatus	Narrow-banded Sand-swimmer				Queensland
	nocturnal	invert.	90	15.3	*** sand shark
intermedius	Northern Narrow-banded Skink				Western Australia
			73	7.8	* sand shark
isolepis	Northern Bar-lipped Skink				Northern Australia
	cathemeral	invert.	67	3.25	** nocturnal ground lizard and litter cryptozoid
musivus	Mosaic Desert Skink				Western Australia
			55	3.16	* litter cryptozoid
pallidus	Western Sand-swimming Skink				Western Australia
			64	5.2	** sand shark
pardalis	Lowlands Bar-lipped Skink				New Guinea
	cathemeral	invert.	67	3.17	*** litter cryptozoid
phantasmus	Ghost Skink				Queensland
			76	8.8	** sand shark
richardsonii	Broad-banded Sand-swimmer				Australia
	nocturnal	invert.	95	18.1	** sand shark
timoriensis	Timor Skink				Indonesia
			78	9.7	

Eroticoscincus sp.	English name				Range core
	Activity	Diet	Size mm	Size g	Ecological business model
graciloides	Elf Skink				Queensland
	diurnal	invert.	34	0.50	** litter cryptozoid

v. 64%). This is surprising given the large size of most jurisdictions (Indonesia, Papua New Guinea, Australian states) and the genus's hidden and substrate-specific lifestyles (digging lizards are often highly localized). Island species are exactly average (57% v. 57%). None has been introduced (v. 2.1%), and none is considered imperiled. This absence of imperilment is also anomalous for digging species, but these species are presumably protected by their comparatively extensive ranges. The extensive ranges for digging species implies substantial gene flow among populations. Complete activity range estimates would be most welcome for these anomalous diggers. In all other documented respects *Eremiascincus* are typical skinks.

Eroticoscincus

Eroticoscincus graciloides is the sole member of its scincid genus and is found in southeastern coastal Queensland. It is a litter cryptozoid with a morphology that reminds me of a *Sphaerodactylus*, though somewhat more slender in build (–0.96 v. –0.46 for skinks, –0.16 for *Sphaerodactylus*). Its tail is short (1.0 v. 1.33 × SVL) and breakable (v. 73%). It is elongate (v. 35%) and shows seasonal color (v. 6%), but the sexes are otherwise similar. It relies on crypsis (v. 9%) for its primary predator concealment. It is found in riparian habitats (v. 23%). It is oviparous (v. 64%), and its eggs develop slowly (0.63 v. –0.40 std. resid.), equating to 68 days. Communal oviposition occurs (v. 2.6%). It is found on Fraser Island (v. 57%). It has not been introduced (v. 2.1%), and it has not been evaluated for conservation status.

Eublepharidae (figs. 4.9A, 4.10A, 4.10B)

The 36 eublepharid geckos are in six genera: *Aeluroscalabotes*, *Coleonyx*, *Eublepharis*, *Goniurosaurus*, *Hemitheconyx*, and *Holodactylus*, which do not group geographically. There are eublepharids in the New World (Utah south to Costa Rica: *Coleonyx*), Asia (Syria east to Japan and Indonesia: *Aeluroscalabotes*, *Eublepharis*, *Goniurosaurus*), and Africa (Mauritania east to the Horn: *Hemitheconyx*, *Holodactylus*). With the mild exception of *Goniurosaurus* and *Holodactylus*, the eublepharids are relatively well known (35% v. 27% cells filled; family v. lizards). Most eublepharids are nocturnal ground lizards (86% v. 8%), but several are ground-foraging tree lizards (7% v. 9%), and one each is a crack hermit (3.4% v. 2.6%) and nocturnal canopy walker (3.4% v. 2.1%). Eublepharids frequently exhibit TSD (at least seven species show evidence of it; at least one species lacks TSD).

They are substantially larger than typical lizards (13.2 g v. 5.3 g) and slightly more robust than normal for their mass (shape index 0.34 v. 0.0 std. resid.). Tails are short (0.76 v. 1.37 × SVL) and mostly breakable (83% v. 57%). A few species have prehensile tails (8.3% v. 5.0%). They lack crests (v. 5%), expanded toepads, and elongation (v. 11%). Most genera have secretory pores (67% v. 15%). Darkening color change is found only in *Eublepharis macularius* (family 2.8% v. 10.7% lizards), and advanced color change is absent (v. 6%).

Mean SSD score (on scale of –2 to 2) is negligible (–0.09 v. 0.36) due to a nearly equal balance of m>f (especially *Eublepharis*) and m<f (especially *Goniurosaurus*) dimorphisms. Sexual dichromatism is scattered throughout the family (11% v. 21%), and seasonal color is found at typical frequency (5.6% v. 6.5%). Blue tails are not found in eublepharids (v. 2.3%). Sexual dimorphism in tail length is rarely strong in lizards (mean ratio 1.07), but eublepharids are an exception (1.21). An especially notable example is *Hemitheconyx caudicinctus*, for which the ratio is 1.85 (m:f). In aggregate, sex differences are moderately strong (1.08 v. 0.75 on 0 to 6 scale).

All documented eublepharids are invertebrate eaters (v. 73%), but they capture these through an unusually rich mixture of foraging modes: sit-and-wait (44% v. 46%), cruise (56% v. 7%), and active (44% v. 58%) foraging. These percentages add to more than 100% because many eublepharid species exhibit multiple modes. Foraging at electric lights and home-range areas have not been reported.

Eublepharids are overwhelming nocturnal (97% v. 25%), with the remainder crepuscular (3.1% v. 1.8%). In keeping with their nocturnality, vocalizations are common. Unexpectedly, spontaneous calling is much less common (8.3% v. 2.5%) than are stress vocalizations (25% v. 2.2%). The only other family to exhibit this unexplained pattern is the mostly nocturnal arboreal Carphodactylidae (33% stress calling to 13% spontaneous calling).

Eublepharids are not reported to bask (v. 18%), and their mean active body temperature is low (27.3°C v. 32°C), likely reflecting their nocturnality.

They are overrepresented in the class of lizards living solitarily (25% v. 10%) and underrepresented among those with socially tolerant males (0% v. 15%). All males are reported to defend space (v. 85%). Male hierarchies are slightly overrepresented (17% v. 7.7%).

Most eublepharids are terrestrial (90% v. 48%), with a few arboreal (7% v. 32%) and cryptozoic microhabitats (3.4% v. 12.1%). In keeping with their terrestriality, burrow refugia frequency is mildly elevated (33% v. 22%). The family is bimodal with regard to rainfall: 40% of species are each found in mesic (1101–3000 mm/y, v. 43%) and arid (151–400 mm/y, v. 30%) belts, with no examples in between. Twenty percent of species are also found in the dry belt (56–150 mm/y, v. 12%). Their elevational distribution is roughly the same as a typical lizard (l/m/h: 100/35/8 v. 82/41/13), even though many other nocturnal families avoid high elevations. The family is underrepresented in littoral (6% v. 17%) and sandy (10% v. 22%) habitats, and it is slightly overrepresented in riparian zones (30% v. 22%). They occur in human habitats at lizard-typical frequencies (17% v. 18.5%).

Despite the number of temperate forms in *Coleonyx* and *Goniurosaurus*, all eublepharids are oviparous (v. 82%), and family members are uniform in clutch counts of two. Thus, mass-adjusted clutch counts are simply a mirror of the species' relative sizes, or arguably a metric of the suboptimality of their clutch sizes. In contrast, the size of young is not obviously constrained and probably reflects selection; mean ontogenetic mass range is low, indicating particularly large young (–0.67 v. 0.0 std. resid.). Incubation periods are slightly short (–0.28 v. 0.0 std. resid.), especially in *Eublepharis* (–0.94 std. resid.). Eggshells are soft (v. 59%), laid in isolation (v. 96%), and adherent at typical frequency (2.8% v. 2.4%). RCM is low (0.177 v. 0.226), as befits a fixed clutch size of two in relatively large lizards. Eublepharids are reproductively notable for their parental care (14% v. 3%), mostly females defending nests. Species that exhibit parental care do not seem to be more or less represented among species that exhibit TSD.

Longevity is elevated (1.16 v. 0.0 std. resid.), perhaps due to low mortality in the aridland eublepharids. The three longest absolute (not mass-adjusted) longevities are *Coleonyx brevis* (34 years, Chihuahuan Desert), *Coleonyx variegatus* (34 years, Great Basin Desert), and *Eublepharis fuscus* (28 years, Sind Desert). Population density information is available only for *Coleonyx variegatus* (0.03 v. 0.54 kg/ha). No eublepharid has been introduced (v. 3.3%). Eublepharids are unfortunately overrepresented in both the ranks of the imperiled (60%

v. 49%) and the IUCN's most endangered lists (30% v. 14%).

In addition to supplying many specimens for the pet trade, Eublepharidae are often burdened with false legends about their venom or malicious powers. Just one example (see genus accounts for more): *Coleonyx variegatus* is greatly feared by at least some residents of Baja California who report that contact with the lizard's feet (even through clothing) will cause the human's flesh to die and slough away. A cow bitten by this gecko is alleged to have died immediately. Saliva of this gecko is believed to cause rashes, and it is thought that contact with the gecko may cause a fatal lung disease. This gecko is even alleged to retain necrotic powers after its death, causing its silhouette to be burned into a human's flesh if the human lies down on the ground where the gecko is buried. A remarkable list of powers for so delicate a creature.

Eublepharis

Eublepharis (six spp.) are the Old World eublepharids analogous to the New World *Coleonyx*. They are similarly shaped and colored, somewhat generously deemed leopard-like in their spots. *Eublepharis* are found from Syria across the southwest Asian deserts to Turkmenistan and India. *Eublepharis* are nocturnal ground lizards and large sized (36.4 g v. 13.2 g) but otherwise morphologically representative. *Eublepharis* are more often capable of basic color change (17% v. 2.8%).

Males are normally larger than females (1.0 v. –0.09 on –2 to 2 SSD scale). Other sex differences include sexual dichromatism (17% v. 11%), secretory pores (100% v. 66%), and tail length (m:f ratio 0.92 v. 1.21); typically this ratio is the reverse of SSD, as it is here. In aggregate, the sex differences rollup is modestly stronger than for the family (1.50 v. 1.08 on 0 to 6 scale).

All *Eublepharis* report only cruise foraging (nocturnal ground lizards rarely move quickly). *E. turcmenicus* oscillates before it moves, which it does on outstretched limbs. Twenty percent of *Eublepharis* are crepuscular (v. 3%), but like *Coleonyx*, there are many exceptions to their usual activity pattern. Activity in *macularius* may be stimulated by rain or high humidity. *Eublepharis* is the only eublepharid genus to report high frequencies of species that call spontaneously (33% v. 8%), but they also have a notably higher frequency of stress calling (50% v. 25%). See the *Coleonyx* account for some ruminations on the implications of such high rates of stress calling. No *Eublepharis* is said to be conspicuously wary (0% v. 19%); they move slowly. Despite residing in some of the world's hottest deserts, their mean active body temperature is a cool 26.5°C (v. 27.3°C).

No clear pattern emerges from the two species described socially: *macularius* occurs in groups and is hierarchical; *turcmenicus* is solitary and both sexes defend space. However, both *macularius* and *turcmenicus* defecate in traditional latrines. *E. macularius* marks with precloacal secretions. These marks do not demarcate a resource territory boundary, but the current position of the marking animal. Perhaps the secretions are designed to overtop the scent marks of subordinate lizards. There is much to be learned about *Eublepharis* sociality.

Burrow refuging is a common practice in *Eublepharis* (50% v. 17%). Although all documented species occur in the arid zone (151–400 mm/y), they are absent from sandy areas (0% v. 10%), possibly because they prefer hard burrows for protection or burrow stability. Burrows in *macularius* are believed to be permanent. *Eublepharis* are overrepresented in littoral (20% v. 7%) and human (40% v. 17%) habitats but slightly underrepresented in riparian zones (20% v. 30%).

E. macularius is notable for TSD (males at middle temperatures). This phenomenon may be present in related species. Incubation is short, even by eublepharid stan-

Eublepharis spp.	English name				Range core
			Size		
	Activity	Diet	mm	g	Ecological business model
angramainyu	Western Leopard Gecko				Iraq
	nocturnal	invert.	141	63	**** nocturnal ground lizard
fuscus	West Indian Leopard Gecko				India
	crepuscular	invert.	151	79	**** nocturnal ground lizard
hardwickii	East Indian Leopard Gecko				India
	nocturnal	invert.	89	14.2	*** nocturnal ground lizard
macularius	Leopard Gecko				Pakistan
	nocturnal	invert.	125	41.6	**** nocturnal ground lizard
satpuraensis	Satpura Hills Gecko				India
			102	21.9	* nocturnal ground lizard
turcmenicus	Turkmenistan Fat-tailed Gecko				Iran
	nocturnal	invert.	120	36.8	**** nocturnal ground lizard

dards (−0.94 v. −0.28 std. resid.), and females often guard eggs (33% v. 14%). In *macularius* the female may guard the clutch until they have hatched. What egg predator, disease, or environmental stressor is uniquely deflected by female attendance in this region of the world? Or is this potentially an artifact of residing in permanent burrows?

Forty percent of *Eublepharis* are considered imperiled (v. 60%), though none is on an IUCN list of the most endangered species (v. 30%). In other respects these are typical eublepharids. Like many other eublepharids, *Eublepharis* have been persecuted for their alleged powers. *E. hardwickii* is considered highly venomous, and the strength of its venom is believed to be discernible by how high it climbs after it bites its victim. The indigenous common name translates, inaccurately on all counts, as "snake that brings death." *E. macularius* is said to kill humans immediately on contact, without requiring a bite or sting; thus it is killed whenever detected. Nonetheless, Masroor (2012) believes that the pet trade is a bigger threat than is wanton human persecution.

Eugongylus spp.	English name				Range core
			Size		
	Activity	Diet	mm	g	Ecological business model
albofasciolatus	Barred Recluse Skink				Pacific Islands
	cathemeral	omnivore	149	29.0	** refugia-anchored diurnal gr. liz.
mentovarius	Halmahera Recluse Skink				Indonesia
			93	17.0	
rufescens	Bar-lipped Sheen Skink				New Guinea
	cathemeral	invert.	129	19.4	** refugia-anchored diurnal gr. liz.
sulaensis	Sula Recluse Skink				Indonesia
			109	28.8	
unilineatus	One-lined Recluse Skink				New Guinea
			94	8.1	

Eugongylus

Eugongylus is a minor radiation (five spp.) of large (18.6 g v. 3.25 g), mostly elongate (60% v. 35%), cathemeral (100% v. 2.4%), burrow-refuging (100% v. 24%) skinks found on the islands (100% v. 57%) of Australasia. The known species are refugia-anchored diurnal ground lizards (v. 30%).

Their build is the exact average for lizards, but comparatively robust for skinks (0.0 v. −0.46 std. resid.). Tails are average in length (1.30 v. 1.33 × SVL) and breakability (60% v. 73%). Sex differences are known only for *albofasciolatus*, for which males are larger than females (1.0 v. −0.31 on −2 to 2 scale); other sex differences are unknown.

The diet of *albofasciolatus* is diverse, including fruit, arthropods, snails, and vertebrates (omnivores 50% v. 13%). The diet of *rufescens* is more conventionally that of an invertebrate eater (v. 84%). Foraging mode, electric light use, and home-range area are unknown. At least 40% of species are conspicuously wary (v. 11%). Direct basking is not reported (v. 21%). *E. rufescens* is found in urban areas (v. 14%).

Reproduction is by eggs (v. 64%), and clutch counts are moderately small (−0.46 v. 0.06 std. resid.).

Endemism is about average (40% v. 64%). None has been introduced (v. 2.1%), nor are any considered imperiled (v. 47%). In other respects *Eugongylus* are much like other skinks.

Eulamprus

Eulamprus is a minor radiation (five spp.) of large (11.3 g v. 3.25 g), nonelongate (0% v. 35% elongate), diurnal (100% v. 74%), direct basking (100% v. 215), conspicuously wary (80% v. 11%), burrow-refuging (100% v. 24%), riparian (100% v. 23%), viviparous (100% v. 36%) skinks found in Australia. The known species are refugia-anchored diurnal ground lizards (v. 30%). They may be semi-aquatic (20% v. 2%) or terrestrial (80% v. 48%). They are comparatively well known (filled cells 40% v. 25%).

They are more robust in build than typical skinks (shape index 0.34 v. −0.46 std. resid.), but otherwise similar. Tail breakage (60% v. 73%) has an interesting consequence in *quoyii*; this species dives into water and swims away when it is threatened and has an intact tail. If freshly autotomized, a threatened *quoyii* will either run away or crawl on the bottom of water courses. Sex differences are unknown. Diets are diverse (omnivores 60% v. 13%), including fruit (*heatwolei*, *kosciuskoi*), leaves (*quoyii*), snails (*quoyii*, *tympanum*), fish (*heatwolei*, *quoyii*), and other ectothermic vertebrates (*heatwolei*, *kosciuskoi*, *quoyii*, *tympanum*). All are active foragers (v. 92%). For primary predator concealment they rely on retreat to vegetation (100% v. 25%). Their mean active body temperature (four spp.) is low (26.3°C v. 31.1°C), presumably reflecting the cooling power of water and high elevation (*kosciuskoi*). The alpine species (*kosciuskoi*) is able to withstand supercooling, and it will intentionally enter very cold water. Information on social relations is lacking, but *quoyii* males will defend space. *Eulamprus* are overrepresented in littoral (40% v. 21%) and human (20% v. 14%) habitats, but not sandy areas (0% v. 25%).

Litter sizes are slightly low (−0.35 v. 0.06 std. resid.), and the ontogenetic mass range suggests that neonates are slightly large (−0.16 v. 0.12 std. resid.). Parental care is known (20% v. 3.2%): neonatal *quoyii* remain for a time in proximity to their mother.

Longevity is about average (0.09 v. 0.17 std. resid.), equating to 3 years in *kosciuskoi*, 6 years in *leuraensis*, and 15 years in *tympanum*. Endemism is low (20% v. 64%). Few species occur on islands (20% v. 57%). None has been introduced (v. 2.1%) and all are considered imperiled (v. 47%), including 20% (v. 14%) on the IUCN's lists of most endangered. In other respects *Eulamprus* are much like other skinks.

Euleptes

Euleptes europaea is a moderately large (1.58 g v. 1.07 g), enlarged toepad, crepuscular (v. 6%), crack hermit (v. 6%) or trunk/wall sphaerodactylid in reduced predation areas from the central Mediterranean (France, Italy, Tunisia). It is found on about 100 islands but is notably missing from the mainland adjacent to many of those. *E. europaea* is the smallest European gecko; males are smaller than females (−1.00 v. 0.04 on −2 to 2 scale). The sexes also differ in color (v. 35%). It may cruise forage (v. 21%). *Euleptes* calls spontaneously (v. 2%) and gives stress calls (v. 2%), but it is not conspicuously wary (v. 7%). It basks directly (v. 6.5%). Individuals may live in groups (v. 17%), and males tolerate the proximity of other males (v. 14%). Its altitudinal range is low to medium (in this case, 0–1510 m). It is associated with littoral (v. 31%), sandy (v. 15%), and human (v. 33%) habitats, but is absent from riparian zones (v. 19%). Incubation is lengthy (0.78 v. 0.44 std. resid.). Eggs are adherent (v. 5.6%) and laid communally (v. 11%). Parental care is not reported (v. 1.4%). Longevity is extraordinary (2.14 v. −0.15 std. resid.), equating to 22 years in this tiny animal. Population density is relatively high (0.67 kg/ha v. 0.32 kg/ha) based on two credible estimates. It has not been introduced (v. 5%), but it is imperiled (v. 41%), though it is not on one of the IUCN's lists of the most endangered (v. 23%). In other respects *Euleptes* is a typical sphaerodactylid.

Eulamprus spp.	English name	Range core	Activity	Diet	Size mm	Size g	Ecological business model
heatwolei	Yellow-bellied Water Skink	New South Wales	diurnal	omnivore	84	12.1	*** refugia-anchored diurnal gr. liz.
kosciuskoi	Alpine Water Skink	Eastern Australia	diurnal	omnivore	73	7.5	*** refugia-anchored diurnal gr. liz.
leuraensis	Blue Mountains Swamp Skink	New South Wales	diurnal	invert.	72	7.3	*** refugia-anchored diurnal gr. liz.
quoyii	Eastern Water Skink	Eastern Australia	diurnal	omnivore	101	22.3	*** refugia-anchored diurnal gr. liz.
tympanum	Southern Water Skink	New South Wales	diurnal	invert.	84	12.1	*** refugia-anchored diurnal gr. liz.

Euleptes sp.	English name	Range core	Activity	Diet	Size mm	Size g	Ecological business model
europaea	European Leaf-toed Gecko	Central Mediterranean	crepuscular	invert.	40	1.58	**** crack hermit or trunk/wall lizard

Eumeces

The genus *Eumeces* once included hundreds of skinks, but as presently understood is just a minor radiation (five spp.) of large (18 g v. 3.25 g), oviparous (100% v. 64%) refugia-anchored diurnal ground lizards (100% v. 30%) found from Morocco to Russia, Kyrgyzstan, and India.

They are robust in build (shape index 0.32 v. −0.46 std. resid.) with only a single species (*schneideri*) considered marginally elongate (20% v. 35%). This species may locomote without use of limbs. It is the only *Eumeces* species documented to break tails (20% v. 73%). In a strong departure from skink norms, 40% (v. 0.1%) exhibit color change; none are advanced changers. SSD (*schneideri*) is m>f (0.50 v. −0.31 on −2 to 2 scale). This species is also sexually dichromatic (20% v. 7.6%). Thus the sex rollup score (on scale of 0 to 6) is higher than the skink norm (0.60 v. 0.28).

Diets are both conventional (invertebrates 50% v. 84%) and omnivorous (50% v. 13%), including fruit, arthropods, snails, and vertebrates. Foraging behavior is not known. *E. schneideri* (20% v. 11%) is conspicuously wary. For primary predator defense the only documented species (*algeriensis*, *schneideri*) rely on retreat to a crevice or burrow (100% v. 43%). All documented males (*algeriensis*, *schneideri*) defend space (v. 94%), but only during the mating season.

Two-thirds of documented species will use burrows as daily refugia (v. 24%). The elevational profile is about normal (l/m/h: 100/50/0 v. 89/35/6). They are overrepresented in sandy areas (80% v. 25%) and underrepresented in human habitats (0% v. 14%).

All are oviparous (v. 64%) with clutches that match their size (0.26 v. 0.06 std. resid.). The hatchlings are exceptionally large (ontogenetic mass range –1.14 v. 0.12 std. resid.). Incubation is remarkably short (–0.89 v. –0.40 std. resid.), equating to 48 days (*schneideri*) to 60 days (*blythianus*). Parental care is notable (40% v. 3.2%), as the females guard their clutches throughout incubation in both *algeriensis* and *schneideri*.

Longevity is relatively long (0.95 v. 0.17 std. resid.), corresponding to 9.3 years (*blythianus*), 22 years (*schneideri*), and 28.8 years (*algeriensis*). Endemism (40% v. 64%) and island presence (20% v. 57%) are low. None has been introduced (v. 2.1%). A typical proportion is imperiled (50% v. 47%), although none is on any of the IUCN's lists of the most endangered. In other respects *Eumeces* are much like other skinks.

In parts of its range *schneideri* are persecuted for allegedly stealing milk by sucking on goats' teats. How did this belief originate? Is it because *schneideri* may locomote without use of limbs and therefore must be evil in some way?

Eumeces spp.	English name		Size		Range core
	Activity	Diet	mm	g	Ecological business model
algeriensis	Algeria Orange-tailed Skink				Morocco
	diurnal	omnivore	146	73	* refugia-anchored diurnal gr. liz.
blythianus	Orange-tailed Skink				Pakistan
	diurnal	invert.	115	33.8	** refugia-anchored diurnal gr. liz.
cholistanensis	Cholistan Sand Skink				Pakistan
			83	11.8	
indothalensis	Thal Mole Skink				Pakistan
		invert.	52	2.66	
schneideri	Flecked Skink				Old World deserts
	diurnal	omnivore	139	24.1	*** refugia-anchored diurnal gr. liz.

Eumecia spp.	English name		Size		Range core
	Activity	Diet	mm	g	Ecological business model
anchietae	Western Serpentiform Skink				Central Africa
	diurnal	invert.	191	57	* litter cryptozoid
johnstoni	Malawian Serpentiform Skink				Malawi
			200	65	

Eumecia

Eumecia is a two-species genus of very large (61 g v. 3.3 g), diurnal (100% v. 74%), terrestrial (100% v. 48%), functionally limbless skinks found in central Africa (Democratic Republic of the Congo and Kenya to Malawi). One (*anchietae*) is tentatively classified as a litter cryptozoid (v. 28%), as it is often concealed in the thatch of dense grasses.

Their build is slightly heavier than family average (shape index –0.17 v. –0.46 std. resid.), with average-length tails (1.34 v. 1.33 × SVL) that are breakable (100% v. 73%). Both species are elongate (100% v. 35%). No sex differences are known. *E. johnstoni* is said to occur at high, though unspecified, elevation.

E. anchietae is viviparous and supports the growing embryos with mammal-like placentation. *E. johnstoni* is endemic (50% v. 64%). Neither occurs on islands (v. 57%) or has been introduced (v. 2.1%). They have not been evaluated for conservation status. In other respects *Eumecia* are probably typical skinks.

Eurolophosaurus

Eurolophosaurus is a small tropidurid genus (three spp.) of comparatively small (9.4 g v. 17.5 g), poorly known (filled cells 18% v. 23%) terrestrial (100% v. 74%) lizards endemic to Brazil. One (*nanuzae*) is suspected of being a refugia-anchored diurnal ground lizard; the others are too poorly known to guess their ecological business model. They are of average build (shape index 1.01 v. 1.10 std. resid.) and possess short tails (1.10 v. 1.56 × SVL) that are not breakable (v. 13% breakable). Two of three exhibit sexual dichromatism (v. 28%), but other sex differences are absent. They are found at low elevation (100% v. 65%) in sandy areas (50% v. 23%). Their clutch counts are about average (−0.47 v. −0.56 std. resid.). They are not found on islands (v. 10%) and have not been introduced (v. 1.5%). One of two evaluated species (*nanuzae*) is imperiled (50% v. 23%), but it is not on one of the IUCN's lists of the most endangered (v. 3%). In other respects *Eurolophosaurus* are typical tropidurids.

Eurydactylodes

Eurydactylodes is a small (four spp.) genus of slow-moving arboreal diplodactylids endemic to New Caledonia. Two are judged to be diurnal canopy walkers (50% v. 7%) and the other two nocturnal canopy walkers (50% v. 25%), though the diel cycles of these cryptic lizards are not well established and they may all be somewhat cathemeral. They are small for diplodactylids (2.61 g v. 4.15 g) with somewhat longer tails (1.06 v. 0.85 × SVL) that are prehensile (v. 26%) and shed only reluctantly (v. 91%). The tails not only have adhesive lamellae, but are also capable of squirting a noxious fluid—no wonder they are reluctant to part with them. They have secretory pores (v. 34%). *Eurydactylodes* also sport a yellow mouth lining, for an unknown reason. At least one species is capable of advanced color change (v. 1.3%). All documented species rely on crypsis (v. 88%).

Only secretory pores are reported among the possible sex differences. Diet is known for only *vieillardi*, which is judged to be an invertebrate eater, though it will take fruit in captivity. Basking is not documented (v. 17%). They are not associated with littoral (v. 17%), sandy (v. 24%), riparian (v. 10%), or human (v. 5%) habitats.

All are oviparous (v. 82%), notable for their very large eggs (up to 16 mm long in lizards of less than 3 g). This large egg size is reflected in the unusually low ontogenetic mass range (−1.10 v. −0.57 std. resid.), indicating very large hatchlings. Contradictory accounts of shell hardness are reported, but Bauer and Sadlier (2000) are definitive in stating that the eggs are anomalous among diplodactylids in being hard-shelled. Perhaps this confusion arose from excessive taxonomic generalization by others or because hard-shelled eggs may be soft when first laid.

No *Eurydactylodes* has been introduced (v. 2%). All are imperiled (v. 81%), and two of four species are among the most endangered by IUCN classification. In other respects they are typical diplodactylids.

Eurolophosaurus spp.	English name / Activity	Diet	Size mm	Size g	Range core / Ecological business model
amathites	Amathites Lizard		58	8.2	Brazil
divaricatus	Crested Sand Lizard		75	16.5	Brazil
nanuzae	Menezes' Lizard	invert.	52	6.2	Brazil; * refugia-anchored diurnal gr. liz.

Eurydactylodes spp.	English name / Activity	Diet	Size mm	Size g	Range core / Ecological business model
agricolae	Bauer's Chameleon Gecko; diurnal		52	2.82	New Caledonia; *** diurnal canopy walker
occidentalis	Western Chameleon Gecko; diurnal		47	2.07	New Caledonia; ** diurnal canopy walker
symmetricus	Large-scaled Chameleon Gecko; cathemeral		52	2.82	New Caledonia; ** nocturnal canopy walker
vieillardi	Vieillard's Chameleon Gecko; cathemeral	invert.	52	2.82	New Caledonia; ** nocturnal canopy walker or diurnal canopy walker

Eurylepis

Eurylepis is a two-species genus of large (7.4 g v. 3.3 g), svelte (shape index −1.20 v. −0.46 std. resid.), extremely elongate (100% v. 35%), cryptozoic (100% v. 30%) skinks found in central Asia deserts (Yemen to Turkmenistan and India). One (*taeniolata*) is tentatively classified as a litter cryptozoid (v. 28%). *E. poonaensis* is known from only a few specimens.

Tails are short (1.04 v. 1.33 × SVL),

Eurylepis spp.	English name				Range core
			Size		
	Activity	Diet	mm	g	Ecological business model
poonaensis	Poona Skink				India
	diurnal	invert.	94	4.65	
taeniolata	Yellow-bellied Mole Skink				Iran
	cathemeral	invert.	137	11.8	** litter cryptozoid or refugia-anchored diurnal gr. liz.

Euspondylus spp.	English name				Range core
			Size		
	Activity	Diet	mm	g	Ecological business model
acutirostris	Sharp-snouted Sun Tegu				Venezuela
			49	2.61	
auyanensis	Auyan Tepui Lizard				Venezuela
			52	3.14	
caideni	Caiden's Sun Tegu				Peru
			68	7.5	
guentheri	Guenther's Sun Tegu				Ecuador
			77	11.0	
josyi	Josy's Sun Tegu				Peru
			67	7.0	
maculatus	Spotted Sun Tegu				Peru
			63	5.8	
monsfumus	Humo Mountain Sun Tegu				Venezuela
			42	1.56	
nellycarrillae	Nelly's Sun Tegu				Peru
			52	3.14	
paxcorpus	Peace Corps Sun Tegu				Peru
			50	2.73	
simonsii	Simons' Sun Tegu				Peru
			44	1.74	

thick, and sometimes (*taeniolata*) breakable (50% v. 73%). Juvenile *taeniolata* tails may be bright blue (v. 4%). Males are smaller than females (*taeniolata*), but no other sex differences are known. *E. taeniolata* is conspicuously wary (50% v. 11%). Both species may use burrows for daily refugia (v. 24%). Though not found in littoral areas (v. 21%), *taeniolata* is found in riparian areas (50% v. 23%), sandy areas (50% v. 25%), and human habitats (50% v. 14%).

Both species are oviparous (100% v. 64%); *taeniolata* has a clutch of 4.5 (–0.09 v. 0.06 std. resid.).

Endemism is low (50% v. 64%). None is found on an island (v. 57%). *E. taeniolata* has been introduced (once) to Jordan (50% v. 2.1%). Only *poonaensis* has been evaluated for conservation status; it is considered imperiled (v. 47%), including placement on the IUCN's lists of the most endangered. *E. taeniolata* is collected for an aphrodisiac. In other respects *Eurylepis* are probably typical skinks.

Euspondylus

Euspondylus is a 10-species genus of relatively large (3.80 g v. 2.35 g) gymnophthalmids in northwestern South America. This is a particularly poorly known group (filled cells 14% v. 19%); the ecological business model is not known for any species, and microhabitat (arboreal) is confidently assigned in only one (*acutirostris*). Diet and diel activity pattern are unknown for any species; thus christening them as "sun tegus" is doubly presumptuous. One species (*paxcorpus*) likely has secretory pores (v. 29%). They are found at particularly high elevations (low elevation: 33% v. 51%; high elevation: 56% v. 33%); the maximum documented elevation is 3800 m (*paxcorpus*). Although none has been judged imperiled (v. 25%), this may be largely a reflection of our ignorance. In other respects *Euspondylus* are probably typical gymnophthalmids.

Eutropis

Eutropis is a major radiation (34 spp.) of slightly large (4.57 g v. 3.25 g), mostly diurnal (85% v. 74%) or cathemeral (11.5% v. 9.0%), mostly terrestrial (90% v. 48%) skinks of diverse ecologies found in southern Asia within the following polygon: Pakistan to Afghanistan to China to Taiwan to Palau to New Guinea to Java to Sumatra to the Maldives to Pakistan. The modal *Eutropis* is a refugia-anchored diurnal ground lizard (75% v. 30%). Others are litter cryptozoids (21% v. 28%) or nocturnal canopy walkers (3.6% v. 0.1%). Many cryptic species occur, and misidentifications plague the literature.

Eutropis are a little heavier in build than an average skink (shape index −0.09 v. −0.46 std. resid.), with longish tails (1.59 v. 1.33 × SVL) that generally break (85% v. 73%). A proportion of species are marginally elongate (21% v. 35%). Blue tails are absent (v. 4%).

Males are typically the larger sex (0.50 v. −0.31 on −2 to 2 scale). The sexes often differ in color (21% v. 7.6%) or in seasonal color (18% v. 6%). Curiously, this genus has the Scincidae's second-highest ratio of m:f relative tail length (1.18 v. 1.03 ratio), for obscure reasons. The generalization that m:f relative tail length is the inverse of m:f SSD is not supported in Scincidae (see family entry). This ratio tends to be high (table 4.1, trait 17) in subterranean foragers (1.24), immobility experts (1.22), and crack hermits (1.12), of which *Eutropis* has none. The *Eutropis* taxa are litter cryptozoids (1.02) and refugia-anchored diurnal ground lizards (1.07), which have low average ratios. Higher ratios are typically found in smaller species (table 3.1, trait 7), which *Eutropis* is not, so what is unique about *Eutropis*? The overall sex differences rollup score for the genus is about average (0.41 v. 0.28 on 0 to 6 scale).

Diet is mostly invertebrates (91% v. 84%); the remaining species are omnivores (9.5% v. 12.6%). Some forage intertidally (e.g., *multifasciata*). One well-studied species (*multifasciata*) appears to be a dietary opportunist, as it includes not only leaves, fruit and all types of invertebrates (including snails) but also vertebrates and carrion. All species are said to be exclusively active foragers (100% v. 92%). *E. multifasciata* is said to forage infrequently, even in suitable weather; that is, only a small percentage of the population is active at any

Eutropis spp.	English name		Size		Range core
	Activity	Diet	mm	g	Ecological business model
allapallensis	Allapalli Grass Skink				India
	diurnal	invert.	45	1.65	** refugia-anchored diurnal gr. liz.
andamanensis	Andaman Black-spotted Skink				India
	diurnal	invert.	104	24.2	** refugia-anchored diurnal gr. liz.
ashwamedhi	Ashwamedhi Writhing Skink				India
	diurnal	invert.	32	0.56	* litter cryptozoid
austini	Gannoruwa Forest Lizard				Sri Lanka
	diurnal		61	4.45	*** litter cryptozoid
beddomei	Beddome's Striped Skink				India
	diurnal	invert.	53	2.80	*** refugia-anchored diurnal gr. liz.
bibronii	Seashore Dune Skink				India
	diurnal	invert.	46	1.11	*** refugia-anchored diurnal gr. liz.
bontocensis	Luzon Montane Mabuya				Philippines
			52	2.66	
carinata	Keeled Common Skink				India
	diurnal	invert.	110	12.7	*** refugia-anchored diurnal gr. liz.
chapaensis	Chapa Ground Skink				Vietnam
	nocturnal		67	5.8	* nocturnal canopy walker
clivicola	Inger's Sun Skink				India
	diurnal	invert.	54	3.02	* refugia-anchored diurnal gr. liz.
cumingi	Cuming's Sun Skink				Philippines
			48	2.02	
darevskii	Darevsky's Grass Skink				Vietnam
			46	1.74	
dissimilis	Striped Grass Skink				Pakistan
	cathemeral	invert.	75	4.41	**** refugia-anchored diurnal gr. liz.
englei	Six-striped Mabuya				Philippines
			59	4.01	
floweri	Taylor's Skink				Sri Lanka
	diurnal		51	2.41	*** litter cryptozoid or refugia-anchored diurnal gr. liz. (geog. var.)
gansi	Gans' Grass Skink				India
		invert.	54	3.02	*** litter cryptozoid
grandis	Buton Ground Skink				Indonesia
			112	30.6	

Eutropis spp.	English name	Activity	Diet	Size mm	Size g	Range core	Ecological business model
greeri	Galle Skink	diurnal		60	4.23	Sri Lanka	*** litter cryptozoid
indeprensa	Incomprehensible Skink	diurnal		57	3.57	Borneo	* refugia-anchored diurnal gr. liz.
innotata	Bronzy-olive Skink	diurnal	invert.	49	2.22	India	** litter cryptozoid
longicaudata	Long-tailed Sun Skink	diurnal	omnivore	110	12.4	Southeast Asia	**** refugia-anchored diurnal gr. liz.
macrophthalma	Large-eyed Ground Skink			87	13.5	Indonesia	
macularia	Variable Sun Skink	cathemeral	invert.	59	2.29	Southern Asia	**** refugia-anchored diurnal gr. liz.
madaraszi	Sri Lanka Spotted Skink	diurnal		54	2.99	Sri Lanka	** refugia-anchored diurnal gr. liz.
multicarinata	Multi-keeled Sun Skink	diurnal	invert.	70	6.9	Indonesia	*** refugia-anchored diurnal gr. liz.
multifasciata	Common Sun Skink	cathemeral	omnivore	104	10.9	Southern East Asia	**** refugia-anchored diurnal gr. liz.
nagarjunensis	Nagarjuni Sun Skink	diurnal	invert.	62	4.50	India	**** refugia-anchored diurnal gr. liz.
quadratilobus	Bhutan Skink			55	3.00	Bhutan	** refugia-anchored diurnal gr. liz.
quadricarinata	Four-keeled Ground Skink	diurnal	invert.	46	1.74	India	** refugia-anchored diurnal gr. liz.
rudis	Rough-headed Sundaland Skink	diurnal	invert.	97	19.5	Southeast Asia	** refugia-anchored diurnal gr. liz.
rugifera	Rough-scaled Sun Skink	diurnal	invert.	64	5.2	Malaysia	** refugia-anchored diurnal gr. liz.
tammanna	Tammanna Point Skink	diurnal	invert.	48	2.08	Sri Lanka	* refugia-anchored diurnal gr. liz.
trivittata	Three-banded Grass Skink	diurnal	invert.	95	8.4	India	** refugia-anchored diurnal gr. liz.

given moment. This suggests that foraging success rates often exceed the lizards' energetic needs. Electric light use (*macularia*) is comparatively common (3.8% v. 0.4%). Home-range area is known only from *multifasciata*, for which it is large (0.97 v. –0.07 std. resid.), equating to about 1000 m^2.

Wariness is present at average frequency (9% v. 11%). Primary predator concealment tactics represented are sand diving (50% v. 22%) and retreat to a crevice or burrow (50% v. 43%). Direct basking is common (27% v. 21%). Mean active body temperature is average (31.3°C v. 31.1°C). All documented males defend space (v. 94%). *E. multifasciata* defends space against other species of lizard, but the consistency of interspecific defense is not established, nor is whether this defense persists outside of the mating season. Are the "other" species of lizards potential predators on their eggs or young?

Rainfall regimes are reasonably well documented (six spp.) and are solely in the mesic belt: 1101–3000 mm/y (100% v. 40%). The elevational profile is on the low elevation side of normal (l/m/h: 96/23/0 v. 89/35/6). *Eutropis* are near normal in representation in littoral (20% v. 21%), sandy (17% v. 25%), and riparian (30% v. 23%) areas and slightly overrepresented in human habitats (30% v. 14%). Human habitations may be hard to avoid in the core range of India.

Reproduction is primarily but not exclusively by eggs (87% v. 64%), and clutch/litter counts are normal (0.03 v. 0.06 std. resid.). Young are comparatively small (ontogenetic mass range 0.53 v. 0.12 std. resid.). At least two species (*carinata*, *longicaudata*) are known or suspected to switch reproductive mode (geographically?). Incubation is notably short (–0.80 v. –0.40 std. resid.), based on

Eutropis spp.	English name				Range core
			Size		
	Activity	Diet	mm	g	Ecological business model
tytleri	Andaman Grass Skink				India
	diurnal	invert.	154	87	** refugia-anchored diurnal gr. liz.

Exila sp.	English name				Range core
			Size		
	Activity	Diet	mm	g	Ecological business model
nigropalmata	Amazonia Gracile Skink				Bolivia
			52	1.63	

Feylinia spp.	English name				Range core
			Size		
	Activity	Diet	mm	g	Ecological business model
boulengeri	Boulenger's Feylinia				Gabon
			81	1.52	
currori	Western Forest Legless Skink				Cameroon
	diurnal	invert.	210	13.0	**** subterranean forager
elegans	Elegant Legless Skink				Gabon
			111	3.13	*** subterranean forager
grandisquamis	Mueller's Legless Skink				Congo
			109	3.04	*** subterranean forager
macrolepis	Boettger's Legless Skink				Congo
			77	1.34	*** subterranean forager
polylepis	Many-scaled Legless Skink				São Tomé and Príncipe
			124	4.03	** subterranean forager

longicaudata (36 days), *macularia* (22 days), and *multicarinata* (38 days). The consistency of these short intervals suggests a region-wide selective force (short suitable intervals before or after the monsoon?). Eggshells are hard in 40% (v. 9%) of species, perhaps indicative of hydric stresses. Parental care occurs at normal frequency (2.9% v. 3.2%). In *macularia* both parents guard the eggs. On Orchid Island (only) *longicaudata* mothers defend their clutches by attacking small oviphagous snakes.

Longevity is short (–0.61 v. 0.17 std. resid.), based on *longicaudata* (3.1 years), *multicarinata* (3 years), and *multifasciata* (6 years). Perhaps it is not surprising that mortality is high in mainland Asia. Endemism is of normal frequency (62% v. 64%). The proportion of species on islands is also typical (65% v. 57%). Introduced species are unusually common (5.9% v. 2.1%). At least one species (*multifasciata*, two introductions) has been observed in association with human transport. *Eutropis* are imperiled at the usual level (53% v. 47%), although there is a surfeit of species that appear on the IUCN's lists of the most endangered (20% v. 14%). *E. dissimilis* is killed on sight in some areas, where it is believed to be a venomous snake masquerading with legs. In some areas *multifasciata* is believed venomous. In other areas it is eaten fried as a cure for itchiness. In other respects *Eutropis* are typical skinks.

Exila

Exila nigropalmata is the sole, very poorly known (filled cells 16% v. 25%) member of its scincid genus and is found in central South America (Peru, Bolivia, Brazil). It is known from only a few specimens. Aside from being viviparous (v. 36%), it is known primarily from its morphology. It is small (1.63 v. 3.25 g), slender (shape index –0.74 v. –0.46 std. resid.), and slightly long-tailed (1.55 v. 1.33 × SVL); the tail is breakable (v. 73%). It is elongate (v. 35%). and probably has a blue tail at some stage (v. 4%). It does not occur on islands (v. 57%), nor is it endemic (v. 64%). It has not been introduced (v. 2.1%), and it has not been evaluated for conservation status.

Feylinia

Feylinia (Scincidae) is a moderate radiation (six spp.) of very poorly documented (filled cells 18% v. 25%), average-sized (3.2 g v. 3.3 g), limbless (several with reduced eyes), subterranean foragers (100% v. 12%) found in equatorial Africa (Cameroon to Kenya, south to Tanzania to Angola, including São Tomé and Príncipe).

For a skink they are exceedingly slender (shape index –2.24 v. –0.46 std. resid.). Tails are short (0.43 v. 1.33 × SVL). One species' tail (*currori*) is known to break (17% v. 73%). Their elevational profile is about normal (l/m/h: 100/25/0 v. 89/35/6). Two documented species (*currori*, *elegans*) are riparian (50% v. 23%). The hearing of *currori* may be good enough to locate termites by their noisy chewing.

Endemism is low (33% v. 64%). Half of the species occur on islands (50% v. 57%). None is known to have been introduced (v. 2.1%). None has been evaluated for conservation status by the IUCN. In other respects *Feylinia* are probably typical legless skinks.

Fojia

Fojia bumui is the sole member of its scincid genus, and it is endemic to Papua New Guinea. Though the common name suggests it may be semi-aquatic, it is actually fairly terrestrial, albeit restricted to riparian zones (v. 23%). It is a small-to-average-sized (2.54 v. 3.25 g) refugia-anchored diurnal ground lizard (v. 30%) that sleeps above ground on vegetation overhanging streams. Sleeping in elevated vegetation is unusual for skinks (v. 4.9%).

It is somewhat tuberculate and a bit more robust than an average skink (shape index −0.17 v. −0.46 std. resid.) and has a slightly longer tail (1.75 v. 1.33 × SVL) that is not reported to break (v. 73%). Males have subdermal glands (not secretory pores, exactly) and differ in color from females (v. 7.6%). The *Fojia* sex differences rollup score (2.0 v. 0.28 on 0 to 6 scale) is the second-highest in Scincidae.

It eats a varied diet of invertebrates (100% v. 84%), which are obtained by sit-and-wait foraging (v. 16%). Home-range area is average (0.01 v. −0.07 std. resid.), equating to about 30 m². Both sexes are spatially intolerant (v. 94% for males).

Fojia is oviparous (v. 64%) and has a clutch of two (−0.14 v. 0.06 std. resid.). It has not been introduced (v. 2.1%), and it has not been evaluated for conservation status.

Furcifer (4.3F)

These 22 species are classic large, colorful chameleons of Madagascar and the nearby Comoro Islands, prime representatives of the "typical" group (see table 5.2 in the Chamaeleonidae entry). They average a bit larger than the mean for the typical group (24.7 g v. 13.5 g) and have a more male-biased SSD (1.50 v. 0.41 on −2 to 2 scale). Greater average SSD is expected in larger species, but *Furcifer* are unusual in the degree of the size discrepancy in individual species. Over half of the species for which SSD is reported exhibit males possessing more than twice the weight of their mates. For example, in *pardalis* the mean adult male weighs about 57 g and the mean adult female is 20 g.

Greater absolute size could also influence the higher proportion (see table 3.2, trait 1) of crested

Fojia sp.	English name				Range core
			Size		
	Activity	Diet	mm	g	Ecological business model
bumui	Rough Stream Lizard				New Guinea
	diurnal	invert.	52	2.54	*** refugia-anchored diurnal gr. liz.

Furcifer spp.	English name				Range core
			Size		
	Activity	Diet	mm	g	Ecological business model
angeli	Gunsight Chameleon				Madagascar
	diurnal	invert.	124	40.8	**** diurnal canopy walker
antimena	Palisade Chameleon				Madagascar
	diurnal	invert.	131	47.4	**** diurnal canopy walker
balteatus	Arc Chameleon				Madagascar
	diurnal	invert.	134	51	**** diurnal canopy walker
belalandaensis	White-lined Chameleon				Madagascar
	diurnal	invert.	95	20.2	**** diurnal canopy walker
bifidus	Bifid Chameleon				Madagascar
	diurnal	invert.	152	71	**** diurnal canopy walker
campani	Campan's Chameleon				Madagascar
	diurnal	invert.	59	5.6	*** diurnal canopy walker
cephalolepis	Grande Comore Chameleon				Comoros
	diurnal	invert.	64	7.1	**** diurnal canopy walker
labordi	Resurrection Chameleon				Madagascar
	diurnal	invert.	93	19.2	**** diurnal canopy walker
lateralis	Carpet Chameleon				Madagascar
	diurnal	invert.	93	19.0	*** diurnal canopy walker
major	Large Side-striped Chameleon				Madagascar
	diurnal		109	28.9	*** diurnal canopy walker
minor	Gaudy Chameleon				Madagascar
	diurnal	invert.	81	13.0	*** diurnal canopy walker

Furcifer spp.	English name				Range core
			Size		
	Activity	Diet	mm	g	Ecological business model
nicosiai	Kirindy Chameleon				Madagascar
	diurnal	invert.	113	32.0	**** diurnal canopy walker
oustaleti	Oustalet's Chameleon				Madagascar
	diurnal	invert.	202	152	**** diurnal canopy walker
pardalis	Panther Chameleon				Madagascar
	diurnal	invert.	91	19.0	**** diurnal canopy walker
petteri	Petter's Chameleon				Madagascar
	diurnal	invert.	74	10.2	**** diurnal canopy walker
polleni	Mayotte Chameleon				Comoros
	diurnal	invert.	84	14.3	**** diurnal canopy walker
rhinoceratus	Rhinoceros Chameleon				Madagascar
	diurnal	invert.	112	30.9	**** diurnal canopy walker
timoni	Timon's Chameleon				Madagascar
	diurnal	invert.	78	11.9	*** diurnal canopy walker
tuzetae	Befandriana Chameleon				Madagascar
	diurnal	invert.	133	49.5	**** diurnal canopy walker
verrucosus	Giant Spiny Chameleon				Madagascar
	diurnal	invert.	188	125	**** diurnal canopy walker
viridis	Ambinanitelo Green Chameleon				Madagascar
	diurnal		95	20.2	*** diurnal canopy walker
willsii	Wills' Canopy Chameleon				Madagascar
	diurnal	invert.	71	9.1	**** diurnal canopy walker

species (59% v. 36% in the typical group), but size should not affect the population density (nine species quantified: 0.66 v. 0.18 kg/ha), the very high clutch count (3.18 v. 2.81 std. resid.), the ontogenetic size range (1.16 v. 0.63 std. resid.), or the lengthy size-adjusted incubation duration (3.22 v. 2.43 std. resid.), all contrasted with the "typical" group. In each of these traits, *Furcifer* exhibit a slightly hypertrophied state of the "typical" condition. For the other traits, they fit the "typical" standard quite closely: diurnal canopy walkers (100% v. 98%), shape index (0.65 v. 0.58 std. resid.), tail length (1.10 v. 1.09 × SVL), and arboreality (100% v. 99%). Their knowledge base, prehensile tails, absence of tail breakage, advanced color change, invertebrate diet, diurnality, reliance on crypsis, frequency of basking, male intolerance and defense of space, modal occurrence in the mesic rainfall belt, longevity, frequency of imperilment (100%), and severe imperilment (18% v. 25%) are all close matches to the chamaeleonid norm. Although all documented *Furcifer* diets are invertebrate eaters in the wild, captive *pardalis* will eat plants.

Furcifer differ from most chamaeleonids in their very high frequency of sexual dichromatism (82% v. 53%), elevated seasonal color (27% v. 18%), relatively shorter male tails (1.07 v. 1.13 × SVL, which is often associated with greater m>f SSD), and a slightly greater sex differences rollup (2.95 v. 2.16 on 0 to 6 scale). Three species (*lateralis*, *minor*, and *oustaleti*) are notable because to human eyes the females are the more colorful sex. *Furcifer* further differ from chamaeleonids in exhibiting a slightly elevated frequency of conspicuous wariness (4.5% v. 0.5%), a higher than average frequency of burrow use (14.3% v. 4.2%), a greater concentration in the lowest elevation belt (95% v. 76%), a reduced presence in the mid-elevation (30% v. 64%) and high-elevation (5% v. 21%) belts, and an increased presence around human habitations (59% v. 23%). Andreone et al. (2005) assert that all *Furcifer* prefer disturbed habitats, though several species (*antimena*, *balteatus*, *rhinoceratus*) are reported to avoid brightly lit areas and may continue activity in the dark.

A large and colorful morphology and presence at low elevations and around humans would be expected to translate into a higher rate of species introduction. *Furcifer* exhibits one of the higher rates of introduction (13.6% v. 5.4%), but no species has been introduced more than twice (0% v. 1.5%). All *Furcifer* are endemic to their country (v. 76%), and all occur on islands (v. 48%).

All *Furcifer* are oviparous (v. 80%). Their mean RCM is slightly lower than the chamaeleonid norm (0.310 v. 0.444), but it is based on only one species (*oustaleti*). *F. labordi* is reproductively notable for exhibiting semelparity in both sexes (breed once and die), a very rare condition in lizards, though several *Furcifer* are very short-lived. Female *pardalis* are also semelparous, but the males may live to breed again. Neither sex of *lateralis* normally live for more than one year. It is possible that there is some geographic variation in the semelparous trait of *labordi*. For the majority of the year, in the best-studied population, *labordi* exists only underground in developing eggs, as adults die after oviposition. Its breeding season is extended, and the incubation period averages 270 days, but growth to maturation is very rapid (ca. 60 days). In another few months, the adults have bred, oviposited, and expired. Copulation in *labordi* is very physical, possibly even forced, though the IUCN (redlist.org, accessed spring 2017) calls it "agnostic."

Gallotia

Gallotia (eight spp.) is a genus of diurnal, terrestrial, mostly giant herbivores (Lacertidae) found in the Canary Islands off the west coast of Africa. As is usual with giant herbivores on islands, many species have been greatly depleted by humans or their commensals, and considerable conservation work on *Gallotia* has been necessary, with the benign side effect of filling many cells in the data matrix (49% v. 33%).

The nomenclatural history of *G. caesaris* is heartbreaking enough to warrant retelling. Two Germans discovered this species more or less simultaneously in 1914: Caesar Boettger (1888–1976) and Phillip Lehrs (1888–1956). In order to get the jump on his rival, Lehrs rushed to London to show his specimens to George Albert Boulenger at the British Museum. By the time he got there, World War I had unexpectedly broken out, and British authorities seized Lehrs as an enemy alien. He was incarcerated for the duration of the war. After the war, Lehrs dropped out of herpetology but persisted in requesting the return of his specimens from the British. The wheels of bureaucracy grind slowly sometimes and Lehrs's request was ultimately approved, but only after the outbreak of World War II, at which time the approval was rescinded. Lehrs did not get his specimens back, but he is recognized as the describer of *Gallotia caesaris*.

Although most are giant herbivores (63% v. 2.4%), many *Gallotia* species exhibit a shift toward herbivory ontogenetically, and the nonherbivorous species are refugia-anchored diurnal ground lizards (38% v. 57%). The diversity in ecological business models is also present in body size, which ranges from 9 g to 332 g (mean 49 g v. 5.1 g), making this genus second only in body size within the family to *Timon*. *G. bravoana* was larger prehistorically, as documented by subfossil remains; in many lizard species with subfossil remains, the subfossils often indicate that larger individuals existed prior to modern stressors such as introduced species or degraded land cover.

Shape index is a robust 0.80 (v. 0.27 std. resid.). Tails are comparatively short (1.69 v. 2.06 × SVL) and breakable (100% v. 92%). *Gallotia* lack secretory pores (v. 4%) and have no color change (v. 2.2%)

The genus's SSD is m>f (1.0 v. 0.49 on −2 to 2 scale). The heads of male *simonyi*, *stehlini*, and possibly *atlantica* are proportionately bigger than those of females. Most species also differ in color (75% v. 33%); seasonal color differences are average (13% v. 15%). The sex differences in color are modulated by the social status of the lizard; subordinate males have some elements of female color in *caesaris* and *galloti*. In *galloti* sex differences extend to voice (vocalizations are rare in diurnal lizards: 3.7%); in this case, the call of subordinate males is different from that of dominant males. The sex differences rollup score highlights these diverse differences (1.75 v. 0.86 on 0 to 6 scale).

All species are either herbivores (43% v. 1.7%) or omnivores (57% v. 19%). *G. galloti* are fond enough of fruit that this species is considered an

Gallotia spp.	English name		Size		Range core
	Activity	Diet	mm	g	Ecological business model
atlantica	Atlantic Lizard				Canary Islands (Spain)
	diurnal	omnivore	79	11.3	** refugia-anchored diurnal gr. liz.
auaritae	Palma Giant Lizard				Canary Islands (Spain)
			248	332	* giant herbivore
bravoana	Gomera Giant Lizard				Canary Islands (Spain)
	diurnal	herbivore	148	73	**** giant herbivore
caesaris	Boettger's Lizard				Canary Islands (Spain)
	diurnal	omnivore	73	9.0	** refugia-anchored diurnal gr. liz.
galloti	Tenerife Lizard				Canary Islands (Spain)
	diurnal	omnivore	104	25.4	*** refugia-anchored diurnal gr. liz.
intermedia	Tenerife Speckled Lizard				Canary Islands (Spain)
	diurnal	herbivore	124	42.8	*** giant herbivore
simonyi	Hierro Giant Lizard				Canary Islands (Spain)
	diurnal	herbivore	165	93	**** giant herbivore
stehlini	Gran Canaria Giant Lizard				Canary Islands (Spain)
	diurnal	omnivore	180	131	**** giant herbivore

agricultural pest on tomatoes. Home-range size is small (–0.72 v. 0.27 std. resid.), based on *galloti* (57 m²).

Spontaneous vocalizations are common (88% v. 2.8%), stress calls less so (13% v. 1.5%). The only documented predator concealment tactic is sand diving (v. 7%) in *atlantica*. Thermoregulatory behavior follows the lacertid template. Males of four documented species defend space during the breeding season (100% v. 90%). None is tolerant (v. 41%) or hierarchical (v. 25%).

The occupied habitats are in the dry (56–150 mm/y) and arid (151–400 mm/y) belts at low (100% v. 87%), medium (57% v. 61%), and high elevation (14% v. 20%). *G. galloti* goes from sea level to 3718 m. They are overrepresented in littoral (38% v. 22%) and human (38% v. 16%) habitats and underrepresented in sandy (13% v. 44%) and riparian (13% v. 28%) ones.

Reproduction is similar to most other lacertids, as are clutch counts (–0.07 v. 0.13 std. resid.), though their incubation is short (–0.73 v. –0.36 std. resid.). The ontogenetic size range is conspicuously high, indicating particularly small hatchlings (0.99 v. –0.05 std. resid.). Mean RCM is a discomfiting 0.435 (v. 0.267 ratio). These latter two benchmarks do not comport well; a normal clutch count of very small hatchlings should not produce such an elevated RCM ratio.

Average mass-adjusted longevity is high (0.80 v. 0.19 std. resid.), corresponding to 45 years in *simonyi*. Population density is remarkably well known ($n = 20$, based on five spp.). Species mean densities range from 2.0 kg/ha (*simonyi*) to a staggering 124 kg/ha (*stehlini*; genus mean 18.4 kg/ha v. 2.07 kg/ha). But all is not well in the islands: *auaritae* is likely extinct; the total population of *intermedia* numbers only in the hundreds; *simonyi* numbers only about 300, with all individuals confined to a single cliff; *bravoana* is found on two cliff sites, but in aggregate these number only about 90 lizards.

All species are endemic (v. 35%) and found on islands (v. 26%). Almost all are imperiled (88% v. 56%), and fully half (50% v. 15.5%) are on the IUCN's lists of the most endangered. The primary threats to these island populations are the usual: cats, dogs, and rats.

Gambelia

Gambelia (three spp.) are large (40.1 g v. 37.5), diurnal terrestrial crotaphytids of western North America. They forage by sitting and waiting, but change refugia, to which they are anchored, often. They cumulatively search over very large home ranges (1.42 v. 0.72 std. resid.). In many other aspects, they are indistinguishable from *Crotaphytus*, the only other crotaphytid genus. See the *Crotaphytus* entry for a detailed comparison of the two genera.

Gambelia are notably less socially pugnacious than are *Crotaphytus*. See the *Crotaphytus* entry for documentation of the comparatively tolerant lifestyle of *Gambelia*.

Gambelia spp.	English name				Range core
			Size		
	Activity	Diet	mm	g	Ecological business model
copeii	Cope's Leopard Lizard				Mexico
	diurnal	invert.	105	44	*** refugia-anchored diurnal gr. liz.
sila	Blunt-nosed Leopard Lizard				California
	diurnal	invert.	100	40.8	*** refugia-anchored diurnal gr. liz.
wislizenii	Long-nosed Leopard Lizard				Arizona
	diurnal	omnivore	107	36.0	**** refugia-anchored diurnal gr. liz.

Garthia

Garthia is a two-species genus of small (1.2 g v. 4.7 g) nocturnal ground phyllodactylids from the southern Andes (Bolivia to Chile). One species is comparatively well documented (*gaudichaudii*, 46% v. 27% cells filled); the other is poorly documented (*penai*, 24% cells filled). The well-documented species is reported to lack expanded toepads; the other may lack them also.

Morphologically, *Garthia* differs from other phyllodactylids primary in its small size, lack of reported color change (v. 11%), and lack of toe pads (v. 8% absence). Sex differences are not reported. Both species are nocturnal (v. 78%). At least *gaudichaudii* calls spontaneously (v. 11%). They are not reported to bask directly (v. 10%), and *gaudichaudii*'s low mean active body temperature (26.3°C v. 26.0°C) supports this. It is reported that *gaudichaudii* is warmer when it is inactive (daytime). *Garthia* is found at the dry end of the rainfall spectrum: hyperdry through arid (in the case of *gaudichaudii*, 13–322 mm/y). *G. gaudichaudii* occurs on beaches (littoral, sand) and under exfoliating rock (0–1100 m elevation), whereas *penai* is found higher (1500–2400 m) and associated only with rock and stones.

Garthia reproduction is generally the same as the family; clutch count

Garthia spp.	English name				Range core
			Size		
	Activity	Diet	mm	g	Ecological business model
gaudichaudii	Chilean Marked Gecko				Chile
	nocturnal	invert.	38	1.49	*** nocturnal ground lizard
penai	Coquimbo Marked Gecko				Chile
	nocturnal	invert.	33	0.98	*** nocturnal ground lizard

Gastropholis spp.	English name				Range core
			Size		
	Activity	Diet	mm	g	Ecological business model
echinata	Spiny-tailed Lizard				Cameroon
	diurnal	invert.	81	12.2	** trunk/wall lizard
prasina	Green Keel-bellied Lizard				Tanzania
	diurnal	invert.	87	15.1	** diurnal canopy walker
tropidopholis	Congo Keel-bellied Lizard				Congo
	diurnal	invert.	92	18.0	** ground-foraging tree lizard
vittata	Striped Keel-bellied Lizard				Tanzania
	diurnal	invert.	86	14.9	* ground-foraging tree lizard

is moderately high (–0.11 v. –0.57 std. resid.), and young are comparatively small (ontogenetic mass range 0.61 v. 0.11 std. resid.). *G. gaudichaudii* females may lay communally. Longevity is comparatively long (0.33 v. 0.15 std. resid.), corresponding to 5.2 years in *gaudichaudii*. No *Garthia* has been found on islands (v. 40%), has been introduced (v. 7.4%), or is imperiled (v. 29%). In other respects *Garthia* are similar to other phyllodactylids.

Gastropholis

Gastropholis (four spp.) is a small genus of large (14.9 g v. 5.1 g) diurnal arboreal lacertids found in equatorial Africa from Liberia to Mozambique. They are rarely seen and very poorly known (filled cells 24% v. 33%); they have diverse ecological business models: ground-foraging tree lizards (50% v. 8%), diurnal canopy walkers (25% v. 1.4%), and trunk/wall lizards (25% v. 2.4%). Note that the common lacertid ecologies are not present in *Gastropholis*. In morphology they are typical lacertids with the exception that they often have prehensile tails (50% v. 0.6%), the only lacertids to do so. Although their shape index is on the high side for a lacertid (0.53 v. 0.27 std. resid.), they convey the visual impression of a fairly slender lizard with spindly limbs. At least two species are primarily green in color, no doubt reflective of their arboreality.

Sex differences are limited to dichromatism (25% v. 33%). Two species are said to be conspicuously wary (50% v. 11%), though the other two are so rarely seen that they may be also. Their primary predator concealment tactic is retreat to crevices/burrows, in this case mostly tree holes. They are documented only from low elevation (v. 87%) and are associated with riparian zones (50% v. 28%). Reproduction is typical. Longevity is undocumented. Endemism is average (25% v. 35%). No species is found on an island (v. 26%). None has been introduced (v. 4.6%) or evaluated for conservation status (v. 56% imperiled). In other respects *Gastropholis* are similar to other lacertids.

Geckolepis

Geckolepis (five spp.) are typical-sized (3.35 g v. 3.31 g) nocturnal arboreal gekkonids from Madagascar and surrounding islands. Documented species are trunk/wall lizards (v. 45%). Though their common name highlights the pattern of their scalation, the most striking feature of their scales is the relatively huge size of the scales and the lizards' penchant for shedding the scales as a defense mechanism. Conversely, their tails (at least in *polylepis*) are shed only reluctantly. For *polylepis* at least, there is an association with water-holding palms, such as *Palmyra*, which might help them with staying hydrated until their skin grows back. Secretory pores are not reported (v. 46%). No other sex differences are known. *Geckolepis* has a relatively small ontogenetic size range (–1.49, the second lowest in Gekkonidae, v. –0.19 std. resid.), indicating exceptionally large young, perhaps to reduce water balance stresses on the young.

Despite their nocturnality and 50% occurrence around humans (v. 27%), they are not known to vocalize. *G. polylepis* is found in groups (unspecified); both socially documented species (*polylepis*, *typica*) have intolerant males. Size-adjusted incubation duration is short (–0.40 v. 0.17 std. resid.; ca. 44 days). In keeping with their trunk/wall habits, most species lay adherent eggs (60% v. 10.5%) and many lay communally (40% v. 8.6%). Reproduction is undistinguished from other gekkonids, though eggs are relatively large (min-

Geckolepis spp.	English name				Range core
			Size		
	Activity	Diet	mm	g	Ecological business model
humbloti	Comoros Fishscale Gecko				Comoros
	nocturnal		56	3.82	
maculata	Spotted Fishscale Gecko				Madagascar
	nocturnal		52	3.07	** trunk/wall lizard
megalepis	Grand Fishscale Gecko				Madagascar
			59	4.53	
polylepis	Spangled Fishscale Gecko				Madagascar
	nocturnal		46	2.22	*** trunk/wall lizard
typica	Grandidier's Fishscale Gecko				Madagascar
	nocturnal		54	3.56	** trunk/wall lizard

imum dimension 11 mm). Longevity is comparatively long (0.90 v. 0.15 std. resid.; 10 years in *typica*). None has been introduced (v. 3.7%); none is considered imperiled (v. 47%). In other respects *Geckolepis* are typical gekkonids.

Gehyra (figs. 2.6B, 4.9B)

Gehyra is a large genus (44 spp.) of the Gekkonidae, with a wide, generally tropical, distribution around the New Guinea/northern Australian core (full native range: Sri Lanka to Japan and French Polynesia). These are notable jurisdictions where *Gehyra* can be found (listed north to south):

Thailand	4 spp.
Vietnam	3
Philippines	1
Indonesia	8
Queensland	10
Northern Territory	13
Western Australia	14
South Australia	5
New South Wales	3

Notable islands where *Gehyra* can be found (listed north to south) include the following:

Borneo	1
Sumatra	2
New Guinea	10

The ecological role of *Gehyra* is primarily that of trunk/wall lizard (73% v. 45%); some diversify by moving downwards: ground-foraging tree lizard (5% v. 5%), nocturnal ground lizard (11% v. 27%); some by moving upwards: nocturnal canopy walker (11% v. 9%).

They are average-sized lizards but large for gekkonids (4.57 g v. 3.31 g). One species (*membranacruralis*) has a prehensile tail (2.3% v. 1.8%) with adhesive lamellae on the tail tip; several have skin that tears easily (*insulensis, mutilata, oceanica*). Secretory pore frequency is low for a gekkonid (32% v. 46%), but basic color change (darkening) is unusually common (43% v. 15%). All species possess expanded toe pads (v. 30%).

SSD is mild, but more m>f than a typical gekkonid (0.21 v. −0.03 on −2 to 2 scale), in keeping with their slightly larger size. Other sex differences are not known, resulting in a reduced sex differences rollup compared to the family (0.39 v. 0.63 on 0 to 6 scale).

Although diet is overwhelmingly the gekkonid standard invertebrates, two large species (*oceanica, vorax*) also consume fruit, nectar, and vertebrates (omnivores 13% v. 4%). They are all considered sit-and-wait foragers (v. 81%), and electric light use is relatively common (6.8% v. 3.8%). Home ranges have not been quantified.

Most are nocturnal (93% v. 73%) and vocalizations are correspondingly common (9.1% v. 6.5% for spontaneous calling). All documented species use crevices and burrows as their primary form of predator concealment; none sleeps in burrows (v. 10.4%). They display the gekkonid standard active body temperature (27.8°C v. 27.6°C).

They do not appear to be social. None of the documented species occurs in groups (v. 25%), and all males are reported to defend space (v. 87%). One-third are documented to live solitarily as opposed to in pairs (v. 7.7%).

Although gekkonids are bimodal in their distribution along a rainfall gradient, *Gehyra* straddle the middle with 40% documented in the arid (151–400 mm/y), intermediate (401–1100 mm/y), and mesic (1101–3000 mm/y) rainfall bands. They are overrepresented in the littoral zone (20% v. 11%) but underrepresented in sandy areas (3% v. 11%).

Reproduction is not much different from other gekkonids, though they seem slightly less inclined to have adherent eggs (4.5% v. 10.5%). The size of hatchlings is lower than an average lizard (ontogenetic range 0.19 v. −0.19 std. resid.). This contrasts with another similar-sized gekkonid that sheds skin to escape predators (*Geckolepis*) in which the young are exceptionally large.

Longevity in *Gehyra* is much longer than that of other gekkonids (1.32 v. 0.14 std. resid.); this corresponds to an absolute maximum longevity of 26 years in *variegata*. Credible population density estimates exist for four species (means in parentheses): *insulensis* (1.5 kg/ha), *mutilata* (0.07

Gehyra spp.	English name				Range core
			Size		
	Activity	Diet	mm	g	Ecological business model
angusticaudata	Chonburi Gecko				Thailand
	nocturnal		50	2.84	
australis	Northern Dtella				Northern Australia
	nocturnal	invert.	67	6.3	** trunk/wall lizard
baliola	Short-tailed Dtella				New Guinea
	nocturnal		84	11.8	** trunk/wall lizard
barea	Banda Island Dtella				Indonesia
	nocturnal		84	11.9	
borroloola	Borroloola Dtella				Queensland
	nocturnal		61	4.90	* ground-foraging tree lizard
brevipalmata	Palau Gecko				Palau
	nocturnal	invert.	73	7.9	** trunk/wall lizard
butleri	Butler's Four-clawed Gecko				Malaysia
	nocturnal		32	0.84	
catenata	Chain-backed Dtella				Queensland
	nocturnal		52	3.21	** nocturnal canopy walker
dubia	Dubious Dtella				Queensland
	nocturnal	invert.	55	3.70	*** trunk/wall lizard
einasleighensis	Einasleigh Rock Dtella				Queensland
	nocturnal		39	1.39	** nocturnal ground lizard
fehlmanni	Fehlmann's Four-clawed Gecko				Thailand
	nocturnal		46	2.22	* nocturnal ground lizard
georgpotthasti	Vanuatu Giant Gecko				Polynesia
	nocturnal	invert.	109	24.0	*** trunk/wall lizard
girloorloo	Kimberley Karst Gecko				Western Australia
			44	1.94	*** ground-foraging tree lizard
insulensis	Pacific Stump-toed Gecko				New Guinea
	nocturnal	invert.	46	1.70	**** nocturnal canopy walker or trunk/wall lizard (where commensal)
interstitialis	Oudeman's Dtella				New Guinea
	nocturnal		76	8.9	
kimberleyi	Robust Termitaria Gecko				Western Australia
	nocturnal		38	1.32	
koira	Koira Dtella				Western Australia
	nocturnal		81	10.6	*** trunk/wall lizard
lacerata	Kanchanaburi Four-clawed Gecko				Thailand
	nocturnal	invert.	50	2.88	
lazelli	Lazell's Dtella				New South Wales
	nocturnal		46	2.18	*** trunk/wall lizard

kg/ha), *oceanica* (5.1 kg/ha), *variegata* (0.9 kg/ha). This spread presents no compelling pattern other than the higher values for island species, though the genus mean (0.84 kg/ha) well exceeds that of the family (0.37 kg/ha). Introductions are exceptionally common in *Gehyra* (any: 6.8% v. 3.7%; >2 introductions: 6.8% v. 2.1%). Seventeen percent of evaluated species are considered imperiled (v. 47%), though only 8.3% (v. 11.3%) are judged by the IUCN to be in the most endangered categories. In other respects *Gehyra* are typical gekkonids.

G. vorax was once eaten by people.

Gehyra spp.	English name				Range core
			Size		
	Activity	Diet	mm	g	Ecological business model
leopoldi	Leopold Dtella				New Guinea
	nocturnal		41	1.61	* nocturnal canopy walker
marginata	Ternate Dtella				Indonesia
	nocturnal		111	25.3	** trunk/wall lizard
membranacruralis	Port Moresby Dtella				New Guinea
	nocturnal		103	20.7	** trunk/wall lizard
minuta	Dwarf Dtella				Northern Territory
	nocturnal		43	1.90	** trunk/wall lizard or immobility expert
montium	Centralian Dtella				Western Australia
	cathemeral		48	2.45	** nocturnal ground lizard
moritzi	Craig's Dtella				Northern Territory
	nocturnal		44	2.03	** trunk/wall lizard
multiporosa	Multi-pored Dtella				Western Australia
	nocturnal		51	3.03	**** trunk/wall lizard
mutilata	Stump-toed Gecko				Southern East Asia
	nocturnal	invert.	54	3.06	**** nocturnal canopy walker
nana	Northern Spotted Rock Dtella				Northern Australia
	cathemeral		48	2.49	* trunk/wall lizard
occidentalis	Kimberley Plateau Dtella				Western Australia
	nocturnal	invert.	65	5.83	** trunk/wall lizard
oceanica	Oceania Gecko				Pacific Islands
	nocturnal	omnivore	81	11.0	**** trunk/wall lizard
pamela	Arnhem Land Watercourse Dtella				Northern Territory
	nocturnal		64	5.5	** trunk/wall lizard
papuana	Papua Four-clawed Gecko				New Guinea
	nocturnal		74	8.4	
pilbara	Pilbara Dtella				Western Australia
	nocturnal	invert.	50	2.80	*** nocturnal ground lizard
pulingka	Rocky Hill Gecko				South Australia
	nocturnal		44	2.03	*** trunk/wall lizard
punctata	Spotted Dtella				Western Australia
	nocturnal	invert.	52	3.07	*** trunk/wall lizard
purpurascens	Purplish Dtella				Australia
	cathemeral	invert.	56	3.83	*** trunk/wall lizard
robusta	Robust Dtella				Queensland
	nocturnal		68	6.6	** trunk/wall lizard
rohan	Orange Eye-ring Giant Gecko				New Guinea
	nocturnal		117	29.1	**** trunk/wall lizard
serraticauda	Fakfak Gecko				Indonesia
	nocturnal		95	16.6	* trunk/wall lizard

Gehyra spp.	English name / Activity	Diet	Size mm	Size g	Range core / Ecological business model
spheniscus	Small Wedge-toed Gecko				Western Australia
	nocturnal		45	2.05	** trunk/wall lizard
variegata	Western Tree Dtella				Western Australia
	nocturnal	invert.	55	3.75	*** trunk/wall lizard or crack hermit (geog. var.)
versicolor	Eastern Tree Dtella				Eastern Australia
	nocturnal		48	2.52	*** trunk/wall lizard
vorax	Fiji Giant Gecko				Fiji
	nocturnal	omnivore	131	40.5	* trunk/wall lizard
xenopus	Crocodile-faced Dtella				Western Australia
	nocturnal	invert.	67	6.2	** trunk/wall lizard

Gekko spp.	English name / Activity	Diet	Size mm	Size g	Range core / Ecological business model
aaronbaueri	Aaron Bauer's Gecko				Laos
			67	6.2	
adleri	Adler's Gecko				Vietnam
			63	5.3	
albofasciolatus	White-lined Mystery Gecko				Thailand
			127	37.1	
athymus	Smooth-scaled Narrow-disked Gecko				Philippines
	crepuscular		95	16.6	** ground-foraging tree lizard
auriverrucosus	Shanxi Gecko				China
			59	4.38	
badenii	Vietnam Golden Gecko				Vietnam
	nocturnal		109	24.4	* nocturnal ground lizard or trunk/wall lizard
boehmei	Boehme's Gecko				Laos
			84	11.9	
bonkowskii	Bonkowski's Gecko				Laos
			59	4.38	
canaensis	Ca Na Marbled Gecko				Vietnam
			87	13.1	
canhi	Canh's Gecko				Vietnam
			80	10.3	
carusadensis	Luzon Karst Gecko				Philippines
			79	9.8	** trunk/wall lizard
chinensis	Gray's Chinese Gecko				China
	nocturnal	invert.	67	6.2	** trunk/wall lizard

Gekko (fig. 4.14D)

Gekko is a large radiation (58 spp.) of the Gekkonidae found primarily in the Oriental biogeographic realm (86%); including likely introductions, it has a minor presence in all others except the Ethiopian. Native distributions range west-east from Sri Lanka to Vanuatu and north-south from China and Japan to Papua New Guinea. The countries with the most species are: China (17), Vietnam (13), and the Philippines (13).

The dominating ecological business model is trunk/wall lizard (88% v. 45%). Minor patterns are farther down the trunk: ground-foraging tree lizards (6% v. 9%) and nocturnal ground lizards (6% v. 8%). Perhaps equally pertinent is that the ecological business models of 25 of 58, or 43%, of *Gekko* could not be estimated from the available data, and only 7% were assigned with full confidence.

On average these are fairly large gekkonids (9.1 g v. 3.31 g), ranging from *similignum* and *wenxianensis* (3.07 g) to the stentorian *gecko* (60 g) and green-eyed *smithii* (62 g). All have expanded toepads (v. 70%). Otherwise they fit a basic gekkonid morphology. Under optimal conditions the lamellae of one 50 g Tokay Gecko (*G. gecko*) can support 210 kg (the mass of three adult humans).

Consonant with their large size, they have larger males (1.0 v. −0.03 on −2 to 2 scale). Aside from limited dichromatism (1.7% v. 6.6%), they are not particularly distinguished between the sexes, and the sex rollup score is a pedestrian 0.55 (v. 0.63 on 0 to 6 scale).

As with most geckos they are primarily invertebrate eaters (82% v. 95%), but they include a notable proportion of omnivores (18% v. 4%). Tokay Geckos (*G. gecko*) are among the omnivores (12% plants). All are said to sit-and-wait forage (v. 81%), which they do fairly frequently at electric lights (5.2% v. 3.8%). *G. gecko* is reported to switch to active foraging when prey are scarce; they are vigilant and may pounce on prey that pass their daytime refuge, but most locomotion is nocturnal (similar in *smithii*). Home-range size is not reported for any *Gekko*.

Most are nocturnal (85% v. 73%), though they are the second-highest genus within Gekkonidae for crepuscular species (5% v. 1%, behind only the monotypic *Narudasia*). They are overrepresented among the gekkonids for calling spontaneously (10.3% v. 6.5%) but not stress calling (1.7% v. 3.7%). The spontaneous call of the male *gecko* is detectable by humans 100 m distant; the female's call is softer. The call is given throughout the year at all times of day, but peak calling is at dawn and dusk during the mating season. For predator concealment *Gekko* do not rely on crypsis (0% v. 50%); instead they are uniquely dependent on hiding in vegetation (100% v. 19%). This fits with the predominance of arboreal microhabitats (94% v. 68%). Despite the generally nocturnal activity pattern, *Gekko* bask directly in a slightly greater proportion than other gekkonids (7% v. 5%); basking may take place at the mouth of their daytime refugia.

All documented species have males that are intolerant of other males (v. 87%). In *monarchus*, groups of up to 50 lizards are reported, but even in this species the males are territorial and will chase off lizards, even those of other species.

All documented *Gekko* are found exclusively in the mesic (1101–3000 mm/y) rainfall belt (v. 66%). They are underrepresented at mid (13% v. 26%) and high (0% v. 2%) elevations. They are overrepresented in littoral (16% v. 11%) and human (45% v. 27%) habitats. They are not known to occur in sandy areas (v. 11%).

Gekko spp.	English name				Range core
			Size		
	Activity	Diet	mm	g	Ecological business model
coi	Sibuyan Gecko				Philippines
			69	7.0	** trunk/wall lizard
crombota	Babuyan Claro Gecko				Philippines
	nocturnal		94	15.9	** trunk/wall lizard
ernstkelleri	Panay Wall Gecko				Philippines
	nocturnal		75	8.7	** trunk/wall lizard
gecko	Tokay Gecko				Southern East Asia
	cathemeral	omnivore	152	60	**** trunk/wall lizard
gigante	Gigante Island Gecko				Philippines
			84	11.9	** trunk/wall lizard
grossmanni	Vietnam Marbled Gecko				Vietnam
	nocturnal		74	8.2	
guishanicus	Guishan Gecko				Taiwan
			57	4.03	
hokouensis	Hokou Gecko				East Asia
	nocturnal	invert.	65	5.8	*** trunk/wall lizard
japonicus	Japan Gecko				East Asia
	nocturnal	invert.	64	5.6	*** trunk/wall lizard
kikuchii	Orchid Island Gecko				Taiwan
			67	6.2	*** trunk/wall lizard
kwangsiensis	Kwangsi Gecko				China
			59	4.53	
lauhachindai	Lauhachinda's Cave Gecko				Thailand
			79	10.1	* trunk/wall lizard
liboensis	Libo Gecko				China
			74	8.4	* trunk/wall lizard
melli	Mell's Gecko				China
			70	7.2	
mindorensis	Mindoro Narrow-disked Gecko				Philippines
	nocturnal		72	7.8	* ground-foraging tree lizard
monarchus	Spotted House Gecko				Borneo
	nocturnal	invert.	87	13.1	**** trunk/wall lizard
nutaphandi	Nutaphand's Red-eyed Gecko				Thailand
			93	15.6	** trunk/wall lizard
palawanensis	Palawan Gecko				Philippines
			57	3.96	** trunk/wall lizard
palmatus	Palmated Gecko				Vietnam
	nocturnal		75	8.5	* trunk/wall lizard
petricolus	Sandstone Gecko				Southeast Asia
	nocturnal		82	10.9	*** trunk/wall lizard

Gekko spp.	English name				Range core
			Size		
	Activity	Diet	mm	g	Ecological business model
porosus	Batan Narrow-disked Gecko				Philippines
			79	9.8	** trunk/wall lizard
reevesii	Black Tokay				Vietnam
			133	41.8	
remotus	Palau Ghost Gecko				Palau
	nocturnal	omnivore	102	20.4	*** trunk/wall lizard
romblon	Romblon Gecko				Philippines
			73	8.0	*** trunk/wall lizard
rossi	Ross' Calayan Gecko				Philippines
			87	12.8	
russelltraini	Russell Train's Marbled Gecko				Vietnam
			69	6.8	
scabridus	Yunnan Gecko				China
	nocturnal	invert.	64	5.7	*** trunk/wall lizard
scientiadventura	Seven-spotted Gecko				Vietnam
	nocturnal		62	5.0	* trunk/wall lizard
sengchanthavongi	Sengchanthavong's Gecko				Laos
			64	5.7	
shibatai	Takara Gecko				Japan
			63	5.3	
siamensis	Thai Green-eyed Gecko				Thailand
	nocturnal	invert.	117	29.1	*** trunk/wall lizard
similignum	Five Finger Mountain Gecko				China
			52	3.07	
smithii	Smith's Green-eyed Gecko				Malaysia
	cathemeral	invert.	154	62	**** trunk/wall lizard
subpalmatus	Coastal China Gecko				China
			61	4.8	
swinhonis	North China Gecko				China
			62	5.2	* nocturnal ground lizard
taibaiensis	Mingtao's Gecko				China
			59	4.53	
takouensis	Ta Kou Marbled Gecko				Vietnam
			86	12.5	
tawaensis	Tawa Gecko				Japan
			60	4.68	** trunk/wall lizard
thakhekensis	Thakhek Gecko				Laos
			66	6.0	** trunk/wall lizard
truongi	Truong's Gecko				Vietnam
			78	9.6	

Their reproduction follows gekkonid patterns with two exceptions attributable to their arboreality: adherent eggs are relatively common (21% v. 10.5%), as is communal oviposition (12% v. 8.6%). Large size may play a role in the dominance of parental care (6.9% v. 1.9%). Females of the largest species (*smithii*) consume spent eggshells, defend the brood, and attend progeny up to and after hatching. In the equally massive Tokay Gecko (*G. gecko*), both parents do some guarding of brood, and the hatchlings may remain with the parents until maturity. In places where the larger *G. gecko* are present, *ernstkelleri* (8.7 g) is said to defend its eggs against even humans; evolutionary history is more likely to include a degree of predation by *gecko*. Both parents defend the clutch and consume dead eggs in *petricolus* (11 g). The apparently fixed clutch count of two results in *Gekko* having a particularly small mass-adjusted clutch count (−0.78 v. −0.26 std. resid.); this is more or less offset by the exceptionally large size of the resultant young (ontogenetic size range −0.49 v. −0.19 std. resid.).

Longevity is recorded for five species of *Gekko* and is relatively long when size adjusted (0.41 v. 0.15 std. resid.); the Tokay Gecko holds the absolute record (23 years). Population density has been credibly quantified for only the Tokay (0.81 kg/ha), which is greater than of the average gekkonid (0.37 kg/ha). Curiously, *gecko* is thought to do much better on islands off the coast of mainland Malaysia and Borneo; the factors responsible for population reduction on the mainland are unclear, but predation on young and defending parents is plausible. *Gekko* have disproportionately been introduced (6.9% v. 3.7%), often more than twice (3.4% v. 2.1%), presumably because their large size, gaudy colors, and bold habits make

Gekko spp.	English name				Range core
			Size		
	Activity	Diet	mm	g	Ecological business model
verreauxi	Andamans Giant Gecko				India
			120	31.6	**** trunk/wall lizard
vertebralis	Kodakara Gecko				Japan
			60	4.68	
vietnamensis	Vietnam Gecko				Vietnam
			74	8.4	
vittatus	Melanesia Ghost Gecko				Solomon Islands
	nocturnal	invert.	110	24.6	** trunk/wall lizard
wenxianensis	Qinling Mountains Gecko				China
			52	3.07	
yakuensis	Yaku Island Gecko				Japan
	nocturnal	invert.	61	4.84	*** trunk/wall lizard

them attractive to the pet trade. *G. gecko* has also been translocated for human food, despite their false reputation for being venomous. Both *gecko* and *swinhonis* are collected for medicinals, including for HIV. They are underrepresented among the imperiled species (31% v. 47%) and the IUCN's most endangered categories (6.3% v. 11.3%). In other respects they are typical gekkonids.

Gekkonidae (figs. 2.6A, 2.6B, 4.5B, 4.7A, 4.7D, 4.9B, 4.9C, 4.9D, 4.10C, 4.10D, 4.14A, 4.14C, 4.14D)

At 1120 species, Gekkonidae is second only to Scincidae (1613 spp.) in size and diversity. They are found in every biogeographic realm, though their center of diversity appears to be the Old World, where they are nearly continuous in distribution at low and moderate latitudes. Only a few central Asian species extend beyond 45° latitude. They are overrepresented in the Ethiopian (38% v. 21%) and Oriental (43% v. 18%) biogeographic realms.

It is easier to specify the ecological business models that they do not exhibit; there are no gekkonids that are giant herbivores (v. 1.6%), sand sharks (v. 1.9%), subterranean foragers (v. 7.8%), or wide-ranging chemosensors (v. 7.4%). As a very general rule, gekkonids are not very mobile (81% v. 46% are sit-and-wait), plant eating (95% v. 73% are invertebrate eaters), diurnal (73% v. 25% are nocturnal), terrestrial (68% v. 32% are arboreal), or cryptozoic (0.9% v. 12.1% are cryptozoic).

Gekkonidae is the largest of the seven families that make up the Gekkota (chapter 2, "Taxonomic Composition"). Members of six of the seven Gekkota families (the outlier is Pygopodidae) are popularly known as geckos. Geckos sensu lato (Gekkota, often excepting Pygopodidae) need to be distinguished from the geckos sensu stricto (Gekkonidae). The 57 genera making up the Gekkonidae are *Afroedura*, *Afrogecko*, *Agamura*, *Ailuronyx*, *Alsophylax*, *Altiphylax*, *Blaesodactylus*, *Bunopus*, *Calodactylodes*, *Chondrodactylus*, *Christinus*, *Cnemaspis*, *Colopus*, *Crossobamon*, *Cryptactites*, *Cyrtodactylus*, *Cyrtopodion*, *Dixonius*, *Ebenavia*, *Elasmodactylus*, *Geckolepis*, *Gehyra*, *Gekko*, *Goggia*, *Hemidactylus*, *Hemiphyllodactylus*, *Heteronotia*, *Homopholis*, *Kolekanos*, *Lepidodactylus*, *Luperosaurus*, *Lygodactylus*, *Matoatoa*, *Mediodactylus*, *Microgecko*, *Nactus*, *Narudasia*, *Pachydactylus*, *Paragehyra*, *Paroedura*, *Parsigecko*, *Perochirus*, *Phelsuma*, *Pseudoceramodactylus*, *Pseudogekko*, *Ptenopus*, *Ptychozoon*, *Ramigekko*, *Rhinogekko*, *Rhoptropella*, *Rhoptropus*, *Stenodactylus*, *Tenuidactylus*, *Trachydactylus*, *Tropiocolotes*, *Urocotyledon*, and *Uroplatus*. Three of these genera are notable as being in the five largest lizard genera (table 2.2): *Cyrtodactylus* (232 spp.), *Hemidactylus* (144 spp.), and *Cnemaspis* (130 spp.). They are exceeded only by *Anolis* (416 spp.) and *Liolaemus* (258 spp.).

The largest group of gekkonids is trunk/wall lizards (45% v. 13%). The second-largest ecological component is nocturnal ground lizards (27% v. 7.7%), which are notably overrepresented in four other gecko (sensu lato) families as well: Carphodactylidae, Diplodactylidae, Eublepharidae, and Phyllodactylidae. Other ecological business models overrepresented within the Gekkonidae are crack hermits (3.2% v. 2.6%), gliders (1.5% v. 1.2%), immobility experts (4.5% v. 3.2%), and nocturnal canopy walkers (8.9% v. 2.1%). Although several well-known human commensals are nocturnal canopy walkers in nature (e.g., *Hemidactylus*, *Lepidodactylus*), it is notable that these canopy walkers are not typical gekkonids. Ecological business models that are also represented in the Gekkonidae, but are underrepresented, include diurnal canopy walkers (1.0% v. 6.4%), ground-foraging tree lizards (5.3% v. 9.2%), litter cryptozoids (0.5% v. 11.6%), and refugia-anchored diurnal ground lizards (3.0% v. 24.5%). It is notable that litter cryptozoids are so scarce in gekkonids, whereas they dominate the closely related, but arguably more basal, Sphaerodactylidae (53%).

The average gekkonid is small

(3.3 g v. 5.3 g), and the size range is fairly small (0.4–76 g) for so large a family. Ninety percent of gekkonids are in the size range 0.76–17.1 g. Gekkonids are of average build (shape index 0.10 v. 0.0 std. resid.). Their tails are on the short side (1.07 v. 1.37 × SVL). Most tails break (87% v. 57%), and gekkonid tails are not usually prehensile (1.8% v. 5.0%). This seems unusual for a largely arboreal clade of small lizards, but it is better understood when one focuses on the prevalence of trunk/wall lizards rather than canopy walkers. Unlike a twig, a trunk cannot easily be grasped with a prehensile tail.

Gekkonids are crestless (v. 5.3%) and almost never elongate (0.9% v. 11.4%), but pores are a dominant feature (46% v. 15%), and basic color change is well represented (14.6% v. 10.7%), whereas advanced color change is rare (2.1% v. 6.0%). Many gekkonids are dark during the day (for thermoregulation or crypsis) and blanched at night. Blue tails are rare in gekkonids (0.6% v. 2.3%). It is notable that on the few occasions when blue tails are found, they are present in males rather than the females and juveniles that dominate blue tails in other lizard families. Many gekkonids, especially the arboreal species, exhibit enlarged toe tips or "toepads." The absence of enlarged toepads is associated with terrestriality or rock climbing and often follows taxonomic lines, with toepads being generally absent in *Agamura*, *Alsophylax*, *Altiphylax*, *Bunopus*, *Cnemaspis*, *Colopus*, *Crossobamon*, *Cyrtodactylus*, *Cyrtopodion*, *Heteronotia*, *Mediodactylus*, *Microgecko*, *Nactus*, *Narudasia*, *Parsigecko*, *Ptenopus*, *Rhinogekko*, *Stenodactylus*, *Tenuidactylus*, and *Trachydactylus*. There are individual species that are exceptions; see genus accounts. About 42% of Gekkonidae are padless, as are about 45% of the legged Gekkota. Gamble et al. (2012) present a formal analysis of the evolution of toepads in geckos; they cite evidence of adhesive toepads evolving on 11 different occasions and being lost at least nine times.

With the exception of secretory pores, sex differences in gekkonids are muted. SSD of the m>f variety almost exactly balances the m<f version, such that mean SSD score (on −2 to 2 scale) is a piddling −0.03 (v. 0.36). Sexual dichromatism (6.6% v. 20.7%) and seasonal color (0.8% v. 6.5%) are uncommon, and mean sexual difference in relative tail length is very close to the lizard standard (1.05 v. 1.07 × SVL). Accordingly, the sex rollup score is an underwhelming 0.63 (v. 0.75, on a 0 to 6 scale).

Although most gekkonids are invertebrate eaters, there are a few carnivores (1.1% v. 3.2%), herbivores (0.3% v. 4.1%), and omnivores (3.8% v. 15.4%). What is entirely missing is ant specialists (0% v. 4.4%). This is especially striking because ant specialists are not missing from the rest of the Gekkota clade. Phyllodactylidae, the closest sister family, based on phylogeny in Pianka et al. (2017), has 1.8% ant specialists, and the next closest family, Sphaerodactylidae, has 16.4%. Taking either the closest Gekkotans or all Gekkotans as the contrast group, Fisher's exact probability test for a 2 × 2 contingency table of counts of ant specialists and non-ant specialists highlights the distinctiveness of Gekkonidae ($p < 0.0001$ in both cases). Gekkonids are ready consumers of ants (18.3% of species list ants in their diet), but zero percent so specialize. As discussed in chapter 3, the abundance of nocturnal ants is roughly equivalent to that during the day, but the predominantly nocturnal gekkonids do not specialize in ant eating.

As noted in the first paragraph of this entry, gekkonids are primarily sit-and-wait foragers, but they are notably overrepresented in cruise foraging as well (23.8% v. 6.6%). Naturally, they are correspondingly underrepresented in active foraging (14% v. 58%). Their use of electric lighting is high (4.3% v. 2.0%), as are all of the closely related Gekkota except Eublepharidae (Phyllodactylidae are 9.2%; Sphaerodactylidae are 6.9%). The distal half of the Gekkota (Carphodactylidae, Pygopodidae, Diplodactylidae) are uniformly zero percent users of electric lighting. Home ranges in gekkonids are moderately small (−0.28 v. 0.0 std. resid.), though the documentation is spare and not very clean (i.e., reported sizes likely to be influenced by unnatural environments).

Although most gekkonids are nocturnal, all patterns are represented: cathemeral is overrepresented (10.8% v. 6.8%); crepuscular (1.0% v. 1.8%) and diurnal (15% v. 67%) are less frequent. Along with the nocturnality comes vocalizations: 6.5% (v. 2.5%) for spontaneous; 3.7% (v. 2.2%) for stress calls. Compared to other geckos, gekkonids are less prone to calling spontaneously: carphodactylids (13.3%), eublepharids (8.3%), phyllodactylids (11.1%), and especially pygopodids (67%) are all higher. Only sphaerodactylids are notably lower (1.9%). All of the gecko families except Sphaerodactylidae (2.3%) also have higher frequencies of stress callers: Carphodactylidae (33%), Diplodactylidae (5.4%), Eublepharidae (25%), Phyllodactylidae (5.9%), Pygopodidae (28%).

The frequency of gekkonid species that are conspicuously wary is lower than an average lizard (4.4% v. 5.8%). It is also lower than eublepharids (19%), pygopodids (8.7%), and sphaerodactylids (7.0%). This lack of wariness is likely connected to, but may not be fully explained by, the gekkonids' reliance on crypsis and nocturnality. Based on the general trends (table 3.7, trait 35), 3.2% of nocturnal species are wary, a bit lower than the 4.4% observed in gekkonids. In general (table 3.7, trait 36), 61% of nocturnal species rely on crypsis, but only 50% of gekkonids do so. Untangling the web of causation will be complex.

The mix of predator concealment tactics exhibited by gekkonids is not very different from that of all lizards. Crypsis (50% v. 43%) and escape to crevices or burrows (31% v. 25%) are slightly overrepresented, and flight (5% v. 11%), sand (0% v. 6%), and vegetation (19% v. 23%) are mildly underrepresented.

Gekkonids are unsurprisingly underrepresented in species that bask directly (4.9% v. 17.9%). The mean active body temperature is lower than a typical (mostly diurnal) lizard (27.6°C v. 31.1°C).

With the exception of male hierarchical frequency (0% v. 7.7%), the social patterns seen in gekkonids are typical of all lizards. The lack of male hierarchical behavior may be related to the visual way in which herpetologists quantify hierarchical behavior. If hierarchical behavior is mediated by pore secretions, it would be unlikely to be noticed by current techniques. In this regard it is notable that the two lowest diel categories for hierarchical behavior (table 3.7, trait 34) are cathemeral (2.9%) and nocturnal (3.8%), the diel patterns prevalent in gekkonids.

Gekkonids are found in all of the six rainfall belts, at frequencies very similar to those of all lizards, with the mild exception of the mesic (1101–3000 mm/y) belt, where gekkonids are overrepresented (66% v. 43%). Gekkonids diverge more from all lizards in elevation; as with many nocturnal groups, they are overrepresented below 1200 m (92% v. 82%) and underrepresented above: 1200–2400 m, 26% v. 41%; >2400 m, 2% v. 13%. Gekkonids like it warm, moist, and low.

Gekkonids are mildly underrepresented in littoral zones (11% v. 16.6%), sand areas (11% v. 21.6%), and riparian zones (17% v. 22.2%), but they are overrepresented in human habitats (27% v. 18.5%). Undoubtedly the advantages of foraging under electric lights have an influence on their attraction to human habitats, but other factors may play a role. For example, most gekkonids are trunk/wall lizards, for which buildings are a massive expansion of the preferred two-dimensional vertical microhabitat.

Gekkonids are all oviparous (v. 82%), with most clutches seemingly fixed almost exclusively at two eggs (mean 1.9 v. 4.0). Only two gekkonid genera are exceptions: *Calodactylodes* (6.3) and *Rhinogekko* (5.0). Expressed as mass-adjusted standardized residuals, gekkonid clutches are slightly small (–0.26 v. 0.0). Do the gekkonids make up for the small counts by increasing frequency of clutches, increasing the size of hatchlings, or both? Unfortunately, the data on clutch frequency are inadequate for general conclusions, but ontogenetic size range indicates that the relative size of young are very slightly larger than average lizards (–0.19 v. 0.0 std. resid.). Incubation duration in gekkonids is typical or slightly long (0.17 v. 0.0 std. resid.). Exceptions (*Christinus, Colopus, Cyrtodactylus, Homopholis*) are discussed in those genus accounts. Gekkonid eggshells are almost always hard (but see *Cyrtodactylus, Paroedura*). Adherent eggs are overrepresented (10.5% v. 2.4%) as they often are in trunk/wall lizards (table 4.1, trait 67). Communal oviposition is also overrepresented (8.6% v. 4.2%). It too is associated with trunk/wall lizards (table 4.1, trait 68). Parental care is underrepresented in Gekkonidae (1.9% v. 2.9%), which is not the expectation based on trunk/wall lizards (3.8% in table 4.1, trait 69). RCM is low (0.14 v. 0.23 ratio), as expected for a small fixed clutch count.

Longevity is about average (0.15 v. 0.0 std. resid.), as is population density (0.37 kg/ha v. 0.54 kg/ha), with exceptions for some island species (see *Hemidactylus*). Island species are not unusually frequent in Gekkonidae overall (44.5% v. 40.3%), but endemism is high (71% v. 64%). Introductions are about average (3.7% v. 3.3% for any introductions), but repeat offenders tend to be more common in the Gekkonidae (>2 introductions: 2.1% v. 1.1%). Gekkonids are imperiled (47% v. 49%) and on the IUCN's lists of most endangered at an average rate (11.3% v. 13.6%).

Gelanesaurus

Gelanesaurus (two spp.) are riparian, diurnal, ground gymnophthalmids found in Ecuador and Colombia.

Gelanesaurus spp.	English name				Range core
			Size		
	Activity	Diet	mm	g	Ecological business model
cochranae	Cochran's Streamside Lizard				Ecuador
			69	7.8	* refugia-anchored diurnal gr. liz.
flavogularis	Yellow-throated Streamside Lizard				Ecuador
	diurnal		62	5.6	* refugia-anchored diurnal gr. liz.

They are larger than typical gymnophthalmids (6.6 g v. 2.4 g) and heavier in build (shape index 0.34 v. −0.32 std. resid.). They have shorter tails (1.40 v. 1.72 × SVL) that are breakable (v. 28%). One species (*flavogularis*) is sexually dichromatic (v. 13%), but other sex differences are not reported. Reproduction is typical for gymnophthalmids, except that the eggs are relatively few in number (−0.58 v. −0.11 std. resid.), perhaps due to the family's apparent restriction to clutches of around two. Neither species is considered imperiled. In other respects *Gelanesaurus* are probably typical gymnophthalmids.

Geocalamus

Geocalamus is small genus (two spp.) of average-sized (2.66 g v. 2.73 g), average build (shape index −4.0 v. −4.0 std. resid.) amphisbaenids from East Africa (Kenya, Tanzania). Endemism is average (50% v. 66%). None occurs on islands (v. 12%). With the few data available, they are undistinguished amphisbaenids.

Geomyersia

Geomyersia is a two-species genus of tiny (0.59 g v. 3.3 g), svelte (shape index −0.78 v. −0.46 std. resid.) litter cryptozoid (100% v. 28%) skinks found on islands northeast of New Guinea (Manus to Shortland Islands).

Tails are short (1.0 v. 1.33 × SVL) and breakable (50% v. 73%). At least *glabra* is elongate (v. 35%). One species (*glabra*) is found in riparian zones (50% v. 23%). At least *glabra* is oviparous (v. 64%) with an unexceptional clutch of one (0.21 v. 0.06 std. resid). Neither species has been introduced (v. 2.1%). One (*glabra*) is imperiled (50% v. 47%); neither is on one of the IUCN's lists of the most endangered (v. 14%). In other respects *Geomyersia* are probably typical skinks.

Geocalamus spp.	English name				Range core
			Size		
	Activity	Diet	mm	g	Ecological business model
acutus	Voi Wedge-snouted Worm Lizard				Kenya
			186	2.83	**** subterranean forager
modestus	Mpwapwa Wedge-snouted Worm Lizard				Tanzania
	diurnal	invert.	181	2.49	**** subterranean forager

Geomyersia spp.	English name				Range core
			Size		
	Activity	Diet	mm	g	Ecological business model
coggeri	Manus Island Skink				New Guinea
			34	0.65	* litter cryptozoid
glabra	Solomon Minute Skink				Papua New Guinea
	nocturnal	invert.	35	0.54	* litter cryptozoid

Geoscincus sp.	English name				Range core
			Size		
	Activity	Diet	mm	g	Ecological business model
haraldmeieri	Meier's Skink				New Caledonia
			89	15.0	

Geoscincus

Geoscincus haraldmeieri is the sole member of its scincid genus, and endemic to New Caledonia. Sadly, the species was described in 1976 on the basis of two male specimens found under one log in a low-elevation forest that was subsequently eliminated. In the subsequent half century, no further specimens have come to light; it may be extinct. It was large (15 g v. 3.3 g).

It was a bit more robust than an average skink (shape index 0.45 v. −0.46 std. resid.) and had a breakable tail (v. 73%). If it still exists, it is imperiled (v. 47%). It is presently listed as critically endangered by the IUCN (v. 14%).

Gerrhonotus

Gerrhonotus is a moderate (seven spp.) genus of small-bodied (7.9 g v. 10.1 g), often-elongate (43% v. 31%) terrestrial anguids from Mexico and adjacent Texas. All known species are refugia-anchored diurnal ground lizards (v. 34%), though many are unassigned.

Their mostly breakable (71% v. 87%) tails are sometimes prehensile (43% v. 31%) and are slightly longer than the anguid norm (1.84 v. 166 × SVL).

All two documented *Gerrhonotus* (*infernalis*, *ophiurus*) are cruise foragers (v. 29%), though active foraging has also been reported in *infernalis*. The same two species have been documented for predator concealment behavior, and both exhibit crypsis (consistent with cruise foraging, but not typical for anguids, which are

Gerrhonotus spp.	English name		Size		Range core
	Activity	Diet	mm	g	Ecological business model
farri	Tamaulipas Alligator Lizard				Mexico
	diurnal		87	9.8	* refugia-anchored diurnal gr. liz.
infernalis	Texas Alligator Lizard				Mexico
	cathemeral	invert.	157	16.7	** refugia-anchored diurnal gr. liz.
lazcanoi	Nuevo Leon Alligator Lizard				Mexico
	diurnal		59	2.56	
liocephalus	Plateau Alligator Lizard				Mexico
	diurnal	invert.	154	16.0	** refugia-anchored diurnal gr. liz.
lugoi	Lugo's Alligator Lizard				Mexico
	crepuscular		77	6.2	
ophiurus	Snake Alligator Lizard				Mexico
	diurnal	invert.	152	15.4	
parvus	Pygmy Alligator Lizard				Mexico
	diurnal		62	2.98	

71% active foragers). Movement in *infernalis* is conspicuously slow. *Gerrhonotus* are not documented to occur in littoral (v. 9%), sandy (v. 15%), or human (v. 11%) habitats.

Gerrhonotus are mostly (67%) oviparous (v. 39%), with a slightly elevated clutch count (1.64, the highest mean in Anguidae, v. 0.85 std. resid.), and a remarkable 57% parental care (v. 17%). All documented females attend their clutches. Male *infernalis* are reported to find females by their scent trails; this species is not documented to be in possession of secretory pores, which would presumably be a source of the scent followed by this trailing behavior. What is the source of the odors that males use?

Longevity is short for an anguid (–0.29 v. 0.83 std. resid.). Population density estimates are unavailable. No *Gerrhonotus* is an island species (v. 17%), nor has any been introduced (v. 3.8%). Eighty percent (v. 63%) are considered imperiled, perhaps a reflection of the high level of endemism (71% v. 65%); only two species extend beyond Mexico marginally into Texas. Twenty percent (v. 28%) are on the IUCN's lists of the most endangered.

Gerrhosauridae

The 37 gerrhosaurids are large (16 g v. 5.3 g), active-foraging (100% v. 58%), diurnal (91% v. 67%), terrestrial (89% v. 48%) lizards endemic to Africa and Madagascar. From a North America perspective, these lizards are strange because they combine a predominantly sedentary tie to refugia (59% are refugia-anchored diurnal ground lizards v. 25%) with active foraging (100%). The most conspicuous active foragers in North America tend to be wide-ranging chemosensors. The second-most frequent ecological business model in Gerrhosauridae is wide-ranging chemosensors (27% v. 7.4%). However, like many skinks and gymnophthalmids, gerrhosaurids use a wide variety of primary predator concealment tactics (see below) while actively foraging. Regrettably, there are no useful data for assessing which, if any, gerrhosaurids have large home-range sizes, and therefore there is no direct evidence of which species could accurately be described as "wide-ranging." Three minor ecological business models in Gerrhosauridae are diurnal canopy walker (6% v. 6%), ground-foraging tree lizard (6% v. 9%), and litter cryptozoid (3% v. 12%).

Seven genera are found within Gerrhosauridae at present: *Broadleysaurus*, *Cordylosaurus*, *Gerrhosaurus*, *Matobosaurus*, *Tetradactylus*, *Tracheloptychus*, and *Zonosaurus*. The latter two are Madagascar endemics. The others are found in southern Africa. *Tetradactylus* are elongate and have variable amounts of limb reduction. All gerrhosaurids have a lateral skin fold similar to Anguidae. The overall body plan of the fully legged gerrhosaurids is similar to that of a particularly corpulent skink or anguid. Shape index values in Gerrhosauridae run from the highly elongate *Tetradactylus* (–1.25 std. resid.) to the beefy *Broadleysaurus* (1.26 std. resid.) and average only slightly above average (0.38 v. 0.0 std. resid.). Relative tail length follows the converse pattern, with *Broadleysaurus* having the shortest tails (1.00 × SVL) and *Tetradactylus* the longest (2.86 × SVL). The family averages 1.92 × SVL (v. 1.37). Tail breakage is limited only in *Tetradactylus* (0% v. the family average of 73%, v. the lizard average of 57%). No tails are prehensile (v. 5%).

Gerrhosauridae lack crests (v. 5.3%). Secretory pores are found only in *Broadleysaurus*, *Cordylosaurus*, and one *Gerrhosaurus*; the family average is 8% (v. 15%). Elongation is limited to *Tetradactylus* (family average 19% v. 11%). Color change is entirely absent (v. 10.7%), and blue tails occur at lizard-typical frequency (2.7% v. 2.3%).

Sexual differences in size are virtually undocumented in Gerrhosauridae. The sexes differ in color at lizard-typical rates (16% v. 21%), but they are among the highest families in exhibiting seasonal color (27% v. 6.5%). In the frequency of seasonal color they are behind only Cordylidae and the small family Crotaphytidae. These other seasonally colorful families hint at a southern African or temperate influence. Considering the three large genera, *Gerrhosaurus* (seven spp.) are nearly universally seasonal in coloration, whereas no *Tetradactylus* (seven spp.) is; the largest genus (*Zonosaurus*, 17 spp.) is mixed.

Gerrhosauridae is unusually rife with omnivores (56% v. 15%), mostly fruit eaters rather than folivores; the remainder are invertebrate eaters (44% v. 73%). Neither sit-and-wait (v. 46%) or cruise (v. 6.6%) foraging is reported for Gerrhosauridae. None has been reported to feed at electric lights (v. 1.4%).

Almost all gerrhosaurids are diurnal (91% v. 67%); the remainder are crepuscular (8.8% v. 1.8%). Many bask directly (24% v. 18%) and many (22% v. 8.2%) are characterized as conspicuously wary. The predator concealment tactics that stand out are flight (27% v. 11%) and sand burial (18% v. 6%). But the other tactics are all present: crypsis (9% v. 43%), crevice-burrows (27% v. 25%), and vegetation (18% v. 23%).

Social behavior is essentially undocumented in gerrhosaurids, but pairing seems to be the default; absent are solitary living (0% v. 10%) and group living (0% v. 20%). Male spatial defense is uniform (100% v. 85%); hierarchies are not reported (v. 7.7%).

Only 11% of gerrhosaurids are arboreal (v. 32%); all of these are *Zonosaurus*. The others reside on the ground (89% v. 48%) and many refuge in burrows (56% v. 22%). They are underrepresented in littoral (6% v. 17%) and human (11% v. 18.5%) habitats, but slightly overrepresented in sandy areas (25% v. 22%) and riparian zones (25% v. 22%).

All are oviparous (v. 82%), with clutch counts that are slightly light (−0.20 v. 0.0 std. resid.). The genera *Broadleysaurus* and *Matobosaurus* have anomalously small clutches (see those genera). Gerrhosaurid eggshells are soft and incubation duration is typical (−0.05 v. 0.0 std. resid.). Adherent eggs (0% v. 2.4%) and parental care (0% v. 2.9%) are not reported, but documentation of reproduction in Gerrhosauridae is minimal. Communal oviposition is reported at typical frequency (5.4% v. 4.2%). RCM is undocumented.

Longevity is unremarkable (−0.09 v. 0.0 std. resid.), but that masks suggestive differences among genera, following the usual axis: *Broadleysaurus* are long-lived (0.89 std. resid., equating to 24 years); *Tetradactylus* are notably short-lived (−1.51 std. resid., equating to 1.3 years, though based on only *T. africanus*). Population density data are unavailable. Endemism is about average (62% v. 64%). Island living is common (57% v. 40%), mostly due to the Madagascan species. None has been introduced (v. 3.3%), and 50% are judged imperiled (v. 49%). Eleven percent (v. 14%) are listed in the most severe IUCN categories of endangerment.

Gerrhosaurus

Gerrhosaurus is a seven-species genus of large (51 g v. 16 g) diurnal terrestrial gerrhosaurids from southern Africa (Gabon to Somalia and south). Most are refugia-anchored diurnal ground lizards (83% v. 59%), though none is assigned with full confidence. One species (*skoogi*) is tentatively considered to be a wide-ranging chemosensor (17% v.

Gerrhosaurus spp.	English name				Range core
			Size		
	Activity	Diet	mm	g	Ecological business model
auritus	Kalahari Plated Lizard				Zambia
	diurnal	omnivore	163	110	*** refugia-anchored diurnal gr. liz.
bulsi	Laurent's Plated Lizard				Congo
			152	89	
flavigularis	Yellow-throated Plated Lizard				Southern Africa
	diurnal	omnivore	110	32.4	** refugia-anchored diurnal gr. liz.
multilineatus	Keeled Plated Lizard				Congo
	diurnal	omnivore	123	4.2	*** refugia-anchored diurnal gr. liz.
nigrolineatus	Black-lined Plated Lizard				Southern Africa
	diurnal	omnivore	127	50	*** refugia-anchored diurnal gr. liz.
skoogi	Desert Plated Lizard				Angola
	diurnal	omnivore	129	53	** wide-ranging chemosensor
typicus	Namaqua Plated Lizard				South Africa
	crepuscular	invert.	98	22.6	*** refugia-anchored diurnal gr. liz.

27%); it emerges from sand to forage on about one day in three. Morphologically *Gerrhosaurus* resemble anguids and are robust by gerrhosaurid standards (shape index 0.92 v. 0.38 std. resid.). Secretory pores are somewhat more common than in the family (14% v. 8%). They stand out in the preponderance of seasonal color (86% v. 27%), exhibited by both sexes in several species. Overall sex differences are mildly elevated (1.43 v. 1.05 on 0 to 6 scale), the highest for any gerrhosaurid genus. They are also family-leading in the frequency of omnivory (83% v. 56%). Their attachment to refugia is likely reflected in the family-leading proportion of species that are conspicuously wary (71% v. 22%), though their known predator defenses are a mix of crypsis (33% v. 9%), flight (33% v. 27%), and crevices and burrows (33% v. 27%). Sand swimming is reported for *skoogi*, though it is not clear that sand diving is a primary predator concealment tactic for this species. Almost all species utilize burrows as refugia (83% v. 56%). They are slightly overrepresented in sandy areas (50% v. 25%) and human habitats (17% v. 11%). Reproduction is normal for lizards of this size (mass-adjusted clutch −0.01 v. −0.20 std. resid.). *Gerrhosaurus* are imperiled at typical rates (50% v. 50%) but missing from any lists of the most endangered (v. 11%). In other respects *Gerrhosaurus* are typical gerrhosaurids.

Glaphyromorphus

Glaphyromorphus (Scincidae) is a moderate radiation (11 spp.) of small (2.18 v. 3.25 g), mostly elongate (82% v. 35%), cathemeral (100% v. 9%) litter cryptozoids (100% v. 28%) found in northern Australia and adjacent parts of Papua New Guinea and Indonesia. Many have proportionately thick tails.

Compared to the average skink they are slender (shape index −0.91 v. −0.46 std. resid.). They have breakable tails (100% v. 73%) that are slightly long (1.51 v. 1.33 × SVL). Males are smaller than females (−1.0 v. −0.31 on −2 to 2 scale), but other sex differences are not known (sex rollup score 0.09 v. 0.28 on 0 to 6 scale). *Glaphyromorphus* eat invertebrates (100% v. 84%), but additional details of their foraging are unavailable. Some are conspicuously wary (18% v. 11%). Direct basking is uncommon (9% v. 21%). At least three of the species with unknown diel cycles are possibly nocturnal. *G. nigricaudis* is known to refuge in soil cracks. *Glaphyromorphus* are overrepresented in littoral zones (44% v. 21%), sandy areas (56% v. 25%), riparian zones (67% v. 23%), and human habitats (22% v. 14%).

Most species are oviparous (89% v. 64%). *G. nigricaudis* may express both reproductive modes, and when laying eggs deposits them at a very advanced developmental stage. They have an ordinary skink clutch count (0.26 v. 0.06 std. resid.). RCM is elevated (0.320 v. 0.236, though that value is based solely on *darwiniensis*).

Nonqualified density estimates suggest that at least two species in this genus are present at very low densities. Endemism is slightly elevated (73% v. 64%). Island species are slightly underrepresented (46% v. 57%). None has been introduced (v. 2.1%). Half are imperiled (50% v. 47%), though none is on any of the

Glaphyromorphus spp.	English name		Size		Range core
	Activity	Diet	mm	g	Ecological business model
clandestinus	Mount Elliot Mulch-skink				Queensland
			65	1.86	* litter cryptozoid
cracens	Slender Mulch-skink				Queensland
		invert.	52	1.08	** litter cryptozoid
crassicaudus	Cape York Mulch-skink				New Guinea
		invert.	50	0.98	** litter cryptozoid
darwiniensis	Northern Mulch-skink				Western Australia
		invert.	53	1.11	** litter cryptozoid
fuscicaudis	Brown-tailed Bar-lipped Skink				Queensland
		invert.	78	4.87	
mjobergi	Atherton Tableland Skink				Queensland
	cathemeral	invert.	83	3.46	*** litter cryptozoid
nigricaudis	Black-tailed Bar-lipped Skink				New Guinea
	cathemeral	invert.	78	4.83	** litter cryptozoid
nyanchupinta	McIlwraith Bar-lipped Skink				Queensland
			48	2.02	* litter cryptozoid
othelarrni	Cape Melville Bar-lipped Skink				Queensland
			76	8.8	* litter cryptozoid
pumilus	Dwarf Mulch-skink				Queensland
	cathemeral	invert.	51	1.03	** litter cryptozoid
punctulatus	Fine-spotted Mulch-skink				Queensland
	cathemeral	invert.	61	1.60	** litter cryptozoid

IUCN's lists of the most endangered (v. 14%). In all other documented respects *Glaphyromorphus* are typical skinks.

Glaucomastix

Glaucomastix (Teiidae) is a small genus (four spp.) of small-sized (8.6 g v. 22.1 g), poorly known (filled cells 20% v. 33%), blue-tailed (100% v. 15%), flight-dependent (100% v. 77%), omnivorous (100% v. 33%) wide-ranging chemosensors (100% v. 97%) endemic to Brazil. They are associated with littoral (100% v. 38%) and sandy (100% v. 47%) areas.

Their shape index is average (0.60 v. 0.88 std. resid.). Males are larger than females (1.00 v. 0.89 on –2 to 2 scale). They bask directly (100% v. 58%). They are terrestrial (100% v. 98%). They are found at low elevation (89–995 m). Clutch count is a little high (0.20 v. –0.51 std. resid.). Based on *littoralis*, longevity is slightly short (–0.89 v. –0.74 std. resid.). No species is found on islands (v. 37%). None has been introduced (v. 8%), and none has been evaluated for conservation status. In other respects *Glaucomastix* are typical teiids.

Glaucomastix spp.	English name		Size		Range core
	Activity	Diet	mm	g	Ecological business model
abaetensis	Bahia Sand Dune Lizard				Brazil
	diurnal	omnivore	61	6.4	*** wide-ranging chemosensor
cyanura	Bluish-tailed Lizard				Brazil
	diurnal		68	9.0	*** wide-ranging chemosensor
littoralis	Coastal Sand Dune Lizard				Brazil
	diurnal		68	9.0	*** wide-ranging chemosensor
venetacauda	Piaui Caatinga Lizard				Brazil
	diurnal		72	10.5	*** wide-ranging chemosensor

Goggia

Goggia is an eight-species genus of very small (0.82 g v. 3.31 g) nocturnal ground gekkonids from southern Africa. One species may be a crack hermit (v. 3.2%). Their shape is relatively slender (shape index –0.12 v. 0.10 std. resid.). All have secretory pores (v. 46%) and enlarged toepads (v. 70%), but color change is not reported (v. 15%). Sex differences are not reported and most accounts note the frequent grouping of one male and one female, suggesting monogamy. Although most are terrestrial, the large species (*microlepidota*) is scansorial and could be considered arboreal (though often on rock walls). They are overrepresented at mid elevation (40% v. 26%) and underrepresented at low (40% v. 92%). They are not found in human habitats (v. 27%). *Goggia* reproduction is typical for gekkonids. The lizards are so small that the fixed clutch of two is relatively larger than the lizard average (0.30 v. –0.26 std. resid.). Adherent eggs are not reported (v. 10.5%), as adhesion is not needed for terrestrial species. One species (*essexi*) may lay communally (v. 8.6%). No *Goggia* has been introduced (v. 3.7%). The large species (*microlepidota*) is deemed to be near threatened (v. 47% imperiled). In other respects *Goggia* are typical gekkonids.

Goggia spp.	English name		Size		Range core
	Activity	Diet	mm	g	Ecological business model
braacki	Braack's Dwarf Leaf-toed Gecko				South Africa
	nocturnal		31	0.75	* nocturnal ground lizard
essexi	Essex's Dwarf Leaf-toed Gecko				South Africa
	nocturnal		26	0.48	* nocturnal ground lizard
gemmula	Richtersveld Dwarf Leaf-toed Gecko				South Africa
	nocturnal	invert.	27	0.51	*** nocturnal ground lizard
hewitti	Hewitt's Dwarf Leaf-toed Gecko				South Africa
	nocturnal		33	0.90	*** nocturnal ground lizard
hexapora	Cederberg Dwarf Leaf-toed Gecko				South Africa
	nocturnal		32	0.80	* nocturnal ground lizard
lineata	Striped Dwarf Leaf-toed Gecko				Namibia
	nocturnal	invert.	30	0.68	*** nocturnal ground lizard
microlepidota	Small-scaled Dwarf Leaf-toed Gecko				South Africa
	nocturnal		56	3.92	* crack hermit
rupicola	Namaqua Dwarf Leaf-toed Gecko				South Africa
	nocturnal	invert.	28	0.59	** nocturnal ground lizard

Gonatodes

Gonatodes is a large genus (30 spp.) of slightly large (1.86 g v. 1.07 g) Sphaerodactylidae. They are mostly padless (93%), mostly diurnal (90% v. 62%), mostly arboreal (67% v. 86%), and mostly ground-foraging tree lizards (52% v. 12%). They are found primarily in northern South America (nine spp. in Venezuela), from Bolivia northward to Colombia and St. Vincent. One species (*albogularis*) is widely distributed in the West Indies and Central America, from Mexico to Panama, Cuba, and Hispaniola, with an outlier in Martinique. Those that are not ground-foraging tree lizards are trunk/wall lizards (29% v. 12%), litter cryptozoids (10% v. 53%), and crack hermits (5% v. 6%). Humans have introduced them many times, and it is now difficult to discern which localities are native.

Some are capable of basic color change (13% v. 6.5%), and one (*ceciliae*) shows advanced color change (3% v. 1.4%). Skin often tears easily. The sexes are about equally likely to be larger (0.17 v. 0.04 on –2 to 2 scale). Sexual dichromatism is common (50% v. 35%); seasonal color is overrepresented (13% v. 3%). Reminiscent of *Pristurus*, *G. concinnatus* waves its tail in a figure eight pattern. *G. vittatus* may hold its tail elevated and curled forward, scorpioid-fashion, another pattern seen in various sphaerodactylid genera.

All eat invertebrates (v. 82%), obtained by a combination of sit-and-wait (80%) and active (20% v. 17%) foraging. Electric light use is average (6.7% v. 6.9%). Home-range area is small (–1.00 v. 0.06 std. resid.), based on one poor sample of *ocellatus* and a qualitative estimate of *albogularis*. The latter species is said to have males that defend space and to have "territories" of 10–20 m^2, but it is also said that large trees support 30–40 individuals. These assessments do not easily align.

Although most species are diurnal, a few are cathemeral (3% v. 15%), crepuscular (3% v. 6%), or nocturnal (3% v. 17%). Many have contradictory assessments of diel activity. Double the usual percentage are

Gonatodes spp.	English name		Size		Range core
	Activity	Diet	mm	g	Ecological business model
albogularis	White-throated Gecko				West Indies
	diurnal	invert.	41	1.73	*** trunk/wall lizard
alexandermendesi	Mendes' Gecko				Guyana
	diurnal		45	2.18	* ground-foraging tree lizard
annularis	Annulated Gecko				Venezuela
	diurnal		47	2.47	**** trunk/wall lizard
antillensis	Antilles Gecko				Curacao
	nocturnal	invert.	36	1.12	*** nocturnal ground lizard
astralis	Astral Gecko				Venezuela
	diurnal		42	1.80	
atricucullaris	Cajamarca Gecko				Peru
	diurnal		33	0.90	
caudiscutatus	Shield-headed Gecko				Ecuador
	diurnal		40	1.63	* litter cryptozoid
ceciliae	Brilliant South American Gecko				Venezuela
	diurnal	invert.	57	4.60	** ground-foraging tree lizard
concinnatus	Collared Forest Gecko				Peru
	diurnal	invert.	45	2.20	*** trunk/wall lizard
daudini	Union Island Gecko				St. Vincent
	crepuscular		30	0.65	** crack hermit
eladioi	Para Gecko				Brazil
	diurnal		34	0.96	** ground-foraging tree lizard
falconensis	Falcon Gecko				Venezuela
	diurnal		54	3.80	
hasemani	Haseman's Gecko				Peru
	diurnal	invert.	41	1.70	** ground-foraging tree lizard
humeralis	Orange-spotted Gecko				Venezuela
	diurnal	invert.	40	1.54	*** trunk/wall lizard
infernalis	Inferno Gecko				Venezuela
	diurnal		56	4.27	
lichenosus	Perija Lichen Gecko				Venezuela
	diurnal		41	1.73	* trunk/wall lizard
ligiae	Llanos Gecko				Peru
	diurnal		41	1.66	
nascimentoi	Nascimento's Gecko				Brazil
	diurnal		49	2.97	
naufragus	Blanquilla Gecko				Venezuela
	diurnal		29	0.60	
ocellatus	Eyespot Gecko				Trinidad and Tobago
	diurnal	invert.	45	2.27	*** ground-foraging tree lizard

Gonatodes spp.	English name	Activity	Diet	Size mm	Size g	Range core	Ecological business model
petersi	Peters' Gecko	diurnal		40	1.53	Venezuela	** ground-foraging tree lizard
purpurogularis	Purple-throated Gecko	diurnal		45	2.27	Venezuela	
riveroi	Rivero-Blanco's Gecko	diurnal		43	2.02	Colombia	* trunk/wall lizard
rozei	Roze's Gecko	diurnal		52	3.37	Venezuela	*** ground-foraging tree lizard
seigliei	Monagas Gecko	diurnal		41	1.71	Venezuela	** ground-foraging tree lizard
superciliaris	Sarisarinama Forest Gecko	diurnal		44	2.17	Venezuela	*** litter cryptozoid
taniae	Aragua Gecko	diurnal		47	2.49	Venezuela	** ground-foraging tree lizard
tapajonicus	Para Gecko	diurnal		49	2.85	Brazil	* ground-foraging tree lizard
timidus	Iwokrama Forest Gecko	diurnal		47	2.49	Guyana	
vittatus	Striped Neotropical Day Gecko	cathemeral	invert.	33	0.86	Trinidad and Tobago	*** ground-foraging tree lizard

Gongylomorphus sp.	English name	Activity	Diet	Size mm	Size g	Range core	Ecological business model
bojerii	Bojer's Skink	diurnal	omnivore	60	5.6	Mauritius	*** refugia-anchored diurnal gr. liz.

conspicuously wary (13% v. 7%). Primary predator concealment tactics are retreat to vegetation (60% v. 27%) or retreat to a crevice or burrow (40% v. 43%). *G. humeralis* may sleep on vegetation over water. Direct basking (10% v. 6.5%) is as common, as is thermal conformance (10% v. 4%). Mean active body temperature is low (28.7°C v. 30.4°C). *G. ceciliae* dies in sunshine at 28°C. The proportion of species with male spatial defense is relatively low (67% v. 86%).

These geckos are found at the wet end of the rainfall spectrum: intermediate (20% v. 30%), mesic (60% v. 45%), and wet (20% v. 25%). Their elevational profile is about normal (l/m/h: 100/13/4 v. 96/20/4). They are underrepresented in littoral areas (13% v. 31%), overrepresented in human habitat (48% v. 33%), and about average in sandy areas (9% v. 15%) and riparian zones (22% v. 19%).

Reproduction closely matches family characteristics; parental care has not been observed (v. 1.4%).

Longevity is notably long (1.45 v. −0.15 std. resid.), corresponding to 13 years in *albogularis*. Population density, based on four credible samples of three species, is average (0.29 kg/ha v. 0.32 kg/ha). Endemism is average (63% v. 65%). Most species do not occur on islands (27% v. 64% on islands). Introductions are very high (13.3% v. 5.1%), including a notable list of species introduced more than twice (6.7% v. 2.3%). Few are imperiled (11% v. 41%), though those 11% (v. 23%) are all on one of the IUCN's lists of the most endangered. In other respects *Gonatodes* are typical sphaerodactylids.

Gongylomorphus

Gongylomorphus bojerii is the sole member of its scincid genus and is currently endemic to Mauritius; it was extirpated from the nearby French island of Reunion. It is a refugia-anchored diurnal ground lizard (v. 30%). It is comparatively well known (filled cells 42% v. 25%). It is a slightly large skink (5.6 g v. 3.3 g) of solid build (shape index 0.40 v. −0.46 std. resid.). Its tail is of average length (1.18 v. 1.33 × SVL) and unbreakable (v. 73% breakable). Males are larger than females (1.0 v. −0.31 on −2 to 2 scale). No other sex difference is noted. It is omnivorous, including eating fruit, ants, ectothermal vertebrates, and carrion. It was once observed smacking an ant nest and devouring the emerging defenders.

It basks directly (v. 21%). It is found in littoral areas (v. 21%). It is oviparous (v. 64%) with a small clutch count (−0.45 v. 0.06 std. resid.). This fits with the slightly large size of hatchlings (ontogenetic mass range −0.26 v. 0.12 std. resid.). It lays at least three clutches per year. It lays communally (v. 2.6%) and sometimes the eggs are adherent when laid (v. 0.1%). The contingent nature of this adherence likely indicates that it is under the control of the mother, as it is in other lizards.

Longevity is short (–0.39 v. 0.17 std. resid.), equating to 4.0 years. It has not been introduced except in the special sense of inter-island conservation translocations to repopulate extirpated populations. It is presently listed as critically endangered by the IUCN (v. 14%).

Goniurosaurus (fig. 4.10B)

Goniurosaurus is a moderate (17 spp.) genus of the Eublepharidae from the eastern edge of Asia: Vietnam, China, and Japan. The genus's mix of ecological business models roughly matches the family: nocturnal ground lizards (73% v. 86%), ground-foraging tree lizards (18% v. 7%), and crack hermits (9% v. 3%). They are about the same size (12 g v. 13.1 g) and build (shape index 0.31 v. 0.34 std. resid.) as the family. They are strikingly similar in color pattern, if slightly more inclined to banding. Morphologically they differ primarily in a lower frequency of secretory pores (47% v. 67%) and an absence of color change (0% v. 2.8%).

In sex differences, they diverge more from the family template. *Goniurosaurus* are characterized by m<f size dimorphism (–0.83 v. –0.09 on –2 to 2 scale). They are less inclined to have sex differences in color (6% v. 11%), and they do not exhibit seasonal color (0% v. 6%). In the ratio of m:f relative tail lengths, they show negligible sex differences (1.03 v. 1.21 ratio). In total, the sex differences rollup is substantially lower in *Goniurosaurus* (0.59 v. 1.08 on 0 to 6 scale).

Goniurosaurus differ from typical eublepharids in lacking reports of cruise foraging (0% v. 56%), but like typical eublepharids they have roughly equal representation of sit-and-wait (100% v. 44%) and active (75% v. 44%) foraging modes. The percentages for *Goniurosaurus* sum to more than 100% because many species exhibit multiple foraging modes.

All evaluated species are nocturnal (100% v. 97%), though the several cave species might be an exception. Unlike typical eublepharids, *Goniurosaurus* vocalize neither spontaneously (0% v. 8%) nor under duress (0% v. 25%). Their mean active body temperature is a bit low (25.8°C v. 27.3°C), likely reflecting the many temperate species. They are slightly

Goniurosaurus spp.	English name		Size		Range core
	Activity	Diet	mm	g	Ecological business model
araneus	Vietnam Leopard Gecko				Vietnam
	nocturnal		103	22.7	*** nocturnal ground lizard
bawanglingensis	Bawangling Leopard Lizard				China
	nocturnal		84	11.4	*** nocturnal ground lizard
catbaensis	Cat Ba Tiger Gecko				Vietnam
	nocturnal		90	14.5	** ground-foraging tree lizard
hainanensis	Hainan Leopard Gecko				China
	nocturnal		99	19.5	
huuliensis	Huu Lian Leopard Gecko				Vietnam
	nocturnal		94	16.5	** ground-foraging tree lizard
kadoorieorum	Kadoories' Cave Gecko				China
			94	16.5	
kuroiwae	Ryukyu Ground Gecko				Japan
	nocturnal	invert.	79	9.6	*** nocturnal ground lizard
kwangsiensis	Guangxi Cave Gecko				China
			87	13.1	
liboensis	Libo Leopard Gecko				China
			88	13.4	
lichtenfelderi	Lichtenfelder's Leopard Gecko				China
	nocturnal		95	17.3	*** crack hermit
luii	Chinese Leopard Gecko				China
	nocturnal	carnivore	119	36.1	*** nocturnal ground lizard
orientalis	Tonaki Island Leopard Gecko				Japan
	nocturnal		69	6.2	*** nocturnal ground lizard
splendens	Tokuno Island Leopard Gecko				Japan
	nocturnal		64	4.87	*** nocturnal ground lizard
toyamai	Iheya Island Leopard Gecko				Japan
	nocturnal		69	6.2	** nocturnal ground lizard
yamashinae	Kume Island Leopard Gecko				Japan
	nocturnal		72	7.1	** nocturnal ground lizard
yingdeensis	Yingde Leopard Gecko				China
	nocturnal		78	9.1	
zhelongi	Zhe-long's Leopard Gecko				China
	nocturnal		76	8.3	

more likely to be active under cover (cryptozoic) than a typical eublepharid (8.3% v. 3.4%). No species with social documentation has hierarchical males (v. 17%). *Goniurosaurus* is the part of the Eublepharidae that occurs in mesic (1101–3000 mm/y) habitats (100% v. 40%). They are less represented at medium elevation than are typical eublepharids (11% v. 35%). Perhaps due to less information, they are underrepresented in littoral zones (0% v. 7%), sandy areas (0% v. 10%), riparian zones (25% v. 30%), and human habitats (8% v. 17%).

Reproduction hews to eublepharid norms, though parental care is slightly less often reported (6% v. 14%), and four species are suspected to exhibit TSD. The suspicion is based on skewed sex ratios of captive litters rather than direct tests of TSD. Young of *Goniurosaurus* are especially large (ontogenetic mass range −0.97 v. −0.67 std. resid.). Documented longevity is slightly shorter (0.53 v. 1.16 std. resid.). Endemism is high (82% v. 53%). All evaluated species are considered to be imperiled (v. 60%), as all are on the IUCN's lists of the most endangered (v. 30%). In other respects *Goniurosaurus* are typical eublepharids.

Gonocephalus

Gonocephalus is a moderate (16 spp.) genus of diurnal (100% v. 98%), arboreal (100% v. 43%) agamids found in southeastern Asia. The common name of anglehead alludes to their relatively large, sharply angled heads. Although poorly known, about half are most likely ground-foraging tree lizards (46% v. 19%) and the others trunk/wall lizards (54%, the third highest in Agamidae, v. 7%). In other words, they all occupy tree trunks and some may also be found episodically on the ground. They are fairly uniform in size and average over twice the weight of a typical agamid (38.9 g v. 15.6 g; range 12–71 g). They have a heavier build (shape index 1.16 v. 0.80 std. resid.), but this large size is not associated with a broader diet. All eat invertebrates (v. 46%), with an agamid-typical frequency of ant specialization (17% v. 20%).

Tails are a bit longer (2.03 v. 1.86 × SVL) and unbreakable (v. 17% breakable). Virtually all have crests (94% v. 38%), but none is known to have secretory pores (v. 13%). They are about twice as likely as the average agamid to darken or have advanced color

Gonocephalus spp.	English name				Range core
			Size		
	Activity	Diet	mm	g	Ecological business model
abbotti	Abbott's Anglehead Lizard				Thailand
	diurnal		126	68	*** trunk/wall lizard
bellii	Bell's Anglehead Lizard				Malaysia
	diurnal		115	51	**** trunk/wall lizard
beyschlagi	Sumatra Forest Lizard				Indonesia
	diurnal		99	32.8	
bornensis	Borneo Forest Lizard				Malaysia
	diurnal	ants	107	41.7	**** trunk/wall lizard
chamaeleontinus	Chameleon Anglehead Lizard				Indonesia
	diurnal	invert.	125	66	**** ground-foraging tree lizard
doriae	Doria's Anglehead Lizard				Indonesia
	diurnal	invert.	104	37.4	
grandis	Giant Anglehead Lizard				Malaysia
	diurnal	invert.	122	61	*** ground-foraging tree lizard
interruptus	Mindanao Anglehead				Philippines
			77	15.0	* trunk/wall lizard
klossi	Kloss' Forest Lizard				Indonesia
	diurnal		127	71	* ground-foraging tree lizard
kuhlii	Kuhl's Forest Lizard				Indonesia
	diurnal		82	18.2	** trunk/wall lizard
lacunosus	Sikulikap Anglehead Dragon				Indonesia
	diurnal		117	53.7	* ground-foraging tree lizard
liogaster	Flame Eye-ring Anglehead Lizard				Borneo
	diurnal	invert.	113	48.8	** ground-foraging tree lizard or trunk/wall lizard
megalepis	Yellow-spotted Sumatra Anglehead				Indonesia
	diurnal		112	46.9	
mjobergi	Murud Mountain Anglehead Lizard				Borneo
	diurnal		72	12.2	* trunk/wall lizard
semperi	White-spotted Anglehead				Philippines
			91	24.8	* trunk/wall lizard
sophiae	Negros Forest Lizard				Philippines
	diurnal	invert.	117	53.7	* ground-foraging tree lizard

change (38% v. 15% for advanced). Males are larger than females (1.00 v. 0.53 on –2 to 2 scale). They possess typical agamid sex differences in color (56% v. 53%), except that seasonal color has not been reported (v. 20%; underreporting possible). Sexual selection appears strongly manifest (sex differences rollup 1.75 v. 1.61 on 0 to 6 scale), as expected for trunk dwelling agamids.

Unlike typical agamids, all documented species are direct baskers (v. 24%). Almost all (93% v. 83%) are found at low elevation (33% extend up to mid elevation), and 94% (v. 36%) are found on islands. All are oviparous (v. 98%); clutch count is low (–0.70 v. 0.20 std. resid.). Based on the mass-adjusted ontogenetic range of sizes, *Gonocephalus* young are especially small (1.23 v. 0.69 std. resid.). The average hatchling is about 1/200 (v. 1/100) of the maximum size. Incubation period is long (0.17 v. –0.55 std. resid.). Thus, the young are few in number, small in size, and slow to develop.

Low reproductive output may be offset by higher longevity, which unfortunately is not known for this genus. No *Gonocephalus* has been introduced (v. 1.7%) or is imperiled (37%).

Gowidon

Gowidon is a tiny genus (two spp.) of large (32 g v. 16 g) semiarboreal agamids, found in Australia, New Guinea, and some nearby islands. Though the two species look very similar—brownish-black with a whitish labial wash that fades out along the upper flanks—and possess a similar long-headed but muscular build (shape index 1.1 v. 0.8 std. resid.) with long tails (3.3, second highest in Agamidae, v. 1.9 × SVL), m>f SSD (1.0 v. 0.53 on –2 to 2 scale), sexual dichromatism (100% v. 53%), color change (100% v. 22%), and moderate crests (100% v. 38%; larger in males), Cogger (2014) judges that the two species belong in different genera.

The small one (*longirostris*) forages somewhat more terrestrially (50% v. 56%), and the large one (*temporalis*) forages more arboreally (50% v. 43%). Both are usually found in proximity to water (riparian 100% v. 21%) in diverse habitats (including dunes: 100% v. 26%) and microhabitats. Both are fast sprinters and may run bipedally; at least *longirostris* is fast enough while running to overtake and capture flying insects. *G. temporalis* is found in wetter areas (ca. 1600 mm/y) and aestivates during the dry season; *longirostris* likes it dry (200–300 mm/y) and has not been reported to aestivate. *G. longirostris* appears to prefer it somewhat warmer (34.5°C v. 24.8°C *temporalis*); the genus mean is low (29.6°C v. 34.8°C). *G. temporalis* tolerates urban areas (genus 50% v. 16%) and is not reported to be conspicuously wary (genus 50% v. 6%); it exhibits a higher RCM (RCM = 0.37) and achieves high densities (2.4 kg/ha, the second-highest genus in Agamidae, v. 0.16 kg/ha), whereas *longirostris* is said to be conspicuously wary, does not occupy urban areas, and has a light RCM (0.15–0.18).

Both bask directly (v. 24%). Both retreat to vegetation as their primary predator concealment tactic (100%, the highest in Agamidae, v. 17%). Both are oviparous (v. 98%). Their mass-adjusted clutch count is low (–0.45 v. 0.20 std. resid.). Their ontogenetic size range is high (0.97 v. 0.69), indicating slightly small young. If they have young that are relatively small and few, one would expect a light RCM ratio, which is observed (0.267 v. 0.336). Curiously, the species with the higher RCM ratio (*temporalis*) has the smaller mass-adjusted clutch count (–0.57 v. –0.34 for *longirostris*). Neither has been introduced (v. 1.7%). They have not been formally evaluated for conservation status.

Gowidon spp.	English name		Size		Range core
	Activity	Diet	mm	g	Ecological business model
longirostris	Long-nosed Dragon				Australia
	diurnal	invert.	89	23.3	** refugia-anchored diurnal gr. liz.
temporalis	Swamplands Lashtail				New Guinea
	diurnal	invert.	110	44.9	*** ground-foraging tree lizard

Graciliscincus

Graciliscincus shonae is the sole member of its scincid genus, a very small (0.75 g v. 3.3 g), elongate (v. 35%) cathemeral (v. 9%) or crepuscular (v. 2.4%) litter cryptozoid (v. 28%) endemic to the island of New Caledonia.

It is slender (shape index –0.85 v. –0.46 std. resid.), with an average tail (1.25 v. 1.33 × SVL) that breaks (v. 73%). Sex differences are not known. It is found in mesic forests at elevations of 150–900 m. It may refuge in soil cracks or other appropriated burrows. It has not been introduced (v. 2.1%) and is considered imperiled (vulnerable according to the IUCN), but it is not on any of the IUCN's lists of the most endangered (v. 14%). In other respects it is likely to be similar to most skinks (e.g., oviparous).

Graciliscincus sp.	English name				Range core
			Size		
	Activity	Diet	mm	g	Ecological business model
shonae	Gracile Burrowing Skink				New Caledonia
	cathemeral		39	0.75	** litter cryptozoid

Grandidierina spp.	English name				Range core
			Size		
	Activity	Diet	mm	g	Ecological business model
fierinensis	Common Burrowing Blind Skink				Madagascar
			61	2.47	**** subterranean forager
lineata	Lined Burrowing Blind Skink				Madagascar
			66	0.95	**** subterranean forager
petiti	Tiny Burrowing Blind Skink				Madagascar
			49	1.40	**** subterranean forager
rubrocaudata	Redtail Burrowing Blind Skink				Madagascar
			73	1.20	**** subterranean forager

Gymnodactylus spp.	English name				Range core
			Size		
	Activity	Diet	mm	g	Ecological business model
amarali	Amaral's Naked-toed Gecko				Brazil
	cathemeral	invert.	43	1.89	* nocturnal ground lizard
darwinii	Darwin's Naked-toed Gecko				Brazil
	nocturnal		47	2.75	
geckoides	Naked-toed Gecko				Brazil
	cathemeral	invert.	43	2.00	
guttulatus	Brazil Naked-toed Gecko				Brazil
	nocturnal		45	2.42	
vanzolini	Vanzolini's Naked-toed Gecko				Brazil
	nocturnal		47	2.70	

Grandidierina

Grandidierina (Scincidae) is a minor radiation (four spp.) of small (1.41 g v. 3.25 g), functionally limbless, vestigial-eyed subterranean foragers (v. 12%) endemic to the island of Madagascar.

They are exceptionally slender for skinks (shape index –1.45 v. –0.46 std. resid.), with tails that are relatively long for subterranean foragers (0.88 v. 1.33 for skinks and 0.37 × SVL for subterranean forager lizards). Some tails may be breakable (50% v. 73%). They are found at low elevation (100% v. 89%) and are associated with littoral areas (50% v. 21%), sandy areas (100% v. 25%), and riparian zones (50% v. 23%). Their reproduction is unknown. They have not been introduced (v. 2.1%). One species (*petiti*, near threatened) is considered imperiled, but none is on one of the IUCN's lists of the most endangered (v. 14%). In other respects *Grandidierina* are much like other subterranean skinks.

Gymnodactylus

Gymnodactylus (five spp.) is a genus of small (2.3 v. 4.72 g) Brazilian nocturnal/cathemeral ground phyllodactylids. Only one species (*amarali*) is known well enough for a tentative assignment of ecological business model (nocturnal ground lizard).

They are not only small but lightly built (shape index 0.15 v. 0.37 std. resid.). They are not reported to exhibit color change (v. 11%). All species lack enlarged toepads (v. 8%).

SSD is in the m<f direction (–1.0 v. 0.32 on –2 to 2 scale), at least in the one documented species (*amarali*). Cathemeral species are overrepresented (40% v. 19%). No species is documented to call spontaneously (v. 11%), though underreporting is possible. Both documented species use crypsis as their primary predator concealment tactic (v. 43%). For a nocturnal/cathemeral group, *Gymnodactylus* has a relatively high mean active body temperature of 31.5°C (v. 26.0°C), albeit based solely on *amarali* and *geckoides*.

Gymnodactylus occupies habitats in the middle of the rainfall spectrum (in this case 400– 2000 mm/y) at low elevation (0–422 m). They are not associated with any of the tabulated habitat specializations.

Reproduction is near normal. Clutch count is slightly elevated (–0.17 v. –0.57 std. resid.). RCM ratio is 0.089 (v. 0.086). All species are endemic to the very large country of Brazil (v. 67%). None occurs on an island (v. 40%). None has been introduced (v. 7.4%), and none has been evaluated for conservation status. In other respects *Gymnodactylus* follow the family characteristics.

Gymnophthalmidae

The 237 gymnophthalmids are small (2.4 g v. 5.3 g), active foraging (100% v. 58%), diurnal (79% v. 67%) Neotropical (100% v. 29%) lizards. Most descriptions of this primarily South American group also include the word "mysterious"; gymnophthalmids are very poorly documented (cells filled: 19% v. 27%). That makes gymnophthalmids second only to Amphisbaenidae in our level of ignorance. It is unsurprising that subterranean amphisbaenids are poorly understood, but what makes the gymnophthalmids so inscrutable? A few (14% v. 8%) of the gymnophthalmids are subterranean as well, but only a few. Seven percent of gymnophthalmids are considered to be conspicuously wary, but that hardly differs from the lizard mean of 8%. Gymnophthalmids are unusually small, an information liability (table 3.1, trait 1), but the association is relatively weak ($r^2 = 0.07$). Gymnophthalmids are found in a part of the world (most are in South America) where herpetologists are relatively sparse. They are unusually well represented in densely foliated habitats at mid (52% v. 41%) and high (33% v. 13%) elevations in mesic areas (77% v. 43%), situations not conducive to observational study. Their reason for occupying high elevation may be the same as their apparent preference for mesic areas: they seem to like it moist. Perhaps gymnophthalmids are rare? Mean documented gymnophthalmid density is 0.19 kg/ha, not far from the lizard mean of 0.54 kg/ha, and ranked right in the middle of the families (23 of 45).

Perhaps gymnophthalmids are of an ecological business model that is hard to observe? There is no single ecological business model that explains the mystery of gymnophthalmids. The mix of business models in gymnophthalmids is very broad; nine of 14 classes are represented (in descending order of frequency): litter cryptozoids (55% v. 12%), refugia-anchored diurnal ground lizards (19% v. 25%), subterranean foragers (14.3% v. 7.8%), ground-foraging tree lizard (4.1% v. 9.2%), nocturnal ground lizards (3.1% v. 7.7%), wide-ranging chemosensors (2.0% v. 7.4%), diurnal canopy walker (1.0% v. 6.4%), and sand sharks (1.0% v. 1.9%). It is notable that only 98 of 237 (41%) gymnophthalmids have an ecological business model even tentatively assigned. But that could be a reflection of our ignorance rather than the cause of it. Litter cryptozoids are the second-most frequently overlooked lizards (table 4.1, trait 2), behind only subterranean foragers, and the modal gymnophthalmid is a litter cryptozoid.

One of the strongest clues to the mystery of gymnophthalmids was offered to me by an eminent gymnophthalmid taxonomist and collector (Tiffany Doan, pers. comm., 2017), when she pointed out to me that among the higher elevation gymnophthalmids *she has never seen one active* (my emphasis). She finds them by exposing them in their refugia. Perhaps they rarely leave their refugia? If so, one would expect that many gymnophthalmids to be crack hermits, but the extant assignments do not support that (0% v. 2.6%). This highlights a probable weakness in the assignment of ecological business models. One cannot readily make assignments when the level of ignorance is high. It is also likely that many of the 139 gymnophthalmids of unknown business model are crack hermits, which may be even harder to observe.

This is an opportunity to point out two particularly weak fields in the natural history data tabulated: "conspicuously wary" and "population density." It is my subjective impression that observers describe a lizard as conspicuously wary when the lizard flees at a distance greater than the observer expects. This means that herpetologists' expectations may play as large a role as does the lizard's flight distance. Herpetologists' expectations are likely to be family- or genus- specific. Herpetologists may expect gymnophthalmids to be mysterious and therefore do not comment on it when their assumption is supported. Perhaps as important as that weakness is the requirement that the herpetologist studying a gymnophthalmid sees the lizard fleeing at a greater than expected distance. What if the fleeing lizard found cover before the human looked in its direction? From watching *Iguana* from a blind, I found that lizard flight distances can often exceed herpetologists' sighting distances (Rodda 1992). Perhaps the gymnophthalmid was hidden before the herpetologist arrived? That would fit with crack hermits and litter cryptozoids. In summary, "conspicuously wary" is a fraught metric, though to date it is the best available. Objective, quantitative, unbiased flight distances would be an improvement, but it is hard to obtain flight distance measurements that do not preferentially collect shorter flights.

Population density is another tricky metric, because it is normally calculated only on particularly dense species in particularly dense populations. That only two species of gymnophthalmids (each sampled once) have credible population density estimates may hint at low densities in the group, or it may simply tell us that population ecologists have not been much inclined to study South American cloud forests. Reptile ecologists desperately need a more random collection of field data.

In summary, my subjective impression is that gymnophthalmids are poorly documented because they are often (1) small, (2) found in herpetologist-sparse countries, (3) found in vegetation-obstructed habitats, (4) reflect hidden or low visibility life styles, (5) wary, and (6) at low population density. If it turns out that gymnophthalmid activity ranges are small (no data yet), they may be amenable to productive study in mesocosms. Surely, they are amenable to field study in the right place and time with the right combination of foul-weather gear, good optics, and patience.

Although gymnophthalmids are rarely observed in the wild, taxonomists have found a remarkable diversity (46) of genera (14 monotypic): *Acratosaura, Adercosaurus, Alexandresaurus, Amapasaurus, Anadia, Anotosaura, Arthrosaura, Bachia, Calyptommatus, Caparaonia, Cercosaura, Colobodactylus, Colobosaura, Colobosauroides, Dryadosaura, Echinosaura, Ecpleopus, Euspondylus, Gelanesaurus, Gymnophthalmus, Heterodactylus, Iphisa, Kaieteurosaurus, Leposoma, Loxopholis, Macropholidus, Marinussaurus, Micrablepharus, Neusticurus, Nothobachia, Pantepuisaurus, Petracola, Pholidobolus, Placosoma, Potamites, Procellosaurinus, Proctoporus, Psilophthalmus, Rhachisaurus, Riama, Riolama, Rondonops, Scriptosaura, Stenolepis, Tretioscincus,* and *Vanzosaura*. Because the gymnophthalmids were once lumped with the larger lizards of the Teiidae, they have often been informally referred to as microteiids. See the Alopoglossidae account for similarities with that family.

Gymnophthalmid bodies differ from typical lizards in their slenderness (shape index −0.32 v. 0.0 std. resid.). Tails differ in their greater length (1.72 v. 1.37 × SVL), reduced breakage (28% v. 57%), and low frequency of prehensile ability (1.7% v. 5.0%). Crests are absent (v. 5.3%). Secretory pores are frequent (29% v. 15%). Elongation is common (41% v. 11%). Color change is not known (v. 11%). Blue tails are present at lizard-typical frequency (2.5% v. 2. 3%).

Sex differences are relatively scarce (0.49 v. 0.75 on 0 to 6 sex rollup score), being due mainly to sex differences in pores. Gymnophthalmids are more inclined to an m<f size difference (mean −0.12 v. 0.36 on −2 to 2 scale), as fits their overall size and perhaps their preference for crevice refugia. Sex differences in color are comparatively muted (13% v. 21% for permanent color; 1.7% v. 6.5% for seasonal color). Gymnophthalmids do not differ from average lizards in m:f ratio of relative tail length (1.05 v. 1.07 ratio).

There are no gymnophthalmids that are herbivores, ant specialists, or omnivores. Carnivores are slightly underrepresented (2.6% v. 3.2%), likely a reflection of small size. The remainder are invertebrate eaters (v. 73%). Gymnophthalmids differ from typical lizards in their aversion to sit-and-wait (0% v. 46%) and cruise (3.8% v. 6.6%) foraging. One does not immediately think of active foraging (96% v. 58%) in animals so rarely seen. Electric light use is not reported (v. 1.4%). Home-range areas are not known.

All four diel activity patterns are present in gymnophthalmids, with cathemeral (13.8% v. 6.8%) and diurnal (79% v. 67%) patterns overrepresented. Nocturnality is the most underrepresented (6.2% v. 24.7%). Vocalizations are not reported (v. 2.5% spontaneous, v. 2.2% stress).

The only reported primary predator concealment tactics are vegetation (73% v. 23%) and crevices-burrows (36% v. 25%). The absence of crypsis (v. 43%) is notable. Direct basking is underreported (8% v. 18%), and thermal conformance is a bit overrepresented (4.6% v. 2.6%). Mean active body temperature is on the low side (28.8°C v. 30.8°C), likely reflecting their high elevation.

Although social relations are hard to document in species that are rarely seen, it is notable that male social tolerance of other males is an extraordinary 88% (v. 15%), exceeded only by two small families (Alopoglossidae and Anniellidae). Only 12.5% (v. 85%) of males defend space.

Gymnophthalmids are found only at the wetter end of the rainfall spectrum (>400 mm/y) and are overrepresented in the mesic band (1101–3000 mm/y), as noted above. They are underrepresented in littoral (3% v. 17%) and sandy (14% v. 22%) areas, and they are slightly overrepresented in riparian areas (29% v. 22%).

Despite their high altitude, all gymnophthalmids are believed oviparous (v. 82%), with clutch count as expected based on body size (−0.11 v. 0.0 std. resid.). As the different genera all seem to have average clutches of two, despite some range in body size, there is some form of phylogenetic constraint on number of eggs. Incubation duration is long (1.11 v. 0.0 std. resid.), presumably a reflection low incubation temperature at high altitude. The mass-adjusted ontogenetic range of masses is notably large in gymnophthalmids (0.62 v. 0.0 std. resid.), albeit based on only 10 of 237 species. This high value indicates relatively small hatchlings. Eggshells are uniformly soft (v. 59%), rarely adherent (0.4% v. 2.4%), and laid communally at typical rates (3.4% v. 4.2%). Parental care is not reported (v. 2.9%), though underreporting is to be expected. RCM is unremarkable (0.192 v. 0.226 ratio).

Longevity is short (−0.61 v. 0.0 std. resid.), albeit based solely on *Loxopholis guianense* (2 years). Gymnophthalmids rarely occur on islands (5% v. 40%). Only the Caribbean *Gymnophthalmus underwoodi* and

Tretioscincus bifasciatus have been introduced (to islands). The former has been introduced > 20 times, often prehistorically. Gymnophthalmids are less often said to be imperiled (25% v. 49%), but are just as likely to be found on an IUCN list of the most endangered (11.5% v. 13.6%). So much to learn, so little time.

Gymnophthalmus

Gymnophthalmus is a seven-species genus of very small (0.79 g v. 2.35 g), diurnal, litter cryptozoid gymnophthalmids ranging from Belize in Central America and Guadeloupe in the Lesser Antilles to Brazil. For gymnophthalmids, this is a particularly well-known group (filled cells 29% v. 19%), though no ecological business models could be assigned with full confidence, and almost nothing is known about activity patterns. They are particularly slender (shape index −0.73 v. −0.32 std. resid.). Tail length is relatively short (1.50 v. 1.72 × SVL) and breakage is common (43% v. 28%). One species (*speciosus*) is documented to have (very minimal) secretory pores (14% v. 29%). Most species are elongate (71% v. 41%). Males are typically smaller than females (−1.0 v. −0.12 on −2 to 2 scale), and the sexes often differ in color (43% v. 13%). Some of the male coloration may actually be seasonal (so noted on two species), and the females and juveniles of three species have a red/orange tail tip. In one case this color has been hypothesized to mimic the color of dangerous sympatric centipedes, but the association may be more widely applicable. Another species (*underwoodi*) has a blue tail (14% v. 2.5%). Wariness is characteristic of this genus (29% v. 7%), and all documented species use retreat to a burrow or crevice as their primary predator concealment tactic. Basking is common (43% v. 8%). Burrow refugia are overrepresented (33% v. 5%). All documented species (v. 77%) are found in the mesic rainfall belt (1101–3000 mm/y) at low elevation (100% v. 51%; at medium elevation: 17% v. 52%). They are overrepresented in littoral (67% v. 3%), sandy (67% v. 14%), and human (50% v. 15%) habitats.

Reproduction is typical, though clutch count is high for such small lizards (0.38 v. −0.11 std. resid.), and incubation is short (0.43 v. 1.11 std. resid.), based on *underwoodi*. *G. underwoodi* is a parthenogenetic unisexual; the others are bisexual. Ontogenetic size range is documented only for *speciosus*, which has an extraordinarily wide range (3.73 v. 0.62 std. resid.), indicating particularly tiny hatchlings. Endemism is notably low in *Gymnophthalmus* (29% v. 68%), and island species are common (57% v. 5%). One species (*underwoodi*) has been introduced more than 20 times (v. 0.4% of gymnophthalmids introduced more than twice). They are imperiled at typical frequency (25% v. 25%) and are overrepresented on the IUCN's most endangered lists (25% v. 11.5%). In other respects *Gymnophthalmus* are probably typical gymnophthalmids.

Haackgreerius

Haackgreerius miopus is the sole member of its scincid genus, a medium-sized (4.5 g v. 3.3 g) lizard endemic to Somalia. It is among the least known lizards (filled cells 11% v. 25%). It is not found on an island (v. 57%), has not been introduced

Gymnophthalmus spp.	English name				Range core
			Size		
	Activity	Diet	mm	g	Ecological business model
cryptus	Ventuari Spectacled Lizard				Venezuela
	diurnal		29	0.47	** litter cryptozoid
leucomystax	White-lipped Spectacled Lizard				Brazil
			40	0.80	
lineatus	Striped Spectacled Lizard				Venezuela
			39	0.70	
pleii	Martinique Spectacled Lizard				Lesser Antilles
	diurnal	invert.	46	2.04	** litter cryptozoid
speciosus	Golden Spectacled Lizard				Central America
	diurnal	invert.	39	0.71	** litter cryptozoid
underwoodi	Underwood's Spectacled Lizard				Venezuela
	diurnal	invert.	37	0.62	** litter cryptozoid
vanzoi	Vanzo's Spectacled Lizard				Brazil
			40	0.79	

Haackgreerius sp.	English name				Range core
			Size		
	Activity	Diet	mm	g	Ecological business model
miopus	Somalia Skink				Somalia
			62	4.50	

Haemodracon spp.	English name				Range core
			Size		
	Activity	Diet	mm	g	Ecological business model
riebeckii	Riebeck's Leaf-toed Gecko				Yemen
	nocturnal		109	33.3	** trunk/wall lizard
trachyrhinus	Rough-snouted Socotran Leaf-toed Gecko				Yemen
	nocturnal		49	3.08	*** ground-foraging tree lizard

Hakaria sp.	English name				Range core
			Size		
	Activity	Diet	mm	g	Ecological business model
simonyi	Socotra Slender Skink				Yemen
	nocturnal		52	1.63	*** nocturnal ground lizard

Harpesaurus spp.	English name				Range core
			Size		
	Activity	Diet	mm	g	Ecological business model
beccarii	Sumatra Nose-horned Lizard				Indonesia
	diurnal	invert.	71	11.5	* diurnal canopy walker
borneensis	Borneo Nose-horned Lizard				Borneo
	diurnal		52	4.28	*** trunk/wall lizard
ensicauda	Nias Nose-horned Lizard				Indonesia
			52	4.47	
modigliani	Riau Nose-horned Lizard				Indonesia
	diurnal		69	10.4	
tricinctus	Java Nose-horned Lizard				Java
			55	5.3	

(v. 2.1%), and has not been evaluated for conservation status. In other respects it is likely to be similar to other skinks.

Haemodracon

Haemodracon is a two-species genus of large (10.1 g v. 4.7 g) nocturnal phyllodactylids endemic to Yemen. The large one is a trunk/wall lizard (v. 44%); the small one a ground-foraging tree lizard (v. 16%). Tails are a bit short (0.90 v. 1.03 × SVL) and breakable (v. 92%). Color change and sex differences are not known. Both occur at low elevation (aggregate range 0–1370), and they are associated with sandy and riparian areas. Reproduction is normal with the exception that the incubation period is long (1.54 v. 0.18 std. resid.), equating to a very wide range in *riebecki*: 118–359 days, suggesting that this species sometimes defers eclosion until the following year. *H. trachyrhinus* has a more conventional 76-day incubation. Clutch count is low, even by phyllodactylid standards (−0.88 v. −0.57 std. resid.). Young are slightly large (ontogenetic mass range −0.05 v. 0.11 std. resid.). Both species are endemic (v. 67%) and occur on islands (v. 40%), have not been introduced (v. 7.4%), and are not considered imperiled (v. 29%). In other respects *Haemodracon* are similar to other phyllodactylids.

Hakaria

Hakaria simonyi is the sole member of its scincid genus, a small (1.63 g v. 3.3 g), extremely elongate (v. 35%), nocturnal ground lizard (v. 2.3%) endemic to the Yemeni island of Socotra. It is a poorly known lizard (filled cells 21% v. 25%). It is only slightly more slender than the average skink (−0.74 v. −0.46). It has an ordinary length tail (1.20 v. 1.33 × SVL), and the tail is breakable (v. 73%). Sex differences are not known. It is conspicuously wary (v. 11%). It occurs over a wide range of elevations (10–1463 m) and is associated with riparian zones (v. 23%). Reproduction is unknown. It has not been introduced (v. 2.1%). Being listed by the IUCN as near threatened, it is imperiled (v. 47%) but not on the IUCN's lists of most endangered (v. 14%). In other respects it is likely to be similar to other skinks.

Harpesaurus

Harpesaurus is a five-species genus of small (6.6 g v. 15.6 g), arboreal (100% v. 43%) agamids famed for their fleshy conical rostral appendages, which are double in *beccarii*. The sexes differ in height of the crests (crests 100% v. 38%), but sexual dimorphism has not been reported in the rostral horns. The genus's natural history is very poorly known (23% v. 30% of cells filled), and only *borneensis* has a moderately convincing business model assignment (trunk/wall lizard: 50% v. 7%). All live in the lowland forests on islands of Indonesia and Malaysia. *H. beccarii* is said to move slowly, chameleon-fashion, on small-diameter perches; *borneensis* is said to prefer tree trunks but has a prehensile tail (20%, the family's third-highest rate, v. 1.4%), which fits much better with canopy walking (50% v. 7%). The diet of only one species (*beccarii*) has been reported (invertebrate eater: 100% v. 46%). Most are believed to be oviparous (75% v. 98%); *borneensis* has been reported to be both egg laying and live bearing, but clutch count (a low −0.35 v. 0.20 std. resid.) is documented only for *borneensis*. None has been introduced (v. 1.7%) or is imperiled (v. 37%).

Harrisoniascincus sp.	English name				Range core
			Size		
	Activity	Diet	mm	g	Ecological business model
zia	Rainforest Cool Skink				Queensland
	diurnal	invert.	53	1.67	* litter cryptozoid or refugia-anchored diurnal gr. liz.

Heliobolus spp.	English name				Range core
			Size		
	Activity	Diet	mm	g	Ecological business model
lugubris	Bushveld Lizard				Botswana
	diurnal	invert.	56	4.11	** wide-ranging chemosensor
neumanni	Neumann's Sand Lizard				East Africa
	diurnal		47	2.51	
nitida	Red-tailed Sand Lizard				Cameroon
	diurnal	invert.	54	3.68	* wide-ranging chemosensor
spekii	Speke's Sand Lizard				East Africa
	diurnal	invert.	52	3.38	** refugia-anchored diurnal gr. liz.

Harrisoniascincus

Harrisoniascincus zia is the sole member of its scincid genus, a small (1.67 g v. 3.3 g), elongate (v. 35%) litter cryptozoid (v. 28%) found in moist coastal forests of eastern Australia. It is relatively well known (filled cells 36% v. 25%). It is only slightly more slender than the average skink (−0.77 v. −0.46). It has an ordinary length tail (1.40 v. 1.33 × SVL), which is breakable (v. 73%). Males show seasonal color (v. 6.2%). It is conspicuously wary (v. 11%), yet it basks directly (v. 21%). It is found at "high" (undefined) elevation and is associated with riparian zones (v. 23%). Reproduction is by numerous (5.4) eggs (0.91 v. 0.06 std. resid.), which the female thrusts into mulch cover using her mouth. It has not been introduced (v. 2.1%), nor does it occur on islands (v. 57%). It has not been evaluated for conservation status. In other respects it is likely to be similar to other skinks.

Heliobolus

Heliobolus (four spp.) is a small genus of smallish (3.37 g v. 5.1 g) diurnal terrestrial lacertids found in equatorial Africa, from Mali to Somalia and South Africa. They are very active and most often are wide-ranging chemosensors (67% v. 25%), resembling New World *Cnemidophorus* or *Aspidoscelis* in demeanor, color, and shape (shape index 0.18 v. 0.27 std. resid.). Their tails are long (2.39 v. 2.06 × SVL) and breakable (100% v. 92%). In *nitida* the sexes differ in color (25% v. 33%); in *lugubris* both sexes show seasonal color (25% v. 15%). Overall they are average in sex differences (1.08 v. 1.07 on 0 to 6 scale). Although *spekii* is listed as sit-and-wait (33% v. 23%), three species are considered to be active foragers, including *lugubris*, the best-studied species. This species is in the 92nd percentile for PTM and the 85th percentile for MPM. All three evaluated species rely mainly on flight (v. 22%) as their primary predator concealment tactic, and *nitida* is considered conspicuously wary (v. 11%). Mean active body temperature is a sizzling 37.5°C (v. 33.3°C).

Although all species are believed to be generalized invertebrate eaters, *lugubris* eats as much as 90% termites. They are found in the dry (56–150 mm/y) and arid zones (151–400 mm/y) at low and medium elevation (*spekii* goes up to 2000 m). Their notable habitat association is with sand (100% v. 44%). Reproduction is average (e.g., mass-adjusted clutch count 0.14 v. 0.13 std. resid.), with the exceptions that their hatchlings are unusually small (ontogenetic mass range 0.64 v. −0.05 std. resid.) and mean RCM is light (0.171 v. 0.267), as befits lizards that rely on flight for predator concealment. No species is endemic (v. 35%). None is found on an island (v. 26%), and none has been introduced (v. 4.6%). They have not been evaluated for conservation status. In other respects *Heliobolus* are similar to other lacertids. *H. lugubris* is famous for the behavioral and morphological mimicry by juveniles of the acid-squirting beetle, the "oogpister."

Hellenolacerta

Hellenolacerta graeca is the sole species in its genus, a largish (8.1 g v. 5.1 g), diurnal, ground-foraging tree lacertid found in the Peloponnese

Peninsula of Greece. It is reasonably well known (filled cells 49% v. 33%), in part because it occurs in a diversity of habitats including houses.

In morphology it is a typical lacertid, though somewhat flattened in cross section. Its tail is a bit long (2.25 v. 2.06 × SVL), breakable, and bright blue in juveniles. The sexes differ in size (m>f) and color. Sex differences aggregate to 3.0 v. 0.86 on 0 to 6 scale. Like most lacertids it eats invertebrates (v. 77%). Unlike most lacertids the males are socially tolerant (v. 41%). *H. graeca* occurs over the elevation range 0–1600 m. It is associated with riparian zones (v. 28%) and human habitats (v. 16%).

Its reproduction is fairly typical for lacertids, though clutch count is on the low side (–0.21 v. 0.13 std. resid.), and hatchlings are notably large (ontogenetic mass range –0.43 v. –0.05 std. resid.). Despite its high latitude, its incubation period is short (44 days or –0.66 v. –0.36 std. resid.). The species is found on islands (v. 26%), and has not been introduced (v. 4.6%). It is an object of conservation concern (v. 56%), but is not on the IUCN lists of the most endangered (v. 15.5%). In other respects *Hellenolacerta* is similar to other lacertids.

Hellenolacerta sp.	English name				Range core
			Size		
	Activity	Diet	mm	g	Ecological business model
graeca	Greek Rock Lizard				Greece
	diurnal	invert.	70	8.1	**** ground-foraging tree lizard

Helodermatidae and *Heloderma*

The family Helodermatidae contains but a single genus, *Heloderma* (two spp.), both species of which are widely known due to their unique venom, size, and natural history. Collectively they range from Guatemala to Nevada. These elongate lizards are almost snakelike in their reliance on infrequent large meals (nestling mammals and birds), which they find by active chemosensory exploration over long distances. For example, *horridum* steadily walks about 25 km annually, but it does so during only 120 hours of activity; 95% of the time it is in shelters. *H. suspectum* spends 96% of its time in shelters and spends only about 70 hours annually foraging. A typical *suspectum* foraging foray covers 215 m and lasts 50 minutes.

Juvenile *suspectum* can consume up to 50% of their body mass in a meal, adults up to 35%. Such huge meals allow them to get by on as few as three to four meals a year. *Heloderma* are not only snakelike in their meal size and possession of venom, they are also snakelike in their body size: the family mean of 809 g exceeds the lizard average by more than 150-fold (5 g). Unlike most snakes, *Heloderma* store massive amounts of energy in their short (0.62 v. 1.37 × SVL), thick tails. Their valuable tails are not breakable: 0% (v. 57%). Accordingly, Helodermatidae are relatively heavy for their length: their shape index is 1.38 (v. 0.0 std. resid.). Also unlike snakes, the venom of *Heloderma* appears to have evolved, and to be used, primarily defensively. The venom delivery system is relatively inefficient at injecting venom into prey, so they do not bite and release as do so many vipers. In any event, nestling birds and mammals are not particularly difficult prey to subdue with a strong bite alone. Instead, the massive jaws of *Heloderma* are applied with continuous force when biting defensively, and the slow seepage of venom into the target suffices to generate memorable pain and thereby "train" the victim not to engage *Heloderma* in future encounters. Most of the features of this paragraph have no analogue in most lizards, except in the sister taxon Varanidae.

Heloderma have no crests (v. 5.3%). Their tails are not prehensile (v. 5.0%). They lack secretory pores (v. 15%). They have no ability to change color (v. 10.7%). And they do not have blue tails (v. 2.3%).

Male *Heloderma* are larger than females (genus 1.00 v. 0.36 on scale of –2 to 2). *Heloderma* sexes do not differ in color (v. 20.7%), nor do they show seasonal color (v. 6.5%). It is not clear that seasonal color would be meaningful, at least in *suspectum*, because their annual activity is largely constrained to the season when nestling birds and mammals are available, which naturally coincides with the lizards' breeding season. Overall, the sex differences of *Heloderma* add up

Heloderma spp.	English name				Range core
			Size		
	Activity	Diet	mm	g	Ecological business model
horridum	Beaded Lizard				Mexico
	crepuscular	carnivore	375	1090	**** wide-ranging chemosensor
suspectum	Gila Monster				Arizona
	cathemeral	carnivore	320	600	**** wide-ranging chemosensor

to 1.50 (v. 0.75 on 0 to 6 scale), commensurate with these lizards' large size.

The large home ranges of *Heloderma* are not commensurate with their body size, but rather they are much larger than that (1.66, by far the highest family, v. 0.0 std. resid.), or about 216,000 m² in *horridum* and 560,000 m² in *suspectum*. *Very* wide-ranging chemosensors. Varanidae are a distant second at 1.01 std. resid., though three varanid species exceed the mass-adjusted home range of *suspectum*. For the tiny portion of the year when helodermatids are foraging, the PTM for *suspectum* is about 0.215, which is the 70th percentile of lizard PTMs.

Although the species differ in nominal diel activity pattern, both species will sometimes be active at all times of day. *H. horridum* is said to be conspicuously wary, *suspectum* less so. Both species bask directly (v. 18%). Their mean active body temperature is 28.3°C (v. 31.1°C).

Males in both species are hierarchical (v. 7.7%). Both species burrow (v. 22%) and occur in the arid rainfall belt (56–150 mm/y: v. 30%) at low (v. 82%) and medium (v. 41%) elevation. They are overrepresented in littoral (50% v. 17%), sandy (50% v. 22%), and riparian (50% v. 22%) habitats, but not human habitats (0% v. 18.5%).

They are oviparous (v. 82%) with a mean clutch count of 6.7 (v. 4.0). The apparently large clutch count is very small when adjusted for their large mass (−1.19 v. 0.0 std. resid.). Similarly, their ontogenetic size range is exceptionally small (−1.87 v. 0.0 std. resid.), indicating that their hatchlings are exceptionally large. Their very long incubation period (1.04 v. 0.0 std. resid.) is lengthened by *suspectum*, for which the eggs laid late in the year may overwinter and hatch the following year (up to 300 days).

Longevity is high (0.86 v. 0.0 std. resid.), equating to 38 years in *horridum* and 30 years in *suspectum*. Conversely, population density is low (0.02 kg/ha v. 0.54 kg/ha), based on four estimates for *suspectum*. Neither species resides on islands (v. 40%); neither has been introduced (v. 3.3%). Both are considered imperiled (v. 49%), though neither is on one of the IUCN's lists of the most endangered species.

A unique addition to the world's saurofauna.

Hemicordylus

Hemicordylus is a tiny genus of average-sized (10.9 g v. 10.8 g) cordylids endemic to South Africa. With two species, one recently described and especially poorly known, species frequencies are relatively meaningless. However, *Hemicordylus* fit in with the typical cordylids in business model, shape index (0.46 v. 0.35 std. resid.), presence of secretory pores (50% v. 69%), some sexual dichromatism (50% v. 37%)—but seasonal color not reported (0% v. 28%)—diet of invertebrates (100% v. 70%), sit-and-wait foraging (100% v. 94%), diurnality (100% v. 99%), direct basking (100% v. 74%), a combination of arboreal and terrestrial microhabitats, viviparity (100% v. 75%), and conservation risk (100% v. 79%), though they are not listed by the IUCN in the most endangered categories (0% v. 4.8%). *Hemicordylus* differ in lacking breakable tails (v. 60% breakable), residing in littoral zones (v. 9%), small clutches (−0.74 v. −0.48 std. resid.) of large young (ontogenetic size range −0.93 v. −0.37 std. resid.), and having short longevity for their size (−0.95 v. 0.02 std. resid.), based solely on a maximum recorded lifespan of 3.3 years in *capensis*. In other respects they comport with typical cordylids.

Hemicordylus spp.	English name				Range core
			Size		
	Activity	Diet	mm	g	Ecological business model
capensis	Graceful Crag Lizard				South Africa
	diurnal	invert.	87	16.3	** trunk/wall lizard
nebulosus	Dwarf Crag Lizard				South Africa
	diurnal		69	7.2	** crack hermit

Hemidactylus (figs. 2.6A, 4.5B, 4.10C)

Hemidactylus is a major radiation of lizards (fourth-largest genus: table 2.2) and the second-largest gekkonid genus (144 spp., after 232 spp. *Cyrtodactylus*). They are native throughout the world, except for the Nearctic (where they have been introduced), though the highest species richnesses are in aridlands of eastern Africa and adjoining Asia. Arrayed from west to east, the core countries and their native *Hemidactylus* species richnesses are as follows:

Ethiopia	20 spp.
Kenya	18
Somalia	30
Yemen	23
Saudi Arabia	8
Oman	16
Iran	5
Pakistan	10
India	33
Sri Lanka	10

These geckos are familiar to most people because they often occur in houses and are frequently introduced (9.0% v. 3.7%). Nonetheless, 22 species (15%) are known from only a few specimens. The most common ecological business models are trunk/wall lizard (57% v. 45%) and noc-

Hemidactylus spp.	English name	Activity	Diet	Size mm	Size g	Range core	Ecological business model
aaronbaueri	Maharashtra Cliff Gecko	nocturnal		101	19.5	India	*** trunk/wall lizard
acanthopholis	Tamil Nadu Rock Gecko	nocturnal		87	13.1	India	
adensis	Aden Gecko	nocturnal		39	1.45	Yemen	
agrius	Brazil Countryside Gecko	nocturnal		57	3.96	Brazil	
albituberculatus	White-tubercled Half-toed Gecko	nocturnal	invert.	65	5.8	Nigeria	*** trunk/wall lizard or nocturnal ground lizard
albivertebralis	White-lined Half-toed Gecko	nocturnal	invert.	54	3.41	Ghana	** trunk/wall lizard
albofasciatus	White-striped Viper Gecko	nocturnal		35	1.07	India	*** nocturnal ground lizard
albopunctatus	White-spotted Gecko	nocturnal		42	1.77	Ethiopia	
alfarraji	Al Farraj's Gecko			51	2.95	Saudi Arabia	
alkiyumii	Ali Kiyumi's Gecko	cathemeral		63	5.32	Yemen	** trunk/wall lizard
anamallensis	Anaimalai Hill Gecko	nocturnal	invert.	64	4.36	India	** trunk/wall lizard
angulatus	Western Half-toed House Gecko	nocturnal	invert.	55	3.73	Western Africa	*** trunk/wall lizard
ansorgii	Ansorge's Half-toed Gecko	nocturnal	invert.	54	3.43	Nigeria	** trunk/wall lizard
aporus	Annobon Gecko	nocturnal		48	2.52	Equatorial Guinea	
aquilonius	Northern Gecko	nocturnal	invert.	47	2.41	India	**** trunk/wall lizard
arnoldi	Arnold's Gecko	nocturnal		68	6.6	Somalia	
asirensis	Asir Mountains Gecko			46	2.22	Saudi Arabia	
awashensis	Awash Gecko	nocturnal		49	2.62	Ethiopia	
barbierii	Turkana Gecko	nocturnal		45	2.08	Kenya	* nocturnal ground lizard
barodanus	Enigmatic Gecko	nocturnal		65	5.8	Ethiopia	

turnal ground lizard (32% v. 27%). Minor elements include ground-foraging tree lizards (8.2% v. 5.3%), gliders (1.2% v. 1.5%), and nocturnal canopy walkers (2.4% v. 8.9%). Morphologically they are very close to the gekkonid norm, though only 0.7% of species lack enlarged toepads (v. 30%).

Sexual dichromatism is present but uncommon (1.4% v. 6.6%); seasonal color is absent (v. 0.8%). Aggregate sex differences (0.74 v. 0.63 on 0 to 6 scale) are those typical of gekkonids.

Foraging modes follow the gekkonid pattern, with the possible exception of a slight overrepresentation of cruising (33% v. 24%). Electric light use is very high (9.0% v. 3.8%), and many species appear to have altered their ecological business model to take advantage of the high prey densities present around electric lights. Size-adjusted home-range area is a bit higher than the gekkonid mean (0.18 v. –0.28 std. resid.). All *Hemidactylus* are nominally invertebrate eaters, though some fruit is taken by *brookii*, nectar by *frenatus*; in houses *brookii*, *frenatus*, *parvimaculatus*, and *platyurus* will eat coffee, milk, syrups, and cooked rice.

The array of diel activity patterns follows the gekkonid mold almost exactly, with the glaring exception that fully diurnal species are absent (v. 15%). Many of the nominally nocturnal species are alert but partially concealed during the day; they will often take prey that wander nearby. Vocalization behavior follows the gekkonid pattern; spontaneous calling is 7.6% (v. 6.5%). Many observers have commented on the loudness of some callers, especially *brookii* and *frenatus*. Males of this species do most of the calling—primarily around dusk and dawn, but also sporadically throughout the day—and have distinct vocalizations associated with advertisement, intermale encounters, and stress calls. In contrast, paired male and female *imbricatus* call to each other antiphonally.

Conspicuous wariness is scarce (0.7% v. 4.4%). Predator concealment is most reliant on crevices and burrows (55% v. 31%), and most remaining species rely on crypsis (36% v. 50%). A few Indian species use tail motions to mimic viper behavior (e.g., *albofasciatus*, *imbricatus*), presumably engendering hesitation in potential predators. Thermoregulatory behavior follows the gekkonid template, and the mean active body temperature is an exact match to the family (27.6°C). Use of microhabitats follows the family mix (e.g., 70% arboreal v. 68%).

Social relations, to the extent they are documented, follow the gekkonid pattern, with most species having males that defend space (90% v. 87%).

Although the highest *Hemidactylus* species richnesses are in arid areas, documented moisture regimes match those of gekkonids in the wetter half of the spectrum: 85% of species are found in mesic (1101–3000 mm/y) areas; more arid sites—most in the Sahel region and Horn of Africa—are likely to be underdocumented. The elevational distribution of *Hemidactylus* species is better documented and matches the family. *Hemidactylus* also matches the family distribution with regard to littoral (13% v. 11%), sandy (15% v. 11%), and riparian (21% v.17%) zones. It differs greatly in the high frequency of *Hemidactylus* in human habitats (45% v. 27%). This urban/commensal proclivity probably contributes to the high frequency of *Hemidactylus* introductions.

Hemidactylus hews to the family pattern in reproduction, with the only exception being a relatively shortened incubation period, perhaps due to high temperatures: −0.19

Hemidactylus spp.	English name				Range core
			Size		
	Activity	Diet	mm	g	Ecological business model
bavazzanoi	Bavazzano's Gecko				Kenya
	nocturnal	invert.	38	1.32	* nocturnal canopy walker
bayonii	Quanza River Gecko				Angola
	nocturnal		41	1.68	
beninensis	Benin Half-toed Gecko				Benin
	nocturnal	invert.	67	6.18	*** nocturnal ground lizard
biokoensis	Bioko Gecko				Equatorial Guinea
	nocturnal		67	6.4	
bouvieri	Cape Verde Gecko				Cape Verde Islands
	nocturnal		42	1.78	*** trunk/wall lizard and nocturnal ground lizard (ssp. differ)
bowringii	Asia House Gecko				East Asia
	nocturnal	invert.	48	2.46	*** nocturnal ground lizard or trunk/wall lizard (around people)
brasilianus	Amaral's Brazilian Gecko				Brazil
	nocturnal		53	3.33	* trunk/wall lizard
brookii	Brooke's House Gecko				Southern Asia
	cathemeral	invert.	54	3.55	**** nocturnal ground lizard or trunk/wall lizard (if elect. lights)
chipkali	Central India Leaf-toed Gecko				India
	nocturnal		62	5.2	*** trunk/wall lizard
citernii	Speedy Gecko				Kenya
	nocturnal		35	1.07	
coalescens	Equatorial Gecko				Congo
	nocturnal		65	5.84	
craspedotus	Frilly House Gecko				Malaysia
	cathemeral	invert.	56	3.82	**** glider, edging to immobility expert
curlei	Parker's Gecko				Ethiopia
	nocturnal		40	1.54	* nocturnal ground lizard
dawudazraqi	Modry's Gecko				Syria
	nocturnal		45	2.11	
depressus	Kandyan Gecko				Sri Lanka
	cathemeral	invert.	73	8.0	**** trunk/wall lizard
dracaenacolus	Socotran Dracaena Gecko				Yemen
	nocturnal		59	4.38	*** trunk/wall lizard
echinus	Spiny Forest Gecko				Congo
	nocturnal		59	4.53	
endophis	Enigmatic Oman Gecko				Oman
			52	3.07	

Hemidactylus spp.	English name				Range core
			Size		
	Activity	Diet	mm	g	Ecological business model
eniangii	Eniang's Gecko				Nigeria
	nocturnal		54	3.43	
fasciatus	Banded Forest Gecko				Nigeria
	nocturnal	invert.	75	8.5	** trunk/wall lizard
festivus	Radiant Gecko				Yemen
	nocturnal		48	2.52	* nocturnal ground lizard
flaviviridis	Yellow-bellied House Gecko				Iran
	cathemeral	invert.	71	7.4	**** trunk/wall lizard
forbesii	Abd al Kuri Gecko				Yemen
	nocturnal		69	6.8	
foudaii	Elba Gecko				Egypt
	nocturnal		41	1.61	*** nocturnal ground lizard
fragilis	Delicate Somali Gecko				Ethiopia
			36	1.14	
frenatus	Cosmopolitan House Gecko				Pan-tropical
	cathemeral	invert.	51.	2.60	**** trunk/wall lizard and nocturnal canopy walker (geog.)
funaiolii	Archer's Post Gecko				Kenya
	nocturnal		45	2.14	
garnotii	Fox Gecko				Indo-Pacific
	nocturnal	invert.	56	3.89	**** nocturnal canopy walker
giganteus	Giant Southern Tree Gecko				India
	nocturnal	invert.	99	18.4	** trunk/wall lizard
gleadowi	Gleadow's House Gecko				Pakistan
	nocturnal		41	1.61	
gracilis	Graceful Leaf-toed Gecko				India
	nocturnal	invert.	36	1.13	** nocturnal ground lizard
granchii	Granchi's Gecko				Somalia
	nocturnal		54	3.43	
graniticolus	Granite Gecko				India
	nocturnal		89	13.7	** trunk/wall lizard
granosus	Sinai Gecko				Egypt
	nocturnal		47	2.41	
granti	Grant's Gecko				Yemen
	nocturnal		65	5.8	*** nocturnal ground lizard
greeffii	São Tomé Gecko				São Tomé and Príncipe
	nocturnal		71	7.4	** trunk/wall lizard
gujaratensis	Gujarat Gecko				India
	nocturnal		56	3.82	*** trunk/wall lizard
hajarensis	Hajar Mountains Gecko				Oman
	nocturnal		55	3.69	* nocturnal ground lizard

v. 0.17 std. resid. *H. squamulatus* may oviposit in termite mounds. *Hemidactylus* is somewhat unusual within Gekkonidae for having young that are smaller than lizard average (ontogenetic mass range 0.14 v. –0.19 std. resid.) while having clutch counts that are family typical (i.e., below lizard average: –0.34 v. –0.26 std. resid.). Thus if they compensate at all for the small number of progeny per clutch, they must do so by increasing clutch frequency, which is insufficiently documented at present.

Longevity is slightly short (–0.02 v. 0.15 std. resid.). Population density is notably high: 3.34 kg/ha v. 0.37 kg/ha (Gekkonidae), though credible data exist for only one island species (*frenatus*). Thus this genus mean is not representative of the genus, which is underrepresented on islands (33% v. 45%). *Hemidactylus* has some imperiled species (22%) but fewer than does the family (47%). They are similarly underrepresented among the ranks of the IUCN's most endangered lists (6.0% v. 11.3%). In other respects *Hemidactylus* are typical gekkonids.

There is a substantial literature documenting the displacement of *Hemidactylus* species by more recently arrived *Hemidactylus*; the proximate mechanism for displacement may be some combination of intra-guild predation and competition, but causation has not been resolved in most if not all displacements.

Although humans share quarters with *Hemidactylus* in much of the tropical world, some individuals in various cultures express phobias or attribute omens to the proximity of these lizards. In Cyprus some residents believe that skin rashes arise from falling feces of *turcicus*. In Sri Lanka *hunae* are killed on sight, in the belief that they cause disease.

Hemidactylus spp.	English name				Range core
			Size		
	Activity	Diet	mm	g	Ecological business model
hemchandrai	Navega Gecko				India
			71	7.4	
homoeolepis	South Arabia Gecko				Yemen
	nocturnal		38	1.34	*** nocturnal ground lizard
hunae	Spotted Giant Gecko				Sri Lanka
	crepuscular	invert.	101	19.3	**** trunk/wall lizard
imbricatus	Carrot-tail Viper Gecko				India
	nocturnal	invert.	49	2.73	**** nocturnal ground lizard
inexpectatus	Hammar-an-Nafur Gecko				Oman
	nocturnal		39	1.43	* trunk/wall lizard
inintellectus	Socotra Rock Gecko				Yemen
	nocturnal		50	2.76	** trunk/wall lizard
isolepis	Scaly Gecko				Ethiopia
	nocturnal	invert.	40	1.56	*** nocturnal ground lizard
ituriensis	Ituri Gecko				Congo
	nocturnal		73	8.0	
jubensis	Ethiopia Gecko				Ethiopia
	nocturnal		59	4.53	
jumailiae	Al-Jumaily's Gecko				Yemen
	nocturnal		48	2.52	
kamdemtohami	Kamdem Toham's Forest Gecko				Cameroon
	nocturnal		60	4.68	
karenorum	Myanmar Spotted Gecko				Myanmar
	nocturnal	invert.	50	2.84	* nocturnal ground lizard
klauberi	Klauber's Gecko				Somalia
	nocturnal		37	1.25	
kundaensis	Kunda Half-toed Gecko				Guinea
	nocturnal	invert.	49	2.67	** trunk/wall lizard
kushmorensis	Kushmore House Gecko				Pakistan
			46	2.22	
kyaboboensis	Kyabobo Gecko				Ghana
	nocturnal		70	7.2	
laevis	Golis Mountains Gecko				Somalia
	nocturnal		37	1.25	
lamaensis	Lama Forest Gecko				Benin
	nocturnal		46	2.22	*** nocturnal ground lizard
lankae	Sri Lanka Leaf-toed Gecko				Sri Lanka
	nocturnal	invert.	70	7.1	**** nocturnal ground lizard
laticaudatus	Andersson's Gecko				Ethiopia
	nocturnal		52	3.18	*** trunk/wall lizard

Hemidactylus spp.	English name				Range core
			Size		
	Activity	Diet	mm	g	Ecological business model
lavadeserticus	Syria House Gecko				Syria
	nocturnal		46	2.18	
lemurinus	Dhofar Ghost Gecko				Yemen
	nocturnal		59	4.53	*** trunk/wall lizard
leschenaultii	Indian Bark Gecko				India
	cathemeral	invert.	78	9.6	**** trunk/wall lizard
longicephalus	Long-headed Gecko				Congo
	nocturnal		61	4.77	* ground-foraging tree lizard
lopezjuradoi	Fogo Island Gecko				Cape Verde Islands
	nocturnal		38	1.32	
luqueorum	Jebel Akhdar Gecko				Oman
	nocturnal		74	8.2	*** trunk/wall lizard
mabouia	Tropical House Gecko				Africa and West Indies
	cathemeral	invert.	59	4.5	*** trunk/wall lizard
macropholis	Large-scaled Gecko				Ethiopia
	nocturnal		60	4.73	*** nocturnal ground lizard
maculatus	Giant Spotted Gecko				India
	nocturnal	invert.	99	18.7	*** trunk/wall lizard
makolowodei	Makolowode's Gecko				Cameroon
	nocturnal	invert.	81	10.6	
mandebensis	Bab-el-Mandeb Gecko				Yemen
	nocturnal		39	1.43	
masirahensis	Masirah Gecko				Oman
	nocturnal		37	1.21	* trunk/wall lizard
matschiei	Matschie's Half-toed Gecko				Nigeria
	nocturnal	invert.	28	0.59	* nocturnal ground lizard
megalops	Parker's Somali Gecko				Somalia
	nocturnal		35	1.07	
mercatorius	Farquhar Island Gecko				Madagascar
	nocturnal	invert.	50	2.86	*** trunk/wall lizard
mindiae	Mount Sinai Gecko				Egypt
	nocturnal		50	2.84	** trunk/wall lizard
minutus	Small Smooth Arabian Gecko				Yemen
	nocturnal		32	0.80	
modestus	Tana River Gecko				Kenya
	nocturnal	invert.	42	1.69	
montanus	Mountain Gecko				Yemen
			58	4.24	
mrimaensis	Kaya Gecko				Kenya
	cathemeral		45	2.12	
muriceus	Prickly Half-toed Gecko				Nigeria
	nocturnal	invert.	51	2.95	* ground-foraging tree lizard

Hemidactylus spp.	English name				Range core
			Size		
	Activity	Diet	mm	g	Ecological business model
murrayi	Murray's House Gecko				Myanmar
	cathemeral		56	3.82	
newtoni	Newton's Gecko				Equatorial Guinea
	nocturnal		45	2.11	
ophiolepis	Snake-scaled Gecko				Ethiopia
	nocturnal		42	1.69	
ophiolepoides	Lanza's Gecko				Ethiopia
	nocturnal		45	2.12	
oxyrhinus	Sharp-nosed Gecko				Yemen
	nocturnal		45	2.12	** nocturnal ground lizard
palaichthus	Trinidad Gecko				Venezuela
	cathemeral		45	2.08	* ground-foraging tree lizard
parvimaculatus	Spotted House Gecko				Indian Ocean Islands
	nocturnal	invert.	49	2.64	**** trunk/wall lizard
pauciporosus	Galgala Gecko				Somalia
			55	3.63	*** trunk/wall lizard
paucituberculatus	Dhofar Marbled Gecko				Oman
	nocturnal		35	1.01	** nocturnal ground lizard
persicus	Persian Gecko				Iran
	nocturnal	invert.	59	4.39	*** nocturnal ground lizard or trunk/wall lizard (elect. light)
pieresii	Gannoruwa Forest Gecko				Sri Lanka
	nocturnal		66	6.0	
platycephalus	Baobab Gecko				East Africa
	nocturnal	invert.	83	12.0	*** trunk/wall lizard
platyurus	Asian Flat-tailed House Gecko				Southeast Asia
	cathemeral	invert.	56	3.92	**** trunk/wall lizard
prashadi	Prashad's Gecko				India
	nocturnal	invert.	77	9.4	*** trunk/wall lizard
principensis	Príncipe Gecko				São Tomé and Príncipe
	nocturnal		57	3.96	** trunk/wall lizard
pseudomuriceus	Azagny Half-toed Gecko				Cameroon
	cathemeral	invert.	49	2.73	** ground-foraging tree lizard
puccionii	Zanzibar Gecko				Somalia
	nocturnal		38	1.32	
pumilio	Pygmy Socotran Gecko				Yemen
	nocturnal		30	0.69	*** nocturnal ground lizard
reticulatus	Reticulate Gecko				India
	nocturnal	invert.	52	3.18	**** nocturnal ground lizard
richardsonii	Richardson's Forest Gecko				Congo
	nocturnal	invert.	66	6.1	** trunk/wall lizard

Hemidactylus spp.	English name				Range core
			Size		
	Activity	Diet	mm	g	Ecological business model
robustus	Red Sea Gecko				Saudi Arabia
	nocturnal	invert.	49	2.71	** ground-foraging tree lizard or trunk/wall lizard (elect. light)
romeshkanicus	Romeshkan Gecko				Iran
	nocturnal		59	4.53	
ruspolii	Prince Ruspoli's Gecko				Ethiopia
	nocturnal		48	2.52	** trunk/wall lizard
saba	Saba Gecko				Yemen
	cathemeral		52	3.07	
sataraensis	Satara Gecko				India
	nocturnal	invert.	43	1.81	** nocturnal ground lizard
scabriceps	Scaly Gecko				India
	nocturnal	invert.	41	1.26	
shihraensis	Al-Shihr Gecko				Yemen
	nocturnal		44	1.96	
sinaitus	African Horn Gecko				Sudan
	nocturnal		50	2.84	* ground-foraging tree lizard
smithi	Smith's Gecko				Ethiopia
	nocturnal		50	2.84	
somalicus	Northern Somalia Gecko				Ethiopia
	nocturnal		40	1.54	
squamulatus	Nyika Gecko				East Africa
	nocturnal	invert.	44	1.94	*** nocturnal ground lizard
stejnegeri	Stejneger's House Gecko				South China Sea Islands
	nocturnal	invert.	52	3.18	** trunk/wall lizard
tanganicus	Dutumi Gecko				Tanzania
	nocturnal		67	6.2	
tasmani	Tasman's Tropical House Gecko				Zimbabwe
	nocturnal		67	6.3	** trunk/wall lizard
taylori	Taylor's Somali Gecko				Somalia
	nocturnal		66	6.0	
tenkatei	Roti House Gecko				Malaysia
	nocturnal		54	3.43	**** trunk/wall lizard
thayene	Shwe-Settaw Gecko				Myanmar
	nocturnal		41	1.61	
treutleri	Golconda Fort Gecko				India
	nocturnal	invert.	59	4.53	** trunk/wall lizard
triedrus	Termite Hill Gecko				Southern Asia
	nocturnal	invert.	77	9.1	**** nocturnal ground lizard
tropidolepis	Mocquard's Gecko				Horn of Africa
	nocturnal	invert.	43	1.91	* nocturnal ground lizard

Hemidactylus spp.	English name				Range core
			Size		
	Activity	Diet	mm	g	Ecological business model
turcicus	Mediterranean Gecko				Eurasia
	cathemeral	invert.	50	2.73	**** ground-foraging tree lizard or trunk/wall lizard (elect. light)
ulii	Joger's Gecko				Yemen
	nocturnal		39	1.37	
vietnamensis	Vietnam House Gecko				Vietnam
	nocturnal		51	2.95	*** trunk/wall lizard
yajurvedi	Kanker Rock Gecko				India
			79	10.0	
yerburii	Yerbury's Red Sea Gecko				Yemen
	nocturnal		63	5.5	** trunk/wall lizard

Hemiergis

Hemiergis is a moderate radiation (seven spp.) of small (1.6 g v. 3.3 g), elongate (v. 35%) cryptozoic (86% v. 30%) or subterranean (14% v. 11%) skinks found in Australia. Most are litter cryptozoids (83% v. 28%); the remainder subterranean foragers (17% v. 12%). Keep in mind that in my application of litter cryptozoid, "litter" may include dense matrices of spinifex grass such as *Triodia*. The common name of *H. millewae* references this association with *Triodia* grass. *Hemiergis* are moderately well known (filled cells 33% v. 25%).

They are slender (shape index –1.17 v. –0.46 std. resid.). Their tails are of average length (1.47 v. 1.33 × SVL) and breakable (100% v. 73%), but not blue (v. 4%).

Males are smaller than females (–1.0 v. –0.31 on –2 to 2 scale), which is consistent with the notion that females are peritoneal-space-limited for embryos (see reduced ontogenetic mass range and litter size three paragraphs below). Other sex differences are not known. Their sex differences rollup score (on 0 to 6 scale) is average (0.29 v. 0.28).

Diets are known for only three species, and each is different (invertebrates: 33% v. 84%; ant specialist: 33% v. 1.6%; omnivore: 33% v. 13%). The omnivore (*peronii*) eats a broad spectrum of foods (fruit, arthropods, snails, ectothermic vertebrates). Some have argued that *peronii*'s appreciable ingestion of seeds is accidental from burrowing, but it is hard to imagine a litter burrower that is unable to keep its mouth shut while burrowing if it wanted to. Foraging mode, electric light use, and home-range area have not been described.

Primary predator defense has been documented for only *initialis* (crypsis: v. 74%). Mean active body temperature (21.6°C v. 31.1°C) is known only for *decresiensis*, a relatively high latitude cryptozoic form. This remarkably low mean active body temperature is however consistent with this lizard's preference of 17°C in a lab temperature gradient. Unsurprisingly, *Hemiergis* have not been reported to bask (v. 21%). *H. peronii* is the only genus member with annual rainfall documentation: 300–1270 mm/y (arid to mesic). The genus is nominally overrepresented in littoral (43% v. 21%), sandy (43% v. 25%), and human (29% v. 14%) habitats

Hemiergis spp.	English name				Range core
			Size		
	Activity	Diet	mm	g	Ecological business model
decresiensis	Three-toed Earless Skink				Southeastern Australia
	diurnal	ants	63	1.73	** litter cryptozoid
gracilipes	Southwestern Mulch-skink				Western Australia
		invert.	75	4.31	
initialis	Southwestern Earless Skink				Southwestern Australia
			44	0.71	** litter cryptozoid
millewae	Triodia Earless Skink				Australia
			52	1.09	** litter cryptozoid
peronii	Lowlands Earless Skink				Southern Australia
		omnivore	65	1.90	**litter cryptozoid
quadrilineata	Two-toed Earless Skink				Western Australia
			62	1.65	* litter cryptozoid
talbingoensis	Eastern Three-toed Earless Skink				New South Wales
			58	1.40	* subterranean forager

and is about average in riparian zones (14% v. 23%).

All are viviparous (v. 36%). They have a slightly elevated litter count (0.33 v. 0.06 std. resid.). Neonates are slightly large (ontogenetic mass range −0.11 v. 0.12 std. resid.). *H. peronii* is known to be inseminated in the fall prior to ovulation; thus the resultant embryos are carried until nearly a year after insemination.

Longevity is short (−0.67 v. 0.17 std. resid.), which equates to 2.4 years for both *decresiensis* and *talbingoensis* (taxonomic confusion between these is likely). Additional data collection may well expand this estimate. Endemism is low (29% v. 64%). Island species are about average (43% v. 57%). None has been introduced (v. 2.1%). Most are imperiled (75% v. 47%), but none is on one of the IUCN's lists of most endangered (v. 14%). In other respects *Hemiergis* are typical skinks.

Hemiphyllodactylus

Hemiphyllodactylus is a 19-species genus of the Gekkonidae centered in Southeast Asia and extending north to China, east to French Polynesia, south to New Caledonia, and west to India. Unfortunately, more is known about its range than its habits. Note that, in the table below, ecological business models were vaguely assignable in only 6 of 19 species (one is firmly known). The diets are crudely known for only four species. Nocturnality (100% v. 73%) is well established.

With only one exception (*insularis*), these are small geckos (genus mean 1.63 g v. 3.31 g family mean) with a slender build (shape index −0.10 v. 0.10 std. resid.). They are the gekkonid genus with the highest frequency of elongate species (elongation documented in 32% of spp. v. 0.9%). Relative tail length is comparatively short (0.88 v. 1.07 × SVL). Tail breakage is reported in only about half of species (58% v. 87%), perhaps because at least 5.3% (v. 1.8%) have prehensile tails. Basic color change is present in 5% (v. 15%). All have enlarged toe pads (v. 70%).

SSD in this genus reflects males that are comparatively smaller than in the average gekkonid (−0.36 v. −0.03 on −2 to 2 scale). Sex differences in color are slightly more frequent (10.5% v. 6.6%). Accordingly, the aggregate sex rollup score, on scale of 0 to 6, is slightly higher for *Hemiphyllodactylus* (1.0 v. 0.63).

Hemiphyllodactylus are fairly arboreal (75% v. 68%) and are overrepresented at medium elevations (47% v. 26%), in littoral zones (20% v. 11%), and in human habitats (60% v. 27%). Given the latter figure, one would expect that ecologists would know more about them. Evidently *Hemiphyllodactylus* are retiring in habits, and they are very small, arboreal, and nocturnal.

Their reproduction is typical of gekkonids, though they are underrepresented among communal ovipositors (0% v. 8.6%) and overrepresented among gekkonids with parental care (5.3% v. 1.9%). This parental care frequency is high for these ecological business models (table 4.1, trait 69).

Based on one estimate for *typus*, the genus is not unusual in its population density (0.21 kg/ha v. 0.37 kg/ha). The same species has been introduced extensively in the Indian and Pacific Ocean islands (5.3% v. 3.7%), likely aided by its inconspicuousness, durable eggs, and small size. *Hemiphyllodactylus* are underrepresented among the gekkonid imperiled species (25% v. 47%), but these rare species are among the IUCN's most severely endangered (25% v. 11.3%). In other respects *Hemiphyllodactylus* are typical gekkonids. With further natural history data this portrait of *Hemiphyllodactylus* may require considerable revision.

Hemiphyllodactylus spp.	English name		Size		Range core
	Activity	Diet	mm	g	Ecological business model
aurantiacus	Southern Ghats Slender Gecko				India
	nocturnal	invert.	38	1.08	* ground-foraging tree lizard
banaensis	Ba Na Dwarf Gecko				Vietnam
	nocturnal		46	2.22	
bintik	Malaysia Spotted Dwarf Gecko				Malaysia
	nocturnal		36	1.10	
changningensis	Changning Slender Gecko				China
	nocturnal		41	1.60	
chiangmaiensis	Chiang Mai Dwarf Gecko				Thailand
	nocturnal		39	1.41	
cicak	Pinang Island Slender Gecko				Malaysia
			32	0.81	
engganoensis	Enggano Dwarf Gecko				Indonesia
	nocturnal		36	1.14	
ganoklonis	Palau Slender Gecko				Palau
	nocturnal	invert.	33	0.69	* nocturnal canopy walker
harterti	Bintang Slender Gecko				Malaysia
	nocturnal		42	1.69	
huishuiensis	Huishui County Gecko				China
			46	2.22	
insularis	Philippine Slender Gecko				Philippines
	nocturnal		102	15.9	* nocturnal canopy walker
kiziriani	Kizirian's Slender Gecko				Laos
	nocturnal		39	1.38	
larutensis	Larut Dwarf Gecko				Malaysia
	nocturnal		47	1.84	*** ground-foraging tree lizard
margarethae	Sumatra Slender Gecko				Indonesia
	nocturnal		41	1.65	
tehtarik	Tebu Mountain Slender-toed Gecko				Malaysia
	nocturnal		38	1.35	
titiwangsaensis	Titiwangsa Slender Gecko				Malaysia
	nocturnal		47	2.37	**** trunk/wall lizard
typus	Indo-Pacific Slender Gecko				Indo-Pacific Islands
	nocturnal	invert.	41	1.31	** nocturnal canopy walker or ground-foraging tree lizard
yunnanensis	Asian Slender Gecko				Thailand
	nocturnal	invert.	44	1.53	
zugi	Zug's Slender Gecko				Vietnam
	nocturnal		43	1.82	

Hemitheconyx

Hemitheconyx is a two-species genus of large (29.5 g v. 13.1 g) eublepharid geckos found in Sahelian regions of Africa. Morphologically they closely resemble other nocturnal ground lizards in the Eublepharidae, though their unregenerated tails are noticeably more ringed and shorter (0.42 v. 0.76 × SVL). No tails are prehensile (v. 8.3%). *Hemitheconyx* lack color change (v. 3%).

Based on their phylogeny, size, and other documented aspects, they are suspected to be m>f in their SSD, though documentation is inadequate. One species (*caudicinctus*) is sexually dichromatic (v. 11%). The ratio of m:f relative tail lengths is an amazing 1.85 in *caudicinctus*, though this ratio may be inflated by inadequate sampling and a very small denominator (female relative tail length is reported to be 0.40). These sex differences sum to the highest sex rollup score (0 to 6 scale) in the Eublepharidae (2.50 v. 1.08).

Although both species are listed as nocturnal, there are diurnality reports for *caudicinctus*. *H. caudicinctus* may be active only at the season when rains begin. This is a short annual activity period, and one can appreciate the possibility that suitable rainy weather might stimulate daytime activity. This species is reported to vocalize, without making a distinction between spontaneous and stress vocalizations. *H. taylori* gives stress calls. Both types of calls are overrepresented in *Hemitheconyx* (50% v. 8.3% for spontaneous; 50% v. 25% for stress).

Both species use burrows as daily refugia (v. 33%). *Hemitheconyx* are relatively well known (42% v. 35% of cells filled), so lack of information is probably not responsible for the genus's absence from reports of presence in littoral zones (0% v. 7%),

Hemitheconyx spp.	English name				Range core
			Size		
	Activity	Diet	mm	g	Ecological business model
caudicinctus	Fat-tailed Gecko				Ghana
	nocturnal	invert.	117	34.1	**** nocturnal ground lizard
taylori	Somaliland Fat-tailed Gecko				Ethiopia
	nocturnal		107	25.6	** nocturnal ground lizard

Heremites spp.	English name				Range core
			Size		
	Activity	Diet	mm	g	Ecological business model
auratus	Golden Grass Skink				Iran
	diurnal	invert.	84	6.0	** ground-foraging tree lizard
septemtaeniatus	Southern Grass Skink				Iraq
	diurnal	invert.	79	5.1	
vittata	Bridled Skink				Mediterranean
	diurnal	invert.	73	4.10	**** refugia-anchored diurnal gr. liz.

sandy areas (0% v. 10%), riparian zones (0% v. 30%), and human habitats (0% v. 17%).

Reproduction follows eublepharid standards; *taylori* shows nest defense (50% v. 14% parental care). The size of young *Hemitheconyx* is notable because they are proportionately smaller than average lizards, whereas all other eublepharids are larger (ontogenetic mass range 0.22 v. –0.67 std. resid.). Longevity is shorter than is typical for eublepharids (0.86 v. 1.16 std. resid.). No *Hemitheconyx* is judged imperiled (v. 60%). In other respects they are typical eublepharids.

Heremites

Heremites is a small genus (three spp.) of medium-large (5.0 g v. 3.3g), elongate (100% v. 35%), diurnal (100% v. 75%), terrestrial (100% v. 38%), viviparous (100% v. 36%) skinks found in the western Palearctic (100% v. 6%). The occupied polygon is Algeria to Greece to Armenia to Uzbekistan to Pakistan to Oman to Ethiopia to Algeria. Ecological business models are hard to generalize; only two species are assigned, and those are inconsistent. Note that *auratus* is not always found in trees—some occupied habitats have none—but uses elevated perches on rock as well.

Tails are about average in length (1.50 v. 1.33 × SVL) and are breakable (100% v. 73%) but not blue (v. 4%).

Based solely on *vittata*, males are smaller than females (–1.0 v. –0.31 on –2 to 2 scale), suggesting reduced selection pressure for male dominance. *H. vittata* sexes differ in color (33% v. 7.6%), but no *Heremites* shows seasonal color (v. 6.2%). The sex differences rollup score is elevated (1.00 v. 0.28 on 0 to 6 scale), but based on an inadequate sample.

All eat invertebrates (v. 84%), taken by active foraging (v. 74%). Home-range area is small (–0.76 v. –0.07 std. resid.), which is based on female *vittata* (9 m²). All species are conspicuously wary (v. 11%). Primary predator concealment tactics are retreat to either crevice-burrows (50% v. 43%) or vegetation (50% v. 25%). Direct basking is the rule (67% v. 21%). Mean active body temperature is warmish (32.0°C v. 31.1°C). *H. auratus* lives in groups (v. 16%); *vittata* is spatially intolerant (v. 94%) during the breeding season. Most species use burrows for daily refugia (67% v. 24%).

H. auratus and *vittata* are reported over the elevations 0–2500 m (taxonomic confusion likely). The genus is nominally overrepresented in sandy areas (67% v. 25%), riparian zones (100% v. 23%), and human habitats (67% v. 14%). They are about average in littoral areas (33% v. 21%).

H. vittata has been described as both oviparous and viviparous. Litter counts for the genus are about what would be expected for their body size (0.21 v. 0.06 std. resid.). Neonates are notably small (ontogenetic mass range 1.18 v. 0.12 std. resid.).

No species is endemic to its country (v. 64%). All occur on some island (v. 57%). Two have been introduced (67% v. 2.1%), and *auratus* has been introduced three times (33% v. 0.7%): Eritrea, Oman, and United Arab Emirates. Two-thirds are imperiled (v. 47%), though none is on one of the IUCN's lists of the most endangered (v. 14%). In other respects *Heremites* are like typical skinks.

Hesperoedura

Hesperoedura reticulata is the sole representative of this diplodactylid genus. It is found in southwestern Australia on the trunks of many smooth-barked eucalypts. A bit larger than an average diplodactylid (5.5 g v. 4.2g), with a nonprehensile but breakable tail that is a little shorter than average (0.70 v. 0.85 × SVL), *reticulata* has an unremarkable appearance. Sex differences are not reported. Though nocturnal, it is not reported to vocalize (v. 5.4%).

Its mean body temperature is low (21.5°C v. 24.3°C). Males defend space (v. 77%). It is not associated with littoral (v. 17%), sandy (v. 24%), riparian (v. 10%), or human (v. 5%) habitats. Population density matches the diplodactylid mean (0.49 kg/ha v. 0.44 kg/ha). It has not been evaluated for conservation status. In other respects it is a typical diplodactylid.

Hesperoedura sp.	English name				Range core
			Size		
	Activity	Diet	mm	g	Ecological business model
reticulata	Reticulated Velvet Gecko				Western Australia
	nocturnal	invert.	65	5.5	*** trunk/wall lizard

Heterodactylus spp.	English name				Range core
			Size		
	Activity	Diet	mm	g	Ecological business model
imbricatus	Rio de Janeiro Elongo-lizard				Brazil
		invert.	88	17.5	
lundii	Lund's Elongo-lizard				Brazil
			52	1.10	
septentrionalis	Diamantina Elongo-lizard				Brazil
			47	0.83	

Heterodactylus

Heterodactylus is a three-species genus of extremely elongate cryptozoic gymnophthalmids endemic to Brazil. They are poorly known, even by gymnophthalmid standards (filled cells 13% v. 19%). Their ecological business models are unknown. They are of average size (2.51 g v. 2.35 g) but more slender build (shape index −0.62 v. −0.32 std. resid.). They eat invertebrates and are found at medium elevation. They have not been evaluated for conservation status. In other respects *Heterodactylus* are probably typical gymnophthalmids.

Heteronotia spp.	English name				Range core
			Size		
	Activity	Diet	mm	g	Ecological business model
atra	Black Pilbara Gecko				Western Australia
	nocturnal		54	3.56	* nocturnal ground lizard
binoei	Bynoe's Prickly Gecko				Australia
	nocturnal	carnivore	47	2.38	**** nocturnal ground lizard
fasciolata	Pale-headed Gecko				Northern Territory
	nocturnal		50	2.84	*** nocturnal ground lizard
planiceps	Banded Prickly Gecko				Western Australia
	nocturnal		47	2.40	**** nocturnal ground lizard
spelea	Desert Cave Gecko				Western Australia
	nocturnal		51	2.93	*** nocturnal ground lizard

Heteronotia

Heteronotia (five spp.) are slightly small (2.79 g v. 3.31 g) nocturnal ground lizards from Australia (Gekkonidae). Their tails are relatively long (1.37 v. 1.07 × SVL) but not prehensile (v. 1.8%). They lack enlarged toepads (v. 30%), secretory pores (v. 46%), and any color change (v. 15%). Males are smaller than females in the only documented species (*binoei*), −1.00 v. −0.03 on −2 to 2 scale, but do not differ in other traits (sex rollup score, on 0 to 6 scale: 0.20 v. 0.63). *H. binoei* is also the only species documented for social interactions: males of this species defend a refuge but not a larger area. They are found in areas with 202–584 mm/y of rain. Parthenogenesis is found in this species and may be routine in some populations. Hatchlings are smaller than the lizard average (ontogenetic range 0.22 v. −0.19 std. resid.), despite the small clutch count (apparently fixed at two: −0.12 v. −0.26 std. resid.). Hatchlings may remain in proximity to their parents, but reproduction is otherwise typical for gekkonids. *Heteronotia* have not been introduced (v. 3.7%) or evaluated for conservation status (v. 47% imperiled). In other respects *Heteronotia* are typical gekkonids.

Holaspis

Holaspis (two spp.) is an unusual lacertid genus of diurnal gliders (subtle flaps on digits, tail, and body), found in equatorial Africa from Guinea to Mozambique. No other gliders are known within the Lacertidae. They are small (2.26 g v. 5.1 g) and lightly built (shape index 0.06 v. 0.27 std. resid.) with a relatively short tail (1.30 v. 2.06 × SVL) that is breakable (v. 92%). One wonders if they are able to control their glide path without their tails. In at least *guentheri* the tails are bright blue in juveniles.

Sex differences are poorly documented or absent (0.0 v. 0.86 on 0 to 6 scale). *H. guentheri* is conspicuously wary (v. 11%). Both bask directly (v. 25%). *H. guentheri* is reported to be capable of gliding across canopy gaps of 10 m. Social relations are not documented in *Holaspis*; it is not yet possible to determine if they share the unusual sexual and spatial traits of other diurnal gliders (e.g., *Draco*). They are found at low elevation (10–800 m); *guentheri* is associated with riparian zones (v. 28%). Reproduction is oviparous with a moderately small clutch (–0.15 v. 0.13 std. resid.), presumably to limit aerodynamic weight. Incubation period is long (0.19 v. –0.36 std. resid.), potentially because eggs are deposited at an early stage of development to free the mother of weight as soon as possible. This incubation period corresponds to 68 days in these small lizards. Multiple broods per year would be expected, but are not yet reported for this genus.

Neither species is endemic (v. 35%); neither occurs on islands (v. 26%). Neither has been introduced (v. 4.6%) or evaluated for conservation status. In other respects they are probably typical lacertids.

Holaspis spp.	English name				Range core
			Size		
	Activity	Diet	mm	g	Ecological business model
guentheri	Western Blue-tailed Gliding Lizard				Nigeria
	diurnal	invert.	47	2.41	*** glider
laevis	Eastern Blue-tailed Gliding Lizard				Mozambique
	diurnal	invert.	45	2.12	**** glider

Holbrookia

Holbrookia (five spp.) is a genus of small (7.0 g v. 19.8 g), relatively well-known (filled cells 54% v. 38%), refugia-anchored diurnal ground phrynosomatids from South Dakota to Mexico. They stand out from other fleet-of-foot phrynosomatids (i.e., not *Phrynosoma*) in being exceptionally wary (60% v. 22%), highly heliophilic (direct bask 80% v. 40%), and frequent users of nighttime burrows (67% v. 29%). Several species gravitate to areas with long sight lines.

They are average in build (shape index 0.80 v. 0.91 std. resid.), prone to tail breakage (40% v. 21%), less inclined to possess secretory pores (20% v. 32%), and bereft of advanced color change (0% v. 4.5%) or blue tails (0% v. 1.3%). Most are m>f in SSD (0.60 v. 0.27 on –2 to 2 scale) and showing sexual dichromatism (80% v. 58%) and seasonal color (80% v. 25%). These roll up to 3.0 v. 1.68 on the 0 to 6 sex differences score.

Though uniform in primary diet class (invertebrate eating), they show some diversity in foraging mode: 100% (v. 98%) sit-and-wait foraging, 20% (v. 3%) cruise foraging, and 20% (v. 8%) active foraging. These add to more than 100% because single species forage in multiple ways. One species (*propinqua*) has been measured for MPM, and within this species different MPM measurements range from 39th percentile to 60th percentile, supporting the generalization of foraging flexibility. Home-range area is moderately large for phrynosomatids (0.73 v. 0.57 std. resid.), equating to 364 m^2 (*maculata*) or 233 m^2 (*propinqua*).

Within the genus there is notable diversity in primary predator concealment tactics, with 25% frequencies each registered for crypsis (v. 32%), flight (v. 14%), sand diving (v. 12%), and hiding in vegetation (v. 24%). Mean active body temperature is hot (37.1°C v. 34.2°C).

Holbrookia straddles the middle of the rainfall spectrum; 67% (v. 48%) for arid (151–400 mm/y) and 100% (v. 52%) for intermediate (401–1100

Holbrookia spp.	English name				Range core
			Size		
	Activity	Diet	mm	g	Ecological business model
approximans	Speckled Earless Lizard				Mexico
	diurnal	invert.	61	8.4	** refugia-anchored diurnal gr. liz.
elegans	Elegant Earless Lizard				Mexico
	diurnal	invert.	58	7.4	** refugia-anchored diurnal gr. liz.
lacerata	Spot-tailed Earless Lizard				Mexico
	diurnal	invert.	58	7.3	** refugia-anchored diurnal gr. liz.
maculata	Lesser Earless Lizard				Kansas
	diurnal	invert.	59	6.5	*** refugia-anchored diurnal gr. liz.
propinqua	Keeled Earless Lizard				Mexico
	diurnal	invert.	53	5.7	*** refugia-anchored diurnal gr. liz.

mm/y). The genus is overrepresented in littoral (40% v. 16%), sandy (100% v. 32%), and riparian (60% v. 40%) habitats and absent from human habitats (v. 18%).

Reproduction is notable only for the light RCM (0.165 v. 0.300 ratio), which is not associated with a particularly svelte shape index (0.80 std. resid.). Clutch count is near normal (0.36 v. 0.68 std. resid.), and hatchling size is only slightly small (ontogenetic mass range 0.15 v. −0.26 std. resid.). Longevity is a little short (−0.46 v. −0.25), though based on only *maculata* (5 years) and *propinqua* (3 years). Population density is near average (0.21 kg/ha v. 0.26 kg/ha), based on *maculata* (0.11 kg/ha) and *propinqua* (0.39 kg/ha). The frequency of imperiled species is about average (25% v. 37%), though none is on one of the IUCN's lists of the most endangered (v. 5.4%). In other respects *Holbrookia* follow the phrynosomatid pattern.

Holcosus spp.	English name		Size		Range core
	Activity	Diet	mm	g	Ecological business model
anomalus	Echternacht's Ameiva				Colombia
	diurnal		88	20.0	** wide-ranging chemosensor
bridgesii	Bridge's Ameiva				Colombia
			95	25.4	* wide-ranging chemosensor
chaitzami	Chaitzam's Ameiva				Mexico
	diurnal		77	13.4	** wide-ranging chemosensor
festivus	Central American Ameiva				Central America
	diurnal	invert.	99	28.5	**** wide-ranging chemosensor
leptophrys	Delicate Ameiva				Panama
	diurnal	invert.	105	34.7	**** wide-ranging chemosensor
niceforoi	Dunn's Ameiva				Colombia
	diurnal		98	28.0	** wide-ranging chemosensor
orcesi	Peters' Ameiva				Ecuador
			85	18.0	* wide-ranging chemosensor
quadrilineatus	Four-lined Ameiva				Costa Rica
	diurnal	invert.	69	9.5	**** wide-ranging chemosensor
septemlineatus	Seven-lined Ameiva				Ecuador
			106	35.4	* wide-ranging chemosensor
undulatus	Rainbow Ameiva				Central America
	diurnal	invert.	85	18.0	**** wide-ranging chemosensor

Holcosus

Holcosus (Teiidae) is a moderate genus (10 spp.) of average-sized (21.5 g v. 22.1 g), space-defending (100% v. 45%), non-flight-dependent (0% v. 77%), secretory-pore-bearing (30%, the highest value in Teiidae, v. 9%), invertebrate-eating (100% v. 62%) wide-ranging chemosensors (100% v. 97%). The genus is found primarily in Central America and extends to southern Mexico, Colombia, and Ecuador. They stand out for having the family's highest frequency of seasonal color (40% v. 10%) and the second-largest home ranges (0.46 v. 0.19 std. resid.). Overall, however, they seem more spatially defensive than other teiids, many of which are very tolerant and also have large home ranges.

They are of average build (0.87 v. 0.88 std. resid.), tail length (2.18 v. 2.30 × SVL), and tail breakability (20% v. 21.3%). Tails are often blue (30% v. 15%), but the blue color of hatchling tails is quickly lost. There is no color change (v. 1.3% for any change).

SSD is average (0.70 v. 0.89 on −2 to 2 scale), as is sexual dichromatism (40% v. 35%) and seasonal color (10% v. 9.6%). The sex differences rollup score is above average (1.40 v. 1.19 on 0 to 6 scale).

An average proportion are conspicuously wary (10% v. 13%). Primary predator concealment is either retreat to a burrow (50% v. 12%) or retreat to vegetation (50% v. 7%). Direct basking is common (40% v. 58%). Mean active body temperature is average (36.5°C v. 36.9°C). Locomotor activity has been quantified in *festivus*, *leptophrys*, and *quadrilineatus*: about 65th percentile for MPM and 78th for PTM. All species use burrows as daily refugia (v. 82%). The only documented rainfall band is wet (100% v. 7%). The documented elevations are (l/m/h) 100/75/0 (v. 95/32/6). They are not found in sandy areas (0% v. 47%) but are about average in the other habitat specialties: littoral (25% v. 38%), riparian (50% v. 46%), and human (38% v. 29%) habitats.

Clutch counts are about average (−0.62 v. −0.51 std. resid.); young are large (ontogenetic mass range −0.56 v. 0.10 std. resid.).

Longevity is short, even by teiid standards (−1.52 v. −0.74 std. resid.),

Holodactylus spp.	English name				Range core
			Size		
	Activity	Diet	mm	g	Ecological business model
africanus	African Clawed Gecko				East Africa
	nocturnal	invert.	71	6.6	**** nocturnal ground lizard
cornii	Somalia Clawed Gecko				Somalia
	nocturnal		77	8.5	

Homonota spp.	English name				Range core
			Size		
	Activity	Diet	mm	g	Ecological business model
andicola	Cei's Marked Gecko				Argentina
	nocturnal		42	1.93	** nocturnal ground lizard
borellii	Borelli's Marked Gecko				Argentina
	nocturnal	invert.	39	1.66	*** nocturnal ground lizard
darwinii	Darwin's Marked Gecko				Chile
	nocturnal	invert.	42	1.99	*** nocturnal ground lizard
fasciata	South America Marked Gecko				Bolivia
	nocturnal	invert.	54	3.60	*** nocturnal ground lizard
rupicola	Rock-loving Marked Gecko				Paraguay
	nocturnal		35	1.18	
taragui	Taragui Marked Gecko				Argentina
	nocturnal		39	1.65	** nocturnal ground lizard
underwoodi	Underwood's Marked Gecko				Argentina
	nocturnal	invert.	48	2.95	*** nocturnal ground lizard
uruguayensis	Uruguay Marked Gecko				Uruguay
	diurnal	invert.	45	2.46	
whitii	Argentina Marked Gecko				Argentina
	crepuscular	invert.	40	1.75	
williamsii	Williams' Marked Gecko				Argentina
	nocturnal		44	2.25	

based on *quadrilineatus* (2 years) and *undulatus* (2 years). Population density of 0.26 kg/ha (v. 1.06 kg/ha) is moderately low, based on *festivus* (*n*=3) and *quadrilineatus* (*n*=1). Endemism is low (30% v. 50%). Island forms are underrepresented (20% v. 37%). None has been introduced (v. 8.3%), and none is imperiled (v. 41%). In other respects *Holcosus* are typical teiids.

Holodactylus

Holodactylus is a two-species genus of small (7.5 g v. 13.1 g) eublepharid geckos found in East Africa. Morphologically they closely resemble other nocturnal ground lizards in the Eublepharidae, with large heads and the family's shortest tails (0.33 v. 0.76 × SVL). No tails are prehensile (v. 8.3%). They lack color change (v. 3%) and secretory pores (v. 67%). Curiously, they have the family's smallest reported ratio of maximum mass divided by minimum mass (−2.74 v. −0.67 std. resid., based on *africanus*). Of the 1090 species in the database with this mass-adjusted value, *africanus* is the sixth from lowest. This indicates that the young are exceptionally large.

Holodactylus are not sexually dichromatic (v. 11%), nor do the sexes differ in relative tail length (1.03 v. 1.21 × SVL).

H. africanus is a cruise forager (v. 56%). In other respects, they are typical eublepharids.

Homonota

Homonota is a moderate radiation (10 spp.) of uniformly small (2.0 g v. 4.7 g) nocturnal ground phyllodactylids from southern South America (Bolivia and Brazil southward). At least 40% of the species lack expanded toepads, an absence often associated with terrestriality. Most have moderately large heads, reminiscent of eublepharids.

Tails are a bit long (1.15 v. 1.03 × SVL). Color change is represented at typical frequency (10% v. 11%). SSD tends to be m<f (−0.50 v. 0.32 on −2 to 2 scale). Other sex differences are not reported. Although mostly nocturnal (80% v. 78%), diurnality (10% v. 0.8%) and crepuscularity (10% v. 2.3%) are also known. Three species have contradictory reports of their diel cycle and crepuscularity may be underreported. This diversity likely indicates some flexibility in diel activity pattern, a trait also seen in the ecologically similar sphaerodactylids. Both spontaneous calling (10% v. 11%) and stress calls (10% v. 6%) are present at typical levels.

Although most species are terrestrial (89% v. 47%), *uruguayensis* is cryptozoic (11% v. 0.9%). Two species (*fasciata*, *uruguayensis*) may spend their inactive periods in burrows. *Homonota* are found in the arid

(151–400 mm/y) portion of the rainfall spectrum, at typical elevations (mostly low, range 0–2500 m). They are absent from littoral areas (v. 20%) but otherwise present in tabulated habitats at normal frequencies.

Reproduction is normal with slightly large clutches (–0.30 v. –0.57 std. resid.); young are larger than family average (ontogenetic mass range –0.47 v. 0.11 std. resid.). Communal oviposition is slightly elevated (30% v. 16%). Parental care is not reported (v. 3.7%). Longevity is high (0.63 v. 0.15 std. resid.), equating to 5.6 years in *fasciata*, *underwoodi*, and *whitii* and a remarkable 17 years in *darwinii*.

Endemism is normal (60% v. 67%). No species occurs on islands (v. 40%). *H. fasciata* may have been introduced to Martinique; although not currently present there, the type specimen was reputed to be from Martinique. Seventeen percent of *Homonota* are considered imperiled (v. 29%), and 17% (v. 7%) of evaluated species are on one of the IUCN's lists of the most endangered (*taragui* is critically endangered). In other respects *Homonota* are typical phyllodactylids.

Homopholis spp.	English name				Range core
			Size		
	Activity	Diet	mm	g	Ecological business model
arnoldi	Bazaruto Gecko				Zimbabwe
	nocturnal	invert.	102	20.3	** trunk/wall lizard
fasciata	Banded Velvet Gecko				East Africa
	nocturnal	invert.	63	5.4	** trunk/wall lizard
mulleri	Muller's Velvet Gecko				South Africa
	nocturnal		68	6.5	* trunk/wall lizard
walbergii	Walberg's Velvet Gecko				Southern Africa
	cathemeral	invert.	95	18.0	** trunk/wall lizard

Homopholis

Homopholis (four spp.) are large (10.7 g v. 3.31 g) trunk/wall gekkonids from eastern and southern Africa (Ethiopia to South Africa, plus *walbergii* is reported for Nigeria). Their tails are relatively short (0.67 v. 1.07 × SVL) and not prehensile (v. 1.8%). They possess enlarged toepads (v. 70%), secretory pores (50% v. 46%), and may show color change (25% v. 15%). Males are smaller than females in maximum size, but dimorphism is explicitly stated for only *walbergii*: –0.50 v. –0.03 on –2 to 2 scale. The sexes differ in color in *walbergii* (25% v. 6.6%). Males have notably longer tails than females in *fasciata*, which is consistent with m<f SSD. The aggregate sex rollup score (on 0 to 6 scale) highlights these multiple differences: 1.25 v. 0.63. Despite all species being nocturnal or cathemeral, spontaneous vocalizations are not reported (v. 6.5%) but stress vocalizations are (25% v. 3.7%). Despite all species being arboreal, *fasciata* may refuge in appropriated burrows (v. 10%). Reproduction is typical for gekkonids, except that mass-adjusted incubation duration is long (1.02 v. 0.17 std. resid.); this equates to 219 days (max. 299 days) in *walbergii*, suggesting developmental adjustment to match eclosion to the best season a year later. In addition, young are notably undersized (ontogenetic size range is high: 1.00 v. –0.19 std. resid.). Maximum longevity in this species is only 5.8 years (–0.26 v. 0.15 std. resid.). *Homopholis* have not been introduced (v. 3.7%); half of evaluated species (v. 47%) are considered imperiled, though none (v. 11.3%) is on the IUCN's lists of most endangered species. In other respects *Homopholis* are typical gekkonids.

Hoplocercidae

The 19 hoplocercids are large (31 g v. 5.3 g), sit-and-wait foraging (100% v. 46%), mostly diurnal (92% v. 67%), mostly arboreal (82% v. 32%), cryptic (100% v. 43%) Neotropical (100% v. 29%) lizards (Panama to Brazil). They seem to evade the attentions of most herpetologists through simple immobility. Hoplocercids are not very well documented (cells filled: 22% v. 27%), for knowledge ranked 40th of 45 families. Hoplocercidae consists of *Enyalioides*, *Hoplocercus*, and *Morunasaurus*. Most of the cataloged ecological diversity is found in *Enyalioides*, which has the family's only diurnal canopy walkers (10% v. 6.4%), ground-foraging tree lizards (50% v. 9.2%), and immobility experts (10% v. 3.2%). Refugia-anchored diurnal ground lizards (30% v. 24.5%) are found only in *Morunasaurus*. No ecological business model can be assigned to the monotypic *Hoplocercus*, but its heavily armored clublike tail and paucity of aboveground observations suggests it may be a crack hermit.

Hoplocercid bodies differ from typical lizards in their solidity (shape index 0.94 v. 0.0 std. resid.) and crests (63% v. 5%). Tails are average in length (1.34 v. 1.37 × SVL), about half are reported to break (47% v. 57%), and none is prehensile (0% v. 5.0%). Secretory pore frequency is average (10.5% v. 15%), as is color change (16% v. 11%). Elongation is absent (0% v. 11%), as are blue tails (v. 2.3%).

Sex differences are relatively scarce

(0.58 v. 0.75 on 0 to 6 sex rollup score), being due mainly to sex differences in size (1.00 v. 0.36 on −2 to 2 scale) and dichromatism (42% v. 21%). Seasonal color is not reported (v. 6.5%). The m:f ratio of relative tail lengths is slightly male-biased (1.14 v. 1.07), but that estimate is based on only three species in one genus (*Enyalioides*).

All documented species eat invertebrates (v. 73%). Those that are not diurnal are crepuscular (7.7% v. 1.8%). Vocalizations are not reported (v. 2.5%). Direct basking is slightly underrepresented (10.5% v. 18%) and thermal conformance slightly overrepresented (5.3% v. 2.6%). Those hoplocercids that are not arboreal are terrestrial (18% v. 48%). Burrow use is documented at typical frequencies (29% v. 22%).

All documented species are in the mesic (1101–3000 mm/y) rainfall band (v. 43%), with a slight overemphasis on low elevations (94% v. 82%; medium elevation, 28% v. 41%; not found high, v. 13%). Hoplocercids are underrepresented in littoral (0% v. 17%), sandy (7% v. 22%), and human (0% v. 18.5%) habitats and overrepresented in riparian zones (40% v. 22%).

All species are oviparous (v. 82%), with a small number of eggs per clutch (−0.69 v. 0.0 std. resid.). Ontogenetic size range is unknown. Eggs are not known to be adherent (v. 2.4%), laid communally (v. 4.2%), or attended by parents (v. 2.9%). There are no data on incubation duration, longevity, or population density.

Endemism is comparatively low (37% v. 64%). Island species are infrequent (5% v. 40%), and no hoplocercids are known to have been introduced (v. 3.3%). They are not as likely as other lizards to be considered imperiled (30% v. 49%), but they appear on the IUCN's lists of the most endangered at typical rates (10% v. 13.6%).

Hoplocercus

Hoplocercus spinosus is the sole member of a hoplocercid genus endemic to Brazil and Bolivia. The species is too poorly known to guess its ecological business model. It is often found in dug burrows, with its weaponized tail outermost, but it is unclear whether it forages from the burrow (in which case it might be a crack hermit) or some other microhabitat. The species is comparatively small (16.4 g v. 31.3 g), and its unbreakable, spiny tail is very short (0.43 v. 1.34 × SVL). Crests are absent (v. 63% present), and secretory pores are present (v. 10.5% present). Rapid color change is absent (v. 11%). The sexes differ in color permanently (v. 42%) but not seasonally (v. 0%). Its diel activity pattern is listed as crepuscular but it also basks (v. 10.5%), and unequivocal observations of its activity seem to be lacking. It is found in sandy areas (v. 7%). Clutch count is a bit high (−0.35 v. −0.69 std. resid.). This species has not been evaluated for conservation status.

Hoplocercus sp.	English name				Range core
			Size		
	Activity	Diet	mm	g	Ecological business model
spinosus	Weapontail Woodlizard				Brazil
	crepuscular	invert.	84	16.4	

Hoplodactylus spp.	English name				Range core
			Size		
	Activity	Diet	mm	g	Ecological business model
delcourti	Kawekaweau				New Zealand
	nocturnal		273	442	
duvaucelii	Duvaucel's Gecko				New Zealand
	nocturnal	omnivore	136	52	*** nocturnal ground lizard

Hoplodactylus

Hoplodactylus is a diplodactylid genus of two phantoms. One species (*delcourti*) is known from only a single dry-skin specimen, which languished unrecognized for decades or more in a Marseille museum. The museum skin turned out to be the world's largest gecko; it is now believed extinct, as no subsequent evidence—either fossil or living material—has come to light from any locality. It was probably extirpated by rodents introduced to New Zealand. Its common name is the aboriginal (Maori) name given to a recently extinct arboreal lizard, presumably the extinct *H. delcourti*. The other phantom (*duvaucelii*) was described from India and attributed to a French naturalist, Alfred Duvaucel (1792–1824), which was a very misleading clue, as Duvaucel could not possibly have collected it from the wild, as he went nowhere close to New Zealand, the lizard's actual home. But *duvaucelii* disappeared from the main islands of New Zealand very early, a victim of introduced rodents. It survives today only on offshore rodent-free islets from which the rodents have either been eliminated or never reached. It is New Zealand's largest extant gecko.

H. duvaucelii (52 g) is a gentle giant that moves relatively little (maximum displacement over three years of 140 m); eats fruit, nectar, and arthropods; peacefully shares quarters with smaller geckos, as well as a mate and offspring; does not defend space; produces a small litter (mean 1.6 young) every other year or so; and produces enough progeny to replace itself by living for as long as 50 years (mass-adjusted 2.16 v. 0.90 std. resid.). Presumably the much larger *delcourti* (roughly eight times the mass) followed the same or a more languorous lifestyle. It is small wonder that they fell victim to the aggressive depredations of hyperactive introduced rodents.

Information on the natural history of *Hoplodactylus* is thus limited to what is known about *duvaucelii*. The likely modal length of *delcourti* was estimated on the basis of the one specimen, and the likely mass of both from lizards was estimated from lizards of similar build. Others have estimated maximum mass of *delcourti* at as much as 2 kg. Regardless of the actual mass, *Hoplodactylus* unquestionably has the highest mean mass of any diplodactylid genus (152 g v. 4.15 g). That is also the highest mean mass of any gecko genus. Compared to diplodactylids, *Hoplodactylus* are stout (shape index 1.04 v. 0.05 std. resid.). *H. duvaucelii* has secretory pores (v. 34%), and males are larger than females. Sexual dichromatism is not reported (v. 21%) and would not be likely in a nocturnal lizard (table 3.4, trait 16).

Duvaucel's Gecko gives stress calls (v. 5.4%) and hides in vegetation (v. 8%). It will bask if sunlight impinges on its refuge entrance (v. 17%), which may be a burrow (v. 22%). It is found in the intermediate rainfall zone (401–1100 mm/y; v. 33%). It is found in littoral areas (v. 17%), but not in sandy areas (v. 24%), riparian zones (v. 10%), or human habitats (v. 5%). Progeny are small compared to average lizards (ontogenetic mass range 0.54 v. 0.0 std. resid.), which is especially unusual within Diplodactylidae (–0.57 std. resid.). Its mean population density (based on three estimates) is a remarkable 23.3 kg/ha (v. 0.44 kg/ha). Life is good in the absence of mammals. Duvaucel's Gecko has been translocated to repopulate islands purged of introduced mammals. Both species are deemed imperiled (v. 81%), though only *delcourti* is placed in IUCN's most endangered categories (v. 29%). In other respects they are typical diplodactylids.

Hyalosaurus sp.	English name				Range core
			Size		
	Activity	Diet	mm	g	Ecological business model
koellikeri	Morocco Glass Lizard				Morocco
	cathemeral		149	11.6	** refugia-anchored diurnal gr. liz.

Hydrosaurus spp.	English name				Range core
			Size		
	Activity	Diet	mm	g	Ecological business model
amboinensis	Sailfin Lizard				Indonesia
	diurnal		259	643	**** giant herbivore
pustulatus	Philippine Sailfin Lizard				Philippines
	diurnal	herbivore	215	360	**** giant herbivore
weberi	Weber's Sailfin Lizard				Indonesia
	diurnal	herbivore	245	540	**** giant herbivore

Hyalosaurus

Hyalosaurus koellikeri is the sole representative of a genus of virtually limbless, terrestrial anguids endemic to Morocco. It is probably a refugia-anchored diurnal ground lizard (v. 34%). As expected for a limbless anguid, it is particularly slender (shape index –1.46 v. –0.60 std. resid.). The average-length tail (1.44 v. 1.66 × SVL) is not prehensile (v. 31%). The sexes are not known to differ in size, color, or morphology. Unusual for Anguidae, it is a cathemeral (v. 7%) cruise forager (v. 7%) that basks directly (v. 18%). It relies on crevices and burrows for predator concealment (v. 25%). Habitat characteristics are sandy areas (v. 15%) and riparian zones (v. 19%) but not human habitats (v. 11%).

It is oviparous (v. 39%). Although clutch count is not known, the ontogenetic mass range indicates that young are especially large (–0.97 v. 0.62 std. resid.). The documented young are 1/20 of the maximum size (v. 1/72).

Longevity is moderate for an anguid (0.47 v. 0.83 std. resid.). It is not found on islands (v. 17%), has not been introduced (v. 3.8%), and is not of conservation concern (v. 63%). In other respects it is a typical anguid.

Hydrosaurus

Hydrosaurus is a small genus (three spp.) of bizarre, very large (500 g, the highest mean in Agamidae, v. 15.6 g), heavily built (shape index 1.91 v. 0.80 std. resid.), direct basking (100% v. 24%), long-tailed (2.35 v. 1.86 × SVL) semiaquatic (100%, tied for first place, v. 0.9%) agamids found at low elevation (100% v. 83%) in the islands (100% v. 36%) of southeastern Asia. All three species are giant herbivores endemic to the Maluku Islands, Indonesia. Instead of sexual dimorphism in crest height, adult males of all species may develop sail fins on their tails, though both sexes have normal agamid nuchal and dorsal crests (100% v. 38%). *H. amboinensis* has fringed toes, probably for propulsion in water or on mud. Snout or neck spines are present on *pustulatus* and *weberi*. Little is known about their behavior or ecology. Juveniles of *pustulatus* may run bipedally on the surface of the water. *H. pustulatus* is said to live in groups (33% v. 12%), defend their nest in captivity (parental care 33% v. 3.5%), and exhibit sex differences in color (33% v. 53%). They are associated with littoral (100% v. 10%), riparian (100% v. 21%), and human (100% v. 16%) habitats but not sandy areas (0% v. 26%).

All are oviparous (100% v. 98%); their clutch counts are extraordinarily small (6.5–7.3) for their size (–0.93 v. 0.20 std. resid.). Furthermore, the young are hatched especially large (ontogenetic range –1.01 v. 0.69 std. resid.). The incubation duration of *Hydrosaurus* is conspicuously short (75 days), even by agamid standards (–1.22 v. –0.55 std. resid.); short incubations are typical of giant herbivores (table 4.1, trait 66).

Longevity is known for all three species (8.8–24 years) and is comparatively long (–0.04 v. –0.46 std. resid.). All are endemic (100% v. 49%). They have not been introduced (v. 1.7%). The flesh of *pustulatus* is highly prized for human food (this fate is likely for all three), and perhaps not coincidentally, it is imperiled (100% v. 37%); *pustulatus* is listed as vulnerable by the IUCN.

Hypsicalotes

Hypsicalotes is a single-species genus of arboreal (v. 43%) agamids found in the vicinity of Mount Kinabalu in northeastern Borneo. There are contradictory claims as to whether its range extends farther. It appears to survive by crypsis (v. 51%), including advanced color change (v. 15%), and it moves only slowly, primarily on trunks. Both sexes have typical agamid crests (v. 38%), and in addition, males have a large dewlap with an anterior fringe of pointed scales. Little else is known (22% v. 30% of cells filled).

Hypsilurus

Hypsilurus is a moderate (17 spp.) genus of large (63 g v. 16 g), stout (shape index 1.30 v. 0.80 std. resid.), arboreal (100% v. 43%) agamids long confused with *Gonocephalus*; note the similarities in common

Hypsicalotes sp.	English name		Size		Range core
	Activity	Diet	mm	g	Ecological business model
kinabaluensis	Kinabalu Crested Dragon				Borneo
	diurnal		113	48.8	** trunk/wall lizard

Hypsilurus spp.	English name		Size		Range core
	Activity	Diet	mm	g	Ecological business model
auritus	West Papua Forest Dragon				New Guinea
	diurnal		102	35.8	
binotatus	Two-marked Forest Dragon				New Guinea
	diurnal	herbivore	152	123	* giant herbivore
bruijnii	Bruijni Forest Dragon				New Guinea
	diurnal		112	46.9	
capreolatus	Parkop Dragon				New Guinea
	diurnal		100	33.0	
geelvinkianus	Geelvink Forest Dragon				New Guinea
	diurnal		81	17.3	
godeffroyi	Papua Angle-headed Lizard				Palau
	diurnal	omnivore	126	69	
hikidanus	Hikida's Forest Dragon				New Guinea
			122	62	
longi	Bismarck Angle-headed Lizard				New Guinea
		omnivore	177	197	
macrolepis	Solomons Tree Dragon				Solomon Islands
	diurnal	invert.	113	48.7	** trunk/wall lizard
magnus	Big Angle-headed Lizard				New Guinea
	diurnal		175	190	
modestus	Modest Forest Dragon				New Guinea
	cathemeral	omnivore	86	20.8	** ground-foraging tree lizard

names. They are all found on New Guinea or nearby, which probably helps explain why they are so poorly known (19% v. 30% of cells filled). Most (76% v. 38%) are crested and have long tails (3.1, tied for highest in Agamidae, v. 1.9 × SVL). Diets, microhabitats, and ecological business models are not confidently assigned. Three of the four whose diets have been sampled were omnivores (75% v. 26%). *H. godeffroyi* is the only species for which incubation duration has been reported, and it is notably short (49 days) even by agamid standards (–1.18 v. –0.55 std. resid.). The small documented clutch count of *godeffroyi* (2.3 eggs) is even more exceptional (–1.36 v. 0.20 std. resid.). Endemism is low (35% v. 49%). Island species proportion is high (100% v. 36%). Five species have been evaluated for conservation risk, and one of these (*macrolepis*) was judged to be imperiled (20% v. 37%: IUCN, near threatened).

Hypsilurus spp.	English name	Activity	Diet	Size mm	Size g	Range core	Ecological business model
nigrigularis	Black-lipped Forest Dragon	diurnal		174	185	Indonesia	
ornatus	Ornate Angle-headed Lizard			120	59.0	New Guinea	
papuensis	Papua Forest Dragon	diurnal		164	155	New Guinea	
schoedei	Vogt's Forest Dragon	diurnal		101	34.3	New Guinea	
schultzewestrumi	Nondugl Angle-headed Lizard	diurnal		128	72.8	New Guinea	
tenuicephalus	Narrow Angle-headed Lizard			117	55	Indonesia	

Iberolacerta

Iberolacerta (eight spp.) is a genus of average-size (5.7 g v. 5.1 g) diurnal, mostly terrestrial lacertids found in mesic areas (1101–3000 mm/y) from Portugal to Slovenia. Eighty percent of classifiable species are refugia-anchored diurnal ground lizards (v. 57%); the other (*horvathi*) is a ground-foraging tree lizard (20% v. 8%). Morphologically they are undistinguished lacertids, save that half (v. 13%) have blue tails as juveniles. Known sex differences are limited to dichromatism in four species (50% v. 33%) and seasonal color in *monticola* (13% v. 15%). Thus the sex differences rollup score (on scale of 0 to 6) is average (0.88 v. 0.86). They all eat invertebrates (v. 77%), by a mixture of sit-and-wait (50% v. 23%) and active (100% v. 92%) foraging. *I. monticola* moves often: it is ranked in the 94th percentile of MPM and 66th percentile in PTM. This species is said to partition diet between the sexes, with males focusing on beetles and females on softer prey. *I. horvathi* may leap to capture flying insects. Home-range area is small (–0.14 v. 0.27 std. resid.), based on *monticola*. This species may occur near the alpine tree line (0–2593 m) and yet has a mean active body temperature of 31.8°C and may suspend activity at midday to prevent hyperthermia in

Iberolacerta spp.	English name	Activity	Diet	Size mm	Size g	Range core	Ecological business model
aranica	Aran Rock Lizard	diurnal		57	4.37	Iberian Peninsula	* refugia-anchored diurnal gr. liz.
aurelioi	Aurelio's Rock Lizard	diurnal		55	4.01	Iberian Peninsula	** refugia-anchored diurnal gr. liz.
bonnali	Pyrenees Rock Lizard	diurnal		59	4.71	France	*** refugia-anchored diurnal gr. liz.
cyreni	Cyren's Rock Lizard	diurnal	invert.	72	8.7	Spain	
galani	Galan's Rock Lizard	diurnal		70	8.0	Spain	
horvathi	Horvath's Rock Lizard	diurnal	invert.	57	4.43	Central Europe	* ground-foraging tree lizard
martinezricai	Pena de Francia Rock Lizard			65	6.6	Spain	
monticola	Iberian Mountain Lizard	diurnal	invert.	66	6.8	Iberian Peninsula	** refugia-anchored diurnal gr. liz.

summer. Based on two species, the genus average mean body temperature is 30.6°C (v. 33.3°C). Direct basking is widely practiced (38% v. 25%) among the eight species.

Male *Iberolacerta* are almost as tolerant as the family norm (33% v. 41%), and male defense of space is only slightly elevated (67% v. 59%). Male hierarchy formation is also slightly high (33% v. 25%).

The elevations where *Iberolacerta* are found ranges 0–3060 m, with all species (v. 61%) found in the intermediate elevation band (1200–2400 m) and three species (*aurelioi*, *bonnali*, and *horvathi* found in the alpine. Except for sandy habitats (0% v. 44%), they are found in the usual lacertid habitat mix.

Reproduction is normal (clutch count 0.25 v. 0.13 std. resid.). *I. monticola* deposits eggs communally (v. 1.9%). Longevity is long (1.39 v. 0.19 std. resid.), based on *aurelioi* (17 years) and *monticola* (15 years). Endemism is typical (38% v. 35%); only one species (*monticola*) occurs on islands (13% v. 26%), and no species has been introduced (v. 4.6%). All evaluated species are considered imperiled (v. 56%), and four species are on the IUCN's lists of the most endangered (50% v. 15.5%). *I. martinezricai* is critically endangered due to small population size (<250 in 2016) and limited distribution (<10 km²), and *aranica*, *aurelioi*, and *cyreni* are classified as endangered.

Ichnotropis

Ichnotropis (six spp.) is a genus of average-size (4.0 g v. 5.1 g) diurnal, terrestrial lacertids found in southern Africa (Democratic Republic of the Congo to South Africa). Some or perhaps all are wide-ranging chemosensors, though ignorance is high in this group (filled cells 21% v. 33%). Three species are deeply mysterious: *chapini* is known from only three specimens; *tanganicana* is known from only a single tailless—probably immature—specimen collected over a century ago from an unspecified location; *grandiceps* has not been collected "in many years." *Ichnotropis* are morphologically indistinct, though *bivittata* and *capensis* have striking lateral jet-black lines and ventrolateral swaths of brilliant yellow and red seasonally in males (25% v. 15% seasonal color). Some species (e.g., *capensis*) have very short lifespans, such that juveniles and adults are rarely present at the same time (most adults die prior to hatching of the next generation). If *capensis* is representative (1.3 years), the size-adjusted longevity is −1.73 (v. 0.19) std. resid. Males in this species are reportedly tolerant and do not defend space. Reproduction is fairly typical for lacertids, but the clutch count is high (0.74 v. 0.13 std. resid.), and incubation is relatively long (−0.02 v. −0.36 std. resid.). Ontogenetic mass range is moderately high, indicating small hatchlings (0.35, the third highest in the family, v. −0.05 std. resid.). No species occurs on an island (v. 26%), nor has any been introduced (v. 4.6%). None is considered imperiled (v. 56%), despite the mysterious half of the genus.

Ichnotropis spp.	English name		Size		Range core
	Activity	Diet	mm	g	Ecological business model
bivittata	Angolan Rough-scaled Lizard				Congo
	diurnal	invert.	65	6.5	* wide-ranging chemosensor
capensis	Cape Rough-scaled Lizard				Southern Africa
	diurnal	invert.	56	4.10	** wide-ranging chemosensor
chapini	Equatorial Rough-scaled Lizard				Congo
	diurnal		51	3.11	
grandiceps	Caprivi Rough-scaled Lizard				Botswana
	diurnal		61	5.2	
microlepidota	Marx's Rough-scaled Lizard				Angola
	diurnal		47	2.40	
tanganicana	Tanzanian Rough-scaled Lizard				Tanzania
	diurnal		54	3.80	

Iguana (fig. 2.6C)

Iguana is a two-species genus of giant (1618 g v. 609 g) herbivores found in the mainland Neotropics and Lesser Antilles (Iguanidae). Thanks to their large size and ready availability in the pet trade, they have been studied extensively as representatives of lizards (filled cells 72% v. 43%), though they are not representative in size, diet, clutch count, or ecology. In many ways they are similar to other iguanids, except that they can be very successful in the trees on the mainland.

Iguana are stout in build (shape index 2.39 v. 2.11 std. resid.) with relatively long tails (2.35 v. 1.53 × SVL) that are readily breakable only in juveniles (v. 38%). Secretory pores are present (100% v. 38%), as are crests (v. 31%), which are greatly enlarged in dominant males. Many populations show basic color change (v. 21%), but not advanced color change (v. 2.4%). Color changing ability seems to decline with age. Despite their diurnality, they have elliptical pupils. Juveniles of both species routinely run bipedally, and all sizes swim rather than walk when possible.

Males are larger than females and some males are many times heavier than their mates (mean 1.00 v. 1.30 on –2 to 2 scale). The sexes are also dichromatic (v. 38%) and may show seasonal color (v. 12%). Males have larger heads (jowls), gular flaps, femoral pores, and nuchal, dorsal, and caudal crests. In aggregate, *Iguana* are the top-rated iguanid genus for sex differences (3.5 v. 1.9 on 0 to 6 scale).

One often reads that *Iguana* are omnivorous because *I. iguana* will eat meat in captivity. The only animal parts that have been found in wild *I. iguana* stomachs—occasional caterpillar or snail—are likely to have been inadvertently consumed with the iguana's quotidian consumption of leaves. *I. delicatissima* takes occasional carrion or immobile prey. Wild *Iguana* will travel some distance to acquire ripe mangoes and other sugary fruits, and they eagerly consume nectar-rich flowers—especially red ones such as hibiscus—but their steady diet is of leaves. Most examples of folivorous lizards are from remote islands, as is *delicatissima*, or desert areas largely devoid of mammalian predators. *I. iguana*, in contrast, thrives in the Neotropical mainland, at least in the absence of cats and dogs (and possibly aided by Pleistocene megafauna extinctions). Both species spend only a minute fraction of their time foraging, estimated at 0.2% in *delicatissima*. In contrast, they spend 10 times as much time courting and mating.

Estimation of the home range of *Iguana* is complicated by the linear distribution of *I. iguana* habitat (see below) and the very different movement patterns of the two sexes. Except for males that are defending mating territories, *Iguana* have moderately large activity ranges, up to 14,000 m^2. However, when expressed in sex-averaged mass-adjusted form, the home-range area is a moderate –0.91 (v. –1.06 std. resid.).

Due to vulnerability while on the ground, *Iguana* are conspicuously wary, with flight distances approaching or exceeding 100 m, unless they are protected from carnivores or habituated to protective humans. Nesting *I. iguana* on the mainland are a major food for jaguars, for example, and both sexes are eaten by humans in many areas, but adults are usually able to escape by retreating upwards into inaccessible vegetation or by plunging into water bodies to escape predators. As a consequence of this predilection for escape into water, *I. iguana* prefer to live near water courses, and large areas of otherwise suitable upland habitat may be unoccupied by adults. Juveniles favor dense copses of thorny plants, regardless of relationship to watercourses.

Both species bask directly. The mean active body temperature is a moderate 33.8°C (v. 34.3°C).

Iguana are not reported to be solitary (v. 8%). Groups are common (v. 42%). Outside of the mating season adults will sleep in physical contact with other iguanas. Juveniles are particularly prone to clustering. During the mating season males space out, but females may sleep in contact or very close proximity. Males defend space (v. 90%) and may be hierarchical (100% v. 25%). In some areas *Iguana* use burrows as daily refugia (100% v. 44%). One notable difference between the species is that *delicatissima* routinely refuges in (dug) burrows, whereas *iguana* do so rarely. *Iguana* are found in a remarkably broad range of rainfall conditions: *delicatissima*: 900–4000 mm/y; *iguana* 350–4600 mm/y. However, they are found only at low elevation (0–1000 m). They appear partial to littoral (100% v. 41%), sandy (50% v. 34%), riparian (100% v. 22%), and human (100% v. 24%) habitats.

Their reproduction is typical oviparity, but with remarkably large clutch counts (2.81 v. 0.45 std. resid.). The mainland species (*iguana*) has the larger mean (30.1 eggs) and maximum (90 eggs) clutch count; the island form (*delicatissima*) may suffer less juvenile mortality and has a lower mean (16 eggs) and maximum (30 eggs). Presumably, the higher predation rate on the mainland is responsible for the larger mainland clutches. In ontogenetic size range, however, they are not distinctive (–0.48 v. –0.55 std. resid.).

Incubation is very short (–1.33 v. –0.76 std. resid.), as the seasonal window suitable for incubation may be short when eggs are oviposited in sand bars during the dry season. Oviposition may be concentrated in limited areas of suitable soil (communal oviposition, 50% v. 6%). Parental care, in the form of female nest defense, is common (50% v. 31%). Female *I. iguana* may travel several kilometers—they will swim if possible—to find a suitable site and take advantage of the soil-loosening activities of their ovipositing peers or other ovipositing species such as crocodiles. *Iguana* may also obtain predator-surveillance benefits from communal nesting.

Iguana reproduction does not appear to be typical of lizards, though it is possibly typical of iguanids, in the complexity of the social aspects of mate selection. Although males may be physiologically competent at a small size (165 g) and early age (1.3 years), they have little reproductive success until they are large enough to dominate their male neighbors. Large males occupy mating territories and enjoy high levels

Iguana spp.	English name				Range core
			Size		
	Activity	Diet	mm	g	Ecological business model
delicatissima	Lesser Antilles Iguana				Lesser Antilles
	diurnal	herbivore	357	1940	**** giant herbivore
iguana	Green Iguana				Neotropics
	diurnal	herbivore	310	1350	**** giant herbivore

of polygyny. Not surprisingly, subordinate males attempt to mate when they can, obtaining copulations through force on females that leave the dominant male's territory, and very small adult males can intrude into male territories—unrecognized as males because they have female coloration and avoid head bobs—and sneak copulations when the dominant male is looking the other way. The resulting sperm competition has apparently led to very large sperm amounts being transferred by the dominant male when he copulates, which apparently causes a temporary shortage of sperm if the male has to service a large harem of highly synchronous females. Accordingly, females jockey for status in the most-subscribed territories, leading to female dominance hierarchies during the copulation season. The larger females have their choice of sleeping perches and obtain copulations when they are most receptive. Small females either wait their turn in popular breeding territories, or move to the territories of males that are less busy and usually smaller. Breeding iguanas lead socially complicated lives.

Longevity is average (0.35 v. 0.16 std. resid.), corresponding to 25–28 years. Population density is an extraordinary 96 kg/ha (v. 12.4 kg/ha). This value is not a one-off; it is based on 20 estimates for *delicatissima* (mean 129.6 kg/ha) and four estimates for *I. iguana* (mean 70.4 kg/ha). Because the latter species tends to reside in lines along the edge of watercourses and therefore has a more linear than areal home range, reliable estimates of population density have proven more difficult. Nonetheless, where unexploited by humans *Iguana* attains one of the highest, if not the highest, biomass densities among true lizards. Over much of its range it has been conspicuously depleted by human and human-augmented (especially dogs and cats) overexploitation. As for many iguanids, local human males assert that consumption of this species—primarily its eggs—cures impotence. Interestingly, it is not eaten in all parts of its range; respondents often note that only unsavory people from the next village, next town, or next country eat iguanas. Apparently impotence is a problem that only other people have. Or, the promised therapeutic benefits do not replicate well.

Endemism is moderate (50% v. 69%). Both species have been introduced (v. 33%), and *I. iguana* is the fourth-most introduced lizard (at least 84 times: table 2.4). Both species are considered to be imperiled (v. 97%) in at least some parts of their range. The genus *Iguana* is composed of only two species, which appear to have nearly identical ecological relations with the natural world. However, their relations with the human world could not be more divergent. *I. delicatissima* is among the IUCN's most endangered species (listed as endangered). Nonetheless, knowledgeable herpetologists argue that it is not adequately recognized for the severity of its conservation risk and should be classified as critically endangered. In general, *Iguana* have greatly declined or disappeared where humans have released dogs, cats, or mongoose. In a further twist of fate, one of the two species, *I. iguana*, has proven to be a prime product for the pet trade, leading it to be sold, and often established, around the globe, whereas *delicatissima* has suffered throughout its Lesser Antillean range and, in several cases, has been displaced by *I. iguana* populations introduced through the release of pets. *I. delicatissima* was originally found on about 30 islands (nine countries) but has been extirpated or lost through hybridization on about half the islands.

Iguanidae (figs. 2.6C, 4.4A, 4.4B, 4.4C, 4.4E, 4.4F)

The 42 iguanids are diurnal (100% v. 67%), giant (609 g v. 5.3 g) herbivores (83% v. 1.6% for giant herbivore business model as percentage of all lizards) found throughout the Neotropics (64% v. 29%) and southern Nearctic (33% v. 6%), plus Fiji. They are the best-known large family (cells filled: 43% v. 27%), just edging out Varanidae (42%) for that honor. As noted in chapter 2, the family Iguanidae was once more inclusive, incorporating what are now 12 families and 1173 species of lizards. This broader use of "iguanas" continues to be reflected in common names and some narratives. Furthermore, lizards worldwide are sometimes colloquially called iguanas or some corruption of that name, such as goannas. Thus care is needed to clarify what is meant by "iguana." This chapter addresses the 42 species of Iguanidae (sensu stricto): *Amblyrhynchus*, *Brachylophus*, *Conolophus*, *Ctenosaura*, *Cyclura*, *Dipsosaurus*, *Iguana*, and *Sauromalus*. Although most are giant herbivores, 12% (v. 9%) are ground-foraging tree lizards, and 5% (v. 6%) are diurnal canopy walkers.

Iguanids are the second-heaviest family (exceeded by Helodermatidae) and the family with the most robust shape index (2.11 v. 0.0 std. resid.). Iguanids have slightly long tails (1.53 v. 1.37 × SVL) that often show an ontogenetic shift from highly breakable to relatively unbreakable. On average, 38% are said to be breakable (v. 57%), but there are major differences among genera (e.g., *Ctenosaura* = 0%; *Sauromalus* = 100%). Tails are not prehensile (v. 5%). Crests are common (31% v. 5%), as are secretory pores (38% v. 15%), with major differences among genera (e.g., *Conolophus* = 0%; *Sauromalus* = 100%). Short-term color change is overrepresented (21% v. 11%), but advanced color change is not (2.4% v. 6.0%). Blue tails are not reported (v. 2.3%).

In keeping with their large size (table 3.3, trait 1), iguanids are strongly m>f in SSD (1.30 v. 0.36 on −2 to 2 scale). The sexes frequently differ in color (38% v. 21%), and seasonal color change is also overrepresented (12% v. 6.5%). Male:female ratio of relative tail lengths is female-biased (0.94 v. 1.07 ratio), as it so often is for species that are m>f in SSD. Collectively these sex differences produce a sex difference rollup score (on scale of 0 to 6) of 1.88 (v. 0.75), ranking Iguanidae as the second-highest large family (after Chamaeleonidae at 2.15).

Iguanid species are mostly herbivorous (84% v. 4%); the remainder (*Ctenosaura* only) are omnivorous (16% v. 15%). In addition, many iguanid individuals are more omnivorous as juveniles and more herbivorous as adults. All three foraging modes are well represented: sit-and-wait foraging (29% v. 46%), cruise foraging (29% v. 6.6%), and active foraging (71% v. 58%). A source of confusion here is that traditionally all iguanas (sensu lato) were glibly lumped as obligate sit-and-wait, even though herbivory is a tough proposition for lizards that must wait for the food to come to them. The more recent literature tends to add a disclaimer for herbivores, but the literature as a whole is characterized by fairly loose definitions of foraging mode, especially in the case of iguanids. One suspects that long intervals devoted to immobile digestion have been misinterpreted as sitting and waiting. Electric light use is not reported for Iguanidae (v. 1.4%), though it is hard to imagine an 800 g lizard scaling a wall in search of hibiscus flowers flitting around electric lights. Home-range areas for iguanids average quite small (−1.06 v. 0.0 std. resid.), though longer-term monitoring may show much larger areas in some species. In my experience (Rodda 1992), the daily movements of iguanids may be quite small, even though the seasonal home range may be vast, as dispersed seasonal fruit are incorporated into the diet. Male iguanids may be more sedentary in order to maintain occupancy of preferred breeding sites. A fuller understanding of home range may require stratification by size, season, and sex.

As noted above, all iguanas are diurnal, and thus one would not expect vocalization to be frequent (0% v. 2.5%). Curiously, stress vocalization is overrepresented (4.8% v. 2.2%). Iguanids may be easily stressed, as 29% (v. 8%) are judged to be conspicuously wary. It is perhaps relevant to point out that this metric is specifically wariness to humans, many of which eat iguanids. *I. iguana* can be very discriminating as regards wariness to different large animals. Adult *iguana* studied in central Venezuela (Rodda 1992) are oblivious to cattle and herons but flee in panic at the sight of humans or foxes. Flight is primarily to crevices or burrows or water (64% v. 25%) or retreat to vegetation (36% v. 23%). Nine percent—all of which are *Brachylophus*—rely on crypsis.

Fifty-five percent of iguanid species (v. 18%) bask directly; none is a thermoconformer (v. 2.6%). Mean active body temperature is a toasty 34.3°C (v. 31.1°C), presumably to expedite the otherwise slow digestion of leafy vegetables.

Iguanids are solitary at a typical frequency (8% v. 10%), but social groups are overrepresented (42% v. 20%). Only 10% of males are considered socially tolerant (v. 15%) and 90% of males defend space (v. 85%), though perhaps only breeding space. Male hierarchies are notably common (25% v. 7.7%), and female hierarchies are relatively common, though not here quantified.

The modal iguanid is terrestrial (62% v. 48%); a minority are arboreal (36% v. 32%). A few others are semi-aquatic (2.4% v. 1.1%), a habit that offers many species a degree of protection from predators and is associated with large size (table 3.9, trait 2).

Iguanids are found in every rainfall band except the hyperdry: dry (56–150 mm/y: 21% v. 12%), arid (151–400 mm/y: 14% v. 30%), intermediate (401–1100 mm/y: 50% v. 27%), mesic (1101–3000 mm/y: 50% v. 43%), and wet (>3000 mm/y: 21% v. 9%). That they are overrepresented in every occupied band except the intermediate probably in part reflects the richer information available for iguanids. The near universality of their distribution may also reflect the geographic reliability of primary production as a food source. The elevational distribution of iguanids is skewed toward lower elevations: low (100% v. 82%), medium (19% v. 41%), and high (0% v. 13%). This apparent preference for warm lowlands no doubt helps with digestion, but it may also reflect the thermal inertia operating against warming a large lizard after a cold night at high elevation.

Iguanids are not only widely spread throughout the rainfall bands, they are also widely spread in habitat types: littoral zones (41% v. 17%), sandy areas (34% v. 22%), riparian zones (22% v. 22%), and human habitats (24% v. 18.5%). The latter figure could be taken to infer that iguanids get along well with humans and their commensals, but another possibility is that in heavily forested regions, basking opportunities are enhanced in the clearings provided by humans, and human plants are highly digestible.

All iguanids are oviparous (v. 82%), with a conspicuously large raw clutch count (13.0 v. 4.0), the highest of any family. When mass adjusted however, the clutch count, still large, ranks seventh (0.45 v. 0.0 std. resid.). This mean conceals vast differences among genera: the genus *Iguana* is nearly three standard deviations above the lizard mean (2.81 v. 0.0 std. resid., or up to 90 eggs; mean of 30.1 in *I. iguana*), whereas *Amblyrhynchus* has a clutch that runs to the opposite extreme (−2.76 v. 0.0 std. resid., or mean of 2.4 eggs). The size of hatchlings shows somewhat reciprocal variation. The family mean for ontogenetic mass range is an unremarkable −0.55 (v. 0.0) std. resid., indicating moderately large hatchlings. But *Amblyrhynchus* (−1.47) and *Brachylophus* (−1.93) have very large hatchlings, whereas those of *Ctenosaura* (0.43) are small. See the genus accounts for further discussion of the major differences within the Iguanidae.

Incubation is notably short (−0.76 v. 0.0 std. resid.), ranging from −1.33 in *Iguana* to 1.90 in *Brachylophus*. Parental care is very high (31% v. 3%). These three measures create an impression that iguanids are, perhaps unconsciously, in a hurry to generate lots of progeny quickly. This implies that at least juvenile mortality is high. At least some of the adults are large at least in part because it provides them with a degree of protection against meso-predators; other advantages include digestive efficiency and sexual domination. Giant herbivores may still be vulnerable to large predators; for example, breeding female *I. iguana* are seasonally a major food for jaguars. But there are more small predators than top ones. Although this general argument applies well to iguanids as a group, there are major differences among genera in the allocation of reproductive effort: *Amblyrhynchus*, an island species that has a mean clutch count of 2.4 is larger in body size than *Iguana*, a mainland lizard that has a mean clutch count of 23.1. Incubation duration may also reflect divergent predator pressure: low in island *Brachylophus*, high in mainland *Iguana*.

RCM is normal (0.231 v. 0.226 ratio). Communal oviposition is overrepresented (14% v. 4.2%). It is unclear if there are any social aspects to communal oviposition in iguanids; for example, crowd-sourcing combined with natal philopatry—a preference for returning to the site of one's birth, which is a demonstrably successful nest site—might provide insight to avoid sites that are suitably moist or warm at the time of oviposition but flood or are shaded later in incubation. Communal oviposition may also arise from independent but instantaneous assessments; for example, when suitable substrates are scarce, as is soft, sunny soil in otherwise forested habitats. The complex seasonal progression of sun spots on the forest floor may be hard for an ovipositing iguana to predict, as are water levels, lending value to deferring to crowd-sourcing as a cue for optimal oviposition sites. Competition for these crowd-sourced oviposition sites is probably the stimulus for so much parental care, usually nest guarding, in iguanids (31% v. 3%). Unguarded nests may have their eggs dug up and exposed by subsequent nesters.

Longevity is mildly lengthy in iguanids (0.16 v. 0.0 std. resid.), ranging from −0.35 in mainland *Ctenosaura* to 1.33 in island *Conolophus*. Population density is dramatically elevated in Iguanidae (12.4 kg/ha v. 0.54 kg/ha). The latter may reflect both the greater food supply associated with eating lower on the food web, and the relative presence of iguanids on islands where predators are reduced (see below) and populations are amendable to quantification. The highest mean genus density is *Iguana* (96 kg/ha, two of three estimates from islands); the lowest from (mainland) *Ctenosaura* (0.3 kg/ha).

Sixty-nine percent of iguanids are endemic (v. 64%). Seventy-nine percent are found on islands (v. 40%). Iguanids are frequently introduced (33% v. 3%), including introductions of a species more than twice (9.5% v. 1.1%). Despite the high densities and frequent successful introductions (some for conservation purposes), virtually all iguanids are imperiled (97% v. 49%), including placement on the IUCN's lists of the most endangered (51% v. 14%). Endangerment is often due to introduction of mammalian predators to islands formerly free of these lizard predators. Rats, mongoose, dogs, cats, and humans are among the key lizard eaters. Unsupported claims of human virility enhancement by iguanid egg/follicle consumption complicate the conservation of these innocuous lizards.

Insulasaurus

Insulasaurus is a minor radiation (four spp.) of smallish (2.48 g v. 3.25 g), short-snouted, diurnal (100% v. 74%), terrestrial (50% v. 48%) or cryptozoic (50% v. 30%) skinks endemic (100% v. 64%) to the Philippines. One is recognizably a litter cryptozoid (50% v. 28%); the other classifiable species is a refugia-anchored diurnal ground lizard (50% v.30%).

They are less slender than the average skink (shape index −0.16 v. −0.46 std. resid.). The sexes are not known to differ in size, but *traanorum* sexes differ in color (25% v. 7.6%). *I. arborens* is conspicuously wary (25% v.11%). This species is a thermoconformer (25% v. 1.5%). Two of three species are found in riparian areas (67% v. 23%). *I. arborens* is oviparous (v. 64%) and has slightly small clutches (−0.37 v. 0.06 std. resid.). Population density information is limited to two samples of *arborens* (0.16 kg/ha v. 0.35 kg/ha), suggesting a fairly normal density for large islands. No *Insulasaurus* have been introduced (v. 2.1%). One-third are imperiled (33% v. 47%), though none is listed on one of the IUCN's lists of the most endangered. In other respects *Insulasaurus* are much like other skinks.

Intellagama

Intellagama is a single-species genus of very large (340 g v. 15.6 g), omnivorous (v. 26%), riparian (v. 21%) agamid found in New Guinea and along the eastern seaboard of Australia. It often sleeps on branches over water, into which it will readily plunge from any height, with a startling splash. It has a long tail (2.5 v. 1.9 × SVL) for an agamid, which may break (v. 17%), and an agamid-typical array of crests (v. 38%) and sexual dichromatism (v. 53%). It basks directly (v. 24%). It eats frogs and fish in addition to opportunistic consumption of all manner of plants and invertebrates. If will forage in the rocky intertidal as well as upland and riparian sites (v. 21%). Its clutch size averages nine, a little small for its large size (−0.24 v. 0.20 std. resid.). Female hatchlings are produced at both particularly low and particularly high incubation temperatures. Offspring are somewhat small (ontogenetic size range 1.09 v. 0.69 std. resid.). One population study found a low density (0.02 kg/ha v. 0.16 kg/ha). For its size, it lives a long time (0.88 v. −0.46 std. resid.), corresponding to 28 years. It has been introduced into South Australia (100% v. 1.7%). It has not been evaluated for conservation status.

Insulasaurus spp.	English name				Range core
			Size		
	Activity	Diet	mm	g	Ecological business model
arborens	Negros Forest Sphenomorph				Philippines
	diurnal	invert.	56	3.27	*** litter cryptozoid
traanorum	Palawan Mountain Sphenomorph				Philippines
			49	2.12	
victoria	Victoria Mountains Lizard				Philippines
			43	1.43	** refugia-anchored diurnal gr. liz.
wrighti	Wright's Mystery Sphenomorph				Philippines
			59	3.84	

Intellagama sp.	English name				Range core
			Size		
	Activity	Diet	mm	g	Ecological business model
lesueurii	Eastern Water Dragon				Queensland
	diurnal	omnivore	211	340	** ground-foraging tree lizard

Iphisa sp.	English name				Range core
			Size		
	Activity	Diet	mm	g	Ecological business model
elegans	Glossy Shade Lizard				Western Amazon
	cathemeral	invert.	52	1.80	** litter cryptozoid

Iphisa

Iphisa elegans is the sole member of this well-known (filled cells 51% v. 19%) genus of elongate Amazonian gymnophthalmids. It is a litter cryptozoid (v. 55%). It is slightly small (1.80 g v. 2.35 g) and of a slender build (shape index −0.62 v. −0.32 std. resid.). Its tail is short (1.20 v. 1.72 × SVL) and breakable (v. 28%). It may locomote without use of its legs. It possesses secretory pores (v. 29%). Males are probably larger than females, and male relative tail length is short relative to female relative tail length (0.93 v. 1.05 ratio). It also shows sexual dichromatism; hence the sex differences rollup score (on scale of 0 to 6) is relatively high (2.00 v. 0.49). Like most gymnophthalmids, it eats invertebrates and is an active forager. Unlike most gymnophthalmids, it is active at all times of day (v. 14%) and conspicuously wary (v. 7%). It is a thermoconformer with an average active temperature of about 28.0°C (v. 28.8°C). Males are socially tolerant (v. 88%). It is partial to sandy areas (v. 14%) and riparian zones (v. 29%), but not littoral (v. 3%) or human (v. 15%) habitats. Clutches are typical, though the hatchlings may remain in the shell fully developed until the eggs are disturbed, at which point they hatch explosively and the hatchlings run away. Its ontogenetic size range is small (−1.64 v. 0.62 std. resid.), indicating particularly large hatchlings. It is not endemic (v. 68%). This species is not known to have been introduced (v. 0.8%), nor has it been evaluated for conservation status.

Iranolacerta

Iranolacerta (two spp.) is a genus of poorly known (filled cells 23% v. 33%) diurnal terrestrial lacertids found in Iran and Azerbaijan, often at high elevation (up to 3200 m). *I. zagrosica* is very tentatively identified as a refugia-anchored diurnal ground lizard. Both species are the same average size (6.05 g v. 5.10 g). Their tails are short (1.68 v. 2.06 × SVL) and breakable (v. 92%). The sexes differ only in color (v. 33%). As a primary predator concealment tactic, they retreat to crevice-burrows (v. 50%). Reproduction is not documented. They are not found on islands (v. 26%), nor have they been introduced (v. 4.6%). They are not imperiled (v. 56%). In other respects they are probably typical lacertids.

Iranolacerta spp.	English name				Range core
			Size		
	Activity	Diet	mm	g	Ecological business model
brandtii	Brandt's Lizard				Iran
	diurnal		64	6.1	
zagrosica	Zagros Mountains Lacerta				Iran
			64	6.1	* refugia-anchored diurnal gr. liz.

Isopachys

Isopachys is a minor radiation (four spp.) of average-sized (3.15 g v. 3.25 g), limbless and reduced-eyed subterranean foragers (100% v. 12%) endemic to Thailand (75% v. 64%); *borealis* extends into Myanmar. They are very slender (shape index −2.22 v. −0.46 std. resid.). Tails are short by skink standards (0.71 v. 1.33 × SVL) and breakable (100% v. 73%). No sex differences are known. All eat invertebrates (100% v. 84%). They are found at low elevation (100% v. 89%) and are associated with sandy places (75% v. 25%). All documented species are viviparous (v. 36%), though litter sizes are unknown. The pressure for viviparity in a subterranean tropical lowland lizard is obscure. One would think that pushing through the soil while pregnant would be unpleasant, but perhaps they haul their embryos up to warm shallow soil during the day. Many skinks seem to readily switch their reproductive mode, but phylogeny may be more constraining in some contexts than in others. No species occurs on islands (v. 57%). None has been introduced (v. 2.1%). None is considered imperiled (v. 47%). In other respects *Isopachys* are much like other subterranean skinks.

Isopachys spp.	English name				Range core
			Size		
	Activity	Diet	mm	g	Ecological business model
anguinoides	Prachuap Snake Skink				Thailand
			77	1.37	**** subterranean forager
borealis	Northern Limbless Skink				Thailand
			136	5.0	**** subterranean forager
gyldenstolpei	Count Gyldenstolpe's Limbless Skink				Thailand
		invert.	156	8.0	**** subterranean forager
roulei	Chonburi Snake Skink				Thailand
			87	1.81	**** subterranean forager

Janetaescincus

Janetaescincus are two species of small (0.85 g v. 3.25 g), functionally limbless, nocturnal (100% v. 15%) subterranean foraging (100% v. 12%) skinks endemic to the Seychelles Islands. For subterranean foragers they are relatively robust (shape index −0.86 v. −0.46 std. resid.). For comparison, the mean shape index for all subterranean forager lizards is −2.75 std. resid. (table 4.1, trait 4). No sex differences are known. All eat invertebrates (100% v. 84%). They are found at low elevation (100% v. 89%; higher elevations are unavailable in the Seychelles). Both species are oviparous (v. 64%), though clutch counts are unknown. None has been introduced (v. 2.1%). Both are considered imperiled (v. 47%) and are on the IUCN's lists of the most endangered (100% v. 14%). In other

Janetaescincus spp.	English name				Range core
			Size		
	Activity	Diet	mm	g	Ecological business model
braueri	Brauer's Skink				Seychelles
	nocturnal	invert.	47	1.24	*** subterranean forager
veseyfitzgeraldi	Vesey's Skink				Seychelles
	nocturnal	invert.	36	0.58	*** subterranean forager or litter cryptozoid

Japalura spp.	English name				Range core
			Size		
	Activity	Diet	mm	g	Ecological business model
andersoniana	Anderson's Mountain Lizard				Himalayas
	diurnal	invert.	63	8.0	**** refugia-anchored diurnal gr. liz.
batangensis	Batang Mountain Lizard				China
			62	7.4	* refugia-anchored diurnal gr. liz.
brevicauda	Snow Mountain Lizard				China
			55	5.3	
brevipes	Short-legged Mountain Lizard				Taiwan
			62	7.4	* ground-foraging tree lizard
chapaensis	Chapa Mountain Lizard				Vietnam
			61	7.3	
dasi	Agaupani Mountain Lizard				Nepal
	diurnal		56	5.5	*** refugia-anchored diurnal gr. liz.
dymondi	Dymond's Mountain Lizard				China
	diurnal		79	16.3	* ground-foraging tree lizard
fasciata	Banded Mountain Lizard				China
			68	10.1	
flaviceps	Szechwan Mountain Lizard				China
	diurnal		75	13.8	* refugia-anchored diurnal gr. liz.
grahami	Yibin Mountain Lizard				China
			46	2.98	
hamptoni	Hampton's Mountain Lizard				Myanmar
	diurnal	invert.	65	8.8	** refugia-anchored diurnal gr. liz.
iadina	Emerald Mountain Lizard				China
			56	5.5	
kumaonensis	Kumaon Mountain Lizard				Himalayas
	diurnal	invert.	57	5.9	** refugia-anchored diurnal gr. liz.
laeviventris	Smooth-bellied Mountain Lizard				China
			61	7.2	

respects *Janetaescincus* are much like other subterranean skinks.

Japalura

Japalura is a large (32 spp.) radiation of long-tailed (2.1 v. 1.9 × SVL), long-legged, small (7.7 g v. 15.6 g; range 3.0–17.4 g) agamids in continental Asia. Only 13% (v. 36%) of species occur on any island. The long legs convey the impression of a lizard about to leap; *planidorsata* is said to hop rather than run. In all probability there are many unnamed species to be discovered, as the group shows many confusing color polymorphisms and sex differences in color (44% v. 53%), though they are morphologically fairly conservative. Most have a definite nuchal crest (50% v. 38%) and a line of pointier scales mid-dorsally, but with little other decisive structural ornamentation. The exception is *vela,* which has a low sailfin dorsal crest (extends to base of legs). As a group they are poorly known (25% v. 30% of cells filled), and the diel activity time is not established for over half of the species. Those that are known are all diurnal (v. 98%), sit-and-wait (v. 88%) invertebrate eaters (v. 48%). All seem to be good climbers, but about half are categorized as refugia-anchored diurnal ground lizards (50% v. 43%), a third ground-foraging tree lizards (35% v. 19%), and the remainder diurnal canopy walkers (10% v. 7%).

The eyes move independently in *tricarinata*. Dewlaps are reported in a few species; those in *variegata* enlarge during the breeding season. Tails rarely break (3% v. 17%). Sex size differences are minimal (0.25 v. 0.53 on –2 to 2 scale), and the sex rollup score (0–6) averages 0.9 (v. 1.6). Mass-adjusted home-range areas are small (–0.86 v. 0.35 std. resid.). Though none is known to call spontaneously, 9% (v. 1.4%, second highest of any

Japalura spp.	English name		Size		Range core
	Activity	Diet	mm	g	Ecological business model
luei	Lue's Mountain Lizard				Taiwan
			63	8.0	* ground-foraging tree lizard
major	Large Mountain Lizard				India
	diurnal	invert.	77	14.6	*** refugia-anchored diurnal gr. liz.
makii	Maki's Mountain Lizard				Taiwan
			70	11.3	* ground-foraging tree lizard
micangshanensis	Shaanxi Mountain Lizard				China
			61	7.2	* ground-foraging tree lizard
ngoclinensis	Ngoc Linh Mountain Lizard				Vietnam
	diurnal		59	6.4	
otai	Ota's Mountain Lizard				India
			50	3.87	
planidorsata	Smooth-scaled Mountain Lizard				India
	diurnal	invert.	47	3.27	**** refugia-anchored diurnal gr. liz.
polygonata	Ryukyu Mountain Lizard				Japan
			63	8.0	*** trunk/wall lizard
sagittifera	Arrow-headed Mountain Lizard				India
			54	4.83	
splendida	Splendid Mountain Lizard				China
		invert.	81	17.3	
swinhonis	Taiwan Mountain Lizard				Taiwan
	diurnal	invert.	71	11.4	* diurnal canopy walker
tricarinata	Three-keeled Mountain Lizard				Himalayas
	diurnal	invert.	52	4.28	**** ground-foraging tree lizard
varcoae	China Mountain Lizard				China
	diurnal		70	11.1	
variegata	Variegated Mountain Lizard				South Asia
	diurnal	invert.	81	17.4	**** refugia-anchored diurnal gr. liz.
vela	Sail Mountain Lizard				China
	diurnal		59	6.4	** refugia-anchored diurnal gr. liz.
yulongensis	Yulong Snow Mountain Lizard				China
			52	4.47	
yunnanensis	Yunnan Mountain Lizard				China
	diurnal		72	12.2	* diurnal canopy walker
zhaoermii	Zhao's Mountain Lizard				China
			70	11.1	* ground-foraging tree lizard

agamid genus) give stress calls. Only 3% (v. 6%) are judged conspicuously wary, though underreporting seems likely. Basking is common (60% v. 24%). Seventy-five percent (v. 51%) rely on crypsis, and the others (25% v. 29%) flee to a crevice or burrow. The mean recorded body temperature is 18.5°C (v. 34.8°C), no doubt reflecting the many high-altitude forms in the Himalayas (39% at high elevation v. 14%). The highest documented elevation is 3650 m in *major*. None is associated with either littoral (v. 10%) or sandy (v. 26%) habitats, and only 8% (v. 16%) are found around humans. Female *J. kumaonensis* have an extraordinary mating rejection posture, flipping onto their back with the male still attached; this species might be m<f in SSD, possibly facilitating this maneuver. Size-adjusted clutch count is high (0.61 v. 0.20 std. resid.). For an average-size lizard (5.9 g), *kumaonensis* has an unusually large clutch (median of 14.4 eggs). *J. variegata* (mean 17.4 g) females may defend their nest sites (parental care 3% v. 3.5%).

Longevity is average (–0.41 v. –0.46 std. resid.), but that is based on a single species (*kumaonensis*). Population density (3.5 kg/ha v. 0.16 kg/ha) is the highest mean for an agamid genus, but it is known from only one study of one species (*swinhonis*). Nearly three-quarters of the species (72% v. 49%) are endemic, but none has been introduced (v. 1.7%). One-quarter (25% v. 37%) are imperiled, though none is on one of the IUCN's lists of the most endangered (v. 5%).

Jarujinia

Jarujinia bipedalis is the sole member of its scincid genus, a smallish (2.45 g v. 3.3 g), functionally (tiny forelimbs) limbless (v. 35%) skink endemic to Thailand. It is very poorly known (filled cells 14% v. 25%). It is more

Jarujinia sp.	English name				Range core
			Size		
	Activity	Diet	mm	g	Ecological business model
bipedalis	Ratchaburi Legless Skink				Thailand
			72	2.45	

slender than the average skink (shape index −1.22 v. −0.46). It has a long tail (2.12 v. 1.33 × SVL) that is not known to be breakable (v. 73% breakable). Sex differences are not known. It is found at low elevation (v. 89%). It has not been introduced (v. 2.1%), nor does it occur on islands (v. 57%). It has not been evaluated for conservation status. In other respects it is likely to be similar to other limbless skinks.

Kaestlea

The genus *Kaestlea* is a minor radiation (five spp.) of small-sized (1.82 g v. 3.25 g), diurnal (100% v. 74%), terrestrial (100% v. 48%), oviparous (100% v. 64%) skinks endemic to India. They are some mix of litter cryptozoids (60% v. 28%) and refugia-anchored diurnal ground lizards (40% v. 30%). Their tails are long (1.98 v. 1.33 × SVL), and at least 60% are breakable (v. 73%). The tails of juvenile *laterimaculata* are bright blue (v. 4%). Most species are elongate (60% v. 35%). All are active foragers (100% v. 92%) and invertebrate eaters (100% v. 84%, though poorly documented). Electric light use is not reported (v. 0.4%) and home-range areas are unknown. They are associated with intermediate elevations (100% v. 35%) and are also present at low (60% v. 89%) and high elevations (40% v. 6%; up to 2600 m for *bilineata*). *K. beddomii* is associated with riparian zones (v. 23%). *K. bilineata* is oviparous (v. 64%), but clutch size is unknown. No species is found on an island (v. 57%). None has been introduced (v. 2.1%). Only *laterimaculata* is imperiled (IUCN, vulnerable), due to restricted distribution (only three localities). None is on one of the IUCN's lists of most endangered (v. 14%). In other respects *Kaestlea* are much like other skinks.

Kaestlea spp.	English name				Range core
			Size		
	Activity	Diet	mm	g	Ecological business model
beddomii	Beddome's Ground Skink				India
	diurnal	invert.	51	2.43	*** litter cryptozoid
bilineata	Nilgiri Hills Ground Skink				India
	diurnal	invert.	56	1.95	* refugia-anchored diurnal gr. liz.
laterimaculata	Side-spotted Ground Skink				India
	diurnal	invert.	45	1.64	* litter cryptozoid
palnica	Palni Hills Ground Skink				India
	diurnal	invert.	49	1.40	*** refugia-anchored diurnal gr. liz.
travancorica	Travancore Ground Skink				India
	diurnal	invert.	54	1.82	** litter cryptozoid

Kaieteurosaurus

Kaieteurosaurus hindsi is the sole member of a gymnophthalmid genus known from only a single wary elongate specimen found in litter in Guyana during the day. The specimen is small (0.85 g v. 2.35 g) and of a slender build (shape index −0.84 v. −0.32 std. resid.). It has secretory pores (v. 29%). It was found at 420 m elevation in a sandy forest. It is endemic (v. 68%). This species is not known to have been introduced (v. 0.8%), nor has it been evaluated for conservation status.

Kaieteurosaurus sp.	English name				Range core
			Size		
	Activity	Diet	mm	g	Ecological business model
hindsi	Kaieteur Shade Lizard				Guyana
	diurnal?		41	0.85	* litter cryptozoid

Kanakysaurus

Kanakysaurus are two species of large (5.3 g v. 3.25 g), nocturnal ground skinks (100% v. 2%) endemic to New Caledonia. Although weakly assigned to the nocturnal ground lizard ecological business model, both species are capable of daytime activity (cathemeral 100% v. 9%) but remain under cover at that time (cryptozoic 100% v. 30%). The two species are similar and hybridize. They are relatively robust (shape index 0.08 v. −0.46 std. resid.). The sexes of *zebratus* differ in color (50% v. 7.6%). They are found at low elevation (100% v. 89%). *K. viviparus* is associated with riparian zones (50% v.23%). Both species are viviparous

Kanakysaurus spp.	English name		Size		Range core
	Activity	Diet	mm	g	Ecological business model
viviparus	Live-bearing New Caledonia Skink				New Caledonia
	cathemeral		69	6.4	* nocturnal ground lizard
zebratus	Zebra-striped New Caledonia Skink				New Caledonia
	cathemeral		61	4.33	* nocturnal ground lizard

Karusasaurus spp.	English name		Size		Range core
	Activity	Diet	mm	g	Ecological business model
jordani	Northern Karusa Lizard				Namibia
	diurnal		106	33.0	*** refugia-anchored diurnal gr. liz.
polyzonus	Southern Karusa Lizard				Namibia
	diurnal	invert.	98	25.0	*** refugia-anchored diurnal gr. liz.

(v. 36%), though litter sizes are unknown. None has been introduced (v. 2.1%). Both are considered imperiled (v. 47%) and are on the IUCN's lists of the most endangered (100% v. 14%). In other respects *Kanakysaurus* are much like other skinks.

Karusasaurus

Karusasaurus is a tiny genus (two spp.) of large-sized (28.8 g v. 10.8 g) cordylids from Namibia and South Africa. With two species, species frequencies are relatively meaningless, but they fit it with the typical cordylids in business model (refugia-anchored diurnal ground lizards 100% v. 29%), tail length (1.3 v. 1.6 × SVL), presence of secretory pores (100% v. 69%), diet (invertebrates 100% v. 70%), sit-and-wait foraging (100% v. 94%), diurnality (100% v. 99%), some conspicuous wariness (50% v. 27%), dependence on crevice refugia (100% v. 20%), direct basking (100% v. 74%), terrestrial microhabitats (100% v. 76%), viviparity (100% v. 75%), endemism (50% v. 53%), and conservation risk (100% v. 79%), though they are not listed by the IUCN in their most endangered categories (v. 4.8%). *Karusasaurus* differ in shape index (0.87 v. 0.35 std. resid.), lacking breakable tails (0% v. 60%), absence of reported dichromatism (0% v. 37%), residence in littoral zones (50% v. 9%), residence in sandy areas (50% v. 3%), residence solely at low elevation (100% v. 55%), litter size (−0.97 v. −0.48 std. resid.), large size of young (ontogenetic range −1.45 v. −0.37 std. resid.), and somewhat greater longevity (0.30 v. 0.02 std. resid.). Over a moderately long life they produce neonates that are both exceptionally numerous and large. In other respects they comport with typical cordylids.

Kentropyx

Kentropyx (Teiidae) is a moderate genus (nine spp.) of slightly small (15.6 g v. 22.1 g), flight-dependent (100% v. 77%), socially tolerant (males 75% v. 55% tolerant), wide-ranging chemosensors (100% v. 97%). The genus is found in much of South America (except Chile and Uruguay) plus Trinidad and Tobago. A disjunct population in Barbados may be introduced. They stand out for leading the family in color change (11% v. 1.3%). They are found in mesic (100% v. 30%) forests at low elevation (100% v. 95%). In seasonally inundated habitats, at least *altamazonica* ascends trees rather than migrating to follow the edge of the rising waters. Perhaps it is this scansorial challenge that gives the genus the family's highest proportions of eggs that are adherent (11% v. 0.6%), and the family's second-highest proportion of eggs laid communally (11% v. 2.6%). The genus's mean RCM ratio, based on five species, is the family's highest at 0.193 (v. 0.163).

They are of average build (0.77 v. 0.88 std. resid.), tail length (2.22 v. 2.30 × SVL), and tail breakability (33% v. 21.3%). Tails are never blue (0% v. 15%). Secretory pores are absent (0% v. 9%).

Average SSD is nearly neutral (0.30 v. 0.89 on −2 to 2 scale). Sexual dichromatism is average (33% v. 35%), and seasonal color is elevated (22% v. 10%). In combination with their family-leading level of color change, these high levels of color communication suggest a genus relatively reliant on visual senses.

Half of the species are invertebrate eaters (50% v. 62%); the remainder are omnivores (50% v. 33%). In relatively small lizards, omnivory occurs often in habitats in which primary productivity, and presumably invertebrate abundance, is low, but this situation does not apply to *Kentropyx*. The small home range (−1.48 v. 0.19 std. resid.) of the genus's only documented species (*striata*) is consistent with the presumption that in such mesic areas, productivity would be relatively high. Perhaps bird predation selects for relatively restrained locomotion in the forests where *Ken-*

Kentropyx spp.	English name				Range core
			Size		
	Activity	Diet	mm	g	Ecological business model
altamazonica	Cocha Whiptail				Venezuela
	diurnal	omnivore	87	19.5	*** wide-ranging chemosensor
borckiana	Guyana Clonal Whiptail				Guiana Shield
			82	15.8	* wide-ranging chemosensor
calcarata	Striped Forest Whiptail				western Amazon
	diurnal	invert.	93	23.4	**** wide-ranging chemosensor
lagartija	Little Whiptail				Argentina
			72	10.5	* wide-ranging chemosensor
paulensis	Boettger's Whiptail				Bolivia
			68	9.1	* wide-ranging chemosensor
pelviceps	Forest Whiptail				western Amazon
	diurnal	omnivore	101	30.9	**** wide-ranging chemosensor
striata	Rain Whiptail				Venezuela
	diurnal	invert.	92	23.0	*** wide-ranging chemosensor
vanzoi	Mato Grosso Whiptail				Bolivia
			60	6.0	* wide-ranging chemosensor
viridistriga	Green-striped Whiptail				Paraguay
	diurnal		86	18.5	* wide-ranging chemosensor

tropyx live. If so, the characterization of these species as "wide-ranging" may be misleading. *K. striata* may not be representative, however, as it is in only the 20th percentile for MPM, notably lower than the genus's other documented species are (*pelviceps*, 66th; *calcarata*, 60th). *K. pelviceps* is the 82nd percentile for PTM; the other species are undocumented in this attribute. Given the wide range of movement values reported, caution is in order for characterizing the movements of *Kentropyx*.

Direct basking is low (44% v. 58%), perhaps due to the mild temperatures in lowland mesic forests. They have an average mean active body temperature (35.6°C v. 36.9°C). They are not reported in littoral areas (v. 38%) and are notably overrepresented in riparian ones (86% v. 46%). They are average to slightly underrepresented in sandy areas (29% v. 47%) and human habitats (14% v. 29%).

K. borckiana is parthenogenetic; the other species are undocumented or bisexual. Mean clutch count is comparatively high (–0.29 v. –0.51 std. resid.) and combines with the mildly small size of hatchlings (ontogenetic mass range 0.33 v. 0.10 std. resid.) to produce the family-leading RCM ratio noted in the first paragraph. Mass-adjusted incubation is lengthy (0.30, the family's third highest, v. –0.38 std. resid.), perhaps reflecting cool temperatures in the shady soil of mesic forest.

Endemism is low (11% v. 50%). Island species are infrequent (22% v. 37%). None has been introduced (v. 8%), and none is imperiled (v. 41%). In other respects *Kentropyx* are typical teiids.

Kinyongia

Kinyongia are the sit-and-wait montane chameleons of East Africa, 20 species representative of the "typical" chameleon group (see Chamaeleonidae entry: table 5.2). *Kinyongia* are an excellent match to the typical group for business model (diurnal canopy walker 100% v. 98%), mass (9.9 g v. 13.5 g), shape index (0.53 v. 0.57 std. resid.), relative tail length (1.31 v. 1.09 × SVL), crests (40% v. 36%), and arboreal microhabitat (100% v. 99%). In clutch count *Kinyongia* is close to the "typical" value, but somewhat intermediate (1.89 v. 2.81 std. resid.). There are no population density estimates for *Kinyongia*. For SSD *Kinyongia* is like *Furcifer* in being more male-biased than the average "typical" chameleon (1.22 v. 0.41 for "typicals" on –2 to 2 scale), and the mass-adjusted incubation duration is extraordinary in *Kinyongia* (4.01 v. 2.43 for "typicals" std. resid.), more so even than *Furcifer* (3.22 std. resid.). In neither of these discrepancies do these typical chameleons resemble the dwarf chameleons. The lengthy incubation of *Kinyongia* seems worthy of a targeted investigation. The relatively high elevation of *Kinyongia* may increase cloud cover and thereby lower soil temperatures, slowing development. If soil temperature is limiting, it may affect other lizards living in the same places. *Kinyongia* shares some geographic range with *Chamaeleo* and *Trioceros*. *Chamaeleo* has a notably variable incubation duration; *Chamaeleo*'s mean ratio of maximum incubation duration to minimum is 2.2, compared to the lizard mean of 1.4. But *Chamaeleo*'s incubation duration is not a particularly lengthy one (1.56 v. 2.43 std. resid. for typical chameleons), whereas *Kinyongia* has a notably lengthy one (4.01 v. 2.43 std. resid.) without having a particularly variable incubation duration (mean

Kinyongia spp.	English name				Range core
			Size		
	Activity	Diet	mm	g	Ecological business model
adolfifriderici	Ituri Dwarf Chameleon				Congo
	diurnal	invert.	51	3.70	**** diurnal canopy walker
asheorum	Mount Nyiro Bearded Chameleon				Kenya
	diurnal	invert.	75	10.7	**** diurnal canopy walker
boehmei	Taita Hills Chameleon				Kenya
	diurnal	invert.	72	9.6	**** diurnal canopy walker
carpenteri	Rwenzori Helmeted Chameleon				Congo
	diurnal	invert.	92	18.2	**** diurnal canopy walker
excubitor	Mount Kenya Sentinel Chameleon				Kenya
	diurnal	invert.	64	7.1	**** diurnal canopy walker
fischeri	Nguru Blade-horned Chameleon				Tanzania
	diurnal	invert.	124	41.0	**** diurnal canopy walker
gyrolepis	Circular-scaled Chameleon				Congo
	diurnal	invert.	63	6.7	*** diurnal canopy walker forced by deforestation to immobility expert
magomberae	Magombera Single-horned Chameleon				Tanzania
	diurnal	invert.	61	6.1	** diurnal canopy walker
matschiei	Giant East Usambara Two-horned Chameleon				Tanzania
	diurnal	invert.	130	46.8	*** diurnal canopy walker
msuyae	Mdandu Chameleon				Tanzania
	diurnal		61	6.0	** diurnal canopy walker
multituberculata	West Usambara Two-horned Chameleon				Tanzania
	diurnal	invert.	101	23.5	*** diurnal canopy walker
mulyai	Mount Nzawa Hornless Chameleon				Congo
			50	3.5	*** diurnal canopy walker
oxyrhina	Uluguru One-horned Chameleon				Tanzania
	diurnal	invert.	57	5.2	*** diurnal canopy walker
tavetana	Kilimanjaro Two-horned Chameleon				Tanzania
	diurnal	invert.	77	11.6	*** diurnal canopy walker
tenuis	Usambara Flap-nosed Chameleon				Kenya
	diurnal	invert.	59	5.7	*** diurnal canopy walker
uluguruensis	Uluguru Two-horned Chameleon				Tanzania
	diurnal	invert.	72	9.4	*** diurnal canopy walker
uthmoelleri	Hanang Hornless Chameleon				Tanzania
	diurnal	invert.	72	9.6	*** diurnal canopy walker
vanheygeni	Poroto Single-horned Chameleon				Tanzania
	diurnal	invert.	57	5.0	*** diurnal canopy walker
vosseleri	Vosseler's Two-horned Chameleon				Tanzania
	diurnal	invert.	91	17.9	*** diurnal canopy walker

ratio 1.1 v. 1.4 for lizards). *Trioceros* is neither lengthy nor variable. Chameleons have the slowest relative egg development of any large family; only the high latitude Sphenodontidae and Leiosauridae are slower. Is it part of the chameleon "Bauplan" to take advantage of the low nest mortality to complete as much development as possible prior to emergence and exposure to high juvenile mortality? Or, is there a strong seasonal advantage to hatching at a particular time of year, dictating slow or seasonally suspended development? Is slow development a plesiomorphic trait retained and difficult to reprogram among all chameleons? It is not characteristic of the arguably basal dwarf chameleons.

Kinyongia differ from chamaeleonids in their reliance on sit-and-wait foraging (100% v. 47%). They are not reported to cruise (0% v. 58%). Only 5% of *Kinyongia* are reported to bask directly (v. 19%). As noted above, *Kinyongia* are underrepresented at low elevation (60% v. 76%) and overrepresented at medium (95% v. 64%) and high (35% v. 21%) elevations. They are also underrepresented in relation to littoral (0% v. 6%), sandy (0% v. 3%), and human (11% v. 23%) habitats. Subjectively, most species seem to be tied to the canopies of large trees or tall forest. Documentation is not available on the rainfall zones in which they are found.

All documented *Kinyongia* are oviparous (v. 80%), which is a little surprising given their high elevation, slow development, and the apparent ease with which other chameleons, even those in the same geographic area (e.g., *Trioceros*), adopt viviparity. *Kinyongia* are short-lived but only slightly more so than other chameleons (−1.26 v. −0.83 std. resid.), and the *Kinyongia* sample is inadequate ($n = 2$ species).

Kinyongia are not found on islands

Kinyongia spp.	English name				Range core
			Size		
	Activity	Diet	mm	g	Ecological business model
xenorhina	Rwenzori Plate-nosed Chameleon				Congo
	diurnal	invert.	83	14.0	*** diurnal canopy walker

Kolekanos sp.	English name				Range core
			Size		
	Activity	Diet	mm	g	Ecological business model
plumicaudus	Feather-tailed Gecko				Angola
			43	1.82	* trunk/wall lizard

(v. 48%), nor have they been introduced (v. 5.4%). In all other respects they are similar to other chamaeleonids. Regrettably this includes imperilment (100% v. 95%), including the most severe IUCN categories (32% v. 25%), largely attributable to restricted geographic ranges.

Kolekanos

Kolekanos plumicaudus is the only member of this unusual gekkonid genus. It is probably a trunk/wall lizard, and it is known only from a small remote desert area of Angola. This lizard's notoriety is due to its extraordinary prehensile tail, which is very wide and flattened—like the lizard's body—and terminated in a feather-like "plume" of dense, soft spines. This small (1.82 g v. 3.31 g) lizard is associated with rock crevices and, though presumed to be nocturnal, it may be exposed under overhangs during the day. The sexes do not differ in color and both lack pores. It has not been introduced (v. 3.7%) or evaluated for conservation status. In other respects it is a typical gekkonid.

Lacerta

Until recently, *Lacerta* was a much larger genus, incorporating, among others, the 13 genera that exhibit the "——lacerta" suffix (e.g., Atlantolacerta, Iberolacerta, etc.). *Lacerta* (sensu stricto) is now an eight-species genus of large (24.2 g v. 5.1 g), diurnal, greenish terrestrial lacertids found throughout the Palearctic biogeographic realm (temperate Eurasia, from Great Britain to Mongolia). Eighty-six percent of classifiable species are refugia-anchored diurnal ground lizards (v. 57%); the other (*media*) is a ground-foraging tree lizard (15% v. 8%). Due to occurrence of many species in Europe, the genus is exceptionally well known (filled cells 60% v. 33%).

Morphologically they are relatively generalized lacertids, with a robust build (shape index 0.64 v. 0.27 std. resid.) and slightly short (1.87 v. 2.06 × SVL) breakable (v. 92%) tails that are not blue (v. 13% blue). Tail autotomy declines ontogenetically in *trilineata*. Secretory pores are common (25% v. 4%). Color change is absent (v. 2.2%).

SSD is subdued for such large lacertids (0.42 v. 0.49 on −2 to 2 scale). Sexual dichromatism is universal (100% v. 33%) and seasonal color differences are widespread (75% v. 15%). In *bilineata*, female seasonal color arises after mating and matches male breeding color, likely signaling the female's unavailability for mating. These sex differences combine into a rollup score (3.0 v. 0.86 on 0 to 6 scale) that is not exceeded within the family (tied with *Hellenolacerta* and *Zootoca*).

Omnivores have a large presence in this genus (43% v. 19%); though underrepresented, invertebrate eaters constitute the majority (57% v. 77%). Much of the vegetable matter eaten by the omnivores seems to be fruit rather than leaves. Both sit-and-wait (50% v. 23%) and active (100% v. 92%) foraging are reported. This is the only lacertid genus with electric light use (12.5% v. 0.3%). Home-range area averages the same as the family (0.30 v. 0.27 std. resid.).

Half of the species are considered conspicuously wary (50% v. 11%). The commonest primary predator concealment tactics are crevices-burrows (43% v. 50%) and vegetation (43% v. 42%), though one species (*agilis*) sometimes practices crypsis (14% v. 2.8%). Most species bask directly (75% v. 25%). Mean active body temperature is low (29.9°C v. 33.3°C), perhaps a reflection of latitude.

Social relations within *Lacerta* are diverse. Some species have tolerant males (20% v. 41%); most defend space (80% v. 59%), but male hierarchies are also widespread (60% v. 25%). Male *agilis* may defend habitual pathways independent of their own activity areas. Male *agilis* may also defend against other lacertids. Are they so amped on hormones that they can't tell the difference, or are they trying to monopolize resources shared with the other species? It is unclear whether this diversity reflects greater attention to *Lacerta*, and therefore greater recognition of multiple social arrangements, or if *Lacerta* lizards are simply more variable or perhaps more dependent on context such as density and season. Social complexity is also a hallmark of reproductive behavior; for ex-

Lacerta spp.	English name				Range core
			Size		
	Activity	Diet	mm	g	Ecological business model
agilis	Sand Lizard				Eurasia
	diurnal	invert.	78	11.0	**** refugia-anchored diurnal gr. liz.
bilineata	Western Green Lacerta				Southern Europe
	diurnal	omnivore	103	25.0	** refugia-anchored diurnal gr. liz.
media	Giant Green Lizard				Iraq
	diurnal	invert.	116	35.0	** ground-foraging tree lizard
pamphylica	South Anatolia Coast Lizard				Turkey
	diurnal		97	21.1	** refugia-anchored diurnal gr. liz.
schreiberi	Iberian Emerald Lizard				Iberian Peninsula
	diurnal	invert.	106	27.4	** refugia-anchored diurnal gr. liz.
strigata	Caspian Green Lizard				Black-Caspian Seas region
	diurnal	invert.	110	30.6	
trilineata	Balkan Emerald Lizard				Balkans
	diurnal	omnivore	96	20.0	*** refugia-anchored diurnal gr. liz. or trunk/wall lizard (elect. lights)
viridis	European Green Lizard				Eastern Europe
	diurnal	omnivore	115	34.4	** refugia-anchored diurnal gr. liz.

ample, *viridis* has complex mating strategies involving both territorial and nonterritorial males. *L. agilis* and *schreiberi* males may mate guard in addition to space defense.

Eighty percent of *Lacerta* use burrows as refugia (v. 48%). On the rainfall spectrum they have a single mode in the intermediate (401–1100 mm/y) belt (75% v. 32%) and range only from the arid (151–400 mm/y) to the mesic (1101–3000 mm/y). All species are found at low and medium elevation (i.e., within 0–2400 m). They are overrepresented in all four specialty habitat types, presumably due to greater information on *Lacerta*: littoral zones (38% v. 22%), sandy areas (50% v. 44%), riparian zones (88% v. 28%), and human habitats (88% v. 16%).

Reproduction is solely by eggs (100% v. 97%), and their size-adjusted clutch counts are the highest of any genus in the family (1.03 v. 0.13 std. resid.). *Lacerta* have the second-widest ontogenetic range of sizes, indicating exceptionally small hatchlings (1.01 v. –0.05 std. resid). Presumably they package their reproductive effort into many small eggs because predation losses are high and relatively independent of hatchling size. Incubation is average in duration (–0.34 v. –0.36 std. resid.). Parental care is overrepresented (25% v. 1.9%) and takes the form of female clutch attendance (*agilis*, *viridis*). RCM is mildly elevated (0.305 v. 0.267).

Longevity in *Lacerta* is average (0.51 v. 0.19 std. resid.), equating to a maximum of 14 years (*media*). Population density is also about average for lacertids (1.60 kg/ha v. 2.07 kg/ha), based on three species (*agilis*, *bilineata*, and *viridis*) with 33 credible samples among them. Seven of the 33 estimates were from islands (mean 0.04 kg/ha); the mainland densities (2.03 kg/ha) are relatively high densities for continental lizards. Endemism is low for these Eurasia-roaming species (13% v. 35%). For example, *agilis* is found in 33 countries. Many species are found on islands (63% v. 26%). Many have been introduced (12.5% v. 4.6%), though none has been introduced more than twice (v. 1.2%). *Lacerta* are imperiled at typical rates (63% v. 56%) and placed on the most endangered lists of the IUCN as expected (12.5% v. 15.5%). *L. viridis* is widely and incorrectly thought to be venomous.

Lacertaspis

The genus *Lacertaspis* is a minor scincid radiation (five spp.) of smallish (2.48 g v. 3.25 g) terrestrial (100% v. 48%) skinks too poorly known (filled cells 18% v. 25%) to assign ecological business models. All species are found in Cameroon, though the genus extends south to the Democratic Republic of the Congo. Their morphology is undistinguished and no sex differences are known. Their elevational profile (l/m/h: 80/60/40 v. 89/35/6) is skewed toward higher elevations (*chriswildi* reaches 2800 m). They are absent from littoral (v. 21%) and sandy (v. 25%) areas. Reproduction is not documented. Endemism overall is low (40% v. 64%). Two species are found on islands (40% v. 57%). None has been introduced (v. 2.1%). They have not been evaluated for conservation status. In other documented respects *Lacertaspis* are much like other skinks.

Lacertaspis spp.	English name				Range core
			Size		
	Activity	Diet	mm	g	Ecological business model
chriswildi	Chris Wild's Skink				Cameroon
			42	1.27	
gemmiventris	Yellow-bellied Mountain Skink				Cameroon
			67	5.9	
lepesmei	Lepesme's Mountain Skink				Cameroon
			51	2.45	
reichenowi	Reichenow's Skink				Congo
			46	1.78	
rohdei	Rohde's Skink				Cameroon
			54	2.90	

Lacertidae

The 323 lacertids are diurnal (95% v. 67%), terrestrial (90% v. 48%), fully limbed active foragers (92% v. 58%) of normal size (5.1 g v. 5.3 g) found in the Old World (through Wallace's Line). They are sometimes considered to be loose analogues of the New World's Teiidae (Zug et al. 2001). Teiids are diurnal (100%), terrestrial (98%), fully limbed active foragers (100%), but they are much larger (22g). More significantly, teiids and lacertids have a different relationship to refugia. Whereas teiids are almost universally wide-ranging chemosensors (97%), lacertids show a great diversity of roles and the modal lacertid is a refugia-anchored diurnal ground lizard (57% v. 25%). Whereas lacertids nearly universally break their tails (92% v. 57%), few teiids do so (21%). Whereas the primary predator concealment tactic in 50% of lacertids is crevices or burrows (v. 25% of lizards), only 12.3% of teiids are so classified. This burrow/crevice difference is made starker by the observation that whereas 48% (v. 22%) of lacertids refuge in burrows, 82% of teiids do so; so teiids understand the merits of getting underground. Whereas 42% of lacertids have their primary predator concealment tactic to be hiding in vegetation (v. 23% of lizards), only 7% of teiids do so.

The number of secondary traits that are a good match between lacertids and teiids is substantial: crests are 0% v. 5.3% in lacertids (0% in teiids); crypsis is rarely practiced by lacertids (3% v. 43%), as it is in teiids (1.8%); arboreality is rare in lacertids (9% v. 32%), as it is in teiids (0%); cryptozoic behavior is almost unpracticed by lacertids (0.7% v. 12.1%), as it is in teiids (0%); and sandy habitats are well frequented by lacertids (44% v. 22%), as they are by teiids (47%). Two important traits found in lacertids are equivocal in teiids: male hierarchies (25% v. 8%) are intermediate (16%) in teiids, and population density is high in lacertids (2.07 k/ha v. 0.54 kg/ha) but intermediate in teiids (1.06 kg/ha). Thus lacertids are somewhat similar to teiids, but dissimilar in important ways as well.

In addition to their modal ecological business model (refugia-anchored diurnal ground lizards), lacertids are also wide-ranging chemosensors (24.5% v. 7.4%), ground-foraging tree lizards (8.2% v. 9.2%), giant herbivores (2.4% v. 1.6%), litter cryptozoids (2.4% v. 11.6%), trunk/wall lizards (2.4% v. 13.0%), diurnal canopy walkers (1.4% v. 6.4%), gliders (1.0% v. 1.2%), and sand sharks (1.0% v. 1.9%). The diversity of ecological business models in Lacertidae is notable and not much removed from the ecological diversity of lizards as a whole. The genera included in Lacertidae are *Acanthodactylus*, *Adolfus*, *Algyroides*, *Anatololacerta*, *Apathya*, *Archaeolacerta*, *Atlantolacerta*, *Australolacerta*, *Congolacerta*, *Dalmatolacerta*, *Darevskia*, *Dinarolacerta*, *Eremias*, *Gallotia*, *Gastropholis*, *Heliobolus*, *Hellenolacerta*, *Holaspis*, *Iberolacerta*, *Ichnotropis*, *Iranolacerta*, *Lacerta*, *Latastia*, *Meroles*, *Mesalina*, *Nucras*, *Omanosaura*, *Ophisops*, *Parvilacerta*, *Pedioplanis*, *Philochortus*, *Phoenicolacerta*, *Podarcis*, *Poromera*, *Psammodromus*, *Pseuderemias*, *Scelarcis*, *Takydromus*, *Teira*, *Timon*, *Tropidosaura*, *Vhembelacerta*, and *Zootoca*.

Lacertids are slightly more heavily built than is an average lizard (shape index 0.27 v. 0.0 std. resid.). Tail length is long (2.06 v. 1.37 × SVL). A few tails are prehensile (0.6% v. 5.0%). Secretory pores are comparatively rare (4.0% v. 14.9%). Anatomists will retort that many lacertids show pores of some stripe, but these pores are often tiny and not sexually dimorphic (my tentative criteria to demonstrate pore activity). This assessment of pores is admittedly weak, as is the scientific community's understanding of their role; much better quantification of pores and secretions would be highly desirable. Only one lacertid is elongate (*Nucras lalandii*) (0.3% v. 11.4%). Basic color change is rare in Lacertidae (2.2% v. 10.7%); advanced color change even rarer (0.9% v. 6.0%).

Lacertids are a mix of m>f and m<f SSD that closely matches that of lizards (0.49 v. 0.36 on −2 to 2 scale). Sexual dichromatism is elevated in Lacertidae (33% v. 21%), as is

seasonal color (14.6% v. 6.5%). Blue tails are far more frequent in lacertids (13.3% v. 2.3%). The ratio of m:f relative tails lengths matches lizards exactly (1.07 v. 1.07). Overall, the sex differences rollup (on scale of 0 to 6) is only slightly elevated in lacertids (0.86 v. 0.75).

Lacertids show all possible diet classes. Perhaps due to their average size they are light on carnivores (0.6% v. 3.2%) and herbivores (1.7% v. 4.1%), about average for invertebrate eaters (77% v. 73%), and a trifle heavy on omnivory (19% v. 15%). There are comparatively few ant specialists among the Lacertidae (1.7% v. 4.4%). As noted above, they are dominated by active-foraging tactics, though sit-and-wait finds a place (23% v. 46%), as does cruise foraging (2.4% v. 6.6%). Electric light use is represented by only one species (0.4% v. 2.0%), *Lacerta trilineata*. Home-range area is modestly large (0.27 v. 0.0 std. resid.), though based on only 14 of 323 species.

Three diel activity patterns are present in Lacertidae; no nocturnal species have been reported (v. 25%). Species active at any time of day (cathemeral) are rare (0.4% v. 6.8%), but crepuscular species are overrepresented (4.4% v. 1.8%). For a group that is overwhelmingly diurnal, spontaneously calling is reported surprisingly often (2.8% v. 2.5%), as are stress calls (1.5% v. 2.2%).

Wariness (10.8% v. 8.2%) is present at typical levels. Most primary predator concealment tactics were outlined above. Other tactics include flight (22.2% v. 10.8%) and sand diving (6.9% v. 5.8%).

Direct basking is slightly more frequent than among lizards (25% v. 18%); thermoconforming is hardly present (0.6% v. 2.6%). Mean active body temperature is a warm 33.3°C (v. 30.8°C). At least six lizard species are known to alternately lift diametrically opposed limbs, apparently to reduce heat gain from hot substrates; all six are in Lacertidae (see the accounts *Acanthodactylus, Meroles, Mesalina*).

Solitary lacertids are not reported (0% v. 10%), and social groups are overrepresented (25% v. 20%). Forty-one percent of males are considered socially tolerant (v. 15%), and 59% of males defend space (v. 85%), though perhaps only breeding space. Male hierarchies are notably common (25% v. 7.7%), as fits in with the high density and elevated frequency of group living.

Lacertids are found in every rainfall band, though slightly underrepresented in the wettest classes: hyperdry (4% v. 3%), dry (56–150 mm/y: 30% v. 12%), arid (151–400 mm/y: 40% v. 30%), intermediate (401–1100 mm/y: 32% v. 27%), mesic (1101–3000 mm/y: 34% v. 43%), and wet (>3000 mm/y: 6% v. 9%). The elevational distribution of lacertids is skewed slightly toward higher elevations: low (87% v. 82%), medium (61% v. 41%), and high (20% v. 13%). Lacertids are not only widely spread throughout the rainfall bands, they are also widely spread in specialty habitat types: littoral areas (22% v. 17%), sandy areas (44% v. 22%), riparian zones (28% v. 22%), and human habitats (16% v. 18.5%).

Almost all lacertids are oviparous (97% v. 82%; some of the others express both modes), with an average mass-adjusted mean clutch count (0.13 v. 0.0 std. resid.) and slightly short incubation period (−0.36 v. 0.0 std. resid.). Hatchlings are of average size, as indicated by Lacertidae's normal ontogenetic range size (−0.05 v. 0.0 std. resid.). Eggshells are hard in a minority of species (14% v. 41%). Adherent eggs are unknown (v. 2.4%); communal oviposition is rare (1.9% v. 4.2%). A number of lacertids are parthenogenetic (7 of 31 in *Darevskia*). Parental care is comparatively rare (1.9% v. 2.9%). RCM is moderately heavy (0.267 v. 0.226), a curiosity for an active forager.

Longevity is short in lacertids (−0.32 v. 0.0 std. resid.). The longest absolute longevity in Lacertidae is 45 years in *Gallotia simonyi*, a giant herbivore of the Canary Islands. The second highest (28 years) is *Timon lepidus*, a large (120 g) but more typical lacertid. As mentioned above, population density in Lacertidae is mildly elevated (2.07 kg/ha v. 0.54 kg/ha). The high value is notable in this case because of the lack of herbivores and paucity of island species and the relatively good documentation (234 estimates of 28 species).

The Lacertidae are dominated by species from the Palearctic biogeographic realm (65% v. 9%); they are also overrepresented in the Ethiopian realm (35% v. 21%). Although they reach Indo-Pacific Rim nations in several parts of Asia, they are not numerous there (Oriental: 8% v. 18%). Thirty-five percent of lacertids are endemic (v. 64%). Twenty-six percent are found on islands (v. 40%). Several have been introduced (4.6% v. 3.3%), including introductions of a species more than twice (1.2% v. 1.1%). Most of the introductions derive from Europe, which is heavily involved in international commerce, including herpetoculture. Despite the high densities, lacertids are overrepresented among the imperiled (56% v. 49%), including placement on the IUCN's lists of the most endangered (15.5% v. 13.6%).

Lacertoides

Lacertoides pardalis is the sole member of its scincid genus, a large (11.5 g v. 3.3 g), diurnal (v. 73%), suspected crack hermit (v. 1.7%) endemic to New Caledonia. It is more robust than the average skink (shape index

Lacertoides sp.	English name				Range core
			Size		
	Activity	Diet	mm	g	Ecological business model
pardalis	New Caledonian Leopard Skink				New Caledonia
	diurnal		82	11.5	* crack hermit

Laemanctus spp.	English name				Range core
			Size		
	Activity	Diet	mm	g	Ecological business model
longipes	Smooth Casque-headed Basilisk				Northern Central America
	diurnal	invert.	120	52	*** diurnal canopy walker
serratus	Serrate Casque-headed Basilisk				Guatemala
	cathemeral	invert.	133	72	* ground-foraging tree lizard

0.36 v. −0.46). It has a long tail (2.25 v. 1.33 × SVL) that is not known to be breakable (v. 73% breakable). Sex differences are not known. It basks directly (v. 21%) and uses burrows or crevices (v. 24%) for its daily refuge. In fact, it has not been observed outside of its burrow entrances. It is found at low (in this case, 250–900 m) elevation (v. 89%). Reproduction is undocumented. It has not been introduced (v. 2.1%), but it is imperiled (v. 57%). It is not on any of the IUCN's lists of the most endangered. In other respects it is likely to be similar to other skinks.

Laemanctus

Laemanctus is a small (two spp.) genus of diurnal arboreal corytophanids that have "casques"—looks somewhat like a bicycle racer's helmet—on their heads rather than sailfins (*Basiliscus*) or "helmets"—nuchal crests bifurcating anteriorly (*Corytophanes*). They are of average size for corytophanids (61 g v. 67 g) and average build for corytophanids (shape index 1.17 v. 1.20 std. resid.). They are found in Central America and are either diurnal canopy walkers (*longipes*) or ground-foraging tree lizards (*serratus*). Both sexes are predominantly green and have very long tails (3.35 v. 2.75 × SVL) that do not readily break. At least one species shows color change (50% v. 22%). *Laemanctus* have not been reported to be sexually dichromatic (0% v. 22%) or to show seasonal color (0% v. 11%).

Both *Laemanctus* are invertebrate eaters (v. 50%), reliant on crypsis and sit-and-wait foraging (v. 80%). Social and thermal behavior are undocumented. They have not been documented in littoral zones (v. 11%), sandy areas (v. 22%), riparian zones (v. 44%), or human habitats (v. 33%).

Both species are oviparous (v. 89%), with a small clutch count (−0.47 v. −0.22) and the family's largest young (ontogenetic range 0.27 v. 1.16 std. resid.), corresponding to hatching at 1/103 of maximum size rather than 1/230. Incubation period is slightly short (−0.83 v. −0.52 std. resid.). The eggs of one species (*longipes*) are documented to hatch in response to disturbance: all development is completed and the hatchlings remain inside the shell unless disturbed by an egg predator, in which case they emerge explosively and immediately run away. Parental care is not reported (v. 11%). Longevity is average (−0.07 v. −0.28 std. resid.). No *Laemanctus* occurs on an island (v. 22%) or has been introduced (v. 33%). *L. serratus* is specially protected in Mexico. In other respects they are typical corytophanids.

Lamprolepis

Lamprolepis is a small genus (three spp.) of medium-large (7.4 g v. 3.3g), diurnal (100% v. 75%), arboreal (100% v. 10%) skinks found in Indonesia, Borneo, New Guinea, Australia, and throughout the Pacific. The genus is easily divided into the island endemic forms (*leucosticta* is found only on Java; *nieuwenhuisii* is found only on Borneo) and *smaragdina*, which is found on at least 312 named Indo-Pacific islands (but not Java or Borneo). The latter is a trunk/wall lizard (33% v. 3%); the endemics are thought to be diurnal canopy walkers (67% v. 2%). Most information is derived from *smaragdina* (filled cells mean for genus: 38% v. 25%).

Lamprolepis are morphologically similar to *Dasia*, with which they were once joined but are now understood to be an example of convergence rather than shared ancestry. *Lamprolepis* are solidly built (shape index 0.19 v. −0.46 std. resid.) and have ordinary length (1.44 v. 1.33 × SVL) breakable tails (v. 73%) that are not blue (v. 4% blue).

Despite their large size, *Lamprolepis* tend to have smaller males than females (based on *smaragdina*). That species also exhibits sexual dichromatism (33% v. 7.6%) and seasonal color (33% v. 6%). The m:f ratio of relative tail lengths is notably skewed toward longer female tails (0.90 v. 1.03 ratio), a fairly unusual condition (within the 16 documented skink genera, only two are lower).

Lamprolepis spp.	English name		Size		Range core
	Activity	Diet	mm	g	Ecological business model
leucosticta	White-spotted Tree Skink				Indonesia
	diurnal		62	4.67	* diurnal canopy walker
nieuwenhuisii	Nieuwenhuis' Tree Skink				Borneo
	diurnal		61	4.33	* diurnal canopy walkers
smaragdina	Emerald Treeskink				New Guinea
	diurnal	omnivore	99	20.0	**** trunk/wall lizard

Note that the lizard generalization of an inverse relationship between SSD and m:f relative tail length ratio (table 3.3, trait 8) does not apply to skinks. The sex differences rollup score (on a scale of 0 to 6) is notably high for *Lamprolepis* (1.00 v. 0.28).

The diet is known only for *smaragdina*, which is fairly broad minded: leaves, fruit, nectar, soft invertebrates, arthropods, ants, wasps, beetles, orthopterans, other insects, ectothermic vertebrates, and, in captivity at least, immobile endotherms. It forages actively (78th percentile of MPM; 59th percentile of PTM), including at electric lights (33% v. 0.4%). Although its home range has not been quantified in planar area, it generally incorporates one to two (occasionally three) trees for both sexes, primarily the trunks. It basks directly (v. 21%) and has a normal mean active body temperature (30.5°C v. 31.1°C).

Social relations are known only for *smaragdina*, which may or may not defend space: the uncertainty may reflect geographic variation in this trait. At low density, spatial defense may be hard to detect (see density comments below).

L. smaragdina is found in mesic and wet areas (in this case 1900–4800 mm/y). The genus is found at low (67% v. 89%) and intermediate (33% v. 35%) elevations. It is associated with human habitats (33% v. 14%).

It is oviparous (v. 64%) and has a notably small clutch for its size (–0.93 v. 0.06 std. resid.). It may lay once or twice a year, so the annual reproductive output is small in number. The hatchlings are relatively small (ontogenetic mass range 0.38 v. 0.12 std. resid.). Therefore, the young are both small in size and few in number. Perhaps juvenile mortality is relatively low. Incubation in *smaragdina* is short (–0.62 v. –0.40 std. resid.), equating to 59 days. Hatchlings may emerge explosively; *smaragdina* hatchlings may emerge nearly instantaneously in response to disturbance.

Longevity is known only for *smaragdina*, which is short-lived (5 years, or –0.48 v. 0.17 std. resid.). This suggests that lifetime reproductive output is fairly low. Population density, based on seven Philippines plus three Marianas estimates for *smaragdina*, is high (1.15 kg/ha v. 0.35 kg/ha). There is substantial variation (0.1 to 2.4 kg/ha) among the estimates; therefore, territorial social pressure may be highly variable. *L. smaragdina* has been introduced a number of times in the Pacific; estimates of number of introductions vary widely. No *Lamprolepis* are imperiled (v. 47%). In other respects *Lamprolepis* are like typical skinks.

Lampropholis

Lampropholis is a moderate radiation (11 spp.) of small (1.20 v. 3.25 g), mostly elongate (64% v. 35%), diurnal (100% v. 74%), direct basking (100% v. 21%) terrestrial (100% v. 28%) skinks found in coastal eastern Australia. Most are refugia-anchored diurnal ground lizards (82% v. 30%); the remainder are litter cryptozoids (18% v. 28%).

Compared to the average skink they are slightly more slender (shape index –0.68 v. –0.46 std. resid.). They have breakable tails (100% v. 73%) that are slightly long (1.48 v. 1.33 × SVL).

In a minority of species the sexes differ in color (18% v. 7.6%, with geographic variation), and seasonal color is found in 9% (v. 6%). Overall their sex differences rollup score (on scale of 0 to 6) is normal (0.27 v. 0.28).

All are generalized invertebrate eaters (v. 84%). Only *delicata* has been characterized by foraging mode; it exhibits both sit-and-wait (100% v. 16%) and active (100% v. 92%) foraging. Electric light use has not been observed (v. 0.4%), and home-range area is not recorded. Eighteen percent (v. 11%) of species are considered conspicuously wary.

Mean active body temperature is cool (27.8°C v. 31.1°C), albeit based solely on *guichenoti*. Although all species are "coastal" in general distribution, this includes species that rise to 1455 m in the coastal ranges. Elevational profile is (l/m/h) 100/100/0 v. 89/35/6. They are lacking from sandy areas (0% v. 25%), but they are otherwise normal in habitat distribution. Soil cracks are used for daily refugia by some *delicata*. *L. guichenoti* may bask communally. Its winter hibernacula may include as many as 50 lizards.

All are oviparous (v. 64%). Clutch count is moderately elevated (0.54 v. 0.06 std. resid.). The hatchlings are the fourth smallest in Scincidae (ontogenetic mass range 1.58 v. 0.12 std. resid.). RCM is the third highest in skinks (0.388 v. 0.236). This suggests strong selection pressure for fecundity and may be connected to the small size of offspring. Incubation is long for a skink but normal for a lizard of this size (0.02 v. –0.40 std. resid.), corresponding to 47–50 days. At the end of incubation, some *delicata* eggs hatch explosively. That is, embryos complete development and wait in the egg unless they are disturbed, at which point they blast out of their shell and run away in a few seconds. Communal oviposition is rife in *Lampropholis* (36% v. 2.6%), an unusual condition in refugia-anchored diurnal ground lizards (3.5%: table 4.1, trait 68). Parental care is unknown (v. 3.2%). Could a mother *Lampropholis* find her eggs in a communal aggregation? Mating balls are uncommon in lizards (they seem to be more common in snakes); mating balls form when a high number of males are physically scrambling to mate simultaneously with a receptive female. In *guichenoti*, as many as 18 lizards have been counted in a mating ball. As in snakes, this is probably related to the high number of individuals in a single hibernaculum. A male waiting outside the hibernaculum has a shot at mating with every female as she emerges, and the male faces an equally impressive array of sexual competitors. The end result, in many cases, is a mating ball.

Longevity is based on *delicata* (2 years) and *guichenoti* (5 years) and is short for a lizard of this size (–0.20 v. 0.17 std. resid.). Credible population density estimates are not available for *Lampropholis*. Endemism is a little high (73% v. 64%). Island species frequency is low (18% v. 57%). Many *Lampropholis* have been introduced (9.1% v. 2.1%), including those species introduced more than twice (9.1% v. 0.7%), in this case eight times. The repeat offender is *delicata,* which has famously disrupted lizard assemblages in New Zealand, Hawaii, and Australia's Lord Howe Island. All *Lampropholis* are imperiled (v. 47%), but none is on one of the IUCN's lists of most endangered (v. 14%). In all other documented respects *Lampropholis* are typical skinks.

Lampropholis spp.	English name				Range core
			Size		
	Activity	Diet	mm	g	Ecological business model
adonis	Diamond-shielded Sunskink				Queensland
	diurnal	invert.	46	1.14	* litter cryptozoid
amicula	Friendly Sunskink				Queensland
	diurnal	invert.	33	0.44	** refugia-anchored diurnal gr. liz.
caligula	Montane Sunskink				New South Wales
	diurnal	invert.	49	1.33	* litter cryptozoid
coggeri	Rainforest Sunskink				Queensland
	diurnal	invert.	40	1.10	* refugia-anchored diurnal gr. liz.
colossus	Bunya Sunskink				Queensland
	diurnal	invert.	51	2.41	* refugia-anchored diurnal gr. liz.
couperi	Plain-backed Sunskink				Queensland
	diurnal	invert.	45	1.68	* refugia-anchored diurnal gr. liz.
delicata	Dark-flecked Garden Sunskink				Eastern Australia
	diurnal	invert.	45	1.05	*** refugia-anchored diurnal gr. liz.
elongata	Elongate Sunskink				New South Wales
	diurnal	invert.	49	0.95	* refugia-anchored diurnal gr. liz.
guichenoti	Pale-flecked Garden Sunskink				Eastern Australia
	diurnal	invert.	44	1.04	** refugia-anchored diurnal gr. liz.
mirabilis	Saxicoline Sunskink				Queensland
	diurnal	invert.	47	1.87	** refugia-anchored diurnal gr. liz.
robertsi	Grey-bellied Sunskink				Queensland
	diurnal	invert.	48	1.25	** refugia-anchored diurnal gr. liz.

Lankascincus

Lankascincus is a moderate radiation (10 spp.) of small-sized (1.57 g v. 3.3 g), mostly elongate (60% v. 35%), litter cryptozoid (100% v. 28%) skinks endemic to Sri Lanka.

They tend to have short snouts but are generally shaped like a typical skink (shape index –0.57 v. –0.46 std. resid.). Tails are exactly average in length (1.33 v. 1.33 × SVL), and most tails are breakable (60% v. 73%). The somewhat scansorial *taprobanensis* has a prehensile tail (10% v. 0.7%). Blue tails are not reported (v. 0.4%)

The sexes often differ in color (40% v. 7.6%), and males of *fallax* are seasonally different in color (10% v. 6.2%).

All appear to be generalized invertebrate eaters (v. 84%). There is no information on electric light use, foraging mode, or home-range area.

Diel activity patterns are equally split between diurnal (50% v. 74%) and crepuscular (50% v. 2.4%). Direct basking is uncommon (10% v. 21%). Social relationships are poorly documented, with only *fallax* males clearly expressed (they defend space, 100% v. 94%).

Described habitats cluster in the mesic portion of the rainfall spectrum (intermediate, 25% v. 34%; mesic, 100% v. 40%). *Lankascincus* are overrepresented in the medium elevational band (67% v. 35%) and in human habitats (29% v. 14%). They are not reported to occur in littoral (v. 21%) or sandy (v. 25%) areas. They inhabit riparian zones at a normal frequency (29% v. 23%).

All species are oviparous (v. 64%) with a normal clutch count (0.03 v. 0.06 std. resid.).

No *Lankascincus* have been introduced (v. 2.1%). All evaluated species are imperiled (100% v. 47%), including 40% (v. 14%) on lists of the IUCN for the most endangered species. In other respects *Lankascincus* are typical skinks.

Lanthanotidae and *Lanthanotus*

Lanthanotus borneensis, a large (64 g v. 5.33 g) semiaquatic lizard endemic to the island of Borneo, is the sole species in Lanthanotidae. As it has rarely been observed in the wild, it is poorly known behaviorally and ecologically, though well studied morphologically (filled cells 48% v. 27%). The majority of undisturbed wild sightings have taken place on the ground at night in light rain, suggesting a nocturnal ground lizard that routinely flees to the protection of darkness or nearly streams.

Reports of maximum SVL (up to 400 mm, a suspiciously round number) are inconsistent with a mean SVL from field measurements of 188 mm, but there is no doubt that this lizard can reach large sizes in captivity. That its modal length is less than its mean length is an unusual pattern found primarily in very large lizards, especially the closely related varanids. It is moderately robust (shape index 0.32 v. 0.00 std. resid.), elongate (v. 11%), and has a relatively short tail (1.09 v. 1.37 × SVL), which is somewhat prehensile (v. 5%) and breaks but does not regenerate. Blue tails are not found in *Lanthanotus* (v. 2.3%). *Lanthanotus* also lacks any crests (v. 5.3%), secretory pores (v. 15%), or color change (v. 11%). It does shed its skin in one piece, as do snakes. Like snakes, its lower jaw is hinged in the middle.

Based on measurements of field-caught lizards, the ratio of mean adult m:f SVLs is 1.16, an SSD consistent with this lizard's size (1.0 v.0.36 on –2 to 2 scale). Also consistent with the m>f difference is body length, and mass is a m<f difference in m:f ratio of relative tail length (0.94 v. 1.07). The sexes do not differ in constant color (v. 21%) or seasonal colors (v. 6.5%), but males do possess blunter jaws, broader heads, and a broader tail base. The head shape differences suggest that intermale competition for mates may take place, or less likely, there are sexual differences in diet or prey capture technique. Added together these sex differences result in a sex rollup score of 3.0

Lankascincus spp.	English name / Activity	Diet	Size mm	Size g	Range core / Ecological business model
deignani	Deignan's Lankaskink				Sri Lanka
	diurnal	invert.	51	1.56	*** litter cryptozoid
deraniyagalae	Deraniyagala's Lankaskink				Sri Lanka
	diurnal	invert.	53	1.70	**** litter cryptozoid
dorsicatenatus	Catenated Lankaskink				Sri Lanka
			42	1.34	*** litter cryptozoid
fallax	Common Lankaskink				Sri Lanka
	crepuscular	invert.	50	1.47	*** litter cryptozoid
gansi	Gans' Lankaskink				Sri Lanka
	crepuscular	invert.	50	2.36	**** litter cryptozoid
greeri	Greer's Lankaskink				Sri Lanka
			52	2.54	
munindradasai	Munindradasa's Lankaskink				Sri Lanka
			41	1.17	** litter cryptozoid
sripadensis	Sripada Range Lankaskink				Sri Lanka
			54	1.77	** litter cryptozoid
taprobanensis	Smooth Lankaskink				Sri Lanka
	diurnal	invert.	50	1.44	*** litter cryptozoid
taylori	Taylor's Lankaskink				Sri Lanka
	crepuscular	invert.	43	0.96	*** litter cryptozoid

(v. 0.75 on 0 to 6 scale), ranked third among all families, exceeded only by Sphenodontidae and Crotaphytidae. The heavy scaring of adult males, and only adult males, is taken to affirm the likelihood of intermale intolerance.

Lanthanotus eats soft invertebrates, arthropods, snails, and fish. The relative proportions of these are not established, nor is the movement pattern through which they are obtained. Home-range area is unknown.

Although *Lanthanotus* is not known to call spontaneously (v. 2.5%), it does give a cry when distressed (v. 2.2%). It is conspicuously wary (v. 8.2%) and cryptic (v. 43%). Its mean active body temperature is a cool 26.3°C (v. 31.3°C), strongly tied to the water temperature in the streams where it is found. In addition to seeking temporary refuge in water, it is reported to spend days in rock crevices, under vegetation or rocks, and in unspecified burrows. The documented elevational band is extremely narrow and low (178–197 m).

Reproduction is by soft-shelled eggs in a small clutch of about four; mass-adjusted count is very low (−0.89 v. 0.0 std. resid.). The eggs are about 30 mm in diameter.

Longevity is relatively short (−0.26 v. 0.0 std. resid.), which equates to 7.8 years. Population density is unknown. This lizard is not endemic to a country; the occupied portion of the island of Borneo is divided between Malaysia and Indonesia. It has not been introduced (v. 3.3%). It is imperiled (v. 49%), though not yet evaluated by the IUCN.

Lanthanotus sp.	English name				Range core
			Size		
	Activity	Diet	mm	g	Ecological business model
borneensis	Borneo Earless Monitor				Borneo
	nocturnal	invert.	170	64	** nocturnal ground lizard

Larutia spp.	English name				Range core
			Size		
	Activity	Diet	mm	g	Ecological business model
larutensis	Larut Hills Larut Skink				Malaysia
	diurnal	invert.	146	13.9	
miodactyla	Titiwanga Skink				Malaysia
			117	8.1	
nubisilvicola	Cloud Forest Skink				Thailand
			94	4.73	* litter cryptozoid
penangensis	Pinang Island Larut Skink				Malaysia
			100	2.50	* litter cryptozoid
puehensis	Berumput Two-toed Skink				Malaysia
			110	6.9	
seribuatensis	Tulai Two-lined Skink				Malaysia
			92	4.39	*** sand shark
sumatrensis	Sumatra Two-toed Skink				Indonesia
			135	11.5	
trifasciata	Three-collared Skink				Malaysia
			188	25.9	* litter cryptozoid

Larutia (fig. 4.12A)

Larutia is a moderate radiation (eight spp.) of large (7.7 g v. 3.3 g), functionally limbless skinks from Southeast Asia (from Thailand south to Borneo and Sumatra). Most are too poorly known to assign to ecological business models, but of the four that are, 75% are litter cryptozoids (v. 28%) and one is a sand shark (25% v. 7%).

They are very slender (shape index −1.31 v. −0.46 std. resid.), but unlike most subterranean forms they have long tails (1.44 v. 1.33 × SVL), and most tails are breakable (75% v. 73%). Sex differences are not reported.

Diet is known only for *larutensis* (generalized invertebrate eater, 100% v. 84%). This species is the only one with a diel activity assignment (diurnal, 100% v. 74%). Three species are conspicuously wary (38% v. 11%), as is so often the case in litter cryptozoids (table 4.1, trait 35). Direct basking is noted (13% v. 21%). Surface activity often follows rain (*larutensis, miodactyla, seribuatensis, trifasciata*).

Their elevational profile is near normal (l/m/h: 88/50/0 v. 89/35/6). They are found in littoral areas (13% v.21%) but are absent from sandy areas (v. 25%), riparian zones (v. 23%), and human habitats (v. 14%). One species (*trifasciata*) is associated with strawberry fields, perhaps because regularly plowing loosens the soil (temporarily, not forever).

All are oviparous (v. 64%), but clutches are very small (−0.89 v. 0.06 std. resid.).

All species are endemic (v. 64%). Half are found on islands (v. 57%). None has been introduced (v. 2.1%), and none is imperiled (v. 47%). In other respects *Larutia* are typical skinks.

Latastia

Latastia (10 spp.) is a genus of diurnal ground lacertids centered in the Sahel (Senegal to Eritrea) and extending from Egypt to Zimbabwe. The one species (*longicaudata*) for which an ecological business model can be suggested is a refugia-anchored diurnal ground lizard, but the level of ignorance on this genus is so high (filled cells 19% v. 33%) that the assignment is very tentative. The illustrated species have a pale background color, numerous longitudinal dark lines, and a long whiplike tail, redolent of *Aspidoscelis* or *Cnemidophorus*. These are lizards of average size (5.24 g v. 5.10 g) and build (shape index 0.28 v. 0.27 std. resid.) with long tails (2.46 v. 2.06 × SVL) that are breakable (90% v. 92%). The best-known species (*longicaudata*) is sexually dichromatic (10% v. 33%), and males have seasonal color in *johnstonii*. *L. longicaudata* is reported to rely primarily on flight for its primary predator concealment. It refuges at night in a burrow that it digs. All species are found at low elevation (v. 87%), 80% (v. 61%) at medium elevation, and none high (v. 20%). They are associated with sandy habitats (57% v. 44%). They are believed to be oviparous. Fifty percent of species are endemic (v. 35%), 10% are found on islands (v. 26%), none has been introduced (v. 4.6%), and 25% (v. 56%) are imperiled, though none is on any of the IUCN's lists of the most endangered.

Latastia spp.	English name	Activity	Diet	Size mm	Size g	Range core	Ecological business model
boscai	Bosca's Long-tailed Lizard	diurnal		51	3.11	Horn of Africa	
caeruleopunctata	Parker's Long-tailed Lizard	diurnal		69	7.6	Ethiopia	
cherchii	Cherch's Long-tailed Lizard	diurnal		63	5.9	Somalia	
doriai	Doria's Long-tailed Lizard	diurnal		72	8.5	Horn of Africa	
johnstonii	Malawi Long-tailed Lizard	diurnal	invert.	54	3.64	Zambia	
longicaudata	Common Long-tailed Lizard	diurnal	invert.	88	15.7	Central Africa	** refugia-anchored diurnal gr. liz.
ornata	Ornate Long-tailed Lizard	diurnal		65	6.5	Guinea-Bissau	
petersiana	Peter's Long-tailed Lizard	diurnal		77	10.7	Somalia	
siebenrocki	Siebenrock's Long-tailed Lizard	diurnal		45	2.17	Guinea	
taylori	Taylor's Long-tailed Lizard	diurnal		40	1.55	Somalia	

Laudakia

Laudakia is a moderate (10 spp.) genus of large (35 g v. 15.6 g), portly (shape index 1.13 v. 0.80 std. resid.) terrestrial agamids of central and eastern Asia, often at high altitude (44% v. 14%) (*tuberculata* up to 5000 m). All whose diet is known are either herbivores (60% v. 8%) or omnivores (40% v. 26%), which may account for their rotund shape. All species seem to be refugia anchored and to bask directly (v. 24%). Rock crevices are listed as the daily retreat for all documented species, though they seem to be habitat generalists.

They do not appear to mind sharing habitat with humans (29% v. 16%). Being concentrated by geologically suitable microhabitats, several species are said to form social groups and exhibit male defense of mating space (100% v. 97%). M>F SSD (1.0 v. 0.53 on −2 to 2 scale) appears to prevail.

Females of two species (*agrorensis*, *tuberculata*) defend their oviposition site, *tuberculata* for up to 13 days (parental care 20% v. 3.5%). A claim that *tuberculata* lays hard-shelled eggs (v. 2%) has been challenged. Mass-adjusted clutch count is high (0.77 v. 0.20 std. resid.). *L. tuberculata* is the only species for which incubation period has been reported (45 days), and it is notably short even by agamid standards (−1.04 v. −0.55 std. resid.). Short incubation periods are characteristic of giant herbivores such as *L. tuberculata* (table 4.1, trait 66).

Based on two species, longevity appears to be relatively short by lizard standards (mean = 0) and long by family standards (−0.14 v. −0.46 std. resid.). None has been introduced (v. 1.7%) or is imperiled (v. 37%). The herbivory of *tuberculata* is alleged to be of sufficient economic importance to farmers to have justified the

Laudakia spp.	English name				Range core
			Size		
	Activity	Diet	mm	g	Ecological business model
agrorensis	Agrore Agama				Pakistan
	diurnal	omni-vores	90	24.2	** refugia-anchored diurnal gr. liz.
dayana	Haridwar Agama				India
			66	9.1	
melanura	Black Rock Agama				Pakistan
	diurnal	herbivore	113	48.8	**** giant herbivore
nupta	Spiny-headed Rock Agama				Iran
	diurnal	herbivore	132	80	**** giant herbivore
nuristanica	Nuristan Agama				Pakistan
			106	39.8	* refugia-anchored diurnal gr. liz.
pakistanica	North Pakistan Agama				India
	diurnal	omnivore	117	53.7	* refugia-anchored diurnal gr. liz.
papenfussi	Papenfuss' Rock Agama				China
			98	31.4	
sacra	Anan's Rock Agama				China
			115	52	* ground-foraging tree lizard
tuberculata	Kashmir Rock Agama				Himalayas
	diurnal	herbivore	99	32.8	*** giant herbivore
wui	Wu's Rock Agama				China
			94	27.3	

effort of growing crops within lizard exclosures, the fencing consisting of thorny shrubs.

Leiocephalidae and *Leiocephalus*

The family Leiocephalidae consists of a single genus: *Leiocephalus*. *Leiocephalus* is a major radiation (31 spp.) of large (12.1 g v. 5.33 g), mostly omnivorous (81% v. 15%), diurnal (v. 67%), terrestrial (96% v. 48%) lizards found in the West Indies (primarily the Greater Antilles and Bahamas). The common name used for the genus highlights the behavior of many species of curling their tails over their head, scorpion-fashion. The name can be misleading however, in that several species do not do this (e.g., *pratensis*, *raviceps*), many species show great geographic variability in this trait, and the context in which it is used varies. For example, in *personatus* and *schreibersii* it is used almost exclusively by males in intermale communication. *L. carinatus* also use it primarily for intermale assertion, but females sometimes display it too.

Most *Leiocephalus* are refugia-anchored diurnal ground lizards (86% v. 25%), but some are ground-foraging tree lizards (9.5% v. 9.2%), and a few are trunk/wall lizards (4.8% v. 13%). Note that all three of these ecological business models are based on short-distance retreat to concealment, and because sightlines are ample, the exhibitors are often territorial. It is notable in this regard that *Leiocephalus* stands out for the dominance of retreat to crevice-burrow as the primary predator concealment tactic (83%, third-highest family, v. 25%). See also the dominance of male territoriality (below).

Although *Leiocephalus* are conspicuously adaptable in behavior and ecology, they are relatively consistent in morphology. Most are somewhat short-snouted and heavily built (shape index 0.62 v. 0.0 std. resid.). Their tails are about average (1.59 v. 1.37 × SVL); many tails are breakable (26% v. 57% documented). No tails are prehensile (v. 5%). Crests are present in 6.5% of *Leiocephalus* (v. 5.3%). They do not have secretory pores (v. 15%), blue tails (v. 2.3%), or short-term color change (v. 11%), though color patterns often show much geographic variability (see below).

Many *Leiocephalus* have m>f SSD, including a number of species for which the males weigh more than twice as much as adult females (*inaguae*, *macropus*, *schreibersii*, and probably some of the extinct species). The mean SSD score (on scale of –2 to 2) is 1.17 (v. 0.36), the highest family mean for moderate-sized lizards (exceeded only by giants: Iguanidae, Sphenodontidae, Varanidae). This is not paired with especially short male tails, though the m:f ratio of relative tail lengths is below average (0.97 v. 1.07 ratio). Sexual dichromatism is present in a majority of species, and probably some extinct ones: 55% v. 21%. It shows much geographic variability. A frequently variable character is orange on the underside of tails (*lunatus*, *melanochlorus*, *personatus*, *schreibersii*). Seasonal color is however not present (0% v. 6.5%). These combine into a sex differences rollup that is high but not exceptional (1.19 v. 0.75 on 0 to 6 scale).

Diet is another area in which

Leiocephalus spp.	English name	Activity	Diet	Size mm	Size g	Range core	Ecological business model
anonymous	Artibonite Curlytail Lizard			95	24.1	Haiti	
apertoculcus	San Rafael Curlytail			117	46.6	Dominican Republic	
barahonensis	Orange-bellied Curlytail	diurnal	omnivore	61	5.8	Hispaniola	*** refugia-anchored diurnal gr. liz.
carinatus	Northern Curly-tailed Lizard	diurnal	omnivore	98	26.4	Cuba	*** refugia-anchored diurnal gr. liz.
cubensis	Cuban Curlytail Lizard	diurnal	omnivore	84	15.9	Cuba	** refugia-anchored diurnal gr. liz.
cuneus	Cave Curlytail Lizard		invert.	152	111	Guadeloupe	
endomychus	Central Haitian Curlytail			59	5.01	Haiti	** trunk/wall lizard
eremitus	Navassa Curlytail Lizard			55	3.80	Navassa Island	
etheridgei	Etheridge's Extinct Lizard			92	21.3	Puerto Rico	
greenwayi	Plana Cay Curlytail Lizard	diurnal	invert.	63	6.4	Bahamas	** refugia-anchored diurnal gr. liz.
herminieri	Martinique Curlytail Lizard			109	40	Martinique	
inaguae	Inagua Curlytail Lizard	diurnal	omnivore	75	11.3	Bahamas	*** refugia-anchored diurnal gr. liz.
jamaicensis	Jamaican Curlytail			131	68	Jamaica	
loxogrammus	Rum Cay Curlytail Lizard	diurnal	omnivore	71	9.3	Bahamas	*** refugia-anchored diurnal gr. liz.
lunatus	Santo Domingo Curlytail Lizard	diurnal		54	3.79	Dominican Republic	*** refugia-anchored diurnal gr. liz.
macropus	Monte Verde Curlytail Lizard	diurnal	omnivore	70	9.1	Cuba	** refugia-anchored diurnal gr. liz.
melanochlorus	Jeremie Curlytail Lizard	diurnal		102	30.5	Haiti	* refugia-anchored diurnal gr. liz.

Leiocephalus shows its variability or opportunism. Only 19% (v. 73%) are invertebrate eaters. The diet of the 81% omnivores is conspicuously variable by sex and geography (island). *L. carinatus* and *punctatus* are conspicuously more herbivorous on small islands. *L. macropus* (m≫f) males eat more plant matter (82%) than do females (58%). The same sex difference shows up in the more subtly m>f *raviceps*, though the proportions and difference are smaller (39% v. 15%). The invertebrate material eaten by these two species is often primarily ants (geographically variable). The invertebrate material eaten by the omnivores *cubensis*, *schreibersii*, *semilineatus*, and *stictigaster* is also dominated by ants. For more general discussion of connections between the consumption of plants and ants see the omnivore section in chapter three. A final "dietary" note is that some large male *schreibersii* are not territorial but are cannibalistic, a distinction that may be lost on the victims.

All *Leiocephalus* are said to be sit-and-wait foragers (100% v. 46%), and only 25% (v. 58%) are also said to be active foragers. However, the omnivores (81%) must actively forage for plants, so this characterization cannot be fully accurate. Home-range area is small (–0.43 v. 0.0 std. resid.) based on *schreibersii* (males about 77 m^2; females about 37 m^2).

Thirteen percent (v. 8%) are conspicuously wary, which fits in with a penchant for dashing to the safety of a refuge. As mentioned above, the primary predator concealment tactic of *Leiocephalus* is retreat to a crevice-burrow, but small proportions of species also rely on sand (17% v. 6%) and vegetation (17% v. 23%).

Many bask directly (42% v. 18%), though thermoconforming is also overrepresented (6.5% v. 2.6%).

Mean active body temperature is a toasty 34.9°C (v. 31.1°C).

Eleven species have been characterized for male social relations and all defend space (v. 85%).

Given the West Indian habitat, one might expect the presence of *Leiocephalus* in human and littoral habitats; their overrepresentation in all tabulated habitat types is consistent with their ecological and behavioral flexibility: littoral zones (72% v. 17%), sandy areas (60% v. 22%), riparian zones (32% v. 22%), and human habitats (44% v. 18.5%).

Reproduction is solely oviparous and mean clutch count is low (−0.47 v. 0.0 std. resid.). Young are about average in relative size (ontogenetic mass range −0.12 v. 0.0 std. resid.). Incubation is short (−0.41 v. 0.0 std. resid.), corresponding to 53–78 days (mostly 60–63 days). *L. schreibersii* adults are fairly tolerant of hatchlings, which are sometimes seen perching on adult females, and natal dispersal may not be immediate after hatching. Given the inclination of large male *schreibersii* to cannibalism, this tolerance of hatchlings may be restricted to kin.

Longevity is average (0.09 v. 0.0 std. resid.), as is population density (0.57 kg/ha v. 0.54 kg/ha). The density comparison is based on only three credible samples of two species (*schreibersii*, *semilineatus*). One might expect higher densities given their island location and low trophic level. Endemism is high despite the tiny size of many Caribbean countries (77% v. 64%). All species live on islands (v. 40%). Almost 10% of *Leiocephalus* have been introduced (9.7% v. 3.3%), including *carinatus*, which has been introduced at least nine times (more than twice: 3.2% v. 1.1%). Seventy percent are considered imperiled (v. 49%), and 44% are on the IUCN's lists of the most endangered (v. 14%). The species that are probably extinct include *anonymous*, *apertoculcus*, *cuneus*, *endomychus*, *eremitus*, *etheridgei*, *herminieri*, *jamaicensis*, *onaneyi*, *partitus*, *pratensis*, and *rhutidira*. Living on islands shortens the demographic feedback time from detrimental introductions and hastens the day when untenable predation eliminates species.

Leiocephalus spp.	English name	Activity	Diet	Size mm	Size g	Range core	Ecological business model
onaneyi	Sierra Curlytail Lizard	diurnal		66	7.4	Cuba	
partitus	Guanica Extinct Curlytail			99	27.2	Puerto Rico	
personatus	Haiti Curlytail Lizard	diurnal	omnivore	67	7.9	Hispaniola	*** refugia-anchored diurnal gr. liz.
pratensis	Atalaye Curlytail Lizard	diurnal		55	4.1	Haiti	** refugia-anchored diurnal gr. liz.
psammodromus	Bastion Bay Curlytail Lizard	diurnal	omnivore	84	16.1	Turks and Caicos Islands	** refugia-anchored diurnal gr. liz.
punctatus	Spotted Curlytail Lizard	diurnal	omnivore	96	25.1	Bahamas	** ground-foraging tree lizard
raviceps	Mountain Curlytail Lizard	diurnal	omnivore	58	4.96	Cuba	*** refugia-anchored diurnal gr. liz.
rhutidira	Lapierre Curlytail Lizard			57	4.48	Haiti	
schreibersii	Red-sided Curly-tailed Lizard	diurnal	omnivore	77	12.2	Hispaniola	*** refugia-anchored diurnal gr. liz.
semilineatus	Thomazeau Curlytail Lizard	diurnal	omnivore	45	2.17	Hispaniola	** refugia-anchored diurnal gr. liz.
sixtoi	Sixto's Curlytail Lizard	diurnal	invert.	72	9.9	Dominican Republic	*** refugia-anchored diurnal gr. liz.
stictigaster	Cuba Striped Curlytail	diurnal	omnivore	69	8.6	Cuba	** refugia-anchored diurnal gr. liz.
varius	Cayman Curly-tailed Lizard	diurnal		88	18.7	Cayman Islands	** refugia-anchored diurnal gr. liz.
vinculum	Cochran's Curlytail Lizard	diurnal		64	6.8	Haiti	*** ground-foraging tree lizard

Leiolepis

Leiolepis is a moderate (nine spp.) genus of speedy, crestless (0% v. 38% crested), burrow-fleeing (100% v. 29%), terrestrial (100% v. 56%) agamids, often with orange flank markings. They are large (67 g v. 15.6 g), portly (shape index 1.32 v. 0.80 std. resid.), lowland (100% v. 83%) residents of continental Southeast Asia and some nearby islands. All whose diet is known are either herbivores (17% v. 8%) or omnivores (83% v. 26%), which may account for their bulky shape. All species seem to be refugia-anchored diurnal ground lizards (100% v. 43%) and to bask directly (100% v. 24%). Males of all species defend space (100% v. 97%), presumably to maintain possession of their burrows. They are habitat and dietary generalists, and they are overrepresented in littoral (56% v. 10%) and sandy (67% v. 26%) habitats.

Leiolepis spp.	English name				Range core
			Size		
	Activity	Diet	mm	g	Ecological business model
belliana	Butterfly Lizard				Southeast Asia
	diurnal	omnivore	137	88	*** refugia-anchored diurnal gr. liz.
boehmei	Boehme's Butterfly Lizard				Thailand
	diurnal		99	32.8	*** refugia-anchored diurnal gr. liz.
guentherpetersi	Peters' Butterfly Lizard				Vietnam
	diurnal	omnivore	121	60	** refugia-anchored diurnal gr. liz.
guttata	Spotted Butterfly Lizard				Vietnam
	diurnal	omnivore	180	207	** refugia-anchored diurnal gr. liz.
ngovantrii	Ngo Van Tri's Lady Butterfly Lizard				Vietnam
	diurnal		100	33.5	*** refugia-anchored diurnal gr. liz.
peguensis	Pegu Butterfly Lizard				Myanmar
	diurnal	omnivore	125	67	**** refugia-anchored diurnal gr. liz.
reevesii	Reeve's Butterfly Lizard				Southeast Asia
	diurnal	omnivore	132	79	**** refugia-anchored diurnal gr. liz.
rubritaeniata	Red-banded Butterfly Lizard				Southeast Asia
	diurnal		132	80	* refugia-anchored diurnal gr. liz.
triploida	Triploid Butterfly Lizard				Thailand
	diurnal	herbivore	115	52	**** refugia-anchored diurnal gr. liz.

The genus is evenly split between parthenogenetic and bisexual species (50% v. 0.8% parthenogenetic, v. 0.27% of Iguania). Male seasonal color, sexual dichromatism (33% v. 53%), and m>f sexual dimorphism (1.00 v. 0.53 on –2 to 2 scale) are present but only mildly developed in three of five bisexuals (undefined in the all-female species). Mass-adjusted clutch count is very small (–1.10 v. 0.20 std. resid.). Hatchling size is normal (ontogenetic size range 0.66 v. 0.69 std. resid.). Incubation period is reported only for *L. guttata*, for which it is notably short (64 days), even by agamid standards (–1.19 v. –0.55 std. resid.). Rapid incubation is characteristic of herbivores and omnivores (table 3.5, trait 46).

Longevity is slightly short (–0.55 v. –0.46 std. resid.), which is not characteristic of herbivores and omnivores (table 3.5, trait 49). The only well-studied species is *belliana*, which runs bipedally when pressed (especially juveniles), seems to be monogamous, and plugs its burrow overnight; the female oviposits in her burrow and allows the hatchlings to remain in her burrow for several months before dispersal (parental care 11% v. 3.5%). Oviposition in *belliana* occurs during the dry season and is timed for hatching to occur at the onset of the rains, presumably to facilitate growth of the young. *L. belliana* is the only species that has been introduced (100% v. 1.7%: Florida), and the only one (v. 37%) generating conservation concern, as a consequence of human consumption for food in Thailand.

Leiolopisma

The genus *Leiolopisma* was once a grab bag of otherwise unassignable skinks, and though most species have been removed from the genus, it continues to have some very strange occupants. Three of the taxonomically, but not literally, surviving species are from the Mascarene Islands (around Mauritius, Indian Ocean) and are probably a natural group. The Ono-i-Lau Ground Skink is from Fiji and is comparatively tiny; it can be expected to someday be removed from the genus, simply on the basis of the vast geographic and likely genetic distance. The Didosaurus and Reunion Skink (both from the Mascarene Islands) are known from only subfossil material and are presumed extinct. What a great tragedy to lose the world's largest skink, the enormous carnivorous Didosaurus. But the strangest of all is the Snafu Skink, which is known from only one type specimen that subsequently disappeared. The type was said to be from New Zealand, but this is almost universally disbelieved. Furthermore, the type specimen is said to match neither the original description nor the describer's redescription. Given that the type specimen is lost, it is hard to understand how it could be known to contradict the descriptions; apparently someone convincing made this judgment before the type was lost. In any event, the type specimen was said to match *Lipinia vulcania*, but apparently no one is willing to synonymize *Leiolopisma fasciolare* with *Lipinia vulcania* given the absence of concrete evidence. For all practical purposes the genus now has two living members, a charismatic 80 g refugia-anchored diurnal ground lizard from Mauritius outliers (*telfairi*) and a 3 g litter cryptozoid from a Fiji outlier (*alazon*).

Leiolopisma spp.	English name				Range core
	Activity	Diet	Size mm	Size g	Ecological business model
alazon	Ono-i-Lau Ground Lizard				Fiji
	diurnal		56	3.29	* litter cryptozoid
ceciliae	Reunion Skink				Reunion (French Overseas Terr.)
			134	56	
fasciolare	Snafu Skink				New Zealand
mauritiana	Didosaurus				Mauritius
	nocturnal	carnivore	252	426	** nocturnal ground lizard
telfairii	Telfair's Skink				Mauritius
	cathemeral	omnivore	136	80	** refugia-anchored diurnal gr. liz.

Leiolopisma is a minor radiation (five spp.) of mostly large or huge (mean 50 g v. 3.25 g) terrestrial (100% v. 48%) skinks either from the Mascarene Islands or Fiji. These are comparatively robust skinks (shape index 0.91 v. −0.46 std. resid.). Tails are short (1.08 v. 1.33 × SVL) and neither breakable (v. 73%) nor blue (v. 4%). *Leiolopisma* are not elongate (v. 35% elongate).

Males are generally larger than females (0.75 v. −0.31 on −2 to 2 scale). *L. telfairi* is sexual dichromatic (20% v. 7.6%). The sex differences rollup score (on scale of 0 to 6) is thus somewhat elevated (0.60 v. 0.28).

Half of the known species are carnivores (50% v. 1%); half are omnivores (v. 13%). The best-studied species (*telfairi*) is dietarily unrestrained, as so often happens on small islands: fruit, nectar, arthropods, orthopterans, other insects, ectothermic vertebrates, endothermic vertebrates that are immobile (eggs, nestlings), and carrion. It will climb into the lap of visitors to beg food. Foraging modes, electric light use, and home-range areas are undocumented.

Diel activity patterns show no theme: 33% cathemeral, 33% diurnal, 33% nocturnal. One wonders how they inferred the nocturnality of a species (*mauritiana*) known only from subfossil material. The early human chroniclers of Mauritius were hungry and yet made no mention of the then-extant Didosaurus, possibly because it was nocturnal (hard to detect prior to the discovery of electricity). *Leiolopisma* are known from only low elevation (v. 89%).

All are oviparous (v. 64%); clutch count is moderately large (0.46 v. 0.06 std. resid.), based on *telfairi*. Based on *telfairi*, the ontogenetic mass range is also elevated, indicating relatively small young (0.66 v. 0.12 std. resid.). Incubation is short (−1.08 v. −0.40 std. resid.), based on *telfairi* (56 days).

Longevity is relatively long (0.59 v. 0.17 std. resid.), equating to 16 years (*telfairi*). All species are endemic (v. 64%). None has been introduced (v. 2.1%, except as conservation repatriations). All are imperiled (v. 47%), including placement on the IUCN's lists of the most endangered (67% v. 14%). In other respects *Leiolopisma* are much like other skinks.

Leiosauridae

The 33 leiosaurids are large (16.4 g v. 5.3 g), poorly known (21% v. 27% filled cells), cryptic (100% v. 43% as primary predator concealment tactic) southern South American (100% Neotropical v. 29%: specifically Ecuador to Brazil) lizards formerly placed in the Iguanidae. They are frequently found at medium (57% v. 41%) or high elevation (36% v. 13%), and yet nearly all (95% v. 82%) are oviparous, though contradictory reports exist for several species. Furthermore, despite their large size and iguanid lineage, all documented sexual size differences are of m<f (−1.0 v. 0.36 on

−2 to 2 scale), suggesting exceptional pressure for either large clutch counts or large eggs. Their mean clutch count (8.0 v. 4.0 eggs) is moderately large for their size (0.68 v. 0.0 std. resid.). The ontogenetic mass range is small (−0.35 v. 0.0 std. resid.), indicating relatively large hatchlings. Perhaps because of comparatively large egg size and high elevation, their eggs incubate more slowly than those of any other lizard family (2.81 v. 0.0 std. resid.), equating to 181–270 days, which is exceeded only by the Tuatara.

Leiosauridae are diverse in their microhabitats and ecological business models. The modal pattern is that of a refugia-anchored diurnal ground lizard (63% v. 25%). Less frequent are diurnal canopy walkers (16% v. 6%), ground-foraging tree lizards (10.5% v. 9.2%), and immobility experts (10.5% v. 3.2%). The family is about evenly split between arboreal (45% v. 32%) and terrestrial forms (50% v. 48%), with a small minority cryptozoic (5% v. 12%). They are overwhelmingly invertebrate eaters (95% v. 73%). One is an omnivore (*Pristidactylus achalensis*). Six modest-sized (2–10 spp.) genera make up the Leiosauridae: *Anisolepis*, *Diplolaemus*, *Enyalius*, *Leiosaurus*, *Pristidactylus*, and *Urostrophus*.

Leiosaurids are robustly built (shape index 0.76 v. 0.0 std. resid.), with many species featuring slender limbs and tails, which accentuate their gestalt as large headed and, to a human, infantile in proportions. Tails are relatively long (1.66 v. 1.37 × SVL), especially in *Anisolepis* (2.79 × SVL). Tail breakage (3% v. 57%) is found only in *Pristidactylus*, and prehensile tails are overrepresented (9.1% v. 5%) but found only in *Anisolepis* and *Urostrophus*. Crests are absent (v. 5.3%), as are secretory pores (v. 15%) and elongation (v. 11%). Basic color change is found (3% v. 11%) but advanced color change is not (v. 6%). Blue tails are absent (v. 2. 3%).

Sex differences in color are found at lizard-typical frequencies (15% v. 21%). Seasonal color is not reported (v. 6.5%). Sex differences in relative tail length (1.20 v. 1.07 ratio) are known for only *Anisolepis* (1.16 ratio) and *Urostrophus* (1.27 ratio), consistent with the general lizard pattern of m<f SSD (table 3.3, trait 8). With the exception of *Anisolepis* (2.0 on 0 to 6 scale), the family has relatively muted sex differences (family mean 0.39 v. 0.75 on 0 to 6 scale).

Foraging mode information is available only for *Enyalius*, for which it is relatively uninformative: one-third sit-and-wait (v. 46%) and two-thirds active (v. 58%). Electric light use is not reported (v. 1.4%), nor are home-range areas.

Diurnal (86% v. 67%) and crepuscular behavior (14% v. 1.8%) are overreported; no leiosaurids are known to be cathemeral (v. 6.8%) or nocturnal (v. 25%). Direct basking is underreported (3% v. 18%), but so is thermoconformance (0% v. 2.6%), suggesting a paucity of observations rather than a thermoregulatory trend. A few males are believed to defend space (v. 85%), but other information on social relations is lacking. Two-thirds of documented species refuge in burrows (v. 22%), at least at some time or by some individuals.

One *Enyalius* (*brasiliensis*) is found in the mesic rainfall belt (1140 mm/y in this case). Leiosaurids are underrepresented in littoral areas (0% v. 17%), sandy areas (16% v. 22%), riparian zones (4% v. 22%), and human habitats (8% v. 18.5%), again suggesting an information shortage.

As noted above, leiosaurids are conspicuously slow in egg incubation, but their RCM (reported only for *Enyalius bibronii*) is only 0.152 (v. 0.226). More information is needed to ascertain whether this represents a common condition in the family; it does not seem compatible with the pattern of large clutches and large hatchlings.

Reported longevity is comparatively short (−0.39 v. 0.0 std. resid.), though based on only *Pristidactylus achalensis* (11 years) and *P. torquatus* (3 years). More information is needed. Endemism is typical (70% v. 64%). No species is found on islands (v. 40%), nor have any been introduced (v. 3.3%). Few are considered imperiled (14% v. 49%), and none is on any of the IUCN's lists of the most endangered (v. 14%).

Leiosaurus

Leiosaurus is a four-species genus of moderately large (21.1 g v. 16.4 g) crepuscular (100% v. 14%), terrestrial (100% v. 50%), cryptic (100% v. 100%) leiosaurids found in Argentina (one species extends into Brazil and Paraguay). All are refugia-anchored diurnal ground lizards (v. 63%).

Leiosaurus are large headed and robust to pudgy (shape index 0.86 v. 0.76) and in possession of relatively skinny, exceptionally short tails (1.10 v. 1.66 × SVL) that are neither breakable (v. 3%) nor prehensile (v. 9%). The species are not known to change color (v. 3%), and the sexes do not differ in color (v. 15%).

They are overrepresented in sandy areas (50% v. 16%). They have exceptionally large clutches (1.93 v. 0.68 std. resid.). Endemism is high (75% v. 70%). None is imperiled (v. 14%). In other respects *Leiosaurus* are probably typical leiosaurids.

Leiosaurus spp.	English name	Activity	Diet	Size mm	Size g	Range core	Ecological business model
bellii	Bell's Iguana	crepuscular	invert.	88	19.7	Argentina	** refugia-anchored diurnal gr. liz.
catamarcensis	Catamarca Iguana	crepuscular	invert.	95	25.6	Argentina	* refugia-anchored diurnal gr. liz.
jaguaris	Jaguar Iguana		invert.	89	20.2	Argentina	* refugia-anchored diurnal gr. liz.
paronae	Peracca's Iguana		invert.	88	19.7	Paraguay	** refugia-anchored diurnal gr. liz.

Lepidoblepharis spp.	English name	Activity	Diet	Size mm	Size g	Range core	Ecological business model
buchwaldi	Buchwald's Scaly-eyed Gecko	diurnal		29	0.59	Ecuador	
colombianus	Colombian Scaly-eyed Gecko	diurnal		42	1.87	Colombia	
conolepis	Spiky Scaly-eyed Gecko	diurnal		41	1.69	Ecuador	**** litter cryptozoid
duolepis	Antioquia Scaly-eyed Gecko	diurnal		42	1.87	Colombia	*** litter cryptozoid
emberawoundule	Burbayar Scaly-eyed Gecko			31	0.74	Panama	
festae	Amazon Scaly-eyed Gecko	diurnal		43	1.97	Ecuador	* litter cryptozoid
grandis	Grand Scaly-eyed Gecko	diurnal		49	2.97	Ecuador	
heyerorum	Vanzolini's Scaly-eyed Gecko	cathemeral		34	1.02	Brazil	* litter cryptozoid
hoogmoedi	Hoogmoed's Scaly-eyed Gecko	diurnal		32	0.81	Brazil	** litter cryptozoid
intermedius	Lowland Scaly-eyed Gecko	cathemeral		33	0.90	Colombia	** litter cryptozoid
microlepis	Tiny Scaly-eyed Gecko	diurnal		27	0.51	Colombia	** litter cryptozoid
miyatai	Miyata's Scaly-eyed Gecko	diurnal		26	0.44	Colombia	*** litter cryptozoid

Lepidoblepharis

Lepidoblepharis is a large genus (21 spp.) of slightly small (0.87 g v. 1.07 g), poorly known (filled cells 20% v. 27%) sphaerodactylids from southern Central America (Nicaragua to Panama) and northwestern South America (Colombia to Peru, Brazil, and French Guiana). They are padless, mostly diurnal (88% v. 62%) litter cryptozoids (100% v. 53%). There are few sex differences other than dichromatism, which is underrepresented (19% v. 35%). All documented species are sit-and-wait (100% v. 83%) foragers. Those that are not diurnal are cathemeral (12% v. 15%). *Lepidoblepharis* are found only in the wettest two rainfall bands: mesic (67% v. 34%) and wet (33% v. 25%). *L. xanthostigma* is found in such wet areas (up to 4600 mm/y) that they may breed preferentially in the dry season. Their elevational profile is near average (l/m/h: 88/31/0 v. 96/20/4). They are overrepresented in riparian zones (29% v. 19%) but underrepresented in littoral (6% v. 31%), sandy (0% v. 15%), and human (24% v. 33%) habitats.

Reproduction is normal, though communal oviposition is infrequent (5% v. 11%), and adherent eggs (v. 6%) and parental care (v. 1.4%) are not documented. RCM ratio is known only from *xanthostigma*, for which it is minimal even by sphaerodactylid standards (0.046 v. 0.095).

Population density is slightly low (0.15 kg/ha v. 0.32 kg/ha), based on one credible sample of *sanctaemartae* and 10 of *xanthostigma*. Endemism is average (67% v. 65%). Island species are uncommon (19% v. 64%), and introductions not reported (0% v. 5%). Twenty-five percent of *Lepidoblepharis* are considered imperiled (v. 41%), all of which are placed on one of the IUCN's lists of the most endangered (v. 23%). In other

Lepidoblepharis spp.	English name				Range core
	Activity	Diet	Size mm	Size g	Ecological business model
montecanoensis	Paraguanan Ground Gecko				Venezuela
	diurnal		24	0.37	* litter cryptozoid
nukak	Nukak Scaly-eyed Gecko				Colombia
			30	0.70	
peraccae	Peracca's Scaly-eyed Gecko				Colombia
	diurnal		30	0.69	* litter cryptozoid
rufigularis	Orange-throated Scaly-eyed Gecko				Panama
			27	0.51	
ruthveni	Ruthven's Scaly-eyed Gecko				Colombia
	diurnal		32	0.84	*** litter cryptozoid
sanctaemartae	Panama Scaly-eyed Gecko				Colombia
	diurnal	invert.	27	0.51	*** litter cryptozoid
victormartinezi	Victor's Scaly-eyed Gecko				Panama
			29	0.59	
williamsi	Williams' Scaly-eyed Gecko				Colombia
	diurnal		31	0.74	*** litter cryptozoid
xanthostigma	Litter Scaly-eyed Gecko				Panama
	diurnal	invert.	37	1.30	**** litter cryptozoid

respects *Lepidoblepharis* are typical sphaerodactylids.

Lepidodactylus (fig. 4.9C)

Lepidodactylus is a large genus (33 spp.) of the Gekkonidae, with a wide, generally tropical distribution on islands bordering or within the western Pacific area, especially Melanesia. *Lepidodactylus* may occur on islands too small to support any other lizard (ca. 0.1 ha.). All *Lepidodactylus* species reside on at least one island (100% v. 45%). Most species are nocturnal canopy walkers (79% v. 9%); the remainder are trunk/wall lizards (21% v. 45%), and electric lighting may propel canopy walkers onto lit walls. *Lepidodactylus* are small (1.79 g v. 3.31 g), with uniformly breakable tails (v. 87%) that are not prehensile (v. 1.8% prehensile in family). Secretory pores are present in typical frequency (39% v. 46%). Basic color change (darkening) is underrepresented (6% v. 15%). All species have expanded toe pads (v. 70%). Sexual size differences emphasize males that are smaller than females (–0.50 v. –0.03 on –2 to 2 scale). Sexual dichromatism is not reported (v. 6.6%).

Foraging mode has been reported (sit-and-wait) for only one species (*lugubris*). This generalized invertebrate-eating species regularly consumes nectar and fruit; it presumably does not wait for the fruit to arrive on its own. Electric light use is frequent (6.1% v. 3.8%). Home ranges are small (–0.77 v. –0.28 std. resid.), based on *lugubris*. As expected for nocturnal arboreal lizards, vocalization is reported (6.1% v. 3.8%), likely underreported. Use of sandy areas (4% v. 11%) and riparian areas (7% v. 17%) is underrepresented, whereas use of human habitat is overrepresented (36% v. 27%). Many species are associated with *Pandanus*, palm trees, or epiphytic plants (all of these may be moisture sources).

Reproduction follows gekkonid standards, except that parthenogenesis is found in *lugubris*. Female *lugubris* mount each other, possibly to stimulate ovulation. *L. ranauensis* (1.97 g) has unusually small eggs (3.6–4.5 mm). The genus averages relatively small young (ontogenetic mass range elevated: 0.32 v. –0.19 std. resid.), a pattern seen in a number of successful gekkonid genera (e.g., *Hemidactylus*).

Credible population density estimates exist for three species: *christiani*: 0.04 kg/ha ($n = 1$); *herrei*: 0.04 kg/ha ($n = 4$); *lugubris*: 1.97 kg/ha ($n = 23$). These compare to a gekkonid mean of 0.37 kg/ha. Introductions have been relatively frequent in *Lepidodactylus*, likely associated with the parthenogenesis of *lugubris* and presence of several species in human habitats and around electric lights (6.1% v. 3.7% for any introductions; 3.0% v. 2.1% for >2 introductions). *Lepidodactylus* are underrepresented among species at conservation risk (24% v. 47%) and those on the IUCN's lists of the most endangered (8.0% v. 11.3%). In other respects *Lepidodactylus* are typical gekkonids.

The species name of *lugubris* (in English, lugubrious, meaning looking or sounding sad or dismal) is responsible for the common name—Mourning Gecko—and has generated much speculation. The clicking call of this species has often and creatively been claimed to sound "mournful"; it is possible, but undocumented, that the describers had appearance rather than voice in mind when they described it as *lugubris*. Gaulke (2011) suggests the darker markings in some way imitate the black arm bands worn by some mourners. Koch (2012) surmises that the impetus was the subdued markings of this species, which is in keeping with numerous other "lugubris" species names used in the nineteenth century for darkly marked but silent species.

Lepidodactylus spp.	English name		Size		Range core
	Activity	Diet	mm	g	Ecological business model
aureolineatus	Golden Scaly-toed Gecko				Philippines
	nocturnal		41	1.63	** nocturnal canopy walker
balioburius	Batan Scaly-toed Gecko				Philippines
	nocturnal		37	1.23	* nocturnal canopy walker
browni	Bootless Bay Gecko				New Guinea
	nocturnal		43	1.80	** nocturnal canopy walker
buleli	Vanuatu Ant-nest Gecko				Vanuatu
	nocturnal		37	1.24	* nocturnal canopy walker
christiani	Negros Broad-tailed Gecko				Philippines
	nocturnal	invert.	42	1.74	*** nocturnal canopy walker
euaensis	'Eua Forest Gecko				Tonga
	nocturnal		45	2.12	* trunk/wall lizard
flavioculuris	Yellow-eyed Gecko				Solomon Islands
	nocturnal		46	2.28	* nocturnal canopy walker
gardineri	Rotuma Forest Gecko				Fiji
	nocturnal	invert.	49	2.60	*** nocturnal canopy walker
guppyi	Solomon Forest Gecko				Solomon Islands
	nocturnal		46	2.20	*** nocturnal canopy walker
herrei	White-lined Smooth-scaled Gecko				Philippines
	nocturnal	invert.	41	1.66	*** nocturnal canopy walker
intermedius	Komodo Scaly-toed Gecko				Indonesia
	nocturnal		41	1.62	** nocturnal canopy walker
labialis	Mindanao Scaly-toed Gecko				Philippines
	nocturnal		54	3.56	*** nocturnal canopy walker
listeri	Christmas Island Chained Gecko				Christmas Island (Australia)
	nocturnal	invert.	51	2.70	*** trunk/wall lizard
lombocensis	Lombok Scaly-toed Gecko				Indonesia
	nocturnal		37	1.23	* nocturnal canopy walker
lugubris	Mourning Gecko				Indo-Pacific Islands
	nocturnal	invert.	42	1.45	**** nocturnal canopy walker or trunk/wall lizard (elect. light)
magnus	Mountain Forest Gecko				New Guinea
	nocturnal		60	4.68	** trunk/wall lizard
manni	Fiji Forest Gecko				Fiji
	nocturnal		44	1.95	** trunk/wall lizard
moestus	Micronesian Flat-tailed Gecko				Micronesia
	nocturnal		36	1.17	*** nocturnal canopy walker or trunk/wall lizard (elect. light)
mutahi	Mutahi Gecko				New Guinea
	nocturnal		45	2.10	** nocturnal canopy walker

Lepidodactylus spp.	English name				Range core
			Size		
	Activity	Diet	mm	g	Ecological business model
novaeguineae	New Guinea Tree Gecko				New Guinea
	nocturnal		42	1.69	* trunk/wall lizard
oligoporus	Mortlock Forest Gecko				Micronesia
	nocturnal	invert.	43	1.84	*** nocturnal canopy walker
oorti	Oort's Forest Gecko				Indonesia
	nocturnal	invert.	51	2.99	* nocturnal canopy walker
orientalis	Savanna Scaly-toed Gecko				New Guinea
	nocturnal		38	1.30	** nocturnal canopy walker
paurolepis	Palau Barred Gecko				Palau
	nocturnal		38	1.33	** nocturnal canopy walker
planicaudus	Small Broad-tailed Smooth-scaled Gecko				Philippines
	nocturnal		39	1.43	**** nocturnal canopy walker
pulcher	Wild Island Gecko				New Guinea
	nocturnal		44	2.00	
pumilus	Slender Chained Gecko				New Guinea
	nocturnal		40	1.55	* nocturnal canopy walker
ranauensis	Kinabalu Mourning Gecko				Malaysia
	nocturnal	invert.	44	1.92	** trunk/wall lizard (at least where commensal)
shebae	Guadalcanal Gecko				Solomon Islands
	nocturnal		35	1.08	
tepukapili	Tuvalu Forest Gecko				Tuvalu
	nocturnal		44	1.96	
vanuatuensis	Vanuatu Forest Gecko				Vanuatu
	nocturnal	invert.	41	1.65	*** nocturnal canopy walker
woodfordi	Melanesia Bisexual Mourning Gecko				New Guinea
	nocturnal		39	1.39	*** nocturnal canopy walker
yami	Lanyu Scaly-toed Gecko				China
	nocturnal		40	1.47	

Lepidophyma

Lepidophyma is a moderate radiation (19 spp.) of xantusiids with large shiny head scales and small dull rugose or tuberculate scales on the rest of their body, as if only the head would be routinely exposed. They are moderately large (6.5 g v. 4.0 g), cruise-foraging (100% v. 33%), cathemeral (50% v. 47%) or nocturnal (50% v. 27%) crack hermits (82% v. 89%) found in Mexico and Central America. Those which are not crack hermits are trunk/wall lizards (18% v. 8%). The ecological business model assignments have low confidence (filled cells 23% v. 26%). Most reported diel activity patterns have weak, conflicting data.

They are typical xantusiids in morphology, albeit with a slightly lower frequency of secretory pores (16% v. 24%), likely underreported. They hew to the family pattern in being m<f in SSD (–1.0 v. –1.0 on –2 to 2 scale) and having few other sex differences (sex differences rollup 0.22 v. 0.39 on 0 to 6 scale). Diets are poorly documented, but most are invertebrate eaters (67% v. 71%), and at least one is an omnivore (33% v. 21%). Wariness is underrepresented (5% v. 15%) and may be underreported. The only thermal behavior reported is thermoconformance (5.3% v. 5.9%). They are about evenly split among arboreal (30% v. 14%), cryptozoic (30% v. 62%), and terrestrial species (40% v. 24%). Their elevational profile emphasizes intermediate elevations: (l/m/h) 82/65/0 (v. 82/45/9). They are associated with riparian (17% v. 16%) and human (22% v. 13%) habitats.

Litter count is about average (0.17 v. –0.07 std. resid.). Neonates are notably small (ontogenetic mass range 0.71 v. –0.19 std. resid.). Parthenogenesis, either occasional (*flavimaculatum*, *tuxtlae*) or routine (*reticulatum*, *sylvaticum*?), appears in *Lepidophyma*.

Longevity is short (0.23 v. 0.93 std. resid.), based on five species. Population density is based on just two credible samples of *flavimaculatum* and is very low (0.02 kg/ha v. 0.49 kg/ha). The majority of species are from the North American part of Mexico (63% v. 76%). Endemism is exactly average (79% v. 79%). No species occurs on an island (v. 6%). None has been introduced (v. 3%). *Lepidophyma* are imperiled at an average family frequency (35% v. 31%), including 5.9% (v. 3.8%) on one of the IUCN's lists of the most endangered. In other respects *Lepidophyma* are typical xantusiids. See the *Xantusia* entry for comparisons with that genus.

Lepidophyma spp.	English name				Range core
			Size		
	Activity	Diet	mm	g	Ecological business model
chicoasensis	Sumidero Tropical Night Lizard				Mexico
			93	16.0	
cuicateca	Cuicatlan Tropical Night Lizard				Mexico
			47	2.04	
dontomasi	MacDougall's Tropical Night Lizard				Mexico
			49	2.33	
flavimaculatum	Yellow-spotted Tropical Night Lizard				Central America
	cathemeral	invert.	105	23.3	** crack hermit
gaigeae	Gaige's Tropical Night Lizard				Mexico
	nocturnal	invert.	61	4.33	* crack hermit
lineri	Liner's Tropical Night Lizard				Mexico
			58	3.77	
lipetzi	Lipetz's Tropical Night Lizard				Mexico
			49	2.23	
lowei	Lowe's Tropical Night Lizard				Mexico
			52	2.77	
mayae	Mayan Tropical Night Lizard				Guatemala
			66	5.7	
micropholis	Cave Tropical Night Lizard				Mexico
	cathemeral	invert.	89	13.8	* crack hermit
occulor	Japlan Tropical Night Lizard				Mexico
		invert.	104	22.2	** trunk/wall lizard
pajapanensis	Pajapan Tropical Night Lizard				Mexico
			69	6.4	** trunk/wall lizard
radula	Yautepec Tropical Night Lizard				Mexico
			47	2.04	
reticulatum	Costa Rica Tropical Night Lizard				Costa Rica
			81	10.7	* crack hermit
smithii	Smith's Tropical Night Lizard				Guatemala
	cathemeral	omnivore	92	15.4	* crack hermit
sylvaticum	Madrean Tropical Night Lizard				Mexico
			82	10.8	* crack hermit
tarascae	Tarascan Tropical Night Lizard				Mexico
	nocturnal		76	8.6	* crack hermit
tuxtlae	Tuxtla Tropical Night Lizard				Mexico
	nocturnal	omnivore	79	9.8	* crack hermit
zongolica	Zongolica Mountains Tropical Night Lizard				Mexico
			61	4.39	* crack hermit

Lepidothyris

Lepidothyris is a small genus (three spp.) of large (33.8 g v. 3.3g) mysterious skinks found in central Africa from Guinea east to Kenya and south to Angola and Zambia. Our ignorance of these lizards (filled cells 22% v. 25%) is in part responsible for the sense of mystery, but the behavior of the lizards must be extraordinarily secretive for such large, gaudy creatures to have evaded herpetological and local awareness. For example, the large, cathemeral *fernandi* is boldly marked with red and black flanks, and has been trapped by Bill Branch (2011) on the forest floor in Gabon and Liberia, but Branch has never seen one moving in nature, and they are unknown to local forest hunters. It is known to bask directly on most days. Is this a (burrow) crack hermit that only occasionally switches burrows? The other species are much less well known. Too few data are available to assign ecological business models.

They are heavier in build than average skinks (shape index 0.39 v. −0.46 std. resid.). *L. fernandi* is marginally elongate (v. 35%).

Males are larger than females (1.0 v. −0.31 of −2 to 2 scale), as expected for their size. Sexual dichromatism is found in *fernandi* (33% v. 7.6%), but no species shows seasonal color (v. 6%). Overall sex difference rollup score (on scale of 0 to 6) is somewhat elevated (0.67 v. 0.28).

Diet of *fernandi* is very broad, ranging from fruit to ectothermic vertebrates. Only cruise foraging is reported (v. 2%), which is unusual for a skink. Electric light use and home-range areas are not known. *L. fernandi* is conspicuously wary (33% v. 11%).

They are found at low and medium elevation (in this case 0–2100 m, based on *fernandi*). They are associated with riparian (100% v. 23%) and human (v. 14%) habitats, but not littoral (0% v. 21%) or sandy (v. 23%) areas.

They are oviparous (100% v. 64%) and have an exceptionally high clutch count (1.06 v. 0.06 std. resid.), again based on *fernandi*. The ontogenetic mass range of *fernandi* is even more exceptional (2.27 v. 0.12 std. resid.), indicating relatively minute hatchlings. Incubation is short (−0.68 v. −0.40 std. resid.), which equates to 59 days (*fernandi*).

Longevity is also short (−0.54 v. 0.17 std. resid.), equating to 5 years (*fernandi*), but it seems prudent to await better sampling before drawing inferences from that datum. Endemism is low (0% v. 64%). One species is found on an island (*fernandi* on Bioko Island: 33% v. 57%). None has been introduced (v. 2.1%) or evaluated for conservation status. In other respects *Lepidothyris* are probably typical skinks.

Lepidothyris spp.	English name		Size		Range core
	Activity	Diet	mm	g	Ecological business model
fernandi	Red-flanked Skink				Nigeria
	cathemeral	invert.	140	24.5	
hinkeli	Hinkel's Skink				Congo
			114	32.6	
striatus	Striped Congo Skink				Congo
			129	48.6	

Leposoma

Leposoma is a six-species genus of exceptionally poorly known (filled cells 13% v. 19%), small (0.82 g v. 2.35 g) gymnophthalmids endemic to Brazil. One species (*scincoides*) is thought to be a litter cryptozoid. At least one species has secretory pores (v. 29%). Two species are elongate (v. 41%). Two species are sexually dichromatic (v. 13%). They have not been evaluated for conservation status. In other respects *Leposoma* are probably typical gymnophthalmids.

Leposoma spp.	English name		Size		Range core
	Activity	Diet	mm	g	Ecological business model
annectans	Ruibal's Shade Lizard				Brazil
			37	0.58	
baturitensis	Baturite Shade Lizard				Brazil
			35	0.85	
nanodactylus	Bahia Shade Lizard				Brazil
			34	0.44	
puk	Relict Forest Shade Lizard				Brazil
			39	1.22	
scincoides	Skink-like Shade Lizard				Brazil
			42	1.55	** litter cryptozoid
sinepollex	Four-fingered Shade Lizard				Brazil
			34	0.76	

Leptoseps

Leptoseps are two species of tiny (0.55 g v. 3.25 g), nearly unknown (each species known from only two specimens: filled cells 19% v. 25%), extremely elongate (100% v. 35%) skinks found in Southeast Asia (one Thailand endemic, one Vietnam endemic). They are likely to be subterranean foragers, for which their elongate bodies enable them to locomote without using limbs. Their limbs are complete, but diminutive. Their bodies are slender (shape index −1.29 v. −0.46 std. resid.) but with full-length (1.27 v. 1.33 × SVL), breakable (100% v. 73%) tails terminating in a spine. Neither has been introduced (v. 2.1%) or evaluated for conservation status. In other respects *Leptoseps* are much like other limbless skinks.

Leptoseps spp.	English name		Size		Range core
	Activity	Diet	mm	g	Ecological business model
osellai	Osella's Skink				Thailand
			39	0.52	
poilani	Poilane's Skink				Vietnam
			40	0.57	

Leptosiaphos

Leptosiaphos is a moderate radiation (18 spp.) of small-sized (1.75 g v. 3.25 g), short-snouted, thick-tailed, litter cryptozoid (100% v. 28%) skinks found, often at altitude, in equatorial Africa (within the polygon of Nigeria to South Sudan to Tanzania to Angola to Nigeria). They are poorly known (filled cells 20% v. 25%). Although the common names highlight the variation in number of digits, the animating theme is fully limbed, but often-elongate (50% v. 35%), active-foraging (100% v. 92%) lizards that search the litter for invertebrates (100% v. 84%).

The only documented sex difference is in color of *blochmanni* and *meleagris* (11% v. 7.6%). Thus their overall sex differences rollup score (on scale of 0 to 6) is low: 0.11 (v. 0.28).

Most are diurnal (86% v. 74%), and the remainder are crepuscular (14% v. 2.4%). There might be seasonal or elevational variation in the boundary between those two conditions. About 6% are conspicuously wary (v. 11%). Direct basking is 5.6% (v. 21%). The majority of species use a burrow as their daily refuge (57% v. 24%). The one rainfall-documented species (*aloysiisabaudiae*) is in the mesic belt (1622 mm/y in this case). Elevation is more widely documented (12 spp.), and *Leptosiaphos* are overrepresented at intermediate (67% v. 35%) and high (25% v. 6%) elevations. They are not reported in littoral (v. 21%) or sandy (v. 25%) habitats, but they are normally frequent in riparian (29 v. 23%) and human (7% v. 14%) habitats.

All documented species are oviparous (v. 64%) and have an ordinary clutch count (0.06 v. 0.06 std. resid.). Communal oviposition (11%) is unusually common for skinks (2.6%) or litter cryptozoids (4.7%: table 4.1, trait 68).

Leptosiaphos spp.	English name				Range core
			Size		
	Activity	Diet	mm	g	Ecological business model
aloysiisabaudiae	Uganda Five-toed Skink				Cameroon
	diurnal	invert.	42	0.89	
amieti	Amiet's Mountain Skink				Cameroon
			46	1.74	
blochmanni	Kivu Three-toed Skink				Rwanda
	diurnal	invert.	49	1.34	* litter cryptozoid
dewittei	De Witte's Five-toed Skink				Congo
			52	2.54	
dungeri	Dunger's Four-fingered Skink				Nigeria
			45	1.05	* litter cryptozoid
fuhni	Fuhn's Skink				Cameroon
			42	1.27	
graueri	Rwanda Five-toed Skink				Congo
	diurnal	invert.	59	2.27	** litter cryptozoid
hackarsi	Virunga Four-toed Skink				Congo
	diurnal	invert.	53	1.69	* litter cryptozoid
hylophilus	Laurent's Five-toed Skink				Congo
			36	0.82	
ianthinoxantha	Yellow and Violet-bellied Mountain Skink				Cameroon
			55	3.12	
kilimensis	Kilimanjaro Five-toed Skink				Tanzania
	crepuscular	invert.	65	2.96	*** litter cryptozoid
koutoui	Koutou's Mountain Skink				Cameroon
			44	0.74	*** litter cryptozoid
luberoensis	De Witte's Lubero Skink				Congo
			49	2.22	
meleagris	Ruwenzori Four-toed Skink				Congo
	diurnal	invert.	63	2.72	** litter cryptozoid
pauliani	Paulian's Mountain Skink				Cameroon
			47	1.92	
rhodurus	Red Five-toed Skink				Congo
			69	6.63	
rhomboidalis	Udzungwa Five-toed Skink				Tanzania
	diurnal	invert.	48	1.29	* litter cryptozoid
vigintiserierum	Twenty-lined Mountain Skink				Cameroon
			45	1.65	* litter cryptozoid

Endemism is about average (50% v. 64%). Island species are rare (6% v. 57%). None has been introduced (v. 2.1%). Relatively few are imperiled (29% v. 47%), and they constitute a depressingly normal complement on the IUCN's lists of most endangered (14% v. 14%). In other documented respects *Leptosiaphos* are probably typical skinks.

Lerista

Lerista is the third largest skink genus (95 spp., table 2.2). Most species are small-sized (1.3 g v. 3.3 g; range 0.38–4.5 g), nocturnal (98% v. 15%), elongate (75% v. 35%), elongation undefined—because absence of any limb pair invalidates the elongation metric—or limbless (2%), invertebrate eating (99% v. 84%), subterranean (27% v. 11%) or cryptozoic (71% v. 30%) dwellers in sandy areas (71% v. 25%) of Australia. The modal *Lerista* is a sand shark (67% v. 6.8%); the remainder subterranean foragers (33% v. 12%). Although there are an intimidating number of *Lerista* species—straining the inventiveness of bestowers of common names—many species are substrate-specific and highly localized in distribution. At any one locality, species richness of *Lerista* may be low.

Lerista spp.	English name / Activity	Diet	Size mm	Size g	Range core / Ecological business model
aericeps	Desert Plain Slider				Queensland
	cathemeral	invert.	49	0.94	**** sand shark
allanae	Greater Robust Fine-lined Slider				Queensland
	nocturnal	invert.	80	3.15	*** subterranean forager
allochira	Cape Range Slider				Western Australia
	nocturnal	invert.	34	0.38	
ameles	Limbless Fine-lined Slider				Queensland
		invert.	54	0.60	
amicorum	Fortescue Slider				Western Australia
	nocturnal	invert.	44	0.74	
apoda	Dampier Land Limbless Slider				Western Australia
	nocturnal	invert.	70	1.08	**** subterranean forager
arenicola	Bight Slider				Southwestern Australia
	nocturnal	invert.	56	1.99	**** sand shark
axillaris	Stripe-sided Robust Slider				Western Australia
	nocturnal	invert.	79	3.06	
baynesi	Baynes' Slider				Southwestern Australia
	nocturnal	invert.	77	2.86	* sand shark
bipes	Northwestern Slider				Australia
	nocturnal	invert.	55	1.27	*** subterranean forager
borealis	Inland Kimberley Slider				Western Australia
	nocturnal	invert.	55	1.27	**** sand shark
bougainvillii	Southeastern Slider				Southeastern Australia
	nocturnal	invert.	59	1.51	**** sand shark
bunglebungle	Bunglebungle Robust Slider				Western Australia
	nocturnal	invert.	55	1.26	
carpentariae	Carpentaria Fine-lined Slider				Northern Territory
	nocturnal	invert.	64	1.81	*** subterranean forager
chordae	Fine-lined Slider				Queensland
	nocturnal	invert.	43	1.42	
christinae	Bold-striped Slider				Western Australia
	nocturnal	invert.	37	0.65	**** sand shark

Naturally these burrowers are slender (shape index −1.19 v. −0.46 std. resid.), with tails that are slightly shorter (0.94 v. 1.33 × SVL) and breakable (99% v. 73%). Tails are occasionally (*zietzi*) blue (1.1% v. 4.0%); *amicorum* and *xanthura* have bright yellow tails and a number (*christinae, cinerea, dorsalis, elegans, flammicauda, fragilis, ingrami, orientalis, separanda, vittata, wilkinsi*) have tails that may be red or orange.

Female fecundity is also constrained by small size, which is likely reflected in the pattern of females exceeding males in size (−1.0 v. −0.31 on −2 to 2 scale). Because SSD is reported for only three species (*baynesi, bougainvillii, dorsalis*), it is difficult to know if the sexes are equal in the other 92 species, or those species are simply unquantified. The absence of other reported sex differences, in combination with the assumption that the absence of SSD information equates to sexual size monomorphism, leads to a very low sex difference rollup (0.03 v. 0.28 on 0 to 6 scale).

Lerista are overwhelmingly (99%) consumers of invertebrates; ants and termites of all life stages are nearly universal in their diets. Only one species (*timida*) is not assessed as an invertebrate eater; 19% of its diet by mass (n = 10) is vertebrates, though this includes the lizard's own skin). This species is problematic, as the leading authority (Cogger 2014) distinguishes it from *muelleri*, but neither the original dietary source nor Reptile Database (2017) do. A reasonable generalization is that all *Lerista* are fundamentally invertebrate eaters. More uniquely for a skink, *Lerista* are about equally split between species deemed to be sit-and-wait (80% v. 16%) and active (70% v. 92%) foragers. Home-range areas have not been quantified for *Lerista*.

It is surprising that 98% of *Lerista* are nocturnal, because it would seem that light is of limited importance to subterranean foragers or sand sharks. However, the emergence onto the surface by either group, especially if emergence is sudden, would draw the unwanted attention of diurnal predators. The visual relevance of this phenomenon is underscored by the many *Lerista* with brightly colored and presumably sacrificial tails. The soaring surface temperature of the sand may be also prohibit emergence during daytime. Some degree of daily vertical migration in the substrate might also be construed by humans as indicating nocturnal behavior; if activity occurs at depth during the day, it might be undetected. A few *Lerista* are also noted to be alert and ambushing prey from the mouth of their burrow during the day (e.g., *aericeps*, *xanthura*), but surface-active only at night or at night after recent rain (e.g., *punctatovittata*). The few non-nocturnal *Lerista* are active at any time (2.3% v. 9%) rather than being diurnal or crepuscular.

Only a small fraction of *Lerista* are deemed conspicuously wary (2.1% v. 11%), but what is the flight distance for a subsurface lizard, and how would it be detected by a human? Presumably any *Lerista* unearthed during the day would immediately dive back into the sand, an inference supported by the few data on primary predator concealment (sand diving 100% v. 22%). Mean active body temperature is a very ordinary 31.3°C (v. 31.1°C).

Lerista spp.	English name				Range core
			Size		
	Activity	Diet	mm	g	Ecological business model
cinerea	Vine-thicket Fine-lined Slider				Queensland
	nocturnal	invert.	64	1.79	*** subterranean forager
clara	Clara Slider				Western Australia
	nocturnal	invert.	44	0.74	* sand shark
colliveri	Nubbined Fine-lined Slider				Queensland
	nocturnal	invert.	77	2.84	
connivens	Blinking Broad-blazed Slider				Western Australia
	nocturnal	invert.	75	2.71	*** subterranean forager
desertorum	Central Deserts Robust Slider				Western Australia
	nocturnal	invert.	79	3.05	**** sand shark
distinguenda	Southwestern Orange-tailed Slider				Western Australia
	nocturnal	invert.	45	0.77	**** sand shark
dorsalis	Southern Slider				Southwestern Australia
	nocturnal	invert.	57	1.39	**** sand shark
edwardsae	Edwards' Slider				South Australia
	nocturnal	invert.	80	3.15	* sand shark
elegans	Elegant Slider				Western Australia
	nocturnal	invert.	37	0.60	**** sand shark
elongata	Wide-striped Mulch-slider				South Australia
	nocturnal	invert.	55	1.22	** sand shark
emmotti	Noonbah Robust Slider				Queensland
	nocturnal	invert.	91	4.33	**** sand shark
eupoda	West Coast Mulga Slider				Western Australia
	nocturnal	invert.	79	3.02	**** sand shark
flammicauda	Pilbara Flame-tailed Slider				Western Australia
	nocturnal	invert.	51	1.50	**** sand shark
fragilis	Eastern Mulch-slider				Queensland
	nocturnal	invert.	52	1.11	**** sand shark
frosti	Centralian Slider				Western Australia
	nocturnal	invert.	56	1.32	**** sand shark
gascoynensis	Gascoyne Broad-blazed Slider				Western Australia
	nocturnal	invert.	63	1.74	*** subterranean forager
gerrardii	Bold-striped Robust Slider				Western Australia
	nocturnal	invert.	76	2.75	**** sand shark
greeri	Southeastern Kimberley Sandslider				Western Australia
	nocturnal	invert.	59	1.47	*** subterranean forager
griffini	Stout Sandslider				Western Australia
	nocturnal	invert.	60	1.54	*** subterranean forager
haroldi	Gnaraloo Mulch-slider				Western Australia
	nocturnal	invert.	39	0.53	**** sand shark

Lerista spp.	English name		Size		Range core
	Activity	Diet	mm	g	Ecological business model
hobsoni	Pentland Slider				Queensland
			56	1.28	* subterranean forager
humphriesi	Taper-tailed West-coast Slider				Western Australia
	nocturnal	invert.	57	1.38	* sand shark
ingrami	McIvor River Slider				Queensland
	nocturnal	invert.	35	0.55	
ips	Robust Duneslider				Western Australia
	nocturnal	invert.	62	1.71	**** subterranean forager
jacksoni	Jackson's Slider				Western Australia
	nocturnal	invert.	41	0.60	* sand shark
kalumburu	Kalumburu Kimberley Slider				Western Australia
	nocturnal	invert.	54	1.20	**** sand shark
karlschmidti	Lesser Robust Fine-lined Slider				Queensland
	nocturnal	invert.	63	1.73	*** subterranean forager
kendricki	Dark Broad-blazed Slider				Western Australia
	nocturnal	invert.	61	1.60	
kennedyensis	Kennedy Range Broad-blazed Slider				Western Australia
	nocturnal	invert.	53	1.16	
kingii	King's Slider				Western Australia
	nocturnal	invert.	42	0.87	* sand shark
labialis	Southern Sandslider				Australia
	nocturnal	invert.	53	1.73	*** subterranean forager
lineata	Perth Slider				Western Australia
	nocturnal	invert.	47	0.80	**** sand shark
lineopunctulata	Dotted-lined Robust Slider				Western Australia
	nocturnal	invert.	93	4.52	*** subterranean forager
macropisthopus	Unpatterned Robust Slider				Western Australia
	nocturnal	invert.	86	3.72	**** sand shark
maculosa	Spotted Slider				Western Australia
	nocturnal	invert.	39	0.53	* sand shark
micra	Little Slider				Western Australia
	nocturnal	invert.	42	0.91	* sand shark
microtus	Southwestern Slider				Southwestern Australia
	nocturnal	invert.	52	1.62	**** sand shark
muelleri	Wood Mulch-slider				Australia
	nocturnal	invert.	44	1.04	**** sand shark
neander	Pilbara Robust Slider				Western Australia
	nocturnal	invert.	77	2.86	**** sand shark
nevinae	Nevin's Slider				Western Australia
	nocturnal	invert.	39	0.74	** sand shark

Social relations are reported for only one *Lerista* (*bougainvillii*), for which the males are intolerant (v. 94%).

Lerista are found at the drier middle of the rainfall spectrum: 71% arid (v. 40%) and 29% intermediate (v. 34%). The few with elevational documentation (four spp.) are all low (v. 89%). Though mostly associated with sandy habitats, they are slightly overrepresented in littoral areas (38% v. 21%) and underrepresented in human habitats (1% v. 14%). They are about average in riparian zones (16.5% v. 23%).

All reproductively documented *Lerista* are oviparous, except for *arenicola*, *bougainvillii*, and *microtis*. *L. arenicola* and *microtis* are contested and may exhibit both reproductive modes. *L. bougainvillii* is definitely both, with the interesting association of insularity: they are oviparous at most sites on the mainland and viviparous on islands south of Australia (generally cool). Reportedly both hard and soft eggshells are known in *bougainvillii*; the other species follow the skink standard (soft). Clutch counts are commensurate with *Lerista*'s small size (0.17 v. 0.06 std. resid.). Ontogenetic mass range is slightly depressed (−0.39 v. 0.12 std. resid.), indicating moderately large hatchlings. Incubation duration is average (−0.41 v. −0.40 std. resid.). RCM ratio is a normal 0.203 (v. 0.236).

Longevity is notably long (0.97 v. 0.17 std. resid.) if poorly documented, equating to 8.3 years in *bipes* (only sp. documented). Population density is low (0.11 kg/ha v. 0.35 kg/ha), based on a single estimate each of *elegans*, *lineata*, and *punctatovittata*. Presumably the productivity of *Lerista*-occupied dunes is generally low. Endemism is high (74% v. 64%), perhaps reflecting both soil-type specialization and the large size of Australian states. Island species are low (25% v. 57%). No *Lerista* have been introduced (v. 2.1%). Most are imperiled (65% v. 47%); about 9% are on one of the IUCN's lists of the most endangered (8.7% v. 14.1%). In other respects *Lerista* are similar to average skinks.

Lerista spp.	English name				Range core
			Size		
	Activity	Diet	mm	g	Ecological business model
nichollsi	Inland Broad-blazed Slider				Western Australia
	nocturnal	invert.	60	1.56	*** subterranean forager
occulta	Carnarvon Slider				Western Australia
	nocturnal	invert.	38	0.70	* sand shark
onsloviana	Onslow Broad-blazed Slider				Western Australia
	nocturnal	invert.	64	1.81	** sand shark
orientalis	Northeastern Orange-tailed Slider				Northern Australia
	nocturnal	invert.	45	0.78	**** sand shark
petersoni	Pale Broad-blazed Slider				Western Australia
	nocturnal	invert.	68	2.13	*** subterranean forager
picturata	Southern Robust Slider				Southwestern Australia
	nocturnal	invert.	79	3.07	**** sand shark
planiventralis	Keeled Slider				Western Australia
	cathemeral	invert.	60	1.55	**** sand shark
praefrontalis	Yampi Sandslider				Western Australia
			61	1.64	* sand shark
praepedita	Blunt-tailed West-coast Slider				Western Australia
	nocturnal	invert.	59	1.49	**** subterranean forager
punctatovittata	Eastern Robust Slider				Eastern Australia
	nocturnal	invert.	87	3.86	**** sand shark
puncticauda	Dotty-tailed Robust Slider				Western Australia
	nocturnal	invert.	77	2.89	* sand shark
quadrivincula	Four-chained Slider				Western Australia
			49	0.93	* sand shark
robusta	Broad-eyed Sandslider				Western Australia
	nocturnal	invert.	58	1.42	* sand shark
rochfordensis	Rochford Slider				Queensland
	nocturnal	invert.	68	2.12	
rolfei	Rolfe's Slider				Western Australia
	nocturnal	invert.	44	0.72	* sand shark
separanda	Dampierland Plain Slider				Western Australia
	nocturnal	invert.	32	0.41	**** sand shark
simillima	Fitzroy Sandslider				Western Australia
	nocturnal	invert.	51	1.05	
speciosa	Pale-striped Mulch-slider				South Australia
			49	1.36	* sand shark
stictopleura	Spotted Broad-blazed Slider				Western Australia
	nocturnal	invert.	53	1.16	*** subterranean forager
storri	Chillagoe Fine-lined Slider				Queensland
	nocturnal	invert.	65	1.88	*** subterranean forager

Lerista spp.	English name				Range core
			Size		
	Activity	Diet	mm	g	Ecological business model
stylis	Arnhem Coast Fine-lined Slider				Northern Territory
	nocturnal	invert.	63	1.76	*** subterranean forager
taeniata	Ribbon Slider				Western Australia
	nocturnal	invert.	41	0.61	* subterranean forager
talpina	Mole Slider				Western Australia
	nocturnal	invert.	37	0.48	
terdigitata	Robust Mulch-slider				Southwestern Australia
	nocturnal	invert.	61	1.58	**** sand shark
timida	Tenuous Mulch-slider				Australia
		carnivore	40	1.10	** sand shark
tridactyla	Dark-backed Mulch-slider				Western Australia
	nocturnal	invert.	52	1.09	* sand shark
uniduo	Slender Broad-blazed Slider				Western Australia
	nocturnal	invert.	53	1.14	* sand shark
vanderduysi	Blackbraes Slider				Queensland
			56	1.28	* subterranean forager
varia	Shark Bay Broad-blazed Slider				Western Australia
	nocturnal	invert.	73	2.48	*** subterranean forager
verhmens	Three-fingered Pilbara Slider				Western Australia
	nocturnal	invert.	45	1.07	* sand shark
vermicularis	Slender Duneslider				Western Australia
	nocturnal	invert.	40	0.57	**** subterranean forager
viduata	Ravensthorpe Range Slider				Western Australia
	nocturnal	invert.	43	0.68	* sand shark
vittata	Side-striped Fine-lined Slider				Queensland
	nocturnal	invert.	64	1.80	*** subterranean forager
walkeri	Coastal Kimberley Slider				Western Australia
	nocturnal	invert.	58	1.41	**** sand shark
wilkinsi	Two-toed Fine-lined Slider				Queensland
	nocturnal	invert.	61	1.62	**** subterranean forager
xanthura	Yellow-tailed Plain Slider				Australia
	nocturnal	invert.	46	0.79	**** sand shark
yuna	Yuna Broad-blazed Slider				Western Australia
	nocturnal	invert.	61	1.58	* sand shark
zietzi	Pilbara Blue-tailed Slider				Western Australia
	nocturnal	invert.	50	0.96	**** sand shark
zonulata	Girdled Slider				Queensland
	nocturnal	invert.	46	0.80	* subterranean forager

Lialis

Lialis is a two-species genus of large (15.2 g v. 5.2 g), cathemeral (100% v. 29%), carnivorous (100% v. 9%) pygopodids from Australia and New Guinea. These comparatively well-known (filled cells 44% v. 31%) lizards specialize in ambushing skinks from a concealed location in the vegetation. *L. burtonis* can consume proportionately huge skinks, having a prey-size ratio reminiscent of viper feeding. It has some unique morphological specializations to go along with this specialized diet. Like ophidian crab eaters, it has hinged recurved teeth (the teeth "lay down" when unable to penetrate and then snag on sutures in the skinks as the mandibles are slid forward). As in the bolyerid snakes, *burtonis* has a mid-mandibular articulation that allows it to encircle protruding hard, cylindrical prey. Like many small snakes, it may use caudal luring to entice curious skinks to approach within striking range. No wonder they are popularly known as snake-lizards.

Lialis spp.	English name				Range core
			Size		
	Activity	Diet	mm	g	Ecological business model
burtonis	Burton's Snake-lizard				Australia
	cathemeral	carnivore	206	13.6	*** litter cryptozoid
jicari	Jiear's Legless Lizard				New Guinea
	cathemeral	carnivore	240	16.8	

Liburnascincus spp.	English name				Range core
			Size		
	Activity	Diet	mm	g	Ecological business model
artemis	Bamboo Rainbow-skink				Queensland
	diurnal		51	2.43	* refugia-anchored diurnal gr. liz.
coensis	Coen Rainbow-skink				Queensland
	diurnal		54	2.88	* refugia-anchored diurnal gr. liz.
mundivensis	Outcrop Rainbow-skink				Queensland
	diurnal		54	2.88	* refugia-anchored diurnal gr. liz.
scirtetis	Black Mountain Rainbow-skink				Queensland
	diurnal	invert.	61	4.43	*** refugia-anchored diurnal gr. liz.

Lialis are slender even by pygopodid standards (shape index –2.35 v. –1.32 std. resid.). They have much shorter tails than other pygopodids (1.30 v. 2.65 × SVL). They do not exhibit color change (v. 4.3%) or seasonal color (v. 2.2%).

Both species exhibit sit-and-wait foraging (100% v. 64%), but *burtonis* may also actively forage (50% v. 64%). Both species vocalize spontaneously (v. 67%). *L. burtonis* basks directly (50% v. 17%) and has a mean active body temperature of 29.0°C (v. 28.3°C). It lives in arid areas (in this case 202–312 mm/y). The elevational range of *jicari* is 0–1600 m. One species (50%) occurs in sandy areas (v. 50%), riparian zones (v. 21%), and human habitats (v. 9%), but not littoral habitats (0% v. 30%).

Lialis clutches are comparatively small relative to their size (–0.84 v. –0.44 std. resid.). *L. burtonis* females may defend their clutch (v. 2.2% parental care), and *jicari* lays communally (v. 6.5%). RCM is light (0.137 v. 0.220), unlike many extreme ambushers.

One population density estimate has been made (*burtonis*): 0.25 kg/ha (v. 0.02 kg/ha). Neither species is endemic (v. 52%), nor has either been introduced (v. 0%). Both species occur on islands (v. 26%). Neither is imperiled (v. 88%). In other respects *Lialis* are similar to other pygopodids.

Liburnascincus

Liburnascincus is a minor radiation (four spp.) of average-sized (3.07 g v. 3.25 g), heliophilic (direct basking 100% v. 21%), saxicolous, refugia-anchored diurnal ground skinks (100% v. 30%) endemic (100% v. 64%) to Queensland. If their common names bear some similarity to skinks of the genus *Carlia*, that commonality reflects their former taxonomic grouping. Ecologically they differ from *Carlia* primarily in being dependent on rock outcrops and crevices.

They are slightly more robust in build than average skinks (shape index –0.09 v. –0.46 std. resid.). They are never elongate (v. 35%). No sex differences have been reported. They are active foragers (100% v. 92%) on invertebrates (100% v. 84%). They are found at low elevation (v. 89%) and are associated with riparian zones (50% v. 23%). They are oviparous (100% v. 64%) with a clutch of two (–0.25 v. 0.06 std. resid.). They make up for the small clutch count by laying repeatedly in a season; adequate quantification is not yet available for broods per year.

They are not reported to be on islands (v. 57%), nor to have been introduced (v. 2.1%). All are imperiled (v. 47%), but none is on any of the IUCN's lists of the most endangered (v. 14%). In other respects *Liburnascincus* are much like other skinks.

Liolaemidae

Liolaemidae is the sixth-largest lizard family (306 spp., table 2.1), yet it has only three genera. The three genera are ecologically distinct: *Ctenoblepharys* (one sp., a mysterious Peruvian coastal desert dweller), *Liolaemus* (258 spp., table 2.2, the second-largest genus of lizards: of diverse diets and reproduction, but 95% are refugia-anchored diurnal ground lizards), and *Phymaturus* (47 spp. of corpulent viviparous herbivores). They share geography: most are in Argentina or Chile (northernmost species reaches Peru; easternmost is in Brazil). They are poorly known (22% v. 27% of cells filled). They are overwhelmingly refugia-anchored diurnal ground lizards (96% v. 24.5%); the remainder (all restricted to *Liolaemus*) are ground-foraging tree lizards (1.6% v. 9.2%), sand sharks (1.6% v. 1.9%), and trunk/wall lizards (1.2% v. 13%).

Ecologically, Liolaemidae are markedly distinct for their diet, foraging mode, diurnality, microhabitat, primary predator concealment strategy, rainfall band, elevation, and reproductive mode. Diet: 26% (v. 4%) are herbivores; 35% (v. 15%) are omnivores. Foraging mode: 84% (v. 46%) are sit-and-wait. Diurnality: 99% (v. 67%). Microhabitat: terrestrial 97% (v. 48%). Primary predator concealment strategy: 46% (v. 25%) are crevice-burrow hiding. Rainfall band: 67% (the second-highest family, v. 12%) reside in the dry belt (56–150 mm/y). Elevation: 44% (the highest family, v. 13%) are found above 2400 m (four species reach 5000 m). Reproductive mode: a mere 33% (v. 82%) are oviparous.

They are moderately large (9.4 g v. 5.3 g) and robust (shape index 0.76 v. 0.0 std. resid.), but not otherwise morphologically specialized. Their ontogenetic range of $\log_{10}$ masses is 1.4 (v. 1.6), a notably low value for such large lizards; do the young need to be relatively larger in this hydrically stressful arid environment? Relative tail length is average 1.41 (v. 1.37 × SVL). Tail breakage is reported to be only 22% (v. 57%); it is possible that some of the shortfall is due to underreporting in this poorly known group. No tails are prehensile (v. 5.0%). No species has crests (v. 5.3%) or is elongate (v. 11%). Secretory pores are 22% (v. 15%). Color change is uncommon (any: 2.0% v. 10.7%; advanced: 1.3% v. 6.0%). Blue tails are rare (0.3% v. 2.3%).

Overall sex differences are about average (0.62 v. 0.75 on 0 to 6 scale). These consist of sex differences in body size (0.45 v. 0.36 on –2 to 2 scale), a value commensurate with their moderately large body size (table 3.3, trait 1); sex differences in color (29% v. 21%); seasonal color (2.3% v. 6.5%); and an average m:f ratio of relative tail lengths (1.05 v. 1.07).

Although many liolaemids eat plants, 34.5% (v. 73%) are known to be generalized invertebrate eaters and 4.6% (v. 4.4%) to be ant specialists. Active foragers are a minority (16% v. 58%), though that proportion could be expanded to include the many plant eaters. Electric light foraging is rare (0.3% v. 2.0%).

As noted above, almost all liolaemids are diurnal; the remaining 0.7% are crepuscular (v. 1.8%). Despite the absence of nocturnal or cathemeral species, 1% (v. 2.5%) call spontaneously and 0.3% (v. 2.2%) give stress calls. Only 2.0% (v. 8.2%) are conspicuously wary. Minority predator avoidance tactics are crypsis (4.2% v. 43%), flight (0.0% v. 10.8%), sand (29% v. 6%), and hiding in vegetation (29% v. 23%). Direct basking is underrepresented (7.8% v. 18%), perhaps at least partially due to underreporting. Thermal conforming is not noted (0% v. 2.6%). Despite the high elevations, mean active body temperature is warm (33.7°C v. 31.1°C).

Sociality is poorly documented, but no species is characterized as solitary (v. 10%); 50% (v. 20%) live in groups of more than two. No males are said to be socially tolerant (v. 15%); males of all documented species defend space (v. 85%). No males form hierarchies (v. 7.7%), perhaps a reflection of the low population densities.

The dominance of terrestriality mentioned above is reinforced by the paucity of species depending on alternate microhabitats: arboreal (2.4% v. 32%), cryptozoic (0.4% v. 12%), subterranean (0% v. 7%), and semiaquatic (0% v. 1.1%). Use of burrows as daily refuges is normal (20% v. 22%). Liolaemidae are underrepresented at low elevation (47% v. 82%) and are found at medium elevation at typical frequency (41% v. 41%). They are overrepresented in sandy areas (32% v. 22%) and underrepresented in the other specialized habitats: littoral (6% v. 17%), riparian (10% v. 22%), and human (2% v. 18.5%) habitats.

Although the majority of Liolaemidae are viviparous, the family's mean clutch/litter count is decidedly average (–0.16 v. 0.0 std. resid.); the anomalies that exist are all in *Phymaturus*, and most other reproductive metrics are near normal. For example, incubation period is only slightly short (–0.28 v. 0.0), despite the cooler temperatures associated with high elevation. The relative size of young is notably large (ontogenetic mass range –0.59 v. 0.0 std. resid.). Parental care is barely reported (0.3% v. 2.9%), but under-documentation is probable.

Longevity is about normal (0.13 v. 0.0 std. resid.). Endemism is high (82% v. 64%), and this may focus attention on the many substrate specializations, or at least the many dispersal barriers, that may be responsible for the remarkable radiation of the liolaemids. They do not seem to have developed the high alpha diversity (i.e., number of species in a site) of anoles. Instead they show a very large number of species whose distribution is limited to a small area. That is, high endemism and rapid species replacement across geography.

Only one liolaemid has been introduced; *Liolaemus wiegmannii* to the mid-Atlantic island of St. Helena. Twenty-one percent (v. 49%) of liolaemids are imperiled, and 4.5% (v. 13.6%) are listed on the IUCN's lists of the most endangered.

Liolaemus

Liolaemus (Liolaemidae) is the second-largest genus of lizards (258 spp., table 2.2) and is found in the southern Neotropics (Peru and Brazil, south to Tierra del Fuego). Most are found in Argentina (165 spp.) or Chile (109 spp.). Peru (21 spp.) and Bolivia (18 spp.) are the only other countries with more than five species. Though many species occur in more or less horizontal habitats, the radiation of *Liolaemus* seems associated with the Andes, presumably due to the heterogeneity in soil and rock cover types generated by the uplift. The topography has not produced an equivalent radiation in ecological business models, as 95% of *Liolaemus* are classified—most only tentatively—as refugia-anchored diurnal ground lizards. The other ecological business models (ground-foraging tree lizards, sand sharks, trunk/wall lizards) each constitute less than 2%.

In part because *Liolaemus* constitute such a large proportion (84%) of the family, the salient characteristics of the family are also those of the genus: omnivory (genus v. family: 41% v. 35%), sit-and-wait foraging (93%

Liolaemus spp.	English name				Range core
			Size		
	Activity	Diet	mm	g	Ecological business model
abaucan	Palo Blanco Lizard				Argentina
	diurnal		56	5.4	* refugia-anchored diurnal gr. liz.
abdalai	Abdala's Lizard				Argentina
	diurnal		48	3.41	* refugia-anchored diurnal gr. liz.
acostai	Acosta's False Strange Lizard				Argentina
	diurnal	ants	64	8.4	* refugia-anchored diurnal gr. liz.
albiceps	White-headed Cordilleran Lizard				Argentina
	diurnal	herbivore	79	15.63	* refugia-anchored diurnal gr. liz.
alticolor	Brilliant Cordilleran Lizard				Bolivia
	diurnal	omnivore	49	3.54	* refugia-anchored diurnal gr. liz.
andinus	Andes Cordilleran Lizard				Argentina
	diurnal	omnivore	80	16.2	* refugia-anchored diurnal gr. liz.
annectens	Caylloma Cordilleran Lizard				Peru
	diurnal		71	11.3	* refugia-anchored diurnal gr. liz.
anomalus	Strange Foothills Lizard				Argentina
	crepuscular	invert.	77	14.8	* refugia-anchored diurnal gr. liz.
antumalguen	Sunwife Lizard				Argentina
	diurnal	invert.	87	21.0	* refugia-anchored diurnal gr. liz.
aparicioi	Effen's Cordilleran Lizard				Bolivia
	diurnal		54	4.79	* refugia-anchored diurnal gr. liz.
arambarensis	Arambare Lizard				Brazil
	diurnal	omnivore	52	4.22	** refugia-anchored diurnal gr. liz.
araucaniensis	Villarrica Volcano Lizard				Argentina
	diurnal	omnivore	58	6.0	* refugia-anchored diurnal gr. liz.

Liolaemus spp.	English name				Range core
			Size		
	Activity	Diet	mm	g	Ecological business model
archeforus	Lakes Region Lizard				Argentina
	diurnal	omnivore	76	13.8	* refugia-anchored diurnal gr. liz.
atacamensis	Atacama Variable Lizard				Chile
	diurnal		65	8.4	* refugia-anchored diurnal gr. liz.
audituvelatus	Covered-ear Lizard				Chile
	diurnal		52	4.41	* refugia-anchored diurnal gr. liz.
austromendocinus	South Mendoza Lizard				Argentina
	diurnal	invert.	84	18.9	* refugia-anchored diurnal gr. liz.
avilai	Avila's Lizard				Argentina
	diurnal		53	4.6	* refugia-anchored diurnal gr. liz.
azarai	Azara's Lizard				Argentina
	diurnal		48	3.39	* refugia-anchored diurnal gr. liz.
baguali	Bagual Mountains Lizard				Argentina
	diurnal	invert.	72	11.7	* refugia-anchored diurnal gr. liz.
bellii	Bell's Cordilleran Lizard				Chile
	diurnal	omnivore	67	9.7	* refugia-anchored diurnal gr. liz.
bibronii	Bibron's Montane Lizard				Argentina
	diurnal	ants	52	3.90	** refugia-anchored diurnal gr. liz.
bitaeniatus	Two-striped Montane Lizard				Argentina
	diurnal	invert.	54	4.79	* refugia-anchored diurnal gr. liz.
boulengeri	Boulenger's Five-striped Lizard				Argentina
	diurnal	invert.	62	7.3	* refugia-anchored diurnal gr. liz.
buergeri	Buerger's Andes Lizard				Argentina
	diurnal	omnivore	80	16.4	* refugia-anchored diurnal gr. liz.
burmeisteri	Burmeister's Montane Lizard				Argentina
	diurnal		70	10.9	* refugia-anchored diurnal gr. liz.
calchaqui	Punta de Agua Lizard				Argentina
	diurnal	herbivore	52	4.41	* refugia-anchored diurnal gr. liz.
canqueli	Canquel Plateau Lizard				Argentina
	diurnal	invert.	81	17.0	** refugia-anchored diurnal gr. liz.

v. 84%), diurnality (99% v. 99%), crevice-burrow reliance as a primary predator concealment tactic (40% v. 46%), terrestriality (97% v. 97%), frequency in the dry rainfall band (60% v. 67%), frequency at high elevation (47% v. 44%), and oviparity (41% v. 33%). The one anomaly is herbivory, for which the genus is notably underrepresented in comparison to the family (10.6% v. 26.4%). However, note the high level of omnivory (see above). Note also that *silvanae* eats foliose lichens.

Liolaemus are average-sized liolaemids (8.1 g v. 9.4 g) and of a liolaemid-standard heavy build (shape index 0.69 v. 0.76 std. resid.). Tails are of average length (1.48 v. 1.41 × SVL) and breakage (24% v. 22%). About a quarter of species have secretory pores (25% v. 22%).

Sexual size differences are slightly m>f, amplified somewhat in *Liolaemus* (0.60 v. 0.45 on –2 to 2 scale), but dichromatism is slightly more subdued in frequency (25% v. 29%). Individual species may of course have strong sexual dichromatism: *ornatus* is so strongly marked that females were originally described as a separate species. Male:female relative tail length ratio is identical to the family (1.05 v. 1.05). Accordingly, the sex differences rollup is indistinguishable (0.60 v. 0.62 on 0 to 6 scale).

Vocalizations are predictably rare in these diurnal lizards but not unknown: *chiliensis* gives stress calls and *zabalai* gives unspecified calls. Wariness, predator concealment tactics, and thermoregulatory behavior are those typical of the family. This includes an ability to raise body temperature to well above ambient (e.g., *signifer* and *silvanae* may be warm enough to be active at air temperatures around freezing). Group living is not reported for *Liolaemus* (v. 50%), but spatial defense by males is universal (v. 100%).

Although the modal *Liolaemus* occurs in the dry rainfall band, as noted above, there is a second mode at the wetter end of the spectrum: (intermediate, or 401–1100 mm/y: 40% v. 33%, and mesic, or 1101–3000 mm/y: 40% v. 33%). The elevational distribution is about equal (l/m/h: 46/43/47 v. 47/41/44). Surprisingly, seven *Liolaemus* are found above 4800 m. Habitat representation is as for the family. Several species (*ditadai, halonastes, salinicola, scapularis*) are specialists for saline lake margins (see also *Ctenophorus* account) or are at least tolerant of salt-encrusted soils.

Clutch/litter count is normal (0.05 v. –0.16 std. resid.). Young are of normal size (ontogenetic mass range –0.41 v. –0.59 std. resid.). Parental care is rare (0.4% v. 0.3%), but interestingly, young of *huacahuasicus* may remain with their parents for up to two years.

Longevity is slightly short (–0.16 v. 0.13 std. resid.). Population density is known for only *ruibali* (1.1 kg/ha), *darwinii* (0.9 kg/ha), *lutzae* (0.3 kg/ha), *arambarensis* (0.1 kg/ha), and *multimaculatus* (0.02 kg/ha), values that seem in line with lizards from relatively unproductive continental areas. All five of these species are associated with sandy areas (v. 37% for genus and 32% for family) and thus may be unrepresentative. Endemism in *Liolaemus* is typical of the family (79% v. 82%). Only one species (*wiegmannii*) has been introduced (0.4% v. 0.3%), as noted in the family account. *Liolaemus* are imperiled at the low family rate (18% v. 21%) for any imperilment; 5.2% v. 4.5% for placement on the IUCN's lists of the most endangered. The species are at risk primarily because of highly restricted distributions.

Several *Liolaemus* are reputedly of

Liolaemus spp.	English name				Range core
			Size		
	Activity	Diet	mm	g	Ecological business model
caparensis	Rock Plateau Lizard				Argentina
	diurnal		52	4.23	* refugia-anchored diurnal gr. liz.
capillitas	Minas Capillitas Cordilleran Lizard				Argentina
	diurnal	ants	76	13.9	** trunk/wall lizard
carlosgarini	Garin's Lizard				Chile
	diurnal		59	6.5	
casamiquelai	Casamiquela's Lizard				Argentina
	diurnal	omnivore	85	19.9	** refugia-anchored diurnal gr. liz.
cazianiae	Caziani's Cordilleran Lizard				Argentina
	diurnal	herbivore	64	8.1	* refugia-anchored diurnal gr. liz.
chacabucoense	Chacabuco Pampas Lizard				Argentina
	diurnal	invert.	56	5.4	* refugia-anchored diurnal gr. liz.
chacoensis	Chaco Sand Lizard				Bolivia
	diurnal	ants	46	2.90	** refugia-anchored diurnal gr. liz.
chaltin	Chaltin Alpine Lizard				Argentina
	diurnal	invert.	52	4.23	* refugia-anchored diurnal gr. liz.
chavin	Chavin Alpine Lizard				Peru
	diurnal		57	5.9	* refugia-anchored diurnal gr. liz.
chehuachekenk	Calcatapul Mountains Lizard				Argentina
	diurnal		84	18.9	
chiliensis	Weeping Lizard				Argentina
	diurnal	invert.	81	17.1	* refugia-anchored diurnal gr. liz.
chillanensis	Chillan Cordilleran Lizard				Argentina
	diurnal	omnivore	66	9.2	* refugia-anchored diurnal gr. liz.
chlorostictus	Jujuy Cordilleran Lizard				Argentina
	diurnal	omnivore	70	10.9	* refugia-anchored diurnal gr. liz.
choique	Choique Pass Lizard				Argentina
	diurnal	invert.	74	13.2	* refugia-anchored diurnal gr. liz.
chungara	Chungara Lake Lizard				Chile
	diurnal		49	3.71	* refugia-anchored diurnal gr. liz.
cinereus	San Guillermo Lizard				Argentina
	diurnal		54	5.0	* refugia-anchored diurnal gr. liz.

Liolaemus spp.	English name				Range core
			Size		
	Activity	Diet	mm	g	Ecological business model
coeruleus	Blue-flanked Montane Lizard				Argentina
	diurnal	omnivore	58	6.2	* refugia-anchored diurnal gr. liz.
confusus	Bewildering Lizard				Chile
	diurnal		56	5.4	
constanzae	Emerald-flecked Lizard				Argentina
	diurnal	omnivore	63	7.7	* refugia-anchored diurnal gr. liz.
crandalli	Mahuida Volcano Lizard				Argentina
	diurnal	omnivore	76	13.9	* refugia-anchored diurnal gr. liz.
cranwelli	Cranwell's Sand Lizard				Bolivia
	diurnal		54	4.84	
crepuscularis	Twilight Lizard				Argentina
	diurnal	herbivore	57	5.6	
cristiani	Aquamarine-flanked Lizard				Chile
	diurnal	omnivore	67	9.6	* refugia-anchored diurnal gr. liz.
curicensis	Curicen Montane Lizard				Chile
	diurnal	invert.	59	6.5	* refugia-anchored diurnal gr. liz.
curis	Hot Springs Lizard				Chile
	diurnal		83	18.4	
cuyanus	Hot Dune Lizard				Argentina
	diurnal	omnivore	85	19.8	* refugia-anchored diurnal gr. liz.
cuyumhue	Neuquén Dunes Lizard				Argentina
	diurnal	invert.	54	4.79	* refugia-anchored diurnal gr. liz.
cyaneinotatus	Blue-shouldered Lizard				Argentina
	diurnal		51	4.05	
cyanogaster	Blue-bellied Patagonia Lizard				Argentina
	diurnal	invert.	60	6.80	* ground-foraging tree lizard
darwinii	Darwin's Dunefield Lizard				Argentina
	diurnal	ants	57	5.5	** refugia-anchored diurnal gr. liz.
diaguita	Diaguita Lizard				Argentina
	diurnal	invert.	45	2.76	* refugia-anchored diurnal gr. liz.
dicktracyi	Tracy's Montane Lizard				Argentina
	diurnal	invert.	75	13.5	* refugia-anchored diurnal gr. liz.
disjunctus	Far Northern Cordilleran Lizard				Peru
	diurnal		75	13.5	

medicinal value (*melanogaster*, *orientalis*, *ornatus*). One might expect such resources to be treated with great reverence; *poconchilensis*, on the other hand, is considered to be venomous. Curiously, three *Liolaemus* feature in the commemoration of Easter (*albiceps*, *orientalis*, *ornatus*: note overlap with medicinal list). At least two of these, *orientalis* and *ornatus*, are ritually killed during Holy Week to absolve sins.

L. dorbignyi honors collector/naturalist Alcide Dessalines d'Orbigny (1802–1857). From the biological diversity of fossil assemblages, d'Orbigny inferred that rock had been laid down over millions of years, the first such refutation of the biblically inspired notion that the Earth was of historical creation. In the modern era one finds it difficult to appreciate what an enormous conceptual leap this inference required.

Liolaemus spp.	English name				Range core
			Size		
	Activity	Diet	mm	g	Ecological business model
ditadai	Di Tada's Lizard				Argentina
	diurnal	invert.	77	14.8	* refugia-anchored diurnal gr. liz.
donosobarrosi	Donoso-Barros' Foothills Lizard				Argentina
	diurnal		56	5.4	* refugia-anchored diurnal gr. liz.
dorbignyi	Dorbigny's Montane Lizard				Argentina
	diurnal	herbivore	82	17.9	* refugia-anchored diurnal gr. liz.
duellmani	Duellman's Andes Lizard				Argentina
	diurnal		69	10.3	* refugia-anchored diurnal gr. liz.
dumerili	Duméril's Desert Shrub Lizard				Argentina
	diurnal		67	9.6	* refugia-anchored diurnal gr. liz.
eleodori	Eleodor's Cordilleran Lizard				Chile
	diurnal		68	9.9	* refugia-anchored diurnal gr. liz.
elongatus	Elongate Rock Lizard				Argentina
	diurnal	invert.	75	13.3	* refugia-anchored diurnal gr. liz.
erguetae	Altiplano Lizard				Bolivia
	diurnal		57	5.9	
erroneus	Erroneous Alpine Lizard				Chile
	diurnal		57	5.6	
escarchadosi	Escarchados Lizard				Argentina
	diurnal		74	13.2	* refugia-anchored diurnal gr. liz.
espinozai	Espinoza's Sand Lizard				Argentina
	diurnal	ants	54	5.0	* refugia-anchored diurnal gr. liz.
etheridgei	Etheridge's Colorful Lizard				Peru
	diurnal	invert.	59	6.6	* refugia-anchored diurnal gr. liz.
exploratorum	Moreno's Ghost Lizard				Argentina
	diurnal		49	3.54	
fabiani	Fabian's Lizard				Chile
	diurnal		71	11.4	
famatinae	Famatina Mountains Lizard				Argentina
	diurnal	omnivore	50	3.93	* refugia-anchored diurnal gr. liz.
fittkaui	Fittkau's Lizard				Bolivia
	diurnal		54	4.79	

Liolaemus spp.	English name				Range core
			Size		
	Activity	Diet	mm	g	Ecological business model
fitzgeraldi	Fitzgerald's Cordilleran Lizard				Argentina
	diurnal		56	5.4	** refugia-anchored diurnal gr. liz.
fitzingerii	Fitzinger's Montane Lizard				Argentina
	diurnal	omnivore	88	22.1	* refugia-anchored diurnal gr. liz.
flavipiceus	Yellow and Black Lizard				Chile
	diurnal		78	15.2	* refugia-anchored diurnal gr. liz.
forsteri	Forster's Bolivia Lizard				Bolivia
	diurnal		83	18.4	
foxi	Fox's Desert Lizard				Chile
	diurnal		69	10.3	* refugia-anchored diurnal gr. liz.
frassinettii	Frassinetti's Coastrange Lizard				Chile
	diurnal		74	13.2	
fuscus	Patagonia Brown Tree Lizard				Chile
	diurnal	invert.	47	3.21	* trunk/wall lizard
gallardoi	Gallardo's Montane Lizard				Argentina
	diurnal	invert.	75	13.5	* refugia-anchored diurnal gr. liz.
goetschi	Goetsch's Sand Lizard				Argentina
	diurnal		62	7.6	* refugia-anchored diurnal gr. liz.
gracielae	Laguna Brava Lizard				Argentina
	diurnal		64	8.4	* refugia-anchored diurnal gr. liz.
gracilis	Collector Darwin Lizard				Argentina
	diurnal	ants	56	4.60	** refugia-anchored diurnal gr. liz.
gravenhorstii	Stony Desert Lizard				Chile
	diurnal		53	4.47	** refugia-anchored diurnal gr. liz.
griseus	Argentina Gray Alpine Lizard				Argentina
	diurnal		59	6.6	** refugia-anchored diurnal gr. liz.
grosseorum	Grosse Family Lizard				Argentina
	diurnal	invert.	49	3.71	** refugia-anchored diurnal gr. liz.
gununakuna	Green Flash Lizard				Argentina
	diurnal		76	13.9	
hajeki	Hajek's Lizard				Chile
	diurnal	invert.	61	7.1	* refugia-anchored diurnal gr. liz.

Liolaemus spp.	English name				Range core
			Size		
	Activity	Diet	mm	g	Ecological business model
halonastes	Salt Crust Cordilleran Lizard				Argentina
	diurnal		61	7.1	* refugia-anchored diurnal gr. liz.
hatcheri	Belgrano Lake Lizard				Argentina
	diurnal	omnivore	62	7.3	* refugia-anchored diurnal gr. liz.
heliodermis	Black-headed Montane Lizard				Argentina
	diurnal		69	10.6	* refugia-anchored diurnal gr. liz.
hellmichi	Hellmich's Montane Lizard				Chile
	diurnal	invert.	52	4.41	* refugia-anchored diurnal gr. liz.
hermannunezi	Nunez's Cordilleran Lizard				Chile
	diurnal		56	5.5	* refugia-anchored diurnal gr. liz.
huacahuasicus	Huaca-Huasi Lakes Lizard				Argentina
	diurnal	herbivore	62	7.6	* refugia-anchored diurnal gr. liz.
huayra	Windy Rocks Lizard				Argentina
	diurnal	herbivore	77	14.4	* refugia-anchored diurnal gr. liz.
igneus	Puna Reddish Lizard				Chile
	diurnal		76	14.1	* refugia-anchored diurnal gr. liz.
inacayali	False Boulenger's Lizard				Argentina
	diurnal	invert.	63	7.9	* refugia-anchored diurnal gr. liz.
incaicus	Calca Montane Lizard				Peru
	diurnal		51	4.05	* refugia-anchored diurnal gr. liz.
insolitus	Aquamarine-flecked Peru Lizard				Peru
	diurnal		56	5.4	
inti	Sun-banded Lizard				Argentina
	diurnal	herbivore	74	12.8	* refugia-anchored diurnal gr. liz.
irregularis	Irregular Alpine Lizard				Argentina
	diurnal	herbivore	79	15.6	* refugia-anchored diurnal gr. liz.
isabelae	Isabella's Cordilleran Lizard				Chile
	diurnal		71	11.3	
jamesi	James' Jararanco				Bolivia
	diurnal	omnivore	81	17.0	* refugia-anchored diurnal gr. liz.
janequeoae	Janequeo's Lizard				Chile
	diurnal		62	7.3	

Liolaemus spp.	English name				Range core
			Size		
	Activity	Diet	mm	g	Ecological business model
josei	Jose's Sanddune Lizard				Argentina
	diurnal		62	7.3	* refugia-anchored diurnal gr. liz.
juanortizi	Ortiz's Cordilleran Lizard				Chile
	diurnal	omnivore	77	14.4	* refugia-anchored diurnal gr. liz.
kingii	King's Patagonia Lizard				Argentina
	diurnal	omnivore	81	17.0	* refugia-anchored diurnal gr. liz.
kolengh	Ceballos Lizard				Argentina
	diurnal	invert.	55	5.1	* refugia-anchored diurnal gr. liz.
koslowskyi	Koslowsky's Montane Lizard				Argentina
	diurnal		62	7.5	* refugia-anchored diurnal gr. liz.
kriegi	Krieg's Montane Lizard				Argentina
	diurnal	omnivore	92	25.0	* refugia-anchored diurnal gr. liz.
kuhlmanni	Kuhlmann's Montane Lizard				Chile
	diurnal	invert.	71	11.3	* refugia-anchored diurnal gr. liz.
laurenti	Laurent's Sand Lizard				Argentina
	diurnal		51	4.14	* refugia-anchored diurnal gr. liz.
lavillai	Lavilla's Cordilleran Lizard				Argentina
	diurnal		56	5.4	** refugia-anchored diurnal gr. liz.
leftrarui	Leftraru's Lizard				Chile
	diurnal	invert.	68	9.9	
lemniscatus	Wreath Desert Lizard				Chile
	diurnal	invert.	48	3.41	* refugia-anchored diurnal gr. liz.
lentus	Confusion Pampas Lizard				Argentina
	diurnal		59	6.6	
leopardinus	Leopard Rock Lizard				Chile
	diurnal	omnivore	76	14.1	* refugia-anchored diurnal gr. liz.
lineomaculatus	Decorated Riparian Lizard				Argentina
	diurnal	omnivore	57	5.7	* refugia-anchored diurnal gr. liz.
loboi	Lobo's Pampas Lizard				Argentina
	diurnal		62	7.3	
lonquimayensis	Lonquimay Volcano Lizard				Chile
	diurnal		59	6.5	

Liolaemus spp.	English name					Range core
			Size			
	Activity	Diet	mm	g	Ecological business model	
lopezi	Lopez's Altiplano Lizard					Chile
	diurnal		57	5.9	* refugia-anchored diurnal gr. liz.	
lorenzmuelleri	Mueller's Montane Lizard					Chile
	diurnal		86	20.5	* refugia-anchored diurnal gr. liz.	
lutzae	Lutz's Sand Lizard					Brazil
	diurnal	omnivore	69	10.6	* refugia-anchored diurnal gr. liz.	
magellanicus	Straits of Magellan Lizard					Chile
	diurnal	omnivore	58	6.1	* refugia-anchored diurnal gr. liz.	
maldonadae	Maldonado's Montane Lizard					Chile
	diurnal		69	10.6		
manueli	Dragon of Torres					Chile
	diurnal		53	4.60	* refugia-anchored diurnal gr. liz.	
mapuche	Scapular-spotted Lizard					Argentina
	diurnal		69	10.6	* refugia-anchored diurnal gr. liz.	
martorii	Martori's Sanddune Lizard					Argentina
	diurnal		64	8.4	* refugia-anchored diurnal gr. liz.	
melaniceps	Chungungo Lizard					Chile
	diurnal	omnivore	84	18.9	* refugia-anchored diurnal gr. liz.	
melanogaster	Ayachuco Alpine Lizard					Peru
	diurnal		74	13.2	* refugia-anchored diurnal gr. liz.	
melanopleurus	Philippi's Desert Lizard					Chile
	diurnal		60	6.8		
melanops	Black-headed Sand Lizard					Argentina
	diurnal		80	16.5	* refugia-anchored diurnal gr. liz.	
millcayac	Millcayac Lizard					Argentina
	diurnal	invert.	65	8.7	* refugia-anchored diurnal gr. liz.	
molinai	Molina's Alpine Lizard					Chile
	diurnal	herbivore	60	6.8	** refugia-anchored diurnal gr. liz.	
montanezi	San Guillermo Sand Lizard					Argentina
	diurnal		55	5.2	* refugia-anchored diurnal gr. liz.	

Liolaemus spp.	English name				Range core
			Size		
	Activity	Diet	mm	g	Ecological business model
montanus	Manchao Mountain Lizard				Argentina
	diurnal	invert.	63	7.9	* refugia-anchored diurnal gr. liz.
monticola	Peak Montane Lizard				Chile
	diurnal	ants	66	9.0	
moradoensis	Valdes Andes Lizard				Chile
	diurnal		59	6.4	
morandae	Morando's Patagonia Lizard				Argentina
	diurnal		54	4.79	* refugia-anchored diurnal gr. liz.
morenoi	Collon Cura Sand Lizard				Argentina
	diurnal	invert.	72	12.0	* refugia-anchored diurnal gr. liz.
multicolor	Many-colored Cordilleran Lizard				Argentina
	diurnal	omnivore	66	9.0	* refugia-anchored diurnal gr. liz.
multimaculatus	Many-spotted Dune Lizard				Argentina
	diurnal	invert.	57	5.7	*sand shark
neuquensis	Neuquén Montane Lizard				Argentina
	diurnal	omnivore	63	6.7	* refugia-anchored diurnal gr. liz.
nigriceps	Black-headed Puna Lizard				Argentina
	diurnal	omnivore	79	16.1	
nigrocoeruleus	Copiapó Lizard				Chile
	diurnal	omnivore	49	3.60	** refugia-anchored diurnal gr. liz.
nigromaculatus	Black-spotted Tree Lizard				Chile
	diurnal	omnivore	73	12.6	* ground-foraging tree lizard
nigroviridis	Black-green Montane Lizard				Chile
	diurnal	omnivore	68	10.1	* refugia-anchored diurnal gr. liz.
nitidus	Shining Desert Lizard				Chile
	diurnal	omnivore	85	19.8	* refugia-anchored diurnal gr. liz.
occipitalis	Skull Sand Lizard				Brazil
	diurnal		58	6.2	
olongasta	Lovely Sand Lizard				Argentina
	diurnal		56	5.4	* sand shark
omorfi	Beautiful Lizard				Chile
	diurnal	omnivore	55	5.2	* refugia-anchored diurnal gr. liz.
orientalis	Eastern Cordilleran Lizard				Argentina
	diurnal	herbivore	81	17.0	* refugia-anchored diurnal gr. liz.

Liolaemus spp.	English name				Range core
			Size		
	Activity	Diet	mm	g	Ecological business model
orko	Fiambala Mountains Lizard				Argentina
	diurnal		60	6.8	** refugia-anchored diurnal gr. liz.
ornatus	Ornate Dimorphic Lizard				Peru
	diurnal	omnivore	60	6.8	* refugia-anchored diurnal gr. liz.
ortizii	Ortiz's Andean Lizard				Peru
	diurnal		60	6.8	* refugia-anchored diurnal gr. liz.
pacha	Harvest Lizard				Argentina
	diurnal	invert.	58	6.1	* refugia-anchored diurnal gr. liz.
pachacutec	Pachacutec's Empire Lizard				Peru
	diurnal		47	3.08	* refugia-anchored diurnal gr. liz.
pachecoi	Frontier Cordilleran Lizard				Chile
	diurnal		75	13.5	
pagaburoi	Pagaburo's Bunchgrass Lizard				Argentina
	diurnal	invert.	49	3.62	
pantherinus	Panther Rock Lizard				Peru
	diurnal	herbivore	59	6.6	** refugia-anchored diurnal gr. liz.
parthenos	Maiden Dune Lizard				Argentina
	diurnal		54	5.0	*** refugia-anchored diurnal gr. liz.
parvus	Small Slender Montane Lizard				Argentina
	diurnal		61	7.1	* refugia-anchored diurnal gr. liz.
patriciaiturrae	Pedernales Cordilleran Lizard				Chile
	diurnal		79	15.6	
paulinae	Paulina's Montane Lizard				Chile
	diurnal	invert.	53	4.60	* refugia-anchored diurnal gr. liz.
petrophilus	Rock-loving Slender Lizard				Argentina
	diurnal	omnivore	77	14.5	** refugia-anchored diurnal gr. liz.
pictus	Painted Tree Lizard				Argentina
	diurnal	omnivore	71	11.1	* refugia-anchored diurnal gr. liz.
pipanaco	Pipanaco Lizard				Argentina
	diurnal	invert.	67	9.3	* refugia-anchored diurnal gr. liz.
platei	Braided Rock Lizard				Chile
	diurnal	invert.	57	5.6	* refugia-anchored diurnal gr. liz.

Liolaemus spp.	English name				Range core
			Size		
	Activity	Diet	mm	g	Ecological business model
pleopholis	Chucuyo Altiplano Lizard				Chile
	diurnal		64	8.4	* refugia-anchored diurnal gr. liz.
poconchilensis	Poconchile Desert Lizard				Chile
	diurnal	invert.	50	3.87	* refugia-anchored diurnal gr. liz.
poecilochromus	Variegated Bunchgrass Lizard				Argentina
	diurnal	herbivore	69	10.5	** refugia-anchored diurnal gr. liz.
polystictus	Huancavelica Lizard				Peru
	diurnal		71	11.3	** refugia-anchored diurnal gr. liz.
porosus	Holey Lizard				Argentina
	diurnal		62	7.3	
pseudoanomalus	False Strange Lizard				Argentina
	crepuscular	invert.	72	11.9	* refugia-anchored diurnal gr. liz.
pseudolemniscatus	False Ribboned Lizard				Chile
	diurnal	invert.	47	3.23	* refugia-anchored diurnal gr. liz.
puelche	Puelche Sand Lizard				Argentina
	diurnal		73	12.4	* refugia-anchored diurnal gr. liz.
pulcherrimus	Beautiful Mudana Lizard				Argentina
	diurnal	herbivore	64	8.1	* refugia-anchored diurnal gr. liz.
puna	Striped Puna Lizard				Bolivia
	diurnal		49	3.71	* refugia-anchored diurnal gr. liz.
punmahuida	Cloudy Mountain Lizard				Argentina
	diurnal	omnivore	79	15.6	* refugia-anchored diurnal gr. liz.
puritamensis	San Pedro Cordilleran Lizard				Bolivia
	diurnal	omnivore	82	17.5	* refugia-anchored diurnal gr. liz.
purul	Parquet Lizard				Argentina
	diurnal	invert.	64	8.4	* refugia-anchored diurnal gr. liz.
pyriphlogos	Flame-sided Cordilleran Lizard				Argentina
	diurnal		50	3.87	* refugia-anchored diurnal gr. liz.
quilmes	Quilmes Lizard				Argentina
	diurnal		73	12.4	* refugia-anchored diurnal gr. liz.

Liolaemus spp.	English name				Range core
			Size		
	Activity	Diet	mm	g	Ecological business model
rabinoi	Rabino's Sanddune Lizard				Argentina
	diurnal	invert.	58	6.1	* sand shark
ramirezae	Ramirez's Montane Lizard				Argentina
	diurnal	omnivore	51	4.02	* refugia-anchored diurnal gr. liz.
ramonensis	Ramon Mountain Lizard				Chile
	diurnal	omnivore	81	16.8	* refugia-anchored diurnal gr. liz.
riodamas	Ladies River Lizard				Chile
	diurnal		75	13.4	
riojanus	La Rioja Sanddune Lizard				Argentina
	diurnal	invert.	58	5.9	* sand shark
robertmertensi	Robert Mertens' Lizard				Argentina
	diurnal	invert.	54	5.0	* refugia-anchored diurnal gr. liz.
robertoi	Donoso-Barros' Montane Lizard				Chile
	diurnal		61	7.1	
robustus	Junin Lizard				Peru
	diurnal		70	10.9	
rosenmanni	Rosenmann's Desert Lizard				Chile
	diurnal		65	8.7	
rothi	Roth's Rock Lizard				Argentina
	diurnal	invert.	84	18.9	* refugia-anchored diurnal gr. liz.
ruibali	Ruibal's Montane Lizard				Argentina
	diurnal	omnivore	52	4.10	* refugia-anchored diurnal gr. liz.
sagei	Sage's Blue-kerchiefed Lizard				Argentina
	diurnal		81	17.0	* refugia-anchored diurnal gr. liz.
salinicola	Salt-loving Foothills Lizard				Argentina
	diurnal		68	9.9	* refugia-anchored diurnal gr. liz.
sanjuanensis	San Juan Bunchgrass Lizard				Argentina
	diurnal	invert.	56	5.4	* refugia-anchored diurnal gr. liz.
sarmientoi	Sarmiento's Arroyo Lizard				Argentina
	diurnal	omnivore	76	14.2	* refugia-anchored diurnal gr. liz.
saxatilis	Comechingones Lizard				Argentina
	diurnal	invert.	52	4.41	* refugia-anchored diurnal gr. liz.
scapularis	Salt Pan Lizard				Argentina
	diurnal	invert.	60	6.8	** refugia-anchored diurnal gr. liz.

Liolaemus spp.	English name / Activity	Diet	Size mm	Size g	Range core / Ecological business model
schmidti	Schmidt's Atacama Lizard				Bolivia
	diurnal		57	5.9	
schroederi	Schroeder's Montane Lizard				Chile
	diurnal	invert.	57	5.8	* refugia-anchored diurnal gr. liz.
scolaroi	Scolaro's Cordilleran Lizard				Argentina
	diurnal		58	6.2	
scorialis	Slag Lizard				Chile
	diurnal		59	6.6	
scrocchii	Scrocchi's Lizard				Argentina
	diurnal	omnivore	77	14.8	* refugia-anchored diurnal gr. liz.
senguer	Senguer River Lizard				Argentina
	diurnal		54	4.79	* refugia-anchored diurnal gr. liz.
septentrionalis	South Chile Lizard				Chile
	diurnal		56	5.3	
shehuen	Uniform Pampas Lizard				Argentina
	diurnal	invert.	81	17.0	* refugia-anchored diurnal gr. liz.
shitan	Black-hearted Lizard				Argentina
	diurnal	invert.	80	16.5	* refugia-anchored diurnal gr. liz.
signifer	Zodiac Cordilleran Lizard				Bolivia
	diurnal	omnivore	76	13.9	** refugia-anchored diurnal gr. liz.
silvai	Silva's Desert Lizard				Chile
	diurnal	invert.	59	6.3	* refugia-anchored diurnal gr. liz.
silvanae	Patagonia Alpine Lizard				Argentina
	diurnal	herbivore	67	9.4	** refugia-anchored diurnal gr. liz.
sitesi	Sites' Pampas Lizard				Argentina
	diurnal	invert.	67	9.3	* refugia-anchored diurnal gr. liz.
smaug	Golden Dragon Lizard				Argentina
	diurnal	omnivore	60	6.8	* refugia-anchored diurnal gr. liz.
somuncurae	Somuncura Plateau Lizard				Argentina
	diurnal	omnivore	72	11.7	* refugia-anchored diurnal gr. liz.
stolzmanni	Stolzmann's Lizard				Chile
	diurnal	omnivore	53	4.60	** refugia-anchored diurnal gr. liz.

Liolaemus spp.	English name				Range core
			Size		
	Activity	Diet	mm	g	Ecological business model
tacnae	Tacna Lizard				Peru
	diurnal		48	3.44	
tacora	Tacora Lizard				Chile
	diurnal		86	20.3	* refugia-anchored diurnal gr. liz.
talampaya	Talampaya Lizard				Argentina
	diurnal	invert.	71	11.3	* refugia-anchored diurnal gr. liz.
tandiliensis	Tandilia Mountains Lizard				Argentina
	diurnal	invert.	52	4.24	* refugia-anchored diurnal gr. liz.
tari	Windy Plateau Lizard				Argentina
	diurnal		82	17.9	* refugia-anchored diurnal gr. liz.
tehuelche	Tehuelche Lizard				Argentina
	diurnal	omnivore	63	8.0	* refugia-anchored diurnal gr. liz.
telsen	Somuncura Rock Lizard				Argentina
	diurnal		63	7.9	* refugia-anchored diurnal gr. liz.
tenuis	Emerald Rock Lizard				Chile
	diurnal	invert.	54	4.86	*** ground-foraging tree lizard
thermarum	Azufre Hot Springs Lizard				Argentina
	diurnal		70	10.9	
thomasi	Thomas' Cordilleran Lizard				Peru
	diurnal		65	8.7	* refugia-anchored diurnal gr. liz.
tolhuaca	Tolhuaca Lizard				Chile
	diurnal		64	8.1	* refugia-anchored diurnal gr. liz.
torresi	Dragon of Torres-Mura				Chile
	diurnal		56	5.4	
tregenzai	Boreal Patagonia Lizard				Argentina
	diurnal		74	12.8	
tristis	Sad Montane Lizard				Argentina
	diurnal	omnivore	71	11.3	* refugia-anchored diurnal gr. liz.
tromen	Tromen Volcano Lizard				Argentina
	diurnal		69	10.3	** refugia-anchored diurnal gr. liz.
tulkas	Speedy Montane Lizard				Argentina
	diurnal		59	6.6	* trunk/wall lizard
ubaghsi	Ubaghs' Leopard Lizard				Chile
	diurnal		77	14.8	

Liolaemus spp.	English name				Range core
			Size		
	Activity	Diet	mm	g	Ecological business model
umbrifer	Shaded Alpine Lizard				Argentina
	diurnal		73	12.4	* refugia-anchored diurnal gr. liz.
uniformis	Chepical Lagoon Lizard				Chile
	diurnal		73	12.4	*** refugia-anchored diurnal gr. liz.
uptoni	Upton's Steppe Lizard				Argentina
	diurnal		72	11.7	* refugia-anchored diurnal gr. liz.
uspallatensis	Uspallata Montane Lizard				Argentina
	diurnal	omnivore	56	5.4	* refugia-anchored diurnal gr. liz.
valdesianus	Valdes Leopard Lizard				Chile
	diurnal	omnivore	81	16.8	* refugia-anchored diurnal gr. liz.
vallecurensis	Cura Valley Puna Lizard				Argentina
	diurnal		64	8.4	* refugia-anchored diurnal gr. liz.
variegatus	Cochabamba Andes Lizard				Bolivia
	diurnal		54	4.90	
velosoi	Iman Mountain Lizard				Chile
	diurnal		52	4.23	
villaricensis	Volcan Villarica Lizard				Chile
	diurnal	omnivore	67	9.3	
vulcanus	Fire Montane Lizard				Argentina
	diurnal	omnivore	79	15.6	* refugia-anchored diurnal gr. liz.
walkeri	Peru Puna Lizard				Peru
	diurnal		55	5.2	
wari	Ayacucho Bunchgrass Lizard				Peru
	diurnal		53	4.60	* refugia-anchored diurnal gr. liz.
wiegmanni	Wiegmann's Dune Lizard				Argentina
	diurnal	omnivore	52	4.42	* refugia-anchored diurnal gr. liz.
williamsi	Galeras Pampas Lizard				Peru
	diurnal		64	8.4	* refugia-anchored diurnal gr. liz.
xanthoviridis	Yellow-green Beach Lizard				Argentina
	diurnal	invert.	82	17.5	* refugia-anchored diurnal gr. liz.
yalguaraz	Yalguaraz Pampas Lizard				Argentina
	diurnal		52	4.41	

Liolaemus spp.	English name				Range core
			Size		
	Activity	Diet	mm	g	Ecological business model
yanalcu	Sooty Cordilleran Lizard				Argentina
	diurnal	omnivore	53	4.60	* refugia-anchored diurnal gr. liz.
yatel	Petrified Forest Lizard				Argentina
	diurnal	omnivore	53	4.60	* refugia-anchored diurnal gr. liz.
zabalai	Biobio Lizard				Chile
	diurnal		75	13.5	
zapallarensis	Zapallar Lizard				Chile
	diurnal	invert.	76	14.2	* refugia-anchored diurnal gr. liz.
zullyae	Zeballos River Lizard				Chile
	diurnal	invert.	73	12.4	* refugia-anchored diurnal gr. liz.

Liopholis

Liopholis is a moderate radiation (11 spp.) of variably large-sized (16.5 g v. 3.25 g; range 5.8 to 119 g), short-snouted (unquantified), mostly omnivorous (79% v. 13%), burrow-refuging (100% v. 24%), viviparous (100% v. 36%) skinks resident in central or southern Australia (including Tasmania). The majority are refugia-anchored diurnal ground lizards (55% v. 30%); the remainder are crack hermits (36% v. 1.7%) or a giant herbivore (9% v. 1%). Color polymorphism is notably common (*margaretae, multiscutata, pulchra, whitii*) in this group of ecologically untypical skinks. In addition, *Liopholis* often dig and occupy a complex burrow network, housing an entire family. At least four of the familial burrowers have traditional "latrines" where all burrow residents defecate. These are usually just outside the burrow entrance (*kintorei, slateri, whitii*), but may also be just inside the entrance (*inornata*). This is associated with a striking frequency of neonates remaining with the mother (or both parents) after parturition (*kintorei, modesta, striata, whitii*). The latter species may need up to 10 days to complete parturition, a luxury presumably affordable only within the safety of a good burrow system. Is family ownership of a burrow system made possible by neonates remaining there (clan defense), or is neonatal lack of dispersal made possible by the burrow "fortress"? Sinn et al. (2008) showed that *whitii* neonates that remained within the fortress were less likely to be cannibalized by neighboring adults.

Liopholis are more heavily built than average skinks (shape index 0.41 v. −0.46 std. resid.). One species (*slateri*) is elongate (v. 35%).

The diet of *Liopholis* emphasizes omnivory, with the proportion of plants in the diet varying widely within the genus and most species also gorging on termites when the opportunity arises. *L. inornata* may eat a large proportion of ants (39%). Unusually for skinks, this genus is about evenly split between sit-and-wait (80% v. 16%) and active (60% v. 92%) foraging. Electric light use and home-range areas are not known.

Although the modal *Liopholis* is diurnal (60% v. 74%) and the remainder cathemeral (40% v. 9%), there is much variation within species. *L. kintorei* and *slateri* have been described as all four diel patterns. Prey may be seized as they pass the burrow entrance during the lizards' nominal inactive period (*inornata, kintorei*). Direct basking is very common (73% v. 21%), if only from the safety of a burrow entrance. Mean active body temperature is 29.4°C (v. 31.1°C), which may be realistic for their environment.

Three species have annual rainfall data, all in the arid belt (151–400 mm/y, 100% v. 40%). *Liopholis* are overrepresented in the intermediate (100% v. 35%) and high (33% v. 6%) elevation belts. They are overrepresented in sandy areas (64% v. 25%) and are absent from riparian (v. 23%) and human (v. 14%) habitats; they are representative in littoral zones (27% v. 21%).

As noted in the first paragraph, *Liopholis* are universally viviparous and often express parental care (36% v. 3%). The average litter count is small (−0.59 v. 0.06 std. resid.), and the neonates are born at a comparatively large size (ontogenetic mass range −0.88 v. 0.12 std. resid.). Only *inornata* is reported to sometimes breed more than once a year (five spp. are specified as once per year). RCM matches the skink norm (0.233 v. 0.236).

Longevity is slightly long (0.37 v. 0.17), based on five species. Population densities are unavailable. Endemism is low (9% v. 64%), despite the vast size of Australian states. Few live on islands (27% v. 57%). None has been introduced (v. 2.1%). Most are imperiled (88% v. 47%), including 12.5% (v. 14%) on the IUCN's lists of the most endangered. In all other documented respects *Liopholis* are typical skinks.

Liopholis spp.	English name		Size		Range core
	Activity	Diet	mm	g	Ecological business model
guthega	Guthega Skink				New South Wales
	diurnal		94	17.9	* refugia-anchored diurnal gr. liz.
inornata	Desert Skink				Australia
	cathemeral	omnivore	72	7.5	*** crack hermit
kintorei	Great Desert Skink				Western Australia
	cathemeral	herbivore	170	119	**** giant herbivore
margaretae	Centralian Ranges Rock-skink				Central Australia
	diurnal	omnivore	94	17.5	* crack hermit
modesta	Eastern Ranges Rock-skink				Queensland
	diurnal	omnivore	96	18.7	** refugia-anchored diurnal gr. liz.
montana	Montane Rock-skink				New South Wales
			93	17	* refugia-anchored diurnal gr. liz.
multiscutata	Southern Sand-skink				Southern Australia
	diurnal	omnivore	80	10.6	** refugia-anchored diurnal gr. liz. or crack hermit
pulchra	Southwestern Rock-skink				Western Australia
	diurnal	omnivore	95	18.2	** refugia-anchored diurnal gr. liz.
slateri	Slater's Desert Skink				Central Australia
	cathemeral	invert.	83	5.8	** crack hermit
striata	Nocturnal Desert Skink				Western Australia
	cathemeral	omnivore	91	16.1	** crack hermit or nocturnal gr. liz.
whitii	White's Skink				Eastern Australia
	diurnal	omnivore	90	15.3	** refugia-anchored diurnal gr. liz.

Lioscincus spp.	English name		Size		Range core
	Activity	Diet	mm	g	Ecological business model
steindachneri	White-lipped Forest Skink				New Caledonia
	diurnal		90	15.4	
vivae	Vivienne's Skink				New Caledonia
	diurnal		49	2.11	** refugia-anchored diurnal gr. liz.

Lioscincus

Lioscincus consists of two species of moderately large (5.7 g v. 3.3 g), diurnal (100% v. 74%) ground skinks endemic to New Caledonia. Only one (*vivae*) is known well enough to infer a likely ecological business model: refugia-anchored diurnal ground lizard.

Lioscincus are more heavily built than average skinks (shape index 0.11 v. –0.46 std. resid.), and they have longer tails (1.95 v. 1.33 × SVL) that are not reported to be breakable (v. 73%). The sexes differ in color in *vivae*, but no other sex differences are known. *L. steindachneri* is conspicuously wary (50% v. 11%); *L. vivae* basks directly (v. 21%). *L. steindachneri* is associated with riparian areas (v. 23%) and may be semi-aquatic. It has been found to possess about six advanced follicles, but the reproductive mode is not yet firmly established. Clutch/litter count, based on the follicles, is normal (0.19 v. 0.06 std. resid.). No *Lioscincus* has been introduced (v. 2.1%); both are imperiled (v. 47%), including placement on the IUCN's lists of the most endangered (v. 14%: *steindachneri*, endangered; *vivae*, critically endangered). In other respects *Lioscincus* are much like other skinks.

Lipinia

Lipinia is a major radiation (28 spp.) of small-sized (1.18 g v. 3.25 g), diurnal (100% v. 74%), mostly arboreal (73% v. 10%) skinks of diverse ecologies found in southeastern Asia and Australasia. They are found within the following polygon: Myanmar and Andaman/Nicobar Islands to Vietnam to Mariana Islands to French Polynesia to New Guinea to Java to the Andaman/Nicobar Islands. New Guinea is a particular hot spot: nine spp. Almost all *Lipinia* (96% v. 57%) are found on islands. All eat either ants (22% v. 1.6%) or are generalized invertebrate eaters (78% v. 84%). The modal *Lipinia* is a diurnal canopy walker (33% v. 1.8%) or a trunk/wall lizard (33% v. 3%). Others are litter cryptozoids (17% v. 28%), refugia-anchored diurnal ground lizards (8% v. 30%), or ground-foraging tree lizards (8% v. 7%). Some of the species have almost never been seen moving (*leptosoma*, *quadrivittata*), suggesting that they might be undetected crack hermits.

Lipinia match skinks well in most dimensions of morphology. They are slightly less inclined to be elongate (25% v. 35%). Although they do not have blue tails (v. 4%), several species have notably gaudy tails: red/orange (*longiceps*, *macrotympanum*, *pulchra*, *vittigera*) or yellow (*leptosoma*).

L. noctua is the most widespread and best-studied species. It has m<f SSD (–1.0 v. –0.31 on –2 to 2 scale). Sexual dichromatism is also known (*pulchella*: 3.6% v. 7.6%). If the poor information available is taken at face value, overall sex differences are very weak in *Lipinia* (0.07 v. 0.28 on 0 to 6 scale).

Dietary data are very poor, but all specific stomach contents include ants. All are active foragers (100% v. 92%), though this is unnecessary and unusual for ant eaters. *Lipinia* match the family for conspicuous wariness (11% v. 11%). The only predator-concealment-documented species (*vittigera*) retreats to vegetation as its primary tactic (100% v. 25%). Social relations are equally undocumented, but *noctua* males defend space (100% v. 94%).

The elevational profile of *Lipinia* is normal (l/m/h: 92/25/0 v. 89/35/6). They are underrepresented in riparian (8% v. 25%) and human habitats (4% v. 14%), but more or less normal in littoral areas (17% v. 21%) and sandy areas (13% v. 25%).

The genus is nearly equally repre-

Lipinia spp.	English name	Activity	Diet	Size mm	Size g	Range core	Ecological business model
albodorsalis	White-backed Slender Skink			48	2.02	New Guinea	
auriculata	Bronze Slender Tree Skink	diurnal	invert.	46	1.14	Philippines	*** diurnal canopy walker
cheesmanae	Cyclops Slender Tree Skink			37	0.89	New Guinea	
inexpectata	Borneo Slender Tree Skink			39	0.99	Borneo	
infralineolata	Lined Slender Tree Skink			44	1.57	Indonesia	
leptosoma	Pandanus Moth Skink	diurnal	invert.	42	1.33	Palau	*** diurnal canopy walker
longiceps	Long Slender Tree Skink			40	1.13	New Guinea	* trunk/wall lizard
macrotympanum	Bay Islands Striped Skink	diurnal		42	0.90	India	* litter cryptozoid or refugia-anchored diurnal gr. liz.
miangensis	Miang Island Skink			37	0.89	Indonesia	
nitens	Sarawak Striped Skink	diurnal	ants	33	0.48	Malaysia	* refugia-anchored diurnal gr. liz.
noctua	Moth Skink	diurnal	ants	41	1.20	Pacific Islands	** ground-foraging tree lizard
nototaenia	Setekwa Slender Tree Skink			44	1.49	New Guinea	
occidentalis	Cenderawasih Slender Tree Skink			38	0.95	New Guinea	
pulchella	Beautiful Lipinia	diurnal	invert.	41	1.26	Philippines	*** trunk/wall lizard
pulchra	Small-legged Slender Tree Skink			39	1.01	New Guinea	
quadrivittata	Four-striped Slender Tree Skink		invert.	41	0.86	Southern Asia	* diurnal canopy walker
rabori	Black Slender Tree Skink	diurnal	invert.	47	1.83	Philippines	*** trunk/wall lizard
relicta	Mentawai Striped Skink			49	2.22	Indonesia	
rouxi	Roux's Slender Tree Skink			35	0.72	New Guinea	

Lipinia spp.	English name				Range core
			Size		
	Activity	Diet	mm	g	Ecological business model
sekayuensis	Sekayu Striped Skink				Malaysia
			40	1.09	
semperi	Semper's Slender Tree Skink				Philippines
			44	1.56	
septentrionalis	Sepik Slender Tree Skink				New Guinea
			40	1.13	
subvittata	Short-striped Tree Skink				Indonesia
			49	1.40	* diurnal canopy walker
surda	Malaysian Striped Skink				Thailand
	diurnal	invert.	47	1.24	* litter cryptozoid
venemai	Brongersma's Slender Tree Skink				Indonesia
			52	2.60	
vittigera	Yellow-striped Tree Skink				Malaysia
	diurnal	invert.	40	1.09	** trunk/wall lizard or ground-foraging tree lizard
vulcania	Vulcan Slender Tree Skink				Philippines
			40	0.80	
zamboangensis	Rusty Slender Tree Skink				Philippines
			41	1.23	

sented by oviparous (46% v. 64%) and viviparous species (54% v. 36%). Several species may be capable of both modes (different areas): *pulchella, relicta, vittigera*. Clutch/litter count is normal (0.09 v. 0.06 std. resid.). They are less normal in the exceptionally small size of the young (ontogenetic mass range 0.71 v. 0.12 std. resid.).

Population density is slightly low (0.19 kg/ha v. 0.25 kg/ha) despite their insularity. It is based on one sample each of *auriculata* (0.04 kg/ha), *noctua* (0.69 kg/ha), and *rabori* (0.25 kg/ha). Endemism is about average (57% v. 64%). Introduction frequency is high (3.6% v. 2.1%), and the responsible species (*noctua*) has been introduced at least nine times (3.6% v. 0.7%). *Lipinia* are not prone to imperilment (12% v. 47%), and none is on one of the IUCN's lists of the most endangered species (v. 14%). In other respects *Lipinia* are typical skinks.

Lissolepis

Lissolepis consists of two species of large (30.4 g v. 3.3 g), comparatively well known (filled cells 38% v. 25%), viviparous (100% v. 36%), semiaquatic (100% v. 2%) refugia-anchored diurnal ground skinks (100% v. 30%) in southern Australia.

Lissolepis are relatively stout (shape index 0.66 v. –0.46 std. resid.), and their tails are moderately long (1.69 v. 1.33 × SVL) and breakable (100% v. 73%). Neither species is elongate (0% v. 35%). No sex differences are known.

Their omnivorous diets are apparently opportunistic and include leaves, fruit, snails, and ectothermal vertebrates. The contribution of plants to the diet of *coventryi* is about 27% of volume, and plants are found in 66% of individuals. Foraging mode has not been assessed but must include active foraging for the plants. Electric light use is not known (v. 0.4%). Home-range area is very small (–1.02 v. –0.07 std. resid.), based on *coventryi* (35 m^2).

Contradictory reports of diel activity are present for both species; they may both be cathemeral. Both are conspicuously wary (v. 11%). Primary predator concealment tactic is some mix of crypsis (50% v. 9%) and retreat to vegetation (50% v. 25%). Both species bask directly (v. 21%). Both species are solitary (v. 8%). Males of both species defend space (v. 94%). Both species refuge in burrows (v. 24%). *L. coventryi* deposits fecal waste in a "latrine" outside of its burrow, suggesting long-term use of the same burrow.

They are both found in littoral (v. 21%) and riparian (v. 23%) habitats, but they are not reported in sandy (v. 25%) or human (v. 14%) habitats.

Their litter size is remarkably small (–0.91 v. 0.06 std. resid.), and the neonates are remarkably large (ontogenetic mass range –0.77 v. 0.12 std. resid.). Furthermore, their RCM (0.370 v. 0.236) is the fourth largest in Scincidae, suggesting that females strain to carry such large fetuses. Parturition requires 2–3 days in *coventryi*. Presumably this lengthy process is made possible by the security of its burrow. One might expect m<f SSD under such circumstances, but SSD is not reported.

Longevity is high (0.59 v. 0.17 std. resid.) and in this case is based on both species (*coventryi*, 9 years; *luctuosa*, 18.2 years). Population density is undocumented. Endemism is about average (50% v. 64%). Neither species is found on an island (v. 57%) nor has either been introduced (v. 2.1%). Both species are imperiled (v. 47%), but neither is on one of the IUCN's lists of most endangered (v. 14%). In other respects *Lissolepis* are much like other skinks.

Lissolepis spp.	English name		Size		Range core
	Activity	Diet	mm	g	Ecological business model
coventryi	Eastern Mourning Skink				Southeastern Australia
	diurnal	omnivore	111	30.1	**** refugia-anchored diurnal gr. liz.
luctuosa	Western Mourning Skink				Southwestern Australia
	cathemeral	omnivore	112	30.8	** refugia-anchored diurnal gr. liz.

Lobulia

Lobulia is a moderate radiation (six spp.) of medium-sized (2.84 g v. 3.25 g), burrow-refuging (100% v. 24%), high-altitude (83% v. 6%), sometimes alpine (often mossy areas), viviparous (100% v. 36%), refugia-anchored diurnal ground skinks (100% v. 30%) endemic to New Guinea.

Lobulia fit the generalized skink morphology well. Their m<f SSD (–0.50 v. –0.31 on –2 to 2 scale) is consistent with this pattern, but only weakly so. The sexes may differ in color (*elegans*: 17% v. 7.6%). Overall, the sex differences rollup is about average (0.50 v. 0.28 on 0 to 6 scale). Dietary information is too fragmentary to draw inferences. Direct basking is especially common (83% v. 21%). Mean active body temperature 26.6°C (v. 31.1°C) is commensurate with their high elevation.

They are found in mesic (100% v. 40%) and wet (100% v. 7%) areas. They are underrepresented at low elevation (17% v. 89%) and mildly overrepresented at intermediate elevations (50% v. 35%). They are overrepresented in riparian areas (40% v. 23%) and absent from littoral (v. 21%), sandy (v. 25%), and human (v. 14%) habitats.

The litter count is normal (–0.03 v. 0.06 std. resid.), and their ontogenetic mass range is only slightly low (–0.23 v. 0.12 std. resid.). RCM ratios are unavailable.

Endemism is high (83% v. 64%). All are island species (v. 57%). None has been introduced (v. 2.1%). One (*glacialis*) is considered at risk due to restricted range (20% v. 47%), but it is not on one of the IUCN's lists of the most endangered (v. 14%). In other respects *Lobulia* are probably typical skinks.

Lobulia spp.	English name		Size		Range core
	Activity	Diet	mm	g	Ecological business model
alpina	Alpine Moss Skink				New Guinea
	diurnal		62	4.49	*** refugia-anchored diurnal gr. liz.
brongersmai	Brongersma's Moss Skink				New Guinea
			52	2.54	
elegans	Elegant Moss Skink				New Guinea
	diurnal		58	3.65	** refugia-anchored diurnal gr. liz.
glacialis	Glacial Moss Skink				Indonesia
	diurnal		51	1.51	** refugia-anchored diurnal gr. liz.
stellaris	Star Mountains Moss Skink				New Guinea
	diurnal	invert.	57	3.41	** refugia-anchored diurnal gr. liz.
subalpina	Subalpine Moss Skink				New Guinea
	diurnal		60	2.43	*** refugia-anchored diurnal gr. liz.

Lophocalotes

Lophocalotes is a single-species genus of moderate-size (13.8 g v. 15.6 g) agamids found in montane shrubs in west-central Sumatra. It sports moderate nuchal and dorsal crests (v. 38%), which may be larger in males. To the extent that it is known, it seems to be a typical arboreal agamid.

Lophocalotes spp.	English name		Size		Range core
	Activity	Diet	mm	g	Ecological business model
ludekingi	Sumatra Crested Dragon				Indonesia
	diurnal	invert.	75	13.8	* diurnal canopy walker

Lophognathus

Lophognathus is a small genus (three spp.) of semiarboreal agamids that bear a striking similarity to *Gowidon*. Like *Gowidon*, they are found in Australia and nearby islands, are moderately large (31 g v. 15.6 g) and are brownish-black with a whitish labial wash that fades out along the upper flanks. *Lophognathus* and *Gowidon* possess a similar muscular build (shape index 1.09 v. 0.80 std. resid.) with long tails (2.9 v. 1.9 × SVL). Other similarities include direct basking (100% v. 24%), m>f SSD (0.50 v. 0.53 on –2 to 2 scale), sexual dichromatism (67% v. 53%), basic color change (67% v. 22%), and moderate crests (67% v. 38%) that are larger in males. Both genera are concentrated along water courses (riparian 100% v. 21%) and otherwise found in diverse habitats (from dunes and mangrove to urban). Both are fast sprinters and may run bipedally. *L. gilberti* waves its forearms, giving rise to the local name of Ta-ta Lizard. *Lophognathus* clutch count is slightly below average (–0.02 v. 0.20 std. resid.), and RCM ratio is light (0.282 v. 0.336). Endemism is low (33% v. 49%). Island species frequency is high (67% v. 36%). None has been introduced (v. 1.7%). Half are imperiled (v. 37%), though none is listed on one of the IUCN's lists of the most endangered (v. 5%).

Lophognathus spp.	English name				Range core
			Size		
	Activity	Diet	mm	g	Ecological business model
burnsi	Burns' Dragon				Queensland
	diurnal		115	52	** ground-foraging tree lizard
gilberti	Gilbert's Dragon				Western Australia
	diurnal	carnivore	101	34.0	*** refugia-anchored diurnal gr. liz.
maculilabris	Timor Laut Dragon				Indonesia
			79	16.3	

Lophosaurus

Lophosaurus is a small genus (three spp.) of large (85 g v. 15.6 g), robust (shape index 1.39 v. 0.80 std. resid.), semiarboreal Australasian agamids with large spines capping their dramatic nuchal and dorsal crests (v. 38%) and often their dewlaps. They are at home in a wide variety of habitats and microhabitats at low elevation (v. 83%). They are reported to both sit-and-wait (100% v. 88%) and actively (100% v. 17%) forage. Unlike most agamids, they are thermoconformers (100% v. 2%), with a low body temperature when active (22.4°C v. 34.8°C). Clutch count is remarkably low for a lizard of this size (–1.15 v. 0.20 std. resid.), but the young are not especially large (ontogenetic mass range 0.66 v. 0.69 std. resid.). Longevity (0.07 v. –0.46 std. resid.) is high for an agamid. Detailed observations are available only for *spinipes*, the most arboreal of the three species. *L. spinipes* has a moderately small home range (300–700 m^2; 0.06 v. 0.35 std. resid.), and both sexes defend space (mating space?); it lives in groups (100% v. 12%) and oviposits communally (33% v. 0.4%), possibly because sunlit patches are scarce in its environment. These features are somewhat unusual for a species that prefers small-diameter perches and relies on crypsis for predator concealment (100% v. 51%). Endemism is low (33% v. 49%). Island species frequency is average (33% v. 36%). None has been introduced (v. 1.7%). None is imperiled (v. 37%).

Lophosaurus spp.	English name				Range core
			Size		
	Activity	Diet	mm	g	Ecological business model
boydii	Boyd's Forest Dragon				Queensland
	diurnal	invert.	142	98	* ground-foraging tree lizard
dilophus	Indonesian Forest Dragon				New Guinea
	diurnal	omnivore	166	163	* ground-foraging tree lizard
spinipes	Southern Angle-headed Dragon				Queensland
	diurnal	invert.	104	38.0	*** diurnal canopy walker

Loveridgea

Loveridgea is a minor genus (two spp.) of exceptionally small (0.79 v. 2.73 g; shape index –4.61 v. 4.0 std. resid.) amphisbaenids endemic (v. 66%) to Tanzania. At least one species (*ionidesii*) has secretory pores (v. 3.4%), is viviparous (v. 14%), and has a litter of two (0.51 v. –0.12 std. resid.). Despite the numerically small count (two), this is a large litter for such a small lizard. Neither species is considered imperiled (v. 23%). Based on the few data available they are otherwise undistinguished amphisbaenids.

Loveridgea spp.	English name				Range core
			Size		
	Activity	Diet	mm	g	Ecological business model
ionidesii	Liwale Round-snouted Worm Lizard				Tanzania
	diurnal	invert.	120	0.48	**** subterranean forager
phylofiniens	Ujiji Round-snouted Worm Lizard				Tanzania
			153	1.28	**** subterranean forager

Loxopholis spp.	English name				Range core
			Size		
	Activity	Diet	mm	g	Ecological business model
caparensis	Caparo Shade Lizard				Venezuela
			39	1.21	
ferreirai	Ferreira's Shade Lizard				Brazil
			34	0.77	
guianense	Guiana Shade Lizard				Guiana Shield
	cathemeral	invert.	37	0.58	**** litter cryptozoid
guianensis	Guiana Creek Lizard				Guyana
			54	3.58	
hexalepis	Six-scaled Shade Lizard				Colombia
			39	1.17	*** litter cryptozoid
hoogmoedi	Hoogmoed's Creek Lizard				Guyana
			52	3.04	
ioanna	Joanna's Shade Lizard				Colombia
			39	1.17	
osvaldoi	Osvaldo's Shade Lizard				Brazil
			34	0.44	* refugia-anchored diurnal gr. liz.
parietalis	Forest Shade Lizard				Ecuador
	cathemeral	invert.	37	1.02	*** litter cryptozoid
percarinatum	Mueller's Shade Lizard				Venezuela
	cathemeral	invert.	34	0.75	*** litter cryptozoid
rugiceps	Panama Shade Lizard				Panama
			40	1.34	*** litter cryptozoid
snethlageae	Snethlage's Shade Lizard				Brazil
	diurnal		33	0.69	*** litter cryptozoid
southi	South's Red-bellied Lizard				Panama
	diurnal	invert.	49	2.49	*** litter cryptozoid

Loxopholis

Loxopholis is a 13-species genus of small (1.15 g v. 2.35 g) litter cryptozoid (88% v. 55%) gymnophthalmids from the mainland Neotropics (Costa Rica to Brazil). Secretory pores are comparatively common (46% v. 29%); elongation is relatively rare (15% v. 41%). The sexes often differ in color (31% v. 12.7%), and some are seasonally colorful (7.7% v. 1.7%). Other sex differences are not reported. The majority are active at any time of day (60% v. 14%); the remainder are diurnal (40% v. 79%). All documented species take refuge in vegetation (v. 73%). Thermoconformers are notably common (23% v. 5%). All males are socially tolerant (v. 88%). *Loxopholis* are associated with riparian areas (40% v. 29%). Reproduction is typical, except that *percarinatum* is parthenogenetic. Clutch counts are relatively high (0.25 v. −0.11 std. resid.). Hatchlings are slightly large (ontogenetic size range 0.39 v. 0.62 std. resid.). RCM is on the light side (0.148 v. 0.192 ratio). There is one population density estimate for *southi* (0.05 kg/ha v. 0.19 kg/ha). Endemism is relatively low (38% v. 68%). No *Loxopholis* are listed as species of conservation concern (v. 25%). In other respects *Loxopholis* are typical gymnophthalmids.

Lubuya

Lubuya ivensii is the sole member of its poorly known (filled cells 12% v. 25%) genus, a large (14.3 g v. 3.3 g), elongate (v. 35%) skink found in the southern part of equatorial Africa: Angola, Democratic Republic of the Congo, Zambia. Reproduction is viviparous (v. 36%). The species is not endemic (v. 64%) or found on an island (v. 57%), nor has it been introduced (v. 2.1%). It has not been evaluated for conservation status. In other respects it is likely to be similar to other skinks.

Lubuya sp.	English name				Range core
			Size		
	Activity	Diet	mm	g	Ecological business model
ivensii	Ivens' Skink				Congo
			115	14.3	

Lucasium spp.	English name				Range core
			Size		
	Activity	Diet	mm	g	Ecological business model
alboguttatum	White-spotted Ground Gecko				Western Australia
	nocturnal	invert.	53	2.86	*** nocturnal ground lizard
bungabinna	Southern Sandplain Gecko				Southwestern Australia
	nocturnal		50	2.46	*** nocturnal ground lizard
byrnei	Gibber Gecko				Eastern Australia
	nocturnal		52	2.78	*** nocturnal ground lizard
damaeum	Beaded Gecko				Australia
	nocturnal	invert.	51	2.67	*** nocturnal ground lizard
immaculatum	Pale-striped Ground Gecko				Australia
	nocturnal		74	8.2	** nocturnal ground lizard
maini	Main's Ground Gecko				Western Australia
	nocturnal	invert.	48	2.22	** nocturnal ground lizard or crack hermit
occultum	Yellow-snouted Ground Gecko				Northern Territory
	nocturnal	invert.	39	1.19	*** nocturnal ground lizard
squarrosum	Mottled Ground Gecko				Western Australia
	nocturnal		52	2.70	*** nocturnal ground lizard
steindachneri	Steindachner's Ground Gecko				Queensland
	nocturnal	invert.	54	3.02	*** nocturnal ground lizard
stenodactylum	Crowned Gecko				Australia
	nocturnal	invert.	54	3.06	*** nocturnal ground lizard
wombeyi	Wombey's Ground Gecko				Western Australia
	nocturnal		46	1.94	*** nocturnal ground lizard

Lucasium

Lucasium (11 spp.) are cryptic, mottled, small (2.69 g v. 4.15 g) nocturnal ground lizards (v. 35%) of the Australian outback. These geckos are fairly typical diplodactylids in shape and proportions, though most lack secretory pores (9% v. 34%), color change (v. 5%), expanded toepads (v. 16%), and sexual dichromatism (v. 6%). Nine percent (v. 5%) are documented to call spontaneously. All males defend space (v. 77%), and three-quarters (v. 22%) use burrows as daily refugia. They are found in the arid (151–400 mm/y) zone (v. 67%) and are overrepresented in sandy areas (64% v. 24%) but not disproportionate in other special habitats (littoral, riparian, human). They are oviparous (v. 82%) and have typical clutch characteristics, but light RCM ratios (0.16 v. 0.23). Hatchling size is close to that of average lizards and is therefore smaller than typical diplodactylids (ontogenetic mass range –0.14 v. –0.57 std. resid.). Their longevity is closer to a typical lizard than a typical diplodactylid (0.28 v. 0.90 std. resid.; mean of 0.0 for lizards). Population density is a bit sparser than a typical diplodactylid (0.15 kg/ha v. 0.44 kg/ha), perhaps reflecting the low primary productivity of their habitat. No *Lucasium* is found on islands (v. 46%) or been introduced (v. 2%). Half of evaluated species are imperiled (v. 81%) and on the IUCN's list of most endangered (v. 29%). In other respects these are typical diplodactylids.

Luperosaurus

Luperosaurus is a 13-species genus of the Gekkonidae found primarily in the islands of Southeast Asia (Philippines, Indonesia, Malaysia, Singapore). Most species are nocturnal canopy walkers (50% v. 9%); others are in the continuum from trunk/wall lizard (10% v. 45%) to immobility expert (20% v. 4.5%) to glider (20% v. 1.5%). All are nocturnal (v. 73%). Almost all species of *Luperosaurus* have restricted distributions, and most are known from only a few specimens. In 2011 Grismer reported that only one species was known from more than four specimens.

They are of average gekkonid size (3.41 g v. 3.31 g), with slightly shorter tails (0.76 v. 1.07 × SVL), and a high proportion of prehensile tails (7.7% v. 1.8%) and secretory pores (62% v. 46%). A number are elongate (7.7% v. 0.9%) and have basic (7.7% v. 14.6%) and advanced (7.7% v. 2.1%) color change. In this case advanced color change is probably associated with reliance on crypsis (100% v. 50%), which comports well with the focus on immobility and gliding. Gliding

Luperosaurus spp.	English name				Range core
			Size		
	Activity	Diet	mm	g	Ecological business model
angliit	Little Brother Gecko				Philippines
	nocturnal		56	3.82	* nocturnal canopy walker
brooksii	Brooks' Camouflage Gecko				Indonesia
	nocturnal		52	3.07	
browni	Brown's Camouflage Gecko				Malaysia
	nocturnal		57	3.25	
corfieldi	Corfield's Gecko				Philippines
	nocturnal		77	9.4	** immobility expert
cumingii	Cuming's Flap-legged Gecko				Philippines
	nocturnal		72	7.6	** nocturnal canopy walker
gulat	Gunob Gecko				Philippines
	nocturnal		67	6.4	* trunk/wall lizard
iskandari	Pak Djoko's Flap-legged Gecko				Indonesia
	nocturnal		59	4.38	* nocturnal canopy walker
joloensis	Jolo Flap-legged Gecko				Philippines
	nocturnal		35	1.07	* nocturnal canopy walker
kubli	Mount Lataan Gecko				Philippines
	nocturnal		84	11.9	*** glider
macgregori	MacGregor's Flap-legged Gecko				Philippines
	nocturnal		52	3.07	* nocturnal canopy walker
palawanensis	Palawan Flap-legged Gecko				Philippines
	nocturnal		47	2.31	* immobility expert or nocturnal canopy walker
sorok	Crocker Range Camouflage Gecko				Malaysia
	nocturnal		34	0.99	**** glider
yasumai	Yasuma's Camouflage Gecko				Indonesia
	nocturnal		37	1.25	

flaps may be exaptations of camouflage flaps. If color change is used to promote crypsis, it is less useful as a social signal, and in *Luperosaurus* there are no examples of sexual dichromatism (v. 6.6%) or seasonal color (v. 0.8%). *Luperosaurus* are not known to forage at electric lights (v. 3.8%). They call spontaneously at gekkonid levels (7.7% v. 6.5%), but are not known to issue distress calls (v. 3.7%). Given the paucity of specimens, it is probable that behavior such as vocalization is underreported. Many are conspicuously wary (15% v. 4.4%).

Most are arboreal (91% v. 68%); a few are terrestrial (9% v. 31%). Some are found in littoral areas (17% v. 11%), but they are underrepresented in sandy areas (0% v. 11%), riparian zones (0% v. 17%), and human habitats (8% v. 27%).

Reproduction follows the gekkonid template, though incubation durations are notably short (–0.74 v. 0.17 std. resid.); this is based solely on the 42-day period of *corfieldi*.

All species are island species (v. 45%). None has been introduced (v. 3.7%). *Luperosaurus* are underrepresented among imperiled species (29% v. 47%), though all of these are listed by the IUCN as among the most endangered (29% v. 11%). In other respects *Luperosaurus* are typical gekkonids.

Lygisaurus

Lygisaurus is a moderate radiation (13 spp.) of very small (0.73 g v. 3.3 g), diurnal (100% v. 74%), oviparous (100% v. 64%), terrestrial (33% v. 48%) or cryptozoic (67% v. 30%) skinks centered on Queensland and extending to New Guinea, the Northern Territory, and New South Wales. Most are litter cryptozoids (80% v. 28%); the remainder are ground-foraging tree lizards (10% v. 7%) or refugia-anchored diurnal ground lizards (10% v. 30%). All of the litter cryptozoid assignments were made with low confidence, and other aspects—lack of elongation, absence of blue tails—suggest that some of these might actually be refugia-anchored diurnal ground lizards. Though they bask directly (69% v. 21%), they often stick to shaded areas while foraging. They are closely related to, and often grouped with, *Carlia*.

Morphologically *Lygisaurus* match skinks well, except for an absence of elongation (0% v. 35%). Blue tails are absent (v. 4%), but several species have bright red/orange tails (*foliorum*, *malleolus*, *parrhasius*). The only sex difference noted is seasonal color, which is expressed in the majority of species (54% v. 6%). In *aeratus* the male seasonal color includes a red tail.

Like most skinks, they are generalized invertebrate eaters (100% v. 84%) that forage actively (100% v. 92%). Home-range area is unknown.

The elevational distribution of *Lygisaurus* is normal (l/m/h: 100/25/0

Lygisaurus spp.	English name	Activity	Diet	Size mm	Size g	Range core	Ecological business model
abscondita	Mt. Surprise Litter-skink			30	0.46	Queensland	* litter cryptozoid
aeratus	Large-disced Litter-skink	diurnal		37	0.87	Queensland	** litter cryptozoid
curtus	Papua Litter-skink			40	1.15	New Guinea	
foliorum	Tree-based Litter-skink		invert.	37	0.89	Queensland	** litter cryptozoid
laevis	Rainforest Edge Litter-skink	diurnal		36	0.77	Queensland	* refugia-anchored diurnal gr. liz.
macfarlani	Translucent Litter-skink	diurnal	invert.	37	0.85	Queensland	** litter cryptozoid
malleolus	Firedart Litter-skink			31	0.50	Queensland	** litter cryptozoid
novaeguineae	New Guinea Four-fingered Skink	diurnal		34	0.69	New Guinea	
parrhasius	Fire-tailed Rainbow Skink	diurnal		34	0.67	Queensland	
rococo	Chillagoe Litter-skink	diurnal		40	1.11	Queensland	** sand shark
sesbrauna	Eastern Cape Litter-skink			33	0.59	Queensland	** litter cryptozoid
tanneri	Endeavor River Litter-skink			35	0.70	Queensland	** litter cryptozoid
zuma	Sun-loving Litter-skink	diurnal		33	0.59	Queensland	* litter cryptozoid

v. 89/35/6). They are overrepresented in riparian habitat (54% v. 23%), representative in littoral (15% v. 21%) and human (15% v. 14%) habitats, and absent from sandy areas (v. 25%).

Clutch count is two (0.34 v. 0.06 std. resid.) and is apparently fixed at that value. Multiple broods per year are not reported. RCM is about average (0.280 v. 0.236).

Longevity and population density are unknown. Endemism is average (77% v. 64%). Island species are low (39% v. 57%). None has been introduced (v. 2.1%). *Lygisaurus* are imperiled at a typical rate (40% v. 47%), though none is on one of the IUCN's lists of the most endangered. In all other documented respects *Lygisaurus* are typical skinks.

Lygodactylus

Lygodactylus is a major radiation (64 spp.) of small (0.94 g v. 3.31 g) gekkonids. However, this group of mostly drab species marks a sharp contrast with the rest of the family, as they are overwhelmingly diurnal (98% v. 15%), and they are found predominantly in the Ethiopian biogeographic realm, including Madagascar (97% v. 38%) with the remaining 3% (v. 1%) in the Neotropics. The majority are trunk/wall lizards (57% v. 45%), but notably include 21% refugia-anchored diurnal ground lizards, which are rare in gekkonids (3%). Other ecological business models include diurnal canopy walker, also very rare in gekkonids (11% v. 1%), ground-foraging tree lizards (9% v. 5%), and immobility experts (2% v. 4.5%). The preponderance of trunk dwellers is a little surprising to me, in that their very small size produces short legs, which are more often associated with smaller diameter perches.

In addition to being small, *Lygodactylus* are slender (shape index −0.10 v. 0.10 std. resid.), without prehensile tails (v. 1.8%), but mostly possessing secretory pores (80% v. 46%). Curiously, at least 10 species possess adhesive lamellae on the tip of their tails, in at least some cases allowing them to be fully supported by the tail tip alone. This is unusual for lizards that lack prehensile tails, but perhaps the combination (adhesive tips and non-prehensile tail) works well on a tree trunk. Possession of any color change is infrequent compared to gekkonids (4.7% v. 14.6%), but advanced color change is more frequent than typical gekkonids (3.1% v. 2.1%). All *Lygodactylus* have enlarged toe pads (v. 70%).

Lygodactylus are slightly more m<f dimorphic than typical gekkonids (−0.33 v. −0.03 on −2 to 2 scale), and 20% (v. 7%) exhibit dichromatism. A few even exhibit seasonal color, which is very rare in gekkonids (3.1% v. 0.8%). Accordingly, the sex rollup score for *Lygodactylus* is relatively high (1.11 v. 0.63 on 0 to 6 scale).

All documented *Lygodactylus* are invertebrate eaters (v. 95%). None is an ant specialist, as is true for the family. However, 56% of the *Lygodac-*

tylus with dietary data consume ants, whereas only 18% of Gekkonids do so. This may reflect both the small size and diurnality of *Lygodactylus* (see ant-eating discussion associated with table 3.5). Only five species have been scored for foraging mode; all were described as sit-and-wait (100% v. 81%); one was also said to be an active forager (20% v. 14%). Two species (*capensis*, *luteopicturatus*) are reported to forage under electric lights (3.1% v. 3.8%). No *Lygodactylus* is reported to vocalize spontaneously (v. 6.5%), a silent condition typical of diurnal lizards, but 1.6% (v. 3.7%) issue stress vocalizations. Wariness is at family-typical levels (4.7% v. 4.4%). Predator concealment tactics largely mirror the family's tactics, with greater reliance on concealment in vegetation (50% v. 19%). For concealment in arboreal vegetation, one might anticipate green body coloration, which is common in the one other large genus of diurnal gekkonids (*Phelsuma*); brilliant green bodies do not, however, occur in *Lygodactylus*. Perhaps their drab colors match the bark of large diameter perches on which they rest, rather than the green leaves within which *Phelsuma* hide. Direct basking is elevated (11% v. 5%), as is the mean active body temperature (35.0°C v. 27.6°C) of *Lygodactylus*, the hottest mean of any gekkonid genus. *Phelsuma*, in contrast, are not particularly warm (mean 29.0°C); perhaps the brown color of *Lygodactylus* expedites warming in these tiny geckos. Social relations follow the family pattern; spatial defense of prospective mates by intolerant males.

Both *Lygodactylus* and Gekkonidae are bimodal with relation to annual rainfall, but *Lygodactylus* exhibits their primary mode in the arid belt (67% v. 27%). *Lygodactylus* is unusually common at medium elevations (47% v. 26%), as might be expected in

Lygodactylus spp.	English name				Range core
			Size		
	Activity	Diet	mm	g	Ecological business model
angolensis	Angola Dwarf Gecko				Southern Africa
	diurnal	invert.	31	0.78	*** trunk/wall lizard
angularis	Angulate Dwarf Gecko				Zambia
	diurnal	invert.	41	1.62	*** trunk/wall lizard
arnoulti	Arnoult's Dwarf Gecko				Madagascar
	diurnal		33	0.90	* refugia-anchored diurnal gr. liz.
bernardi	Bernard's Dwarf Gecko				Zimbabwe
	diurnal		36	1.13	* refugia-anchored diurnal gr. liz.
bivittis	Tiny-scaled Dwarf Gecko				Madagascar
	diurnal	invert.	35	1.07	* trunk/wall lizard
blancae	Lake Itasy Gecko				Madagascar
	diurnal	invert.	34	1.01	*** trunk/wall lizard
blanci	Mount Ibity Dwarf Gecko				Madagascar
	diurnal		34	0.96	* refugia-anchored diurnal gr. liz.
bradfieldi	Bradfield's Dwarf Gecko				Botswana
	diurnal	invert.	30	0.67	** diurnal canopy walker
broadleyi	Broadley's Dwarf Gecko				Tanzania
	diurnal	invert.	25	0.40	
capensis	Cape Dwarf Gecko				Southern Africa
	diurnal	invert.	36	1.15	*** ground-foraging tree lizard
chobiensis	Chobe Dwarf Gecko				Botswana
	diurnal	invert.	35	1.07	*** trunk/wall lizard
conradti	Conradt's Dwarf Gecko				Tanzania
	diurnal	invert.	27	0.49	
conraui	Conrau's Dwarf Gecko				Nigeria
	diurnal	invert.	35	1.05	** trunk/wall lizard
decaryi	Angel's Dwarf Gecko				Madagascar
	diurnal		27	0.52	* trunk/wall lizard
depressus	Africa Spotted Dwarf Gecko				Central Africa
	diurnal		37	1.20	** trunk/wall lizard
expectatus	Ambilobe Dwarf Gecko				Madagascar
	diurnal		29	0.65	** trunk/wall lizard
fischeri	Fischer's Dwarf Gecko				Cameroon
	diurnal	invert.	39	1.46	* diurnal canopy walker
grandisonae	Grandison's Dwarf Gecko				Ethiopia
	diurnal	invert.	29	0.63	
graniticolus	Granite Dwarf Gecko				South Africa
	diurnal		36	1.13	* refugia-anchored diurnal gr. liz.

Lygodactylus spp.	English name				Range core
			Size		
	Activity	Diet	mm	g	Ecological business model
gravis	Usambara Dwarf Gecko				Tanzania
	diurnal	invert.	39	1.46	* ground-foraging tree lizard
grotei	Grote's Dwarf Gecko				Mozambique
	diurnal		33	0.92	
guibei	Western Dwarf Gecko				Madagascar
	diurnal		38	1.32	** trunk/wall lizard
gutturalis	Chevron-throated Dwarf Gecko				Central Africa
	diurnal	invert.	38	1.33	*** trunk/wall lizard
heterurus	Boettger's Dwarf Gecko				Madagascar
	diurnal		27	0.52	* trunk/wall lizard
inexpectatus	Dar es Salaam Dwarf Gecko				Tanzania
	diurnal		33	0.88	
insularis	Insular Dwarf Gecko				Madagascar
	diurnal		28	0.57	
intermedius	Andringitra Dwarf Gecko				Madagascar
	diurnal		32	0.82	
keniensis	Kenya Dwarf Gecko				Ethiopia
	diurnal	invert.	35	1.07	**** trunk/wall lizard
kimhowelli	Kim Howell's Dwarf Gecko				Tanzania
	diurnal	invert.	35	1.04	** trunk/wall lizard
klemmeri	Klemmer's Dwarf Gecko				Madagascar
	diurnal		29	0.65	** trunk/wall lizard
klugei	Kluge's Dwarf Gecko				Brazil
	diurnal	invert.	29	0.65	** diurnal canopy walker
laterimaculatus	Side-spotted Dwarf Gecko				Tanzania
	diurnal		30	0.67	** diurnal canopy walker
lawrencei	Lawrence's Dwarf Gecko				Angola
	diurnal		34	0.94	* trunk/wall lizard
luteopicturatus	Yellow-headed Dwarf Gecko				Tanzania
	diurnal	invert.	40	1.50	** trunk/wall lizard
madagascariensis	Madagascar Dwarf Gecko				Madagascar
	diurnal		36	1.13	* trunk/wall lizard
manni	Mann's Dwarf Gecko				Tanzania
	diurnal	invert.	38	1.33	** diurnal canopy walker
methueni	Methuen's Dwarf Gecko				South Africa
	diurnal		38	1.31	*** trunk/wall lizard
miops	Guenther's Dwarf Gecko				Madagascar
	diurnal		34	0.95	* trunk/wall lizard
mirabilis	Ankaratra Dwarf Gecko				Madagascar
	diurnal		30	0.71	** refugia-anchored diurnal gr. liz.

comparatively high-altitude Africa. *Lygodactylus* are underrepresented in sandy (3% v. 11%) and riparian (3% v. 17%) areas, but overrepresented in human habitat (39% v. 27%). Perhaps they avoid some mesopredators by proximity to humans.

All documented reproductive conditions match the Gekkonidae. *Lygodactylus* is so small that the apparently obligatory clutch count of two is above the mass-adjusted lizard average (0.20 v. –0.26 std. resid.). Furthermore, *Lygodactylus* deviates from the family pattern in having young that are smaller than the lizard average (ontogenetic size range 0.36 v. –0.19 std. resid.), a pattern also seen in other gekkonid radiations: *Hemidactylus* and *Lepidodactylus*. *Lygodactylus capensis* has an unusually wide range of incubation durations (60–177 days), perhaps reflecting the wide range of latitudes occupied.

Longevity is short (–0.43 v. 0.15 std. resid.), but this is based on only four species. *Lygodactylus* have been introduced at a gekkonid-typical rate (3.1% v. 3.7%), though no species has been introduced more than twice (v. 2.1%). *Lygodactylus* are imperiled at a gekkonid-typical rate as well: 49% (v. 47%) for any statement of concern; 13.5% (v. 11.3%) for IUCN listing in the most endangered categories. In other respects *Lygodactylus* are typical gekkonids.

Lygodactylus spp.	English name	Activity	Diet	Size mm	Size g	Range core	Ecological business model
mombasicus	Mombasa Dwarf Gecko	diurnal	invert.	37	1.23	Tanzania	** trunk/wall lizard
montanus	Mount Ivohibe Gecko	diurnal		37	1.19	Madagascar	** refugia-anchored diurnal gr. liz.
nigropunctatus	Black-spotted Dwarf Gecko	diurnal		33	0.92	South Africa	** ground-foraging tree lizard
ocellatus	Ocellated Dwarf Gecko	diurnal		33	0.88	South Africa	* refugia-anchored diurnal gr. liz.
ornatus	Madagascar Ornate Dwarf Gecko	diurnal		29	0.64	Madagascar	* refugia-anchored diurnal gr. liz.
pauliani	Paulian's Dwarf Gecko	diurnal		35	1.07	Madagascar	* refugia-anchored diurnal gr. liz.
picturatus	White-headed Dwarf Gecko	diurnal	invert.	38	1.32	Central Africa	**** trunk/wall lizard
pictus	Robust Dwarf Gecko	diurnal		38	1.27	Madagascar	** trunk/wall lizard
rarus	Thin Dwarf Gecko	diurnal		36	1.13	Madagascar	** refugia-anchored diurnal gr. liz.
regulus	Prince Dwarf Gecko	diurnal		39	1.37	Mozambique	
rex	King Dwarf Gecko	diurnal		49	2.62	Mozambique	
roavolana	Two-moon Dwarf Gecko	diurnal		26	0.48	Madagascar	** ground-foraging tree lizard
scheffleri	Scheffler's Dwarf Gecko	diurnal		28	0.58	Tanzania	
scorteccii	Scortecci's Dwarf Gecko	diurnal	invert.	35	1.07	Ethiopia	
somalicus	Somali Dwarf Gecko	diurnal	invert.	33	0.89	Ethiopia	**** trunk/wall lizard
stevensoni	Stevenson's Dwarf Gecko	diurnal		34	0.97	South Africa	* ground foraging tree lizard
thomensis	Annobon Dwarf Gecko	diurnal	invert.	36	1.11	Liberia	* trunk/wall lizard
tolampyae	Grandidier's Dwarf Gecko	diurnal		35	1.03	Madagascar	*** trunk/wall lizard

Lygodactylus spp.	English name				Range core
			Size		
	Activity	Diet	mm	g	Ecological business model
tuberosus	Bumpy Dwarf Gecko				Madagascar
	diurnal		37	1.20	*** trunk/wall lizard
verticillatus	Mocquard's Dwarf Gecko				Madagascar
	diurnal		29	0.63	** trunk/wall lizard
viscatus	Copal Dwarf Gecko				Tanzania
	diurnal	invert.	29	0.62	* immobility expert
waterbergensis	Waterberg Dwarf Gecko				South Africa
	diurnal		37	1.22	* refugia-anchored diurnal gr. liz.
wetzeli	South American Dwarf Gecko				Paraguay
	diurnal		30	0.68	* trunk/wall lizard
williamsi	Turquoise Dwarf Gecko				Tanzania
	crepuscular	invert.	35	1.07	** diurnal canopy walker
wojnowskii	Mount Kenya Dwarf Gecko				Kenya
	diurnal		37	1.19	* trunk/wall lizard

Lygosoma (fig. 4.8A)

Lygosoma is a major radiation (30 spp.) of slightly small (2.60 g v. 3.25 g), mostly elongate (77% v. 35%), mostly litter cryptozoic (79% v. 28%) skinks found in two Old World clusters: equatorial Africa (from Ghana to Somalia and Mozambique) and southern Asia (from Pakistan/Nepal/Sri Lanka to China/Vietnam/Philippines/Borneo). Those that are not litter cryptozoids are most likely sand sharks (10.5% v. 6.8%), nocturnal ground lizards (5.3% v. 2.3%), or subterranean foragers (5.3% v. 11.8%). Although no *Lygosoma* are totally limbless, two-thirds are extremely elongate. That is, the gap separating adpressed limbs exceeds 25% of the axilla-groin distance.

Lygosoma are a little lighter in build than an average skink (shape index −0.88 v. −0.46 std.). Blue tails are not present in *Lygosoma* (v. 4%); *punctata* juveniles may have brilliant red tails.

SSD is characterized in only one *Lygosoma* (*quadrupes*, m<f). Sexual dichromatism (3.3% v. 7.6%) and seasonal color (3.3% v. 6.2%) are both present in *bowringii*. *L. bowringii* is particularly well studied where it occurs as an invader on Australia's Christmas Island, hinting that other forms of color sex differences might be present elsewhere and not yet detected.

All *Lygosoma* are invertebrate eaters (100% v. 84%) and active foragers (100% v. 92%). Termites are often mentioned as a prey item.

All four diel activity patterns are present in *Lygosoma*; crepuscular is notably overrepresented (15.4% v. 2.4%); diurnal is mildly underrepresented (61.5% v. 74%). Several species are noted to locomote without use of limbs (especially when pressed), and that inactivity may extend over many days. Three species have been described in relation to primary predator concealment tactic, and all three rely on sand swimming for escape (100% v. 22%). Reports of direct basking are rare (6.7% v. 21%). Mean active body temperature is a sizzling 38.0°C (v. 31.1°C), though this is based on one species (*bowringii*).

Rainfall-documented *Lygosoma* are toward the wet end of the skink spectrum: they are slightly overrepresented in the intermediate band (50% v. 34%) and much overrepresented in the mesic (100% v. 40%). Their elevational profile is about average (l/m/h: 95/23/0 v. 89/35/6). They are slightly overrepresented in human habitats (25% v. 14%) and slightly underrepresented in sandy areas (13% v. 25%). They are a close match for littoral (21% v. 21%) and riparian (21% v. 23%) areas.

All are oviparous (v. 64%) with average clutch counts (0.03 v. 0.06 std. resid.). The wide ontogenetic mass range (1.04 v. 0.12 std. resid.) indicates small hatchlings. Incubation is about average (−0.34 v. −0.40 std. resid.). At least one species (*corpulentum*) may have multiple broods in a year.

Endemism is about average (53% v. 64%). Island species are few (37% v. 57%). Introductions are common (6.7% v. 2.1%), though none has been introduced more than twice (v. 0.7%). *L. bowringii* has been introduced to the Philippines as well as Christmas Island; *pembanum* has been introduced in Kenya. No *Lygosoma* is considered imperiled (v. 47%). In other respects *Lygosoma* are typical skinks.

Lygosoma spp.	English name				Range core
			Size		
	Activity	Diet	mm	g	Ecological business model
albopunctata	White-spotted Supple Skink				Southern Asia
	diurnal	invert.	57	1.34	*** litter cryptozoid
angeli	Angel's Supple Skink				Southeast Asia
		invert.	89	4.14	** sand shark or litter cryptozoid
anguinum	Myanmar Supple Skink				Myanmar
		invert.	52	1.06	*** litter cryptozoid
bampfyldei	Black-naped Masked Skink				Borneo
			111	7.0	
boehmei	Boehme's Supple Skink				Vietnam
	nocturnal		74	2.63	*** nocturnal ground lizard
bowringii	Bowring's Supple Skink				Southeast Asia
	crepuscular	invert.	51	1.02	**** litter cryptozoid
chaperi	Serpentiform Skink				Ghana
			56	1.95	
corpulentum	Fat Skink				Southeast Asia
	diurnal	invert.	149	14.5	** sand shark or litter cryptozoid
frontoparietale	Taylor's Writhing Skink				Thailand
			39	0.71	
goaensis	Goa Writhing Skink				India
	diurnal	invert.	47	1.92	** litter cryptozoid
guentheri	Guenther's Writhing Skink				India
	diurnal	invert.	88	3.98	*** litter cryptozoid
haroldyoungi	Banded Supple Skink				Thailand
	diurnal	invert.	115	7.7	** subterranean forager or sand shark
herberti	Peninsular Supple Skink				Thailand
	diurnal		57	1.38	*** litter cryptozoid
isodactylum	Even-toed Supple Skink				Southeast Asia
			93	4.56	* litter cryptozoid
koratense	Ratchasima Supple Skink				Thailand
			95	4.82	* litter cryptozoid
laeviceps	Common Writhing Skink				Ethiopia
			70	6.9	
lanceolatum	Bazaruto Writhing Skink				Mozambique
			72	7.5	* litter cryptozoid
lineata	Lined Writhing Skink				India
	crepuscular	invert.	50	0.99	** litter cryptozoid
lineolatum	Lined Supple Skink				Myanmar
		invert.	54	1.21	*** litter cryptozoid

Lygosoma spp.	English name				Range core
			Size		
	Activity	Diet	mm	g	Ecological business model
mafianum	Mafia Writhing Skink				Tanzania
			71	7.1	
opisthorhodum	Indragiri Supple Skink				Indonesia
			76	4.53	
pembanum	Pemba Island Writhing Skink				Tanzania
	nocturnal		75	8.6	
popae	Mt. Popa Writhing Skink				Myanmar
			53	2.78	
pruthi	Pruthi's Writhing Skink				India
	diurnal	invert.	57	3.57	
punctata	Punctate Supple Skink				Southern Asia
	diurnal	invert.	78	2.92	*** litter cryptozoid
quadrupes	Short-limbed Supple Skink				Southeast Asia
	cathemeral	invert.	78	2.95	
singha	Sri Lanka Supple Skink				Sri Lanka
			42	0.65	
tabonorum	Palawan Supple Skink				Philippines
			65	1.94	* litter cryptozoid
veunsaiensis	Veun Sai Forest Skink				Cambodia
			33	0.36	** litter cryptozoid
vosmaeri	Vosmaer's Writhing Skink				India
		invert.	52	1.06	* litter cryptozoid

Lyriocephalus sp.	English name				Range core
			Size		
	Activity	Diet	mm	g	Ecological business model
scutatus	Hump-nosed Lizard				Sri Lanka
	diurnal	omnivore	141	98	** ground-foraging tree lizard

Lyriocephalus

Lyriocephalus is a single-species genus; it is a large (98 g v. 15.6 g) Sri Lankan agamid with a spectacular warning display. This gaudy—when choosing to be—arboreal lizard has a large globular yellow nose ornament, a huge bright yellow dewlap (larger in males), a cerulean belly, and a prominent—at least in adult males—yellow-green serrated nuchal-dorsal crest (v. 38%). When agitated it faces its antagonist and opens its enormous mouth to reveal a bright, pure red interior.

Overall shape is robust (1.44 v. 0.80 std. resid.). It lacks secretory pores (v. 13%). Its tail is short (1.0 v. 1.86 × SVL). The sexes differ in color (v. 53%), and it can change its color (v. 15% advanced color change). The latter is likely associated with both its cryptic predator concealment (v. 51%) and its strong sexual dichromatism. Its sex differences rollup score is the fourth highest in Agamidae (3.00 v. 1.61 on 0 to 6 scale). It moves slowly and may death feign. It does not bask (v. 24% basking directly). It will dig up earthworms and grubs, suggesting an unusual—for an agamid—reliance on chemical cues for prey discovery. It is found in mesic (v. 20%) forests and human habitats (v. 16%). Its clutch count of 4.5 eggs is small for its size (−0.93 v. 0.20 std. resid.). Its ontogenetic size range is also small (−0.06 v. 0.69 std. resid.), indicating that the young are not only few in number but large in size. Its longevity is notably short (−1.40 v. −0.46 std. resid.). It has not been introduced (v. 1.7%). It is listed as near threatened by the IUCN (v. 37% imperiled).

Mabuya

Mabuya is a moderate radiation (13 spp.) of large (10.3 g v. 3.3 g), diurnal (100% v. 74%). viviparous (100% v. 36%) skinks of diverse ecologies found in the vicinity of the Caribbean (Colombia and Venezuela through the Lesser Antilles and Hispaniola). This genus has recently undergone a radical shrinkage. The following genera were wholly or in part contained in *Mabuya* (those with an asterisk were removed recently): *Alinea**, *Aspronema*, *Ateuchosaurus*, *Brasiliscincus**, *Capitellum**, *Chioninia*, *Copeoglossum**, *Emoia*, *Eumeces*, *Eutropis**, *Heremites*, *Lubuya*, *Manciola*, *Marisora**, *Notomabuya*, *Orosaura**, *Panopa**, *Plestiodon*, *Psychosaura**, *Spondylurus**, *Toenayar*, *Trachylepis**, and *Varzea**. The remaining *Mabuya* species are primarily insular forms, many of which are likely extinct (*desiradae*, *grandisterrae*, *hispaniolae*, *mabouya*, *montserratae*), victims of mongoose and other introduced species. The six living species are not very well known (filled cells 21% v. 25%). The modal surviving species is a ground-foraging tree lizard (50% v. 7%); others are refugia-anchored diurnal ground lizards (33% v. 30%) or crack hermits (17% v. 1.7%). It would be interesting to know if the extirpated forms were a random selection from these ecological business models, or if lizards in certain microhabitats (e.g., terrestrial species) were more vulnerable to extinction.

Mabuya are more robust that average skinks (shape index 0.31 v. −0.46 std. resid.). Their tails are a bit longer (1.61 v. 1.33 × SVL) but just as breakable (92% v. 73%). They are not elongate (v. 35%) and do not have blue tails (v. 4%).

Five species have documentation of SSD and all are m<f (−1.00 v. −0.31 on −2 to 2 scale), a surprising condition for lizards this large (table 3.1, trait 6). No other sex differences are known.

All eat invertebrates (v. 84%), which they acquire by active foraging (v. 92%). Home-range size and electric light use are unknown. Direct basking is present at typical levels (15% v. 21%). Mean active body temperature is slightly warm (33.0°C v. 31.1°C).

The elevational profile of *Mabuya* is normal (l/m/h: 86/29/0 v. 89/35/6). They are overrepresented in littoral (33% v. 21%) and human (56% v. 14%) habitats, and they are absent from sandy (v. 25%) and riparian (v. 23%) areas.

Their litter size is as expected for their body size (−0.09 v. 0.06 std. resid.). There is no evidence for multiple broods per year. The young are large (ontogenetic mass range −0.80 v. 0.12 std. resid.). Large neonates seem to be common in island species of large skinks and might constitute a selection pressure for large female size in this case.

Mabuya spp.	English name					Range core
			Size			
	Activity	Diet	mm	g		Ecological business model
berengerae	San Andres Skink					Colombia
	diurnal		81	10.8		*** crack hermit or diurnal canopy walker
cochonae	Cochons Skink					Guadeloupe
			79	10.2		
desiradae	Desirade Skink					Guadeloupe
	diurnal		79	10.2		* refugia-anchored diurnal gr. liz.
dominicana	Dominica Skink					Dominica
	diurnal		82	11.2		** ground-foraging tree lizard
grandisterrae	Grande-Terre Skink					Guadeloupe
			82	11.5		
guadeloupae	Guadeloupe Skink					Guadeloupe
			85	12.8		
hispaniolae	Hispaniola Two-lined Skink					Dominican Republic
			76	8.8		
mabouya	Greater Martinique Skink					Martinique
	diurnal	invert.	82	11.2		
meridensis	Merida Skink					Venezuela
			64	5.2		** ground-foraging tree lizard
montserratae	Montserrat Skink					Montserrat
			79	10.2		
parviterrae	Petite Terre Skink					Guadeloupe
			83	11.8		
pergravis	Providencia Skink					Colombia
	diurnal		82	11.3		* ground-foraging tree lizard
zuliae	Maracaibo Skink					Venezuela
	diurnal		82	11.2		* refugia-anchored diurnal gr. liz.

Population density has been estimated 12 times for *dominicana* (0.53 kg/ha v. 0.35 kg/ha), a value lower than might be expected for an island, though mammals other than the mongoose are present on Dominica. Endemism is very high (92% v. 64%). Most species are found on islands (85% v. 57%). None has been introduced (v. 2.1%). Sadly, most are imperiled (92% v. 47%), including placement of 25% (v. 14%) on the IUCN's lists of the most endangered. Though likely extinct, *desiradae* and *grandisterrae* had not been evaluated by the IUCN at the time of compilation (mid-2017). In all other documented respects *Mabuya* are typical skinks.

Macropholidus

Macropholidus is a four-species genus of very poorly known (filled cells 12% v. 19%) gymnophthalmids found in Ecuador and Peru. Too little information is available to assign them to

Macropholidus spp.	English name				Range core
			Size		
	Activity	Diet	mm	g	Ecological business model
annectans	Parker's Large-scaled Lizard				Ecuador
			52	1.88	
ataktolepis	Cachil Forest Lizard				Peru
			40	1.31	
huancabambae	Huancabamba Large-scaled Lizard				Peru
			48	2.41	
ruthveni	Ruthven's Large-scaled Lizard				Peru
			42	1.56	

Madascincus spp.	English name				Range core
			Size		
	Activity	Diet	mm	g	Ecological business model
ankodabensis	Ankodabe Skink				Madagascar
			46	1.14	
arenicola	Sand-loving Skink				Madagascar
	nocturnal		68	3.35	* nocturnal ground lizard
igneocaudatus	Red-tailed Skink				Madagascar
			62	2.60	
macrolepis	Rusty Skink				Madagascar
	nocturnal		34	0.49	** litter cryptozoid
melanopleura	Common Madagascar Skink				Madagascar
	diurnal	invert.	50	1.45	**** litter cryptozoid
miafina	Petit Tsingy Skink				Madagascar
			53	2.78	*** litter cryptozoid
minutus	Manongarivo Dwarf Skink				Madagascar
	diurnal		40	0.79	**** litter cryptozoid
mouroundavae	Morondava Skink				Madagascar
	diurnal		62	2.57	**** litter cryptozoid
nanus	Nano Skink				Madagascar
			33	0.48	
polleni	Madagascar Coastal Skink				Madagascar
			75	4.37	**** litter cryptozoid
pyrurus	Ibity Mountain Skink				Madagascar
			48	2.02	* litter cryptozoid
stumpffi	Stumpff's Skink				Madagascar
			91	7.4	

ecological business models. They are somewhat small (1.75 g v. 2.35 g). At least one is elongate. Secretory pores and other sex differences are not reported. They are found at medium elevation (v. 52%). All species are endemic to their countries (v. 68%). They have not been evaluated for conservation risk. In other respects *Macropholidus* are probably typical gymnophthalmids.

Madascincus

Madascincus is a moderate radiation (12 spp.) of small-sized (1.81 g v. 3.25 g), mostly elongate (83% v. 35%), mostly litter cryptozoid (88% v. 28%) skinks endemic to Madagascar. Those that are not litter cryptozoids are nocturnal ground lizards (13% v. 2%).

They are slightly more slender than most skinks (shape index −0.65 v. −0.46 std. resid.), but they are otherwise similar. None has a blue tail (v. 4%), though several have bright red tails (*igneocaudatus*, *miafina*, *pyrurus*). No sex differences are known, though *minutus* is likely m<f in size and has an extraordinarily low m:f ratio of relative tail lengths (0.68 v. 1.03).

Only nocturnal (40% v. 14.5%) and diurnal (60% v. 74%) *Madascincus* are known. Such an even split usually suggests unrecognized cathemeral species, but the narratives do not support that in this case. Data on *arenicola* activity come from pitfall captures, as the species has not been observed active.

One species is quantified for annual rainfall (*polleni*); it is found in the mesic band (in this case 2250 mm/y). The genus's elevational profile is skewed toward intermediate elevations (l/m/h: 75/50/0 v. 89/35/6). *Madascincus* have a normal frequency in sandy areas (33% v. 25%), but they are absent from littoral (v. 21%), riparian (v. 23%), and human (v. 14%) habitats.

Most documented species are oviparous (75% v. 64%); *igneocaudatus* appears to be viviparous in some localities (25% v. 36%). Clutch/litter count is high for such small skinks (0.62 v. 0.06 std. resid.). Incubation duration is normal (–0.36 v. –0.40 std. resid.).

No *Madascincus* has been introduced (v. 2.1%). They are imperiled at the usual rate (40% v. 47%), including 20% (*arenicola*, critically endangered) that are on the IUCN's lists of the most endangered (v. 14%). In all other documented respects *Madascincus* are typical skinks.

Malayodracon sp.	English name				Range core
			Size		
	Activity	Diet	mm	g	Ecological business model
robinsonii	Robinson's Anglehead Lizard				Malaysia
	diurnal	invert.	122	62	** ground-foraging tree lizard or diurnal canopy walker

Manciola sp.	English name				Range core
			Size		
	Activity	Diet	mm	g	Ecological business model
guaporicola	Neotropics Small-handed Skink				Paraguay
	diurnal	invert.	69	3.44	* refugia-anchored diurnal gr. liz.

Mantheyus sp.	English name				Range core
			Size		
	Activity	Diet	mm	g	Ecological business model
phuwuanensis	Phu Wua Lizard				Thailand
	cathemeral	invert.	74	13.0	** refugia-anchored diurnal gr. liz.

Malayodracon

Malayodracon is a single-species genus of a large (62 g v. 15.6 g), robust (shape index 1.3 v. 0.80 std. resid.), crested (v. 38%), arboreal (v. 43%) agamid from peninsular Malaysia. It has a long tail (2.3 v. 1.9 × SVL) and is found on perches of all diameters. The males have slightly larger dewlaps, and the sexes differ in color (v. 53%) and size (m>f; 1.0 v. 0.53 on –2 to 2 scale). These add up to strongly differentiated sexes (3.0 v. 1.6 on 0 to 6 sex differences rollup score). It can change its hue (v. 15% advanced color change), a capability likely to support its cryptic predator concealment (v. 51%). It is found in mossy/lichenous mid-elevation (v. 52%) forests, often near streams (v. 21%), but will also tolerate disturbed habitats (v. 16%). Its clutch count of about seven is low for a lizard of this size (–0.10 v. 0.20 std. resid.). It has not been introduced (v. 1.7%) or evaluated for conservation status.

Manciola

Manciola guaporicola is the sole member of its genus. It is an average-sized (3.4 g v. 3.3 g), elongate (v. 35%), refugia-anchored diurnal ground skink (v. 30%) found in the center of South America: Bolivia, Paraguay, and Brazil. Its tail is comparatively long (1.80 v. 1.33 × SVL) and breakable (v. 73%). Males are smaller than females (–1.0 v. –0.31 on –2 to 2 scale). Other sex differences are not known. It forages actively (v. 92%) for invertebrates (v. 84%). It basks directly (v. 21%) and has a mean active body temperature of about 35°C (v. 31.1°C). For concealment it retreats to vegetation (v. 25%). It is associated with sandy habitats (v. 25%). Reproduction is somewhat uncertain, but both modes have been reported. Clutch/litter count is normal (0.34 v. 0.06 std. resid.). It is not endemic (v. 64%), it is not found on an island (v. 57%), and it has not been introduced (v. 2.1%). It has not been evaluated for conservation status. In other respects it is likely to be similar to other skinks.

Mantheyus

Mantheyus is a single-species genus of average-sized (13.0 g v. 15.6 g), crestless (v. 38%), terrestrial (v. 56%) agamid distinguished at the genus level by possession of haired skin sense organs. It is found in Thailand and Laos. As the lizard wedges itself into cracks or inflates to prevent dislodgement, the skin sense organs may be of value in matching to the contours of the protective crack. It has a moderately long tail (2.1 v. 1.9 × SVL) and is otherwise a fairly typical, moderately dimorphic (dewlap and color, v. 53%) agamid. For an agamid (v. 0.8%) or a refugia-anchored diurnal ground lizard (v. 3.5%: table 4.1, trait 67), this species it is unusual in that the eggs are adhered to rock rather than deposited unattached in the soil. Adherent eggs are characteristic of crack hermits, as are specialized morphologies for anchoring in cracks; the ecological business model assignment should be revisited when more field observations become available. Clutch count is small (–0.26 v. 0.20 std. resid.), and incubation is relatively long (–0.11 v. –0.55 std. resid.). It has not been introduced (v. 1.7%) or evaluated for conservation status.

Marinussaurus

Marinussaurus curupira is the sole member of a poorly known (filled cells 12% v. 19%) gymnophthalmid genus from Brazil. The species is small (1.56 g v. 2.35 g) and of a slender build (shape index –0.61 v. –0.32 std. resid.). It is elongate (v. 41%) and lacks secretory pores (v. 29%). No other information about sex differences was reported. It is endemic (v. 68%). This species is not known to have been introduced (v. 0.8%), nor has it been evaluated for conservation status.

Marisora

Marisora is a moderate radiation (seven spp.) of modestly large (7.1 g v. 3.3 g), viviparous (100% v. 36%), conspicuously wary (57% v. 11%) refugia-anchored diurnal ground (80% v. 30%) skinks found in the Northern Neotropics from Mexico southward through Colombia and Venezuela to Trinidad and Tobago northward to St. Vincent. The remainder are ground-foraging tree lizards (20% v. 7%).

They are less slender than average skinks (shape index 0.07 v. –0.46 std. resid.). Their tails are slightly long (1.64 v. 1.33 × SVL) and breakable (100% v. 73%) but not blue (v. 4%).

Males are smaller than females (–1.0 v. –0.31 on –2 to 2 scale), which is consistent with the notion that females are space limited for embryos (see reduced ontogenetic mass range and litter size below). Other sex differences are not known. Their sex differences rollup score (on 0 to 6 scale) is average (0.29 v. 0.28).

All eat invertebrates (v. 84%), which they obtain by active foraging (v. 92%).

Primary predator concealment is based on crevice-burrows (67% v. 43%) or vegetation (33% v. 25%). Direct basking is common (29% v. 21%), though *brachypoda* exposes only part of its body while basking. This genus is found at the wettest end of the rainfall spectrum: mesic (50% v. 40%), wet (100% v. 7%). They are found primarily at low elevation (l/m/h: 100/33/0 v. 89/35/6). They are overrepresented in littoral (40% v. 21%) and human (100% v. 14%) habitats and absent from riparian zones (v. 23%).

Litter size is normal (0.21 v. 0.06 std. resid.); neonates are large (ontogenetic mass range –0.56 v. 0.12 std. resid.). Parental care is known in *brachypoda* (female attends neonates for a few days: 14% v. 3%).

Endemism is low (29% v. 64%), perhaps in part because so many of the resident countries are themselves small. Island species are overrepresented (71% v. 57%). No *Marisora* has been introduced (v. 2.1%). They are imperiled at the usual rate (43% v. 47%), including placement on the IUCN's lists of the most endangered (14% v. 14%). *M. magnacornae* has not been collected in more than a century. *M. brachypoda* is believed by some local residents to be venomous.

Marinussaurus sp.	English name		Size		Range core
	Activity	Diet	mm	g	Ecological business model
curupira	Marinus' Lizard				Brazil
			49	1.56	

Marisora spp.	English name		Size		Range core
	Activity	Diet	mm	g	Ecological business model
alliacea	Costa Rica Four-lined Skink				Costa Rica
	diurnal		81	10.9	*** refugia-anchored diurnal gr. liz.
aurulae	Lesser Windward Skink				Trinidad and Tobago
			73	7.8	** refugia-anchored diurnal gr. liz.
brachypoda	Shiny Skink				Central America
	diurnal	invert.	66	4.90	** refugia-anchored diurnal gr. liz.
falconensis	Venezuela Coast Skink				Venezuela
	diurnal		90	7.3	* refugia-anchored diurnal gr. liz.
magnacornae	Corn Island Skink				Nicaragua
			64	5.2	
roatanae	Roatan Skink				Honduras
			74	8.0	
unimarginata	Middle America Long-limbed Skink				Costa Rica
	diurnal	invert.	72	7.5	** ground-foraging tree lizard

In other respects *Marisora* are typical skinks.

Marmorosphax

Marmorosphax is a minor radiation (five spp.) of uniformly small (2.56 g v. 3.25 g), cathemeral (100% v. 9%), viviparous (100% v. 36%) litter cryptozoid (100% v. 28%) skinks endemic to New Caledonia.

These are slightly thickened skinks (shape index –0.14 v. –0.46 std. resid.) with normal tails (1.50 v. 1.33 × SVL) that are often breakable (40% v. 73%). None has a blue tail (v. 4%), but the juveniles of *tricolor* may have bright red tails. The only known sex difference is color (40% v. 7.6%). Most are conspicuously wary (60% v. 11%). They are found exclusively at low elevation (v. 89%). None has been introduced (v. 2.1%). Most are imperiled (80% v. 47%), including overrepresentation on the IUCN's lists of the most endangered (40% v. 14%). In other respects *Marmorosphax* are much like other skinks.

Marmorosphax spp.	English name				Range core
			Size		
	Activity	Diet	mm	g	Ecological business model
boulinda	Boulinda Litterskink				New Caledonia
	cathemeral		47	1.92	**** litter cryptozoid
kaala	Mount Kaala Skink				New Caledonia
	cathemeral		49	2.22	*** litter cryptozoid
montana	Montane Marble-throated Skink				New Caledonia
			57	3.57	* litter cryptozoid
taom	Mount Taom Marble-throated Skink				New Caledonia
	cathemeral		49	2.11	** litter cryptozoid
tricolor	Marble-throated Skink				New Caledonia
	cathemeral	invert.	57	3.42	**** litter cryptozoid

Matoatoa spp.	English name				Range core
			Size		
	Activity	Diet	mm	g	Ecological business model
brevipes	Western Ghost Gecko				Madagascar
	nocturnal	invert.	38	1.32	
spannringi	Fiadanana Ghost Gecko				Madagascar
	nocturnal	invert.	51	2.95	** nocturnal canopy walker

Matobosaurus spp.	English name				Range core
			Size		
	Activity	Diet	mm	g	Ecological business model
maltzahni	Western Giant Plated Lizard				Angola
			188	171	
validus	Giant Plated Lizard				Southern Africa
	diurnal	omnivore	215	258	** refugia-anchored diurnal gr. liz.

Matoatoa

Matoatoa (two spp.) are small (1.97 g v. 3.31 g) nocturnal gekkonids from Madagascar. Neither is well known; one is known from a few specimens in a single locality. One is arboreal (trunk/wall); the other uncertain. Both have adhesive lamellae on their tail tip, suggesting arboreality. They are characteristically slender (shape index 0.02 v. 0.10 std. resid.). Secretory pores are not reported (v. 46%), nor is any other sex difference. One lives in rain forest; the other is found in a dry shrubby woodland. Neither species has been introduced; both are considered imperiled (100% v. 47%). In other documented respects *Matoatoa* are typical gekkonids.

Matobosaurus

Matobosaurus (two spp.) are very large (210 g v. 16 g) refugia-anchored diurnal ground gerrhosaurids found in southern Africa. By virtue of their massive shape index (1.32 v. 0.38 std. resid.), they convey an impression of a small-headed *Tupinambis*. Like most gerrhosaurids, they lack secretory pores (0% v. 8%). Few other sex differences are reported, but at least one species shows seasonal color (v. 27%). The one diet-documented species (*validus*) is an omnivore, taking as much as 72% (by occurrence) of its diet in leaves and fruit. It is conspicuously wary (v. 22%). Clutch count is anomalously low (–1.31 v. –0.20 std. resid.), based on *validus*. The eggs of *validus* seem large (24.5 × 46 mm), though a more objective measure, mass-adjusted ontogenetic size range, is not available. Incubation is short (–0.64 v. –0.05 std. resid.), which equates to 95 days. Longevity is long (0.34 v. –0.09 std. resid.), which equates to 17.2 years in *validus*.

Matobosaurus has not been evaluated for conservation status. In other respects *Matobosaurus* are typical gerrhosaurids.

Mediodactylus

Mediodactylus is a 13-species genus of padless Gekkonidae found primarily in southwestern Asia (full range: Italy to China, Ukraine to Jordan). Most are trunk/wall lizards (71% v. 45%); some are nocturnal ground lizards (14% v. 27%) and refugia-anchored diurnal ground lizards (14% v. 3%). They are small (1.56 g v. 3.31 g), of average build (shape index –0.03 v. 0.10 std. resid.), and in possession of slightly long (1.13 v. 1.07 × SVL), breakable tails (92% v. 87%). Most have secretory pores (62% v. 46%). Many are capable of basic color change (23% v. 15%).

Mediodactylus spp.	English name				Range core
			Size		
	Activity	Diet	mm	g	Ecological business model
amictopholis	Mount Hermon Gecko				Syria
	diurnal	invert.	40	1.55	*** refugia-anchored diurnal gr. liz.
aspratilis	Iranian Keel-scaled Gecko				Iran
	nocturnal		28	0.57	
brachykolon	Short-limbed Bent-toed Gecko				Pakistan
	nocturnal		46	2.22	
heterocercus	Blanford's Rough-scaled Gecko				Iraq
	nocturnal		45	2.10	** trunk/wall lizard
heteropholis	Iraq Keel-scaled Gecko				Iran
			39	1.37	
ilamensis	Ilamian Keel-scaled Gecko				Iran
	nocturnal		36	1.10	
kotschyi	Mediterranean Thin-toed Gecko				Mediterranean
	cathemeral	invert.	45	2.12	*** trunk/wall lizard or ground-foraging tree lizard (geog. var.)
narynensis	Naryn Thin-toed Gecko				Kyrgyzstan
	nocturnal		49	2.62	
russowii	Gray Thin-toed Gecko				Central Asia
	cathemeral	invert.	44	2.00	**** trunk/wall lizard or ground-foraging tree lizard
sagittifer	Jaz Murian Bent-toed Gecko				Pakistan
	nocturnal		31	0.77	*** trunk/wall lizard
spinicauda	Kopet Dagh Thin-toed Gecko				Southern Caspian Sea
	nocturnal	invert	41	1.67	*** nocturnal ground lizard
stevenandersoni	Anderson's Bent-toed Gecko				Iran
	nocturnal		36	1.10	
walli	Chitral Gecko				Pakistan
	nocturnal	invert.	54	3.43	** trunk/wall lizard

Commensurate with their small size they exhibit m<f SSD (–1.00 v. –0.03 on –2 to 2 scale). Although they are not permanently dichromatic, 7.7% of species show seasonal color (v. 0.8%). All characterized species forage by sit-and-wait (v. 81%), though *russowii* will forage up to 200 m from its daily retreat in areas where rock crevices are plentiful. This species is also reported to leap off trees or cliffs to subdue prey spotted on the ground beneath them. Electric light use is especially common in this genus (23% v. 4%).

Hours of activity seem flexible, and although most species are considered nocturnal (75% v. 73%), several exemplars have been seen active or heard calling at other times of day. Spontaneous vocalizations are conspicuous (*sagittifer* is audible by humans at 50 m) and taxonomically overrepresented (38.5% v. 6.5%), as are stress calls (15.4% v. 3.7%). Conspicuous wariness is notable (7.7% v. 4.4%).

All documented species eat invertebrates (v. 81%). *M. kotschyi* is notable in that it eats as much as 70% of its diet in beetle larvae, suggesting the use of nonvisual cues for prey detection.

For predator concealment all species rely on retreat to crevices or burrows (v. 31%). Despite their residence in relatively hot deserts, these mostly nocturnal geckos maintain low active body temperatures (mean 23.5°C v. 27.6°C). Calls of *kotschyi* are heard at air temperatures of <10°C.

Most species are found only solitarily (67% v. 7.7%), though all males are thought to defend space (v. 87%). Burrow use for daily refugia is notably common (38% v. 10%), even though most species are primarily arboreal (63% v. 68%). They are overrepresented at medium elevations (67% v. 26%) and overrepresented in human habitats (40% v. 27%).

Reproduction hews to the gekkonid pattern, with communal

oviposition notably frequent (15% v. 8.6%). Species such as *kotschyi*, which has a large latitudinal range, vary in number of clutches per year from one (far north) to five (southern limit of range). The size of young is below lizard average and thereby deviates from the family pattern (ontogenetic mass range 0.50 v. –0.19 std. resid.).

Population density is documented for only *kotschyi* (one estimate: 0.15 kg/ha v. 0.37 kg/ha). Study of *amictopholis* populations has been limited by extensive antipersonnel minefields in its range. *M. kotschyi* has been introduced at least three times (genus 7.7% v. 2.1% family). *Mediodactylus* are imperiled at a family-typical rate (44% v. 47%). Eleven percent (v. 11%) are listed by the IUCN in the most extreme categories of endangerment. In other respects *Mediodactylus* are typical gekkonids.

Medopheos

Medopheos edracanthus is the colorful sole member of its teiid genus of medium-small (13.0 g v. 22.1 g), riparian (100% v. 46%) wide-ranging chemosensors (100% v. 97%) found in Ecuador and Peru.

Medopheos is of medium build (shape index 0.71 v. 0.88 std. resid.) and is sexually dichromatic (v. 35%). It lacks secretory pores (v. 9%), color change (v. 1.3%), seasonal color (v. 10%), and a blue tail (v. 15%). It basks directly (v. 58%) and is found at low and medium elevations (0–2335 m). Its clutch count is unknown. It is not endemic (v. 50%), found on an island (v. 37%), or imperiled (v. 41%). It has not been introduced (v. 8%). In other respects *Medopheos* are typical teiids.

Medopheos sp.	English name		Size		Range core
	Activity	Diet	mm	g	Ecological business model
edracanthus	Bocourt's Ameiva				Peru
	diurnal	invert.	77	13.0	** wide-ranging chemosensor

Melanoseps spp.	English name		Size		Range core
	Activity	Diet	mm	g	Ecological business model
ater	Black Limbless Skink				Central Africa
			141	5.5	** litter cryptozoid
emmrichi	Uluguru Limbless Skink				Tanzania
			135	4.94	* subterranean forager
longicauda	Long-tailed Limbless Skink				Tanzania
			60	0.77	
loveridgei	Loveridge's Limbless Skink				Tanzania
		invert.	111	3.13	
occidentalis	Western Limbless Skink				Equatorial Africa
		invert.	96	4.91	*** subterranean forager
pygmaeus	Pygmy Limbless Skink				Tanzania
			72	1.15	
rondoensis	Rondo Limbless Skink				Tanzania
			99	2.40	
uzungwensis	Udzungwa Mountains Limbless Skink				Tanzania
			159	7.2	

Melanoseps

Melanoseps is a moderate radiation (eight spp.) of average-sized (2.99 g v. 3.3 g), functionally limbless skinks from Africa (Cameroon east to Kenya, south to Mozambique, west to Namibia). Many have reduced eyes as well as reduced or missing limbs. Two of three assignable species are subterranean foragers (67% v. 12%); the other is a litter cryptozoid (33% v. 28%).

They are very slender (shape index –2.10 v. –0.46 std. resid.); they have the short tails characteristic of subterranean forms (0.40 v. 1.33 × SVL), and only one tail is breakable (13% v. 73%). Sex differences are not reported. Diet is generalized invertebrate eater (100% v. 84%). Diel activity is unknown. Their elevational profile is near normal (l/m/h: 80/20/0 v. 89/35/6). All reproductively documented species (*emmrichi, occidentalis)* are viviparous (v. 36%), but only *emmrichi* has a known litter count (–0.14 v. 0.06 std. resid.). For some reason *loveridgei* is presumed to be oviparous. Endemism is exactly average (63% v. 64%). None is found on islands (v. 57%). None has been introduced (v. 2.1%), and half are imperiled (50% v. 47%) and are found on the IUCN's lists of the most endangered (50% v. 14%). In other respects *Melanoseps* are typical limbless skinks.

Menetia

Menetia is a moderate radiation (six spp.), of uniformly tiny (0.32 g v. 3.25 g), diurnal (100% v. 74%), conspicuously wary (100% v. 11%), direct

Menetia spp.	English name				Range core
			Size		
	Activity	Diet	mm	g	Ecological business model
alanae	Top End Dwarf Skink				Northern Territory
	diurnal	invert.	28	0.28	*** litter cryptozoid
amaura	Shark Bay Dwarf Skink				Western Australia
	diurnal	invert.	28	0.28	
concinna	Jabiluka Dwarf Skink				Northern Territory
	diurnal	invert.	31	0.39	**** litter cryptozoid
greyii	Common Dwarf Skink				Australia
	diurnal	invert.	29	0.32	** refugia-anchored diurnal gr. liz.
maini	Northern Dwarf Skink				Northern Australia
	diurnal	invert.	27	0.27	*** litter cryptozoid
surda	Western Dwarf Skink				Western Australia
	diurnal	invert.	32	0.41	*** litter cryptozoid

basking (100% v. 21%), elongate (100% v. 35%), oviparous (100% v. 64%) litter cryptozoid (80% v. 28%) skinks in Australia. The remainder are refugia-anchored diurnal ground lizards (20% v. 30%).

They are relatively slender (shape index –1.03 v. –0.46 std. resid.), though their tail length is normal (1.38 v. 1.33 × SVL) and tails are breakable (100% v. 73%). No information suggesting SSD is available for *Menetia*, but the sexes may differ in color (*amaura* 17% v. 7.6%) or seasonally (*greyii* 17% v. 6.2%).

They are generalized invertebrate eaters (100%), using active foraging (100% v. 92%). Mean active body temperature is warm (33.8°C v. 31.1°C). They are found at the arid side of the skink rainfall spectrum (arid: 100% v. 40%; intermediate: 100% v. 34%). They are known only from low elevation (100% v. 89%). They are most notably overrepresented in littoral (40% v. 21%) and sandy (67% v. 25%) habitats.

Their mean clutch count is high (0.62 v. 0.06 std. resid.). Their hatchlings are notable for small size (ontogenetic mass range 2.25, the third-highest value in Scincidae, v. 0.12 std resid.), which is unusual for a lizard this small in maximum size. As the ontogenetic mass range is based solely on *greyii*, this outlier should be treated with reservations until documented in more species. RCM is minimal (0.159 v. 0.236), based on *greyii*. Incubation is very long (0.90 v. –0.40 std. resid.) at 75 days, also based on *greyii*, which has both parthenogenetic and bisexual populations.

There is one credible population density (*greyii*); it is among the lowest recorded for lizards (0.001 kg/ha v. 0.35 kg/ha). Endemism is about average (67% v. 64%). Insular forms are comparatively rare (33% v. 57%). None has been introduced (v. 2.1%), and none is considered imperiled (v. 47%). In other respects *Menetia* are probably typical skinks.

Meroles

Meroles (eight spp.) is a genus of diurnal ground lacertids in southern Africa (Angola and Tanzania to South Africa). Many are sand dwelling, possessing a pointed or shovel-shaped snout, and marked with a mixture of longitudinal lines alternating with longitudinal rows of spots. Most are wide-ranging chemosensors (75% v. 25%), with individual species that are refugia-anchored diurnal ground lizards (*suborbitalis*) and sand sharks (*anchietae*).

The diet—a surfeit of omnivores (33% v. 19%) with much ingestion of termites—and movement patterns of these lizards are strikingly disrespectful of the usual lizard foraging dogma. Many move often and rapidly, which is no surprise for a family dominated by active foragers, but half of the species in this genus are listed as sit-and-wait (v. 23%), and those sit-and-wait species are often characterized by some of the highest movement rates. Some would argue that this juxtaposition illustrates the flaws of the assumption that each species resides at one or the other pole of the foraging mode continuum. For example, *suborbitalis* (50% termites) is reported to forage solely by sit-and-wait, and yet it is in the 85th percentile for MPM and 60th percentile for PTM. *M. squamulosus* (80% termites) is said to be both sit-and-wait and active foraging, yet it is in the 95th percentile for MPM and 90th percentile for PTM. *M. ctenodactylus* (diet unknown) is also considered to forage by both modes and is in the 97th percentile for MPM, 76th for PTM. Conversely, the solely active-foraging *reticulatus* (arthropod diet not specified) is in the 4th percentile for MPM. *M. cuneirostris* (omnivore including termites) has intermediate MPM and PTM scores, but it sometimes sits by ant trails to rob the ants of the seeds they are carrying. But it moves far too frequently to be considering a cryptic ambush forager like a *Phrynosoma*, and its primary predator concealment tactic is flight. Of course it is not like *Phrynosoma* in

Meroles spp.	English name				Range core
	Activity	Diet	Size mm	Size g	Ecological business model
anchietae	Shovel-snouted Lizard				Angola
	diurnal	omnivore	49	2.72	** sand shark
ctenodactylus	Smith's Desert Lizard				Namibia
	diurnal		76	10.2	** wide-ranging chemosensor
cuneirostris	Wedge-snouted Desert Lizard				Namibia
	crepuscular	omnivore	54	3.80	** wide-ranging chemosensor
knoxii	Knox's Desert Lizard				Namibia
	diurnal	invert.	58	4.68	** wide-ranging chemosensor or refugia-anchored diurnal gr. liz.
micropholidotus	Small-scaled Desert Lizard				Namibia
	diurnal		58	4.68	** wide-ranging chemosensor
reticulatus	Reticulated Desert Lizard				Namibia
	diurnal	invert.	48	2.59	* wide-ranging chemosensor
squamulosus	Common Rough-scaled Lizard				Southern Africa
	diurnal	invert.	63	5.8	*** wide-ranging chemosensor
suborbitalis	Spotted Desert Lizard				Botswana
	diurnal	invert.	59	4.76	** refugia-anchored diurnal gr. liz.

that it does not eat ants, but only the booty the ants are carrying.

Morphologically *Meroles* are typical lacertids (e.g., 4.48 g v. 5.10 g), though they have a notably small ontogenetic range in body sizes (1.1 v. 1.5 ratio of largest to smallest $\log_{10}$ mass). None is reported to sport a blue tail (v. 13%). For normal-sized lizards, they are more m>f sexually dimorphic than usual (1.00 v. 0.49 on –2 to 2 scale). Sexual dichromatism (25% v. 33%) and seasonal color (12.5% v. 14.6%) are at usual frequencies, as is the sex differences rollup (0.63 v. 0.86 on 0 to 6 scale).

Only three species have been characterized for primary predator concealment tactic, and they are all different: sand (*ctenodactylus*), flight (*cuneirostris*), and vegetation (*suborbitalis*).

Direct basking is common (38% v. 25%), and mean active body temperature is high (36.0°C v. 33.3°C). *M. anchietae* raises diametrically opposed limbs alternately to reduce heat gain from the hot desert sands.

The one species for which social group size is documented (*squamulosus)* lives in groups (v. 25%). Male defense behavior is reported for *anchietae* (tolerant or not) and *cuneirostris* (tolerant), suggesting that spatial tolerance may be high in male *Meroles*. During the mating season, male *anchietae* oversee a harem of four to five females, which may consume the males' attentions, leaving little time for spatial defense.

Meroles is at the dry end of the rainfall spectrum: hyperdry (33% v. 4%), dry (33% v. 30%), and arid (67% v. 40%). They are documented only at low elevation (v. 87%). They are overrepresented in sandy areas (100% v. 44%).

Meroles are oviparous with typical clutch counts (0.13 v. 0.13 std. resid.). The mass-adjusted ontogenetic mass range is the lowest of any lacertid genus, indicating exceptionally large young (–0.96 v. –0.05 std. resid., based on $n = 4$ spp.). The genus's other reproductive anomaly is very long incubation (1.87 v. –0.36 std. resid.), corresponding to 165 days in the one species documented (*squamulosus*). This species is very short-lived (1.5 years maximum, or –1.65 v. 0.19 std. resid.) and may need to prolong incubation such that the young emerge at a seasonally optimal time the following year. Some have argued that the phenology of this "annual" lizard is timed to exactly offset that of another annual lizard in the same environment (*Ichnotropis capensis*), such that the two species are not present simultaneously; one is in the egg stage when the other is aboveground, and vice versa.

Endemism in *Meroles* is low (13% v. 35%). Island species are not present (0% v. 26%), none has been introduced (v. 4.6%), and none has been evaluated for conservation status.

Mesalina

Mesalina (14 spp.) is a moderate genus of small-sized (2.43 g v. 5.10 g), slender (shape index 0.08 v. 0.27 std. resid.), diurnal terrestrial lacertids—think svelte *Acanthodactylus*—centered in the Middle East and extending west to Western Sahara, north to Turkmenistan, east to India, and south to Ethiopia. The modal ecological business model is refugia-anchored diurnal ground lizard (63% v. 57%); the remainder are wide-ranging chemosensors (38% v. 25%). Morphologically they are typical lacertids, though many have bluish (often a pale blue) tails primarily in juveniles (36% v. 13%), and some have secretory pores (7% v. 4%).

The sexes do not differ in size (0.0

v. 0.49 on –2 to 2 scale). A few are sexually dichromatic (14% v. 33%); none has seasonal color (v. 15%). Males have comparatively long relative tail length (1.14 v. 1.07 m:f ratio).

Home-range area is large (1.3 v. 0.27 std. resid., based on *guttulata*), especially in association with mate searching by males (up to 9000 m²). Although all species are nominally diurnal, *watsonana* has been seen active at night in the light of gas flares. It is considered to be an annual species, so perhaps it is in a hurry to complete business when light permits. Twenty-one percent are conspicuously wary (v. 11%), and the range of primary predator concealment tactics does not include flight (which is characteristic of active foragers); instead it includes crypsis (17% v. 3%), crevices-burrow (50% v. 50%), and retreat to vegetation (33% v. 42%). Use of burrows for nighttime refugia is high, perhaps universal (86% v. 48%), and at least two species (*adramitana*, *brevirostris*) are notable for plugging the entrance to *unoccupied* burrows. Most lizards plug occupied burrows to protect themselves from discovery by predators. Either the lizard predators in this part of the world remember the locations of open burrows and return to them at night to consume the occupants or other unsavory burrow dwellers move into unplugged burrows when unoccupied.

Thirty-six percent of *Mesalina* bask directly (v. 25%), and their mean active body temperature is an average temperature (33.7°C v. 33.3°C). The sand is hot, however, and *adramitana*, *brevirostris*, and *olivieri* lift diametrically opposed feet alternately to limit heat gain.

Group living is not reported (v. 25%). Male social tolerance is high (50% v. 41%) and spatial defense low (50% v. 59%).

Mesalina spp.	English name				Range core
			Size		
	Activity	Diet	mm	g	Ecological business model
adramitana	Hadramawt Sand Lizard				Arabian Peninsula
	diurnal	invert.	42	1.81	*** wide-ranging chemosensor
ayunensis	Dhofar Sand Lizard				Oman
	diurnal	invert.	41	1.63	
bahaeldini	Mount Sinai Lizard				Egypt
	diurnal		47	2.40	
balfouri	Balfour's Desert Racer				Yemen
	diurnal		46	2.29	
brevirostris	Short-nosed Lizard				Iran
	diurnal	invert.	48	2.63	** refugia-anchored diurnal gr. liz.
ercolinii	Bud Bud Desert Racer				Somalia
	diurnal		57	4.26	
guttulata	Small-spotted Desert Racer				Old World deserts
	diurnal	invert.	52	2.72	*** wide-ranging chemosensor
kuri	Abd al Kuri Desert Lizard				Yemen
	diurnal		50	2.99	
martini	Red Sea Lizard				Sudan
	diurnal		43	1.90	* refugia-anchored diurnal gr. liz.
olivieri	Olivier's Desert Racer				Northern Sahara
	diurnal	invert.	43	1.93	* wide-ranging chemosensor
pasteuri	Pasteur's Desert Racer				Sahara
	diurnal	invert.	44	2.00	** refugia-anchored diurnal gr. liz.
rubropunctata	Red-spotted Desert Racer				Northern Africa
	diurnal	invert.	51	3.20	
simoni	Simon's Desert Racer				Morocco
			44	2.04	* refugia-anchored diurnal gr. liz.
watsonana	Long-tailed Desert Lizard				Pakistan
	diurnal	invert.	53	3.50	* refugia-anchored diurnal gr. liz.

The one rainfall-documented species (*guttulata*) lives in the dry rainfall band (in this case, 76 mm/y). *Mesalina* are overrepresented in littoral (38% v. 22%) and sandy (100% v. 44%) areas.

Reproduction is typical for lacertids (clutch count 0.30 v. 0.13 std. resid.). Hatchling size is high, as indicated by the low ontogenetic mass range (–0.59 v. –0.05 std. resid.). Despite the large young, they have an average RCM (0.295 v. 0.267). Longevity is on the short side (–0.03 v. 0.19 std. resid.), at least for the three documented species (4–5 years).

Several are considered to be "annuals": *watsonana* adults disappear late in the calendar year, suggesting programmed senescence or very low survival. Population density is low (0.12 kg/ha v. 2.07 kg/ha), commensurate with the low primary productivity of these deserts. The genus mean is, however, based solely on *guttulata* ($n = 2$ estimates) and may not be representative.

Endemism is average (43% v. 35%); island species are frequent (57% v. 26%). None has been introduced (v. 4.6%), and 11% are imperiled (v. 56%). None is on any of the IUCN lists of the most endangered (v. 15.5%). In other respects *Mesalina* are typical desert lacertids.

Mesaspis

Mesaspis is a moderate (seven spp.) genus of small-bodied (3.8 g v. 10.1 g), elongate terrestrial anguids from high-altitude mesic areas of Mexico and Central America (Guatemala to Panama). Most are refugia-anchored diurnal ground lizards (60% v. 34%); others are litter cryptozoids (40% v. 6%).

Mesaspis spp.	English name				Range core
			Size		
	Activity	Diet	mm	g	Ecological business model
antauges	Mount Orizaba Alligator Lizard				Mexico
			70	4.56	
cuchumatanus	Cuchumatanes Alligator Lizard				Guatemala
			61	2.78	** refugia-anchored diurnal gr. liz. or litter cryptozoid
gadovii	Gadow's Alligator Lizard				Mexico
			84	8.7	** refugia-anchored diurnal gr. liz.
juarezi	Sierra Juarez Alligator Lizard				Mexico
			67	3.87	
monticola	Highland Alligator Lizard				Costa Rica
	diurnal	invert.	73	2.47	** litter cryptozoid
moreletii	Morelet's Alligator Lizard				Guatemala
	diurnal	invert.	81	3.20	** refugia-anchored diurnal gr. liz.
viridiflava	Dwarf Alligator Lizard				Mexico
			63	3.14	

Sexual dichromatism is mildly overrepresented (43% v. 26%), but seasonal color is not reported (v. 3.8%). To the extent that they are known, all are said to be diurnal (v. 70%) active foragers (v. 71%) on the ground (v. 45%). Direct basking is common (29% v. 15%). They are underrepresented at low (0% v. 47%) and medium elevations (67% v. 77%) and overrepresented at high elevations (100% v. 41%). They are notably absent from riparian zones (0% v. 19%).

All known species are viviparous or both oviparous and viviparous (v. 61%). Litters are slightly small (0.50 v. 0.85 std. resid.), but more noteworthy is the large size of young (1/6 v. 1/72 maximum size), indicated by the reduced ontogenetic size range (–2.12 v. 0.62 std. resid.). Parental care is about average for an anguid (14% v. 17%).

They are short-lived (–0.68 v. 0.83 std. resid.). Endemism is average (71% v. 65%). There are no island species (v. 17%), none has been introduced (v. 3.8%), and 33% (v. 63%) are imperiled, including 17% (v. 28%) on one of the IUCN's lists of the most endangered.

Mesobaena

Mesobaena is small genus (two spp.) of typically small (3.0 v. 2.7 g; shape index –3.92 v. –3.97 std. resid.) amphisbaenids from South America (Colombia to Brazil). Endemism is average (50% v. 66%). Neither is found on islands (v. 12%) or is imperiled (v. 23%). With the few data available, they are undistinguished amphisbaenids.

Mesobaena spp.	English name				Range core
			Size		
	Activity	Diet	mm	g	Ecological business model
huebneri	Inirida Word Lizard				Venezuela
			189	3.00	** subterranean forager
rhachicephala	Blue-eyed Worm Lizard				Brazil
			190	3.05	**** subterranean forager

Mesoscincus

Mesoscincus is a small genus (three spp.) of large (10.3 g v. 3.3g), elongate (67% v. 35%), diurnal (v. 74%) litter cryptozoid (100% v. 28%) skinks from Central America (Mexico to Costa Rica).

Sex differences, diet, and foraging

Mesoscincus spp.	English name		Size		Range core
	Activity	Diet	mm	g	Ecological business model
altamirani	Tepalcatepec Skink				Mexico
			81	10.8	* litter cryptozoid
managuae	Managua Skink				Nicaragua
	diurnal		99	9.3	* litter cryptozoid
schwartzei	Schwartze's Skink				Guatemala
	diurnal		104	10.8	* litter cryptozoid

Micrablepharus spp.	English name		Size		Range core
	Activity	Diet	mm	g	Ecological business model
atticolus	Ant-loving Tinyfoot Lizard				Brazil
	diurnal	invert.	39	0.69	** litter cryptozoid
maximiliani	Maximilian's Tinyfoot Lizard				Paraguay
	diurnal	invert.	40	0.77	** litter cryptozoid

Microgecko spp.	English name		Size		Range core
	Activity	Diet	mm	g	Ecological business model
chabaharensis	Chabahar Dwarf Gecko				Iran
			30	0.68	
depressus	Low Lying Gecko				Pakistan
	nocturnal	invert.	32	0.81	*** nocturnal ground lizard
helenae	Khuzestan Dwarf Gecko				Iran
	nocturnal	invert.	28	0.59	*** nocturnal ground lizard
latifi	Latifi's Dwarf Gecko				Iran
	nocturnal		28	0.57	** nocturnal ground lizard or crack hermit
persicus	Persian Dwarf Gecko				Iran
	nocturnal	invert.	31	0.73	* nocturnal ground lizard

mode are unknown. They are conspicuously wary (67% v. 11%), and they bask directly (67% v. 21%). They are found at low elevation (100% v. 89%) and are overrepresented in human habitat (67% v. 14%). Reproduction is not documented. Endemism is low (33% v. 64%). They do not occur on islands (v. 57%), and they have not been introduced (v. 2.1%). None is considered imperiled (v. 47%). In other respects *Mesoscincus* are probably typical skinks.

Micrablepharus

Micrablepharus (two spp.) are diurnal, marginally elongate, blue-tailed, ground gymnophthalmids found in central South America (Paraguay, Bolivia, Brazil). These litter cryptozoids are much smaller than typical gymnophthalmids (0.73 g v. 2.4 g) and thinner in build (shape index −0.93 v. −0.32 std. resid.). They have typical tails (1.65 v. 1.72 × SVL) that are not reported to be breakable (v. 28%). One species (*maximiliani*) is sexually dimorphic for pores (v. 29%), but other sex differences are not reported. *M. maximiliani* is also reported to be conspicuously wary (v. 7%) and to dig burrows (v. 5%). Active body temperature is a warm 34.5°C (v. 28.8°C). Their generalized invertebrate diet is unremarkable, though *maximiliani* is said to emphasize leaf-hoppers.

Reproduction is typical for gymnophthalmids in raw form, but some anomalies appear when corrected for size. Mass-adjusted clutch count is slightly large (0.38 v. −0.11 std. resid.). Ontogenetic size range is exceptionally small (−2.87 v. 0.62 std. resid.), indicating very large hatchlings. This suggests that *Micrablepharus* are straining to produce young that are both large in size and numerous in number. Unfortunately, this ontogenetic size range is based solely on data from one species (*maximiliani*), though the lengths of *atticolus* imply an even greater ontogenetic range in this species (hatchling unweighed). RCM ratios are normal, however (0.198 v. 0.192). Neither species is considered imperiled. In other respects *Micrablepharus* are probably typical gymnophthalmids.

Microgecko

Microgecko (five spp.) are very small (0.67 g v. 3.31 g), padless nocturnal ground lizards from Iran/Iraq/Pakistan (Gekkonidae). Most are banded and slightly svelte (shape index −0.17 v. 0.10 std. resid.). A few species are known to have secretory pores (20% v. 46%). Other sex differences are not documented. Conspicuous wariness has been noted (20% v. 4.4%), and their primary predator concealment is in vegetation (100% v. 19%). One species (*persicus*) is cryptozoic (25% v. 0.9%). *M. depressus* will appropriate the burrows of other animals (20% v. 10%). Several species are reported to have their surface activity triggered by rain. Four of five species are found in the medium elevation band (80% v. 26%). Riparian areas are favored (50% v. 17%). Clutches are small when uncorrected for mass (1.0 v. 1.9) and large when corrected (0.13 v. –0.26 std. resid.). *Microgecko* have not been introduced (v. 3.7%) nor are they imperiled (v. 47%). In other respects *Microgecko* are typical gekkonids.

Microlophus

Microlophus are the common ground lizards on the west coast of South America, including the ubiquitous "lava" lizards of the Galápagos Islands. *Microlophus* is a large tropidurid genus (23 spp.) of slightly large (20.7 g v. 17.5 g), crevice-burrow retreating (100% v. 29%), mostly refugia-anchored diurnal ground lizards (95% v. 67%). A small minority are ground-foraging tree lizards (5% v. 9%).

They are of average build (shape index 1.12 v. 1.10 std. resid.) and tail length (1.33 v. 1.56 × SVL). Tails are of average fragility (9% v. 13% breakable). Crests are uncommon (4.3% v. 10.3%). Color change is absent (v. 3%).

The males are often much larger than females (1.20 v. 0.95 on –2 to 2 scale), and the m:f relative tail length ratio is correspondingly low (0.98 v. 1.06). Sexual dichromatism is underrepresented (17% v. 28%), and seasonal color frequency is about average (4.3% v. 2.9%). The sex differences rollup score is mildly elevated (1.13 v. 0.80 on 0 to 6 scale).

The diet spectrum of the genus is similar to the family, with a slightly greater emphasis on omnivory (56%

Microlophus spp.	English name				Range core
			Size		
	Activity	Diet	mm	g	Ecological business model
albemarlensis	Galápagos Lava Lizard				Ecuador
	diurnal	invert.	88	25.0	*** refugia-anchored diurnal gr. liz.
atacamensis	Atacama Pacific Iguana				Chile
	diurnal	omnivore	98	33.9	** refugia-anchored diurnal gr. liz.
barringtonensis	Santa Fe Lava Lizard				Ecuador
	diurnal	invert.	77	17.8	*** refugia-anchored diurnal gr. liz.
bivittatus	San Cristobal Lava Lizard				Ecuador
	diurnal	invert.	76	17.1	*** refugia-anchored diurnal gr. liz.
delanonis	Espanola Lava Lizard				Ecuador
	diurnal	omnivore	112	48.4	*** refugia-anchored diurnal gr. liz.
duncanensis	Pinzon Lava Lizard				Ecuador
	diurnal	invert.	78	18.1	*** refugia-anchored diurnal gr. liz.
grayii	Floreana Lava Lizard				Ecuador
	diurnal	invert.	82	20.8	*** refugia-anchored diurnal gr. liz.
habelii	Marchena Lava Lizard				Ecuador
	diurnal	invert.	93	29.3	*** refugia-anchored diurnal gr. liz.
heterolepis	Tacna Pacific Iguana				Chile
	diurnal	omnivore	102	38.1	** refugia-anchored diurnal gr. liz.
indefatigabilis	Santa Cruz Lava Lizard				Ecuador
	diurnal	omnivore	77	17.5	*** refugia-anchored diurnal gr. liz.
jacobii	Santiago Lava Lizard				Ecuador
	diurnal		70	13.7	*** refugia-anchored diurnal gr. liz.
koepckeorum	Red-faced Pacific Lizard				Peru
	diurnal		67	12.2	* refugia-anchored diurnal gr. liz.
occipitalis	Knobbed Pacific Iguana				Peru
	diurnal		63	10.2	* refugia-anchored diurnal gr. liz.
pacificus	Pinta Lava Lizard				Ecuador
	diurnal	omnivore	84	22.3	*** refugia-anchored diurnal gr. liz.
peruvianus	Peru Pacific Lizard				Peru
	diurnal		92	28.2	* refugia-anchored diurnal gr. liz.

Microlophus spp.	English name	Activity	Diet	Size mm	Size g	Range core	Ecological business model
quadrivittatus	Four-banded Pacific Iguana			93	29.7	Chile	
stolzmanni	Stolzmann's Pacific Iguana	diurnal	omnivore	82	21.0	Peru	* refugia-anchored diurnal gr. liz.
tarapacensis	Tarapaca Pacific Iguana	diurnal	herbivore	88	25.3	Chile	
theresiae	Theresia's Pacific Iguana		invert.	86	23.7	Peru	
theresioides	Pica Oasis Pacific Iguana	diurnal	omnivore	82	21.0	Chile	* ground-foraging tree lizard
thoracicus	Tschudi's Pacific Iguana	diurnal	omnivore	73	15.2	Peru	* refugia-anchored diurnal gr. liz.
tigris	Tiger Pacific Iguana	diurnal	omnivore	75	16.2	Peru	* refugia-anchored diurnal gr. liz.
yanezi	Yanez's Pacific Iguana	diurnal	omnivore	62	9.9	Chile	* refugia-anchored diurnal gr. liz.

v. 44%) and an absence of ant specialization (v. 6.3%). Herbivory (5.6% v. 6.3%) and invertebrate eating (39% v. 44%) are average. Electric light use is not known (v. 1.3%). Home-range area is average (−0.38 v. −0.38 std. resid.). Mean active body temperature is undistinguished (30.8°C v. 31.8°C). At least two species (*atacamensis, heterolepis*) forage intertidally. *M. theresioides* may run bipedally.

Their elevational profile (l/m/h: 94/29/12 v. 65/47/36) emphasizes low elevations but includes the full range up to 3000 m. They are overrepresented in littoral areas (43% v. 15%), sandy areas (61%, the second-highest value in Tropiduridae, v. 23%), riparian areas (26% v. 18%), and human habitat (30% v. 14%).

Clutch count is slightly low (−0.70 v. −0.56 std. resid.). Hatchlings are large (ontogenetic mass range −0.99 v. −0.22 std. resid.). RCM is undocumented. Incubation duration is long (0.67 v. 0.40 std. resid.), perhaps reflecting the influence of cool maritime fog and winds in the occupied area. Communal oviposition (17.4% v. 3.7%) and parental care (17.4% v. 3.7%) are overrepresented. Parental care takes the form of transient nest guarding by females.

Endemism is high (83% v. 74%). Many species are found on islands (48% v. 10%). None has been introduced (v. 1.5%), and only 16% (v. 23%) are considered imperiled. None is on one of the IUCN's lists of the most endangered (v. 3.0%). In other respects *Microlophus* are typical tropidurids.

M. theresiae honors Therese Prinzessin von Bayern, collector of the type. Collecting type specimens is not an activity one often associates with princesses.

Mniarogekko

Mniarogekko spp.	English name	Activity	Diet	Size mm	Size g	Range core	Ecological business model
chahoua	Mossy Prehensile-tailed Gecko	nocturnal		114	30.8	New Caledonia	** trunk/wall lizard
jalu	Green Spirit Gecko	nocturnal		109	26.9	New Caledonia	* trunk/wall lizard

Mniarogekko is a two-species genus of large (28.8 g v. 4.14 g), nocturnal (v. 79%), arboreal (v. 64%), diplodactylid geckos endemic to New Caledonia. They are highly variable in their cryptic markings and stouter than the average diplodactylid (shape index 0.59 v. 0.05 std. resid.). Their tails are prehensile (v. 34%) and have adhesive lamellae. Such a valuable tail is shed only with great reluctance, and regrowth is marginal. They have secretory pores (v. 34%), but other sex differences are undocumented. Vocalization is not recorded (v. 5%). They are associated with large trees and often found in riparian zones (v. 10%). Observed social groupings suggest monogamy.

They are oviparous (v. 82%) and lay eggs that are hard when laid but soften during incubation, the opposite of the usual pattern. Incubation is relatively short (−0.36 v. −0.22 std. resid.). Females may guard the clutch. Longevity is a bit high (1.15 v. 0.90 std. resid.), which equates to 20 years for so large a lizard. They have not been introduced (v. 2%). Both species are considered imperiled (v. 81%), though neither is on the IUCN's lists of the most endangered (v. 29%). In other respects they are typical diplodactylids.

Mochlus

Mochlus is a moderate radiation (13 spp.) of reasonably large (6.8 g v. 3.3 g), crepuscular (67% v. 2%) or diurnal (33% v. 74%), oviparous (100% v. 64%) skinks from Africa south of the central Sahara. They are diverse in microhabitat and ecological business model. The modal model is a litter cryptozoid (40% v. 28%) or sand shark (40% v. 7%). The other classifiable ecological business model (*guineensis*) is refugia-anchored diurnal ground lizard.

They have a slightly heavier build than an average skink (shape index –0.11 v. –0.46 std. resid.). Elongation occurs at family-typical frequency (39% v. 35%). Tails are notably short (0.99 v. 1.33 × SVL) and somewhat breakable (39% v. 73%) but not blue (v. 4%). Sex differences are not known.

Diet is generalized invertebrate eater (100% v. 84%). One species (*afer*) uses burrows for its daily refuge. It and *paedocarinatum* are often found inside termite mounds; *paedocarinatum* has not been seen active. The extremely elongate *sundevalli* is a sand swimmer. *Mochlus* is found in a wide region in the middle of the rainfall spectrum (arid to mesic; in this case 375–1300 mm/y). The elevational profile is normal (l/m/h: 100/38/0 v. 89/35/6). They are not known to occupy human habitats (0% v. 14%).

Mean clutch count is about average (0.26 v. 0.06 std. resid.). The ontogenetic mass range of the genus, based solely on *sundevalli*, is the highest in Scincidae (3.13 v. 0.12 std. resid.), indicating exceptionally small hatchlings. This needs replication. Measurements of RCM are unavailable and could clarify this relationship.

Longevity is short (–0.90 v. 0.17 std. resid.) but is based solely on *sundevalli* (2.5 years). Endemism is low (23% v. 64%). Few *Mochlus* live on islands (8% v. 57%). None has been introduced (v. 2.1%), and none is imperiled (v. 47%). In all other documented respects *Mochlus* are typical skinks.

Mochlus spp.	English name				Range core
			Size		
	Activity	Diet	mm	g	Ecological business model
afer	Peters' Writhing Skink				Southern Africa
	crepuscular	invert.	98	9.2	** sand shark
brevicaudis	Short-tailed Writhing Skink				Ghana
		invert.	65	5.4	*** litter cryptozoid
grandisonianum	Lanza's Desert Writhing Skink				Somalia
			52	2.66	
guineensis	Guinea Writhing Skink				Ghana
		invert.	67	3.16	* refugia-anchored diurnal gr. liz.
mabuiiforme	Mabuya-like Writhing Skink				Kenya
	diurnal		169	116	
mocquardi	Mocquard's Writhing Skink				Southern Sahara
			67	5.8	** litter cryptozoid
paedocarinatum	Lanza's Highland Writhing Skink				Ethiopia
			69	6.6	
productum	Boulenger's Writhing Skink				Somalia
			91	15.8	
simonettai	Simonetta's Writhing Skink				Somalia
			72	7.6	
somalicum	Somali Writhing Skink				East Africa
			70	6.9	
sundevallii	Sundevall's Writhing Skink				Southern Africa
	crepuscular	invert.	89	4.08	* sand shark
tanae	Tana River Writhing Skink				Kenya
			79	5.1	
vinciguerrae	Vinciguerra's Writhing Skink				Ethiopia
			55	1.89	

Mokopirirakau

Mokopirirakau is a small (four spp.) genus of secretive, cryptic, large-sized (8.4 g v. 4.15 g), nocturnal arboreal New Zealand diplodactylids. Assignment of ecological business models is uncertain. They all have a distinctive bright orange mouth lining and are otherwise similar to other diplodactylid geckos, except that they

Mokopirirakau spp.	English name				Range core
			Size		
	Activity	Diet	mm	g	Ecological business model
cryptozoicus	Takitimu Gecko				New Zealand
	nocturnal		77	9.3	* crack hermit
granulatus	Forest Gecko				New Zealand
	cathemeral		77	9.3	* nocturnal canopy walker
kahutarae	Black-eyed Gecko				New Zealand
	nocturnal		74	8.3	** nocturnal ground lizard
nebulosus	Cloudy Gecko				New Zealand
	nocturnal		70	6.9	

lack prehensile tails (v. 26% possess them), many lack expanded toepads (v. 16%), most are capable of color change (75% v. 5%), 25% (v. 1.3%) have advanced color changing ability, they will bask directly (v. 17%) if the sun hits their refuge entrance, 75% extend up to medium elevation (v. 11%), they are all viviparous (v. 18%), and their longevity is very high (1.69 v. 0.90 std. resid.; up to 25 years), as is so often the case in New Zealand. They are not associated with littoral areas (v. 17%), sandy areas (v. 24%), riparian zones (v. 10%), or human habitats (v. 5%). They have not been introduced (v. 2%). All evaluated species are imperiled (v. 81%), though none is listed on the IUCN's most endangered lists (v. 29%).

Moloch (fig. 5.3A)

Moloch is a single-species genus of crestless (v. 38%), terrestrial (v. 56%) agamid from Australia, distinguished by its covering of conical spines, and celebrated worldwide for its otherworldly appearance (fig. 5.3A). "Devil" is part of several unwarranted common names for this innocuous species. Perhaps this lizard should also be celebrated for its ability to draw soil moisture and atmospheric moisture into its mouth through capillary action.

It is frequently cited as an example of convergent evolution with North American horned lizards, *Phrynosoma* spp. (table 5.4; figs. 4.7B, 5.3B). *Phrynosoma* are strikingly similar desert ant specialists, though Schwenk (2000) challenged the analogy on the basis that *Moloch* feeds on very large numbers of small innocuous ants, whereas *Phrynosoma* feed on smaller numbers of *Pogonomyrmex* ants, which are larger and come bearing ferocious stings. As there are many other species of desert ant specialists that do not have the spines or the wide bodies of *Moloch/Phrynosoma*, there may be additional factors in the causation of their spininess and shape (see Myrmecovore discussion under "Diet" in chapter 3). Note that the frequency of crests is strongly elevated in ant specialists (table 3.2, trait 10; table 3.5, trait 5), but both *Moloch* and *Phrynosoma* lack crests; potentially the conspicuous spines possessed by both *Moloch* and *Phrynosoma* replace the role of crests in these ant specialists. Both *Moloch* and *Phrynosoma* are extremely cryptic (v. 51%), sit-and-wait (v. 88%) immobility experts (v. 4.5%).

With its short (0.74 v. 1.9 × SVL) tail, large size (43 g v. 15.6 g), squat shape (1.76 v. 0.80 std. resid.), m<f dimorphism (–1.0 v. 0.53 on –2 to 2 scale), lack of dichromatism (v. 53%), possession of advanced color change (v. 15%), and absence of secretory pores (v. 12.6%), *Moloch* is not a typical agamid. With two exceptions, it shares all of these anomalies with *Phrynosoma*: secretory pores are present but their role is unknown in *Phrynosoma*, and their absence in *Moloch* is no better understood. The other exception is that *Phrynosoma* are no squatter in build than the rest of the family and are not as heavily weighted as *Moloch*. *Moloch* also moves with a characteristic slow jerky back-and-forth gait, which *Phrynosoma* do not do. Over a year *Moloch* is said to wander extensively like some *Phrynosoma*, but the only quantification of *Moloch*'s movements was short-term and found very limited movements (~100 m^2; –0.69 v. 0.35 std. resid.). Long-term tracking of *Moloch* is desperately needed. *Moloch* basks directly (v. 24%) and has a high mean body temperature (33.8°C v. 34.8°C), though it exhibits a wide range of body temperatures (27.5°C–40°C). Male *Moloch* are reported to strongly defend a small breeding site (v. 97%), and at least some individuals hibernate or aestivate; therefore there may be a strong seasonal component to *Moloch* movements, site fidelity, and spatial defense. Male *Phrynosoma* are comparatively tolerant of other males, and *Phrynosoma* vary greatly among species in their site fidelity.

Moloch is found in arid (v. 35%) and sandy (v. 26%) habitats; it is not associated with littoral (v. 10%), riparian (v. 21%), or human (v. 16%) habitats; these biases are mirrored in *Phrynosoma*.

One argument that has been offered for the wide bodies of *Moloch* and *Phrynosoma*—an argument seemingly applicable to many other ant specialists, but not fitting their

A B

Fig. 5.3. A famous case of convergent evolution. **A,** *Moloch horridus* (Agamidae). **B,** *Phrynosoma mcallii* (Phrynosomatidae). **A,** M. Whiting; **B,** L. Grismer

Table 5.4. Key natural history differences between *Moloch* and *Phrynosoma*

Trait	*Moloch*	*Phrynosoma*
Density of vegetation	Closed	Open
Shape	Robust	Flattened and wide
Secretory pores	Absent	Greatly elevated (77% v. 32%)
Diet	Many innocuous ants	Fewer well-defended ants
Home range	Small (inadequate sample)	Moderate to huge
Male defense of space	Sites	No defense
Mass-adjusted clutch count	Average (0.0 std. resid.)	Very high (2.12 std. resid.)
Relative clutch mass	Low (0.178 ratio)	High (0.342 ratio)
Mass-adjusted longevity	High (1.05 std. resid.)	Near average (0.17 std. resid.)
Parental care	None (0%)	Elevated (4.5% v. 2.9%)
Incubation duration	Long (0.20 std. resid.)	Short (–0.62 std. resid.)

Moloch sp.	English name				Range core
			Size		
	Activity	Diet	mm	g	Ecological business model
horridus	Moloch				Australia
	diurnal	ants	90	43	**** immobility expert

morphology—is that the peritoneal space is needed for the large volume of low-grade food ingested. It is difficult to quantify the volume of food needed by a particular lizard, but data exist to compare species' peritoneal-space needs for reproduction. *Moloch*'s clutch count of 6.8 is at exactly the lizard average (0.0 v. 0.20 std. resid.), as is its ontogenetic size range (0.0 v. 0.69 std. resid.), indicating that its clutch is neither particularly large in number nor composed of large young. *Moloch* has a notably low RCM ratio (0.178 v. 0.336). *Moloch* does not have the high-energy sprinting lifestyle often believed to drive low weight-loading and therefore low RCM. Low RCM may not be required by the lifestyle of *Moloch*, or perhaps *Moloch* suffer from low energy intake and cannot produce larger clutches, given their diet and prevailing prey availability. These facts largely eliminate the argument that extra peritoneal space is needed for reproduction in *Moloch*. In contrast, *Phrynosoma* are characterized by especially high clutch counts and heavy RCMs; thus the peritoneal-space rationale for wide bodies could apply to both diet and reproduction in *Phrynosoma*. Perhaps the relatively large size of female *Moloch* provides females with extra room for ants, which might be energetically helpful for egg development. Her energy requirements for egg production may exceed his for mate finding, as evidenced by the m<f SSD in both genera. Many factors remain to be explored in the energetics of *Moloch* reproduction. *Moloch* is notably slow in egg incubation duration (111 days), especially when contrasted with agamids (0.20 v. –0.55 std. resid.). In this regard,

Moloch not only fails to fit with agamids, but it also diverges from ant specialists (v. −0.40) and *Phrynosoma* (v. −0.62, table 5.4).

For its size, *Moloch* is long-lived (1.05 v. −0.46 std. resid.), equating to 20 years, whereas *Phrynosoma* are not especially long-lived (table 5.4). Density data for *Moloch* are unavailable. *Moloch* is portrayed as "common," but "common" *Phrynosoma* have low densities when estimated quantitatively. *Moloch* is not endemic (v. 49%) or found on islands (v. 36%). It has not been introduced (v. 1.7%). It is not categorized as at conservation risk (v. 37% imperiled).

Monopeltis

Monopeltis is a major (20 spp.) radiation of large (10.7 g v. 2.7 g; range 1.0–125 g), mildly robust (shape index −3.27 v. −3.97 std. resid.) amphisbaenids in southern and central Africa: Cameroon and Central African Republic to South Africa. Their tails are on the short side of amphisbaenids (0.08 v. 0.11 × SVL); none of the tails are reported to break (v. 15%). Sand is mentioned in 5 of 10 species' habitat descriptions (50% v. 28%). Both species with comments on reproductive mode are believed to be viviparous (v. 14%), with *capensis* having a litter of two, which is notably small for lizards of this size (−0.63 v. −0.12 std. resid.). *M. capensis* is also notable for having a recorded longevity (1.8 years) that, when size adjusted, is −1.5 std. resid., which is substantially longer than other amphisbaenids (v. −3.02 std. resid.). Endemism (45% v. 66%) and insularity (5% v. 12%) are on the low side. The frequency of imperiled species is average (25% v. 23%), though none is on one of the IUCN's lists of the most endangered (v. 13%). In other respects *Monopeltis* are similar to other amphisbaenids.

Monopeltis spp.	English name	Activity	Diet	Size mm	Size g	Range core	Ecological business model
adercae	Lualaba Worm Lizard			373	45.7	Congo	**** subterranean forager
anchietae	Anchieta's Spade-snouted Worm Lizard		invert.	248	8.9	Botswana	**** subterranean forager
capensis	Cape Spade-snouted Worm Lizard	diurnal	invert.	247	8.65	South Africa	**** subterranean forager
decosteri	De Coster's Worm Lizard			217	5.2	Zimbabwe	**** subterranean forager
galeata	Hallowell's Spade-snouted Worm Lizard			361	39.9	Congo	**** subterranean forager
guentheri	Western Congo Worm Lizard			190	3.00	Congo	**** subterranean forager
infuscata	Dusky Spade-snouted Worm Lizard			235	7.1	South Africa	**** subterranean forager
jugularis	Peters' Spade-snouted Worm Lizard			466	111	Equatorial Guinea	**** subterranean forager
kabindae	Kabinda Worm Lizard			144	0.99	Congo	**** subterranean forager
leonhardi	Kalahari Spade-snouted Worm Lizard		invert.	217	5.2	Botswana	**** subterranean forager
luandae	Luanda Worm Lizard			284	15.3	Angola	**** subterranean forager
perplexus	Capelongo Worm Lizard			224	5.9	Angola	**** subterranean forager
remaclei	De Witte's Worm Lizard			423	76	Congo	**** subterranean forager
rhodesiana	Zimbabwe Spade-snouted Worm Lizard			217	5.2	Zimbabwe	**** subterranean forager
scalper	Carved Worm Lizard			216	5.1	Congo	**** subterranean forager
schoutedeni	Middle Congo Worm Lizard			480	125	Congo	**** subterranean forager
sphenorhynchus	Slender Spade-snouted Worm Lizard			255	10.0	Botswana	**** subterranean forager
vanderysti	Vanderyst's Worm Lizard			270	12.5	Congo	**** subterranean forager
welwitschii	Pungo Andongo Worm Lizard			219	5.3	Angola	**** subterranean forager
zambezensis	Zambezi Spade-snouted Worm Lizard			187	2.88	Zimbabwe	**** subterranean forager

Morethia

Morethia are archetypical little brown skinks, often with a darker mid-lateral stripe. Juveniles (and some adults) sport a reddish or bright red tail (63%), which they may wave conspicuously. Most adults show seasonal color (88% v. 6%). They constitute a moderate radiation (eight spp.) of small-sized (1.63 g v. 3.3 g), diurnal (100% v. 74%), direct-basking (100% v. 21%), oviparous (100% v. 64%), ground-dwelling skinks from throughout Australia. Most are refugia-anchored diurnal ground lizards (88% v. 30%); the remaining species (*taeniopleura*) is a litter cryptozoid (13% v. 28%).

Morphologically they are average skinks, though none is elongate (v. 35%) and tails are a bit longer (1.67 v. 1.33 × SVL). They do not have blue tails (v. 4%). Sex differences are limited to the seasonal color mentioned above. All eat invertebrates (v. 84%), but they break the skink mold in their heavy reliance on sit-and-wait foraging (100% v. 16%). Some are active foragers as well (38% v. 92%). *M. obscura* steals prey from ants. Home-range area is unknown.

Many are conspicuously wary (38% v. 11%). All are reported to sand swim, but the primary predator concealment tactic is retreat to a crevice or burrow (100% v. 43%). Mean active body temperature is relatively hot (34.0°C v. 31.1°C). Anomalously for skinks or refugia-anchored diurnal ground lizards (table 4.1, trait 46), the only socially documented species (*boulengeri*) has males that are tolerant (v. 5.6%); none is reported to defend space. They are overrepresented in littoral (63% v. 21%) and sandy (50% v. 25%) habitats, representative in riparian areas (25% v. 23%), and absent from human habitats (0% v. 14%).

Clutch counts are average for their size (0.25 v. 0.06 std. resid.); young are small (ontogenetic mass range 0.88 v. 0.12 std. resid.). Several species breed more than once a year. Incubation is slightly long for a skink (−0.20 v. −0.40 std. resid.), equating to 58 days (*adelaidensis*, *boulengeri*) or 29 days (*obscura*). Communal oviposition is known (*adelaidensis* 13% v. 3%). RCM, based on *boulengeri* and *storri* is high (0.300 v. 0.236).

Population density is high (1.01 kg/ha v. 0.35 kg/ha), based on 12 estimates of *boulengeri*. Endemism is low (13% v. 64%), despite the large size of Australian states. Island species are few (38% v. 57%), even though all *Morethia* have distributions that reach the coast. None has been introduced (v. 2.1%). Half of evaluated species are imperiled (50% v. 47%), though none is on one of the IUCN's lists of the most endangered (v. 14%). In other respects *Morethia* are typical skinks.

Morethia spp.	English name				Range core
			Size		
	Activity	Diet	mm	g	Ecological business model
adelaidensis	Saltbush Morethia Skink				Australia
	diurnal	invert.	50	2.34	** refugia-anchored diurnal gr. liz.
boulengeri	Southeastern Morethia Skink				Australia
	diurnal	invert.	49	2.09	*** refugia-anchored diurnal gr. liz.
butleri	Woodland Morethia Skink				Southwestern Australia
	diurnal	invert.	50	2.34	** refugia-anchored diurnal gr. liz.
lineoocellata	West Coast Morethia Skink				Western Australia
	diurnal	invert.	48	2.02	** refugia-anchored diurnal gr. liz.
obscura	Shrubland Morethia Skink				Australia
	diurnal	invert.	50	2.34	*** refugia-anchored diurnal gr. liz.
ruficauda	Lined Firetail Skink				Western Australia
	diurnal	invert.	39	1.05	*** refugia-anchored diurnal gr. liz.
storri	Top End Firetail Skink				Western Australia
	diurnal	invert.	36	0.83	** refugia-anchored diurnal gr. liz.
taeniopleura	Fire-tailed Skink				Queensland
	diurnal	invert.	40	1.09	* litter cryptozoid

Morunasaurus

Morunasaurus is a three-species genus of refugia-anchored diurnal terrestrial hoplocercids found in northwestern South America and Panama. Although tentatively listed as refugia-anchored diurnal ground lizards, they may actually be crack (in this case, burrow) hermits, rarely leaving the mouth of their burrows. Virtually all of the natural history data for this genus are prefaced with the qualifier "presumed to be."

They are of average size for Hoplocercidae (32.1 g v. 31.3 g). They are not reported to have crests (v. 63%), secretory pores (v. 10.5%), color change (v. 16%), or sex differences. Their tails are somewhat spiny, but not nearly as intimidating as those of *Hoplocercus*. *M. groi* is a thermoconformer (v. 5%). *Morunasaurus* are found at low elevation (v. 94%), often in riparian areas (67% v. 40%). Clutch count is low (–0.92 v. –0.69 std. resid.), based on *annularis*. One species (*peruvianus*) is endemic (v. 37%). None is found on islands (v. 5%). Two species are considered imperiled (v. 30%), and *groi* is listed by IUCN as one of the most endangered species (v. 10%). In other respects *Morunasaurus* are probably typical hoplocercids.

Morunasaurus spp.	English name				Range core
			Size		
	Activity	Diet	mm	g	Ecological business model
annularis	Ringed Woodlizard				Colombia
	diurnal		107	35.6	* refugia-anchored diurnal gr. liz.
groi	Gro's Manticore				Colombia
	diurnal		88	18.7	** refugia-anchored diurnal gr. liz.
peruvianus	Cenepa Woodlizard				Peru
			119	49.4	* refugia-anchored diurnal gr. liz.

Nactus spp.	English name				Range core
			Size		
	Activity	Diet	mm	g	Ecological business model
acutus	Rossel Island Gecko				New Guinea
	nocturnal		36	1.10	
cheverti	Chevert Gecko				Queensland
	nocturnal		53	3.23	*** nocturnal ground lizard
coindemirensis	Lesser Night Gecko				Mauritius
	nocturnal	invert.	35	1.07	*** nocturnal ground lizard
eboracensis	Cape York Bent-toed Gecko				Queensland
	nocturnal		54	3.46	*** nocturnal ground lizard
galgajuga	Black Mountain Gecko				Queensland
	cathemeral		47	2.41	*** nocturnal ground lizard
kunan	Bumblebee Gecko				New Guinea
	nocturnal		50	2.84	
multicarinatus	Melanesia Slender-toed Gecko				Solomon Islands
	nocturnal	invert.	57	3.96	*** nocturnal ground lizard
pelagicus	Pelagic Gecko				Pacific Islands
	nocturnal	invert.	54	3.10	*** nocturnal ground lizard
serpensinsula	Serpent Island Ground Gecko				Mauritius
	nocturnal	invert.	54	3.54	*** nocturnal ground lizard
soniae	Reunion Ground Gecko				Reunion
	nocturnal		42	1.69	
sphaerodactylodes	Sunrise Islands Gecko				New Guinea
	nocturnal		26	0.47	* nocturnal ground lizard

Nactus

Nactus is a 12-species genus of small (1.90 g v. 3.31 g), padless, nocturnal ground lizards (Gekkonidae) found in Australasia and the Mascarene Islands near Madagascar. In regard to the latter, it is noteworthy that prevailing winds and currents in the Mascarene Islands are from the northeast (i.e., originating in Australasia). These generators of drift are presumably responsible for a number of species—the most famous being the coelacanth fish—with disjunct distributions spanning the Indian Ocean. One *Nactus* species (*pelagicus*) is parthenogenetic, no doubt facilitating its colonization of islands throughout the western Pacific.

Nactus mirrors most gekkonids in general form, frequency of secretory pores (50% v. 46%), and incidence of basic color change (17% v. 15%). Sex differences are muted (sex rollup score on scale of 0 to 6: 0.45 v. 0.63), and females tend to be larger than males: –0.50 (v. –0.03 on scale of –2 to 2). Sexual dichromatism (v. 6.6%) and seasonal color are not reported.

Dietary details are known only for *serpensinsula*, a generalized invertebrate eater. *N. pelagicus* is said to forage by sit-and-wait (v. 81%). Spontaneous calling is reported in 17% (v. 7%) and stress calls in 8% (v. 4%). *N. pelagicus* is conspicuously wary (8% v. 4%).

Reproduction is gekkonid-normal, with the wrinkle that *pelagicus* is parthenogenetic.

Population density has been estimated only for *pelagicus*, for which it is typical (0.34 kg/ha v. 0.37 kg/ha). Ninety-two percent of *Nactus* are island species (v. 45%); *pelagicus* has

Nactus spp.	English name	Activity	Diet	Size mm	Size g	Range core	Ecological business model
vankampeni	Van Kampen's Gecko	nocturnal		31	0.72	New Guinea	

Nadzikambia spp.	English name	Activity	Diet	Size mm	Size g	Range core	Ecological business model
baylissi	Mount Mabu Chameleon	diurnal	invert.	60	5.9	Mozambique	*** diurnal canopy walker
mlanjensis	Mulanje Chameleon	diurnal	invert.	69	8.7	Malawi	*** diurnal canopy walker

Namazonurus spp.	English name	Activity	Diet	Size mm	Size g	Range core	Ecological business model
campbelli	Campbell's Nama Lizard	diurnal	invert.	66	6.1	Namibia	*** crack hermit
lawrenci	Lawrence's Nama Lizard	diurnal		68	6.6	South Africa	*** crack hermit
namaquensis	Namaqua Nama Lizard	diurnal	invert.	72	8.3	Namibia	* crack hermit
peersi	Peer's Nama Lizard	diurnal	invert.	73	8.8	South Africa	** crack hermit
pustulatus	Herero Nama Lizard	diurnal	invert.	71	7.9	Namibia	* crack hermit

arguably been introduced (8% v. 3.7%). Conservation status is about the same as gekkonids: 50% (v. 47%) imperiled by some measure; 0% (v. 11%) listed by the IUCN among the most at-risk species. In other respects *Nactus* are typical gekkonids. In one peculiar circumstance, *Nactus* stands out: *serpensinsula* may express tonic immobility for many hours, at least if placed on its back on a soft substrate. Never underestimate the patience of a reptile.

Nadzikambia

Nadzikambia is a tiny genus (two spp.) of small-sized (7.2 g v. 13.5 g) typical (table 5.2) chameleons from Malawi/Mozambique. They are relatively poorly known (29% v. 38% of cells filled), and they are placed in their own genus solely because of genetic distance to other typical chameleons, rather than any ecological or morphological distinctiveness. With only two species—one recently described and especially poorly known—species frequencies are relatively meaningless, but they fit in with the typical chameleons in business model, shape index (0.50 v. 0.57 std. resid.), tail length (1.15 v. 1.09 × SVL), crests, and arboreality. The one species for which clutch count has been recorded has relatively small clutches (0.81 v. 2.81 std. resid.), not far removed from the 0.60 of the dwarf chameleon group. Longevity is short, even by chameleon standards (−1.34 v. −0.83 std. resid.). In other respects they comport well with the remaining chamaeleonids. Unfortunately, all are imperiled (100% v. 95%).

Namazonurus

Namazonurus (five spp.) are diurnal crack hermit cordylids of a modest size (7.5 g v. 10.8 g) and typical build (shape index 0.30 v. 0.35 std. resid.) from Namibia and South Africa. Outwardly, they are very similar in morphology to *Cordylus*. Their tails are slightly shorter than average cordylids (1.00 v. 1.16 × SVL) and do not break (v. 60%). Secretory pores are reported for only one species (v. 69%); the sexes are not dichromatic (v. 37%), nor do they have seasonal color (20% v. 28%). Accordingly, their sex differences rollup score on a scale of 0 to 6 is low (0.20 v. 1.41).

Their diet is invertebrates (v. 70%). Two of five (40% v. 27%) are conspicuously wary. All bask directly (v. 74%). Social life has not been reported for *Namazonurus*. At least one is found in littoral areas (20% v. 9%). All are viviparous (v. 75%), with an average litter count (−0.44 v. −0.48 std. resid.) of exceptionally large young (ontogenetic range −1.86 v. −0.37 std. resid.) and no reports of parental care (v. 7.4%). Exceptionally large young are characteristic of crack hermits (table 4.1, trait 64), but on the same basis one would have higher expectations for parental care frequency (table 4.1, trait 69). Parental care is difficult to detect in the field. Size-adjusted longevity is longer than average (0.67 v. 0.02 std. resid.), commensurate with the business model (v. 0.74 std. resid.). Most species are endemic (80% v. 53%), and all are imperiled (v. 79%), though none is among the IUCN listings of most endangered (v. 4.8%). In other respects they fit the cordylid template.

Nangura

Nangura spinosa is the sole member of its genus. It is a large-sized (13.5 g v. 3.3 g), pebble-skinned, viviparous (v. 36%), riparian (v. 23%), crack hermit skink that lives in colonies (v. 16%) in one tiny area of mainland Queensland. Family burrows might be inherited, as neonates are reported to remain with their mothers for at least nine months.

Nangura is heavier built than an average skink (shape index 0.39 v. −0.46 std. resid.). Its tail is short (0.90 v. 1.33 × SVL) and breakable (v. 73%). Sex differences are not known. Their invertebrate prey are taken by sit-and-wait (v. 16%). *Nangura* has not been introduced (v. 2.1%). It is imperiled (v. 47%), including placement on one of the IUCN's lists of the most endangered (in this case critically endangered). In other respects *Nangura* is likely to be similar to other skinks.

Nangura sp.	English name				Range core
			Size		
	Activity	Diet	mm	g	Ecological business model
spinosa	Nangur Spiny Skink				Queensland
		invert.	87	13.5	** crack hermit

Nannoscincus spp.	English name				Range core
			Size		
	Activity	Diet	mm	g	Ecological business model
exos	North New Caledonia Dwarf Skink				New Caledonia
			36	0.58	* litter cryptozoid
fuscus	Central Chain Dwarf Skink				New Caledonia
					** litter cryptozoid
garrulus	Talkative Dwarf Skink				New Caledonia
			47	1.24	** litter cryptozoid
gracilis	Gracile Dwarf Skink				New Caledonia
		invert.	44	1.04	** litter cryptozoid
greeri	Greer's Dwarf Skink				New Caledonia
			34	0.49	** litter cryptozoid
hanchisteus	Pindai Dwarf Skink				New Caledonia
			34	0.49	** litter cryptozoid
humectus	Foret Plate Dwarf Skink				New Caledonia
			35	0.55	** litter cryptozoid
koniambo	Mount Koniambo Skink				New Caledonia
			34	0.49	*** litter cryptozoid
manautei	Manaute's Dwarf Skink				New Caledonia
			34	0.39	** litter cryptozoid
mariei	Earless Dwarf Skink				New Caledonia
			42	0.91	** litter cryptozoid
rankini	Rankin's Dwarf Skink				New Caledonia
			39	0.71	** litter cryptozoid
slevini	Slevin's Dwarf Skink				New Caledonia
			40	0.57	** litter cryptozoid

Nannoscincus

Nannoscincus is a moderate radiation (12 spp.) of very small-sized (0.65 g v. 3.25 g), elongate (100% v. 35%), oviparous (100% v. 64%), litter cryptozoid (100% v. 28%) skinks endemic to New Caledonia. They are among the very few skinks that vocalize (stress calls 8.3% v. 0.4%).

They are more slender than most skinks (shape index −0.94 v. −0.46 std. resid.) and have shorter tails (0.96 v. 1.33 × SVL). None has a blue tail (v. 4%). Sex differences are unknown. One species (*gracilis*) has dietary information (generalized invertebrate eater). *N. mariei* is conspicuously wary (8% v. 11%). They are absent from littoral (v. 21%), sandy (v. 25%), and human (v. 14%) habitats, but they are present at normal levels in riparian zones (18% v. 23%). *N. mariei* may refuge in soil cracks.

Clutch count is normal (0.24 v. 0.06 std. resid.); although the mean is two, it does not seem to be fixed (range one to three). Multiple broods per year seem likely but have not been documented. No *Nannoscincus* has been introduced (v. 2.1%). All are to some degree imperiled (v. 47%), including 70% of evaluated species landing on one of the IUCN's lists of most endangered (v. 14%). In all other documented respects *Nannoscincus* are typical skinks.

Narudasia

Narudasia festiva is the sole representative of a padless gekkonid genus found in Namibia. It is very small (0.68 g v. 3.31 g) and slender (shape index –0.17 v. 0.10 std. resid.), and some individuals are "festively" marked, often with an orangish tail. Few firm facts are available about its behavior and foraging. Sexual dimorphism is not reported. It is often found around rock, but its foraging microhabitat is not known. As far as is known, it reproduces like other gekkonids. This gecko is so small that the apparently fixed clutch count of two is larger than the lizard average when mass adjusted (0.38 v. –0.26 std. resid.). It has not been introduced, and it has not been formally evaluated for conservation status. In other respects *Narudasia* are typical gekkonids.

Narudasia sp.	English name				Range core
			Size		
	Activity	Diet	mm	g	Ecological business model
festiva	Festive Gecko				Namibia
	crepuscular	invert.	30	0.68	

Naultinus spp.	English name				Range core
			Size		
	Activity	Diet	mm	g	Ecological business model
elegans	Elegant Green Gecko				New Zealand
	diurnal		77	9.3	*** diurnal canopy walker
gemmeus	South Island Jewelled Gecko				New Zealand
	diurnal	omnivore	67	5.9	*** diurnal canopy walker
grayii	Northland Green Gecko				New Zealand
	diurnal		77	9.3	*** diurnal canopy walker
manukanus	Marlborough Green Gecko				New Zealand
	diurnal		68	6.4	** immobility expert
punctatus	Barking Gecko				New Zealand
	diurnal		77	9.3	** diurnal canopy walker or immobility expert
rudis	Rough Gecko				New Zealand
	diurnal		61	4.48	** diurnal canopy walker or immobility expert
stellatus	Starred Gecko				New Zealand
	diurnal		67	5.9	** diurnal canopy walker or immobility expert
tuberculatus	West Coast Green Gecko				New Zealand
	diurnal		70	6.9	** diurnal canopy walker or immobility expert

Naultinus (figs. 4.3D, 4.7C)

Naultinus (eight spp.) are exquisitely cryptic, nearly immobile, green, diurnal, arboreal, diplodactylid geckos hiding in the dense foliage of selected trees in New Zealand. They are medium-sized (7.0 g v. 4.15 g), long-tailed (1.89 v. 0.85 × SVL), and prehensile-tailed (v. 26%) lizards. Their sedentariness has been documented in *manukanus*, for which the mean daily movement of 0.59 m has been measured. They are 100% sit-and-wait foragers, as are all diplodactylids. As with many diplodactylids, males tend to be smaller than females; but unlike most diplodactylids, 88% are sexually dichromatic (v. 6%). Thus they have a sex differences rollup score of 1 (on 0 to 6 scale), which is higher, but not much higher, than the family (0.54).

Two species are recorded to give stress calls (25% v. 5%). They all bask directly (v. 17%). They are underrepresented in sandy areas (0% v. 24%). As is so often the case in New Zealand, they are viviparous (v. 18%) and live for a long time (1.75 v. 0.90 std. resid.), up to 47 years in *stellatus*. Like most diplodactylids, but even more so, the young of *Naultinus* are especially large (ontogenetic range –0.90 v. –0.47 std. resid.). *N. grayii* parents have been seen to defend their neonates (parental care 13% v. 3%). Population density is high (1.34 kg/ha v. 0.44 kg/ha) and is recorded at a mean density of 4 kg/ha in *manukanus*. This species has been translocated to additional offshore islands for conservation purposes. All evaluated species are considered imperiled (v. 81%), though none is on the IUCN's lists of most endangered. In other respects *Naultinus* are typical diplodactylids.

Nebulifera

Nebulifera robusta is the sole representative of this diplodactylid genus. It is found in eastern Australia on the trunks and branches of trees. A bit larger than an average diplodactylid (8.2 g v. 4.2 g), it sports a bulbous and breakable, but nonprehensile, tail. It lacks secretory pores (v. 34%). Other sex differences are not reported. It is cathemeral and gives stress vocalizations (v. 5%). Males defend space (v. 77%) but interact with other individuals on the same tree; their activity areas do not extend beyond a single tree. It is associated with riparian zones (v. 10%) and human habitats (v. 5%). Young are relatively large, even by diplodactylid standards (ontogenetic range –1.24 v. –0.57 std. resid.). Maximum reported longevity is a bit short for a diplodactylid (0.39 v. 0.90 std. resid.). It has not been introduced or evaluated for conservation status. In other respects it is a typical diplodactylid.

Nephrurus

Nephrurus is a moderate (nine spp.) carphodactylid genus of striking terrestrial geckos: large heads, large eyes, pronounced eyebrows, and most species with a conspicuous spherical knob on the end of their very short tails (0.26 v. 0.54 × SVL). Many species have striking white dots, often on their tails. They are of average carphodactylid size (13.9 g v. 14.5 g), are all nocturnal ground lizards (v. 30%), and are distributed throughout Australia. A test of their nocturnal visual acuity would reveal whether they are as oriented toward the visual modality as their morphology suggests they might be, a condition relatively unusual in nocturnal species. Despite their short tails, two-thirds of the species report tail breakage (v. 87%); the three shortest-tailed species (<0.15 × SVL) cannot break their tails. The majority (57% v. 40%) are carnivores (including many of the smallest *Nephrurus*), and the second smallest (*laevissimus*) is an omnivore. Oddly, the largest species with dietary information are invertebrate eaters (29% v. 50%). All carphodactylids are sit-and-wait (100% v. 100%). At surface temperatures of less than 8°C, *wheeleri* digs a small depression, enters it, and uses its forelegs to toss soil onto its back, which presumably allows it to continue surveillance while minimizing heat loss. *Nephrurus* have carphodactylid-typical mean body temperature (23.7°C v. 22.8°C).

Nephrurus vocalizes at carphodactylid-typical frequencies (11% v. 13%), but almost all species give stress calls (89% v. 33%). Does this startle nocturnal predators, or are there kin that can be warned residing within earshot?

Nephrurus are accomplished burrowers. *N. levis* is known to dig a side chamber off underground tunnels dug by other species and plug the entrance to the side branch with sand when in residence. Seventy-eight percent (v. 28%) of *Nephrurus* are associated with sand or dunes.

Nebulifera sp.	English name				Range core
			Size		
	Activity	Diet	mm	g	Ecological business model
robusta	Robust Velvet Gecko				Queensland
	cathemeral	invert.	74	8.2	*** trunk/wall lizard

Nephrurus spp.	English name				Range core
			Size		
	Activity	Diet	mm	g	Ecological business model
amyae	Centralian Knob-tailed Gecko				Northern Territory
	nocturnal		117	26.6	*** nocturnal ground lizard
asper	Rough Knob-tail				Northern Australia
	nocturnal	invert.	106	20.8	*** nocturnal ground lizard
deleani	Pernatty Knob-tail				South Australia
	nocturnal	carnivore	86	12.7	*** nocturnal ground lizard
laevissimus	Smooth Knob-tail				Western Australia
	nocturnal	omnivore	75	9.4	*** nocturnal ground lizard
levis	Three-lined Knob-tail				Australia
	cathemeral	carnivore	83	11.8	*** nocturnal ground lizard
sheai	Northern Knob-tail				Western Australia
	nocturnal	invert.	106	21.0	*** nocturnal ground lizard
stellatus	Stellate Knob-tail				Southwestern Australia
	nocturnal	carnivore	80	10.9	*** nocturnal ground lizard
vertebralis	Midline Knob-tail				Southwestern Australia
	nocturnal	carnivore	74	9.0	*** nocturnal ground lizard
wheeleri	Banded Knob-tail				Western Australia
	nocturnal		85	12.3	*** nocturnal ground lizard

Nephrurus are absent from riparian zones (0% v. 24%); is ground-level riparian vegetation too dense for *Nephrurus* to forage effectively, is the soil too compacted in riparian zones for effective digging, or are there hostile mammals in riparian zones? Only 25% of *Nephrurus* are judged to be at conservation risk (v. 67%). In other respects *Nephrurus* are typical carphodactylids.

Nessia

Nessia (eight spp.) are slightly small (2.28 g v. 3.25 g), highly elongate (38% v. 35%) or functionally limbless skinks endemic to Sri Lanka. Although all may be subterranean foragers (100% v. 12%), some may be litter cryptozoids or sand sharks. They are very slender (shape index –1.58 v. –0.46 std. resid.), and like most subterranean foragers, they have very short tails (0.56 v. 1.33 × SVL) that do not break (v. 73% breakable). *N. bipes* has documented m<f SSD (–1.0 v. –0.31 on –2 to 2 scale). No other sex differences are known. Their diets are those of generalized invertebrate eaters (100% v. 84%), with an emphasis on ants and termites. They are evenly split between nocturnal (50% v. 14.5%) and diurnal (50% v. 74%) species. They are found in the mesic rainfall belt (in this case 1500–2500 mm/y) and predominantly low elevations (l/m/h: 100/14/0 v. 89/35/6). They are noticeably overrepresented in human habitat (50% v. 14%) and absent from riparian zones (0% v. 23%). They are normal in littoral (17% v. 21%) and sandy (33% v. 25%) areas.

All are oviparous (v. 64%) and have clutch counts commensurate with their size (0.03 v. 0.06 std. resid.).

None has been introduced (v. 2.1%). All are imperiled (v. 47%), and 67% (v. 14%) are on the IUCN's lists of the most endangered. In other respects *Nessia* are typical limbless skinks.

Nessia spp.	English name				Range core
			Size		
	Activity	Diet	mm	g	Ecological business model
bipes	Two-legged Snakeskink				Sri Lanka
	diurnal	invert.	71	2.37	** subterranean forager or litter cryptozoid
burtonii	Three-toed Snakeskink				Sri Lanka
	nocturnal	invert.	65	1.89	**** subterranean forager
deraniyagalai	Deraniyagala's Snakeskink				Sri Lanka
			72	1.17	* subterranean forager or sand shark
didactyla	Two-toed Snakeskink				Sri Lanka
			72	2.42	** subterranean forager
hickanala	Shark-head Snakeskink				Sri Lanka
			118	3.59	** subterranean forager
layardi	Layard's Snakeskink				Sri Lanka
			70	1.09	** subterranean forager
monodactyla	One-toed Snakeskink				Sri Lanka
	nocturnal	invert.	76	2.79	**** subterranean forager
sarasinorum	Sarasin's Snakeskink				Sri Lanka
	diurnal	invert.	98	5.3	**** subterranean forager

Neusticurus

Neusticurus is a five-species genus of exceptionally large (16.2 g v. 2.35 g) diurnal, terrestrial, riparian gymnophthalmids found in northern South America. They often sleep in foliage over water but are active as either litter cryptozoids (50% v. 55%) or refugia-anchored diurnal ground lizards (50% v. 19%). Either group may take temporary refuge in water. They are relatively heavily built (shape index 0.65 v. –0.32 std. resid.). They are not elongate (v. 41%). Tails are of typical length (1.86 v. 1.72 × SVL) and often breakable (40% v. 28%). Most have secretory pores (60% v. 29%) and are unusual among gymnophthalmids in having males larger than females (1.00 v. –0.12 on –2 to 2 scale), presumably reflecting their large size (table 3.3, trait 1). Many species are sexually dichromatic (40% v. 13%), though they lack seasonal color (v. 1.7%). Their sex rollup score (on 0 to 6 scale) is high (1.20 v. 0.49).

Their large size makes carnivory (33% v. 2.6%) possible, though the majority are invertebrate eaters (67% v. 97%). Wariness is overrepresented (20% v. 7%) and yet may be underreported. After a swim, even tropical lizards may have a need to warm up; direct basking is overrepresented (40% v. 8%), as is thermal conformance (20% v. 5%). The one documented species, *bicarinatus*, has males that are socially tolerant (v. 88%). Burrow refugia are not reported (v. 5%). They are often found at low elevation (100% v. 51%) and are absent from high elevation (v. 33%). Unsurprisingly, they are overrepresented in riparian habitats

Neusticurus spp.	English name		Size		Range core
	Activity	Diet	mm	g	Ecological business model
bicarinatus	Two-crested Stream Lizard				Guiana Shield
	diurnal	carnivore	90	19.0	** litter cryptozoid
medemi	Medem's Stream Lizard				Colombia
	diurnal	invert.	96	23.2	* refugia-anchored diurnal gr. liz.
racenisi	Roze's Stream Lizard				Venezuela
	diurnal		84	15.2	** refugia-anchored diurnal gr. liz.
rudis	Ruddy Stream Lizard				Venezuela
	diurnal	invert.	77	11.0	** litter cryptozoid
tatei	Tate's Stream Lizard				Venezuela
			84	14.8	

Ninurta sp.	English name		Size		Range core
	Activity	Diet	mm	g	Ecological business model
coeruleopunctatus	Blue-spotted Lizard				South Africa
	diurnal	invert.	61	4.60	* refugia-anchored diurnal gr. liz. or crack hermit

(100% v. 29%) and absent from littoral areas (v. 3%), sandy areas (v. 14%), and human habitat (v. 15%).

Reproduction is typical, though clutches are small (–0.83 v. –0.11 std. resid.) and incubation duration is short (0.51 v. 1.11 std. resid.). No *Neusticurus* is found on islands (v. 5%), and none is considered imperiled (v. 25%). In other respects *Neusticurus* are probably typical gymnophthalmids.

Ninurta

Ninurta coeruleopunctatus is the sole member of this genus of cordylid lizards, endemic to South Africa. It is small (4.6 g v. 10.8 g), with a typical lizard build (shape index 0.1 v. 0.35 std. resid.), is less armored and pointy than *Cordylus*, and has the large jowl bulges found in *Pseudocordylus*. Its tail is not reported to break (v. 60%). It has secretory pores (v. 69%), and the sexes differ in color (v. 37%), though it lacks seasonal color (v. 28%). This species is a sit-and-wait forager, basks directly (v. 74%), digs burrows (v. 10%), resides in littoral areas (v. 9%), is viviparous (v. 75%), has slightly numerous young (0.02 v. –0.48 std. resid.), is short-lived (–0.62 v. 0.02 std. resid., 3.2 years), and has not been evaluated for conservation risk. In other respects it is a typical cordylid.

Niveoscincus

Niveoscincus (eight spp.) are from cool, temperate southern Australia and Tasmania. They are average-sized (3.56 g v. 3.25 g), sometimes marginally elongate (38% v. 35%), diurnal (100% v. 74%), viviparous (100% v. 36%), direct-basking (100% v. 231%) skinks of diverse ecologies that all retreat to a crevice or burrow for predator concealment (100% v. 43%). Most are refugia-anchored diurnal ground lizards (63% v. 30%); two are ground-foraging tree lizards (25% v. 7%), and the remainder (*orocryptus*) is a diurnal canopy walker (13% v. 2%).

They differ morphologically from a typical skink only in the occasional possession (*pretiosus*) of a partially prehensile tail (13% v. 0.7%). Three species have SSD, all m<f (–1.00 v. –0.31 on –2 to 2 scale). One species (*metallicus*) has sexual dichromatism. In aggregate, the genus's sex differences rollup score is near normal (0.50 v. 0.28 on 0 to 6 scale).

They eat generalized invertebrates (100% v. 84%), which they acquire by both sit-and-wait and active foraging (foraging mode documented only in *greeni*). Notable dietary items include nectar (*metallicus*) and carrion (*palfreymani*, *pretiosus*). Based on *microlepidotus*, they have small home ranges (–0.37 v. –0.07 std. resid.), equating to about 12 m^2. Their mean active body temperature is low (27.8°C v. 31.1°C), as is the ambient temperature in much of their range. They tolerate much colder temperatures under some conditions; for example, *greeni* will take refuge from predators in 5°C water. When the ambient temperature soars to 22°C, pregnant *microlepidotus* will seek cooler temperature than nonpregnant ones—their human counterparts will relate, but this is an extraordinary behavior in a reptile. Conversely, *palfreymani* will bask communally when it is cool, despite the readiness of males to challenge the presence of other males in other contexts (spatial defense 100% v. 94%). The alpine/subalpine *orocryptus* may not become active when it is cloudy.

Their elevational profile is near normal (l/m/h: 100/25/0 v. 89/35/6).

Niveoscincus spp.	English name	Activity	Diet	Size mm	Size g	Range core	Ecological business model
coventryi	Southern Forest Cool-skink	diurnal	invert.	45	1.07	New South Wales	* refugia-anchored diurnal gr. liz.
greeni	Alpine Cool-skink	diurnal	invert.	64	5.1	Tasmania	* refugia-anchored diurnal gr. liz.
metallicus	Metallic Cool-skink	diurnal	invert.	57	2.07	Tasmania	*** refugia-anchored diurnal gr. liz.
microlepidotus	Boulder Cool-skink	diurnal	invert.	59	2.28	Tasmania	* ground-foraging tree lizard
ocellatus	Ocellated Cool-skink	diurnal	invert.	70	6.8	Tasmania	* refugia-anchored diurnal gr. liz.
orocryptus	Heath Cool-skink	diurnal	invert.	59	3.87	Tasmania	* diurnal canopy walker
palfreymani	Pedra Branca Cool-skink	diurnal	invert.	80	10.5	Tasmania	** refugia-anchored diurnal gr. liz.
pretiosus	Agile Cool-skink	diurnal	invert.	57	3.60	Tasmania	* ground-foraging tree lizard

Nothobachia sp.	English name	Activity	Diet	Size mm	Size g	Range core	Ecological business model
ablephara	Shovel-nosed Lizard			52	1.06	Brazil	*** subterranean forager

Notomabuya sp.	English name	Activity	Diet	Size mm	Size g	Range core	Ecological business model
frenata	Southern Neotropics Skink	diurnal	invert.	71	7.1	Paraguay	** trunk/wall lizard

They match average skinks in their proportions in littoral, sandy, riparian, and human habitats.

Litter counts are average (0.04 v. 0.06 std. resid.). Young are moderately small (ontogenetic mass range 0.53 v. 0.12 std. resid.). Several species are documented to skip breeding in some years (*greeni*, *microlepidotus*, *orocryptus*).

Longevity is high (0.95 v. 0.17 std. resid.), which is based on 12 years in *ocellatus*. Presumably at least part of their high longevity is attributable to short active seasons at their high latitude. Endemism is a little high (75% v. 64%). Most species are insular (88% v. 57%). None has been introduced (v. 2.1%). All are imperiled (v. 47%), but none is on one of the IUCN's lists of the most endangered (v. 14%). The Pedra Branca Cool-skink occurs solely on a 2.5 ha island (Pedra Branca) south of Tasmania; the global population of this species is around 300 individuals. In other respects *Niveoscincus* are typical skinks.

Nothobachia

Nothobachia ablephara is the sole member of a poorly known (filled cells 17% v. 19%) subterranean gymnophthalmid genus from Brazil. The species is small (1.06 g v. 2.35 g) and extremely slender (shape index –1.29 v. –0.32 std. resid.). It is extremely elongated (v. 41%) and possesses secretory pores (v. 29%). Pores are notably rare in subterranean foragers (table 4.1, trait 9). No other information about sex differences was found. It is endemic to Brazil (v. 68%). This species is not known to have been introduced (v. 0.8%), nor has it been evaluated for conservation status.

Notomabuya

Notomabuya frenata is the sole member of its genus. It is a large-sized (7.1 g v. 3.3 g), viviparous (v. 36%), diurnal (v. 74%) trunk/wall skink in south-central South America: Bolivia, Paraguay, Brazil, and Argentina.

Notomabuya is more heavily built than an average skink (shape index 0.18 v. –0.46 std. resid.). Sex differences are not known. Its invertebrate prey are taken by active foraging (v. 92%). Males defend space (v. 94%). *Notomabuya* occupies human habitats (v. 14%). Its mean litter count is average (0.03 v. 0.06 std. resid.); it breeds continuously, once or twice per year. Young are small (ontogenetic mass range 0.65 v. 0.12 std. resid.). It is not endemic (v. 64%), does not occur on islands (v. 57%), and has not been introduced (v. 2.1%). It is often accidentally transported with produce shipments, so its range may have been inadvertently expanded. It has not been evaluated for conservation status. Some Paraguayan residents consider it poisonous. In other respects *Notomabuya* is likely to be similar to other skinks.

Notoscincus

Notoscincus are two species of very small (0.78 g v. 3.3 g), diurnal (100% v. 74%), direct-basking (100% v. 21%), conspicuously wary (100% v. 11%), oviparous (100% v. 64%), litter-cryptozoic skinks (100% v. 28%) found in northern Australia.

They are even more slender than an average skink (shape index –0.68 v. –0.46 std. resid.). Their breakable (100% v. 73%) tails are a bit longer than average (1.63 v. 1.33 × SVL). Both species are marginally elongate (v. 35% elongate). Sex differences are not known. One is a generalized invertebrate eater (v. 84%). The other is an obligatory ant specialist (v. 1.6%); it captures the ants by both sit-and-wait (v. 16%) and active (v. 92%) foraging. Both species may use burrows for their daily refugia. They are overrepresented in littoral (50% v. 21%), sandy (50% v. 25%), and riparian (100% v. 23%) areas; they are absent from human habitat (v. 14%).

Clutch count is high (0.66 v. 0.06 std. resid.), based on *ornatus*. Endemism is about average (50% v. 64%). *N. ornatus* is found on at least four islands (v. 57%). Neither has been introduced (v. 2.1%). Both are imperiled (v. 47%), but neither is on one of the IUCN's lists of the most endangered (v. 14%). In other respects *Notoscincus* are much like other skinks.

Notoscincus spp.	English name		Size		Range core
	Activity	Diet	mm	g	Ecological business model
butleri	Lined Soil-crevice Skink				Western Australia
	diurnal	invert.	39	1.06	** litter cryptozoid
ornatus	Ornate Soil-crevice Skink				Australia
	diurnal	ants	36	0.57	* litter cryptozoid

Nucras

Nucras (10 spp.) is a genus of short-snouted, large (8.17 g v. 5.10 g), secretive, crepuscular (100% v. 4%), terrestrial lacertids found in southern Africa from Angola and Kenya south to South Africa. Most are wide-ranging chemosensors (71% v. 25%); the others are a refugia-anchored diurnal ground lizard (14% v. 57%) and a sand shark (14% v. 1%).

This genus presents a mystery. Most observers describe these comparatively large and colorful lizards as secretive and rarely seen. Yet the two movement-quantified species are some of the most active of all lizards. For example, *intertexta* is in the 97th percentile for MPM and the 93rd percentile for PTM. *N. tessellata* is the 92nd percentile for MPM and 87th for PTM. How can such incessant and wide-ranging behavior be so readily overlooked? One possibility lies with their unusual diel activity pattern. Crepuscularity is rarely seen in lizards in general (1.8%) and lacertids in particular (4.4%). Almost all lacertids are diurnal (95.3%), but all documented *Nucras* are crepuscular. Another factor may be the penchant these lizards have for refuging for long periods of time, that is, without emerging onto the surface. Useful statistics on the proportion of time each lizard species is active during clement weather are unavailable for the vast majority of species, but *Nucras* has been singled out as notable for lengthy periods of inactivity. That too will limit their detectability by herpetologists. Crepuscularity and infrequent activity help explain their "secretiveness," but why might they be crepuscular and inactive for long periods of time? One possibility for their crepuscularity is their dietary preference for scorpions. Fifty percent of the diet of *tessellata* is scorpions; an unquantified but large portion of *intertexta*'s diet is scorpions. How does a normally diurnal lacertid find nocturnal scorpions? Possibly by unearthing them in burrows or under rocks, although burrowing is not reported for *Nucras*: 0% (v. 48%). My guess is that their crepuscularity allows them to capture early-emerging or late-retreating scorpions. Another dietary item that may reward crepuscularity is termites, which mostly fly at night but remain superabundant the following dawn. Of the seven species of *Nucras* with any dietary data, five specify termite eating. Often, when termites are available, the availability of millions of these nutrient-rich insects enables lizards to gorge and thereafter remain inactive for long periods of time (i.e., "secretive"). For example, *caesicaudata* is said to emerge only when termites swarm. This hypothesis suggests that *Nucras* will be found to have unusually capacious stomachs.

Nucras are morphologically distinguishable from an average lacertid only on the basis of size. Though often said to have particularly long tails (2.04 × SVL), that statement is true in comparison to lizards (1.37 × SVL) but not in relation to lacertids (2.06

Nucras spp.	English name	Activity	Diet	Size mm	Size g	Range core	Ecological business model
boulengeri	Boulenger's Scrub Lizard	crepuscular		58	4.58	Tanzania	
caesicaudata	Blue-tailed Sandveld Lizard	crepuscular	invert.	55	3.90	Zimbabwe	* sand shark
holubi	Holub's Sandveld Lizard			66	6.9	South Africa	
intertexta	Spotted Sandveld Lizard	crepuscular	invert.	76	10.3	Botswana	** wide-ranging chemosensor
lalandii	Delalande's Sandveld Lizard	crepuscular	invert.	90	13.3	South Africa	** wide-ranging chemosensor
livida	Karoo Sandveld Lizard	crepuscular	invert.	73	9.1	Namibia	** refugia-anchored diurnal gr. liz.
ornata	Ornate Sandveld Lizard	crepuscular	omnivore	78	10.8	Southern Africa	* wide-ranging chemosensor
scalaris	Scaled Sandveld Lizard	crepuscular		72	8.8	Angola	
taeniolata	Striped Sandveld Lizard	crepuscular	invert.	76	10.1	Southern Africa	* wide-ranging chemosensor
tessellata	Western Sandveld Lizard	crepuscular	invert.	73	9.0	Botswana	** wide-ranging chemosensor

Oedodera sp.	English name	Activity	Diet	Size mm	Size g	Range core	Ecological business model
marmorata	New Caledonia Marbled Gecko	nocturnal		53	2.94	New Caledonia	** nocturnal canopy walker

× SVL). At least one *Nucras* (*caesicaudata*) has a blue tail (10% v. 14%), but juveniles of *boulengeri* may have bright orange tails. *N. lalandii* is elongate (10% v. 0.3%); it inhabits dense grasslands and may be expected to propel itself without use of legs when pressed.

Of the four species characterized for foraging mode, one (*lalandii*) cruises (25% v. 2.4%); the others are active foragers (75% v. 92%). Two species (*boulengeri*, *ornata*) are conspicuously wary (20% v. 11%). Two of three species quantified for primary predator concealment tactic rely on flight (67% v. 22%); the other (*tessellata*) retreats to a crevice-burrow (33% v. 50%). Basking behavior is normal for lacertids, but based on *intertexta* and *tessellata*, they have a very high mean active body temperature (39.1°C v. 33.3°C). Those two are the only thermally documented species, but they are particularly active, for which high temperature is an asset. They also might be getting especially hot to retain heat for crepuscular activity after direct basking is no longer possible. *Nucras* are found at the arid end of the lacertids' rainfall spectrum (56–400 mm/y). The most notable habitat association of *Nucras* is their preference for sand (75% v. 44%).

Reproduction is normal for lacertids, clutch count is average based on eight species (0.05 v. 0.13 std. resid.), and ontogenetic mass range is only slightly elevated (i.e., young are relatively small) based on four species (0.30 v. –0.05 std. resid.). This suggest that RCM should be slightly light, which it is in the one species (*tessellata*) documented (0.188 v. 0.267 ratio). Longevity is normal (0.01 v. 0.19 std. resid.). Endemism is low (20% v. 35%). No species is found on islands (v. 26%); none has been introduced (v. 4.6%). Half are imperiled (v. 56%), though none is found on any of the IUCN's lists of the most endangered (v. 15.5%).

Oedodera

Oedodera marmorata is the sole representative of this diplodactylid genus from New Caledonia. It is a small (2.9 g v. 4.2 g), poorly known (21% v. 28% of cells filled) nocturnal canopy walker. Its tail is not prehensile (v. 26%), it possesses secretory pores (v. 34%), it is oviparous (v. 82%), and it is listed among the most endangered of the IUCN categories (v. 29%). In other respects it is a typical diplodactylid, as far as is known. It should not be confused with *Oedura marmorata*.

Oedura

Oedura is an Australian genus (14 spp.) of large-sized diplodactylids (10.6 g v. 4.15 g). Most are trunk/wall lizards (69% v. 19.5%), which in this case readily shift from trunks in forested areas to walls in rocky areas. The genus also has crack hermits (8%

Oedura spp.	English name				Range core
			Size		
	Activity	Diet	mm	g	Ecological business model
bella	Gulf Marbled Velvet Gecko				Queensland
			76	8.9	
castelnaui	Northern Velvet Gecko				Queensland
	nocturnal	invert.	86	12.9	*** ground-foraging tree lizard
cincta	Inland Marbled Velvet Gecko				Eastern Australia
			89	14.3	* trunk/wall lizard or ground-foraging tree lizard
coggeri	Northern Spotted Velvet Gecko				Queensland
	nocturnal		64	5.2	*** nocturnal ground lizard
filicipoda	Fringe-toed Velvet Gecko				Western Australia
	nocturnal		93	16.1	*** trunk/wall lizard
fimbria	Western Marbled Velvet Gecko				Western Australia
			88	13.6	*** trunk/wall lizard
gemmata	Dotted Velvet Gecko				Northern Territory
	nocturnal		89	14.1	** trunk/wall lizard
gracilis	Gracile Velvet Gecko				Western Australia
	nocturnal		76	9.0	** trunk/wall lizard
jowalbinna	Quinkan Velvet Gecko				Queensland
	nocturnal		59	4.02	*** trunk/wall lizard
luritja	Mereenie Velvet Gecko				Northern Territory
			80	10.4	* trunk/wall lizard
marmorata	Marbled Velvet Gecko				Northern Territory
	cathemeral	invert.	92	15.6	** crack hermit
monilis	Ocellated Velvet Gecko				Queensland
	cathemeral	invert.	79	10.1	**** ground-foraging tree lizard
murrumanu	Limestone Range Velvet Gecko				Western Australia
	nocturnal		90	14.8	* trunk/wall lizard
tryoni	Southern Spotted Velvet Gecko				Queensland
	nocturnal	herbivore	77	9.2	** trunk/wall lizard

v. 4%), ground-foraging tree lizards (15% v. 7%), and nocturnal ground lizards (8% v. 35%). Eighty percent are nocturnal (v. 79%) and 20% cathemeral (v. 13%).

Oedura mostly lack prehensile tails (7% v. 26%), though *filicipoda* has adhesive lamellae on the end of its prehensile tail. About one-fourth of species have minimal (darkening) color change ability (v. 5%). Documented species are m<f in SSD, despite their moderately large size. Other sex differences are not reported. One of four documented species is herbivorous (9.2 g *tryoni*). All males defend space (v. 77%). Arboreal species (85% v. 64%) outnumber terrestrial ones (15% v. 34%). They are underrepresented in sandy areas (0% v. 24%) and overrepresented in riparian ones (15% v. 10%).

They are oviparous (v. 82%). Young are notably large (ontogenetic range −0.90 v. −0.57 std. resid.). *O. castelnaui* has an unusually variable incubation period (58–145 days). RCM is high (0.304 v. 0.225), based on five species. Both communal oviposition (7% v. 2.7%) and parental care (7% v. 3.4%) are reported for this genus.

Longevity is high, even by diplodactylid standards (1.36 v. 0.90 std. resid.), which translates to 21–22 years in *marmorata* and *tryoni*. Species in this genus have not been introduced (v. 2%), and none has been evaluated for imperiled species (v. 81%). In all other documented respects these are typical diplodactylids. Note the confusingly similar diplodactylid genus *Oedodera*, which also has a *marmorata* species.

Oligosoma

Oligosoma is *the* radiation (41 spp.) of native New Zealand skinks (*lichenigera* is found on some islands half way between Australia and New Zealand). They are relatively large (6.9 g v. 3.25 g), and sizes vary widely (1.1–124 g). The modal *Oligosoma* is a refugia-anchored diurnal ground lizard (60% v. 30%), but 20% (v. 28%) are litter cryptozoids, 14% (v. 2%) are nocturnal ground lizards, and 3% each are crack hermits (v. 1.7%) and ground-foraging tree lizards (v. 7%). They are equally diverse in diet, foraging mode, diel activity pattern, predator concealment tactics, thermal behavior, microhabitat, and habitat. With the exception of one New Zealand species (*suteri*) and the non-New Zealand species (*lichenigera*), however, they are consistently viviparous (93% v. 36%).

Their overall shape is closer to that of an average lizard than an average skink (shape index 0.05 v. −0.46 std.

resid.). Tail breakage is low (37% v. 73%), but perhaps underreporting is involved. Elongation is present but underrepresented in *Oligosoma* (22% v. 35%). They do not have blue tails (v. 4%).

The sexes are not known to differ in size, but they occasionally differ in color (2.4% v. 7.6%) and seasonal color (5% v. 6%). Generally, however, they are sexually monomorphic (sex differences rollup 0.07 v. 0.28 on 0 to 6 scale).

Oligosoma stand out for the high frequency of omnivores (53% v. 13%); the remainder are generalized invertebrate eaters (37% v. 84%). Several dietary items stand out: berries are explicitly mentioned as a favored food for *burganae*, *inconspicuum*, and *maccanni*. Some have claimed that skinks may be an agricultural pest on New Zealand strawberries. Spiders are mentioned as a major food for *burganae*, *maccanni*, *polychroma*, and *zelandicum*. Carrion is noted for *fallai*, *oliveri*, and *smithi*. The Lord Howe and Norfolk Islands Skink was once observed to roll an egg until it broke (and then lap its contents). Several species are known to forage in the intertidal zone (*acrinasum*, *suteri*). All three foraging modes are well represented in *Oligosoma*: sit-and-wait (22% v. 16%), cruise (11% v. 2%), and active foraging (89% v. 92%). Electric light foraging has not been reported. Home-range area is small (−0.52 v. −0.07 std. resid.), based on *burganae* (20 m^2), *grande* (154 m^2), *maccanni* (300 m^2), *oliveri* (15 m^2), *polychroma* (28 m^2), and *zelandicum* (15 m^2).

All four diel activity patterns are represented: cathemeral (22% v. 9%), crepuscular (8% v. 2.4%), diurnal (62% v. 74%), and nocturnal (8% v. 14.5%). At least six of the reported patterns are contested. This may reflect the high latitude of some of these species and the likelihood that

Oligosoma spp.	English name		Size		Range core
	Activity	Diet	mm	g	Ecological business model
acrinasum	Fiordland Skink				New Zealand
	diurnal	invert.	72	7.6	*** refugia-anchored diurnal gr. liz.
aeneum	Copper Skink				New Zealand
	crepuscular	invert.	54	1.76	*** litter cryptozoid
alani	Robust Skink				New Zealand
	nocturnal		112	13.1	*** nocturnal ground lizard
burganae	Burgan Skink				New Zealand
	diurnal	omnivore	57	3.45	** refugia-anchored diurnal gr. liz.
chloronoton	Green Skink				New Zealand
	diurnal		99	20.6	
fallai	Falla's Skink				New Zealand
	cathemeral	omnivore	114	33.0	** refugia-anchored diurnal gr. liz.
grande	Grand Skink				New Zealand
	diurnal	omnivore	96	19.0	*** refugia-anchored diurnal gr. liz.
hardyi	Hardy's Skink				New Zealand
	crepuscular		52	1.64	*** litter cryptozoid
homalonotum	Chevron Skink				New Zealand
	diurnal		114	32.6	
inconspicuum	Cryptic Skink				New Zealand
	diurnal	omnivore	59	3.82	* refugia-anchored diurnal gr. liz.
infrapunctatum	Speckled Skink				New Zealand
	diurnal		95	8.4	* refugia-anchored diurnal gr. liz.
judgei	Barrier Skink				New Zealand
			74	8.3	
levidensum	Slight Skink				New Zealand
	diurnal		46	1.13	** litter cryptozoid
lichenigera	Lord Howe and Norfolk Islands Skink				Australia
	cathemeral	invert.	71	3.83	* crack hermit
lineoocellatum	Spotted Skink				New Zealand
	diurnal	omnivore	74	4.20	** refugia-anchored diurnal gr. liz.
longipes	Northern Long-toed Skink				New Zealand
	diurnal		63	4.85	** refugia-anchored diurnal gr. liz.
maccanni	McCann's Skink				New Zealand
	diurnal	omnivore	60	4.13	*** refugia-anchored diurnal gr. liz.

Oligosoma spp.	English name				Range core
			Size		
	Activity	Diet	mm	g	Ecological business model
macgregori	McGregor's Skink				New Zealand
	crepuscular		94	17.6	** litter cryptozoid
microlepis	Small-scaled Skink				New Zealand
	diurnal		59	4.01	** refugia-anchored diurnal gr. liz.
moco	Moko Skink				New Zealand
	diurnal	invert.	64	5.2	**** refugia-anchored diurnal gr. liz.
nigriplantare	Chathams Skink				New Zealand
	diurnal	invert.	67	6.03	** refugia-anchored diurnal gr. liz.
northlandi	Northland Skink				New Zealand
			124	42.7	
notosaurus	Southern Skink				New Zealand
	diurnal		74	8.0	** refugia-anchored diurnal gr. liz.
oliveri	Marbled Skink				New Zealand
	cathemeral	omnivore	88	14.4	**** nocturnal ground lizard
ornatum	Ornate Skink				New Zealand
	cathemeral		67	3.22	** litter cryptozoid
otagense	Otago Skink				New Zealand
	cathemeral	omnivore	115	34.0	*** refugia-anchored diurnal gr. liz.
pikitanga	Sinbad Skink				New Zealand
	diurnal		81	11.1	
polychroma	Northern Grass Skink				New Zealand
	diurnal	omnivore	60	4.00	** refugia-anchored diurnal gr. liz.
repens	Eyres Skink				New Zealand
			54	2.92	
roimata	Aorangi Skink				New Zealand
	cathemeral	invert.	56	3.29	*** litter cryptozoid
smithi	Shore Skink				New Zealand
	cathemeral	omnivore	64	5.1	** refugia-anchored diurnal gr. liz.
stenotis	Small-eared Skink				New Zealand
	diurnal		63	4.85	** refugia-anchored diurnal gr. liz.
striatum	North Island Striped Skink				New Zealand
	diurnal		64	5.0	** ground-foraging tree lizard
suteri	Egg-laying Skink				New Zealand
	nocturnal	invert.	77	9.0	*** nocturnal ground lizard

they switch activity time depending on the season. Up to eight *oliveri* have been seen simultaneously lapping water from the same pool at night. Twenty-two percent are conspicuously wary (v. 11%). Sixty percent (v. 9%) rely on crypsis as their primary predator concealment tactic; 40% (v. 43%) retreat to a crevice or burrow. The majority (71% v. 21%) bask directly, but 10% (v. 1.5%) are thermoconformers. Mean active body temperature is a very cool 20.8°C (v. 31.1°C). In *whitakeri*, pregnant females are known to bask directly, but other *whitakeri* may not bask. At least 30% (v. 24%) of species use burrows for daily refugia.

Social relations are diverse: 17% (v. 8%) live solitarily; 50% (v. 16%) live in groups of more than family members. Male spatial tolerance is the third highest of any scincid genus (27% v. 5.6%). *O. otagense* may mate guard. Male spatial intolerance is still the modal pattern (73% v. 94%). Male hierarchies have been noted in 9% (v. 5.6%). At least three species (*burganae*, *inconspicuum*, and *polychroma*) have been observed defending space against other skink species, suggesting either a deficiency in species recognition signals or behaviorally mediated interspecific competition.

Oligosoma are found at the wet end of the skink rainfall spectrum (>400 mm/y; intermediate through wet). They are overrepresented in the intermediate (75% v. 34%) and wet (17% v. 7%) bands. Their elevational profile is normal (l/m/h: 94/44/6 v. 89/35/6). Among tabulated habitat types, they are overrepresented only in littoral areas (56% v. 21%). They are average in sandy areas, riparian zones, and human habitats.

Litter counts are slightly under average (–0.16 v. 0.06 std. resid.). Their ontogenetic range of masses is slightly small (−0.38 v. 0.12 std. resid.), indicating comparatively large neonates. See the Scincidae entry for evidence that this might be associated with their viviparity. Incubation is slightly long (0.15 v. −0.40 std. resid.), based on the two oviparous species, presumably a reflection of their cold environments. Communal oviposition is found in *suteri* (2.4% v. 2.6%).

Longevity is long (0.91 v. 0.17 std. resid.), likely associated with short active seasons at high latitude. The longest are *otagense* (44 years), *grande* (30 years), and *alani* (20 years). The two shortest maxima are *moco* (3 years) and *polychroma* (6 years). The population density of *Oligosoma* is not only exceptionally high (2.27 kg/ha v. 0.35 kg/ha; within skinks, second only to the single species documented in *Carlia*) but exceptionally well sampled (11 species, 25 estimates). The three densest species are *oliveri* (45.0 kg/ha; $n = 1$), *polychroma* (11.8 kg/ha; $n = 4$), and *lineoocellatum* (7.0 kg/ha; $n = 1$). The lowest three are *ornatum* (0.03 kg/ha; $n = 1$), *otagense* (0.50 kg/ha; $n = 5$), and *grande* (0.51 kg/ha; $n = 5$). The latter two are endangered species and therefore particularly challenging and worthy to sample.

All species are island (v. 57%) endemics (v. 64%). None has been introduced (v. 2.1%), except for conservation translocations to repatriate extirpated populations. Most *Oligosoma* are imperiled (88% v. 47%), including 7.7% (v. 14%) placed on one of the IUCN's lists of the most endangered. In other respects *Oligosoma* are typical skinks.

The diversity of ecologies within *Oligosoma* suggests a classical divergent radiation rather that speciation triggered by population discontinuities (e.g., many substrate-specific Australian lizards). It may offer some perspective for comparison to better-studied radiations such as those of Caribbean anoles.

Oligosoma spp.	English name		Size		Range core
	Activity	Diet	mm	g	Ecological business model
taumakae	Taumaka Skink				New Zealand
	diurnal		82	11.5	** refugia-anchored diurnal gr. liz.
tekakahu	Te Kakahu Skink				New Zealand
	diurnal		64	5.2	** refugia-anchored diurnal gr. liz.
toka	Nevis Skink				New Zealand
			59	4.02	* refugia-anchored diurnal gr. liz.
townsi	Hauraki Skink				New Zealand
	nocturnal		77	4.77	*** nocturnal ground lizard
waimatense	Scree Skink				New Zealand
	diurnal		99	20.6	** refugia-anchored diurnal gr. liz.
whitakeri	Whitaker's Skink				New Zealand
	cathemeral	invert.	86	13.0	** nocturnal ground lizard
zelandicum	Glossy Brown Skink				New Zealand
	diurnal	invert.	58	3.70	** litter cryptozoid (if litter) or refugia-anchored diurnal gr. liz.

Omanosaura spp.	English name		Size		Range core
	Activity	Diet	mm	g	Ecological business model
cyanura	Blue-tailed Oman Lizard				Oman
	diurnal	invert.	52	3.38	* wide-ranging chemosensor
jayakari	Jayakar's Oman Lizard				Oman
	diurnal	omnivore	145	69	*** wide-ranging chemosensor

Omanosaura

Omanosaura (two spp.) is a genus of large (15.2 g v. 5.1 g), wide-ranging scansorial lacertids in Oman and the United Arab Emirates. Though the average size in the genus is moderately large, one species is small and the other large. Their tails are long (2.57 v. 2.06 × SVL) and breakable (v. 92%); in one species (*cyanura*) tails are blue, especially in juveniles. The larger species (*jayakari*) includes both fruit and leaves in its diet. It forages by both sit-and-wait and active foraging, whereas the smaller species is reported to use only active foraging. Both species range from sea level to fairly high (*cyanura* to 2120 m; *jayakari* to 2750 m). They are associated with rocky and riparian habitats (100% v. 28%). Clutches are slightly small (−0.20 v. 0.13 std. resid.), and young are fairly large (ontogenetic mass range, based on *jayakari* −0.67, the third lowest in Lacertidae, v. −0.05 std. resid.). Longevity is notably long (0.71 v. 0.19 std. resid.), which equates to 17 years in *jayakari*, the only quantified species. Neither species is endemic (v. 35%), found on an island (v. 26%), introduced (v. 4.6%), or imperiled (v. 56%). In other respects they are probably typical lacertids.

Ophidiocephalus

Ophidiocephalus taeniatus is the sole species in a genus of average-sized (4.7 g v. 5.2 g) cathemeral (v. 29%) pygopodids from a tiny area in the very center of Australia (straddles the border between the Northern Territory and South Australia). The location is so remote that no specimens were found between the original description in 1897 and 1978. It is a sand swimmer, but it will also refuge in deep soil cracks. Its tail is short for a pygopodid (1.5 v. 2.7 × SVL). Like most pygopodids it calls spontaneously (v. 67%). It will hide from predators by diving into sand (v. 50%). It basks directly (v. 17%). It does not lay communally (v. 6.5%) or show parental care (v. 2.2%). It is not endemic (v. 52%), does not occur on islands (v. 26%; any islands near Coober Pedy?), and has not been introduced (v. 0%), but it is considered imperiled (v. 88%). It is not, however, on any of the IUCN's lists of the most endangered (v. 6%). In other respects *Ophidiocephalus* is similar to other pygopodids.

Ophiodes

Ophiodes (five spp.) are functionally limbless (stylar hind limb flaps), large (11.0 g v. 6.3 g) terrestrial diploglossids found in southern South America. Their ecological business models are highly uncertain, but two species appear to be refugia-anchored diurnal ground lizards (v. 12%). Because they are elongate (100% v. 71%), they are slender in build (shape index −1.96 v. −0.98 std. resid.). Their longish tails (1.43 v. 1.27 × SVL) are all breakable (v. 98%). *O. intermedius*, *striatus*, and *vertebralis* are colloquially known as glass snakes for the ease of tail loss. No sex differences have been reported (0 v. 0.28 on 0 to 6 scale). All known species are invertebrate eaters (v. 73%), though *vertebralis* is reported to eat some leaves. They are cathemeral (67% v. 25%) or diurnal (33% v. 70%), though *striatus* is suspected of being more crepuscular than diurnal. They are not known to vocalize (v. 0%), bask (v. 8%), or be conspicuously wary (v. 16%). They are overrepresented in littoral (20% v. 7%) and riparian (60% v. 18%) areas. As far as is known they are all viviparous (v. 86%), with large or normal litter counts (1.11 v. 0.86 std. resid.). They stand out for having exceptionally small young (ontogenetic size range 3.14 v. 2.13 std. resid.), the most extreme value in the first-ranked family. Parental care has not been observed (v. 8%). They have not been introduced (v. 0%), nor are they imperiled (v. 51%). In other respects these are typical diploglossids.

Ophidiocephalus sp.	English name				Range core
			Size		
	Activity	Diet	mm	g	Ecological business model
taeniatus	Bronzeback				Central Australia
	cathemeral	invert.	95	4.72	**** litter cryptozoid

Ophiodes spp.	English name				Range core
			Size		
	Activity	Diet	mm	g	Ecological business model
fragilis	Fragile Worm Lizard				Paraguay
			160	14.2	
intermedius	Middle Worm Lizard				Bolivia
	cathemeral	invert.	201	27.2	
luciae	Lucy's Worm Lizard				Paraguay
			122	6.5	
striatus	Striped Worm Lizard				Paraguay
	diurnal	invert.	213	3.40	* refugia-anchored diurnal gr. liz.
vertebralis	Jointed Worm Lizard				Uruguay
	cathemeral	invert.	178	19.2	* refugia-anchored diurnal gr. liz.

Ophiomorus

Ophiomorus is a moderate radiation (11 spp.) of average-sized (3.15 g v. 3.25 g), functionally limbless (91% v. 35% elongate, all with reduced limbs), sand-hiding (100% v. 22%) skinks found from Greece and Israel to Turkmenistan and India. Most are sand sharks (70% v. 7%); the remainder are subterranean foragers (30% v. 12%). Blue tails are common among the sand sharks (27% v. 4%). *Ophiomorus* are nearly equally split between cathemeral (43% v. 9%) and nocturnal (57% v. 15%) diel activity patterns. Somewhat unexpectedly, these nocturnal/cathemeral cryptozoic lizards are mostly viviparous (86% v. 36%), though the reproductive mode of five species is contested. Viviparity is a trait one associates with high-elevation/high-latitude basking lizards that can transport the embryos into the sunshine to expedite development. However, no *Ophiomorus* is reported to bask (v. 21%). This may reflect the high ambient temperatures present in this part of the world during the breeding season (spring–summer). By tunneling to the optimal depth in sand, they may be able to convey the embryos to the optimal developmental temperature, whereas eggs are fixed at the depth where they are laid. The mobile incubator advantage of viviparity may apply to many subterranean species (table 3.9, trait 43).

Like most limbless lizards, *Ophiomorus* are extremely slender (shape index –1.26 v. –0.46 std. resid.). Tail lengths are short (0.85 v. 1.33 × SVL); this length is more in keeping with sand sharks (table 4.1, trait 5) than with subterranean foragers. Tails are documented to break in about half of the species (55% v. 73%); this is consistent with very short subterranean tails often being unbreakable.

Mean SSD is m<f: –0.80 (v. –0.31) on –2 to 2 scale. Larger female size may be associated with the exceptionally large size of the neonates (third paragraph below). No other sex differences are known.

Ophiomorus are generalized invertebrate eaters (100% v. 84%); ants and termites are often mentioned. No information is available on foraging mode or home-range size. More than a quarter of the species are conspicuously wary (27% v. 11%).

Social relations are little known, but males of *persicus*, *raithmai*, and *tridactylus* (the latter contested) are said to defend space (100% v. 94%). All documented species are found solely at low elevation (100% v. 89%). They are overrepresented in sandy areas (60% v. 23%) and absent from riparian (v. 23%) and human (v. 14%) habitats.

Clutch/litter count is slightly elevated (0.38 v. 0.06 std. resid.). Based on *punctatissimus*, *Ophiomorus* has the family's third-smallest ontogenetic ranges of masses (–1.31 v. 0.12 std. resid.). This suggests particularly large neonates. Combined with the evidence on elevated litter count, this suggests size pressure on the females, which is consistent with the observed m<f SSD. Incubation duration is about average (–0.41 v. –0.40 std. resid.). Parental care is not reported (v. 3.2%).

Longevity is very close to the lizard average (–0.02 v. 0.17 std. resid.), equating to 5.8 years (*tridactylus*) and 3.5 years (*latastii*). It is unlikely that the greater longevity of *tridactylus* is responsible for this species being sold as an aphrodisiac and a medicinal. Endemism in the genus is low (27% v. 64%), despite their likely substrate specializations. Island species are very low (9% v. 57%). None has been introduced (v. 2.1%). *Ophiomorus* are imperiled at the usual frequency (43% v. 47%). None is on the IUCN's lists of the most endangered (v. 14%). In all other documented respects *Ophiomorus* are typical skinks.

Ophiomorus spp.	English name				Range core
			Size		
	Activity	Diet	mm	g	Ecological business model
blanfordi	Makran Sand-swimmer				Pakistan
	nocturnal	invert.	78	2.95	** sand shark
brevipes	Irano-Baluchistan Sandfish				Afghanistan
	nocturnal	invert.	81	3.23	**** sand shark
chernovi	Chernov's Snake Skink				Southern Caspian Sea
			81	3.23	
latastii	Lataste's Legless Skink				Syria
	cathemeral	invert.	89	1.88	*** subterranean forager
maranjabensis	Maranjab Snake Skink				Iran
			72	3.89	*** subterranean forager
nuchalis	Plateau Snake Skink				Iran
	nocturnal		90	4.22	* sand shark
persicus	Persia Snake Skink				Iran
			69	2.16	* sand shark
punctatissimus	Greek Snake Skink				Greece
			81	3.23	*** subterranean forager
raithmai	Three-fingered Sandfish				India
	cathemeral	invert.	80	3.15	*** sand shark and subterranean forager
streeti	Street's Sand Skink				Iran
	nocturnal	invert.	88	3.98	* sand shark
tridactylus	Three-toed Sandfish				Pakistan
	cathemeral	invert.	85	3.64	*** sand shark

Ophioscincus

Ophioscincus is a small genus (three spp.) of small-sized (1.40 g v. 3.25 g), functionally limbless, subterranean forager (100% v. 12%) skinks found along the east coast of Australia. As with most subterranean foragers, their shape index is exceptionally low (–1.87 v. –0.46 std resid.). Their tails are short for skinks (0.80 v. 1.33 × SVL), but not as short as an average subterranean forager (0.37 × SVL; table 4.1, trait 5). Their tails are also breakable (100% v. 73%); short subterranean tails are often resistant to breakage. No sex differences are known. All eat generalized invertebrates (100% v. 84%); termites are specified foods in two of three. They are overrepresented in littoral (67% v. 21%) and sandy (67% v. 25%) habitats; they are absent from riparian (v. 23%) and human (v. 14%) habitats. All are oviparous (v. 64%) and have clutch counts commensurate with their size or slightly more numerous (0.27 v. 0.06 std. resid.). Endemism (67% v. 64%) and insularity (67% v. 57%) are at the usual level. None has been introduced (v. 2.1%). All are imperiled (v. 47%), but none is on one of the IUCN's lists of the most endangered (v. 14%). In other respects *Ophioscincus* are probably typical limbless skinks.

Ophioscincus spp.	English name				Range core
			Size		
	Activity	Diet	mm	g	Ecological business model
cooloolensis	Cooloola Snake-skink				Queensland
		invert.	63	0.86	*** subterranean forager
ophioscincus	Yolk-bellied Snake-skink				Queensland
		invert.	83	1.59	**** subterranean forager
truncatus	Short-limbed Snake-skink				Queensland
		invert.	67	2.00	*** subterranean forager

Ophisaurus

Ophisaurus is a small (five spp.) genus of medium-large (19.1 g v. 10.1 g), slender (shape index –1.4 v. –0.6) limbless anguids from low elevations (v. 47%) of the southeastern United States (Nebraska, Wisconsin, and Virginia southward) and Mexico. Their common name reflects the fragility of their tails, which are long (2.27 v. 1.66 × SVL) but not prehensile (v. 31%). *O. compressus* does not lose its tail readily or regenerate it well. Despite their residence in well-studied areas, taxonomic confusion has greatly limited the amount of natural history information available.

The sexes often differ in color (60% v. 26%), but seasonal color is not reported (v. 3.8%). Three of four are m>f in relative size. All are considered to be active foragers (v. 71%), and a remarkable 75% (v. 13%) are crepuscular; the other 25% are diurnal (v. 70%). Sixty percent (v. 18%) are judged to be conspicuously wary, an attribute perhaps associated with their active foraging. Forty percent (v. 15%) bask directly, perhaps dictated by their high latitude. Males are mostly tolerant (75% v. 54%), and only 25% defend space (v. 46%). They are cryptozoic (50% v. 14%) or terrestrial (v. 53%). All dig burrow refugia (v. 36%). They are often found in littoral areas (60% v. 9%), associated with sandy soils (100% v. 15%), and present in human habitats (40% v. 11%). As with other anguids, they prefer mesic areas, and rainfall triggers activity in many *Ophisaurus*.

Ophisaurus are all oviparous (v. 39%), despite their high latitude. Clutches are above average in number of young (1.12 v. 0.85 std. resid.) and smallness of young (ontogenetic range of masses 1.37 v. 0.62 std. resid.). The latter indicates that young average 1/174 of maximum mass (v. 1/72). Sixty percent have exhibited parental care (v. 17%). Unlike many lizards' parental care, these mothers remain with the clutch throughout incubation and, in the case of *attenuatus*, have been observed to raise the temperature of the

Ophisaurus spp.	English name				Range core
			Size		
	Activity	Diet	mm	g	Ecological business model
attenuatus	Slender Glass Lizard				Southeastern United States
	crepuscular	invert.	220	36.0	** wide-ranging chemosensor
ceroni	Ceron's Glass Lizard				Mexico
			161	14.5	
compressus	Island Glass Lizard				Georgia (state)
	crepuscular	invert.	165	15.3	
mimicus	Mimic Glass Lizard				Southeastern United States
	diurnal	invert.	131	7.9	
ventralis	Eastern Glass Lizard				Southeastern United States
	crepuscular	invert.	229	40.1	

clutch through shivering thermogenesis, as do some pythons.

Population density is moderately high by mainland lizard standards (2.36 kg/ha), but low by anguid standards (v. 9.22 kg/ha), though both of these estimates strike me as selectively obtained rather than representative. Sixty percent of species live on at least one island (v. 17%), one (*ventralis*) has been introduced (20% v. 3.8%), and 60% (v. 63%) are imperiled. Twenty percent (v. 28%) are on the IUCN's lists of the most endangered.

Ophisops spp.	English name				Range core
			Size		
	Activity	Diet	mm	g	Ecological business model
beddomei	Beddome's Snake-eyed Lizard				India
	diurnal	invert.	36	1.11	** refugia-anchored diurnal gr. liz.
elbaensis	Mount Elba Snake-eyed Lizard				Egypt
	diurnal		34	0.99	* refugia-anchored diurnal gr. liz.
elegans	Elegant Snake-eyed Lizard				Old World Deserts
	diurnal	invert.	52	3.37	** refugia-anchored diurnal gr. liz.
jerdonii	Jerdon's Snake-eyed Lizard				Pakistan
	diurnal	invert.	44	2.09	*** refugia-anchored diurnal gr. liz.
leschenaultii	Leschenault's Snake-eyed Lizard				India
	diurnal	invert.	46	2.36	**** refugia-anchored diurnal gr. liz.
microlepis	Small-scaled Snake-eyed Lizard				India
	diurnal	invert.	56	4.10	**** refugia-anchored diurnal gr. liz.
minor	Dwarf Snake-eyed Lizard				India
	diurnal		46	2.31	* refugia-anchored diurnal gr. liz.
occidentalis	Mograbin Snake-eyed Lizard				Sahara
	diurnal		46	2.24	* refugia-anchored diurnal gr. liz.

Ophisops

Ophisops (eight spp.) is a lacertid genus of small (2.11 g v. 5.10 g), burrow-using (80% v. 48%), refugia-anchored diurnal ground lizards (100% v. 57%) found in Old World aridlands from Morocco east to Bulgaria, Afghanistan, and India and south to Sri Lanka). Morphologically they are typical lacertids, though they are slightly more likely to have secretory pores (12.5% v. 4%), sexes that differ in color (38% v. 33%), and seasonal color (25% v. 15%). None has a blue tail, though juveniles of *microlepis* may have bright red tails. Like most lacertids they are active foragers (100% v. 92%). *O. elegans* is notable for exhibiting frequent movement: 97th percentile for MPM and 90th percentile for PTM. Some *Ophisops* are conspicuously wary (25% v. 11%). The genus exhibits a diversity of typical lacertid predator concealment tactics: 40% flee to crevices or burrows (v. 50%), 40% flee to vegetation (v. 42%), and 20% give flight (v. 22%).

Direct basking is documented at typical frequency (25% v. 25%), but thermoconformance is overrepresented (13% v. 0.6%), a somewhat unusual condition for active foragers (table 3.6, trait 28). Mean body temperature is known for only *elegans* (30.0°C v. 33.3°C).

O. jerdonii is an outlier in that it is often subterranean or cryptozoic. Adult *jerdonii* are rare after May, suggesting either a long dry-season aestivation or very short longevity. This seems unusual for a lizard that lives from sea level to 3000 m and therefore must adapt to a wide variety of environmental conditions. The genus as a whole does not exhibit any strong signal as regards special habitat preferences.

Clutch count is the fourth highest in Lacertidae (0.65 v. 0.13 std. resid.). This fits well with the small size of offspring (ontogenetic mass range 0.71, the fourth highest in Lacertidae, v. −0.05 std. resid.). Mortality of *Ophisops* is probably high and relatively independent of size. Longevity is a trifle short (5 years) in *elegans*, the only aged species (0.04 v. 0.19 std. resid.). Endemism is low (13% v. 35%). Species occur on islands at the usual rate (38% v. 26%). None has been introduced (v. 4.6%). A low number are imperiled (29% v. 56%). None of the imperiled species are on any of the IUCN's lists of the most endangered (v. 15.5%).

Opluridae

Opluridae is a very small family (eight spp.) of two former iguanid genera found exclusively in Madagascar and nearby islands: *Chalarodon* and *Oplurus*. The salient ecological feature of the family is its omnivory (100% v. 15%). All are also diurnal (v. 67%). Refugia-anchored diurnal ground lizards are overrepresented (50% v. 24.5%), but *Oplurus* also has two ground-foraging tree lizards (25% v. 9%) and two trunk/wall lizards (25% v. 13%). In part because this is a curiously disjunct group of the former Iguanidae—no others are found in Africa, Asia, or Australia—the family has received some research attention (filled cells 34% v. 27%).

Oplurids are large (19.8 g v. 5.3 g) and robust (shape index 0.70 v. 0.0 std. resid.), though this generality conceals important differences among the two genera: *Chalarodon* is small (4.3 g v. 19.8 g) and skinny (shape index −0.07 v. 0.70); *Oplurus* is large (33.0 g v. 19.8 g) and hefty (shape index 0.96 v. 0.70). Oplurids have low frequencies of tail breakage (12.5% v. 57%) but tails of typical length (1.44 v. 1.37 × SVL) that are not prehensile (0% v. 5.0%). Oplurids usually lack crests (25% v. 5.3%); all oplurid crests are found in *Chalarodon*. All species lack secretory pores (v. 15%), elongation (v. 11%), color change (v. 11%), and blue tails (v. 2.3%).

The mean SSD in oplurids is m>f, which might superficially appear related to their large size, but this feature is documented only in *Chalarodon*. Some species, only in *Oplurus*, have sexual dichromatism (family mean 12.5% v. 21%); seasonal color is present in both genera (family mean 25% v. 6.5%). Overall their sex differences rollup is exactly average (0.75 v. 0.75 on 0 to 6 scale).

Foraging mode is documented only in *Oplurus*: 67% (v. 46%) sit-and-wait; 33% (v. 58%) active. Electric light use is not reported (v. 1.4%). Home-range area, documented solely in *Oplurus*, is very small: −1.31 (v. 0.0 std. resid.), equating to 30–100 m^2.

Vocalizations (v. 2.5%) and wariness (v. 8.2%) are not reported. Oplurids rely primarily on retreat to a crevice or burrow for primary predator concealment (83% v. 25%). Some species also use flight (17% v. 11%). Direct basking is well represented (50% v. 18%) and is presumably responsible for the high active body temperature (35.3°C, third-highest family, v. 31.1°C), a remarkable value for largely sedentary lizards. This warrants further documentation, as it may account for the family's restriction to lower elevations (see below). Oplurids, documented only in *Oplurus*, live in groups (100% v. 20%), but this does not stop the males from defending space (100% v. 85%). Hierarchies are not reported (v. 7.7%).

Oplurids are split between arboreality (50% v. 32%, though only in *Oplurus*) and terrestriality (50% v. 48%). In keeping with this terrestriality, 40% (v. 22%) of species sometimes refuge during the night in burrows. Habitat documentation in relation to rainfall is limited to *C. madagascariensis*, which is found only in the intermediate rainfall belt (401–1100 mm/y). The mix of elevations represented is skewed toward lower elevations (l/m/h: 100/20/0 v. 82/41/13). Oplurids are slightly overrepresented in littoral (25% v. 17%), sandy (25% v. 22%), and human habitats (25% v. 18.5%) but absent from riparian zones (v. 22%).

All known oplurids are oviparous (v. 82%) with small clutch counts (−0.62 v. 0.0 std. resid.) and short incubation (−0.74 v. 0.0 std. resid.), albeit the latter based on just two *Oplurus*. Young are unusually small, again based on two *Oplurus*, as indicated by the elevated ontogenetic mass range (0.78 v. 0.0 std. resid.). Thus, this family does not represent a tradeoff between egg size and number, but has few small eggs. Eggs are not adherent (v. 2.4%) but may be communally deposited (12.5% v. 4.2%). Parental care is not reported (v. 2.9%). RCM is known for only *O. cuvieri*, which is low-normal (0.204 v. 0.226).

Longevity is based on just two *Oplurus* but is long (0.51 v. 0.0 std. resid). As expected for Madagascar species, endemism is high (88% v. 64%). *Oplurus cuvieri* is also found on the nearby Comoro Islands; it is unclear whether this distribution is due to natural or human-aided dispersal. No oplurids are imperiled (v. 49%), despite their Madagascar homeland.

Oplurus

Oplurus (six spp.) is a genus of large (33 g v. 19.8 g) Madagascar oplurids of diverse ecological business models; equal numbers (2 spp.) are ground-foraging tree lizards, refugia-anchored diurnal ground lizards, and trunk/wall lizards. They are heavily built (shape index 0.96 v. 0.70 std. resid.), and their normal-length tails (1.39 v. 1.44 × SVL) do not break (v. 12.5%). They lack crests (v. 25%). Sexual dichromatism (17% v. 12.5%) and seasonal color (17% v. 6.5%) are known in *saxicola*. Overall their sex differences rollup is modest (0.33 v. 0.75 on 0 to 6 scale). Their mix of foraging modes matches the family (67% sit-and-wait; 33% active). The largest species (*cuvieri*) is notable for its slow, infrequent movements: 9th percentile for MPM and 6th percentile for PTM. Home-range area is small (−1.31 v. −1.31 std. resid.), based on *cuvieri* (30 m^2 in males) and *quadrimaculatus* (>100 m^2 in males). The only primary predator concealment tactic documented (five spp.) is retreat to a crevice or burrow (v. 83%). Group living is reported in *saxicola*. Male spatial defense (undefined) is reported in four species. Four species are arboreal (v. 50%) and two are terrestrial (v. 50%). One of four documented species may use a burrow as a nightly refuge (25% v. 40%). Elevations follow the family pattern (100% low, 33% v. 20% medium). Use of sandy areas is not reported (v. 25%), but other habitat types follow the family template (25% each for littoral and human habitats; 0% for riparian zones). Clutch count is average (−0.62 v. −0.62 std. resid.). Ontogenetic mass range indicates the same small hatchling size as the family (0.78 v. 0.78 std. resid.). Incubation is short (−0.74 v. −0.74), based on *cyclurus* (72 days) and *quadrimaculatus* (60 days). Longevity is high (0.51 v. 0.51 std. resid.) based on *cuvieri* (21 years) and *cyclurus* (9.3 years). In other respects *Oplurus* follow the family characteristics.

Oplurus spp.	English name				Range core
			Size		
	Activity	Diet	mm	g	Ecological business model
cuvieri	Cuvier's Madagascar Swift				Madagascar
	diurnal	omnivore	128	73	*** ground-foraging tree lizard
cyclurus	Merrem's Madagascar Swift				Madagascar
	diurnal		116	49	** ground-foraging tree lizard
fierinensis	Anzamala Madagascar Swift				Madagascar
	diurnal		83	13.1	** trunk/wall lizard
grandidieri	Grandidier's Madagascar Swift				Madagascar
	diurnal		100	27.3	* trunk/wall lizard
quadrimaculatus	Duméril's Madagascar Swift				Madagascar
	diurnal	omnivore	116	49.7	** refugia-anchored diurnal gr. liz.
saxicola	Marked Madagascar Swift				Madagascar
	diurnal		92	20.1	*** refugia-anchored diurnal gr. liz.

Oriocalotes

Oriocalotes is a single-species genus of small (7.2 g v. 15.6 g) arboreal (v. 43%) agamids from the Himalayan plateau and adjacent uplands: China, India, and possibly Bhutan. It has a weak crest (v. 38%) and spines above its eyes. Regrettably, too little is known about this species to distinguish it ecologically. SSD is m<f (−1.0 v. 0.53 on −2 to 2 scale). Clutch count is average (0.23 v. 0.20 std. resid.). It is not found on an island (v. 36%), and it has not been introduced (v. 1.7%) or evaluated for conservation status.

Oriocalotes sp.	English name				Range core
			Size		
	Activity	Diet	mm	g	Ecological business model
paulus	Khasi Hills Forest Lizard				Himalayas
	diurnal	invert.	61	7.2	

Orosaura

Orosaura nebulosylvestris is the sole member of its genus. It is a large-sized (9.9 g v. 3.3 g), largely unknown (filled cells 15% v. 25%) skink endemic to Venezuela.

Orosaura is more heavily built than an average skink (shape index 0.29 v. −0.46 std. resid.). No sex differences are known. It spans the elevational range 920–2360 m. It does not occur on an island (v. 57%), nor has it been introduced (v. 2.1%). It is not imperiled (v. 47%). In other respects *Orosaura* is likely to be similar to other skinks.

Orosaura sp.	English name				Range core
			Size		
	Activity	Diet	mm	g	Ecological business model
nebulosylvestris	Venezuela Cloud Forest Skink				Venezuela
			79	9.9	

Orraya

Orraya occultus is the sole occupant of this Queensland carphodactylid genus. Despite targeted searches, only a few specimens are known. It is a leaf-tailed gecko with a conspicuously long neck and long digits. The

Orraya sp.	English name				Range core
			Size		
	Activity	Diet	mm	g	Ecological business model
occultus	McIlwraith Leaf-tailed Gecko				Queensland
	nocturnal		96	16.4	** immobility expert

tail is short (0.25 v. 0.54 × SVL). It has secretory pores (v. 10%). It is an immobility expert (v. 47%) associated with large granite boulders in riparian settings (v. 24%). As with many immobility experts, it has a very restricted range. Despite the limited range and apparent rarity, it is not listed as a species of conservation concern (v. 6% imperiled). As far as is known it is otherwise similar to the other family members.

Otocryptis

Otocryptis is minor genus (three spp.) of small (6 g v. 15.6 g), lanky (0.63 v. 0.80 std. resid.), terrestrial (v. 56%) South Asian agamids. They are morphologically distinct in possessing very long spindly legs (hence kangaroo lizards), negligible crests (v. 38%), huge dewlaps (especially *wiegmanni*), and long tails (2.06 v. 1.9 × SVL). All three species are strongly associated with leaf litter, yet they do not have the morphology of litter cryptozoids. A reasonable guess is that they forage above the leaf litter, but they dive into it to conceal themselves when threatened. They are a good example of lizards that do not completely fit into the litter cryptozoid ecological business model, or are misclassified. All three are reported to run bipedally, and hopping seems likely. They are reported to sit-and-wait (v. 88%), but at least one species is omnivorous, requiring at least some active foraging. Two of the three have contradictory accounts of the diel activity cycle, and the third, which is nominally diurnal, is famous for lying spread-eagled on asphalt roads or other hot rocks at night. Thermoregulation in this genus is reported to be direct basking (33% v. 24%), indirect basking, and thermoconforming (33% v. 2%). One species (*wiegmanni*) is reported to break its tail (unusual for agamids), but it is also noted that the tail does not regenerate normally. In summary, these lizards seem fond of breaking rules. Their social relations and reproduction hew to agamid standards.

All three are endemic (v. 49%), none has been introduced (v. 1.7%), and *beddomei* (endangered) and *wiegmanni* (near threatened) are listed as imperiled (100% v. 37%); 50% (v. 5%) are on IUCN lists of the most endangered.

Otocryptis spp.	English name				Range core
			Size		
	Activity	Diet	mm	g	Ecological business model
beddomei	Indian Kangaroo Lizard				India
	cathemeral	invert.	45	2.83	**** litter cryptozoid
nigristigma	Lowland Kangaroo Lizard				Sri Lanka
	diurnal		65	8.8	*** litter cryptozoid
wiegmanni	Sri Lanka Kangaroo Lizard				Sri Lanka
	diurnal	omnivore	65	8.8	** refugia-anchored diurnal gr. liz. or litter cryptozoid

Otosaurus

Otosaurus cumingi is the sole member of its genus. It is a very large (46.1 g v. 3.3 g) litter cryptozoid (v. 28%) skink endemic to the Philippines, from Luzon to Mindanao. Although too little is known to understand this skink's foraging behavior, its size suggests that it might be a predator of normal-sized litter cryptozoids.

Orosaura is more heavily built than an average skink (shape index 0.79 v. −0.46 std. resid.). No sex differences are known. It spans the elevational range 0–1000 m. It is associated with riparian zones (v. 23%). Although its reproductive mode is unknown, it is thought to produce about 2.5 young, which is conspicuously few for its size (−1.15 v. 0.06 std. resid.). It has not been introduced (v. 2.1%) and is not imperiled (v. 47%). In other respects *Otosaurus* is likely to be similar to other skinks.

Otosaurus sp.	English name				Range core
			Size		
	Activity	Diet	mm	g	Ecological business model
cumingi	Cuming's Eared Sphenomorph				Philippines
			127	46.1	** litter cryptozoid

Ouroborus

Ouroborus cataphractus is the sole member of this genus of cordylid lizards and is endemic to South Africa. It is large (22.4 g v. 10.8 g), with a mildly robust build (shape index 0.76 v. 0.35 std. resid.). It is ar-

Ouroborus sp.	English name				Range core
			Size		
	Activity	Diet	mm	g	Ecological business model
cataphractus	Armadillo Lizard				South Africa
	crepuscular	invert.	95	22.4	*** crack hermit

mored and pointy like *Cordylus* with a similar crevice-oriented lifestyle, but it may live in large colonies and has charmed the world by its predilection for biting its own tail and forming a donut when threatened with capture in the open. This habit is presumed to defeat saurophagous snakes that need an end to begin swallowing. Colonies of *O. cataphractus* may number in the hundreds, though it shares its crack with only its mate and young. *Ouroborus* is also reported to feed its neonate (modal litter count of one, equates to –1.26, the smallest in Cordylidae, v. –0.48 std. resid.), an extraordinary form of reptilian parental care. The sexes differ in color (v. 37%), but it lacks seasonal color (v. 28%). It is crepuscular (v. 1.5%), conspicuously wary (v. 27%), resides in low elevation areas (v. 55%), and is long-lived (1.53 v. 0.02 std. resid.; 25 years). Its locomotor activity is subjectively associated with spring rains. In other respects it is a typical cordylid.

Pachycalamus

Pachycalamus brevis is the sole member of its trogonophid genus of medium-large (3.57 g v. 1.78 g), speckled subterranean foragers (100% v. 80%) endemic to Yemen. *Pachycalamus* is of moderate build (shape index –3.01 v. –3.43 std. resid.) and possesses secretory pores (v. 17%). Other sex differences are unknown. It is found at low elevation, in this case 20–700 m, and is associated with sandy (v. 100%) areas. Its clutch/litter count is unknown. It is found on Socotra Island (v. 37% insular). In other respects *Pachycalamus* is a typical trogonophid.

Pachydactylus

Pachydactylus is a major radiation (56 spp.) of small (2.18 g v. 3.31 g) nocturnal gekkonids in southern Africa, from Congo south on the west coast

Pachycalamus sp.	English name				Range core
			Size		
	Activity	Diet	mm	g	Ecological business model
brevis	Yemen Short Worm Lizard				Yemen
			151	3.57	**** subterranean forager

Pachydactylus spp.	English name				Range core
			Size		
	Activity	Diet	mm	g	Ecological business model
acuminatus	Helmeringhausen Gecko				Namibia
	nocturnal		51	2.97	*** nocturnal ground lizard
affinis	Transvaal Gecko				South Africa
	nocturnal		49	2.70	*** nocturnal ground lizard
amoenus	Kamaggas Gecko				South Africa
	nocturnal		35	1.07	* nocturnal ground lizard
angolensis	Angola Large-scaled Gecko				Namibia
			38	1.32	** nocturnal ground lizard
atorquatus	Augrabies Gecko				South Africa
	nocturnal		48	2.52	* nocturnal ground lizard
austeni	Austen's Dune Gecko				South Africa
	nocturnal		39	1.41	*** nocturnal ground lizard
barnardi	Barnard's Gecko				Namibia
	nocturnal		52	3.09	*** nocturnal ground lizard
bicolor	Bicolored Velvety Gecko				Namibia
	nocturnal		38	1.27	*** nocturnal ground lizard
boehmei	Boehme's Namib Gecko				Namibia
	nocturnal		41	1.64	* nocturnal ground lizard
capensis	Cape Gecko				Southern Africa
	nocturnal	invert.	56	3.94	*** nocturnal ground lizard
caraculicus	Angolan Banded Gecko				Namibia
	nocturnal		37	1.21	*** nocturnal ground lizard
carinatus	Richtersveld Gecko				South Africa
	nocturnal		42	1.77	*** nocturnal ground lizard
etultra	Sossus Gecko				Namibia
	nocturnal		41	1.65	* nocturnal ground lizard
fasciatus	Banded Gecko				Namibia
	nocturnal		47	2.36	**** nocturnal ground lizard
formosus	Karoo Gecko				South Africa
	nocturnal		52	3.09	*** nocturnal ground lizard
gaiasensis	Brandberg Gecko				Namibia
	nocturnal		56	3.85	*** nocturnal ground lizard
geitje	Ocellated Gecko				South Africa
	nocturnal	invert.	37	1.20	*** nocturnal ground lizard

Pachydactylus spp.	English name	Activity	Diet	Size mm	Size g	Range core	Ecological business model
goodi	Good's Gecko	nocturnal		45	2.11	South Africa	* nocturnal ground lizard
griffini	Griffin's Gecko	nocturnal		36	1.15	Namibia	* nocturnal ground lizard
haackei	Haacke's Gecko	nocturnal	invert.	73	7.9	Namibia	*** nocturnal ground lizard
katanganus	Katanga Thick-toed Gecko	nocturnal		54	3.39	Congo	* nocturnal ground lizard
kladaroderma	Southern Fragile-skinned Gecko	nocturnal	invert.	70	7.0	South Africa	*** trunk/wall lizard
kobosensis	Baster Gecko	nocturnal		48	2.52	Namibia	* nocturnal ground lizard
labialis	Western Cape Gecko	nocturnal		39	1.37	South Africa	*** nocturnal ground lizard
laevigatus	Fischer's Gecko	cathemeral		74	8.4	Namibia	** trunk/wall lizard
macrolepis	Springbokfontein Gecko	nocturnal		47	2.31	South Africa	** nocturnal ground lizard
maculatus	Large-spotted Gecko	nocturnal	invert.	45	2.14	South Africa	*** nocturnal ground lizard
maraisi	Johan's Gecko	nocturnal		39	1.38	Namibia	** nocturnal ground lizard
mariquensis	Marico Gecko	nocturnal	invert.	48	2.46	Namibia	*** nocturnal ground lizard
mclachlani	McLachlan's Gecko	nocturnal		44	1.92	Namibia	* nocturnal ground lizard
monicae	Monica's Gecko	nocturnal		45	2.1	Namibia	*** nocturnal ground lizard
montanus	Namaqua Mountain Gecko	nocturnal		40	1.56	Namibia	* nocturnal ground lizard
namaquensis	Namaqua Gecko	cathemeral	invert.	72	7.7	South Africa	** nocturnal ground lizard
oculatus	Golden Spotted Gecko	nocturnal	invert.	43	1.91	South Africa	*** nocturnal ground lizard
oreophilus	Kaokoveld Gecko	nocturnal		47	2.41	Angola	*** nocturnal ground lizard
oshaughnessyi	O'Shaughnessy's Gecko	nocturnal		50	2.82	Southern Africa	*** nocturnal ground lizard
otaviensis	Otavi Gecko	nocturnal		40	1.57	Namibia	* nocturnal ground lizard

and Mozambique south on the east. Most are nocturnal ground lizards (91% v. 27%), and the others are trunk/wall lizards (9% v. 45%). Tail breakage (98% v. 87%) and prehensile tails (1.8% v. 1.8%) follow family norms. Toes without expanded toepads are rare (1.8% v. 30%) and secretory pores unreported (v. 46%). The combination of terrestriality and toes possessing expanded toepads is unusual; a test not reported in earlier chapters found a "strong" (*phi* = 0.305) negative association of expanded toepads with terrestriality rather than arboreality (Fisher's exact probability < 0.0001, $n = 1253$). The explanation for this anomaly is likely the reliance of *Pachydactylus* on rock surfaces and rock refugia; climbing smooth rocks is likely aided by possession of expanded toe pads. Color change is present at about half the family frequency (7% v. 15%) and advanced color change at the family norm (1.8% v. 2.1%).

Most *Pachydactylus* have males smaller than females (−0.67 v. −0.03 on −2 to 2 scale), though this is often the only sex difference. The sex rollup score for *Pachydactylus* is thus low: 0.21 v. 0.63 on 0 to 6 scale.

Diet and foraging follow the family pattern. Spontaneous calling occurs at a normal frequency (7.1% v. 6.5%), but stress calls are unknown (0% v. 3.7%). Predator concealment relies primarily on crypsis (100% v. 50%), and no species is reported to be conspicuously wary (v. 4.4%). At least six species are reported to have skin that tears exceptionally easily. Mean active body temperature is a cool 25.3°C (v. 27.6°C), presumably reflecting their latitude and nocturnality. Group living is common (67% v. 25%), although so is spatial defense by males (100% v. 87%).

Pachydactylus are found toward the arid end of the gekkonid rainfall spectrum (dry and arid: 56–400 mm/y). They are conspicuously overrepresented in sandy areas (48% v. 11%) and underrepresented in human habitat (11% v. 27%). Eighty-three percent of species refuge in rock crevices. *P. maculatus* is reported to refuge, communally, in giant snail shells. Why communally? To reduce water loss? *P. rangei* digs burrows up to 500 mm long. *P. vanzyli* also digs burrows, and closes the entrance when it is inside.

Reproduction is similar to that of other gekkonids.

Longevity is relatively short (−0.33 v. 0.15 std. resid.). *P. geitje* has been introduced to the mid-Atlantic island of Ascension (1.8% v. 3.7% insular). A large proportion of *Pachydactylus* are imperiled (64% v. 47%), though none is on any of the IUCN's lists of the most endangered (v. 11.3%). *P. maculatus* is feared as venomous by local residents. In other respects *Pachydactylus* are typical gekkonids.

Pachydactylus spp.	English name				Range core
			Size		
	Activity	Diet	mm	g	Ecological business model
parascutatus	Sesfontein Gecko				Namibia
	nocturnal		46	1.15	*** nocturnal ground lizard
punctatus	Speckled Gecko				Southern Africa
	nocturnal	invert.	35	1.07	*** nocturnal ground lizard
purcelli	Purcell's Gecko				South Africa
	nocturnal		46	2.22	*** nocturnal ground lizard
rangei	Web-footed Gecko				Namibia
	nocturnal	invert.	66	6.1	*** nocturnal ground lizard
reconditus	Cryptic Gecko				Namibia
	crepuscular		40	1.49	** trunk/wall lizard
robertsi	Karasberg Gecko				Namibia
	nocturnal		39	1.38	*** nocturnal ground lizard
rugosus	Rough Thick-toed Gecko				Botswana
	nocturnal	invert.	54	3.47	** trunk/wall lizard
sansteynae	San Steyn's Gecko				Namibia
	nocturnal	invert.	42	1.70	*** nocturnal ground lizard
scherzi	Scherz's Gecko				Namibia
	nocturnal		33	0.86	*** nocturnal ground lizard
scutatus	Common Large-scaled Gecko				Namibia
	nocturnal		39	1.41	*** nocturnal ground lizard
serval	Western Spotted Gecko				Namibia
	nocturnal		41	1.67	*** nocturnal ground lizard
tigrinus	Tiger Gecko				Botswana
	nocturnal		46	2.22	*** nocturnal ground lizard
tsodiloensis	Tsodilo Thick-toed Gecko				Botswana
	nocturnal	invert.	51	2.93	*** nocturnal ground lizard
vansoni	Van Son's Gecko				Mozambique
	nocturnal		48	2.58	*** nocturnal ground lizard
vanzyli	Kaoko Dune Gecko				Angola
	nocturnal	invert.	54	3.54	*** nocturnal ground lizard
visseri	Visser's Gecko				South Africa
	nocturnal		42	1.76	*** nocturnal ground lizard
waterbergensis	Waterberg Gecko				Namibia
	nocturnal		45	2.06	* trunk/wall lizard
weberi	Weber's Gecko				Namibia
	nocturnal	invert.	42	1.78	*** nocturnal ground lizard
werneri	Werner's Gecko				Namibia
	nocturnal		47	2.41	** nocturnal ground lizard

Palleon

Palleon is a tiny (two spp.) genus of smaller than average-sized (1.7 g v. 2.0 g) dwarf chameleons (see table 5.2 in Chamaeleonidae account for group definition) from Madagascar. The species described in 1995, *lolontany*, is known solely from two sleeping females. With only two species, species frequencies are relatively meaningless, but they fit *Palleon* with the dwarf chameleons in size, business model (though evidence is contradictory), shape index (0.33 v. 0.32 std. resid.), tail length (0.50 v. 0.53 × SVL), crests, terrestriality, clutch count (0.40 v. 0.60 std. resid.), and population density (0.03 v. 0.03 kg/ha, though based on only two samples of one species). There is a shockingly large difference between the sexes in relative tail length (m:f ratio: 2.22 v. 1.13), likely due to inadequate sampling. *Palleon* are not reported to bask (v. 19%). Both species occur at medium elevations (v. 64%); one is found in a riparian zone (v. 21%). They are probably both oviparous; one is so documented. In other respects they comport well with the chamaeleonids. Unfortunately, they are both imperiled (100%), as are most members of the family (v. 95%), though none is on one of the IUCN's worst lists (0% v. 25%).

Pamelaescincus

Pamelaescincus gardineri is the sole member of its genus. It is a large (7.3 g v. 3.3 g) litter cryptozoid (v. 28%) skink endemic to the Seychelles Islands, Indian Ocean.

Pamelaescincus is slightly more heavily built than an average skink (shape index 0.21 v. −0.46 std. resid.). This is a relatively uncommon skink that expresses m>f SSD (1.0 v. −0.31 on −2 to 2 scale). No other sex differences are reported, but this sexual size difference suggests that intermale competition is relatively important in this skink. Populations are diurnal/crepuscular on Fregate, La Digue, Mahé, and Praslin, but nocturnal on Cerf, Curieuse, Cousin, and other small islands. Several predators have been fingered as possibly responsible for this curious arrangement; *Trachylepis wrightii*, a larger (40 g) predatory skink is a prime suspect. *Pamelaescincus* spans the elevational range 0–500 m. Population density (two estimates) averages 0.45 kg/ha (v. 0.35 kg/ha), though a more elevated density might be expected on such small remote islands. Introduced crepuscular tenrecs (*Tenrec ecaudatus*) have been held responsible for the low density, but the evidence is not overwhelming. *Pamelaescincus* has been introduced (v. 2.1%). It is not imperiled (v. 47%). In other respects *Pamelaescincus* is likely to be similar to other skinks.

Palleon spp.	English name				Range core
			Size		
	Activity	Diet	mm	g	Ecological business model
lolontany	Tsaratanana Mountains Chameleon				Madagascar
	diurnal	invert.	32	1.12	
nasus	Elongate Leaf Chameleon				Madagascar
	diurnal	invert.	44	2.63	* immobility expert

Pamelaescincus sp.	English name				Range core
			Size		
	Activity	Diet	mm	g	Ecological business model
gardineri	Gardiner's Skink				Seychelles Islands
	cathemeral	invert.	71	7.3	*** litter cryptozoid

Panaspis

Panaspis is a moderate radiation (15 spp.) of small-sized (1.21 g v. 3.25 g), short-snouted, diurnal (100% v. 74%), oviparous (100% v. 64%) skinks found, often at altitude, in Africa south of the Sahara. They express at least three ecological business models: litter cryptozoid (50% v. 28%), refugia-anchored diurnal ground lizard (38% v. 30%), and sand shark (13% v. 7%). Although their common name emphasizes their lack of eyelids, this is a common phenomenon among very small lizards, presumably facilitating water conservation (see "Mass," chapter 3).

Panaspis are morphologically average skinks, except that elongation is somewhat rarer (20% v. 35%). Blue tails are somewhat more frequent (13% v. 4%). *P. maculicollis* may locomote without use of limbs.

Sex differences are surprisingly bold for such small and secretive lizards. SSD is m>f (0.50 v. −0.31 on −2 to 2 scale). Sexual dichromatism is frequent (40% v. 8%), as is seasonal color (13% v. 6%). These combine into a relatively strong sex differences rollup score: 0.60 (v. 0.28 on a scale of 0 to 6).

Diet has been quantified only for *wahlbergi*, which eats generalized invertebrates, with many ants, termites, and beetles. Three *Panaspis* are all active foragers (100% v. 92%). Home-range area is known only for female *nimbaensis* (250 m^2), which is 1.0 (v. –0.07) standardized residuals above lizard mean for such small lizards. Two *Panaspis* have had their basking behavior characterized: *breviceps* conforms; *cabindae* basks indirectly.

Their elevational profile is skewed upwards (l/m/h: 73/45/18 v. 89/35/6), as might reflect the topography of Africa. They are not recorded in sandy areas (v. 25%) and are underrepresented in littoral areas (9% v. 21%), riparian zones (18% v. 23%), and human habitats (9% v. 14%).

Their clutch count is normal (0.24 v. 0.06 std. resid.); hatchlings are exceptionally small (ontogenetic mass range 1.15 v. 0.12 std. resid.). Incubation duration is based on *wahlbergi* (45 days) and is a little fast for skinks (–0.13 v. –0.40 std. resid.). Hatchlings of *maculicollis* may emerge explosively in response to disturbance. Communal oviposition is relatively common (13% v. 2.6%), for reasons that are not apparent. Possibly there is seasonal flooding of oviposition sites and the females cannot identify such sites by the moisture content of the soil at the time of oviposition, leading to reliance on crowd-sourced sites combined with natal philopatry.

Longevity is short (–0.89 v. 0.17 std. resid.), based on *maculicollis* and *nimbaensis* (each 2 years). Sex-specific longevities are given for *wahlbergi*, and these are also short (males 1 years; females 1.5 years). Three credible population estimates are available for *nimbaensis*; these are notably low (0.01 kg/ha v. 0.35 kg/ha). Thus *Panaspis* struggle to achieve high density and do not live long. Endemism is low (33% v. 64%). Island species are underrepresented (27% v. 57%). None has been introduced (v. 2.1%). About a quarter are imperiled (29% v. 47%). A typical percentage is on one of the IUCN's lists of the most endangered (14% v. 14%). In other documented respects *Panaspis* are typical skinks.

Panaspis spp.	English name				Range core
			Size		
	Activity	Diet	mm	g	Ecological business model
africanus	Guinea Snake-eyed Skink				São Tomé and Príncipe
			43	1.41	
annobonensis	Annobon Snake-eyed Skink				Equatorial Guinea
			40	1.13	
breviceps	Short-headed Lidless Skink				Congo
			59	4.01	* refugia-anchored diurnal gr. liz.
burgeoni	Ruwenzori Mountains Lidless Skink				Congo
			50	2.32	
cabindae	Cabinda Lidless Skink				Congo
	diurnal	invert.	38	0.92	* refugia-anchored diurnal gr. liz.
duruarum	African Snake-eyed Skink				Cameroon
			42	0.87	* sand shark
helleri	Heller's Lidless Skink				Congo
			54	2.90	
maculicollis	Spotted-neck Snake-eyed Skink				Botswana
	diurnal		38	0.91	*** litter cryptozoid
megalurus	Blue-tailed Lidless Skink				Tanzania
	diurnal		39	0.75	** litter cryptozoid
nimbaensis	Tristao's Lidless Skink				Guinea
	diurnal	invert.	52	2.53	* litter cryptozoid
seydeli	Seydel's Snake-eyed Skink				Congo
			37	0.84	
tancredi	Ethiopia Lidless Skink				Congo
	diurnal		29	0.42	
togoensis	Togo Lidless Skink				Cameroon
			47	1.89	* litter cryptozoid
wahlbergi	Wahlberg's Snake-eyed Skink				Southern Africa
	diurnal	invert.	49	1.40	** refugia-anchored diurnal gr. liz.
wilsoni	Wilson's Snake-eyed Skink				Sudan
			27	0.33	

Paniegekko

Paniegekko madjo is the sole representative of this diplodactylid genus from New Caledonia. It is medium sized (5.0 g v. 4.2 g) and poorly known (24% v. 28% of cells filled). The ecological business model of this species is not evident. Its breakable (v. 91%), prehensile (v. 26%) tail is on the long side (1.16 v. 0.85 × SVL) and has an adhesive pad on the tip; *Paniegekko* possesses secretory pores

Paniegekko sp.	English name				Range core
			Size		
	Activity	Diet	mm	g	Ecological business model
madjo	High Elevation Bavayia				New Caledonia
	nocturnal		63	4.98	

Panopa spp.	English name				Range core
			Size		
	Activity	Diet	mm	g	Ecological business model
carvalhoi	Amazonia Blue-tailed Skink				Brazil
	diurnal	invert.	58	2.17	** refugia-anchored diurnal gr. liz.
croizati	Turimiquire Blue-tailed Skink				Venezuela
	diurnal		64	5.0	* refugia-anchored diurnal gr. liz.

Pantepuisaurus sp.	English name				Range core
			Size		
	Activity	Diet	mm	g	Ecological business model
rodriguesi	Mount Maringma Lizard				Guyana
			51	2.86	

Papuascincus spp.	English name				Range core
			Size		
	Activity	Diet	mm	g	Ecological business model
buergersi	Sepik River Ground Skink				New Guinea
			52	2.66	
morokanus	Moroka Mountain Skink				New Guinea
	diurnal	ants	47	1.83	* refugia-anchored diurnal gr. liz.
phaeodes	Sepik Basin Skink				New Guinea
			42	1.27	
stanleyanus	Stanley Mountains Skink				New Guinea
	diurnal		52	2.66	** refugia-anchored diurnal gr. liz.

(v. 34%). It is nocturnal (v. 79%), probably arboreal (v. 64%), oviparous (v. 82%), and imperiled (v. 81%), though not listed among the most endangered of the IUCN categories (v. 29%). Though the common name indicates high elevation, this is so only by New Caledonian standards (850–1100 m). In other respects it is a typical diplodactylid, as far as is known.

Panopa

Panopa are two species of average-sized (3.3 g v. 3.3 g), blue-tailed (100% v. 4%), viviparous (100% v. 36%), refugia-anchored diurnal ground (100% v. 30%) skinks found in Venezuela and Brazil.

Panopa match skinks well in morphology; *carvalhoi* is elongate (v. 35%). Sex differences are not known. Specific dietary information is unavailable. *P. croizati* basks directly (v. 21%). *P. carvalhoi* is found in human habitat (v. 14%). Viviparity has no obvious cause (the usual drivers are high elevation and high latitude). Endemism (50% v. 64%) and island species (50% v. 57%) are as expected. Neither has been introduced (v. 2.1%). Neither is considered imperiled (v. 47%). In other respects *Panopa* are much like other skinks.

Pantepuisaurus

Pantepuisaurus rodriguesi is the sole member of a gymnophthalmid genus known poorly (filled cells 14% v. 19%) from 2080 m elevation in Guyana. The species is of average size (2.86 g v. 2.35 g) and of average lizard build (shape index 0.04 v. –0.32 std. resid.). It is not elongate (v. 41%), nor is it known to possess secretory pores (v. 29%). No other information about sex differences is reported. It is endemic (v. 68%). This species is not known to have been introduced (v. 0.8%), nor has it been evaluated for conservation status.

Papuascincus

Papuascincus is a minor radiation (four spp.) of small-sized (2.01 g v. 3.25 g), high elevation (100% v. 6%), poorly known (filled cells 17% v. 25%), refugia-anchored diurnal ground skinks (100% v. 30%) endemic (100% v. 64%) to New Guinea.

Tail lengths are long (1.80 v. 133 × SVL), and tails are not reported to be breakable (0% v. 73%). No tails are blue (v. 4%). No species is elongate (v. 35%). No sex differences are known.

The only dietary item recorded is

ants. Direct basking is noted (25% v. 21%) in *stanleyanus*. The elevational distributions of the two documented species cover all three bands (in this case 700–3440 m). *Papuascincus* is not recorded in littoral (v. 21%) or sandy (v. 25%) areas, but they are overrepresented in riparian (50% v. 23%) and human (50% v. 14%) habitats. They are oviparous (v. 64%) with a slightly small clutch count (−0.16 v. 0.06 std. resid.), based on *stanleyanus*. Based on *stanleyanus*, *Papuascincus* has a low ontogenetic mass range (−0.23 v. 0.12 std. resid), indicating moderately large hatchlings. *P. stanleyanus* breeds about five times a year, suggesting a strategy of small clutches laid often.

Longevity is about average (0.34 v. 0.17 std. resid.), based on *stanleyanus* (6 years). Endemism is about average (75% v. 64%). None has been introduced (v. 2.1%), and none is imperiled (v. 47%). In other respects *Papuascincus* are much like other skinks.

Paracontias spp.	English name		Size		Range core
	Activity	Diet	mm	g	Ecological business model
ampijoroensis	Ampijoroa Skink				Madagascar
			54	0.61	* subterranean forager
brocchii	Stone Legless Skink				Madagascar
			94	2.13	*** litter cryptozoid
fasika	Littoral Forest Legless Skink				Madagascar
			33	0.20	
hafa	Strange Legless Skink				Madagascar
			59	0.73	*** litter cryptozoid
hildebrandti	Hildebrandt's Legless Skink				Madagascar
			45	0.40	*** litter cryptozoid
holomelas	Anzahamaru Legless Skink				Madagascar
			124	4.03	* litter cryptozoid
kankana	Mahasoa Forest Worm Lizard				Madagascar
			52	0.54	
mahamavo	Mahamavo Skink				Madagascar
			54	0.59	
manify	Manarikoba Forest Worm Lizard				Madagascar
			57	0.69	
milloti	Mamoko Island Legless Skink				Madagascar
			39	0.29	
minimus	Minimal Worm Lizard				Madagascar
			63	0.85	
rothschildi	Rothchild's Legless Skink				Madagascar
			49	0.47	
tsararano	Tsararano Worm Lizard				Madagascar
	nocturnal		57	0.67	
vermisaurus	Angozongahy Worm Lizard				Madagascar
			53	0.57	* litter cryptozoid

Paracontias

Paracontias is a moderate radiation (14 spp.) of very small-sized (0.66 g v. 3.25 g), totally limbless (100%), litter cryptozoid (83% v. 28%) or subterranean forager (17% v. 12%) skinks endemic to Madagascar.

These vermiform lizards are naturally more slender than a typical skink (shape index −2.14 v. −0.46 std. resid.). Their tails are short (0.79 v. 1.33 × SVL), though not as short as pure subterranean foragers would be (table 4.1, trait 5). Tail breakage is between that of ordinary skinks and subterranean foragers (21% v. 73%). Blue tails are absent (v. 4%). *P. fasika* is among the world's smallest reptiles. Eyes are vestigial in *minimus*.

No sex differences are known in *Paracontias*. One species (*tsararano*) is deemed nocturnal (v. 14.5%), but the others are unknown. *P. brocchii* is said to be conspicuously wary (7% v. 11%). Two species are classifiable in relation to rainfall spectrum and they are both mesic (v. 40%). The elevational spectrum of *Paracontias* is skewed low (l/m/h: 100/15/0 v. 89/35/6). They are not noticeably different from average skinks in habitat benchmarks, although the frequency of riparian species seems low (7% v. 23%).

Reproductive mode is unknown. None has been introduced (v. 2.1%). An ordinary frequency of species is imperiled (55% v. 47%), including 27% (v. 14%) that are on the IUCN's lists of the most endangered. In other documented respects *Paracontias* are probably typical skinks.

Paradelma

Paradelma orientalis is the sole species in a genus of large (9.0 g v. 5.2 g) nocturnal ground (v. 11%) pygopodids endemic to Queensland. Its dark nape and crown are notably similar

Paradelma sp.	English name				Range core
			Size		
	Activity	Diet	mm	g	Ecological business model
orientalis	Brigalow Scaly-foot				Queensland
	nocturnal	omnivore	153	9.0	** nocturnal ground lizard

Paragehyra spp.	English name				Range core
			Size		
	Activity	Diet	mm	g	Ecological business model
austeni	Andohahela Cave Gecko				Madagascar
			52	3.18	*** trunk/wall lizard
felicitae	Anja Reserve Gecko				Madagascar
	cathemeral		57	3.96	*** trunk/wall lizard
gabriellae	Slender Stream Gecko				Madagascar
	cathemeral	invert.	64	5.5	** trunk/wall lizard
petiti	Angel's Petite Gecko				Madagascar
	nocturnal	invert.	58	4.15	** trunk/wall lizard

to the coloration of juvenile Brown Snakes (*Pseudonaja*). When confronted, it enhances this mimicry by abruptly raising its head and rapidly flicking its tongue. Otherwise, it moves very slowly, as is so often the case in nocturnal ground lizards. Its ferocious defensive behavior belies its gentle diet, which includes fruit and nectar.

It is a bit more slender than a typical pygopodid (shape index −1.86 v. −1.32 std. resid.) and has a shorter tail (1.90 v. 2.65 × SVL). It calls spontaneously (v. 67%). It does not spend its days in a burrow (0% v. 37%). It is not associated with any habitat specializations. Reproduction is typical. It does occur on an island (v. 26%: Boyne Island) and is imperiled (v. 88%). It does not, however, appear on any of the IUCN's lists of the most endangered species (v. 6%). In other respects *Paradelma* is similar to other pygopodids.

Paragehyra

Paragehyra (four spp.) are trunk/wall gekkonids from Madagascar. These particular trunk/wall lizards are all found on walls, often cave walls, where they are frequently active at any time (cathemeral 67% v. 11%) and found primarily in riparian zones (75% v. 17%). *Paragehyra* are slightly larger than typical gekkonids (4.12 g v. 3.31 g), with longer tails (1.28 v. 1.07 × SVL). They superficially resemble *Hemidactylus*, but lack color change (v. 15%) and infrequently have secretory pores (25% v. 46%). Other sex differences are not reported. They are not reported to forage at electric lights (v. 3.8%). Reproduction fits the gekkonid mold, though adherent eggs (50% v. 11%) and communal oviposition (25% v. 9%) are especially common, as befits a gecko living on cave walls. They have not been introduced (v. 3.7%). Many are cave endemics with restricted distributions, leading to a high level of imperilment (100% v. 47%); half of evaluated species appear on the IUCN's list of most endangered species (v. 11%). In other respects *Paragehyra* are typical gekkonids.

Paralaudakia

Paralaudakia is a moderate (eight spp.) genus of crestless (v. 38%), rock-loving, terrestrial (v. 56%) agamids from the steppes of central Asia: Turkey to Kyrgyzstan and Mongolia, south to Iran, India, and China. They are large (48 g v. 15.6 g), portly (shape index 1.22 v. 0.80 std. resid.), mostly omnivorous (75% v. 26%), sit-and-wait foragers (100% v. 88%; except for plant eating) of small home ranges (*caucasia* is ~100 m^2; −0.87 v. 0.35 std. resid.). Their small home ranges likely reflect their attachment to rock crevices and burrows. Tails are unusually breakable (50% v. 17%). Advanced color change is present (25% v. 15%). Where rocks are missing, the lizards are either absent or they dig burrows (100% v. 29%); when opportunity presents itself, many will climb and bask directly (100% v. 24%).

Agamid-typical levels of sexual dichromatism (63% v. 53%), seasonal color (25% v. 20%), and dimorphism (1.00 v. 0.53 on −2 to 2 scale) exist, but records are few. Their elevational profile covers 300–4000 m and is skewed upwards (l/m/h: 71/100/57 v. 83/52/14), likely reflecting the elevations available in their desert plateaus. They are absent from littoral (v. 10%) and sandy (v. 26%) areas.

Males seem to defend only mating sites; female *caucasia* may defend their nest sites (parental care 13% v. 3.5%), which will produce an excess of males if the soil temperature is high. TSD is not otherwise known in the genus. Clutch counts are more numerous than the family average (0.58 v. 0.20 std. resid.), and hatchlings are relatively large (ontogenetic range 0.12 v. 0.69 std. resid.). Incuba-

Paralaudakia spp.	English name				Range core
			Size		
	Activity	Diet	mm	g	Ecological business model
badakhshana	Badakhshana Rock Agama				Central Asia
	diurnal	omnivore	82	18.2	**** refugia-anchored diurnal gr. liz.
bochariensis	Tajik Rock Agama				Tajikistan
			95	28.6	
caucasia	Caucasia Rock Agama				Iran
	diurnal	omnivore	129	73.2	*** giant herbivore
erythrogaster	Redbelly Rock Agama				Afghanistan
	diurnal	invert.	117	54.8	** refugia-anchored diurnal gr. liz.
himalayana	West Himalaya Agama				Central Asia
	diurnal	omnivore	113	48.8	** refugia-anchored diurnal gr. liz.
lehmanni	Turkestan Rock Agama				Central Asia
	diurnal		117	54.3	** ground-foraging tree lizard
microlepis	Small-scaled Rock Agama				Afghanistan
	diurnal		116	52.7	** refugia-anchored diurnal gr. liz.
stoliczkana	Mongolia Rock Agama				China
	diurnal		138	91	* refugia-anchored diurnal gr. liz.

Paralipinia sp.	English name				Range core
			Size		
	Activity	Diet	mm	g	Ecological business model
rara	Double Subdigital-pads Skink				Vietnam
	diurnal		42	0.87	

tion period is notably short in *Paralaudakia* (59 days), even by agamid standards (–0.92 v. –0.55 std. resid.).

Longevity, based on three species, is long (–0.14 v. –0.46 std. resid.). None is endemic (v. 49%), on islands (v. 36%), or listed as imperiled (v. 37%), though some conservation anxiety is associated with *erythrogaster* because of its restricted range.

Paralipinia

Paralipinia rara is the sole member of its genus. It is a very small (0.87 g v. 3.3 g), elongate (v. 35%), arboreal (v. 10%) skink endemic to Vietnam.

Paralipinia is slightly lighter built than an average skink (shape index –0.89 v. –0.46 std. resid.). No sex differences are known. The microhabitat of this species is unclear; the original specimen was found at night 2 m above ground on a tree trunk, but the describers speculate that it might be a canopy species, which would account for its apparent rarity. The diel cycle of this lizard is also unclear. The describers do not indicate whether the lizard was active at the time of its discovery, though few diurnal lizards would be found at night on a tree trunk; in contrast, many diurnal lizards sleep on leaves, in tree holes, or on the distal ends of twigs. Das (2010) calls it diurnal, without indicating the basis for that claim.

This species inhabits riparian areas (v. 23%). It is not an island species (v. 57%), and it has not been introduced (v. 2.1%). It has not been evaluated for conservation status. In other respects *Paralipinia* is likely to be similar to other skinks.

Paroedura

Paroedura is an 18-species genus of nocturnal Madagascan gekkonids, popularly known as ground geckos, though the modal ecological business model is ground-foraging tree lizards (33% v. 5%). Three other ecological business models are nearly equally frequent: trunk/wall lizard (27% v. 45%), nocturnal canopy walker (20% v. 9%), and nocturnal ground lizard (20% v. 27%).

Paroedura are a little larger than a typical gekkonid (4.37g v. 3.31 g) and have short tails (0.80 v. 1.07 × SVL). They lack secretory pores (v. 46%) and color change (v. 15%). All have small toepads enlarged only at the very ends of the digits.

Paroedura are unusual gekkonids in the preponderance of m>f species (mean 1.00 v. –0.03 on –2 to 2 scale), but other sex differences are lacking and therefore the sex rollup score is a negligible 0.06 (v. 0.63 on a scale of 0 to 6).

Only two species have been reported for foraging mode—both sit-and-wait. Explicit dietary information is not available. *Paroedura* have not been documented to forage under electric lights or to have spontaneous vocalizations. Stress vocalizations, in contrast, are elevated

Paroedura spp.	English name				Range core
			Size		
	Activity	Diet	mm	g	Ecological business model
androyensis	Grandidier's Madagascar Ground Gecko				Madagascar
	nocturnal	invert.	43	1.83	*** nocturnal ground lizard
bastardi	Mocquard's Madagascar Ground Gecko				Madagascar
	nocturnal	invert.	67	6.2	*** ground-foraging tree lizard
gracilis	Graceful Madagascar Ground Gecko				Madagascar
	nocturnal	invert.	57	4.10	*** nocturnal canopy walker
homalorhina	Northern Madagascar Ground Gecko				Madagascar
	cathemeral	invert.	62	5.2	** trunk/wall lizard
hordiesi	Hordies' Ground Gecko				Madagascar
	nocturnal		51	2.95	
ibityensis	Ibity Ground Gecko				Madagascar
	cathemeral	invert.	53	3.31	** trunk/wall lizard
karstophila	Namoroka Ground Gecko				Madagascar
	nocturnal	invert.	49	2.62	** ground-foraging tree lizard
lohatsara	Grand-headed Ground Gecko				Madagascar
	nocturnal	invert.	67	6.4	** ground-foraging tree lizard
maingoka	Scorpion Ground Gecko				Madagascar
	nocturnal	invert.	60	4.68	* nocturnal ground lizard
masobe	Madagascar Big-eyed Gecko				Madagascar
	nocturnal	invert.	86	12.5	*** nocturnal canopy walker
oviceps	Nosy Be Cliff Gecko				Madagascar
	nocturnal	invert.	59	4.38	** trunk/wall lizard
picta	Madagascar Panther Gecko				Madagascar
	nocturnal	invert.	75	8.7	*** nocturnal ground lizard
sanctijohannis	Comoro Ground Gecko				Comoros
	nocturnal	invert.	67	6.2	*** ground-foraging tree lizard
stellata	Mayotte Ground Gecko				Comoros
	nocturnal		54	3.43	
stumpffi	Island Ground Gecko				Madagascar
	nocturnal	invert.	63	5.32	** ground-foraging tree lizard
tanjaka	Limestone Wall Gecko				Madagascar
	nocturnal	invert.	82	11.1	*** trunk/wall lizard
vahiny	Newcomer Ground Gecko				Madagascar
	nocturnal	invert.	39	1.46	
vazimba	Little Ghost Tree Gecko				Madagascar
	nocturnal	invert.	45	2.12	*** nocturnal canopy walker

(11% v. 4%). No species is said to be conspicuously wary. Only *gracilis* has been characterized in relation to primary predator concealment tactic; it hides in vegetation. *P. maingoka* waves its tail in scorpion fashion when confronted.

Males exhibit a high level of social tolerance (33% v. 13%) and are less defensive than typical gekkonids (67% v. 87%). Burrow use is relatively common (29% v. 10%). These geckos are found in the mesic portion of the rainfall spectrum (1101–3000 mm/y). They are underrepresented in human habitats (6% v. 27%).

Reproduction is similar to that of other gekkonids, though communal oviposition is less frequent (0% v. 9%). The young of *Paroedura* are exceptionally large (ontogenetic range small: −0.99 v. −0.19 std. resid.).

Reported longevity is short (−0.24 v. 0.15 std. resid.), though based on only one species (*picta*), and higher maxima are likely to accrue with more observations. No *Paroedura* has been introduced (v. 3.7%). Many species (56% v. 47%) are considered imperiled; 25% are on the IUCN's lists of most endangered (v. 11%). *P. masobe* has an appealing "ET"-like appearance due to large binocular eyes, but it rarely survives in captivity. In other respects *Paroedura* are typical gekkonids.

Parsigecko

Parsigecko ziaiei is the sole representative of a padless nocturnal ground lizard genus (Gekkonidae) found at medium elevation (1596–1697 m) in Iran. It is small (1.25 g v. 3.31 g) and slender (shape index −0.03 v. 0.10 std. resid.). It is known from only a few specimens in a single locality. Few firm facts are available about its behavior and foraging. Sexual dimorphism is not reported. As far as is known it reproduces like other gekkonids, albeit with a documented clutch count of one, which fits its small size (mass-adjusted clutch count −0.13 v. −0.26 std. resid.). It has not been introduced, and it has not been formally evaluated for conservation status. In other respects *Parsigecko* is a typical gekkonid.

Parsigecko sp.	English name				Range core
			Size		
	Activity	Diet	mm	g	Ecological business model
ziaiei	Ziaie's Pars-gecko				Iran
	nocturnal		37	1.25	*** nocturnal ground lizard

Parvilacerta

Parvilacerta (two spp.) is a genus of average (4.5 g v. 5.1 g), diurnal, terrestrial, montane lacertids found in Turkey and neighboring countries. Both species are tentatively identified as refugia-anchored diurnal ground lizards (v. 57%). They are found only toward the arid end of the lacertids' rainfall spectrum (56–400 mm/y). They reach high elevations (2400, 2550 m) and are found in grasslands and alpine meadows. The smaller species (*parva*) is associated with stony riparian habitats, and its primary predator concealment is crevices-burrows. Clutch count is moderately elevated (0.43 v. 0.13 std. resid.), and the hatchlings are comparatively large (ontogenetic mass range is low −0.46 v. −0.05 std resid.). One wonders how they can energetically afford to produce many young that are both numerous and large: *fraasii* breeds only once a year, but *parva* may lay three clutches in a season. Incubation is short (−0.55 v. −0.36 std. resid.), equating to 34 days in *parva* and 48 days in *fraasii*. Longevity is relatively long (0.81 v. 0.19 std. resid.), equating to 12 years in the *fraasii*. *P. fraasii* is endemic to Lebanon (50% v. 35%). Neither species occurs on islands (v. 26%), nor has either been introduced (v. 4.6%). Both are imperiled (v. 56%), and *fraasii* is on the IUCN's list of most endangered (endangered in this case). In other respects they are probably typical lacertids.

Parvilacerta spp.	English name				Range core
			Size		
	Activity	Diet	mm	g	Ecological business model
fraasii	Fraas' Lizard				Lebanon
	diurnal		63	6.0	** refugia-anchored diurnal gr. liz.
parva	Lizard of Asia Minor				Turkey
	diurnal	invert.	53	3.45	** refugia-anchored diurnal gr. liz.

Parvoscincus

Parvoscincus is a major radiation (24 spp.) of small-sized (1.51 g v. 3.25 g), diurnal (100% v. 74%), mostly litter-cryptozoic (79% v. 28%) skinks endemic to the Philippines. In addition to terrestrial (41% v. 48%) and cryptozoic (41% v. 30%) microhabitats, a substantial minority of *Parvoscincus* are semiaquatic (18% v. 2%). These are very poorly known lizards (filled cells 19% v. 25%); in only 2 of 24 species is diet class known (generalized invertebrates 100% v. 84%). Diel activity pattern is documented in only three species.

Parvoscincus closely resemble average skinks in morphology. They are somewhat less prone to elonga-

Parvoscincus spp.	English name				Range core
			Size		
	Activity	Diet	mm	g	Ecological business model
abstrusus	Polillo Sphenomorph				Philippines
			42	1.29	
agtorum	Agta Sphenomorph				Philippines
			42	1.26	
arvindiesmosi	Arvin's Sphenomorph				Philippines
			40	1.14	
aurorus	Maria Aurora Sphenomorph				Philippines
			43	1.38	
banahaoensis	Mt. Banahao Sphenomorph				Philippines
			42	1.28	
beyeri	Beyer's Sphenomorph				Philippines
			63	4.74	*** litter cryptozoid
boyingi	Boying's Zambales Mountain Skink				Philippines
			56	3.34	
decipiens	Black-sided Sphenomorph				Philippines
			40	1.09	*** litter cryptozoid

Parvoscincus spp.	English name				Range core
			Size		
	Activity	Diet	mm	g	Ecological business model
duwendorum	Cordillera Aquatic Skink				Philippines
			50	2.00	* refugia-anchored diurnal gr. liz.
hadros	Aurora Mountain Skink				Philippines
			75	8.4	
igorotorum	Igorot Cordillera Mountain Skink				Philippines
			52	2.68	
jimmymcguirei	McGuire's Sphenomorph				Philippines
			44	1.57	
kitangladensis	Mount Kitanglad Skink				Philippines
			52	2.54	
laterimaculatus	Side-spotted Sphenomorph				Philippines
			50	2.25	*** litter cryptozoid
lawtoni	Lawton's Sphenomorph				Philippines
			41	1.24	
leucospilos	White-spotted Sphenomorph				Philippines
			53	2.72	
luzonense	Highland Sphenomorph				Philippines
			44	0.99	** litter cryptozoid
manananggalae	Aurora Aquatic Skink				Philippines
	diurnal		49	2.22	* refugia-anchored diurnal gr. liz.
palaliensis	Palali Mountain Skink				Philippines
			38	0.91	
palawanensis	Palawan Earless Sphenomorph				Philippines
			34	0.52	
sisoni	Sison's Dwarf Sphenomorph				Philippines
	diurnal	invert.	35	0.52	** litter cryptozoid
steerei	Steere's Forest Skink				Philippines
	diurnal	invert.	33	0.46	*** litter cryptozoid
tagapayo	Walter's Tagapayo Skink				Philippines
			30	0.45	*** litter cryptozoid
tikbalangi	Sierra Madres Aquatic Skink				Philippines
			49	2.11	

tion (17% v. 35%). Their tails are a bit shorter (1.08 v. 1.33 × SVL). No sex differences are known. Their elevational profile is skewed toward intermediate elevations (l/m/h: 74/52/0 v. 89/35/6). They are associated with riparian zones (27% v. 23%) but not with the other tabulated habitat types: littoral (0% v. 21%), sandy areas (0% v. 25%), human habitat (0% v. 14%).

They are oviparous (v. 64%) and have a normal clutch count (0.20 v. 0.06 std. resid.). This mass-adjusted count is based on putatively fixed clutch counts of two in *agtorum* and one in *steerei*.

Population density has been credibly estimated twice for *steerei* (0.03 kg/ha v. 0.35 kg/ha). More data on more species is needed to determine if this low density is characteristic of *Parvoscincus*. No *Parvoscincus* have been introduced (v. 2.1%). They are imperiled at an average frequency (45% v. 47%). None is on one of the IUCN's lists of the most endangered (v. 14%). In other respects *Parvoscincus* are typical skinks.

Pedioplanis

Pedioplanis (13 spp.) is a moderate genus of small-sized (2.73 g v. 5.10 g), slightly slender (shape index 0.12 v. 0.27 std. resid.), diurnal terrestrial lacertids found in southwestern Africa. The modal ecological business model is refugia-anchored diurnal ground lizard (67% v. 57%); the remainder are wide-ranging chemosensors (33% v. 25%). As happens so often in the coastal deserts of Angola to South Africa, termites play a large role in the diet of these lizards. The importance of termites is highlighted in six of seven species with dietary data. In two cases, termites are judged to account for more the 50% of the diet. These lizards are mostly longitudinally striped on a tan background color, with alternate lines and rows of spots.

SSD is not documented. Secretory pores are not reported (v. 4%). Sexual dichromatism is present but infrequent (15% v. 33%). Seasonal color is not reported (v. 15%). The sex differences rollup score, on scale of 0 to 6, for *Pedioplanis* is just 0.15 (v. 0.86).

One species is reportedly carnivorous (*husabensis*), though the documentation is weak. Omnivory is significant (25% v. 19%) and likely underreported. The others are typical lacertid invertebrate eaters (63% v.

Pedioplanis spp.	English name				Range core
			Size		
	Activity	Diet	mm	g	Ecological business model
benguellensis	Angolan Sand Lizard				Namibia
	diurnal		46	2.24	* refugia-anchored diurnal gr. liz.
breviceps	Short-headed Sand Lizard				Namibia
	diurnal		41	1.64	* wide-ranging chemosensor
burchelli	Burchell's Sand Lizard				South Africa
	diurnal	invert.	53	3.52	*** refugia-anchored diurnal gr. liz.
gaerdesi	Kaokoveld Sand Lizard				Namibia
	diurnal	omnivore	47	2.42	** refugia-anchored diurnal gr. liz.
haackei	Haacke's Sand Lizard				Angola
			46	2.29	
huntleyi	Huntley's Sand Lizard				Angola
			50	2.99	
husabensis	Husab Sand Lizard				Namibia
	diurnal	carnivore	51	3.16	
inornata	Plain Sand Lizard				South Africa
	diurnal	invert.	47	2.47	* wide-ranging chemosensor
laticeps	Cape Sand Lizard				South Africa
	diurnal		54	3.76	
lineoocellata	Spotted Sand Lizard				Botswana
	diurnal	invert.	56	4.15	*** refugia-anchored diurnal gr. liz.
namaquensis	Namaqua Sand Lizard				Botswana
	diurnal	invert.	50	2.89	*** wide-ranging chemosensor
rubens	Waterberg Sand Lizard				Namibia
	diurnal	invert.	45	2.14	*** refugia-anchored diurnal gr. liz.
undata	Western Sand Lizard				Angola
	diurnal	omnivore	50	2.95	** refugia-anchored diurnal gr. liz.

77%). About half of the species are characterized as sit-and-wait foragers (43% v. 23%), and the other half are active foragers (57% v. 92%). Some of the active foragers have notably high objective movement statistics: *namaquensis* is in the 89th percentile for both MPM and PTM. The corresponding values for *undata* are 76th and 87th. *P. lineoocellata* is described as sit-and-wait, but apparently it does not do so patiently: its corresponding statistics are 60th and 51st.

Mean active body temperature is a toasty 37.4°C (v. 33.3°C). All four species with nightly refuge characterization use burrows (100% v. 48%). The genus's mode for rainfall distribution is in the arid belt (151–400 mm/y). All six elevationally documented species are found at low elevation (maximum 1066 m). Their notable habitat association is sand, though its frequency is average in relation to lacertids (46% v. 44%).

Reproduction is typical of lacertids (clutch count 0.15 v. 0.13 std. resid.), with slightly large hatchlings (ontogenetic mass ratio −0.40 v. −0.05 std. resid.). RCM is low (0.199 v. 0.267), although it is based on only *lineoocellata* and *namaquensis*.

Longevity (based on 5 years in *lineoocellata*) is slightly short (−0.02 v. 0.19 std. resid.). Endemism is high (46% v. 35%). No species occurs on islands (v. 26%) or has been introduced (v. 4.6%). None is considered imperiled (v. 56%). In other respects *Pedioplanis* are typical desert lacertids.

Perochirus

Perochirus (three spp.) are trunk/wall gekkonids from Western Pacific islands, especially Micronesia. Their distribution is scattered, suggestive of spotty survival in the face of introductions of predators. The large species (*scutellatus*) appears to survive on only the most remote atolls of Micronesia. One species (*guentheri*), known from only a few specimens, has an isolated distribution on two islands of Vanuatu. The widespread Micronesian species (*ateles*) is missing from some islands heavily impacted by humans. As far as is known, all reside on tree trunks and some buildings. *Perochirus* are relatively large (9.5 g v. 3.31 g) and robust (shape index 0.35 v. 0.10 std. resid.), with comparatively short tails (0.86 v. 1.07 × SVL). One species (*ateles*) has secretory pores (v. 46%) and darkens (33% v. 15%). All have enlarged toepads (v. 70%).

In keeping with their large size, SSD is m>f for all species (mean SSD score 1.0 on scale of −2 to 2). The sex differences rollup is therefore higher than a typical gekkonid (1.33 v. 0.63 on 0 to 6 scale).

Perochirus spp.	English name				Range core
			Size		
	Activity	Diet	mm	g	Ecological business model
ateles	Micronesia Gecko				Western Pacific Islands
	nocturnal	invert.	69	8.01	*** trunk/wall lizard
guentheri	Vanuatu Saw-tailed Gecko				Vanuatu
	nocturnal	invert.	60	4.68	
scutellatus	Giant Micronesia Gecko				Micronesia
	cathemeral	invert.	106	22.5	*** trunk/wall lizard

Petracola spp.	English name				Range core
			Size		
	Activity	Diet	mm	g	Ecological business model
angustisoma	Bongara Lizard				Peru
			41	1.36	
*labiooculari*s	Huanuco Lizard				Peru
			56	3.90	
ventrimaculatus	Spotted Lightbulb Lizard				Peru
			56	3.90	
waka	Lower Mountain Lizard				Peru
			45	1.88	

Diet is invertebrates; electric light use and home-range area are not quantified. Exposed basking is known for *ateles*, and *scutellatus* is active at all times, at least on islands such as Ulithi Atoll, where it may form groups of up to 20 individuals. *P. ateles* forms groups of four to five. Nonetheless, both of these species are reported to have males that defend space. No species is reported to be conspicuously wary (v. 4.4%), and none is documented to vocalize (v. 6.5%).

The widespread species (*ateles*) is found around humans, at frequencies typical of gekkonids (33% v. 27%). One species is documented to have adherent eggs (33% v. 10.5%), as is common for trunk/wall lizards, but communal oviposition—often found in arboreal species—is not documented (v. 8.6%). The apparently fixed clutch count of two is notably small for a lizard this large (mass-adjusted: −0.64 v. −0.26 std. resid.).

Population density is on the low side for gekkonids (0.19 kg/ha v. 0.37 kg/ha). None has been introduced (v. 3.7%). All are considered imperiled (v. 47%), and even the widespread species in considered endangered by the IUCN (33% v. 11%). In other respects *Perochirus* are typical gekkonids.

Petracola

Petracola is a four-species genus of poorly known (filled cells 16% v. 19%) gymnophthalmids endemic to middle and higher elevations of Peru. Too little information is available to assign them to ecological business models, though the three species characterized by microhabitat are all said to be cryptozoic. They are of medium size (2.50 g v. 2.35 g) and average build (−0.01 v. −0.32 std. resid.). They are not elongate (v. 41%). Secretory pores are present in *angustisoma* (v. 29%), but other sex differences are not reported. *P. angustisoma* is said to be conspicuously wary (v. 7%). Recorded elevations are 1889–3600 m (75% v. 33% at high elevation). All species are endemic (v. 68%). They are not considered to be at conservation risk (v. 25%). In other respects *Petracola* are probably typical gymnophthalmids.

Petrosaurus (fig. 4.11A)

Petrosaurus (four spp.) is a genus of large (30.4 g v. 19.8 g), saxicolous, refugia-anchored diurnal ground phrynosomatids from the area where California meets Mexico. They are robust in build (shape index 1.04 v. 0.91). Tails are relatively long (1.62 v. 1.22 × SVL) but not breakable (0% v. 21%). They are notable for the frequency of secretory pores (75% v. 32%).

SSD is m>f (1.0 v. 0.27 on −2 to 2 scale), and chronic color differences between the sexes show up at a high frequency (75% v. 58%), as do seasonal colors (75% v. 25%). The sex differences rollup score is thus moderately elevated (2.50 v. 1.68 on 0 to 6 scale). Two species are invertebrate eaters (50% v. 62%) and two are omnivores (50% v. 20%). Foraging mode is not documented. Home-range size is small (−0.45 v. 0.57 std. resid.), based on *mearnsi* (ca. 80 m^2). Though nominally diurnal, at least two species (*repens*, *thalassinus*) are noted to be active on some nights.

Two are said to be conspicuously wary (50% v. 22%). Primary predator concealment tactics are reported to be flight (50% v. 14%) and crevice-burrows (50% v. 32%). These three facts seem to fit together well. Because *Petrosaurus* rely on rapid flight to a rock crevice, they are judged to be exceptionally wary. To run rapidly they need to be warm; 75% (v. 40%)

Petrosaurus spp.	English name				Range core
			Size		
	Activity	Diet	mm	g	Ecological business model
mearnsi	Banded Rock Lizard				Mexico
	diurnal	invert.	81	18.2	** refugia-anchored diurnal gr. liz.
repens	Central Baja California Banded Rock Lizard				Mexico
	diurnal	omnivore	102	32.9	** refugia-anchored diurnal gr. liz.
slevini	Slevin's Banded Rock Lizard				Mexico
	diurnal	invert.	85	20.5	** refugia-anchored diurnal gr. liz.
thalassinus	San Lucan Rock Lizard				Mexico
	diurnal	omnivore	134	70	** refugia-anchored diurnal gr. liz.

Phaeoscincus spp.	English name				Range core
			Size		
	Activity	Diet	mm	g	Ecological business model
ouinensis	Mount Ouin Skink				New Caledonia
			38	0.92	
taomensis	Mount Taom Skink				New Caledonia
			47	2.00	

Phasmasaurus spp.	English name				Range core
			Size		
	Activity	Diet	mm	g	Ecological business model
maruia	Maruia Maquis Skink				New Caledonia
	diurnal		53	2.78	*** refugia-anchored diurnal gr. liz.
tillieri	Tillier's Maquis Skink				New Caledonia
	diurnal		55	3.15	** refugia-anchored diurnal gr. liz.

of *Petrosaurus* are reported to bask directly.

All species are found at low elevation (100% v. 73%); half range up in elevation into the medium elevation band (50% v. 76%). They are not found in littoral (v. 16%) or human (v. 18%) habitats, but are overrepresented in riparian zones (100% v. 40%). In such an arid area, exposed rock crevices may be associated with dry stream beds (i.e., riparian).

Clutch count is a little low for their large size (0.33 v. 0.68 std. resid.). Ontogenetic mass range is a little high, indicating small young (0.12 v. −0.26 std. resid.). Incubation (based solely on *thalassinus*: 56 days) is notably short (−1.04 v. −0.58 std. resid.), perhaps reflecting the high summer temperatures in their environment.

Longevity is long (0.91 v. −0.25 std. resid.), though based on only *thalassinus* (20 years). Endemism is high (75% v. 58%). All species occur on at least one island (v. 21%). None has been introduced (v. 3.9%). Twenty-five percent are considered imperiled (v. 37%), though none is found on any of the IUCN's lists of the most endangered (v. 5.4%). The bold patterning and bright colors of *thalassinus* (fig. 4.11A) make it a magnet for the illicit pet trade. In other respects *Petrosaurus* follow phrynosomatid norms.

Phaeoscincus

Phaeoscincus are two species of small-sized (1.3 g v. 3.3 g), terrestrial (100% v. 48%) skinks endemic to New Caledonia. Little is known (filled cells 15% v. 25%). Adult males of *ouinensis* have yet to be detected; *taomensis* is known from only one subadult (modal size estimated from similar New Caledonian skinks).

Morphology is unspecialized. Sex differences are unknown. Known elevations are 1000–1020 m. None has been introduced (v. 2.1%) or evaluated for conservation status. In other documented respects *Phaeoscincus* are much like other skinks.

Phasmasaurus

Phasmasaurus are two species of average-sized (3.0 g v. 3.3 g), direct basking (100% v. 21%), conspicuously wary (100% v. 11%), very long-tailed (2.70 v. 1.33 × SVL), refugia-anchored diurnal ground (100% v. 30%) skinks endemic to New Caledonia.

Phasmasaurus have a slightly more robust build than average skinks (shape index −0.09 v. −0.46 std. resid.). Sex size differences are unknown, but the sexes differ in color in *tillieri* (50% v. 7.6%). Known elevations are 0–1000 m. Reproduction is by eggs in *maruia* and live birth in *tillieri*. One craves more information about this difference and its ramifications in such closely related species living in such similar environments. Clutch/litter size is known only for *maruia* (−0.05 v. 0.06 std. resid.). Neither species has been introduced (v. 2.1%). Both are imperiled (v. 47%); *maruia* (endangered) is on one of the IUCN's lists of the most endangered (50% v. 14%). In other respects *Phasmasaurus* are much like other skinks.

Phelsuma

Phelsuma is a major radiation (52 spp.) of moderate-sized (3.55 g v. 3.31 g) predominantly diurnal gekkonids found in eastern Africa, Madagascar, and other Indian Ocean Islands, including the Andaman Islands. These lizards are often brightly colored over a brilliant green base and are relatively well-known (38% cells filled v. 25%) and charismatic lizards. They are especially familiar in anthropomorphic form to those viewing commercials for the American insurance firm Geico. They may be native only to Indian Ocean Islands, as the populations on mainland Africa may represent introductions. These are among the most often introduced lizards (11.5% v. 3.7%), and nearly 6% of species have been introduced repeatedly (5.8% v. 2.1% for >2 introductions). Ninety percent of *Phelsuma* are trunk/wall lizards (v. 45%); the remainder are a smattering of diurnal canopy walkers (4% v. 1%), ground-foraging tree lizards (4% v. 5.3%), and nocturnal ground lizards (2% v. 27%). The surprise is the paucity of microhabitats other than tree trunks. They do not appear to have diversified their microhabitats with the frequency associated with *Anolis*; no ecological business model is found in more than

Phelsuma spp.	English name		Size		Range core
	Activity	Diet	mm	g	Ecological business model
abbotti	Aldabra Day Gecko				Madagascar
	diurnal	invert.	57	4.09	*** trunk/wall lizard
andamanense	Andaman Day Gecko				India
	diurnal	invert.	55	3.69	**** trunk/wall lizard
antanosy	Littoral Day Gecko				Madagascar
	cathemeral	invert.	44	1.94	**** trunk/wall lizard
astriata	Stripeless Day Gecko				Seychelles
	diurnal	omnivore	52	3.16	**** trunk/wall lizard
barbouri	Barbour's Day Gecko				Madagascar
	diurnal	invert.	55	3.69	*** ground-foraging tree lizard
berghofi	Berghof's Day Gecko				Madagascar
	diurnal	invert.	51	2.95	** ground-foraging tree lizard
borai	Bemaraha Day Gecko				Madagascar
	diurnal		40	1.56	
borbonica	Reunion Forest Day Gecko				Reunion
	diurnal	invert.	60	4.69	**** trunk/wall lizard
breviceps	Short-headed Day Gecko				Madagascar
	diurnal	invert.	48	2.52	*** trunk/wall lizard
cepediana	Blue-tailed Day Gecko				Mauritius
	diurnal	omnivore	54	3.41	**** trunk/wall lizard
comorensis	Grand Comore Day Gecko				Comoros
	diurnal	invert.	51	3.01	*** trunk/wall lizard
dorsivittata	Striped-back Day Gecko				Madagascar
	diurnal	invert.	53	3.31	*** trunk/wall lizard
dubia	Olive Day Gecko				Tanzania
	diurnal	omnivore	56	3.89	**** trunk/wall lizard
edwardnewtoni	Newton's Day Gecko				Mauritius
	diurnal	omnivore	87	12.8	** trunk/wall lizard
flavigularis	Yellow-throated Day Gecko				Madagascar
	diurnal	invert.	59	4.53	*** trunk/wall lizard
gigas	Liénard's Day Gecko				Mauritius
	nocturnal	omnivore	165	76	* nocturnal ground lizard
gouldi	Anja Day Gecko				Madagascar
			42	1.70	** trunk/wall lizard
grandis	Madagascar Giant Day Gecko				Madagascar
	diurnal	omnivore	96	17.0	**** trunk/wall lizard
guentheri	Guenther's Day Gecko				Mauritius
	cathemeral	carnivore	124	34.3	**** trunk/wall lizard
guimbeaui	Orange-spotted Day Gecko				Mauritius
	diurnal	invert.	59	4.53	**** trunk/wall lizard

42% of anoles. There seems to be some specialization for the species of tree whose trunk *Phelsuma* occupy—travelers' palms, bamboo, pandanus, bananas, coconuts, dead trees—but this has not been quantified.

As expected for arboreal lizards that occupy tree trunks rather than smaller perches, they do not have prehensile tails (v. 1.8%). As expected for arboreal lizards rather than terrestrial ones (see *Pachydactylus* account), they possess enlarged toe pads (0% v. 30% padless). Secretory pores are nearly universal in *Phelsuma* (98% v. 46%). Color change is almost as uniform (77% v. 15%), and advanced color change is not unusual (17.3% v. 2.1%). Blue tails are present at un-gekkonid frequencies (9.6% v. 0.6%), perhaps at least in part because bright blue is a recurring color highlight in *Phelsuma* generally.

Phelsuma are more prone to m>f SSD than are comparably sized geckos (0.93 v. −0.03 on −2 to 2 scale). This fits with their diurnality (table 3.7, trait 11) compared to the generally nocturnal gekkonids. Many *Phelsuma* have sex differences in color (31% v. 6.6%), and seasonal color is low but much more frequent than in gekkonids (3.8% v. 0.8%). All these sex differences add up to one of the highest sex rollup scores for gekkonids (1.75 v. 0.63 on 0 to 6 scale).

The largest *Phelsuma* are able to consume vertebrates (2.1% v. 1.1%), though most are invertebrate eaters (83% v. 95%). Omnivores are overrepresented (15% v. 3.8%); none of the seven omnivores is reported to eat leaves—five lap nectar and all seven consume fruit. All documented foraging modes are sit-and-wait (v. 81%); electric light use is common (7.7% v. 3.8%) and probably underreported—electric lights may be uncommon in many species' ranges.

Phelsuma spp.	English name				Range core
			Size		
	Activity	Diet	mm	g	Ecological business model
guttata	Speckled Day Gecko				Madagascar
	diurnal	invert.	58	4.24	**** trunk/wall lizard
hielscheri	Pandanus-loving Day Gecko				Madagascar
	diurnal	invert.	62	5.0	**** trunk/wall lizard
hoeschi	Hoesch's Day Gecko				Madagascar
	cathemeral		36	1.10	*** trunk/wall lizard
inexpectata	Manapany Day Gecko				Reunion
	diurnal	invert.	51	2.95	** trunk/wall lizard
kely	Andranokoditra Day Gecko				Madagascar
	diurnal	invert.	33	0.90	*** diurnal canopy walker
klemmeri	Blue Bamboo Day Gecko				Madagascar
	cathemeral	invert.	40	1.54	*** trunk/wall lizard
kochi	Koch's Madagascar Day Gecko				Madagascar
	diurnal	invert.	86	13.0	**** trunk/wall lizard
laticauda	Gold Dust Day Gecko				Madagascar
	diurnal	invert.	56	3.83	**** trunk/wall lizard
lineata	Yellow-striped Day Gecko				Madagascar
	diurnal	invert.	56	3.78	**** trunk/wall lizard
madagascariensis	Madagascar Day Gecko				Madagascar
	diurnal	invert.	93	15.8	**** trunk/wall lizard
malamakibo	Andohahela Day Gecko				Madagascar
	diurnal	invert.	53	3.31	*** trunk/wall lizard (low elev.) or *** diurnal canopy walker (high elev.)
masohoala	Masoala Day Gecko				Madagascar
	diurnal	invert.	43	1.86	
modesta	Modest Day Gecko				Madagascar
	diurnal	invert.	49	2.68	** trunk/wall lizard
mutabilis	Thick-tailed Day Gecko				Madagascar
	diurnal	invert.	45	2.12	*** trunk/wall lizard
nigristriata	Island Day Gecko				Comoros
	diurnal	invert.	42	1.69	** trunk/wall lizard
ornata	Ornate Day Gecko				Mauritius
	cathemeral	omnivore	50	7.6	*** trunk/wall lizard
parkeri	Pemba Day Gecko				Tanzania
	diurnal	invert.	60	4.69	**** trunk/wall lizard
parva	Tamatave Day Gecko				Madagascar
	diurnal	invert.	35	1.07	*** trunk/wall lizard
pasteuri	Pasteur's Day Gecko				Comoros
	diurnal	invert.	47	2.41	**** trunk/wall lizard

Phelsuma spp.	English name				Range core
			Size		
	Activity	Diet	mm	g	Ecological business model
pronki	Andramasina Day Gecko				Madagascar
	diurnal	invert.	45	2.12	*** trunk/wall lizard
pusilla	Banana Leaf Day Gecko				Madagascar
	diurnal	invert.	36	1.13	*** trunk/wall lizard
quadriocellata	Peacock Day Gecko				Madagascar
	diurnal	invert.	53	3.31	**** trunk/wall lizard
ravenala	Traveler's Palm Day Gecko				Madagascar
	diurnal		53	3.31	**** trunk/wall lizard
robertmertensi	Mertens' Day Gecko				Comoros
	diurnal	invert.	45	2.12	*** trunk/wall lizard
roesleri	Roesler's Day Gecko				Madagascar
	diurnal		35	1.03	*** trunk/wall lizard
rosagularis	Upland Forest Day Gecko				Mauritius
	diurnal	invert.	59	4.38	**** trunk/wall lizard
seippi	Seipp's Day Gecko				Madagascar
	diurnal	invert.	49	2.62	**** trunk/wall lizard
serraticauda	Serrated Day Gecko				Madagascar
	diurnal	invert.	57	3.96	**** trunk/wall lizard
standingi	Standing's Day Gecko				Madagascar
	diurnal	invert.	76	8.9	**** trunk/wall lizard
sundbergi	Seychelles Giant Day Gecko				Seychelles
	diurnal	invert.	84	11.8	*** trunk/wall lizard
vanheygeni	Van Heygen's Day Gecko				Madagascar
	diurnal	invert.	33	0.87	*** diurnal canopy walker
v-nigra	Chevron-headed Day Gecko				Comoros
	diurnal	invert.	47	2.41	*** trunk/wall lizard or diurnal canopy walker (geog. var.)

Mass-adjusted home-range size is nominally a bit larger than typical gekkonids (0.09 v. –0.28 std. resid.), though it is based on just one species (*kochi*: 13 g) and may therefore be unrepresentative.

All diel activity patterns are present in *Phelsuma*: active at all times (cathemeral) is underrepresented (5.9% v. 10.8%), crepuscular is overrepresented (3.9% v. 1.0%), diurnal is greatly overrepresented (88% v. 15%), and nocturnality is correspondingly rare (2% v. 73%). In accord with the latter, spontaneous vocalization is rare (3.8% v. 6.5%), though stress calling is found at typical frequency (3.8% v. 3.7%).

Phelsuma are conspicuously wary at the family frequency (5.8% v. 4.4%), and all documented species rely primarily on crypsis for predator concealment (100% v. 50%). Given how many *Phelsuma* reside on trunks rather than leaves, the dominance of green coloration seems a bit surprising for lizards so reliant on crypsis on mostly brown bark. The many bright blues, yellows, and reds among the green background color suggest that communication with other lizards may have trumped crypsis in many cases. Has anyone checked on palatability of *Phelsuma* to avian predators? The bright colors might be aposematic in some environments. At least *kochi* darkens its body colors in response to avian predator alarm calls (Ito et al. 2013), suggesting it is palatable and prefers to be hidden when in the vicinity of predatory birds.

Direct basking is overrepresented (17.3% v. 4.9%) and not much different from another genus (*Anolis*) of similarly size green arboreal lizards (19.5%). Thermoconforming is overrepresented (3.8% v. 1.4%) but much less than reported for *Anolis* (16.1%). Mean active body temperature is exactly the same as reported for anoles (29.0°C), slightly warmer than for an average gekkonid (27.6°C).

Social tolerance is not characteristic of *Phelsuma*, though solitary living is not reported (v. 7.7%). Group living is comparatively rare (6.7% v. 25.3%), and only 5.6% (v. 13.3%) of species are reported to have tolerant males. Spatial defense by males is nearly universal (94% v. 87%). The nectar-eating *astriata* is reported to defend nectar sources not only against other *Phelsuma*, but against other nectivorous lizard species as well. One older report notes an association of male *guentheri* with communally deposited eggs, suggesting that males might contact receptive females coming to lay. Another possibility is that male attendance represents male parental care, which is not otherwise documented in *Phelsuma*.

Phelsuma are absent from the hyperdry (v. 2%) and dry (v. 13%) precipitation bands (0–150 mm/y). They are mildly overrepresented at the wetter end of the spectrum: in mesic areas (1101–3000 mm/y), 85% v. 66%; in wet areas (>3000 mm/y), 8% v. 2%. *Phelsuma* are overwhelmingly found at low elevations (98% v. 92%), but that likely reflects in part the elevations present on Indian Ocean Islands. They are overrepresented in littoral (33% v. 11%) and human (54% v. 27%) habitats, but underrepresented in sandy areas (2% v. 11%) and riparian zones (9.6% v. 17%). Many species are said to avoid continuous native forest.

Reproduction in *Phelsuma* conforms to gekkonid norms, though TSD is widely reported (at least 36 species) and may be universal. The mean incubation period is somewhat short (–0.17 v. 0.17 std. resid.), and many eggs are adherent (42% v. 10.5%) or deposited communally (19.2% v. 8.6%). The latter two characters are in keeping with *Phelsuma*'s arboreality. In contrast to communal oviposition, at least three species (*flavigularis, madagascariensis, quadriocellata*) are said to split up their clutches and deposit eggs singly. At least two species (*breviceps, flavigularis*) lay eggs in exposed locations (not under bark, in crevices, etc.). *P. breviceps* lays it exposed eggs within protective thorn tangles. Parental care is notably common in *Phelsuma* (11.5% v. 1.9%), typically in the form of females defending the eggs and consuming hatched shell fragments. The latter suggests a potential shortage of calcium in their diets.

Longevity is well above the gekkonid norm (0.79 v. 0.15 std. resid.), though it is hard to know the extent to which this discrepancy is due to the much greater attention given to *Phelsuma* by herpetoculturists. *P. standingi* (22 years) and *grandis* (20 years) are the genus leaders in absolute longevity. The mean population densities are known for *astriata* in the Seychelles (0.44 kg/ha, $n = 7$), *ornata* on a Mauritius outlier (1.7 kg/ha, $n = 3$), and *sundbergi* in the Seychelles (2.1 kg/ha, $n = 8$). These average to 1.16 kg/ha, well above the gekkonid norm (0.37 kg/ha), but not surprising for small island populations. Alas, these high densities (documented however in only three species) are insufficient to protect *Phelsuma* from imperilment (98% v. 47%), including one-quarter of species that are on the IUCN's lists of the most endangered (25% v. 11.3%). Two of the larger species (*edwardnewtoni*, 12.8 g; *gigas*, 76 g) are extinct. In other respects *Phelsuma* are typical gekkonids.

Philochortus

Philochortus spp.	English name				Range core
	Activity	Diet	Size mm	Size g	Ecological business model
hardeggeri	Hardegger's Shield-backed Lizard				Horn of Africa
	diurnal		68	7.3	
intermedius	Shield-backed Ground Lizard				Horn of Africa
	diurnal	invert.	70	8.0	
neumanni	Neumann's Orange-tailed Lizard				Yemen
	diurnal		68	7.3	
phillipsi	Phillips' Shield-backed Lizard				Ethiopia
	diurnal		52	3.38	
spinalis	Peters' Shield-backed Lizard				Southern Sahara
	diurnal		51	3.11	
zolii	Zolio's Shield-backed Ground Lizard				Sahara
	diurnal	invert.	54	3.70	* ground-foraging tree lizard

Philochortus (six spp.) is a genus of average-size (5.1 g v. 5.1 g) diurnal, terrestrial lacertids found in the deserts from Mali east to Saudi Arabia and Egypt and south to Kenya. Ignorance is high in this group (filled cells 21% v. 33%). Only one species has been very tentatively assigned an ecological business model. Most species have dark longitudinal lines on a light brown background; juveniles sometimes have reddish-orange tails. One species (*zolii*) is associated with oases, where it digs burrows for its nightly refuge, which may be true for many *Philochortus*.

Philochortus are morphologically indistinct lacertids, except for their long tails (2.46 v. 2.06 × SVL). Sex differences are not reported. No species is endemic to a single country (v. 35%). None has been introduced (v. 4.6%). One-third of evaluated species are imperiled (v. 56%) and are listed by the IUCN as among the most endangered (*zolii*, endangered).

Phoboscincus

Phoboscincus are two species of very large (80 g v. 3.3 g) elusive skinks found in the New Caledonia region, including the Loyalty Islands. *P. bocourti* was considered possibly extinct, until two specimens were found on a small offshore islet, suggesting that introduced predators have eliminated it from the main islands. Both species have large recurved teeth in the front of their mouth, suggesting a diet of other vertebrates, perhaps slippery prey

Phoboscincus spp.	English name		Size		Range core
	Activity	Diet	mm	g	Ecological business model
bocourti	Bocourt's Giant Skink				New Caledonia
	diurnal		212	77	* ground-foraging tree lizard
garnieri	Garnier's Giant Skink				New Caledonia
	diurnal		152	84	** refugia-anchored diurnal gr. liz.

Phoenicolacerta spp.	English name		Size		Range core
	Activity	Diet	mm	g	Ecological business model
cyanisparsa	Rift Valley Lizard				Turkey
	diurnal		56	4.10	
kulzeri	Mount Hermon Lizard				Syria
	diurnal		72	8.5	* refugia-anchored diurnal gr. liz.
laevis	Agile Lebanon Lizard				Syria
	diurnal	invert.	67	6.1	** wide-ranging chemosensor
troodica	Troodos Lizard				Cyprus
	diurnal	omnivore	57	4.40	** ground-foraging tree lizard

like skinks. Too little information is known about *bocourti* to identify other commonalities. *P. bocourti* is elongate (50% v. 35%).

Phoboscincus have a more robust build than average skinks (shape index 0.44 v. −0.46 std. resid.). Tails are slightly longer (1.50 v. 1.33 × SVL). Sex differences are unknown. Both are likely diurnal (v. 74%) and both bask directly (v. 21%). Both are conspicuously wary (v. 11%). Both retreat to vegetation when pressed (100% v. 25%). *P. garnieri* may use a burrow for a daily retreat (v. 24%). Both are found solely at low elevation (v. 89%). Both are found in the littoral zone (v. 21%); *garnieri* tolerates human habitats (50% v. 14%).

Reproduction is unknown, though it seems likely that they do it. Neither species has been introduced (v. 2.1%). *P. bocourti* is imperiled (50% v. 47%) and, being identified as endangered, is on one of the IUCN's lists of the most endangered (v. 14%). In other respects *Phoboscincus* are much like other skinks.

Phoenicolacerta

Phoenicolacerta (four spp.) is a small genus of average (5.5 g v. 5.1 g) diurnal lacertids found in the Middle East from Turkey to Israel and on Cyprus. They are of diverse microhabitats (half arboreal, half terrestrial) and ecological business models (see table). Their general shape resembles *Lacerta*, with a large head and often bright colors on the chin (shape index 0.25 v. 0.27 std. resid.). Their tails are average (2.11 v. 2.06 × SVL) and breakable (100% v. 92%).

The sexes differ in every tabulated way except pores: m>f size (1.0 v. 0.49 on −2 to 2 scale), sexes differ in color (75% v. 33%), seasonal color (50% v. 15%), blue tails (50% v. 14%), and m:f ratio of relative tail length (1.16 v. 1.07, the highest ratio of any lacertid genus). In aggregate, 2.25 v. 0.86 on 0 to 6 sex differences rollup.

Diet is largely unknown, but is known to include fruit, leaves, invertebrates, and vertebrates. One species (*laevis*) is reported to forage actively (69th percentile MPM; 74th percentile PTM). They show typical lacertid tactics for predator concealment, including retreat to vegetation or a crevice-burrow.

Direct basking is the rule (75% v. 25%). One species occurs in groups (*troodica*); one (*kulzeri*) is said to defend space. They are well represented in littoral (50% v. 22%), sandy (50% v. 44%), riparian (50% v. 28%), and human (75% v. 16%) habitats. Reproduction is typical for lacertids (clutch count 0.08 v. 0.13 std. resid.).

Longevity is short (−0.24 v. 0.19 std. resid.) based on *laevis* (4.6 years). They have not been introduced (v. 4.6%). Half of evaluated species are imperiled (v. 56%); one species (*kulzeri*, endangered) is on an IUCN list of the most endangered (v. 15.5%). In other respects *Phoenicolacerta* are similar to other lacertids.

Pholidobolus

Pholidobolus is a nine-species genus of medium-sized (3.11 g v. 2.35 g) mostly terrestrial gymnophthalmids from northwestern South America and Panama. They are taxonomically distinguished from their sister genus (*Macropholidus*) by the presence of a ventrolateral fold, hence the common name of bellyfold lizards. Secretory pores are comparatively common (46% v. 29%). Most species cannot be assigned an ecological business model due to inadequate natural history information. Tails are a bit long (1.98 v. 1.72 × SVL) and are not reported to be breakable (v. 28%). One species (*ulisesi*) has been characterized for SSD (m<f). The same species has been reported to be sexually dichromatic (11% v. 13%). *P. vertebralis* is conspicuously wary (11% v. 7%). These are generally high elevation species: medium elevation 88% (v. 52%); high elevation 63% (v. 33%). This may account of the extraordinarily long incubation period reported: 216 days, based on *montium* (2.99 v. 1.11 std. resid.). Clutch size is slightly small (−0.24 v. −0.11 std. resid.). Most species are endemic (78% v. 68%). Forty percent (v. 25%) are considered imperiled, though none is on an IUCN list of the most endangered species (v. 11.5%). In other respects *Pholidobolus* are typical gymnophthalmids.

Pholidobolus spp.	English name		Size		Range core
	Activity	Diet	mm	g	Ecological business model
affinis	Wandering Bellyfold Lizard				Ecuador
			51	2.87	
anomalus	Strange Bellyfold Lizard				Peru
			54	3.46	
dicrus	Uzzell's Langelot				Ecuador
			47	2.25	
hillisi	Hillis' Bellyfold Lizard				Ecuador
			49	2.61	
macbrydei	MacBryde's Bellyfold Lizard				Ecuador
			49	2.61	
montium	Mountain Bellyfold Lizard				Colombia
			57	4.07	
prefrontalis	Montanucci's Bellyfold Lizard				Ecuador
			54	3.58	
ulisesi	Huamantanga Lizard				Peru
	diurnal		51	2.80	** litter cryptozoid
vertebralis	Rainbow Langelot				Colombia
	diurnal	invert.	58	4.42	*** wide-ranging chemosensor

Pholidoscelis

Pholidoscelis (Teiidae, formerly *Ameiva*) is a major radiation (20 spp.) of average-sized (25.3 g v. 22.1 g), solitary (100% v. 29%), mostly omnivorous (75% v. 33%), wide-ranging chemosensors (100% v. 97%) found on islands (100% v. 37%) throughout the West Indies, though absent from Grand Cayman and the southern Lesser Antilles.

Shape index (0.92 v. 0.88 std. resid), tail length (2.20 v. 2.30 × SVL), tail breakage (25% v. 21%), and blue tails (15% v. 15%) are average. Secretory pores (0% v. 9%) and color change are absent (0% v. 1.3%).

SSD is strongly m>f (1.40 v. 0.89 on −2 to 2 scale). Sexual dichromatism may be a little more frequent than average (40% v. 35%). Seasonal color is absent (v. 10%).

Their diets are diverse; generalized invertebrate eaters are notably scarce (19% v. 62%), and one species (*griswoldi*) is a carnivore (6% v. 3%). In *corax*, a dozen individuals have been observed to feed from the same plant, without apparent animosity between them. Home-range area is small, especially by teiid standards (−0.18 v. 0.19 std. resid.). At least four species are known to locomote bipedally using only hind limbs, but *maynardi* has the unusual attribute of sometimes locomoting using only forelimbs. The locomotion of *exsul* has been quantified, but its average MPM have ranged from the 26th to the 90th percentile of lizard scores; PTM was more stable (71st to 85th percentile).

Stress calls are given (5%, the

Pholidoscelis spp.	English name		Size		Range core
	Activity	Diet	mm	g	Ecological business model
atrata	Redonda Ameiva				Antigua and Barbuda
	diurnal	omnivore	84	17.1	* wide-ranging chemosensor
auberi	Auber's Ameiva				Cuba
	diurnal	invert.	95	25.1	** wide-ranging chemosensor
chrysolaema	Hispaniola Giant Ameiva				Hispaniola
	diurnal	omnivore	106	35.7	*** wide-ranging chemosensor
cineraceus	Guadeloupe Bank Ameiva				Guadeloupe
	diurnal	omnivore	117	47.8	* wide-ranging chemosensor
corax	Little Scrub Ground Lizard				Anguilla
	diurnal	omnivore	98	28.0	** wide-ranging chemosensor
corvina	Sombrero Ameiva				Anguilla
	diurnal	omnivore	104	34.0	* wide-ranging chemosensor

Pholidoscelis spp.	English name				Range core
			Size		
	Activity	Diet	mm	g	Ecological business model
dorsalis	Jamaica Ameiva				Jamaica
	diurnal	omnivore	93	23.7	* wide-ranging chemosensor
erythrocephala	Red-faced Ground Lizard				St. Kitts and Nevis
	diurnal	omnivore	98	27.9	** wide-ranging chemosensor
exsul	Puerto Rico Bank Ameiva				Puerto Rico
	diurnal	omnivore	123	55	** wide-ranging chemosensor
fuscata	Dominica Ameiva				Dominica
	diurnal	invert.	156	132	* wide-ranging chemosensor
griswoldi	Antigua Ameiva				Antigua and Barbuda
	diurnal	carnivore	122	55	*** wide-ranging chemosensor
lineolata	Pygmy Blue-tailed Ameiva				Hispaniola
	diurnal	omnivore	49	3.28	* wide-ranging chemosensor
major	Terre de Bas Ameiva				Guadeloupe
			150	105	
maynardi	Great Inagua Ameiva				Bahamas
	diurnal	invert.	61	6.4	* wide-ranging chemosensor
plei	Anguilla Bank Ameiva				Lesser Antilles
	diurnal	omnivore	119	51	**** wide-ranging chemosensor
pluvianotata	Montserrat Ameiva				Montserrat
	diurnal		130	67	*** wide-ranging chemosensor
polops	St. Croix Ameiva				US Virgin Islands
	diurnal	omnivore	59	5.7	*** wide-ranging chemosensor
taeniura	Hispaniola Blue-tailed Ameiva				Hispaniola
	diurnal	omnivore	84	17.6	* wide-ranging chemosensor
umbratilis	Shady Ameiva				Dominican Republic
			102	31.9	* wide-ranging chemosensor
wetmorei	Puerto Rico Blue-tailed Ameiva				Puerto Rico
	diurnal		46	2.68	** wide-ranging chemosensor

highest proportion in Teiidae, v. 0.6%). One-fifth are conspicuously wary (20% v. 13%). Direct basking is average (55% v. 58%), and mean active body temperature is near average (35.2°C v. 36.9°C).

In keeping with their solitary ways, only 25% (v. 55%) of males are socially tolerant; 75% (v. 45%) defend space. Half of the species have males or both sexes that develop social hierarchies (50% v. 16%). This is a trait that tends to emerge at high population densities (see below). Mate guarding occurs in *plei*, though females may reject the consortship of lower-ranking males. In general, this genus seems beset by pugnacious territorial holders, an atypical state of affairs for teiids.

They are found in the intermediate rainfall belt (100% v. 50%) at low (100% v. 95%) and occasionally medium (8% v. 32%) elevation. They are overrepresented in littoral (71% v. 38%), sandy (65% v. 47%), and human (59% v. 29%) habitats. They are comparatively scarce in riparian zones (29% v. 46%).

Clutch counts are moderately low (–0.77 v. –0.51 std. resid.), and young are notably small (ontogenetic mass range 0.76 v. 0.10 std. resid.), suggesting that RCM ratio should be low. Data on that are available only for *exsul* and *wetmorei* (mean 0.170 v. 0.163). Incubation duration is about average (–0.47 v. –0.38 std. resid.). Communal oviposition is present (*chrysolaema* 5.0% v. 2.6%). This species is interesting in that females share a nest burrow entrance but do not guard their nest against accidental excavation by later arriving females.

As befits island species, population density is high (3.17 kg/ha v. 1.06 kg/ha), based on 25 credible samples of seven species. Endemism is high (65% v. 50%). Ten percent of *Pholidoscelis* have been introduced (v. 8%), but none has been introduced more than twice (v. 0.6%). Most species are imperiled (73% v. 41%), including a disturbing 40% (v. 12%) that are on one of the IUCN's lists of the most endangered. Mongooses are held responsible for at least four of the population declines. In other respects *Pholidoscelis* are typical teiids.

Phoxophrys

Phoxophrys is a minor (five spp.) genus of small (4.7 g v. 15.6 g), weakly crested (80% v. 38%), large-headed, arboreal (100% v. 43%) agamids from the Greater Sunda Islands of Indonesia and Malaysia. Most seem to be drawn to small-diameter perches, and this diurnal canopy walking activity is reflected in a preference for crypsis (100% v. 51%), slow movement, and a high frequency of advanced color change (40% v. 15%). Most species have a ridge above the eye, leading to the common names referencing an eyebrow. SSD is not reported, and sexual dichromatism is found at an agamid-typical frequency (60% v. 53%). *Phoxophrys* are not associated with littoral (0% v. 10%) or sandy (0% v. 26%) habitats. Clutch counts are low (−0.34 v. 0.20 std. resid.), equating to 2.7 eggs (v. 6.4). Endemism is average (40% v. 49%). All species occur on islands (v. 36%). None has been introduced (v. 1.7%) or evaluated for conservation status. Three species (*cephalum*, *nigrilabris*, *spiniceps*) are noted for facing an oppressor and opening their mouth when stressed, though their mouth lining does not exhibit any startle colors. *P. spiniceps* has conspicuous spines above the eye and rear of cranium.

Phoxophrys spp.	English name		Size		Range core
	Activity	Diet	mm	g	Ecological business model
borneensis	Sabah Eyebrow Lizard				Borneo
			57	5.7	
cephalum	Large-headed Eyebrow Lizard				Borneo
	diurnal		69	10.8	* ground-foraging tree lizard
nigrilabris	Black-lipped Eyebrow Lizard				Borneo
	diurnal	invert.	51	4.10	* trunk/wall lizard or diurnal canopy walker
spiniceps	Sarawak Eyebrow Lizard				Borneo
	diurnal		52	4.47	* diurnal canopy walker
tuberculata	Sumatra Eyebrow Lizard				Indonesia
	diurnal		40	1.97	

Phrynocephalus

Phrynocephalus is a speciose (31 spp.) radiation of small (4.6 g v. 15.6 g), crestless (v. 38%), dry-zone, terrestrial (100% v. 56%) agamids in central Asia, from Turkey and Yemen east through Russia to Mongolia and south to India. With the exception of *mystaceus*, males of which can be sand sharks, they are all known to be refugia-anchored diurnal ground lizards (96% v. 43%).

For agamids they are short-tailed (1.18 v. 1.86 × SVL) and have a low frequency of color change (10% v. 22% for any; 0% v. 15% for advanced), sexual dichromatism (29% v. 53%), and seasonal color (10% v. 20%). *Phrynocephalus* have few sex differences overall (0.65 v. 1.61 on 0–6 scale), a slightly elevated frequency of active foraging (33% v. 17%), and a low reliance on crypsis (24% v. 51%). Some flee (18% v. 7.6%), many dive into sand (35%, the second-highest frequency in Agamidae, v. 6.7%), but few dive into crevice-burrows (18% v. 29%) or vegetation (6% v. 17%). Home-range area is the second highest in Agamidae (1.06 v. 0.35 std. resid.), likely reflecting the low productivity of their environment.

Little is known about their social lives. They are found in the hyperdry (17% v. 17%); they are the highest agamid genus in the dry (83% v. 30%), representative in the arid (33% v. 35%), and absent from intermediate (v. 12%) and wetter locales. They are especially frequent in sand and dunes (82% v. 26%) but absent from human habitats (v. 16%). *Phrynoceph-*

Phrynocephalus spp.	English name		Size		Range core
	Activity	Diet	mm	g	Ecological business model
ahvazicus	Khuzestan Toad-headed Agama				Iran
			46	2.98	
ananjevae	Natalia's Toad-headed Agama				Iran
			45	2.83	
arabicus	Arabia Toad-headed Agama				Saudi Arabia
	diurnal	invert.	51	4.08	** refugia-anchored diurnal gr. liz. or immobility expert
axillaris	Yarkand Toad-headed Agama				China
		invert.	46	3.03	* refugia-anchored diurnal gr. liz.
clarkorum	Afghan Toad-headed Agama				Afghanistan
	diurnal	invert.	42	2.20	** refugia-anchored diurnal gr. liz.
erythrurus	Sagus Kul Lizard				China
			49	3.75	
euptilopus	Alcock's Toad-headed Agama				Afghanistan
	diurnal	invert.	54	5.1	*** refugia-anchored diurnal gr. liz.

alus range up to 5400 m in elevation and into Mongolia in latitude, which may be reflected in the high proportion (22% v. 2%) of viviparous species. Clutches on average are small (–0.16 v. 0.20 std. resid.). Incubation duration is average (–0.56 v. –0.55 std. resid.). Hatchlings/neonates are relatively large, as reflected in the small ontogenetic range (0.41 v. 0.69 std. resid.), corresponding to young averaging 1/44 of maximum mass rather than 1/104.

Longevity is slightly short in relation to lizards (–0.19 v. 0.0 std. resid.) but long in relation to the mostly tropical agamids (v. –0.46). Population density estimates are available for seven species and are low (0.01 v. 0.16 kg/ha), likely reflecting the low productivity of the cold, dry, climate. *P. mystaceus* has been claimed to have a much higher population density (mean of 1.75 kg/ha; maximum of 3.4 kg/ha), but those two estimates were judged to be nonqualified. No *Phrynocephalus* has been introduced (v. 1.7%). About half of the evaluated species are deemed imperiled (54% v. 37%), including 15% (v. 5%) that are on the IUCN's lists of the most endangered.

Phrynocephalus spp.	English name	Activity	Diet	Size mm	Size g	Range core / Ecological business model
forsythii	Xinjiang Toadhead Agama			53	4.54	China
frontalis	Shansi Toadhead Agama		invert.	49	3.57	China * refugia-anchored diurnal gr. liz.
golubewii	Golubev's Toad Agama	diurnal	invert.	58	6.2	Turkmenistan * refugia-anchored diurnal gr. liz.
guttatus	Spotted Toad-headed Agama	diurnal	omnivore	54	4.90	Mongolia ** refugia-anchored diurnal gr. liz.
helioscopus	Sunwatcher	diurnal	omnivore	56	5.4	Central Asia ** refugia-anchored diurnal gr. liz.
interscapularis	Rose-shouldered Toad Agama	diurnal	invert.	36	1.39	Central Asia ** refugia-anchored diurnal gr. liz.
lutensis	Lut Desert Toad-headed Agama			63	8.0	Iran
luteoguttatus	Yellow-speckled Toad-headed Agama	diurnal	omnivore	43	2.44	Afghanistan *** refugia-anchored diurnal gr. liz.
maculatus	Blacktail Toad-headed Agama	diurnal	omnivore	74	13.4	Iran *** refugia-anchored diurnal gr. liz.
mystaceus	Bearded Toad-headed Agama	diurnal	omnivore	88	22.4	Central Asia **** sand shark (male) and *** refugia-anchored diurnal gr. liz. (female)
ornatus	Striped Toad-headed Agama	diurnal	invert.	43	2.44	Afghanistan *** refugia-anchored diurnal gr. liz.
persicus	Persia Toad-headed Agama	diurnal	invert.	50	4.00	Turkey * refugia-anchored diurnal gr. liz.
przewalskii	Przewalski's Toad-headed Agama		invert.	61	7.2	China * refugia-anchored diurnal gr. liz.
putjatai	Guide Toad-headed Agama			74	13.0	China
raddei	Rose-spotted Toad Agama		invert.	51	4.10	Turkmenistan ** refugia-anchored diurnal gr. liz.

Phrynocephalus spp.	English name				Range core
			Size		
	Activity	Diet	mm	g	Ecological business model
reticulatus	Reticulated Toad-headed Agama				Afghanistan
	diurnal	omnivore	49	3.59	**** refugia-anchored diurnal gr. liz.
roborowskii	Roborowski's Toad-headed Agama				China
			67	9.7	
rossikowi	Khentau Toad Agama				Turkmenistan
			45	2.83	* refugia-anchored diurnal gr. liz.
sakoi	Oman Toad-headed Agama				Oman
			52	4.38	
scutellatus	Gray Toad-headed Agama				Afghanistan
	diurnal	ants	49	3.75	**** refugia-anchored diurnal gr. liz.
strauchi	Strauch's Toad-headed Agama				Uzbekistan
	diurnal		45	2.83	** refugia-anchored diurnal gr. liz.
theobaldi	Snow Lizard				Himalayas
	diurnal	omnivore	53	4.73	**** refugia-anchored diurnal gr. liz.
versicolor	Piebald Toad-headed Agama				Mongolia
	diurnal	invert.	53	4.73	** refugia-anchored diurnal gr. liz.
vlangalii	Ching Hai Toad-headed Agama				China
			65	8.7	** refugia-anchored diurnal gr. liz.

Phrynosoma (figs. 4.7B, 5.3B)

Phrynosoma is a major radiation (22 spp.) of average-sized (14.8 g v. 11.2 g) phrynosomatid immobility experts (100% v. 16%). They are found in the western Nearctic from British Columbia south to central Mexico. With their pancake-shaped spiny bodies and placid demeanor, they are the prized possession of nearly every child who discovers one. This enchantment follows herpetologists as they get older (filled cells 42% v. 38%), despite the difficulty finding these cryptic lizards on demand. Ironically, because of their specialized lifestyle, they share few natural history themes with the other family members; *Phrynosoma* traits are generally not those of phrynosomatids.

Phrynosoma approach life as a waiting game, maintaining extreme crypsis and rarely moving except to avoid being stepped on or to change the places where they sit-and-wait. They eat only what walks in front of their faces, which is often ants. Some species of ants can mass and attack their predators, so horned lizards need to be hard for ants to see and some species must change their ambushing position regularly to give time for an aggrieved colony to relax. *Phrynosoma* have a disruptive silhouette unlike any lizard except the ant-eating *Moloch*; one of many hypotheses for this unusual silhouette (see "Diet" in chapter 3 and table 3.2, trait 10) is that this unusual profile may make it hard for ants to detect them. The *Phrynosoma* species that roam over large areas often live in sandy places, which facilitates sleeping where dusk finds them, as they can merely shimmy into the omnipresent sand. This roaming lifestyle does not lend itself to male territorial defense or the large male size necessary for territory acquisition. Instead, the relatively larger females have exceptionally large clutches and promote their offspring's survival with parental care. How does this hypothesized narrative play out in quantified natural history attributes?

The crypsis of *Phrynosoma* is legendary among those that search for them, but that is an unquantified metric. *Phrynosoma* are reported to rely on crypsis as a primary predator concealment tactic at a frequency that is the highest in Phrynosomatidae: 92% v. 32%. Along with this is their penchant for not moving unless they are about to be stepped on, which means their immobility is misconstrued by humans as unwariness (0% v. 22% conspicuously wary). Morphologically, crypsis is manifest in their pancake shape (shape index 1.03 v. 0.91 std. resid.) and their exceedingly short tails (0.40 v. 1.22 × SVL). Along with rudimentary tails goes reduced tail breakage (0% v. 21%). Bright blue tails are out (0% v. 1.3%). Extreme crypsis also implies foregoing sexual dichromatism (4.5% v. 58%) and embracing basic color change (50% v. 20%) to background match. Because intraspecific communication is not visual, *Phrynosoma* might rely on chemical communication (secretory pores 77% v. 32%). Most likely such chemical cues in low density lizards such as *Phrynosoma* facilitate mate finding. Another possibility is that secretory pores are promoting chemical crypsis from the perspective of the lizards' prey; potentially the chemical signals mask detection by intensively chemosensory ants? *Moloch* lacks secretory pores (see table 5.4 in *Moloch* account) and pores are generally avoided in ant specialists (table 3.2, trait 10).

Phrynosoma are among the lizards most specialized for ant eating: 83% (v. 17%). Ants are a major dietary item even in the species that are not ant specialists (*brevirostris*, *ditmarsi*, *goodei*). All ants are not created equal; they differ in the ferocity of their stings and bites, and in their propensity to trigger massed counterattacks. Overall, *Phrynosoma* have large home ranges (1.23 v. 0.57 std. resid.) and such large home ranges have been suggested to lighten the predatory pressure on any single ant colony and thereby minimize massed counterattacks by dangerous ants. *Moloch horridus* eats ants that are not aggressive, and it has only a tiny documented home range (100 m^2; see that account). One curiosity is that some species of *Phrynosoma* are wide-ranging (e.g., *modestum* home range 2728 m^2; *cornutum* 10,800 m^2; *platyrhinos* 14,780 m^2; *mcallii* 10,000–100,000 m^2) and some are not (*douglasii* 764 m^2; *solare* 110 m^2). Do the wide-ranging species wander to minimize retaliation by the ants (e.g., *cornutum*), or to avoid depleting the ants in a small area?

The second paragraph noted that the wide-ranging *Phrynosoma* often live in sandy areas (55% v. 32%). They are also good diggers; despite their unstreamlined shape, 78% (v. 29%) may use burrows as nightly retreats.

A wandering lifestyle does not lend itself to territorial defense, as noted above. *Phrynosoma* are conspicuously divergent from other phrynosomatids in their lack of male territorial defense (0% v. 90%). They are socially tolerant (100% v. 11%). This may be reflected in m<f dimorphism (–1.03 v. 0.27 on –2 to 2 scale). Alternately, one might view the relatively large size of female horned lizards as evidence of strong selection pressure for output of young. *Phrynosoma* have conspicuously large clutch/litter counts (2.12, the highest value in the family, v. 0.68 std. resid.). These aspects highlight how aberrant *Phrynosoma* are from the rest of the Phrynosomatidae. Is the relatively large female size a result of fecundity selection? Does fecundity selection drive the evolution of pancake-shaped bodies (absent in *Moloch*, *Draco*, and many other ant-eating lizards)? My guess is that vegetation provides the necessary concealment for *Moloch*, *Draco*, and many other ant-eating lizards, whereas *Phrynosoma* are often immobile out in the open and therefore are under pressure to avoid casting shadows that would reveal their presence to birds and other predators that search visually. A pancake-shaped body that tapers at the edges of the body and tail (i.e., *Phrynosoma*, *Uroplatus*) minimizes shadows. A pancake-shaped body also allows for larger clutches/litters. This comports well with *Phrynosoma*'s large RCMs (0.342 v. 0.300), the highest in Phrynosomatidae. Note that *Phrynosoma* do not have young of exceptional size (ontogenetic range 0.04 v. –0.26 std. resid.); the small elevation of this value indicates that their young are relatively small.

Parental care (4.5% v. 1.9%) is another way to elevate the recruitment of young. In an unusual case, parental care in *Phrynosoma* takes the form of mothers attacking small snakes (*Salvadora hexalepis*) that eat *Phrynosoma* eggs (Sherbrooke 2017; W. Sherbrooke, pers. comm. 2018). Mother *solare* have been observed rolling freshly deposited eggs in soil to coat them, perhaps to obscure color, prevent adhesion, or alter scent. Young from viviparous species may remain with the mother after parturition and obtain some protection (*hernandesi*). Incubation duration in *Phrynosoma* is average (–0.62 v. –0.58 std. resid.). Differences between *Phrynosoma* and *Moloch* are summarized in table 5.4 (*Moloch* entry).

Increased longevity might be another way to increase lifetime recruitment. Among phrynosomatids it is second highest in *Phrynosoma* (0.17 v. –0.25 std. resid.). This equates to 13.7 years in *asio* but only 5 years in *modestum*. The median maximum age is 8 years (7 of 11 species are 8–9 years).

Because humans are so readily charmed by these innocuous lizards, *Phrynosoma* are often introduced to new places (9.1% v. 3.9%). In part because ants easily stow-away in cargo, they too are often translocated and the ensuing disruption of ant communities has had detrimental impacts on the conservation status of *Phrynosoma* (75% v. 37% are imperiled), though none is on one of IUCN's list of the most endangered (v. 14%), probably because most *Phrynosoma* have large geographic ranges and are capable of eating alternate prey.

The preceding narrative highlights the ways in which *Phrynosoma* differ from other phrynosomatids; in a number of less critical ways *Phrynosoma* appear to be more typical phrynosomatids. For example, advanced color change is rare in both groups (4.5% v. 4.5%). The ratio of m:f relative tail length is moderately higher in *Phrynosoma* (1.24 v. 1.13), perhaps because such short tails do not leave sufficient room for the males' hemipenes. The sex differences rollup score is not highly divergent in *Phrynosoma* (1.82 v. 1.68 on 0 to 6 scale), despite the very different ways in which sex differences are manifest: bright colors in many refugia-anchored phrynosomatids, crypsis in *Phrynosoma*. Both *Phrynosoma* and Phrynosomatidae are heavily reliant on sit-and-wait foraging (80% v. 98%); the difference is that some *Phrynosoma* cruise forage (20%), whereas few other phrynosomatids do (3.3%). All phrynosomatids are strictly diurnal (100% v. 100%), though individuals of many taxa (e.g., *P. asio, cornutum, modestum*) may continue activity after dark in favorable weather. Direct basking is common (41% v. 40%). Body temperatures do not differ (34.1°C v. 34.2°C). Terrestriality is common (95% v. 77%). *Phrynosoma* are found in a broad range of rainfall areas (dry to mesic: 65–2125 mm/y), with a mode near the center of the spectrum (arid: 151–400 mm/y); phrynosomatids have a slightly wider range (hyperdry to mesic) and a slightly wetter mode (intermediate: 401–1100 mm/y). Aridland *Phrynosoma* of several species (*cornutum, platyrhinos*) are capable of collecting dew or rain that falls on their back, transferring it to the mouth by capillary action, a skill also manifest in *Moloch*. Potentially the forest of spines possessed by both *Moloch* and *Phrynosoma* play a role in creating the surface capillaries that channel moisture to their mouths. The elevational profile of *Phrynosoma* barely differs from phrynosomatids (l/m/h: 79/79/26 v. 73/76/36). Their use of littoral (5% v. 16%), riparian (30% v. 40%), and human (5% v. 18%) habitats are also broadly similar.

Reproduction is slightly more often viviparous in *Phrynosoma* (53% v. 35%), possibly a reflection of their higher latitudes. Incubation period is similarly short (−0.62 v. −0.58

Phrynosoma spp.	English name	Activity	Diet	Size mm	Size g	Range core	Ecological business model
asio	Giant Horned Lizard	diurnal	ants	109	40.2	Mexico	*** immobility expert
bauri	Baur's Short-horned Lizard	diurnal	ants	70	12.0	Colorado	*** immobility expert
blainvillii	Blainville's Horned Lizard	diurnal	ants	70	12.1	Mexico	*** immobility expert
braconnieri	Short-tailed Horned Lizard	diurnal	ants	64	9.4	Mexico	*** immobility expert
brevirostris	Great Plains Short-horned Lizard	diurnal	invert.	62	8.9	Rocky Mountain states	*** immobility expert
cerroense	Cedros Island Horned Lizard			70	12.2	Mexico	*** immobility expert
cornutum	Texas Horned Lizard	diurnal	ants	88	22.1	Texas	**** immobility expert
coronatum	Coast Horned Lizard	diurnal	ants	87	22.0	Mexico	*** immobility expert
diminutum	San Luis Valley Short-horned Lizard	diurnal		53	5.9	Colorado	*** immobility expert
ditmarsi	Rock Horned Lizard		invert.	74	14.3	Mexico	*** immobility expert
douglasii	Pygmy Short-horned Lizard	diurnal	ants	64	15.5	Oregon	**** immobility expert
goodei	Goode's Horned Lizard	diurnal	invert.	75	14.7	Mexico	*** immobility expert
hernandesi	Mountain Short-horned Lizard	diurnal	ants	86	21.0	Rocky Mountain states	*** immobility expert
mcallii	Flat-tailed Horned Lizard	diurnal	ants	77	15.9	California	*** immobility expert
modestum	Round-tailed Horned Lizard	diurnal	ants	57	7.1	New Mexico	**** immobility expert
orbiculare	Mountain Horned Lizard	diurnal	ants	81	17.7	Mexico	*** immobility expert
ornatissimum	Desert Short-horned Lizard			88	22.4	Mexico	** immobility expert
platyrhinos	Desert Horned Lizard	diurnal	ants	76	22.7	Western United States	**** immobility expert
sherbrookei	Sherbrooke's Horned Lizard	diurnal	ants	53	5.7	Mexico	*** immobility expert
solare	Regal Horned Lizard	diurnal	ants	93	54	Mexico	*** immobility expert

Phrynosoma spp.	English name				Range core
			Size		
	Activity	Diet	mm	g	Ecological business model
taurus	Bull Horned Lizard				Mexico
	diurnal	ants	76	15.1	** immobility expert
wigginsi	Gulf Coast Horned Lizard				Mexico
	diurnal		66	10.3	** immobility expert

std. resid.). Rapid development may reflect the short summer growing season at high latitudes. Population density is on the low side (0.13 kg/ha v. 0.26 kg/ha), but well within the range of variation in Phrynosomatidae (0.04 kg/ha in *Callisaurus*; 1.10 in *Uta*). Anecdotal accounts support the notion that prior to disruption of native ant communities, *Phrynosoma* were often present at much higher densities (*blainvillii, cornutum*). Although immobility experts are often associated with low population density (table 4.1, trait 71), one does not imagine an ant specialist to be limited by prey availability under natural conditions. Too high a population density would facilitate predators' acquisition of a search image for the lizards, but *Phrynosoma* are rarely so common. Endemism is a bit lower in *Phrynosoma* (14% v. 21%), no doubt reflecting their continental range.

Phrynosoma coronatum are unique in copulating belly to belly (female lying on her back) and the male biting the female under the chin.

Phrynosoma of at least eight species exhibit an extraordinary predator defense, primarily against canids or felids. When caught in the open and unable to retreat, by sudden vasoconstriction they puncture an artery in the corner of their eyes and squirt blood into the open mouths of the approaching predator. *Phrynosoma* blood contains compounds—presumably derived from the noxious chemicals of their ant prey—that are offensive to canids and felids but not to other predators, which is why *Phrynosoma* rarely squirt blood at other predators such as humans.

P. coronatum is believed by residents of Baja California to possess spines capable of burning human flesh; *P. solare* are believed to be deadly.

The explorer David Douglas (1799–1834) is the honoree commemorated in *P. douglasii*. He collected many type specimens (e.g., Douglas Fir trees) in many parts of the world. He died by falling into a pitfall trap, into which he was preceded by a feral bull, which gored him upon his descent. Field work can be nasty, brutish, and short.

Phrynosomatidae (figs. 2.5A, 4.7B, 4.11A, 4.11B, 4.11D, 5.3B)

The Phrynosomatidae are a large North American radiation (154 spp.) of relatively well-studied (filled cells 38% v. 27%) former iguanid genera: *Callisaurus, Cophosaurus, Holbrookia, Petrosaurus, Phrynosoma, Sceloporus, Uma, Urosaurus,* and *Uta.* Many species reside in Mexico, but the family extends from Panama to Canada. Despite the name association, *Phrynosoma* is an ecological outlier within the Phrynosomatidae; most phrynosomatids resemble other genera, such as the diverse (100 spp.) *Sceloporus*. The family is notably diurnal (100% v. 67%), terrestrial (77% v. 48%), Nearctic (88% v. 6%), and ecologically diverse (six ecological business models). Most are refugia-anchored diurnal ground lizards (52% v. 24.5%). Others are ground-foraging tree lizards (20% v. 9.2%), immobility experts (16.3% v. 3.2%), trunk/wall lizards (10.4% v. 13%), diurnal canopy walkers (0.7% v. 6.4%), and litter cryptozoids (0.7% v. 11.6%). The family has the third-highest frequency of sexual dichromatism (58% v. 21%), the third-highest proportion of ant specialists (17% v. 4.4%), the third-highest representation at high elevation (36% v. 13%), and the fourth-highest RCM (0.300 v. 0.226).

Phrynosomatids are large (11.2 g v. 5.3 g) and robust (shape index 0.91 v. 0.0 std. resid.). Tails are on the short side (1.22 v. 1.37 × SVL) with fewer species having breakable tails (21% v. 57%). Phrynosomatids lack prehensile tails (v. 5.0%), crests (v. 5.3%), and elongation (v. 11.4%), but frequently possess secretory pores (32% v. 15%) and color change (19.5% v. 11%). A few exhibit advanced color change (4.5% v. 6.0%) and blue tails (1.3% v. 2.3%).

SSD takes the usual m>f form, but weakly so (0.27 v. 0.36 on −2 to 2 scale). In addition to the very high sexual dichromatism noted above, phrynosomatids have the fourth-highest frequency of seasonal color (25.3% v. 6.5%) and a greater than average m:f ratio of relative tail lengths (1.13 v. 1.07, cf. table 3.3, trait 8). These combine into a notably high sex rollup score of 1.68 (v. 0.75) on a scale of 0 to 6.

In addition to the league-ranked myrmecovory noted above, all diets except carnivory are found in the Phrynosomatidae: herbivory (1.0% v. 4.1%), invertebrate eating (62% v. 73%), and omnivory (20% v. 15.4%). Almost all species are said to forage by sit-and-wait (98% v. 46%), though many observers diverge from my practice by treating foraging mode as a family-invariant trait. My tabulation shows that at least 8.3% (v. 58%) of phrynosomatids exhibit active foraging; the true value must be at least 34% to cover the plant eaters) and 3.3% (v. 6.6%) cruise forage. Electric light foraging is not reported (v. 1.4%). Home-range size is large (0.57 v. 0.0 std. resid.).

Despite their diurnality, 0.6% (v. 2.2%) give stress vocalizations, though none (v. 2.5%) calls spontaneously. Twenty-two percent are said to be conspicuously wary (v. 8.2%). All primary predator concealment tactics are well represented within Phrynosomatidae: crypsis (32% v. 43%), flight (14% v. 11%), sand (12% v. 6%), crevice-burrows (32% v. 25%), and retreat to vegetation (24% v. 22%).

Direct basking is a phrynosomatid staple (40% v. 18%) and thermoconforming is not reported (v. 2.6%). Mean active body temperature is 34.2°C (v. 31.1°C). As many have noted, these often-temperate lizards are active at tropical body temperatures.

Solitary living has not been reported for phrynosomatids (v. 10%); group living is about normal (19% v. 20%). A small minority of phrynosomatid males are socially tolerant (11% v. 15%), and male defense of space is routine (89% v. 85%). Within the routine group showing defense of space, it is notable that male phrynosomatids are in second place for exhibiting year-round defense (4.5% v. 1.4%), and females are in first place (21% v. 3.4%). In part, documentation of year-round phenomena may reflect more thorough study of predominantly North American lizards. A small number of species are documented to exhibit male hierarchies (3.0% v. 7.7%).

Although 77% are terrestrial, 22% (v. 32%) are arboreal, and a small number (0.7% v. 12%) are cryptozoic. Twenty-nine percent (v. 22%) use burrows at some time as their nightly retreats.

Phrynosomatids are overrepresented at the dry end of the rainfall spectrum. Specifically they are overrepresented in the dry (56–150 mm/y: 24% v. 12%), arid (151–400 mm/y: 48% v. 30%), and intermediate (401–1100 mm/y: 52% v. 27%). Correspondingly, they are underrepresented in the mesic (1101–3000 mm/y: 28% v. 43%) and absent from the wet (v. 9%). As noted in the opening paragraph, they are particularly overrepresented at high elevation (36% v. 13%); they are also frequent at mid elevations (76% v. 41%). Correspondingly they are underrepresented at low elevation (73% v. 82%). They are overrepresented in sandy areas (32% v. 22%) and riparian zones (40% v. 22%) and present in littoral zones (16% v. 17%) and human habitats (18% v. 18.5%) at typical frequencies.

Oviparity is universal in all phrynosomatid genera except *Phrynosoma* (47%) and *Sceloporus* (56%), in both cases viviparity is best developed at high latitude and high elevation. The family mean for oviparity (65%) is lower than the lizard mean (82%), presumably because phrynosomatids so often occur at high latitude and high elevation. Clutch/litter count is high (0.68 v. 0.0 std. resid.); the genus *Phrynosoma* is a notable outlier (2.11 std. resid.). The average phrynosomatid young is slightly large, as indicated by the ontogenetic mass range (−0.26 v. 0.0 std. resid.), with some notable outliers. *Uta* are small (1.31 std. resid.) and *Uma* are large (−1.20 std. resid.). Incubation is short (−0.58 v. 0.0 std. resid.) and is negative (less than mass-adjusted lizard average) for each phrynosomatid genus, possibly a reflection of their high active body temperatures (in viviparous species) or the requirement to expedite development during the short high-latitude summer (in high-latitude species). Eggs are soft and nonadherent in all cases. Communal oviposition is known only in *Uta* (0.6% family average v. 4.2%). Parental care is recorded in *Phrynosoma*, *Sceloporus*, and *Urosaurus* (1.9% family average v. 2.9%). As noted in the first paragraph, RCM is high in Phrynosomatidae. Is this a reflection of available peritoneal space? It is highest in *Phrynosoma* (0.342; shape index 1.03) and *Urosaurus* (0.302; shape index 0.64) and lowest in *Callisaurus* (0.154; shape index 0.85) and *Holbrookia* (0.165; shape index 0.80). Note that all of these shape index values are above zero (i.e., above average), but the RCMs straddle the lizard average (0.226). There is no statistical relationship between RCM and shape index in lizards overall ($r^2 = 0.02$, $p = 0.99$), presumably because many lizards' clutch counts are constrained more by phylogeny than peritoneal volume.

Longevity is slightly short in Phrynosomatidae: −0.25 (v. 0.0 std. resid.). Population density in the family is exceptionally well documented, with numerous credible population estimates for every genus except *Petrosaurus*. On average, however, Phrynosomatidae are slightly sparser than typical lizards (0.26 kg/ha v. 0.54 kg/ha), an unsurprising result for continental lizards in relatively unproductive environments.

Endemism is about average (58% v. 64%). Relatively few species occur on islands (21% v. 40%). Introductions are slightly more frequent in this family (3.9% v. 3.3%), though fewer than average have been introduced more than twice (0.6% v. 1.1%). The frequency of imperiled species is moderately low (37% v. 49%), as is the frequency of species on the IUCN's lists of most endangered (5.4% v. 13.6%). The arid lands primarily occupied by Phrynosomatidae have attracted relatively sparse human development.

Phyllodactylidae

The Phyllodactylidae (135 spp.) are an odd mix of nocturnal+cathemeral, arboreal+terrestrial lizards found in two clusters: the entire Neotropics north to Southern California, and the northern African tropics from the Cape Verde Islands to Ghana and Somalia, north to Portugal and Pakistan. Though hardly global, they are entirely absent only from the Australasian and Oriental biogeographic realms. The dominant ecological business models are also somewhat disparate: trunk/wall lizards (44% v. 13%) and nocturnal ground lizards (40% v. 8%). The intervening microhabitat is weakly represented: ground-foraging tree lizards (16% v. 9%), with a crack hermit (1% v. 2.6%) thrown in for good measure. Although ground lizards and trunk/wall lizards seem at first brush to be separated microhabitats, there are eight examples, albeit solely in *Phyllodactylus* and *Ptyodactylus*, of nocturnal ground lizards being diverted by electric lighting into becoming trunk/wall lizards. Why crawl across the mammal-infested ground when one can effortlessly score innumerable moths beating themselves to death against an electric light? Curiously, there are no examples of phyllodactylid ground-foraging tree lizards becoming trunk/wall lizards, though that would seem like an easier transition. The following genera make up Phyllodactylidae: *Asaccus, Garthia, Gymnodactylus, Haemodracon, Homonota, Phyllodactylus, Phyllopezus, Ptyodactylus, Tarentola,* and *Thecadactylus*.

The Phyllodactylidae are notable among lizard families for the number of attributes for which they appear at or near the top of the league tables. For example, they are the third-highest family for nocturnal ground lizards (40% v. 7.7%). They are the second-highest family for trunk/wall lizards (44% v. 13%). They are the top family for electric light use (9.2% v. 1.4%). Phyllodactylids call spontaneously at a higher frequency (11% v. 2.5%) than any families other than pygopodids and carphodactylids. Phyllodactylids have the highest representation in the hyperdry rainfall band (27% v. 3%) and the fourth-highest representation in the dry band (36% v. 12%). They appear in human habitats at the second-highest frequency (40% v. 18.5%). Their eggshells are adherent at a frequency (10.4% v. 2.4%) second only to Gekkonidae, and they are ranked third for communal oviposition (16.3% v. 4.2%).

Phyllodactylids are ordinary in mean size (4.74 g v. 5.33 g) and build (shape index 0.37 v. 0.0). Their tails are short (1.03 v. 1.37 × SVL) and very breakable (92% v. 57%). A few tails are prehensile (2.2% v. 5.0%). No phyllodactylids are crested (v. 5.3%), elongate (v. 11.4%), or in possession of secretory pores (v. 15%). The latter is somewhat unexpected for nocturnal/cathemeral species (table 3.2, traits 17, 20). Although 11% (v. 11%) of phyllodactylids darken, more advanced color change is absent (v. 6.0%), as are blue tails (v. 2.3%) and seasonal color change (v. 6.5%). Phyllodactylids are mostly drab. In general, the terrestrial species lack enlarged toepads and the arboreal species possess them (see phyllodactylid genus accounts). Sexual dichromatism is present but underrepresented (6.7% v. 21%). SSD is weakly m>f (0.32 v. 0.36 on –2 to 2 scale). In contrast, the m:f ratio of relative tail lengths is high (1.14 v. 1.07 ratio), the fifth-highest family. In aggregate, these sex differences are relatively weak: 0.31 (v. 0.75) on 0 to 6 sex differences rollup scale.

All diets except herbivory are found within Phyllodactylidae, but invertebrate eating dominates (93% v. 73%). Carnivory (1.8% v. 3.2%), myrmecovory (1.8% v. 4.4%), and omnivory (3.6% v. 15.4%) are all conspicuously underrepresented. The commitment of Phyllodactylidae to sit-and-wait foraging (94% v. 46%) is notable (fifth place among families). Cruise foraging is also overrepresented (12.5% v. 6.6%), but active foraging is not (31% v. 58%). Home-range area is slightly large (0.34 v. 0.0 std. resid.), but this value is based on a single species (*Thecadactylus rapicauda*) in an aberrant environment (gazebo).

The characterization of diel activity patterns in phyllodactylids is remarkably controversial. At least 44 species (of 135) had contradictory assessments of their activity patterns. Nineteen percent of the total (v. 6.8%) were judged to be active at all times of day (cathemeral), whereas another 14% of the total were categorized as nocturnal-contested. In combination with the 64% of the phyllodactylid sample that were nocturnal-uncontested, the range of nocturnal frequency in the family is 64%–78% (v. 25%). Almost no species are diurnal (0.8% v. 67%) or crepuscular (2.3% v. 1.8%). Thus the modal pattern in Phyllodactylidae is nocturnality, with 19%–33% (v. 6.8%) of species either cathemeral or arguably cathemeral.

Cathemeral behavior is associated with spontaneous calling (table 3.7, trait 31), as is nocturnality to a lesser degree. This is consistent with the notably high level of spontaneous calling in phyllodactylids (11% v. 2.5%). It is also consistent with the high frequency of stress calls (table 3.7, trait 32) in phyllodactylids (5.9% v. 2.2%).

Nocturnal species tend not to be conspicuously wary (table 3.7, trait 35), and phyllodactylids fit this pattern (1.5% v. 8.2% wary). Their boldness around humans may also be manifest in their penchant for electric light feeding (9.2% v. 1.4%) and occupancy of human habitats (40% v. 18.5%).

Their frequencies of primary predator concealment tactics show no flagrant deviations from lizard norms: crypsis (43% v. 43%), flight (14% v. 11%), sand (0% v. 5.8%), crevice-burrows (29% v. 25%), and vegetation (14% v. 23%). The latter underrepresentation is probably due in part to the paucity of vegetation in their Old World ranges—Sahara to Middle Eastern deserts; the absence of sand diving is less likely to follow the same attribution.

Direct basking is infrequent (9.6% v. 18%), likely reflecting their nocturnality or difficulty detecting indirect basking at night. Their low active body temperature (26.0°C v. 31.1°C) is probably also due to nocturnality.

Little is known about the social lives of phyllodactylids, but the easily quantified traits do not greatly diverge from lizard norms: solitary living (10% v. 10%), group living (20% v. 20%), male social tolerance (0% v. 15%), male spatial defense (100% v. 85%), and male hierarchy formation (0% v. 15%). The overall impression is of trenchant male intolerance.

Arboreality is overrepresented in the Phyllodactylidae (52% v. 32%); terrestriality is about average (47% v. 48%). The other microhabitats get negligible use. Reliance on burrows for daily retreats is notably underrepresented (5.7% v. 22%). As noted in the second paragraph, phyllodactylids are conspicuously overrepresented at the dry end of the rainfall spectrum; their elevational profile emphasizes lower elevations (l/m/h: 93/34/6 v. 82/41/13). As also mentioned above, Phyllodactylidae are notably overrepresented in human habitats, but occur in the other tabulated habitats at about average frequencies: littoral (20% v. 17%), sandy (25% v. 22%), and riparian (16% v. 22%) zones.

All are oviparous (v. 82%). Reproduction in phyllodactylids follows a pattern of few eggs deposited often rather than large clutches laid infrequently (clutch count −0.57 v. 0.0 std. resid.). Young are of normal size (ontogenetic size range 0.11 v. 0.00 std. resid.). RCM is therefore low (0.086 v. 0.226). As with many geckos, eggshells are hard (100% v. 41%). Incubation duration (0.18 v. 0.0 std. resid.) is normal. Parental care (3.7% v. 2.9%) may be slightly elevated.

Longevity is about normal (0.15 v. 0.0 std. resid.). Population density is known for only a single species (*Phyllodactylus wirshingi*, *n*=1 sample; 0.8 kg/ha v. 0.5 kg/ha) and is therefore not generalizable. The frequencies of endemism (67% v. 64%) and island species (40% v. 40%) are normal. Introductions, however, are about double the lizard frequency (7.4% v. 3.3%). Species introduced more than twice are also elevated (1.5% v. 1.1%). These introduction rankings are the sixth highest among lizard families. They are consistent with both anecdotal evidence (e.g., a *Tarentola* from Spain endured transport to Scotland in a load of bananas and survived there for years) and the strong correlates tabulated in table 3.14: electric light use, human habitats, and population density. Two of three strong correlates characterize the phyllodactylid condition; the third is consistent with the weak population evidence available. Although most introductions of Phyllodactylidae have been to places beyond the focus of Western scientists (e.g., Cape Verde Islands), a few have entered North America (*Tarentola annularis* in Florida; *T. mauritanica* in Florida and California) or Europe (*T. mauritanica*). This family may have high potential for further introductions, perhaps especially *Tarentola*.

A low number of Phyllodactylidae are imperiled (29% v. 49%) or are listed on one of the IUCN's lists of most endangered (7.1% v. 13.6%).

Phyllodactylus

Phyllodactylus is a major radiation (52 spp.) of moderately small (3.54 g v. 4.74 g), mostly nocturnal phyllodactylids from the New World, from the west coast of California south to the Neotropics. Their Neotropical distribution includes all mainland areas as well as the West Indies. Their modal ecological business model is trunk/wall lizard (49% v. 44%), though several species are noted to be so primarily or exclusively where attracted to feeding under electric lights. Many species are originally or presently nocturnal ground lizards (39% v. 40%). A small minority are ground-foraging tree lizards (13% v. 16%).

Phyllodactylus closely resemble the morphology of the phyllodactylids, except that no species have prehensile tails (v. 2.2%). Virtually all species have expanded toe pads, at least on the rear digits. Pads are terminal on the digits. *P. insularis* has a reflective layer in its retina, producing an eye shine. Sex differences are very minor (0.15 v. 0.31 on 0 to 6 scale), consisting primarily of rare sexual dichromatism (1.9% v. 6.7%).

Almost all *Phyllodactylus* are invertebrate eaters (95% v. 73%), but a few are omnivores (5.3% v. 3.6%). Several species feed intertidally, at least opportunistically; *angustidigitus* does so exclusively. The only foraging mode documented for the genus is sit-and-wait (v. 94%). Electric light use is slightly more frequent than in the

Phyllodactylus spp.	English name				Range core
			Size		
	Activity	Diet	mm	g	Ecological business model
angustidigitus	Narrow Leaf-toed Gecko				Peru
	nocturnal	invert.	49	3.20	**** nocturnal ground lizard
barringtonensis	Santa Fe Leaf-toed Gecko				Ecuador
	nocturnal		40	1.68	*** nocturnal ground lizard
baurii	Pinta Leaf-toed Gecko				Ecuador
	nocturnal		46	2.56	*** nocturnal ground lizard
bordai	Guerreran Leaf-toed Gecko				Mexico
	crepuscular		51	3.47	** trunk/wall lizard
bugastrolepis	Santa Catalina Leaf-toed Gecko				Mexico
	nocturnal	invert.	54	4.27	** nocturnal ground lizard
clinatus	Mount Illescas Leaf-toed Gecko				Peru
	nocturnal		37	1.41	
darwini	Darwin's Leaf-toed Gecko				Ecuador
	nocturnal		61	5.98	** nocturnal ground lizard or trunk/wall lizard (where lit)
davisi	Davis' Leaf-toed Gecko				Mexico
	cathemeral		59	5.34	
delcampoi	Del Campo's Leaf-toed Gecko				Mexico
	nocturnal		74	10.4	** trunk/wall lizard
delsolari	Del Solar's Leaf-toed Gecko				Peru
	nocturnal	invert.	67	8.0	*** trunk/wall lizard
dixoni	Dixon's Leaf-toed Gecko				Venezuela
	nocturnal		64	6.8	** trunk/wall lizard
duellmani	Duellman's Pygmy Leaf-toed Gecko				Mexico
	nocturnal		40	1.75	
duncanensis	Pinzon Leaf-toed Gecko				Ecuador
	nocturnal		42	1.98	* nocturnal ground lizard
galapagensis	Galapagos Leaf-toed Gecko				Ecuador
	nocturnal		42	1.98	*** nocturnal ground lizard or trunk/wall lizard (where lit)
gerrhopygus	South America Leaf-toed Gecko				Chile
	nocturnal	omnivore	46	2.62	** nocturnal ground lizard or trunk/wall lizard (where lit)
gilberti	Wolf Leaf-toed Gecko				Ecuador
	nocturnal		49	3.22	*** nocturnal ground lizard
gorii	Espanola Leaf-toed Gecko				Ecuador
			39	1.57	*** nocturnal ground lizard
heterurus	Tarapaca Leaf-toed Gecko				Chile
	nocturnal	invert.	39	1.57	
homolepidurus	Sonora Leaf-toed Gecko				Mexico
	cathemeral	invert.	59	5.5	** trunk/wall lizard

family (11.8% v. 9.2%). Home-range area is not documented.

Most *Phyllodactylus* are nocturnal (90% v. 78%), but an appreciable minority are active at all times (cathemeral) 7.8% (v. 19%) or crepuscular (2.0% v. 2.3%). Several of the nominally nocturnal species appear to keep more flexible hours than implied by that categorization. Despite their nocturnality/cathemerality, few call spontaneously (3.8% v. 11.1%). More give stress calls (5.8% v. 5.9%). Conspicuous wariness is observed at typical frequency (1.9% v. 1.5%).

Predator concealment tactics are not documented for *Phyllodactylus*. Direct basking is rare (5.8% v. 9.6%). Mean active body temperature is very low (22.8°C v. 26.0°C), based on four species. Annual rainfall data are available for only *palmeus* (2162–2570 mm/y). Elevational profile is normal (l/m/h: 97/27/3 v. 93/34/6). Littoral habitats (33% v.20%) are overrepresented; sandy areas (20% v. 25%), riparian zones (11% v. 16%), and human habitats (47% v. 40%) are about average.

Reproduction is very close to the phyllodactylid norm (clutch count –0.42 v. –0.57 std. resid.). Incubation is a trifle long (0.39 v. 0.18 std. resid.). Adherent eggs (5.8% v. 10.4%), communally deposited clutches (7.7% v. 16.3%), and parental care (1.9% v. 3.7%) are underrepresented.

Longevity is notably short (–0.52 v. 0.15 std. resid., based on four species), corresponding to 1.7 years (*microphyllus*) to 6.4 years (*reissii*). Population density is based on only one sample of one species (*wirshingi*, 0.8 kg/ha), which is computationally average for the family as there are no other estimates. Endemism is high (79% v. 67%). A disproportionate number of species occur on islands (52% v. 40%). Relatively few have been introduced (3.8% v. 7.4%), none more than twice (v. 1.5%). *Phyllodac-*

tylus are imperiled at an elevated level (44% v. 29%), including 11% (v. 7%) listed by the IUCN within the ranks of the most endangered species.

It is difficult to be terrified by a tiny gecko, but it can be done; *gerrhopygus* is believed by some local residents to be venomous. Some local residents believe that *homolepidurus* will cause necrosis of the nearest human flesh, even if the flesh is separated from the gecko by a layer of clothing.

Phyllodactylus spp.	English name				Range core
			Size		
	Activity	Diet	mm	g	Ecological business model
inaequalis	Peru Leaf-toed Gecko				Chile
	nocturnal	invert.	39	1.55	
insularis	Island Leaf-toed Gecko				Honduras
	nocturnal		59	5.4	* trunk/wall lizard
interandinus	Andes Leaf-toed Gecko				Peru
	nocturnal		42	2.01	
johnwrighti	Rio Huancabamba Leaf-toed Gecko				Peru
	nocturnal		42	1.94	
julieni	Aruba Leaf-toed Gecko				Aruba
	nocturnal		50	3.36	** trunk/wall lizard (where lit)
kofordi	Coastal Leaf-toed Gecko				Peru
	nocturnal	invert.	40	1.75	*** nocturnal ground lizard
lanei	Lane's Leaf-toed Gecko				Mexico
	nocturnal		65	7.3	*** trunk/wall lizard
leei	San Cristobal Leaf-toed Gecko				Ecuador
	nocturnal		42	1.93	*** nocturnal ground lizard or trunk/wall lizard (where lit)
leoni	Leon River Leaf-toed Gecko				Ecuador
	nocturnal		49	3.08	
lepidopygus	Western Leaf-toed Gecko				Peru
	nocturnal		46	2.60	*** trunk/wall lizard
magister	Noble's Leaf-toed Gecko				Peru
	nocturnal		59	5.3	*** trunk/wall lizard
martini	Dutch Leaf-toed Gecko				Curacao
	nocturnal		48	2.95	*** trunk/wall lizard (where lit)
microphyllus	Central Leaf-toed Gecko				Peru
	nocturnal	invert.	48	2.98	**** nocturnal ground lizard
muralis	Oaxaca Leaf-toed Gecko				Mexico
	nocturnal		55	4.44	
nocticolus	Peninsula Leaf-toed Gecko				Mexico
	cathemeral	invert.	52	3.65	** trunk/wall lizard
pachamama	Mother Earth Gecko				Peru
	nocturnal	invert.	64	7.0	** trunk/wall lizard
palmeus	Honduras Leaf-toed Gecko				Honduras
	nocturnal		64	6.8	*** trunk/wall lizard
papenfussi	Papenfuss' Leaf-toed Gecko				Mexico
	nocturnal		36	1.22	
paralepis	Guanaja Leaf-toed Gecko				Honduras
	nocturnal		59	5.5	
partidus	Partida Norte Leaf-toed Gecko				Mexico
	nocturnal		57	4.96	* nocturnal ground lizard

Phyllodactylus spp.	English name				Range core
			Size		
	Activity	Diet	mm	g	Ecological business model
paucituberculatus	Rio Marquez Valley Gecko				Mexico
	nocturnal		59	5.5	* trunk/wall lizard
pulcher	Barbados Leaf-toed Gecko				Barbados
	nocturnal	invert.	57	4.96	*** trunk/wall lizard
pumilus	Ecuador Leaf-toed Gecko				Ecuador
	nocturnal		46	2.58	
reissii	Guayaquil Leaf-toed Gecko				Peru
	nocturnal	invert.	60	5.8	*** trunk/wall lizard
rutteni	Venezuela Leaf-toed Gecko				Venezuela
	nocturnal		46	2.58	
sentosus	Lima Leaf-toed Gecko				Peru
	nocturnal		49	3.22	** nocturnal ground lizard
thompsoni	Thompson's Leaf-toed Gecko				Peru
	nocturnal	invert.	39	1.66	** ground-foraging tree lizard
transversalis	Malpelo Leaf-toed Gecko				Colombia
	nocturnal	invert.	60	5.7	*** ground-foraging tree lizard
tuberculatus	Yellow-bellied Leaf-toed Gecko				Central America
	cathemeral	invert.	73	10.1	*** trunk/wall lizard
unctus	San Lucan Leaf-toed Gecko				Mexico
	nocturnal	invert.	56	4.53	** ground-foraging tree lizard
ventralis	Margarita Leaf-toed Gecko				Venezuela
	nocturnal	invert.	62	6.2	*** trunk/wall lizard
wirshingi	Puerto Rico Leaf-toed Gecko				Hispaniola
	nocturnal		57	4.78	** ground-foraging tree lizard
xanti	Cape Leaf-toed Gecko				Mexico
	nocturnal	invert.	54	4.11	*** ground-foraging tree lizard

Phyllopezus spp.	English name				Range core
			Size		
	Activity	Diet	mm	g	Ecological business model
lutzae	Bogert's Gecko				Brazil
	nocturnal		59	5.3	
maranjonensis	Chachapoyas Giant Gecko				Peru
	nocturnal	omnivore	92	19.7	*** trunk/wall lizard
periosus	Peraiba Gecko				Brazil
	nocturnal	invert.	103	28.0	* trunk/wall lizard
pollicaris	Chaco Leaf-footed Gecko				Paraguay
	nocturnal	invert.	82	14.0	* trunk/wall lizard

Phyllopezus

Phyllopezus (four spp.) is a genus of large (14.3 v. 4.72 g), nocturnal (100% v. 78%), trunk/wall phyllodactylids found in southern South America, from Peru to Argentina and Brazil. They are solidly built for phyllodactylids (shape index 0.62 v. 0.37 std. resid.), bearing tails of normal length (1.10 v. 1.03 × SVL) and breakability (100% v. 92%). Basic color change is present (25% v. 11%). Sex differences are not reported. Although most species are nominally invertebrate eaters (67% v. 93%), *maranjonensis* is an omnivore (v. 3.6%) and *periosus* consumes nectar. Reproduction follows the family template, but clutches are notably low in number of eggs (−0.95 v. −0.57 std. resid.), and young are large in size (ontogenetic mass range −0.21 v. 0.11 std. resid.); communal oviposition is reported for *pollicaris*. The same species has a typical RCM (0.106 v. 0.086). No *Phyllopezus* is found on an island (v. 40%). None has been introduced (v. 7.4%), and none has been evaluated for conservation status. In other respects *Phyllopezus* follow the family characteristics.

Phyllurus

Phyllurus is a moderate (nine spp.) carphodactylid genus of extremely cryptic geckos, often with a very broad leaf-shaped tail (0.76 v. 0.54 × SVL). With such an elaborate tail, one might think that they would part with it reluctantly if at all, but tails of all species are believed to be breakable (100% v. 87%). *Phyllurus* are found in very restricted distributions on the east coast of Australia and have a diversity of ecological business models: immobility expert (67% v. 47%), trunk/wall lizard (22% v. 7%), and crack hermit (11% v. 13%). Note that the one crack hermit (*platurus*) forages inside its crack during the day but ventures abroad at night. *Phyllurus* differs from average carphodactylids in smaller mean size (11.4 g v. 14.5 g) and absence of pores (0% v. 10%). They are all invertebrate eaters (v. 50%). Their mix of daily activity patterns mirrors that of the family. Unlike other family members, they are not known to live in groups (v. 33%).

Almost all are arboreal (89% v. 53%). Their clutch count is average, as fits their size and the fixed clutch of two (–0.79 v. –0.82). The size of young is closer to the lizard average than to the large size typical of carphodactylids (–0.26 v. –1.21 std. resid.). Incubation duration is normal (0.17 v. –0.07 std. resid.). Longevity is long (1.40 v. 0.42 std. resid.), notably different from the superficially similar *Saltuarius* (–1.83 std. resid.). In other respects, they conform to carphodactylid patterns. All evaluated species are of conservation concern (v. 67%), and one species (*gulbaru*) is on the IUCN lists of the most endangered (13% v. 13%).

Phyllurus spp.	English name				Range core
			Size		
	Activity	Diet	mm	g	Ecological business model
amnicola	Mount Elliot Broad-tailed Gecko				Queensland
	nocturnal		99	18.0	* immobility expert
caudiannulatus	Ringed Thin-tail Gecko				Queensland
	nocturnal		84	12.0	*** trunk/wall lizard
championae	Conners' Range Broad-tailed Gecko				Queensland
	nocturnal		69	7.6	* immobility expert
gulbaru	Palluma Thin-tailed Gecko				Queensland
	nocturnal		77	9.8	* immobility expert
isis	Mount Jukes Broad-tailed Gecko				Queensland
	nocturnal		83	11.9	* immobility expert
kabikabi	Gympie Broad-tailed Gecko				Queensland
	nocturnal		73	8.6	* immobility expert
nepthys	Peppered-belly Broad-tailed Gecko				Queensland
	nocturnal		90	14.3	* trunk/wall lizard
ossa	Mount Ossa Broad-tailed Gecko				Queensland
	nocturnal		80	10.9	* immobility expert
platurus	Southern Broad-tailed Gecko				New South Wales
	cathemeral	invert.	87	13.0	*** crack hermit

Phymaturus

Phymaturus is a major radiation (47 spp.) of large (21.4 g v. 9.4 g), herbivorous (94% v. 26%), group-living (100% v. 50%), viviparous (100% v. 67%) liolaemids, found exclusively in Argentina and Chile. All are refugia-anchored diurnal ground lizards (v. 96%), and most are conspicuously corpulent (mean shape index 1.13 v. 0.76 std. resid.). North American herpetologists will be reminded of *Sauromalus*. It is easy to imagine this genus on an evolutionary pathway to becoming giant herbivores. At present they represent the 65th to 92nd percentiles of lizard mass (including Varanidae and Iguanidae).

Tails are short for liolaemids (1.06 v. 1.41 × SVL) and somewhat less frequently breakable (13% v. 22%). Secretory pores are comparatively scarce (4.3% v. 22%). Color change is not reported (v. 2%).

In contrast to other liolaemids, *Phymaturus* are m<f in average SSD (–0.17 v. 0.45 on –2 to 2 scale). This is unusual for large lizards and may reflect their penchant for wedging in tapering rock cracks, which would seem to reward small size, a factor that may also supply a countervailing selection pressure to becoming giant herbivores. Female shrinkage might be constrained by fecundity selection. Sexual dichromatism is not so constrained and is much

Phymaturus spp.	English name				Range core
			Size		
	Activity	Diet	mm	g	Ecological business model
aguanegra	Black Water Lizard				Argentina
	diurnal	herbivore	80	19.3	* refugia-anchored diurnal gr. liz.
aguedae	Agueda's Rock Lizard				Chile
			77	16.6	* refugia-anchored diurnal gr. liz.
alicahuense	Valparaiso Lizard				Chile
			84	23.0	* refugia-anchored diurnal gr. liz.
antofagastensis	Antofagasta Mountains Lizard				Argentina
	diurnal	herbivore	84	22.8	*** refugia-anchored diurnal gr. liz.

Phymaturus spp.	English name	Activity	Diet	Size mm	Size g	Range core	Ecological business model
bibronii	Ovalle Highlands Lizard			83	21.7	Chile	* refugia-anchored diurnal gr. liz.
cacivioi	Cacivio's Lizard			85	23.4	Argentina	* refugia-anchored diurnal gr. liz.
calcogaster	Chubut Rock Lizard	diurnal	herbivore	75	15.6	Argentina	*** refugia-anchored diurnal gr. liz.
camilae	Camila's Patagonia Rock Lizard	diurnal	herbivore	85	23.5	Argentina	** refugia-anchored diurnal gr. liz.
castillensis	Castillo Mountains Rock Lizard	diurnal	herbivore	77	16.6	Argentina	*** refugia-anchored diurnal gr. liz.
ceii	Cei's Rock Lizard	diurnal	herbivore	81	19.6	Argentina	*** refugia-anchored diurnal gr. liz.
curivilcun	Black Lizard	diurnal	herbivore	84	22.7	Argentina	** refugia-anchored diurnal gr. liz.
damasense	Las Damas River Lizard			90	28.6	Chile	* refugia-anchored diurnal gr. liz.
darwini	Darwin's Rock Lizard			85	23.6	Chile	* refugia-anchored diurnal gr. liz.
delheyi	Delhey's Patagonia Lizard	diurnal	herbivore	81	20.4	Argentina	* refugia-anchored diurnal gr. liz.
denotatus	Fading-shoulder Rock Lizard	diurnal	herbivore	89	27.8	Argentina	* refugia-anchored diurnal gr. liz.
desuetus	Rio Negro Rock Lizard	diurnal		61	7.8	Argentina	* refugia-anchored diurnal gr. liz.
dorsimaculatus	Copahue Lizard	diurnal	herbivore	76	16.1	Argentina	*** refugia-anchored diurnal gr. liz.
etheridgei	Etheridge's Rock Lizard	diurnal		79	18.5	Argentina	* refugia-anchored diurnal gr. liz.
excelsus	Eye of Water Lizard	diurnal	herbivore	74	14.7	Argentina	*** refugia-anchored diurnal gr. liz.

more frequent in *Phymaturus* (55% v. 29%). Seasonal color is present at family-typical levels (2.1% v. 2.3%). Blue tails are not present, in keeping with the reduced frequency of tail autotomy. Many *Phymaturus* have armored or spiny tails with which they block access to their refuge, a feature risky and costly to discard. Add these sexual differences together and one obtains a mildly elevated sex differences rollup (0.72 v. 0.62 on 0 to 6 scale).

Phymaturus are almost universally herbivorous, as noted above; the remainder are omnivorous (6.1% v. 34.5%). Half of documented species are reported to be sit-and-wait foragers (50% v. 84%), and the other half are reported to be active foragers (50% v. 16%). This cannot be correct, as plant eaters (100% of *Phymaturus*) are by definition active foragers; predators on invertebrates (6% of *Phymaturus*) may rely exclusively on sit-and-wait. Electric light use is not reported, nor is it expected for plant eaters.

Despite their uniform diurnality, 4.3% of *Phymaturus* (*dorsimaculatus*, *vociferator*) vocalize. The context of this calling is not documented. All documented *Phymaturus* rely on crevices or burrows (v. 46%) for their primary predator concealment. Only 4.3% are documented to bask directly (v. 7.8%), despite their uniform viviparity. Perhaps this basking frequency reflects underreporting; the well-accepted rationale for viviparity at high elevation is that it advances progeny development via maternal basking. Mean active body temperature is a typical 33.5°C (v. 33.7°C), suggesting that *Phymaturus* are somehow acquiring high temperatures despite their high altitude. Is indirect basking present and unreported?

For nightly retreats only 3.3% of *Phymaturus* species are documented to reside in burrows. This is a surprising limitation given the places in which they may live. For contrast, 75% of *Sauromalus* and 67% of *Cyclura*—two genera with similar lifestyles—use burrows. Perhaps the digits of *Phymaturus* are poorly suited to excavation. Perhaps *Phymaturus* face nighttime predators that can readily excavate them from soil burrows: mustelids such as tayra or grison? It may be notable that another mainland group of smallish giant herbivores, *Ctenosaura*, also report a low frequency (17%) of burrow retreats.

In comparison to other Liolaemidae, *Phymaturus* are found only at the drier end of the rainfall spectrum (dry and arid). Note that rainfall documentation is limited to a single species (*curivilcun*), but it comports with my impression of the genus. Documentation of elevation is much better and does not differ appreciably from that of the family (l/m/h: 55/30/30 v. 47/41/44). Frequencies of habitat specializations in all categories are lower for *Phymaturus* than that for the family, suggesting possible underreporting: littoral (0% v. 6%), sandy (5% v. 32%), riparian (7.5% v. 10%), and human (0% v. 2%) habitats. The infrequency in sandy areas is consonant with their disinclination to use easily excavated refugia.

Litter counts are available for 12 species and the individual species ranges all include two (most range two to two), suggesting a possible phylogenetic constraint. For their size, this litter count is very low (−1.07 v. −0.16 std. resid.). Neonates are, however, large (mass-adjusted ontogenetic range of body masses: −1.08 v. −0.59 std. resid.). In many giant herbivores found on islands (see Iguanidae accounts) one observes large young, perhaps as a reflection of a large-size advantage of herbivorous lizards; *Phymaturus* is somewhat unusual in exhibiting the

Phymaturus spp.	English name				Range core
			Size		
	Activity	Diet	mm	g	Ecological business model
extrilidus	Dauntless Rock Lizard				Argentina
	diurnal	herbivore	84	22.4	*** refugia-anchored diurnal gr. liz.
felixi	White-flecked Rock Lizard				Argentina
	diurnal	herbivore	70	12.4	*** refugia-anchored diurnal gr. liz.
indistinctus	Las Pulgas Rock Lizard				Argentina
	diurnal	herbivore	88	26.2	*** refugia-anchored diurnal gr. liz.
laurenti	Laurent's Rock Lizard				Argentina
	diurnal	herbivore	82	21.1	*** refugia-anchored diurnal gr. liz.
mallimaccii	Thorny-tailed Rock Lizard				Argentina
	diurnal	herbivore	86	24.6	*** refugia-anchored diurnal gr. liz.
manuelae	Many-colored Rock Lizard				Argentina
	diurnal		86	24.5	* refugia-anchored diurnal gr. liz.
maulense	Altos de Lircay Rock Lizard				Chile
			87	25.0	* refugia-anchored diurnal gr. liz.
nevadoi	Snowy Mountains Rock Lizard				Argentina
	diurnal	herbivore	85	23.3	*** refugia-anchored diurnal gr. liz.
palluma	High Mountains Rock Lizard				Argentina
	diurnal	omnivore	93	27.7	*** refugia-anchored diurnal gr. liz.
patagonicus	Argentina Chuckwalla				Argentina
	diurnal	herbivore	89	27.3	*** refugia-anchored diurnal gr. liz.
payuniae	Payun Rock Lizard				Argentina
	diurnal	herbivore	82	20.9	*** refugia-anchored diurnal gr. liz.
punae	San Guillermo Rock Lizard				Argentina
	diurnal	herbivore	91	30.0	*** refugia-anchored diurnal gr. liz.
querque	Laguna Blanca Spinytail				Argentina
	diurnal	herbivore	84	22.4	*** refugia-anchored diurnal gr. liz.
rahuensis	Rahue Rock Lizard				Argentina
	diurnal		81	19.8	* refugia-anchored diurnal gr. liz.
roigorum	Five Roigs Rock Lizard				Argentina
	diurnal	herbivore	92	31.1	*** refugia-anchored diurnal gr. liz.

Phymaturus spp.	English name				Range core
			Size		
	Activity	Diet	mm	g	Ecological business model
sinervoi	Sinervo's Rock Lizard				Argentina
	diurnal	herbivore	87	25.0	*** refugia-anchored diurnal gr. liz.
sitesi	Sites' Rock Lizard				Argentina
	diurnal	herbivore	79	18.3	*** refugia-anchored diurnal gr. liz.
somuncurensis	Somuncura Plateau Rock Lizard				Argentina
	diurnal	herbivore	88	26.2	*** refugia-anchored diurnal gr. liz.
spectabilis	Climber Rock Lizard				Argentina
	diurnal	herbivore	82	21.1	*** refugia-anchored diurnal gr. liz.
spurcus	Lackluster Rock Lizard				Argentina
	diurnal	herbivore	76	16.1	*** refugia-anchored diurnal gr. liz.
tenebrosus	Gloomy Rock Lizard				Argentina
	diurnal	herbivore	91	29.4	*** refugia-anchored diurnal gr. liz.
tromen	Tromen Rock Lizard				Argentina
	diurnal		84	23.0	* refugia-anchored diurnal gr. liz.
verdugo	Peteroa Volcano Rock Lizard				Argentina
	diurnal		95	34.2	* refugia-anchored diurnal gr. liz.
videlai	Videla's Rock Lizard				Argentina
	diurnal		77	17.1	* refugia-anchored diurnal gr. liz.
vociferator	Calling Rock Lizard				Argentina
	diurnal	herbivore	83	21.7	** refugia-anchored diurnal gr. liz.
williamsi	Williams' Rock Lizard				Argentina
	diurnal	herbivore	84	23.0	*** refugia-anchored diurnal gr. liz.
yachanana	Iron Range Rock Lizard				Argentina
	diurnal	omnivore	77	17.1	*** refugia-anchored diurnal gr. liz.
zapalensis	Zapala Rock Lizard				Argentina
	diurnal	herbivore	85	23.7	*** refugia-anchored diurnal gr. liz.

same large neonatal size in a continental lizard, though one also sees this phenomenon in the high-latitude continental genera *Dipsosaurus* and *Sauromalus*. Digestive efficiency may be elevated by large size, even if size-dependent predation is a minor or neutral selective force on the mainland.

Longevity is on the long side in *Phymaturus* (0.71 v. 0.13 std. resid.), but this is based on only two species: *palluma* (12 years), *tenebrosus* (16 years). A remarkable 96% (v. 82%) of *Phymaturus* are endemic to their countries. They have not been introduced (v. 0.3%). About a third are imperiled (31% v. 21%), though none is listed on any of the IUCN's lists of most endangered (v. 4.5%). In other respects *Phymaturus* are typical liolaemids.

Physignathus

Physignathus is a single-species agamid genus of giant (223 g v. 15.6 g) herbivores (v. 6%) from lowland Southeast Asia, ranging from China and Vietnam west to Thailand and perhaps Myanmar. It has a long tail (2.6 v. 1.9 × SVL); secretory pores (v. 13%); dramatic nuchal, dorsal, and tail crests (v. 38%) that are larger in males; and males that are more than twofold the mass of females (2.0 v. 0.53 on –2 to 2 scale), differ in color (v. 53%), and have relatively larger heads. In aggregate, *Physignathus* has the most sex differences in Agamidae: 4.0 (v. 1.6) on 0 to 6 scale. The species has advanced color change (hue; v. 15%), may run bipedally, and is usually found in the vicinity of water (semiaquatic v. 0.9%; riparian v. 21%). Males defend space (v. 97%). The lizards sleep on branches overhanging the water, and they are conspicuously wary (v. 6%). I am guessing that this species is large enough and herbivorous enough to be highly desirable to local cooks. The proportion of plants in its diet increases with size and age, but opinions are contradictory on what that proportion is (temporal and geographic variation seem likely). Clutch count is about average (0.11 v. 0.20 std. resid.), though young are small (0.91 v. 0.69 std. resid.), reflecting a young/maximum mass ratio of 1/288 (v. 1/104). Incubation is

Physignathus sp.	English name		Size		Range core
	Activity	Diet	mm	g	Ecological business model
cocincinus	Indochinese Water Dragon				Southeast Asia
	diurnal	omnivore	184	223	**** giant herbivore

Pinoyscincus spp.	English name		Size		Range core
	Activity	Diet	mm	g	Ecological business model
abdictus	Jagor-like Sphenomorph				Philippines
			82	11.4	*** litter cryptozoid
coxi	Cox's Sphenomorph				Philippines
	diurnal		74	8.1	** litter cryptozoid
jagori	Jagor's Sphenomorph				Philippines
	diurnal	invert.	84	11.2	** refugia-anchored diurnal gr. liz.
llanosi	Samar-Leyte Sphenomorph				Philippines
			77	9.4	*** refugia-anchored diurnal gr. liz.
mindanensis	Southeastern Sphenomorph				Philippines
			49	1.35	*** litter cryptozoid

short (−0.89 v. −0.55). Mass-adjusted longevity is long (0.44 v. −0.46 std. resid.). Endemism is low (0% v. 49%); *Physignathus* is not native on any islands (0% v. 36%). It has been introduced on Taiwan and in Malaysia (v. 1.7% introduced). It was judged as vulnerable in the 1992 Vietnamese Red Data Book, but was not so listed by the IUCN (v. 5% on one of the IUCN's lists of most endangered).

Pinoyscincus

Pinoyscincus (formerly in *Sphenomorphus*) is a minor radiation (five spp.) of moderately large (6.8 g v. 3.25 g), diurnal (100% v. 74%), mostly riparian (80% v. 23%) litter cryptozoid (60% v. 28%) or refugia-anchored diurnal ground (40% v. 30%) skinks endemic to the Philippines.

These are slightly beefier skinks (shape index 0.09 v. −0.46 std. resid.), with normal tails (1.50 v. 1.33 × SVL) that are breakable (100% v. 73%). A few are elongate (20% v. 35%). None has a blue tail (v. 4%). The largest difference between *Pinoyscincus* and an average skink is that *Pinoyscincus* has a notably large ontogenetic mass range (0.88 v. 0.12 std. resid.), suggesting particularly small hatchlings.

Pinoyscincus is an unusual skink genus in that m>f SSD is the rule (1.0 v. −0.31 on −2 to 2 scale). Seasonal color is also present (*jagori* 20% v. 6%). This produces a mildly elevated sex differences rollup score on scale of 0 to 6: 0.60 (v. 0.28).

Diet has been quantified only for *jagori*, which is classified as a generalized invertebrate eater (v. 84%) but also eats fruit and ectothermic vertebrates. There are no data on foraging mode, electric light use, or home-range area.

Elevational profile is about normal (l/m/h: 100/40/0 v. 89/35/6). They are about average in presence in human habitat (20% v. 14%) but are absent from littoral (v. 21%) and sandy (v. 25%) areas.

All are oviparous (v. 64%); their mean clutch count is small (−0.59 v. 0.06 std. resid.). This is unexpected given the notably small size of hatchlings. An alternate way to produce many small hatchlings per year is to breed many times per year, but the data to judge that are not available (*jagori* averages about 1.5 clutches per year).

Longevity is short (−0.87 v. 0.17 std. resid.), based on *coxi* (3 years). Population density is low (0.01 kg/ha v. 0.35 kg/ha) based on one sample of *jagori*. All species are insular (v. 57%) and endemic (v. 64%). None has been introduced (v. 2.1%). A normal percentage of species are imperiled (40% v. 47%), and none are found on one of the IUCN's lists of most endangered (v. 14%). In other respects *Pinoyscincus* are much like other skinks.

Placosoma

Placosoma is a four-species genus of poorly known (filled cells 16% v. 19%) gymnophthalmids endemic to Brazil. Too little information is available to assign most to ecological business models, though *limaverdorum* is likely a litter cryptozoid. They are medium sized (2.25 g v. 2.35 g) and of average build (−0.02 v. −0.32 std. resid.). They are not elongate (v. 41%). Secretory pores are present in *limaverdorum* (v. 29%), but other sex differences are not reported. Reproduction appears to be typical for gymnophthalmids. All species are endemic (v. 68%). They are not considered to be at conservation risk (v. 25%). In other respects *Placosoma* are probably typical gymnophthalmids.

Placosoma spp.	English name	Activity	Diet	Size mm	Size g	Range core	Ecological business model
cipoense	Cunha's Brazilian Lizard			59	4.79	Brazil	
cordylinum	Tschudi's Brazilian Lizard			41	1.41	Brazil	
glabellum	Peters' Brazil Lizard			47	2.25	Brazil	
limaverdorum	Ceara Litter Lizard			43	1.70	Brazil	*** litter cryptozoid

Platysaurus spp.	English name	Activity	Diet	Size mm	Size g	Range core	Ecological business model
attenboroughi	Attenborough's Flat Lizard	diurnal		71	7.9	South Africa	
broadleyi	Augrabies Flat Lizard	diurnal	omnivore	71	7.9	South Africa	*** trunk/wall lizard
capensis	Cape Flat Lizard	diurnal		71	8.0	Namibia	*** trunk/wall lizard
guttatus	Dwarf Flat Lizard	diurnal	omnivore	84	14.5	Southern Africa	*** ground-foraging tree lizard
imperator	Emperor Flat Lizard	diurnal	omnivore	110	38.2	Zimbabwe	** refugia-anchored diurnal gr. liz.
intermedius	Common Flat Lizard	diurnal	omnivore	96	23.3	Zimbabwe	** trunk/wall lizard
lebomboensis	Lebombo Flat Lizard	diurnal		63	5.2	South Africa	** trunk/wall lizard
maculatus	Spotted Flat Lizard	diurnal	invert.	64	5.4	Mozambique	** trunk/wall lizard
minor	Waterberg Flat Lizard	diurnal	omnivore	62	4.75	South Africa	** trunk/wall lizard
mitchelli	Mitchell's Flat Lizard	diurnal	omnivore	47	1.74	Malawi	
monotropis	Orange-throated Flat Lizard	diurnal		66	6.1	South Africa	** trunk/wall lizard
ocellatus	Ocellated Flat Lizard	diurnal	omnivore	77	10.7	Mozambique	*** ground-foraging tree lizard
orientalis	Sekukhune Flat Lizard	diurnal	invert.	73	8.7	South Africa	

Platysaurus

Platysaurus (16 spp.) differ from most cordylids in that their scales are relatively smooth, their bodies and limbs are flattened to facilitate entry into low cracks, and the males seem irrationally exuberant when it comes to sexual colors, especially seasonal embellishments. They are virtually restricted to rock outcrops, from South Africa north to Tanzania on the east and Namibia on the west. The breakdown of their ecological business models is as follows: 62% (v. 14%) are trunk/wall lizards, 23% (v. 29%) are refugia-anchored diurnal ground lizards, and 15% (v. 6%) are ground-foraging tree (or cliff) lizards. Collectively they are colloquially known as flat lizards.

On average, flat lizards are small for cordylids (7.9 g v. 10.8 g); they are of typical build (shape index 0.32 v. 0.35 std. resid.) with normal tails (1.23 v. 1.16 × SVL) that break (94% v. 60%). They usually have secretory pores (88% v. 69%).

Males are larger than females (1.0 on –2 to 2 scale), are almost always colored differently (94% v. 37%), and have seasonal color (81% v. 28%). In aggregate, this produces a high sex differences rollup score (2.81 v. 1.41 on 0 to 6 scale).

Flat lizards have diverse diets, despite their small size: only 36% (v. 70%) are invertebrate eaters; 64% (v. 30%) are omnivores. Does their restriction to the vicinity of exfoliating rock walls limit their dietary options? They may be forced to eat anything within reach. All are treated as sit-and-wait foragers, presumably excluding the search for plants, and 50% (v. 19%) are also considered active foragers. Many species are heliophilic, and at least 81% (v. 74%) bask directly.

Evidently they feel secure in their ability to whisk into narrow cracks,

Platysaurus spp.	English name		Size		Range core
	Activity	Diet	mm	g	Ecological business model
pungweensis	Pungwe Flat Lizard				Mozambique
	diurnal	invert.	78	10.9	** refugia-anchored diurnal gr. liz.
relictus	Soutpansberg Flat Lizard				South Africa
	diurnal		61	4.62	*** trunk/wall lizard
torquatus	Striped Flat Lizard				Zimbabwe
	diurnal	invert.	65	5.8	** refugia-anchored diurnal gr. liz.

as only 6% are considered conspicuously wary (v. 26.5%). Forty percent (v. 29%) live in groups, and all species have males (v. 100%) that defend space, presumably protecting at least ownership of individual cracks. Although 30 *intermedius* have been found sharing a crack, it was not reported whether this was during the mating season or whether the group included multiple mature males. Two-thirds (v. 43%) of species have males that form hierarchies.

About half of *Platysaurus* climb walls or trees (v. 17%), 27% are cryptozoic (v. 8%), and 20% are terrestrial (v. 76%). They have not been noted in littoral areas (v. 9%), sandy areas (v. 3%), or in human habitats (v. 0%).

All are oviparous (v. 25%). They have clutches of average count (−0.60 v. −0.48 std. resid.) but young that are exceptionally small (ontogenetic range 1.48 v. −0.37 std. resid.). This ontogenetic mass range indicates that the average young is 1/176 of maximum size versus 1/43 for the average cordylid. Communal oviposition is routine (88% v. 21%). Parental care has not been observed (v. 7.4%). Longevity is high for cordylids (0.72 v. 0.02 std. resid.): 15 years for *intermedius* and *guttatus*.

As with other cordylids, *Platysaurus* have not been introduced, though some species are in the pet trade (e.g., *imperator*). Forty percent (v. 79%) are judged to be imperiled, though none is IUCN listed as particularly endangered (v. 4.8%). In other respects these are typical cordylids.

Plestiodon (fig. 4.12B)

Plestiodon is a major radiation (47 spp.) of mildly large (5.0 g v. 3.25 g), often blue-tailed (62% v. 35%; a few others are bright red or pink), mostly diurnal (89% v. 74%), mostly cryptozoic (39% v. 30%) or terrestrial (46% v. 48%) skinks found primarily in temperate north latitudes with some spread into the adjacent tropics. The 28 species found in the Nearctic (60% v. 2%) are located from British Columbia and Manitoba to Connecticut and south to southern Mexico, plus Bermuda. *P. sumichrasti* is found in the Neotropics (2% v. 5%), from Mexico to Honduras. The 12 Palearctic *Plestiodon* (26% v. 6%) are found in temperate Japan, China, and North Korea. The seven Oriental *Plestiodon* (15% v. 24%) are found in areas adjacent to the Palearctic species: on the mainland from southern China to Vietnam and Thailand, and offshore islands such as Taiwan. The modal *Plestiodon* is a litter cryptozoid (54% v. 28%), and more than a third are refugia-anchored diurnal ground lizards (39% v. 30%). The remainder are ground-foraging tree lizards (5% v. 7%) and sand sharks (2.6% v. 6.8%).

Some species undergo dramatic ontogenetic color change in addition to routine ontogenetic disappearance of blue tails. For example, the juveniles of *obsoletus* differ so much from adults that the two were treated as separate genera for more than 50 years, even though the two original descriptions appeared in the same volume of the same journal.

Plestiodon are much like typical skinks in morphology. Their shape index is a little heavier: −0.11 (v. −0.46 std. resid.). Their tails are a bit longer (1.50 v. 1.33 × SVL) and somewhat less breakable (45% v. 73%). The fat content of *Plestiodon* tails may be very high—47% in *gilberti* and *skiltonianus*—but this precious resource is losable in those species. Carmichael and Williams (2004) claim that the blue tails of *inexpectatus* might be toxic to some predators. Elongation is very close to average (32% v. 35%).

Net SSD is very close to neutral: −0.07 v. −0.31 on −2 to 2 scale. Nineteen percent (v. 7.6%) show dichromatism; 32% (v. 6%) show seasonal color. These factors combine to a relatively high 0.81 (v. 0.28) on the 0 to 6 sex differences rollup score.

There are no purely carnivorous or herbivorous *Plestiodon*, but all other dietary classes are present: invertebrate eaters (89% v. 84%), ant specialists (3.6% v. 1.6%), and omnivores (7.1% v. 12.6%). The omnivores may have a very broad dietary spectrum; for example, *laticeps* eats everything from fruit and snails to ectothermic and immobile endothermic prey. The Bermuda *Plestiodon* (*longirostris*) is unique in shifting ontogenetically from insects to carrion spillage from seabird nests. Active foraging predominates (70% v. 92%), but sit-and-wait foraging is also reasonably common (30% v. 16%). Electric light use is not reported. Home-range area (based on four spp.) is high (0.54 v. −0.07 std. resid.), equating to

Plestiodon spp.	English name	Activity	Diet	Size mm	Size g	Range core	Ecological business model
anthracinus	Coal Skink	diurnal	invert.	58	3.68	Southeastern United States	**** litter cryptozoid
barbouri	Barbour's Eyelid Skink			59	4.01	Japan	** refugia-anchored diurnal gr. liz.
bilineatus	Two-lined Durango Litterskink	diurnal		55	1.90	Mexico	**** litter cryptozoid
brevirostris	Short-nosed Skink	diurnal	invert.	58	2.17	Mexico	*** litter cryptozoid
callicephalus	Mountain Skink	diurnal	invert.	62	2.57	Mexico	**** litter cryptozoid
capito	Gail's Eyelid Skink			67	5.8	China	
chinensis	Chinese Blue-tailed Skink	diurnal	invert.	105	25.3	China	**** refugia-anchored diurnal gr. liz.
colimensis	Colima Skink	diurnal		56	3.29	Mexico	** litter cryptozoid
copei	Cope's Skink	diurnal	invert.	63	4.93	Mexico	** refugia-anchored diurnal gr. liz.
coreensis	Smith's Skink			94	17.9	North Korea	
dicei	Tamaulipas Litterskink	diurnal		46	1.15	Mexico	*** litter cryptozoid
dugesii	Duges' Skink	diurnal		59	3.86	Mexico	** refugia-anchored diurnal gr. liz.
egregius	Mole Skink	diurnal	invert.	50	1.43	Georgia	** litter cryptozoid or subterranean forager
elegans	Shanghai Elegant Skink	diurnal	omnivore	79	10.2	East Asia	
fasciatus	Common Five-lined Skink	diurnal	invert.	68	6.3	Eastern United States	*** litter cryptozoid or ground-foraging tree lizard (geog. var.)
finitimus	Tokyo Skink			74	8.0	Japan	* refugia-anchored diurnal gr. liz.
gilberti	Gilbert's Skink	diurnal	invert.	88	6.8	Arizona	** litter cryptozoid

95–800 m². Males may have much larger home ranges during the season when they are searching for mates (e.g., *fasciatus*, *obsoletus*).

All diel patterns except nocturnal are present in *Plestiodon*: cathemeral (9% v. 9%), crepuscular (2.9% v. 2.4%), and diurnal (89% v. 74%). For many species there is an association between activity and recent rainfall. *P. laticeps* is notable as one of the most incessant foragers (94th percentile for PTM), though it ranks in the lower half (47th percentile) for MPM. *Plestiodon* are notably prone to retreating from humans in haste; 36% (v. 11%) are conspicuously wary. The forms of primary predator concealment *Plestiodon* use do not include pure flight, however (0% v. 4%). Instead, they retreat to sand (12.5% v. 22%), crevices or burrows (75% v. 43%), and vegetation (12.5% v. 25%). *P. reynoldsi* (fig. 4.12B) may also death feign. An ordinary proportion bask directly (17% v. 21%). None thermoconforms (v. 1.5%). Mean active body temperature is average 31.2°C (v. 31.1°C).

Group living is not reported in *Plestiodon*. Instead, all males defend space (v. 94%) and 7% (v. 5.6%) form male hierarchies.

Plestiodon occupy climate space in the middle of the rainfall spectrum: arid (63% v. 40%), intermediate (75% v. 34%), and mesic (38% v. 40%). They are absent from the hyperdry (<55 mm/y) and dry (151–400 mm/y) as well as the wet (>3000 mm/y). Their elevational profile is skewed toward medium and higher elevations (l/m/h: 73/60/27 v. 89/35/6). They are overrepresented in riparian areas (43% v. 23%) but found in typical frequencies in littoral (14% v. 21%), sandy (23% v. 25%), and human (17% v. 14%) habitats.

They are mostly oviparous (81% v. 64%) and have large clutches/litters (0.81 v. 0.06 std. resid.). The large clutches are consistent with the high RCM ratio (0.351 v. 0.236). Hatchlings/neonates average small (ontogenetic mass range 0.74 v. 0.12 std. resid.). Two species are solely viviparous (*brevirostris*: high elevation Mexico; *copei*: Mexico, elevation undocumented); whereas *bilineatus* (high elevation Mexico), *callicephalus* (medium elevation Mexico), *gilberti* (California to Mexico), *latiscutatus* (Japan), and *lynxe* (high elevation Mexico) have been reported to exhibit both reproductive modes. Incubation period is slightly short (−0.56 v. −0.40 std. resid.). Communal oviposition is known (4.3% v. 2.6%), though with conflicting evidence. Some males are believed to mate guard (*laticeps*).

What really stands out in the reproduction of *Plestiodon*, however, is the remarkably high level of parental care (53% v. 3%). Among polytypic scincid genera this level of parental care is exceeded only by *Tiliqua* (57%); 139 of 155 scincid genera have no documented parental care. In most cases the *Plestiodon* female guards the clutch, usually throughout incubation: *anthracinus*, *callicephalus*, *capito*, *chinensis*, *egregius*, *elegans*, *fasciatus*, *inexpectatus*, *japonicus*, *kishinouyei*, *laticeps*, *latiscutatus*, *multivirgatus*, *obsoletus*, *quadrilineatus*, *reynoldsi*, *septentrionalis*, *skiltonianus*, *stimpsonii*, *sumichrasti*, and *tetragrammus*. In most of these, and in the viviparous species, the hatchlings/neonates remain with the mother for a few days after hatching/birth. Experiments with *egregius* have demonstrated that the female protects the clutch from mold, at least in part by regular rotation of the eggs, as in birds. The female may rewet the eggs during incubation, such as in *fasciatus*. Brood frequencies in *Plestiodon* are generally only once per year and may be less in *latiscutatus*, *obsoletus*, *reynoldsi*, and *tetragrammus*. A few

Plestiodon spp.	English name				Range core
			Size		
	Activity	Diet	mm	g	Ecological business model
indubitus	Questionable Litterskink				Mexico
	diurnal		55	1.91	*** litter cryptozoid
inexpectatus	Southeastern Five-lined Skink				Southeastern United States
	diurnal	invert.	71	7.0	** ground-foraging tree lizard
japonicus	Southern Japan Five-lined Skink				Japan
	diurnal	invert.	74	8.0	*** refugia-anchored diurnal gr. liz.
kishinouyei	Southern Ryukyus Japan Skink				Japan
	cathemeral	invert.	132	53	** refugia-anchored diurnal gr. liz.
kuchinoshimensis	Kuchin Island Skink				Japan
	diurnal		69	6.5	** refugia-anchored diurnal gr. liz.
lagunensis	San Lucan Skink				Mexico
	diurnal	invert.	58	3.65	** litter cryptozoid
laticeps	Broad-headed Skink				Eastern United States
	diurnal	omnivore	105	25.2	*** ground-foraging tree lizard
latiscutatus	Okada's Five-lined Skink				Japan
			77	9.3	*** refugia-anchored diurnal gr. liz.
liui	Hikida's Skink				China
			57	3.57	
longirostris	Bermuda Skink				Bermuda
		invert.	66	5.6	* refugia-anchored diurnal gr. liz.
lynxe	Oak Forest Skink				Mexico
	diurnal	invert.	63	2.70	*** litter cryptozoid
marginatus	Tokara Island Skink				Japan
	diurnal	invert.	81	10.8	** refugia-anchored diurnal gr. liz.
multilineatus	Chihuahua Skink				Mexico
	diurnal	invert.	59	4.01	* refugia-anchored diurnal gr. liz.
multivirgatus	Many-lined Skink				New Mexico
	diurnal	ants	63	2.66	** litter cryptozoid
nietoi	Guerrero Mountain Skink				Mexico
			64	5.2	
obsoletus	Great Plains Skink				Southwestern United States
	cathemeral	invert.	110	12.7	*** refugia-anchored diurnal gr. liz.
ochoterenae	Guerrero Skink				Mexico
	diurnal		52	2.55	

Plestiodon spp.	English name				Range core
	Activity	Diet	Size mm	Size g	Ecological business model
oshimensis	Amami Island Skink				Japan
			80	10.5	* refugia-anchored diurnal gr. liz.
parviauriculatus	Northern Pygmy Skink				Mexico
	diurnal	invert.	47	1.24	*** litter cryptozoid
parvulus	Southern Pygmy Skink				Mexico
			52	2.54	** litter cryptozoid
popei	Pope's Skink				China
			50	2.30	
quadrilineatus	Four-lined Blue-tailed Skink				Southeast Asia
	diurnal	invert.	69	3.55	**** refugia-anchored diurnal gr. liz.
reynoldsi	Florida Sand Skink				Florida
	cathemeral	invert.	54	1.20	**** sand shark
septentrionalis	Prairie Skink				Nebraska
	crepuscular	invert.	71	3.81	*** litter cryptozoid
skiltonianus	Western Skink				Western United States
	diurnal	invert.	67	5.8	** litter cryptozoid
stimpsonii	Yaeyama Seven-lined Skink				Japan
	diurnal	invert.	67	5.8	* litter cryptozoid
sumichrasti	Sumichrast's Skink				Northern Central America
	diurnal	invert.	81	10.8	** litter cryptozoid
tamdaoensis	Tam Dao Blue-tailed Skink				Vietnam
	diurnal		102	22.6	* litter cryptozoid
tetragrammus	Four-lined Skink				Mexico
	diurnal	invert.	64	2.88	*** litter cryptozoid
tunganus	Tung River Skink				China
			67	6.0	

Pletholax sp.	English name				Range core
	Activity	Diet	Size mm	Size g	Ecological business model
gracilis	Keeled Legless Lizard				Western Australia
	diurnal	invert.	76	3.45	* litter cryptozoid

species, such as *egregius* and *multivirgatus*, occasionally double-clutch. That high frequencies of scincid parental care are known in two places laden with herpetologists—Australia (*Tiliqua*) and North America (*Plestiodon*)—raises the possibility that parental care might be more prevalent but unrecognized elsewhere in Scincidae.

Longevity is moderately long in *Plestiodon*: 0.39 (v. 0.17 std. resid.), equating to 10 years in several species (*fasciatus, laticeps, reynoldsi*) and 27 years in the Bermuda species (*longirostris*). There are credible population estimates for nine samples of three species of *Plestiodon* (0.96 kg/ha v. 0.35 kg/ha), but this mean density may conceal more than it reveals. The sand shark *reynoldsi* ($n = 4$; 0.32 kg/ha) is in the middle, the continental *obsoletus* ($n = 3$; 0.07 kg/ha) is low, and the Japanese small-island *latiscutatus* ($n = 2$; 37.5 kg/ha) is much denser. Most *Plestiodon* are not on islands (43% v. 57%). Endemism is average (64% v. 64%). One species has been introduced (*japonicus* to Hachijo Island: 2.1% v. 2.1%), but none has been introduced more than twice (v. 0.7%). *Plestiodon* are imperiled at the usual rate (52% v. 47%), but only *longirostris* is on one of the IUCN's lists (in this case critically endangered) of the most endangered species (3.4% v. 14%).

Pletholax

Pletholax gracilis is the sole species in a genus of smallish (3.45 g v. 5.16 g), diurnal (v. 61%), litter cryptozoid pygopodids from Western Australia. It is a sand swimmer and it may leap. Its diet has been difficult to ascertain, as the few detected specimens have mostly possessed no visible stomach contents. It laps plant juices and may lap the juices of captured arthropods rather than swallowing their hard exoskeletons, as does *Pygopus lepidopodus* and *P. nigriceps*.

The tail of *Pletholax* is slightly long (3.30 v. 2.65 × SVL). Known sex differences are limited to male seasonal color (v. 2.2%). Though diurnal, it follows the pygopodid template in vocalizing spontaneously (v. 67%). It is conspicuously wary (v. 8.7%). For predator concealment it dives into sand (v. 50%). It basks directly (v. 17%). One collection site had a mean annual rainfall of 860 mm/y (intermediate). It is found in littoral (v. 30%), sandy (v. 50%), and human (v. 9%) habitats. Reproduction is pygopodid typical.

It is endemic to Western Australia (v. 52%) and not found on an island (v. 26%). It is considered imperiled (v. 88%), but it is not on any of the IUCN's lists of the most endangered (v. 6%). In other respects *Pletholax* is similar to other pygopodids.

Plica

Plica (Tropiduridae) is a moderate genus (eight spp.) of large (29.7 g v. 22.1 g), colorful trunk/wall lizards (100% v. 13%) found in northern South America (excluding Paraguay, Argentina, Chile, and Uruguay) plus Trinidad and Tobago. They are the leading genus in Tropiduridae for advanced color change (12.5% v. 0.6%), seasonal color (12.5% v. 2.9%), and group living (67% v. 36%). Of the traits for which they are the second highest, some notables are crests (63% v. 10%), ant specialization (33% v. 4.4%), retreat to vegetation (50% v. 36%), direct basking (25% v. 16%), presence at intermediate elevations (33% v. 47%), presence in the mesic rainfall band (100% v. 89%), and frequency of introduced species (12.5% v. 1.5%).

They are of average build (shape index 1.18 v. 1.10 std. resid.) but longer tail length (1.93 v. 1.56 × SVL). Tails are usually breakable (63% v. 13%). Two species are capable of darkening (25% v. 3%).

SSD is m>f at the family average (1.00 v. 0.95 on –2 to 2 scale). Sexual dichromatism is slightly elevated (38% v. 28%), as is seasonal color (first paragraph). M:F relative tail length ratio is low (0.98 v. 1.06), as it often is with m>f SSD. On balance, *Plica* has about the same overall sex differences rollup as the family (0.88 v. 0.80 on 0 to 6 scale).

Diet is exclusively invertebrates, though at least *umbra* specializes in ants. Crypsis (50% v. 43%) and retreat to vegetation (50% v. 36%) are the primary predator concealment tactics. Direct basking is average (25% v. 16%). Mean active body temperature is low (27.0°C v. 31.8°C), based on *umbra*. Despite the high frequency of group living, all documented *Plica* males defend space (v. 100%).

Their elevational profile is skewed low: (l/m/h) 100/33/0 (v. 65/47/36). *Plica* are absent from littoral (v. 15%) and sandy (v. 23%) areas and overrepresented in riparian (40% v. 18%) and human (40%, the second-highest value in Tropiduridae, v. 14%) habitats.

Clutch count is exceptionally low (–1.02 v. –0.56 std. resid.) and young are notably large (ontogenetic mass range –0.60 v. –0.22 std. resid.). The former apparently dominates the overall size of the clutch, as RCM ratio is low (0.118 v. 0.156). Incubation duration is short for a tropidurid (–0.07 v. 0.40 std. resid.), based on *P. plica*.

The same species has a maximum recorded longevity of 5.2 years, which is long for a tropidurid of this size (–0.79 v. –1.17 std. resid.), though short for a lizard (v. 0.0 std. resid.). Endemism is somewhat low (50% v. 74%). Island species are average (12.5% v. 10%). Introductions are very high (first paragraph). No *Plica* has been evaluated for conservation status. In other respects *Plica* are typical tropidurids.

Plica spp.	English name				Range core
			Size		
	Activity	Diet	mm	g	Ecological business model
caribeana	Caribbean Treerunner				Venezuela
	diurnal	invert.	98	33.8	**** trunk/wall lizard
kathleenae	Kathleen's Treerunner				Guyana
	Diurnal		98	33.9	
lumaria	Guaiquinima Tepui Treerunner				Venezuela
	diurnal		81	20.1	
medemi	Medem's Treerunner				Colombia
	diurnal		89	26.4	
pansticta	Corocoro Mountain Treerunner				Venezuela
	diurnal		94	30.6	** trunk/wall lizard
plica	Collared Treerunner				Guyana
	diurnal	invert.	127	68	**** trunk/wall lizard
rayi	Ray's Treerunner				Venezuela
	diurnal		85	23.1	** trunk/wall lizard
umbra	Blue-lipped Treerunner				Venezuela
	diurnal	ants	81	20.4	*** trunk/wall lizard

Podarcis

Podarcis is a major radiation (23 spp.) of average-sized (5.7 g v. 5.1 g) refugia-anchored diurnal ground (96% v. 57%) lacertids found primarily on the European side of the Mediterranean, but extending from Netherlands to the Ukraine and south to Turkey and Morocco to Tunisia. Their heliophilic rock-loving habits lead to fragmented native ranges, which trigger conservation concern,

Podarcis spp.	English name				Range core
			Size		
	Activity	Diet	mm	g	Ecological business model
bocagei	Bocage's Wall Lizard				Iberian Peninsula
	diurnal	invert	59	4.73	**** refugia-anchored diurnal gr. liz.
carbonelli	Carbonell's Wall Lizard				Iberian Peninsula
	diurnal	invert.	54	3.71	* refugia-anchored diurnal gr. liz.
cretensis	Crete Wall Lizard				Greece
			58	4.67	** refugia-anchored diurnal gr. liz.
erhardii	Cyclades Wall Lizard				Balkans
	diurnal	invert.	66	6.6	*** refugia-anchored diurnal gr. liz.
filfolensis	Malta Wall Lizard				Malta
	diurnal	omnivore	67	7.1	** refugia-anchored diurnal gr. liz.
gaigeae	Skyros Wall Lizard				Greece
	diurnal	omnivore	72	8.8	** refugia-anchored diurnal gr. liz.
guadarramae	Galicia Wall Lizard				Spain
	diurnal	invert.	59	4.79	*** refugia-anchored diurnal gr. liz.
hispanicus	Iberian Wall Lizard				Iberian Peninsula
	diurnal	omnivore	59	4.81	*** refugia-anchored diurnal gr. liz.
levendis	Pori Wall Lizard				Greece
			65	6.5	* refugia-anchored diurnal gr. liz.
lilfordi	Lilford's Wall Lizard				Spain
	diurnal	omnivore	64	6.2	** refugia-anchored diurnal gr. liz.
liolepis	Columbretes Wall Lizard				Iberian Peninsula
	diurnal	omnivore	63	5.8	*** refugia-anchored diurnal gr. liz.
melisellensis	Dalmatian Wall Lizard				Balkans
	diurnal	omnivore	60	5.2	*** refugia-anchored diurnal gr. liz.
milensis	Milos Wall Lizard				Greece
	diurnal	invert.	58	4.48	* refugia-anchored diurnal gr. liz.
muralis	Common Wall Lizard				Europe
	diurnal	omnivore	62	4.50	*** refugia-anchored diurnal gr. liz.
peloponnesiacus	Peloponnese Wall Lizard				Greece
	diurnal	invert.	70	8.0	** ground-foraging tree lizard

especially for species with limited ranges. Their bright colors and physical proximity to commerce has likely promoted species introductions—at least 27 introductions documented—which trigger conservation concern. In combination with proximity to Western scientists, these factors have resulted in a particularly well-studied genus (filled cells 51% v. 33%).

Morphologically these are fairly typical lacertids, with the exception that they are unusually capable of color change (8.7% v. 2.2%), including advanced color change (4.3% v. 0.9%). Blue tails are overrepresented (22% v. 14%). The sexes almost always differ in color (91% v. 33%) and frequently show seasonal color (26% v. 15%). In a case of seasonal colors combining with introduced species to cause trouble, *siculus* males in introduced populations are known to attack other lizard species that show green coloration. The green coloration may be a badge for breeding *siculus* males. *Podarcis* are more likely than average lacertids to be m>f sexually dimorphic in size (0.86 v. 0.49 on –2 to 2 scale); combined with other sex differences, this leads to an exceptionally high sex differences rollup—only three genera are higher in Lacertidae—of 2.13 (v. 0.86 on 0 to 6 scale). Many *Podarcis* males have large heads (literally): *bocagei, guadarramae, hispanicus, peloponnesiacus, siculus*, and *vaucheri*.

Only half of *Podarcis* species are invertebrate eaters (v. 77%); the remaining 50% are omnivores (v. 19%), which is unusual in such small lizards (table 3.5, trait 2). All *Podarcis* are considered to be active foragers (v. 92%). At least five species show a pattern of high MPM, while showing a percentile for PTM that is at least 20 percentiles lower: *hispanicus* (96/70), *melisellensis* (91/64), *muralis* (95/68), *peloponnesiacus* (89/60), *tiliguerta* (84/54). This spread is not so large in *lilfordi* (86th and 81st, respectively), but the general pattern suggests that there is a component to the movement of *Podarcis* that accentuates MPM over PTM; apparently they dart very expeditiously or for very short distances. Home-range areas have not been documented, in part because home ranges are so sensitive to the geometry of supporting rock crevices and are therefore hard to generalize. Only 4% (v. 11%) are conspicuously wary. They exhibit strong reliance on retreat to crevice-burrows (44% v. 50%) and vegetation (44% v. 42%). Eleven percent dive into sand (v. 7%).

Fifty-two percent bask directly (v. 25%), and their mean active body temperature is average for lacertids (33.2°C v. 33.3°C).

Male social tolerance is low (15% v. 41%); male spatial defense is high (85% v. 59%). This fits better with their association with refugia (table 4.1, trait 47). Male hierarchies are present (15% v. 25%), associated with high-density living (see population density, below). In the mating season, at least *carbonelli* and *muralis* have both territorial and satellite males.

Thirty-one percent of *Podarcis* species include burrows among their nightly retreats (v. 48%). Presumably it might be higher if so many were not living in protective rock crevices. Their modal habitat in relation to rainfall is the intermediate band (401–1100 mm/y; 75% v. 32%); they are absent from areas with <150 mm/y. They are most often found at low elevation (100% v. 87%), but their elevational profile does not differ much from that of lacertids (100/65/15 v. 87/61/20). They are overrepresented in littoral (68% v. 22%) and human (50% v. 16%) habitats and exhibit average frequency in sandy areas (50% v. 44%) and riparian habitats (27% v. 28%).

Podarcis spp.	English name		Size		Range core
	Activity	Diet	mm	g	Ecological business model
pityusensis	Ibiza Wall Lizard				Spain
	diurnal	omnivore	70	8.1	** refugia-anchored diurnal gr. liz.
raffonei	Aeolian Wall Lizard				Italy
	diurnal	invert.	70	8.0	* refugia-anchored diurnal gr. liz.
siculus	Italian Wall Lizard				Balkans
	diurnal	omnivore	69	7.6	**** refugia-anchored diurnal gr. liz.
tauricus	Balkan Wall Lizard				Balkans
	diurnal	omnivore	67	7.1	** refugia-anchored diurnal gr. liz.
tiliguerta	Tyrrhenian Wall Lizard				Italy
	diurnal	invert.	66	6.6	*** refugia-anchored diurnal gr. liz.
vaucheri	West Mediterranean Wall Lizard				Morocco
	diurnal	invert.	57	4.46	** refugia-anchored diurnal gr. liz.
virescens	Starter Castle Wall Lizard				Iberian Peninsula
	diurnal		54	3.66	** refugia-anchored diurnal gr. liz.
waglerianus	Sicilian Wall Lizard				Italy
	diurnal	invert.	59	4.90	*** refugia-anchored diurnal gr. liz.

Podarcis are much like Lacertidae in reproduction (clutch count −0.05 v. 0.13 std. resid.). Notable differences include communal oviposition (8.7% v. 1.9%) and slightly elevated RCM (0.301 v. 0.267).

Longevity is average (0.18 v. 0.19 std. resid.). Population density is high (3.18 kg/ha v. 2.07 kg/ha) based on a remarkable coverage of 10 species estimated a total of 115 times. The species' means range from 0.3 kg/ha (*muralis*) to 15.9 kg/ha (*lilfordi*, *siculus*). This excludes 48 samples of 15 species that were judged to be less credible, often due to measuring small concentrations of lizards at rock features.

Endemism in *Podarcis* is normal in frequency (43% v. 35%). Almost all species occur on islands (91% v. 26%). *Podarcis* have been widely introduced (26% v. 4.6%), including excess species introduced more than twice (8.7% v. 1.2%). Almost all species are objects of conservation concern (95% v. 56%), and 19% (v. 15.5%) are on the IUCN's lists of the most endangered. In other respects *Podarcis* appear to be typical lacertids.

P. erhardii is thought to be tolerated in the vicinity of some bird nests because it captures flies that deposit larvae that are parasites on the bird nestlings. *P. lilfordi* is one of the few lizards that is an important pollinator of selected plant species (Traveset and Riera 2005).

Pogona spp.	English name				Range core
	Activity	Diet	Size mm	Size g	Ecological business model
barbata	Bearded Dragon				Eastern Australia
	diurnal	omnivore	190	245	*** refugia-anchored diurnal gr. liz. or ground-foraging tree lizard (geog.)
henrylawsoni	Down's Bearded Dragon				Queensland
	diurnal	omnivore	126	69	* refugia-anchored diurnal gr. liz.
microlepidota	Kimberly Bearded Dragon				Western Australia
	diurnal	omnivore	146	109	
minima	Western Bearded Dragon				Western Australia
	diurnal	omnivore	98	31	* ground-foraging tree lizard
minor	Dwarf Bearded Dragon				Western Australia
	diurnal	omnivore	132	79.5	** ground-foraging tree lizard
mitchelli	Northwest Bearded Dragon				Western Australia
	diurnal	omnivore	144	104	** ground-foraging tree lizard
nullarbor	Nullarbor Bearded Dragon				Southwestern Australia
	diurnal	omnivore	123	63	* ground-foraging tree lizard
vitticeps	Central Bearded Dragon				Eastern Australia
	diurnal	omnivore	187	232	*** ground-foraging tree lizard

Pogona

Pogona is a moderate (eight spp.) genus of large (96 g v. 15.6 g), robust (shape index 1.4 v. 0.8 std. resid.), crestless (v. 38% crested) but spiny, semiarboreal agamids from Australia. Spines are particularly prominent at the rear of the head and on the body laterally. The spines set near the angle of the jaw, and the ability of some species to inflate their throat, are responsible for the appellation "bearded." *Pogona* routinely open their mouths as part of their threat display, which is not entirely a bluff, as their prodigious bite is combined, as least in *barbata*, with toxins in the saliva. Most species show some color change (75% v. 22% any; 24% v. 15% advanced), but *barbata* is unusual in that the regions of the body are specialized for different color change objectives: neck and belly regions are reserved for social color change, whereas back and sides are reserved for thermoregulatory color change.

Sexual differences are those typical for agamids, including dimorphism (0.50 v. 0.53 on –2 to 2 scale), dichromatism (63% v. 53%), and seasonal color (13% v. 20%). All documented species bask (v. 24%) but also rely on crypsis for predator concealment (100% v. 51%). Mean body temperature is 33.4°C (v. 34.8°C). One of three documented species shows tolerance among males (33% v. 3%). Climbing ability is widespread, but accounts are contradictory as to the degree of arboreality. *Pogona* are found primarily (67% v. 35%) in the arid zone (151–400 mm/y) and secondarily (33% v. 12%) in the intermediate rainfall zone (401–1100 mm/y). Half of the species (v. 26%) are found associated with sand or dunes.

Clutches are of numerous (13.7 v. 6.4) eggs (1.48 v. 0.20 std. resid.), but RCM is on the light side (0.293 v. 0.336). The ontogenetic mass range is notably high (1.67 v. 0.69 std. resid.), indicating relatively small offspring (1/380 of maximum mass v. 1/100). This suggests that RCM ratio may be depressed by the small size of offspring produced. Incubation duration is short (–0.73 v. –0.55 std. resid.). *P. vitticeps* is unusual in that TSD occurs but only at the highest temperatures, where production of high-quality females overrides the genetic sex determination prevalent under normal incubation conditions.

Mass-adjusted longevity is about average for lizards (–0.11 v. 0.0 std. resid.) but a bit long for agamids (v. –0.46 std. resid.). Population density is unsurprising (0.33 v. 0.16 kg/ha). Endemism is low (38% v. 49%). Only a quarter of species occur on islands (25% v. 36%). None has been introduced (v. 1.7%). Half are considered imperiled (v. 37%), but none is found on any of the IUCN's lists of the most endangered (v. 5%).

Polychrotidae and *Polychrus*

Polychrotidae and *Polychrus* is a moderate radiation (seven spp.) of large (42.6 g v. 5.3 g) diurnal canopy walkers (v. 6%) found in forests of the mainland Neotropics from Honduras to Brazil, excluding Chile, but possibly including Uruguay. Although Polychrotidae now includes only *Polychrus*, it once included *Anolis*, and the general morphological similarity between the two is apparent. If you think of a very large anole with little or no neck constriction, you get some image of a *Polychrus*. Like anoles, they possess dewlaps, the color of which, at least in the case of *gutturosus*, can be changed quickly. In recognition of their preference for small-diameter perches, they are sometimes given the name of bush anoles. They are

Polychrus spp.	English name				Range core
			Size		
	Activity	Diet	mm	g	Ecological business model
acutirostris	Sharp-nosed Monkey Lizard				Paraguay
	diurnal	omnivore	113	41.9	*** diurnal canopy walker
femoralis	Werner's Bush Anole				Peru
	diurnal		87	17.8	*** diurnal canopy walker
gutturosus	Neotropical Chameleon				Panama
	diurnal	omnivore	131	68	**** diurnal canopy walker
jacquelinae	Jacqueline's Bush Anole				Peru
			112	41.3	** diurnal canopy walker
liogaster	Boulenger's Bush Anole				Bolivia
	diurnal	invert.	118	48.5	* diurnal canopy walker
marmoratus	Common Monkey Lizard				Venezuela
	diurnal	omnivore	120	52	**** diurnal canopy walker
peruvianus	Peru Bush Anole				Peru
			118	48.5	* diurnal canopy walkers

also associated with chameleons because their eyes are turreted and move independently, their toes may be zygodactylous (two digits opposing the other toes), they move slowly and may rock back and forth in a swaying motion, and they may be motionless for hours. Their tails are only marginally prehensile, however (43% v. 5%).

They are more massively built (especially anteriorly) than an average lizard (shape index 1.04 v. 0.0 std. resid.). Their tails are much longer (2.41 v. 1.37 × SVL) and, like chameleons, are not breakable (v. 57%). Their uniformly advanced color changing ability bears some resemblance to chameleons and far exceeds familial norms (43% v. 11% basic change; 43% v. 6% advanced color change). Secretory pores are present in low but typical frequency (14% v. 15%). They are never elongate (v. 11%). Unlike chameleons, they are never crested (v. 5%).

All four SSD-documented species have m<f SSD. Their mean SSD score is –1.0 v. 0.36 on –2 to 2 scale, which is unusual for such large lizards. The sexes routinely differ in color at about twice the family frequency (43% v. 21%), as they also do for seasonal color (14% v. 6.5%). They do not have blue tails (v. 2.3%). The m:f ratio of relative tail length is about average (1.10 v. 1.07). Collectively this produces a sex differences rollup that is appreciably elevated but not extraordinary (1.43 v. 0.75 on 0 to 6 scale).

Most are omnivores (75% v. 15%); the remainder, which may take minor amounts of plant matter, are invertebrate eaters (25% v. 73%). All species are said to both sit-and-wait (100% v. 46%) and cruise forage (100% v. 7%); *gutturosus* is also said to forage actively (33% v. 58%). Electric light use is not documented (v. 2.0%), and home-range size has not been quantified.

They do not vocalize (v. 2.5%), which would mar the crypsis upon which all species rely for the primary predator concealment (100%, the highest family, v. 43%). Twenty-nine percent bask directly (v. 18%), and their mean active body temperature is a toasty 35.0°C (v. 31.1°C). All are judged to be solitary (v. 10%), though no males are known to defend space (v. 85%). The latter characteristics distinguish them sharply from both anoles and chameleons.

They are found at the wet end of the rainfall spectrum (mesic: 67% v. 43%; wet: 33% v. 9%). Their elevational profile is about average but they are not found high (l/m/h: 80/40/0 v. 82/41/13). They are not reported to occur in littoral (0% v. 17%) or sandy (0% v. 22%) areas, and they coexist with humans at average frequencies (14% v. 18.5%). They are much smitten by riparian areas, however (57% v. 22%).

They are uniformly oviparous (v. 82%), with the second-largest clutches of any family (1.26 v. 0.0 std. resid.), behind only Chamaeleonidae. This sharply distinguishes them from anoles (–0.61 std. resid.). The hatchlings are the fifth-smallest family (ontogenetic mass range 0.73 v. 0.0 std. resid.). The copulation duration of *gutturosus* is measured in hours, during which they switch sides and hemipenes, suggesting significant sperm competition. Their incubation period is a bit longer than average (0.31 v. 0.0 std. resid.) but nothing like the 1.89 (std. resid.) of chameleons. Their RCMs (mean 0.275) are only a little above the lizard average (0.226), way below the chameleon one (0.444), and way above the anole mean RCM (0.092). Endemism is low (14% v. 64%). A few are found on islands (14% v. 40%). None is known to have been introduced (v. 3.3%), and none is considered to be imperiled (v. 49%).

P. marmoratus is believed by some local residents to be highly venomous.

Poromera sp.	English name				Range core
			Size		
	Activity	Diet	mm	g	Ecological business model
fordii	Swamp Lizard				Congo
	diurnal		56	4.10	** refugia-anchored diurnal gr. liz.

Potamites spp.	English name				Range core
			Size		
	Activity	Diet	mm	g	Ecological business model
apodemus	Uzzell's Streamside Lizard				Costa Rica
	crepuscular		56	3.90	** refugia-anchored diurnal gr. liz.
ecpleopus	Cope's Streamside Lizard				Colombia
	diurnal	invert.	63	5.9	** refugia-anchored diurnal gr. liz.
erythrocularis	Orange-eyering Streamside Lizard				Peru
			69	8.0	
juruazensis	Juruai Streamside Lizard				Peru
	diurnal		48	2.31	** refugia-anchored diurnal gr. liz.
montanicola	Montane Streamside Lizard				Peru
			59	4.60	
ocellatus	Ocellated Streamside Lizard				Bolivia
			64	6.0	
strangulatus	Big-scaled Streamside Lizard				Peru
			79	12.5	* refugia-anchored diurnal gr. liz.

Poromera

Poromera fordii is the sole species in a genus of average (4.1 g v. 5.1 g), refugia-anchored diurnal ground lacertids found in equatorial Africa, from Cameroon to the Democratic Republic of Congo. It is very poorly known (filled cells 21% v. 33%) but is understood to be closely allied with water courses, especially in *Raphia* swamps. Tail length is comparatively short (1.50 v. 2.06 × SVL) and breakable. Sex differences are not reported. Its known elevational range is 10–1300 m, in both primary and secondary forest. Reproduction traits are not documented. It is found on Bioko Island, but has not been introduced. It has not been evaluated for conservation status. In other respects *Poromera* is similar to other lacertids.

Potamites

Potamites is a seven-species genus of slightly large (5.47 g v. 2.35 g) semiaquatic gymnophthalmids from Bolivia north to Costa Rica. Although all classifiable species are refugia-anchored diurnal ground lizards, all assignments are only weakly supported. *Potamites* have comparatively large heads, often larger in males, and a mildly robust body build (shape index 0.27 v. −0.32 std. resid.). Secretory pores are comparatively common (71% v. 29%). They are not elongate (v. 41%). SSD tends to the m>f side (0.50 v. −0.12 on −2 to 2 scale), though m<f species are also reported. Sexual dichromatism is relatively common (29% v. 13%). Seasonal color is not reported (v. 1.7%). The ratio of m:f relative tail lengths is documented for only one species (*ecpleopus*), for which it is higher than typical (1.16 v. 1.05 ratio). High ratios often connote m<f SSD, which is common in gymnophthalmids but does not apply in this case: *ecpleopus* is m>f. The sex rollup score for *Potamites* is 1.29 (v. 0.49 on 0 to 6 scale), the fourth highest of 46 genera in Gymnophthalmidae.

Although *apodemus* is characterized as crepuscular, this datum is poorly supported; the other documented species are diurnal. Primary predator concealment tactic for *Potamites* is both retreat to crevices and burrows and concealment in vegetation. Mean active body temperature is a mild 27.5°C (v. 28.8°C). One species (*apodemus*) is tentatively said to defend space, in part because only one individual is found per stream. The other documented species are considered to be socially tolerant.

Potamites are found in areas characterized as mesic or wet, in this case 3000–4600 mm/y. They are found over a wide range of elevations (0–2098 m). They are absent from littoral areas (v. 3%) and human habitat (v. 15%), and they are overrepresented in sandy areas (50% v. 14%) and, of course, riparian zones (100% v. 29%).

Reproduction is typical. The ontogenetic range of sizes is notably large (1.47 v. 0.62 std. resid.), albeit based on only one species (*ecpleopus*). This indicates unusually small hatchlings.

Prasinohaema spp.	English name		Size		Range core
	Activity	Diet	mm	g	Ecological business model
flavipes	Common Green Tree Skink				New Guinea
	diurnal	invert.	74	8.3	** diurnal canopy walker
parkeri	Parker's Green Tree Skink				New Guinea
			47	1.92	
prehensicauda	Prehensile Green Tree Skink				New Guinea
	diurnal	invert.	63	4.85	* diurnal canopy walker
semoni	Semon's Green Tree Skink				New Guinea
			62	4.67	
virens	Green-blooded Vineskink				New Guinea
	diurnal	invert.	53	2.72	*** diurnal canopy walker

One species (*ecpleopus*) lays communally (v. 3.4%).

Endemism is relatively low (43% v. 68%). None is found on islands (v. 5%). No *Potamites* is deemed imperiled (v. 25%). In other respects *Potamites* are typical gymnophthalmids.

Prasinohaema

Prasinohaema is a minor radiation (five spp.) of average-sized (3.97 g v. 3.25 g), diurnal canopy walker (100% v. 1.8%) skinks found in New Guinea, New Guinea outliers, and the Solomon Islands. Their signature attribute is green blood, which contains high quantities of biliverdin, a digestive agent that is toxic to many vertebrates at the level found in *Prasinohaema*. The green color is evident in the skin of the lizards, their open mouths, and even their eggs (even if not eaten with ham). As inclusion of this blood compound has evolved four times, it presumably conveys some protection to the lizards against predators, diseases, or parasites, but the issue is not yet resolved. *Prasinohaema* are also distinguished by a prehensile tail (100% v. 0.7%) that bears adhesive or glandular lamellae on the tail tip. These canopy skinks are cryptic, move slowly, and can climb glass.

These are slightly robust skinks (shape index 0.01 v. −0.46 std. resid.), with normal tails (1.35 v. 1.33 × SVL) that are not breakable (0% v. 73%). Sex differences are not known.

All documented species forage actively (100% v. 92%) for generalized invertebrates (100% v. 84%). Their elevational profile is skewed upwards (l/m/h: 67/67/33 v. 89/35/6). They are overrepresented in human habitats (33% v. 14%) but absent from littoral (v. 21%), sandy (v. 25%), and riparian (v. 23%) areas.

Both oviparity (*virens* v. 64%) and viviparity (*flavipes*, *prehensicauda*, and *semoni*; v. 36%) are known. Clutch/litter size is low (−0.17 v. 0.06 std. resid.) and may be fixed at two eggs/embryos.

Endemism is normal (60% v. 64%). All species occur on islands (v. 57%). None has been introduced (v. 2.1%), and none is imperiled (v. 47%). In other respects *Prasinohaema* are much like other skinks.

Pristidactylus

Pristidactylus (10 spp.) is a moderate genus of average-sized (14.8 g v. 16.4 g), refugia-anchored diurnal ground (100% v. 63%) leiosaurids from Argentina, Chile, and possibly Peru.

Pristidactylus are of typical build (shape index 0.72 v. 0.76) with large heads and average tails (1.50 v. 1.66 × SVL) that break in at least one species (10% v. 3%) but are not prehensile (v. 9%). One species (*torquatus*) exhibits basic color change (10% v. 3%). Documented sex differences are exclusively dichromatism (30% v. 15%). Only four species have dietary information, but those diets are more diverse than is usual for leiosaurids: one (*achalensis*) is omnivorous—leaves in addition to invertebrates—and *casuhatiensis* specializes in snails, an unusual diet in general and exceptionally so in its habitat, which has very sparse vegetation.

Stress calls are known in this genus (10% v. 3%), the only genus so recognized in *Leiosauridae*. Direct basking is also reported (10% v. 3%), again the only genus so recognized in the family. Most genus members are terrestrial (83% v. 50%), though *torquatus* is arboreal (17% v. 45%). Half of documented species refuge in burrows (50% v. 67%). *Pristidactylus* are underrepresented at low elevation (40% v. 71%) and overrepresented at medium (80% v. 57%). They are somewhat associated with sandy areas (33% v. 16%).

Their average clutch count is well below the family norm (−0.31 v. 0.68 std. resid.) for reasons that are not known. Their exceptionally long incubation period (3.62 v. 2.81 std. resid.) should be viewed with the understanding that this value is based on a single species (*torquatus*) for which the incubation duration is annotated as being uncertain but very long. Presumably this species' eggs are set to hatch the following year.

Conversely, longevity is notably short (−0.39 v. −0.39), although only one species (*Pristidactylus torquatus*) in the family has a known longevity (3 years). Additional data will no doubt lengthen this maximum.

Ninety percent of *Pristidactylus* are

Pristidactylus spp.	English name				Range core
			Size		
	Activity	Diet	mm	g	Ecological business model
achalensis	Argentina Steppe Iguana				Argentina
	diurnal	omnivore	106	36.6	* refugia-anchored diurnal gr. liz.
alvaroi	Alvaro's Steppe Iguana				Chile
			73	10.4	
araucanus	Gallardo's Steppe Iguana				Argentina
			83	16.1	
casuhatiensis	Casuhatien Steppe Iguana				Argentina
	diurnal	invert.	62	5.9	*** refugia-anchored diurnal gr. liz.
fasciatus	Banded Steppe Iguana				Argentina
	diurnal		81	14.8	** refugia-anchored diurnal gr. liz.
nigroiugulus	Black-throated Steppe Iguana				Argentina
			83	19.7	
scapulatus	Burmeister's Steppe Iguana				Argentina
		invert.	92	23.4	** refugia-anchored diurnal gr. liz.
torquatus	Philippi's Forest Iguana				Chile
		invert.	88	19.7	
valeriae	Donoso's Steppe Iguana				Chile
	diurnal		67	7.9	
volcanensis	Volcano Steppe Iguana				Chile
			79	13.5	

endemic (v. 70%). None is imperiled (v. 14%). In other respects *Pristidactylus* are typical leiosaurids.

Pristurus

Pristurus is a large genus (25 spp.) of slightly oversized (1.37 g v. 1.07 g), padless sphaerodactylids that break the sphaerodactylid mold. They are found in Old World deserts from Egypt south to Kenya and east to Iran and possibly Pakistan, with one outlier species in Mauritania. The richest concentration is in the southern Arabia Peninsula: 12 are in Yemen. Males are often crested (40%, the highest value in the family, v. 5%). Their crest extends onto their tail, which they curl and uncurl and wave sideways and up and down, hence the "semaphore" common name. In this bleak environment they have diversified: five ecological business models are found within the genus, including refugia-anchored diurnal ground lizard (55%, highest within the family, v. 7%), diurnal canopy walkers (14%, the highest within the family, v. 1.8%), trunk/wall lizards (18% v. 12%). ground-foraging tree lizards (9% v. 12%), and immobility experts (4.5%, the second highest within the family, v. 1.2%). Advanced color change is relatively common in *Pristurus* (4.0%, the second highest in the family, v. 1.4%). Tails are the second longest in the family (1.33 v. 1.07 × SVL). The males are often much larger than females (1.5, the highest mean in the family, v. 0.04 on –2 to 2 scale). The hallmark diet of the genus is ant specialization (63%, the highest value in the family, v. 16%); of the dietary nonspecialists, 60% eat ants and the remaining 40% specify termites. All three foraging tactics are used, but their reliance on active foraging (43% v. 17%) is the highest in the family. As a primary predator concealment tactic, they emphasize flight (35%, the second-highest value in the family, v. 23%). To quicken their flight, they maintain the highest mean active body temperature in the family (33.1°C v. 30.4°C). Their presence at high elevation is the second highest in the family (16% v. 4%). Their incubation duration is the second longest in the family (0.75 v. 0.44 std. resid.). They may run bipedally (*collaris, crucifer, ornithocephalus*). These benchmarks in the league tables do not always combine into a coherent whole, but they clearly indicate that *Pristurus* is not an average sphaerodactylid. Indeed, in some ways it seems more like a non-gecko, such as a lacertid.

In other respects, *Pristurus* are notable without being anomalous. Basic color change is only slightly elevated (8% v. 6.5%). Sexual dichromatism is comparatively subdued (16% v. 35%). Seasonal color is absent (v. 2.8%). Female tails are relatively longer than male tails (m:f ratio 0.95 v. 1.02). Invertebrate eating is found in the *Pristurus* that are not ant specialists (38% v. 82%). Sit-and-wait foraging frequency is average (86% v. 83% of species). Cruise foraging is only slightly elevated (29% v. 21%). Cathemeral (17% v. 15%) and diurnal (83% v. 62%) habits are mildly elevated. Spontaneous calling is unknown (v. 1.9%), but stress calling is high (8% v. 2.3%). *Pristurus* are not conspicuously wary (v. 7%). When not relying on flight, *Pristurus* may use crypsis (18% v. 17%), retreat to a crevice or burrow (29% v. 43%), or retreat to vegetation (18% v. 27%). Direct basking is high (28% v. 6.5%), as is thermoconforming (8% v. 4%). The only social behavior observation is that males defend space (100% v. 86%). As implied above, their elevational profile is skewed upwards (l/m/h: 89/58/16 v. 96/20/4). Perhaps it is due to habitat availability in their part of the world, but *Pristurus* are overrepresented in littoral areas (36% v. 31%), sandy areas (50% v. 15%), and riparian zones (41% v. 19%). They are about average in frequency of species using human habitat (27% v. 33%).

Reproductive traits match the family, except for incubation duration (highlighted above). Note also that incubation duration can be highly variable, for example, 45–188 days in *sokotranus*. There are no reports of adherent eggs (v. 5.6%), communal oviposition (v. 11%), or parental care (v. 1.4%).

Longevity is high (0.45 v. –0.15 std. resid.), equating to 4.9 years in *rupestris*. Endemism is average (60% v. 65%). Island species are underrepresented (48% v. 64%) but not as much as might be expected for desert lizards. Introductions are about average (4.0% v. 5.1%), but none has been introduced more than twice (v. 2.3%). Few are imperiled (5% v. 41%), and none is on one of the IUCN's lists of the most endangered (v. 23%). In other respects *Pristurus* are typical sphaerodactylids.

Pristurus spp.	English name				Range core
			Size		
	Activity	Diet	mm	g	Ecological business model
abdelkuri	Abd el Kuri Semaphore Gecko				Yemen
	diurnal	ants	36	1.15	** refugia-anchored diurnal gr. liz.
adrarensis	Adrar Semaphore Gecko				Mauritania
	diurnal	invert.	32	0.82	** refugia-anchored diurnal gr. liz.
carteri	Carter's Semaphore Gecko				Saudi Arabia
	cathemeral	ants	62	5.8	** immobility expert
celerrimus	Oman Semaphore Gecko				Oman
	diurnal	ants	38	1.36	** trunk/wall lizard
collaris	Collared Semaphore Gecko				Yemen
	diurnal	ants	53	3.58	
crucifer	Cross-marked Semaphore Gecko				Red Sea margins
	cathemeral	invert.	35	1.06	*** refugia-anchored diurnal gr. liz.
flavipunctatus	Rueppell's Semaphore Gecko				Red Sea margins
	diurnal	ants	38	1.32	** trunk/wall lizard
gallagheri	Gallagher's Semaphore Gecko				Oman
	diurnal	invert.	38	1.36	*** trunk/wall lizard
guichardi	Guichard's Semaphore Gecko				Yemen
	diurnal		36	1.16	** diurnal canopy walker
insignis	Unsigned Semaphore Gecko				Yemen
	diurnal		55	3.99	*** trunk/wall lizard
insignoides	Haggier Massif Semaphore Gecko				Yemen
	diurnal		47	2.60	** diurnal canopy walker
longipes	Peters' Semaphore Gecko				Yemen
			38	1.36	
mazbah	Mazbah Semaphore Gecko				Yemen
			37	1.29	
minimus	Least Semaphore Gecko				Oman
	cathemeral	invert.	31	0.74	** refugia-anchored diurnal gr. liz.
obsti	Obst's Semaphore Gecko				Yemen
	diurnal		36	1.18	** diurnal canopy walker
ornithocephalus	Bird-headed Semaphore Gecko				Yemen
	diurnal	ants	52	3.51	** refugia-anchored diurnal gr. liz.
phillipsii	Phillips' Semaphore Gecko				Ethiopia
	diurnal	ants	30	0.71	* refugia-anchored diurnal gr. liz. or litter cryptozoid
popovi	Popov's Semaphore Gecko				Yemen
	diurnal	ants	36	1.18	* refugia-anchored diurnal gr. liz. or trunk/wall lizard (urban)

Pristurus spp.	English name				Range core
			Size		
	Activity	Diet	mm	g	Ecological business model
rupestris	Blanford's Semaphore Gecko				Saudi Arabia
	cathemeral	ants	31	0.75	**** ground-foraging tree lizard or trunk/wall lizard (urban)
saada	Saada Semaphore Gecko				Yemen
	diurnal	invert.	39	1.44	** refugia-anchored diurnal gr. liz.
samhaensis	Samha Semaphore Gecko				Yemen
	diurnal		37	1.25	** refugia-anchored diurnal gr. liz.
schneideri	Schneider's Semaphore Gecko				Yemen
	diurnal		29	0.61	* refugia-anchored diurnal gr. liz.
simonettai	Coastal Semaphore Gecko				Somalia
	diurnal		29	0.59	* refugia-anchored diurnal gr. liz.
sokotranus	Socotra Semaphore Gecko				Yemen
	diurnal	ants	38	1.40	** ground-foraging tree lizard
somalicus	Somalia Semaphore Gecko				Ethiopia
	diurnal	invert.	37	1.29	** refugia-anchored diurnal gr. liz.

Proablepharus spp.	English name				Range core
			Size		
	Activity	Diet	mm	g	Ecological business model
barrylyoni	Lyon's Soil-crevice Skink				Queensland
			48	1.25	* crack hermit
kinghorni	Red-tailed Soil-crevice Skink				Queensland
	diurnal	invert.	44	1.00	** crack hermit
naranjicaudus	Orange-tailed Soil-crevice Skink				Western Australia
			44	1.03	* litter cryptozoid
reginae	Western Soil-crevice Skink				Western Australia
	diurnal	invert.	40	1.09	** litter cryptozoid
tenuis	Northern Soil-crevice Skink				Northern Australia
	diurnal	invert.	32	0.41	** litter cryptozoid

Proablepharus

Proablepharus is a minor radiation (five spp.) of very small (0.90 g v. 3.25 g), elongate (100% v. 35%) diurnal (100% v. 74%), sit-and-wait (100% v. 16%), invertebrate-eating (100% v. 84%) skinks found in inland Australia. Like many small lizards, they do not have movable eyelids (possibly a water conservation measure). Most are tentatively judged to be litter cryptozoids (60% v. 28%); the remainder are crack hermits (40% v. 1.7%).

These are slender skinks (shape index −0.79 v. −0.46 std. resid.), with slightly long tails (1.66 v. 1.33 × SVL) that are breakable (100% v. 73%). Although blue tails are not known (v. 4%), 80% of *Proablepharus*, especially young juveniles, exhibit red or orange tails. Sex size differences are not known, but both sexual dichromatism (40% v. 7.6%) and seasonal color (40% v. 6%) are common. These combine to a substantial but not overwhelming sex differences rollup score (on scale of 0 to 6) of 0.80 (v. 0.28).

Sixty percent of *Proablepharus* are conspicuously wary (v. 11%). An equal percentage bask directly (v. 21%), but all *Proablepharus* are cryptozoic (100% v. 30%). Rain stimulates surface activity in at least three species (*kinghorni*, *reginae*, *tenuis*). Three species (*barrylyoni*, *kinghorni*, *naranjicaudus*) refuge during the dry season in clay soils that produce deep cracks when dried out. They are overrepresented in sandy habitats (40% v. 25%) and absent from human habitats (v. 14%) but at normal frequency in littoral (20% v. 21%) and riparian (20% v. 23%) areas.

They are all oviparous (v. 64%) with slightly large clutches (0.32 v. 0.06 std. resid.) of one or two.

Population density has been estimated once for *reginae* (0.04 kg/ha v. 0.35 kg/ha), a low density that may reflect the low primary productivity of their outback environments. Endemism is low (20% v. 64%), despite their tie to highly fragmented specialty soils. Island species are only slightly low (40% v. 57%), despite their generally inland locations. None has been introduced (v. 2.1%). All are imperiled (v. 47%), though none is on one of the IUCN's lists of the most endangered (v. 14%). In other respects *Proablepharus* are much like other skinks.

Procellosaurinus spp.	English name	Activity	Diet	Size mm	Size g	Range core	Ecological business model
erythrocercus	Rodrigues' Red Lizard			30	0.29	Brazil	** subterranean forager or sand shark
tetradactylus	Rodrigues' Four-fingered Lizard			29	0.45	Brazil	

Proctoporus spp.	English name	Activity	Diet	Size mm	Size g	Range core	Ecological business model
bolivianus	Bolivia Lightbulb Lizard			53	1.96	Peru	** litter cryptozoid
carabaya	Cusco Lightbulb Lizard			57	4.24	Peru	
cephalolineatus	Head-lined Lizard			46	2.04	Venezuela	*** litter cryptozoid
chasqui	Chasqui Lightbulb Lizard	diurnal		62	5.6	Peru	** refugia-anchored diurnal gr. liz.
guentheri	Guenther's Lightbulb Lizard			43	1.65	Bolivia	** litter cryptozoid
iridescens	Iridescent Lightbulb Lizard			55	3.74	Peru	
kiziriani	Kizirian's Lightbulb Lizard			52	3.00	Peru	
lacertus	Stejneger's Lightbulb Lizard			54	3.43	Peru	
laudahnae	Monika's Lightbulb Lizard			56	3.90	Peru	
machupicchu	Machu Picchu Andean Lizard			43	1.62	Peru	
oreades	Nymph Sun Tegu			53	3.28	Peru	
pachyurus	Tschudi's Lightbulb Lizard			51	2.86	Peru	** litter cryptozoid
rahmi	Rahm's Sun Tegu			67	4.18	Peru	
spinalis	Boulenger's Sun Tegu			59	4.79	Peru	
sucullucu	Urubamba Lightbulb Lizard	diurnal		49	2.61	Peru	** litter cryptozoid

Procellosaurinus

Procellosaurinus (two spp.) are tiny (0.36 g v. 2.4 g), elongate, red-tailed ground gymnophthalmids endemic to Brazil. Both species are probably subterranean foragers, sand sharks, or litter cryptozoids. They are thin in build (shape index –0.93 v. –0.32 std. resid.). Sex differences, diel activity pattern, diet, predator avoidance tactics, and reproduction are not reported. They are both endemic to their country (v. 68%). Neither has been evaluated for conservation status. In other respects *Procellosaurinus* may be typical gymnophthalmids.

Proctoporus

Proctoporus (17 spp.) is a large genus of diurnal, cryptozoic, mostly high elevation gymnophthalmids centered in Peru and extending south to Argentina and north to Trinidad and Tobago. Most species are too poorly known (filled cells 18% v. 19%) to assign ecological business models. *Proctoporus* are of average size (2.89 v. 2 35 g) and build (shape index –0.04 v. –0.32 std. resid.), with tails of average length (1.74 v. 1.72 × SVL); at least one species has tails documented to be breakable (6% v. 28%). Secretory pores are present (24% v. 29%), and some species are elongate (18% v. 41%). SSD has been characterized only in *xestus*, for which it is m<f. The same species is reported to be sexually dichromatic (6% v. 13%). Sex difference in relative tail length has been characterized only in *machupicchu*, for which the ratio is 1.11 (v. 1.05), consistent with a tendency to have m<f SSD. Social relations are known only for *bolivianus*, which has tolerant males.

This genus reaches elevations of at least 4098 m, with a near-record—exceeded only by *Riolama*—88% (v. 33%) in the highest elevation band (>2400 m). Only 19% of species are found below 1200 m (v. 51%).

Proctoporus spp.	English name				Range core
			Size		
	Activity	Diet	mm	g	Ecological business model
unsaacae	Unsaac Lightbulb Lizard				Peru
	diurnal		47	1.29	** litter cryptozoid
xestus	River Lightbulb Lizard				Peru
	diurnal		51	2.86	

Proscelotes spp.	English name				Range core
			Size		
	Activity	Diet	mm	g	Ecological business model
aenea	Montane Skink				Mozambique
			54	1.82	
arnoldi	Arnold's Skink				Zimbabwe
	diurnal	invert.	73	4.05	
eggeli	Usambara Five-toed Fossorial Skink				Tanzania
		invert.	82	5.66	** litter cryptozoid

They are notably underrepresented in riparian habitat (0% v. 29%) and overrepresented in human habitat (33% v. 15%). Reproduction is poorly known but typical for gymnophthalmids (clutch of two; no viviparous species despite the cold elevation). Many species are endemic (82% v. 68%). Only 14% of species are considered imperiled (v. 25%), but all 14% are on the IUCN's lists of the most endangered (v. 11.5%). In other respects *Proctoporus* are typical gymnophthalmids.

Proscelotes

Proscelotes is a small genus (three spp.) of average-sized (3.47 g v. 3.25 g), elongate (100% v. 35%), cryptozoic (100% v. 30%) skinks found in southeastern Africa, from Tanzania to Zimbabwe. *P. eggeli* may be a litter cryptozoid.

Although two species have reduced limbs (*aenea*, *arnoldi*), the genus does not have the extreme attenuation associated with true subterranean foragers (shape index −0.63 v. −0.46 std. resid.); the mean shape index for all subterranean forager lizards (table 4.1, trait 4) is −2.75. The tails of *Proscelotes* break (100% v. 73% v. 38% mean for subterranean foragers). Furthermore, blue tails are found in juvenile *arnoldi* (33% v. 4% v. 0.3% for subterranean foragers: table 4.1, trait 16). No sex differences are reported.

Proscelotes eat invertebrates (100% v. 84%). One species (*arnoldi*) is diurnal (v. 30%). *P. eggeli* is found at an elevation of 1600 m (medium). Reproduction is known only in *arnoldi*, which may exhibit either viviparity or oviparity. The mean clutch/litter count for this species is high (0.46 v. 0.06 std. resid.). Endemism is normal (67% v. 64%). No species occurs on islands (v. 57%). None has been introduced (v. 2.1%). None has been evaluated for conservation status. In other respects *Proscelotes* are probably typical skinks.

Psammodromus

Psammodromus (six spp.) is a genus of small (2.82 g v. 5.1 g) refugia-anchored diurnal ground lacertids found on both sides of the Strait of Gibraltar, from Portugal to Italy in Europe and Morocco to Tunisia in Africa.

They are lightly built (shape index 0.12 v. 0.27 std. resid.) with slightly short tails (1.87 v. 2.06 × SVL) that are usually breakable (67% v. 92%). Secretory pores are present (17% v. 4%). *Psammodromus* are usually m<f in SSD (−0.50 v. 0.49 on −2 to 2 scale). Some have sexual dichromatism (17% v. 33%), but most have seasonal color (50%+ v. 15%). The seasonal color frequency is here labeled as 50%+ because in addition to the three species with unequivocal seasonal coloration, *occidentalis* is said to "usually" have nuptial coloration. "Nuptial" coloration is ordinarily regarded as seasonal, but if *occidentalis* routinely shows it, it might better be characterized as sexual dichromatism. In sum, *Psammodromus* are often colorful and show above-average sex differences (1.17 v. 0.86 on 0 to 6 scale).

Diet is known for three species, of which *algirus* is reported to have plants 23% by items. All three species are active foragers (v. 92%). Consistent with the tendency of lacertids to have higher MPM than PTM percentiles, *hispanicus* has mean percentiles of 99th (MPM) and 72nd (PTM). The corresponding values for *algirus* are 81st and 67th, respectively.

Most give stress calls (67% documented v. 1.5%); 33% give spontaneous vocalizations (v. 22%), both comparatively rare traits (1.6%) in diurnal lizards (table 3.7, traits 31, 32). For predator concealment *Psammodromus* exhibit a lacertid-typical wide range of tactics: flight (33% v. 22%), crevice-burrow (17% v. 50%), and vegetation (50% v. 25%). Direct basking is common (50% v. 25%), and they have a moderate mean active body temperature (31.1°C v.

Psammodromus spp.	English name				Range core
			Size		
	Activity	Diet	mm	g	Ecological business model
algirus	Large Psammodromus				Mediterranean
	diurnal	omnivore	75	9.6	*** refugia-anchored diurnal gr. liz.
blanci	Blanc's Sand Racer				Algeria
	diurnal		42	1.81	
edwarsianus	East Iberian Psammodromus				Iberian Peninsula
	diurnal	invert.	43	1.90	* refugia-anchored diurnal gr. liz.
hispanicus	Spanish Psammodromus				Iberian Peninsula
	diurnal	invert.	48	2.58	** refugia-anchored diurnal gr. liz.
microdactylus	Small-fingered Psammodromus				Morocco
			54	3.64	
occidentalis	Madrid Psammodromus				Iberian Peninsula
	diurnal		41	1.63	

Psammophilus spp.	English name				Range core
			Size		
	Activity	Diet	mm	g	Ecological business model
blanfordanus	Peninsular Dwarf Rock Agama				India
	diurnal	invert.	84	19.2	*** refugia-anchored diurnal gr. liz.
dorsalis	South Indian Rock Agama				India
	diurnal	invert.	109	44.2	**** refugia-anchored diurnal gr. liz.

33.3°C). The only documentation of social relations is that males defend space (v. 59%).

Compared to a typical lacertid, *Psammodromus* is shifted slightly toward the middle of the rainfall spectrum: arid (33% v. 40%), intermediate (50% v. 32%), mesic (17% v. 34%). All species are found at both low (v. 87%) and medium (v. 61%) elevation. A usual frequency (25% v. 20%) is found high (>2400 m). *Psammodromus* are overrepresented in littoral (40% v. 22%), sandy (60% v. 44%), and human (40% v. 16%) habitats. They are not especially associated with riparian areas (20% v. 28%).

Their reproduction is lacertid-typical (clutch count 0.21 v. 0.13 std resid.). Possible mate guarding is reported for *algirus*. *P. blanci* has one of the shortest-known lizard copulation times: 10 seconds. Perhaps they are eager to resume their incessant MPM.

Longevity is comparatively short (−0.18 v. 0.19 std. resid.), based on *algirus* (7 years) and *hispanicus* (3 years). Many *hispanicus* are said to die after their first reproduction, indicating a trend toward semelparity. Endemism is low in *Psammodromus* (17% v. 35%), as is representation on islands (17% v. 26%). Nonetheless, they have often been introduced (16.7% v. 4.6%). All species are judged imperiled (v. 56%) and 25% (v. 15.5%) are on one of the IUCN's lists of most endangered. There have been few observations of *microdactylus* since the 1930s.

Psammophilus

Psammophilus is a pair of large (29 g v. 15.6 g) semiterrestrial agamid species found in India. These lizards are exquisitely cryptic (50% v. 51%) against their favored rocks most of the year, but breeding males are as gaudy as they were once cryptic: *blanfordanus* males may acquire a crimson jawline and head/body crest against a jet-black background, whereas *dorsalis* males become bright yellow on the back and jawline. Males are larger than females (1.0 v. 0.53 on −2 to 2 scale). Unusual for agamids, their tails readily break (100% v. 17%). As is typical for agamids, broken tails do not regenerate. Both species have tiny nuchal crests (v. 38%). Both species are conspicuously wary (v. 6%). Both are sit-and-wait foragers (v. 88%); *dorsalis* is documented to move comparatively little (MPM and PTM around 10th percentile). Males perch higher than females at all times of year and higher yet in the mating season. Feeding appears to be opportunistic and may include many ants as well as occasional vertebrates. They are found in mesic or wet areas (100% v. 7%) over a wide elevational range (0–2100 m). Clutch counts are average (0.17 v. 0.20 std. resid.). Longevity is short (−1.00 v. −0.46 std. resid.). Both species are endemic (100% v. 49%). Reported densities are high, but areal density is dependent on exactly which rock crevices are included, precluding objective comparison with other lizards. Neither species has been found on an island (v. 36%), and neither has been introduced (v. 1.7%). Neither species is considered imperiled (v. 37%).

Pseudemoia

Pseudemoia is a moderate radiation (six spp.) of slightly small (2.46 g v. 3.25 g), diurnal (100% v. 74%), direct basking (100% v. 21%), viviparous (100% v. 64%) skinks found in southern (temperate) Australia. Half are litter cryptozoids (v. 28%); others are refugia-anchored diurnal ground lizards (v. 30%) or trunk/wall lizards (17% v. 3%). They are morphologically similar to typical skinks.

Males are smaller than females (−1.0 v. −0.31 on −2 to 2 scale), as is often the case with smaller skinks. Although none is sexually dichromatic (v. 7.6%), almost all show seasonal color (83% v. 6%). Thus their sex rollup score (on scale of 0 to 6) is 1.5 (v. 0.28), the fifth-highest value among 155 skink genera.

The dietary mix is similar to skinks: invertebrate eaters (83% v. 84%) and omnivores (17% v. 13%). All are active foragers (v. 92%). *P. rawlinsoni* is conspicuously wary (17% v. 11%). Mean active body temperature is close to the skink average (30.1°C v. 31.1°C).

P. spenceri lives in groups (v. 16%). No other social information is available.

They are well represented in littoral (50% v. 21%) and riparian (50% v. 23%) situations, absent from human habitat (v. 14%), and representative in sandy areas (17% v. 25%).

Litter size is moderately large (0.41 v. 0.06 std. resid.). The young are slightly small (ontogenetic mass range 0.34 v. 0.12 std. resid.). Longevity is about average (0.05 v. 0.17 std. resid.), based on *entrecasteauxii* (5 years). Endemism is low (17% v. 64%). Two-thirds of species occur on islands (v. 57%). None has been introduced (v. 2.1%). All are imperiled (v. 47%), but none is on one of the IUCN's lists of the most endangered (v. 14%). In other respects *Pseudemoia* are probably typical skinks.

Pseudemoia spp.	English name				Range core
			Size		
	Activity	Diet	mm	g	Ecological business model
baudini	Great Bight Cool-skink				Southwestern Australia
	diurnal	invert.	47	1.24	** litter cryptozoid
cryodroma	Alpine Bog-skink				Victoria
	diurnal	invert.	53	1.70	* refugia-anchored diurnal gr. liz.
entrecasteauxii	Tussock Cool-skink				Southeastern Australia
	diurnal	omnivore	56	3.29	* litter cryptozoid
pagenstecheri	Tussock Skink				New South Wales
	diurnal	invert.	58	3.62	* refugia-anchored diurnal gr. liz.
rawlinsoni	Swampland Cool-skink				New South Wales
	diurnal	invert.	52	2.58	** litter cryptozoid
spenceri	Trunk-climbing Cool-skink				New South Wales
	diurnal	invert.	57	3.43	*** trunk/wall lizard

Pseuderemias spp.	English name				Range core
			Size		
	Activity	Diet	mm	g	Ecological business model
brenneri	Brenner's Sand Racer				Horn of Africa
	diurnal		47	2.51	
erythrosticta	Boulenger's Sand Racer				Somalia
	diurnal		47	2.40	
mucronata	Blanford's Sand Racer				Red Sea
	diurnal		47	2.40	
savagei	Kandala Sand Racer				Somalia
	diurnal		43	1.90	
septemstriata	Seven-striped Sand Racer				Somalia
	diurnal		45	2.19	
smithii	Smith's Sand Racer				Ethiopia
	diurnal		43	1.90	
striatus	Peters' Sand Racer				Horn of Africa
	diurnal		43	1.90	

Pseuderemias

Pseuderemias (seven spp.) is a very poorly known (filled cells 17% v. 33%) genus of small (2.16 g v. 5.10 g) diurnal terrestrial lacertids, found along the east coast of northern Africa from Egypt to Kenya. They are too poorly known to state the diet or ecological business model for any species. Morphologically they are lightly built lacertids (shape index 0.06 v. 0.27) that have long tails (2.44 v. 2.06 × SVL) that are uniformly breakable (v. 92%). Sex differences are not documented. They are associated with sandy areas (100% v. 44%).

They are likely oviparous (v. 97%). Endemism is at typical frequency (29% v. 35%). None is found on an island (v. 26%), and none has been introduced (v. 4.6%). None is considered imperiled (v. 56%). In other respects they are probably typical lacertids.

Pseudoacontias

Pseudoacontias is a minor radiation (four spp.) of large-sized (11.7 g v. 3.3 g), functionally or literally limbless skinks endemic to Madagascar. As is so often the case with subterranean foragers (67% v. 12%), natural history information is scarce (filled cells 20% v. 25%) and insufficient to specify even the diel cycle or diet of any of the species. One species is tentatively classified as a litter cryptozoid (33% v. 28%).

These attenuate lizards have very low shape index values (–1.71 v. –0.46 std. resid.). Their tails are very short (0.39 v. 1.33 × SVL) and unbreakable (v. 73% breakable). Sex differences are not known. *P. angelorum* is conspicuously wary (25% v. 11%). They are found at low elevation (in this case <650 m). Reproductive mode is unknown. None has been introduced (v. 2.1%). All are imperiled (v. 47%), and 67% of the evaluated species are on the IUCN's lists of the most endangered (v. 14%). In other respects *Pseudoacontias* are much like other limbless skinks.

Pseudoacontias spp.	English name				Range core
			Size		
	Activity	Diet	mm	g	Ecological business model
angelorum	Marojejy Legless Skink				Madagascar
			157	16.7	* subterranean forager
madagascariensis	Madagascar Legless Skink				Madagascar
			152	6.5	
menamainty	Berara Forest Legless Skink				Madagascar
			169	20.1	** subterranean forager
unicolor	Mauve Legless Skink				Madagascar
			172	8.6	** litter cryptozoid

Pseudocalotes (fig. 4.3C)

Pseudocalotes is a major (21 spp.) radiation of moderate-sized (12.8 g v. 15. 6 g), often crested (76% v. 38%), arboreal (79% v. 43%) agamids in southern and southeastern Asia, found from India to China and south to Indonesia. They are poorly known (24% of cells filled v. 34%). Only two have their ecological business model assigned with full confidence, though their behavior suggests that most may be diurnal canopy walkers (75% v. 7%). None has been studied for their diet. Ten species are known from only a few specimens. They have long tails (2.04 v. 1.86 × SVL). They are not very sexually dimorphic (0.17 v. 0.53 on –2 to 2 scale) or sexually dichromatic (24% v. 53%), though 29% (v. 22%) exhibit some color change (19% v. 15% advanced). None has been reported to feed at electric lights (v. 0.2%). In a close fit to their generally retiring nature, both reported species rely on crypsis (100% v. 51%). Two species are reported to move notably slowly. Two species are reported to bask directly

Pseudocalotes spp.	English name				Range core
			Size		
	Activity	Diet	mm	g	Ecological business model
andamanensis	Tamil Nadu Forest Lizard				India
	diurnal	invert.	77	14.7	** diurnal canopy walker
austeniana	Abor Hills Agama				India
			93	26.7	
brevipes	Short-footed Long-headed Lizard				China
	diurnal		65	8.8	
cybelidermus	Purple-throated False Garden Lizard				Indonesia
			77	14.6	
dringi	Dring's Long-headed Lizard				Malaysia
	diurnal		59	6.6	** diurnal canopy walker
drogon	Drogon's False Garden Lizard				Malaysia
			74	13.4	
flavigula	Yellow-throated Long-headed Lizard				Malaysia
	diurnal		67	9.5	* trunk/wall lizard
floweri	Flower's Long-headed Lizard				Southeast Asia
	diurnal		80	16.8	
guttalineatus	Dash-lined False Garden Lizard				Indonesia
			77	15.0	

Pseudocalotes spp.	English name				Range core
			Size		
	Activity	Diet	mm	g	Ecological business model
kakhienensis	Kakhien Hills Spiny Lizard				Myanmar
	diurnal	invert.	100	33.5	* diurnal canopy walker
khaonanensis	Khao Nan Long-headed Lizard				Thailand
	diurnal		84	19.8	** diurnal canopy walker
kingdonwardi	Kingdon-Ward's Forest Lizard				Myanmar
	diurnal	invert.	83	18.7	
larutensis	Bukit Larut Long-headed Lizard				Malaysia
	diurnal		70	10.9	**** diurnal canopy walker
microlepis	Indochinese Long-headed Lizard				Myanmar
	diurnal	invert.	79	16.3	**** ground-foraging tree lizard
poilani	Laos Long-headed Lizard				Laos
	diurnal		74	13.0	
rhaegal	Rhaegal's False Garden Lizard				Malaysia
			70	11.1	
rhammanotus	Stitched-back False Garden Lizard				Indonesia
			58	6.2	
saravacensis	Lanjak-Entimau Lizard				Malaysia
	diurnal		68	10.1	
tympanistriga	Javamatra Long-headed Lizard				Indonesia
	diurnal	invert.	67	9.8	* diurnal canopy walker
viserion	Viserion's False Garden Lizard				Malaysia
	diurnal		66	9.1	
ziegleri	Ziegler's Tree Lizard				Vietnam
			65	8.8	

Pseudoceramodactylus sp.	English name				Range core
			Size		
	Activity	Diet	mm	g	Ecological business model
khobarensis	Gulf Short-fingered Gecko				Saudi Arabia
	nocturnal		59	4.38	**** nocturnal ground lizard

Pseudocophotis spp.	English name				Range core
			Size		
	Activity	Diet	mm	g	Ecological business model
kontumensis	Kon Tum Monkey-tailed Lizard				Vietnam
	diurnal	invert.	72	12.2	* diurnal canopy walker
sumatrana	Javamatra Monkey-tailed Lizard				Indonesia
			72	12.2	

(v. 25%). Clutch count is conspicuously small (−0.50 v. 0.20 std. resid.). Incubation is long for agamids (−0.19 v. −0.55 std. resid.). Endemism is high (71% v. 49%). Island species are as expected (29% v. 36%). None has been introduced (v. 1.7%), and none is listed as imperiled (v. 37%).

Pseudoceramodactylus

Pseudoceramodactylus khobarensis is the sole representative of a padless nocturnal ground lizard genus (Gekkonidae) found at low elevation (0–100 m) in vegetationless salt pans on the Arabian Peninsula. A better-studied species with a similar ecology is *Ctenophorus maculosus* (Agamidae). *P. khobarensis* is presumed to burrow under the salt crust, and the gecko possesses unique digits that are described as not fringed but "puffy" and with 10–15 staggered rows of sharply pointed scales. These scales are thought to prevent caking by salty silt. *P. khobarensis* is of average size (4.38 g v. 3.31 g) with a comparatively short tail (0.78 v. 1.07 × SVL). It has highly protuberant nostrils and lacks secretory pores (v. 46%). The only reported sex difference is in size: m<f. It sits-and-waits. Mean body temperature is about 30.0°C (v. 27.6°C). It is associated with littoral (v. 11%) and sandy (v. 11%) areas but not riparian zones (v. 17%) or human habitats (v. 27%). Its reproduction is typical of gekkonids. It has not been introduced (v. 3.8%) nor is it imperiled (v. 47%). In other respects *Pseudoceramodactylus* is a typical gekkonid.

Pseudocophotis

Pseudocophotis is a tiny (two spp.) genus of average-sized (12.2 g v. 15.6 g), moderately crested (v. 38%), prehensile-tailed (100% v. 1.4%), arboreal (v. 43%) agamids found in southeastern Asia, from Vietnam and

Indonesia. Very little is known about these lizards; *kontumensis* is said to often perch low over streams (riparian 50% v. 21%). Clutch count is high (0.55 v. 0.20 std. resid.). Both species are endemic (100% v. 49%). One is on an island (50% v. 36%). Neither has been introduced (v. 1.7%) or evaluated for conservation status.

Pseudocordylus

Pseudocordylus is a moderate (six spp.) genus of large-sized (26.4 v. 10.8 g), medium-build (shape index 0.83 v. 0.35 std. resid.) cordylids from southern Africa. Their bodies are not quite as spiny as *Cordylus*, and they are more colorful, especially in the breeding season, but like *Cordylus*, they are restricted to the vicinity of their rock refugia. Tail length is not documented, but breakage is reported to be absent (v. 60%). Some species have secretory pores (33% v. 69%).

Pseudocordylus spp.	English name				Range core
			Size		
	Activity	Diet	mm	g	Ecological business model
langi	Lang's Crag Lizard				South Africa
	diurnal	omnivore	82	13.0	*** refugia-anchored diurnal gr. liz.
melanotus	Common Crag Lizard				South Africa
	diurnal	omnivore	111	38.9	*** refugia-anchored diurnal gr. liz.
microlepidotus	Cape Crag Lizard				South Africa
	diurnal	invert.	124	57.9	*** refugia-anchored diurnal gr. liz.
spinosus	Spiny Crag Lizard				South Africa
	diurnal	invert.	77	10.5	** refugia-anchored diurnal gr. liz.
subviridis	Drakensberg Crag Lizard				South Africa
	diurnal		99	26.4	*** refugia-anchored diurnal gr. liz.
transvaalensis	Northern Crag Lizard				South Africa
	diurnal		113	42.1	*** refugia-anchored diurnal gr. liz.

Pseudocordylus are more often m>f dimorphic in size between the sexes (0.80 v. 0.36 on –2 to 2 scale), dichromatic in permanent color (67% v. 37%), and seasonally colorful (67% v. 28%). These combine to a substantially greater sex differences rollup score (2.50 v. 1.41 on 0 to 6 scale).

About half of the species are invertebrate eaters (50% v. 73%) and half omnivores (50% v. 30%). All forage by sit-and-wait, except when pursuing plants, and often scan from a rocky prominence. *P. melanotus* movements have been quantified by MPM (lowest percentile) and PTM (second-lowest percentile). Home-range size is unknown. One species (*spinosus*) is reported to be conspicuously wary (17% v. 27%). Groups are common (83% v. 74%), but colonies of *Pseudocordylus* tend to be less dense than *Cordylus*. Males defend space around their crevice, at least during the breeding season. *P. transvaalensis* males may acquire a harem during the breeding season. Some species report crack sharing outside of breeding. Many use burrows (40% v. 10%). They are overrepresented at high elevation (67% v. 21%) and have not been reported in littoral (0% v. 9%), sandy (0% v. 3%), or human (0% v. 0%) habitats.

All *Pseudocordylus* are viviparous (v. 75%), and they have a typical litter count (–0.78 v. –0.48 std. resid.). Young are slightly smaller than average (ontogenetic range 0.20 v. –0.37 std. resid.). This equates to a neonate that is 1/82 of maximum size versus 1/43. One species (*transvaalensis*) is reported to attend the young (17% v. 7.4%). Longevity is short (–1.30 v. 0.02 std. resid., based on five of six spp.). There are no population density estimates. Half of *Pseudocordylus* species are considered imperiled (v. 79%), though none is listed by the IUCN in the most endangered categories (v. 4.8%). In other respects they are typical cordylids.

Pseudogekko

Pseudogekko is an eight-species genus of little-known (23% v. 25% of cells filled) average-sized (3.12 g v. 2.75 g) nocturnal canopy gekkonids from the Philippines. Despite their predilection for small-diameter perches, none is reported to have prehensile tails (v. 1.8%). Secretory pores are common (63% v. 46%), but no other sex differences are documented. One species (*brevipes*) is elongate (13% v. 0.9%). Color change is not reported (v. 15%). A statement about foraging mode is available for one species (*brevipes*: sit-and-wait). This same species vocalizes (v. 6.5%), a call described as soft, infrequent "twittering" (probably even fewer than 140 characters). Stress calling is not documented (0% v. 3.7%). One species (*brevipes*) is found in human habitats (17% v.27%); none is reported in littoral (v. 11%), sandy (v. 11%), or riparian (v. 17%) habitats. Reproduction is typical for gekkonids. Population density is known from only one

Pseudogekko spp.	English name				Range core
			Size		
	Activity	Diet	mm	g	Ecological business model
atiorum	Central Visayan False Gecko				Philippines
			47	2.41	
brevipes	Orange-spotted Smooth-scaled Gecko				Philippines
	nocturnal	invert.	48	2.00	** trunk/wall lizard
chavacano	Zamboanga False Gecko				Philippines
			49	2.73	
compresicorpus	Cylindrical-bodied Gecko				Philippines
	nocturnal		52	3.18	** nocturnal canopy walker
ditoy	Leyte Diminutive False Gecko				Philippines
			47	2.41	** nocturnal canopy walker
isapa	Romblon Province False Gecko				Philippines
			54	3.56	* nocturnal canopy walker
pungkaypinit	Southern Philippines False Gecko				Philippines
			66	6.1	** nocturnal canopy walker
smaragdinus	Green Smooth-scaled Gecko				Philippines
	nocturnal		57	4.13	** nocturnal canopy walker

Pseudogonatodes spp.	English name				Range core
			Size		
	Activity	Diet	mm	g	Ecological business model
barbouri	Barbour's Clawed Gecko				Peru
	diurnal		24	0.35	
furvus	Colombia Clawed Gecko				Colombia
	diurnal		42	1.78	
gasconi	Gascon's Clawed Gecko				Brazil
	diurnal		27	0.47	
guianensis	Amazon Clawed Gecko				Venezuela
	cathemeral	invert.	28	0.57	*** litter cryptozoid
lunulatus	Venezuela Clawed Gecko				Venezuela
	diurnal		30	0.66	**** litter cryptozoid
manessi	Maness' Clawed Gecko				Venezuela
	diurnal		37	1.22	**** litter cryptozoid
peruvianus	Peru Clawed Gecko				Peru
	diurnal		32	0.84	

sample of one species (*brevipes*) and is low for a gekkonid (0.02 kg/ha v. 0.37 kg/ha). *Pseudogekko* have not been introduced (v. 3.8%); one of three evaluated species is imperiled (v. 47%), though one suspects that lack of information is a significant impediment to evaluating conservation status. In other respects *Pseudogekko* are typical gekkonids.

Pseudogonatodes

Pseudogonatodes is a seven-species genus of uniformly very small (0.73 g v. 1.07 g), padless, litter cryptozoid (100% v. 53%) sphaerodactylids found in northwestern South America, from Colombia east to French Guiana and south to Peru.

They are poorly known (filled cells 21% v. 27%). Tails are short (0.87 v. 1.07 × SVL). Only one species (*guianensis*) has documentation of SSD and it is contradictory. Some species show sexual dichromatism (14% v. 35%). *P. guianensis* males have relatively shorter tails (m:f ratio 0.92 v. 1.02). The modal *Pseudogonatodes* is diurnal (86% v. 62%); the remainder are cathemeral (14% v. 15%). All documented species conceal themselves from predators by retreating to vegetation (v. 27%). Males defend space (100% v. 86%). They are absent from littoral (v. 31%) and human (v. 33%) habitats and representative in sandy areas (25% v. 15%) and riparian zones (25% v. 19%). The skin of riparian *guianensis* is refractory to wetting, such that this tiny lizard can float indefinitely on the water's surface tension. My guess is that this feature evolved not for floating but for preventing chronic skin wetness that would facilitate the growth of pathogens in its chronically damp leaf litter environment. It also tears its skin easily, presumably for predator escape.

Reproductive values are normal; the RCM mean (based on *guianensis*) is a little high (0.117 v. 0.095). Endemism is normal (57% v. 65%). No species occurs on islands (v. 64%). None has been introduced (v. 5%), and none is imperiled (v. 41%). In other respects *Pseudogonatodes* are typical sphaerodactylids.

Pseudopus

Pseudopus is a single-species genus of a virtually limbless, terrestrial anguid found from the Adriatic to Kazakhstan. It is a well-known (64% v. 27% of cells filled), wide-ranging chemosensor (v. 13%). Although small (4.3 g v. 10.1g) and slender (shape index –1.69 v. –0.6), it is nonetheless carnivorous (v. 7%). Its tail breaks with difficulty and is a bit short (1.43 v. 1.66 × SVL) and not prehensile (v. 31%). The species is diurnal (v. 67%) and flees (v. 11%), rather than hides, from predators. It basks directly (v. 18%) and, unlike many wide-ranging chemosensors, is not particularly wary (v. 18% in anguids). It has an anguid-typical body temperature of around 24.5°C. It is terrestrial (v. 45%) and will use burrows (v. 36%). It is found at low and medium elevations, in littoral areas (v. 9%), and around human habitats (v. 11%).

It is oviparous (v. 39%), and its clutch is laid in an advanced stage of development. The young are small (ontogenetic range 2.08 v. 0.62 std. resid.). The female may attend the clutch until hatching (100% v. 17% parental care in anguids). It has been introduced to Austria and Slovenia (v. 3.8%) and is imperiled (v. 63%) in portions of its range. It is not on one of the IUCN's lists of the most endangered (v. 28%).

Pseudopus sp.	English name				Range core
			Size		
	Activity	Diet	mm	g	Ecological business model
apodus	European Glass Lizard				Eurasia
	diurnal	carnivore	106	4.3	*** wide-ranging chemosensor

Pseudothecadactylus

Pseudothecadactylus is a small (three spp.) genus of secretive, large-sized (18.4 g v. 4.15 g), nocturnal, arboreal—cave wall or tree trunk—northern Australian diplodactylids. All three exhibit some degree of flexibility in time of activity, but they prefer low light or light rainfall when foraging in the open. At least *lindneri* may forage at midday under overhangs or other deeply shaded sites. They vocalize both spontaneously and triggered by noises (v. 5%) during the day, which may reveal their whereabouts to approaching humans who would not otherwise be aware of them. At least one species—*australis*—has a bright red eyeshine, enabling headlamp-wearing humans to locate them at night. The mean size of *australis* is reported to be larger than the maximum size, provoking some uncertainty as to the size of the largest species in the genus. All three species have an adhesive pad under the tip of their more or less prehensile tails (v. 26%).

As befits their large size, they exhibit somewhat greater sex differences than typical diplodactylids (1.33 v. 0.54 on 0 to 6 scale). Although two of three species are considered invertebrate eaters, all three will take vertebrates at least opportunistically. Two of three are judged to be cathemeral (v. 13%). None is said to be conspicuously wary (v. 3%). They are overrepresented in sandy areas (33% v. 24%) but not reported in littoral (v. 17%), riparian (v. 10%), or human (v. 5%) habitats.

All are oviparous (v. 82%); the egg size of only one species is reported (*lindneri*) and those eggs are extraordinarily elongate (7 × 32 mm). This large egg size matches their ontogenetic mass range, which indicates exceptionally large young in this genus (–1.67, the lowest value in Diplodactylidae, v. –0.47 std. resid.). *Pseudothecadactylus* young average 1/13 of maximum size, versus the family ratio of 1/23. Incubation is unusually short (–0.46 v. –0.22 std. resid.). Longevity is high, even for a diplodactylid (1.23 v. 0.90 std. resid.), equating to 18 years in *lindneri*.

Pseudothecadactylus species have not been introduced (v. 2%) or evaluated for conservation concern. In other respects they are typical diplodactylids.

Pseudothecadactylus spp.	English name				Range core
			Size		
	Activity	Diet	mm	g	Ecological business model
australis	Giant Tree Gecko				Queensland
	nocturnal	invert.	106	24.2	*** trunk/wall lizard
cavaticus	Western Giant Cave Gecko				Western Australia
	cathemeral	invert.	94	17.1	** trunk/wall lizard
lindneri	Lindner's Giant Cave Gecko				Northern Territory
	cathemeral	carnivore	91	15.1	*** trunk/wall lizard

Pseudotrapelus

Pseudotrapelus is a moderate (six spp.) genus of average-sized (17.6 g v. 15.6 g), average-shape (shape index 0.9 v. 0.8 std. resid.), poorly known (20% v. 30% of cells filled), crestless (v. 38%), terrestrial (v. 56%) agamids of the Middle East (from Libya to Sudan to Oman to Syria). They have relatively large heads (larger yet in male *sinaitus*), breakable tails (v. 17%), secretory pores (v. 13%; larger in males), and, for the documented species, m>f SSD (1.00 v. 0.53 on –2 to 2 scale), sexual dichromatism (v.

Pseudotrapelus spp.	English name		Size		Range core
	Activity	Diet	mm	g	Ecological business model
aqabensis	Aqaba Agama				Jordan
			74	13.0	
chlodnickii	Dammed Agama				Sudan
	diurnal		93	26.7	
dhofarensis	Dhofar Agama				Oman
			82	18.2	
jensvindumi	Vindum's Agama				Oman
			75	13.8	
neumanni	Neumann's Agama				Yemen
	diurnal		81	17.3	** refugia-anchored diurnal gr. liz.
sinaitus	Sinai Agama				Saudi Arabia
	diurnal	omnivore	84	19.6	*** refugia-anchored diurnal gr. liz.

Psilophthalmus sp.	English name		Size		Range core
	Activity	Diet	mm	g	Ecological business model
paeminosus	Slithery Sand Lizard				Brazil
			34	0.44	** sand shark

Psychosaura spp.	English name		Size		Range core
	Activity	Diet	mm	g	Ecological business model
agmosticha	Caatinga Sharp-nosed Skink				Brazil
			70	6.8	* immobility expert
macrorhyncha	Atlantic Forest Sharp-nosed Skink				Brazil
		invert.	69	6.4	* immobility expert

53%), seasonal color (v. 20%), and advanced color change (v. 15%). Their omnivory (v. 26%) is likely associated with the low primary productivity of their habitat. Reasonable documentation is available only for *sinaitus*, which is active in the midday heat (median body temperature 39.5°C v. 34.8°C), runs to a crevice or burrow to evade predators (v. 29%), has both sexes defend space during the breeding season (male spatial defense v. 97%), and is found over a wide range in elevations (0–2500 m) in extremely arid zones (rainfall not quantified). *Pseudotrapelus* are associated with sandy (33% v. 26%) and riparian areas (33% v. 21%).

Egg count is about average (0.35 v. 0.20 std. resid.). Mother *sinaitus* are reported to defend both eggs and hatchlings (parental care 17% v. 3.5%). Mass-adjusted longevity is slightly short compared to all lizards (–0.20 v. 0.0 std. resid.) but slightly long in comparison to agamids (v. –0.46 std. resid.). Endemism is average (50% v. 49%). Island species are infrequent (17% v. 36%). No species has been introduced (v. 1.7%), and none is on any lists of conservation concern.

Psilophthalmus

Psilophthalmus paeminosus is the sole member of a gymnophthalmid genus of tiny (0.44 g v. 2.35 g), elongate sand sharks endemic to Brazil. The species is of slender build (shape index –1.12 v. –0.32 std. resid.). It possesses secretory pores (v. 29%), but other sex differences are not reported. The lizard-average frequency of secretory pores in sand sharks is low (3%: table 4.1, trait 9). *Psilophthalmus* occurs at low elevation, 250–450 m, in association with shrubby dunes. Reproduction is typical for gymnophthalmids, though clutch count is elevated (0.53 v. –0.11 std. resid.) and RCM is on the low side (0.141 v. 0.192 ratio). This species is considered imperiled (v. 25%), but it is not on the IUCN's lists of the most endangered species (v. 11.5%).

Psychosaura

Psychosaura are two species of moderately large (6.6 g v. 3.3 g), elusive skinks that hide in arboreal (100% v. 10%) bromeliads in southern Brazil. They are tentatively categorized as immobility experts (100% v. 0.2%), but they could be crack hermits if the report of foraging mode is in error. They are reported to be active foragers (100% v. 92%) and viviparous (100% v. 36%).

They are more robust that average skinks (shape index 0.16 v. –0.46 std. resid.). Tails are breakable, but the length is not known. The ontogenetic range of $\log_{10}$ masses is somewhat low (27-fold v. 39-fold; –0.28 v. 0.12 std. resid.), indicating comparatively large offspring. Sex differences are unknown. Their diet is invertebrates (v. 84%). Thermal and social behaviors are unknown. They are found at low elevation, in this case below 450 m. Litter counts are normal (–0.03 v. 0.06 std. resid.). Both species are endemic to Brazil (v. 64%). One is found on an island (50% v. 57%)—Queimada Grande. Neither has been introduced (v. 2.1%), nor has either been evaluated for conservation status. In other respects *Psychosaura* are much like other skinks.

Ptenopus spp.	English name				Range core
			Size		
	Activity	Diet	mm	g	Ecological business model
carpi	Carp's Barking Gecko				Namibia
	nocturnal		52	3.17	*** nocturnal ground lizard
garrulus	Common Barking Gecko				Botswana
	nocturnal	invert.	51	2.96	**** nocturnal ground lizard
kochi	Koch's Barking Gecko				Namibia
	nocturnal		55	3.65	**** nocturnal ground lizard

Ptenopus

Ptenopus (three spp.) are short-snouted, loudly vocalizing, burrow-using, nocturnal ground gekkonids from southwestern Africa, south of an east-west line from Namibia to Zimbabwe. The males call loudly enough to engender human complaints, but even more remarkably, they turn their extensive burrows over to the females that agree to mate with them. Thus the burrows appear to be a nuptial gift to females and therefore an energetic contribution to their reproductive success. Apparently something to crow about. Stress calls are not reported (v. 3.7%). Calling is predominantly at dusk, though some calls are heard throughout the day; though surface-active at night, all species appear to be at least periodically alert around the clock within their burrows. In the event of swarming termites, a rich food item, they may be active on the surface during the day. These easily located lizards have garnered consider study (42% v. 25% of cells filled) despite their remote geographic location.

Body size is typical (3.25 g v. 3.31 g) and tails are short (0.74 v. 1.07 × SVL). Secretory pores are absent (v. 46%), as are expanded toe pads (v. 30%), color change (v. 15%), and SSD (0.0 v. –0.03 on –2 to 2 scale). Conversely, sexual dichromatism is common (67% v. 7%), as is seasonal color (33% v. 6.5%). This reliance on color for sexual differentiation is unexpected for a nocturnal gecko and suggests that at least some sexual interactions occur with enough light to support color vision. Based entirely on these color differences, their sex rollup score is high (1.33 v. 0.63 on 0 to 6 scale).

Diet and foraging have been characterized only for *garrulus*, which is an invertebrate eater—though leaves are sometimes taken—that forages by both sit-and-wait and active methods. The majority of its intake is termites (ca. 62%). Electric light use (v. 3.8%) and home-range size are not reported.

Apart from mating and burrow hand-over, males are solitary and defend their burrows. They are found in low elevation areas that are dry (56–150 mm/y: 100% v. 13%) or arid (151–400 mm/y: 100% v. 27%). They favor sandy (100% v. 11%) and riparian (67% v. 17%) areas, and they are not found in human habitats (0% v. 27%).

Clutches are thought to be single eggs laid an average of 1.2–1.5 times per year. For such small clutches, RCM is naturally low (0.092 v. 0.140 ratio). Young are exceptionally large in *Ptenopus*: ontogenetic mass range is small (–0.72 v. –0.19 std. resid.).

Longevity is comparatively high (0.57 v. 0.15 std. resid.), based solely on *garrulus* (7.4 absolute years). No *Ptenopus* have been introduced (v. 3.7%), and none is considered imperiled (v. 47%). In other documented respects *Ptenopus* are typical gekkonids.

Ptychoglossus

Ptychoglossus is a moderate (15 spp.) radiation of average-sized (2.4 g v. 2.5%), average-build (shape index –0.11 v. –0.06 std. resid.), litter-cryptozoic (100% v. 100%) alopoglossids from the region where Central America meets South America. They occur from Costa Rica to Peru and Bolivia, east to the Guiana Shield. They differ in only a few respects from the other alopoglossids (genus *Alopoglossus*): unlike *Alopoglossus*, they all break their tails (100% v. 14% for *Alopoglossus*), few have secretory pores (13% v. 43% in *Alopoglossus*), 20% are elongate (v. 0% in *Alopoglossus*), and the one SSD-documented species (*brevifrontalis*) is m>f (v. the opposite in *Alopoglossus*). *P. plicatus* may oviposit in ant nests. Endemism is about average (53% v. 45%). Unlike *Alopoglossus*, some species occur on islands (6.7% v. 0% in *Alopoglossus*). None has been introduced (v. 0%). Species are imperiled at an average rate (30% v. 29%). Unlike *Alopoglossus*, 10% of *Ptychoglossus* are on one of the IUCN's lists of the most endangered (v. 7%).

Ptychoglossus spp.	English name		Size		Range core
	Activity	Diet	mm	g	Ecological business model
bicolor	Werner's Largescale Lizard				Colombia
			54	3.58	** litter cryptozoid
bilineatus	Two-lined Largescale Lizard				Colombia
			49	2.49	
brevifrontalis	Boulenger's Largescale Lizard				Venezuela
	diurnal	invert.	52	1.82	*** litter cryptozoid
danieli	Daniel's Largescale Lizard				Colombia
			50	2.73	*** litter cryptozoid
eurylepis	Largescale Lizard				Colombia
			39	1.17	
festae	Peracca's Largescale Lizard				Panama
	diurnal		50	1.64	*** litter cryptozoid
gorgonae	Rueda's Largescale Lizard				Colombia
	cathemeral		65	6.5	*** litter cryptozoid
grandisquamatus	Variable Largescale Lizard				Colombia
			42	1.56	
kugleri	Kugler's Largescale Lizard				Venezuela
	crepuscular		50	2.73	*** litter cryptozoid
myersi	Darien Keeled Litter Lizard				Colombia
	diurnal		47	1.29	**** litter cryptozoid
nicefori	Loveridge's Largescale Lizard				Venezuela
			47	2.25	* litter cryptozoid
plicatus	Keeled Leaf Litter Lizard				Panama
	diurnal	invert.	58	4.34	*** litter cryptozoid
romaleos	Burly Largescale Lizard				Colombia
	diurnal		48	2.37	*** litter cryptozoid
stenolepis	San Antonio Mountain Largescale Lizard				Colombia
			52	3.14	
vallensis	Cauca Valley Largescale Lizard				Colombia
	diurnal		48	2.37	*** litter cryptozoid

Ptychozoon

Ptychozoon is a nine-species genus of large (8.0 g v. 3.31 g) nocturnal gliders (Gekkonidae) found in southeastern Asia, from India to China and south to the Philippines and Indonesia. Tails are short (0.89 v. 1.07 × SVL), often prehensile (22% v. 1.8%), and like the rest of the body and limbs, tend to be elaborated with a segmented lateral fringe composed of "lappets." As with many gliders, extreme crypsis is also aided by the lateral fringe, and *Ptychozoon* relies primarily on crypsis for predator concealment. They may sleep during day or night in exposed locations that precisely match their dorsal coloration. This match is aided by basic (44% v. 15%) and advanced (22% v. 1.8%) color change, among the highest in gekkonids. Secretory pores are very common (78% v. 46%), and toes are expanded into webbed pads (100% v. 70%), probably to support gliding as well as adhesion. Their prehensile tails also do double duty, not only providing a steering mechanism while gliding, but also used for intra- and intersexual communication.

Males are larger than females (1.0 v. −0.03 on −2 to 2 scale), but sexual dichromatism is lacking—presumably it would conflict with crypsis—yielding a gekkonid-typical sex differences rollup (0.78 v. 0.63 on 0 to 6 scale).

Electric light use is not reported (v. 3.8%). In keeping with their nocturnality, spontaneous vocalizations are present (11% v. 6.5%), but stress calls are not reported (v. 3.7%). One-third of species are said to be conspicuously wary (v. 4.4%); the others may be concerned but are wedded to immobile crypsis (100% v. 50%) as a primary predator concealment. Both direct basking (11% v. 5%) and thermoconformance (11% v. 1.4%) are known.

Ptychozoon are overrepresented in littoral (29% v. 11%) and human (71% v. 27%) habitats and unrepresented in sandy habitats (0% v. 11%). There is a weak general correlation between padlessness in geckos and occurrence in sandy areas (Fisher's Exact [one-tailed] = 0.04, *phi* = 0.049; not reported elsewhere), supporting the supposition that toepads, such as in *Ptychozoon*, might be negatively impacted by sand.

Reproduction is typical of gekkonids, with adherent eggs (44% v. 10.5%) and communal oviposition (11% v. 2%) overrepresented, as is typical of arboreal species. Parental care is reported for female *lionotum*, which defend eggs and consume spent shells. The young are unusually large (ontogenetic size range small: −0.76 v. −0.19 std. resid.).

P. horsfieldii may be introduced on Okinawa, Japan. One in three evalu-

Ptychozoon spp.	English name				Range core
			Size		
	Activity	Diet	mm	g	Ecological business model
bannaense	Banna Gliding Gecko				China
			72	7.8	**** glider
horsfieldii	Horsfield's Gliding Gecko				Malaysia
	nocturnal	invert.	67	6.2	**** glider
intermedium	Philippines Gliding Gecko				Philippines
	nocturnal		81	10.6	**** glider
kaengkrachanense	Kaeng Krachan Gliding Gecko				Thailand
	nocturnal		71	7.4	**** glider
kuhli	Kuhl's Gliding Gecko				Malaysia
	nocturnal	invert.	86	12.5	**** glider
lionotum	Smooth-backed Gliding Gecko				Southern Asia
	cathemeral	invert.	81	10.6	**** glider
nicobarensis	Nicobar Gliding Gecko				India
	nocturnal		81	10.6	**** glider
rhacophorus	Kinabalu Gliding Gecko				Malaysia
	nocturnal	invert.	63	5.3	**** glider
trinotaterra	Three-banded Gliding Gecko				Thailand
	nocturnal		60	4.68	**** glider

Ptyctolaemus spp.	English name				Range core
			Size		
	Activity	Diet	mm	g	Ecological business model
collicristatus	Mount Victoria Fan-throated Lizard				Myanmar
	diurnal		74	13.4	** ground-foraging tree lizard
gularis	Green Fan-throated Lizard				Myanmar
	cathemeral	invert.	72	11.8	**** ground-foraging tree lizard

ated species is considered imperiled (v. 47%), though none (v. 11.3%) is classified within the IUCN's most endangered categories. In other respects *Ptychozoon* are typical gekkonids.

Ptyctolaemus

Ptyctolaemus is a tiny (two spp.) genus of average-sized (12.6 g v. 15.6 g), low-crested (100% v. 38%) ground-foraging tree (100% v. 19%) agamids found in southern Asia, from Bangladesh to China and Myanmar. As suggested by the common names, they have colorful extensible dewlaps that are larger in males. Their tails are a bit long (2.12 v. 1.86 × SVL). Males are larger than females in *collicristatus* (1.00 v. 0.53 on –2 to 2 scale). *P. gularis* death feigns, occurs in riparian zones (50% v. 21%), and has extraordinarily large clutches (2.52, the second-highest value in Agamidae, v. 0.20 std. resid.), equating to a midpoint clutch of 14.5 eggs. Both species occur in towns (v. 16%). *P. gularis* is listed as endangered in the Indian Red List (v. 37% imperiled). Neither is listed on one of the IUCN's lists of the most endangered (v. 5%). In other respects *Ptyctolaemus* are average agamids.

Ptyodactylus

Ptyodactylus is a minor radiation (eight spp.) of large (9.9 g v. 4.7 g), scansorial (i.e., climbing) phyllodactylids in Old World deserts from Western Sahara and Ghana east to Somalia, Syria, and Pakistan. They are the best-known genus in the family (filled cells 39% v. 27%). Ecologically the genus is about half ground-foraging tree lizards (57% v. 16%) and half trunk/wall lizards (43% v. 44%).

Morphologically they are typical phyllodactylids, with the exception that half of the species exhibit basic color change (50% v. 11%). All have expanded toe pads (v. 92%). SSD is m>f (1.0 v. 0.32 on –2 to 2 scale). Half of the species show sexual dichromatism (v. 6.7%). The m:f ratio of relative tail lengths is unusually equal (1.02 v. 1.14 ratio). In aggregate, sex differences are moderately large (1.0 v. 0.31 on 0 to 6 scale).

They are all invertebrate eaters, though one fruit was found once in a *hasselquistii*. Species differ in their resistance to scorpion venom: *guttatus* is resistant, *puiseuxi* is not. Both sit-and-wait (100% v. 94%) and cruise (50% v. 13%) foraging are recorded. Electric light use is the highest in phyllodactylids (40% v. 9%). Home-range size is unknown.

Most species are cathemeral (80% v. 19%); the remainder are crepuscular (20% v. 2.3%). One imagines that the very harsh seasonal temperatures in this part of the world promote flexible activity schedules. In keeping with, or perhaps amplifying from, their cathemerality, spontaneous calling frequency is very high (63%, the highest in Phyllodactylidae, v.

Ptyodactylus spp.	English name				Range core
			Size		
	Activity	Diet	mm	g	Ecological business model
dhofarensis	Dhofar Fan-footed Gecko				Oman
			77	12.0	* ground-foraging tree lizard
guttatus	Spotted Fan-footed Gecko				Saudi Arabia
	cathemeral	invert.	74	10.5	** ground-foraging tree lizard
hasselquistii	Yellow Fan-footed Gecko				Saudi Arabia
	cathemeral	invert.	79	12.9	* trunk/wall lizard
homolepis	Pakistan Fan-fingered Gecko				Pakistan
		invert.	88	17.6	
orlovi	Tanuf Fan-footed Gecko				Oman
			69	8.7	* ground-foraging tree lizard
oudrii	Oudri's Fan-footed Gecko				Algeria
	crepuscular	invert.	51	3.00	** trunk/wall lizard
puiseuxi	Levant Fan-fingered Gecko				Jordan
	cathemeral	invert.	70	9.0	** ground-foraging tree lizard or trunk/wall lizard (where lit)
ragazzii	Ragazzi's Fan-toed Gecko				North Africa
	cathemeral	invert.	81	13.7	*** trunk/wall lizard

Pygmaeascincus spp.	English name				Range core
			Size		
	Activity	Diet	mm	g	Ecological business model
koshlandae	Fine-browed Dwarf Skink				Queensland
	diurnal	invert.	28	0.29	*** litter cryptozoid
sadlieri	Magnetic Island Dwarf Skink				Queensland
	diurnal	invert.	29	0.31	* litter cryptozoid
timlowi	Dwarf Litter-skink				Queensland
	diurnal	invert.	29	0.31	* litter cryptozoid

11%), and stress calls are known from 25% (the highest in Phyllodactylidae, v. 6%). Conspicuous wariness is high for phyllodactylids (13%, the highest in Phyllodactylidae, v. 1.5%). All documented species (*guttatus*, *oudrii*) use retreat to crevices or burrows as their primary predator concealment tactic (v. 29%). Direct basking is remarkably high for a phyllodactylid (63%, by far the highest in Phyllodactylidae, v. 10%), and mean active body temperature (32.0°C v. 26.0°C) is the hottest of any phyllodactylid genus.

A typical frequency of *Ptyodactylus* live in groups (20% v. 20%), but males of all species still defend space (100% v. 100%).

The elevational profile of *Ptyodactylus* is the usual (l/m/h: 100/33/17 v. 93/34/6). *Ptyodactylus* are slightly overrepresented in sandy (50% v. 25%), riparian (33% v. 19%), and human (83% v. 40%) habitats; they are absent from littoral zones (v. 20%).

Reproduction is as expected (clutch count −0.64 v. −0.57 std. resid.), except that eggs are often adherent (63% v. 10%), laid communally (50% v. 16%), and attended by both parents (parental care 25%, the highest in Phyllodactylidae, v. 3.7%). Young are slightly large (ontogenetic mass range −0.19 v. 0.11 std. resid.). In the case of *oudrii*, egg survival increases by occurrence of the eggs in communal nests. In *hasselquistii*, parental attendance of offspring may extend beyond hatching.

Maximum longevity is the highest among phyllodactylid genera (0.83 v. 0.15 std. resid.) and is well documented (five spp.), equating to 10–11 years in four species and 14.6 years in *hasselquistii*. Population density is unknown. Endemism is low (25% v. 67%). Few species occur on islands (13% v. 40%). None has been introduced (v. 7.4%). Half of evaluated species are considered imperiled (v. 29%), though none is listed by the IUCN on one of its lists of the most endangered. In other respects *Ptyodactylus* are typical phyllodactylids.

Pygmaeascincus

Pygmaeascincus is a small genus (three spp.) of uniformly tiny (0.31 g v. 3.25 g), elongate (100% v. 35%), diurnal (100% v. 74%), oviparous (100% v. 64%), litter-cryptozoic (100% v. 28%) skinks endemic to Queensland.

They are slender (shape index −1.06 v. −0.46 std. resid.) with ordinary length tails (1.20 v. 1.33 × SVL) that are breakable (100% v. 73%). The only documented sex difference is seasonal color (*timlowi*: 33% v. 6%). Dietary specifics are not available. All are conspicuously wary (v. 11%). They are associated with riparian habitats (33% v. 23%) but absent from littoral (v. 21%), sandy (v. 25%), and human (v. 14%) habitats. *P. sadlieri* occurs on Magnetic Island (33% v. 57%). None has been introduced (v. 2.1%). All are imperiled (v. 47%), though none is listed

Pygomeles spp.	English name / Activity	Diet	Size mm	Size g	Range core / Ecological business model
braconnieri	Braconnier's Short Skink				Madagascar
			131	20.5	*** subterranean forager
petteri	Petter's Short Skink				Madagascar
			125	4.14	** subterranean forager
trivittatus	Three-striped Slender Skink				Madagascar
			118	15.2	

on one of the IUCN's lists of the most endangered (v. 14%). In other respects *Pygmaeascincus* are probably typical skinks.

Pygomeles

Pygomeles is a small scincid genus (three spp.) of relatively large (10.9 g v. 3.25 g), elongate (67% v. 35%) or limbless subterranean foragers (100% v. 12%) endemic to Madagascar.

They are slender (shape index –0.99 v. –0.46 std. resid.) with moderately short tails (0.61 v. 1.33 × SVL) that are mostly breakable (67% v. 73%). Sex differences are unknown. They are found at low elevation (<300 m). Both reproductive modes are known for *trivittatus*; the other species are unknown. Clutch/litter count for *trivittatus* is low (–0.59 v. 0.06 std. resid.). As with virtually all Madagascar species, all are island species (v. 57%) and endemic (v. 64%). None has been introduced (v. 2.1%). One of two evaluated species is imperiled (v. 47%), and *petteri* is listed as endangered on the IUCN's list of the most endangered (v. 14%). In other respects *Pygomeles* are probably typical limbless skinks.

Pygopodidae (fig. 4.8C)

The 46 pygopodids are an Australian plus Papua New Guinean family of average-sized (5.0 g v. 5.2 g) functionally limbless lizards that are mostly diurnal (61% v. 66%) and cryptozoic (71% v. 12%). Most are litter cryptozoids (68% v. 12%), a categorization that for most species implies that the lizards are routinely concealed by fallen leaves. In the case of pygopodids, however, the habitual concealment is often within dense clumps of stiff bunch grasses known collectively as spinifex, though several plant genera are involved. Those pygopodids that are not litter cryptozoids are subterranean foragers (19% v. 8%), nocturnal ground lizards (11% v.8%), or crack hermits (2.7% v. 2.6%). The genera involved are *Aprasia*, *Delma*, *Lialis*, *Ophidiocephalus*, *Paradelma*, *Pletholax*, and *Pygopus*. The pygopodids are closely related to geckos, despite their divergent morphology and ecology.

That Pygopodidae are unique is highlighted by their notable rank in comparison to other lizard families on a number of key attributes. For example, their frequency of litter cryptozoid ecological business model, mentioned above, is ranked behind only the small Neotropical family Alopoglossidae. Their cryptozoic microhabitat, also mentioned above, is behind only the Alopoglossidae and the small family Xenosauridae. Their serpentine body form has an average shape index of –1.32 (v. 0.0), the lowest value of any aboveground family. Their tails are remarkably long (2.65 v. 1.37 × SVL), second only to the tiny North America family Corytophanidae. The long-tailed trait is the opposite of the condition normally associated with limbless lizards, which are mostly subterranean foragers and have extremely short tails. Their ontogenetic range of masses is –2.12 std. resid. (v. 0.0), the lowest family mean, easily eclipsing the –1.87 of Helodermatidae, indicating that young pygopodids are comparatively huge. This low ontogenetic range of sizes is not due to a small sample, as it is found in all three of the documented genera. The mean SSD of pygopodids is strongly m<f (–0.96 on –2 to 2 scale), marginally higher than only the –1.0 scores of the small families Polychrotidae, Leiosauridae, and Xantusiidae. Pygopodids are remarkably vocal; their frequency of spontaneous calling (67% v. 3%) ranks them well ahead of all other families (compare to second-place Carphodactylidae at 13%). Pygopodids are also prone to concealing themselves from predators in loose sand (50% v. 6%), nearly double the frequency of second-place Liolaemidae (29%). One attribute of pygopodids is not tabulated in any league tables but deserves special mention due to its rarity in lizards: invertebrate-eating pygopodids of several genera—some species in at least *Pletholax* and *Pygopus*—consume their prey by lapping the crushed or dismembered arthropods' juices rather than swallowing the exoskeleton.

Although pygopodids are functionally limbless, most retain tiny hind limb flaps that play a role in mating and moving through dense grass clumps at low speed. Their long tails are critical to propulsion through the dense grass but are nonetheless shed with alacrity (96% v. 57%); they regrow broken tails very quickly. Pygopodids lack prehensile tails (v. 5.0%), crests (v. 5.3%), secretory pores (v. 15%), elongation (v. 11%), blue tails (v. 2.3%), and advanced color change (v. 6%). A few *Delma* can change darkness (4.3% of family v. 10.7% of lizards).

Overall sex differences are few despite the strong size difference (sex rollup score 0.50 v. 0.75 on 0 to 6 scale). The only other recorded sex difference is seasonal color in *Pletholax* (2.2% of family v. 6.5% of lizards).

Herbivory is lacking (v. 4.1%), but all other diet classes are found in Pygopodidae: carnivory (8.7% v. 3.2%), invertebrate eating (65% v. 73%), ant specialists (8.7% v. 4.4%), and omnivory (17% v. 15%). Both sit-and-wait (66% v. 46%) and active (64% v. 58%) foraging are overrepresented in pygopodids. Cruise foraging (v. 7%) and electric light use (v. 2.0%) are not known. Home-range sizes have not been estimated.

Nocturnality is underrepresented (10% v. 25%) in Pygopodidae, and activity at all times of day is overrepresented (29% v. 7%). Diurnality (61% v. 67%) and crepuscularity (0% v. 1.8%) are about as frequent as would be expected. Spontaneous calling was highlighted in the second paragraph; stress calling is also highly elevated (28% v. 2.2%), the fourth-highest family mean. Conversely, conspicuous wariness is about average (8.7% v. 8.2%).

Sand diving as a primary predator concealment tactic was mentioned above. The other tactics except crypsis (0% v. 43%) and hiding in crevices or burrows (0% v. 25%) are present in expected frequencies: flight (25% v. 11%) and hiding in vegetation (25% v. 23%). Direct basking is also normal in frequency (17% v. 18%); thermoconformance is not reported (v. 2.6%). The mean active body temperatures of pygopodids are about normal except for *Pygopus* (24.5°C). The family mean is 28.3°C (v. 31.1°C).

Arboreality (2.4% v. 32%) and terrestriality (20% v. 48%) are underrepresented in pygopodids. Cryptozoic behavior is extraordinary, as highlighted in the introductory paragraphs. The others (subterranean, semiaquatic) are about average. Burrow use for daily refugia is slightly elevated (37% v. 22%). Pygopodids reside at the middle of the rainfall spectrum: 67% arid (v. 30%) and 33% intermediate (v. 27%). Elevations are not adequately documented. Pygopodids are underrepresented in human habitats (9% v. 19%) and overrepresented in littoral (30% v. 17%) and sandy (50% v. 22%) areas. They are found in riparian areas at expected frequency (21% v. 22%).

All pygopodids are oviparous (v. 82%). With the exception of *Lialis burtonis* (one to three eggs), all pygopodids appear to have a fixed clutch count of two, a reflection perhaps of their gecko heritage. For their size this clutch count is small but not exceptional (−0.44 v. 0.0 std. resid.). Incubation period is very close to average (−0.04 v. 0.0). Eggshells are soft (v. 41% hard) and nonadherent (v. 2.4%) but may be deposited communally (6.5% v. 4.2%). Parental care is known in *Delma* (family mean 2.2% v. 2.9% for lizards). RCM is notably normal (0.220 v. 0.226).

Population density is low (0.02 kg/ha v. 0.54 kg/ha), albeit based on few samples ($n = 1$ each of two species: *Delma nasuta, Lialis burtonis*). Endemism is about average (52% v. 64%). Island species are underrepresented (26% v. 40%), not a surprise in the outback. Introductions are not known (v. 3.3%). Most pygopodids are considered imperiled (88% v. 49%), but few are at serious risk globally; 6% v. 14% are on the IUCN's lists of the most endangered. Australian conservationists are diligent at highlighting species whose limited distributions may put them at risk, hence many species are considered imperiled but not globally endangered.

Pygopus

Pygopus (five spp.) is a genus of large (10.0 v. 5.2 g), functionally limbless, nocturnal/cathemeral surface pygopodids from Australia. Most are nocturnal ground lizards (75% v. 11%); one (*lepidopodus*) may be a litter cryptozoid (25% v. 68%). All have a somewhat snakelike coloration (likely mimicry of *Pseudonaja nuchalis*), and they behave like snakes when they are cornered. This mimicry extends to even the autotomized tail of *nigriceps*, which writhes as a *Pseudonaja* locomotes.

Pygopus are slender, even by pygopodid standards (shape index −1.97 v. −1.32 std. resid.). Their tails are

Pygopus spp.	English name				Range core
			Size		
	Activity	Diet	mm	g	Ecological business model
lepidopodus	Common Scalyfoot				Australia
	cathemeral	omnivore	196	12.7	* litter cryptozoid
nigriceps	Western Hooded Scalyfoot				Australia
	nocturnal	invert.	166	10.1	*** nocturnal ground lizard
robertsi	Cape York Scalyfoot				Queensland
		invert.	143	8.2	
schraderi	Eastern Hooded Scalyfoot				Eastern Australia
	nocturnal	invert.	168	10.3	*** nocturnal ground lizard
steelescotti	Northern Hooded Scalyfoot				Northern Australia
	nocturnal	invert.	159	9.5	*** nocturnal ground lizard

comparatively short (1.91 v. 2.65 × SVL). They do not show color change (v. 4.3%).

Most species eat invertebrates (80% v. 65%); the remainder are omnivores (20% v. 17%). *P. lepidopodus* specializes on spiders (68% of diet), but it laps the spider's juices rather than swallowing the exoskeleton. This affection for juice may account for its ability in captivity to thrive on a diet of bananas. *P. nigriceps* also laps arthropod juices rather than swallowing the exoskeleton, but it specializes (ca. 33%) on scorpions. All documented species are active foragers (v. 64%), but sit-and-wait foraging is common also (50% v. 64%). Home-range area is unknown.

The nominally nocturnal species (75% v. 10%) may be more crepuscular in cooler weather. No *Pygopus* is reported to call spontaneously (v. 67%), but all give stress calls (v. 28%). None is conspicuously wary (v. 9%). The only reported predator concealment tactic is flight (100% v. 25%). Mean active body temperature is cool (24.5°C v. 28.3°C), presumably a reflection of their nocturnality. *P. lepidopodus* is reported to burrow by twisting rather than pushing. It is also said to carry displaced soil in its mouth. *Pygopus* are found in the middle of the rainfall spectrum (arid: 50% v. 67%; intermediate: 50% v. 33%). They do not appear in human (v. 9%) or riparian (v. 21%) habitats, but are in littoral (20% v. 30%) and sandy (60% v. 50%) ones at typical frequencies.

Reproduction is typical, but communal oviposition is especially common (40% v. 6.5%). Clutches are comparatively small (−0.69 v. −0.44 std. resid.), and young are slightly large (ontogenetic mass range −2.32 v. −2.12 std. resid.). RCM is on the light side (0.176 v. 0.220 ratio).

Longevity is about average (0.15 v. 0.24 std. resid.), albeit based on *nigriceps* alone (7 years). Endemism in *Pygopus* is low (20% v. 52%), despite the large size of Australian states. No species is found on an island (v. 26%). All documented species are imperiled (v. 88%), but none is on one of the IUCN's lists of the most endangered (v. 6%). In other respects *Pygopus* follow the family characteristics.

Quedenfeldtia

Quedenfeldtia is a two-species genus of larger (1.52 g v. 1.07 g), padless, diurnal (100% v. 62%) trunk/wall (100% v. 12%) sphaerodactylids found in Morocco and Western Sahara. Tails are long (1.43 v. 1.07 × SVL). Secretory pores (*trachyblepharus*) may be present, though it is unclear if they are actually secretory (50% v. 0.5%). Color change may be advanced, but possibly seasonal, in *trachyblepharus* (50% v. 1.4%). Unquestionably the sexes differ in color (100% v. 35%) and show seasonal color (50% v. 3%). Therefore their sex differences rollup score is elevated (1.50, the second highest in the family, v. 0.54 on 0 to 6 scale). Primary predator concealment is retreat to a crevice or burrow (100% v. 43%). Social relations are reported contradictorily; they are simultaneously claimed to be colonial, territorial, and intolerant and also to share crevices. They bask directly (100% v. 6.5%), *trachyblepharus* has been seen doing so at air temperatures as low as 2.2°C. Both species occur in all elevational bands: 0–4000 m in both species. They are overrepresented in all tabulated habitat specialties, except human habitat (0% v. 33%): littoral (50% v. 31%), sandy (100% v. 15%), and riparian (50% v. 19%) zones.

Reproduction is normal, though young are exceptionally small (ontogenetic mass range 0.94 v. −0.12 std. resid.). Eggs are adherent (100% v. 6%) and laid communally (50% v. 11%). Endemism is normal (50% v. 65%). They are not found on islands (v. 64%), nor has either been introduced (v. 5%). They are imperiled at a normal frequency (50% v. 41%). None is found on any of the IUCN's lists of the most endangered (v. 223%). In other respects *Quedenfeldtia* are typical sphaerodactylids.

Ramigekko

Ramigekko swartbergensis is the sole representative of a genus (Gekkonidae) of South African crack hermits. It is moderately large (6.2 g v. 3.31 g) and robustly built (shape index 0.23

Quedenfeldtia spp.	English name				Range core
			Size		
	Activity	Diet	mm	g	Ecological business model
moerens	Morocco Day Gecko				Morocco
	diurnal		38	1.36	** trunk/wall lizard
trachyblepharus	Atlas Day Gecko				Morocco
	diurnal	invert.	41	1.70	** trunk/wall lizard

Ramigekko sp.	English name				Range core
			Size		
	Activity	Diet	mm	g	Ecological business model
swartbergensis	Swartberg Gecko				South Africa
	nocturnal		66	6.2	*** crack hermit

v. 0.10 std. resid.). It is not reported to possess secretory pores (v. 46%), color change (v. 15%), SSD, or sexual dichromatism (v. 6.6%). Both sexes are socially tolerant, and social groupings have been noted. They live at medium elevation (ca. 1850 m) on vertical rock walls, refuging in crevices. They have not been introduced (v. 3.7%) and are considered biologically at risk (v. 47%) due to their very limited range. In other respects *Ramigekko* are typical gekkonids.

Rankinia sp.	English name				Range core
			Size		
	Activity	Diet	mm	g	Ecological business model
diemensis	Mountain Dragon				New South Wales
	diurnal	ants	66	9.2	* immobility expert

Rhachisaurus sp.	English name				Range core
			Size		
	Activity	Diet	mm	g	Ecological business model
brachylepis	Dixon's Earless Lizard				Brazil
			53	3.28	

Rankinia

Rankinia is a single-species genus of a well-known (42% v. 30% of cells filled), small (9.2 g v. 15.6 g), crestless (v. 38% crested), terrestrial (v. 56%) ant-specialist (v. 20%) agamid from southeastern Australia including Tasmania. It is an immobility expert (v. 4.5%). It is unusual in having adult males only about half of the mass of adult females (–1.00 v. 0.53 on –2 to 2 scale), and it lacks other sex differences. The m<f size difference is characteristic of immobility experts, but the absence of other sex differences is not (table 4.1, traits 14–15). The large female mass is perhaps associated with the extraordinarily high RCM found in this species (mean 0.480, individuals as high as 0.860, v. 0.336). This RCM value is the third highest in Agamidae, suggesting great selective pressure for fecundity. The young are slightly larger than average (ontogenetic mass range 0.54 v. 0.69 std. resid., equating to 1/59 v. 1/104). Clutch count, however, is only average (0.24 v. 0.20 std. resid.). Oddly, this high-latitude species is oviparous. Perhaps egg size is unusually large (egg size not reported). Longevity is the third highest for an agamid (0.63 v. –0.46 std. resid.), which equates to 10 years.

As with so many immobility experts, it is reported to practice crypsis (v. 51%) and sit-and-wait foraging (v. 88%), and it is not considered to be conspicuously wary (v. 6%). It basks directly (v. 24%) and is associated with sandy areas (26%). It has not been introduced (v. 36%). *Rankinia diemensis* is of conservation concern; one population in Victoria is classified as critically endangered by that state.

Rhachisaurus

Rhachisaurus brachylepis is the sole member of a very poorly known (filled cells 12% v. 19%) gymnophthalmid genus endemic to Brazil. The species is too poorly known to guess its ecological business model. The species is medium sized (3.28 g v. 2.35 g) and of mildly robust build (shape index 0.10 v. –0.32 std. resid.). It is not known to possess secretory pores (v. 29%); other sex differences, diet, diel activity pattern, predator avoidance tactics, social relations, and reproduction are also unknown. This species has not been evaluated for conservation status.

Rhacodactylus

After several sections of the genus were cleaved, *Rhacodactylus* is now a small (four spp.) genus of large (54 g v. 4.15 g), nocturnal arboreal New Caledonia diplodactylids. *R. leachianus* is the largest extant gecko (195 g) and is therefore popular in the pet trade, for which numerous "designer morphs" have been bred. *R. leachianus* has a thick neck and massive torso (shape index 1.11 std. resid.) that matches its size, as do the other *Rhacodactylus* to a lesser degree (genus mean shape index 0.76 v. 0.05 std. resid. for family).

Rhacodactylus are known better than many smaller species (41% v. 28% of cells filled). Three of four species are trunk/wall lizards (v. 19.5%), and the other may be a nocturnal canopy walker (v. 25%). Tails are short (0.75 v. 0.85 × SVL), breakable (v. 91%), and prehensile (75% v. 26%). The tail of *leachianus* seems especially small (0.41 × SVL) in relation to its massive bulk. A low crest of spines is found in *auriculatus* (v. 1.3%). Color change (darkening) is found in *leachianus* (v. 5.4%).

Sexual dichromatism is found in half of the species (v. 6%), but overall sex differences are family-typical (0.75 v. 0.54 on 0 to 6 scale).

Rhacodactylus is atypical in diet: all evaluated species are omnivores (v. 15%). Even when offered vertebrates in captivity, *Rhacodactylus* are prone to eat vegetable matter, especially fig fruits. *R. trachyrhynchus* is presumed to be omnivorous, though only vege-

Rhacodactylus spp.	English name				Range core
			Size		
	Activity	Diet	mm	g	Ecological business model
auriculatus	Gargoyle Gecko				New Caledonia
	cathemeral	omnivore	108	26.2	*** nocturnal canopy walker
leachianus	New Caledonia Giant Gecko				New Caledonia
	cathemeral	omnivore	209	19 5	*** trunk/wall lizard
trachycephalus	Lesser Rough-snouted Giant Gecko				New Caledonia
	nocturnal		109	26.9	** trunk/wall lizard
trachyrhynchus	Rough-snouted Giant Gecko				New Caledonia
	nocturnal	omnivore	145	64	* trunk/wall lizard

table matter is documented in its diet.

Half of species are cathemeral (v. 13%), and half are nocturnal (v. 79%). Daytime activity on the large island of New Caledonia seems to be limited to basking, and there may be large geographic differences in activity cycle (more flexibility on small islands). Half of the species are documented to vocalize spontaneously (v. 5%) and when stressed (v. 5%).

Social tolerance is not well understood in the wild. *R. trachycephalus* is reported to aggregate in tree holes, with individuals clinging to each other, perhaps to conserve moisture or heat. In contrast, *leachianus* males are said to defend their tree holes against other males, and *leachianus* females are said to expel other females. In at least two species the primary grouping seems to be monogamous pairs, possibly with young. All bask (v. 17%).

They are partial to riparian areas (25% v. 10%) but are not recorded in littoral (v. 17%), sandy (v. 24%), or human (v. 5%) habitats.

Rhacodactylus are reproductively heterogeneous. Two species (*trachycephalus, trachyrhynchus*) are viviparous (v. 18%), and in both there is evidence of females attending the neonates (v. 3.4% parental care in family). *R. leachianus* is oviparous (though this is contradicted), with parthenogenesis occurring on occasion and skewed hatchling sex ratios suggesting male-high TSD. *R. auriculatus* is also alleged to show TSD, but reports as to which sex is produced at higher temperatures are contradictory. Gestation length has not been quantified but is reported to be "long" in the viviparous *trachyrhynchus*; incubation is typical in the oviparous forms (–0.26 v. –0.22 std. resid.).

Longevity is about average for a diplodactylid (0.71 v. 0.90 std. resid.), although this size-adjusted value equates to 30 years in *leachianus*. *Rhacodactylus* have not been introduced (v. 2%). Three species are imperiled (v. 81%), and two are listed on the IUCN's lists of most endangered (v. 29%). In other respects these are typical diplodactylids.

Rhampholeon

Rhampholeon are the primary terrestrial chameleons of East Africa, though *spectrum* extends west to Nigeria. The 19 species are representatives of the "dwarf" chameleon group (see Chamaeleonidae entry, table 5.2). They are small (3.2 g v. 2.0 g for the dwarf group). Ninety-five percent are immobility experts (v. 98% for dwarf group). Their shape index is also a close fit to the dwarf group (0.38 v. 0.32 std. resid.), as is their relative tail length (0.38 v. 0.53 × SVL), absence of crests (v. 0%), SSD score on –2 to 2 scale (–0.56 v. –0.70), arboreality (10.5% v. 17.3%), clutch count (0.79 v. 0.60 std. resid.), ontogenetic size range (–0.06 v. –0.16 std. resid.), and size-adjusted incubation duration (0.85 v. 0.44 std. resid.). In keeping with their lifestyle of terrestrial immobility, at least two species mimic dead leaves (*boulengeri, uluguruensis*).

The majority of species are reported to have prehensile tails (63% v. 88% family). Color change is subdued in this genus (any or advanced: 79% v. 92% family). Seasonal color is present only at low levels (5.3% v. 18.2%). The ratio of m:f relative tail lengths is larger than in chamaeleonids (1.26 v. 1.13), perhaps because the males' tails are otherwise too short to comfortably accommodate their hemipenes. Many species in *Rhampholeon* will vibrate when stressed either by social tension or by capture by a herpetologist. All males defend space (v. 93%). They are overrepresented at medium elevations (95% v. 64%) but are not found at high elevation (v. 21%). They are not reported for littoral (v. 6%), sandy (v. 3%), or human (v. 23%) habitats.

All are oviparous (v. 80%). *R. gorongosae* males, which are much smaller than their mates, are reported to perch on the female for extended periods, presumably to mate guard. *R. marshalli* eggs are deposited at an advanced stage of development (embryos are 70% of the egg volume at deposition). A similar phenomenon may episodically affect *spectrum*, which is anomalous within this genus for having a wide range of incubation durations (28–185 days). The other *Rhampholeon* average a 1.2 ratio

Rhampholeon spp.	English name	Activity	Diet	Size mm	Size g	Range core	Ecological business model
acuminatus	Nguru Pygmy Chameleon	diurnal	invert.	50	3.63	Tanzania	*** diurnal canopy walker
beraduccii	Mahenge Pygmy Chameleon	diurnal	invert.	25	0.57	Tanzania	** immobility expert
boulengeri	Boulenger's Pygmy Chameleon	diurnal	invert.	51	3.70	Central Africa	*** immobility expert
bruessoworum	Mount Inago Pygmy Chameleon			43	2.46	Mozambique	** immobility expert
chapmanorum	Chapmans' Pygmy Chameleon	diurnal	invert.	47	2.98	Malawi	*** immobility expert
gorongosae	Mount Gorongosa Pygmy Chameleon	diurnal	invert.	60	5.86	Mozambique	*** immobility expert
hattinghi	Mount Nzawa Pygmy Chameleon	diurnal		49	3.43	Congo	*** immobility expert
marshalli	Marshall's Pygmy Chameleon	diurnal	invert.	60	5.8	Mozambique	*** immobility expert
maspictus	Mount Mabu Pygmy Chameleon	diurnal	invert.	56	4.85	Mozambique	** immobility expert
moyeri	Udzungwa Pygmy Chameleon	diurnal	invert.	47	2.98	Tanzania	*** immobility expert
nchisiensis	Nchisi Pygmy Chameleon	diurnal	invert.	51	3.77	Zambia	** immobility expert
nebulauctor	Mount Chiperone Pygmy Chameleon	diurnal	invert.	44	2.62	Mozambique	*** immobility expert
platyceps	Mount Mulanje Pygmy Chameleon	diurnal	invert.	59	5.7	Mozambique	*** immobility expert
spectrum	Western Pygmy Chameleon	diurnal	invert.	49	3.50	Nigeria	*** immobility expert
spinosus	Rosette-nosed Chameleon	diurnal	invert.	46	2.86	Tanzania	*** immobility expert
temporalis	East Usambara Pygmy Chameleon	diurnal	invert.	45	2.74	Tanzania	*** immobility expert
tilburyi	Mount Namuli Pygmy Chameleon	diurnal		49	3.36	Mozambique	*** immobility expert
uluguruensis	Uluguru Pygmy Chameleon	diurnal	invert.	45	2.74	Tanzania	*** immobility expert
viridis	Pare Pygmy Chameleon	diurnal	invert.	43	2.41	Tanzania	** immobility expert

of longest duration/shortest duration (v. 1.4 lizard average).

Only about 10% occur on islands (v. 48%). None has been introduced (v. 5.4%). They are slightly underrepresented in the ranks of the mildly imperiled (74% v. 95%), though 47% are included in the most severe IUCN categories (v. 25%).

Rhineuridae and *Rhineura*

Rhineura floridana is the only species in the amphisbaenian family Rhineuridae. It is a moderately large-sized (17.9 g v. 5.3 g lizards), relatively slender (shape index –2.6 v. 0.0 std. resid.; –3.97 for Amphisbaenidae), limbless subterranean forager from Florida and Georgia. Its adaptation to subterranean existence extends to reduced eyes. Tails are very short (0.06 × SVL v. 0.11 for amphisbaenids) and are not known to break. It is partial to sandy areas (v. 22%). *Rhineura* are oviparous, with medium-sized clutches (2.0 v. 3.2 for amphisbaenids). It is not endemic (v. 64%), does not occur on an island (v. 40%), has not been introduced (v. 3.3%), and is not considered imperiled (v. 49%). In other respects information is lacking or undistinguished from other amphisbaenians.

Rhineura sp.	English name				Range core
			Size		
	Activity	Diet	mm	g	Ecological business model
floridana	Florida Wormlizard				Florida
		invert.	261	17.9	**** subterranean forager

Rhinogekko spp.	English name				Range core
			Size		
	Activity	Diet	mm	g	Ecological business model
femoralis	Kharan Spider Gecko				Pakistan
	nocturnal	invert.	54	3.43	*** nocturnal ground lizard
misonnei	Misonne's Swollen-nose Gecko				Pakistan
		invert.	67	6.4	* nocturnal ground lizard

Rhoptropella sp.	English name				Range core
			Size		
	Activity	Diet	mm	g	Ecological business model
ocellata	Namaqua Day Gecko				Namibia
	diurnal	invert.	39	1.37	*** refugia-anchored diurnal ground lizard

Rhinogekko

Rhinogekko (two spp.) are padless nocturnal ground gekkonids from Iran and Pakistan. They possess small nose "horns" that are globular in *femoralis* and tubular in *misonnei*. It is possible these nasal appendages aid in keeping sand out. This is a reasonable inference given the habitually violent winds that wrack the sandy wastes of species such as *misonnei*, which was first discovered in a site more than 100 km distant from any plant! Presumably it subsists on wind-blown insects.

Rhinogekko are of medium-large size (4.68 g v. 3.31 g) and build (shape index 0.16 v. 0.10 std. resid.). Both have secretory pores (v. 46%), but lack color change (v. 15%) and sexual dichromatism (v. 6.6%). They are not known to feed under electric lights (v. 3.8%). They rely on crypsis for predator concealment (v. 50%). They are overrepresented only in sandy habitats (50% v. 11%). Reproduction is documented only in *femoralis*, which for a gekkonid has the shockingly high clutch count of five (up to eight in some individuals). Mass-adjusted clutch count is the highest in the family (0.52 v. −0.26 std. resid.). No *Rhinogekko* have been introduced (v. 3.7%) or evaluated for conservation status. In other respects *Rhinogekko* are typical gekkonids.

Rhoptropella

Rhoptropella ocellata is an active-foraging refugia-anchored diurnal ground lizard from Namibia and South Africa, more like a typical lizard than the gekkonid it is. It matches the gekkonid template closely only in reproduction (oviparous, clutch of two). It is small (1.37 g v. 3.31 g) and slender (shape index −0.08 v. 0.10 std. resid.). It has secretory pores (v. 87%) but lacks color changing ability (v. 15%). Incubation duration is extended (0.48 v. 0.17 std. resid.). It has not been introduced (v. 3.7%); it is considered to be at conservation risk due to its restricted distribution (v. 47%). It is not on any of the IUCN's lists of the most endangered. In other respects *Rhoptropella* is a typical gekkonid.

Rhoptropus

Rhoptropus is a seven-species genus of typical-sized (3.08 g v. 3.31 g) diurnal gekkonids from Angola and Namibia. The majority are refugia-anchored diurnal ground lizards (67% v. 3%); the remainder trunk/wall lizards (33% v. 45%). Morphologically they match typical gekkonids, though they appear to lack color changing ability (0% v. 15%). Though m:f SSD has not been reported directly, the m:f ratio of relative tail lengths is notably male biased (1.15 v. 1.05), suggesting that SSD is likely to be m<f. All species are sit-and-wait. *R. afer* is notable for moving so little that conventional movement statistics (MPM, PTM) have not been measurable, though its sprint speed (10.8 kph) is the highest recorded from a gecko. Spontaneous vocalization is surprisingly frequent for a diurnal species (29% v. 6.5%), but stress calls have not been documented (0% v. 3.7%). No species has been described as conspicuously wary (0% v. 4.4%). Direct basking is overrepresented (14.3% v. 4.9%). Males of all documented species defend space (v. 87%). The only habitat type they are associated with is sandy areas (17% v. 11%). Adherent (43% v. 10.5%) and communally deposited eggs (29% v. 9%) are notably frequent and are found in the genus members that are refugia-anchored diurnal ground lizards, an unusual combination (table 4.1, traits 67, 68). *Rhoptropus* have not been introduced (v. 3.7%), nor have they been

Rhoptropus spp.	English name	Activity	Diet	Size mm	Size g	Range core	Ecological business model
afer	Namib Day Gecko	diurnal	invert.	49	2.73	Namibia	*** refugia-anchored diurnal gr. liz.
barnardi	Barnard's Namib Day Gecko	diurnal	invert.	42	1.74	Angola	** refugia-anchored diurnal gr. liz.
biporosus	Kaokoveld Namib Day Gecko	diurnal	invert.	47	2.31	Namibia	** refugia-anchored diurnal gr. liz.
boultoni	Boulton's Namib Day Gecko	diurnal	invert.	62	5.1	Angola	*** trunk/wall lizard
bradfieldi	Bradfield's Namib Day Gecko	diurnal	invert.	58	4.20	Namibia	*** refugia-anchored diurnal gr. liz.
diporus	Brandberg Namib Day Gecko	diurnal	invert.	51	2.95	Namibia	** trunk/wall lizard
taeniostictus	Angola Slender Day Gecko	diurnal	invert.	56	3.82	Angola	

Rhynchoedura spp.	English name	Activity	Diet	Size mm	Size g	Range core	Ecological business model
angusta	Border Beaked Gecko	nocturnal		48	2.10	Queensland	*** nocturnal ground lizard
eyrensis	Eyre Basin Beaked Gecko	nocturnal		49	2.25	Queensland	*** nocturnal ground lizard
mentalis	Brigalow Beaked Gecko	nocturnal		47	2.06	Queensland	*** nocturnal ground lizard
ormsbyi	Eastern Beaked Gecko	nocturnal	invert.	47	2.06	Queensland	*** nocturnal ground lizard
ornata	Western Beaked Gecko	nocturnal	invert.	50	2.38	Australia	*** nocturnal ground lizard
sexapora	Northern Beaked Gecko	nocturnal		47	1.99	Western Australia	*** nocturnal ground lizard

evaluated for conservation status. In other respects *Rhoptropus* are typical gekkonids.

Rhynchoedura

Rhynchoedura (six spp.) are uniformly small (2.14 g v. 4.15 g) nocturnal ground lizards (v. 35%) (Diplodactylidae) of the Australian outback. Their slightly short (0.69 v. 0.85 × SVL) tails all break (v. 91%) and are not prehensile (v. 26%). They lack secretory pores (v. 34%); documentation of other sex differences is absent. Several lack enlarged toepads. All are invertebrate eaters (v. 80%), though this is based on only two species, both of which appear to be termite specialists (*ormsbyi*, *ornata*). Despite their nocturnality, and perhaps in light of their terrestriality, they do not call (v. 5%). They are not known to bask (v. 17%). Despite their lack of basking, they have a higher than usual body temperature (mean 27.8°C v. 24.3°C), presumably reflecting high ambient temperatures in their environment. All are known to use burrow refugia (v. 22%). They are overrepresented in sandy areas (40% v. 24%) and not recorded in littoral (v. 17%), riparian (v. 10%), or human (v. 5%) habitats.

They are oviparous (v. 82%), with family-typical reproduction. Longevity is short (−0.59 v. 0.90 std. resid.), based on one species (*ornata*) at 2.8 years, an inadequate sample. Population density is similarly based on one estimate of one species (*ornata*), the value (0.067 kg/ha v. 0.44 kg/ha) likely reflecting low productivity of this desert environment. One species has been evaluated for conservation risk (*ornata*) and is considered imperiled (v. 81%), though none is on any of the IUCN's lists of most endangered (v. 29%). In other respects these are typical diplodactylids.

Riama

Riama is a large genus (30 spp.) of mostly elongate (73% v. 41%), mostly litter-cryptozoic (71% v. 55%) Gymnophthalmidae, found often at medium (80% v. 52%) or higher (53% v. 33%) elevations in the northern Andes from Ecuador to Venezuela and Trinidad and Tobago. They are slightly larger than typical gymnophthalmids (3.10 g v. 2.35 g) but of normal build (shape index −0.27 v. −0.32 std. resid.) and relative tail length (1.62 v. 1.72 × SVL). Only 6.7% (v. 28%) are reported to have breakable tails, but underreporting seems likely. Similarly, secretory pores (13% v. 29%) may be more prevalent. Sexual dichromatism is reported for only *shrevei* and *vieta* (7% v. 13%). Diet (invertebrates) is known for only *unicolor*. Diel activity pattern is reported for only three species and each is different (only crepuscular is not claimed). No species is said to be conspicuously wary (v. 7%). *R. oculata* climbs, but six other species are found primarily on the ground (*shrevei*) or associated with litter (71% v. 45%). Reproduction is fairly typical, though communal oviposition is slightly overrepresented (6.7% v. 3.4%). Ninety percent of species are endemic (v. 68%). Nearly half are considered imperiled (47% v. 25%), and they are greatly overrepresented among those listed by the IUCN as most endangered (27% v. 11.5%). In other respects *Riama* are typical gymnophthalmids.

The "lightbulb" appellation is applied to these and similar gymnophthalmids that may have ocelli on the flanks, especially of adult males. The "lightbulb" part got started with *R. shrevei*, which was reported to have glowing ocelli. Subsequent observations suggest that the observer was in fact seeing a bright reflection of their own light. As far as is known, the centers of the ocelli are highly reflective, but not luminous. Keep looking though.

Riama spp.	English name				Range core
			Size		
	Activity	Diet	mm	g	Ecological business model
achlyens	Highway Lightbulb Lizard				Venezuela
			72	9.1	*** litter cryptozoid
afrania	Afranio's Lightbulb Lizard				Colombia
			70	4.96	
anatoloros	East Slope Lightbulb Lizard				Ecuador
			54	2.15	
aurea	El Oro Lightbulb Lizard				Ecuador
			50	1.64	
balneator	Tungurahua Lightbulb Lizard				Ecuador
			56	2.34	
cashcaensis	Cashca Lightbulb Lizard				Ecuador
			62	3.23	
colomaromani	Coloma-Roman's Lightbulb Lizard				Ecuador
			69	4.80	
columbiana	Colombia Lightbulb Lizard				Colombia
			64	6.3	
crypta	Covert Lightbulb Lizard				Ecuador
			59	2.76	
hyposticta	Boulenger's Lightbulb Lizard				Ecuador
			68	4.48	
inanis	Portuguesa Lightbulb Lizard				Venezuela
			47	1.29	*** litter cryptozoid
kiziriani	Azuay Lightbulb Lizard				Ecuador
			53	1.97	
labionis	Large-lipped Lightbulb Lizard				Ecuador
			52	1.88	
laevis	Shiny Lightbulb Lizard				Colombia
			61	5.2	*** litter cryptozoid
luctuosa	Aragua Lightbulb Lizard				Venezuela
			75	10.4	
meleagris	Brown Lightbulb Lizard				Colombia
			67	4.2	
oculata	Ocellated Lightbulb Lizard				Ecuador
	diurnal		72	5.4	*** ground-foraging tree lizard
orcesi	Orces' Lightbulb Lizard				Ecuador
			54	2.15	
petrorum	Three Peters' Lightbulb Lizard				Ecuador
			64	3.61	
raneyi	Raney's Lightbulb Lizard				Ecuador
			68	4.48	

Riama spp.	English name / Activity	Diet	Size mm	Size g	Range core / Ecological business model
rhodogaster	Paria Lightbulb Lizard				Venezuela
			43	1.67	
shrevei	Shreve's Lightbulb Lizard				Trinidad and Tobago
	nocturnal		43	1.65	*** nocturnal ground lizard
simotera	O'Shaughnessy's Lightbulb Lizard				Colombia
			66	4.03	
stellae	Stella's Lightbulb Lizard				Colombia
			45	1.93	
stigmatoral	Spotted-belly Lightbulb Lizard				Ecuador
			66	4.03	
striata	Striped Lightbulb Lizard				Colombia
			58	4.46	
unicolor	Unicolored Lightbulb Lizard				Ecuador
	cathemeral	invert.	58	2.65	**** litter cryptozoid
vespertina	Western Lightbulb Lizard				Ecuador
			58	2.65	
vieta	Wrinkled Lightbulb Lizard				Ecuador
			47	1.29	* litter cryptozoid
yumborum	Yumbo Lightbulb Lizard				Ecuador
			52	1.80	

Rieppeleon spp.	English name / Activity	Diet	Size mm	Size g	Range core / Ecological business model
brachyurus	Zomba Pygmy Chameleon				Mozambique
	diurnal	invert.	49	3.36	*** immobility expert
brevicaudatus	Bearded Pygmy Chameleon				Tanzania
	diurnal	invert.	56	4.90	*** immobility expert
kerstenii	Pygmy Grass Chameleon				East Africa
	crepuscular	invert.	52	4.06	** immobility expert

Rieppeleon

Rieppeleon is small genus (three spp.) of small (4.1 g v. 8.1 g), short-tailed, dwarf chameleons from East Africa. They match the dwarf chameleon template (see Chamaeleonidae, table 5.2) in ecological business model (100% immobility expert v. 98% for dwarf group), shape index (0.41 v. 0.32 std. resid. for dwarf group, etc.), tail length (0.33 v. 0.53 × SVL), absence of crests (v. 0%), SSD on −2 to 2 scale (−0.67 v. −0.70), terrestrial or cryptozoic microhabitat (100% v. 83%), clutch count (1.02 v. 0.60 std. resid.), and incubation duration (0.01 v. 0.44 std. resid.). There are no data for ontogenetic size range or population density of *Rieppeleon*. In other words, they fit the dwarf chameleon template in all known respects.

Rieppeleon differs from other chamaeleonids in the low frequency of prehensile tails (33% v. 88% for family), no doubt because the tails are too short to grasp anything, a slightly higher frequency of sexual dichromatism (67% v. 53% family, etc.) and seasonal color (33% v. 18%), larger m:f relative tail length ratio (1.26 v. 1.13: note similarity to *Rhampholeon*), a slightly higher overall sex differences rollup (3.0 v. 2.2 on 0 to 6 scale), and uniform oviparity (v. 80%).

Rieppeleon species are widely distributed; none is endemic (v. 76%). None has been introduced (v. 5.4%). None is considered imperiled (v. 95%). In all other respects *Rieppeleon* are indistinguishable from other chamaeleonids. They are conspicuously similar to *Rhampholeon* in that they also produce a buzzing vibration when handled, and—like *Rhampholeon gorongosae—kerstenii* males will ride on their larger mating partners for several days, presumably as a form of mate guarding.

Riolama

Riolama is a four-species genus of poorly known (filled cells 16% v. 19%) terrestrial gymnophthalmids found on tepuis (high elevation mesas) in Venezuela and adjacent Guyana. None is found in the low elevation belt (v. 51%); 75% are found in medium elevations (1200–2400 m: v. 52%); all are found high (v. 33%). Too little information is available to assign most to ecological business models. None is elongate (v. 41%). They are of average size (2.24 g v. 2.35 g) and build (shape index −0.04 v. −0.32 std. resid.). They have long tails (2.10 v. 1.72 × SVL) that are uniformly breakable (v. 28%). All have secretory pores (v. 29%), but no other sex differences are reported.

Riolama spp.	English name				Range core
			Size		
	Activity	Diet	mm	g	Ecological business model
inopinata	Murispan-tepui Summit Lizard				Venezuela
	diurnal		40	1.30	*** refugia-anchored diurnal gr. liz.
leucosticta	White-spotted Summit Lizard				Venezuela
			47	2.25	
luridiventris	Marahuaca Tepui Lizard				Venezuela
			52	3.00	
uzzelli	Uzzell's Tepui Lizard				Venezuela
			51	2.86	

Ristella spp.	English name				Range core
			Size		
	Activity	Diet	mm	g	Ecological business model
beddomii	Beddome's Cat Skink				India
	diurnal	invert.	38	0.68	**** litter cryptozoid
guentheri	Guenther's Cat Skink				India
		invert.	38	0.68	* litter cryptozoid
rurkii	Rurk's Cat Skink				India
	diurnal	invert.	43	1.41	
travancorica	Travancore Cat Skink				India
	diurnal	invert.	38	0.68	*** litter cryptozoid

Rondonops spp.	English name				Range core
			Size		
	Activity	Diet	mm	g	Ecological business model
biscutatus	Good Pepper Lizard				Brazil
	cathemeral		54	3.51	*** litter cryptozoid
xanthomystax	Rio Abacaxis Yellow-lipped Lizard				Brazil
			57	4.24	*** litter cryptozoid

Reproduction, diet, and population biology are undocumented. Three-quarters of species are endemic to their countries (v. 68%). None is considered imperiled (v. 25%). In other respects *Riolama* are probably typical gymnophthalmids.

Ristella

Ristella is a minor radiation (four spp.) of very small (0.82 g v. 3.3 g), diurnal (100% v. 73%), riparian (67% v. 23%), mostly elongate (75% v. 35%) litter-cryptozoic (100% v. 28%) skinks endemic to India. Their claws are completely retractable within sheaths, giving rise to the listed common names.

They are slightly more slender than average skinks (shape index –0.76 v. –0.46 std. resid.) but otherwise morphologically average. No sex differences are known. There are no specifics on diet or foraging. They occur over the elevational range 0–2000 m and may be disproportionately common at medium elevations (75% v. 35%). They are not recorded from littoral (v. 21%), sandy (v. 25%), or human (v. 14%) habitats.

They are oviparous (100% v. 64%) and have moderately high clutch counts (0.51 v. 0.06 std. resid.). They oviposit during the wet season—southwestern monsoons. Multiple broods per year are not known.

No species occurs on islands (v. 57%). None has been introduced (v. 2.1%). None is considered imperiled (v. 47%). In other respects *Ristella* are much like other skinks.

Rondonops

Rondonops (two spp.) are litter cryptozoid gymnophthalmids endemic to Brazil. They are known better than typical gymnophthalmids (filled cells 30% v. 19%). The lizards are on the large side of the family for mass (3.86 g v. 2.35 g) and build (shape index 0.16 v. –0.32 std. resid.). They have long tails (2.18 v. 1.72 × SVL). One species (50%) is reported to have breakable tails (v. 28%). Both have secretory pores (v. 41%). One is reported to be m>f in SSD; the other is not clearly dimorphic (genus 0.50 v. –0.12 on –2 to 2 scale). Direct basking is reported (*xanthomystax*). Both are found in riparian zones (v. 29%). *Rondonops* are oviparous, but the clutch count is reported for both species to be only one (–0.57 v. –0.11 std. resid.). *R. biscutatus* is thought to breed only during the dry season, an unusual circumstance likely precipitated by flooding of oviposition sites. Neither species is considered imperiled (v. 25%). In other respects *Rondonops* are probably typical gymnophthalmids.

Saara

Saara is small genus (three spp.) of very large (282 g, the third highest in Agamidae, v. 15.6 g), beefy (shape index 1.74 v. 0.80 std. resid.), terrestrial (v. 56%), herbivorous (v. 8%) agamids from the deserts of southwest Asia, extending from Iraq and Afghanistan to India. They are all considered giant herbivores (v. 6%). Unlike many agamids, these burrow-refuging lizards (v. 40%) have relatively short and narrow heads, wide bodies, and short (0.78 v. 1.86 × SVL), armored, spiny tails, presumably to discourage eviction from their burrows. They resemble lizards of the genus *Uromastyx* (to which they were once assigned). All three species dig their own burrows; they do this primarily during the rainy season, as the ground is otherwise too hard to dig. Predators will find it very difficult to dig a *Saara* out of its burrow except during the rains. *S. loricata* is reported to remain within 25 m of its burrow and to be active at an astonishingly high mean body temperature (44.0°C, the highest in Agamidae, v. 34.8°C). Both *hardwickii* and *loricata* are reported to reside in groups of up to 100 individuals in adjacent burrows (group living 50% v. 12%). *S. hardwickii* is said to remain inside its burrow most of the time and to block off the entrance to the burrow when it is occupied. Among agamids, they are the fifth highest in wariness (67% v. 6%).

Females oviposit large clutches (0.82 v. 0.20 std. resid.), equating to 12.8 eggs, in side-chambers of their burrows, and allow the hatchlings to remain with them for 1–3 months (parental care 33% v. 3.5%). Burrows are clearly central to their existence, and understandably *Saara* defend them. Males are larger than females in the two documented species (1.00 v. 0.53 on –2 to 2 scale), and the sexes differ in color in the third (33% v. 53% dichromatic). *S. hardwickii* is reported to death feign; *loricata* can darken (basic color change 33% v. 22%). Although all three species are primarily diurnal, *hardwickii* becomes crepuscular (v. 0.5%) in the hottest months. They are found in dry and arid zones (56–400 mm/y). Longevity is short even for an agamid (–0.75 v. –0.46 std. resid.). None is endemic (v. 49%), is found on an island (v. 36%), or has been introduced (v. 1.7%). All are imperiled (v. 37%), but none is on one of the IUCN's lists of the most endangered (v. 5%).

However much humans may admire their domestic engineering and solicitous parental care, it has not kept humans from collecting *Saara* for their fat, which is reputed to be an aphrodisiac, their skin (leather), and their bodies (meat). To keep the meat fresh without the annoyance of the lizards running away, *Saara* have historically been stored alive with their backs broken. All three species are monitored for trade volume.

Saara spp.	English name				Range core
			Size		
	Activity	Diet	mm	g	Ecological business model
asmussi	Seistan Spinytailed Ground Lizard				Afghanistan
	diurnal	herbivore	201	294	**** giant herbivore
hardwickii	Hardwicke's Spinytailed Lizard				Pakistan
	diurnal	herbivore	180	207	**** giant herbivore
loricata	Mesopotamian Spinytailed Lizard				Iran
	diurnal	herbivore	216	367	**** giant herbivore

Saiphos sp.	English name				Range core
			Size		
	Activity	Diet	mm	g	Ecological business model
equalis	Three-toed Skink				New South Wales
		invert.	74	2.57	** subterranean forager

Saiphos

Saiphos equalis is the sole member of its genus, a smallish (2.57 g v. 3.3 g), elongate (v. 35%), functionally limbless subterranean forager (v. 12%) skink endemic to coastal eastern Australia in Queensland and New South Wales.

S. equalis is more lightly built than an average skink (shape index –1.24 v. –0.46 std. resid.). It has a relatively long tail (1.80 v. 1.33 × SVL) that is breakable (v. 73%). It may sand swim. It is found in urban areas (v. 14% human habitat). It may reproduce by either eggs or live birth. The number of young is average (0.14 v. 0.06 std. resid.). When eggs are laid, they are in an advanced stage of development and take an average of only five days to hatch (–1.11 v. –0.40 std. resid.). It is even possible for a single clutch/litter to include both hatchlings and neonates (Laird et al. 2019). *Saiphos* is not endemic (v. 64%). It is not found on an island (v. 57%). It has not been introduced (v. 2.1%), nor has it been evaluated for conservation status. In other respects *Saiphos* is likely to be similar to other skinks.

Salea spp.	English name				Range core
			Size		
	Activity	Diet	mm	g	Ecological business model
anamallayana	Anaimalai Spiny Lizard				India
	diurnal	invert.	89	23.0	** ground-foraging tree lizard
gularis	Mirzapore Garden Lizard				India
			96	29.3	
horsfieldii	Nilgiris Spiny Lizard				India
	diurnal	invert.	77	15.0	*** diurnal canopy walker

Saltuarius spp.	English name				Range core
			Size		
	Activity	Diet	mm	g	Ecological business model
cornutus	Northern Leaf-tailed Gecko				Queensland
	nocturnal		136	37.5	**** immobility expert
eximius	Cape Melville Leaf-tailed Gecko				Queensland
			95	16.2	**** immobility expert
kateae	Upper Clarence Leaf-tailed Gecko				New South Wales
	nocturnal		97	17.1	**** immobility expert
moritzi	Moritz's Leaf-tailed Gecko				New South Wales
	nocturnal		97	17.0	**** immobility expert
salebrosus	Rough-throated Leaf-tailed Gecko				Queensland
	nocturnal		129	33.0	**** immobility expert
swaini	Southern Leaf-tailed Gecko				Queensland
	nocturnal		116	25.9	**** immobility expert
wyberba	Granite Belt Leaf-tailed Gecko				Queensland
	nocturnal		97	17.0	**** immobility expert

Salea

Salea is small genus (three spp.) of medium-sized (21.6 g v. 15.6 g), robust (shape index 0.99 v. 0.80 std. resid.), arboreal (v. 43%) agamids endemic (100% v. 49%) to India. *S. gularis*, described in 1853 from a now untraceable locality is virtually unknown. The other two species are fairly typical arboreal agamids, with coarse scales, moderate crests (v. 38%), large heads, long tails (2.23 v. 1.86 × SVL), frequent advanced color change (33% v. 14.7%), m>f SSD (1.00 v. 0.53 on –2 to 2 scale), sexual dichromatism (33% v. 53%), crypsis (100% v. 51%), and sit-and-wait foraging (100% v. 88%). *S. anamallayana* is found in mesic areas (1101–300 mm/y) at medium elevation (v. 52%). *Salea* tolerate human habitats (100% v. 16%). Clutches are very small (–0.47 v. 0.20 std. resid.). None occurs on an island (v. 36%), has been introduced (v. 1.7%), or is imperiled (v. 37%).

Saltuarius

Saltuarius is a moderate (seven spp.) carphodactylid genus of extremely cryptic geckos, often with a very broad leaf-shaped tail (0.57 v. 0.54 × SVL). With such an elaborate tail, one might think that they would part with it only reluctantly if at all, but all species have breakable tails (100% v. 87%). *S. swaini* tails are reported to leap in a 0.5 m diameter circle for about 2 minutes after autotomy, suggesting that even in death the tails do yeoman service for their former owners. As with *Phyllurus*, *Saltuarius* are found in very restricted distributions on the east coast of Australia. *Saltuarius* differ from *Phyllurus* in their larger size (22.1 g v. 14.5 g family; 11.4 g in *Phyllurus*) and possession of secretory pores (29% v. 10% family; absent in *Phyllurus*). *Saltuarius* are all arboreal (v. 53%), nocturnal (v. 86%) immobility experts (v. 47%). They are notably underrepresented in riparian zones (0% v. 24%); one would think that the dense vegetation of riparian zones would be advantageous for a cryptic immobility expert—are there critical predators that thrive in the riparian areas? *Saltuarius* clutches are apparently fixed at two, which is few for their size (–1.20 v. –0.82 std. resid.); conversely, the young are large, as is typical of carphodactylids (ontogenetic size range –1.32 v. –1.21 std. resid.). Either they are very short-lived (–1.83 v. 0.42 std. resid.) or herpetoculturists have yet to work out long-term survival in captivity (record is only 2 years, for *cornutus*). In other respects they conform to carphodactylid patterns. To date, none has been listed as of conservation concern (v. 67%).

Salvator

Salvator is a small teiid genus (three spp.) of huge (1410 g v. 22.1 g), moderately well-known (filled cells 38% v. 33%, though *duseni* is virtually unknown), omnivorous (100% v. 33%) wide-ranging chemosensors (100% v. 97%) that lay an extraordinary number of eggs (2.37, the highest value in Teiidae, v. –0.51 std. resid.). They are found in southern South America, from Bolivia to Argentina, Uruguay, and Brazil. This genus, formerly in *Tupinambis*, constitutes part of the group of giant lizards popularly known as tegus. The best known, *merianae*, has some ability to produce heat metabolically.

Body shape is exceptionally robust (shape index 2.10 v. 0.88 std. resid.). Tails are short (1.86 v. 2.30 × SVL) and breakable (67% v. 21%) but not blue (v. 15%). *Salvator* are incapable of color change (v. 1.3%). Both sexes have pores that are large; the sexes differ only in size (m>f; 1.00 v. 0.89 on –2 to 2 scale).

They are open-minded omnivores, eating items such as leaves, fruit, nectar, arthropods, snails, orthopterans, other insects, ectothermic vertebrates, immobile endothermic vertebrates, mobile endothermic vertebrates, fish, and carrion. Flowers are an important part of their diet, as least seasonally. Despite their massive size, they may run bipedally, at least as juveniles. Home-range area and quantitative movement statistics are not known. *S. rufescens* is conspicuously wary (33% v. 13%). Their primary predator concealment tactic is retreat to vegetation (50% v. 7%) or retreat to a burrow (50% v. 12%). Two of three species are documented to bask directly (67% v. 58%). All documented species use burrows as daily refuges.

Males of all socially documented species are tolerant (v. 55%).

They are found exclusively at low elevation (100% v. 95%), and are notably absent from littoral areas (v. 38%), but occur at representative frequencies in sandy areas (50% v. 47%), riparian zones (50% v. 46%), and human habitat (50% v. 29%).

In keeping with their high clutch counts, young are notably small (ontogenetic mass range 0.59 v. 0.10 std. resid.). Mass-adjusted incubation is the shortest in Teiidae (–1.19 v. –0.38 std. resid.). Despite the small investment in an individual hatchling, at least two-thirds (v. 3%) of *Salvator* species show parental care of the clutch: *merianae* and *rufescens* females guard the clutch, at least initially, and may appropriate other species' burrows for the purpose of oviposition. *S. duseni* is undocumented for reproduction.

Longevity is slightly long by teiid standards (–0.42 v. –0.74 std. resid.), equating to 20 years in *merianae*. Population density is very high for mainland species: 3.25 kg/ha (v. 1.06 kg/ha), based on six credible samples of *merianae*. All species are widespread; none is endemic (0% v. 50%). One species (*merianae*) occurs on an island (33% v. 37%), and it has been introduced at least twice (33% v. 8%), to Fernando de Noronha and Florida. All evaluated species are imperiled (100% v. 41%), though none is on one of the IUCN's lists of the most endangered (v. 12%). In other respects *Salvator* are typical teiids.

Two species are collected in large numbers: *merianae* for medicinal, leather, and food purposes; *rufescens* for food and leather. Most of the international skin trade in tegus is *S. merianae* mislabeled as *Tupinambis teguixin*.

Salvator spp.	English name				Range core
			Size		
	Activity	Diet	mm	g	Ecological business model
duseni	Dusén's Tegu				Paraguay
	diurnal	omnivore	302	925	** wide-ranging chemosensor
merianae	Argentine Giant Tegu				Paraguay
	diurnal	omnivore	332	1245	**** wide-ranging chemosensor
rufescens	Red Tegu				Paraguay
	diurnal	omnivore	412	2430	**** wide-ranging chemosensor

Saproscincus

Saproscincus is a moderate radiation (12 spp.) of small-sized (1.55 g v. 3.25 g), mostly diurnal (82% v. 74%), terrestrial (91% v. 48%) or cryptozoic (9% v. 30%), oviparous (100% v. 64%), communally ovipositing (83% v. 2.6%) skinks found in coastal eastern Australia. The core group is litter cryptozoids (73% v. 28%), but they have diversified not only during the day to refugia-anchored diurnal ground lizards (18% v. 30%), but also into the night as nocturnal ground lizards (9% v. 2.3%). There may be some species, tentatively assigned to litter cryptozoids, which are actually refugia-anchored diurnal ground lizards. In keeping with these more open microhabitats, they have sexual dichromatism (33% v. 7.6%) and seasonal color (17% v. 6%). Some have other bright colors; *tetradactylus* has a conspicuous orange iris.

Saproscincus differ morphologically from average skinks only in their reduced frequency of elongation (17% v. 35%).

In addition to the color mentioned in the first paragraph, *Saproscincus* differ between the sexes in size: m<f (–1.0 v. –0.31 on –2 to 2 scale). Their sex rollup score is above average (0.58 v. 0.28 on 0 to 6 scale).

They eat invertebrates (100% v. 84%) and appear to be generalized dietary opportunists. Like most skinks, they are active foragers (100% v. 92%). They are conspicuously wary at an average frequency (17% v. 11%).

Saproscincus spp.	English name	Activity	Diet	Size mm	Size g	Range core	Ecological business model
basiliscus	Basilisk Shadeskink	diurnal	invert.	44	1.58	Queensland	* refugia-anchored diurnal gr. liz.
challengeri	Orange-tailed Shadeskink	diurnal	invert.	52	1.57	Queensland	** litter cryptozoid
czechurai	Wedge-snouted Shadeskink	diurnal	invert.	37	0.85	Queensland	**** litter cryptozoid
eungellensis	Eungella Shadeskink	diurnal		60	4.24	Queensland	** litter cryptozoid
hannahae	Hannah's Shadeskink	diurnal	invert.	41	1.17	Queensland	** litter cryptozoid
lewisi	Cooktown Shadeskink	diurnal	invert.	40	1.15	Queensland	** litter cryptozoid
mustelinus	Weasel Skink	crepuscular	invert.	52	1.58	New South Wales	** litter cryptozoid
oriarus	Heath Shadeskink	nocturnal	invert.	42	1.29	Queensland	*** nocturnal ground lizard
rosei	Rose's Shadeskink	diurnal	invert.	58	3.77	Queensland	** litter cryptozoid
saltus	Melville Range Shadeskink			40	1.13	Queensland	
spectabilis	Pale-lipped Shadeskink	diurnal	invert.	56	3.23	Queensland	** litter cryptozoid
tetradactylus	Four-fingered Shadeskink	diurnal	invert.	32	0.57	Queensland	* refugia-anchored diurnal gr. liz.

Direct basking is relatively uncommon (8% v. 21%), but they are also not found to thermoconform (0% v. 1.5%). They are overrepresented in riparian (36% v. 23%) habitat, absent from sandy areas (v. 25%), and about average in littoral (27% v. 21%) and human (9% v. 14%) habitats.

Their clutch count is commensurate with their body size (0.12 v. 0.06 std. resid.). Their remarkable frequency of communal oviposition (83%, first paragraph) is especially surprising in relation to their ecological business model (table 4.1, trait 68). For average litter cryptozoids, it is 4.7%; for refugia-anchored diurnal ground lizards, 3.5%. What unique factors impinge on *Saproscincus* to produce or sustain this association with communal oviposition? Few sun-warmed nest sites in the forest? Are suitable soils geographically rare or clumped? *Saproscincus* also show an elevated frequency of parental care (8.3% v. 3.2%); this takes the form of female clutch attendance (*mustelinus*).

Endemism is very close to the skink average (58% v. 64%). Island species are comparatively rare (17% v. 57%) despite their coastal distributions. One species (*mustelinus*) has been introduced (8.3% v. 2.1%), but only once (0% v. 0.7% for >2 introductions). Many *Saproscincus* are imperiled (67% v. 47%), but none is on one of the IUCN's lists of the most endangered (v. 14%). In other documented respects *Saproscincus* are typical skinks.

Sarada

Sarada has a place of honor in the history of science. Charles Darwin used the sexually dimorphic and dichromatic features of the genus to illustrate his theory of sexual selection. Unlike his theory of natural selection, which was met with forehead-slapping concurrence among scientists, sexual selection was not widely embraced by his scientific contemporaries and continued to be controversial into the late twentieth century. Virtually all aspects of his sexual selection theory have now become scientific consensus, but curiously this genus is little better known than it was in Darwin's time. For example, there is only one clutch size estimate (15) for one of the species, and there is no dietary or population information for any of them.

Sarada is a small genus (three spp.) of relatively small (6.1 g v. 15.6 g), terrestrial (v. 56%) agamids endemic (v. 49%) to India. All are categorized as refugia-anchored diurnal ground lizards (100% v. 43%), though with limited confidence in the assignment (2 on 0–4 scale). They are partial to refuging in soil cracks, at least during the dry season. Males of all three species have dewlaps and are larger than the females (1.00 v. 0.53 on –2 to 2 scale). As Darwin knew, *Sarada* have strong sex differences overall (2.33 v. 1.61 on 0 to 6 scale). Both sexes have breakable tails (v. 17%) and lack short-term color change (v. 22%). Their elevational distribu-

Sarada spp.	English name				Range core
			Size		
	Activity	Diet	mm	g	Ecological business model
darwini	Darwin's Large Fan-throated Lizard				India
	diurnal		55	5.1	** refugia-anchored diurnal gr. liz.
deccanensis	Deccan Fan-throated Lizard				India
	diurnal		56	5.6	** refugia-anchored diurnal gr. liz
superba	Superb Large Fan-throated Lizard				India
	diurnal		63	8.0	** refugia-anchored diurnal gr. liz.

Saurodactylus spp.	English name				Range core
			Size		
	Activity	Diet	mm	g	Ecological business model
brosseti	Brosset's Lizard-toed Gecko				Morocco
	cathemeral	invert.	32	0.79	** crack hermit
fasciatus	Banded Dappled Gecko				Morocco
	crepuscular	invert.	37	1.22	** crack hermit
mauritanicus	Morocco Lizard-fingered Gecko				Morocco
	cathemeral	invert.	34	0.96	** nocturnal ground lizard (night) or crack hermit (day)

tion is skewed low (l/m/h: 100/33/0 v. 83/52/14). They are not associated with littoral (v. 10%), sandy (v. 26%), riparian (v. 21%), or human (v. 16%) habitats. The one mass-adjusted clutch count is the highest genus average in Agamidae (2.80 v. 0.20 std. resid.). They do not occur on islands (v. 36%). None has been introduced (v. 1.7%) or evaluated for conservation status.

Saurodactylus

Saurodactylus is a three-species genus of normal-sized (0.97 g v. 1.07 g), padless, cathemeral (67% v. 15%) or crepuscular (33% v. 6%) sphaerodactylids found in Spain south to Western Sahara and including Algeria and Morocco. The modal *Saurodactylus* is a crack hermit (67% v. 6%), but one may be a nocturnal ground lizard (33% v. 6.5%). All three species may be exposed at night but forage within their crevices during the day. Food may be scarce for them; *mauritanicus* is reported to be so enthused by the sight of prey that it will bite nearby conspecifics. Perhaps this represents the sudden emergence of an impulse to defend foraging space when a choice food item approaches the defender.

Tails are slightly long (1.24 v. 1.07 × SVL). Male tails are proportionately longer than in females (m:f ratio 1.13 v. 1.02). The sexes differ in color (67% v. 35%). Males defend space (*fasciatus*), but the sequestered space may be no more than their home crack.

Elevational profile is normal (l/m/h: 100/33/0 v. 96/20/4). The only habitat association is with sandy areas (67% v. 15%).

Reproduction is normal, though incubation is on the short side (0.06 v. 0.44 std. resid.), perhaps reflecting the high temperatures in the Saharan soil.

Longevity is a little long (0.08 v. −0.15 std. resid.), equating to 3.7 years in *brosseti*. None has been introduced (v. 5%). A normal proportion is imperiled (33% v. 41%). None is on any of the IUCN's lists of the most endangered (v. 23%). In other respects *Saurodactylus* are typical sphaerodactylids.

Sauromalus

Sauromalus (five spp.) is a genus of diurnal terrestrial giant herbivores (Iguanidae) found in the southwestern United States and adjacent northwestern Mexico. The one widespread species (*ater*) and four island endemics hybridize and yet show distinct geographic color patterns, yielding great confusion in nomenclature. They all share the distinctive trait of inflating themselves in narrow crevices when pursued. As the lizards so wedged are not removable, commercial collectors sometimes resort to physical destruction of the rock crevices to extract their quarry.

Sauromalus are relatively small iguanids (333 g v. 609 g); their shape index is on the low side (1.91 v. 2.11 std. resid.). Their tails are short (1.06 v. 1.53 × SVL) and breakable, primarily in juveniles (100% v. 38%). Secretory pores are present (100% v. 38%), but crests are absent (v. 31%). Color change ability (21% v. 21%) declines ontogenetically, and advanced color change is absent (v. 2.4%).

Sex differences in *Sauromalus* follow the iguanid template, though size differences are subdued (0.75 v. 1.30 on −2 to 2 scale); sexual dichromatism is lower (20% v. 38%), seasonal color is absent (v. 11.9%), and the sex rollup score (on 0 to 6 scale) is 1.60 (v. 1.88).

All species are herbivores (v. 84%), with a preference for flowers. They

Sauromalus spp.	English name		Size		Range core
	Activity	Diet	mm	g	Ecological business model
ater	Common Chuckwalla				California
	diurnal	herbivore	162	138	**** giant herbivore
hispidus	Spiny Chuckwalla				Mexico
	diurnal	herbivore	243	620	**** giant herbivore
klauberi	Santa Catalina Chuckwalla				Mexico
	diurnal	herbivore	148	142	**** giant herbivore
slevini	Montserrat Chuckwalla				Mexico
	diurnal	herbivore	188	290	**** giant herbivore
varius	San Esteban Island Chuckwalla				Mexico
	diurnal	herbivore	300	1160	**** giant herbivore

Scelarcis sp.	English name		Size		Range core
	Activity	Diet	mm	g	Ecological business model
perspicillata	Morocco Rock Lizard				Algeria
	diurnal	omnivore	52	3.23	*** trunk/wall lizard

have been described as sit-and-wait (50% v. 29%), cruise (50% v. 29%), and active (50% v. 71%) foragers, though sitting and waiting for flowers to arrive would strain the patience of even a chuckwalla. Home-range size is large for an iguanid (0.21 v. –1.06 std. resid.), as large as 19,000 m^2 in male *ater*.

Sauromalus are not reported to give stress calls (v. 4.8%), but they are conspicuously wary (60% v. 29%), as befits a lizard needing only to sprint to an impregnable crevice. They bask directly (100% v. 55%).

Solitariness is not reported (v. 8%), and group living universal (100% v. 42%). Some males are said to defend space (33%v. 90%); hierarchies are absent (0% v. 25%).

Burrow use is common (75% v. 44%). *Sauromalus* exploit a drier environment than typical iguanids, with their mode in the dry (56–150 mm/y; 100% v. 21%) and presence in the arid (50% v. 14%) and intermediate (401–1101 mm/y: 50% v. 50%). Their elevational profile is normal (100%) at low and 67% (v. 19%) at medium (does not reach over 1316 m). They are overrepresented in littoral (60% v. 41%) and riparian (60% v. 22%) habitats and underrepresented in sandy areas (20% v. 34%) and human habitats (0% v. 24%).

Reproduction is similar to most other mainland iguanids, with an adjusted clutch count that is the third highest in the family (1.64 v. 0.45 std. resid.). Ontogenetic size range is comparatively low (–0.93 v. –0.55 std. resid.), indicating larger hatchlings. *Sauromalus* incubation period is on the short side (–0.91 v. –0.76 std. resid.). Communal oviposition is not reported (v. 14%), and parental care is nest defense (40% v. 31%). In the widespread *ater* parental care extends beyond the time of hatching: hatchlings may associate with their mother.

Longevity is long (0.55 v. 0.16 std. resid.), equating to 65 years in *ater*. Population density is low for an iguanid, but high for a temperate lizard (5.5 kg/ha v. 12.4 kg/ha); the *Sauromalus* data set is based on samples of one *varius*, three *hispidus*, and eight *ater* populations. No *Sauromalus* has been introduced (v. 33%), and all are considered imperiled (v. 97%). None is listed on the IUCN's lists of most endangered (v. 51%). At least three species are eaten by local residents. In other respects *Sauromalus* are typical iguanids.

Scelarcis

Scelarcis perspicillata is the sole species in a genus of smallish (3.2 g v. 5.1 g), diurnal trunk/wall lacertids found in Algeria and Morocco. It is at home on arboreal vegetation and on cliffs and constructed walls. It is agile enough to capture insects in flight, though it also eats fruit. In build it is slightly lighter than an average lacertid (shape index 0.13 v. 0.27 std. resid.). Tail length is comparatively short (1.50 v. 2.06 × SVL), and tails are breakable. Juveniles may have blue tails. Males are larger than females (1.0 v. 0.49 on –2 to 2 scale). For predator concealment it relies primarily on flight (v. 22%). It basks directly (v. 25%). Its elevational range is notably broad (30–2650 m), and it is associated with both riparian (v. 28%) and human (v. 16%) habitats. The mean clutch count is on the small side of lacertids (–0.16 v. 0.13 std. resid.), and the young are slightly large (ontogenetic mass range –0.52 v. –0.05 std. resid.). The eggs are notably elongate: minimum and maximum dimensions are 6 and 24.4 mm. Longevity is about average (0.05 v. 0.19 std. resid.), equating to 5 years. *Scelarcis* does not naturally occur on islands (v. 26%) but has been introduced (v. 4.6%) to Menorca. It is not considered imperiled (v. 56%). In other respects

Scelarcis is probably similar to other lacertids.

Sceloporus (fig. 4.11B)

Sceloporus is the largest phrynosomatid genus (100 spp.; table 2.2) and the archetypical refugia-anchored diurnal lizard of the Nearctic. The vast majority of species are found in the Nearctic biogeographic realm (83% v. 88%); the family extends into Central America as far as Costa Rica. Because *Sceloporus* makes up about two-thirds of the Phrynosomatidae (100 out of 154 spp.), mean values for the genus are often similar to mean values for the family. *Sceloporus* are medium sized (11.1 g v. 11.2 g) and have radiated into a number of ecological business models: refugia-anchored diurnal ground lizard (54% v. 52%), ground-foraging tree lizard (29% v. 20%), trunk/wall lizard (16% v. 10%), and litter cryptozoid (1.2% v. 0.7%). The essence of *Sceloporus* is its ecological generality. Like a typical lizard, most *Sceloporus* eat generalized invertebrates (82% v. 62% family), have small home ranges (0.28 v. 0.57 std. resid.), and protect themselves by running to crevice-burrows (39% v. 32%) or vegetation (36% v. 24%). For overnight refuge they do not often rely on burrows (16% v. 29%), and they are infrequently found in sandy habitats (18% v. 32%). *Sceloporus* houses the quintessential generalist lizards; Lemos-Espinal and Dixon (2013) note that *Sceloporus grammicus* is "the most adaptable lizard in Mexico."

Their build fits the phrynosomatid template (shape index 0.92 v. 0.91 std. resid.). Tails are of moderate length (1.31 v. 1.22 × SVL) and sometimes break (25% v. 21%). They sometimes have secretory pores (18% v. 32%) or color change (9% v. 20%). Advanced color change is present but rare (2% v. 4.5%). Blue tails are present at typical frequency (2.0% v. 1.3%).

Sceloporus spp.	English name		Size		Range core
	Activity	Diet	mm	g	Ecological business model
acanthinus	Bocourt's Spiny Lizard				Guatemala
	diurnal		75	14.6	** trunk/wall lizard
adleri	Adler's Spiny Lizard				Mexico
			62	8.7	** ground-foraging tree lizard
aeneus	Black-bellied Bunchgrass Lizard				Mexico
	diurnal	omnivore	49	4.8	*** refugia-anchored diurnal gr. liz.
albiventris	Western White-bellied Spiny Lizard				Mexico
	diurnal	invert.	82	18.3	** trunk/wall lizard
anahuacus	Anahuacan Bunchgrass Lizard				Mexico
	diurnal	invert.	51	5.1	** refugia-anchored diurnal gr. liz.
angustus	Santa Cruz Spiny Lizard				Mexico
	diurnal	herbivore	75	14.7	** ground-foraging tree lizard
arenicolus	Dunes Sagebrush Lizard				New Mexico
	diurnal	invert.	57	7.0	*** refugia-anchored diurnal gr. liz.
asper	Sardonic Spiny Lizard				Mexico
	diurnal		67	10.9	
aurantius	Southern Occidental Bunchgrass Lizard				Mexico
	diurnal		49	4.68	** refugia-anchored diurnal gr. liz.
aureolus	Southern Crevice Spiny Lizard				Mexico
			84	19.5	
becki	Island Fence Lizard				California
	diurnal		62	8.8	* refugia-anchored diurnal gr. liz.
bicanthalis	Trans-volcanic Bunchgrass Lizard				Mexico
	diurnal	invert.	50	5.0	* refugia-anchored diurnal gr. liz.
bimaculosus	Twin-spotted Spiny Lizard				Mexico
	diurnal	omnivore	93	26.0	** refugia-anchored diurnal gr. liz.
bulleri	Buller's Spiny Lizard				Mexico
	diurnal		94	27.1	** ground-foraging tree lizard
caeruleus	Blue Ornate Spiny Lizard				Mexico
			70	12.2	
carinatus	Keeled Spiny Lizard				Mexico
	diurnal		67	10.6	** ground-foraging tree lizard
cautus	Shy Spiny Lizard				Mexico
	diurnal	invert.	73	13.5	** refugia-anchored diurnal gr. liz.

Sex differences are those typical of phrynosomatids: limited m>f SSD (0.41 v. 0.27 on –2 to 2 scale), moderate sexual dichromatism (67% v. 58%), limited seasonal color (16% v. 25%), and bias in m:f relative tail length ratio (1.06 v. 1.13). The sex differences rollup is subdued compared to the family (1.39 v. 1.68 on 0 to 6 scale).

All diets except carnivory are present, but mostly at low levels: herbivory (2% v. 1%), ant specialists (2% v. 17%), and omnivores (14% v. 20%). Many observers comment on the opportunism in the diet of *Sceloporus*. It is notable that though ant specialization is very rare (2%) in *Sceloporus*, ant consumption is widespread: 49% of species. Sit-and-wait foraging is the rule for *Sceloporus* (100% v. 98%). Active foraging is reported for 5.1% of species (v. 8.3%); cruise foraging is not noted (v. 3.3%).

Although all species are nominally diurnal, at least seven species have been noted to occasionally remain active after dark: *angustus*, *clarkii*, *hunsakeri*, *magister*, *nelsoni*, *orcutti*, and *zosteromus*. This opportunistic nocturnality has been noted throughout the Phrynosomatidae. Occasional bipedalism is also widely reported in the genus: *malachiticus*, *siniferus*, *undulatus*, and *woodi*.

A normal frequency of wariness is characteristic of *Sceloporus* (23% v. 22%). Underrepresented primary predator concealment tactics are crypsis (12% v. 32%), flight (9% v. 14%), and sand diving (3% v. 12%). Direct basking is reported at typical levels (33% v. 40%). Mean active body temperature is an ordinary 32.4°C (v. 34.2°C). No species is said to be a thermoconformer but many tolerate a wide range of temperatures; *consobrinus* and *ornatus* have been noted to be active adjacent to snowbanks.

Sceloporus spp.	English name				Range core
			Size		
	Activity	Diet	mm	g	Ecological business model
chaneyi	Chaney's Bunchgrass Lizard				Mexico
	diurnal		48	4.43	** refugia-anchored diurnal gr. liz.
chrysostictus	Yucatán Spiny Lizard				Guatemala
	diurnal	invert.	53	5.8	** ground-foraging tree lizard
clarkii	Clark's Spiny Lizard				Arizona
	diurnal	invert.	103	49.2	*** trunk/wall lizard
consobrinus	Prairie Lizard				New Mexico
	diurnal	invert.	64	9.6	**** refugia-anchored diurnal gr. liz.
couchii	Couch's Spiny Lizard				Mexico
	diurnal		55	6.4	* refugia-anchored diurnal gr. liz.
cowlesi	Southwestern Fence Lizard				New Mexico
	diurnal	invert.	65	9.8	** trunk/wall lizard
cozumelae	Cozumel Spiny Lizard				Mexico
	diurnal	invert.	50	4.86	** refugia-anchored diurnal gr. liz.
cryptus	Juarez Mountains Spiny Lizard				Mexico
			57	6.9	
cupreus	Upland Long-tailed Spiny Lizard				Mexico
			47	4.09	
cyanogenys	Blue Spiny Lizard				Mexico
	diurnal	invert.	108	39.1	*** refugia-anchored diurnal gr. liz.
cyanostictus	Blue-spotted Spiny Lizard				Mexico
	diurnal	invert.	69	11.8	**** refugia-anchored diurnal gr. liz.
druckercolini	Graceful Mountain Tree Lizard				Mexico
			87	22.0	
dugesii	Duges' Spiny Lizard				Mexico
	diurnal	invert.	72	13.2	** refugia-anchored diurnal gr. liz.
edbelli	Bell's Spiny Lizard				Mexico
	diurnal		71	12.6	*** ground-foraging tree lizard
edwardtaylori	Taylor's Spiny Lizard				Mexico
	diurnal		86	21.0	** refugia-anchored diurnal gr. liz.
exsul	Queretaro Desert Spiny Lizard				Mexico
			56	6.6	
formosus	Emerald Spiny Lizard				Mexico
	diurnal		71	12.5	** refugia-anchored diurnal gr. liz.

Sceloporus spp.	English name	Range core	Activity	Diet	Size mm	Size g	Ecological business model
gadoviae	Gadow's Spiny Lizard	Mexico			59	7.6	** refugia-anchored diurnal gr. liz.
gadsdeni	Lagoon Spiny Lizard	Mexico	diurnal		67	10.9	*** refugia-anchored diurnal gr. liz.
goldmani	Goldman's Bunchgrass Lizard	Mexico			52	5.6	
graciosus	Common Sagebrush Lizard	Western United States	diurnal	invert.	64	9.5	*** refugia-anchored diurnal gr. liz.
grammicus	Graphic Spiny Lizard	Mexico	diurnal	invert.	59	7.8	*** trunk/wall lizard
grandaevus	Cerralvo Spiny Lizard	Mexico	diurnal	omnivore	67	10.9	** ground-foraging tree lizard
halli	Hall's Spiny Lizard	Mexico	diurnal		43	3.30	
heterolepis	Odd-scaled Spiny Lizard	Mexico	diurnal		65	10.0	* trunk/wall lizard
horridus	Horrible Spiny Lizard	Mexico	diurnal	invert.	88	22.4	* trunk/wall lizard
hunsakeri	Hunsaker's Spiny Lizard	Mexico	diurnal	invert.	72	12.9	* refugia-anchored diurnal gr. liz.
insignis	Coalcoman Black-collared Lizard	Mexico			82	18.7	*** ground-foraging tree lizard
internasalis	Tehuantepec Spiny Lizard	Mexico	diurnal		75	14.8	
jalapae	Xalapa Spiny Lizard	Mexico	diurnal		52	5.5	* refugia-anchored diurnal gr. liz.
jarrovii	Yarrow's Spiny Lizard	Mexico	diurnal	invert.	79	16.9	**** refugia-anchored diurnal gr. liz.
lemosespinali	Lemos-Espinal's Spiny Lizard	Mexico	diurnal	invert.	48	4.43	*** trunk/wall lizard
licki	Cape Spiny Lizard	Mexico	diurnal	invert.	84	19.9	* ground-foraging tree lizard or * refugia-anchored diurnal gr. liz. (geog. var.)
lineatulus	Santa Catalina Spiny Lizard	Mexico	diurnal		89	23.2	*** refugia-anchored diurnal gr. liz.

Group living is slightly elevated (27% v. 19%), presumably because so many species are drawn to rock piles. Only a few species have tolerant males, however (2.3% v. 10.6%). Male spatial defense is nearly universal (98% v. 89%). Male hierarchies develop occasionally (2.3% v. 3.0%).

Arboreal microhabitat use is slightly overrepresented (31% v. 22%). *Sceloporus* are found in dry through mesic rainfall bands at ordinary levels. A slight excess is tabulated in the mesic (1101–3000 mm/y) category (38% v. 28%). The elevational profile is ordinary, with a slight excess at medium elevation (83% v. 76%). Frequencies in the tabulated habitat types match the family, except possibly in human habitats (26% v. 18%).

Oviparity is present in a slight majority of species (56% v. 65%), and clutch/litter count is average (0.58 v. 0.68 std resid.). The size of young is also average (ontogenetic mass range −0.36 v. −0.26 std. resid.). Incubation duration is typical of the family (−0.57 v. −0.58 std. resid.), equating to 68 days in *undulatus* for example. RCM is slightly elevated (0.326 v. 0.300). Though parental care is present at low levels (1.0% v. 1.9%), it is notable that a captive *orcutti* was observed to backfill her nest excavation for up to 25 days after oviposition. Hatchling *undulatus* are sometimes observed to associate with adults.

Longevity is slightly short (−0.49 v. −0.25 std. resid.), equating to 4 years in *undulatus* for example. Many observers comment on high juvenile mortality in *Sceloporus*. Mean population density is about average (0.35 kg/ha v. 0.26 kg/ha). Endemism is a trifle elevated (66% v. 58%). Somewhat less than the family proportion occurs on islands (13% v. 21%). Introductions are slightly low (3.0% v. 3.9%), in keeping with their lack of insularity and their paucity in the pet trade, presumably due to their generally subdued colors. Perhaps because they so often embrace rocky habitats, which are not developers' choice for building or agriculture, imperilment is comparatively low in *Sceloporus*: 23% v. 37%. An ordinary frequency of species is on one of the IUCN's lists of most endangered (4.9% v. 5.4%).

Sceloporus clarkii is believed by some residents of Mexico to be venomous. *Sceloporus uniformis* (fig. 4.11B) is notable for the fear it engenders in some indigenous people who are loath to urinate in the vicinity of this species for fear that it will run up their legs and wedge in their vaginas.

Sceloporus spp.	English name				Range core
			Size		
	Activity	Diet	mm	g	Ecological business model
lunae	Luna's Spiny Lizard				Guatemala
	diurnal		77	15.8	
lundelli	Lundell's Spiny Lizard				Guatemala
	diurnal		85	20.4	**** trunk/wall lizard
macdougalli	MacDougall's Spiny Lizard				Mexico
			71	12.6	
maculosus	Nazas River Spiny Lizard				Mexico
	diurnal		45	3.76	** refugia-anchored diurnal gr. liz.
magister	Desert Spiny Lizard				Arizona
	diurnal	omnivore	102	33.1	*** ground-foraging tree lizard
malachiticus	Green Spiny Lizard				Central America
	diurnal	invert.	77	15.5	**** ground-foraging tree lizard
megalepidurus	Cursorial Spiny Lizard				Mexico
	diurnal		47	4.28	* refugia-anchored diurnal gr. liz.
melanorhinus	Black-nosed Lizard				Mexico
	diurnal		85	20.1	** trunk/wall lizard
merriami	Canyon Lizard				Mexico
	diurnal	ants	51	3.40	*** trunk/wall lizard
minor	Minor Spiny Lizard				Mexico
	diurnal	invert.	72	13.0	**** refugia-anchored diurnal gr. liz.
mucronatus	Central Cleft Lizard				Mexico
	diurnal	invert.	89	23.0	** refugia-anchored diurnal gr. liz.
nelsoni	Nelson's Spiny Lizard				Mexico
	diurnal	invert.	56	6.8	** ground-foraging tree lizard
oberon	Oberon Spiny Lizard				Mexico
	diurnal		79	17.0	**** refugia-anchored diurnal gr. liz.
occidentalis	Western Fence Lizard				California
	diurnal	invert.	68	11.5	*** ground-foraging tree lizard
ochoterenae	Ochoterena's Spiny Lizard				Mexico
	diurnal		54	6.0	
olivaceus	Texas Spiny Lizard				Texas
	diurnal	invert.	93	25.8	*** ground-foraging tree lizard
omiltemanus	Southern Cleft Lizard				Mexico
			84	20.0	
orcutti	Granite Spiny Lizard				Mexico
	diurnal	invert.	95	75	** ground-foraging tree lizard

Sceloporus spp.	English name				Range core
			Size		
	Activity	Diet	mm	g	Ecological business model
ornatus	Ornate Spiny Lizard				Mexico
	diurnal	invert.	79	17.0	** refugia-anchored diurnal gr. liz.
palaciosi	Palacios' Bunchgrass Lizard				Mexico
	diurnal	invert.	58	7.4	* refugia-anchored diurnal gr. liz.
parvus	Blue-bellied Lizard				Mexico
	diurnal	invert.	46	3.92	** refugia-anchored diurnal gr. liz.
poinsettii	Crevice Spiny Lizard				Mexico
	diurnal	omnivore	107	37.7	**** refugia-anchored diurnal gr. liz.
pyrocephalus	Red-headed Spiny Lizard				Mexico
	diurnal		60	8.1	** refugia-anchored diurnal gr. liz.
salvini	Salvin's Spiny Lizard				Mexico
	diurnal		75	14.8	** ground-foraging tree lizard
samcolemani	Coleman's Bunchgrass Lizard				Mexico
	diurnal		53	5.6	* refugia-anchored diurnal gr. liz.
scalaris	Pale-bellied Bunchgrass Lizard				Mexico
	diurnal	invert.	57	7.1	** refugia-anchored diurnal gr. liz.
serrifer	Rough-scaled Lizard				Guatemala
	diurnal	invert.	90	24.0	*** trunk/wall lizard
shannonorum	Shannons' Spiny Lizard				Mexico
			65	10.0	
siniferus	Long-tailed Spiny Lizard				Mexico
	diurnal		58	7.3	
slevini	Slevin's Bunchgrass Lizard				Arizona
	diurnal	invert.	48	4.40	** refugia-anchored diurnal gr. liz.
smaragdinus	Bocourt's Emerald Lizard				Guatemala
	diurnal		67	10.8	*** ground-foraging tree lizard
smithi	Smith's Rose-bellied Lizard				Mexico
	diurnal		58	7.2	* litter cryptozoid
spinosus	Spiny Lizard				Mexico
	diurnal	invert.	102	33.0	* ground-foraging tree lizard
squamosus	Dwarf Spiny Lizard				Central America
	diurnal		52	5.4	** refugia-anchored diurnal gr. liz.
stejnegeri	Stejneger's Spiny Lizard				Mexico
			77	15.4	

Sceloporus spp.	English name				Range core
			Size		
	Activity	Diet	mm	g	Ecological business model
subpictus	Southern Cursorial Lizard				Mexico
			57	6.9	** ground-foraging tree lizard
sugillatus	Zempoala Bar-bellied Lizard				Mexico
	diurnal	invert.	72	13.2	*** refugia-anchored diurnal gr. liz.
taeniocnemis	Guatemalan Emerald Spiny Lizard				Mexico
	diurnal		69	11.5	
tanneri	Tanner's Spiny Lizard				Mexico
	diurnal		71	12.6	*** trunk/wall lizard
teapensis	Rose-bellied Spiny Lizard				Guatemala
	diurnal	invert.	57	6.9	** ground-foraging tree lizard
torquatus	Crevice Swift				Mexico
	diurnal	omnivore	100	31.5	* refugia-anchored diurnal gr. liz. or crack hermit
tristichus	Plateau Fence Lizard				Colorado Plateau
	diurnal		67	11.7	* refugia-anchored diurnal gr. liz.
undulatus	Eastern Fence Lizard				Arkansas
	diurnal	invert.	62	8.6	*** ground-foraging tree lizard
uniformis	Yellow-backed Spiny Lizard				Utah
	diurnal	omnivore	98	30.2	** ground-foraging tree lizard
utiformis	Cope's Largescale Spiny Lizard				Mexico
	diurnal	invert.	66	10.2	*** refugia-anchored diurnal gr. liz.
variabilis	Rose-bellied Lizard				Central America
	diurnal	invert.	61	8.5	**** ground-foraging tree lizard
virgatus	Striped Plateau Lizard				New Mexico
	diurnal	invert.	57	7.1	*** ground-foraging tree lizard
woodi	Florida Scrub Lizard				Florida
	diurnal	invert.	53	5.7	*** refugia-anchored diurnal gr. liz.
zosteromus	Baja California Spiny Lizard				Mexico
	diurnal	invert.	94	26.5	*** refugia-anchored diurnal gr. liz.

Scelotes

Scelotes is a major radiation (22 spp.) of moderately small-sized (2.48 g v. 3.25 g), diurnal (100% v. 74%), elongate (73%) or functionally limbless, cryptozoic (32% v. 30%) or subterranean (68% v. 11%) skinks found mostly in South Africa. The genus is distributed south of a west-east line connecting Namibia to Mozambique, with two outliers: Tanzania and the island of Bioko in Equatorial Guinea. *Scelotes* are reminiscent of Australia's *Lerista* skinks, but they show a wider range of limb development: from five digits on all limbs to

Scelotes spp.	English name	Activity	Diet	Size mm	Size g	Range core	Ecological business model
anguineus	Algoa Dwarf Burrowing Skink		invert.	64	0.88	South Africa	*** subterranean forager
arenicola	Zululand Dwarf Burrowing Skink			70	1.10	South Africa	*** subterranean forager
bidigittatus	Lowveld Dwarf Burrowing Skink			69	3.43	South Africa	*** litter cryptozoid
bipes	Silvery Dwarf Burrowing Skink		invert.	67	3.17	South Africa	** subterranean forager
bourquini	Bourquin's Dwarf Burrowing Skink			84	3.52	South Africa	** subterranean forager
caffer	Cape Dwarf Burrowing Skink			47	1.19	South Africa	** litter cryptozoid
capensis	Western Dwarf Burrowing Skink			50	1.45	South Africa	** litter cryptozoid
duttoni	Dutton's Dwarf Burrowing Skink			55	1.83	Mozambique	* litter cryptozoid
fitzsimonsi	FitzSimons' Dwarf Burrowing Skink			55	0.62	Mozambique	*** subterranean forager
gronovii	Gronovi's Dwarf Burrowing Skink			59	2.23	South Africa	*** subterranean forager
guentheri	Guenther's Dwarf Burrowing Skink			81	5.4	South Africa	
inornatus	Smith's Dwarf Burrowing Skink			73	1.22	South Africa	* subterranean forager
insularis	Bazarutu Dwarf Burrowing Skink			52	0.55	Mozambique	** litter cryptozoid
kasneri	Kasner's Dwarf Burrowing Skink			96	8.7	South Africa	** subterranean forager or litter cryptozoid
limpopoensis	Limpopo Dwarf Burrowing Skink	diurnal		69	3.53	Botswana	*** subterranean forager
mirus	Montane Dwarf Burrowing Skink			69	3.53	South Africa	** refugia-anchored diurnal gr. liz.
montispectus	Bloubergstrand Dwarf Burrowing Skink			105	11.1	South Africa	
mossambicus	Mozambique Dwarf Burrowing Skink			61	2.48	South Africa	** subterranean forager
poensis	Bioko Dwarf Burrowing Skink			65	2.98	Equatorial Guinea	

completely limbless. Even when fully limbed however, the limbs are noticeably diminutive and widely separated (i.e., species is elongate). The modal *Scelotes* is a subterranean forager (63% v. 112%), and most others are litter cryptozoids (26% v. 28%), but there is also a refugia-anchored diurnal ground lizard (5% v. 30%) and a sand shark (5% v. 7%).

As with most subterranean foragers, *Scelotes* are attenuate (shape index –0.98 v. –0.46 std. resid.) and tails are shorter (1.00 v. 1.33 × SVL). Blue tails are average in frequency (4.5% v. 4%). No sex differences are known. They are skewed toward low elevation (l/m/h: 100/21/0 v. 89/35/6). They are overrepresented in littoral (50% v. 21%) and sandy (80% v. 25%) areas, absent from riparian zones (v. 23%), and infrequent in human habitat (5% v. 14%).

All are viviparous (v. 36%); litters are average in count (0.00 v. 0.06 std. resid.), with no reports of multiple broods per year.

Longevity is short (–0.96 v. 0.17 std. resid.), though based solely on *limpopoensis* (2.3 years). There are two credible population density estimates for *bidigittatus* (mean 0.0007 kg/ha v. 0.35 kg/ha). Naturally, the spots chosen to dig up these subterranean lizards were previously assessed as being favorable, so this low density probably exceeds a landscape average. Endemism is average (64% v. 64%). Island species are underrepresented (18% v. 57%). No species has been introduced (v. 2.1%). A usual proportion are imperiled (44% v. 47%), and 12.5% (v. 14%) are on one of the IUCN's lists of the most endangered. In other respects *Scelotes* are typical skinks.

Scelotes spp.	English name				Range core
			Size		
	Activity	Diet	mm	g	Ecological business model
sexlineatus	Striped Dwarf Burrowing Skink				South Africa
			80	5.3	*** subterranean forager
uluguruensis	Uluguru Dwarf Burrowing Skink				Tanzania
		invert.	72	3.97	** subterranean forager
vestigifer	Coastal Dwarf Burrowing Skink				South Africa
			61	4.43	** sand shark

Scincella spp.	English name				Range core
			Size		
	Activity	Diet	mm	g	Ecological business model
apraefrontalis	Huulien Ground Skink				Vietnam
			35	0.74	
assatus	Red Forest Skink				Honduras
	diurnal	invert.	49	1.34	**** litter cryptozoid
barbouri	Barbour's Ground Skink				China
			44	1.49	
boettgeri	Sakishima Ground Skink				Japan
	cathemeral		49	2.22	**** litter cryptozoid
capitanea	Nepal Large Ground Skink				Nepal
			66	5.7	* litter cryptozoid
caudaequinae	Horsehead Ground Skink				Mexico
			44	1.57	
cherriei	Brown Forest Skink				Central America
	diurnal	invert.	53	1.73	**** litter cryptozoid
darevskii	Darevsky's Ground Skink				Vietnam
			73	7.8	* litter cryptozoid
devorator	Grasshopper-eating Forest Skink				Vietnam
	diurnal	invert.	53	1.69	
doriae	Doria's Litterskink				Southeast Asia
	diurnal	invert.	52	1.57	**** litter cryptozoid
forbesora	Hidalgo Ground Skink				Mexico
			53	2.78	
formosensis	Van Denburgh's Ground Skink				Taiwan
			43	1.44	
gemmingeri	Forest Ground Skink				Mexico
			56	1.95	
huanrenensis	Huanren Ground Skink				China
			52	1.59	
incerta	Stuart's Forest Skink				Guatemala
	diurnal		57	2.09	**** litter cryptozoid

Scincella

Scincella constitute a major radiation (34 spp.) of consistently small-sized (1.63 g v. 3.25 g), mostly diurnal (88% v. 74%), mostly cryptozoic (50% v. 30%) or terrestrial (44% v. 48%) skinks that are found in two clusters: Mexico-centered (9 spp.) and East Asia (25 spp.). The Mexican-centered cluster extends from Kansas east to New Jersey and south to Panama. The East Asian cluster ranges from Nepal and India east to China and Japan—including islands such as Taiwan and Hainan—and south to Sulawesi, Indonesia. The most-speciose countries are Mexico (9 spp.), Vietnam (11 spp.), and China (11 spp.). Most species are litter cryptozoids (88% v. 28%); the remainder are refugia-anchored diurnal ground lizards (12% v. 30%).

Scincella are very close to average skinks in morphology. Their tails are slightly longer (1.50 v. 1.33 × SVL), and they are slightly more inclined to be elongate (44% v. 35%). Blue tail frequency is slightly lower (2.9% v. 4.0%), but *assata* has a bright orange tail in juveniles. The large tail of *lateralis* is a fat storage organ: 47%–75% of total fat reserves. This species is known to return later to the site of a tail loss event and consume the tail if it is still present.

Overall sex differences are very similar to those of average skinks (0.29 v. 0.28 on 0 to 6 scale), and they achieve this in most of the same ways. Size dimorphism is more m<f (–0.75 v. –0.31 on –2 to 2 scale). Dichromatism is about the same (6% v. 7.6%). Seasonal color is indistinguishable (6% v. 6%). The m:f ratio of relative tail lengths is slightly higher (1.09 v. 1.03); that is, males have slightly longer tails, which fits with this trait being inversely related to SSD in most lizards (though not skinks: see table 3.3, trait 8, and Scincidae entry in this chapter).

Scincella spp.	English name	Range core	Activity	Diet	Size mm	Size g	Ecological business model
inconspicua	Hidden Ground Skink	Indonesia			49	2.22	* litter cryptozoid
kikaapoa	Kikapue Ground Skink	Mexico			44	1.57	
lateralis	Little Brown Skink	Eastern United States	diurnal	invert.	47	1.19	**** litter cryptozoid
macrotis	Nicobar Ground Skink	India	diurnal	invert.	27	0.30	
melanosticta	Black-spotted Ground Skink	Southeast Asia	diurnal	invert.	52	1.60	**** litter cryptozoid
modesta	Dark Forest Skink	China	diurnal	invert.	48	2.02	*** litter cryptozoid
monticola	Snow Mountain Ground Skink	Vietnam			52	2.54	
ochracea	Tawny Ground Skink	Vietnam			46	1.74	
potanini	Sichuan Ground Skink	China			41	1.23	
przewalskii	Gansu Ground Skink	China			40	1.13	
punctatolineata	Spot-lined Litterskink	Myanmar	diurnal	invert.	38	0.51	**** litter cryptozoid
reevesii	Reeves' Ground Skink	Southeast Asia	diurnal	invert.	49	1.38	**** litter cryptozoid
rufocaudatus	Red-tailed Ground Skink	Southeast Asia	diurnal	invert.	46	1.14	*** litter cryptozoid
rupicola	Rock-dwelling Ground Skink	Vietnam	diurnal		41	1.20	* refugia-anchored diurnal gr. liz.
schmidti	Mt. Washan Skink	China			43	1.40	
silvicola	Taylor's Ground Skink	Mexico	cathemeral	invert.	51	2.43	*** refugia-anchored diurnal gr. liz.
tsinlingensis	Central China Ground Skink	China			59	2.32	
vandenburghi	Tsushima Smooth Skink	Japan	diurnal	invert.	46	1.16	**** litter cryptozoid
victoriana	Mount Victoria Ground Skink	Myanmar	diurnal		64	2.89	

All *Scincella* are generalized invertebrate eaters, though detailed dietary data are available for only four species (*cherriei, lateralis, modesta, reevesii*). Only two species (*cherriei* and *lateralis*) have foraging mode descriptors: cruise (50% v. 2%) and active (100% v. 92%). Based on those same two species, home-range area is large (1.06 v. –0.07 std. resid.), equating to about 88 m^2 (*cherriei*) and 76 m^2 (*lateralis*).

The *Scincella* that are not diurnal are cathemeral (12% v. 9%). Twelve percent (v. 11%) are conspicuously wary. For primary predator defense, they rely on either crypsis (33% v. 9%) or retreat to a crevice or burrow (67% v. 43%). Some species will locomote without use of limbs; at least *lateralis* will seek refuge in water if pressed. Twelve percent bask directly (v. 21%). Based on one tropical species (*cherriei*), they achieve a mean active body temperature that is notably low (26.0°C v. 31.1°C).

Social relations are almost unknown. *S. cherriei* males may or may not defend space (contradictory reports). *S. lateralis* males may defend a site (e.g., burrow) or a larger area; it is unclear if home-range defense in this species is seasonal or year-round. There are no reports of explicit male spatial tolerance or hierarchies in *Scincella*.

S. devorator is arboreal (6% v. 10%), but the other species reside on the ground. Half of the rainfall-documented species are found in every band from arid (151–400 mm/y) to wet (>3000 mm/y). *S. cherriei* is found in areas up to 4600 mm/y. The genus's elevational profile is shifted to higher elevations (l/m/h: 76/57/29 v. 89/35/6). They are underrepresented in littoral (5% v. 21%) and sandy (5% v. 25%) areas, but normal for riparian zones (25% v. 23%) and human habitat (10% v. 14%).

Both reproductive modes are

present, though oviparity dominates (79% v. 64%). The three known viviparous species are *gemmingeri* (low to medium elevation, tropical Mexico), *huanrenensis* (low elevation, temperate China), and *reevesii* (low to high elevation, Southeast Asia). Mean clutch/litter count is slightly high (0.43 v. 0.06 std. resid.). The young are comparatively small (ontogenetic mass range 1.10 v. 0.12 std. resid.). The number of broods per year is not well documented but ranges from one to four. In *lateralis*, the interclutch interval ranges 21–35 days. Incubation period (based on *lateralis*) is about average for lizards and is therefore a bit long for skinks (−0.03 v. −0.40 std. resid.), equating to 48 days. The eggs of *lateralis* are at an advanced stage of development at oviposition, so this species may not be representative. Communal oviposition is found in *lateralis* (3% v. 3%); parental care is not known (v. 3%).

Longevity is average (0.06 v. 0.17 std. resid.), based on *lateralis* (4 years). Credible population estimates are available for *cherriei* (n = 6; 0.44 kg/ha) and *lateralis* (n = 5; 0.70 kg/ha), or roughly the same as the skink average (0.35 kg/ha). Endemism is normal (59% v. 64%). Island species are comparatively scarce (27% v. 57%). None has been introduced (v. 2.1%). Half are imperiled (v. 47%), though none is on one of the IUCN's lists of the most endangered (14%). In other respects *Scincella* are typical skinks.

Scincidae (figs. 2.7, 4.2D, 4.4D, 4.8A, 4.8B, 4.11F, 4.12A, 4.12B, 4.12C, 4.14B)

The 1613 scincids are a global family of small (3.25 g v. 5.33 g) lizards that defy a simple ecological characterization. In part this is due simply to family size; Scincidae, the largest family, has 31% more species than Gekkonidae, which is the second-largest family. The addition of species tends to increase the spectrum of ecologies represented. For example, the only ecological business model missing from the skinks is glider; the 13 others have representatives within skinks. For comparison, gekkonids have only 10 ecological business models; many small families have one. In descending order of frequency, the ecological business models of skinks are refugia-anchored diurnal ground lizard (30% v. 25%), litter cryptozoid (28% v. 12%), subterranean forager (12% v. 8%), ground-foraging tree lizard (7.0% v. 9.2%), sand shark (6.8% v. 1.9%), wide-ranging chemosensor (6.0% v. 7.4%), trunk/wall lizard (3.0% v. 13.0%), nocturnal ground lizard (2.3% v. 7.7%), diurnal canopy walker (1.8% v. 6.4%), crack hermit (1.7% v. 2.6%), giant herbivore (1.7% v. 2.6%), immobility expert (0.2% v. 3.2%), and nocturnal canopy walker (0.1% v. 2.1%). At latitudes of less than 45°, skinks are missing only from the Andes and the center of the Sahara. Skinks are an unusual family in that English has a common name for them, "skinks," that corresponds exactly to the current familial designation, Scincidae. In contrast, "geckos" are found in many families beyond Gekkonidae; "iguanas" are found in many families besides Iguanidae, and so forth. This suggests that skinks are morphologically distinctive, mostly likely the predominance of small size, elongation, and glossy sheen.

Phylogenetically focused works typically stress the advanced olfactory-vomeronasal capabilities of scincids and invariably highlight the family's frequent possession of unornamented and often shiny scales. Yet many scincids have carinated or rugose scales and many other families, even iguanids, have substantial—though often poorly understood—vomeronasal abilities.

The glossiness of scales tells us relatively little about ecology; vomeronasal capabilities are often used to imply a chemosensory lifestyle like that of teiids or varanids. Unlike teiids or varanids, skinks are rarely wide-ranging chemosensors (6.0% v. 7.4% lizards, 97% teiids, and 92% varanids). As characterized in the literature, active foraging is prevalent in scincids (92% v. 58%), but this has no obvious ecological correlates in sex differences, diet, diel activity pattern, behavior, habitat, reproduction, demography, or morphology, except blue tails (4.0% v. 2.3%). The frequency of blue tails in Scincidae is third-ranked, behind only Teiidae (100% active foragers) and Lacertidae (92% active). Skinks are prone to sand diving (22% v. 6%) as a primary predator concealment tactic, a feature arguably connected to active forging, but note that their frequency of sand diving is third-ranked behind often sit-and-wait pygopodids and liolaemids. Skinks do not especially rely on flight (4% v. 11% lizards) as do so many active foragers (e.g., 77% of teiids). Indeed, their modal predator concealment tactic is retreat to a crevice or burrow (43% v. 31% lizards), a tactic often associated with Iguania and sedentary foragers. Consistent with crevice-burrow propensity is that scincids' modal ecological business model is refugia-anchored diurnal ground lizard (30% v. 24.5%). Many active foragers guard mates or tolerate fellow males (e.g., 55% teiids, 59% varanids); scincids do not exhibit much tolerance: 5.6% (v. 15% in lizards). Instead skinks defend space (94% v. 85% lizards). In these respects skinks diverge from active foragers.

What ecological themes permeate the Scincidae? The three ecological business models in which they are most overrepresented are litter cryptozoid (28% v. 12%), sand shark

(6.8% v. 1.9%, a family frequency second only to the tiny Trogonophidae), and subterranean forager (11.8% v. 7.8%). These are all "hidden" lifestyles, even though they involve active foraging. In contrast, the four ecological business models in which they are most underrepresented are very "exposed" ones: immobility expert (0.2% v. 4.5%), diurnal canopy walkers (1.8% v. 6.4%), nocturnal canopy walkers (0.1% v. 2.1%), and trunk/wall lizards (3% v. 13%).

What morphological attributes reflect their hidden lives? Elongation (35% v. 11%), behind only the large families Diploglossidae and Gymnophthalmidae; their mean shape index is correspondingly low: –0.46 (v. 0.0 std. resid.). Elongation is closely associated with subterranean or cryptozoic microhabitats. Absence of crests (0% v. 5.3%) is also associated with hidden lives; near absence of color change (0.1% v. 10.7% for basic color change; 0% v. 6% for advanced color change) is characteristic of nonvisual settings. Sixteen percent of skinks are functionally limbless, and most of these are subterranean foragers (12% v. 8%); the other functionally limbless skinks are mostly litter cryptozoids.

Another way to quantify exposure tolerance is by microhabitats: skinks are overrepresented as cryptozoids (30% v. 12%) and rarely found on elevated perches (10% v. 32% arboreal). Most skinks consume a diet of invertebrates (84% v. 73%)—prey easier to find while hidden. Skinks have few sex differences (sex rollup score 0.28 v. 0.75 on 0 to 6 scale); no need for flashy colors or visual enhancements if one is going to stay hidden. Skinks have a preponderance of species that are m<f (–0.31 v. 0.36 on –2 to 2 scale), even though clutch size seems to be invariant in many skink genera and the opportunity for relatively larger females to expand clutch count would not appear to be available to them. Regardless, most skink males are comparatively small and unadorned. Similarly, skinks do not rely on (visual) crypsis (9.0% v. 43%); that capability finds the most application if one is exposed. In summary, skinks are hidden active foragers, often tied to refugia rather than wandering.

There are a few ecological characteristics that do not fit neatly into this schema. One is scincids' high frequency of viviparity (oviparity 64% v. 82%). Viviparity is commonly associated with high latitude or high elevation (Shine 1985) but is also found in many lowland tropical skinks. Viviparous females gain by their ability to raise the temperature of their developing fetuses by basking, a relatively exposed action. Reproductive mode does not relate directly to foraging mode or microhabitat, but there is a curious association in skinks that is weak or absent in lizards overall; small skinks (<2 g) are 87% oviparous ($n = 339$); skinks >30 g are only 21% oviparous ($n = 34$). The relative size of progeny is not related to reproductive mode in small skinks (<5 g). Large (>5g) viviparous skinks have comparatively large neonates (mean ontogenetic range of $\log_{10}$ masses 42-fold; mean mass-adjusted std. resid. –0.278), whereas large (>5 g) oviparous skinks have comparatively small hatchlings (mean ontogenetic range of $\log_{10}$ masses 60-fold; mean mass-adjusted std. resid. 0.373). Mass-adjusted means compared with a t-test for unequal variances yields $p = 0.0008$ (two-tailed). This suggests that viviparity in Scincidae may facilitate the production of relatively larger neonates, perhaps through maternal feeding of fetuses or improved thermal buffering. The dominance of viviparity in large skinks implies that production of large neonates may be of greater value in larger species; perhaps because large neonates are better positioned to more quickly outgrow their vulnerability. Neonates of small skink species may never attain a size of diminishing vulnerability. See also table 3.10, trait 5, and associated discussion.

A curiosity of skinks is that they do not exhibit the weak negative correlation between SSD and m:f relative tail length found in lizards in general (table 3.3, trait 8). The trend in skinks is arguably opposite; although the trend is of a positive correlation, it is statistically insignificant despite the large sample size, perhaps washed out by the vast ecological diversity of skinks.

Another disparate attribute of skinks is the high proportion of skinks that reside on islands (57% v. 40%). That may simply reflect the part of the world in which they are overrepresented: the Australasian (46% v. 19%) and Oriental (24% v. 18%) biogeographic realms. Another arguably anomalous attribute is the near absence of secretory pores in skinks: 0.4% v. 15%. One might think that chemical communication would be an obvious advantage to lizards living an otherwise hidden lifestyle. Skinks are notably unlikely to be nocturnal (14.5% v. 25%); nocturnality would seem like a good way to be "hidden," and indeed 14.5% of skinks do it, but it is not a scincid staple. Why? Phylogeny? Fitting in with reduced nocturnality (table 3.7, trait 31) and elevated attachment to refugia (table 4.1, trait 33) is the near complete absence of vocalizations (0.1% v. 2.5%). Perhaps a lizard of predictable (refugia-anchored) location does not need to advertise where it is? Are skinks found in populations that are dense enough to preclude concupiscent individuals having to draw mates from a distance?

The mean home-range area of

skinks is almost exactly average (−0.07 v. 0.0 std. resid.). This fits with their attachment to refugia, perhaps less so to their active foraging. Evidently the two tendencies (small home range for the refugia-anchored; large home ranges for active foragers) simply cancel each other out and produce an intermediate value.

Four of four lizard species documented to steal booty from columns of ants are in Scincidae (see accounts of *Cryptoblepharus*, *Meroles*, and *Morethia*); four of five lizards known to be kleptoparasitic on other lizards are in Scincidae (see accounts of *Carlia* and *Emoia*); the other example is *Atlantolacerta andreanskyi*. Is larceny a skink specialty?

The remaining cataloged ecological traits of skinks closely approximate those of other lizards (chapter 2). For example, skinks exhibit an average frequency of the other ecological business models, an average distribution of the other diets, an average habitat, rainfall, and elevation spectra, average longevity, and so forth.

This tendency to be average does not necessarily apply to particular genera (see genus accounts), of which Scincidae has 155 (as of May 2017): *Ablepharus, Acontias, Alinea, Amphiglossus, Anepischetosia, Anomalopus, Aspronema, Asymblepharus, Ateuchosaurus, Barkudia, Bassiana, Bellatorias, Brachymeles, Brasiliscincus, Caesoris, Caledoniscincus, Calyptotis, Capitellum, Carlia, Celatiscincus, Chalcides, Chalcidoseps, Chioninia, Coeranoscincus, Coggeria, Concinnia, Copeoglossum, Cophoscincopus, Corucia, Cryptoblepharus, Ctenotus, Cyclodomorphus, Dasia, Egernia, Emoia, Epibator, Eremiascincus, Eroticoscincus, Eugongylus, Eulamprus, Eumeces, Eumecia, Eurylepis, Eutropis, Exila, Feylinia, Fojia, Geomyersia, Geoscincus, Glaphyromorphus, Gongylomorphus, Graciliscincus, Grandidierina, Haackgreerius, Hakaria, Harrisoniascincus, Hemiergis, Heremites, Insulasaurus, Isopachys, Janetaescincus, Jarujinia, Kaestlea, Kanakysaurus, Lacertaspis, Lacertoides, Lamprolepis, Lampropholis, Lankascincus, Larutia, Leiolopisma, Lepidothyris, Leptoseps, Leptosiaphos, Lerista, Liburnascincus, Liopholis, Lioscincus, Lipinia, Lissolepis, Lobulia, Lubuya, Lygisaurus, Lygosoma, Mabuya, Madascincus, Manciola, Marisora, Marmorosphax, Melanoseps, Menetia, Mesoscincus, Mochlus, Morethia, Nangura, Nannoscincus, Nessia, Niveoscincus, Notomabuya, Notoscincus, Oligosoma, Ophiomorus, Ophioscincus, Orosaura, Otosaurus, Pamelaescincus, Panaspis, Panopa, Papuascincus, Paracontias, Paralipinia, Parvoscincus, Phaeoscincus, Phasmasaurus, Phoboscincus, Pinoyscincus, Plestiodon, Prasinohaema, Proablepharus, Proscelotes, Pseudemoia, Pseudoacontias, Psychosaura, Pygmaeascincus, Pygomeles, Ristella, Saiphos, Saproscincus, Scelotes, Scincella, Scincopus, Scincus, Scolecoseps, Sepsina, Sepsophis, Sigaloseps, Silvascincus, Simiscincus, Sphenomorphus, Spondylurus, Tachygyia, Techmarscincus, Tiliqua, Toenayar, Trachylepis, Tribolonotus, Tropidophorus, Tropidoscincus, Tumbunascincus, Typhlacontias, Typhlosaurus, Tytthoscincus, Varzea, Vietnascincus,* and *Voeltzkowia*.

Individual species naturally show even greater variability than do genera. For example, whereas the mean scincid has a mass of 3.25 g, the family's full range is 0.2 g (1st percentile of mass: *Paracontias fasika*) to 1047 g (99th percentile of mass: *Egernia striolata*). There are many ways to be a successful lizard, and individual skink species exhibit most of them.

Scincopus sp.	English name				Range core
			Size		
	Activity	Diet	mm	g	Ecological business model
fasciatus	Banded Skink				Sahara
	nocturnal		153	31.1	*** nocturnal ground lizard

Scincopus

Scincopus fasciatus is the sole member of its genus. It is a large (31 g v. 3.3 g), elongate (v. 35%), nocturnal ground (v. 2%) skink found throughout the Sahara, from Algeria, Mauritania, and Mali to Sudan and Nigeria. Although nocturnal ground lizards are scarce in skinks, the summer Sahara might be a good place to avoid daytime activity.

Despite their large size, their shape index is only slightly heavier than that of a typical skink (−0.28 v. −0.46 std. resid.). Tail length is very short (0.52 v. 1.33 × SVL) and breakable (v. 73%). The sexes are alike. Diet is unknown, but herbivory is suspected. *S. fasciatus* is conspicuously wary (v. 11%). It thermoconforms (v. 1.5%) or obtains heat indirectly. It relies on burrows (v. 24%) or sand for its daily refuge. It is found at low elevation (v. 89%). The only tabulated habitat association is sand (v. 25%). There are contradictory accounts of its reproductive mode, and no information is available on its clutch or litter count. It is not endemic (v. 64%), does not occur on islands (v. 57%), has not been introduced (v. 2.1%), and is not imperiled (v. 47%). In other respects *Scincopus* is likely to be similar to other skinks.

Scincus

Scincus is a minor radiation (four spp.) of large-sized (19.6 g v. 3.3 g), omnivorous (v. 13%), sand-dwelling (v. 25%), mostly elongate (75% v.

Scincus spp.	English name				Range core
			Size		
	Activity	Diet	mm	g	Ecological business model
albifasciatus	White-banded Sandfish				Mauritania
	diurnal	omnivore	104	24.7	*** sand shark
hemprichii	Yemen Sandfish				Southern Arabia
	nocturnal		122	16.9	* nocturnal ground lizard
mitranus	Arabian Sandfish				Saudi Arabia
	diurnal	omnivore	108	11.9	**** wide-ranging chemosensor and sand shark
scincus	Common Sandfish				Northern Africa
	cathemeral	omnivore	111	29.8	**** sand shark

35%) skinks found in the North African and Arabian deserts, from Western Sahara eastward to Iran and possibly Pakistan. They occupy a variety of ecological business models that are rare in skinks: sand shark (50% v. 7%), nocturnal ground lizard (25% v. 2%), and wide-ranging chemosensor (25% v. 6%).

Their shape index is higher than average (0.11 v. –0.46 std. resid.). Their thick tails are much shorter than average (0.70 v. 1.33 × SVL) but of normal breakability (75% v. 73%). Juvenile *mitranus* have blue tails (25% v. 4%).

Males are larger than females (1.0 v. –0.31 on –2 to 2 scale), but color sex differences are absent. Mean sex rollup score (on scale of 0 to 6) is mildly elevated (0.75 v. 0.28).

All documented species are omnivores; adult *S. scincus* eat more plants during the summer, when plant material constitutes a slim majority of the stomach volume. Naturally, all species are active foragers (v. 92%). Home-range area is likely to be very large in this low-productivity environment, but it has not been quantified. The primary foraging microhabitat of *mitranus* is dune slip faces, over which they may travel >40 m daily—and more if moving between dunes.

Almost all diel activity patterns are present in *Scincus*: cathemeral (25% v. 9%), diurnal (50% v. 74%), and nocturnal (25% v. 15%). Activity may be suspended at midday during the summer. For primary predator concealment, all rely on sand (v. 22%). The viper *Cerastes vipera* does not bite and release when envenomating *S. scincus*; instead, it bites and holds to prevent the dying lizard from running beyond the detection distance of the snake. All *Scincus* are sand swimmers, though *S. scincus* uses it limbs when moving through moist sand. In dry sand, it can disappear into the sand in 0.41 sec. It positions itself in the sand column for foraging and thermoregulating. Direct basking is known for *mitranus* and *scincus* (50% v. 21%). Mean active body temperature for the genus is a toasty 35.5°C (v. 31.1°C). Half of the species are conspicuously wary (v. 11%).

Male *albofasciatus* are socially intolerant during breeding. For daily refuge at least 67% (v. 24%) use burrows. They are found in littoral zones (25% v. 21%) but are absent from riparian areas (v. 23%) and human habitats (v. 14%).

S. mitranus is viviparous (v. 36%); the reproductive mode of *scincus* has been reported both ways and may be geographically variable. The mode of *albifasciatus* is not overtly documented, but an incubation period of 64 days implies oviparity (v. 64%). Those three species are all said to reproduce only once per year; mean clutch/litter count is low (–0.43 v. 0.06 std. resid.). Young may be slightly large (ontogenetic mass range –0.12 v. 0.12 std. resid.). Incubation period is about average (–0.61 v. –0.40 std. resid.). Female *scincus* have the interesting constraint that they will copulate only once per day.

Longevity is about average (0.28 v. 0.17 std. resid.), equating to 10 years in *scincus*. No species are endemic to their country (v. 64%). Two species occur on islands (50% v. 57%); *mitranus* is found on Masirah and Qeshm islands; *scincus* is found on Bahrain and Sir Baniyas. None has been introduced (v. 2.1%). One evaluated species is imperiled (33% v. 47%: *scincus* is vulnerable in Israel), though none is on one of the IUCN's lists of the most endangered (v. 14%). Each of these sand swimmers leaves a detectable furrow when moving at shallow depth; a skilled child is said to be able to catch 10–50 *scincus* in a day, making them vulnerable to overcollecting—for the pet and aphrodisiac trades—in populated areas. *S. albifasciatus* is also collected for the pet and "medicinal" trades. In other respects *Scincus* are much like other skinks.

Scolecoseps

Scolecoseps is a trio (three spp.) of uniformly small (1.91 g v. 3.3 g), limbless (100%), very poorly known (filled cells 16% v. 25%) skinks in east Africa (Tanzania, Mozambique). They are too poorly known to assign diel cycle, diet, or ecological business model. They are extremely slender (shape index –2.21 v. –0.46 std. resid.). Tail length is short by any standard (0.29 v. 1.33 × SVL). Sex differences are unknown. They are found solely at low elevation (100% v. 89%) in littoral areas (100% v. 21%). One species (*acon-*

Scolecoseps spp.	English name				Range core
			Size		
	Activity	Diet	mm	g	Ecological business model
acontias	Legless Sand Skink				Tanzania
			95	2.20	
boulengeri	Boulenger's Legless Skink				Mozambique
			92	2.02	
litipoensis	Litipo Sand Skink				Tanzania
			83	1.61	

Scriptosaura sp.	English name				Range core
			Size		
	Activity	Diet	mm	g	Ecological business model
catimbau	Catimbau Sand Swimmer				Brazil
			47	0.86	

Sepsina spp.	English name				Range core
			Size		
	Activity	Diet	mm	g	Ecological business model
alberti	Albert's Burrowing Skink				Namibia
	diurnal		47	1.19	*** litter cryptozoid
angolensis	Angolan Burrowing Skink				Congo
			74	4.20	*** litter cryptozoid
bayoni	Bayon's Burrowing Skink				Congo
			60	4.17	* litter cryptozoid
copei	Cope's Burrowing Skink				Angola
			67	5.79	* litter cryptozoid
tetradactyla	Four-toed Burrowing Skink				Zambia
		invert.	75	2.69	* subterranean forager

tias) is known to be viviparous (v. 36%). All are endemic (v. 64%). None is found on an island (v. 57%); none has been introduced (v. 2.1%). They are not imperiled (v. 47%), though the IUCN judges *litipoensis* to be very rare. In other respects *Scolecoseps* are probably typical limbless skinks.

Scriptosaura

Scriptosaura catimbau is the sole member of a very poorly known (filled cells 13% v. 19%) functionally limbless gymnophthalmid genus endemic to Brazil. The species is too poorly known to guess its ecological business model. The species is very small (0.86 g v. 2.35 g) and mildly elongate (shape index –1.25 v. –0.32 std. resid.). It possesses secretory pores (v. 29%), but other sex differences, diet, diel activity pattern, predator avoidance tactics, social relations, and reproduction are not reported. This species has not been evaluated for conservation status.

Sepsina

Sepsina is a minor radiation (five spp.) of very poorly known (filled cells 17% v. 25%), average-sized (3.18 g v. 3.25 g), mostly elongate (at least 60% v. 35%), mostly litter cryptozoid (80% v. 28%) skinks found in south central Africa, from the Democratic Republic of the Congo to Tanzania and south to Namibia and Mozambique. At least some appear to be transitioning to subterranean foragers (20% v. 12%). Their tails are somewhat short but not as short as obligate diggers (1.00 v. 1.33 × SVL). Their tails mostly break (60% v. 73%). Blue tails, though perhaps not bright blue, are found in 40% (v. 4%). *S. tetradactyla* is an invertebrate eater, and the others likely are as well (v. 84%). One species (*alberti*) is diurnal (v. 74%). They are found at low elevations (v. 89%) and are associated with sandy areas (33% v. 25%) and riparian zones (33% v. 23%). Reproductive mode is unknown but presumedly oviparous (v. 64%). Endemism is low (40% v. 64%). None is found on an island (v. 57%). None has been introduced (v. 2.1%). Half are imperiled (v. 47%), though none is on one of the IUCN's lists of the most endangered (v. 14%). In other respects *Sepsina* are much like other skinks.

Sepsophis

Sepsophis punctatus is the sole member of its genus. It is a large (6.4 g v. 3.3 g), functionally limbless, subterranean forager (v. 12%) skink endemic to India. Despite this digging lifestyle, it has a fairly long tail (1.00 v. 1.33 × SVL), and it is not extremely attenuate (shape index –1.21 v. –0.46 std. resid.). It is associated with litter and loose soil, and it may be transitional to/from a litter cryptozoid. Sex differences are not reported. It feeds on a variety of invertebrates, includ-

Sepsophis sp.	English name				Range core
			Size		
	Activity	Diet	mm	g	Ecological business model
punctatus	Eastern Ghats Burrowing Skink				India
			107	6.4	** subterranean forager

Shinisaurus sp.	English name				Range core
			Size		
	Activity	Diet	mm	g	Ecological business model
crocodilurus	Chinese Crocodile Lizard				China
	diurnal	carnivore	138	46.6	*** refugia-anchored diurnal gr. liz.

Sigaloseps spp.	English name				Range core
			Size		
	Activity	Diet	mm	g	Ecological business model
balios	Mount Humboldt Skink				New Caledonia
	diurnal		45	1.10	
conditus	Pourina River Skink				New Caledonia
	diurnal		42	1.30	** litter cryptozoid
deplanchei	Deplanche's Shiny Skink				New Caledonia
	cathemeral	invert.	41	0.84	**** litter cryptozoid
ferrugicauda	Russet-tailed Mountain Skink				New Caledonia
			54	1.75	* litter cryptozoid
pisinnus	Ningua Dwarf Skink				New Caledonia
	diurnal		36	0.57	** litter cryptozoid
ruficauda	Red-tailed Shiny Skink				New Caledonia
	cathemeral		49	1.40	** litter cryptozoid

ing termites. It is found only at low elevations (v. 89%); several sources cite an upper elevational limit of 300 m, despite the type locality being at 900 m. It is not found on an island (v. 57%), has not been introduced (v. 2.1%), and has not been evaluated for conservation status. In other respects *Sepsophis* is likely to be similar to other limbless skinks.

Shinisauridae and *Shinisaurus*

Shinisaurus crocodilurus is the sole representative of the family Shinisauridae. It is a diurnal (v. 67%), semiaquatic (v. 1%), carnivorous (v. 3.2%) lizard with low crests (v. 5%) and conspicuous rectilinear plates covering the dorsal surface of its tail, hence the association with crocodiles. Unlike a true crocodile, it has a very blunt snout. It is found in Vietnam and China, where it lives as a refugia-anchored diurnal ground lizard (v. 25%). It is relatively well known (filled cells 49% v. 27%). Its tail is normal in length (1.21 v. 1.37 × SVL) but unbreakable (v. 57% breakable). The primary sex difference is that males have shorter tails (m:f relative tail length ratio 0.88 v. 1.07). It forages actively (v. 58%) at low and medium elevation—in this case 200–1500 m—in riparian areas (v. 1.5%).

It is viviparous (v. 18%) with a normal-sized litter (0.12 v. 0.0 std. resid.) of particularly large young (ontogenetic size range −0.91 v. 0.0 std. resid.). The neonates remain in the vicinity of the mother (parental care v. 3%), at least in captivity. Gestation takes a remarkable 270 days, presumably to ensure that birth takes place at a favorable season.

Longevity is extended (0.32 v. 0.0 std. resid.), corresponding to 11.4 years. It is not endemic (v. 64%), is not found on an island (v. 40%), and has not been introduced (v. 3.3%), but it is considered imperiled (v. 49%), including placement on one of the IUCN's lists (endangered) of the most endangered (v. 14%).

Sigaloseps

Sigaloseps is a moderate radiation (six spp.), of small (1.09 g v. 3.25 g), diurnal (60% v. 74%) or cathemeral (40% v. 9%), mostly elongate (83% v. 35%), oviparous (100% v. 64%) litter cryptozoid (100% v. 28%) skinks endemic to New Caledonia. Aside from their size and absence of blue tails (v. 4%, though *pisinnus* may have a bright orange tail and *ruficauda*'s is red), they are morphologically similar to typical skinks. Known sex differences are limited to dichromatism (50% v. 8%). Dietary details are unavailable, but *deplanchei* eats invertebrates (v. 84%). Two are conspicuously wary (33% v. 11%). One (*deplanchei*) is known to be a thermoconformer (17% v. 1.5%). Their elevational profile is biased toward middle elevations (l/m/h: 80/60/0 v. 89/35/6). *S. deplanchei* is found in riparian zones (20% v. 23%), but they are absent from littoral (v. 21%), sandy (v. 25%), and human (v. 14%) hab-

Silvascincus spp.	English name				Range core
			Size		
	Activity	Diet	mm	g	Ecological business model
murrayi	Murray's Skink				Queensland
	cathemeral	omnivore	88	14.4	* crack hermit
tryoni	Tryon's Skink				Queensland
	diurnal	invert.	91	16.1	* litter cryptozoid

Simiscincus sp.	English name				Range core
			Size		
	Activity	Diet	mm	g	Ecological business model
aurantiacus	Orange-bellied Burrowing Skink				New Caledonia
			70	3.65	** litter cryptozoid

itats. Mean clutch count is normal to slightly high for their size (0.32 v. 0.06 std. resid.). None has been introduced (v. 2.1%). All are imperiled (v. 47%), but none is on one of the IUCN's lists of the most endangered (v. 14%). In other respects *Sigaloseps* are typical skinks.

Silvascincus

Silvascincus are two species of large (15.2 g v. 3.3 g), terrestrial (50% v. 48%) or cryptozoic (50% v. 30%), viviparous (v. 36%) rain forest skinks found in extreme eastern Australia, at the point on the coast where New South Wales meets Queensland. The omnivore (*murrayi*) may be a crack hermit; it is often observed with its head protruding from a wood cavity, though it is also described as sometimes sleeping exposed on rock faces at night. The invertebrate eater (*tryoni*) seems to be a more conventional litter cryptozoid, though both are described as active foragers (v. 92%); *murrayi* also sit-and-waits.

Their build is more robust than a typical skink (0.45 v. –0.46 std. resid.). Their tails are shorter (1.00 v. 1.33 × SVL) and breakable (100% v. 73%). Sex differences are absent. *S. murrayi* basks directly (v. 21%) and lives in groups that could be extended families. Both species will use burrows for daily refugia (100% v. 24%). They are found at low elevation (v. 89%) but are not associated with any of the tabulated habitat categories (littoral, riparian, sandy, or human habitats).

Litter counts are normal to low (–0.41 v. 0.06 std. resid.). They are not endemic (v. 64%), do not occur on islands (v. 57%), have not been introduced (v. 2.1%), and are not imperiled (v. 47%). In other respects *Silvascincus* are much like other skinks.

Simiscincus

Simiscincus aurantiacus is the sole member of its genus, an average (3.65 g v. 3.3 g), elongate (v. 35%) litter cryptozoid (v. 28%) skink endemic to New Caledonia. It is known from a single specimen (filled cells 22% v. 25%) that has a noticeably rounded snout. Aside from its apparent rarity, its claim to fame is that it has the highest number of premaxillary teeth of any skink. Its diet is unknown; does that high tooth count suggest that it eats things that are hard to hold onto, perhaps slugs? It is otherwise morphologically average for a skink. No sex differences are knowable (given the sample size). It was found at low elevation (v. 89%). Obviously, it has not been introduced (v. 2%), and it is imperiled (v. 47%), though not yet on one of the IUCN's lists of the most endangered. In other respects *Simiscincus* is likely to be similar to other skinks.

Sitana

Sitana is a moderate (10 spp.) genus of small (2.9 g v. 15.6 g), mostly crestless (10% v. 38%), terrestrial (v. 56%) agamids of lowland (100% v. 83%) south Asia, from Nepal to Sri Lanka. All are considered refugia-anchored diurnal ground lizards (v. 43%). Their hallmark is the males' large colorful dewlaps. Most documented species are sexually dichromatic (50% v. 53%) and show seasonal color (40% v. 20%), but oddly, only m<f SSD is known (–1.00 v. 0.53 on 2 to 2 scale). Females are about 65% heavier than males in *ponticeriana*, the best studied species. Unlike grass anoles, which they superficially resemble, *Sitana* have large clutches (1.58, the fourth highest in Agamidae, v. 0.20 std. resid., equating to 9.0 eggs), which may exert selection pressure for large female size. As a mass-adjusted proportion of maximum size (1.53, the fifth highest in Agamidae, v. 0.69 std. resid.), the young are comparatively small.

Tails are long (2.26 v. 1.86 × SVL) and are substantially longer in males (mean m:f ratio 1.23, the highest value in Agamidae, v. 1.07). Tails that are documented break (40% v. 17%) but do not regenerate. One species (*sivalensis*) can change the hue of its skin (advanced color change 10% v. 15%), and two (*fusca*, *schleichi*) are able to darken (basic color change 30% v. 22%), though underreporting seems probable. Of the four species

Sitana spp.	English name				Range core
			Size		
	Activity	Diet	mm	g	Ecological business model
bahiri	Bahir's Fan-throated Lizard				Sri Lanka
			45	2.83	* refugia-anchored diurnal gr. liz.
devakai	Devaka's Fan-throated Lizard				India
			42	2.27	
fusca	Dark-brown Fan-throated Lizard				Nepal
	diurnal	invert.	45	2.76	*** refugia-anchored diurnal gr. liz.
laticeps	Broad-headed Fan-throated Lizard				India
			46	2.94	** refugia-anchored diurnal gr. liz.
marudhamneydhal	Tamil Nadu Fan-throated Lizard				India
			49	3.59	
ponticeriana	Cuvier's Fan-throated Lizard				India
	diurnal	omnivore	56	5.38	**** refugia-anchored diurnal gr. liz.
schleichi	Suklaphantah Sitana				Nepal
	diurnal	invert.	37	1.57	*** refugia-anchored diurnal gr. liz.
sivalensis	Siwalik Sitana				Nepal
	diurnal	invert.	42	2.29	*** refugia-anchored diurnal gr. liz.
spinaecephalus	Spiny-headed Fan-throated Lizard				India
	diurnal		48	3.46	** refugia-anchored diurnal gr. liz.
visiri	Palm-leaf Fan-throated Lizard				India
	diurnal		49	3.67	* refugia-anchored diurnal gr. liz.

with diet documentation, three are invertebrate eaters (75% v. 46%), and *ponticeriana* is, at 5 g, a notably small omnivore (25% v. 26%). All forage by sit-and-wait (100% v. 88%). For predator concealment, two of four rely on crypsis (50% v. 51%). Annual rainfall documentation is limited to *ponticeriana*, but it suggests an overrepresentation in the intermediate (100% v. 12%) and mesic (100% v. 20%) part of the spectrum. *Sitana* are well represented in littoral (33% v. 10%) and sandy (56% v. 26%) habitats, about average in riparian zones (11% v. 21%), and absent from human habitat (v. 16%).

The longevity of *ponticeriana* is as much as 6 years, which is lengthy for so small a lizard and especially so for an agamid (0.13 v. −0.46 std. resid.). Population is based on two credible estimates of *ponticeriana* and is slightly high (0.28 kg/ha v. 0.16 kg/ha). Endemism is high (80% v. 49%). *Sitana* have neither been introduced (v. 1.7%) nor listed as imperiled (v. 37%). In other respects they are typical agamids.

Smaug

Smaug is a moderate (eight spp.) cordylid radiation of crack hermits (75% v. 43%) and refugia-anchored diurnal ground lizards (25% v. 29%) in southern Africa. The most famous of the genus is the giant member (*giganteus*), the Sungazer, which gets its name from a characteristic head-up basking and surveillance pose. The Sungazer is notable as the spiniest cordylid, extra prickly perhaps because it is burrow-based rather than dependent on rock crevices. Its ability to range beyond rock outcrops may be facilitated by its large size. *S. warreni* spines are comparatively demure, but most *Smaug* are very well armored.

Mean mass for *Smaug* (43.6 g) is well above the cordylid average (10.8 g), and *Smaug* are more robust (shape index 1.05 v. 0.35 std. resid.). Tails do not normally break (0% v. 60%). Half of the species are considered conspicuously wary (v. 27%). *Smaug* are not known for living in groups (v. 29%), but their social lives are not well documented. Documented elevations are comparatively low (up to 1700 m), though not associated with littoral (v. 9%) or sandy (v. 3%) areas.

All species are viviparous (v. 75%), with a litter count that fits their size (−0.65 v. −0.48 std. resid.) for their family. Neonate size is normal (ontogenetic range −0.51 v. −0.37 std. resid.). Parental care—young remain with mother for a time—is reported for only *giganteus* (v. 7%). Longevity is mildly elevated (0.39 v. 0.02 std. resid.). Only *S. giganteus* is considered imperiled (v. 79%), and none is among those listed by the IUCN as severely endangered (v. 4.8%). *S. giganteus* is at risk from habitat destruction and commercial collecting for the pet trade. In other respects *Smaug* are typical cordylids.

Smaug spp.	English name				Range core
			Size		
	Activity	Diet	mm	g	Ecological business model
barbertonensis	Barberton Dragon Lizard				South Africa
			111	39.0	** refugia-anchored diurnal gr. liz.
breyeri	Waterberg Dragon Lizard				South Africa
	diurnal		112	40.8	**crack hermit
depressus	Zoutpansberg Dragon Lizard				South Africa
			96	23.2	** crack hermit
giganteus	Sungazer				South Africa
	diurnal	omnivore	168	174	*** refugia-anchored diurnal gr. liz.
mossambicus	Gorongosa Dragon Lizard				Mozambique
	diurnal		93	20.4	** crack hermit
regius	Regal Dragon Lizard				Zimbabwe
	diurnal	invert.	119	51	*** crack hermit
vandami	Van Dam's Dragon Lizard				South Africa
	diurnal		112	40.8	*** crack hermit
warreni	Warren's Dragon Lizard				Mozambique
	diurnal	invert.	117	47.8	*** crack hermit

Sphaerodactylidae (fig. 4.8D)

Sphaerodactylidae is the eighth-largest family (214 spp., table 2.1) and a part of the gecko clade (chapter 2). Though sometimes loosely referred to as day geckos, their diurnal frequency (62%) is less than an average lizard (67%). Furthermore, many diel classifications in Sphaerodactylidae are highly uncertain and often contested. The best-known genus is *Sphaerodactylus* (105 spp.), but the family also includes: *Aristelliger, Chatogekko, Coleodactylus, Euleptes, Gonatodes, Lepidoblepharis, Pristurus, Pseudogonatodes, Quedenfeldtia, Saurodactylus*, and *Teratoscincus*. They are very widely distributed, most numerous in the Neotropics (83% v. 29%) and absent from only the Australasian and Oriental biogeographic realms. They are ecologically diverse: 8 of 14 ecological business models, two more than in 416 species of anoles. In descending order of frequency the business models are litter cryptozoids (53% v. 12%), ground-foraging tree lizards (12% v. 9%), trunk/wall lizards (12% v. 13%), refugia-anchored diurnal ground lizards (7% v. 24.5%), nocturnal ground lizards (6.5% v. 7.7%), crack hermits (6% v. 2.6%), diurnal canopy walkers (1.8% v. 6.4%), and immobility experts (1.2% v. 3.2%). Although often characterized as small (1.07 g v. 5.33 g), they span the lizard size distribution from the 0th percentile (0.21 g) to the 93rd (37.9 g).

Morphologically they differ from many other geckos in their lack of expanded toepads (99% padless). The family has the fourth-highest frequency of ant specialists (16.4% v. 4.4%); their consumption of generalized invertebrates is much higher than lizards as a group (82% v. 46%). Their use of electric lights is the third highest (6.9% v. 2.0%). Their modal microhabitat is cryptozoic (41% v. 12%). Their elevational profile is skewed toward low elevations (l/m/h: 96/20/4 v. 82/41/13). They are a regular presence in many human habitats (33% v. 19%). They are universally oviparous (100% v. 82%) and committed to hard-shelled eggs (100% v. 41), tied for the top billing in that category. Those two reproductive attributes may be connected, as one imagines that egg retention—to transition to viviparity—would be difficult or pointless for a hard-shelled egg. All clutches are one or two eggs, which leads to an exceptionally low RCM (0.095 v. 0.226). Sphaerodactylidae is the third-highest family for frequency of adherent eggs (5.6% v. 2.4%). The hard-shelled eggs may be adapted to salt exposure, facilitating natural overwater dispersal. The attached eggs and the adults' predilection for human habitats may combine to promote accidental human transport, as the family has the third-highest frequency of multiple (>2) introductions (2.3% v. 1.1%). These distinctive attributes of Sphaerodactylidae add up to a family that is diverse, opportunistic, often secretive, easily dispersed, and reproductively geared toward small clutches that are deposited often.

Tail length is moderately short (1.07 v. 1.37 × SVL), and most tails are breakable (94% v. 57%) but rarely blue (0.5% v. 2.3%) or prehensile (0.5% v. 5.0%). The rarity of blue tails is unexpected for a group so rife with litter cryptozoids (see *Sphaerodactylus* account). Qualitatively, this family abounds in species that tear their skin easily, presumably to escape from predators. Perhaps tearable skin serves some of the same functions as blue tails. Crests are present (4.7% v. 5.3%) but only in *Pristurus*. Secretory pores (0.5%, only *Quedenfeldtia*, v. 11%) and elongation (0.5%, only *Sphaerodactylus*, v. 11%) are rare. Color change is relatively infrequent

(6.5% v. 11% for any color change; 1.4% v. 6.0% for advanced).

The family has no strong pattern of SSD (0.04 v. 0.36 on –2 to 2 scale), but individual genera have distinct patterns (see those accounts). Sexual dichromatism is common (35% v. 21%), but seasonal color is not (2.8% v. 6.5%). The average sphaerodactylid shows little difference between the sexes in relative tail length (1.02 v. 1.07 ratio). Consequently, their sex rollup score (on scale of 0 to 6) is modest (0.54 v. 0.75).

No sphaerodactylids are carnivores (v. 3.2%) or herbivores (v. 4.1%). Omnivory (1.5% v. 15%) is found only in *Aristelliger*. Sphaerodactylids are diverse in foraging mode: sit-and-wait (83% v. 46%), cruise (21% v. 6.6%), and active (17% v. 58%). Home-range size is very close to the lizard average (0.06 v. 0.0 std. resid.), though sampling of this trait is poor.

The diel patterns shown by sphaerodactylids are diverse. Cathemeral (15% v. 7%) and crepuscular (6.3% v. 1.8%) diel activity patterns are overrepresented; diurnal (62% v. 67%) and nocturnal (17% v. 25%) patterns are underrepresented. The apparent diel opportunism may be related to the cryptozoic lives led by many sphaerodactylids; why hold to a rigid schedule when foraging under cover? Spontaneous calling is low (1.9% v. 2.5%), and stress calling frequency is normal in frequency (2.3% v. 2.2%). An average number of species are described as conspicuously wary (7% v. 8%).

The primary predator concealment tactics used by sphaerodactylids emphasize retreat to a crevice or burrow (43% v. 25%), flight (23% v. 11%), and retreat to vegetation (27% v. 23%). Crypsis is underused (17% v. 43%), and sand diving is absent (v. 6%). As fits their cryptozoic microhabitat, direct basking is underrepresented (6.5% v. 18%); thermoconforming is overrepresented (4.2% v. 2.6%). Mean active body temperature is average (30.4°C v. 31.1°C). Several species in various sphaerodactylid genera are reported to suffer from hyperthermia when handled.

Solitary living is about average (8% v. 10%), as is group living (17% v. 20%), male tolerance (14% v. 15%), and male defense of space (86% v. 85%). Male hierarchies have not been observed (0% v. 7.7%). In their social lives sphaerodactylids do not differ much from other lizards.

The family is well represented in the arboreal stratum (29% v. 32%), underrepresented on the surface (terrestrial 30% v. 48%), and absent from subterranean (v. 7%) and semiaquatic (v. 1.1%) microhabitats. The family is strongly skewed toward the wetter end of the rainfall spectrum, and the dry mode is nearly vacant (5% v. 33%). The modal band is mesic (45% v. 43%); they are overrepresented in the wet (>3000 mm/y: 25% v. 9%) and average in the intermediate band (30% v. 27%). They are overrepresented in littoral habitats (31% v. 17%) and about average in sandy areas (15% v. 22%) and riparian zones (19% v. 22%).

Mass-adjusted clutch count is near average (–0.11 v. 0.0 std. resid.), suggesting that their apparent restriction to one- or two-egg clutches may match their size rather than be a phylogenetic imposition. Similarly, their ontogenetic size range (–0.12 v. 0.0 std. resid.) denotes hatchlings of an average size. Incubation is long (0.44 v. 0.0 std. resid.). Communal oviposition is frequent (11% v. 4%); parental care is not (1.4% v. 2.9%). Sphaerodactylids are ranked first in frequency of year-round breeding (65% v. 22%), possibly because their low clutch count rewards—or induces—continuous reproduction. Continuous breeding is not an attribute of litter cryptozoids (14% v. 22%), suggesting that the selective pressure for continuous breeding is not the litter environment.

Mass-adjusted longevity is about average (–0.15 v. 0.0 std. resid.), as is population density (0.32 kg/ha v. 0.54 kg/ha). Endemism is normal (65% v. 64%). Island species are frequent (64% v. 40%). Introductions are high (5.1% v. 3.3%), as further explained above. A normal frequency of species is imperiled (41% v. 49%), including an excessive number of species on the IUCN's lists of the most endangered (23% v. 13.6%). Many of the critical species are likely to be island forms suffering from introduced predators and habitat degradation.

Sphaerodactylus (fig. 4.8D)

Sphaerodactylus, the largest genus (105 spp., table 2.2) of the Sphaerodactylidae, has five ecological business models but is dominated by litter cryptozoids (77% v. 53%). The others are ground-foraging tree lizards (8% v. 12%), crack hermits (7% v. 6%), trunk/wall lizards (4.6% v. 12.4%), and nocturnal ground lizards (3.4% v. 6.5%). They are found primarily in the West Indies and secondarily in Central America, a range very similar to the anoles (for comparison, see *Anolis* entry):

Islands (North to South)	
Cuba	20 spp.
Hispaniola	34
Puerto Rico	6
Jamaica	8
Mainland Countries (North to South)	
Mexico	3
Guatemala	3
Honduras	10
Nicaragua	5
Costa Rica	4
Panama	4
Colombia	4

Sphaerodactylus are especially concentrated in Hispaniola, whereas anoles are richest in Colombia. It seems that *Sphaerodactylus* might be a good model system for studying speciation in the West Indies.

Sphaerodactylus are small, even for sphaerodactylids (0.72 g v. 1.07 g) and relatively uniform in size (range 0.21–1.44 g, with the exception of one outlier, the gigantic, 2.17 g, Cocos Island *pacificus*). The smallest *Sphaerodactylus* (*ariasae*, *parthenopion*) are among the world's smallest amniotes (but note *Amphisbaena talisiae* mass). Not coincidentally, the genus also has a representative (fig. 4.8D) that was recorded as the world's densest amniote (*macrolepis*: 52,800 individuals/ha). When density is measured in kg/ha, however, Aldabra Tortoises move to the head of the amniote line.

Aside from their small size and high density, *Sphaerodactylus* head the league tables in relatively minor and inchoate ways. For example, their frequency of blue tails (1.0% v. 0.5%) is the highest in the family. They have the family's second-highest frequency of ant specialists (4.3% v. 16.4%). They have the greatest reliance on crypsis (100% v. 17%). They are the most-represented sphaerodactylid genus in the intermediate rainfall band (56% v. 30%). They have the second-highest frequency of parental care (1.9% v. 1.4%).

Sphaerodactylus are padless (100% v. 99%); tails are a little short (0.99 v. 1.07 × SVL). One species (*ramsdeni*) is elongate (1.0% v. 0.5%). In addition to the blue tails, at least one *Sphaerodactylus* (*elegans*) sports a bright red tail in juveniles. Basic color change is present but underrepresented (2.9% v. 6.5%). Males are often smaller than females (–0.40 v. 0.04 on –2 to 2 scale), as is often the case with the smaller species. Sexual dichromatism is common (42% v. 35%). Seasonal color is rare (1.0% v. 2.8%). The m:f ratio of relative tail lengths indicates

Sphaerodactylus spp.	English name	Activity	Diet	Size mm	Size g	Range core	Ecological business model
alphus	Guanaja Ocellated Sphaero			39	1.42	Honduras	
altavelensis	Alto Velo Sphaero	diurnal	invert.	30	0.69	Hispaniola	*** litter cryptozoid
argivus	Cayman Sphaero			27	0.52	Cayman Islands	** litter cryptozoid
argus	Ocellated Sphaero	diurnal	invert.	31	0.74	Nicaragua	*** litter cryptozoid
ariasae	Jaragua Sphaero			20	0.21	Dominican Republic	*** litter cryptozoid
armasi	Guantanamo Sphaero			32	0.84	Cuba	
armstrongi	Armstrong's Sphaero			32	0.79	Hispaniola	* litter cryptozoid
asterulus	Haiti Sphaero		invert.	32	0.79	Haiti	* litter cryptozoid
beattyi	Saint Croix Sphaero			31	0.74	US Virgin Islands	*** litter cryptozoid
becki	Beck's Sphaero	cathemeral		32	0.79	Navassa Island	** litter cryptozoid
bromeliarum	El Yunque Sphaero	cathemeral	invert.	30	0.64	Cuba	
caicosensis	Caicos Sphaero			32	0.84	Turks and Caicos Islands	** litter cryptozoid
callocricus	Callous Sphaero			29	0.64	Dominican Republic	** litter cryptozoid
celicara	Baracoan Eyespot Sphaero			32	0.84	Cuba	* ground-foraging tree lizard
cinereus	Gray Sphaero	crepuscular		34	0.97	Haiti	** trunk/wall lizard (elect. lights) or litter cryptozoid
clenchi	Peninsula Sphaero	crepuscular	invert.	34	0.96	Dominican Republic	*** litter cryptozoid
cochranae	Cochran's Sphaero	nocturnal		31	0.74	Dominican Republic	** crack hermit
continentalis	Continental Sphaero			30	0.65	Guatemala	
copei	Cope's Sphaero			39	1.44	Hispaniola	** litter cryptozoid

Sphaerodactylus spp.	English name				Range core
			Size		
	Activity	Diet	mm	g	Ecological business model
corticola	Rum Cay Sphaero				Bahamas
	nocturnal	invert.	37	1.29	** nocturnal ground lizard or litter cryptozoid
cricoderus	Turquino Collared Sphaero				Cuba
			27	0.51	** litter cryptozoid
cryphius	Bahoruco Mountains Sphaero				Dominican Republic
			25	0.40	* litter cryptozoid
dacnicolor	Eastern Jamaica Sphaero				Jamaica
			32	0.84	*** litter cryptozoid
darlingtoni	Darlington's Sphaero				Dominican Republic
	diurnal		30	0.69	
difficilis	Difficult Sphaero				Hispaniola
	diurnal	invert.	32	0.81	*** litter cryptozoid
dimorphicus	Santiago de Cuba Sphaero				Cuba
			33	0.92	* litter cryptozoid
docimus	Cabo Cruz Banded Sphaero				Cuba
			31	0.74	
dunni	Dunn's Sphaero				Honduras
	diurnal		29	0.64	*** litter cryptozoid
elasmorhynchus	Marche Leon Sphaero				Haiti
			22	0.25	
elegans	Ashy Sphaero				Hispaniola
	crepuscular	invert.	34	0.94	*** trunk/wall lizard or ground-foraging tree lizard
elegantulus	Antigua Sphaero				Antigua and Barbuda
	nocturnal	invert.	30	0.69	** litter cryptozoid
epiurus	Hispaniola Spotted-tail Sphaero				Dominican Republic
			27	0.51	* trunk/wall lizard
exsul	Swan Islands Sphaero				Honduras
			27	0.47	
fantasticus	Fantastic Sphaero				Guadeloupe
	crepuscular	invert.	30	0.69	** litter cryptozoid
gaigeae	Chevronated Sphaero				Puerto Rico
			25	0.41	*** litter cryptozoid
gilvitorques	Jamaica Sphaero				Jamaica
			29	0.59	
glaucus	Sleek Sphaero				Guatemala
	cathemeral	invert.	30	0.65	* ground-foraging tree lizard
goniorhynchus	Cabarita Sphaero				Jamaica
			32	0.84	*** litter cryptozoid

that females have the relatively longer tails (0.96 v. 1.02 ratio).

Generalized invertebrate consumption is not quite universal (96% v. 82%) because of the ant specialists (see above). Electric light use is about average (5.7% v. 6.9%). Home-range areas are unknown, but net displacement over 10 months is 3–5 m in *vincenti*.

All four diel activity patterns are present in substantial proportions; only diurnal is underrepresented: cathemeral (20% v. 15%), crepuscular (15% v. 6%), diurnal (34% v. 62%), and nocturnal (32% v. 17%). Cryptozoic life may allow flexibility in activity patterns. Mean active body temperature is average (30.0°C v. 30.4°C). Group living is common (50% v. 17%), although the definition of this may be vague for litter dwellers. Male social tolerance is high (21% v. 14%). Male spatial defense is about average (79% v. 86%).

The elevational profile of the genus is low (l/m/h: 100/5/0 v. 96/20/4). They are about average in representation in littoral (40% v. 31%), sandy (4% v. 15%), riparian (12% v. 19%), and human (35% v. 33%) habitats. Reproduction is normal.

Longevity is short (−0.79 v. −0.15 std. resid.), ranging from 1.3 years (*savagei*) to 4.0 years (*vincenti*). The longevity pattern is presently based on only eight species and will no doubt increase in magnitude as more species are sampled. Average population density is high (0.62 kg/ha v. 0.32 kg/ha), based on 14 credible samples of six species. Endemism is average (74% v. 65%). Most species occur on at least one island (95% v. 64%). Introductions are about average for the family (4.8% v. 5.1%), including 2.9% (v. 2.3%) of species introduced more than twice. Most *Sphaerodactylus* are imperiled (60% v. 41%), and 36% (v. 23%) are on one of the IUCN's lists of the most endangered. This high level of endangerment likely reflects the severe damage that has been done to West Indian Islands by the dense human development and the many introduced predators. Many species have distributions limited to a single or small number of islands.

In 1919 in Cuba, *S. elegans* was reported to often lurk behind pictures of "the Virgin"; is that because such images predominated on Cuban walls in 1919, or did the lizard express a preference? In Costa Rica *millepunctatus* is widely believed to be venomous, though there is disagreement as to whether it injects the venom through biting or stinging with its tail tip. Scorpion-like envenomation is also greatly feared from *glaucus* (Guatemala).

Sphaerodactylus spp.	English name				Range core
			Size		
	Activity	Diet	mm	g	Ecological business model
graptolaemus	Costa Rico Sphaero				Costa Rica
	cathemeral		32	0.84	** litter cryptozoid or ground-foraging tree lizard
guanajae	Guanaja Sphaero				Honduras
			28	0.56	
heliconiae	Santa Marta Sphaero				Colombia
			32	0.80	
homolepis	Yellow-tailed Sphaero				Costa Rica
	diurnal		33	0.90	
inaguae	Inagua Sphaero				Bahamas
	nocturnal	invert.	30	0.69	*** trunk/wall lizard
intermedius	Mantanzas Sphaero				Cuba
			35	1.08	* litter cryptozoid
kirbyi	Grenada Bank Sphaero				Grenada
	diurnal		26	0.44	*** litter cryptozoid
klauberi	Klauber's Sphaero				Puerto Rico
	diurnal	invert.	36	1.15	*** litter cryptozoid
ladae	Martin Garcia Sphaero				Dominican Republic
	nocturnal		29	0.59	* crack hermit
lazelli	Cap-Haitien Sphaero				Haiti
			32	0.79	
leonardovaldesi	Roatan Sphaero				Honduras
			30	0.69	* litter cryptozoid
leucaster	Eastern Sphaero				Dominican Republic
	nocturnal		32	0.79	* nocturnal ground lizard
levinsi	Desecheo Sphaero				Puerto Rico
			29	0.64	* litter cryptozoid
lineolatus	Panama Sphaero				Colombia
			32	0.84	*** litter cryptozoid
macrolepis	Puerto Rico Sphaero				Puerto Rico
	crepuscular	invert.	25	0.40	**** litter cryptozoid
mariguanae	Mayaguana Sphaero				Bahamas
			39	1.44	* litter cryptozoid
microlepis	Little-scaled Sphaero				St. Lucia
		invert.	34	0.96	*** litter cryptozoid
micropithecus	Monito Sphaero				Puerto Rico
			35	1.08	** litter cryptozoid
millepunctatus	Spotted Sphaero				Central America
	diurnal	ants	35	1.04	** ground-foraging tree lizard
molei	Tobago Sphaero				Venezuela
	diurnal	invert.	29	0.64	** ground-foraging tree lizard

Sphaerodactylus spp.	English name				Range core
			Size		
	Activity	Diet	mm	g	Ecological business model
monensis	Mona Sphaero				Puerto Rico
	cathemeral	invert.	30	0.71	*** litter cryptozoid
nicholsi	Nichols' Sphaero				Puerto Rico
	diurnal	invert.	24	0.34	**** litter cryptozoid
nigropunctatus	Black-spotted Sphaero				Cuba
			37	1.25	** litter cryptozoid
notatus	Reef Gecko				Cuba
	crepuscular	invert.	29	0.62	*** litter cryptozoid
nycteropus	Morne Dubois Sphaero				Haiti
			24	0.37	* litter cryptozoid
ocoae	Peravia Sphaero				Dominican Republic
			36	1.15	** litter cryptozoid
oliveri	Juventud Sphaero				Cuba
	diurnal		32	0.83	** litter cryptozoid
omoglaux	Fond Parisien Sphaero				Haiti
			24	0.34	** litter cryptozoid
oxyrhinus	Jamaica Sharp-nosed Sphaero				Jamaica
			32	0.81	** crack hermit
pacificus	Cocos Sphaero				Costa Rica
	nocturnal	invert.	44	2.17	*** litter cryptozoid or nocturnal ground lizard
parkeri	Parker's Sphaero				Jamaica
	nocturnal		34	1.02	** litter cryptozoid
parthenopion	Virgin Gorda Sphaero				British Virgin Islands
			20	0.21	** litter cryptozoid
parvus	Anguilla Bank Sphaero				Lesser Antilles
	nocturnal	invert.	31	0.72	* crack hermit
perissodactylius	Dominican Sphaero				Dominican Republic
			26	0.44	** litter cryptozoid
phyzacinus	Saintes Sphaero				Guadeloupe
			27	0.51	** litter cryptozoid
pimienta	Pepper Sphaero				Cuba
			35	1.04	* litter cryptozoid
plummeri	Barahona Big-scaled Sphaero				Dominican Republic
	diurnal		25	0.40	** litter cryptozoid
poindexteri	Poindexter's Sphaero				Honduras
			31	0.72	* litter cryptozoid
ramsdeni	Ramsden's Sphaero				Cuba
			33	0.71	* crack hermit
randi	Pedernales Sphaero				Dominican Republic
			31	0.76	** litter cryptozoid

Sphaerodactylus spp.	English name				Range core
			Size		
	Activity	Diet	mm	g	Ecological business model
rhabdotus	Valle de Neiba Sphaero				Dominican Republic
			32	0.84	** litter cryptozoid
richardi	Zapata Big-scaled Sphaero				Cuba
			31	0.72	** litter cryptozoid
richardsonii	Richardson's Sphaero				Jamaica
			38	1.36	** litter cryptozoid
roosevelti	Roosevelt's Sphaero				Puerto Rico
	nocturnal		36	1.17	** litter cryptozoid
rosaurae	Bay Island Sphaero				Honduras
	cathemeral		37	1.29	* ground-foraging tree lizard
ruibali	Ruibal's Sphaero				Cuba
			32	0.84	** litter cryptozoid
sabanus	Saba Sphaero				St. Kitts and Nevis
	diurnal		31	0.74	**** litter cryptozoid
samanensis	Samana Bay Sphaero				Dominican Republic
			30	0.69	*** litter cryptozoid
savagei	Altagracia Speckled Sphaero				Dominican Republic
	nocturnal		32	0.85	** litter cryptozoid
scaber	Camaguey Sphaero				Cuba
			32	0.80	** litter cryptozoid
scapularis	Boulenger's Sphaero				Colombia
			33	0.90	
schuberti	Neiba Agave Sphaero				Dominican Republic
			33	0.90	** litter cryptozoid
schwartzi	Schwartz's Sphaero				Cuba
			24	0.34	** litter cryptozoid
semasiops	Cockpit Eyespot Sphaero				Jamaica
			30	0.65	*** crack hermit
shrevei	Shreve's Sphaero				Haiti
			31	0.74	
siboney	Siboney's Sphaero				Cuba
			32	0.85	** litter cryptozoid
sommeri	Terre Nueve Sphaero				Haiti
	nocturnal		34	1.02	*** nocturnal ground lizard
sputator	Leeward Islands Sphaero				Lesser Antilles
	cathemeral		33	0.87	** ground-foraging tree lizard or litter cryptozoid
storeyae	Canarreos Sphaero				Cuba
			31	0.76	** litter cryptozoid
streptophorus	Hispaniola Sphaero				Hispaniola
			28	0.55	*** litter cryptozoid

Sphaerodactylus spp.	English name / Activity	Diet	Size mm	Size g	Range core / Ecological business model
thompsoni	Barahona Limestone Sphaero				Hispaniola
	nocturnal		33	0.90	** ground-foraging tree lizard or litter cryptozoid
torrei	Cuba Broad-banded Sphaero				Cuba
			37	1.21	** litter cryptozoid
townsendi	Townsend's Sphaero				Puerto Rico
			28	0.53	*** litter cryptozoid
underwoodi	Turk's Island Sphaero				Turks and Caicos Islands
			32	0.84	
vincenti	Vincent's Sphaero				Lesser Antilles
	cathemeral	invert.	29	0.60	*** litter cryptozoid or crack hermit
williamsi	Haiti Striped Sphaero				Haiti
			25	0.40	
zygaena	Dame-Marie Sphaero				Haiti
			32	0.84	

Sphenodontidae and *Sphenodon* (figs. 2.5B, 4.11C)

Sphenodon punctatus, or Tuatara, is the sole living member of its genus, family, suborder, and order (Rhynchocephalia, the outgroup to Squamata). For a brief time a second species was recognized, but all populations are now judged to be part of the same gene pool. Because of its unique phylogenetic position, it has been much studied and is very well known (filled cells 71% v. 27%; see especially Cree 2014). Ecologically, it differs little from larger lizards (*Sphenodon* is 200 g, the 98th percentile of lizard mass, v. 5.3 g), save that it occurs in relatively cool locations (air temperature range: 7°C–20.5°C) in temperate New Zealand and has evolved a life history that keeps pace with the low temperatures and predator-free environment in which it is found. Though the longevity record for this species is currently about 83 years (i.e., as long as good records have been kept), most Tuatara researchers believe that much longer periods will be documented with additional time. Its longevity already exceeds that of any true lizard.

The pace of its life is commensurate with its longevity; a captive Tuatara was reported in a century-old government report to have held its position, as outlined by pencil on the floor, for six weeks. Tuatara use a number of burrows, but the mean distance moved between burrows is <1 m. Territory locations may be maintained for 40–50 years. Ingestion of a large food item (e.g., sparrow) may take 2 hours. Egg development usually takes 4–5 years, gestation—ovulation to oviposition—about 275 days, and incubation about 405 days. Tuatara oviposit about every fourth year. Survival is excellent on the predator-free islands: 8 of 24 marked in 1957 were recaptured in 1988. While one should never underestimate the patience of a reptile, Tuatara convey a strong impression that they are physiologically marching to the beat of a distinctly languid drummer.

Tuatara were present throughout New Zealand at the time of initial human colonization, but human commensals (especially rats) quickly eliminated them from the larger New Zealand islands, the New Zealand "mainland." Most field data have been collected from a few small outlier islands in Cook Strait (41° S), which separates the large New Zealand islands. However, the New Zealanders ("Kiwis") take good care of their national icons and are repatriating Tuataras to many of their original locations after introduced predators have been eradicated. These new Tuatara populations will eventually provide field data on natural populations in much warmer areas (down to 34° S).

Tuatara have proportionately large heads, but the species otherwise resembles an average lizard. Its shape index (1.60, second-highest family, v. 0.0 std. resid.) reflects their bulky build. Their breakable tails are slightly short (1.10 v. 1.37 × SVL) but not prehensile (v. 5%) or blue (v. 2.3%). They possess crests (v. 5.3%) but lack secretory pores (v. 15%), elongation (v. 11%), and any color change (v. 11%).

Males are much larger than females (2.00 v. 0.36 on –2 to 2 scale) and have larger crests, larger and more angular heads, and more slender abdomens. The sexes differ in color (v. 21%) but do not show seasonal color (v. 6.5%). These sex differences roll up to 4.00 (v. 0.36 on 0 to 6 scale), higher than any lizard. As the strongly dimorphic Tuatara is the only living representative of the outgroup to lizards, one might be tempted to infer that the basal lizard was strongly differentiated between the sexes. Given the lability of such traits, and the 240–250 million years that the Tuatara has been on its own

Sphenodon sp.	English name		Size		Range core
	Activity	Diet	mm	g	Ecological business model
punctatus	Tuatara				New Zealand
	cathemeral	omnivore	180	200	*** refugia-anchored diurnal gr. liz.

evolutionary trajectory, this inference strikes me as poorly supported. This inferential weakness arises repeatedly in inferring the basal condition of many lizard traits.

Many extant Tuatara are found on islands used by nesting seabirds; on average, 45% of the carbon in a male Tuatara is derived from seabirds, their nestlings, commensals, offal, and so on. Tuatara are especially fond of beetles but will also eat leaves, fruit, soft invertebrates, arthropods, snails, ants, orthopterans, other insects, ectothermic vertebrates, and immobile (eggs, nestlings) endothermic vertebrates. They forage by both sit-and-wait (v. 46%) and active (v. 58%) movements, though their home-range area is remarkably small (–1.99, second-lowest family, v. 0.0 std. resid.). This equates to a home range of about 14 m^2 for females and 16–87 m^2 for males. Total daily movements are less than 63 m. The Tuatara does not use electric lights.

The Tuatara is active at any time of day (v. 6.8%), though adult locomotion for foraging is predominantly at night, especially when their burrows are surrounded by more open areas. Therefore this species is not an archetypical refugia-anchored "diurnal" ground lizard; in assigning an ecological business model, I chose to emphasize its very strong attachment to its refuge rather than its conditional attachment to nocturnality. However, Tuatara are arguably crack hermits with nocturnal forays or nocturnal ground lizards with strong attachment to their refugia. The absence of natural predators on adult Tuatara leaves ambiguous many natural history traits of this species; for example, which is their primary predator concealment tactic where there is no regular predator? Would Tuatara's foraging forays be smaller if a predator were present? This unique insular non-lizard shows that lizard ecological business models are not always discrete.

Sphenodon does not call spontaneously (v. 2.5%), but it does give stress calls (v. 2.2%). Juveniles are conspicuously wary (v. 8.2%). Tuatara bask directly (v. 18%). Mean active body temperature is very low (17.8°C v. 31.1°C). Activity is reduced in the winter, but the Tuatara is not found in areas that freeze hard.

The Tuatara is generally solitary (v. 10%); it defends (v. 85%) only the burrow it is in. Its primary predator concealment tactic, at least from humans, is retreat to a burrow (v. 25%).

The Tuatara is found in the intermediate rainfall band (v. 27%) at low (0–722 m) elevation (v. 82%). It occurs in the littoral zone (v. 17%) but not in sandy areas (v. 22%), riparian areas (v. 22%), or human habitats (v. 18.5%).

Many aspects of Tuatara reproduction are novel (see first paragraph). *Sphenodon* lacks an intromittent organ, and therefore they oppose cloacae for up to 120 minutes. The male holds the female with his arms and does not use a neck or shoulder bite. Mass-adjusted clutch count is slightly low (–0.19 v. 0.0 std. resid.), equating to about 8.4 eggs. Sex determination is temperature-sensitive, male offspring dominating at higher temperatures. The eggs are soft (v. 41% hard) and not adherent (v. 2.4% adherent) but laid communally (v. 4.2%). The reasons for communal oviposition are believed to be common attraction to areas of friable warm soil. Females migrant a short distance—about 84 m, but up to 200 m—to oviposit, and they defend their clutch for a few days, usually 3 but up to 14 days. Incubation period is without peer in lizards (5.66 v. 0.0 std. resid.). The resultant young are notably small (ontogenetic mass range 1.99, the second highest in "lizards," v. 0.0 std. resid.), equating to hatchlings that weigh 1/625 (v. 1/44) of maximum mass. Adult Tuatara, especially the very large adult male Tuatara, consume juveniles; no wonder juveniles are conspicuously wary.

In addition to unmatched longevity, the population density achieved by Tuatara (mean of 259 kg/ha v. 0.54 kg/ha, based on 15 qualified estimates) is without saurian peer (and likely without squamate peer, though I have not tabulated snake densities). Lack of predation, low metabolic requirements, and extra energy inputs from seabirds appear to be the major contributing factors. The Tuatara has not been introduced (v. 3.3%), except to repatriate it to localities from which it was previously extirpated. It is considered to be imperiled (v. 49%), but it is not on any of the IUCN's lists of the most endangered (v. 14%). In other respects *Sphenodon* is ecologically similar to lizards.

Sphenomorphus (fig. 2.7)

Sphenomorphus is the largest genus (110 spp., table 2.2) of the Scincidae, and it is widespread in Southeast Asia and Australasia, especially New Guinea, but anywhere within the following polygon: Nepal to China,

Sphenomorphus spp.	English name				Range core
			Size		
	Activity	Diet	mm	g	Ecological business model
acutus	Pointy-headed Sphenomorph				Philippines
	diurnal		65	5.6	*** ground-foraging tree lizard
aignanus	Misima Forest Skink				New Guinea
			67	5.8	
alfredi	Alfred's Forest Skink				Malaysia
			33	0.46	
annectens	Pug-nosed Skink				New Guinea
			44	1.49	
anomalopus	Long-toed Forest Skink				Indonesia
			59	4.01	
anotus	Masba Creek Earless Skink				New Guinea
			33	0.46	** litter cryptozoid
apalpebratus	Wide-eyed Litterskink				India
	diurnal		39	0.75	** litter cryptozoid
bacboensis	Bacbo Skink				Vietnam
			39	1.07	
bignelli	Slot Skink				Solomon Islands
	diurnal	invert.	34	0.52	** litter cryptozoid
brunneus	Plain Brown Skink				New Guinea
	nocturnal		72	3.86	**** litter cryptozoid
buenloicus	Buen Loi Forest Skink				Vietnam
	diurnal	invert.	50	1.45	*** litter cryptozoid
buettikoferi	Buettikofer's Forest Skink				Indonesia
			34	0.69	
cameronicus	Cameron Highlands Forest Skink				Malaysia
	diurnal		59	2.32	** litter cryptozoid
capitolythos	Kopstein's Headstone Skink				Indonesia
			67	3.26	
celebense	Sulawesi Skink				Indonesia
			52	2.66	
cinereus	Ash-colored Skink				New Guinea
			84	6.1	* litter cryptozoid
concinnatus	Elegant Forest Skink				Solomon Islands
	diurnal	invert.	62	4.49	**** litter cryptozoid
consobrinus	Bacan Skink				Indonesia
			37	0.84	
cophias	Tahan Mountain Forest Skink				Malaysia
			42	0.65	
courcyanum	Medog Skink				Himalayas
	diurnal	invert.	45	1.09	** litter cryptozoid

Taiwan, Palau, Solomon Islands, New Guinea, Indonesia, Sri Lanka; one species (*rarus*) occurs in Panama. In the recent past there were many more species in the genus, including some or all that are now in *Calyptotis*, *Concinnia*, *Ctenotus*, *Eremiascincus*, *Eulamprus*, *Glaphyromorphus*, *Hemiergis*, *Insulasaurus*, *Lankascincus*, *Lipinia*, *Lygosoma*, *Oligosoma*, *Otosaurus*, *Parvoscincus*, *Pinoyscincus*, *Scincella*, *Silvascincus*, *Tropidophorus*, *Tumbunascincus*, and *Tytthoscincus*. This litany suggests that further splitting is to be expected and that species found to be missing from the following tabulation of *Sphenomorphus* are likely to be found in one of the above genera.

Sphenomorphus is large and diverse, and there are few ecological characteristics that unite the genus and distinguish it from the family. They are slightly small (2.6 g v. 3.3 g) but include a wide range of sizes (0.38–79 g). They forage for invertebrates (100% v. 84%) mostly on the ground: terrestrial (57% v. 48%) or cryptozoic (33% v. 30%). They are found at the wet end of the rainfall spectrum: mesic (50% v. 40%), wet (50% v. 7%). They are notably overrepresented in two ecological business models: litter cryptozoids (67% v. 28%) and ground-foraging tree lizards (15% v. 7%). Four other ecological business models are also found within *Sphenomorphus*: refugia-anchored diurnal ground lizard (12.5% v. 30%), crack hermit (2.1% v. 1.7%), nocturnal ground lizard (2.1% v. 2.3%), and subterranean forager (2.1% v. 12%). Some species lead a plethodontid-like existence, residing primarily under rocks near streams. The genus is poorly known (filled cells 21% v. 25%); there is some indication of diet in only 18% of the species. A statement regarding their diel activity pattern is available for just 35%.

Sphenomorphus have no morphological features to distinguish them from other skinks, except that blue tails are not reported (v. 4%), though male *dussumieri* may have red tails. The recorded sex differences are almost exactly balanced between m>f and m<f size differences (−0.07 v. −0.31 on −2 to 2 scale). Dichromatism matches the family closely (5.5% v. 7.6%). Seasonal color is comparatively rare in *Sphenomorphus* (2.7% v. 6.2%). The m:f relative tail length ratio within *Sphenomorphus* is the sixth highest among Scincidae, but it is not all that aberrant (1.08 v. 1.03 ratio). The overall sex differences rollup score (on scale of 0 to 6) is therefore a paltry 0.12 (v. 0.28).

Cruise foraging is small but overrepresented (10% v. 2%). Sit-and-wait is slightly underrepresented (10% v. 16%); active foraging is universal (100% v. 92%). *S. maculatus* eats large enough meals that it will be inactive for several days after gorging. Electric light use is unknown (v. 0.4%). Home-range area has not been estimated for any species in the genus. The mix of diel activity patterns is not far removed from the family average, but crepuscular is overrepresented (10% v. 2.4%) and nocturnal is underrepresented (7.5% v. 14.5%). Conspicuous wariness is uncommon (5.5% v. 11%). The overrepresented primary predator concealment tactics are retreat to vegetation (43% v. 25%) and crypsis (14% v. 9%). Flight (v. 4%) and sand diving (v. 22%) are not present. Direct basking is rare (4.5% v. 21%); thermoconformance is uncommon but overrepresented (3.6% v. 1.5%). Mean active body temperature is about average (32.0°C v. 31.1°C). *S. sabanus* has an unusual resting posture: dangling from a small branch by its hind claws. Is this twig mimicry?

There is very little information on social relations: group living (0%

Sphenomorphus spp.	English name				Range core
			Size		
	Activity	Diet	mm	g	Ecological business model
cranei	Crane's Skink				Solomon Islands
	crepuscular	invert.	62	2.64	**** litter cryptozoid
crassus	Thick Forest Skink				Malaysia
			68	3.35	* refugia-anchored diurnal gr. liz.
cryptotis	Depressed-ear Forest Skink				Vietnam
	diurnal		69	6.4	* refugia-anchored diurnal gr. liz.
cyanolaemus	Blue-throated Forest Skink				Borneo
	diurnal		52	2.60	**** ground-foraging tree lizard
darlingtoni	Darlington's Forest Skink				New Guinea
			55	1.89	
derooyae	De Rooij's Forest Skink				New Guinea
			70	3.65	
diwata	Diwata Mountains Sphenomorph				Philippines
			53	2.80	*** litter cryptozoid
dussumieri	Dussumier's Forest Skink				India
	diurnal	invert.	55	3.15	
fasciatus	Banded Sphenomorph				Borneo
			92	7.7	*** litter cryptozoid
forbesi	Forbes' Forest Skink				New Guinea
			52	1.57	*** litter cryptozoid
fragilis	Hall Sound Skink				New Guinea
	cathemeral		48	1.29	** nocturnal ground lizard (juveniles are litter cryptozoids)
fragosus	Fragmented Skink				New Guinea
			60	4.18	
fuscolineatus	Brown-lined Forest Skink				New Guinea
			52	2.54	*** litter cryptozoid
grandisonae	Grandison's Forest Skink				Thailand
			31	0.38	* litter cryptozoid
granulatus	Albert Edward Skink				New Guinea
			59	4.01	
haasi	Haas' Forest Skink				Borneo
	diurnal		50	2.32	** ground-foraging tree lizard
helenae	Notaburi Forest Skink				Thailand
			70	3.50	
incognitus	South China Sea Sphenomorph				China
	diurnal	invert.	85	12.7	**** litter cryptozoid

Sphenomorphus spp.	English name	Activity	Diet	Size mm	Size g	Range core	Ecological business model
indicus	Indian Forest Skink	diurnal	invert.	76	4.51	East Asia	*** refugia-anchored diurnal gr. liz. or litter cryptozoid (geog. var.)
jobiensis	New Guinea Ground Skink	crepuscular	invert.	92	16.6	New Guinea	*** refugia-anchored diurnal gr. liz.
kinabaluensis	Kinabalu Forest Skink	diurnal	invert.	52	2.54	Malaysia	*** litter cryptozoid
latifasciatus	Japen Sphenomorph			149	78	Indonesia	
leptofasciatus	Central Highlands Skink	crepuscular		71	3.76	New Guinea	** litter cryptozoid
lineopunctulatus	Spot-lined Forest Skink			69	6.6	Southeast Asia	
longicaudatus	Lorentz River Long-tailed Skink			77	2.82	New Guinea	**** litter cryptozoid
loriae	Mid-mountain Forest Skink			67	3.16	New Guinea	
louisiadensis	Rossel Forest Skink			44	1.57	New Guinea	
maculatus	Spotted Forest Skink	diurnal	invert.	56	3.30	Southern East Asia	**** refugia-anchored diurnal gr. liz.
maculicollus	White-throated Forest Skink			44	1.57	Malaysia	
maindroni	New Guinea Mystery Skink			63	4.74	New Guinea	
malaisei	Demotivating Skink			41	0.85	Myanmar	
malayanum	Malay Forest Skink			56	1.95	Indonesia	
megalops	Annandale's Litterskink	diurnal	invert.	48	1.95	Sri Lanka	* litter cryptozoid
melanopogon	Lesser Sundas Forest Skink	cathemeral	invert.	74	8.0	Indonesia	
meyeri	Wokan Skink			112	31	New Guinea	
microtympanus	Small-eared Forest Skink			42	0.87	New Guinea	
mimicus	Dwarf Forest Skink	diurnal		39	1.07	Thailand	** litter cryptozoid

v. 16%), male social tolerance (0% v. 5.6%), and male hierarchies (0% v. 5.6%) are unknown. The only documented pattern is male spatial defense (100% v. 94%).

There are a few arboreal *Sphenomorphus* (6.6% v. 10%) and even one semiaquatic species (*cryptotis* 1.3% v. 2.1%). This species readily enters water and sleeps on vegetation overhanging water, but the microhabitat distribution of its foraging forays is not known. Few *Sphenomorphus* sleep in burrows (3.0% v. 24.2%). The elevational profile of *Sphenomorphus* is slightly skewed toward medium elevations (l/m/h: 85/49/2 v. 89/35/6). They are scarce in littoral (5% v. 21%) and sandy (1% v. 25%) areas but representative in riparian (24% v. 23%) and human (15% v. 14%) habitats.

The distribution of reproductive modes within *Sphenomorphus* is skewed toward oviparity (84% v. 64%); viviparity is present exclusively or optionally in seven species (6% v. 36%). Incubation in the viviparity-optional species (e.g., *fragilis*) may be very short, as the eggs are in an advanced stage of development when laid. *S. cyanolaemus* and *kinabaluensis* oviposit in ant nests. Mean clutch/litter count is normal (0.12 v. 0.06 std. resid.). Young are moderately small (ontogenetic mass range 0.49 v. 0.12 std. resid.). Communal oviposition is very rare (0.9% v. 2.6%). Female *maculatus* are receptive over a period of 2–3 days, during which time they mate with one to three males for 1–2 minutes each.

Longevity information is available only for *indicus* (6 years) and *maculatus* (1 year). In the unlikely event that those are representative of the genus, *Sphenomorphus* are comparatively short-lived (–0.91 v. 0.17 std. resid.). Endemism is marginally low (55% v. 64%). Insular species are high (80% v. 57%). One *Sphenomorphus* (*indi-*

cus) has been introduced (0.9% v. 2.1%: to China); no species has been introduced more than once. A very low percentage is imperiled (7% v. 47%), though one wonders what role ignorance of their status plays in this assessment. No species is on one of the IUCN's lists of the most endangered (v. 14%).

Sphenomorphus spp.	English name	Activity	Diet	Size mm	Size g	Range core	Ecological business model
mimikanum	Mimika Forest Skink			74	8.0	Indonesia	
minutus	Diminutive Forest Skink			39	0.71	New Guinea	
misolense	Misool Island Skink			42	1.34	Indonesia	
modigliani	Sipora Island Skink			39	1.01	Indonesia	
muelleri	Mueller's Forest Skink	nocturnal		156	16.5	New Guinea	
multisquamatus	Graph-paper Forest Skink	diurnal	invert.	60	4.18	Borneo	* litter cryptozoid
murudensis	Mount Murud Forest Skink	diurnal		45	1.65	Borneo	
necopinatus	Bogor Forest Skink	diurnal		41	1.20	Indonesia	
neuhaussi	Sattelberg Forest Skink			105	11.1	New Guinea	
nigriventris	Hellwig Mountains Skink			74	4.19	New Guinea	*** litter cryptozoid
nigrolabris	Black-lipped Forest Skink			77	9.4	Indonesia	
nigrolineatus	Black-lined Forest Skink			63	2.72	New Guinea	
oligolepis	Small-scaled Forest Skink			49	1.34	New Guinea	
orientalis	Mount Victoria Skink			48	2.02	Myanmar	
papuae	Mount Lamington Forest Skink			67	5.8	New Guinea	
praesignis	Blotched Forest Skink	diurnal		96	18.8	Thailand	** crack hermit
pratti	Pratt's Forest Skink		invert.	77	4.77	New Guinea	* subterranean forager
puncticentralis	Black-spotted Forest Skink	diurnal		42	1.27	Indonesia	
rarus	Guabo River Skink			47	1.83	Panama	*** litter cryptozoid
sabanus	Sabah Forest Skink	diurnal	invert.	51	2.49	Borneo	*** ground-foraging tree lizard

Sphenomorphus spp.	English name				Range core
			Size		
	Activity	Diet	mm	g	Ecological business model
sananus	Sula Islands Skink				Indonesia
			32	0.56	
sanctus	Yellow-striped Forest Skink				Indonesia
	diurnal		51	2.51	
sarasinorum	Sarasin's Skink				Indonesia
			64	5.0	
schlegeli	Komodo Island Skink				Indonesia
			33	0.60	
schultzei	Schultze's Forest Skink				New Guinea
			43	0.95	* litter cryptozoid
scotophilus	Selangor Forest Skink				Malaysia
	diurnal	invert.	53	2.83	**** ground-foraging tree lizard
scutatus	Palau Ground Skink				Palau
			39	0.74	*** litter cryptozoid
senja	Alpenglow Skink				Malaysia
			56	1.95	
sheai	Shea's Forest Skink				Vietnam
			35	0.70	* litter cryptozoid
shelfordi	Mount Penrisen Skink				Borneo
			59	4.01	
simus	Common Lowland Skink				New Guinea
	diurnal		51	2.43	* ground-foraging tree lizard
solomonis	Solomon Ground Skink				New Guinea
	cathemeral	invert.	59	2.23	**** litter cryptozoid
stellatus	Starred Forest Skink				Southeast Asia
	diurnal	invert.	67	5.8	*** ground-foraging tree lizard
striolatus	Flores Forest Skink				Indonesia
			49	2.11	
sungaicolus	Malay Water Skink				Malaysia
			74	4.19	* refugia-anchored diurnal gr. liz.
taiwanensis	Mount Hohuan Skink				Taiwan
			56	3.26	
tanahtinggi	Montane Forest Skink				Malaysia
			55	3.15	
tanneri	Tanner's Skink				Solomon Islands
			43	0.96	** litter cryptozoid
taylori	Taylor's Solomon Skink				New Guinea
			134	56	

Sphenomorphus spp.	English name	Activity	Diet	Size mm	Size g	Range core	Ecological business model
tenuiculus	Narrow-necked Forest Skink			52	2.66	Borneo	
tersus	Nakhon Si Thammarat Forest Skink	cathemeral		78	9.7	Malaysia	
tetradactylus	Four-fingered Skink	nocturnal		35	0.41	Vietnam	
tonkinensis	Tonkin Forest Skink			47	1.92	Vietnam	
transversus	Bougainville Barred Skink			58	3.71	Solomon Islands	
tridigitus	Three-toed Forest Skink			37	0.63	Vietnam	
tritaeniatus	Three-striped Forest Skink			43	1.41	Vietnam	
tropidonotus	Sulawesi Swamp Skink			53	2.78	Indonesia	
undulatus	Wavy-backed Forest Skink	crepuscular		49	1.35	New Guinea	* litter cryptozoid
vanheurni	Ijang Mountains Skink	diurnal		55	3.15	Indonesia	
variegatus	Vivid Skink			57	3.55	Borneo	** litter cryptozoid
wollastoni	Kamoro Forest Skink			74	4.19	Indonesia	
woodfordi	Woodford's Ground Skink	diurnal	invert.	106	11.3	New Guinea	*** litter cryptozoid
zimmeri	Mekongga Mountains Skink			60	4.17	Indonesia	

Spondylurus

Spondylurus is a moderate radiation (17 spp.) of large-sized (6.8 g v. 3.25 g), diurnal (100% v. 74%) skinks living in the West Indies, especially the Puerto Rican Bank. Almost all populations were treated as *sloanii* until Hedges and Conn (2012) split them. Unfortunately, many of the new species were already extirpated by mongoose or other human introductions at the time of their recognition. The following species are known from only a few specimens, and in most cases there are no recent observations: *anegadae*, *haitiae*, *lineolatus*, *martinae*, *monae*, *nitidus*, *spilonotus*, and most populations of *turksae*. Those species are therefore not represented in many of the following generalizations. Most *Spondylurus* are refugia-anchored diurnal ground lizards (75% v. 30%). The remainder are ground-foraging tree lizards (17% v. 7%) or litter cryptozoids (8% v. 28%).

Spondylurus differ from average

Spondylurus spp.	English name	Activity	Diet	Size		Range core	Ecological business model
				mm	g		
anegadae	Anegada Skink	diurnal		59	4.01	British Virgin Islands	
caicosae	Caicos Island Skink	diurnal		65	5.4	Turks and Caicos Islands	** ground-foraging tree lizard
culebrae	Culebra Skink			79	10.2	Puerto Rico	** refugia-anchored diurnal gr. liz.
fulgidus	Jamaica Skink	diurnal	carnivore	70	6.9	Jamaica	** ground-foraging tree lizard
haitiae	Hispaniola Four-lined Skink			70	6.8	Haiti	
lineolatus	Hispaniola Ten-lined Skink			55	1.89	Hispaniola	
macleani	Carrot Rock Skink			67	6.0	British Virgin Islands	* refugia-anchored diurnal gr. liz.
magnacruzae	Greater St. Croix Skink			86	13.2	US Virgin Islands	* refugia-anchored diurnal gr. liz.
martinae	Saint Martin Skink			69	6.4	St. Martin	
monae	Mona Skink			72	7.3	Puerto Rico	** refugia-anchored diurnal gr. liz.
monitae	Monito Skink			77	9.4	Puerto Rico	* refugia-anchored diurnal gr. liz.
nitidus	Puerto Rico Skink	diurnal	invert.	78	9.7	Puerto Rico	*** refugia-anchored diurnal gr. liz.
powelli	Anguilla Bank Skink			61	4.33	Anguilla	*** refugia-anchored diurnal gr. liz.
semitaeniatus	Lesser Virgin Islands Skink	diurnal		69	6.4	Virgin Islands	*** refugia-anchored diurnal gr. liz.
sloanii	Virgin Islands Bronze Skink	diurnal	omnivore	80	10.5	Virgin Islands	** refugia-anchored diurnal gr. liz.
spilonotus	Greater Virgin Islands Skink			86	13.2	US Virgin Islands	
turksae	Turks Islands Skink	diurnal		66	5.6	Turks and Caicos Islands	** litter cryptozoid

skinks in their build (shape index 0.15 v. –0.46 std. resid.), lack of elongation (6% v. 35%), and high frequency of blue tails (12% v. 4%). Blue tails are found in adult *lineolatus*, which is an unusual life stage for blue tails and may reflect mimicry of the syntopic blue-tailed *Ameiva*. All documented sexual size differences are m<f (–1.0 v. –0.31 on –2 to 2 scale). This suggests selection pressure for high fecundity, which is not manifest in litter count (see below). They do not show sexual dichromatism (v. 8%) or seasonal color (v. 6%). Their sex differences rollup score (on 0 to 6 scale) is mildly elevated (0.59 v. 0.28).

Spondylurus are diverse in diet: carnivores (33% v. 1%), invertebrate eaters (33% v. 84%), and omnivores (33% v. 13%). Foraging mode, electric light use, and home-range area are not known. The only species documented for primary predator avoidance tactic (*fulgidus*) retreats to crevices or burrows (v. 43%). The MPM tabulation of foraging *semitaeniatus* puts them in the 74th percentile of all lizards; for PTM they are also relatively active (61st percentile). Direct basking is common (35% v. 21%), as is thermoconformance (6% v. 1.5%). Three species (*lineolatus*, *monae*, *powelli*) have rainfall records; these are clustered in the intermediate band (100% v. 34%), and one species extends into the mesic band (33% v. 40%). They are overrepresented in littoral (33% v. 21%) and human (27% v. 14%) habitats, as may reflect habitat opportunities of those islands in general. *Spondylurus* are underrepresented in sandy areas (7% v. 25%) and riparian habitat (0% v. 23%).

All are viviparous (v. 36%) and have a mean litter count that is slightly low (–0.25 v. 0.06 std. resid.). As noted above, this does not seem to support the hypothesis that the females of *Spondylurus* are under great pressure to produce many offspring, though perhaps the offspring are exceptionally large at birth. There is no record of multiple litters per year. For *magnacruzae*, Hedges and Conn (2012) note that "two of the four specimens were collected while copulating." Presumably, it was the lizards that were copulating.

Endemism is high (71% v. 64%). None has been introduced (v. 2.1%). All are imperiled (v. 47%), and 88% (v. 14%) are on one of the IUCN's lists of the most endangered. In other respects *Spondylurus* are typical skinks.

Stellagama sp.	English name				Range core
			Size		
	Activity	Diet	mm	g	Ecological business model
stellio	Hardun Rock Agama				Mediterranean
	diurnal	omnivore	110	45	**** refugia-anchored diurnal gr. liz.

Stellagama

Stellagama is a single-species genus of stout (shape index 1.21 v. 0.80 std. resid.), large (45 g v. 15.6 g), crestless (v. 38% crested), terrestrial (v. 56%) agamid surrounding the eastern Mediterranean, from Greece and North Macedonia to Jordan, Saudi Arabia, and Egypt. It is a refugia-anchored diurnal ground lizard (v. 43%) and a particularly well-studied one (cells filled 68% v. 30%). It is sexually dichromatic (v. 53%), males show seasonal color (v. 20%), and both sexes can change hue (v. 15% advanced color change). SSD is somewhat variable, but males are usually larger, have larger canine teeth, longer hind limbs, larger heads, and larger secretory pores (v. 13%). These sex differences sum to the fourth-highest rollup score in Agamidae: 3.00 (v. 1.61 on 0 to 6 scale).

They can run bipedally but have difficulty climbing smooth vertical surfaces such as palm tree trunks. They have typical agamid tail lengths (1.47 v. 1.86 × SVL), and the tails break (v. 17%) but only with difficulty. Females are reported to be conspicuously wary, but males are not. Males defend their crevice-burrow refuge against other males (v. 97%), and in some cases may be busy, as colonies of up to 120 individuals have been reported (group living v. 12%). At high densities, territories transition to hierarchies (v. 18%); at low densities, male activity ranges do not overlap. Males may be accompanied by more than one female, though activity areas are not especially large (250 m^2 or –0.30 v. 0.35 std. resid.). Hardun Rock Agamas bask and have an average body temperature of around 36.5°C (v. 34.8°C) but may suspend midday activity during the hottest months to keep their temperature below 39°C. The species occurs from sea level to over 2600 m and occupies a wide variety of habitats, including littoral areas (v. 10%), dunes (v. 26%), riparian areas (v. 21%), and human habitats (v. 16%). *Stellagama* is an opportunistic feeder, including leaves, fruit, snails, ants, wasps, and vertebrates (especially hymenopterans).

The Hardun Rock Agama's clutch count is slightly high (0.37 v. 0.20 std. resid.), and young are exceptionally small (ontogenetic size range 2.41, the highest in Agamidae, v. 0.69 std. resid.). The wide ontogenetic size range indicates that the average young is 1/470 of the maximum size (v. 1/100 for Agamidae). Incubation duration is about average (–0.39 v. –0.55 std. resid.).

For an agamid it is long-lived (0.21 v. –0.46 std. resid.). It is native on many islands (v. 36%). *Stellagama* has been introduced on Malta and several Greek islands (v. 0.6% for >2 introductions). Though listed in the European Union's Bern Convention (100% v. 37% imperiled), it is not globally imperiled (v. 5%) or on one of the IUCN's lists of the most endangered.

Stenocercus

Stenocercus is the largest tropidurid genus (68 spp.). Most species are medium-sized (14.7 g v. 17.5 g; range 5.5–28 g) refugia-anchored diurnal ground lizards (66% v. 67%). However, they have the family's highest proportion of immobility experts (20.7% v. 7.6%) and litter cryptozoids (6.9% v. 2.5%). The remainder are ground-foraging tree lizards (3.4% v. 8.9%) and trunk/wall lizards (3.4% v. 12.7%). The majority rely on crypsis (60% v. 43%) for primary predator concealment, the second-highest proportion in the family. The disruptive colorations of *aculeatus*, *doellojuradoi*, and *fimbriatus* bear a remarkable resemblance to fallen leaves. *Stenocercus* is the family's top-rated genus for use of burrows for daily refugia (15% v. 5%). *Stenocercus* are found at relatively high elevations in the Andes (up to 4034 m) and the genus is top-ranked genus for medium (64% v. 47%) and high elevations (59% v. 36%). The species are poorly known (filled cells 19% v. 23%); only 43% of species have enough information to guess at a likely ecological business model. *Stenocercus* are found in the Andean countries of South America (plus Brazil, Paraguay, Uruguay, minus Chile). Peru is the most species rich (41 spp.), Ecuador is a distant second (17 spp.), and vast Brazil is third (10 spp.).

Stenocercus spp.	English name		Size		Range core
	Activity	Diet	mm	g	Ecological business model
aculeatus	O'Shaughnessy's Whorltail				Peru
	diurnal		87	24.2	*** immobility expert
albolineatus	Rescued Whorltail				Brazil
	diurnal	invert.	76	16.9	* immobility expert
amydrorhytus	Ancash Whorltail				Peru
			58	8.2	* immobility expert
angel	Angel Whorltail				Colombia
			72	14.4	
angulifer	False O'Shaughnessy's Whorltail				Ecuador
			78	18.2	
apurimacus	Fritts' Whorltail				Peru
	diurnal		69	13.3	** refugia-anchored diurnal gr. liz.
arndti	La Granja Whorltail				Peru
			74	15.6	
azureus	Blue Whorltail				Uruguay
	diurnal	invert.	69	12.9	
boettgeri	Boettger's Whorltail				Peru
	diurnal		87	24.2	** refugia-anchored diurnal gr. liz.
bolivarensis	Bolivar Whorltail				Colombia
			74	15.6	
cadlei	Cadle's Whorltail				Ecuador
			70	13.7	
caducus	Cope's Whorltail				Bolivia
	diurnal	invert.	76	17.1	
carrioni	Parker's Whorltail				Ecuador
			62	9.9	
chinchaoensis	Chinchao Whorltail				Peru
			71	14.0	
chlorostictus	Chorro Whorltail				Peru
			63	10.2	
chota	Imbabura Whorltail				Ecuador
			79	18.7	
chrysopygus	Golden Whorltail				Peru
	diurnal		65	11.1	** refugia-anchored diurnal gr. liz.
crassicaudatus	Spiny Whorltail				Peru
			77	17.8	
cupreus	Copper Whorltail				Peru
	diurnal		65	11.2	** refugia-anchored diurnal gr. liz.

Stenocercus are average in build (shape index 1.07 v. 1.10 std. resid.), tail length (1.67 v. 1.56 × SVL), and tail breakability (8.8% v. 13.2%). Crests (8.8% v. 13.2%) and basic color change (1.5% v. 2.9%) are present but subdued.

SSD is strong, but the reverse of the usual tropidurid condition; it is m<f (−0.70, the lowest value in the family, v. 0.95 on −2 to 2 scale). Sexual dichromatism is about average (22% v. 28%). Seasonal color is absent (v. 2.9%). M:F relative tail length ratio is also the reverse of the usual condition (m:f 1.12 v. 1.06 ratio). Overall sex differences rollup score is low (0.31 v. 0.80 on 0 to 6 scale), as it often is in m<f SSD species. This diminished SSD is probably related to the genus's reliance on crypsis (see first paragraph).

Invertebrate eating is the rule in *Stenocercus* (89% v. 44%); the remaining species are omnivores (11% v. 44%). Wariness is about average (2.9% v. 3.7%). The species that do not rely on crypsis are evenly divided between retreat to crevice-burrows (20% v. 29%) and retreat to vegetation (20% v. 36%). Direct basking is exactly average (16.2% v. 16.2%). *Stenocercus* is a largely terrestrial group (87% v. 74%), especially if one includes the cryptozoids (3.2% v. 1.2%). The remainder are arboreal (10% v. 25%). Their rainfall distribution is much like the family: intermediate (25% v. 22%) and mesic (75% v. 89%). They are underrepresented in all tabulated habitat specialties: littoral (5% v. 15%), sandy (5% v. 23%), riparian (12% v. 18%), and human (2% v. 14%) habitats.

Clutch count is about average (−0.47 v. −0.56 std. resid.). The hatchlings are relatively large (ontogenetic mass range −0.81 v. −0.22 std. resid.). Incubation duration is exactly average (0.40 v. 0.40 std. resid.). Most documented species lay eggs just once a year, though the reproductive cycles in a population may not be synchronous, and some laying may occur in most months.

No credible population estimates are available, though a noncredible estimate is very low (0.03 kg/ha v. 1.38 kg/ha). Endemism is exactly average (74% v. 74%). No species occurs on an island (v. 10%), nor has any been introduced (v. 1.5%). A normal proportion is imperiled (24% v. 23%), with a mild excess of species on one of the IUCN's lists of the most endangered (5.4% v. 3.0%). In other respects *Stenocercus* are similar to average tropidurids.

S. simonsii honors collector Perry O. Simons (1869–1901) who was murdered by his guide in the Andes. *S. marmoratus* is sometimes killed for absolution of all sins; this practice is especially likely to be carried out during Holy Week. This could come in handy, in case there are any herpetologists in need of absolution.

Stenocercus spp.	English name		Size		Range core
	Activity	Diet	mm	g	Ecological business model
doellojuradoi	La Rioja Whorltail				Argentina
	diurnal	invert.	67	11.9	* refugia-anchored diurnal gr. liz.
dumerilii	Andre's Whorltail				Brazil
	diurnal	invert.	91	28.0	* immobility expert
empetrus	Rock Whorltail				Peru
	diurnal		83	21.6	** refugia-anchored diurnal gr. liz.
erythrogaster	Red-bellied Whorltail				Colombia
	diurnal		74	16.0	*** litter cryptozoid
eunetopsis	Udima Whorltail				Peru
			69	12.9	
festae	Peracca's Whorltail				Ecuador
			82	21.1	
fimbriatus	Western Leaf Whorltail				Brazil
	diurnal		74	16.0	**** immobility expert
formosus	Formosa Whorltail				Peru
			73	15.2	
frittsi	Huancavelica Whorltail				Peru
			66	11.5	
guentheri	Guenther's Whorltail				Colombia
			78	18.2	
haenschi	Haensch's Whorltail				Ecuador
			64	10.5	
huancabambae	San Jose Whorltail				Peru
			80	19.6	
humeralis	Patterned Whorltail				Peru
			89	26.4	* refugia-anchored diurnal gr. liz.
imitator	Clandestine Whorltail				Peru
	diurnal		81	20.1	* refugia-anchored diurnal gr. liz.
iridescens	Iridescent Whorltail				Ecuador
			80	19.6	
ivitus	Ivy Whorltail				Peru
	diurnal		56	7.4	
johaberfellneri	Huarmey Whorltail				Peru
			50	5.5	
lache	Devil's Pulpit Whorltail				Colombia
			72	14.8	* refugia-anchored diurnal gr. liz.

Stenocercus spp.	English name	Activity	Diet	Size mm	Size g	Range core	Ecological business model
latebrosus	Chail Forest Whorltail	diurnal		64	10.5	Peru	* refugia-anchored diurnal gr. liz.
limitaris	Alamor Whorltail			79	18.7	Peru	
marmoratus	Spotted Whorltail	diurnal		69	12.9	Argentina	** refugia-anchored diurnal gr. liz.
melanopygus	Dark Whorltail			70	13.7	Peru	
modestus	Modest Whorltail	diurnal	invert.	61	9.3	Peru	* refugia-anchored diurnal gr. liz.
nigromaculatus	Black-spotted Whorltail			65	11.2	Peru	
nubicola	Cloud Forest Whorltail	diurnal		61	9.3	Peru	
ochoai	Ochoa Whorltail	diurnal		77	17.8	Peru	*** refugia-anchored diurnal gr. liz.
omari	Libertad Lizard	diurnal		59	8.4	Peru	*** refugia-anchored diurnal gr. liz.
orientalis	Eastern Whorltail	diurnal		66	11.5	Peru	*** refugia-anchored diurnal gr. liz.
ornatissimus	Lesser Ornate Whorltail	diurnal	omnivore	63	10.2	Peru	** refugia-anchored diurnal gr. liz.
ornatus	Gray's Whorltail			70	13.7	Peru	
pectinatus	Argentina Whorltail	diurnal	invert.	67	11.9	Argentina	* litter cryptozoid
percultus	Carnival Whorltail			86	23.7	Peru	
praeornatus	Greater Ornate Whorltail			81	20.1	Peru	*** refugia-anchored diurnal gr. liz.
prionotus	Botanical Garden Whorltail			76	16.9	Peru	
puyango	Lowland Whorltail	diurnal		92	28.2	Peru	

Stenocercus spp.	English name				Range core
			Size		
	Activity	Diet	mm	g	Ecological business model
quinarius	Five-crested Whorltail				Brazil
			74	15.6	
rhodomelas	Red-black Whorltail				Ecuador
			79	19.1	
roseiventris	Rosy Whorltail				Argentina
	diurnal	invert.	85	23.1	** refugia-anchored diurnal gr. liz.
santander	Santander Whorltail				Colombia
			78	18.2	
scapularis	Boulenger's Whorltail				Peru
			75	16.5	
simonsii	Simons' Whorltail				Ecuador
			72	14.8	
sinesaccus	West Brazil Whorltail				Brazil
	diurnal		72	14.9	* immobility expert
squarrosus	Saw-crested Whorltail				Brazil
			72	14.8	
stigmosus	Cajamarca Whorltail				Peru
			58	8.1	
torquatus	East Slope Whorltail				Peru
			69	13.3	** trunk/wall lizard
trachycephalus	Duméril's' Whorltail				Colombia
	diurnal		76	16.9	** refugia-anchored diurnal gr. liz.
tricristatus	Three-crested Whorltail				Brazil
			72	14.8	
variabilis	Variable Whorltail				Peru
			77	17.3	
varius	Mist Whorltail				Ecuador
			70	13.7	*** ground-foraging tree lizard

Stenodactylus

Stenodactylus is a 13-species genus of the typical-sized (3.02 g v. 3.31 g) nocturnal ground gekkonids from the Sahara and deserts of the Middle East, from Senegal and Morocco east to Turkey, Kenya, and Iran. They have a number of unique behaviors. They tend to forage slowly on stiffly erect limbs over long distances—hundreds of meters—without returning to their prior refugium. They dig a burrow and use their tails to sweep loose sand over the entrance to close it as they enter. This behavior is thought to protect them from crow predation. *Stenodactylus* are also sensitive to the odors of predatory snakes; they will abandon a burrow in which the scent of a snake predator appears, as least in *doriae*. Closing their burrow might preclude snake detection of the burrow. Their use of tails for burrow closure is evidently critical, as most species appear to have partially or mostly lost the ability to shed their short tails (0.80 v. 1.07 × SVL). Their locomotor activity has been associated with bright moonlight, which is the reverse of the association in many nocturnal lizards. Subjectively, these lizards have very large eyes, suggesting a reliance on visual cues despite their nocturnality. A majority of *Stenodactylus* have been described as both sit-and-wait and cruise foragers; the latter is anomalous for gekkonids (83% v. 24%).

Secretory pores are uncommon (7.7% v. 46%), perhaps to deprive predatory snakes of information about their whereabouts. Most *Stenodactylus* lack enlarged toepads (85% v. 30%). Males are smaller than females in all documented species (mean SSD –1.14 v. –0.03 on –2 to 2 scale). Sexual dichromatism is absent (v. 6.6%) though male seasonal color is especially common (7.7% v. 0.6%), which is highly unusual in nocturnal lizards (table 3.4, trait 16). One suspects that sexual encounters in this species (*petrii*) occur with enough light for color vision; one observer anomalously characterized *petrii* as crepuscular. Electric light use has not been reported; electric lights may be fairly hard to come by in the Sahara. Spontaneous (15% v. 6.5%) and stress calling (15% v. 3.7%) are relatively common, even more than is typical for nocturnal lizards (table 3.7, traits 31, 32). Primary predator concealment is mostly by vegetation (75% v. 19%).

Direct basking is relatively common for a gekkonid (7.7% v. 4.9%), though documented only in *Stenodactylus sthenodactylus*, which is day active in winter. Mean active body temperature is a cool 26.6°C (v. 27.6°C). Though not absolutely cool, or especially cool in relation to other

Stenodactylus spp.	English name		Size		Range core
	Activity	Diet	mm	g	Ecological business model
affinis	Murray's Comb-fingered Gecko				Iran
	nocturnal		52	3.18	* nocturnal ground lizard
arabicus	Arabian Sand Gecko				Saudi Arabia
	nocturnal		40	1.47	**** nocturnal ground lizard
doriae	Middle Eastern Sand Gecko				Iraq
	nocturnal	invert.	66	6.0	*** nocturnal ground lizard
grandiceps	Large-headed Thin-toed Gecko				Syria
	nocturnal	invert.	52	3.19	*** nocturnal ground lizard
leptocosymbotes	Southern Big-headed Gecko				Oman
	nocturnal		67	6.2	*** nocturnal ground lizard
mauritanicus	Mauritania Gecko				Sahara
	nocturnal		43	1.86	* nocturnal ground lizard
petrii	Petrie's Gecko				Sahara
	nocturnal	invert.	56	3.93	**** nocturnal ground lizard
pulcher	Beautiful Sand Gecko				Southern Arabia
	nocturnal		45	2.12	** nocturnal ground lizard
sharqiyahensis	Sharqiyah Sands Gecko				Oman
	nocturnal		34	1.01	
slevini	Slevin's Big-headed Gecko				Saudi Arabia
	nocturnal		72	7.8	**** nocturnal ground lizard
stenurus	Western Maghreb Sand Gecko				Northern Sahara
	nocturnal		49	2.73	* nocturnal ground lizard
sthenodactylus	Elegant Gecko				Red Sea
	nocturnal	invert.	51	2.99	**** nocturnal ground lizard
yemenensis	Yemen Sand Gecko				Southern Arabia
	nocturnal		52	3.18	* nocturnal ground lizard

geckos, it is cool in comparison to summer temperatures in the Sahara.

Male social relations in *Stenodactylus* are surprisingly tolerant (67% v. 13%). Male spatial defense is noted only in a minority of species (33% v. 87%). Given the frequency of roaming foraging, and the ready availability of sand for burrow construction, these lizards may have no reason to contest turf.

Stenodactylus are overrepresented in littoral areas (30% v. 11%) and especially in sandy habitats (90% v. 11%); they are underrepresented in riparian zones (10% v. 17%) and absent from human habitats (v. 27%).

Their reproduction is similar to most gekkonids. Young are notably large (ontogenetic size range low: –0.55 v. –0.19 std. resid.).

Mass-adjusted longevity is short (–0.41 v. 0.15 std. resid.), possibly reflecting the paucity of data on these remote lizards. Three species are noticeably long-lived—*petrii*, 13.0 years; *sthenodactylus*, 9.1 years; *doriae*, 8.9 years—but several others are said to be less than one year, possibly a failure of herpetoculture. *Stenodactylus*

Stenolepis sp.	English name				Range core
			Size		
	Activity	Diet	mm	g	Ecological business model
ridleyi	Igarassu Forest Lizard				Brazil
			42	1.47	

Strobilurus sp.	English name				Range core
			Size		
	Activity	Diet	mm	g	Ecological business model
torquatus	Thorn-tailed Iguana				Brazil
			90	26.7	* trunk/wall lizard

have not been introduced (v. 3.7%). A typical proportion are considered imperiled (38% v. 47%), though none is listed among the most endangered by the IUCN (v. 11.3%). In other respects *Stenodactylus* are typical gekkonids. Note the comparatively mellifluous common name for *Stenodactylus sthenodactylus*.

Stenolepis

Stenolepis ridleyi is the sole member of a very poorly known (filled cells 11% v. 19%) gymnophthalmid genus endemic to Brazil. This genus has the dubious distinction of being the least-documented genus in the least-documented aboveground family. The species is too poorly known to guess its ecological business model. The species is small (1.47 g v. 2.35 g). Sex differences, diet, diel activity pattern, predator avoidance tactics, social relations, and reproduction are not reported. This species has not been evaluated for conservation status.

Strobilurus

Strobilurus torquatus is the sole member of its tropidurid genus of large (26.7 g v. 17.5 g) trunk/wall lizards (100% v. 13%) endemic to Brazil. It is of average build (shape index 1.15 v. 1.10 std. resid.). Males are larger than females (1.0 v. 0.95 on –2 to 2 scale). It is not recorded in littoral (v. 15%), sandy (v. 23%), riparian (v. 18%), or human (v. 14%) habitats. Its clutch count is high (0.34 v. –0.56 std. resid.), equating to 7.4 eggs. It is not found on an island (v. 10%), it has not been introduced (v. 1.5%), and it has not been evaluated for conservation status. In other respects *Strobilurus* is probably a typical tropidurid.

Strophurus

Strophurus is a moderate (19 spp.) radiation of small (3.3 g v. 4.2 g), nocturnal (89% v. 79%), arboreal (94% v. 64%) diplodactylids in Australia. Their signature possession is a tail that, for most species, is able to exude, or even squirt, noxious sticky gels. Such tails are often accompanied by orange spines or markings that are likely a warning (aposematic) to potential predators such as *Lialis burtonis* (which rejects *Strophurus*). All species rely on crypsis (v. 88%); those species that hide in spinifex grasses (*Plectrachne, Triodia*) are typically striped and colloquially known as phasmid geckos; those that hide in other vegetation are mottled. Originally a "phasmid" was only a stick insect that walks in a slow rocking back and forth motion that mimics the wind-driven motion of vegetation. By adopting this same motion, these lizards can apparently also progress (slowly) without triggering the neurological motion detectors in visual predators or perhaps prey. Most are nocturnal canopy walkers (56% v. 25%), albeit often at very low perch heights; the others are mostly ground-foraging tree lizards (31% v. 7%), with one each of immobility expert (v. 1.6%) and litter cryptozoid (v. 0.8%) in spinifex tangles.

Strophurus are reasonably well known (34% v. 28% of cells filled), short-tailed (0.68 v. 0.85 × SVL), occasionally subtlety crested (5.3% v. 1.3%), and lacking in color changing ability (v. 5.4%). *Strophurus* tend to have small spines or tubercles, especially on the tail; three species have several small but distinct spines above their eyes (*ciliaris, krisalys, strophurus*). These supraocular spines are present in both sexes; their function is unknown. When *Strophurus* have color, it is likely some variant of orange, typically either in spines or in their irises. Some species (e.g., *taenicauda*) partially shut their eyes when endeavoring to be especially cryptic, presumably to conceal their bright irises or alter their lizard-like gestalt.

Reported sex differences are limited to secretory pores (42% v. 34%) and m<f SSD (–0.88 v. –0.65 on –2 to 2 scale). Accordingly, their sex rollup score (on 0 to 6 scale) is modest (0.74 v. 0.54).

With one exception, all *Strophurus* eat primarily invertebrates; the exception (*assimilis*) is recorded feeding only on nectar, but examination of stomach contents would likely reveal a broader and most probably invertebrate diet. Few or no vertebrates feed exclusively on nectar, as it lacks many essential nutrients.

Strophurus emits spontaneous

Strophurus spp.	English name				Range core
			Size		
	Activity	Diet	mm	g	Ecological business model
assimilis	Goldfields Spiny-tailed Gecko				Western Australia
	nocturnal	herbivore	66	5.7	
ciliaris	Spiny-tailed Gecko				Australia
	cathemeral	invert.	72	7.6	*** ground-foraging tree lizard
congoo	Congoo Gecko				Queensland
			45	1.72	
elderi	Australian Jewelled Gecko				Australia
	nocturnal	invert.	43	1.55	*** nocturnal canopy walker
horneri	Arnhem Phasmid Gecko				Northern Territory
	nocturnal		38	1.06	** litter cryptozoid
intermedius	Southern Spiny-tailed Gecko				Australia
	nocturnal	invert.	64	5.3	*** ground-foraging tree lizard
jeanae	Southern Phasmid Gecko				Western Australia
	nocturnal		45	1.83	*** nocturnal canopy walker
krisalys	Kristin's Spiny-tailed Gecko				Queensland
	nocturnal		65	5.8	** nocturnal canopy walker
mcmillani	Short-tailed Striped Gecko				Western Australia
	nocturnal		47	2.02	* nocturnal canopy walker
michaelseni	Robust Striped Gecko				Western Australia
	nocturnal	invert.	54	3.14	* nocturnal canopy walker
rankini	Rankin's Spiny-tailed Gecko				Western Australia
	nocturnal	invert.	55	3.20	* nocturnal canopy walker
robinsoni	Robinson's Spinifex Gecko				Western Australia
	nocturnal		52	2.72	** nocturnal canopy walker
spinigerus	Southwestern Spiny-tailed Gecko				Western Australia
	nocturnal	invert.	64	5.1	** nocturnal canopy walker
strophurus	Western Spiny-tailed Gecko				Western Australia
	nocturnal	invert.	64	5.1	** ground-foraging tree lizard
taeniatus	Phasmid Striped Gecko				Northern Australia
	nocturnal	invert.	47	2.01	
taenicauda	Golden-tailed Gecko				Queensland
	nocturnal	invert.	61	4.54	*** immobility expert
wellingtonae	Western Shield Spiny-tailed Gecko				Western Australia
	nocturnal	invert.	72	7.4	* ground-foraging tree lizard
williamsi	Eastern Spiny-tailed Gecko				Queensland
	cathemeral		59	4.10	** nocturnal canopy walker
wilsoni	Mount Augustus Spiny-tailed Gecko				Western Australia
	nocturnal		52	2.74	** ground-foraging tree lizard

vocalizations at the family-typical frequency (5.3% v. 5.4%), but stress calls are not recorded (v. 5.4%). Perhaps they rely on their gooey tails for predator repulsion instead. All species rely on crypsis as their primary predator concealment (v. 88%). Although their mean body temperature when active is a typical 25.1°C (v. 24.3°C), they have a notably wide range of active body temperatures, typically 20°C–31.5°C. The ability of these nocturnal lizards to change their temperature is likely limited; they seem to thermoconform as a consequence, though no report specifically addressed thermoconformance. All males defend space (v. 77%). No *Strophurus* is documented to burrow (v. 22%). They are slightly overrepresented in sandy areas (32% v. 24%) and underrepresented in riparian zones (0% v. 10%).

They are oviparous (v. 82%) with typical clutch count and incubation. One species (*ciliaris*) is reported to have adherent eggs (v. 0.7%), but communal oviposition (v. 2.7%) and parental care (v. 3.4%) are absent. Their RCM ratios are light (0.143 v. 0.225 ratio).

Longevity is short, especially for a diplodactylid (–0.34 v. 0.90 std. resid). Population density is relatively low (0.13 kg/ha v. 0.44 kg/ha), likely reflecting the low productivity of their desert environment. *Strophurus* have not been introduced (v. 2%), and 40% (v. 81%) of species are considered imperiled, though none is on an IUCN list of the most endangered (v. 29%). In other respects *Strophurus* are typical diplodactylids.

Tachygyia

Tachygyia microlepis is the sole member of its genus. It is a very large (58 g v. 3.3 g) skink endemic to the Pacific Island nation of Tonga. It is known from only two 1827 specimens (filled cells 14% v. 25%), which were not documented in relation to natural history. It is presumed extinct, due to the usual perpetrators of island extinctions. Its likely shape index (0.87 v. −0.46 std. resid.) suggests a fairly robust skink. Its tail length is slightly short (1.10 v. 1.33 × SVL). It was terrestrial. Endemism was high (100% v. 64%), as was insularity (v. 57%). It has not been introduced (v. 2.1%), and of course, it is imperiled, if it even survives (v. 47%). It is on one of the IUCN lists of the most endangered (extinct; v. 14%). In other respects *Tachygyia* was likely to be similar to other skinks.

Tachygyia sp.	English name	Range core	Activity	Diet	Size mm	Size g	Ecological business model
microlepis	Small-scaled Giant Skink	Tonga			136	58	

Takydromus

Takydromus is a major radiation (22 spp.) of slightly small (4.0 g v. 5.1 g) but very long-tailed (3.22 v. 2.06 × SVL) diurnal scansorial lacertids found in eastern Asia from Russia and Japan south to Indonesia and west to India. The long-tailed theme reaches its apex in *sexlineatus*, whose tail attains 5.6 × SVL, the longest of any lizard. Most *Takydromus* move through a dense matrix of grass and other small diameter vegetation, using their long tails for serpentine propulsion. Many tails have sharp edges, which promotes control when pushing through fine vegetation. The modal *Takydromus* is a wide-ranging chemosensor (v. 25%); others are diurnal canopy walkers (17% v. 1.4%) or refugia-anchored diurnal ground lizards (8% v. 57%).

Despite the importance of their tails, tails are breakable (91% v.92%). Color change is found in *smaragdinus* (4.5% v. 2.2%), including advanced color change (4.5% v. 0.9%).

SSD is strongly in the m<f category (−0.58 v. 0.49 on −2 to 2 scale). Sexual dichromatism is present but underrepresented (18% v. 33%), as is seasonal color (9% v. 15%). Blue tails are absent (v. 14%).

Most *Takydromus* are invertebrate eaters (88% v. 77%); *sauteri* is a fruit-eating omnivore (13% v. 19%). All are active foragers (v. 92%). Home-range area is moderate (0.5 v. 0.27 std. resid.). All use retreat to vegetation as their primary predator concealment tactic (v. 42%). Direct basking occurs at the usual frequency (23% v. 25%).

All males are thought to be tolerant (100% v. 41%), a rare condition in lizards (v. 15%), but one consistent with the paucity of m>f size dimorphism and the high frequency of wide-ranging chemosensors.

About a third of *Takydromus* are considered arboreal (29% v. 9%); the remainder are basically terrestrial (71% v. 90%), though most *Takydromus* are at ease propelling themselves through vegetation that is near but not on the ground. They do not burrow (v. 48%).

Two species (*kuehnei*, *sauteri*) are documented with regard to rainfall (2500 mm/y, 3000 mm/y), both in

Takydromus spp.	English name	Range core	Activity	Diet	Size mm	Size g	Ecological business model
amurensis	Amur Grass Lizard	East Asia	diurnal		69	7.7	** refugia-anchored diurnal gr. liz.
dorsalis	Sakishima Grass Lizard	Japan	diurnal		62	5.5	*** diurnal canopy walker
formosanus	West Taiwan Grass Lizard	Taiwan		invert.	45	2.10	
hani	Southeast Asian Green Climber	Vietnam	diurnal		66	6.7	** wide-ranging chemosensor
haughtonianus	Goalpara Grass Lizard	India	diurnal	invert.	52	3.38	
hsuehshanensis	Mount Hsiao-hsueh Lizard	Taiwan			60	5.1	
intermedius	Stejneger's Rock Racer	China			54	3.66	
khasiensis	Khasi Hills Long-tailed Lizard	India	diurnal	invert.	59	4.87	
kuehnei	Thompson's Grass Lizard	China	diurnal	invert.	53	3.51	*** wide-ranging chemosensor
luyeanus	Southeast Taiwan Grass Lizard	Taiwan			48	2.63	** wide-ranging chemosensor

Takydromus spp.	English name				Range core
	Activity	Diet	Size mm	Size g	Ecological business model
madaensis	Ma Da Grass Lizard				Vietnam
	diurnal		47	2.40	
sauteri	Koshun Grass Lizard				Taiwan
	diurnal	omnivore	66	6.74	*** diurnal canopy walker (low)
septentrionalis	China Grass Lizard				China
			64	6.0	
sexlineatus	Long-tailed Climber				Southern East Asia
	diurnal	invert.	57	4.45	** wide-ranging chemosensor
sikkimensis	Sikkim Grass Lizard				India
			52	3.38	
smaragdinus	Green Grass Lizard				Japan
	diurnal		52	3.27	** wide-ranging chemosensor
stejnegeri	Pescadores Grass Lizard				Taiwan
			54	3.64	* wide-ranging chemosensor
sylvaticus	Chongan Ground Lizard				China
			51	3.11	
tachydromoides	Japanese Grass Lizard				Japan
	diurnal	invert.	56	4.08	**** wide-ranging chemosensor
toyamai	Miyako Grass Lizard				Japan
	diurnal		52	3.36	** wide-ranging chemosensor
viridipunctatus	North Taiwan Grass Lizard				Taiwan
	diurnal		51	3.12	** wide-ranging chemosensor
wolteri	Mountain Grass Lizard				China
	diurnal	invert.	54	3.63	

the mesic band. Elevational distribution is normal. They are absent from sandy areas (v. 44%) but otherwise undistinguished in terms of habitat.

Reproduction is fairly normal, with clutch count small (−0.15 v. 0.13 std. resid.), perhaps to facilitate their active foraging on spindly perches. This is supported by the small size of the young (ontogenetic mass range 0.65 v. −0.05 std. resid.). Consequently, one would expect a light RCM, which is not supported (0.295 v. 0.267), based solely on *tachydromoides*. Incubation is short (−0.64 v. −0.36), equating to between 29 days (*wolteri*) and 46 days (*sexlineatus*)

Longevity is known from only two species (*sexlineatus*, 1.5 years; *tachydromoides*, 9 years), averaging −0.42 (v. 0.19 std. resid.). Population density, based solely on *tachydromoides* is average (1.84 kg/ha v. 2.07 kg/ha). Endemism is high (68% v. 35%). Many species are found on islands (64% v. 26%). One-quarter are imperiled (25% v. 56%) and placed on the IUCN's lists of the most endangered (25% v. 15.5%). In other respects *Takydromus* are typical lacertids.

Tarentola

Tarentola is a major radiation (31 spp.) of large-headed, mostly tuberculate, moderately large (8.5g v. 4.7 g), nocturnal/cathemeral phyllodactylids from two widely separated tropical/subtropical areas: Jamaica and Cuba through the Bahamas and the Old World aridlands from the Cape Verde Islands through Israel, extending to Spain and Greece on the north and Nigeria and Central African Republic on the south. The modal *Tarentola* is a nocturnal ground lizard (57% v. 40%); the remainder are evenly split between ground-foraging tree lizards (20% v.15%) and trunk/wall lizards (23% v. 44%).

They are morphologically typical phyllodactylids, though *chazaliae* may lack toepads (v. 8%). This species also wraps its prehensile tail alongside itself like a cat when it rests. Sex differences are moderate (0.52 v. 0.32 on 0 to 6 scale). SSD is mostly m>f (1.0 v. 0.32 on −2 to 2 scale), and the m:f ratio of relative tail length (1.34 v. 1.14) is the highest of any phyllodactylid genus, based on only *annularis* and *chazaliae*. Color differences between the sexes are not reported.

T. gigas (44 g) is large enough to eat vertebrates (v. 1.8%), specializing in the eggs and young of nesting seabirds on the Cape Verde Islands. *T. annularis* is an ant specialist despite its large size (27g). The others are conventional invertebrate eaters (85% v. 93%), though *mauritanica* is also reported to take fruit. The genus is unusual among phyllodactylids in that large proportions of the genus are associated with each of the foraging modes: sit-and-wait (85% v. 93%), cruise (17% v. 13%), and active foraging (67% v. 31%). *T. deserti* is said to stalk prey from 1 m. Electric light use is common (9.7% v. 9.2%). Home-range sizes are unknown, although *annularis* is notable for occupying the same site for weeks at a time; the ants must be slow learners.

Most *Tarentola* are nocturnal (84% v. 78%); the remainder are cathemeral (16% v. 19%). Note that five of 26 nominally nocturnal species have been reported to be exposed or active in the daytime. As expected from cathemeral/nocturnal lizards, spontaneous calling (12.9% v. 11.1%) and stress vocalizations (6.5% v. 5.9%) are moderately common. No *Tarentola* is conspicuously wary (v. 1.5%). Primary predator concealment tactics are a mix of flight (50% v. 14%) and crypsis (50% v. 43%). Direct basking is fairly common for phyllodactylids: 13% v. 10%. Mean active body temperature is a very cool 23.6°C (v. 26.0°C), based on *chazaliae* and *mauritanica*.

Both solitariness (50% v. 10%) and group living (50% v. 20%) are unusually common in *Tarentola*.

Tarentola are found in rainfall regimes from hyperdry (<55 mm/y) to mesic (1101–3000 mm/y). The elevational distribution is normal (l/m/h: 100/43/7 v. 93/34/6). They are over-represented in sandy areas (45% v. 25%).

TSD is known from at least *boettgeri* and *mauritanica*. *T. annularis* and *deserti* males show the notable adaptation of attempting to copulate with each female on both sides (i.e., using each hemipene). This is presumably a reflection of intense sperm competition; *fascicularis* females are known to storm sperm over winter. Reproduction is otherwise typical for phyllodactylids, with the possible exception of elevated parental care (6.5% v. 3.7%): *annularis* and *fascicularis* females guard their eggs. Clutch count is mildly low (−0.78 v. −0.57 std. resid.). Young are slightly small (ontogenetic mass range 0.38 v. 0.11 std. resid.).

Longevity is about average (−0.05

Tarentola spp.	English name				Range core
			Size		
	Activity	Diet	mm	g	Ecological business model
albertschwartzi	Jamaica Giant Gecko				Jamaica
	nocturnal		107	31.4	
americana	Cuba Giant Gecko				Cuba
	nocturnal	invert.	95	22.1	**** trunk/wall lizard
angustimentalis	East Canary Gecko				Spain
	nocturnal	invert.	65	7.3	*** nocturnal ground lizard
annularis	White-spotted Wall Gecko				western Africa
	cathemeral	ants	101	26.6	**** ground-foraging tree lizard
bischoffi	Selvagens Gecko				Portugal
	nocturnal		52	3.80	* nocturnal ground lizard or ground-foraging tree lizard
boavistensis	Boa Vista Wall Gecko				Cape Verde Islands
	nocturnal		65	7.1	** nocturnal ground lizard
bocagei	Bocage's Cape Verde Gecko				Cape Verde Islands
	nocturnal		57	4.82	** nocturnal ground lizard
boehmei	Boehme's Gecko				Western Sahara
	nocturnal		76	11.6	** trunk/wall lizard
boettgeri	Gran Canaria Gecko				Spain
	nocturnal		61	5.9	* nocturnal ground lizard
caboverdiana	Santa Antao Gecko				Cape Verde Islands
	nocturnal		62	6.1	* nocturnal ground lizard
chazaliae	Helmet Gecko				Morocco
	cathemeral	invert.	66	7.4	**** nocturnal ground lizard
crombiei	Crombie's Gecko				Cuba
	nocturnal	invert.	51	3.00	** trunk/wall lizard
darwini	Darwin's Wall Gecko				Cape Verde Islands
	nocturnal		67	7.7	*** nocturnal ground lizard
delalandii	Tenerife Gecko				Spain
	cathemeral		63	6.7	** nocturnal ground lizard
deserti	Desert Wall Gecko				Northern Sahara
	nocturnal	invert.	79	12.9	** ground-foraging tree lizard
ephippiata	African Wall Gecko				Africa
	nocturnal	invert.	91	19.1	*** trunk/wall lizard
fascicularis	Wolfgang's Moorish Gecko				Libya
	cathemeral	invert.	72	9.9	** ground-foraging tree lizard
fogoensis	Fogo Wall Lizard				Cape Verde Islands
	nocturnal		59	5.3	** nocturnal ground lizard
gigas	Giant Wall Gecko				Cape Verde Islands
	nocturnal	carnivore	120	43.9	*** nocturnal ground lizard

Tarentola spp.	English name	Activity	Diet	Size mm	Size g	Range core	Ecological business model
gomerensis	Gomera Gecko	nocturnal		64	6.8	Spain	* ground-foraging tree lizard
maioensis	Maio Wall Gecko	nocturnal		60	5.7	Cape Verde Islands	*** nocturnal ground lizard
mauritanica	Moorish Wall Gecko	cathemeral	invert.	68	8.3	Western Mediterranean	**** trunk/wall lizard
mindiae	Qattara Gecko	nocturnal		67	8.0	Egypt	* trunk/wall lizard
neglecta	Algeria Wall Gecko	nocturnal	invert.	58	5.1	Northern Sahara	* ground-foraging tree lizard
nicolauensis	Sao Nicolau Wall Gecko	nocturnal		59	5.4	Cape Verde Islands	*** nocturnal ground lizard
parvicarinata	West African Wall Gecko	nocturnal	invert.	79	12.6	Senegal	** ground-foraging tree lizard
pastoria	Pastoria Wall Gecko	nocturnal	invert.	83	15.0	Guinea	*** trunk/wall lizard
protogigas	Bravacim Wall Gecko	nocturnal		75	11.1	Cape Verde Islands	** nocturnal ground lizard
raziana	Desertas Islands Wall Gecko	nocturnal		47	2.70	Cape Verde Islands	** nocturnal ground lizard
rudis	Santimaria Wall Gecko	nocturnal		68	8.3	Cape Verde Islands	** nocturnal ground lizard
substituta	São Vicente Wall Gecko	nocturnal		54	4.14	Cape Verde Islands	*** nocturnal ground lizard

v. 0.15 std. resid.), based on remarkably complete data (13 spp.). The longest-lived, *annularis*, lives up to 19.8 years; the shortest maximum (2.4 years) is shared by five species. No population density estimates are available, though 40–50 individuals of *annularis* have been observed on single huts in Sudan. Endemism is normal (61% v. 67%). The majority of species occur on islands (71% v. 40%). Introductions have been especially common (19.4% v. 7.4%), including 6.5% (v. 1.5%) of species that have been introduced more than twice, *mauritanica* at least 10 times. Only 19% (v. 29%) of *Tarentola* are considered imperiled, and 3.7% (v. 7.1%) are on the IUCN's lists of the most endangered. In other documented respects *Tarentola* are typical phyllodactylids.

The genus is named in honor of Taranto, the southern Italian harbor where this species occurs, and which also gave its name to the tarantula spider and the tarantella dance.

Techmarscincus

Techmarscincus jigurru is the sole member of its genus, a moderately large (6.1 g v. 3.3 g), saxicolous, oviparous (v. 64%) refugia-anchored diurnal ground (v. 30%) skink endemic to a single mountaintop (Bartle Frere; 1440–1620 m occupied) on the coast of Queensland. Its head and neck are noticeably narrower than its body, its tail is long (1.90 v. 1.33 × SVL), and its shape index is mildly robust (0.16 v. −0.46 std. resid.), but otherwise it appears as a typical skink. It is a generalized invertebrate eater (v. 84%). It basks directly (v. 21%). Its clutch of around four eggs is normal for its size (0.04 v. 0.06 std. resid.). It is called a cool-skink, not because of its supreme self-confidence but because it lives among rocks in forests at elevations that are comparatively cool.

It has not been introduced (v. 2.1%). It is imperiled (v. 47%) but is not on one of the IUCN's lists of the most endangered (v. 14%). In other respects *Techmarscincus* is likely to be similar to other skinks.

Techmarscincus sp.	English name	Activity	Diet	Size mm	Size g	Range core	Ecological business model
jigurru	Bartle Frere Cool-skink	diurnal	invert.	67	6.1	Queensland	** refugia-anchored diurnal gr. liz.

Teiidae (figs. 4.11E, 4.15A, 4.15B)

The 156 teiids (table 2.1) are the archetypical fast-moving (see genus accounts), active-foraging (100% v. 58%), heliophilic (58% v. 18% are documented direct baskers) terrestrial (98%, the fourth-highest family, v. 48%) lizards. They are all diurnal (100% v. 67%). Almost all are wide-ranging chemosensors (97% v. 7%). Most species are Neotropical (78% v. 29%); the remainder are Nearctic (24% v. 6%). Few species extend beyond 40° latitude. With their incessant movements, they are easily observed (filled cells 33% v. 27%). The downside to such high levels of locomotor activity is elevated exposure to predation. Quantification of predation is always hard, but it is noteworthy that many teiids spend long periods of time hidden, even when the weather is suitable, and they have very exacting standards for suitable weather. Thus, their conspicuousness need not translate directly into high mortality per unit calendar time. It would be helpful to have more estimates of mortality per foraging hour, but values so close to zero are hard to estimate accurately. Lifetime mortality risk may be reflected in their short size-adjusted longevity (–0.74, the fourth-lowest value, v. 0.0 std. resid.). The families with even lower known mass-adjusted longevity are Chamaeleonidae, another family with many exposed, incessant if slow movers, and three inadequately documented families of subterranean foragers (Amphisbaenidae, Bipedidae, and Trogonophidae). Therefore, it is likely that teiids and chamaeleonids pay a high cost for their frequent movements. The energetic gain from their foraging bouts may also be high, but quantification of this is lacking at the family level.

In what other ways do teiids stand out? Their tails are long (2.30, the fourth-highest family, v. 1.37 × SVL) and often blue (14.7%, the highest family frequency, v. 2.3%). As the movements of most species destroy any hope of remaining cryptic, they might as well have a blue tail to divert attacks. The long tail probably aids with sprint speed and bipedalism, a combination which is common in teiids. One wonders if high sprint speed also characterized the many bipedal dinosaurs (e.g., *Tyrannosaurus rex*). Teiids are hot (mean active body temperature 36.9°C, the second-highest value, v. 31.1°C), the faster to sprint with. They often rely on flight (77%, the second-highest family, v. 11%) for primary predator concealment. Burrows are often used for a daily refuge (82%, the fourth highest, v. 22%). One way in which they do not stand out is diet: all five diet classes are well represented within Teiidae (see below). Their active foraging mode does not apparently constrain their trophic level.

Those that are not wide-ranging chemosensors are refugia-anchored diurnal ground lizards (2.6% v. 24.5%) or giant herbivores (0.6% v. 1.6%). Teiids are moderately large (22.1 g v. 5.3 g) and stoutly built (shape index 0.88 v. 0.00 std. resid.). Tails are often unbreakable (21% v. 57% breakable). Presumably, all of the blue tails are breakable. Secretory pores are mildly underrepresented (9.0% v. 15%), but this is at least partially due to sexual monomorphism in pores. Teiids often have large pores that do not differ between the sexes, and thus are poorly characterized by the pore metric used in this work. Crests (v. 5.3%) and elongation (v. 11%) are absent. Color change is rare (1.3% v. 10.7% for any color change; 0.6% v. 6.0% for advanced color change); teiids are not trying to be cryptic by color matching. The teiid genera are *Ameiva, Ameivula, Aspidoscelis, Aurivela, Callopistes, Cnemidophorus, Contomastix, Crocodilurus, Dicrodon, Dracaena, Glaucomastix, Holcosus, Kentropyx, Medopheos, Pholidoscelis, Salvator, Teius*, and *Tupinambis*.

Teiids may be colorful for mating reasons: sexual dichromatism is overrepresented (35% v. 21%), as is seasonal color (9.6% v. 6.5%). SSD (m>f) is also elevated (0.89 v. 0.36 on –2 to 2 scale). Their sex differences rollup score (on scale of 0 to 6) is therefore high (1.19 v. 0.75).

Omnivory is overrepresented in teiids (33% v. 15%). Carnivory (3.3% v. 3.2%) is about average. The others are present but underrepresented: herbivory (1.1% v. 4.1%), invertebrate eating (62% v. 73%), and ant specialization (1.1% v. 4.4%). Sit-and-wait (v. 46%) and cruise foraging (v. 6.6%) are absent. Electric light use is not known (v. 2.0%).

Home-range area is slightly elevated (0.19 v. 0.0 std. resid.), but the average teiid's movements do not cover the vast areas that one might imagine. There are nine families with larger mass-adjusted home ranges, and individual genera have comparatively small areas (*Ameiva, Cnemidophorus, Kentropyx, Pholidoscelis*, and *Teius* have mass-adjusted home ranges that are smaller than the lizard norm). This observation challenges the stereotype that teiids are "wide-ranging"; perhaps it is better to consider them to be frequently moving rather than wide-ranging. Many more data are needed on the home ranges of lizards to clarify the connections between foraging mode, regularity of movements, and home-range area.

Although archetypical teiids sprint away from predators, appreciable minorities express all of the other primary predator defenses: crypsis (1.8% v. 43%), sand diving (1.8% v. 5.8%), retreat to a crevice or burrow (12.3% v. 25%), and hiding in vegetation (7.0% v. 23%). Conspicuous wariness is overrepresented (12.8% v. 8.2%), though wariness is difficult to assess in teiids, as they most likely were darting away before the herpetologist arrived.

Social behaviors are not well understood in Teiidae. Solitariness is strongly overrepresented (29% v. 10%). Group living is slightly overrepresented (29% v. 20%). Male social tolerance is the fifth highest (55% v. 15%). Male spatial defense is low (45% v. 85%), but it is much above zero. Male dominance hierarchies are strongly overrepresented (16% v. 7.7%).

The annual rainfall spectrum of the family is fairly close to that of all lizards in the drier half of the spectrum. A second mesic mode is not evident (mesic: 30% v. 43%; wet: 7% v. 9%). Their elevational profile is skewed toward lower elevations (l/m/h: 95/32/6 v. 82/41/13). They are overrepresented in littoral (38% v. 17%), sandy (47% v. 22%), riparian (46% v. 22%), and human (29% v. 18.5%) habitats.

All teiids lay eggs (v. 82% oviparous). Many are famous for their parthenogenesis (see genus accounts). Clutch counts are small (−0.51 v. 0.0 std. resid.), and the size of the hatchlings is normal to slightly small (ontogenetic mass range slight elevated 0.10 v. 0.0 std. resid.). These combine to produce a low RCM (0.163 v. 0.226), which may be the point. It is hard to sprint when burdened with many eggs. Incubation is short (−0.38 v. 0.0 std. resid.). They may lay often, to make up for the small clutches, but the data to properly evaluate this possibility are available only for selected species. Eggshells are soft (v. 41% hard), rarely adherent (0.6% v. 2.4%), and only occasionally laid communally (2.6% v. 4.2%). Parental care is expressed at a typical frequency (3.2% v. 2.9%).

Population density has been credibly estimated for 25 species; those estimates average somewhat high (1.06 kg/ha v. 0.54 kg/ha). Thus, the high mortality implied by their short longevities is not obviously depressing their population density. Endemism is about average (50% v. 64%). Island species are about average (37% v. 40%). Teiids have often been introduced (8.3% v. 3.3%); presumably, these are mostly intentional introductions, as frenetic lizards do not go unnoticed as stowaways. Multiple (>2) introductions are underrepresented (0.6% v. 1.1%). Teiids are imperiled at an average rate (41% v. 49%), including placement on one of the IUCN's lists of the most endangered (12.1% v. 13.6%). In other respects teiids are similar to other lizards.

Although other families (e.g., Lacertidae, Varanidae) have been likened to teiids in their frequent movements, reliance on chemosensation, and heliophilia, those families are notably different in important ways. For example, lacertids (see Lacertidae entry) are much more anchored to known refugia. Varanids (see Varanidae entry) are more inclined to travel or seek refuge in trees rather than sprint to safety. The teiid lifestyle is one successful model but not one that other lizards have copied exactly.

Teira sp.	English name				Range core
			Size		
	Activity	Diet	mm	g	Ecological business model
dugesii	Madeira Lizard				Portugal
	diurnal	omnivore	67	7.1	** wide-ranging chemosensor

Teira

Teira dugesii is the sole species in a lacertid genus of larger-sized (7.1 g v. 5.1 g) wide-ranging chemosensors found in the Portuguese islands off northwestern Africa. It was originally resident in the Madeira Islands, possibly carried by humans to the Selvagens—this lizard has been discovered on ships at sea—and assuredly transported by humans to the Azores. It achieves particularly high densities within storm petrel colonies, no doubt benefiting from spilled prey and the attracted flies. It will also take leaves and fruit and is considered an agricultural pest by grape growers.

It has a typical lacertid form, though it is a bit heavier (shape index 0.37 v. 0.27 std. resid.), and its tail is a bit shorter (1.70 v. 2.06 × SVL). It has advanced color change (v. 0.9%), and the sexes differ in color (v. 33%). It basks directly (v. 25%) but does not use burrow refugia (v. 48%). It is very broad-minded when it comes to rainfall regimes, occurring from areas of <300 mm/y to >3000 mm/y, at elevations 0–1862 m. It resides in the littoral zone (v. 22%) as well as human habitats (v. 16%). Clutch count is small (−0.42 v. 0.13 std. resid.), and young are relatively large (ontogenetic mass range −0.58 v. −0.05 std. resid.), but reproduction is otherwise undistinguished. Longevity is very high (1.30 v. 0.19 std. resid.), which equates to 16 years. Although there are no credible population density estimates, an unsupported density estimate of 200 kg/ha may reflect potential density in very small areas. Nonetheless, it is considered imperiled (v. 56%), though it does not appear on any of the IUCN's lists of most endangered species (v. 15.5%). In other respects *Teira* is probably similar to other lacertids.

Teius

Teius is a small teiid genus (three spp.) of slightly large (28.3 g v. 22.1 g), moderately well-known (filled cells 40% v. 33%), sand-fleeing (100%, highest value in Teiidae, v. 1.8%), socially tolerant (100% v. 55%) wide-ranging chemosensors (100% v. 97%)

found in southern South America, from Bolivia to Argentina, Uruguay, and Brazil. They resemble striped whiptails with a bright green wash on the head and anterior back.

Body shape is average (shape index 0.85 v. 0.88 std. resid.). Tails are average (2.15 v. 2.30 × SVL) and breakable (100% v. 21%) but not blue (v. 15%). *Teius* are incapable of color change (v. 1.3%) and lack secretory pores (v. 9%).

The sexes differ mildly in size (0.50 v. 0.89 on –2 to 2 scale), color (33% v. 35%), and seasonal color (33% v. 10%).

Two of three species are invertebrate eaters (see species listing: 67% v. 62%); *teyou*, the largest species, is an ant specialist (33% v. 1.1%). Home-range area is small (–0.25 v. 0.19 std. resid.), based on *oculatus*. Direct basking is standard (67% v. 58%); mean active body temperature is a sizzling 39.0°C (v. 36.9°C), based on *oculatus*. All species use burrows as a daily refuge (v. 82%). *T. teyou* may run bipedally and is an arm waver—while foraging it raises one arm and waves it at intervals.

They are found exclusively at low elevation (100% v. 95%) and are of average frequency in littoral areas (33% v. 38%), but they are overrepresented in sandy areas (67% v. 47%), riparian zones (67% v. 46%), and human habitat (67% v. 29%).

Reproduction is fairly standard, save that *suquiensis* is parthenogenetic. Clutch count is average (–0.40 v. –0.51 std. resid.), young are slightly large (ontogenetic mass range –0.17 v. 0.10 std. resid.), and incubation is short (–0.65 v. –0.38 std. resid.). *T. oculatus* and *teyou* may oviposit in ant mounds.

Longevity is not documented. Population density is high (2.5 kg/ha v. 1.06 kg/ha), albeit based solely on one credible sample of *oculatus*. Endemism is low (33% v. 50%). No species occurs on islands (v. 38%) or has been introduced (v. 8%). They have not been evaluated for conservation status. In other respects *Teius* are typical teiids.

Teius spp.	English name				Range core
			Size		
	Activity	Diet	mm	g	Ecological business model
oculatus	Eyewitness Whiptail				Paraguay
	diurnal	invert.	97	24.0	*** wide-ranging chemosensor
suquiensis	Cordoba Whiptail				Argentina
	diurnal	invert.	99	25.9	** wide-ranging chemosensor
teyou	Four-toed Whiptail				Paraguay
	diurnal	ants	110	36.4	*** wide-ranging chemosensor

Tenuidactylus spp.	English name				Range core
			Size		
	Activity	Diet	mm	g	Ecological business model
bogdanovi	Central Asia Thin-toed Gecko				Uzbekistan
			61	4.84	
caspius	Caspian Thin-toed Gecko				Central Asia
	cathemeral	invert.	61	4.82	*** trunk/wall lizard or nocturnal ground lizard
dadunensis	Xinjiang Gecko				China
			48	2.52	
elongatus	Yangihissar Gecko				China
	nocturnal		50	2.74	** nocturnal ground lizard
fedtschenkoi	Turkestan Rock Gecko				Afghanistan
	cathemeral	omnivore	62	5.0	** crack hermit or trunk/wall lizard
longipes	Nikolsky's Long-toed Gecko				Afghanistan
			59	4.42	
turcmenicus	Turkmenian Thin-toed Gecko				Afghanistan
	nocturnal	invert.	65	5.9	** trunk/wall lizard or ground-foraging tree lizard
voraginosus	Southwest Thin-toed Gecko				Afghanistan
	cathemeral	invert.	58	4.30	** trunk/wall lizard

Tenuidactylus

Tenuidactylus is an eight-species genus of padless, cathemeral/nocturnal gekkonids found in central Asia, from Azerbaijan to Mongolia and southward to Iran, China, and possibly India. Though nominally well known (filled cells 35% v. 25%), there is much uncertainty regarding their diel activity pattern and either uncertainty or geographic variation in their microhabitat choice. The result is that none can be assigned to an ecological business model with certainty. Most are likely trunk/wall lizards (60% v. 45%); others are nocturnal ground lizards (20% v. 27%) and crack hermits (20% v. 3.2%).

They are slightly larger than average gekkonids (4.16 g v. 3.31 g) and have slightly longer tails (1.31 v. 1.07 × SVL). Secretory pores are common and may be universal (75% v. 46%). As befits their slightly large size, they are more likely than typical gekkonids to have m>f dimorphism (0.75 v. −0.03 on −2 to 2 scale). Though the sexes are not known to differ in fixed color, seasonal color is present (12.5% v. 0.8%). Overall, the sex differences are stronger than for a typical gekkonid (1.13 v. 0.63 on 0 to 6 scale).

Although omnivory (25% v. 3.8%) is the categorization for one of four species with dietary information (*fedtschenkoi*), the ingestion of plant material in this species seems to be variable and is unquantified. Foraging mode is equally uncertain; only in *caspius* (sit-and-wait) and *turcmenicus* (sit-and-wait, active) has foraging pattern been described.

Consistent with cathemeral/nocturnal activity, spontaneous calling (12.5% v. 6.5%) and stress calling (38% v. 3.7%) are strongly overrepresented. None is said to be conspicuously wary (v. 4.4%). Direct basking is frequent (25% v. 5%).

Solitary living (33% v. 8%) and group living (33% v. 25%) are overrepresented, but male social tolerance is not (0% v. 13%). All documented males defend space (100% v. 87%).

These lizards are found at the dry end of the rainfall spectrum (0–400 mm/y). They are not reported from littoral areas (v. 11%), though that likely reflects the paucity of oceans in central Asia. Conversely, they are overrepresented in sandy areas (17% v. 11%), riparian zones (50% v. 17%), and human habitat (67% v. 27%).

Reproduction is typical, though incubation duration tends to be short (−0.16 v. 0.17 std. resid.). Communal oviposition (25% v. 8.6%) is surprisingly common, though both cases (*caspius, fedtschenkoi*) are from lizards that at least sometimes reside in vertical microhabitats (trunk/wall lizards or crack hermits), where communal oviposition is commonest (table 4.1, trait 68). Young are especially small (ontogenetic mass range large: 0.53, the second highest in Gekkonidae, v. −0.19 std. resid.). Thus, clutches of *Tenuidactylus* are few in number and small in offspring size. They breed multiple times per season, though the number of times is not well documented.

Two species have been introduced (*caspius* in Armenia and the Republic of Georgia; *fedtschenkoi* in at least Kyrgyzstan). Most evaluated species are imperiled (80% v. 47%), though none is one any of the IUCN lists of the most endangered (v. 11.3%). In other respects *Tenuidactylus* are typical gekkonids.

Teratoscincus

Teratoscincus is a six-species genus of large-sized (8.5 g, the heaviest genus average in the family, v. 1.07 g) nocturnal ground lizards (100%, the highest value in Sphaerodactylidae, v. 6.5%) found in Central Asian deserts, from Qatar to Pakistan, Mongolia, and China. They have large heads, large eyes, short snouts, short tails (0.62, the shortest value in the family, v. 1.07 × SVL), and fragile skin. Though this combination is apt to induce baby noises in gecko aficionados, local residents often fear them. *T. scincus* is believed to be fatal on contact. *T. keyserlingii* has a bright eyeshine. All species seem to move slowly on stiffly erect limbs. At least two species can stridulate softly by chaffing their rough skin against itself, a noise that may be audible to conspecifics. In at least some cases, conspecifics have been alerted to seek cover by the sound made by an autotomized *Teratoscincus* tail.

Despite their remote location, this genus is one of the best known (filled cells 44% v. 27%). Their shape index is the second highest in the family (0.19 v. −0.08 std. resid.). Two species are able to darken (33% v. 6.5%). SSD is absent (0.0 v. 0.04 on −2 to 2 scale). The sexes can be compared in relative tail length in two species: *keyserlingii* (m<f) and *microlepis* (m>f).

Teratoscincus spp.	English name		Size		Range core
	Activity	Diet	mm	g	Ecological business model
bedriagai	Bedriaga's Wonder Gecko				Iran
	nocturnal	invert.	61	4.11	*** nocturnal ground lizard
keyserlingii	Fragile Wonder Gecko				Iran
	nocturnal	invert.	95	17.3	**** nocturnal ground lizard
microlepis	Small-scaled Wonder Gecko				Pakistan
	nocturnal	invert.	66	5.3	*** nocturnal ground lizard
przewalskii	Przewalski's Wonder Gecko				China
	nocturnal	invert.	78	9.02	**** nocturnal ground lizard
roborowskii	Roborowski's Wonder Gecko				China
	nocturnal		79	9.3	*** nocturnal ground lizard
scincus	Frog-eyed Wonder Gecko				Iran
	nocturnal	invert.	85	12.0	**** nocturnal ground lizard

Tetradactylus spp.	English name / Activity	Diet	Size mm	Size g	Range core / Ecological business model
africanus	Eastern Long-tailed Seps				South Africa
	diurnal		66	1.97	* wide-ranging chemosensor
breyeri	Breyer's Long-tailed Seps				South Africa
	diurnal		56	1.33	* wide-ranging chemosensor
eastwoodae	Eastwood's Long-tailed Seps				South Africa
	diurnal		50	0.97	
ellenbergeri	Ellenberger's Long-tailed Seps				Central Africa
	diurnal	invert.	61	1.60	* wide-ranging chemosensor
seps	Short-legged Seps				South Africa
	diurnal	invert.	56	1.30	** litter cryptozoid
tetradactylus	Common Long-tailed Seps				South Africa
	diurnal		58	1.41	* wide-ranging chemosensor
udzungwensis	Udzungwa Long-tailed Seps				Tanzania
			57	1.38	* wide-ranging chemosensor

All documented species are generalized invertebrate eaters (v. 82%). Half sit-and-wait (50% v. 83%) and all cruise (100% v. 21%). Home-range area has been quantified only for *przewalskii* (about 1175 m^2): 1.1, the highest value in the family, v. 0.06 std. resid.

In keeping with their nocturnality, vocalizations are relatively common: spontaneous (17% v. 2%); stress (17%, the second-highest frequency in family, v. 2%). For predator concealment, one-third (v. 17%) rely on crypsis, and two-thirds (v. 43%) retreat to a crevice or burrow. *T. keyserlingii* and *scincus* plug their burrows when occupied. Their strongest habitat association is with sandy areas (100%, tied for highest in the family, v. 15%). They are not associated with riparian (v. 19%) or human (v. 33%) habitats. They are about average in littoral zones (20% v. 31%).

Teratoscincus seem to have broken out of the family's commitment to one-egg clutches (mean 1.8, which is –0.73, the second-lowest value in family, v. –0.11 std. resid. when mass adjusted). These notably small clutches are of young that are conspicuously large (ontogenetic mass range –1.1, the lowest value in family, v. –0.12 std. resid.). Incubation is short (–0.05 v. 0.44 std. resid.). Eggs are not adherent (v. 5.6%), but they are sometimes deposited in a communal site (17% v. 11%). Parental care (female nest defense in *scincus*) is also observed (17%, the highest value in family, v. 1.4%).

Longevity is high (0.60 v. –0.15 std. resid.), based on three species (6–13 years). Endemism is low (17% v. 65%). No species occurs on islands (v. 64%), nor has any been introduced (v. 5.1%). Two-thirds are imperiled (v. 41%), though none is on one of the IUCN's lists of the most endangered (v. 23%). In other respects *Teratoscincus* are typical sphaerodactylids.

Tetradactylus

Tetradactylus is a seven-species genus of small (1.40 g v. 16 g) elongate, limb-reduced, diurnal terrestrial gerrhosaurids from southern Africa, ranging from the Democratic Republic of the Congo to Tanzania and South Africa. They are poorly known (filled cells 26% v. 29%), and ecological business models are uncertain. Most appear to be wide-ranging chemosensors (83% v. 27%), and one is most likely a litter cryptozoid (17% v. 3%). Their tails are mostly very long (mean 2.86 v. 1.92 × SVL), with the longest-tailed species (*ellenbergeri*) averaging 3.4 × SVL. This lengthening is accompanied by reduction in body width, such that the mean shape index of *Tetradactylus* is a very slender –1.25 (v. 0.38 std. resid.). All species have small limbs (some no more than flaps) and are therefore elongate (100% v. 19%). No sex differences are known. Given their size, it is not surprising that all eat invertebrates (v. 44%). Wariness is about as expected for the family (14% v. 24%). Half of documented species use burrow refugia (50% v. 56%). They are found at relatively high elevation, underrepresented at low (60% v. 93%), and overrepresented at medium elevation (60% v. 32%). Their mix of habitats follows the family pattern, except that they are not reported for human habitats (v. 11%). As far as is known *Tetradactylus* reproduce like other gerrhosaurids, though clutch size is the largest in the family (0.21 v. –0.20 std. resid.), and communal oviposition sites are somewhat more common (14% v. 5%). Longevity is short (–1.51 v. –0.09 std. resid.), though one suspects that additional data and experience will raise the 1.3-year maximum, which is based solely on *africanus*. Although the genus is widespread, 71% of species are endemic (v. 62%). None is found on an island (v. 57%). They are imperiled at frequencies slightly higher than the family (57% v. 50%) for any concern; 29% v. 11% for the IUCN's most endangered categories. *T. east-*

Thaumatorhynchus sp.	English name				Range core
			Size		
	Activity	Diet	mm	g	Ecological business model
brooksi	Sumatra Unicorn Lizard				Indonesia
	diurnal		52	4.47	

Thecadactylus spp.	English name				Range core
			Size		
	Activity	Diet	mm	g	Ecological business model
oskrobapreinorum	Saint Martin Turnip-tailed Gecko				Saint Martin
	nocturnal		80	1.3.	*** trunk/wall lizard
rapicauda	Turnip-tailed Gecko				Neotropics
	nocturnal	invert.	100	25.5	*** trunk/wall lizard and ground-foraging tree lizard
solimoensis	Solimoes River Gecko				Colombia
	nocturnal	invert.	103	28.0	*** trunk/wall lizard

woodae is deemed extinct. In other respects *Tetradactylus* are typical gerrhosaurids.

Thaumatorhynchus

Thaumatorhynchus is a single-species genus of little-known (filled cells 21% v. 34%), minimally crested (v. 38%), arboreal (v. 43%) oviparous (v. 98%) agamid endemic to the Indonesian island of Sumatra. Its hallmark is a flexible 10 mm long rostral horn, apparently larger in males. Little else is known about this strange lizard. It has not been introduced (v. 1.7%) or evaluated for conservation status.

Thecadactylus

Thecadactylus (three spp.) is a genus of large (21.2 g v. 4.74 g) nocturnal (100% v. 78%) trunk/wall phyllodactylids found in the Neotropics from Mexico to Brazil and in the Lesser Antilles but not the Greater Antilles. *T. rapicauda* has the largest distribution of any New World gecko. *Thecadactylus* have a robust build (0.73 v. 0.37 shape index) and short tails (0.70 v. 1.03 × SVL). Original tails, which are often lost, may have a swollen base and a pointed tip, which has aided the emergence of myths about these lizards. In various places, *rapicauda* are associated with scorpions. In northern South America some residents not only believe this species to be poisonous if ingested or contacted but also venomous, by virtue of the inferred tail stinger, and capable of human fatalities. The larger *Thecadactylus* species may be dangerous to smaller lizards; *rapicauda* may eat small ectothermic vertebrates as well as leaves. This species is the largest gecko in the New World. It will tear its skin readily to escape capture. It can parachute among trees with assistance from its extensively webbed toes. It has a reflective layer in its retina and therefore an eye shine, a feature that may be present in the other *Thecadactylus* but is not yet so documented. *T. rapicauda*'s tail is somewhat prehensile (v. 2.2%). It can darken (v. 11% basic color change).

Male *Thecadactylus* are smaller than females in the only SSD-documented species (*rapicauda* –1.0 v. 0.32 on –2 to 2 scale). Other morphological characteristics of *Thecadactylus* match the family.

Both sit-and-wait (100% v. 94%) and active (50% v. 31%) foraging are recorded for *Thecadactylus*. Electric lights are used by *rapicauda* (33% v. 9%). Home-range size for the family is based solely on *rapicauda* (0.34 std. resid.), equating to 570 m^2, though this value is based on monitoring it in a built structure (gazebo). The same species vocalizes throughout the day (33% v. 11%), but it is not reported to give stress calls (v. 5.9%). The primary predator concealment tactic is retreat to vegetation (v. 14%). This species is reported to bask directly and indirectly and to thermally conform. Its mean active body temperature is a normal 26.3°C (v. 26.0°C). This species, the only one documented for rainfall regime, occurs over a notable range of rainfall: intermediate to wet, in this case 900–4600 mm/y. All three species in the genus are found at low elevation (<1200 m). They are not found in littoral (v. 20%), sandy (v. 25%), or riparian (v. 16%) areas, but they are overrepresented in human habitats (100% v. 40%).

Reproduction is by eggs; the clutch count is very low (–1.33 v. –0.57 std. resid.). Young are notably large (ontogenetic mass range –0.34 v. 0.11 std. resid.). Incubation is extremely protracted (2.26 v. 0.18 std. resid.), corresponding to 206 days in *rapicauda*. Adherent (33% v. 10%) and communally deposited eggs (33% v. 16%) are common. Parental care is not known (v. 3.7%). RCM is light (0.059 v. 0.086), further documenting that this genus's egg-laying strategy appears to be "often" (not quantified) rather than "many" or "large."

Longevity information is based

solely on *rapicauda* (6.9 years), which is short when size adjusted (–0.14 v. 0.15 std. resid.). Credible estimates of population density are not available. No species is endemic (v. 67%). Two species occur on islands (67% v. 40%). One species (*rapicauda*) has been introduced (33% v. 7.4%), to St. Croix and possibly St. Thomas in the US Virgin Islands. No *Thecadactylus* is considered imperiled (v. 29%). In other respects *Thecadactylus* follow the family characteristics.

Tiliqua (fig. 4.4D)

Tiliqua (seven spp.) are large (40 g v. 3.3 g), slender (shape index –1.13 v. –0.46 std. resid.), elongate (100% v. 35%), blue-tongued (86%), short-tailed (0.50 v. 1.33 × SVL), mostly diurnal (71% v. 74%), viviparous (100% v. 36%) mostly giant herbivores (86% v. 1%) living in Australia, Papua New Guinea, and Indonesia. That these are remarkable skinks is attested by the number of attributes in which they lead the skink league tables: filled cells (52% v. 25%; first in Scincidae), giant herbivores (see above; second only to *Corucia*), prehensile tails (14% v. 0.7%; behind only *Corucia, Prasinohaema*), m:f relative tail length (1.25 v. 1.03 ratio; first in Scincidae), sex differences rollup (2.14 v. 0.28 on 0 to 6 scale; first in Scincidae), home-range area (1.20 v. –0.07 std. resid.; first in Scincidae), male hierarchies (67% v. 5.6%; behind only *n*=1 sample of *Asymblepharus*), and parental care (57% v. 3%; behind monotypic *Corucia, Tumbunascincus*). They are fourth in the proportion of species introduced (27% v. 2.1%; after only genera with two or three species). *Tiliqua* have many attributes in common with the herbivorous *Corucia* and *Cyclodomorphus*, the other genus of blue-tongued skinks. At the conclusion of this account is a tabulation of the similarities and dissimilarities between the blue-tongue genera (*Tiliqua* and *Cyclodomorphus*).

As noted above, most *Tiliqua* are giant herbivores. The anomaly (*adelaidensis*) is an equally remarkable crack hermit that lives almost exclusively in the appropriated burrows of trap-door spiders. This anomalously small *Tiliqua* (4 g; the others are at least an order of magnitude larger) is fiercely solitary in its burrow and captures prey walking by the well-concealed entrance. Thus, their heads face outward, and their tails are not armored and potentially facing an intruding predator, as is characteristic of skinks such as *Egernia*. Bizarrely for a terrestrial/cryptozoic lizard, *adelaidensis* has a prehensile tail, lacking in the rest of the genus and present in only 2.3% of crack hermits. This tail nonetheless will break (v. 73%). Prehensile tails are generally unbreakable and are associated with arboreality (table 4.1, trait 7). To what avail is a prehensile tail in a burrow-dwelling crack hermit? Perhaps their tails are entangled in roots or wedged in the dense soil if needed to prevent extraction by a predator, possibly by extinct species of monitor lizards. *T. adelaidensis* lacks a blue tongue. Perhaps it is too small to terrorize any predator with the threat of biting (see *Cyclodomorphus* entry, however, for a possible case of blue-tongue mimicry).

A short tail is not expected based on size or ecological business model, but it implies that *Tiliqua* are unlikely to flee as a primary predator concealment tactic. Indeed, the only such tactics reported for *Tiliqua* are retreat to a crevice-burrow (50% v. 43%) or retreat to vegetation (50% v. 25%). Long-distance flight would also be associated with long limbs, which these elongate lizards do not possess.

As noted above, *Tiliqua* are distinctively different between the sexes. The big surprise is that such large lizards do not follow the general pattern (table 3.3, trait 1) of m>f. Instead most males are decisively smaller than their mates (–0.75 v. –0.31 on –2 to 2 scale). This suggests that either selective pressures on the female for high fecundity prevail (see reproduction comments below), or that intermale competition for mates is relatively subdued (see male social relations comments below). A second big sex-differences surprise is the extraordinarily high m:f ratio of relative tail lengths (1.25). There are no clear associations with this attribute; do females have especially short tails or males especially long ones? What would create selection pressure for such features? Although m>f tail length ratios are associated with m<f SSD in lizards generally, this relationship does not hold in Scincidae, though perhaps *Tiliqua* is the scincid exception that does hold. It is notable that the relative head size of males shifts both ways in relation to females: male heads are bigger/wider/longer in *adelaidensis*, *gigas*, *nigrolutea*, and *scincoides*, but they are smaller/narrower/shorter in *occipitalis* and *rugosa*. These differences are not apparently associated with dietary differences or intermale competition. Are they part of a beauty pageant that highlights the context of their blue tongues?

The dietary scope of *Tiliqua* is wide. Most giant herbivores specialize on leaves, but *Tiliqua* have few such inhibitions. For example, the largest species, *scincoides*, includes not only leaves/flowers and fruit but also fungi, snails, fish, immobile endotherms, and carrion. Fungus eating is documented in five of seven *Tiliqua*. Perhaps the low primary productivity of their environment rewards broad-mindedness in diet, but *scincoides* is found in comparatively mesic eastern coastal areas of Australia. *T. multifasciata* is found in

Tiliqua spp.	English name				Range core
			Size		
	Activity	Diet	mm	g	Ecological business model
adelaidensis	Pygmy Blue-tongue				South Australia
	diurnal	omnivore	90	4.20	**** crack hermit
gigas	New Guinea Blue-tongue				New Guinea
	diurnal	omnivore	258	57	** giant herbivore
multifasciata	Centralian Blue-tongued Skink				Australia
	cathemeral	omnivore	240	47.5	**** giant herbivore
nigrolutea	Blotched Blue-tongued Lizard				New South Wales
	diurnal	omnivore	272	65	*** giant herbivore
occipitalis	Western Blue-tongued Lizard				Australia
	diurnal	omnivore	257	56	** giant herbivore
rugosa	Shingleback				Australia
	diurnal	herbivore	259	58	**** giant herbivore
scincoides	Eastern Blue-tongued Lizard				Australia
	cathemeral	omnivore	284	72	*** giant herbivore

the more arid parts of Australia, and its diet is about the same, though fish, fungi, and endotherms are not mentioned. All species are reported to forage by both sit-and-wait (100% v. 16%) and active (100% v. 92%) foraging. Sit-and-wait is not characteristic of skinks. Carrion-eating in *rugosa* extends to consumption of feces of emus and kangaroos, two herbivores that probably pass some digestible vegetable matter and may provide beneficial microflora.

The remarkably large home-range areas of *Tiliqua* demand explanation. Care needs to be taken with this statistic, in that the quantified *Tiliqua* home ranges are based on *rugosa* and *scincoides*. The home range of *adelaidensis* is so small as to frustrate attempts to measure it. *T. rugosa* has the largest range, and except for the low productivity of the environment, it is hard to see why it roams over such large areas (680–20,400 m^2). Although slow moving, *rugosa* keeps on moving; its PTM is in the 55th percentile. Males move about twice as far as females, as is also the case in *scincoides*, suggesting that mate searching is part of their restlessness. Although considered monogamous over multiple years (see below), *rugosa*'s mate attendance is maintained for only the mating period each year (6–8 weeks). Are they checking out the prospects during the remainder of the year? The majority of adults remained within 100 m (a circle with a 100 m radius is 31,416 m^2) of their capture point over periods of up to six years; it seems likely that mate searching was a factor motivating those that exceeded this displacement distance. *T. scincoides* is also a roamer (5100–12,700 m^2); its social life is less well understood but demonstrates that large home-range areas are not limited to *rugosa*.

The diel activity patterns of *Tiliqua* are primarily diurnal but also cathemeral (29% v. 9%). This likely reflects the difficulties of walking across superheated substrates in the Australian summer. About a third of species are consistently wary (39% v. 11%). Direct basking is universal (100% v. 21%), perhaps a requirement for a large lizard that has chilled during the cold desert night. Mean active body temperature is warm (32.9°C v. 31.1°C). *T. nigrolutea* did not contribute body temperature data but is reported to be active at low temperatures.

All species live more or less solitarily or with a family (100% v. 8%). Group living is not reported (v. 16%). All males defend space, as do most skinks (100% v. 94%). However, hierarchies among males are common, as noted above. Hierarchies often arise when populations are dense (see density discussion below). Most *rugosa* return to their long-term mates over multiple years (67% of males and 74% of females); one pair has been documented to reunite annually for 27 years (Leu et al. 2015 in Whiting and While 2017).

Most species make use of burrows for daily refugia (86% v. 24%). This may reflect the features available in their environment. *Tiliqua* are found in the middle of the rainfall spectrum (only arid to mesic) and are overrepresented in the intermediate rainfall belt (401–1100 mm/y: 75% v. 34%). They are overrepresented in littoral (57% v. 21%), sandy (57% v. 25%), and human (57% v. 14%) habitats. They are about average in frequency in riparian zones (29% v. 23%). *T. scincoides* are actively averse to roads, which no doubt assists their survival and shows them in a more intelligent light than many reputedly smarter mammals.

Litter counts are moderately low (−0.27 v. 0.06 std. resid.). *Tiliqua* have a slightly small ontogenetic mass range (−0.13 v. 0.12 std. resid.), indicating slightly large offspring. Relative "clutch" mass is moderately above the skink average (0.295 v. 0.236 ratio), suggesting that females are moderately heavily laden when full term. Average litter frequencies are less than one per year in five of seven species (only tiny *adelaidensis* regularly breeds once per year). The range of litters per year in *nigrolutea* is less than once every other year: 0.38–0.45. This RCM ratio is consistent with the m<f SSD, implying fecundity selection on female body size prevails over male selection for intermale combat success or hierarchical dominance. Conversely, the low litter counts suggest that the selective pressure may not be simply for number of offspring but number of relatively large offspring.

The high frequency of relatively brief parental care in *Tiliqua* is perhaps related to their high longevity (see next paragraph), low frequency of breeding (preceding paragraph), viviparity, and the need for producing large young that survive. Parental care takes the form of female attendance of the young for a few days. In *rugosa*, both neonates and mother consume the placenta, and both can recognize each other in tests comparing them to unrelated conspecifics. Such tests have not been run in *scincoides* to the best of my knowledge, but in that species only the neonates consume the placenta. In *nigrolutea* the mother consumes the placenta and thereby assists the young in freeing themselves.

Longevity is markedly high (0.90 v. 0.17 std. resid.) and well based (six of seven species). The longest living are *rugosa* (>50 years) and *scincoides* (30 years); the shortest are *occipitalis* (7.4 years) and *adelaidensis* (8 years). Population density is also high (1.11 kg/ha v. 0.35 kg/ha), albeit based on just *adelaidensis* ($n = 1$; 1.0 kg/ha) and *rugosa* ($n = 10$; 1.23 kg/ha). These high values suggest that population density may be contributing to the emergence of male dominance hierarchies. The high densities are perhaps more notable in light of the low primary productivity in *adelaidensis*/*rugosa* habitat, but high densities are also characteristic (table 4.1, trait 71) of giant herbivores such as *rugosa*.

Endemism is low (14% v. 64%). Island species frequency is average (57% v. 57%). Two species have been introduced (first paragraph): *rugosa* to Tasmania and *scincoides* to New South Wales and South Australia (Kangaroo Island). Humans are more prone to move large edible species. *Tiliqua*'s edibility and digging (soil specialization) requirements may have contributed to their high frequency of imperilment (100% v. 47%), including placement on one of the IUCN's lists of the most endangered (33% v. 14%).

T. adelaidensis offers an interesting object lesson in the perils of concluding that the absence of information is information on absence. Although a few specimens had been collected prior to 1959, the absence of recent detections had led the silverbacks of Australian herpetology to conclude by 1990 that *adelaidensis* was probably extinct. And then one turned up in the stomach of a road-killed snake. The search was rejuvenated, and it was eventually discovered that the historical information on the hiding place of this lizard—rock or wood cavities—was erroneous or at the very least misleading. *T. adelaidensis* live in open grassland, away from wood or rock, in the soil cavities dug and then abandoned by trap-door spiders. Although their range is now very restricted, due to habitat loss from urban development and plowing of fields (and therefore the spider burrows), they are locally common.

A comparison between the smaller blue-tongued skinks (*Cyclodomorphus*, mean 8.7 g) and the larger blue-tongued skinks (*Tiliqua*, 40 g) is intriguing because those two genera are likely sister taxa and yet differ in their size and ecological business models. As noted above, *Tiliqua* are primarily giant herbivores and secondarily crack hermits, whereas *Cyclodomorphus* are neither, but they are litter cryptozoids (33%), nocturnal ground lizards (33%), refugia-anchored diurnal ground lizards (22%), and ground-foraging tree lizards (11%). Which of their traits match their phylogeny and which their ecology? In the following comparisons, each genus is contrasted to Scincidae and the two genera are considered a match if their deviation from the skink norm is in the same direction and approximate magnitude.

The two blue-tongued genera are notably similar in morphology (shape index, elongation, prehensile tails), some sex differences (size dimorphism, lack of seasonal color, sex differences rollup), omnivorous diet, cathemerality, wariness, some aspects of habitat (littoral, human habitat), reproductive mode, longevity, and population density. They are not a good match in relation to tail length (*Cyclodomorphus* are slightly short; *Tiliqua* are really short), diurnality (*Cyclodomorphus* strongly avoid diurnality; *Tiliqua* are average), and endemism (*Cyclodomorphus* are nearly average; *Tiliqua* are notably multijurisdictional). The two genera show strongly contrasting trends with regard to foraging mode (*Cyclodomorphus* avoid sit-and-wait; *Tiliqua* embrace it), ontogenetic mass range (high in *Cyclodomorphus*; low in *Tiliqua*), mass-adjusted litter count (high in *Cyclodomorphus*; low in *Tiliqua*), and introductions (none in *Cyclodomorphus*; elevated frequencies in *Tiliqua*). Some of the many traits that appear to be phylogenetically conserved may be at least in part a response to similar body sizes (both genera are above the skink mean size) and similar environments (their geographic ranges overlap extensively): general diet, diel activity pattern, habitat preferences, longevity, and population density. Nonetheless it is conspicuous how similar the natural histories of these two ecologically disparate and yet closely related genera are.

Timon

Timon spp.	English name		Size		Range core
	Activity	Diet	mm	g	Ecological business model
kurdistanicus	Kurdistan Lizard				Iraq
	diurnal		115	34.6	* wide-ranging chemosensor (juveniles are litter cryptozoids)
lepidus	Ocellated Lizard				Iberian Peninsula
	diurnal	omnivore	175	120	*** wide-ranging chemosensor
nevadensis	Sierra Nevada Lizard				Spain
			165	100	
pater	North African Ocellated Lizard				Algeria
	diurnal	invert.	136	56	
princeps	Zagros Lizard				Iraq
	diurnal		112	31.8	
tangitanus	Morocco Ocellated Lizard				Morocco
	diurnal	invert.	123	42.2	

Timon (six spp.) is a genus of very large (56 g, first in Lacertidae, v. 5.1 g) wide-ranging chemosensors found on both sides of the Strait of Gibraltar, from Portugal to Italy and Morocco to Tunisia, as well as a disjunct set of species found in the Middle East, from Syria and Turkey to Iran. Along with a muscular build (shape index 0.85 v. 0.27 std. resid.), all species seem to be marked with colorful lateral or dorsal ocelli. These animals are important in the pet trade and much information has been obtained from the herpetoculture of *lepidus*.

Their tails are average in length (1.99 v. 2.06 × SVL) and breakable (67% v. 92%). They are m>f in SSD (1.0 v. 0.49 on –2 to 2 scale). Most have sexual dichromatism (83% v. 33%) and some have seasonal color (17% v. 15%). In sum, *Timon* exhibit above-average sex differences (1.50 v. 0.86 on 0 to 6 scale).

Diet is known for three species, of which *lepidus* is reported to eat plants (reported in different studies as 80% by occurrence or 10% by mass). Foraging mode is not reported. Home-range area is small (0.10 v. 0.27 std. resid.), based solely on *lepidus*. This species is notable for the noise it makes while foraging in undergrowth. Curiously, this is the only *Timon* that is said to be conspicuously wary (v. 11%). *Timon* hews to the lacertid pattern of using multiple predator concealment tactics, primarily retreat to a crevice-burrow (67% v. 50%) or hiding in vegetation (33% v. 42%). One species (*lepidus*) is documented to bask directly (17% v. 25%).

Social relations in *Timon* are documented for only one species (again *lepidus*), for which reports indicate that it is tolerant, intolerant, defends a mate, and defends a harem. Apparently they are flexible in their social relations. *T. lepidus* is flexible also in rainfall regime (400 to >3000 mm/y) and elevation (0–2500 m). The habitats of five species are described and they do not notably diverge from the lacertid pattern in relation to littoral (20% v.22%), sandy (60% v. 44%), riparian (20% v. 28%), or human (20% v. 16%) habitats.

Reproduction is normal except for clutch count (0.82, the highest in Lacertidae, v. 0.13 std. resid.) and parental care (*lepidus* females defend their clutches, at least in captivity). The mean clutch count is 10.3 eggs. RCM is on the high side (0.305 v. 0.267), as usual based on *lepidus*.

Longevity is high (0.77 v. 0.19 std. resid.), based on three species, with *lepidus* notably high at 28 years. Population density is mildly elevated, based on three estimates of *lepidus* (3.2 kg/ha v. 2.07 kg/ha). Endemism in the genus is low (17% v. 35%). Island species frequency is normal (33% v. 26%). No *Timon* has been introduced (v. 4.6%), and only 25% are considered imperiled (v. 56%). None is placed on one of the IUCN's lists of the most endangered species (v. 15.5%).

Toenayar

Toenayar sp.	English name		Size		Range core
	Activity	Diet	mm	g	Ecological business model
novemcarinata	Nine-keeled Sun Skink				Malaysia
	diurnal	invert.	81	10.8	** refugia-anchored diurnal gr. liz.

Toenayar novemcarinata, formerly in *Eutropis*, is the sole member of its genus. It is a moderately large (10.8 g v. 3.3 g) refugia-anchored diurnal ground (v. 30%) skink found from Thailand and potentially India to Brunei. Aside from a slightly heavier build (shape index 0.32 v. –0.46 std.

resid.), it is morphologically a typical skink. Sex differences are unknown. It is found at low and medium elevations (in this case 0–2000 m). Its reproduction is not documented. It is not endemic (v. 64%), but it is found on islands (v. 57%). It has not been introduced (v. 2.1%), nor is it considered to be imperiled (v. 47%). In other respects *Toenayar* is likely to be similar to other skinks.

Toropuku

Toropuku stephensi is the sole representative of this diplodactylid genus from New Zealand. It is a medium-sized (7.4 g v. 4.2 g) and a reasonably well-known (35% v. 28% of cells filled) nocturnal canopy walker that may bask at the entrance to its refuge. Its tail is a bit long for a diplodactylid (1.19 v. 0.85 × SVL) and breakable (v. 91%). It has secretory pores (v. 34%), but other sex differences are not documented. It is conspicuously wary (v. 2.7%). It is found in littoral areas (v. 17%) but not in sandy areas (v. 24%), riparian zones (v. 10%), or human habitats (v. 5%). As with most New Zealand lizards, it is viviparous (v. 18%) and has high longevity (1.29 v. 0.90 std. resid.). It has not been introduced (v. 2%), and it is considered imperiled (v. 81%), though it is not on the IUCN's lists of the most endangered (v. 29%). In other respects it is a typical diplodactylid.

Tracheloptychus

Tracheloptychus (two spp.) are medium-sized (11.2 g v. 16 g) wide-ranging chemosensor gerrhosaurids found in Madagascar. New World readers can imagine a particularly corpulent *Ameiva*. Their breakable tails (v. 73%) are on the short side (1.55 v. 1.92 × SVL). *T. petersi* is sexually dichromatic. They do not climb, but *petersi* digs burrows for refugia (v. 56%). They are found in littoral zones (50% v. 6%), sand areas (100% v. 25%), and riparian habitats (50% v. 25%) but not human habitat (v. 11%). Clutch count is low (−0.40 v. −0.20 std. resid.), and incubation is a trifle slow (0.13 v. −0.05 std. resid.). Both species are endemic (v. 62%). *T. petersi* is listed by the IUCN as vulnerable. In other respects *Tracheloptychus* are typical gerrhosaurids.

Trachydactylus

Trachydactylus (two spp.) are padless nocturnal ground gekkonids found in Yemen and Oman. As far as is known, they cruise slowly around the desert at night on stiff extended limbs. They are slightly small (2.67 g v. 3.31 g) with short tails (0.78 v. 1.07 × SVL). They have secretory pores (100% v. 46%) but lack color change (v. 15%). They show no SSD, sexual dichromatism, or seasonal color. Their diets are unknown. Commensurate with their nocturnality, they call spontaneously (50% v. 6.5%) and under duress (100% v. 3.7%). They are not conspicuously wary (v. 4.4%). Though they are not reported to bask and are active at night, they have one of the Gekkonidae's highest mean body temperatures (35.0°C v. 27.6°C). Of the tabulated habitat types, they are associated only with riparian zones (100% v. 17%). They are not explicitly associated with sand despite its abundance in their range (0% v. 11%). Reproduction is typical for gekkonids; adherent (0% v. 10.5%) or communally laid (0% v. 8.6%) eggs are not reported. They have not been introduced (v. 3.7%), nor have they been evaluated for conservation status. In other respects *Trachydactylus* are typical gekkonids.

Toropuku sp.	English name				Range core
			Size		
	Activity	Diet	mm	g	Ecological business model
stephensi	Southern Striped Gecko				New Zealand
	nocturnal		72	7.4	** nocturnal canopy walker

Tracheloptychus spp.	English name				Range core
			Size		
	Activity	Diet	mm	g	Ecological business model
madagascariensis	Madagascar Keeled Stout Lizard				Madagascar
	diurnal		74	9.5	** wide-ranging chemosensor
petersi	Peters' Keeled Stout Lizard				Madagascar
	diurnal		82	13.1	** wide-ranging chemosensor

Trachydactylus spp.	English name				Range core
			Size		
	Activity	Diet	mm	g	Ecological business model
hajarensis	Banded Ground Gecko				Oman
	nocturnal		45	2.12	*** nocturnal ground lizard
spatalurus	Spacious Ground Gecko				Yemen
	nocturnal		53	3.37	*** nocturnal ground lizard

Trachylepis

Trachylepis is the fourth-largest skink genus (78 spp., table 2.2); many of its species were recently in *Euprepis* or *Mabuya*. Most species are medium-large (6.9 g v. 3.3 g; range 1.27–136 g), nonelongate (v. 35% elongate), mostly diurnal (96% v. 74%) skinks of diverse ecologies. Ninety-seven percent (v. 20%) are found in the Ethiopian biogeographic realm. Those found outside of Africa are found on the South Atlantic island of Fernando de Noronha (*atlantica*) and doubtfully Brazil (*maculata*). *T. maculata* is known from only a single 1839 type specimen, the geographic origin of which is uncertain, possibly the West Indies instead of Brazil. The vast majority are refugia-anchored diurnal ground lizards (75% v. 30%). The remainder are ground-foraging tree lizards (20% v. 7%), trunk/wall lizards (3.4% v. 3.0%), and diurnal canopy walkers (1.7% v. 1.8%). If refugia-anchored diurnal ground lizard is the starting condition, the others have moved progressively and diminishingly further upwards into arboreal microhabitats. Curiously, they have not moved downward into the second-most-common skink business model: litter cryptozoid. Given their skink ancestry, it is not surprising that they also have not moved into less mobile roles (e.g., crack hermit) or into nocturnal business models. Nonetheless, they have embraced sedentary (sit-and-wait) foraging to a degree uncommon in skinks (41% v. 16%).

They are more robust than a typical skink (shape index 0.18 v. −0.46 std. resid.). Tail length (1.59 v. 1.33 × SVL), breakability (89% v. 73%), and blue coloration (5% v. 4%) are normal.

Unlike typical smaller skinks, *Trachylepis* males tend to be larger than females (0.56 v. −0.31 on −2 to

Trachylepis spp.	English name				Range core
			Size		
	Activity	Diet	mm	g	Ecological business model
acutilabris	Wedge-snouted Skink				Southwestern Africa
	diurnal	invert.	51	2.49	** refugia-anchored diurnal gr. liz.
adamastor	Adamastor Skink				São Tomé and Príncipe
			89	15.0	
affinis	Striped Forest Skink				Congo
	diurnal		66	5.7	*** refugia-anchored diurnal gr. liz.
albilabris	Orange-throated Skink				Cameroon
	diurnal	invert.	65	5.7	*** refugia-anchored diurnal gr. liz.
albotaeniata	Pemba Island Skink				Tanzania
			56	3.29	
angolensis	Angolan Skink				Angola
			66	5.6	
atlantica	Fernando de Noronha Skink				Brazil
	diurnal	omnivore	84	12.1	**** refugia-anchored diurnal gr. liz.
aureopunctata	Gold-dotted Skink				Madagascar
	diurnal		68	6.2	*** refugia-anchored diurnal gr. liz.
bayonii	Bayon's Skink				Tanzania
	diurnal	invert.	67	6.0	*** refugia-anchored diurnal gr. liz.
bensonii	Benson's Skink				Guinea
	diurnal	invert.	52	2.66	*** refugia-anchored diurnal gr. liz.
betsileana	Betsileo Skink				Madagascar
			136	58	
binotata	Ovambo Tree Skink				Angola
	diurnal	invert.	97	19.3	*** ground-foraging tree lizard
bocagii	Tropical Skink				Zambia
			62	4.50	* refugia-anchored diurnal gr. liz.
boettgeri	Boettger's Skink				Madagascar
			49	2.15	* refugia-anchored diurnal gr. liz.
boulengeri	Boulenger's Skink				Tanzania
	diurnal	invert.	79	10.0	** ground-foraging tree lizard (north) or ** refugia-anchored diurnal gr. liz. (south)
brauni	Ukinga Mountain Skink				Tanzania
	diurnal		57	3.45	** refugia-anchored diurnal gr. liz.

2 scale). *Trachylepis* also embrace sexual dichromatism (13% v. 7.6%) and seasonal color (9% v. 6%) at slightly elevated levels. Overall, however, their sex differences rollup score (on scale of 0 to 6) is indistinguishable (0.32 v. 0.28).

All dietary classes except herbivore are represented in *Trachylepis*: carnivore (2.8% v. 1.0%), invertebrate eating (75% v. 84%), ant specialization (2.8% v. 1.6%), and omnivores (19% v. 13%). Although sit-and-wait foraging is notably common in *Trachylepis*, active foraging is also common (71% v. 92%). Electric light use is not known (v. 0.4%). Home-range area is slightly small (−0.19 v. −0.07 std. resid.). Unfortunately, this home-range statistic is based on only two species (*sechellensis*, *wrightii*), both found in the Seychelles Islands. Both are specialists for harvesting eggs and carrion from seabird colonies and may not be representative (27 m^2, 78 m^2). *T. sechellensis* opens bird eggs by pushing them off branches; *wrightii* opens the bird eggs by pushing them off cliffs.

The few *Trachylepis* that are not diurnal are active at any time (cathemeral 2% v. 9%) or crepuscular (2% v. 2.4%). For primary predator concealment tactic, they are notably overrepresented in the flight category (33% v. 4%) and underrepresented in sand diving (11% v. 22%) and retreat to crevice-burrows (22% v. 43%). This penchant for high-energy "flight" contrasts with the genus's equally notable embrace of sedentary (sit-and-wait) foraging. Direct basking is present at the usual frequency (23% v. 21%). Mean active body temperature is hot (33.9°C v. 31.1°C).

Social relations are poorly documented in *Trachylepis*, but group living is notably common (75% v. 16%). This is consistent with socially tolerant males (12.5% v. 5.6%), the

Trachylepis spp.	English name				Range core
			Size		
	Activity	Diet	mm	g	Ecological business model
brevicollis	Short-necked Skink				East Africa
	cathemeral	invert.	123	41.8	*** refugia-anchored diurnal gr. liz.
buettneri	Buettner's Skink				Central Africa
	diurnal	invert.	71	7.2	** diurnal canopy walker
capensis	Cape Skink				Botswana
	diurnal	ants	96	19.0	** refugia-anchored diurnal gr. liz.
casuarinae	Casuarina Island Skink				Mozambique
			76	8.8	
chimbana	Chimba Skink				Angola
	diurnal		61	4.31	** refugia-anchored diurnal gr. liz.
comorensis	Comoro Skink				Comoros
			89	15.0	*** refugia-anchored diurnal gr. liz.
cristinae	Abd Al Kuri Skink				Yemen
			91	15.8	* refugia-anchored diurnal gr. liz.
depressa	Eastern Coastal Skink				Mozambique
	diurnal		69	6.5	** ground-foraging tree lizard
dichroma	Two-colored Skink				Tanzania
	diurnal	omnivore	95	18.3	* refugia-anchored diurnal gr. liz.
dumasi	Passing Zone Skink				Madagascar
	diurnal		49	2.11	** refugia-anchored diurnal gr. liz.
elegans	Elegant Skink				Madagascar
	diurnal	invert.	55	3.09	*** refugia-anchored diurnal gr. liz.
ferrarai	Ferrara's Skink				Somalia
			63	4.85	
gravenhorstii	Gravenhorst's Skink				Madagascar
	diurnal		71	7.0	*** refugia-anchored diurnal gr. liz.
hemmingi	Somali Skink				Somalia
			67	6.0	
hildebrandtii	Hildebrandt's Skink				Ethiopia
			68	6.2	
hoeschi	Hoesch's Skink				Namibia
	diurnal	invert.	72	7.5	*** refugia-anchored diurnal gr. liz.

Trachylepis spp.	English name				Range core
			Size		
	Activity	Diet	mm	g	Ecological business model
homalocephala	Red-sided Skink				South Africa
	diurnal		73	7.8	** refugia-anchored diurnal gr. liz.
infralineata	Europa Skink				Comoros
			87	13.5	
irregularis	Alpine-meadow Skink				Kenya
	diurnal	invert.	70	7.0	*** refugia-anchored diurnal gr. liz.
lacertiformis	Bronze Rock Skink				Zimbabwe
	crepuscular	invert.	47	1.94	*** refugia-anchored diurnal gr. liz.
laevis	Angola Blue-tailed Skink				Namibia
	diurnal	invert.	53	2.82	*** refugia-anchored diurnal gr. liz.
lavarambo	Nosy Be Skink				Madagascar
			53	2.78	
loluiensis	Lolui Island Skink				Uganda
	diurnal	invert.	78	9.7	**** refugia-anchored diurnal gr. liz.
maculata	Demerara Skink				Brazil
			71	7.1	
maculilabris	Speckle-lipped Skink				Tropical Africa
	diurnal	invert.	74	8.1	*** ground-foraging tree lizard
madagascariensis	Malagasy Skink				Madagascar
	diurnal		65	5.4	** refugia-anchored diurnal gr. liz.
makolowodei	Makolowode's Skink				Cameroon
			96	18.8	
margaritifera	Five-lined Rainbow Skink				Southeastern Africa
	diurnal	invert.	87	3.8	*** refugia-anchored diurnal gr. liz.
megalura	Grass-top Skink				Central Africa
	diurnal	invert.	64	5.2	*** refugia-anchored diurnal gr. liz.
mekuana	Bamboutos Mountain Grass Skink				Cameroon
			55	3.15	* refugia-anchored diurnal gr. liz.
mlanjensis	Mulanje Skink				Malawi
			65	5.4	
nancycoutuae	Isalo Skink				Madagascar
			42	1.27	*** refugia-anchored diurnal gr. liz.

fourth-highest genus rate in Scincidae. Male spatial defense is correspondingly low (88% v. 94%). Male hierarchies are not reported (v. 5.6%). This tolerance suggests that population densities are relatively low (see that section, below).

Burrows are used as daily refugia at a typical rate (30% v. 24%). *Trachylepis* are bimodal with regard to the rainfall spectrum, with the primary mode in the dry end (second-highest rate in Scincidae for dry rainfall belt). They are overrepresented in dry (56–150 mm/y: 33% v. 4%) and arid (151–400 mm/y: 67% v. 40%) zones. Their elevational profile is skewed slightly toward higher elevations (l/m/h: 84/51/12 v. 89/35/6). They are slightly underrepresented in littoral habitats (11% v. 21%) and overrepresented in human ones (26% v. 14%).

Over half of the species are viviparous (54% v. 36%). Seven species are either documented as being capable of switching reproductive mode, or they have contradictory accounts suggesting possible geographic variation in reproductive mode. Clutch/litter counts are moderately elevated (0.40 v. 0.06 std. resid.). Young are notably small (ontogenetic mass range 0.95 v. 0.12 std. resid.). Incubation is slightly short (−0.64 v. −0.40 std. resid.). One species (*madagascariensis*) has hard eggshells (v. 9%). One species (*quinquetaeniata*) has a very curious oviposition practice of laying eggs in communal sites (1.3% v. 2.6%) but limiting the number of eggs in a communal site to about five. This is strange because the midpoint of mean clutch counts for this species is 7.4, indicating that females divide their clutches among sites but share sites. Evidently, they do not want to put all of their eggs into one basket. One wonders about the synchrony of oviposition at shared sites. In some species, egg predation is amplified for recently laid eggs, which are presumably more odiferous due to fresh oviductal secretions, and full-term eggs, which are perhaps laden with metabolic products. If that were the case in this species, one would think mixing eggs of differing developmental stages would be detrimental. The females may instead choose to share oviposition sites only with eggs at a similar stage of development. This species is found in 32 countries, and introduced into Egypt, Florida, and possibly India; there should be ample opportunities to study its unique ovipositing behavior. Parental care is absent in *Trachylepis* (v. 3.2%). RCM is normal (0.214 v. 0.236 ratio).

Longevity is very short (–0.80 v. 0.17 std. resid.). This value is based on 16 species (15 on mainland) and therefore is not merely a small-sample artifact. The shortest absolute longevity is 1.5 years (*buettneri*); the longest is 9 years (*dichroma*). This suggests that predation on skinks is relatively heavy in Africa. Population density is exactly average (0.35 kg/ha v. 0.35 kg/ha). This misleading mean is based on 10 samples of three species, and these samples show a striking disparity between the island species (*sechellensis* 5.5 kg/ha) and the mainland species (*buettneri* 0.095 kg/ha, *maculilabris* 0.079 kg/ha). The mean population density on mainland Africa ($n = 2$ species!) is thus a very low 0.087 kg/ha, which seems consistent with the inference from longevity that predation may be high on mainland skinks.

Endemism is slightly low (46% v. 64%). Island species are slightly infrequent (44% v. 57%). Introductions are notably high (5.1% v. 2.1%), including more than two introductions of species (2.6% v. 0.7%). Few *Trachylepis* are imperiled (27% v. 47%) and none is on one of the IUCN's lists of the most endangered (v. 14%). In

Trachylepis spp.	English name		Size		Range core
	Activity	Diet	mm	g	Ecological business model
nganghae	Ngang-Ha Mountain Grass Skink				Cameroon
			51	2.43	** refugia-anchored diurnal gr. liz.
occidentalis	Western Three-striped Skink				Botswana
	diurnal	invert.	89	14.6	** refugia-anchored diurnal gr. liz.
ozorii	Pagalu Skink				Equatorial Guinea
			79	10.0	
pendeana	Pende River Skink				Central African Republic
			67	6.0	** refugia-anchored diurnal gr. liz.
perrotetii	Orange-flanked Skink				Western Africa
	diurnal	invert.	113	32.4	** refugia-anchored diurnal gr. liz.
planifrons	African Tree Skink				Congo
	diurnal	invert.	89	14.8	*** ground-foraging tree lizard
polytropis	Large Forest Skink				Nigeria
	diurnal	invert.	96	18.8	** refugia-anchored diurnal gr. liz.
principensis	Príncipe Skink				São Tomé and Príncipe
			72	7.6	
pulcherrima	Beautiful Skink				Congo
			96	18.8	
punctatissima	Montane Speckled Skink				Southern Africa
	diurnal	invert.	80	10.4	*** ground-foraging tree lizard
punctulata	Speckled Sand Skink				Southern Africa
	diurnal	invert.	47	1.93	** refugia-anchored diurnal gr. liz.
quinquetaeniata	African Five-lined Skink				Africa
	diurnal	omnivore	106	25.8	*** ground-foraging tree lizard
rodenburgi	Rodenburg's Skink				Nigeria
	diurnal	invert.	63	4.85	*** refugia-anchored diurnal gr. liz.
sechellensis	Seychelles Skink				Seychelles Islands
	diurnal	omnivore	83	11.9	*** ground-foraging tree lizard
socotrana	Socotra Islands Skink				Yemen
	diurnal		81	11.1	** refugia-anchored diurnal gr. liz.
sparsa	Karasburg Tree Skink				Botswana
	diurnal	invert.	79	10.0	** ground-foraging tree lizard
spilogaster	Kalahari Tree Skink				Namibia
	diurnal	invert.	72	7.5	*** trunk/wall lizard

Trachylepis spp.	English name				Range core
			Size		
	Activity	Diet	mm	g	Ecological business model
striata	Africa Striped Skink				Southern Africa
	diurnal	omnivore	85	12.5	** ground-foraging tree lizard
sulcata	Western Rock Skink				Namibia
	diurnal	omnivore	74	8.0	*** refugia-anchored diurnal gr. liz.
tandrefana	West Malagasy Skink				Madagascar
			51	2.43	*** refugia-anchored diurnal gr. liz.
tavaratra	North Cape Skink				Madagascar
			54	2.90	
tessellata	Tessellated Skink				Oman
	diurnal		81	10.8	** trunk/wall lizard
thomensis	São Tomé Skink				São Tomé and Príncipe
			79	10.2	
varia	Variable Skink				Southern Africa
	diurnal	invert.	57	3.48	** ground-foraging tree lizard
variegata	Variegated Skink				Namibia
	diurnal	omnivore	47	1.86	*** refugia-anchored diurnal gr. liz.
vato	Boulder Skink				Madagascar
	diurnal		49	2.11	** refugia-anchored diurnal gr. liz.
vezo	Starry Night Skink				Madagascar
			48	2.02	** refugia-anchored diurnal gr. liz.
volamenaloha	Copper-headed Skink				Madagascar
			47	1.83	** refugia-anchored diurnal gr. liz.
wingati	Wingate's Skink				Ethiopia
	diurnal		81	10.8	
wrightii	Wright's Skink				Seychelles Islands
	diurnal	carnivore	121	40.0	*** ground-foraging tree lizard

other respects *Trachylepis* are similar to average skinks.

Trapelus

Trapelus is a moderate (13 spp.) radiation of average-sized (21.8 g), mostly crestless, large-headed terrestrial agamids in northern Africa and southwestern Asia, from Morocco and Mauritania to Russia and China, south to Chad, Yemen, and India. They are all refugia-anchored diurnal ground lizards (v. 43%). Half are omnivores (50% v. 26%), the other half invertebrate eaters (50% v. 46%). They have relatively short tails (1.33 v. 1.86 × SVL) and an elevated proportion of species with reluctantly breakable tails (31% v. 17%), secretory pores (31% v. 13%), color change (46% v. 22% for any change; 38% v. 13% for advanced), sexual dichromatism (69% v. 53%), and seasonal color (77% v. 20%). Many males are seasonally gaudy. Some species (e.g., *agilis*) are capable of nearly instantaneous color change. Dimorphic features such as gular sacs are common but inconsistent among individuals and populations. Though they have many sexual differences (2.9 v. 1.6 on 0–6 sex differences rollup scale), they are not especially sexually dimorphic in body size (0.33 v. 0.53 on –2 to 2 scale).

Somewhat unusual for agamid genera, they have been described with all three foraging modes: sit-and-wait (83% v. 88%), cruise (17% v. 6%), and active (50% v. 17%) foraging. In their low productivity environment—mostly deserts, including the central Sahara—they may need to be opportunistic foragers. For an agamid genus, they are diverse in predator concealment tactics: crypsis (44% v. 51%), crevice-burrow (22% v. 29%), and hide in vegetation (39% v. 34%). Their social lives have not been reported. They live in the hyperdry zone (<55 mm/y: 25% v. 17%), dry zone (56–150 mm/y: 75% v. 30%), and arid zone (151–400 mm/y: 50% v. 35%).

Reproductive traits closely match those of Agamidae, except for a high frequency of hard-shelled eggs (17% v. 2%) and a high frequency of adherent eggs (8% v. 0.8%). Perhaps the hard-shelled eggs in this case protect against desiccation.

Their mass-adjusted longevity is slightly short compared to agamids (–0.67 v. –0.46 std. resid.). Endemism is very low (8% v. 49%). Island species are relatively infrequent (15% v. 36%). They have not been introduced (v. 1.7%), but they do suffer from agamid-typical levels of imperilment (38% v. 37%). None is on one

Trapelus spp.	English name	Activity	Diet	Size mm	Size g	Range core	Ecological business model
agilis	Brilliant Ground Agama	diurnal	invert.	84	19.0	Iran	**** refugia-anchored diurnal gr. liz.
agnetae	Desert Agama	diurnal	omnivore	80	16.5	Saudi Arabia	*** refugia-anchored diurnal gr. liz.
boehmei	Morocco Agama	diurnal		91	24.8	Algeria	*** refugia-anchored diurnal gr. liz.
flavimaculatus	Yellow-spotted Agama	diurnal	invert.	102	35.8	Saudi Arabia	
jayakari	Muscat Agama			119	56.9	Southern Arabia	
megalonyx	Afghan Ground Agama	diurnal	invert.	75	13.6	Pakistan	**** refugia-anchored diurnal gr. liz.
mutabilis	Changeable Agama	diurnal	omnivore	77	15.1	Northern Africa	*** refugia-anchored diurnal gr. liz.
rubrigularis	Red-throated Agama	diurnal	invert.	77	15.0	Pakistan	**** refugia-anchored diurnal gr. liz.
ruderatus	Horn-scaled Steppe Agama	diurnal	omnivore	90	23.7	Iraq	** refugia-anchored diurnal gr. liz.
sanguinolentus	Steppe Agama	diurnal	omnivore	114	50.4	Central Asia	** refugia-anchored diurnal gr. liz.
savignii	Savigny's Agama	diurnal	omnivore	90	24.1	Egypt	** refugia-anchored diurnal gr. liz.
schmitzi	Schmitz' Agama			60	6.9	Sahara	
tournevillei	Erg Agama	diurnal	invert.	79	15.9	Northern Sahara	** refugia-anchored diurnal gr. liz.

of the IUCN's lists of most endangered (v. 5%).

Tretioscincus

Tretioscincus is a three-species genus of diurnal gymnophthalmids found in northern South America and a few nearby islands. The confusing genus name accurately suggests some evolutionary convergence with characters often found in skinks. They are comparatively well known by gymnophthalmid standards (filled cells 31% v. 19%) and share dark flanks, dorsolateral cream lines, and bright blue tails (v. 2.5%). Remarkably, the blue tails may be missing in juveniles, at least in *agilis* and *oriximinensis*. Their ecological business models are litter cryptozoid (67% v. 55%) and ground-foraging tree lizard (33% v. 4%). They are of average size (2.22 g v. 2.35 g) and build (shape index –0.20 v. –0.32 std. resid.). Tails are typical (1.59 v. 1.72 × SVL) and breakable (v. 28%). They possess secretory pores (v. 29%), and may be elongate (33% v. 41%). They are sexually monomorphic in size (0.0 v. –0.12 on –2 to 2 scale), but *oriximinensis* males have unusually long tails compared to females (1.16 v. 1.05 ratio of m:f).

T. bifasciatus is reported to be conspicuously wary (33% v. 7%). *Tretioscincus* retreat to vegetation or burrow/crevices to conceal themselves from predators. Two-thirds bask directly (v. 8%). Males of the one documented species (*agilis*) are socially tolerant. All three species are found in human habitats (v. 15%). One species (*bifasciatus*) is found on islands (v. 5%) and has been introduced to two other islands (v. 5%). None is considered imperiled (v. 25%). In other respects *Tretioscincus* are probably typical gymnophthalmids.

Tretioscincus spp.	English name				Range core
			Size		
	Activity	Diet	mm	g	Ecological business model
agilis	Smooth False Skink				Venezuela
	diurnal	invert.	46	2.11	**** ground-foraging tree lizard
bifasciatus	Striped False Skink				Venezuela
	diurnal		54	2.08	*** litter cryptozoid
oriximinensis	Oriximina False Skink				Colombia
	diurnal	invert.	49	2.51	*** litter cryptozoid

Tribolonotus spp.	English name				Range core
			Size		
	Activity	Diet	mm	g	Ecological business model
annectens	New Britain Spiny Skink				New Guinea
			45	1.65	
blanchardi	Blanchard's Crocodile Skink				New Guinea
	diurnal		36	0.83	** litter cryptozoid
brongersmai	Manus Spiny Skink				New Guinea
			54	2.93	
gracilis	Mascara Spiny Skink				New Guinea
	diurnal		83	11.8	** litter cryptozoid
novaeguineae	New Guinea Spiny Skink				New Guinea
	diurnal		87	13.8	** litter cryptozoid
ponceleti	Poncelet's Crocodile Skink				New Guinea
	nocturnal	invert.	108	27.2	** sand shark
pseudoponceleti	Western Crocodile Skink				New Guinea
			66	5.6	
schmidti	Schmidt's Crocodile Skink				Solomon Islands
	diurnal	invert.	37	0.85	* crack hermit

Tribolonotus

Tribolonotus (eight spp.) are atypical skinks possessing spiny or rugose scales. This feature has no obvious ecological correlate, though perhaps it inhibits scale wetting in an excessively wet environment. *Tribolonotus* are of normal size (4.14 g v. 3.25 g), diel pattern (diurnal 80% v. 74%), and range of ecological business models: litter cryptozoid (60% v. 28%), crack hermit (20% v. 2%), and sand shark (20% v. 7%). They are found on New Guinea and surrounding islands in Indonesia and the Solomon Islands. They have an extraordinary frequency of secretory pores (88% v. 0.4%, the highest value in Scincidae), but there is no obvious connection between this attribute and their spiny scales. Despite their modest body size, they have the family's fifth-highest SSD score (1.0 v. –0.31 on –2 to 2 scale). They do not bask (v. 21%) but do thermoconform (25% v. 1.5%) and can die in the hand from overheating. They call spontaneously (25% v. 0.1%, the highest rate in Scincidae). They have high levels of parental care, especially for oviparous skinks (12.5% v. 3.2%). Thus, there are many unique attributes associated with this genus but no apparent narrative that unites these features.

Morphologically *Tribolonotus* have large heads and robust bodies (shape index 0.01 v. –0.46 std. resid.). Tail breakage is low (13% v. 73%) and no tails are blue (v. 4%).

The unusual dominance of m>f SSD is not paired with other sex differences aside from an extraordinarily low m:f ratio of relative tail lengths (0.83 v. 1.03 ratio, unfortunately based solely on *brongersmai*). Their sex differences rollup score is high (1.0 v. 0.28 on 0 to 6 scale). Perhaps spiny skin is a mate-attracting feature in *Tribolonotus*.

Diet is poorly documented but unremarkable (invertebrate eating 100% v. 84%). Electric light use and home-range area are unknown. Wariness is unremarkable (13% v. 11%). They are found primarily at low elevation (l/m/h: 100/20/0 v. 89/35/6). They are absent from littoral (v. 21%), sandy (v. 25%), and human (v. 14%) habitats.

Except for *schmidti* they are all oviparous (88% v. 64%) and have a fixed litter/clutch count of one, which is conspicuously low for a skink of this size (–0.65 v. 0.06 std. resid.). They almost undoubtedly lay more than one clutch per year, but this is not documented. Incubation is short (–0.79 v. –0.40 std. resid.) based on *gracilis* (43 days). The female of this species tends both the eggs (perhaps to remove mold) and the young (duration not specified).

Longevity of *ponceleti* is 8 years (0.03 v. 0.17 std. resid.). Endemism is low (38% v. 64%). All species live on islands (v. 57%). None has been introduced (v. 2.1%) or is imperiled (v. 47%). In other respects *Tribolonotus* are typical skinks.

Trioceros

Trioceros is the largest chamaeleonid genus (40 spp.). It is composed of "typical" chameleons (see table 5.2 in Chamaeleonidae) from central Africa, many of which sport flamboyant rostral horns. With the exception of incubation duration (see below) and ontogenetic size range (see below), they fit all aspects of the "typical" chameleon template: ecological business model (100% v. 98% diurnal canopy walkers), mass (12.6 g v. 13.5 g), shape index (0.56 v. 0.57 std. resid.), crests (30% v. 36%), relative tail length (1.01 v. 1.09 × SVL), SSD on a scale of −2 to 2 (0.27 v. 0.41), arboreality (100% v. 99%), and clutch count (2.1 v. 2.8 std. resid.). There are no population density estimates for *Trioceros*.

Incubation duration is anomalously short (1.13 v. 2.43 std. resid. for "typical" chameleons, 1.89 for all chamaeleonids, and 0.44 for dwarf chameleons). Ontogenetic size range is between the chamaeleonid and lizard norms (0.28 v. 0.53 for family or 0.0 for lizards). The low value indicates relatively large offspring compared to "typical" chameleons and relatively small offspring compared to dwarf chameleons (v. −0.16 std. resid.). Their geographic location near the equator may serve to expedite egg development in the low elevation forms that are oviparous. In contrast, viviparous species may have long gestations: the mean of the gestation midpoints is 171 days, and three species gestate up to 300 days (*fuelleborni*, *jacksonii*, and *werneri*). Thus, the oviparous minority of *Trioceros* may be matched to warm environments.

Note also that although the mean

Trioceros spp.	English name				Range core
			Size		
	Activity	Diet	mm	g	Ecological business model
affinis	Ethiopian Highland Chameleon				Ethiopia
	diurnal	invert.	67	8.0	**** diurnal canopy walker
balebicornutus	Bale Two-horned Chameleon				Ethiopia
	diurnal	invert.	64	7.1	**** diurnal canopy walker
bitaeniatus	Two-striped Chameleon				Central Africa
	diurnal	invert.	74	10.2	**** diurnal canopy walker
camerunensis	Cameroon Dwarf Chameleon				Cameroon
	diurnal	invert.	77	11.4	*** diurnal canopy walker
chapini	Chapin's Chameleon				Congo
	diurnal	invert.	67	7.8	*** diurnal canopy walker
conirostratus	Sudanese Unicorn Chameleon				Kenya
	diurnal	invert.	57	5.2	*** diurnal canopy walker
cristatus	Crested Chameleon				Central Africa
	diurnal	invert.	124	40.8	**** diurnal canopy walker
deremensis	Usambara Three-horned Chameleon				Tanzania
	diurnal	invert.	123	40.4	**** diurnal canopy walker
ellioti	Montane Side-striped Chameleon				Central Africa
	diurnal	invert.	79	12.3	**** diurnal canopy walker
feae	Bioko Montane Chameleon				Equatorial Guinea
	diurnal	invert.	84	14.3	**** diurnal canopy walker
fuelleborni	Poroto Three-horned Chameleon				Tanzania
	diurnal	invert.	91	17.8	**** diurnal canopy walker
goetzei	Goetze's Whistling Chameleon				Zambia
	diurnal	invert.	78	11.7	*** diurnal canopy walker
hanangensis	Mount Hanang Chameleon				Tanzania
	diurnal	invert.	63	6.7	*** diurnal canopy walker
harennae	Harenna Forest Chameleon				Ethiopia
	diurnal	invert.	62	6.5	**** diurnal canopy walker
hoehnelii	Kenya High-casqued Chameleon				Kenya
	diurnal	invert.	79	12.1	**** diurnal canopy walker
incornutus	Ukinga Hornless Chameleon				Tanzania
	diurnal	invert.	72	9.6	**** diurnal canopy walker
ituriensis	Ituri Forest Chameleon				Congo
	diurnal	invert.	98	21.7	*** diurnal canopy walker
jacksonii	Jackson's Chameleon				Tanzania
	diurnal	invert.	118	35.6	**** diurnal canopy walker
johnstoni	Rwenzori Three-horned Chameleon				Congo
	diurnal	invert.	99	22.5	**** diurnal canopy walker
kinangopensis	Aberdare Mountains Chameleon				Kenya
	diurnal	invert.	54	4.48	*** diurnal canopy walker

Trioceros spp.	English name	Range core	Activity	Diet	Size mm	Size g	Ecological business model
kinetensis	Mount Kineti Chameleon	South Sudan	diurnal	invert.	59	5.5	**** diurnal canopy walker
laterispinis	Spiny-sided Chameleon	Tanzania	diurnal	invert.	66	7.5	**** diurnal canopy walker
marsabitensis	Marsabit One-horned Chameleon	Kenya	diurnal	invert.	71	9.2	**** diurnal canopy walker
melleri	Giant One-horned Chameleon	Mozambique	diurnal	invert.	211	171	**** diurnal canopy walker
montium	Cameroon Sailfin Chameleon	Cameroon	diurnal	invert.	88	16.5	*** diurnal canopy walker
narraioca	Mount Kulal Chameleon	Kenya	diurnal	invert.	71	9.2	**** diurnal canopy walker
ntunte	Mount Nyiru Chameleon	Kenya	diurnal	invert.	66	7.5	**** diurnal canopy walker
nyirit	Cherangani Hills Chameleon	Kenya	diurnal	invert.	74	10.3	**** diurnal canopy walker
oweni	Owen's Three-horned Chameleon	Congo	diurnal	invert.	119	37.0	**** diurnal canopy walker
perreti	Southern Peacock Chameleon	Cameroon			71	9.2	**** diurnal canopy walker
pfefferi	Pfeffer's Chameleon	Cameroon	diurnal	invert.	77	11.6	**** diurnal canopy walker
quadricornis	Four-horned Chameleon	Nigeria	diurnal	invert.	117	35.2	**** diurnal canopy walker
rudis	Rwenzori Bearded Chameleon	Congo	diurnal	invert.	62	6.5	*** diurnal canopy walker
schoutedeni	Schouteden's Dwarf Chameleon	Congo	diurnal	invert.	70	9.0	** diurnal canopy walker
schubotzi	Mount Kenya Dwarf Chameleon	Kenya	diurnal	invert.	69	8.5	*** diurnal canopy walker
serratus	Serrated Chameleon	Nigeria			75	10.7	**** diurnal canopy walker
sternfeldi	Tanzanian Dwarf Chameleon	Tanzania	diurnal	invert.	72	9.7	*** diurnal canopy walker
tempeli	Udzungwa Double-bearded Chameleon	Tanzania	diurnal	invert.	82	13.4	*** diurnal canopy walker
werneri	Werner's Three-horned Chameleon	Tanzania	diurnal	invert.	88	16.2	**** diurnal canopy walker
wiedersheimi	Northern Peacock Chameleon	Nigeria	diurnal	invert.	82	13.4	**** diurnal canopy walker

score on a scale of –2 to 2 for SSD (0.27) is low, this reflects a close frequency match between m>f and m<f species rather than an absence of SSD. In general, m>f SSD increases with size, but that is not the case in *Trioceros*, for which the largest species (*melleri*) has minimal SSD and also notably lacks sex differences in rostral horns and hemipenial bulges.

Several species are said to reside in aggregations (*affinis*, *bitaeniatus*). Three species are reported to be monogamous (*bitaeniatus*, *hoehnelii*, *jacksonii*); *hoehnelii* pairs are said to remain together for multiple years. This species is also reported to occasionally snag their tongues on rose bush thorns and in due course expire from the incapacitation. The power of chameleon tongues is illustrated by *montium*, for which a 100 g lizard was measured to generate a 43 g pull with its tongue. *T. jacksonii* uses it tongue in another novel way; it is reported to wipe saliva on branches and then snap up the flies that are attracted to the saliva deposit. The reporting authors do not claim that this feat was premeditated. *T. goetzei* is noted for the whistle-like noise it emits when captured, presumably for the startle effect. This sound is not given spontaneously.

In most other respects *Trioceros* are similar to most chamaeleonids; there are some exceptions: seasonal color is less frequent (5.0% v. 18.2%), home-range area is relatively large (0.07 v. –1.12 std. resid., though based on only one sample of one species—*jacksonii*), direct basking is relatively infrequent (10.0 v. 18.7%) despite several species living in very cold, even snowy, high-elevation environments. As well, *Trioceros* are underrepresented at low elevation (53% v. 76%) and overrepresented at medium (78% v. 64%) and high (60% v. 21%) elevations. They are underrepresented in riparian zones (13% v. 21%) and comparatively frequent in human habitats (43% v. 23%). Viviparity in *Trioceros* is exceptionally frequent (62% v. 30%: high elevation?) and longevity is relatively long for a chameleon (–0.13 v. –0.83 std. resid.). Although 98% of *Trioceros* are considered imperiled (v. 95% of chameleons), only 8% are in the most extreme IUCN categories (v. 25%). Factors commonly mentioned with regard to conservation vulnerability include heavy collecting for the pet trade and fragmented and restricted distributions.

Trogonophidae

The six trogonophids are very small (1.78 g v. 5.33 g), mostly subterranean foraging (80% v. 7.8%), nocturnal (100%, tied for first place with Lanthanotidae, v. 25%) lizards found in the western Mediterranean, from Spain to Morocco, Algeria, and Tunisia, and in the Red Sea area, from Somalia to the Arabian Peninsula (Yemen to Kuwait). Those which are not subterranean foragers are likely sand sharks (20%, the highest family, v. 1.9%). No family of lizards has more than 20% sand sharks; Scincidae is second with 6.8%, suggesting that this ecological business model is not phylogenetically canalized but arises opportunistically in numerous families (seven) when suitable substrates are available. Although the lizard mean for nocturnality is 25%, among subterranean foragers it is 11.7% (table 4.1, trait 32), emphasizing how unusual Trogonophidae is in its 100% nocturnality. It is tempting, but erroneous, to pronounce the family name with a double *i*.

Another trait in which Trogonophidae is an outlier is longevity (1.27, behind only Sphenodontidae, v. 0.0 std. resid.), based on 14 years in 4.3 g *Trogonophis wiegmanni*. Longevity is typically short in subterranean foragers (–0.92 v. 0.0 std. resid. for lizards; table 4.1, trait 70), highlighting the uniqueness of Trogonophidae. Extended longevity may be more widespread than is documented in subterranean foragers, as so few data have been collected on them. Another noteworthy attribute of Trogonophidae is their extraordinary RCM ratio (0.500 v. 0.226), based on *wiegmanni*. This statistic does not flatter the concept that RCM is responsive to locomotor needs; imagine pushing through soil while so enormously pregnant, the highest mean of any family. In contrast, subterranean foragers overall average about the same RCM as average lizards (0.252 v. 0.226; table 4.1, trait 65).

Another remarkable attribute of this family is their very low mean active body temperature, exceeded only by Anniellidae, Sphenodontidae, and Xenosauridae: 22.0°C (v. 30.8°C). This temperature is based on a laboratory gradient study of *wiegmanni* and needs field confirmation and sampling of additional species. It is much lower than other subterranean foragers (27.6°C; table 4.1, trait 43).

The Trogonophidae genera are *Agamodon*, *Diplometopon*, *Pachycalamus*, and *Trogonophis*. Collectively they are poorly known (filled cells 22% v. 27%). Their average shape index is –3.43 (v. 0.0 std. resid.), the second lowest (behind Anniellidae). Their tails are a mere 0.07 v. 1.37 × SVL, longer only than Cadeidae and Rhineuridae. Such short tails do not break (v. 57% breakable). Secretory pores are relatively common in Trogonophidae (17% v. 15% in lizards) and compared to 3.9% in subterranean foragers (table 4.1, trait 9). No other sex differences are known, leading to one of the lowest sex differences rollup scores (0.17 v. 0.75 on 0 to 6 scale).

Diet is known for only *wiegmanni*, which is an ant specialist. Direct basking is known in Trogonophidae (17% v. 18%), based on *wiegmanni*, which also basks indirectly. *T. wiegmanni* is found in the intermediate rainfall band (v. 27%). The family's elevational profile is skewed toward low elevation (l/m/h: 100/33/0 v. 82/41/13). They are overrepresented in littoral (33% v. 17%) and sandy (100% v. 22%) areas and absent in riparian (v. 22%) and human (v. 18.5%) habitats.

Equal numbers of trogonophids are oviparous (50% v. 82%) and viviparous, but more data are needed as this inference is based on only two species. Clutch/litter count, based on *wiegmanni*, is about average (0.25 v. 0.0 std. resid.). The young, based on *wiegmanni*, are about average in size (ontogenetic mass range 0.30 v. 0.0 std. resid.). Parental care is known (17% v. 3%) in *wiegmanni*: neonates, which are up to 50% of the length of their mother, may accompany her. To what end? Use her tunnel?

Endemism is high (33% v. 9%). Island species frequency is about average (33% v. 40%). Trogonophids have not been introduced (0% v. 3.3%), nor are they imperiled (v. 49%).

Trogonophis

Trogonophis wiegmanni is the sole member of its trogonophid genus of large (4.34 g v. 1.78 g), handsomely yellow-and-purple speckled subterranean foragers (100% v. 80%) with reduced eyes. It is found in countries bordering the western Mediterranean: Spain, Morocco, Algeria, and Tunisia. It is by far the best-known trogonophid (filled cells 53% v. 22%), and many "facts" about trogonophids are based solely on this species. Surface forays occur in midsummer at night; the purpose of these nocturnal movements is unknown, but periodic exposure—note basking (below)—may account for retention of color in this otherwise hidden lizard. For a trogonophid, *Trogonophis* is of moderate build (shape index –2.96 v. –3.43 std. resid.) and lacks secretory pores (v. 17%). Other sex differences are unknown. It specializes on consumption of immature ants and termites. It may bask directly (v. 17%) and has the low active body temperature (22°C) described in the Trogonophidae account. It is found in the intermediate rainfall belt (in this case 600 mm/y) at low and medium elevation (in this case 0–1900 m) and is associated with sandy areas (v. 100%). It has been described as both oviparous and viviparous, though the latter is more credible. Its clutch/litter count is 4.3 (0.25 v. 0.25 std. resid.), and it breeds about every other year. RCM ratio is an intimidating 0.500 (v. 0.252 for average subterranean foragers: table 4.1, trait 65). The neonates, which are up to 50% of the length of their mother, follow her for a time (see Trogonophidae account). It is not endemic (v. 67%) but is found on many islands (v. 37% insular). In other respects, *Trogonophis* is a typical trogonophid. Irrigation of fields either attracts this species or draws it toward the surface, where many are killed by modern plows. There is some concern that this process has negative consequences for the species, but this lizard is not considered imperiled (v. 0%).

Trogonophis sp.	English name		Size		Range core
	Activity	Diet	mm	g	Ecological business model
wiegmanni	Checkerboard Worm Lizard				Algeria
	nocturnal	ants	161	4.34	**** subterranean forager

Tropidophorus spp.	English name		Size		Range core
	Activity	Diet	mm	g	Ecological business model
assamensis	Northeastern Water Skink				India
	cathemeral	invert.	58	3.71	* nocturnal ground lizard
baconi	Bacon's Water Skink				Indonesia
			95	18.3	
baviensis	Ba Vi Water Skink				Vietnam
			75	8.6	
beccarii	Beccari's Water Skink				Borneo
	diurnal	invert.	81	10.9	** refugia-anchored diurnal gr. liz.
berdmorei	Berdmore's Water Skink				Southeast Asia
	cathemeral	invert.	76	4.62	** refugia-anchored diurnal gr. liz.
boehmei	Boehme's Water Skink				Vietnam
			84	12.1	
brookei	Brooke's Water Skink				Borneo
	diurnal		88	14.2	*** refugia-anchored diurnal gr. liz.
cocincinensis	Southeast Asia Water Skink				Southeast Asia
	diurnal		82	11.2	** refugia-anchored diurnal gr. liz.
davaoensis	Davao Waterside Skink				Philippines
		invert.	79	9.9	** refugia-anchored diurnal gr. liz.
grayi	Spiny Waterside Skink				Philippines
	diurnal	invert.	93	17.0	* crack hermit
guangxiensis	Darning Shan Water Skink				China
			55	3.15	

Tropidophorus

Tropidophorus is a major radiation (28 spp.) of moderately large (6.7 g v. 3.3 g), mostly semiaquatic (65% v. 2%, third highest in Scincidae), mostly cryptic (50% v. 9%) skinks found in southern Asia: India east to China and south through the Philippines to Indonesia. They are distinctly unskink-like in possessing conspicuously striated or keeled rather than smooth scales. Most are refugia-anchored diurnal ground lizards (63% v. 30%); the remainder are nocturnal ground lizards (16% v. 2.3%), crack hermits (10.5% v. 1.7%), ground-foraging tree lizards (5.3% v. 7.0%), or trunk/wall lizards (5.3% v. 3.0%). Some *Tropidophorus* forage in the water—consume water insects—whereas most other semiaquatic lizards take refuge in water but forage mainly on land.

They are proportioned as an average lizard (shape index 0.01 v. –0.46 std. resid.). They lack blue tails (v. 4%) and are elongate at an average frequency (29% v. 35%). They match skinks in SSD (–0.33 v. –0.31 on –2 to 2 scale). Sexual dichromatism is present (3.6% v. 7.6%), but seasonal color is not (v. 6.2%). Thus their sex differences rollup score (on scale of 0 to 6) is subdued (0.11 v. 0.28). There is a curious difference between the sleeping sites of the sexes in *brookei*: males sleep in rock crevices, but females and juveniles sleep on vegetation overhanging streams.

Both diet (invertebrates 100% v. 84%) and foraging mode (active 100% v. 92%) are unremarkable. Electric light use and home-range areas are unknown.

More than regular skinks, they are prone to activity at all times of day (24% v. 9%). They are not crepuscular (v. 2.4%), but diurnality (59% v. 74%) and nocturnality (18% v. 15%) frequencies are normal. *T. mocquardii* gives stress calls, which is very uncommon in skinks (3.6% v. 0.4%). No species is said to be conspicuously wary (v. 11%). There is a heavy reliance on crypsis (as noted in first paragraph) and retreat to crevice-burrows (50% v. 43%). Direct basking is rare (3.6% v. 21%). Social affairs are undocumented. Elevational distribution is normal (l/m/h: 88/27/0 v. 89/35/6). Naturally, they are often found in riparian habitats (80% v. 23%); they are absent from littoral (v. 21%), sandy (v. 25%), and human (v. 14%) habitats.

All are viviparous (v. 36%) and have slightly large litter counts (0.37 v. 0.06 std. resid.). The young are about average or possibly a bit large (ontogenetic mass range –0.06 v. 0.12 std. resid.). Parental care is suspected in *baviensis* (3.6% v. 3.2%).

Longevity is about average (0.21 v. 0.17 std. resid.), as is endemism (50% v. 64%) and insularity (46% v. 57%). None has been introduced (v. 2.1%), and none is imperiled (v. 47%).

In other respects *Tropidophorus* are typical skinks.

Tropidophorus spp.	English name	Activity	Diet	Size mm	Size g	Range core	Ecological business model
hainanus	Hainan Water Skink	diurnal	invert.	48	2.02	Vietnam	** refugia-anchored diurnal gr. liz.
hangnam	Spiny-tailed Water Skink	nocturnal	invert.	82	11.2	Thailand	**** nocturnal ground lizard
iniquus	Kajan Headwaters Skink			78	9.7	Indonesia	
laotus	Laos Water Skink	cathemeral		74	4.19	Thailand	* refugia-anchored diurnal gr. liz.
latiscutatus	Phu Wua Water Skink	nocturnal		82	5.7	Thailand	** nocturnal ground lizard
matsuii	Matsui's Water Skink			77	4.65	Thailand	* refugia-anchored diurnal gr. liz.
microlepis	Small-scaled Water Skink	cathemeral	invert.	69	3.45	Southeast Asia	** refugia-anchored diurnal gr. liz. or nocturnal ground lizard
micropus	Stink Water Skink	diurnal		38	0.68	Indonesia	** refugia-anchored diurnal gr. liz.
misaminius	Misamis Waterside Skink			89	15.0	Philippines	* crack hermit
mocquardii	Mocquard's Water Skink	diurnal		77	9.4	Malaysia	* ground-foraging tree lizard
murphyi	Murphy's Water Skink	nocturnal		78	9.7	Vietnam	*** trunk/wall lizard
noggei	Nogge's Water Skink		invert.	93	17.3	Vietnam	
partelloi	Partello's Waterside Skink			100	21.6	Philippines	
perplexus	Perplexing Water Skink	diurnal		62	4.5	Borneo	
robinsoni	Robinson's Water Skink	diurnal		63	4.85	Thailand	
sinicus	China Water Skink		invert.	60	2.39	Vietnam	**** refugia-anchored diurnal gr. liz.
thai	Thai Water Skink	diurnal		70	3.65	Thailand	** refugia-anchored diurnal gr. liz.

T. partelloi honors Colonel Joseph McDowell Trimble Partello, who survived Custer's Last Stand.

Tropidosaura

Tropidosaura (four spp.) is a small genus of undersized (3.4 g v. 5.1 g) refugia-anchored diurnal ground lacertids found in montane (1400–3337 m) South Africa and Lesotho. The sexes differ primarily in color (25% v. 33%), especially seasonally (75% v. 15%). Most are conspicuously wary (75% v. 11%). At least half bask directly (50% v. 25%). *T. cottrelli* may dig a burrow for its nightly refuge (33% v. 48%). They are not associated with littoral (0% v. 22%), sandy (0% v. 44%), or human (0% v.16%) habitats. Riparian use is normal (25% v. 28%). Clutch count is normal (0.33 v. 0.13 std. resid.), though incubation is very short (−0.62 v. −0.36 std. resid.), albeit based on only *montana* (34 days). This is surprising given the high elevation, suggesting that developing eggs are retained for a time to expedite their early development through basking. No species occurs on islands (v. 26%); none has been introduced (v. 4.6%). One (*cottrelli*) is imperiled (25% v. 56%). None is listed on one of the IUCN's lists of the most endangered (v. 15.5%). In other respects *Tropidosaura* are similar to other lacertids.

Tropidosaura spp.	English name				Range core
			Size		
	Activity	Diet	mm	g	Ecological business model
cottrelli	Cottrell's Mountain Lizard				South Africa
	diurnal	invert.	58	4.51	*** refugia-anchored diurnal gr. liz.
essexi	Essex's Mountain Lizard				South Africa
	diurnal		46	2.24	*** refugia-anchored diurnal gr. liz.
gularis	Cape Mountain Lizard				South Africa
	diurnal	invert.	54	3.73	*** refugia-anchored diurnal gr. liz.
montana	Common Mountain Lizard				South Africa
	crepuscular	invert.	53	3.49	*** refugia-anchored diurnal gr. liz.

Tropidoscincus spp.	English name				Range core
			Size		
	Activity	Diet	mm	g	Ecological business model
aubrianus	Aubry's Whiptailed Skink				New Caledonia
	diurnal		95	18.3	** refugia-anchored diurnal gr. liz.
boreus	Northern Whiptailed Skink				New Caledonia
	diurnal		77	9.4	*** refugia-anchored diurnal gr. liz.
variabilis	Southern Whiptailed Skink				New Caledonia
	diurnal	invert.	65	5.4	*** refugia-anchored diurnal gr. liz.

Tropidoscincus

Tropidoscincus is a small genus (three spp.) of large-sized (9.8 g v. 3.3 g), long-tailed (2.27 v. 1.33 × SVL), sexually dichromatic (100% v. 8%), oviparous (100% v. 64%) refugia-anchored diurnal ground (100% v. 30%) skinks endemic to New Caledonia. They are stockier than an average skink (shape index 0.30 v. −0.46 std. resid.). The sexes differ only in color. All bask directly (v. 21%). They are found in littoral areas (100% v. 21%) but not reported from sandy areas (v. 25%), riparian zones (v. 23%), or human habitat (14%). Clutch size is small for their size (−0.53 v. 0.06 std. resid.). They have not been introduced (v. 2.1%). One-third are imperiled (v. 47%), but none is on one of the IUCN's lists of the most endangered (v. 14%). In other respects *Tropidoscincus* are probably typical skinks.

Tropiduridae

The tropidurids are remarkably unremarkable lizards. They are not in the top or bottom three families in ranked listings of any tabulated trait. Thus, they are distinctly average. They are notably large (17.5 g v. 5.33 g), taxonomically successful (136 spp.), ecologically diverse (six business models), and devoted to sit-and-wait foraging (100% v. 46%). They are diurnal (100% v. 67%), oviparous (100% v. 82%), committed to male spatial defense (100% v. 85%), found in mesic areas (89% v. 43%), very short-lived (−1.17 v. 0.0 std. resid.), and they exhibit low RCM ratios (0.156 v. 0.226). Most species are refugia-anchored diurnal ground lizards (67% v. 24.5%). The others are trunk/wall lizards (13% v. 13%), ground-foraging tree lizards (9% v. 9%), immobility experts (7.6% v. 3.2%), litter cryptozoids (2.5% v. 11.6%), and diurnal canopy walkers (1.3% v. 6.4%). The constituent genera are *Eurolophosaurus*, *Microlophus*, *Plica*, *Stenocercus*, *Strobilurus*, *Tropidurus*, *Uracentron*, and *Uranoscodon*. They are found throughout South America, plus Trinidad and Tobago. They are poorly documented (filled cells 23% v. 27%).

They are robust in build (shape index 1.10 v. 0.0 std. resid.), with moderate-length tails (1.56 v. 1.37 × SVL) that are mostly unbreakable (13% v. 57% breakable) and never prehensile (v. 5%) or blue (v. 2.3%). Crests are slightly overrepresented (10% v. 5%); color changing ability is underrepresented (2.9% v. 10.7% for basic; 0.7% v. 6.0% for advanced). They do not exhibit secretory pores (v. 15%) or elongation (v. 11%).

As with most larger lizards, they exhibit m>f SSD (0.95 v. 0.36 on −2 to 2 scale). Sexual dichromatism is about average (28% v. 21%); seasonal color is underrepresented (2.9% v. 6.5%). Sexual dimorphism in relative tail length is absent (1.06 v. 1.07 m:f ratio). Accordingly, their sex differences rollup score if also very close to the lizard mean (0.80 v. 0.75 on 0 to 6 scale).

Their diet mix is omnivore-heavy, but close to the lizard profile: carnivory is absent (0% v. 3.2%), herbivory is uncommon (6.3% v. 4.1%), invertebrate eating is underrepresented (44% v. 73%), ant specialization is about average (6.3% v. 4.4%), and omnivory is overrepresented (44% v. 15%). In addition to the sit-and-wait dominance mentioned in the first paragraph, cruise foraging is not reported (v. 6.6%) and active foraging is rare (5.6% v. 58%). Electric light use is near the lizard average (1.3% v. 2.0%). Home-range area is not well documented (five species) and is slightly small (−0.38 v. 0.0 std. resid.).

Tropidurids are comparatively unwary (3.7% v. 8.2%). They do not use flight (v. 11%) or burial in sand (v. 6%) as a primary predator concealment tactic; instead they use crypsis (43% v. 43%), retreat to a burrow or crevice (29% v. 25%), and retreat to vegetation (36% v. 23%). They bask directly at a typical frequency (16% v. 18%) and rarely thermoconform (0.7% v. 2.6%). Their mean active body temperature is very close to the lizard average (31.8°C v. 31.1°C).

Group living is somewhat overrepresented (36% v. 20%), but they are not reported to live solitarily (v. 10%), be tolerant of other males (v. 15%), or develop male social hierarchies (v. 7.7%).

In relation to microhabitats, tropidurids are overrepresented in terrestrial strata (74% v. 48%), underrepresented in cryptozoic (1.2% v. 12.1%) strata, and absent underground. They are about average in arboreal places (25% v. 32%). They rarely use burrows as daily refugia (5.4% v. 22.2%).

They are found only in the intermediate (22% v. 27%) and mesic (89% v. 43%) rainfall belts. Their elevational profile is skewed toward high elevations (l/m/h: 65/47/36 v. 82/41/13). Their presence in the tabulated special habitats are all average: littoral (15% v. 17%), sandy (23% v. 22%), riparian (18% v. 22%), and human (14% v. 18.5%) habitats.

All known tropidurids are bisexual and, as noted in the first paragraph, oviparous. Their clutch counts are mildly low (−0.56 v. 0.0 std. resid.), and young are slightly large in size (ontogenetic mass range −0.22 v. 0.0 std. resid.). The low clutch counts are presumably responsible for the low RCM ratios highlighted in the first paragraph. Many tropidurids lay multiple clutches in a year, but quantification of this trait is insufficient to assess reproductive output. Incubation duration is mildly lengthy (0.40 v. 0.0 std. resid.). All eggs are soft (v. 41% hard) and nonadherent (v. 2.4% adherent). Communal oviposition is about average (3.7% v. 4.2%), as is parental care (3.7% v. 2.9%).

Longevity is short, as highlighted in the first paragraph. Population density has been credibly quantified only six times, for just two species within the genus *Tropidurus*. For those mainland species, it averaged fairly high (1.38 kg/ha v. 0.54 kg/ha). Endemism is high (74% v. 64%); island species are scarce (10% v. 40%). Introductions are relatively low (1.5% v. 3.3%), and no species has been introduced more than twice (v. 1.1%). Relatively few are imperiled (23% v. 49%), and only 3% (v. 13.6%) of the evaluated species are on one of the IUCN's lists of the most endangered.

Tropidurus

Tropidurus (Tropiduridae) is a major radiation (30 spp.) of average-sized (19.5 g v. 17.5 g), mostly omnivorous (71%, the highest in the family, v. 44%) or herbivorous (14%, the highest in the family, v. 6%), mostly refugia-anchored diurnal ground lizards (70% v. 67%) found in all South American countries except Chile and Ecuador. Primary countries are Brazil (20 spp.), Paraguay (eight spp.), and Bolivia (eight spp.). Subsidiary ecological business models are ground-foraging tree lizards (20% v. 9%) and trunk/wall lizards (10% v. 13%). The genus leads the family in m>f SSD (1.20 v. 0.95 on −2 to 2 scale), active foraging (10% v. 5.6%), electric light use (3.3% v. 1.3%), wariness (6.7% v. 3.7%), mean active body temperature (35.3°C v. 31.8°C), presence in the intermediate rainfall belt (100% v. 22%), length of incubation (1.19 v. 0.40 std. resid.), and RCM ratio (0.164 v. 0.156). Overall, these loosely combine to create an impression of a relatively territorial species, though direct documentation of that is thin. *T. hispidus* males defend harems.

Tropidurus spp.	English name	Range core	Activity	Diet	Size		Ecological business model
					mm	g	
arenarius	Huacho Lizard	Peru			74	15.6	
bogerti	Auyen Tepui Lizard	Venezuela	diurnal		75	16.5	*** refugia-anchored diurnal gr. liz.
callathelys	Huanchaca Rock Lizard	Bolivia			79	19.1	** refugia-anchored diurnal gr. liz.
catalanensis	Southern Cone Rock Lizard	Paraguay		omnivore	99	35.0	
chromatops	Paragua Rock Lizard	Bolivia			85	23.3	** refugia-anchored diurnal gr. liz.
cocorobensis	Cocorobo Sand Lizard	Brazil		omnivore	65	10.9	
erythrocephalus	Red-headed Rock Lizard	Brazil		omnivore	70	13.6	
etheridgei	Southern Cone Sand Lizard	Paraguay	diurnal	omnivore	81	19.9	** refugia-anchored diurnal gr. liz.
guarani	Hu Mountain Rock Lizard	Paraguay			88	25.6	** refugia-anchored diurnal gr. liz.
helenae	Piaui Rock Lizard	Brazil			59	8.7	
hispidus	Northern Rock Lizard	Western Amazon	diurnal	omnivore	97	32.7	** ground-foraging tree lizard
hygomi	Hygom's Sand Lizard	Brazil		herbivore	62	9.9	
imbituba	Imbituba Rocky Shore Lizard	Brazil			97	32.5	* refugia-anchored diurnal gr. liz.
insulanus	Para Rock Lizard	Brazil			71	14.2	* refugia-anchored diurnal gr. liz.
itambere	Itambere Rock Lizard	Brazil	diurnal	omnivore	73	15.2	** refugia-anchored diurnal gr. liz.
jaguaribanus	Jaguaribe Valley Rock Lizard	Brazil	diurnal		83	21.6	
lagunablanca	Laguna Blanca Lizard	Paraguay			90	27.0	

Shape index (1.11 v. 1.10 std. resid), tail length (1.47 v. 1.56 × SVL), tail breakage (13% v. 13%), and crest frequency (13% v. 10%) are normal. Color change is absent (0% v. 1.3%).

Sexual dichromatism is rife (40% v. 28%) and, in some cases, dramatic. The sexes differ so much in *melanopleurus*—female cryptic, male jet black with white spangles—that they have been repeatedly described as different species. Seasonal color is overrepresented (6.7% v. 2.9%). These combine to a sex differences rollup score (on a scale of 0 to 6) that is the family's highest (1.60 v. 0.80).

Due to the preponderance of herbivores and omnivores (first paragraph), invertebrate eaters are grossly underrepresented (15% v. 44%). Adult *psammonastes* are partial to eating flowers. *T. torquatus* focus on flowers and fruits. The family's only datum on home-range area comes from this species (−0.38 v. −0.38 std. resid.), a low value compared to other lizards (0.0 std. resid.). Most predator avoidance is by retreat to a crevice or burrow (40% v. 29%) or vegetation (40% v. 36%). The remainder rely on crypsis (20% v. 43%). Direct basking is about average (23% v. 16%).

Group living is the family's second highest (50% v. 36%). The genus's elevational profile is skewed low (l/m/h: 93/21/0 v. 65/47/36). Use of specialty habitats is about the same as the family: littoral (8% v. 15%), sandy (25% v. 23%), riparian (13% v. 18%), and human (13% v. 14%) habitats.

Clutch count is about average (−0.47 v. −0.56 std. resid.). Hatchlings are comparatively small (ontogenetic mass range 0.32 v. −0.22 std. resid.). RCM ratio is slightly high (first paragraph). Incubation duration is more unusually high (first paragraph). Some breeding goes on throughout the year, and *hispidus* has multiple clutches, which may be oviposited in ant mounds.

Tropidurus spp.	English name / Activity	Diet	Size mm	Size g	Range core / Ecological business model
melanopleurus	Highway Stripe Lizard				Bolivia
	diurnal		83	21.5	*** refugia-anchored diurnal gr. liz.
montanus	Southern Mountain Rock Lizard				Brazil
	diurnal	invert.	78	18.0	* ground-foraging tree lizard
mucujensis	Mucuje Rock Lizard				Brazil
			65	10.9	
oreadicus	Brazil Treebase Lizard				Brazil
	diurnal	omnivore	88	25.1	**** ground-foraging tree lizard
pinima	Santo Inacio Rock Lizard				Brazil
		herbivore	70	13.7	*** refugia-anchored diurnal gr. liz.
psammonastes	Xique-Xique Sand Lizard				Brazil
	diurnal	omnivore	86	23.4	* refugia-anchored diurnal gr. liz.
semitaeniatus	Mono-striped Rock Lizard				Brazil
	diurnal	omnivore	78	18.0	*** refugia-anchored diurnal gr. liz.
sertanejo	Caatinga Lizard				Brazil
	diurnal		66	11.8	*** refugia-anchored diurnal gr. liz.
spinulosus	Chaco Spiny Lizard				Paraguay
	diurnal	invert.	97	32.6	*** trunk/wall lizard
tarara	Natural Cerradoes Lizard				Paraguay
			97	33.2	
teyumirim	Ybycui Rock Lizard				Paraguay
			77	17.3	*** refugia-anchored diurnal gr. liz.
torquatus	Amazon Rock Lizard				Brazil
	diurnal	omnivore	94	30.1	**** ground-foraging tree lizard
xanthochilus	Santa Cruz Tree Lizard				Bolivia
	diurnal		99	35.2	** trunk/wall lizard

Population density is based on *itambere* ($n = 4$) and *torquatus* ($n = 2$) and averages 1.38 kg/ha, the only values for the family (v. 0.54 kg/ha for lizards). Longevity is short (–1.52 v. –1.17 std. resid.), based on *hispidus* (2 years) and *torquatus* (3 years). Endemism is average (77% v. 74%). Island species are minor (3.3% v. 10.3%). Introductions are overrepresented (3.3% v. 1.5%). Imperiled species are an average frequency (25% v. 23%), though none is on one of the IUCN's lists of the most endangered (v. 3%). In other respects *Tropidurus* are typical tropidurids.

Tropiocolotes

Tropiocolotes (Gekkonidae) is a 10-species genus of very small (0.73 g v. 3.31 g), padless, nocturnal ground lizards from the Sahara and nearby parts of the Middle East, from Western Sahara to Egypt, Somalia, and Iran. In addition to their small size—the largest species barely exceeds 1 g—they are slender (shape index –0.16 v. 0.10 std. resid.). They tend to have smaller males than females (–0.38 v. –0.03 on –2 to 2 scale), as befits their small size, but sexual dichromatism is not reported (v. 6.6%). Pore frequency is average (40% v. 46%). Cruise foraging is notably common (80% v. 24%), though both sit-and-wait (40% v. 81%) and active (20% v. 14%) foraging are known. Electric light use is not (0% v. 3.8%). Commensurate with their nocturnality, spontaneous vocalization is common (30% v. 6.5%), as is stress calling (30% v. 3.7%). In *nubicus*, the calls from these tiny geckos are audible to humans from 50 m. No species is conspicuously wary (v. 4.4%). Unusually for gekkonids, both documented species (*nattereri*, *steudneri*) rely principally on flight for predator concealment (v. 4.8%), though normal locomotion is conspicuously slow in four of four speed-documented species. The speed needed to escape a nocturnal predator may be less than that associated with daytime flight. Social relations are documented only for *nubicus*, which is said to live in undefined groups. *T. nubicus* is anomalous in that its body touches the substrate when locomoting. Most other desert nocturnal ground lizards, including *bisharicus* and *nattereri*, hold their bodies well above the ground on straightened limbs, presumably to increase sight lines. It would be of interest to know whether the body dragging trait of *nubicus* was manifest in both sexes, as only the males of

Tropiocolotes spp.	English name				Range core
			Size		
	Activity	Diet	mm	g	Ecological business model
algericus	Algerian Pygmy Gecko				Algeria
	nocturnal	invert.	36	1.13	**** nocturnal ground lizard
bisharicus	Bishari Pygmy Gecko				Egypt
	nocturnal		32	0.79	**** nocturnal ground lizard
nattereri	Natterer's Pygmy Gecko				Egypt
	nocturnal	invert.	27	0.50	**** nocturnal ground lizard
naybandensis	Nayband Pygmy Gecko				Iran
	nocturnal		33	0.86	
nubicus	Nubian Pygmy Gecko				Egypt
	nocturnal		31	0.77	**** nocturnal ground lizard
scorteccii	Scortecci's Arabian Gecko				Yemen
	nocturnal		28	0.55	**** nocturnal ground lizard
somalicus	Parker's Pygmy Gecko				Horn of Africa
	nocturnal		29	0.65	*** nocturnal ground lizard
steudneri	Steudner's Pygmy Gecko				Sahara
	nocturnal	invert.	30	0.69	**** nocturnal ground lizard
tripolitanus	Tripoli Pygmy Gecko				Sahara
	nocturnal	invert.	32	0.83	*** nocturnal ground lizard
wolfgangboehmei	Arabia Pygmy Gecko				Saudi Arabia
	nocturnal		31	0.70	

Tukutuku sp.	English name				Range core
			Size		
	Activity	Diet	mm	g	Ecological business model
rakiurae	Harlequin Gecko				New Zealand
	cathemeral		60	4.32	** nocturnal ground lizard

this species appear to have secretory pores. The genus is conspicuously overrepresented in sandy areas (88% v. 11%) and riparian zones (38% v. 17%). Clutch count is documented in seven species, for which it is uniformly one, a normal value for lizards of this size (0.08 v. –0.26 std. resid.). Young are comparatively large (ontogenetic size range is small: –0.43 v. –0.19 std. resid.). Up to six clutches may be deposited per year. Incubation duration is somewhat lengthy (0.34 v. 0.17 std. resid.). Longevity is conspicuously short (–0.89 v. 0.15 std. resid.), though documented in only three species. *Tropiocolotes* have not been introduced (v. 3.7%), nor are any considered imperiled (v. 47%). In other respects *Tropiocolotes* are typical gekkonids.

Tukutuku

Tukutuku rakiurae is the sole representative of this diplodactylid genus from chilly Steward Island south of the main islands of New Zealand. It is an average-sized diplodactylid (4.3 g v. 4.2 g) nocturnal—in this case cathemeral—ground lizard (v. 35%) that may bask. It lacks expanded toepads. It has secretory pores (v. 34%), but other sex differences are not documented. It calls spontaneously (v. 5.4%). It is found in riparian zones (v. 10%) but not in littoral areas (17%), sandy areas (v. 24%), or around human habitats (v. 5%). As with most New Zealand lizards, it is viviparous (v. 18%). It has not been introduced (v. 2%), and it is considered imperiled (v. 81%), though it is not on the IUCN's lists of the most endangered (v. 29%). In other respects it is a typical diplodactylid.

Tumbunascincus

Tumbunascincus luteilateralis, formerly in *Concinnia*, is the sole member of its genus. It is a moderately large (14.0 g v. 3.3 g), orange-sided refugia-anchored diurnal ground (v. 30%) skink endemic to the Clarke [mountain] Range in Queensland. Aside from a heavier build (shape index 0.40 v. –0.46 std. resid.), it is morphologically a typical skink. Sex differences are unknown. It gives stress calls (v. 0.4%). It basks directly (v. 21%), and males defend space (v. 94%). It is found at low and possibly medium elevations (in this case documented "above 900 m"). It is viviparous (v. 36%), with an ordinary or a slightly small-sized litter (–0.19 v. 0.06 std. resid.). Neonates remain for a time in the vicinity of their mother (parental care v. 3.2%). It is endemic (v. 64%), but it is not found on an island (v. 57%). It has not been introduced (v. 2.1%), and it is considered to be imperiled (v. 47%) but is not on one of the IUCN's lists of the most endangered (v. 14%). In other respects *Tumbunascincus* is likely to be similar to other skinks.

Tumbunascincus sp.	English name		Size		Range core
	Activity	Diet	mm	g	Ecological business model
luteilateralis	Orange-speckled Forest-skink				Queensland
	cathemeral	invert.	88	14.0	** refugia-anchored diurnal gr. liz.

Tupinambis spp.	English name		Size		Range core
	Activity	Diet	mm	g	Ecological business model
cryptus	Cryptic Golden Tegu				Venezuela
	diurnal	omnivore	268	636	**** wide-ranging chemosensor
cuzcoensis	Cusco Tegu				Peru
	diurnal	invert.	186	204	**** wide-ranging chemosensor
longilineus	Slender Gold Tegu				Brazil
	diurnal	omnivore	183	194	** wide-ranging chemosensor
palustris	São Paulo Swamp Tegu				Brazil
			241	456	* wide-ranging chemosensor
quadrilineatus	Four-lined Tegu				Brazil
			202	265	* wide-ranging chemosensor
teguixin	Common Golden Tegu				South America
	diurnal	omnivore	326	1168	**** wide-ranging chemosensor
zuliensis	Maracaibo Basin Tegu				Venezuela
			204	274	**** wide-ranging chemosensor

Tupinambis

Tupinambis (Teiidae) is a moderate genus (seven spp.) of very large (371 g v. 22.1 g), burrow-concealment-dependent (100% v. 77%), socially tolerant (males 100% v. 55% tolerant), wide-ranging chemosensors (100% v. 97%). One might expect them to defend ownership of their burrows, but they do not. The genus is found in much of South America except Chile and Paraguay, plus Trinidad and Tobago. Most species in this genus were treated as *teguixin* until recently and confused with the larger tegus in the leather trade, which are mostly *Salvator*. *Tupinambis* tegus are poorly known (filled cells 26% v. 33%).

They are of heavy build (1.70 v. 0.88 std. resid.), and have short tails (1.97 v. 2.30 × SVL) and low tail breakability (14% v. 21.3%). Tails are never blue (0% v. 15%). Secretory pores may be present (14% v. 9%). Bipedalism is reported among juveniles of *cryptus* and *teguixin*. The latter species has been known to sleep while floating on vegetation.

Despite their large size, average SSD is more nearly neutral (0.50 v. 0.89 on −2 to 2 scale). Sexual dichromatism (v. 35%) and seasonal color (v. 10%) are absent. *T. cryptus* is unique in having male tails noticeably smaller relative to SVL than those of females (0.79 v. 1.04 m:f ratio for family). The genus's overall sex differences rollup is subdued (0.57 v. 1.19 on 0 to 6 scale).

Most species are omnivores (75% v. 33%), the other diet-known species (*cuzcoensis*) is an invertebrate eater (25% v. 62%). *T. teguixin* will raid honeybee hives for the honey. Home-range sizes are unknown. Wariness is elevated (29% v. 13%). Direct basking is underreported (43% v. 58%). They are found in low elevation (100% v. 95%) mesic areas (100% v. 30%). They are habitat generalists found at about average proportions in littoral areas (33% v. 38%), riparian zones (67% v. 46%), sandy areas (33% v. 47%), and human habitats (67% v. 29%).

Clutch counts (−0.40 v. −0.51 std. resid.) and the size of hatchlings (ontogenetic mass range 0.16 v. 0.10 std. resid.) are about normal. Several species, at least *cryptus* and *teguixin*, will oviposit in termite mounds. Communal oviposition is also reported (14% v. 2.6%). Females will defend their nest site at least temporarily (parental care 14% v. 3.2%) and may help young escape from termite mounds. Rainy weather may also soften the mounds. Incubation period, based on *teguixin*, is lengthy (0.63 v. −0.38 std. resid.), possibly in part due to low insolation in forested areas.

Longevity is comparatively long for a teiid (−0.19 v. −0.74 std. resid.), based on *teguixin* (16.1 years). Population density is unknown. Endemism is low (29% v. 50%). One species (*cryptus*: six islands) occurs on an island (14% v. 37%); none has been introduced (v. 8%). All are imperiled (v. 41%), but none is listed on one of the IUCN's lists of the most endangered species (v. 12%). In other respects *Tupinambis* are typical teiids.

Tupinambis are relatively heavily exploited by humans: *cryptus* are taken for food and medicinals; *teguixin* for food, leather, pets, and medicinals.

Tympanocryptis

Tympanocryptis is a moderate (15 spp.) radiation of uniformly small (3.25–6.7 g v. 15.6 g), rounded, crestless (v. 38% crested), terrestrial (93% v. 56%) agamids from the aridlands of outback Australia. They are poorly known (filled cells 24% v. 34%), with most research effort going to identifying cryptic (i.e., look-alike) species. Not only are the taxa cryptic, but the lizards are also. All documented species rely on crypsis (v. 51%), and many are characterized as pebble mimics because they look, lie alongside, and often act very similar to the untold billions of rounded, wind-polished rocks that define the "gibber" deserts in Australia. A few occur in other habitats, such as grassland. Eighty-two percent (v. 4.5%) are characterized as immobility experts; the remainder are refugia-anchored diurnal ground lizards (18% v. 43%). They may all be invertebrate eaters (v. 46%), but supportive documentation is at hand for only two species.

Most have secretory pores (80% v. 13%). Females tend to be larger than males (–0.38 v. 0.53 on scale of –2 to 2), only 27% are sexually dichromatic (v. 53%), none is reported to show seasonal color (v. 20%), and none has been found to change color (v. 22%). None is said to be conspicuously wary (v. 6%), most likely because they freeze rather than run. When inactive, many species retreat to soil cracks, if available, or hide under rocks. One species, *intima*, has been found sleeping on the surface of the ground. Although desert dwelling, only 7% of species are found in sandy areas (v. 26%).

Mass-adjusted clutch count is notably high (1.20 v. 0.20 std. resid.), and the young are notably large (ontogenetic range is low: –0.14 v. 0.69 std. resid). This equates to young of *Tympanocryptis* averaging 1/26 v. 1/100 of the maximum size. Acting like a pebble in the Australian outback appears to be a winning strategy, as they produce young that are both large and numerous. Presumably mortality is correspondingly high and size-dependent. Parental care has not been seen (v. 3.5%).

Longevity is slightly above average for lizards (0.18 std. resid.), which is notably high for agamids (v. –0.46 std. resid.). Endemism is average (53% v. 49%); insularity is rare (7% v.

Tympanocryptis spp.	English name				Range core
			Size		
	Activity	Diet	mm	g	Ecological business model
centralis	Centralian Earless Dragon				Australia
	diurnal		53	4.54	
cephalus	Coastal Pebble-mimic Dragon				Western Australia
	cathemeral		54	4.96	***immobility expert
condaminensis	Condamine Earless Dragon				Queensland
			54	4.96	
diabolicus	Hammersley Pebble-mimic Dragon				Western Australia
	diurnal		54	5.1	*** immobility expert
fortescuensis	Fortescue Marsh Pebble-mimic				Western Australia
	diurnal		56	5.6	*** immobility expert
gigas	Gascoyne Pebble-mimic				Western Australia
	diurnal		57	5.9	** immobility expert
houstoni	Nullarbor Earless Dragon				Southwestern Australia
	diurnal		56	5.5	
intima	Gibber Earless Dragon				Eastern Australia
	diurnal	invert.	54	4.93	* immobility expert
lineata	Lined Earless Dragon				Australia
	diurnal	invert.	57	5.8	* immobility expert
pentalineata	Five-lined Earless Dragon				Queensland
	diurnal		52	4.28	* immobility expert
pinguicolla	Grassland Earless Dragon				New South Wales
	diurnal		52	4.44	** refugia-anchored diurnal gr. liz.
pseudopsephos	Goldfields Pebble-mimic Dragon				Western Australia
	diurnal		50	3.93	*** immobility expert
tetraporophora	Eyrean Earless Dragon				Australia
	diurnal		60	6.7	** refugia-anchored diurnal gr. liz.
uniformis	Even-scaled Earless Dragon				Western Australia
	diurnal		47	3.25	* immobility expert
wilsoni	Roma Earless Dragon				Queensland
	diurnal		48	3.43	

Typhlacontias spp.	English name	Activity	Diet	Size mm	Size g	Range core	Ecological business model
brevipes	FitzSimons' Burrowing Skink	nocturnal	invert.	99	5.4	Namibia	*** subterranean forager or sand shark
gracilis	Roux's Burrowing Skink			71	1.12	Botswana	** subterranean forager
johnsonii	Western Burrowing Skink	nocturnal		90	1.95	Angola	*** subterranean forager or sand shark
kataviensis	Katavi Burrowing Skink			64	0.88	Tanzania	* subterranean forager
punctatissimus	Speckled Burrowing Skink			70	2.23	Angola	*** subterranean forager or sand shark
rohani	Kalahari Burrowing Skink	nocturnal	invert.	72	1.18	Botswana	*** subterranean forager or sand shark
rudebecki	Rudebeck's Burrowing Sink			68	1.02	Angola	* subterranean forager

Typhlosaurus spp.	English name	Activity	Diet	Size mm	Size g	Range core	Ecological business model
braini	Brain's Blind Legless Skink	nocturnal	invert.	161	7.4	Namibia	**** subterranean forager
caecus	Cuvier's Blind Legless Skink			165	7.9	South Africa	**** subterranean forager
lomiae	Lomi's Blind Legless Skink			96	2.27	South Africa	**** subterranean forager
meyeri	Meyer's Blind Legless Skink			140	5.3	Namibia	**** subterranean forager
vermis	Boulenger's Blind Legless Skink			212	14.0	South Africa	**** subterranean forager

36%). None has been introduced (v. 1.7%). Two of three evaluated species are considered imperiled (v. 37%), primarily due to loss of native habitat rather than small ranges. None appears on any of the IUCN's lists of the most endangered (v. 5%).

Typhlacontias

Typhlacontias (seven spp.) are small-sized (1.6 g v. 3.3 g), slender (shape index –1.90 v. –0.46 std. resid.), functionally or literally limbless (100%), short-tailed (0.45 v. 1.33 × SVL), nocturnal (100% v. 15%) subterranean forager (100% v. 12%) skinks living in tropical southern Africa, from Angola east to Tanzania and south to Namibia and Botswana. Termites feature in their generalized invertebrate diet (100% v. 84%). They are associated with sandy areas (100% v. 25%) and riparian zones (40% v. 23%). Despite their lowland tropical environment and the lack of evidence that these subterranean lizards bask to expedite offspring development, they are viviparous (100% v. 36%), with an unremarkable litter count (0.19 v. 0.06 std. resid.). Endemism is low (43% v. 64%). They are not found on islands (v. 57%) and have not been introduced (v. 2.1%). The one conservation-evaluated species (*kataviensis*) is imperiled (100% v. 47%), including placement on one of the IUCN's lists of the most endangered (100% v. 14%). In other respects *Typhlacontias* are probably typical limbless skinks.

Typhlosaurus

Typhlosaurus (five spp.) are moderately large (6.3 g v. 3.3 g), slender (shape index –2.27 v. –0.46 std. resid.), literally limbless and with reduced eyes (100%), nocturnal (100% v. 15%) subterranean foragers (100% v. 12%) living in temperate southern Africa, in Namibia and South Africa. In comparison to the similar and geographically proximate *Typhlacontias* skinks, *Typhlosaurus* skinks are larger, found at higher latitudes, and are further along the evolutionary progression to subterranean specialization. Sex differences are unknown. Termites feature in their generalized invertebrate diet (100% v. 84%). They are restricted to low elevations (0–900 m in this case). They are associated with sandy areas (100% v. 25%) and are representative in littoral zones (20% v. 21%). Despite their lowland environment and the lack of evidence that these subterranean lizards bask to expedite offspring development, they are viviparous

(100% v. 36%), with a moderately low mass-adjusted litter count (–0.55 v. 0.06 std. resid.). Endemism is high (80% v. 64%). They are not found on islands (v. 57%) and have not been introduced (v. 2.1%). One evaluated species (*lomiae*) is imperiled (33% v. 47%) but is not on one of the IUCN's lists of the most endangered (0% v. 14%). In other respects *Typhlosaurus* are probably typical limbless skinks.

Tytthoscincus (fig. 4.8B)

Tytthoscincus is a major radiation (17 spp.) of very small (0.63 g v. 3.25 g), elongate (71% v. 35%), litter cryptozoid (89% v. 28%) skinks living in Malaysia and possibly also Thailand and the islands of Indonesia and the southern Philippines. One species (*temengorensis*) may be a nocturnal ground lizard (11% v. 2.3%).

Tytthoscincus spp.	English name	Activity	Diet	Size mm	Size g	Range core	Ecological business model
aesculeticola	Mount Lamaku Litter-skink	diurnal		40	0.79	Malaysia	* litter cryptozoid
atrigularis	Zamboanga Litter-skink			35	0.52	Philippines	
batupanggah	Curse-stone Litter-skink			33	0.61	Malaysia	* litter cryptozoid
biparietalis	Sulu Litter Skink			35	0.52	Philippines	** litter cryptozoid
bukitensis	Fraser's Hill Forest Skink			41	0.60	Malaysia	** litter cryptozoid
butleri	Butler's Forest Skink	diurnal		41	0.60	Malaysia	** litter cryptozoid
hallieri	Hallier's Forest Skink			50	2.32	Borneo	* litter cryptozoid
ishaki	Tioman Island Forest Skink	crepuscular		45	0.77	Malaysia	**** litter cryptozoid
langkawiensis	Langkawi Islands Forest Skink			36	0.43	Malaysia	
leproauricularis	Scale-eared Tiny Leaf-litter Skink			33	0.62	Malaysia	
panchorensis	Bukit Panchor Forest Skink			35	0.53	Malaysia	
parvus	Luhu Litter Skink			35	0.74	Indonesia	
perhentianensis	Perhentian Island Forest Skink			31	0.30	Malaysia	
sibuensis	Sibu Island Forest Skink			29	0.31	Malaysia	
temengorensis	Temengor Forest Skink	nocturnal		35	0.40	Malaysia	* nocturnal ground lizard
temmincki	Temminck's Forest Skink	diurnal		47	1.2	Indonesia	**** litter cryptozoid
textus	Soundara Forest Skink			39	1.07	Indonesia	

Their body plan is somewhat slender for a skink (shape index –0.92 v. –0.46 std. resid.). The sexes are known to differ only in color (6% v. 8%). Their diet is unknown, but they forage actively (100% v. 92%). They have a diversity of diel activity patterns: crepuscular (20% v. 2%), diurnal (60% v. 74%), and nocturnal (20% v. 14.5%). Their frequency of conspicuous wariness is normal (12% v. 11%). The only reported basking behavior is thermoconformance (12% v. 1.5%). Their elevational profile is normal (l/m/h: 87/40/0 v. 89/35/6). They are not associated with any of the tabulated habitat specializations, but they are near representative in littoral (7% v. 21%) and riparian (14% v. 23%) habitats. They are absent from sandy areas (v. 25%) and human habitats (v. 14%).

All are oviparous (v. 64%). Their clutch count fits their size (0.21 v. 0.06 std. resid.). Incubation is long for a skink (0.13 v. –0.40 std. resid.), though based solely on *temmincki* (56 days).

Endemism is high (88% v. 64%). Most species are found on an island (77% v. 57%). None has been introduced (v. 2.1%). About half are imperiled (50% v. 47%), including placement on one of the IUCN's lists of the most endangered (50% v. 14%). In other respects *Tytthoscincus* are typical skinks.

Uma

Uma (six spp.) is a genus of large (21.1 g v. 11.2 g) Nearctic (Nevada

Uma spp.	English name				Range core
			Size		
	Activity	Diet	mm	g	Ecological business model
cowlesi	Yuma Desert Fringe-toed Lizards				Mexico
	diurnal	omnivore	91	32.6	** refugia-anchored diurnal gr. liz.
exsul	Exiled Fringe-toed Lizard				Mexico
	diurnal	invert.	74	13.9	** refugia-anchored diurnal gr. liz.
inornata	Coachella Fringe-toed Lizard				California
	diurnal	omnivore	95	27.8	*** refugia-anchored diurnal gr. liz.
notata	Colorado Desert Fringe-toed Lizard				California
	diurnal	omnivore	92	25.5	** refugia-anchored diurnal gr. liz.
paraphygas	Chihuahua Fringe-toed Lizard				Mexico
	diurnal	omnivore	69	11.5	** refugia-anchored diurnal gr. liz.
scoparia	Mohave Fringe-toed Lizard				California
	diurnal	omnivore	90	23.9	** refugia-anchored diurnal gr. liz.

to Mexico) refugia-anchored diurnal ground phrynosomatids that specialize in sandy or dune areas (100% v. 32%). A fringe of conical toe scales aids propulsion in loose sand. Some species (e.g., *paraphygas*) are renowned for their ability to bury into sand while running at full speed. Being located near many fine universities, they are comparatively well known (filled cells 45% v. 38%). They are normally proportioned for phrynosomatids (shape index 1.04 v. 0.91 std. resid.). Their tails are of average length (1.21 v. 1.22 × SVL) and breakability (17% v. 21%). Secretory pores are present but not especially common (17% v. 32%). Color change is absent (v. 20%). SSD is m>f (1.0 v. 0.27 on –2 to 2 scale). Sexual dichromatism is subdued (33% v. 58%), and seasonal color is elevated (67% v. 25%). Blue tails are not reported (v. 1.3%). Overall, their sex differences rollup is mildly elevated (2.17 v. 1.68 on 0 to 6 scale).

Considering the scarcity of vegetation in their habitat, one might expect a dietary reliance on animal matter, but omnivory is the rule (83% v. 20%). Only one species (*exsul*) foregoes plant food (17% v. 62% invertebrate eaters). All species are reported to sit-and-wait (v. 98%), but only 25% to forage actively (v. 8%), despite the necessity of doing so to acquire plant food.

Wariness is mildly elevated (33% v. 22%). Diving into sand is the most frequent primary predator concealment tactic (67% v. 12%). Minority tactics include flight (17% v. 14%) and retreat to crevices and burrows (17% v. 32%). Mean active body temperature is a very warm 37.6°C (v. 34.2°C). All documented species have males that defend space (v. 89%), though the object of their belligerence is unclear: mate sequestration or food? Sixty percent of species may retreat to a burrow at night (v. 29%).

They are found in the dry and arid rainfall belts (specifically 94–230 mm/y). Occupied elevations range 0–1400 m.

All are oviparous (v. 65%) and have relatively small clutches (–0.57 v. 0.68 std. resid.). Young are large (ontogenetic mass range –1.20 v. –0.26 std. resid.). Incubation is short (–0.63 v. –0.58 std. resid.), equating to 50 days in *paraphygas* and 66 days in *cowlesi*. RCM is the lightest among phrynosomatid genera: 0.134 (v. 0.300).

Longevity is moderate (0.05 v. –0.25 std. resid.), corresponding to about 8 years in *inornata* and *notata*. Population density is average (0.30 kg/ha v. 0.26 kg/ha), as is endemism (50% v. 58%). None has been introduced (v. 3.9%), but most are imperiled (83% v. 37%), including a third that are on one of the IUCN's lists of the most endangered (33% v. 5.4%). In California, large portions of suitable habitat have been lost to off-highway vehicles and urban development. In other respects *Uma* follow the family characteristics.

Underwoodisaurus

Underwoodisaurus is small genus (two spp.) of average-sized (12.6 g v. 14.5 g) terrestrial carphodactylids found in mainland Australia. Their white spots are reminiscent of *Nephrurus*. They are crack hermits (v. 13%), and they leave their cracks only at night, if then. *U. milii* is known to ambush from the crack entrance day and night. It is reported to be in groups, but it is unclear if group members share foraging positions or simply cluster in geographically clustered rock outcrops. Tails are relatively long (0.80 v. 0.54 × SVL), and both species break their tails (v. 87%). They maintain a low body temperature (20.3°C v. 22.8°C), at least in *milii*. Clutch count is the normal two (–0.79 v. –0.82 std. resid.), and their ontogenetic range of sizes is higher

Underwoodisaurus spp.	English name				Range core
			Size		
	Activity	Diet	mm	g	Ecological business model
milii	Thick-tailed Gecko				Australia
	cathemeral	invert.	87	13.1	*** crack hermit
seorsus	Pilbara Barking Gecko				Western Australia
	nocturnal		84	12.2	** crack hermit

Uracentron spp.	English name				Range core
			Size		
	Activity	Diet	mm	g	Ecological business model
azureum	Green Thornytail				Brazil
	diurnal	ants	72	14.8	* diurnal canopy walker
flaviceps	Amazon Thornytail				Western Amazon
	diurnal	ants	98	34.3	**** trunk/wall lizard

than the carphodactylid average (−0.77 v. −1.21 std. resid.), indicating relatively smaller young, though the young are still very large by lizard standards. *U. milii* is reported to oviposit communally (v. 13%), and its maximum longevity is 18 years (1.27 v. 0.42 std. resid.). They have not been evaluated for conservation status.

Uracentron

Uracentron is a small tropidurid genus (two spp.) of medium-sized (22.5 g v. 17.5 g), well-known (filled cells 49% v. 23%), arboreal (100% v. 25%) ant specialists (100%, the highest in family, v. 6.3%) found in northwestern South America, from Colombia to Peru, Brazil, and the Guiana Shield. They have short unbreakable tails (0.56 v. 1.56 × SVL) that are heavily armored with formidable spines. They differ primarily in perch-diameter preference; one is a diurnal canopy walker (50% v. 1.3%); the other is a trunk/wall lizard (50% v. 12.7%). They lead the family's rank tables for basic color change (50% v. 3%), retreat to crevice-burrow (100% v. 36%), retreat to vegetation (100% v. 16%), frequency in littoral habitats (50% v. 15%), frequency in human habitats (50% v. 14%), communal oviposition (50% v. 3.7%), parental care (50% v. 3.7%), and island species (50% v. 10%). *U. flaviceps* is socially structured around possession of a tree hole. When inside its hole, *flaviceps* presumably blocks entry with its spiny tail; males defend the best holes and acquire a harem of females that will join in the cozy refuge. Groups as large as 16—up to 10 females plus lingering offspring—are associated with single defended holes. It would not be surprising if *azureum* used their spiny tails for similar group living, though tree holes and group size are likely to be smaller in their smaller diameter branches.

Uracentron do not have crests (v. 10%) or advanced color change (v. 0.7%). Presumably crests are not a viable ornament when crammed in a tree hole.

Males are larger than females (1.00 v. 0.95 on −2 to 2 scale) and more colorful (*flaviceps*). Male *flaviceps* also have longer heads and limbs, as fits a strongly territorial social system. Overall, the genus's sex differences rollup score is elevated (1.50, the second highest in family, v. 0.80 on 0 to 6 scale).

Mean active body temperature is average (31.0°C v. 31.8°C). They are found in mesic areas (in this case, 2400–3000 mm/y). The PTM for *azureum* is the 83rd percentile. Is this a measurement of territory-patrolling males, or is their foraging mode active or cruising? It does not comport well with sit-and-wait, as described for *flaviceps*. Both species are found at low elevation, 0–500 m in this case. They are overrepresented in riparian zones (50% v. 18%) due to *azureum*.

Clutch count is the second lowest in the family (−1.00 v. −0.56 std. resid.), and hatchlings are slightly large (ontogenetic mass range −0.51 v. −0.22 std. resid.), but data on RCM ratio and incubation duration are lacking.

Population density and longevity are unknown. Neither species is endemic (v. 74%), nor has either been introduced (v. 1.5%). They have not been evaluated for conservation status. In other respects *Uracentron* are typical tropidurids.

Uranoscodon

Uranoscodon superciliosus is the sole member of its tropidurid genus of large (60 g v. 17.5 g), short-snouted, low-crested (v. 10%) ground-foraging tree lizards (100% v. 13%) found in lowland (0–250 m) regions of the upper Amazon Basin, from Bolivia, Peru, Colombia, Venezuela, the Guiana Shield, and of course Brazil. The Mophead Iguana is the ultimate sit-and-wait predator (0th percentile of MPM), watching for invertebrate movement from the safety of vegetation overlooking the edge of seasonally changing water bodies. It seems likely that some of its prey are exposed by rising waters. It gleans most prey from the surface of the land or the surface of the water. It moves its home range to track the seasonal migration of the water's edge. When pressed, it will run bipedally, including across the surface of the water. Fringed toes aid this form of locomotion. It sleeps on arboreal vegetation overhanging water and will dive into the water to escape predators and herpetologists. It relies on crypsis (v. 43%) as its primary predator concealment tactic.

Uranoscodon is the best-known genus in the family (filled cells 51% v. 23%). Its mean mass is the highest of any genus in the family. It has the stoutest build (shape index 1.29. v. 1.10 std. resid.). Its breakable (v. 13%) tail is the longest (2.10 v. 1.56 × SVL).

The Mophead Iguana is a thermoconformer (v. 0.7%) and has a low mean active body temperature (28.0°C v. 31.8°C), presumably because moving around to track sun spots would compromises its extreme immobility and crypsis.

Its clutch count is the highest in the family (0.46 v. −0.56 std. resid.), and the young are the smallest (ontogenetic mass range 0.33, the highest in Tropiduridae, v. −0.22 std. resid.). Incubation duration is short (−0.13 v. 0.40 std. resid.). These benchmarks suggest high random, but not size dependent, mortality, which is supported by its relatively short longevity (−0.86 v. −1.17 std. resid.), though that longevity (4.8 years) is absolutely long for a tropidurid.

It is not endemic (v. 74%), does not occur on a named island (v. 10%), and has not been introduced (v. 1.5%). It has not been evaluated for conservation status. In other respects *Uranoscodon* is probably a typical tropidurid.

Uranoscodon sp.	English name				Range core
			Size		
	Activity	Diet	mm	g	Ecological business model
superciliosus	Mophead Iguana				Northern South America
	diurnal	invert	121	60	*** ground-foraging tree lizard

Urocotyledon spp.	English name				Range core
			Size		
	Activity	Diet	mm	g	Ecological business model
inexpectata	Seychelles Sucker-tailed Gecko				Seychelles Islands
	nocturnal	invert.	38	1.32	*** crack hermit
palmata	Equatorial Tree Gecko				Congo
	nocturnal		51	2.95	* crack hermit
rasmusseni	Rasmussen's Gecko				Tanzania
	nocturnal		39	1.46	
weileri	Weiler's Tree Gecko				Cameroon
	nocturnal		42	1.69	
wolterstorffi	Uluguru Tail-pad Gecko				Tanzania
	nocturnal	invert.	45	2.08	* immobility expert

Urocotyledon

Urocotyledon is a five-species genus of small (1.82 g v. 3.31 g), arboreal, nocturnal tropical African gekkonids found in Cameroon and from Gabon to Tanzania (also the Seychelles Islands). The genus is morphologically distinctive for uniformly possessing adhesive lamellae on the tip of the tail. Although arboreal, none is a typical canopy or trunk dweller, but all specialize in immobility, either by refuging in a crack (67% v. 3.2%) or lying exposed and immobile (33% v. 4.5%). Their color and behavior belie a commitment to crypsis (100% v. 50%). The genus is poorly known (filled cells 21% v. 25%), and two species are insufficiently known to characterize.

Tails are relatively short (0.82 v. 1.07 × SVL) but not listed as prehensile, despite the adhesive tail lamellae. The do not have secretory pores (v. 46%) or color change (v. 15%). The sexes do not differ in color (v. 6.6%), and size differences, if any, are undocumented. Males have much longer tails than females (1.46 v. 1.05 ratio) in the one species documented (*palmata*). Based on that species and others with m>f relative tail length ratio, the genus is likely to have predominantly m<f SSD.

Urocotyledon have not been introduced (v. 3.7%). One in three evaluated species is considered imperiled (v. 47%), though none is on one of the IUCN's lists of the most endangered species. In other respects *Urocotyledon* are typical gekkonids.

Uromastyx

Uromastyx is a moderate (15 spp.) radiation of large (166 g v. 15.6 g), robust (shape index 1.59 v. 0.80 std. resid.), crestless (v. 38% crested), terrestrial (v. 56%) agamids from the aridlands of North Africa east to Iran. They are found across the Sahel from Mauritania to Somalia, Yemen, Oman, and Iran, north to the

Uromastyx spp.	English name	Activity	Diet	Size mm	Size g	Range core	Ecological business model
acanthinura	North Africa Mastigure	diurnal	herbivore	256	622	Northern Sahara	**** giant herbivore
aegyptia	Egypt Mastigure	diurnal	herbivore	314	1170	Saudi Arabia	**** giant herbivore
alfredschmidti	Schmidt's Mastigure	diurnal	herbivore	174	185	Northern Sahara	**** giant herbivore
benti	Bent's Spiny-tailed Lizard	diurnal	herbivore	149	116	Yemen	**** giant herbivore
dispar	South Sahara Mastigure	diurnal	herbivore	196	270	Sahara	**** giant herbivore
geyri	Geyr's Mastigure	diurnal	herbivore	171	178	Algeria	**** giant herbivore
macfadyeni	Somalia Mastigure	diurnal		97	30.0	Somalia	** refugia-anchored diurnal gr. liz.
nigriventris	Morocco Spiny-tailed Lizard	diurnal	omnivore	160	145	Morocco	** refugia-anchored diurnal gr. liz.
occidentalis	Western Giant Dob	diurnal		229	439	Western Sahara	** giant herbivore
ocellata	Ocellated Mastigure	diurnal	herbivore	134	84	Horn of Africa	**** giant herbivore
ornata	Ornate Mastigure	diurnal	herbivore	154	128	Saudi Arabia	**** giant herbivore
princeps	Princely Mastigure	diurnal	omnivore	138	91	Somalia	** refugia-anchored diurnal gr. liz.
shobraki	Yemen Spiny-tailed Mastigure			158	138	Yemen	** giant herbivore
thomasi	Oman Spiny-tailed Lizard	diurnal	herbivore	145	106	Oman	**** giant herbivore
yemenensis	South Arabia Spiny-tailed Lizard			141	98	Southern Arabia	* refugia-anchored diurnal gr. liz.

Mediterranean and Syria. About 73% are giant herbivores (v. 6%), and the others are refugia-anchored diurnal ground lizards (27% v. 43%). These lizards occur in strikingly hostile environments: soil temperatures may reach 75°C at midday, some sites have never recorded rain. Several species are active in the middle of the day; others shift activity to cooler hours in midsummer. One (*acanthinura*) was found to commute 1500 m from its burrow to browse on a single shriveled plant. This species' gut passage rate is reported to be 30 days! At low temperatures (30°C, which is evidently considered to be low in this area), they can survive a year without eating. They seem to be in an unending race to procreate, as they have large clutches (1.30 v. 0.20 std. resid.), breathtaking RCM (mean 0.618, the second highest in Agamidae, v. 0.336), and live up to 33 years (mass-adjusted mean longevity 0.34 v. −0.36 std. resid.). Young are comparatively small (ontogenetic range −0.01 v. 0.69 std. resid.). Four species are reported to have female nest guarding (*acanthinura*, *aegyptia*, *ocellata*, *ornata*), most for 4–6 weeks (parental care 27% v. 3.5%). What are they guarding against? Many giant herbivores guard their eggs to keep other females from inadvertently digging them up, but that seems unlikely under the circumstances. Loose soil would not seem to be in short supply. Ants? Small carnivores? Oviphagous snakes? Desiccation?

On average, tails are short (0.68 v. 1.86 × SVL). Forty percent (v. 22%) exhibit darkening color change. Mean SSD is 1.0 (m>f) v. 0.53 on −2 to 2 scale, despite the evident pressure on females for high fecundity. Most species are herbivores (82% v. 8%); the remainder are omnivores (18% v. 26%). All are active foragers (v. 17%). Presumably, this predilection for actively pursuing low trophic level food reflects the extremely low productivity of their environments. Home-range areas are large for the lizards' size (0.50 v. 0.35 std. resid.). Only 20% of species are considered conspicuously wary (v. 6%). Most (88% v. 29%) rely on crevices or burrows for predator concealment. All bask directly (v. 24%). Although all live by themselves, at least in their burrow, several species are found in colonial groups. *U. aegyptia* is found in groups of up to four; *acanthinura* in groups of up to 20. Two-thirds (v. 17%) of documented species occur in the hyperdry (<55 mm/y) and most of the others (58% v. 30%) in the dry (56–150 mm/y). They occur at all elevations from the littoral (14% v. 10%) up to 3200 m; the single species *yemenensis* covers that range by itself. Half are found in sand or dunes (50% v. 26%). None is found around humans (v. 16%). One species (*acanthinura*) is reported to have adherent eggs (7% v. 0.8%), which is unusual for a terrestrial nester.

Perhaps there is some payoff for living in such a harsh environment: longevity is long (0.34 v. −0.46 std. resid.), and population density is average (0.14 kg/ha v. 0.16 kg/ha), based on two estimates of *nigriventris*. Endemism is low (27% v. 49%), as is insular species frequency (20% v. 37%). They have not been introduced (v. 1.7%).

All evaluated species are considered at conservation risk (100% v. 37%), and despite their residence in areas nearly unpopulated by humans, most evaluated species suffer from direct human take: *acanthinura* is overharvested for the pet trade, food (especially tail and ovaries), shark bait, and medicinals. Skins are used as milk bottles for infants (legs form nipples), and older children are given live animals, with the lizards' mouths sewed shut, for living dolls. *U. aegyptia* is used for food. *U. alfredschmidti* is used for food and medicinals, and it suffers from competition with livestock for forage. *U. benti* is in high demand for the pet trade. *U. ocellata* is believed suffering from habitat loss and collection for the pet trade. *U. ornata* is associated only with overharvesting for the pet trade. For a *Uromastyx*, it seems living in the Saharan outback is not getting away from it all.

Uroplatus

Uroplatus is a 15-species genus of the quintessentially cryptic Madagascar gekkonids. They are large (11 g v. 3.31 g) and robust (0.31 v. 0.10 shape index), many with spectacularly flattened or widened tails, some of which bear an uncanny resemblance to leaves, replete with visually convincing "insect damage." Almost all species are immobility experts (87% v. 4.5%); the few others are nocturnal canopy walkers (13% v. 9%). Nearly a third of cells (32% v. 25%) are filled for this group, but much of the information derives from herpetoculture rather than field observations, so many facts regarding natural behavior and ecology are unknown.

The tails are among the shortest in Gekkonidae (0.53 v. 1.07 × SVL), and they are not prehensile (v. 1.8%). Secretory pores are absent (v. 46% present). Color changing ability is always the "advanced" sort (pattern, color: 20% v. 2.1%). All species exhibit ex-

Uroplatus spp.	English name				Range core
			Size		
	Activity	Diet	mm	g	Ecological business model
alluaudi	North Madagascar Leaf-tailed Gecko				Madagascar
	nocturnal	invert.	66	6.0	*** immobility expert
ebenaui	Nosy Be Leaf-tailed Gecko				Madagascar
	nocturnal	invert.	56	3.89	**** immobility expert
fiera	Fiery Leaf-toed Gecko				Madagascar
			57	4.10	**** immobility expert
fimbriatus	Common Leaf-tailed Gecko				Madagascar
	nocturnal	invert.	149	57	**** immobility expert
finiavana	Drip-tailed Leaf-toed Gecko				Madagascar
	nocturnal	invert.	56	3.92	**** immobility expert
giganteus	Giant Madagascar Leaf-tailed Gecko				Madagascar
	nocturnal	invert.	152	61	**** immobility expert
guentheri	Guenther's Leaf-tailed Gecko				Madagascar
	nocturnal	invert.	66	6.0	*** immobility expert
henkeli	Large-bodied Madagascar Leaf-tailed Gecko				Madagascar
	nocturnal	invert.	124	34.3	**** immobility expert
lineatus	Striped Leaf-tailed Gecko				Madagascar
	nocturnal	invert.	133	42.0	*** immobility expert
malahelo	Velvety Leaf-tailed Gecko				Madagascar
	nocturnal	invert.	66	6.0	** nocturnal canopy walker
malama	Fringeless Leaf-tailed Gecko				Madagascar
	nocturnal	invert.	65	5.8	** nocturnal canopy walker
phantasticus	Madagascar Leaf-tailed Gecko				Madagascar
	nocturnal	invert.	65	5.8	**** immobility expert
pietschmanni	Crenulated Leaf-toed Gecko				Madagascar
	nocturnal	invert.	67	6.4	**** immobility expert
sameiti	White-mouthed Leaf-tailed Gecko				Madagascar
	nocturnal	invert.	92	15.3	**** immobility expert
sikorae	Black-mouthed Lichen Gecko				Madagascar
	nocturnal	invert.	97	17.7	**** immobility expert

panded toe pads (v. 70%). *U. fimbriatus* has the largest number of teeth of any gecko (>300 teeth).

Sexual size differences are not documented, and other sex differences hew to the gekkonid norm.

Foraging mode (cruise) has been characterized in only one species (*fimbriatus*). Electric light use is not reported (v. 3.8%). Spontaneous vocalizations have not been reported (v. 6.5%) but stress calling has (13% v. 3.7%). *Uroplatus* exhibits a high level of social tolerance among males (67% v. 13%), and only *lineatus* is known to defend space (33% v. 87%).

The rainfall band is known for only one species (*henkeli*), which is found in the mesic zone (1101–3000 mm/y).

Reproduction in *Uroplatus* generally follows the gekkonid pattern, though eggs are spherical rather than elongate in at least five species (*ebenaui, guentheri, lineatus, malama, phantasticus*). The apparently fixed egg count of two is small for such large geckos (−0.75 v. −0.26 std. resid.). The young are exceptionally large (ontogenetic mass range is small: −0.80 v. −0.19 std. resid.). Female *henkeli* attend their clutch in captivity (6.7% v. 1.9% parental care in gekkonids).

Longevity is short (−0.71 v. 0.15 std. resid.), despite much herpetoculture effort to extend it. No species has been introduced (v. 3.7%), but all are considered to be at conservation risk (v. 47%). The proportion on the IUCN's lists of most endangered is double that of gekkonids in general (21% v. 11%), likely reflecting both the limited distributions of many species and the intense collecting pressure for the pet trade. In other respects *Uroplatus* are typical gekkonids.

Urosaurus

Urosaurus (eight spp.) is a genus of small (4.7 g v. 11.2 g), southwestern Nearctic (Wyoming to southern Mexico) phrynosomatids of diverse—mostly scansorial—ecologies: ground-foraging tree lizards (43% v. 20%), refugia-anchored diurnal ground lizards (29% v. 52%), diurnal canopy walkers (14% v. 0.7%), and trunk/wall lizards (14% v. 10%). *U. bicarinatus* ranges into the northernmost Neotropics.

Comparatively slender in build (shape index 0.64 v. 0.91 std. resid.), they have moderately long tails (1.88 v. 1.22 × SVL), secretory pores (38% v. 32%), and color change (basic 50% v. 20%; advanced 25% v. 4.5%). They are more m>f sexually size dimorphic than the average phrynosomatid (0.80 v. 0.27 on −2 to 2 scale) despite their small size. Sexual dichromatic species are in the majority (63% v. 58%), and seasonal color is present in *ornatus* (13% v. 25%). Accordingly, the sex differences rollup score is mildly elevated (1.88 v. 1.68 on 0 to 6 scale).

Most species are conventional invertebrate eaters (86% v. 62%), though *nigricaudis* is an ant specialist (15% v. 17%). Along with the phrynosomatid standard of sit-and-wait foraging (100% v. 98%), one-quarter of *Urosaurus* also forage actively (25% v. 4.5%). *U. auriculatus* has been characterized as being a "random" forager, perhaps indicating that it changes direction frequently. There are objective movement statistics for *ornatus*, suggesting intermediate foraging behavior (MPM 65th percentile; PTM 35th percentile). Mean home-range area for *Urosaurus* is close to the phrynosomatid norm (0.46 v. 0.57 std. resid.), though it is based solely on *ornatus* (ca. 93 m^2).

Although all species are viewed as being diurnal, *nigricaudus* may sometimes be active at night. Wariness is

Urosaurus spp.	English name				Range core
			Size		
	Activity	Diet	mm	g	Ecological business model
auriculatus	Socorro Island Tree Lizard				Mexico
	diurnal	invert.	62	8.9	* refugia-anchored diurnal gr. liz.
bicarinatus	Tropical Tree Lizard				Mexico
	diurnal	invert.	51	5.2	*** trunk/wall lizard
clarionensis	Clarion Island Tree Lizard				Mexico
	diurnal	invert.	54	5.9	** refugia-anchored diurnal gr. liz.
gadovi	Gadow's Tree Lizard				Mexico
			47	4.26	
graciosus	Long-tailed Brush Lizard				Arizona
	diurnal	invert.	53	3.40	**** diurnal canopy walker
lahtelai	Baja California Brush Lizard				Mexico
	diurnal	invert.	51	5.2	*** ground-foraging tree lizard
nigricaudus	Black-tailed Brush Lizard				Mexico
	diurnal	ants	48	4.36	** ground-foraging tree lizard
ornatus	Ornate Tree Lizard				New Mexico
	diurnal	invert.	48	2.60	**** ground-foraging tree lizard or refugia-anchored diurnal gr. liz. (geog. var.)

reported at a slightly reduced level (13% v. 22%). The primary predator concealment tactics of crypsis (40% v. 32%) and retreat to crevice-burrows (40% v. 32%) are slightly overrepresented; flight (0% v. 14%) and sand (0% v. 12%) are not known, and vegetation concealment is about average (20% v. 24%). Direct basking is common (50% v. 40%), and mean active body temperature is slightly elevated (35.2°C v. 34.2°C).

Urosaurus occupy the arid end of the phrynosomatid rainfall spectrum (not found in mesic or wet), with a slight overrepresentation in the intermediate band (401–1100 mm/y: 67% v. 52%). All species are found at low elevation (100% v. 73%), but 29% (v. 36%) of species extend to high elevations (*ornatus* to 2770 m). *Urosaurus* are about average for representation in littoral (25% v. 16%), riparian (38% v. 40%), and human (25% v. 18%) habitats. They may be underrepresented in sandy areas (13% v. 32%).

Reproduction has been a focus of much attention due to throat/dewlap color polymorphism in *ornatus* (e.g., Hamilton and Sullivan 2005) associated with male mating tactics. However, in broad outline the reproduction of *Urosaurus* is notable primarily for exclusive oviparity (100% v. 65%) and parental care (12.5% v. 1.9%); female *ornatus* defend their clutches. In other respects *Urosaurus* reproduction is as expected for phrynosomatid lizards of this size: clutch count (0.83 v. 0.68 std. resid.), incubation (–0.34 v. –0.58 std. resid., equating to 40–52 days), ontogenetic mass range (0.12 v. –0.26 std. resid.), and RCM 0.302 (v. 0.300).

Longevity is normal (–0.25 v. –0.25 std. resid.), equating to 3–5 years. Population density (based solely on *ornatus*) averages 0.33 kg/ha (v. 0.26 kg/ha), slightly greater than 0.13 kg/ha *Phrynosoma*, and the same as 0.35 kg/ha *Sceloporus*. Endemism is normal (63% v. 58%). Half of all *Urosaurus* are found on islands (50% v. 21%). None has been introduced (v. 3.9%). One-quarter of evaluated species are deemed imperiled (v. 37%), and *auriculatus* (endangered) is on one of the IUCN's lists of the most endangered (12.5% v. 5.4%). In other respects *Urosaurus* follow the family characteristics.

Urostrophus spp.	English name				Range core
			Size		
	Activity	Diet	mm	g	Ecological business model
gallardoi	Gallardo's Crypsis Iguana				Bolivia
	diurnal	invert.	65	7.1	** immobility expert
vautieri	Brazil Steppe Iguana				Brazil
	diurnal		76	12.0	*** immobility expert

Urostrophus

Urostrophus is the best-documented leiosaurid genus (filled cells 31% v. 21%) but perhaps only because it is depauperate: two species, both diurnal (v. 86%), arboreal (v. 45%) immobility experts. They are found in Argentina, Bolivia, and Brazil. They are morphologically indistinct lizards: (9.2 g v. 16.4 g), shape index (0.55 v. 0.76 std. resid.), tail length (1.47 v. 1.66 × SVL), lacking tail breakability (0% v. 3%), possessing prehensile tails (100% v. 9%), and lacking color change (0% v. 3%). The only documented sex difference is an elevated m:f ratio of relative tail length (1.27 v. 1.20).

Only *gallardoi* is documented for elevational range (250–700 m). Clutch count is average (0.85 v. 0.68 std. resid.). *U. vauteri* is endemic (50% v. 70%). Neither species has been evaluated for conservation status. In other respects *Urostrophus* are probably typical leiosaurids.

Uta (fig. 4.11D)

Uta (seven spp.) is a genus of small (5.8 g v. 11.2 g), spotted southwestern Nearctic (Idaho to Mexico) refugia-anchored diurnal ground phrynosomatids. They are comparatively well studied (filled cells 42% v. 38%), especially *stansburiana* (81%).

They are of slightly lighter build than average phrynosomatids (shape index 0.77 v. 0.91 std. resid.). Their tails are only slightly long (1.43 v. 1.22 × SVL) and are of average breakability (14% v. 21%). Most have secretory pores (57% v. 32%). Color change is at an elevated frequency (29% v. 19.5% basic; 14% v. 4.5% advanced). They do not have bright blue tails (v. 1.3%), though the tail of *nolascensis* could be described as green with turquoise flecking. The malachite-green tail of *squamata* could be construed as bluish.

They are more m>f in SSD than average phrynosomatids (0.75 v. 0.27 on –2 to 2 scale) despite their smaller size. Almost all species are sexually dichromatic (86% v. 58%) and show seasonal color (86% v. 25%). All five species that show seasonal color show seasonal color in females (*palmeri* in both sexes); this female color is associated with gravidity and deflects the amorous attentions of aggressive males. These sex differences combine for a relatively high sex differences rollup score of 2.71 (v. 1.68 on 0 to 6 scale).

Unlike an average phrynosomatid,

Uta go for omnivory (71% v. 20%), despite their small size. The ratio of plant to animal food appears to be highly opportunistic in the omnivorous species. The others are conventional invertebrate eaters (29% v. 62%). No foraging mode except for sit-and-wait is reported for *Uta*. Home range is comparatively large (1.10 v. 0.57 std. resid.) based on *stansburiana* (up to 2190 m^2 in males, but averaging about 366 m^2).

All species are diurnal, but *stansburiana* has been found on the surface at night (not entirely clear whether the observed lizard was sleeping or active). *Uta* are conspicuously wary at a normal level (29% v. 22%), and all species rely primarily on scampering to a crevice or burrow (v. 32%) when a predator appears. They are fervent direct baskers (71% v. 40%) and attain a normal mean active body temperature (34.0°C v. 34.2°C).

All males defend space (v. 89%), and 25% of species are documented to sometimes form male hierarchies (v. 3%). Forty percent of species (v. 29%) may use burrows as nocturnal retreats.

Only *stansburiana* is documented for rainfall range, which is an extraordinarily wide 28–1170 mm/y (hyperdry through mesic). Its elevational range is similarly expansive: 0–2750 m. The other two elevationally documented species (*nolascensis*, *palmeri*) are found only low (<1200 m). Though not usually found in human habitats (v. 18%), *Uta* are overrepresented in littoral areas (86% v. 16%) and about average in sandy areas (43% v. 32%) and riparian zones (43% v. 40%). *U. encantadae* is especially active in the littoral zone, feeding often on intertidal isopods. It has nasal salt glands to address the consequent high salt loading. A similar situation, and nasal salt glands, prevails in *tumidarostra*, which is said to prefer salt water to fresh water for drinking. Nasal salt glands may be expected in *lowei* and *palmeri*, the other intertidal foragers.

Like *Urosaurus*, *Uta* has been closely examined for its reproduction because of polymorphic throat color "badges" associated with differing levels of testosterone and territorial assertiveness (e.g., Corl et al. 2010). Mate guarding has been observed in several species at least in some color morphs. Like *Urosaurus*, *Uta* are oviparous (100% v. 65%) and lay smallish clutches (−0.01 v. 0.68 std. resid.) that are incubated for slightly longer than average (−0.25 v. −0.58 std. resid.). Although only two species have been documented for ontogenetic mass range (*encantadae*, *stansburiana*), the mean is the highest in the family (1.31 v. −0.26 std. resid.), indicating exceptionally small hatchlings. The incubation period is documented for only *encantadae* (46 days) and *stansburiana* (63 days). Unlike *Urosaurus*, *Uta* lack parental care (v. 1.9%), have light RCM (0.195 v. 0.300), and may lay communally (14% v. 0.6%).

Longevity is about average for lizards but slightly long for phrynosomatids (0.03 v. −0.25 std. resid.), equating to 4.8 years in *stansburiana*, the only documented species. Population density is very high in *Uta* compared to average phrynosomatids (1.10 kg/ha v. 0.26 kg/ha). Closer examination of the record indicates a strong divergence between the well-documented (n=54 estimates) mainland *stansburiana* (0.15 kg/ha) and the well-documented (n=8 estimates) island *palmeri* (8.0 kg/ha).

Uta spp.	English name				Range core
			Size		
	Activity	Diet	mm	g	Ecological business model
encantadae	Enchanted Side-blotched Lizard				Mexico
		omnivore	52	5.6	*** refugia-anchored diurnal gr. liz.
lowei	Dead Side-blotched Lizard				Mexico
	diurnal	omnivore	57	6.9	*** refugia-anchored diurnal gr. liz.
nolascensis	San Pedro Nolasco Side-blotched Lizard				Mexico
	diurnal	omnivore	48	4.46	*** refugia-anchored diurnal gr. liz.
palmeri	San Pedro Martir Side-blotched Lizard				Mexico
	diurnal	omnivore	67	10.8	*** refugia-anchored diurnal gr. liz.
squamata	Santa Catalina Side-blotched Lizard				Mexico
	diurnal	omnivore	49	4.73	*** refugia-anchored diurnal gr. liz.
stansburiana	Common Side-blotched Lizard				Southwestern United States
	diurnal	invert.	46	2.90	**** refugia-anchored diurnal gr. liz.
tumidarostra	Swollen-nosed Side-blotched Lizard				Mexico
	diurnal	invert.	62	8.9	*** refugia-anchored diurnal gr. liz.

U. palmeri lives on an island where seabirds nest; it feeds on seabird food spillage (fish) and the attracted flies during the booby nesting season and is omnivorous at other times and during El Niño years when boobies may not nest.

Endemism is high in *Uta* (86% v. 58%). All species reside on at least one island (v. 21%). *U. stansburiana* has been introduced to a California locality (14% v. 3.9%). Most species are imperiled (86% v. 37%), but none is listed on one of the IUCN's lists of most endangered species (v. 5.4%). In other respects *Uta* follow the family characteristics.

Uvidicolus

Uvidicolus sphyrurus is the sole species in this genus of carphodactylids. It is found on the border between Queensland and New South Wales. It is small for a carphodactylid (6.4 g v. 14.5 g). It will ambush prey from its crevice at any time of day; it is a crack hermit (v. 13%). Its clutch count of two, invariant in the family, gives this small species a slightly larger relative count (−0.50 v. −0.82 std. resid.), and its ontogenetic size range is very low (−2.12 v. −1.21 std. resid.), indicating exceptionally large young (1/6.6 v. 1/18.6 for carphodactylids and 1/44 for lizards). Maximum longevity is at least 10 years (0.73 v. 0.42 std. resid.). In other respects it typifies carphodactylids. It is considered near threatened by the IUCN and is listed by both states (Queensland and New South Wales) and the Australian government.

Uvidicolus sp.	English name		Size		Range core
	Activity	Diet	mm	g	Ecological business model
sphyrurus	Border Thick-tailed Gecko				New South Wales
	cathemeral	invert.	64	6.4	*** crack hermit

Vanzosaura

Vanzosaura is a three-species genus of diurnal terrestrial gymnophthalmids found in Brazil and (*rubricauda*) adjoining southern countries of South America. *V. rubricauda* is the source of almost all information on the genus; it has longitudinal striping and a bright red tail. It may inhabit ant nests. *Vanzosaura* are small (0.79 g v. 2.35 g). Tail lengths are typical (1.63 v. 1.72 × SVL), and tails are breakable (v. 28%). They possess secretory pores (67% v. 29%) and may be elongate (33% v. 41%). Sexual size difference has not been established, but *rubricauda* is sexually dichromatic. It also has males that are socially tolerant (v. 88%). Reproduction is typical, though clutch counts are high (0.41 v. −0.11 std. resid.) and RCM (*rubricauda*) is comparatively light (0.141 v. 0.192 ratio). These two reproductive facts seem inconsistent, but related objective measures (e.g., ontogenetic size range) are unavailable to resolve this. No species is found on an island (v. 5%), and none has been evaluated for conservation status. In other respects *Vanzosaura* are probably typical gymnophthalmids.

Vanzosaura spp.	English name		Size		Range core
	Activity	Diet	mm	g	Ecological business model
multiscutata	Multi-striped Spectacled Lizard				Brazil
			37	0.97	
rubricauda	Red-tailed Spectacled Lizard				Paraguay
	diurnal	invert.	38	0.63	** refugia-anchored diurnal gr. liz.
savanicola	Savanna Spectacled Lizard				Brazil
			35	0.82	

Varanidae and *Varanus* (figs. 2.6D, 4.15C, 4.15D)

Varanus, the only genus (79 spp.) of the Varanidae, has one dominant ecological business model, the wide-ranging chemosensor (92.3%, the third-highest family, v. 7.4%), and the remainder are refugia-anchored diurnal ground lizards (7.7% v. 24.5%). "Third-highest family" seems to be the ranking for Varanidae, and it also applies to mass (455 g v. 5.3 g; note the wide range within the genus: 8.2–4820 g), SSD (1.20 v. 0.36 on −2 to 2 scale), proportion of carnivores (79% v. 3%), proportion of semiaquatic species (9.6% v. 1.1%), proportion of species in the hyperdry rainfall belt (6.7% v. 2.9%), and proportion of imperiled species (100%, trade is regulated for all *Varanus,* v. 49%). The rankings mold is broken only for mass-adjusted home-range size (1.01 v. 0.0 std. resid.), for which Varanidae is the second highest. These features can be summarized as large, m>f, carnivores that roam widely, including residence in very dry and aquatic environments. Varanidae are reasonably well known (filled cells 42% v. 27%).

In Australia they are generally called goannas, but outside of Australia *Varanus* are known as monitors. The name "monitor" comes from an unlikely legend that upon seeing a crocodile, a varanid will whistle, thereby warning people nearby. Humans do not always respond to this generosity in kind: *rudicollis* is thought to spit venom. *V. griseus* is said to cause human males to become impotent if the lizard runs between the human's legs. Both *griseus* and *bengalensis* are said to cause human disasters if the lizard sees the teeth of the human. Are toothy smiles unlikely in varanid-infested parts of the world? On a more factual front, some *Varanus* such as *flavescens*, *griseus*, *komodoensis*, *scalaris*, and *varius*, and possibly all species, have saliva that is somewhat venomous.

A note on mass: traditionally, lizard species are compared by their maximum recorded size (length or mass); I have opted to characterize the mass of the individual most likely to occur (mode). The better-studied species of *Varanus* have been discovered to have anomalously left skewed mass distributions, and my mass value reflects these empirical data, whereas most uncharacterized varanids were estimated with a generic algorithm that assumes the lizard-typical right-skewed mass distribution. Thus, poorly-studied varanids will appear to have relatively larger sizes than their better-studied congeners. Regardless, most varanids are very, very large lizards. The largest *komodoensis* actually weighed by Auffenberg (1981) was 54 kg. A maximum of 100 kg is plausible, but the mass of this species at modal SVL (540 mm) is about 2.4 kg.

Varanus are found virtually everywhere throughout Africa (except Madagascar), southern Asia, and Australasia. The family's core area is northern Australia and nearby islands. Example jurisdictions with

Varanus spp.	English name	Range core	Activity	Diet	Size mm	Size g	Ecological business model
acanthurus	Ridge-tailed Monitor	Australia	diurnal	carnivore	195	127	*** refugia-anchored diurnal gr. liz.
albigularis	White-throated Monitor	Southern Africa	diurnal	carnivore	560	3880	**** wide-ranging chemosensor
auffenbergi	Auffenberg's Monitor	Indonesia	diurnal	carnivore	164	73	** wide-ranging chemosensor
bangonorum	Bangon Monitor Lizard	Philippines	diurnal		288	450	** wide-ranging chemosensor
baritji	Black-spotted Spiny-tailed Monitor	Northern Territory	diurnal	invert.	197	132	**** refugia-anchored diurnal gr. liz.
beccarii	Black Tree Monitor	Indonesia	diurnal	carnivore	261	330	** wide-ranging chemosensor
bengalensis	Bengal Monitor	Southern Asia	diurnal	carnivore	280	410	**** wide-ranging chemosensor
bitatawa	San Ildefonso Monitor	Philippines	diurnal	omnivore	598	4820	*** wide-ranging chemosensor
boehmei	Golden Speckled Tree Monitor	Indonesia	diurnal	carnivore	240	250	*** wide-ranging chemosensor
bogerti	Bogert's Tree Monitor	New Guinea	diurnal		235	233	*** wide-ranging chemosensor
brevicaudus	Short-tailed Pygmy Monitor	Australia	diurnal	carnivore	95	9.0	**** refugia-anchored diurnal gr. liz.
bushi	Bush's Monitor	Western Australia	diurnal	carnivore	120	19.0	*** wide-ranging chemosensor
caerulivirens	Turquoise Monitor	Indonesia	diurnal	carnivore	335	736	*** wide-ranging chemosensor
caudolineatus	Stripe-tailed Monitor	Western Australia	diurnal	invert.	117	21.5	*** wide-ranging chemosensor
cerambonensis	Ambon Mangrove Monitor	Indonesia	diurnal	carnivore	347	830	*** wide-ranging chemosensor
cumingi	Mindanao Water Monitor	Philippines	diurnal	invert.	550	3575	*** wide-ranging chemosensor
dalubhasa	Enteng's Monitor Lizard	Philippines	diurnal		370	1015	*** wide-ranging chemosensor
doreanus	Blue-tailed Monitor	New Guinea	diurnal	carnivore	364	962	*** wide-ranging chemosensor

Varanus richness, from west to east, are as follows:

Nigeria	2 spp.
Dem. Rep. Congo	3
South Africa	2
Sudan	2
Saudi Arabia	2
Iran	3
India	5
Philippines	12
New Guinea (one island)	9
Indonesia	28
Western Australia	21
Northern Territory	19
Queensland	21
South Australia	9
New South Wales	4

Varanids have long necks but a generally robust build (shape index 1.12 v. 0.0 std. resid.). Tail lengths are near average length (1.52 v. 1.37 × SVL) but are rarely breakable (1.3% v. 57%). A substantial minority of varanid species have prehensile tails (19% v. 5.0%). No varanids have crests (v. 5.3%) or advanced color change (v. 6.0%), but a few have secretory pores (1.3% v. 15%), are elongate (6.3% v. 11.4%), show basic color change (5.1% v. 10.7%), or have blue tails (1.3% v. 2.3%). *V. semiremex* has nasal salt-secreting glands.

As noted in the first paragraph, *Varanus* are strongly m>f in SSD, but other sex differences are fairly muted: dichromatism (2.5% v. 20.7%), seasonal color (2.5% v. 6.5%). The ratio of m:f relative tail lengths is about average (1.04 v. 1.07). The genus's sex differences rollup score (on scale of 0 to 6) is an unremarkable 0.91 (v. 0.75).

In addition to the carnivory highlighted in the first paragraph, a few *Varanus* are herbivores (1.5% v. 4.1%), invertebrate eaters (16.2% v. 73%), and omnivores (2.9% v. 15.4%). Many food items not normally taken by lizards are taken by varanids, for example, *flavescens* eats not only frogs but frog egg masses. *V. nuchalis* and probably *varius* are able to ingest Cane Toads (*Rhinella horri-*

Varanus spp.	English name	Activity	Diet	Size mm	Size g	Range core	Ecological business model
dumerilii	Harlequin Monitor	cathemeral	carnivore	418	1515	Malaysia	*** wide-ranging chemosensor
eremius	Pygmy Desert Monitor	diurnal	carnivore	141	32.0	Australia	**** wide-ranging chemosensor
exanthematicus	African Savannah Monitor	diurnal	invert.	435	1711	Central Africa	**** wide-ranging chemosensor
finschi	Finsch's Monitor	diurnal	carnivore	332	715	New Guinea	*** wide-ranging chemosensor
flavescens	Yellow Monitor	diurnal	carnivore	350	840	South Asia	**** wide-ranging chemosensor
giganteus	Perentie	diurnal	carnivore	655	4320	Australia	**** wide-ranging chemosensor
gilleni	Pygmy Mulga Monitor	diurnal	carnivore	136	28.9	Australia	*** wide-ranging chemosensor
glauerti	Kimberley Rock Monitor	diurnal	carnivore	186	110	Western Australia	**** wide-ranging chemosensor
glebopalma	Black-palmed Monitor	cathemeral	carnivore	276	392	Northern Australia	**** wide-ranging chemosensor
gouldii	Gould's Goanna	diurnal	carnivore	300	350	Australia	**** wide-ranging chemosensor
griseus	Desert Monitor	diurnal	carnivore	476	2300	Old World Deserts	**** wide-ranging chemosensor
hamersleyensis	Southern Pilbara Rock Goanna	diurnal		124	29.3	Western Australia	*** wide-ranging chemosensor
indicus	Mangrove Monitor	diurnal	carnivore	310	482	New Guinea	**** wide-ranging chemosensor
jobiensis	Peach-throated Monitor	diurnal	carnivore	356	900	New Guinea	*** wide-ranging chemosensor
juxtindicus	Rennell Island Monitor	diurnal	invert.	369	1010	Solomon Islands	*** wide-ranging chemosensor
keithhornei	Canopy Monitor	diurnal	invert.	231	222	Queensland	*** wide-ranging chemosensor
kingorum	Long-tailed Rock Monitor	diurnal	invert.	95	12.6	Western Australia	*** wide-ranging chemosensor

Varanus spp.	English name				Range core
			Size		
	Activity	Diet	mm	g	Ecological business model
komodoensis	Komodo Dragon				Indonesia
	diurnal	carnivore	540	2400	**** wide-ranging chemo-sensor
kordensis	Biak Emerald Monitor				Indonesia
	diurnal	invert.	229	215	*** wide-ranging chemosensor
lirungensis	Talaud Mangrove Monitor				Indonesia
	diurnal		263	336	*** wide-ranging chemosensor
mabitang	Panay Monitor				Philippines
	diurnal	herbivore	522	3087	*** wide-ranging chemosensor
macraei	MacRae's Monitor				Indonesia
	diurnal	carnivore	282	424	*** wide-ranging chemosensor
marmoratus	Marbled Water Monitor				Philippines
	cathemeral	carnivore	522	3095	**** wide-ranging chemo-sensor
melinus	Quince Monitor				Indonesia
	cathemeral	carnivore	391	1215	*** wide-ranging chemosensor
mertensi	Mertens' Water Monitor				Northern Australia
	diurnal	carnivore	374	510	*** wide-ranging chemosensor
mitchelli	Mitchell's Water Monitor				Northern Australia
	diurnal	carnivore	246	175	*** wide-ranging chemosensor
nebulosus	Clouded Monitor				Southeast Asia
	diurnal	carnivore	484	2430	**** wide-ranging chemo-sensor
nesterovi	Nesterov's Desert Monitor				Iran
			373	1050	*** wide-ranging chemosensor
niloticus	Nile Monitor				Africa
	diurnal	carnivore	660	3565	**** wide-ranging chemo-sensor
nuchalis	Rough-necked Water Monitor				Philippines
	cathemeral	carnivore	510	2850	**** wide-ranging chemo-sensor
obor	Sago Monitor				Indonesia
	diurnal		342	785	*** wide-ranging chemosensor
olivaceus	Gray's Monitor				Philippines
	diurnal	omnivore	575	4000	**** wide-ranging chemo-sensor
palawanensis	Palawan Islands Monitor				Philippines
	diurnal		524	3125	*** wide-ranging chemosensor
panoptes	Yellow-spotted Monitor				Northern Territory
	diurnal	carnivore	474	1060	**** wide-ranging chemo-sensor

bilis). *V. giganteus* and *varius* will eat small kangaroos; *giganteus* is said to ambush passing gulls from concealment under parked cars. *V. komodoensis* obtain deer for about half of their intake as adults. However, there is reason to doubt the popular Pakistani claim that hungry *griseus* will attack camels.

Though predominantly active foragers (97% v. 58%), nearly one-quarter also sit-and-wait (22% v. 46%). None cruise (v. 6.6%). King and Rhodes (1982) argued that the foraging patterns of *Varanus* differed greatly by sex—females sedentary/males roam—to which they attributed the conspicuous male-bias in varanid specimens. Subsequent tabulations of home-range sizes indicate that some species are greatly divergent—*tristis* male home ranges exceed those of females by a factor of 10—whereas others seem closer to parity (*bengalensis*) in foraging habitats. In all recorded cases, however, males roam further. *V. tristis* is reported to flush small lizards out of leaf litter by disturbing the litter with its tail. Surface-walking *albigularis* is said to be able to find subterranean mammals by smell. *V. komodoensis* may travel 3 km upwind to find carrion. The large home range of *Varanus* was noted in the first paragraph, but that average concealed considerable interspecific variation. *V. albigularis* home ranges are very large, the maximum one documented was 28,000,000 m^2 (2.7 v. 0.0 std. resid.). Individual *albigularis* and *varius* may walk 3–6 km per day. In contrast, *brevicaudus* has net movements of 14–25 m over periods of up to 751 days. The activity area of *flavescens* is much smaller than that for an average lizard of its size (–1.09 v. 0.0 std. resid.). But tiny *eremius* may travel >1 km per day. Single forays by *griseus* may extend 10 km (home range 1.6 v. 0.0 std. resid.). Many species may run bipedally, especially as juveniles. Death feigning is widespread (at least *albigularis, exanthematicus, mabitang, niloticus, rudicollis, salvator*) and may represent loss of consciousness.

The vast majority of *Varanus* are diurnal (91% v. 67%), but a representative proportion are active at all times of day (7.7% v. 6.8%) or are crepuscular (1.3% v. 1.8%). One species (*bengalensis*) calls spontaneously during the breeding period (1.3% v. 2.5%), but none gives stress calls (v. 2.2%). Wariness occurs at a high level (32% v. 8%), presumably because so many *Varanus* are collected commercially.

Varanus are adaptable with regard to primary predator concealment, eschewing only sand diving: crypsis (10% v. 43%), flight (15% v. 11%), retreat to a crevice or burrow (35% v. 25%), and retreat to vegetation (55% v. 23%). The majority bask directly (51% v. 18%) and a few thermoconform (3.8% v. 2.6%). Their mean active body temperature is a warm 33.7°C (v. 30.8°C). The maximum recorded field body temperature of *tristis* was 47.3°C!

Group living is average in frequency (22% v. 20%); male social tolerance is the fourth highest among families (59% v. 15%). Most other species have males that defend space (41% v. 85%). Male hierarchies are double the average frequency (15% v. 7.7%). Nine species have been observed for group composition, which is disproportionately m+f (78% v. 69%); does this suggest mate guarding or monogamy?

The modal *Varanus* is terrestrial (55% v. 48%), but arboreality is also common (34% v. 32%). One species (*exanthematicus*) is cryptozoic (1.4% v. 12%), despite being a wide-ranging chemosensor. Almost all of its prey are taken from underneath the leaf litter or deeper in the soil. The majority of species use burrows as a daily

Varanus spp.	English name				Range core
			Size		
	Activity	Diet	mm	g	Ecological business model
pilbarensis	Pilbara Rock Monitor				Western Australia
	cathemeral	carnivore	135	38.6	*** refugia-anchored diurnal gr. liz.
prasinus	Emerald Monitor				New Guinea
	diurnal	carnivore	243	262	*** wide-ranging chemosensor
primordius	Northern Ridge-tailed Monitor				Northern Territory
	diurnal	carnivore	171	84	*** refugia-anchored diurnal gr. liz.
rainerguentheri	Guenther's Mangrove Monitor				Indonesia
	diurnal	carnivore	224	200	*** wide-ranging chemosensor
rasmusseni	Tawi Tawi Monitor				Philippines
	diurnal		395	1255	*** wide-ranging chemosensor
reisingeri	Yellow Tree Monitor				Indonesia
	diurnal	carnivore	209	160	* wide-ranging chemosensor
rosenbergi	Heath Monitor				Australia
	diurnal	carnivore	423	1560	**** wide-ranging chemosensor
rudicollis	Rough-necked Monitor				Malaysia
	diurnal	invert.	378	1090	**** wide-ranging chemosensor
salvadorii	Crocodile Monitor				New Guinea
	diurnal	carnivore	596	4750	**** wide-ranging chemosensor
salvator	Water Monitor				Southeast Asia
	diurnal	carnivore	540	2880	**** wide-ranging chemosensor
samarensis	Samar Water Monitor				Philippines
	diurnal		450	1900	*** wide-ranging chemosensor
scalaris	Western Spotted Tree Monitor				Western Australia
	diurnal	carnivore	200	140	*** wide-ranging chemosensor
semiremex	Rusty Monitor				Queensland
	diurnal	carnivore	220	187	*** wide-ranging chemosensor
semotus	Mussau Monitor				New Guinea
	diurnal	carnivore	295	490	*** wide-ranging chemosensor
similis	Northern Spotted Tree Monitor				Queensland
	diurnal	carnivore	190	118	**** wide-ranging chemosensor
sparnus	Dampier Peninsula Goanna				Western Australia
	diurnal		92	8.2	*** wide-ranging chemosensor
spenceri	Spencer's Monitor				Queensland
	cathemeral	carnivore	433	1700	*** wide-ranging chemosensor

Varanus spp.	English name				Range core
			Size		
	Activity	Diet	mm	g	Ecological business model
spinulosus	Spiny-necked Monitor				New Guinea
	diurnal	carnivore	315	600	*** wide-ranging chemosensor
storri	Storr's Monitor				Northern Australia
	diurnal	invert.	116	23.7	*** refugia-anchored diurnal gr. liz.
telenesetes	Rossel Island Monitor				New Guinea
	diurnal	carnivore	164	74	
timorensis	Timor Monitor				Indonesia
	diurnal	carnivore	244	263	*** wide-ranging chemosensor
togianus	Togian Islands Water Monitor				Indonesia
	diurnal	carnivore	309	570	*** wide-ranging chemosensor
tristis	Black-headed Monitor				Australia
	diurnal	carnivore	229	214	**** wide-ranging chemosensor
varius	Lace Monitor				Eastern Australia
	diurnal	carnivore	541	3475	**** wide-ranging chemosensor
yemenensis	Yemen Monitor				Red Sea margins
	diurnal	invert.	452	1940	*** wide-ranging chemosensor
yuwonoi	Tri-colored Monitor				Indonesia
	diurnal	carnivore	389	1200	*** wide-ranging chemosensor
zugorum	Zugs' Monitor				Indonesia
	diurnal	carnivore	320	500	** wide-ranging chemosensor

refugium (51% v. 22%). *V. acanthurus*, *caudolineatus*, *niloticus*, and *varius* are known to refuge inside termite mounds; *albigularis* may refuge inside hanging beehives. *Varanus* follow the lizard pattern of exhibiting bimodality in annual rainfall belts: there are no *Varanus* documented to use the intermediate band, but 73% use the next drier zone (v. 30%, arid belt) and 27% (v. 43%) use the next wetter zone (mesic). The other bands are all used by about 7% of *Varanus* (see first paragraph). The elevational profile of *Varanus* is strongly skewed toward low elevations (l/m/h: 100/19/0 v. 82/41/13), possibly to avoid competition from more cold-adapted mammals. *Varanus* are of average representation in sandy areas (22% v. 22%) and human habitat (19% v. 19%) and are overrepresented in littoral areas (30% v. 17%) and riparian areas (41% v. 22%).

All *Varanus* are oviparous (v. 82%). Female receptivity periods seem to be short, and mate guarding is commonly reported; multiple males may attend receptive females in some species (e.g., *similis*). *V. varius* attempts to mate with each female on both sides (i.e., use both hemipenes), suggesting sperm competition. Clutch count is slightly small (−0.36 v. 0.0 std. resid.); the young are of average size (ontogenetic mass ratio 0.02 v. 0.0 std. resid.). Not surprisingly, RCM ratio is therefore near average as well (0.282 v. 0.226). Incubation duration is long (0.71 v. 0.0 std. resid.) and may be variable to synchronize eclosion to the arrival of rains. Number of broods per year is highly variable and can be as high as eight (*kingorum*). Eggshells are universally soft (v. 41% hard). They are adherent at an average frequency (2.5% v. 2.4%). Communal oviposition is slightly elevated (5.1% v. 4.2%) and parental care is very common (17.7% v. 2.9%). Parental care generally takes the form of female defense of nest sites or female excavation of young at the time of hatching, especially if eggs are deposited in termite mounds. At least 12 species oviposit in termite mounds, and many, but not all of these species, return to excavate the hatchlings. At least two species (*komodoensis*, *marmoratus*) oviposit in megapode mounds.

Longevity is average (−0.06 v. 0.0 std. resid.). Population density is low (0.12 kg/ha v. 0.54 kg/ha), based on 14 credible estimates of six species (eight estimates are of *bengalensis*). Endemism is slightly low (52% v. 64%) but will probably rise as more island populations are elevated to full species. Most species occur on islands (72% v. 40%). Introductions are elevated (5.1% v.3.3% for any introductions; 1.3% v. 1.1% for >2 introductions per species). *V. indicus* was introduced, with minimal success, as a control agent for coconut crop pests. Some species are accomplished dispersers on their own; *salvator* was the first vertebrate species to recolonize Krakatoa after the landmark 1883 irruption. As highlighted in the first paragraph, all *Varanus* are deemed vulnerable to commercial depletion and are therefore considered imperiled (v. 49%), but only one species (*mabitang*, endangered) is on one of the IUCN's lists of the most endangered (v. 13.6%). Among the species collected for leather, eggs, and meat are *bengalensis, bitatawa, flavescens, griseus, indicus, nebulosus, niloticus*, and *salvator*.

Varzea spp.	English name				Range core
			Size		
	Activity	Diet	mm	g	Ecological business model
altamazonica	Upper Amazon Floodplain Skink				Peru
	diurnal	invert.	85	12.5	** refugia-anchored diurnal gr. liz.
bistriata	Lower Amazon Floodplain Skink				Brazil
	diurnal	invert.	87	6.6	*** ground-foraging tree lizard

Vhembelacerta sp.	English name				Range core
			Size		
	Activity	Diet	mm	g	Ecological business model
rupicola	Soutpansberg Rock Lizard				South Africa
	diurnal		46	2.24	**** refugia-anchored diurnal gr. liz.

Varzea

Varzea are two species of large (9.1g v. 3.3 g), blue-tailed (100% v. 4%), diurnal (100% v. 74%), viviparous (v. 36%) rain forest skinks found in South America, from French Guiana to Brazil and Bolivia. One is semiarboreal (ground-foraging tree lizard 50% v. 7%); the other is terrestrial (refugia-anchored diurnal ground lizard 50% v. 30%).

Their build is more robust than a typical skink (–0.09 v. –0.46 std. resid.). Their tails are normal (1.43 v. 1.33 × SVL) and breakable (100% v. 73%). One (*bistriata*) is elongate (v. 35%).

Males are smaller than females (–1.0 v. –0.31 on –2 to 2 scale), which is not expected for such large lizards (table 3.3, trait 1). The only other sex difference is that the m:f relative tail length ratio is strongly skewed toward longer male tails (1.13 v. 1.03), the third-highest value in Scincidae. This is consistent with the weak but significant overall pattern in lizards—table 3.3, trait 8—but not with the contrary but insignificant pattern in Scincidae.

Their generalized invertebrate diet is unremarkable (100% v. 84%), but they notably forage by both sit-and-wait (100% v. 16%) and active foraging (100% v. 92%). Mean active body temperature is normal (32.3°C v. 31.1°C). They are found in mesic areas (1101–3000 mm/y) at mostly low elevations (in this case 0–1800 m). They are associated with riparian (50% v. 23%) and human (50% v. 14%) habitats. The association with human habitats is believed to be mediated by a preference for light gaps.

Despite their lowland tropical environment, both species are viviparous (v. 36%) with size-appropriate litters (0.13 v. 0.06 std. resid.). In this case viviparity might be a product of the rarity of light gaps in the Amazonian forest; the females may carry the developing embryos into the scarce sun flecks rather than leaving them in a shaded nest. The neonates, based on *bistriata*, are of average size, but they place a substantial physical burden on the mother: relative "clutch" mass is 0.581 (v. 0.236). In the case of *bistriata*, the females ovulate and conceive at a very tender age and size, but their remarkably slow gestation (270–365 days) ensures that the females will have grown sufficiently by the time of parturition.

Neither species is endemic (v. 64%); *bistriata* occurs on Marajo Island (50% v. 57%). Neither has been introduced (v. 2.1%) or is imperiled (v. 47%). In other respects *Varzea* are much like other skinks.

Vhembelacerta

Vhembelacerta rupicola is the sole species in a lacertid genus of (2.2 g v. 5.1 g) refugia-anchored diurnal ground lizards endemic to South Africa. It is light in build (shape index 0.04 v. 0.27 std. resid.) with a moderately short tail (1.50 v. 2.06 × SVL) that is breakable (v. 92%). Sex differences are not reported. Its primary predator concealment tactic is retreat to its crevice-burrow (v. 50%). It is documented over the elevation range 800–1600 m. Clutch count is small (–0.09 v. 0.13 std. resid.). It is not found on islands (v. 26%). It has not been introduced (v. 4.6%). It is considered imperiled (v. 56%), but it is not on one of the IUCN's lists of most endangered (v. 15.5%). In other respects *Vhembelacerta* is probably similar to other lacertids.

Vietnascincus

Vietnascincus rugosus is the sole member of its genus, a little-known (filled cells 18% v. 25%), moderately large (6.2 g v. 3.3 g) trunk/wall (v. 3%) skink endemic to Vietnam. It is slightly more robust than an average skink (shape index 0.15 v. –0.46 std. resid.). It is found at moderately low elevation (700 m in this case). Reproduction is unknown. It is not found on an island (v. 57%), nor has it been introduced (v. 2.1%). It is not imperiled (v. 47%). In other respects

Vietnascincus sp.	English name				Range core
			Size		
	Activity	Diet	mm	g	Ecological business model
rugosus	Rough Vietnam Skink				Vietnam
	diurnal		68	6.2	* trunk/wall lizard

Voeltzkowia spp.	English name				Range core
			Size		
	Activity	Diet	mm	g	Ecological business model
mira	Burrowing Blind Skink				Madagascar
			67	0.97	**** subterranean forager
mobydick	Moby's Mermaid Skink				Madagascar
			60	1.55	**** subterranean forager
yamagishii	Yamagishi's Mermaid Skink				Madagascar
			72	3.86	**** subterranean forager

Woodworthia spp.	English name				Range core
			Size		
	Activity	Diet	mm	g	Ecological business model
brunneus	Waitaha Gecko				New Zealand
	nocturnal		67	5.9	** trunk/wall lizard
chrysosireticus	Goldstripe Gecko				New Zealand
	cathemeral		67	6.1	** nocturnal ground lizard
maculatus	Raukawa Gecko				New Zealand
	nocturnal	omnivore	68	6.4	** ground-foraging tree lizard

Vietnascincus is likely to be similar to other skinks.

Voeltzkowia

Voeltzkowia is a trio (three spp.) of small-sized (1.8 g v. 3.3 g), vestigial-eyed (100%), functionally limbless (100%) subterranean forager (100% v. 12%) skinks from Madagascar. *V. mobydick* is unpigmented, as was the legendary whale. They are fairly slender but not as extreme as many subterranean foragers; their shape index is –1.35 std. resid. (v. –0.46 for Scincidae or –2.75 for subterranean foragers: table 4.1, trait 4). Their tails are short compared to skinks (1.02 v. 1.33 × SVL) but long for subterranean foragers (0.37, see table 4.1, trait 5). One (*yamagishii*) tail is breakable (33% v. 73%). They are oviparous (100% v. 64%), with slightly large clutch counts (0.50 v. 0.06 std. resid.). They have not been introduced (v. 2.1%) but are imperiled (100% v. 47%), including placement on the IUCN's lists of the most endangered (100% v. 14%). In other respects *Voeltzkowia* are probably typical limbless skinks.

Woodworthia

Woodworthia is a small (three spp.) genus of reasonably well-known (35% v. 28% of cells filled), moderately large-sized (6.1 g v. 4.15 g) New Zealand diplodactylids of diverse habits. Their tails are a bit long for diplodactylids (1.17 v. 0.85 × SVL). They have secretory pores (v. 34%), but other sex differences are not reported. The diet of only one species is documented (*maculatus*); its stomach contents were dominated by flowers (58% by occurrence). The same species is conspicuously wary (v. 2.7%). All three species are reported to bask directly (v. 17%), despite their generally nocturnal habits.

Living in groups is common (at least two of three spp.), and males of all species are reported to be tolerant of other males (v. 24%). *Woodworthia* is overrepresented in littoral areas (100% v. 17%), sandy areas (33% v. 24%), and around humans (33% v. 5%), but it has not been reported in riparian zones (v. 10%). As with most New Zealand lizards, *Woodworthia* are viviparous (v. 18%). Unlike most diplodactylids, *Woodworthia* young are smaller than lizard average (ontogenetic mass range 0.13, the second-highest value in Diplodactylidae, v. –0.57 std. resid.). Parental care has not been reported (v. 3.4%). Longevity is very high (1.86 v. 0.90 std. resid.), which in the case of *brunneus* and *maculatus* equates to 42 years and 37 years, respectively. Population density is also high, based on two estimates for *maculatus*: 3.9 kg/ha (v. 0.44 kg/ha).

Woodworthia have not been introduced (v. 2%). *W. chrysosireticus* is considered imperiled (v. 81%), though no *Woodworthia* is listed on the IUCN's list of most endangered species (v. 29%). In other respects these are typical diplodactylids.

Xantusia (fig. 4.2A)

Xantusia constitute a moderate radiation (14 spp.) of xantusiids that are strikingly similar to *Lepidophyma* but with smoother body scales. They are smaller (2.3 g v. 6.5g), found further north—Mexico to California, Nevada, and Utah—and are much better known (filled cells 32% v. 23%). The latter may be responsible for the other apparent differences between the genera: *Xantusia* have more diverse reports of foraging modes, more diverse reported diel patterns, are more often said to be conspicuously wary (29% v. 5% in *Lepidophyma*), are more often characterized as cryptozoic (90% v. 30% in *Lepidophyma*), and have greater reported longevity (1.32 v. 0.23 std. resid. in *Lepidophyma*), based on nine *Xantusia* longevities. Subsequent reference values in this entry will list the family's mean value (not *Lepidophyma*). The common names of "night" lizard are misleading for both genera; see Xantusiidae entry.

Xantusia are small-sized (2.3 g v. 4.0 g) crack hermits (100% v. 89%). They are of average build (–0.07 v. 0.05 std. resid.) with shorter tails (1.08 v. 1.21 × SVL). Secretory pores are average in frequency (29% v. 24%). Basic color change is found more often in the genus (14% v. 6%). As with the family, *Xantusia* are m<f in SSD (–1.0 v. –1.0 on –2 to 2 scale). Sexual dichromatism is slightly more frequent (7% v. 3%).

Xantusia has no ant specialists (v. 7%) but has a typical mix of invertebrate eaters (86% v. 71%) and omnivores (14% v. 21%). They are reported to both sit-and-wait (100% v. 67%) and forage actively (50% v. 33%) but not cruise (0% v. 33%). Home-range area is the same as for the family, as it is documented only in *Xantusia* (–1.31 std. resid.).

Cathemeral (38% v. 47%) and diurnal (38% v. 20%) activity are the dominant and coequal patterns in "night" lizards. Crepuscular (13% v. 6.7%) and nocturnal (13% v. 27%) species are also coequal, but minor. Conspicuous wariness is common (29% v. 15%). Direct basking is overrepresented (14% v. 6%); thermoconformance is average (7% v. 6%). Mean active body temperature is slightly cooler than the family (26.8°C v. 27.6°C). Use of burrows for daily refugia is slightly elevated (18% v. 8%). Unlike *Lepidophyma*, which is associated with wet habitats, *Xantusia* is associated with the dry rainfall belt (50% v. 50%). More documentation is needed for both genera. The elevational profile of *Xantusia* is normal (l/m/h: 82/45/9 v. 83/55/3). They are found in littoral (17% v. 6%) and riparian (17% v. 16%) zones but not sandy areas (v. 0%) or human habitat (v. 13%). Woodrats (*Neotoma* sp.) often surround their nests with spiny plant materials (especially cholla cactus joints); *vigilis* has the same habitat modification preference and often cohabits with *Neotoma*. Genetic sampling indicates that aggregations of *vigilis* often consist of family members; neonates may remain with their mothers and may share hibernacula

Xantusia spp.	English name		Size		Range core
	Activity	Diet	mm	g	Ecological business model
arizonae	Arizona Night Lizard				Arizona
		invert.	49	2.21	*** crack hermit
bezyi	Bezy's Night Lizard				Arizona
	diurnal	invert.	56	3.32	*** crack hermit
bolsonae	Bolson Night Lizard				Mexico
			51	2.58	*** crack hermit
extorris	Durango Night Lizard				Mexico
	nocturnal		41	1.28	*** crack hermit
gilberti	Cape Night Lizard				Mexico
	diurnal	invert.	41	1.31	** crack hermit
gracilis	Sandstone Night Lizard				California
			61	4.35	*** crack hermit
henshawi	Granite Night Lizard				Mexico
	cathemeral	invert.	60	4.13	**** crack hermit
jaycolei	Cole's Night Lizard				Mexico
	cathemeral		39	1.17	** crack hermit
riversiana	Island Night Lizard				California
	diurnal	omnivore	85	27.7	**** crack hermit
sanchezi	Zacatecas Night Lizard				Mexico
	cathemeral	invert.	45	1.77	** crack hermit
sherbrookei	Sherbrooke's Night Lizard				Mexico
			35	0.83	** crack hermit
sierrae	Sierra Night Lizard				California
			44	1.66	** crack hermit
vigilis	Desert Night Lizard				California
	cathemeral	invert.	50	2.40	**** crack hermit
wigginsi	Wiggin's Night Lizard				Mexico
			39	1.17	** crack hermit

for several years. It is not a strain to imagine female offspring inheriting the well-feathered beds of their mothers.

Litter count is average (–0.20 v. –0.07 std. resid.). Longevity is notably high (1.32 v. 0.93 std. resid.), highlighted among nine species by *riversiana* (32.9 years). Based on four credible samples of the same species, population density is very high (12.5 kg/ha v. 0.49 kg/ha). This species was once listed as threatened in the United States and therefore received additional study, which probably resulted in the discovery and quantification of the most-dense populations.

The genus is found entirely within the Nearctic biogeographic realm (v. 76%). Endemism is average (79% v. 79%). Island species are about average (7% v. 6%). One introduction of *Xantusia* has been made (*vigilis* to Santa Catalina Island: 7.1% v. 2.9%), though no species has been introduced more than twice (v. 0%). They are underrepresented among imperiled xantusiids (13% v. 31%), and none is on one of the IUCN's lists of the most endangered (v. 6%). In other respects *Xantusia* are typical xantusiids. The flesh of *gilberti* is reported by Seri Amerindians to be toxic.

Xantusiidae (fig. 4.2A)

The 34 xantusiids are moderately small (4.0 g v. 5.3 g), viviparous (100% v. 18%), and mostly crack hermits (89% v. 3%). They are found in Mexico and adjoining areas, with *Xantusia vigilis* extending northward to Nevada and Utah and two species reaching Panama. *Cricosaura* are found in Cuba. Those few that are not crack hermits are trunk/wall lizards (7.7% v. 13%). Although Xantusiidae are not the leading family for any single trait, they are near the top in breakable tails (100%, third highest, v. 57%), cruise foraging (33%, fourth highest, v. 6.6%), activity at any time of day (47%, fourth highest, v. 6.8%), retreat to crevice-burrows when threatened (100%, tied for second, v. 25%), cryptozoic microhabitats (62%, fourth, v. 12%), residence in the dry rainfall belt (50%, third, v. 12%), and residence at the other end of the rainfall spectrum—wet (50%, third, v. 9%). The overarching theme seems to be slow-moving persistent secretiveness in diverse habitats. The Xantusiidae genera are *Cricosaura*, *Lepidophyma*, and *Xantusia*.

The secretiveness of Xantusiidae has resulted in a number of misunderstandings. The common name of "night" lizards accurately described only a small minority of the species. Apparently herpetologists who can't find a lizard during the day tend to assume that it is nocturnal. In several well-studied xantusiids (e.g., *Xantusia riversiana*), the lizards have been found to be active primarily during the day but so secretive in their locomotion that they are rarely seen abroad. They are also sedentary: only 9% of *X. vigilis* changed crack during a six-year study. The difficulty in detecting such homebodies caused *X. riversiana* to be initially listed by the United States as threatened and removed from protection after discovering its true abundance. The IUCN initially deemed this species rare (1986–1996), then vulnerable (1996–2007), and then least concern (2007–2017). Although the US Fish and Wildlife Service proudly declares this species to have been "delisted due to Recovery," it is perhaps more accurate to suggest, as does the trajectory of IUCN listings, that this species was initially deemed to be scarce on the basis of failing to understand its cryptic life style and that subsequent research demonstrated that it was numerous though very well hidden. The earlier listings of species formerly deemed imperiled on the basis of inadequate information are said by the IUCN to be "ingenuine" listings; over time scientific study helps clarify species' ecologies and steer attention to the most imperiled species.

Xantusiids are of average build (shape index 0.05 v. 0.00 std. resid.). Their tails are slightly short (1.21 v. 1.37 × SVL). Secretory pores are common (24% v. 15%), but blue tails (0% v. 2.3%), prehensile tails (0% v. 5%), crests (0% v. 5.3%), elongation (0% v. 11%), basic color change (6% v. 11%), and advanced color change (0% v. 6%) are not.

Males are smaller than females (–1.00 v. 0.36 on –2 to 2 scale), but other sex differences are relatively muted. Sexual dichromatism (2.9% v. 20.7%) is the only other regular sex difference; sex difference rollup score, on scale of 0 to 6 is 0.39 (v. 0.75).

Diets are diverse in xantusiids: invertebrate eating (71% v. 73%), ant specialization (7.1% v. 4.4%), and omnivory (21.4% v. 15.4%). Foraging modes are equally diverse: sit-and-wait (67% v. 46%), cruise (33% v. 6.6%), and active (33% v. 58%) foraging. In keeping with their general secretiveness, electric lights are not used (v. 2.0%). In keeping with their sedentariness, home ranges are very small (–1.31 v. 0.0 std. resid.), based on *Xantusia riversiana* (ca. 17.2 m^2).

All diel activity patterns are present in Xantusiidae, but cathemeral (47% v. 7%) and crepuscular (6.7% v. 1.8%) are overrepresented; nocturnal is average (27% v. 25%), and diurnal is comparatively rare (20% v. 67%). No xantusiid vocalizes (v. 2.5% call spontaneously). They are wary (15% v. 8%). Direct basking is uncommon (6% v. 18%); thermoconformance is as common as basking (6% v. 2.6%). Mean active body temperature is low (27.6°C v. 30.8°C). Social relations are essentially unknown, though all described males defend space (100% v. 85%). Burrows are infrequently used as daily retreats (8% v. 22%).

The striking bimodality of their rainfall belt residence was noted in the first paragraph. Their elevational profile is about average (l/m/h: 83/55/3 v. 82/41/13). They are underrepresented in all tabulated habitat specializations: littoral (6% v. 17%), sandy (0% v. 22%), riparian (16% v. 22%), and human (13% v. 18.5%) habitats.

They are all viviparous, as noted in the first paragraph; litter count is average (−0.07 v. 0.0 std. resid.). Neonates are about average or slightly large (ontogenetic mass range −0.19 v. 0.0 std. resid.). RCM ratio is slightly low (0.193 v. 0.226). Parental care is normal (2.9% v. 2.9%), though see the *Xantusia* account for evidence of substantially delayed juvenile dispersal.

Longevity is long (0.93 v. 0.0 std. resid.), based on 14 species. The longest is 32.9 years for *X. riversiana*. This fits with the hypothesis that immobility enhances survivorship. Mean population density is average (0.49 kg/ha v. 0.54 kg/ha), but the family mean is based on two very disparate density means. *X. riversiana* is very dense, based on four credible samples (12.5 kg/ha), whereas *L. flavimaculatum* goes to the opposite extreme (0.02 kg/ha). As befits sedentary lizards, endemism is high (79% v. 64%). Island species are uncommon (6% v. 40%). They have been introduced at a normal frequency (2.9% v. 3.3%), presumably as stowaways rather than intentional translocations. No species has been introduced more than twice (v. 1.1%). They are somewhat underrepresented among the imperiled (31% v. 49%), especially so among those on the IUCN lists of the most endangered (3.8% v. 13.6%).

Xenagama spp.	English name / Activity	Diet	Size mm	Size g	Range core / Ecological business model
batillifera	Vaillant's Strange Agama				Ethiopia
	diurnal		82	17.8	*** refugia-anchored diurnal gr. liz.
taylori	Taylor's Strange Agama				Ethiopia
			71	11.5	*** refugia-anchored diurnal gr. liz.
wilmsi	Wilms' Strange Agama				Ethiopia
			79	15.9	** refugia-anchored diurnal gr. liz.
zonura	Ethiopian Ridgeback Agama				Ethiopia
			70	11.1	** refugia-anchored diurnal gr. liz.

Xenagama

Xenagama is a small (four spp.) genus of average-sized (13.8 g v. 15.6 g), average build (shape index 0.85 v. 0.80 std. resid.), poorly known (filled cells 21% v. 30%), crestless (v. 38% crested) terrestrial (v. 56%) agamids from Ethiopia and Somalia. All four species are found in both countries and nowhere else. All are refugia-anchored diurnal ground lizards (v. 43%). Most are immediately recognizable by their large heads and short (0.64 v. 1.86 × SVL), spatulate, spiny, armored tails that taper abruptly to a wirelike extension, the "terminal filament." These lizards retreat headfirst into soft soil burrows (100% v. 29%) with their tails facing any rearguard action. All reported species are sexually dichromatic (v. 53%), and half exhibit seasonal color (50% v. 20%). One is found around humans (25% v. 16%); two are associated with sand or dunes (50% v. 26%). None has been reported as introduced (v. 1.7%) or imperiled (v. 37%).

Xenosauridae and *Xenosaurus* (figs. 4.2B, 4.2C)

Xenosaurus, the only genus (11 spp.) of the Xenosauridae, has one ecological business model, the crack hermit (100% v. 2.6%). They are large (17.4 g v. 5.3 g), diurnal (100% v. 67%) viviparous (v. 18%) lizards found in the Mexican uplands (*rackhami* reaches Guatemala). The group is reasonably well known (filled cells 37% v. 27%). Their skin texture is pebbly throughout, hence the "knob-scaled" common names. An illustration of the sedentariness of these crack hermits is that the mean net daily movement distance of *newmanorum* is 39 mm (not meters or centimeters). Daily activity radius of *grandis* is about 1 m.

Xenosaurids are of average build (shape index 0.33 v. 0.0 std. resid.); tails are short (0.90 v. 1.37 × SVL) and unbreakable (v. 57% breakable). Tails are not prehensile (v. 5.0%) or blue (v. 2.3%). They do not have crests (v. 5.3%) or secretory pores (v. 15%); they are not elongate (v. 11%) and do not change color (v. 11%).

Males tend to be smaller than females (−0.64 v. 0.36 on −2 to 2 scale), possibly to fit into cracks with females, as receptive females remain in their home cracks and males must find and join them, at least momentarily. Males tend to have slightly larger heads and longer limbs, but do not exhibit dichromatism (v. 21%) or seasonal color (v. 6.5%).

Most species eat invertebrates (88% v. 73%); *newmanorum* is an omnivore (13% v. 15%). The omnivore must make at least short forays out of its crack to obtain leaves, but it does not apparently eat fruit or flowers that might force it to go further afield. With the exception of *newmanorum* in folivorous mode, all are sit-and-wait foragers (100% v. 46%).

Direct basking occurs at the mouth of crevices (9% v. 18%). Thermoconformance (18%, the highest family, v. 2.6%) is more frequent than basking. Mean active body temperature is very low (21.6°C, third lowest, v. 30.8°C). They are found in the middle of the rainfall spectrum: intermediate (50% v. 27%) and mesic (50% v. 43%). They are found predominantly at intermediate elevations (80% v. 41%) but also occur low (50% v. 82%). Evidently there are many unoccupied cracks at lower and higher elevations, but crack hermits have such low vagility that they may be unable to disperse to those vacancies. They are not associated with any of the tabulated specialty habitat types (littoral, sandy, riparian, or human habitats).

Litter counts are small (–0.51 v. 0.0 std. resid.), perhaps to facilitate squeezing pregnant mothers into their home cracks or because neonates are so large. Ontogenetic mass range is very small (–0.94 v. 0.0 std. resid.), suggesting a selective advantage to larger young. In several species (at least *newmanorum*, *platyceps*, *rectocollaris*) the young remain with their presumed mother in her home crack for periods up to 132 days (parental care 27% v. 3%). This suggests that neonates are relatively defenseless on their own, or cannot safely find a suitable infant-sized crack, and can benefit from female attendance. RCM ratio is 0.238 (v. 0.226), suggesting that the small litter count is traded for larger neonates, rather than shrinking the pregnant mothers' volume.

The only regular lizard family ($n > 7$) with a higher mass-adjusted longevity is Eublepharidae, suggesting that life in the confines of a crack has survival benefits. Xenosauridae longevity averages 0.68 (v. 0.0) std. resid., albeit based solely on *grandis* (12 years). Credible population density estimates ($n = 4$) are available only for that species (4.5 kg/ha v. 0.54 kg/ha). Population densities in crack hermits are difficult to compare, as the density of crack hermits is likely to be set by the area's geology rather than ecology. Most xenosaurids are found in the Nearctic (100% v. 6%); 18% (v. 29%) extend into the Neotropics. Endemism is high (91% v. 64%). Island species are absent (0% v. 40%). None has been introduced (v. 3.3%). Most are imperiled (71% v. 49%), presumably due to limited range and vagility. Over one-quarter (29% v. 14%) are on the IUCN's lists of the most endangered.

Xenosaurus spp.	English name		Size		Range core
	Activity	Diet	mm	g	Ecological business model
agrenon	Oaxaca Knob-scaled Lizard				Mexico
	diurnal		101	19.1	**** crack hermit
arboreus	Arboreal Knob-scaled Lizard				Mexico
			89	12.1	**** crack hermit
grandis	Grand Knob-scaled Lizard				Mexico
	diurnal	invert.	106	22.6	**** crack hermit
mendozai	Queretero Knob-scaled Lizard				Mexico
	diurnal	invert.	94	14.6	**** crack hermit
newmanorum	Newmans' Knob-scaled Lizard				Mexico
	diurnal	omnivore	105	21.8	**** crack hermit
penai	Pena's Knob-scaled Lizard				Mexico
	diurnal	invert.	89	12.2	**** crack hermit
phalaroanthereon	Chin-spotted Knob-scaled Lizard				Mexico
	diurnal	invert.	109	24.7	**** crack hermit
platyceps	Flathead Knob-scaled Lizard				Mexico
	diurnal	invert.	105	21.6	**** crack hermit
rackhami	Rackham's Knob-scaled Lizard				Mexico
	diurnal		101	19.1	**** crack hermit
rectocollaris	Pallid Knob-scaled Lizard				Mexico
	diurnal	invert.	92	14.0	**** crack hermit
tzacualtipantecus	Zacualtipan Knob-scaled Lizard				Mexico
	diurnal	invert.	97	16.0	**** crack hermit

Zonosaurus

Zonosaurus is a 17-species genus of the diurnal Madagascan gerrhosaurids, with most species found on the ground: refugia-anchored diurnal ground lizards 77% (v. 59%). Others are diurnal canopy walkers (12% v. 6%) and ground-foraging tree lizards (12% v. 6%). They resemble mildly hefty skinks, though possessing the lateral fold characteristic of gerrhosaurids. They are a bit larger than average gerrhosaurids (21.3 g v. 16.2 g) and robustly built (shape index 0.68 v. 0.38 std. resid.). Secretory pores are not reported (v. 8%), but sexual dichromatism is reported often (24% v. 16%). Seasonal color is unusually common with reference to all lizards, but relatively rare for gerrhosaurids (12% v. 27% for family and 6.5% for lizards). *Zonosaurus* are a mix of invertebrate eaters (57% v. 44%) and omnivores (43% v. 56%). They follow the family pattern in being mostly diurnal (88% v. 91%); the remainder are crepuscular (12% v. 9%). A few are conspicuously wary (6% v. 22%). The documented primary predator concealment tactics are flight (50% v. 27%) and sand diving (50% v. 18%). Two species are highlighted for the ease with which they shed skin scales (*subunicolor*, *tsingy*), presumably a last-second predator escape tactic. Direct basking is widely documented (41% v. 24%). *Z. laticaudatus* is documented for mean active body temperature at the family mean (34.0°C). Forty percent of species are known to use burrow refugia (v. 56%). All four documented species are found in the mesic rainfall belt (around 2250 mm/y). They are found in all the tabulated habitat types at family-typical levels except littoral areas (0% v. 6%). Clutches are slightly small (−0.42 v. −0.20 std. resid.). Longevity matches the family pattern, though female *laticaudatus* are notable for dividing their clutches at oviposition. Are they bet hedging against chemosensory predators? Half of evaluated species are considered imperiled (v. 50%), though only 6.3% (v. 10.7%) are on the IUCN's lists of the most endangered. In other respects *Zonosaurus* are typical gerrhosaurids.

Zonosaurus spp.	English name				Range core
			Size		
	Activity	Diet	mm	g	Ecological business model
aeneus	Bronze Girdled Lizard				Madagascar
	diurnal		66	6.78	** refugia-anchored diurnal gr. liz.
anelanelany	Strand Girdled Lizard				Madagascar
	diurnal		76	10.4	* refugia-anchored diurnal gr. liz.
bemaraha	Outcrop Girdled Lizard				Madagascar
	diurnal		63	5.9	* refugia-anchored diurnal gr. liz.
boettgeri	Boettger's Girdled Lizard				Madagascar
	diurnal		95	20.9	*** diurnal canopy walker
brygooi	Brygoo's Girdled Lizard				Madagascar
	diurnal		65	6.5	** refugia-anchored diurnal gr. liz.
haraldmeieri	Green Madagascar Girdled Lizard				Madagascar
	diurnal		119	41.3	** refugia-anchored diurnal gr. liz.
karsteni	Karsten's Girdled Lizard				Madagascar
	diurnal	invert.	104	27.8	** refugia-anchored diurnal gr. liz.
laticaudatus	Western Girdled Lizard				Madagascar
	diurnal	omnivore	135	62	** refugia-anchored diurnal gr. liz.
madagascariensis	Madagascar Girdled Lizard				Madagascar
	diurnal	invert.	116	38.3	** refugia-anchored diurnal gr. liz.
maramaintso	Maramaintso Girdled Lizard				Madagascar
	diurnal		95	20.9	* diurnal canopy walker
maximus	Southeastern Girdled Lizard				Madagascar
	diurnal	invert.	200	207	*** refugia-anchored diurnal gr. liz.
ornatus	Ornate Girdled Lizard				Madagascar
	diurnal	invert.	106	29	** ground-foraging tree lizard
quadrilineatus	Four-lined Girdled Lizard				Madagascar
	crepuscular	omnivore	143	74	** refugia-anchored diurnal gr. liz.
rufipes	Red-legged Girdled Lizard				Madagascar
	diurnal		72	9.0	** ground-foraging tree lizard
subunicolor	Azure-chinned Girdled Lizard				Madagascar
	diurnal		71	8.4	** refugia-anchored diurnal gr. liz.
trilineatus	Three-lined Girdled Lizard				Madagascar
	crepuscular	omnivore	120	42.9	** refugia-anchored diurnal gr. liz.

Zonosaurus spp.	English name				Range core
			Size		
	Activity	Diet	mm	g	Ecological business model
tsingy	Tsingy Girdled Lizard				Madagascar
	diurnal		70	8.2	**** refugia-anchored diurnal gr. liz.

Zootoca sp.	English name				Range core
			Size		
	Activity	Diet	mm	g	Ecological business model
vivipara	European Common Lizard				Eurasia
	cathemeral	invert.	62	5.61	*** refugia-anchored diurnal gr. liz.

Zootoca

Zootoca vivipara is the sole species in a lacertid genus of average-sized (5.6 g v. 5.1 g) refugia-anchored diurnal ground lizards found from Great Britain and Spain to Japan. It may be night active in warm weather. Although the species name correctly implies that the species is viviparous in most of its range, it lays eggs in the southern parts of its range. Its tail is a trifle short (1.75 v. 2.06 × SVL). It has secretory pores. Males are smaller than females (–1.0 v. 0.49 on –2 to 2 scale) but have proportionately larger heads. The sexes differ in color (v. 33%). Male tails are shorter in relative length than are females (0.94 v. 1.07 ratio). The sex differences rollup (on scale of 0 to 6) is 3.0 (v. 0.86), tied with *Lacerta* as the highest genus in Lacertidae.

They eat many invertebrates (v. 77%) but avoid ants. They are reported to both sit-and-wait (v. 23%) and actively (v. 92%) forage. Home-range area is large (1.2 v. 0.27 std. resid.), corresponding to an area of about 540 m². Males venture beyond this limit when searching for mates. Movements are very high; 98th percentile for MPM and 80th percentile for PTM, keeping with the lacertid generalization that the percentile rank of MPM appreciable exceeds that of PTM. It hides from predators in vegetation (v. 42%). It basks directly (v. 25%); pregnant females are especially fond of sunning themselves.

Males are socially tolerant (v. 41%), an unusual trait for refugia-anchored species (table 4.1, trait 46).

They may dig burrows for refuge during their inactive period (v. 48%). Their elevational range is 0–3000 m. This wide ecological amplitude is also reflected in their wide geographic range (39 countries) and ability to range north of the Arctic Circle in several places. They also occur in all the specialized habitats evaluated: littoral (v. 22%), sandy (v. 44%), riparian (v. 28%), and human (v. 16%) habitats.

As noted above, the species is both viviparous and oviparous, with a moderately large clutch/litter count (0.64 v. 0.13 std. resid.). More distinctly, it also exhibits the smallest young of any genus in the family (ontogenetic mass range 1.87 v. –0.05 std. resid.). Incubation period is relatively short (–0.51 v. –0.36), equating to 46 days. When they have eggshells, these may be hard (v. 14%), which would seem to deter the evolution of viviparity but has not done so. Females may lay communally (v. 1.9%). Neonates may remain with their mother for a time (parental care v. 1.9%). Their RCM is exceptionally high (0.482 v. 0.267).

Their longevity is high (1.0 v. 0.19 std. resid.), equating to 12 years. Population density has been credibly estimated 45 times, for a mean density that is average for lacertids (1.6 kg/ha v. 2.07 kg/ha). They are found on a few islands (Great Britain, Hokkaido, Ireland, Sakhalin). They have not been introduced (v. 4.6%). They are considered to be of conservation concern in several localities (v. 57% imperiled) but are not globally of concern (v. 15.5% IUCN listings of most endangered). In other respects *Zootoca* is similar to other lacertids.

Zygaspis

Zygaspis is a moderate (eight spp.) genus of small (1.9 v. 2.7 g; shape index −4.2 v. −4.0 std. resid.) amphisbaenids from southern Africa, ranging from the Democratic Republic of the Congo to South Africa. Their tails are the longest for an amphisbaenid genus (0.23 v. 0.11 × SVL), and tail breakage is about average (13% v. 15%). For such small lizards the clutch count of 3.5 is high (0.43 v. −0.12 std. resid.). See the family account for mention of a population density estimate from this group. Endemism is average (63% v. 66%). None occurs on an island (v. 12%). None is imperiled (v. 23%).

Zygaspis spp.	English name	Activity	Diet	Size mm	Size g	Range core	Ecological business model
dolichomenta	Equatorial Round-headed Worm Lizard			171	1.99	Congo	**** subterranean forager
ferox	Ferocious Round-headed Worm Lizard		invert.	185	2.71	Zimbabwe	**** subterranean forager
kafuensis	Kafue Round-headed Worm Lizard			166	1.79	Zambia	**** subterranean forager
maraisi	Cabo Delgado Worm Lizard			157	1.43	Mozambique	**** subterranean forager
nigra	Black Round-headed Worm Lizard		invert.	213	4.83	Namibia	**** subterranean forager
quadrifrons	Kalahari Round-snouted Worm Lizard		invert.	184	2.67	Southern Africa	**** subterranean forager
vandami	Van Dam's Round-headed Worm Lizard		invert.	143	0.98	Mozambique	**** subterranean forager
violacea	Violet Round-headed Worm Lizard			143	0.98	Mozambique	**** subterranean forager

Acknowledgments

Preparing a book-length manuscript requires many things, foremost among them are time and the support of loved ones, for which I especially wish to thank Renée Rondeau. Bob Reed of the United States Geological Survey (USGS) graciously spearheaded the provision of institutional support and made possible the computer and library facilities upon which this project was totally dependent. Three USGS librarians did the heavy lifting of finding and supplying hundreds of obscure papers: Rebecca Bolger, Lily Sheng, and Jenny Stevens. Many individuals suggested or supplied essential and sometimes unpublished literature directly: Aaron Bauer, Alison Cree, Leo Fleishman, Enrique Font, Lee Fitzgerald, Tom Fritts, Harry Greene, Michael McCoid, Shai Meiri, Maria Novosolov, Gad Perry, Eric Pianka, Bob Reed, and Martin Whiting. I could not have done it without them.

Among the giants whose shoulders I have stood upon are those contributing to, and helping to compile, the taxonomy of lizards, especially the contributors to Reptile Database (www.reptile-database.org), curated by Peter Uetz, and the hundreds of contributors to the lizard assessments published on the IUCN Red List of Threatened Species (www.iucnredlist.org), whose contributions are too numerous to cite individually.

Many individuals contributed in small and large ways to my understanding of lizard behavior and ecology; I could not hope to enumerate all of the important contributions. Leá Bonewell, Earl Campbell, Kathy Dean-Bradley, Bjorn Lardner, Tom Fritts, Gad Perry, Bob Reed, Renée Rondeau, and Eli Wostl not only helped shape my insights but also contributed directly to the published and unpublished field work cited herein.

Helpful reviewers, including Kraig Adler, Brian Bock, Leo Fleishman, Tom Fritts, Craig Guyer, Bjorn Lardner, Melia Nafus, Gad Perry, and Martin Whiting flagged many wayward statements. Reviewers are the unsung and uncompensated heroes of science. Thank you so much for volunteering your time and effort.

Any use of trade, product, or firm names is for descriptive purposes only and does not imply endorsement by the US government.

The spectacular photographs portrayed herein are largely the work of Lee Grismer, the world's best lizard photographer. Additional superlative photographs were provided by Chandler Eaglestone, Robert Fisher, Glenn Gerber, Bjorn Lardner, Julio Lemos Espinal, James Reardon, Bob Reed, Renée Rondeau, Susy Sanoja-Sarabia, Martin Whiting, and Eli Wostl.

The exertions and counsel of my editors at Johns Hopkins University Press were irreplaceable: Vince Burke, Tiffany Gasbarrini, Kyle Kretzer, Julie McCarthy, Carrie Love, and Esther Rodriguez.

The editorial contributions of Joe Mitchell were sadly interrupted, but he will be remembered for the warmth and generosity with which he always offered them.

Thank you all.

Appendix A Database Fields and Comments on Their Tabulation

The objective of the database is to characterize the ecology of each lizard species. A reader should be able to grasp each species' basic natural history, to the extent recorded, from a glance at the tabulated fields (size, distribution, diet, etc.). I have omitted locality-dependent attributes, such as the time of year when hibernation or breeding commenced or the plant species that are prevalent in the occupied habitat. I have also avoided traits that extend into related subject areas such as physiology (e.g., thermal preferenda) or development (e.g., relationship of incubation time to temperature). These related features can be explored by reading the cited references, but they are not tabulated in the database.

Following a few introductory remarks on the sampling unit—species—and my approach to citing sources, I list each field topically, adding a parenthetic explanation for the fields with nonintuitive names. Following the topical listing, all fields are listed alphabetically, in **boldface**, followed by a complete description of their derivation. Where germane, I discuss the challenges I encountered in pooling the literature values. The special challenges of estimating species size (appendix B) and population density (appendix C) are described in those appendices. The appendices fully explain the methods outlined in chapter 1.

Species

The unit of interest in this work is the species. Subspecies names are listed, but subspecies are not characterized. Because the natural history data are tabulated by species, it is critical that the application of each species name be in agreement among all the observers whose data were combined, and that it be understood by the readers of the book. The first step was to choose a consensus list of species: I used the taxonomic units as they were understood by Reptile Database (www.reptile-database.org, curated by Peter Uetz) in May 2017. If the name given to a species in a tabulated source subsequently became ambiguous—what are now understood to be multiple species were treated as one at the time of an author's account—I discarded the confounded information unless it could be parsed into the Reptile Database's nomenclature. If the species name used by an earlier author remained valid in current usage, the ambiguity created by a subsequent taxonomic split was addressed whenever recognized, but in some cases could have been overlooked. In most but not all cases, the current name for a subsequently divided taxon can be ascertained from the geographic locality of the source data. About half of the total effort involved in compiling this record of lizard natural histories was expended in parsing such taxonomic uncertainties; unfortunately, many secondary sources were impossible to parse. Myhrvold et al. (2015) intentionally conflated all species with overlapping definitions through time. I have done my best to disentangle the complex web of protean names but no doubt have failed in some cases.

If you are searching for a species name that is not given in chapter 5, review the family narrative to identify other genera in which this species might now be found. For example, you will not find the dactyloid *Norops ahli* as a listed species in either chapter 5 or the Species field in the database. However, the Dactyloidae family entry in chapter 5 will specify that all dactyloid species are in *Anolis*, which is where you will find *Anolis ahli*. Similarly, a computer search of the database's Species field will discover one skink and one dactyloid with the *ahli* species name. The Systematic field of the dactyloid record will report *Anolis ahli*'s episodic placement in *Norops*. Confirm that the lizard in question lives in the designated geographic range; if it does not, explore subsequent taxonomic splits using Reptile Database (www.reptile-database.org).

Be especially wary of species names used for introduced populations, as current species concepts are ambiguous with reference to lineages that have been brought together or separated by human agency. For example, the gene pools of *Anolis ordinatus*, originally from the Bahamas, and *A. sagrei*, originally from Cuba, have merged in their mutual introduced range in Florida. Either name could be used for the Florida population, but the name of the numerically dominant Cuban lineage is used more often.

One difference between this database and many other sources is that I have attempted to explicitly list each trait for each species individually. A key objective of science is to generalize from individual species, and most authors have done so with zeal. There are dangers inherent in this process. For example, suppose that the reproductive mode is known for five of eight species within a genus and all known species are oviparous. Some authors conclude that all members of the genus are oviparous. Revision in the taxonomy of the group may invalidate that generalization, even if it was true for the prior taxonomy. Additional field work may discover exceptions. A species described subsequent to the generalization may contradict it. These possibilities create challenges for those trying to tabulate the traits of particular species from generalized statements.

Absence of a given feature is rarely noted explicitly; the absence of a statement is often intended to be taken as evidence for the absence of a trait. For example, if species one, two, and seven in a seven-species genus are described as sexually dichromatic, one might be expected to infer that species three to six are sexually monochromatic, but this inference is tenuous. Compilers of natural history would greatly benefit from the inclusion of explicit negative statements (e.g., "males and females do not differ in color"); the routine practice of inserting weasel words—such as, the sexes are "similar"—may sometimes be necessary, but they propagate or even generate ambiguity. Are the species "similar" but not the same? Is the difference between the species subtle, or is the source simply unsure?

In generalizing from family or genus accounts I have limited my interpretation to the geographic area covered by the source at the time of its publication. For example, if a 1994 guide for Pakistan states that all Gekkonidae have hard-shelled eggs, I have scored only Pakistani gekkonids that were listed in the 1994 source as having hard-shelled eggs. This is apt to be excessively conservative, as the structure of eggshells is evolutionarily conservative (but single species exceptions are known to occur within a single genus). However, there are generalizations in the literature that are overly general. I have worked diligently to ferret out exceptions, and I have refrained from scoring generalized traits if I distrusted the source or had another reason to be skeptical of the generalization reported. For example, it is my impression that diel activity patterns are far more labile than many taxonomic works credit; therefore, I have been particularly chary of generalizing from vague statements such as "these nocturnal geckos. . . ." A more helpful statement of the same would be: "all geckos in this family are nocturnal." Regrettably, some such strong statements are simply untrue. The best I can do is document the sources I have used. I have endeavored to list as sources for each species account not only sources of explicit species-specific statements but also sources of generalizations.

Note that a species is an aggregation of multiple populations, whereas many of the traits tabulated in the database are individual behaviors: home-range size, diel activity pattern, reproductive mode, and so forth. This raises the possibility that a species may exhibit different traits in different populations—diurnal in New Jersey, nocturnal in Georgia—or under different conditions. Where such sources of variation are understood, they are so described in the narrative field accompanying the relevant datum (e.g., ActivityNarrative, ReproEarlyDevNarrative). Because such variation may exist, one should not assume that the full range of values reported in a given field will be manifest in any single geographic location. For example, a species may be characterized as cathemeral—active at any time—but all those in New Jersey may be strictly diurnal, and those in Georgia may be strictly nocturnal.

Source Citations

Sources for all reproductive data are given in the field ReproSources, those for morphology in SizeSources, and so forth. Source fields indicate that germane information was obtained from the listed sources, even though preceding or subsequent sources may have eclipsed the information so obtained. For example, one paper might report a brood count range of two to four eggs and another paper one to five eggs. The brood count data fields will show the complete range as one to five, but the document that reported two to four will still be listed among the sources contributing to the tabulation of brood count. If both earlier and later editions of a work were consulted, the earlier edition was not ordinarily cited. If I was reasonably certain that the same datum was reported in multiple places, I endeavored to cite only the more accessible, recent, or authoritative work. In particular, population density estimates are often reported in multiple works, often with incomplete overlap in authorship and numerical values; normally I cited only the most authoritative—usually the latest—such work.

In a few cases, I omitted data that were clearly misleading; for example, maturation size inferred from the smaller of n=2 adult females. I have generally omitted references reporting a small number of habitats or sizes of a few specimens.

The book's literature cited, the database's source fields, and the literature cited included with the database list sources in alphabetical order by the call: the citation for "Rand and Dickins 1942" would appear before that for "Rand et al. 1932," as is the practice in *Journal of Herpetology* and similar journals. Thus, all two-author papers precede all three-author papers by the same senior author, because "XXXX and YYYY" always alphabetizes before "XXXX et al." Publication year is relevant to the arrangement of two-author citations only in ordering multiple papers by the exact same two authors: "Smith and Jones 1932" would appear before "Smith and Jones 1933." Similarly, a paper by "Rand, Smith, and Jones 1993" would precede "Rand, Jones, and Smith 1994" in the literature cited, because the call "Rand et al. 1993" appears alphabetically before the call "Rand et al. 1994." Thus all "et al." papers by a given senior author are arranged by publication year, not the alphabetical order of junior authors. Different papers may have the same call; if more than one paper matches the call, check titles for content. If the titles do not clarify, the reader may find it fruitful to investigate both sources. On occasion, the same *apparent* author will be listed twice: for example, "Cox et al. 1998; Cox et al. 2003" rather than "Cox et al. 1998, 2003." A repeating listing such as the first example indicates that two different authors with the same last name are involved. Because there are 16 different Smiths, I included first and middle initials in the calls for papers where Smith was a senior author. Note that some authors with a compound family name are not completely consistent with their bylines, and compound names may or may not be hyphenated (Lemos Espinal = Lemos-Espinal). Some authors reported their optional family name as a middle initial or omitted it entirely in some bylines. I tried to list each author in a consistent way, but some name variants were unavoidable or undetectable.

In many cases where a primary source is cited in a secondary document, the two are given, for example, "Lang 1932 in Loveridge 1947." If the first date is omitted (e.g., "Lang in Loveridge 1947"), the information was most likely given by Lang to Loveridge in a personal communication or by some pathway that we can no longer follow. In many cases, I have credited only the secondary source if the primary source was unpublished, is a transient website, is difficult to obtain (e.g., an obscure government document or a thesis or report from small or remote institution), or if the primary source was an author of the secondary work.

In many cases, a primary source was cited by several secondary sources; in such cases I have recorded only the first two such secondary sources that I encountered, even though it was common for secondary sources to extract differing data from the same primary source. When the originating citation is reported in the source that I read, I have repeated the citation of the originating source in the literature cited, but I cannot vouch for the accuracy of the citation.

When a secondary source cited three or more papers as the basis for a synthetic fact (typically species size or the range of habitat types occupied), I have cited only the secondary source as the basis for the synthetic datum. Be aware that secondary citations have the potential to generate information that is not in the original source cited. Citation demonstrates that the secondary author credits the primary author with offering pertinent information, but it does not ensure that the secondary author did not interpret, combine, or enhance that information through additional processing. A simple example would be a secondary author estimating the mass of a hatchling on the basis of a snout-vent length (SVL) measurement reported by the primary author. The two authors in combination may have produced the data that I have cited; therefore full retracing of the source of a datum may require consultation with both the primary and secondary sources.

I have attempted to cite the most important primary sources from within a synthetic summary, but I have undoubtedly missed a number of key papers. In particular, I had difficulty parsing the original source of facts cited in tabular form in synthetic compendia that extracted myriad statistics for hundreds or even thousands of species from a long list of sources. This problem will arise for users of my database as well. Even in cases where only a single source was listed, I found it difficult to determine whether the conclusion cited was that of the primary author or that of the secondary author interpreting the primary author's observations. In some cases of uncertain attribution, I resorted to citing only the secondary author that I read. I apologize to unintentionally excluded authors; the opportunity to climb on your giant shoulders is fervently appreciated.

Some loss of precision is inevitable in a meta-analysis such as this, which collates sources that often fail to report sample sizes, almost never report metrics of variation, and in any event collected the data in different ways. I endeavored to report numerical values to the number of digits warranted, but mixing sources of variable precision leads to muddled means, and inflated appearances of precision can result from sources that overstate their precision. Be aware that unsexed samples are often larger than sexed samples and may be more precise than the sum of the male and female results combined.

I have not included the citations of original species descriptions that I did not consult; in these cases, the original description source call (e.g., Species described by "Boettger (1834)") is given solely to clarify the taxonomic unit being described.

References and other call lists in this database and text are generally given in alphabetical order; however, for database data that are time sensitive (e.g., attribution of a common name, legal status of protected species, etc.), the references are listed in chronological order, in an effort to give maximal credit to the author who first contributed the datum in question and to anchor a transient condition to a particular date range. For example, a species might have been officially listed as "Threatened" in 1988, but that status changed to "Endangered" in 1994. Citations of this transient condition are organized by date to help understand the chronology. If three or more sources state that a species is classified by a certain protective classification, I may report only the earliest and latest sources.

Ongoing aggregator sources (Reptile Database and the IUCN Red List of Threatened Species: www.iucnredlist.org) are extremely valuable sources of information but do not normally generate data (though both sources sometimes convene experts and compile unpublished observations); they primarily report data and filter it. For time-sensitive information, therefore, these two *ongoing* aggregator sources are called chronologically by the most recent year in which they were consulted. If the last year given is recent (e.g., 2017), the cited condition may prevail beyond the date on which I last consulted the site. For example, a species that is listed as "Critically Endangered (1994–2017)," may still be so (readers are encouraged to consult www.iucnredlist.org for current information). The 2017 date does not denote the time of a declassification but rather the most recent date in which I checked the species status.

There are more than 141,000 calls of 5045 unique references in the database. These constituted a remarkable diversity of sources; the most cited journal—*Journal of Herpetology*—was responsible for only 341 unique papers. The next most referenced source was *Herpetological Review* (288 works). The journal *Ecology* evidently contains few natural history data, as it was responsible for only 81 cited works.

Topical Outline of Fields

In the following topical outline, the left-most headings represent topics, the first-indented headings represent data fields (parenthetically expanded for nonintuitive names), and the second-indented headings are fields derived, extracted, or manipulated from source data fields. A few fields (e.g., SVLModeSpEst) are data fields for species with suitable empirical data or derived estimates if empirical data were unavailable. The order of this topical listing is followed in the downloadable database, and the general order is followed throughout this book. A full explanation of all field names, arranged alphabetically, follows this topical outline.

Database Fields

- Species
- CommonName
- RangeCore
- ActivCode (time of day)
- DietCode
- SVLModeSpEst (length in mm at adult length mode)
- MassModeSpEst (mass in grams at adult length mode)
- BusModel (includes 0–4 asterisks denoting confidence in assignment)
- Taxonomy
    - Genus
    - Family
  - SystematicNarrative
- Subspecies
  - CommonNames
  - EnglishNamesNotes
- Color
  - SexesDifferColor
  - ReproSeasonColor
  - BlueTails
  - ColorChange
    - ColorNum (an ordinal scoring of ColorChange)
- ColorPhotoSources
- ColorSources
- ColorNarrative
- Geographic Range
  - GeogCode (biogeographic realms)
  - NativeCountries
    - SingJurisEndemic
  - NativeIslands
  - IntroCountries
    - IntroCounCt
  - IntroIslands
    - IntroIslCt
    - MaxIntros (the larger of IntroCounCt and IntroIslCt)
  - RangeMapSources
  - RangeNarrative
- Morphology and Size
    - Log_{10}Mass
    - ShapeIndex
  - SVLMaturingMaleRange
  - MassMaturingMaleRange
  - AgeMaturingMaleRange
  - SVLMeanMaleMin
    - SVLMeanMaleMidpoint
  - SVLMeanMaleMax
  - MassMeanMaleRange
  - SVLMaxsizeMaleMax

MassMaxsizeMaleMax
AgeMaxsizeMaleMax
TotalMaxsizeMaleMax
TailLengthMaleRange
  TailEst-Male
SVLMinsizeUnkMin ("Unk" designates samples of unsexed or unspecified sex)
MassMinsizeUnkMin
TotalMinsizeUnkMin
  MaxOverMinMass (= MOMM)
  SizeAdjMOMM
SVLMaturingUnkRange
MassMaturingUnkRange
AgeMaturingUnkRange
SVLMeanUnkMin
  SVLMeanUnkMidpoint
SVLMeanUnkMax
MassMeanUnkRange
SVLMaxsizeUnkMax
MassMaxsizeUnkMax
AgeMaxsizeUnkMax
  Log_{10}Age
  MassAdjLongevity
TotalMaxsizeUnkMax
TailLengthUnkRange
  TailEst-Unk
SVLMaturingFemRange
MassMaturingFemRange
AgeMaturingFemRange
SVLMeanFemMin
  SVLMeanFemMidpoint
SVLMeanFemMax
MassMeanFemRange
SVLMaxsizeFemMax
MassMaxsizeFemMax
AgeMaxsizeFemMax
TotalMaxsizeFemMax
TailLengthFemRange
  TailEst-Fem
  RatioMFSVLMidpoint
  TailEstM-FRatio
Mass-SVLRegression
  Intercept
  Power
TailBreaks
TailPrehensile
SpecialMorph
  Padless (geckos only)
SSD (sexual size dimorphism)
  OrdSSD (ordinal rank of SSD)
SexDimorphStructures
  Crest
  Pores
  SexDiffRollup
  DewHead (dewlap size relative to head; anoles only)
  DewMF (dewlap SSD; anoles only)
Elongate
LimbReduction
EyeReduction
SizeSources
SizeNarrative

Activity and Spacing
  DielNum (ordinal score of ActivCode)
SeasonalInactivity
ConspicuouslyWary
  OrdWary (ordinal rank of wariness)
PrimPredAvoidTact
  Crypsis
  Flight
  Sand
  Crev-burrow
  Veg
ElectricLightUse
Vocalizes
  Sound
  CallSpontane
  Stress
Basks
  Direct
  Indirect
  Conforms
BodyTempMin
BodyTempMax
  BodyTempMid
AirtempMin
AirtempMax
SubstrateTempMin
SubstrateTempMax
InRefuge (composition of social groups)
  OrdInRefuge (ordinal rank of group size)
SpacingMale
  OrdMaleSpace (ordinal rank of male intolerance)
  MaleHierarchy
  MaleContested
SpacingFem
  OrdFemSpace
  FemHierarchy
  FemContested
ForagingMode
  S-WForage (sit-and-wait foragers)

- CruForage (cruise foragers)
- ActForage (active foragers)

MPMMin (lowest value of Moves Per Minute)
MPMMax
- PercMPMMin (percentile rank of MPM, etc.)
- PercMPMMid
- PercMPMMax

PTMMin (lowest value of Proportion of Time Moving)
PTMMax
- PercPTMMin (percentile rank of PTM, etc.)
- PercPTMMid
- PercPTMMax

ActAreaJuvRange
ActAreaMaleRange
ActAreaUnkRange
ActAreaFemRange
- MeanLog$_{10}$ActArea
- ActAreaMassResid (size-free standardized residuals of home-range area)

ActivitySpacingSources
ActivityNarrative

Habitat
- AnnPrecRange
  - PrecipBands
  - PrecHyperdry
  - PrecDry
  - PrecArid
  - PrecIntermed
  - PrecMesic
  - PrecWet
- ElevationRange
  - ElevBands
  - ElevL
  - ElevM
  - ElevH
- SleepRefuge
  - SandSleep
  - RockCrev
  - UnderVegRock
  - Burrow
  - ElevVeg
- ActiveSurface
  - Scansorial
  - MicroCode (modal microhabitat)
  - PrimBM (primary ecological business model)
  - SecBM
  - BMConfidence (confidence in assignment of ecological business model)
- Landscape
  - Littoral
  - DunesSand
  - Rock
  - Trees
  - Riparian
  - HumanHabitat
- HabitatSources
- HabitatPhotos
- HabitatNarrative

Reproduction
- ReproMode
- Parthenogenesis
- TSD (temperature-dependent sex determination)
- OvipSeasLength
  - Seasonal
  - Extended
  - Continuous
- GestationMin
- GestationMax
- IncubMin
  - IncubMidpoint
  - IncubMidMassSR
- IncubMax
- ClutchCountMin
- ClutchCountMeanRange
  - ClutchMidpoint
  - MassAdjClutch
- ClutchCountMax
- ClutchesPerYearMin
- ClutchPerYearMeanRange
- ClutchesPerYearMax
- ICI-Min (shortest duration of inter-clutch interval)
- ICI-Max
- RCMMin
- RCMMeanRange
- RCMMax
  - RCMMidpoint
- EggshellMinsizeMin
- EggshellMaxsizeMax
- EggshellHard
- Adherent
- CommunalOvipSite
- ParentalCare
- ReproSources
- ReproEarlyDevNarrative

Diet
- DietClasses
  - AntsReported
  - PlantsReported
- DietSources
- Diet Narrative

Population
- SubjAbund (subjective abundance)

SubjAbundSourceCt
SubjAbundMean
QSampleSize (*Q* in this case indicates a qualified absolute population density estimate)
QMeanDens (number per hectare)
QMaxDens
QMeanBiomass (kilograms per hectare)
Log_{10}QMeanBiom
QMaxBiomass
Log_{10}QMaxBiomass
NQSampleSize (nonqualified absolute density estimate)
NQMeanDens
NQMaxDens
NQMeanBiomass
Log_{10}NQMeanBiom
NQMaxBiomass
Log_{10}NQMaxBiom
PopSources
ConservStatus
Population Narrative
OrdImperiled (ordinal rank of degree of imperilment)
Overall Data Quality
TextFieldCount
NumFieldCount
FilledCount
PerFilled

Alphabetical Listing of Field Names

ActAreaFemRange The range of mean minimum-convex-polygon activity areas, in m^2, for adult females. For example, if three females were sampled and the individual home ranges for the three females were 10, 15, and 20 m^2, the value reported in the field would be 10–20. If the source merely stated that the mean female home range was 15 m^2, that would be the sole value in the field. I avoided reporting activity area sizes for individual females, but I sometimes included them if there were no other data or if the original source was ambiguous with regard to the number of females sampled. I excluded data if the number of fixes indicated that temporal sampling was insufficient.

ActAreaJuvRange The range of mean minimum-convex-polygon activity areas, in m^2, for immature lizards. I avoided reporting activity area sizes for individual juveniles, but I sometimes included them if there were no other data or if the original source was ambiguous with regard to the number of juveniles sampled. I excluded data if the number of fixes indicated that temporal sampling was insufficient.

ActAreaMaleRange The range of mean minimum-convex-polygon activity areas, in m^2, for adult males. I avoided reporting activity area sizes for individual males, but I sometimes included them if there were no other data or if the original source was ambiguous with regard to the number of males sampled. I excluded data if the number of fixes indicated that temporal sampling was insufficient.

ActAreaMassResid A measure of home-range size that is independent of the lizard's size. For a given species, I regressed the log_{10} of the midpoint of the ActAreaUnkRange or, if that was unavailable, the averaged midpoints of males and females on the Log_{10}Mass of the species. The sample size was 216 species, and the standard error associated with each was 0.994. The highly significant (p = 2.7E–31) relationship was log_{10} activity area = 0.964 × log_{10}Mass + 1.080. The standardized residuals (mean = 0, standard deviation = 1) for each species constituted the ActAreaMassResid.

ActAreaUnkRange The range of mean minimum-convex-polygon activity areas, in m^2, for adults of pooled or unknown sex. I avoided reporting activity area sizes for individual adults, but I sometimes included them if there were no other data or if the original source was ambiguous with regard to the number or maturation condition of the lizards sampled. I excluded data if the number of fixes indicated that temporal sampling was insufficient.

ActForage Blank if no foraging mode reported; "0" if a foraging mode other than active was reported; otherwise "1."

ActivCode Blank if information was lacking with regard to diel activity pattern: "a" for activity possible both day and night (= cathemeral), even if activity was predominantly at one time during a given season or type of weather (e.g., night-active following rain); "c" if activity restricted to dawn and dusk (= crepuscular); "d" if activity limited to daylight hours (= diurnal), even if activity was reduced at midday during hot weather; "n" if activity took place during hours of darkness (= nocturnal), even if activity was often restricted to particular hours of darkness rather than extending uniformly throughout the night. These activity score choices were mutually exclusive.

Many observers infer the diel activity pattern of a species from that of congeners. Based on the reports I tabulated, this is often a dubious inference. If I was aware of this generalization, I accepted it only if it seemed highly likely for the taxon in question. Many diel activity reports were contradicted. If contradictory sources were about evenly matched in credibility, I left this field blank.

I did not infer diel activity from morphology (cf. Schmitz and Motani 2011); for example, the firmly diurnal *Iguana iguana* has elliptical pupils.

Note that diel activity conflates two concepts: (1) that the lizard is alert to social and predatory opportunities (alert), and (2) that the lizard is outside of its refugium and locomoting (locomotor activity, usually visible to a herpetologist). Some observers consider only one of these types to be activity. Note also that many lizards may not have the decisive difference in alertness implied by the human concepts of "awake" and "asleep." Many nominally nocturnal lizards are alert and will capture prey that approach their positions during the day, but they leave the confines of their refuge only at night. I tentatively treat such species as cathemeral. This judgment is ambiguous for the many arboreal and cliff-side lizards that spend their locomotor-inactive period in exposed places. Alertness is particularly difficult to assess in species that lack eyelids.

Many assessments of diel activity pattern are based on informal observations. Many collectors appear to assume that a lizard did not move prior to its detection by the herpetologist, an assumption my hide-based observations document as often erroneous. Some collectors assume that an animal obtained from a concealed location (e.g., under a piece of wood) must be active at the opposite time of day (nocturnal, if the board flipped during daytime). Thus there are likely to be numerous situations in which the diel activity cycle was wrongly inferred (e.g., a lizard hunts by immobility, a lizard refuges in exposed location, a lizard forages under logs, a lizard runs under the log as the herpetologist approached, etc.). In the worst cases, the time of day of capture is treated as the time of activity regardless of the action of the lizard at the time of capture. Based on the number of contradictions reported in the literature, diel activity is one of the least-objectively scored natural history traits, or it is much more variable and susceptible to detection problems than is generally understood. My activity code was targeted at expressing the potential range of observed activity rather than assuming a single time of day for locomotor and alert activity.

Activity at electric lights is not considered for this field.

ActiveSurface A text field that lists all surfaces upon which the lizard may forage. The choices were "subterranean," "on ground," "vertical wall," "commensal," "low vegetation," "arboreal," "marine," and "aquatic." A lizard that refuges in a burrow is not considered to forage underground ("subterranean"), though numerous accounts do not clearly distinguish these possibilities. "On ground" includes foraging both under and above loose surface debris (litter, duff, etc.). "Vertical wall" is any more or less vertical surface that is not vegetation (e.g., a cliff or wall). "Commensal" indicates a lizard's habit of foraging on the outside or inside of *occupied* dwellings. If the lizard will use only unoccupied dwellings, the active surface would be listed as "vertical wall." I distinguish "low vegetation" (<2 m high) from "arboreal" (0 to any height on vegetation) only if the source lists perch heights or other information discriminating perch heights. Perch heights inferred from lizard positions on trees after the trees are felled are unreliable and have been ignored. "Marine" indicates that the lizard will enter salt water to forage, not just to escape predation. "Aquatic" is the freshwater equivalent, though it is often claimed without regard to where prey are captured. I looked for dietary evidence to suggest that foraging, rather than just escape, occurs in water.

This field presents some unique challenges for species that routinely reside in one stratum and forage in another. For example, *Anolis acutus*, of the US Virgin Islands, routinely rests *on tree trunks* but captures most prey on the ground. Savage (2002) argues that for anoles such as this, foraging consists of scanning the environment and then rushing to a prey item once it has been spotted. As the scanning element of this foraging occurs exclusively in trees for such anoles, but the prey capture element often occurs on the ground, the "foraging" surface is a combination of terrestrial and arboreal surfaces. There is no widely used term to describe this common hybrid foraging; see the "ground-foraging tree lizard" business model described in chapters 2 and 4.

Another quandary is presented by saxicolous species that mostly clamber over or among vertical or very steep rock faces or buildings. This habit grades imperceptibly into both terrestrial and arboreal/commensal foraging, and any strict division of this continuum is fraught. Furthermore, saxicoly is often a reflection of the lizard's refuge needs rather than, or in addition to, its foraging surface. My database recognizes "rock" as an important Landscape feature and "rock crevice" as an important SleepRefuge, but it has no explicit recognition beyond "vert wall" for rock as a foraging surface.

ActivityNarrative Any narrative helpful in comprehending activity, spacing, thermoregulation, movement, and social scores. See also ReproEarlyDevNarrative, as some modulators of activity derive from reproductive impulses, sometimes without a clear connection to, or distinction from, reproduction.

ActivitySpacingSources Any sources consulted in assigning activity, spacing, thermoregulation, movement, or social scores.

Adherent A "true" indicates that the species at least sometimes lays eggs that adhere to either their substrate or each other. In some cases, this trait is known to be under the control of the ovipositing female. She can hold the eggs off the substrate until the fluid emitted with the egg dries and thereby loses its adhesion.

AgeMaturingFemRange The age range, in years, over which females may first be able to ovulate.

AgeMaturingMaleRange The age range, in years, over which males may first be able to inseminate females.

AgeMaturingUnkRange The age range, in years, over which unreported sex individuals mature.

AgeMaxsizeFemMax The highest age, in years, reported for females of this species. Although wild individuals are the preferred sample, most data are from captivity.

AgeMaxsizeMaleMax The highest age, in years, reported for males of this species. Although wild individuals are the preferred sample, most data are from captivity.

AgeMaxsizeUnkMax The highest age, in years, reported for individuals of unreported or unknown sex of this species. Although wild individuals are the preferred sample, most data are from captivity.

AirtempMax The highest shaded air temperature recorded while the species was known to be alert or locomoting.

AirtempMin The lowest shaded air temperature recorded while the species was known to be alert or locomoting.

AnnPrecRange The empirical range, in mm/y, for sites occupied by the species. I have not included modeled annual rainfall averages, as my experience with these revealed large errors. Modeled annual rainfall data are likely to become acceptably accurate in the near future.

AntsReported Blank if diet data unavailable; "0" if ants not reported; "1" if adult ants of any caste are eaten in any amount >1% of diet. It would be useful to discriminate worker and soldier ants from defenseless reproductives, but such data are rarely available.

Basks Text field of three nonexclusive choices: "direct," "indirect," "conforms." Indirect basking is behavior that would alter the lizard's temperature without sitting directly in sunshine; for example, clinging the underside of a warm rock or to any surface of a warm rock that is no longer in sunshine.

BlueTails Blank if information on tail color unavailable or tails are not bright blue at any life stage. A single letter conveys the other possible choices: "j" if possessed only by juveniles; "f" if possessed by both juveniles and at least some adult females; "m" if possessed by adult males; "a" (all) if possessed by at least some juveniles, adult males, and adult females; "y" (yes) if blue tails are present in the population but the life stage composition of blue-tailed individuals is inadequately specified. Bright red tails may have many of the same functions of blue tails; they are noted in ColorNarrative.

BMConfidence 0–4, indicating the degree of confidence I had in assigning an ecological business model to the species. In some cases, especially island lizards, multiple ecological business models were assigned with no loss of confidence. In a few cases, different life stages exhibit different ecological business models. However, multiple business models may also reflect uncertainty, which is directly, if subjectively, denoted by this field, even if only a single model is mentioned in BusModel.

BodyTempMax The maximum cloacal temperature, in degrees Celsius, measured from a lizard that is alert or locomoting in nature. Many literature sources do not report either minimum or maximum body temperatures of active lizards but state a mean and standard deviation. For these sources I inserted a mean plus one standard deviation (rounded upwards to the nearest 0.5°C) as a conservative estimate of a maximum.

BodyTempMid The midpoint of the temperature range given by BodyTempMax and BodyTempMin.

BodytempMin The minimum cloacal temperature, in degrees Celsius, measured from a lizard that is alert or locomoting in nature. Many literature sources do not report either minimum or maximum body temperatures of active lizards but state a mean and standard deviation. For these sources I inserted a mean minus one standard deviation (rounded downwards to the nearest 0.5°C) as a conservative estimate of a minimum.

Burrow This field references the use of either dug or appropriated burrows during the inactive period. It does not denote digging for prey capture or use of underground cavities for escaping predation, though numerous literature sources were ambiguous with regard to one or both of those related uses of "burrow." The field is blank if data on daily refugia are unavailable; "0" if data are available but burrows are not mentioned; "1" otherwise.

BusModel A text field that is blank if no ecological business model can be inferred. If one or two ecological business models are likely based on the information available (see chapter 4), the plausible models are mentioned, preceded by one to four asterisks, which convey my confidence in the assignment. Four asterisks indicate full confidence in the assignment if the available field data are accurate; a single asterisk may be no more than a guess based on minimal empirical data or the ecological business model of closely related species that are similar but better studied. This is the field that appears at the end of each species listing in the taxon accounts (chapter 5).

CallSpontane This field is blank if vocalization data are absent; "0" if the lizard's vocalization behavior is known but does not include spontaneous calling; "1" otherwise. This field is derived from the Vocalizes text field. Spontaneous calling can be very difficult to elicit in captivity, and it can be very hard to identify the caller in the field; it is probably substantially underreported.

ClutchCountMax The maximum number of eggs or young recorded from a single clutch or litter. I place a note in ReproEarlyDevNarrative if I suspect that clutch counts of communal ovipositors represent multiple females. There is considerable uncertainty as to what exactly constitutes a clutch; litter count is less ambiguous. Clutch count may be the number of eggs laid within the span of a few days for oviparous species; such eggs may or may not be deposited in the same physical location. Other observers insist that a clutch must be laid in a single place at a single time. For species such as anoles, which generally lay single eggs repeatedly throughout the reproductive season but sometimes retain an earlier egg until such time as a suitable oviposition site presents itself, authors have been somewhat inconsistent in counting which eggs belong in a given "clutch." For such species, the maximum clutch count may include eggs that other authors treat as a separate clutch. The two eggs of an anole "clutch" may be laid a day or two apart, separated by several weeks before another staggered two-egg clutch is deposited, or the alternation of individual eggs may be more evenly spaced. In the absence of information on the pattern of oviposition, clutch count may be a fairly ambiguous concept for anoles.

Confusion arises also because clutch sizes are often inferred from dissections of specimens—even using enlarged follicles, which routinely exceed the number of oviductal eggs—rather than counts from oviposited clutches, which may be lower yet. For species that lay larger clutches, variation among years and sites is conspicuous, presumably in an adaptive way (if energetically stressed, the female produces fewer eggs). This dynamism implies the need for large, geographically dispersed, and long-term samples to adequately characterize the mean clutch count for a given species. Samples of this scope are scarce, and wide spreads in the reported mean clutch sizes are to be expected.

ClutchCountMeanRange The minimum–maximum means reported for clutch or litter counts. Some sources are vague as to whether the values they report are true means or general impressions; most sources omit their sample size.

ClutchCountMin The minimum number of eggs or young recorded from a single clutch or litter. See caveats under ClutchCountMax.

ClutchesPerYearMax The maximum number of clutches or litters per year, often unrealistically obtained from pampered captive females, which may give a count that is biased high. Field determination of clutch frequency can be fairly challenging, especially for species that may skip years. Values <1 (i.e., years are skipped) are likely imprecise. For example, many of the reported values are exactly 0.5 or 1.5, which implies that exactly half of the females nest annually, and the other half skip exactly one year or nest exactly twice in a year. Such values should be read as substantially less precise than arithmetic inference indicates. Many tabulations of species use the placeholder "2," which is undoubtedly biased low, to characterize species for which unquantified multiple clutches per year are known or suspected. One hopes that these various biases cancel each other out, but my reservations about their validity kept me from using these data to compute annual reproductive output. For a more rigorously quantified metric with similar motivation see ICI (inter-clutch interval).

ClutchesPerYearMin The minimum number of clutches or litters per year. See immediately preceding entry.

ClutchMidpoint The estimated mean clutch count, obtained from the midpoint of the clutch or litter count means if possible or the midpoint of the maximum and minimum values if necessary.

ClutchPerYearMeanRange The minimum–maximum range of mean number of clutches or litters per year. These data were rarely available. See caveats under ClutchesPerYearMax.

ColorChange A text field with five mutually exclusive possibilities: "none," "darken" (but no change in pattern or hue), "pattern" (includes darkening, but excludes changes in hue), "color" (includes darkening, patterning, and changes in hue), or "yes" (type of color change unspecified). Counts of the latter were combined with darkening to recognize basic color change; pattern or color were required to credit advanced color change. These three levels (none, basic, advanced) are not sharply defined and can be viewed as arbitrary bands on a continuum of length 0–3. For the reasons outlined under SexesDifferColor, conflicting claims were resolved in favor of the more advanced color changing ability.

ColorNarrative Any text elaboration of data on sexual dichromatism or seasonal, short-term, or ontogenetic color change.

ColorNum An ordinal ranking of color changing ability, as tabulated in ColorChange: blank indicates no data; "0" codes for none; "1" for darken or yes; "2" for pattern; "3" for color.

ColorPhotoSources Sources are listed if their good-quality color photographs are of living specimens. For species for which no such photographs exist, lower quality representations were sometimes noted. This list is not comprehensive for common species, even within the universe of the ColorSources cited. The call includes the page number on which the photograph appears, if paginated; otherwise, the plate or figure number. This field is often a convenient way to quickly locate a photograph or species account.

ColorSources The sources used for quantifying colors.

CommonName Note the similarity to the field immediately following. This field lists one unique English common name in widest or standardized use, if such a name exists. In the event that no such name is available, or if the available name is misleading (e.g., naming a ground-dwelling liolaemid "tree iguana") or widely disregarded, I have chosen one. The unique common names appear in the species listings in chapter 5.

In choosing among English common names, I have endeavored to avoid names that are likely to become awkward as a result of taxonomic revision. For example, common names that include a genus name (such as Common Mabuya) are vulnerable. On the other hand, some scientific names have been loosely adopted by lay users (e.g., "sphaero" for Sphaerodactylid geckos), which provides a useful linguistic dimension for distinguishing among the seemingly bottomless font of gecko species. I have accepted those transformations of genus names that are in wide use and seem unlikely to fall victim to systematic turbulence. I have not attempted to force common names to transliterate the scientific name or reflect the animal's phylogenetic position; we already have a naming system for that. Names that include directions (Eastern Skink) have been avoided if the direction is likely to be misconstrued in parts of the range. I have also endeavored to avoid names that are excessively long, that are difficult to speak in English, or that have ungrammatical constructions (e.g., Blackhead Emo Skink). Descriptors referring to a species' relative abundance ("Common Girdled Lizard") suffer from inappropriateness in parts of their range where they are relatively scarce (and is that "common" in comparison to other girdled lizards or other lizards?). I have often avoided naming a lizard after its describer or patronym, though I have accepted those that have come into common usage. I especially try to avoid using patronymic names for which the honoree's place in history has been lost in the mists of time. I have favored names that describe some unique feature that would be readily evident to the casual observer (e.g., Three-horned Chameleon) or that is intuitively associated with the animal (Aldabra Lizard). For narrow endemics, I favor names that will elicit local pride in local fauna, but I avoid giving a jurisdictional name (e.g., Venezuelan Lizard) if the species also occurs elsewhere (e.g., Colombia), as I have encountered resistance to protecting species that have been named as being from some place other than "ours." I have tried to update geographic references to place names in current usage (wipe away that colonialist aftertaste). I have favored names that capture the essence from are among a cluster of similar permutations in wide usage. I favor names entirely in English; solidarity with local peoples can be conveyed in other ways. I have favored the names popularized by the most recent work for the general public. While cognizant of all of these constraints, I provide a globally unique "common" name; on occasion, the result comes at some expense to simplicity and euphony.

CommonNames This text field lists alphabetically each of the common names that I have encountered, followed parenthetically by the earliest, latest, and other key sources that I encountered. I undoubtedly missed some sources, especially early ones. I annotate names that are factually or orthographically erroneous. See English-

NamesNotes for brief biographic sketches of persons mentioned among the CommonNames. See Reptile Database (www.reptile-database.org) for patronyms used solely for scientific names.

CommunalOvipSite "True" if a record of communal oviposition was found in the literature; "false" otherwise. For small species, this generally means eggs found in physical contact or nearly so in the same cavity. For large species, the nests may be clustered in space but laid in separate excavations or side chambers.

Conforms Blank if thermoregulatory behavior not reported; "0" if thermoregulatory behavior does not include thermoconformance; "1" otherwise.

ConservStatus A text field listing formal classifications in descending geographic order. International classifications—mostly those on the IUCN Red List of Threatened Species (www.iucnredlist.org)—are first, followed by national and subnational listings. The dates following each classification are the first and last years for which I have verified this classification. I have not individually attributed to the authors who conducted the IUCN assessment; my apologies and kudos to the hundreds of herpetological conservationists who have volunteered to screen the thousands of lizard species! I most recently documented most ongoing classifications in 2017. If the reported time span appears to end in 2017, consider the possibility that I last checked it in 2017, and it continued beyond that last-verified date. The IUCN ratings of interest in order of descending risk are as follows: EX, extinct; CR, critically endangered; EN, endangered; VU, vulnerable; and NT, near threatened. I reported lesser statuses (LC, least concern; DD, data deficient, etc.) only when these represent a revision of a species previously deemed of conservation concern. See www.iucnredlist.org for definitions of their various classifications, which have evolved over time (note dates).

A species whose trade is regulated by the Convention on International Trade in Endangered Species (CITES) or the European Union's Bern Convention may also be listed (by those conventions' applicable appendices). National and subnational statuses are given by the terms used locally. Occasionally I reported statuses suggested by individual scientists or credible nongovernmental organizations, such as the Natural Heritage Programs. I usually avoided citing statuses that reflect rarity only at the edge of a range.

ConspicuouslyWary A text field that notes the life stage of species reported to be conspicuously wary. See OrdWary for an ordinal condensation of this information. This field provides a warning that behaviors that can be difficult to observe may be underreported in a species so flagged.

Different scientists have very different flight distance expectations and these expectations depend on taxon and context. One scientist deemed an island anole especially wary if it retreated from humans at distances of slightly over one meter and then ran a mean of 0.3 m further away; in other cases flight thresholds exceed 100 m. Some species flee on the basis of nonvisual cues, such as vibrations (footfalls) or noise, and thus are routinely invisible to the observer. Reported flight propensities are highly context specific, as indicated by the species that fled at 100 m when cold and in an area where persecuted, yet the same species ignores humans walking by in urban areas and will climb onto human laps for food where habitually fed by tourists. The utility of this field is probably greatest for species too small to be preyed upon by humans, as the wariness of small species is unlikely to be the product of experience with predatory humans. In contrast, the largest lizards—iguanas, monitors—vary greatly in their circumspection depending on whether they are locally deemed delectable. Often the large lizard species learn which humans to run from, and they can often discriminate, sometimes on blatantly racist cues, between those that may attack and those that most likely will not.

Continuous Blank if no data are available on the length of the species' breeding season; "0" if broods are known to occur only in certain months; "1" if some oviposition or birthing occurs throughout the year, however modulated.

Crest In my reckoning, a lizard is said to have a crest if a line of spiny dorsal scales would be visible to another lizard from a moderate distance. This excludes differences of scale ornamentation of interest primarily to taxonomists, lateral fringes, and crests visible to lizards only when in contact. This field has a value of "1" if crests are present in either sex; "0" otherwise. This field was originally scored to highlight sexually dimorphic crests. Some small or monomorphic crests may have been overlooked; a greatly expanded series of fields will provide greater resolution and precision on crest size, position, and dimorphism in the next edition of the database.

Crev-burrow This field characterizes the species' primary predator concealment tactic (PrimPredAvoidTact) as retreat to a crevice or burrow (from among the choices of crypsis, flight, sand, crevice-burrow, or vegetation). The field is blank if no data are available on any choice; "1" if the lizard's primary tactic is immediate retreat to a crevice or burrow; “0” otherwise. Lizards that first run out of visual range (“flight”) or initially hide in vegetation (“vegetation”) may subsequently retreat to a crevice or burrow without earning a “1” in this field.

CruForage Blank if information unavailable on foraging mode; “1” if cruise foraging (Butler 2005) reported; “0” otherwise.

Crypsis This field characterizes the species’ primary predator concealment tactic (PrimPredAvoidTact) as crypsis (from among the choices of crypsis, flight, sand, crevice-burrow, or vegetation). The field is blank if data were unavailable on any choice; “1” if the lizard’s primary tactic is immobility; “0” otherwise. Lizards that freeze initially but may flee if they are about to be stepped upon or grabbed may still qualify for crypsis as their primary tactic.

DewHead This field applies only to species with a dewlap and has been scored only for anoles. It expresses the size of the dewlap in relation to the male lizard’s head: “0” = no dewlap; “1” = dewlap rudimentary; “2” = dewlap small; “3” = unspecified size or medium; “4” = dewlap larger than its head; “5” = much larger than the lizard’s head.

DewMF This field compares the relative size of male and female dewlaps and has been used only for anoles. The range matches that of OrdSSD for body sexual size dimorphism, except that no anoles have been documented to have female dewlaps much larger than male dewlaps, hence the range of scores is only –1 to 2 (rather than –2 to 2): “–1” = m<f; “0” = m=f; “1” = m>f; “2” = m>>f.

DielNum The DielNum assumes that nocturnal and diurnal are polar opposites and that cathemeral and crepuscular lie in the middle: “–1” = nocturnal; “0” = crepuscular or cathemeral; “1” = diurnal. Species for which no diel information is available are left blank. As explained in chapter 3 in relation to table 3.7, this assumption is probably invalid and this field was not used in my analysis.

DietNarrative The diet narrative was used to flesh out understanding of the list of food items taken, especially by flagging the relative importance of particular foods. For example, the proportion of diet that is plant material or whether this composition changes seasonally or ontogenetically.

DietClasses A text field built around the following nonexclusive categories, arrayed roughly in trophic rank.

“Leaves” include both foliage and leafy flower parts (e.g., petals), though nectar is undoubtedly ingested along with some flowers and may be the goal of flower ingestion.

“Fruit” includes both hard fruits (seeds) and soft fruits, as well as pollen.

“Nectar” means plant juices of all types, including sap, guttation, nectar, and juices oozing from soft or damaged fruits. These foods are distinguished and treated as nonherbivorous because they require no special dentition, gut symbionts, or physiology to digest—dissolved sugar just passes into the blood. Nectar is not detectable by gross examination of gut samples, is undoubtedly underreported, and may be taken opportunistically by a wide variety of lizards.

“Fungi” means fungi, which are unlikely to be well quantified by any gut samples.

“Soft invert” includes worms, slugs, caterpillars, grubs, and any other invertebrate lacking a hard exoskeleton (also likely to be undersampled).

“Arthropods” is a catchall for any invertebrate not captured by another category; the commonest type is spiders, but other hazardous prey such as scorpions or millipedes are also included. Isopods, crayfish, and less intimidating noninsectan arthropods are also included.

“Snails” indicates any invertebrate with a conspicuously tough shell, requiring special techniques or powerful muscles and teeth to penetrate.

“Ants” denotes adult Formicidae. Eggs and larvae are classified as “soft invert” if so identified in the source document.

“Wasps” denotes non-formicid hymenopteran adults.

“Termites” includes both winged and nonwinged adults; immature stages are “soft invert” if so identified in the source document.

“Beetles” denotes adult coleopterans and life stages that have rigid exoskeletons; other life stages may be scored as “soft inverts.”

“Orthops” includes large, chitinous insects such as crickets, grasshoppers, cockroaches, praying mantises, stick insects, and so on.

“Other insects” include all other Insecta, such as moths, butterflies, dipterans, homopterans, hemipterans, and so on. Note, for example, that Lepidoptera, in the absence of further specificity, could be subterranean, under litter, above ground, arboreal, nonvolant, or volant, emphasizing the ineffectiveness of using taxonomic classifications of food items to ask ecological questions.

“Fish” seems self-explanatory, including fish taken when alive or dead if living status not discernible from gastrointestinal contents. If discernibly dead at the time of ingestion, fish would instead be categorized as carrion.

“Ecto verts” includes all other terrestrial ectothermic

vertebrates, including eggs of ectotherms and examples of cannibalism.

"Endo Immobile" means nestling birds and mammals or bird eggs.

"Endo verts" includes only those mammals and birds that have the ability to escape at the time of ingestion, or unspecified accounts of birds and mammals.

"Carrion" is vertebrate material ingested when its dead condition at time of ingestion is detectable from the information presented.

Only important food components (>1% by volume or weight) are listed, although this assessment is often based on my guess, as the sources were rarely so specific. I have listed plant material—cited as "leaves" if the sources did not specify which type of plant material—even if the source was of the opinion that the plant material was ingested by accident. I did not separately tabulate claims of cannibalism because I think most carnivores will consume unrelated small conspecifics when the opportunity arises; the key question—whether such behavior is demographically important—is beyond the scope of this work.

The categories in which dietary items are usually tabulated are taxonomic: insect orders. Information tabulated by microhabitat stratum—distinguishing aquatic insects from terrestrial insects, for example—would be substantially more useful.

DietCode An exclusive categorization of each species' primary diet in the wild: "c" = carnivore (at least 10% vertebrates); "h" = herbivore (at least 90% plant material); "i" = invertebrate eating (<10% vertebrates or <1% plants); "m" = myrmecovore or ant specialist (>50% ants, at least as some times of year; no implication that diet is restricted to ants); "o" = omnivore (1%–90% plants). Captive foods can be very misleading and were ignored. Many sources could not quantify the proportional breakdown in dietary classes. I used volume when that metric was available.

DietSources Sources used for the characterization of diet.

Direct Blank if thermoregulatory behavior not reported; "0" if thermoregulatory behavior did not include direct basking; "1" otherwise.

DunesSand Blank in the absence of Landscape habitat data; "0" if habitat data did not include the keywords "dunes" or "sand"; "1" otherwise.

EggshellHard A short text field including the words "hard" or "soft." Denotes whether the incubating eggshell is rigid (= calcareous) or not. Even "hard" shells may be a bit supple when oviposited, but they harden within a few minutes. Some lizards manipulate the eggs with their hindlimbs during the drying process, perhaps to alter shape or influence degree of adhesion. Water permeability is loosely correlated with eggshell hardness, but important exceptions have recently been identified (Andrews 2017), and some intermediate hardness eggs have been identified.

EggshellMaxsizeMax A numerical field giving the maximum recorded dimension, in mm, of an egg. Note that a comparison of this with the next field does not define the average shape of an egg, as disparity between the minimum and maximum dimensions can be due to either elliptical eggs of fixed shape or individual or geographic variation in the size of spherical eggs.

EggshellMinsizeMin A numerical field giving the minimum recorded dimension, in mm, of an egg. Note that a comparison of this with the immediately previous field does not define the average shape of an egg, as disparity between the minimum and maximum dimensions can be due to either elliptical eggs of fixed shape or individual or geographic variation in the size of spherical eggs.

ElectricLightUse "True" if feeding at electric lights reported; "false" otherwise. This behavior is considered artificial and does not affect the score for diel activity.

ElevationRange The height above sea level, in m, of known localities. Be alert for unbounded ranges (e.g., –2400); many sources report only the upper elevational limit.

ElevBands One to three letters expressing which of three elevations bands are inhabited: "l" (low) = <1200 m; "m" (med) = 1201–2400 m; "h" (high) = >2400 m.

ElevH Blank if data on elevation absent; "0" if all recorded elevations <2400 m; "1" otherwise.

ElevL Blank if data on elevation absent; "0" if all recorded elevations >1200 m; "1" otherwise.

ElevM Blank if data on elevation absent; "0" if all recorded elevations <1200 m or >2400 m; "1" otherwise.

ElevVeg Blank if data on sleeping site absent; "0" if all sleeping sites on or below ground; "1" otherwise. Note that lizards that sleep in elevated vegetation may be visually concealed (e.g., under bark) or not.

Elongate Blank if evidence of elongation absent; "elongate" if gently adpressed limbs fail to overlap; "marginal" if gently adpressed limbs sometimes overlap and sometimes do not; "extreme" if the adpressed limbs are separated by at least 25% of the interaxillary distance. For the tabulation of elongation (e.g., table 3.2), the three nonblank scores were expressed as a proportion of the species pool. Mild underreporting bias is likely for this variable.

EnglishNamesNotes This text field was used to explain the rationale for nonobvious common names, flag common name errors, and provide a biographical sketch of people honored in a common name. The data for the brief biographical sketches were shamelessly extracted from Kraig Adler's three *Contributions to the History of Herpetology* volumes (1989, 2007, and 2012, issued by the Society for the Study of Amphibians and Reptiles, often in association with World Congresses of Herpetology), Beolens et al. (2011), Shea (2012), and Reptile Database (2007–2017). Kraig Adler also edited some entries in this field. Thank you to those sources and contributors (especially John Applegarth's lovely (2012) index of taxonomic authors). The impressive number of persons named in either scientific or common names testifies to the wealth of effort contributed by thousands of individuals to the aggregation of natural history knowledge. The mini-biographies suggest that this work is hazardous: hundreds died conducting field work. For example, no less than three of them were killed by aggrieved elephants. One fell into a pitfall trap and was gored to death by the feral bull that had preceded him. Disease and accidents felled many more. These human "statistics" are not manifest in economists' accounting, but they testify to the depth of human yearning for knowledge of the natural world.

Extended Blank if there was no information on the seasonality of breeding; "0" if oviposition or birth occurs over periods of <5 or >10 months; "1" otherwise.

EyeReduction Blank if information lacking on eye condition or eyes believed to be normal; "reduced" if the normal lizard eye exhibits a loss of transparency or elements; "vestigial" if the opacity and structure of tissue suggests impaired ability to render photic images.

Family The taxonomic family of the species, as assigned by Reptile Database in May 2017.

FemContested Blank if information lacking on the social relations among females; "0" if consistent information reported on female social relations; otherwise "1."

FemHierarchy Blank if information lacking on the social relations among females; "0" if available information does not include female hierarchies; "1" otherwise.

FilledCount The number of database cells occupied, exclusive of derived fields. A field was considered derived if it recoded the contents of another field; derived fields are identified by being further indented in the outline of fields at the beginning of this appendix. The FilledCount number is a nonconservative estimate, as a cell was treated as occupied regardless of the quality or quantity of information.

Flight This field characterizes the species' primary predator concealment tactic (PrimPredAvoidTact) as flight (from among the choices of crypsis, flight, sand, crevice-burrow, or vegetation). The field is blank if no data were available on any choice; "1" if the lizard's primary tactic was sprinting to beyond the predator's visual range; "0" otherwise. At the point at which the fleeing lizard gets beyond the predator's visual range, it may conceal itself by other means (sand diving, retreat to a crevice, burrow, or vegetation).

ForagingMode If the animal had been reported as foraging by waiting for prey to approach it, "s-w" (sit-and-wait) was noted. If the animal had been seen to move slowly but relatively continuously through the environment, pausing for periods to observe and attack prey that are moving toward it, it was labeled "cruise" (Butler 2005). If the lizard moved more or less continuously while foraging, it was labeled "active." Note that this arrangement does NOT make the traditional assumption that a species exhibits only one mode; the three categories are nonexclusive. In a number of species, all three foraging modes were noted. Assuming a single condition for a species or higher taxon is particularly inappropriate for herbivorous species, which are often described as sit-and-wait, implying a degree of locomotor forbearance that would inexorably lead herbivores to starvation. Moreover, strenuous and trenchant criticisms of these foraging terms have been registered (Perry 1999, 2007; Perry and Pianka 1997).

These designations were provided primarily by the literature sources, or the literature sources described the lizard's foraging in words that approximate the definitions of these terms. Papers prior to 2005 did not use the "cruise" descriptor, which complicated the interpretation of earlier sources. I inferred the application of "cruise" from the behavior reported. Furthermore, some authors appeared to assign foraging mode categories on the basis of taxon: all iguanids were considered to be "s-w," all lacertids "active," and so on. Several authors reporting foraging modes also noted that it was often difficult to distinguish movements motivated by social factors, such as territorial patrolling, from food gathering, which may occur concurrently. Furthermore, I found a number of "active" descriptors that appeared to convey the author's awe at the speed of the lizard's sprint speed fleeing from the herpetologist, rather than the undisturbed lizard's characteristic foraging speed. Despite these reservations, I have presented the qualitative descriptions in an effort to quantify the diversity of foraging behavior reported and to offer an opportunity to match these qualitative descriptors to quantitative measures (see MPM and PTM fields) and compare foraging patterns to nonforaging attributes (e.g., crests).

Genus A field provided for ease of sorting, searching, subsetting, and so forth.

GeogCode One to five letters representing Wallace's biogeographic realms, each coded by their first letter except for Neotropical: "a" = Australasian; "e" = Ethiopian; "n" = Nearctic; "t" = Neotropical; "o" = Oriental; "p" = Palearctic.

GestationMax For viviparous species, this is the maximum number of days between effective copulation and delivery; for oviparous species it is the maximum duration between effective copulation and oviposition.

GestationMin For viviparous species this is the minimum number of days between effective copulation and delivery; for oviparous species it is the minimum duration between effective copulation and oviposition.

HabitatNarrative Any notes useful in fleshing out the classification fields for sleeping site, microhabitat, or habitat.

HabitatPhotos Readily available sources of photographs illustrating a site—as opposed to a habitat type—in which the focal species has been caught. In my experience, these photos are far more helpful for characterizing occupied habitats than are verbal descriptions, which are severely vulnerable to the vagaries of local adaptation of the English language. They are of course less helpful than vegetation plot data, of which there were too few.

HabitatSources Sources consulted in the tabulation of rainfall, elevation, microhabitats, habitats, and sleeping sites.

HumanHabitat Blank if information lacking on habitat/landscape and active surface/microhabitat; "1" if the lizard is reported to be active on "commensal" surfaces or in "urban" landscapes; "0" otherwise.

ICI-Max The inter-clutch interval is the number of days between oviposition or birth events. This field gives the maximum number of days exclusive of the interval between reproductive seasons.

ICI-Min The inter-clutch interval is the number of days between oviposition or birth events. This field gives the minimum number of days.

IncubMax This is the maximum number of days between egg deposition and emergence of the young, preferably in nature. Captive studies can be distorted by aberrant temperature or osmotic conditions, context which is not tabulated in the database but can sometimes be obtained from the cited source. Emergence from wild nests may include many days after hatching but before emergence from the soil. This posthatching interval may include a shedding event and yolk resorption.

IncubMidMassSR The standardized residual of the regression IncubMidpoint on Log_{10}Mass, a measure of incubation duration that is independent of the lizard's size. The sample size was 1050 species, and the standard error associated with each was 49.9 days. The highly significant ($p = 1.8E{-}54$) relationship was incubation duration = 33.14 × $\log_{10}$Mass + 46.7. The standardized residual (mean = 0, standard deviation = 1) for each species constituted the IncubMidMassSR.

IncubMidpoint The average of IncubMin and IncubMax.

IncubMin This is the minimum number of days between egg deposition and emergence of the young, preferably in nature. Captive studies can be distorted by aberrant temperature or osmotic conditions, context which is not tabulated in the database but can sometimes be obtained from the cited source. Emergence from wild nests may include many days after hatching but before emergence from the soil. This posthatching interval may include a shedding event and yolk resorption.

Indirect Blank if thermoregulatory behavior not reported; "0" if thermoregulatory behavior does not include indirect basking (e.g., maximizing contact with a warm substrate); "1" otherwise.

InRefuge A text field that summarizes the size and sex composition of groups of lizards. The most common was a single male with one or more females. The many text descriptors were further aggregated in the field OrdInRefuge. "Refuge" in this case is applied loosely and includes an area where all individuals were socially interacting at close range, often spanning the geographic space of more than one daily refugium. My objective was to quantify single refugia, but the vagueness of many reports precluded such specificity; many authors treated a species as grouped if they were found in rock crevices that were geologically clumped.

Intercept The coefficients listed in the Intercept and Power fields are those which I used and are uniformly of the form $\text{Log}_{10}\text{Mass}$ (in g) = Intercept coefficient + Power coefficient × $\log_{10}$(SVLModeSpEst in mm).

IntroCounCt The number of political jurisdictions in which introduced populations were reported, as listed in IntroCountries. The recognition of nativeness of a population is often uncertain. I followed Kraus (2009) in most cases and noted uncertainties in the Range Narrative or parenthetically in IntroCountries.

IntroCountries These are the political jurisdictions, ordered alphabetically, in which introduced populations were documented. These jurisdictions are countries as understood by the United Nations, plus Canadian, Australian, and American provinces or states. The introduced populations are annotated (sometimes in Range Narrative) if there is information indicating that the population died out.

IntroIslands The named islands on which introduced populations have been found and believed established unless annotated otherwise. The lists are alphabetical and omit diacritics, as do most English-speakers when pronouncing the names. I referred to current usage in Google Earth to update older island names. Identical island names used in more than one part of the world may be parenthetically referenced to their political or geographic location. For example, there are at least six Green islands in the low islands of the West Indies: Green (Grenada), Green (S. Andros), Green (San Salvador), Green (A&B), Green (BVI), and Green (USVI). See the island synonymy in the database. Islands may be named at several grouping levels (island, atoll, group, archipelago). I preferred the lowest level, whenever this option was documented. I attempted to eliminate duplicate reports based on different grouping levels for the same locality, but I cannot guarantee that I always succeeded.

The word used for "island" varies in different parts of the world: "pulau" in Indonesia, "——jima" in Japan, "nosy" in Madagascar, "île" in French speaking colonies, "key" in the Caribbean, and so on. These words are redundant in a list of island names and have generally been excised: Key West, therefore, is given as "West." Where I specify "Isl." after an island group name, that suffix denotes that the name applies to an island *group* or archipelago rather than to an individual island: "Hawaii" is the single (so called "Big") island, whereas "Hawaii Isl." denotes the entire chain/archipelago. Furthermore, "Hawaii" may be found in the country field to designate a political rather than geographic demarcation. Because country lists are segregated from island lists, each list is unambiguous with regard to whether it references an island or political jurisdiction.

IntroIslCt The number of islands in which introduced populations were reported, as listed in IntroIslands.

Landscape Landscape is the term I prefer in place of habitat, as many observers use "habitat" in a loose way to also incorporate the fields I discriminate as SleepRefuge and ActiveSurface.

I scored each lizard species on three aspects of habitat sensu lato: where it stays when it is inactive (SleepRefuge), the surface on which it forages (ActiveSurface), and the structural classes of substrate and vegetation in the surrounding environment (Landscape). The first two of these collectively characterize the microhabitat; the Landscape field is similar to habitat without, however, naming species of dominant plants.

Perhaps the most vexing of all natural history traits considered in this work was that of habitat in the loose sense. Few of the sources chose to clearly distinguish microhabitat from habitat, and many were actually describing the lizard's hiding place—where they uncovered the inactive lizard—as if that equated to the lizard's habitat or microhabitat. Many species were known only from having been collected from beneath some sort of debris, such as coconut husks, old boards, and so on. Did the lizard run under there as the collector approached (in which case the site says something about the availability of hiding places from predators), or is it a daily refugium occupied during the day by a nocturnal species or vice versa if collected at night (though this is rarely explicit), or was the lizard foraging under the debris pile (in which case the underside of the debris pile is the lizard's active surface)? We do not understand a lizard's ecology until we know where and how it forages when it does not feel threatened by a human, and yet this information is clearly understood for only a small minority of mostly diurnal lizard species.

Aside from discriminating inactive refugia from active surfaces, the characterization of habitat sensu stricto (Landscape) is fraught with difficulty because we have so little understanding of which aspects of habitat are important to the lizards. Many sources made reference to the steepness of the terrain, distinguishing hilly areas from flats, from montane, and so forth, but we do not know that lizards care about the overall slope of their neighborhood. I have made an effort to tabulate elevation of occupied sites, not because lizards care about height above sea level, but because higher sites are generally cooler, and lizards care a great deal about temperature. Meiri et al. (2013) found that environmental temperatures were a better predictor of natural history traits than body temperatures, but few environmental temperatures are documented in the literature (see AirtempMin and AirtempMax). Elevation as a proxy for temperature has several serious weaknesses: a given temperature can occur at many different elevations at different latitudes and localities, temperature varies a great deal throughout the year, and we rarely know which temperature is the one of importance to the lizard. Thus we have introduced chameleons living at sea level in Hawaii, whereas in their East African native range they are found only at high elevation. As climate change alters the temperature/elevation relationships on Earth, we will also begin to find that historic elevations are a misleading indicator of climatic suitability.

Characterization of the Landscape (habitat sensu stricto) has traditionally relied on naming the dominant species of vegetation. This information is almost useless to readers not familiar with the botany of the site, and it is impractical to tabulate in a global meta-tabulation such as this. Furthermore, most sources describe habitats in terms relative to other habitats in their region. Thus "dry" or "dense" forests mean something very different in Peru than they mean in New Hampshire or Madagascar. Many sources refer to areas with spiny or sclerophyllous vegetation, but though understandable, those characterizations do not always transfer accurately to other parts of the world. For example, one description reads, ". . . inhabits well-drained sandy fields, rosemary shrub, sandy parking lot edges, and myriad other such habitats." As this is not a burrowing species, the theme of sand may simply reflect the soil type present in that part of the world. The association with well-drained soils implies very different soil conditions in, say, Bangladesh or Mauritania. The reference to a particular plant species ("rosemary") is meaningful to herpetologists familiar with the site but not to those from elsewhere. And I don't know what to make of sandy parking lot edges (are the lots paved?) or what would constitute "other such habitats." How does one aggregate such descriptions?

A further challenge in characterizing habitat arises in places where natural habitats have been replaced by anthropogenic ones. The landscape code "urban" indicates a vegetatively disturbed site, which could be a garden clearing in a forest or in town and is not necessarily urbanized; the active surface code "commensal" indicates that the lizard forages inside or on the outer walls of occupied buildings. Thus I intended the "urban" landscape code in the *absence* of the "commensal" code to indicate that the lizard occurs in vegetatively disturbed zones but does not usually forage on or in occupied buildings. If "commensal" is also noted, the species occurs on or in occupied buildings and is urban in that sense. A common remark is that a species is known to "utilize agricultural landscapes" (such remarks are often validly included to help assess the degree to which agricultural development will degrade natural habitats from the lizard's perspective). Such a statement cannot be parsed to distinguish a lawn from a forest plantation. Without an indication of what crop is involved, and whether the crop is protected by shade plants, such a remark is virtually uninterpretable in terms of Landscape or habitat requirements.

My hunch is that herpetologists will eventually come up with some practical way to characterize the climate experienced by a lizard during its active season and the constraints on it enduring a nonactive season, if any, in terms that are meaningful to lizards. I predict that these climatic constraints will also largely dictate the structure and composition of the vegetation, but that the composition of the vegetation will rarely be of great importance to most lizards, most of which do not eat vegetation. The structure of the habitat and microhabitat, on the other hand, are likely to be of appreciable importance (Scott Newbold and MacMahon 2014). Therefore, until we figure out a practical way to characterize the climate of habitats and microclimate of microhabitats from the lizards' point of view, our best way forward may be to describe the structure and microstructure of occupied environments in relatively universal terms; that is, not relying on regional descriptors. I tried to do this. As inspection of the Landscape field will show, a surprisingly large number of lizards reside across a wide variety of the structural habitat types.

Landscape is my habitat characterization in the narrow sense. It is a text field with 13 nonexclusive categories. The first four flag specialized habitat combinations: littoral, riparian, caves, and dunes. The other terms are all structural and are arrayed from belowground to aboveground. The full range of occupied habitats is documented in this field, without reference to which habitat type is most favored by the species, or whether that feature is obligatory or optional. If there is a strong or specific habitat mode, it may be highlighted in HabitatNarrative.

"Littoral" indicates that the source noted proximity of occupied habitat to seawater, not necessarily limiting that to within the zone showing distinct salt-tolerant vegetation; fog belts may extend inland for dozens of kilometers from the literal coast. The shorelines of very large freshwater lakes (e.g., Black Sea, Lake Superior) are treated as littoral for the purpose of this tabulation, as they engender geomorphological and climatic phenomena redolent of saltwater shorelines.

"Riparian" indicates that the lizard was found in proximity to a stream or watercourse—running or still, intermittent or permanent—not necessarily within the band of vegetation requiring high levels of ground moisture. In the case of desert areas, the riparian zones may not even contain surface water much of the year. In arid areas, one suspects that "riparian" may inadvertently describe a lizard as saxicolous, because rock outcrops are a regular feature of aridland canyons/draws/washes/gullies/intermittent streams.

"Cave" indicates a rock cavity of appreciable size and permanence; it may be fully dark away from the entrance, but it need not be.

"Dunes" indicates loose sand in mounds, with grains that will relocate in high winds, but mounds may also be partially vegetated; mounds need not be anywhere near a coast (e.g., Sahara).

"Sand" indicates a friable substrate loose enough for rapid penetration by digging lizards. It does not require the presence of dunes, but neither does it exclude those; dunes always imply the presence of sand.

"Stones" indicates a soil type that is composed of particles heavy enough and cohesive enough to preclude rapid penetration by a digging lizard but without having particles large enough to generate cavities providing protection from predators.

"Rock" indicates solid rock material large enough to provide predator protection when a lizard is underneath it and a solid surface for climbing by those scansorial lizards equipped to scale it.

"Grass" is any herb in the environment that covers some of the ground but does not have woody stems—bamboo is taxonomically a grass, but structurally a tree. All herbs are treated as "grass" for the purpose of this field. Some lizard species are typically active in grassy meadows—pratincolous—but that characterization refers to active surface rather than habitat; the "grass" indicated under habitat may or may not be continuous or dense enough to be described as a meadow. "Grass" can include herbaceous row crops such as rice, wheat, or corn.

"Shrub" is any low woody vegetation, succulent or otherwise, that fails to qualify as a tree.

"Scattered trees" are woody plants with a dominant bole and a canopy that fails to approach that of its neighbors on all sides. Orchards, some coffee plantations and many other open-canopy agroforests are categorized as "scattered trees."

"Forest" is the default for forested areas, including closed-canopy agroforests.

"Primary forest" indicates that particularly dense canopy is favored for the species; it need not require an area that has never been logged. It is more mesic than the "forest" category. "Rain forest" means different things in different parts of the world, but it will often evoke my categorization of "primary forest," as will any forest that exhibits a closed canopy, high vegetative density, and a relatively shady, moist microclimate.

"Urban" denotes the lizard's presence within disturbed areas, but it does not imply modern European-style urbanization; an African village or remote abandoned sawmill is categorized as urban in this work. Categorization of the Landscape as "urban" does not imply that the lizard occurs on, or in, houses (see the ActiveSurface category of "commensal").

An example of the distinction between ActiveSurface and Landscape can be found in the word "shrub." A lizard that frequently perches on a shrub in a forest will be scored as having "low veg" as its ActiveSurface but "forest" as its Landscape. In contrast, an aridland lizard might typically reside or hunt by sit-and-wait under a shrub, in which case its ActiveSurface is "on ground" but its Landscape is "shrub." Another lizard in the same desert might often perch in shrubs, in which case I would score its ActiveSurface as "low veg" and its Landscape as "shrub." In some areas, perching on fence posts is widely included in habitat descriptions; I have treated this "habitat" as representing a "low veg" ActiveSurface and "scattered trees" (i.e., sun penetration between canopies) as its Landscape.

With regard to the distinction between Landscape (habitat) and ActiveSurface (microhabitat), the commonly used descriptor "saxicolous" was found to be problematic. I was usually unable to determine if saxicolous was a landscape requirement because (1) the lizard needed to sleep in rock crevices (SleepRefuge), (2) the lizard escapes from predators into rock crevices (see PrimPredAvoidTact), (3) the lizard forages on vertical rock walls ("vert wall" in ActiveSurface), (4) the lizard was often located near water courses, where rocks are often exposed, but that association is perhaps coincidental to the lizards need for high humidity, (5) the substrate was karst and limestone rock that permits water to rapidly percolate and causes moisture stress to the vegetation, or perhaps lowers the humidity for the lizard or its predators, parasites, and so on, or (6) other forms of rock on the surface provide a solid structure under which burrows can be extended without collapsing, and so forth. Without knowing how the lizard uses rocks, if at all, the descriptor "saxicolous" is challenging to parse into features of ActiveSurface (microhabitat) and Landscape (habitat).

LimbReduction A text field that describes which limbs are reduced and to what extent. A limb was considered reduced if it possessed fewer than four digits. "Limbless" means that there are no external signs of the limbs; "extreme" indicates that the limbs have one or two, usually very small, digits. These definitions do not comport exactly with those used in some other reviews (e.g., Meiri 2008), or with definitions used by some taxonomists, who appear to assess relative limb reduction by subjective relative length.

Littoral See the Landscape field for a definition of "littoral." This field is blank if Landscape was unpopulated; "1" if littoral was a keyword listed in Landscape; "0" otherwise.

Log$_{10}$Age Log$_{10}$ transformation of AgeMaxsizeUnkMax.

Log$_{10}$Mass Log$_{10}$ transformation of MassModeSpEst.

Log$_{10}$NQMaxBiom The log$_{10}$ transformation of the highest nonqualified biomass (kg/ha) estimate (NQMaxBiomass). See appendix C for criteria used to rate a population density estimate as nonqualified (NQ).

Log$_{10}$NQMeanBiom The log$_{10}$ transformation of the mean nonqualified biomass (kg/ha) estimate (NQMeanBiomass). See appendix C for criteria used to rate a biomass estimate as nonqualified (NQ).

Log$_{10}$QMaxBiomass The log$_{10}$ transformation of the highest qualified biomass (kg/ha) estimate (QMaxBiomass). See appendix C for criteria used to rate a biomass estimate as qualified (Q).

Log$_{10}$QMeanBiom The log$_{10}$ transformation of the mean qualified biomass (kg/ha) estimate (QMeanBiomass). See appendix C for criteria used to rate a biomass estimate as qualified (Q).

MaleContested Blank if information lacking on the social relations among males; "0" if consistent information reported on male social relations; otherwise "1."

MaleHierarchy Blank if information lacking on the social relations among males; "0" if available information did not include male hierarchies; "1" otherwise.

MassAdjClutch The standardized residuals of the regression ClutchMidpoint on Log$_{10}$Mass, a measure of clutch or litter count that is independent of the lizard's size. The sample size was 2973 species, and the standard error associated with each was 3.82 young. The highly significant (p = 7.5E–196) relationship was clutch count = 3.474 × log$_{10}$Mass + 1.138. The standardized residuals (mean = 0, standard deviation = 1) for each species constituted the MassAdjClutch.

MassAdjLongevity The standardized residuals of the regression log$_{10}$ AgeMaxsizeUnkMax on Log$_{10}$Mass, a measure of maximum longevity that is independent of the lizard's size. The sample size was 879 species, and the standard error associated with each was 0.342 years. The highly significant (p = 1.6E–47) relationship was log$_{10}$ longevity = 0.230 × log$_{10}$Mass + 0.564. The standardized residuals (mean = 0, standard deviation = 1) for each species constituted the MassAdjLongevity.

MassMaturingFemRange The mass (g) range of females believed to be maturing. As this measure could be collected from a wide geographic range over time, the metric has the potential to conflate spatial, individual, and temporal variability. A common literature practice was to report the length or mass of the smallest female bearing enlarged follicles. This measure of the maturing range is biased downwards because it does not also report the largest female that failed to commence breeding. This measure is biased upwards because it is a minimum and would necessarily get smaller with a larger sample size. Let us hope that these biases canceled each other out.

MassMaturingMaleRange The mass (g) range of males believed to be maturing. As this measure could be collected from a wide geographic range over time, the metric has the potential to conflate spatial, individual, and temporal variability. See the caveat in the preceding field.

MassMaturingUnkRange The mass (g) range of unsexed or unspecified individuals believed to be maturing. As this measure could be collected from a wide geographic range over time, the metric has the potential to conflate spatial, gender, individual, and temporal variability. See the caveat listed in MassMaturingFemRange.

MassMaxsizeFemMax The weight (g) of the heaviest female.

MassMaxsizeMaleMax The weight (g) of the heaviest male.

MassMaxsizeUnkMax The weight (g) of the heaviest unsexed or unspecified individual.

MassMeanFemRange Ideally, this would be the gram range of empirical mass means corresponding to the SVL of the mean adult female in an unbiased sample of the population. Often, however, a report is ambiguous with regard to whether all immature females were excluded from the mean, whether the mean is of weighed individuals (as opposed to the weight of females at the mean length), whether mass was extrapolated from length, and whether mass was estimated in some manner from a different sample of the population. Tabulations may give the mean of unsexed individuals in place of their target population (females) as the best available proxy, and this proxy is repeated by others as if it were a true sample. Given the number of species in which the reported mean unsexed adult and mean adult female metrics are identical, this conflation must be a regular occurrence.

MassMeanMaleRange Ideally, this would be the gram range of empirical mass means corresponding to the SVL of the mean adult male in an unbiased sample of the population. See the previous field for several caveats.

MassMeanUnkRange Ideally, this would be the gram range of empirical mass means corresponding to the SVL of the mean unsexed or mixed-sex pool of adults in an unbiased sample of the population. See MassMeanFemRange for several caveats.

MassMinsizeUnkMin This is the smallest mass (g) recorded for the species.

MassModeSpEst This is my best estimate of the mass (g) of the adult most likely to be encountered in an unbiased sample. See appendix B for a full explanation.

Mass-SVLRegression This text field includes the equations available for estimating mass from SVL for this species. When species-specific equations were unavailable, I used increasingly general formulae, often those posited by Meiri (2010) or Feldman et al. (2015).

MaxIntros The larger of two fields: IntroCounCt or IntroIslCt.

MaxOverMinMass (MOMM)
The $\log_{10}$ value of the ratio: MassMaxsizeUnkMax / MassMinsizeUnkMin. This was found to have a size signal—it tended to be larger in larger lizards—and so was further adjusted (see SizeAdjMOMM).

MeanLog$_{10}$ActArea The mean activity area of a species was estimated from the midpoint of the adult male and female ranges or the range of unsexed adults, and it was subjected to a $\log_{10}$ transformation. This value was expected and confirmed to have a size signal—larger species have larger home ranges—and was further adjusted (see ActAreaMassResid).

MicroCode Short for microhabitat code, this single letter code was the single best choice from among the following microhabitats: "f" = fossorial; "c" = cryptozoic; "t" = terrestrial; "a" = arboreal; "q" = aquatic; "m" = marine. I attempted to characterize the surface or medium in which the species was most often seen; I excluded surfaces used only as refugia or for immobile thermoregulation. Species that use different strata at different life stages—typically juveniles on the ground and adults above ground—were coded by the stratum of the adult stage. Species that foraged on the ground but hidden under objects (rocks, litter, vegetation) were termed cryptozoic, though that term was probably sometimes reported for species in which foraging behavior was simply not understood (it was found under a rock; little else was known). Some sources used the terms "sub-fossorial" or "semi-fossorial" to mean roughly what I mean by cryptozoic. I used "fossorial" to denote that the animal forages by pushing through soil or sand, whether in tunnels that the lizard creates or in tunnels that it appropriates from other fossorial species. The presence of lizards under logs or rocks is ambiguous, as it may be that the lizard took refuge there as the herpetologist approached (a predator refugium), or that the herpetologist was hunting for lizards during the lizard's inactive period (inactive refugium), and it refuges under logs or rocks when it is asleep or otherwise inactive during its activity period (temporary refugium). It may also mean that the lizard was foraging under logs or rocks (fossorial). "Marine" and "aquatic" are reserved for species that are usually encountered in those media; any species that occasionally takes refuge in fresh or salt water would thus not be characterized as aquatic or marine.

MPMMax The highest published value of mean moves per minute.

MPMMin The lowest published value of mean moves per minute.

NativeCountries These are the political jurisdictions in which native populations have been found, ordered alphabetically. These jurisdictions are countries as understood by the United Nations, plus Canadian, Australian, and American provinces or states. For contested regions of the world, I apologize for any inadvertent offenses; I tried to list the jurisdiction that would issue a recognized research permit.

NativeIslands These are the names of islands on which native populations have been found. See IntroIslands for pointers on island names, which are ordered alphabetically.

NQMaxBiomass The maximum biomass estimate (kg/ha) produced by a population density estimate that was deemed less credible (NQ or nonqualified). See appendix C for the criteria used in this judgment.

NQMaxDens The maximum density estimate (number per hectare) obtained from a study that was deemed less credible (NQ or nonqualified). See appendix C for the criteria used in this judgment.

NQMeanBiomass The mean biomass estimate (kg/ha) produced by population density estimates that were deemed less credible (NQ or nonqualified). See appendix C for the criteria used in this judgment.

NQMeanDens The mean density estimate (number per hectare) produced by population density estimates that were deemed less credible (NQ or nonqualified). See appendix C for the criteria used in this judgment.

NQSampleSize The number of population density estimates that were deemed less credible (NQ or nonqualified). See appendix C for the criteria used in this judgment.

NumFieldCount The number of numerical fields that were filled for the species.

OrdFemSpace An ordinal condensation of the SpacingFem field: blank if SpacingFem blank; "0" if females socially tolerant; "1" if females defended space, formed a hierarchy, or were socially intolerant; "2" if the females defended entire home ranges; "3" if females defended space year-round; "4" if females chased males away in nonbreeding periods.

OrdImperiled An ordinal condensation of ConservStatus: blank if species not evaluated for conservation status; "0" if the status of the species had been investigated and was listed by the IUCN as LC or DD or if it was an equivalent, credible low-risk category or lacked any status; "1" if the species was judged to be at risk by a competent scientist or nongovernmental organization, was imperiled but only in certain portions of its range, or otherwise lacked official range-wide protective status; "2" if imperiled range wide but the threat did not reach IUCN categories EN, CR, or EX; "3" if classified by the IUCN as EN, CR, or EX.

OrdInRefuge An ordinal condensation of InRefuge: blank if no data in InRefuge; "1" if solitary; "2" if only ever found in groups of one adult male, one or more adult females and optionally young (presumed offspring), or if grouping confirmed but composition unknown; "3" if found in a group consisting of more than one adult male; "4" if found in multispecies groups.

OrdMaleSpace An ordinal condensation of the SpacingMale field: blank if SpacingMale blank; "0" if males socially tolerant; "1" if males defended space, formed a hierarchy, or were socially intolerant; "2" if the males defended entire home ranges; "3" if males defended space year-round; "4" if males chased females away in nonbreeding periods.

OrdSSD An ordinal condensation of the text field SSD: "–2" if males averaged smaller than 50% of the mass of their female counterparts; "–1" if males were smaller by a lesser or an unquantified amount; "–0.5" if contradictory data were obtained that supported both m<f and m=f size relationships; "0" if the sexes were found to be indistinguishable in size; "0.5" for contradictory data supporting both m>f and m=f size relationships; "1" if males were larger than females but not twice the mass of females; "2" if males were twice the mass of females. All size comparisons are based solely on adults. Many reports seemed to lack precise quantification or unbiased sampling. See SSD for additional caveats.

OrdWary An ordinal condensation of the text field ConspicuouslyWary: blank if ConspicuouslyWary blank; "0" if explicitly unwary; "1" if wariness characterization was controverted or partial (one sex, certain places, etc.); "2" if consistently judged to be conspicuously wary.

OvipSeasLength A text field with three exclusive choices for the populations' duration of oviposition or birthing: "seas" if less than five months in duration; "extend" if 5–10 months in duration; "cont" if >10 months or year-round, however seasonally modulated the amount.

Padless A nominal variable reflecting width of gecko (sensu lato) toepads: blank if not a gecko; "0" if digits had pads that were noticeably wider than the digit; "1" otherwise.

ParentalCare A text field consisting of a single exclusive letter denoting the gender and duration of parental care: blank if care was not documented; "m" if males provided care prior to hatching or birth; "f" if females provided care prior to hatching or birth; "b" if both sexes provided care prior to hatching or birth; "u" if a gender-unspecified adult provided care prior to hatching or birth; "a" if care was provisioned after hatching or birth. The latter code allows for prehatching or birth parental care. Details may be provided in ReproEarlyDevNarrative.

The most comprehensive review of parental care in lizards is that of Somma (2003), which takes a very inclusive perspective on parental care. Parental care is usually of interest because it implies effort being directed toward organisms that share only half their genes with the individual parent. Thus there is a genetic basis for potential conflicts of interest between the parties, and the vast majority of lizards appear to provide little care for progeny once the eggs or young have been deposited in an appropriate environment. Somma (2003) took the view that situations that *might* imply parental care were to be tabulated, even if the parsimonious explanation is parental self-interest. For example, Somma's "NC" category documented that a parent (usually the mother) occupied a nest depression or built a chamber. While this might be parental care in the strictest of senses, it might equally benefit the mother—protection from elements and predation while ovipositing or birthing—and is so widespread in lizards that I did not tabulate it. It seems to me to be an example of where parent and offspring often have a shared interest, not a conflict.

Similarly, I have ignored Somma's "SA" category, which documented that one or more parents allow the offspring to remain in their vicinity after birth/hatching. In my experience, the vast majority of lizards *in nature* ignore the presence of hatchling/neonates, as they do not consume substantial limited food or habitat resources, nor do they threaten the sexual hegemony of the resident adults. In contrast, maturing consexual juveniles are often expelled, but that is not the situation considered in category "SA." For lizards that in the wild allow hatchlings/neonates to remain in their vicinity (regardless of parentage), this proximity evidently entails negligible fitness cost, and therefore does not demonstrate an appreciable conflict of interest. Only in the rare cases—none directly documented in Somma (2003)—where adults in nature are intolerant of the presence of unrelated neonates/hatchlings, or routinely eat them, would the tolerance in nature of related neonates/hatchlings be noteworthy. See Whiting and While (2017) for several examples from *Egernia*. This targeted tolerance would suggest kin recognition, which is not widely documented in lizards, and would be reported in ReproEarlyDevNarrative. In captivity, many species show aberrant behavior with regard to close confinement with conspecifics; unfortunately, many of the papers cited by Somma (2003) are from captivity, without an appropriate disclaimer to that effect.

For viviparous species, there are two additional Somma categories ("OO" and "AN") that might be examples of parental care or parental self-interest. "OO" codes for parental consumption of stillborn young; "AN" codes for parental consumption of birth fluids. Such consumption benefits both mother and progeny by removing odor cues from the site. Assisting progeny in the emergence from birth membranes shortens the process—a process that is risky for both of them. In both cases, consumption of the tissue benefits the mother by replenishing her severely depleted energy or water stores. In my experience, these actions are widespread and unsurprising in viviparous species of many vertebrates. I did not code them as examples of parental care, as the benefits to the mother are clear, though the offspring no doubt benefit as well.

The largest number of postbirth examples of parental care in Somma (2003) are observations of mother and offspring remaining in close proximity after birth (Somma category "CN"). I find the demonstration of proximity to be ambiguous, as by definition two relatively immobile entities that start from the same geographic point are more likely to remain in close proximity than to instantly become distant from each other. Over time, random movement will increase spatial separation, which is what is routinely observed. Somma (2003) also failed to categorically distinguish observations of this made in the wild versus those made in captivity. In the close confines of captivity, a parent that is defending its own personal space or territory could easily be mistaken for defending its progeny (which, due to captivity, are unable to disperse). Thus some care is needed in the interpretation of these putative manifestations of parental care.

Parthenogenesis A short text field of three mutually exclusive possibilities: "bisexual" if parthenogenesis had been looked for and not found, "rare" if occasional examples of known or suspected parthenogenesis had been logged; "normal" if the species is all-female, or otherwise presents strong evidence of obligate parthenogenesis.

PercMPMMax The percentile rank (1st–100th percentile) of MPMMax.

PercMPMMid The percentile rank (1st–100th percentile) of the midpoint of MPMMin and MPMMax.

PercMPMMin The percentile rank (1st–100th percentile) of MPMMin.

PercPTMMax The percentile rank (1st–100th percentile) of PTMMax.

PercPTMMid The percentile rank (1st–100th percentile) of the midpoint of PTMMin and PTMMax.

PercPTMMin The percentile rank (1st–100th percentile) of PTMMin.

PerFilled NumFieldCount/163 (the number of nonderived fields), expressed as a percentage.

PlantsReported A nominal variable documenting whether plant material (>1%) had been found in diet samples: blank if diet samples unavailable; "0" if plant material absent or <1%; "1" otherwise.

PopSources The sources used for SubjAbund.

Population Narrative Supplementary population information, including legends and myths.

Pores "0" if secretory pores absent; "1" if present. I tabulated pores if they were sexually dimorphic or conspicuous (i.e., secretory). Diminutive sexually monomorphic pores were ignored. As pores are presumed to operate for intraspecific communication, and most infraspecific lizard chemical communication is presumed to be directed at deterring consexuals (i.e., stay away), relatively few secretory pores are likely to be missed by this standard. Unfortunately, knowledge of the use of secretory pores is largely speculative, and this logic may be erroneous in some and possibly many species. An expanded scoring of pores will provide greater precision and resolution on pore attributes in the next edition of the database.

Power The coefficients listed in the Intercept and Power fields are those that I used to estimate MassModeSpEst and are uniformly of the form Log_{10}Mass (in g) = Intercept coefficient + Power coefficient × log_{10}(SVLModeSpEst in mm).

PrecArid Blank if rainfall data lacking for the species; "1" if annual rainfall 151–400 mm/y; "0" otherwise.

PrecDry Blank if rainfall data lacking for the species; "1" if annual rainfall 56–150 mm/y; "0" otherwise.

PrecHyperdry Blank if rainfall data lacking for the species; "1" if annual rainfall <56 mm/y; "0" otherwise.

PrecIntermed Blank if rainfall data lacking for the species; "1" if annual rainfall 401–1100 mm/y; "0" otherwise.

PrecipBands The count of rainfall bands occupied by the species (0 to 6).

PrecMesic Blank if rainfall data lacking for the species; "1" if annual rainfall 1101–3000 mm/y; "0" otherwise.

PrecWet Blank if rainfall data lacking for the species; "1" if annual rainfall >3000 mm/y; "0" otherwise.

PrimBM Acronym of the primary ecological business model. For example, the score for species that are wide-ranging chemosensors is "WRC."

PrimPredAvoidTact A short text field of the following nonexclusive choices: "crypsis," "flight," "sand," "crev-burrow" (crevice-burrow), and "veg" (vegetation). It would be logical for only one of these to be selected as the *primary* predator avoidance tactic, but the literature on this matter is too thin to assemble a relative ranking for most species. Instead, I countenanced any choice reported. Each of these choices was subsequently coded into nominal variables named after the keyword (see these for details of assignments).

PTMMax The highest published value of proportion of time moving. This metric is sometimes alternately reported as percentage of time moving (100-fold the value reported here).

PTMMin The lowest published value of proportion of time moving. This metric is sometimes alternately reported as percentage of time moving (100-fold the value reported here).

QMaxBiomass The maximum biomass estimate (kg/ha) produced by a population density estimate that was deemed credible (Q or qualified). See appendix C for the criteria used in this judgment.

QMaxDens The maximum density (number per hectare) produced by a population density estimate that was deemed credible (Q or qualified). See appendix C for the criteria used in this judgment.

QMeanBiomass The mean biomass estimate (kg/ha) produced from population density estimates that were deemed credible (Q or qualified). See appendix C for the criteria used in this judgment.

Here are some standards to help guide interpretation of this value. Islands that are small enough to *lack* a full suite of predators often engender much higher lizard population densities (Rodda and Dean-Bradley 2002; Novosolov et al. 2013). A biomass density of 0.01 kg/ha is very low for a continental species and indicates extreme rarity for a lizard on a small island. A biomass density of 0.1 kg/ha is low for a continental species and extraordinarily low for a lizard on a small island. A biomass density of 1 kg/ha is high for a continental species but typical for a lizard population on a small island; a biomass density of 4 kg/ha is about average for a lizard on a small island but very high for a continental species; a biomass density of 15 kg/ha is high for a lizard on a small island and extraordinary for a continental species (= mean biomass density of elephants in the Serengeti). Maximum reported lizard biomass densities are on islands (ca. 150 kg/ha). Island Tuatara are off the scale (mean 259 kg/ha).

QMeanDens The mean density estimate (number per hectare) produced by population density estimates that were deemed credible (Q or qualified). See appendix C for the criteria used in this judgment.

QSampleSize The number of qualified density estimates that were summarized in the Q (qualified) density estimates. See appendix C for the criteria used in this judgment.

RangeCore A geographic name or short phrase characterizing the species' range core or distribution.

RangeMapSources A field listing sources that provided a credible range map. Sources, such as this database, that record or map only political jurisdictions are not a source of a credible range map. As the objective of this work did not include distribution mapping, sources that were consulted for lists of political jurisdictions or island records, but did not provide a credible range map, are not cited in this field. Consult Reptile Database (www.reptile-database.org) or Roll et al. (2017) for geographic distributions.

RangeNarrative A narrative field to explain controverted distributions and uncertainties regarding the naturalness of possibly introduced populations.

RatioMFSVLMidpoint The m:f ratio of adult midpoints of snout-vent length (SVL) estimates by sex: SVLMeanMaleMidpoint/SVLMeanFemMidpoint). Values were available for only a small minority of species, most of which clustered near 1.0 (sexual monomorphism). Many of the species with sexual size dimorphism (SSD) reported did not have data in the SVLMeanMaleMidpoint and SVLMeanFemMidpoint fields, suggesting that sex-specific measurements are more often collected when the size difference between the sexes is uncertain.

RCMMax The highest reported ratio: (aggregate clutch or litter mass) / (female postpartum or postoviposition mass).

RCMMeanRange The range of reported mean ratios: (aggregate clutch or litter mass) / (female postpartum or postoviposition mass).

RCMMidpoint The midpoint of the range of reported mean ratios (aggregate clutch or litter mass) / (female postpartum or postoviposition mass), if mean values available. If mean values are unavailable, the midpoint of RCMMin and RCMMax.

RCMMin The lowest reported ratio (aggregate clutch or litter mass) / (female postpartum or postoviposition mass).

ReproEarlyDevNarrative A narrative field elaborating on the reproduction fields. See also the ActivityNarrative, as some elements of reproductive behavior are not always immediately recognized as motivated by mating concerns.

ReproMode A short text field distinguishing oviparous from pooled viviparous/ovoviviparous species: the choices were "ovip," "vivip," or "both." "Both" may reflect contradictory accounts as well as species that are capable of both reproductive modes. I made no attempt to ascertain the degree to which nutrients are provided directly by the mother in "viviparous" species.

ReproSeasonColor A short text field indicating that males ("m"), females ("f"), or both ("b") sexes exhibit seasonal color when reproducing. Female color seems to most often be associated with ovulation hormones, which peak at a time when the female's sexual readiness is turning a sharp corner from available to disinterested. Often a female's "nuptial" redness—for it is routinely red or orange—is stimulating to males as it begins but rapidly turns to repulsion as the color deepens and is prolonged. It may serve to reduce the harassment experienced by females who are already inseminated.

ReproSources The sources used for reproductive data.

Riparian A nominal variable expressing whether the "riparian" keyword was tabulated in Landscape: blank if Landscape blank; "1" if "riparian" registered; "0" otherwise.

Rock A nominal variable expressing whether the "rock" or "cave" keywords were tabulated in Landscape: blank if Landscape blank; "1" if "rock" or "cave" registered; "0" otherwise. Note distinction from following field.

RockCrev A nominal variable expressing whether the "rock crevice" keyword was tabulated in SleepRefuge: blank if SleepRefuge blank; "1" if "rock crevice" registered; "0" otherwise.

Sand This field characterizes the species' primary predator concealment tactic (PrimPredAvoidTact) as sand diving (from among the choices of crypsis, flight, sand, crevice-burrow, or vegetation). The field was left blank if no predator avoidance tactic data were available; "1" if the lizard's primary tactic was sand diving; "0" otherwise. Note distinction from following field and from DunesSand.

SandSleep A nominal variable expressing whether the "sand" keyword was tabulated in SleepRefuge: blank if SleepRefuge blank; "1" if "sand" registered; "0" otherwise.

Scansorial A nominal variable expressing whether any of the keywords "arboreal," "low veg," or "vert wall" were tabulated in ActiveSurface: blank if ActiveSurface blank; "1" if any of these three scansorial keywords were registered; "0" otherwise.

Seasonal A nominal variable expressing whether the "seas" keyword was tabulated in OvipSeasLength: blank if OvipSeasLength blank; "1" if "seas" registered; "0" otherwise.

SeasonalInactivity A short text field to describe a seasonal reduction in activity that might occur due to excessive cold, heat, wetness, or dryness. The inactive period did not need to be complete—individuals might emerge on clement days, for example—nor need it necessarily affect all life stages: rapidly growing juveniles are often more ready to risk exposure.

SecBM A secondary ecological business model, if any, by acronym.

SexDiffRollup This composite classification sums (false = "0," true = "1") credits for each of five different measures of sexual difference: (1) SexesDifferColor, (2) ReproSeasonColor, (3) OrdSSD not m=f, (4) the sexes have at least one morphological structure or ornament that distinguishes them (dewlap, crest, secretory pores, etc.), and (5) TailEstM-FRatio is either <0.9 or >1.1 (i.e., a difference >10%) if based on means, or a difference of >~20% was considered plausible if based on individual relative tail lengths. One extra point was awarded if the body mass difference between mean adult males and mean adult females exceeded twofold (i.e., the OrdSSD score was –2 or 2); however if RatioMFSVLMidpoint was in the opposite direction from OrdSSD, only the consensus of those metrics was counted. Note that RatioMFSVLMidpoint is not exactly the same metric as OrdSSD, so being of opposite sign does not denote a contradiction, only that the signals are mixed.

This composite "rollup" score was meant to combine features that reflect visual cues, olfactory cues, and body mass. It was intended to reflect the magnitude of the overall difference between the sexes, but it does not capture the direction of the difference; for that, inspect the individual scores. The rollup was accentuated for species that differed in many respects and minimal for those species that differ, however greatly, in only one dimension. This rollup measure is vulnerable to misinterpretation when data are lacking for the component subscores. A highly dimorphic species will likely be given only a modest rollup score if it was poorly known. A high rollup score indicates strong sex differences regardless of data completeness. A low rollup score is strongly indicative only for well-known species. A blank score denotes a unisexual (all-female) species.

Note that there are additional database traits that may have tabulated sex differences—maturation age, longevity, blue tails, spatial defense, activity range size, active surface—which are not rolled up in this index.

SexDimorphStructures A long text field that describes differences detectable by another lizard at a distance; that is, it omits many microscopic characters of interest to taxonomists. In general, I have omitted hemipenial bulges, as these might not be visible in an approaching lizard. A text phrase in this field that a sex possesses structure "X" indicates that the other sex does not.

SexesDifferColor True ("y"), false ("n"), or blank. I include ventral colors (omitted by some reporters) and omit juvenile colors. Colors that exist only on a structure that is lacking in the opposite sex are ambiguous. My criterion was, is there any color information that would communicate a strange lizard's sex to another lizard? As this question is not of primary importance to the taxonomists who write the majority of the color descriptions, and as color may also be lost or obscured due to specimen preservation or due to short-term (metachrosis), ontogenetic, seasonal, thermal, or geographic variation, the literature is replete with examples color differences that were initially overlooked. Therefore, in the case of controverted claims, I erred on the side of recognizing sexual dichromatism. Nonetheless, this field is vulnerable to bias from underreporting.

ShapeIndex The standardized residual of Log_{10}Mass regressed on log_{10} (SVLModeSpEst), a measure of shape that is independent of lizard size. A negative value indicates a species that is lighter than an average lizard of that length would be; a positive value indicates an unusually heavy one. The sample size was 6525 species and the standard error associated with each was 0.339. The highly significant ($p = 0$) relationship was Log_{10}Mass = 2.399 × log_{10} (SVLModeSpEst) – 3.653.

SingJurisEndemic "0" if the species' range spans more than one of the political jurisdictions tabulated (see NativeCountries); "1" otherwise.

SizeAdjMOMM The standardized residual of MaxOverMinMass regressed on Log_{10}Mass, a measure of ontogenetic mass range that is independent of lizard size. A negative value indicates a species with a reduced ontogenetic mass range (i.e., young are comparatively large); a positive value indicates an expanded ontogenetic mass range (i.e., young are comparatively small). The sample size was 1089 species and the standard error associated with each was 0.333. The highly significant (p = 2.1E–139) relationship was MaxOverMinMass = 0.409 × Log_{10}Mass + 1.195.

SizeNarrative A text field challenging or explicating unusual values (mostly data reliability questions) in morphological data.

SizeSources Sources consulted for morphological data.

SleepRefuge A text field composed of the following keywords arranged on a quasi-continuum from below ground to canopy top. I did not include a daily refuge for subterranean lizards, as that information was not available.

"Sand" indicates that the aboveground active lizard buried itself in loose sand when inactive.

"Rock crevice" indicates that the inactive lizard refuged in rock crevices—rock on both sides—as opposed to "under veg-rock."

"Under veg-rock" denotes that soil, sand, or humus formed one side of the refuge cavity, which could be a crevice that is natural, dug by another animal, or dug by the focal individual. The upper surface of a refugium labeled "under veg-rock" can be any natural or unnatural material. In the Caribbean, it is usually human trash. Cavities dug "under veg-rock" are usually small and are generally ambiguous as regards their excavation. Cavities that are clearly excavated by a vertebrate generally fall into one of the following three categories.

"Appropr burrow" (= appropriated burrow) indicates that a burrow, with or without a capping structure such as wood or rock, was dug by another species but appropriated by the lizard in question.

"Dug burrow" specifies a burrow excavated by the resident lizard. The burrow categories sometimes meld into the single "under veg-rock" category, in that the digger of the burrow often selected a rigid structure to support the roof of the burrow entrance. Thus the space underlying the rock or vegetation may or may not be excavated by either the resident or some other animal. Is such a refuge a burrow or simply the underside of a cover object?

"Unspec burrow" indicates that the creator of the burrow was not specified.

"Elev veg" indicates that the lizard sleeps in either an exposed or concealed arboreal location.

Sound A nominal variable indicating whether the lizard was known to make any sound beyond a hiss. "0" if sound production was not recorded; "1" otherwise.

SpacingFem A short text field describing what is known about non-nest-site spatial relations among females. Most choices are "tolerant," "defend," "hierarchy," "defend mating" (season), and so on. For females that oviposit outside of their normal home range, it is easy to distinguish routine female spatial defense from nest-site defense—the latter is reported only in ReproEarlyDevNarrative. For species that oviposit within their home range, distinguishing the two relies on understanding the female individual's reproductive phenology, and casual field observations may conflate the two.

SpacingMale A short text field describing what is known about spatial relations among males, such as "tolerant," "defend," "hierarchy," "defend mating" (season), and so on. See preceding field.

SpecialMorph A short text field describing unusual morphological attributes possessed by both sexes.

Species The species name recognized by Reptile Database (www.reptile-database.org) in May 2017.

SSD A short text statement of the size relationship between the sexes. Typical values were "m>f," "m≫f," "m<=f," and so on. An inequality sign repeated indicates differences of large magnitude (mass ratio > 2.0). The presence of different operands (e.g., <>) or a question mark indicates contradictory assessments. The entry "na" indicates an all-female species, for which SSD in not applicable. Note that population size means from the wild reflect not only physiological growth processes but also size-specific mortality, both of which may differ between the sexes.

Stress A nominal variable indicating the production of stress vocalizations: "0" indicates that the lizard did not produce stress vocalizations, although other information was available on vocalizations; "1" if stress calls were known.

SubjAbund A short text field offering five nonexclusive qualitative population density estimates: "very rare," "rare," "moderate," "common," "very common." These ranks do not have any objective meaning and could be replaced with an ordinal 1–5 scale. A given species may have several different density estimates by one or more observers, reflecting geographic and habitat variation (or simple chance). The range of all observers' estimates was reflected in this field.

In my experience, observers often gave their judgment of a species' density within the context of what was expected of similar or closely related species. As noted in chapter 4, these qualitative estimates bear no statistical relationship to quantitative measures of density, but they may be of value for herpetologists who can intuit the unstated context in which the original estimate was couched. Although most observers seemed to judge abundance on the basis of the taxon involved, some compared sighting rates to those of other species in the area; for these authors, "common" may indicate that the focal taxon is seen more often than syntopic lizard species.

SubjAbundMean The numeric mean of the SubjAbund scores, on a scale of 1–5.

SubjAbundSourceCt The number of literature sources contributing to SubjAbundMean.

Subspecies An alphabetical list of the subspecies recognized by Reptile Database (www.reptile-database.org) in May 2017.

SubstrateTempMax The maximum field-measured substrate temperature (°C) traversed by an active lizard.

SubstrateTempMin The minimum field-measured substrate temperature (°C) traversed by an active lizard.

SVLMaturingFemRange The range of snout-vent lengths (mm) reported for females in the process of maturing. Often a source will simply report the smallest female captured with developing follicles. This value is sample-size sensitive and takes no account of larger but immature females. These biases are in opposite directions, and I hope they cancel each other out. None of my SVL measurements included head ornamentation (primarily on chameleons) that protruded beyond the unornamented snout.

SVLMaturingMaleRange The range of snout-vent lengths (mm) reported for males in the process of maturing. Often a source will simply report the smallest male captured with enlarged testes. This value is sample-size sensitive and takes no account of larger but immature males. These biases are in opposite directions, and I hope they cancel each other out. None of my SVL measurements included head ornamentation (primarily on chameleons) that protruded beyond the unornamented snout.

SVLMaturingUnkRange The range of snout-vent lengths (mm) reported for lizards of unknown or unspecified sex in the process of maturing. Often a source will simply report the smallest individual captured with signs of maturity. This value is sample-size sensitive and takes no account of larger but immature animals. These biases are in opposite directions, and I hope they cancel each other out. None of my SVL measurements include head ornamentation (primarily on chameleons) that protrudes beyond the unornamented snout.

SVLMaxsizeFemMax The snout-vent length (mm) of the largest female measured. None of my SVL measurements included head ornamentation (primarily on chameleons) that protruded beyond the unornamented snout.

SVLMaxsizeMaleMax The snout-vent length (mm) of the largest male measured. None of my SVL measurements included head ornamentation (primarily on chameleons) that protruded beyond the unornamented snout.

SVLMaxsizeUnkMax The snout-vent length (mm) of the largest individual measured. None of my SVL measurements included head ornamentation (primarily on chameleons) that protruded beyond the unornamented snout.

SVLMeanFemMax The largest mean snout-vent length (mm) reported for adult females. Many sources failed to report whether immature females may have been included in the sample; young of the year were routinely excluded. None of my SVL measurements included head ornamentation (primarily on chameleons) that protruded beyond the unornamented snout.

SVLMeanFemMidpoint The mean of SVLMeanFemMin and SVLMeanFemMax.

SVLMeanFemMin The smallest mean snout-vent length (mm) reported for adult females. Many sources failed to report whether immature females may have been included in the sample; young of the year were routinely excluded.

SVLMeanMaleMax The largest mean snout-vent length (mm) reported for adult males. Many sources failed to report whether immature males may have been included in the sample; young of the year were routinely excluded. None of my SVL measurements included head ornamentation (primarily on chameleons) that protruded beyond the unornamented snout.

SVLMeanMaleMidpoint The mean of SVLMeanMaleMin and SVLMeanMaleMax. None of my SVL measurements included head ornamentation (primarily on chameleons) that protruded beyond the unornamented snout.

SVLMeanMaleMin The smallest mean snout-vent length (mm) reported for adult males. Many sources failed to report whether immature males may have been included in the sample; young of the year were routinely excluded.

SVLMeanUnkMax The largest mean snout-vent length (mm) reported for adults of unknown or unspecified sex. Many sources failed to report whether immatures may have been included in the sample; young of the year were routinely excluded. None of my SVL measurements included head ornamentation (primarily on chameleons) that protruded beyond the unornamented snout.

SVLMeanUnkMidpoint The mean of SVLMeanUnkMin and SVLMeanUnkMax.

SVLMeanUnkMin The smallest mean snout-vent length (mm) reported for adults of unknown or unspecified sex. Many sources failed to report whether immatures may have been included in the sample; young of the year were routinely excluded.

SVLMinsizeUnkMin The snout-vent length (mm) of the smallest individual measured.

SVLModeSpEst The empirical or estimated snout-vent length (mm) of the adult length mode. See appendix B for an explanation of this estimation.

S-WForage A nominal variable indicating the species' exhibition of sit-and-wait foraging. Blank if foraging mode information lacking; "1" if sit-and-wait reported; "0" otherwise.

SystematicNarrative To assist the reader in sorting through nomenclatural chaos, five types of data were included in the SystematicNarrative field. The first sentence, in the form "Species described by ______ (description citation) in [genus][species]" gives the citation of the paper that originally described the form, whether as a species or subspecies, along with the original genus or species name if that name is not still in use. The most common reason for the obsolescence of the original name is that the species was subsequently transferred to another genus. The second most common reason is that the species was originally described as a subspecies of another species and subsequently raised in rank to become a full-fledged species of its own. The first sentence should make this history evident.

Not uncommonly, a species has wandered around in a taxonomic pilgrimage. Thus the second sentence, if present, lists the names of all genera in which the species has resided over the last 100 years. Thus if you are looking for the skink *Eumeces reynoldsi* and can't find it under that name, search for skinks with the species name "*reynolds*"—endings of species names need match the gender of the genus name and therefore may change when a species is moved from one genus to another; thus a search on the full species name (e.g., "*reynoldsi*") may fail to find it. Omit the ending from your search string and you will usually find it if the species name is still in use. In this case, the only name that matches the string "*reynolds*" is *Plestiodon reynoldsi*, which the taxonomic field's second sentence helpfully notes was once in *Eumeces*. Had you known this species as *Neoseps reynoldsi*, you would also have found that association in the first sentence, which notes that Stejneger, the original describer, described it in *Neoseps*.

If the species name with which you are familiar is no longer in use, or is now in a restricted use and therefore no longer applies to the lizard you have in mind, some help may be found in the third (optional) sentence of the Systematic field, which is in the form "Confused with ______ (species names)." For example, suppose you are searching for *Zygaspis longicauda*. That name is no longer in use, and you won't find it in a simple search of the taxon field. If you search the taxon field for "*longicaud*" within Amphisbaenidae, you turn up only *Dalophia longicauda*, which was originally in *Monopeltis*, not *Zygaspis* (that is, there is no mention in the Systematic field of *Zygaspis*). Maybe that's not the amphisbaenian you're looking for, even though its geographic range is in the same area of southern Africa. But if you search the Systematic field for "*longicaud*" you turn up a record for *Zygaspis violacea*, which in the third part of the Systematic field has a note that it was confused with *longicauda*. The "confused with" sentence can be very helpful at disentangling nomenclatural meanderings.

Confusion about species names can arise because of uncertainty about the features that define a species, because two different people independently described the same species, or because species are so similar in appearance that the key attribute that distinguishes them is or was overlooked. Sometimes all three of these same problems arise together. But in other cases the confusion is in poor identification rather than taxonomic uncertainty. In the latter case, an additional optional sentence may shed some light: "Similar to ______ (species name)." This sentence should caution the reader that some of the pertinent natural history data may be erroneously associated with the other species name.

Finally, the last standard part of the Systematic field lists forms that were once listed as subspecies of the species in question. This helps to warn the reader that a split may have cleaved off the population that is actually of interest. For example, suppose one is interested in a Malagasy lizard by the name of *Cryptoblepharus boutonii*. There is a species with that name, but as noted in the Systematic field, it now includes only populations on the islands around Mauritius, as dozens of former populations—often named subspecies—were elevated to full species rank. As it happens, the Malagasy subspecies *cognatus* and *voeltzkowi* were among those raises to full species rank. So any *C. boutonii* from Madagascar are now known as either *C. cognatus* or *C. voeltzkowi*. The fifth standard element of the Systematic field can often point you in the direction of discovering a new species name from the subspecies names it lists as formerly included. Too many such names have been elevated from subspecies of *C. boutonii* to list them all individually, but the fifth element warns you to search for them if your population is not from Mauritius.

Additional elements are included in the taxonomic field as needed. The most common is a warning that the taxon is known from only a few specimens or a single sex, which implies that all of the derived natural history data may have an insufficient sample size, and that the taxon may be invalid, parthenogenetic, or extinct. A bisexual taxon so qualified is likely to be extremely rare if it is still extant.

TailBreaks "True" if the species' tail is documented to break with any frequency (e.g., >1%), regardless of whether it has a fracture plane or regrows; "false" otherwise. Underreporting bias is likely for this variable.

TailEst-Fem My best estimate of the mean relative tail length ratio (relative to SVL) for adult females. I have provided these single best estimates by gender based on the following descending order of preference:

1. If two or more empirical mean estimates exist, use the midpoint of the reported range.
2. If only a single empirical mean is known, use it.
3. If only two or more individual relative tail length estimates are available, use their midpoint.
4. If only one individual relative tail length estimate is available, use it.

TailEst-Male My best estimate of the mean relative tail length ratio (relative to SVL) for adult males. See TailEst-Fem for the algorithm used to estimate this value.

TailEst-Unk My best estimate of the mean relative tail length ratio (relative to SVL) for lizards of unknown or unspecified sex. See TailEst-Fem for the basic algorithm for estimation, which is supplemented in the special case of TailEst-Unk, with the following two supplementary lowest-priority options:

1. If only a single individual relative tail length estimate or none is available, but there are values for *TailEst-Male* and *TailEst-Fem*, average the sex-specific values.
2. If either *TailEst-Male* and *TailEst-Fem* are unavailable, and no empirical value is reported for unknown gender individuals, use the genus average for *TailEst-Unk*.

TailEstM-FRatio The ratio of TailEst-Male to TailEst-Fem.

TailLengthFemRange Tail length ranges are in a special form: absolute minimum–lowest mean–highest mean–absolute maximum. All values are of length of unbroken tails relative to adult female SVL; both tails and SVLs measured in mm. A tilde (~) precedes any value that represents an individual measurement; a tilde is omitted if the reported relative tail length is a mean. Thus the string is in the form ~# - # - # - ~# (e.g., "~0.5-0.95-1.20-~1.41"). If a reported "mean" is suspect (i.e., does not comport with other measurements, or it appears to be excessively round; e.g., = exactly 1.0 or 0.5, etc.), I may flag it as if it were an individual measurement.

For a few species that have no better data, I have sometimes used string and a photograph taken perpendicular to the body axis to estimate the approximate relative tail length. Such photographic estimates are qualified with a "p" suffix. These photographic approximations were removed whenever supplanted by an empirical value.

TailLengthMaleRange See TailLengthFemRange. A string of numbers in the form ~# - # - # - ~# for the relative (to SVL) length of unbroken tails in adult males. A tilde precedes any value that represents an individual measurement or is otherwise suspect; a tilde was omitted if the reported relative tail length (expressed in proportion to the individual male's SVL; both measurements in mm) was a mean.

TailLengthUnkRange See TailLengthFemRange. A string of numbers in the form ~# - # - # - ~# for the relative (to SVL) length of unbroken tails in populations of unknown or unspecified sex. A tilde precedes any value that represents an individual measurement or is otherwise suspect; a tilde was omitted if the reported relative tail length (expressed in proportion to the individual's SVL; both measurements in mm) was a mean.

TailPrehensile A short text field consisting of "yes," "no," or "limited" or left blank.

TextFieldCount The number of filled text fields, exclusive of derived fields.

TotalMaxsizeFemMax The total length (in mm) of the largest female measured. This can be approximately converted to SVL using the TailEst-Fem ratio.

TotalMaxsizeMaleMax The total length (in mm) of the largest male measured. None of my length measurements included head ornamentation that protruded beyond the unornamented snout. Total can be approximately converted to SVL using the TailEst-Male ratio.

TotalMaxsizeUnkMax The total length (in mm) of the largest measured individual of unknown or unspecified sex. This can be approximately converted to SVL using the TailEst-Unk ratio.

TotalMinsizeUnkMin The total length (in mm) of the smallest measured individual. This cannot be approximately converted to SVL using the TailEst-Unk ratio because tail lengths of neonates and hatchlings are often very different from the adult proportions specified in TailEst-Unk. As the field SVLMinsizeUnkMin may have been taken from a different specimen than TotalMinsizeUnkMin, it is best not to infer a neonatal/hatchling TailEst ratio from those two fields.

Trees This nominal variable indicates whether trees were reported to be part of the species' Landscape. Blank if Landscape information lacking; "1" if trees present (as "scattered trees," "forest," or "primary forest"); "0" otherwise.

TSD A short text field indicating whether temperature-dependent sex determination (TSD) had been indicated or firmly documented, and if so, which sex was produced at which temperature: "yes," "no," "male high," "fem mid," and so on.

UnderVegRock A nominal variable recording whether the species was known to spend its inactive period under vegetation or rocks. Blank if SleepRefuge was undocumented; "1" if "under veg-rock" keyword was used; "0" otherwise.

Veg This field characterizes the species' primary predator concealment tactic (PrimPredAvoidTact) as retreat to vegetation (from among the choices of crypsis, flight, sand, crevice-burrow, and vegetation). The field is blank if no data were available on any choice; "1" if the lizard's primary tactic was retreat to vegetation; "0" otherwise.

Vocalizes A short text field that describes vocalizations, exclusive of hissing, by time of day, gender, context, and so on.

Appendix B Methods for Quantifying Lizard Sizes

Size is a critical ecological attribute: larger species eat smaller ones, choose from a different array of prey, and fear a differing suite of predators. The energy flow through an ecosystem is often conceptualized in transfers of stored energy (biomass) from one trophic level to another. In social encounters, the top predictor of dominance is almost always size. The probability of anthropogenic extinction is tied to size (Slavenko et al. 2016).

To quantify species' relative sizes, one would like to have a single size metric with which to compare all species of lizards. The most commonly used size metric is the maximum snout-vent length (SVL), which is available for virtually all species (Feldman et al. 2015). There are two conspicuous difficulties with this measure. Some lizards are as much as 400 times heavier than others of the same SVL, and maxima are inherently dependent on the number of lizards measured (which vary from $n = 1$ to $n =$ tens of thousands). An example of the first is that 95 mm SVL *Sceloporus orcutti* (75 g) is simply not comparable in size to 95 mm SVL *Cynisca degrysi* (0.19 g). The first problem is widely addressed by registering size as body mass. This is hampered by the absence of empirical weights for many species. For the species for which I did not have empirical masses, I estimated masses using species-specific mass-SVL regressions whenever possible, and progressively more general mass-SVL regressions when necessary. I encourage lizard ecologists to weigh and report the mass of the lizards they sample. Until that is done, we will have to estimate mass from reported lengths for many species.

Overcoming the second problem, of varying sample size, is more convoluted. If the distribution of individual lengths within a species hewed to the normal distribution, the obvious choice would be to use mean SVL and convert that to mass. Regrettably, SVL distributions are almost never normal, and evidence is accumulating that typical sampling protocols grossly undersample smaller lizards (Rodda et al. 2015), invalidating many extant samples. This problem can be somewhat ameliorated by restricting a sample to adults. This requires knowledge of the size of each sex as they mature, information that for many species is poorly documented at best. The size of the smallest mature female is widely known, but that is not an unbiased estimator of mean maturational size in either sex, and it is a minimum, which gets progressively smaller as the sample size is increased.

Another complication is that mean adult size is dependent not only on lizard growth trajectory but also on size-dependent mortality. In most cases this problem is minor, but it becomes potentially troublesome in species that have anthropogenic mortality in addition to natural mortality. Anthropogenic mortality can arise directly if the largest adults are most keenly exploited for a product such as leather. This may result in a distribution with a long right tail, as few individuals survive to large size. Anthropogenic mortality can arise indirectly if a species introduced by humans preys selectively on lizards of a particular size. This is most notable among giant Caribbean island iguanids for which introduced predators such as rats, mongoose, and mice can eliminate or greatly curtail juvenile survival, though the large adults are relatively invulnerable to these small pesky mammals. Over time, the population, if it survives at all, tends to become dominated by large (i.e., less vulnerable) adults. Another way to think of this outcome is the "left-skewing" of the species size distribution. The left tail is long and thin; most surviving individuals pile up near their upper size limit.

Because juvenile mortality is generally higher than adult mortality and growth declines with time, typical valid SVL distributions of most lizards are left-skewed (Rodda et al. 2015); mean SVL is located at some nonobvious point on the left shoulder of the distribution, and the median is located higher up on the same shoulder. The mode, by definition, is at the peak of the size distribution. Although typical SVL distributions are left-skewed, some very large, heavily exploited lizards (e.g., *Varanus*, Auffenberg 1981, 1988, 1994) are strongly right-skewed, indicating that the modes are smaller than the medians and means, and that the medians and means are at some nondescript point on the right shoulder of the distributions. In either case (right or left skew), maximum SVL moves to the right as additional lizards are sampled, and mean population SVL shifts to the left as sampling improves and more of the hard-to-sample juveniles are included. Therefore, both population mean and maximum SVL are somewhat sensitive to sample size. We cannot control sample size, which varies at least five orders of magnitude among lizard species. We can minimize the influence of juvenile undersampling by excluding juveniles, and we can eliminate both juvenile undersampling and sample-size bias by focusing on the adult mode.

I propose that adult modal SVL be the preferred metric for linear measurements. Although, like all measures, it is hard to estimate with very small samples, it does not change in response to increasing sample size, it is invulnerable to the undersampling of juveniles, and it represents the size of the lizard individual that you are most likely to encounter. In that sense, it is a maximum likelihood estimate. It is closer to the estimate often reported in field guides for "typical" size—often inaccurately described as average SVL. I have attempted to estimate modal adult SVL for all lizard species, and modal adult SVL is the value presented in the genus account tables.

There is no unequivocal best method for estimating modal adult SVL from extant size measurements. I collated length and mass data for 34 lizard species (table B.1) of greatly varying size (1st to 100th percentiles of lizard masses), for which plausibly unbiased sampling had been conducted and for which means, modes, and medians of both length and mass could be extracted; 17 of these datasets were unpublished data from total removal sampling that we conducted; see Rodda et al. (2001) for method. Because most field studies focus on larger species, lizards in the top five percentiles of size were greatly overrepresented (12 spp.) among the species used for calibration (table B.1). Fortunately, the absolute size of extrapolation errors of small species are themselves very small.

Table B.1. Species used to calibrate relationships among various size metrics. Only the mean adult snout-vent length (SVL), modal SVL, and mean mass at modal SVL are listed here. See cited sources for additional information. Study site names are either political jurisdictions or a single island, unless followed by "Isl.," denoting an archipelago. The percentile of mass field is provided to draw attention to the overrepresentation of data on very large species.

Species	Study site	Source	Adult mean SVL	Modal SVL	Mass at modal SVL	Percentile of mass
Anolis acutus	St. Croix Isl.	Ruibal and Philobosian 1974	45.6	43	2.3	38
Anolis carolinensis	Mariana Isl.	Rodda et al. unpub.	59.3	66	5.95	60
Calumma nasutum	Madagascar	Brady and Griffiths 2000	47	47	2.3	39
Calumma oshaughnessyi	Madagascar	Brady and Griffiths 2000	130	125	50	95
Carlia ailanpalai	Mariana Isl.	Rodda et al. unpub.	53.3	56	4.21	47
Cryptoblepharus poecilopleurus	Mariana Isl.	Rodda et al. unpub.	41.7	42	1.31	14
Cyclura carinata	Bahama Isl.	Iverson 1979	250	235	500	99
Cyclura cornuta	Mona Isl.	Wiewandt 1977	495	480	5100	100
Cyclura pinguis	Anegada Isl.	Carey 1975	501	530	7500	100
Emoia atrocostata	Mariana Isl.	Rodda et al. unpub.	64.5	63	4.8	51
Emoia caeruleocauda	Mariana Isl.	Rodda et al. unpub.	42.78	49	2.01	24
Emoia slevini	Mariana Isl.	Rodda et al. unpub.	70	70	7.6	65
Gehyra insulensis	Mariana Isl.	Rodda et al. unpub.	46.7	46	1.7	20
Gehyra oceanica	Mariana Isl.	Rodda et al. unpub.	79.3	81	10.93	74
Gongylomorphus bojerii	Mauritius Isl.	Jones 1993	64	63	5.4	56
Hemidactylus frenatus	Mariana Isl.	Rodda et al. unpub.	49.7	51	2.65	31
Iguana iguana	Flamenco Isl.	Dugan 1980	313	310	1350	99
Lamprolepis smaragdina	Mariana Isl.	Rodda et al. unpub.	98	99	20.1	86
Leiolopisma telfairii	Round Isl.	Jones 1993	133	136	76	97
Lepidodactylus lugubris	Mariana Isl.	Rodda et al. unpub.	41.4	42	1.45	16
Lipinia noctua	Mariana Isl.	Rodda et al. unpub.	43.1	41	1.13	12
Nactus pelagicus	Mariana Isl.	Rodda et al. unpub.	55.6	54	3.1	37

Table B.1 continued

Species	Study site	Source	Adult mean SVL	Modal SVL	Mass at modal SVL	Percentile of mass
Oligosoma whitakeri	Mercury Isl.	Towns 1994	85	86	14.5	78
Perochirus ateles	Mariana Isl.	Rodda et al. unpub.	57.5	54	4	67
Podarcis liolepis	Columbrete Grande Isl.	Castilla and Bauwens 1991	63	63	5.8	57
Sauromalus ater	California	Johnson 1965	160	162	162	97
Sceloporus occidentalis	California	Davis 1967	69	68	12.6	75
Sceloporus tristichus	Arizona	Tinkle 1982	61	62	8.8	76
Sphaerodactylus macrolepis	Guana Isl.	Rodda et al. unpub.	22.9	24.5	0.371	1
Uromastyx nigriventris	Morocco	Grenot 1976	220	200	280	98
Varanus albigularis	Namibia	Phillips 1995	575	540	3700	100
Varanus indicus	Guam Isl.	Wikramanayake and Dryden 1988	367	310	453	99
Varanus indicus	Mariana Isl.	Rodda et al. unpub.	397	275	290	99
Varanus palawanensis	Calauit Isl.	Gaulke 1989	550	520	2450	100

I found that for these 34 species, modal SVL could be predicted with high accuracy ($r^2 > 0.99$) from knowledge of mean adult SVL (modal SVL = 0.9585 × mean adult SVL + 1.3 mm), and with fair accuracy ($r^2 = 0.93$) from maximum SVL (modal SVL = 0.7135 × maximum SVL + 9.5 mm). Extrapolating from total length provided some information ($r^2 = 0.82$): modal SVL = 0.2939 × maximum total length + 29.4 mm. Maximum values will not work as well for species for which the maxima are based on small samples. Furthermore, extrapolations from total length will work poorly for species with unusually long or short tails. Accordingly, I did not base any estimates of modal SVL on total length. Instead, I obtained my estimate of modal SVL from the averaged estimates from mean and maximum SVL.

If all mean SVL measurements in the full lizard database were precise, I would have preferred to estimate modal SVL solely from the means, but I found that a number of the secondary literature estimates of "mean" SVL were in fact closer to the concept of "typical" (i.e., modal) length, especially for poorly documented species. Use of such values in place of true means produced an upward bias in the estimate of modal SVL. The upward bias was, however, limited by averaging that estimate with one based on maximum SVL, which tended to be an underestimate for poorly documented species, and therefore produced an offsetting downward bias. In the case of well-quantified species, there should be little bias from either estimate or their average.

I found that I could also estimate body mass at modal size directly from linear measurements, especially from adult mean SVL ($r^2 = 0.97$): most likely mass = 0.0000277 × mean adult mass + 2.9962mm. The estimates from maximum SVL were slightly less convincing ($r^2 = 0.94$): most likely mass = 0.0000266 × maximum SVL + 2.8669 mm. The weakness of estimating mass directly from length is that unusually elongate or squat species produce large errors not adequately vetted among the species in the modeled data set. Comparison of estimates with known masses on especially elongate lizards convinced me not to directly estimate body mass from any single regression on mean or maximum SVL.

Instead, I relied on species-, genus-, or family-specific mass-SVL regressions in the literature, specified in the Mass-SVLRegression field in each species' database record (see appendix A). These regressions were mostly those developed by Meiri (2010) and extended by Novosolov et al. (2013) and Feldman et al. (2015). Those teams from Meiri's lab approached the problem of body size variation both taxonomically—by lizard family—and morphologically—segregating limbless/limb-reduced forms. My tabulation of limb reduction used slightly different thresholds and I recognized an intermediate level of elongation not classified by Meiri's group, but the combination of morphology and family produced plausible results with the slight modification that, whereas Meiri divided most families into limbed and limbless forms, I divided them into four tiers: (1) fully legged, (2) elongate, (3) functionally limbless, and (4) fully limbless. I used the Meiri group's equations for categories 1, 3, and 4, and generated an intermediate regression for the elongate forms (category 2). In the case of Gekkonidae and Varanidae, I found empirical data with which to calibrate the positions of the intermediate curves; in the other cases, I placed the intermediate curves in a correspondingly intermediate position. In each case, the elongate species are so labeled in the Elongation field and the corresponding regressions quantified in the Mass-SVLRegression field. In the cases for which I had empirical modal SVL and mass data, those were used in the place of regression-derived estimates, as denoted in the SizeNarrative field (appendix A). If the reported mass data were suggestive but not compelling estimates of mass at modal adult SVL, I shaded the regression estimate of mass at modal adult SVL by rounding in the direction suggested by the empirical mass values.

Appendix C Standards for Qualification of Absolute Population Densities

In my experience (Rodda 2012, 2012), some of the absolute population density estimates that appear in the literature do not do an adequate job of adjusting for imperfect detectability, sample area uncertainty, and other possible biases. Some are raw counts treated as densities. Because "absolute" population density estimates are intended to be used to compare across taxon, time, and space, both false positives and false negatives arise in comparing values obtained using grossly disparate or untenable assumptions. Turner (1977) identified a number of reports that were particularly suspect; I have followed his lead and identified subsequent papers with similar limitations. The suspect sources are denoted as "unqualified" in the database and are tabulated separately under the "NQ" fields in the main database.

The three most-frequently invoked suspect assumptions are as follows:

1. All individuals in the population were seen or trapped—perfect detection. A milder variant of this assumption is that all individuals are assumed to be equally likely to be seen or trapped—absence of detection heterogeneity.
2. The area sampled is defined by the size of the area searched or trapped and is equivalent in density to average habitat for this species—representative habitat.
3. During the course of sampling, the sampled population was not augmented with the arrival of individuals from outside of the study area—closure.

Assumptions 1 and 3 can be addressed with appropriate analytical models, and estimates that did so were categorized as qualified. Assumption 2 can be addressed with appropriate placement of sample plots, in which case the obtained estimates were classified as qualified (listed under the "Q" fields in the main database).

Detectability quantification is now understood to be a central requirement for population density estimation (Turner 1977; Mazerolle et al. 2007), but it was not always so, and the untenable assumption of perfect detectability continues to be made, especially in less formal population estimates. The number of unique individuals in a mark-recapture sample—MKA, or minimum number known alive—was almost never judged valid. It might be a reasonable approximation if the time interval was very short and per-individual detectability was near 1. Imperfect detectability can be addressed by a number of mark-recapture and repeated sample protocols (Rodda 2012). An invalid assumption of perfect detection was the foremost reason for disqualifying an absolute population density estimate. The strongest case for accepting an assumption of perfect detection occurred in certain chameleons that sleep in conspicuous arboreal positions in sparsely vegetated woodland.

Detection heterogeneity is less often addressed. An example would be juveniles that behave differently from adults, typically being much harder to detect. Demure females are sometimes much harder to detect than blustering territorial or hierarchical males. Even if a simple mark-recapture protocol is used to estimate species-specific detectability, detection heterogeneity can produce gross errors in density estimation because the detection data stream is dominated by detections of the more detectable stratum of the population, and the detectable stratum dominates the estimation of overall detectability. Many accounts mention the apparent rarity of one sex or size class, a red flag that crucial heterogeneity in detectability is present and must be addressed. Some analytical models do a good job of addressing capture or detection heterogeneity; such analyses were deemed qualified.

The assumption that sampled habitat was representative was often tacit. The most problematic cases occurred when the study site was implicitly or explicitly chosen because there was a high density of the target species present. The best studies chose the study site on the basis of representative habitat characteristics and without advanced knowledge of the target species density.

A common approach was to sample the lizards in an urban area that was convenient to the scientists' residence or office, or because the lizards were easy to see in the simplified urban environment. With few exceptions, such studies were disqualified, as they cannot be anchored to a surface area that would be equivalent to that used by another scientist. A more subtle form of the same problem arises with species that are arrayed linearly along water courses or are clustered at rock outcrops or masonry structures. In each of these cases, there is almost no objective algorithm to match the study conditions to those undertaken by another scientist. How would one compare the effective size of rock outcrops, even if the two populations being compared were of the same species? As we wish to compare among species, even a count of the number and size of suitable cracks—if that were knowable—would not allow a credible comparison. I rejected virtually all studies based on unique habitats that would not be considered spatially representative.

The closure assumption (3, above) can be addressed by some computational models, and those studies which did so were deemed qualified. One less-satisfying analytical approach is to posit a "superpopulation" that resides in an unknown area that is larger than the trapping grid or study plot. I did not accept such superpopulation analyses unless the size of the area occupied by the superpopulation was plausibly estimated, for example by radiotelemetric documentation of lizard movements. For lizard species of low detectability, a common gambit is to pool captures over all or most of an active season and treat these as a single sample. Many studies used a simple Lincoln-Peterson index based on a whole season's captures; this model assumes closure and a season-long sample is unlikely to be closed. I accepted these only if there was strong evidence that appreciable numbers of nonresident lizards had not entered the population during the sampling period.

All mark-recapture traps or capture grids capture some individuals whose home range is mostly outside of the formal study area. There are various analytical methods available for estimating the effective area of a mark-recapture study (Otis et al. 1978; Parmenter et al. 2003; Efford 2004; Efford et al. 2004, 2005), and any that incorporated these techniques were deemed qualified.

To help me with the tabulation of analytical assumptions, I asked 11 questions of each plausible study; the answers to these are tabulated in the fields Q1–Q11 in the DENSITIES worksheet of the downloadable database. That is, the answer to question 1 below appears in the Q1 field in the database and so on. They are roughly ranked from most favorable (a) to least so (z). Note that some answers were obtained by personal communication with the author and may not be evident in the published work.

1. In terms of population density, was the system selected for study representative of typical habitat in the area? Or were animals concentrated by water, localized refugia, habitat, or some other feature?
   a. System selected on the basis of its population biology representativeness
   b. System studied and reported for reasons unrelated to the population density, which was presumed to be typical
   c. Area chosen for study was typical habitat, but this habitat is very localized, with animals concentrated near water, refugia, and so on
   d. System chosen for study or reported because of high densities; reason for high density may include concentration at optimal localities
   z. Not stated
2. The area actually sampled was ______ in size.
   a. >5 ha
   b. 1–5 ha
   c. 0.1–0.99 ha
   d. <0.1 ha
   z. not stated
3. The minimum number of animals (marked and unmarked) sampled during a sampling event that produced a population estimate was
   a. >50
   b. 10–50
   c. <10
   z. not stated
4. The demographic strata of the population sample were
   a. the entire population
   b. all residents, including some juveniles
   c. both sexes of resident adults
   d. one sex of resident adults
   z. not stated
5. In relation to the annual population cycle, the described sampling was
   a. averaged or averageable over the cycle
   b. near the trough (i.e., just before emergence of young)
   c. near the peak (i.e., just after the emergence of young)
   d. at a known date, but relation to cycle unknown
   z. date not reported
6. The animals were detected by
   a. multiple methods
      i. direct nondisruptive observation, usually visually

    ii. by exposure in refugia (e.g., hand capture or rock flipping)
    iii. random movement trapping (pitfalls or isolated traps)
    iv. enhanced trapping (baited, drift fences, etc.)
  z. not stated
7. The fraction of animals that were undetected was estimated by
  a. total sampling at conclusion of study
  b. mark-recapture
  c. removal sampling
  d. validated calibration runs
  e. distance sampling (100% detection on centerline)
  z. not stated
8. The analytical assumptions of the method were
  a. tested and found true
  b. tested and found mildly wanting
  c. tested and found to be significantly wanting
  d. not tested, but thought to be true
  e. not tested, but likely to fail
  z. not stated
9. Sampling was stratified by size classes to address heterogeneity
  a. known to be unnecessary
  b. done
  c. assumed to be unnecessary
  d. recognized to be necessary, but not effectively accomplished
  z. not stated
10. The boundaries of the study area were defined on the basis of
  a. physical barriers that prevented population movement
  b. population flux estimated with open population model
  c. home ranges known; divided between inside and outside on that basis
  d. assumed half a row of home ranges beyond outer sampling points
  e. habitat discontinuities discouraged population movement
  f. assumed closure on the basis of good statistical evidence
  g. assumed closure on the basis of anecdotal evidence
  h. assumed closure for an unstated reason
  z. not stated
11. Were empirical estimates validated in this system or a similar one (external validity)?
  a. a successful validation test was performed or cited
  b. some validation evidence claimed or reported
  z. validation not discussed

The other fields in the DENSITIES database are all self-explanatory, with the exception that Island (y/n) is "yes" if the island studied was smaller than Madagascar.

Literature Cited

Adler, K. 1989. *Contributions to the History of Herpetology, Vol. 1.* Society for the Study of Amphibians and Reptiles.

Adler, K. 2007. *Contributions to the History of Herpetology; Vol. 2.* St. Louis, MO, USA: Society for the Study of Amphibians and Reptiles.

Adler, K. 2012. *Contributions to the History of Herpetology, Vol. 3*: Society for the Study of Amphibians and Reptiles.

Alberts, A. C. 1989. Ultraviolet visual sensitivity in Desert Iguanas: Implications for pheromone detection. *Animal Behaviour* 38:129–137.

Andreone, F., F. M. Guarino, and J. E. Randrianirina. 2005. Life history traits, age profile and conservation of the Panther Chameleon, *Furcifer pardalis* (Cuvier 1829) at Nosy Be, NW Madagascar. *Tropical Zoology* 18:209–225.

Andrews, R. M. 1976. Growth rate in island and mainland anoline lizards. *Copeia* 1976:477–482.

Andrews, R. M. 1979. Evolution of life histories: A comparison of *Anolis* lizards from matched island and mainland habitats. *Breviora* (454):1–51.

Andrews, R. M. 2000. Evolution of viviparity in squamate reptiles (*Sceloporus* spp.): A variant of the cold-climate model. *Journal of Zoology (London)* 250:243–253.

Andrews, R. M. 2017. Novel eggshell of the New Caledonian diplodactylid gecko species *Correlophus ciliatus* (= *Rhacodactylus ciliatus*). *Journal of Herpetology* 51:173–177.

Applegarth, J. S. 2012. Index of authors in taxonomic herpetology, 3rd ed. In *Contributions to the History of Herpetology, Vol. 3*, edited by K. Adler, pp. 387–469. Society for the Study of Amphibians and Reptiles.

Auffenberg, W. 1981. *The Behavioral Ecology of the Komodo Monitor.* Gainesville, FL, USA: University Presses of Florida.

Auffenberg, W. 1988. *Gray's Monitor Lizard.* Gainesville, FL, USA: University Presses of Florida.

Auffenberg, W. 1994. *The Bengal Monitor.* Gainesville, FL, USA: University Press of Florida.

Bauer, A. M., and R. A. Sadlier. 2000. *The Herpetofauna of New Caledonia.* Society for the Study of Amphibians and Reptiles, in cooperation with Institut de recherche pour le développement.

Bauwens, D., and R. Díaz-Uriarte. 1997. Covariation in life-history traits in lacertid lizards: A comparative study. *American Naturalist* 149:91–111.

Bell, W. J. 1991. *Searching Behaviour.* London, UK: Chapman and Hall.

Beolens, B., M. Watkins, and M. Grayson. 2011. *The Eponym Dictionary of Reptiles.* Baltimore, MD, USA: Johns Hopkins University Press.

Beuttell, K., and J. B. Losos. 1999. Ecological morphology of Caribbean anoles. *Herpetological Monographs* 13:1–28.

Blackburn, D. G. 1982. Evolutionary origins of viviparity in the Reptilia. I. Sauria. *Amphibia-Reptilia* 3:185–205.

Bogert, C. M. 1959. How reptiles regulate their body temperature. *Scientific American* 200:105–120.

Bolger, D. T., and T. J. Case. 1992. Intra- and interspecific interference behaviour among sexual and asexual geckos. *Animal Behaviour* 44:21–30.

Brady, L. D., and R. A. Griffiths. 2000. *Status Assessment of Chameleons in Madagascar.* Gland, Switzerland: International Union Conservation Nature Species Survival Commission.

Branch, W. R. 1998. *Field Guide to the Snakes and Other Reptiles of Southern Africa*, 3rd ed. Sanibel Island, FL, USA: Ralph Curtis Books Publ.

Branch, W. R. 2011. *Lepidothyrys fernandi harlani* (Hallowell, 1844). Harlan's Fire Skink. *African Herp News* 54:21–22.

Brandstaetter, F. 1992. Observations on the territory-marking behaviour of the gecko *Eublepharis macularius*. *Hamadryad* 17:17–20.

Burghardt, G. M. 1977. Of iguanas and dinosaurs: Social behavior and communication in neonate reptiles. *American Zoologist* 17:177–190.

Burghardt, G. M., H. W. Greene, and A. S. Rand. 1977. Social behavior in hatchling green iguanas: Life in a reptile rookery. *Science* 195:689–691.

Butler, M. A. 2005. Foraging mode of the chameleon, *Bradypodion pumilum*: A challenge to the sit-and-wait versus active forager paradigm? *Biological Journal of the Linnean Society* 84:797–808.

Carey, W. M. 1975. The Rock Iguana, *Cyclura pinguis*, on Anegada, British Virgin Islands, with notes on *Cyclura ricordi* and *Cyclura cornuta* on Hispaniola. *Bulletin of the Florida State Museum, Biological Sciences* 19:189–233.

Carmichael, P., and W. Williams. 2004. *Florida's Fabulous Reptiles and Amphibians: Snakes, Lizards, Alligators, Frogs, and Turtles.* Tampa, FL, USA: World Publications.

Castilla, A. M., and D. Bauwens. 1991. Observations on the natural history, present status, and conservation of the insular lizard *Podarcis hispanica atrata* on the Columbretes Archipelago, Spain. *Biological Conservation* 58:69–84.

Church, G., and C.-S. Lim. 1961. The distribution of three species of house gecko in Bandung (Java). *Herpetologica* 17:199–201.

Clark, D. L., J. M. Macedonia, J. W. Rowe, K. Kamp, and C. A. Valle. 2017. Responses of Galapagos Lava Lizards (*Microlophus bivittatus*) to manipulation of female nuptial coloration on lizard robots. *Herpetologica* 73:323–330.

Cogger, H. G. 2014. *Reptiles and Amphibians of Australia*, 7th Ed. Collingwood, NSW, Australia: CSIRO Publications.

Cooper, W. E., Jr., and L. J. Vitt. 2002. Distribution, extent, and evolution of plant consumption by lizards. *Journal of Zoology (London)* 257:485–517.

Corl, A., A. R. Davis, S. R. Kuchta, and B. Sinervo. 2010. Selective loss of polymorphic mating types is associated with rapid phenotypic evolution during morphic speciation. *Proceedings of the National Academy of Sciences USA* 107:4254–4259.

Costa, G. C., L. J. Vitt, E. R. Pianka, D. O. Mesquita, and G. R. Colli. 2008. Optimal foraging constrains macroecological patterns: Body size and dietary niche breadth in lizards. *Global Ecology and Biogeography* 17:670–677.

Cowles, R. B., and C. M. Bogert. 1944. A preliminary study of the thermal requirements of desert reptiles. *Bulletin of the American Museum of Natural History* 83:261–296.

Cox, R. M., S. L. Skelly, and H. B. John-Alder. 2003. A comparative test of adaptive hypotheses for sexual size dimorphism in lizards. *Evolution* 57:1653–1669.

Cree, A. 2014. *Tuatara: Biology and Conservation of a Venerable Survivor*. Christchurch, New Zealand: Canterbury University Press.

Darwin, C. 1871. *The Descent of Man and Selection in Relation to Sex*. New York, NY, USA: Appleton.

Das, I. 2010. *A Field Guide to Reptiles of Southeast Asia*. Singapore: New Holland Publications.

Davis, J. 1967. Growth and size of the Western Fence Lizard (*Sceloporus occidentalis*). *Copeia* 1967:721–731.

Dixon, J. R., and J. A. Lemos-Espinal. 2010. *Amphibians and Reptiles of the State of Querétaro, Mexico*. Mexico: Texas A&M, Universidad Nacional Autónoma de México, Comisión Nacional para el Conocimiento y Uso de la Biodiversidad.

Dugan, B. A. 1980. A field study of the social structure, mating system and display behavior of the Green Iguana (*Iguana iguana*), PhD dissertation, University of Tennessee, Knoxville, TN, USA.

Dunham, A. E. 1980. An experimental study of interspecific competition between the iguanid lizards *Sceloporus merriami* and *Urosaurus ornatus*. *Ecological Monographs* 50:309–330.

Dunham, A. E., D. W. Tinkle, and J. W. Gibbons. 1978. Body size in island lizards: A cautionary tale. *Ecology* 59:1230 1238.

Efford, M. G. 2004. Density estimation in live-trapping studies. *Oikos* 106:598–610.

Efford, M. G., D. K. Dawson, and C. S. Robbins. 2004. DENSITY: Software for analyzing capture-recapture data from passive detector arrays. *Animal Biodiversity and Conservation* 27:217–228.

Efford, M. G., B. Warburton, M. C. Coleman, and R. J. Barker. 2005. A field test of two methods for density estimation. *Wildlife Society Bulletin* 33:731–738.

Ehmann, H. 1992. *Encyclopedia of Australian Animals: Reptiles*. Pymble, NSW, Australia: Angus and Robertson.

Feldman, A., N. Sabath, R. A. Pyron, I. Mayrose, and S. Meiri. 2015. Body sizes and diversification rates of lizards, snakes, amphisbaenians and the Tuatara. *Global Ecology and Biogeography* 25:187–197.

Ferguson, G. A. 1966. *Statistical Analysis in Psychology and Education*. New York, NY, USA: McGraw-Hill.

Fisher, R. N., A. V. Suarez, and T. J. Case. 2002. Spatial patterns in the abundance of the Coastal Horned Lizard. *Conservation Biology* 16:205–215.

Fitch, H. S. 1968. Temperature and behavior of some equatorial lizards. *Herpetologica* 24:35–38.

Fleishman, L. J. 1992. The influence of the sensory system and the environment on motion patterns in the visual displays of anoline lizards and other vertebrates. *American Naturalist* 139:S36–S61.

Gamble, T., E. Greenbaum, T. R. Jackman, and A. M. Bauer. 2015. Into the light: Diurnality has evolved multiple times in geckos. *Biological Journal of the Linnean Society* 115:896–910.

Gamble, T., E. Greenbaum, A. P. Russell, T. R. Jackman, and A. M. Bauer. 2012. Repeated origins and loss of toepads in geckos. *PLoS ONE* 7 (6):e39429.

Gaulke, M. 1989. Zur Biologie des Bindenwaranes, unter Berucksichtigung der paläogeographischen Verbeitung und der phylogenetischen Entwicklung der Varanidae. *Courier Forschungsinstitut Senckenberg* 112:1–229.

Gaulke, M. 2011. *The Herpetofauna of Panay Island, Philippines: An Illustrated Field Guide*. Frankfurt am Main, Germany: Edition Chimaira.

Greene, H. W. 1988. Antipredator mechanisms in reptiles. In *Biology of the Reptilia, vol. 16, Ecology B, Defense and Life History*, edited by C. Gans and R. B. Huey, pp. 1–152. Ann Arbor MI, USA: Branta Books.

Greer, A. E. 1983. On the adaptive significance of the reptilian spectacle: The evidence from scincid, teiid, and lacertid lizards. In *Advances in Herpetology and Evolutionary Biology*, edited by A. G. J. Rhodin and K. I. Miyata, pp. 213–221. Cambridge, MA, USA: Museum of Comparative Zoology.

Greer, A. E. 2001. Distribution of maximum snout-vent length among species of scincid lizards. *Journal of Herpetology* 35:383–395.

Grenot, C. 1976. *Ecophysiologie du Lézard Saharien* Uromastix acanthinurus *Bell, 1825 (Agamidae Herbivore)*. Paris, France: École Normale Supérieure Publications du Laboratoire de Zoologie.

Grismer, L. L. 2011. *Lizards of Peninsular Malaysia, Singapore, and Their Adjacent Archipelagos*. Frankfurt am Main, Germany: Edition Chimaira.

Grismer, L. L., P. L. Wood, Jr., S. Anuar, A. Riyanto, N. Ahmad, A. M. Mohd, . . . O. S. G. Pauwels. 2014. Systematics and natural history of Southeast Asia rock geckos (genus *Cnemaspis* Strauch, 1887) with descriptions of eight new species from Malaysia, Thailand, and Indonesia. *Zootaxa* 3880:1–147.

Hamilton, P. S., and B. K. Sullivan. 2005. Female mate attraction in Ornate Tree Lizards, *Urosaurus ornatus*: A multivariate analysis. *Animal Behaviour* 69:219–224.

Harrison, A., and S. Poe. 2012. Evolution of an ornament, the dewlap, in females of the lizard genus *Anolis*. *Biological Journal of the Linnean Society* 106:191–201.

Heatwole, H., and J. Taylor. 1987. *Ecology of Reptiles*. Chipping Norton, NSW, Australia: Surrey Beatty.

Hedges, S. B., and C. E. Conn. 2012. A new skink fauna from Caribbean islands (Squamata, Mabuyidae, Mabuyinae). *Zootaxa* 3288:1–244.

Henderson, R. W., and R. Powell. 2009. *Natural History of West Indian Reptiles and Amphibians*. Gainesville, FL, USA: University Press of Florida.

Henke, S. E. 1998. The effect of multiple search items and item abundance on the efficiency of human searchers. *Journal of Herpetology* 32:112–115.

Herrel, A. 2007. Herbivory and foraging mode in lizards. In *Lizard Ecology: The Evolutionary Consequences of Foraging Mode*, edited by S. M. Reilly, L. D. McBrayer, and D. B. Miles, pp. 209–236. Cambridge, UK: Cambridge University Press.

Hill, K. A. 1994. *An Introduction to Ground Search Management for the Police Authority*. International Association of Chiefs of Police.

Hölldobler, B., and E. O. Wilson. 1990. *The Ants*. Cambridge, MA, USA: Belknap Press.

Huey, R. B., and A. F. Bennett. 1987. Phylogenetic studies of coadaptation: Preferred temperatures versus optimal performance temperatures of lizards. *Evolution* 41 (5):1098–1115.

Huey, R. B., and E. R. Pianka. 1981. Ecological consequences of foraging mode. *Ecology* 62:991–999.

Huey, R. B., and M. Slatkin. 1976. Cost and benefits of lizard thermoregulation. *Quarterly Review of Biology* 51:363–384.

Huey, R. B., P. H. Niewiarowski, J. H. Kaufmann, and J. C. Herron. 1989. Thermal biology of nocturnal endotherms: Is sprint performance of geckos maximal at low body temperatures? *Physiological Zoology* 62:488–504.

Hutchinson, G. E. 1959. Homage to Santa Rosalia, or why are there so many kinds of animals? *American Naturalist* 93:145–149.

Hutchinson, G. E. 1978. *An Introduction to Population Ecology*. New Haven, CT, USA: Yale University Press.

Inger, R. F., and R. K. Colwell. 1977. Organization of contiguous communities of amphibians and reptiles in Thailand. *Ecological Monographs* 47:229–253.

Ingram, T., A. Harrison, D. L. Mahler, M. d. R. Castañeda, R. E. Glor, A. Herrel, . . . J. B. Losos. 2016. Comparative tests of the role of dewlap size in *Anolis* lizard speciation. *Proceedings of the Royal Society of London B* 283:20162199.

Irschick, D. J., L. J. Vitt, P. A. Zani, and J. B. Losos. 1997. A comparison of evolutionary radiations in mainland and Caribbean *Anolis* lizards. *Ecology* 78:2191–2203.

Ito, R., I. Ikeuchi, and A. Mori. 2013. A day gecko darkens its body color in response to avian alarm calls. *Current Herpetology* 32:26–33.

IUCN. 2017. *The IUCN Red List of Threatened Species*. http://www.iucnredlist.org.

Iverson, J. B. 1979. Behavior and ecology of the Rock Iguana *Cyclura carinata*. *Bulletin of the Florida State Museum, Biological Sciences* 24:175–358.

Jackson, J. F., W. F. Ingram, III, and H. W. Campbell. 1976. The dorsal pigmentation pattern of snakes as an antipredator strategy: A multivariate approach. *American Naturalist* 110:1029–1053.

Johnson, M. A., M. Leal, L. Rodríguez Schettino, A. Chamizo Lara, L. J. Revell, and J. B. Losos. 2008. A phylogenetic perspective on foraging mode evolution and habitat use in West Indian *Anolis* Lizards. *Animal Behaviour* 75:555–563.

Johnson, S. R. 1965. An ecological study of the Chuckwalla, *Sauromalus obesus* Baird, in the western Mojave Desert. *American Midland Naturalist* 73:1–29.

Jones, C. G. 1993. The ecology and conservation of Mauritian skinks. *Proceedings of the Royal Society of Arts and Letters of Mauritius* 5:71–95.

King, D. R., and L. Rhodes. 1982. Sex ratio and breeding season of *Varanus acanthurus*. *Copeia* 1982:784–787.

Klaver, C. J. J., and W. Böhme. 1986. Phylogeny and classification of the Chamaeleonidae (Sauria) with special reference to hemipenis morphology. *Bonner Zoologische Monographie* 22:1–64.

Koch, A. 2012. *Discovery, Diversity, and Distribution of the Amphibians and Reptiles of Sulawesi and Its Offshore Islands*. Frankfurt am Main, Germany: Edition Chimaira.

Kraus, F. 2009. *Alien Reptiles and Amphibians: A Scientific Compendium and Analysis*. New York, NY, USA: Springer Series in Invasion Ecology 4.

Laird, M. K., M. B. Thompson, and C. M. Whittington. 2019. Facultative oviparity in a viviparous skink (*Saiphos equalis*). *Biology Letters* 15:20180827.

Leão, S. M., E. R. Pianka, and N. Pelegrin. 2018. Is there evidence for population regulation in amphibians and reptiles? *Journal of Herpetology* 52:28–33.

Lemos-Espinal, J. A., and J. R. Dixon. 2013. *Amphibians and Reptiles of San Luis Potosí*. Eagle Mountain, UT, USA: Eagle Mountain Publishing.

Lemos-Espinal, J. A., G. R. Smith, and R. E. Ballinger. 2003. Variation in growth and demography of a knob-scaled lizard (*Xenosaurus newmanorum*: Xenosauridae) from a seasonal tropical environment in Mexico. *Biotropica* 35:240–249.

Leu, S. T., D. Burzacott, M. J. Whiting, and C. M. Bull. 2015. Mate familiarity drives long-term monogamy in a lizard: Evidence from a thirty-year field study. *Ethology* 121:720–728.

Losos, J. B. 1992. The evolution of convergent structure in Caribbean *Anolis* communities. *Systematic Biology* 41:403–420.

Losos, J. B. 2009. *Lizards in an Evolutionary Tree: Ecology and Adaptive Radiation of Anoles*. Berkeley, CA, USA: University of California Press.

Losos, J. B., and L.-R. Chu. 1998. Examination of factors potentially affecting dewlap size in Caribbean anoles. *Copeia* 1998:430–438.

MacArthur, R. H. 1972. *Geographical Ecology*. New York, NY, USA: Harper and Row.

Macrini, T. E., D. J. Irschick, and J. B. Losos. 2003. Ecomorphological differences in toepad characteristics between mainland and island anoles. *Journal of Herpetology* 37:52–58.

Masroor, R. 2012. *A Contribution to the Herpetology of Northern Pakistan: The Amphibians and Reptiles of Margalla Hills National Park and Surrounding Regions*. Salt Lake City, UT, USA: Society for the Study of Amphibians and Reptiles.

Mazerolle, M. J., L. L. Bailey, W. L. Kendall, J. A. Royle, S. J. Converse, and J. D. Nichols. 2007. Making great leaps forward: Accounting for detectability in herpetological field studies. *Journal of Herpetology* 41:672–689.

McDiarmid, R. W., M. S. Foster, C. Guyer, J. W. Gibbons, and N. Chernoff. 2012. *Reptile Biodiversity: Standard Methods for Inventory and Monitoring*. Berkeley, CA, USA: University of California Press.

Meiri, S. 2008. Evolution and ecology of lizard body sizes. *Global Ecology and Biogeography* 17:724–734.

Meiri, S. 2010. Length-weight allometries in lizards. *Journal of Zoology* 281:218–226.

Meiri, S., A. M. Bauer, L. Chirio, G. R. Colli, I. Das, T. M. Doan, . . . R. Van Damme. 2013. Are lizards feeling the heat? A tale of ecology and evolution under two temperatures. *Global Ecology and Biogeography* 22:834–845.

Mesquita, D. O., G. R. Colli, G. C. Costa, T. s. B. Costa, D. B. Shepard, L. J. Vitt, and E. R. Pianka. 2015. Life history data of lizards of the world. *Ecology* 96:594–594.

Minton, S. A., Jr. 1966. A contribution to the herpetology of West Pakistan. *Bulletin of the American Museum of Natural History* 134:29–184.

Mitchell, J. C., R. E. Jung Brown, and B. Bartholomew. 2008. *Urban Herpetology*. Salt Lake City, UT, USA: Society for the Study of Amphibians and Reptiles.

Myhrvold, N. P., E. Baldridge, B. Chan, D. Sivam, D. L. Freeman, and S. K. Morgan Ernest. 2015. An amniote life-history database to perform comparative analyses with birds, mammals, and reptiles. *Ecology* 96:3109–3109.

Novosolov, M., P. Raia, and S. Meiri. 2013. The island syndrome in lizards. *Global Ecology and Biogeography* 22:184–191.

Otis, D. L., K. P. Burnham, G. C. White, and D. R. Anderson. 1978. Statistical inference from capture data on closed animal populations. *Wildlife Monographs* (62):1–135.

Packard, G. C., C. R. Tracy, and J. J. Roth. 1977. The physiological ecology of reptilian eggs and embryos, and the evolution of viviparity within the class Reptilia. *Biological Review* 52:71–105.

Parmenter, R. R., T. L. Yates, D. R. Anderson, K. P. Burnham, J. L. Dunnum, A. B. Franklin, . . . G. C. White. 2003. Small-mammal density estimation: A field comparison of grid-based vs. web-based density estimators. *Ecological Monographs* 73:1–26.

Pearl, J., and D. Mackenzie. 2018. *The Book of Why: The New Science of Cause and Effect*. New York NY, USA: Basic Books.

Perry, G. 1999. The evolution of search modes: Ecological versus phylogenetic perspectives. *American Naturalist* 153:98–109.

Perry, G. 2007. Movement patterns in lizards: Measurement, modality, and behavioral correlates. In *Lizard Ecology: The Evolutionary Consequences of Foraging Mode*, edited by S. M. Reilly, L. D. McBrayer, and D. B. Miles, pp. 13–48. Cambridge, UK: Cambridge University Press.

Perry, G., and T. Garland, Jr. 2002. Lizard home ranges revisited: Effects of sex, body size, diet, habitat, and phylogeny. *Ecology* 83:1870–1885.

Perry, G., and E. R. Pianka. 1997. Animal foraging: Past, present, and future. *Trends in Ecology & Evolution* 12:360–364.

Phillips, J. A. 1995. Movement patterns and density of *Varanus albigularis*. *Journal of Herpetology* 29:407–416.

Pianka, E. R., and L. J. Vitt. 2003. *Lizards: Windows to the Evolution of Diversity*. Berkeley, CA, USA: University of California Press.

Pianka, E. R., L. J. Vitt, N. Pelegrin, D. B. Fitzgerald, and K. O. Winemiller. 2017. Toward a periodic table of niches, or exploring the lizard niche hypervolume. *American Naturalist* 190:601–616.

Pocheville, A. 2015. The ecological niche: History and recent controversies. In *Handbook of Evolutionary Thinking in the Sciences*, edited by T. Heams, P. Huneman, G. Lecointre, and M. Silberstein, pp. 547–586. Dortrecht, Netherlands: Springer Science+Business Media.

Poulin, B., G. Lefebvre, R. Ibáñez, C. Jaramillo, C. Hernández, and A. S. Rand. 2001. Avian predation upon lizards and frogs in a neotropical forest understorey. *Journal of Tropical Ecology* 17:21–40.

Rabinowitz, D. 1981. Seven forms of rarity. In *The Biological Aspects of Rare Plant Conservation*, edited by H. Synge, pp. 205–217. Chichester, UK: Wiley.

Rabinowitz, D., S. Cairns, and T. Dillon. 1986. Seven forms of rarity and their frequency in the flora of the British Isles. In *Conservation Biology: The Science of Scarcity and Diversity*, edited by M. E. Soulé, pp. 182–204. Sunderland, MA, USA: Sinauer.

Rand, A. S., and H. W. Greene. 1982. Latitude and climate in the phenology of reproduction in the Green Iguana, *Iguana iguana*. In *Iguanas of the World: Their Behavior, Ecology, and Conservation*, edited by G. M. Burghardt and A. S. Rand, pp. 142–149. Park Ridge, NJ, USA: Noyes Publications.

Reichling, H. 1957. Transpiration und Vorzugstemperatur mitteleuropäischer Reptilien und Amphibien. Zoologische Jahrbücher (Zoologie) 67:1–64.

Reilly, S. M., L. D. McBrayer, and D. B. Miles. 2007. *Lizard Ecology*. Cambridge, UK: Cambridge University Press.

Reptile Database. 2019. Uetz, P., P. Freed, and J. Hošek, eds. http://www.reptile-database.org.

Rodda, G. H. 1992. The mating behavior of Iguana iguana. *Smithsonian Contributions to Zoology* 534:1–40.

Rodda, G. H. 2012. Population size and demographics. In *Reptile Biodiversity: Standard Methods for Inventory and Monitoring*, edited by R. W. McDiarmid, M. S. Foster, C. Guyer, J. W. Gibbons, and N. Chernoff, pp. 283–322. Berkeley, CA, USA: University of California Press.

Rodda, G. H. 2012. Statistical properties of techniques and validation. In *Reptile Biodiversity: Standard Methods for Inventory and Monitoring*, edited by R. W. McDiarmid, M. S. Foster, C. Guyer, J. W. Gibbons, and N. Chernoff, pp. 197–203. Berkeley, CA, USA: University of California Press.

Rodda, G. H., and K. Dean-Bradley. 2002. Excess density compensation of island herpetofaunal assemblages. *Journal of Biogeography* 29:1–10.

Rodda, G. H., G. Perry, R. J. Rondeau, and J. D. Lazell, Jr. 2001. The densest terrestrial vertebrate. *Journal of Tropical Ecology* 17:331–338.

Rodda, G. H., A. A. Yackel Adams, E. W. Campbell, and T. H. Fritts. 2015. General herpetological collecting is size-biased for five Pacific lizards. *Journal of Herpetology* 49:507–512.

Rodda, G. H., K. Dean-Bradley, E. W. Campbell, T. H. Fritts, B. Lardner, A. A. Yackel Adams, and R. N. Reed. 2015. Stability of detectability over 17 years at a single site and other lizard detection comparisons from Guam. *Journal of Herpetology* 49:513–521.

Roll, U., A. Feldman, M. Novosolov, A. Allison, A. M. Bauer, R. Bernard, . . . S. Meiri. 2017. The global distribution of tetrapods reveals a need for targeted reptile conservation. *Nature Ecology and Evolution* 1:1677–1682.

Root, R. B. 1967. The niche exploitation pattern of the Blue-Gray Gnatcatcher. *Ecological Monographs* 37:317–350.

Rose, B. 1981. Factors affecting activity in *Sceloporus virgatus*. *Ecology* 62:706–716.

Roughgarden, J. 1995. *Anolis Lizards of the Caribbean: Ecology, Evolution, and Plate Tectonics*. New York, NY, USA: Oxford University Press.

Ruibal, R., and R. Philibosian. 1974. The population ecology of the lizard *Anolis acutus*. *Ecology* 55:525–537.

Savage, J. M. 2002. *The Amphibians and Reptiles of Costa Rica*. Chicago, IL, USA: University of Chicago Press.

Scharf, I., A. Feldman, M. Novosolov, D. Pincheira-Donoso, I. Das, M. Böhm, . . . S. Meiri. 2015. Late bloomers and baby boomers: Ecological drivers of longevity in squamates and the Tuatara. *Global Ecology and Biogeography* 24:396–405.

Schmitz, L., and R. Motani. 2011. Nocturnality in dinosaurs inferred from scleral ring and orbit morphology. *Science* 332:705–708.

Schoener, T. W. 1979. Inferring the properties of predation and other injury-producing agents from injury frequencies. *Ecology* 60:1110–1115.

Schoener, T. W., and A. Schoener. 1980. Ecological and demographic correlates of injury rates in some Bahamian *Anolis* Lizards. *Copeia* 1980:839–850.

Schwenk, K. 2000. Feeding in lepidosaurs. In *Feeding*, edited by K. Schwenk, pp. 175–291. San Diego, CA, USA: Academic Press.

Scott Newbold, T. A., and J. A. MacMahon. 2014. Determinants of habitat selection by Desert Horned Lizards (*Phrynosoma platyrhinos*): The importance of abiotic factors associated with vegetation structure. *Journal of Herpetology* 48:306–316.

Shea, G. M. 2012. [Review of] The eponym dictionary of reptiles by Beolens et al. *Copeia* 2012:578–590.

Sherbrooke, W. C. 2017. Antipredator nest guarding by female horned lizard (*Phrynosoma*): Iguanian parental care. *Herpetologica* 73:331–337.

Shine, R. 1985. The evolution of viviparity in reptiles: An ecological analysis. In *Biology of the Reptilia, Vol. 15*, edited by C. Gans and F. Billett, pp. 605–694. New York, NY, USA: John Wiley and Sons.

Shine, R. 1995. A new hypothesis for the evolution of viviparity in reptiles. *American Naturalist* 145:809–823.

Simberloff, D., and T. Dayan. 1991. The guild concept and the structure of ecological communities. *Annual Review of Ecology and Systematics* 22:115–143.

Sinn, D. L., G. M. While, and E. Wapstra. 2008. Maternal care in a social lizard: Links between female aggression and offspring fitness. *Animal Behaviour* 76:1249–1257.

Slavenko, A., O. J. S. Tallowin, Y. Itescu, P. Raia, and S. Meiri. 2016. Late Quaternary reptile extinctions: Size matters, insularity dominates. *Global Ecology and Biogeography* 25:1308–1320.

Spawls, S., K. Howell, H. Hinkel, and M. Menegon. 2018. *Field Guide to East African Reptiles*, 2nd ed. London, UK: Bloomsbury.

Somma, L. A. 2003. *Parental Behavior in Lepidosaurian and Testudinian Reptiles. A Literature Survey*. Malabar, FL, USA: Krieger.

Stamps, J. A. 1977. Social behavior and spacing patterns in lizards. In *Biology of the Reptilia, Vol. 7. Ecology and Behaviour*, edited by C. Gans and D. W. Tinkle, pp. 265–334. London, UK: Academic Press.

Thompson, W. L., G. C. White, and C. Gowan. 1998. *Monitoring Vertebrate Populations*. San Diego, CA, USA: Academic Press.

Tinkle, D. W. 1982. Results of experimental density manipulation in an Arizona lizard community. *Ecology* 63:57–65.

Tolley, K., and M. Burger. 2007. *Chameleons of Southern Africa*. Cape Town, South Africa: Struik.

Towns, D. R. 1994. The role of ecological restoration in the conservation of Whitaker's Skink (*Cyclodina whitakeri*), a rare New Zealand lizard (Lacertilia: Scincidae). *New Zealand Journal of Zoology* 21:457–471.

Traveset, A., and N. Riera. 2005. Disruption of a plant-lizard seed dispersal system and its ecological effects on a threatened endemic plant in the Balearic Islands. *Conservation Biology* 19:421–431.

Turner, F. B. 1977. The dynamics of populations of squamates, crocodilians and rhynchocephalians. In *Biology of the Reptilia, Vol. 7*, edited by C. Gans and D. W. Tinkle, pp. 157–264. New York, NY, USA: Academic Press.

Uetz, P., P. Freed, and J. Hošek, eds. 2019. *The Reptile Database*. http://www.reptile-database.org.

Vanhooydonck, B., and R. Van Damme. 1999. Evolutionary relationships between body shape and habitat use in lacertid lizards. *Evolutionary Ecology Research* 1:785–805.

Vitt, L. J. 1981. Lizard reproduction: Habitat specificity and constraints on relative clutch mass. *American Naturalist* 117:506–514.

Vitt, L. J., and E. R. Pianka. 2005. Deep history impacts present-day ecology and biodiversity. *Proceedings of the National Academy of Sciences USA* 102:7777–7778.

Werner, Y. L., H. Takahashi, Y. Yasukawa, and H. Ota. 2006. Factors affecting foraging behaviour, as seen in a nocturnal ground lizard, *Goniurosaurus kuroiwae kuroiwae*. *Journal of Natural History* 40:439–459.

Whiting, M. J., and G. M. While. 2017. Sociality in lizards. In *Comparative social evolution*, edited by D. R. Rubenstein and P. Abbot, pp. 390–426. Cambridge, UK: Cambridge Univ. Press.

Wiewandt, T. A. 1977. Ecology, behavior, and management of the Mona Island Ground Iguana, *Cyclura stejnegeri*, PhD dissertation, Cornell University, Ithaca, NY, USA.

Wikramanayake, E. D., and G. L. Dryden. 1988. The reproductive ecology of *Varanus indicus* on Guam. *Herpetologica* 44:338–344.

Williams, E. E. 1972. The origin of faunas. Evolution of lizard congeners in a complex island fauna: A trial analysis. *Evolutionary Biology* 6:47–89.

Wilson, E. O. 1975. *Sociobiology: The New Synthesis*. Cambridge, MA, USA: Belknap Press.

Zug, G. R., L. J. Vitt, and J. P. Caldwell. 2001. *Herpetology: An Introductory Biology of Amphibians and Reptiles*, 2nd Ed. San Diego, CA, USA: Academic Press.

Index

Note to readers: For common species names and subject terms used extensively in this work (crest, home range, etc.), all table entries and the most important text passages are indexed; look for additional relevant narratives adjacent to the cited tables.